THE ELEMENTS†

Name	Symbol	Atomic Number	Atomic Weight*	Name	Symbol	Atomic Number	Atomic Weight*
Actinium	Ac	89	(227)	Mercury	Hg		
Aluminum	Al	13	26.981539	Molybdenum			.4
Americium	Am	95	(243)	Nielsbohrium			(262)
Antimony	Sb	51	121.75	Neodymium		60	144.24
Argon	Ar	18	39.948	Neon	Ne	10	20.1797
Arsenic	As	33	74.92159	Neptunium	Np	93	237.05
Astatine	At	85	(210)	Nickel	Ni	28	58.69
Barium	Ba	56	137.327	Niobium	Nb	41	92.90638
Berkelium	Bk	97	(247)	Nitrogen	N	7	14.00674
Beryllium	Be	4	9.012182	Nobelium	No	102	(259)
Bismuth	Bi	83	208.98037	Osmium	Os	76	190.2
Boron	B	5	10.811	Oxygen	O	8	15.9994
Bromine	Br	35	79.904	Palladium	Pd	46	106.42
Cadmium	Cd	48	112.411	Phosphorus	P	15	30.973762
Calcium	Ca	20	40.078	Platinum	Pt	78	195.08
Californium	Cf	98	(251)	Plutonium	Pu	94	(244)
Carbon	C	6	12.011	Polonium	Po	84	(209)
Cerium	Ce	58	140.115	Potassium	K	19	39.0983
Cesium	Cs	55	132.90543	Praseodymium	Pr	59	140.90765
Chlorine	Cl	17	35.4527	Promethium	Pm	61	(145)
Chromium	Cr	24	51.9961	Protactinium	Pa	91	231.03588
Cobalt	Co	27	58.93320	Radium	Ra	88	226.03
Copper	Cu	29	63.546	Radon	Rn	86	(222)
Curium	Cm	96	(247)	Rhenium	Re	75	186.207
Dysprosium	Dy	66	162.50	Rhodium	Rh	45	102.90550
Einsteinium	Es	99	(254)	Rubidium	Rb	37	85.4678
Erbium	Er	68	167.26	Ruthenium	Ru	44	101.07
Europium	Eu	63	151.965	Rutherfordium	Rf	104	(261)
Fermium	Fm	100	(257)	Samarium	Sm	62	150.36
Fluorine	F	9	18.9984032	Scandium	Sc	21	44.955910
Francium	Fr	87	(223)	Seaborgium	Sg	106	(263)
Gadolinium	Gd	64	157.25	Selenium	Se	34	78.96
Gallium	Ga	31	69.723	Silicon	Si	14	28.0855
Germanium	Ge	32	72.61	Silver	Ag	47	107.8682
Gold	Au	79	196.96654	Sodium	Na	11	22.989768
Hafnium	Hf	72	178.49	Strontium	Sr	38	87.62
Hahnium	Ha	105	(262)	Sulfur	S	16	32.066
Hassium	Hs	108	(265)	Tantalum	Ta	73	180.9479
Helium	He	2	4.002602	Technetium	Tc	43	(98)
Holmium	Ho	67	164.93032	Tellurium	Te	52	127.60
Hydrogen	H	1	1.00794	Terbium	Tb	65	158.92534
Indium	In	49	114.82	Thallium	Tl	81	204.3833
Iodine	I	53	126.90447	Thorium	Th	90	232.0381
Iridium	Ir	77	192.22	Thulium	Tm	69	168.93421
Iron	Fe	26	55.847	Tin	Sn	50	118.710
Krypton	Kr	36	83.80	Titanium	Ti	22	47.88
Lanthanum	La	57	138.9055	Tungsten	W	74	183.85
Lawrencium	Lr	103	(260)	Uranium	U	92	238.0289
Lead	Pb	82	207.2	Vanadium	V	23	50.9415
Lithium	Li	3	6.941	Xenon	Xe	54	131.29
Lutetium	Lu	71	174.967	Ytterbium	Yb	70	173.04
Magnesium	Mg	12	24.3050	Yttrium	Y	39	88.90585
Manganese	Mn	25	54.93805	Zinc	Zn	30	65.39
Meitnerium	Mt	109	(266)	Zirconium	Zr	40	91.224
Mendelevium	Md	101	(258)				

†Only 109 elements are listed, as there have been no names assigned to elements 110 and 111. Names listed for elements 104–109 are those recommended by the American Chemical Society Committee on Nomenclature.

*Based on relative atomic mass of ^{12}C = 12, 1987 IUPAC values. Values in parentheses are the mass numbers of the isotopes with the longest half-life.

World of Chemistry

SECOND EDITION

2/2/96

Molly–

Thanks for your interest in bringing together chemistry and geology to address global issues.

Best Wishes,

Mel

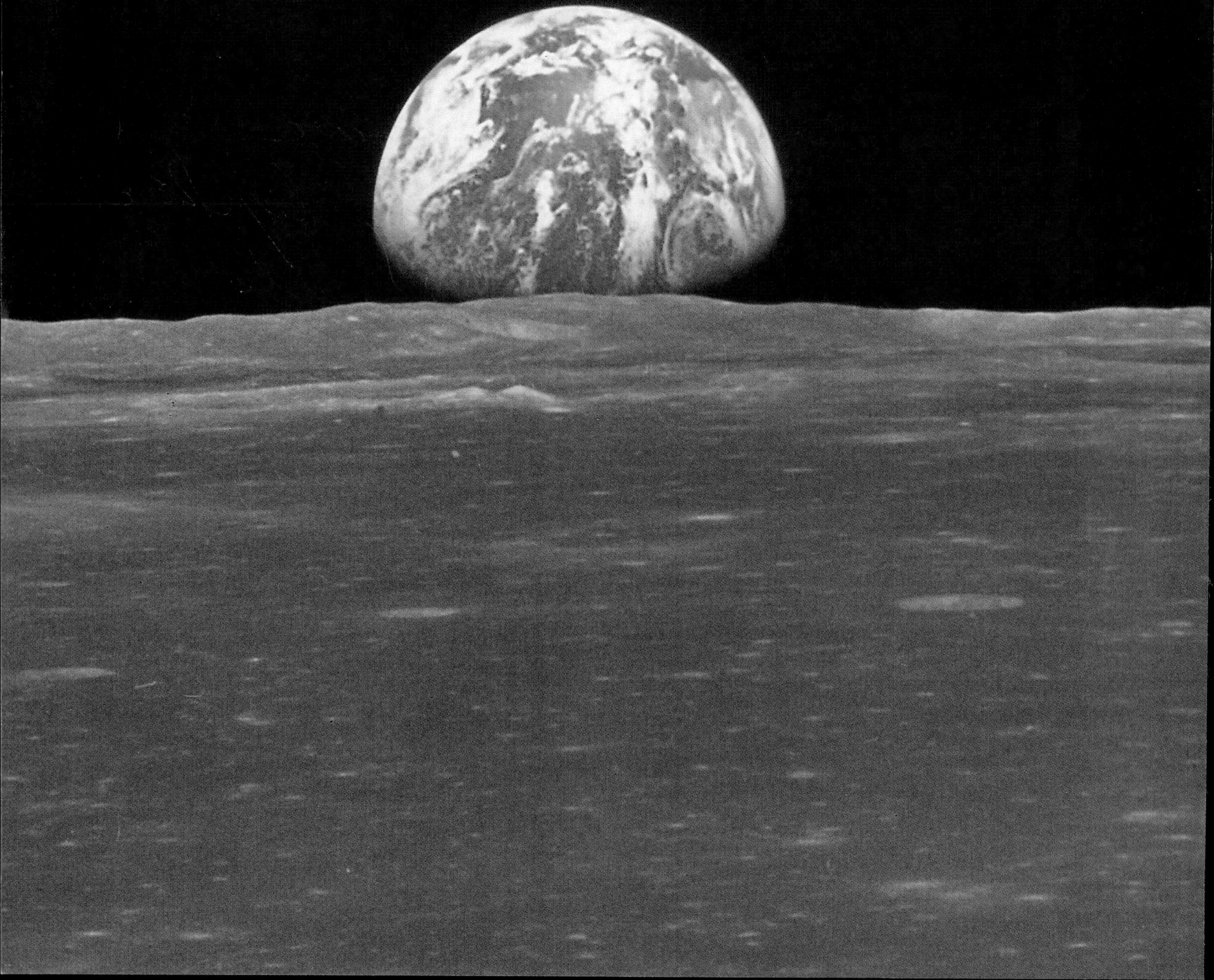

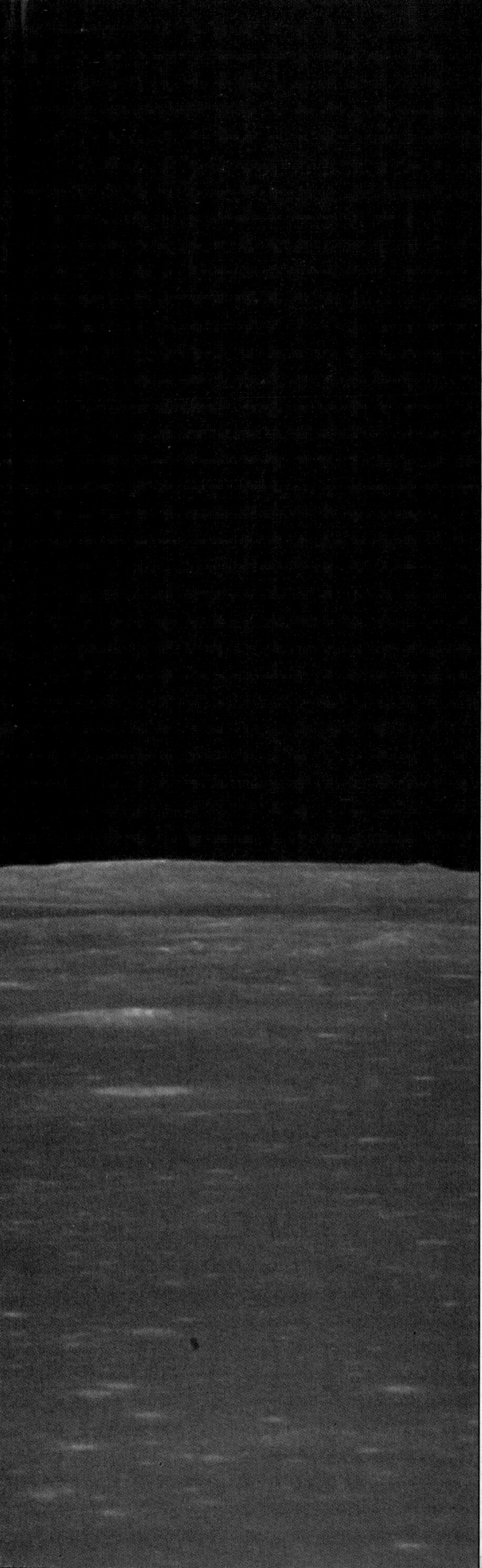

World of Chemistry

SECOND EDITION

Melvin D. Joesten
Harvie Branscomb Distinguished Professor
Vanderbilt University
Nashville, Tennessee

James L. Wood
Adjunct Professor
David Lipscomb University
Nashville, Tennessee

Mary E. Castellion
Contributing Editor
Norwalk, Connecticut

David O. Johnston
David Lipscomb University
Nashville, Tennessee

John T. Netterville
David Lipscomb University
Nashville, Tennessee

Several of the Boxed Features entitled *The World of Chemistry* provided by

Isidore Adler†
University of Maryland

Nava Ben-Zvi
Hebrew University of Jerusalem
and Open University of Israel

Saunders Golden Sunburst Series
SAUNDERS COLLEGE PUBLISHING
Harcourt Brace College Publishers

Fort Worth   Philadelphia   San Diego   New York   Orlando   Austin
San Antonio   Toronto   Montreal   London   Sydney   Tokyo

Text Typeface: New Baskerville
Compositor: Progressive Information Technologies
Vice President/Publisher: John Vondeling
Senior Developmental Editor: Beth Rosato
Project Editor: Maureen Iannuzzi
Copy Editor: Janis Moore
Art Director: Caroline McGowan
Art and Design Coordinator: Kathleen Flanagan
Photo Researcher: Sue C. Howard
Text Designer: Merce Wilczek
Cover Designer: Larry Didona/Caroline McGowan
Text Artwork: George Kelvin/Science Graphics; and Rolin Graphics, Inc.
Layout Artists: Wendy Cummisky and Anne Muldrow
Senior Production Manager: Charlene Squibb
Product Manager: Angus McDonald
Cover Credits: Chemical Earth © George Diebold/The Stock Market, Science Equipment © Tom Tracy/The Stock Market
Frontispiece: NASA

Printed in the United States of America

World of Chemistry, Second Edition
ISBN 0-03-004463-4

Library of Congress Catalog Card Number: 95-071011
5678901234 000 10 987654321

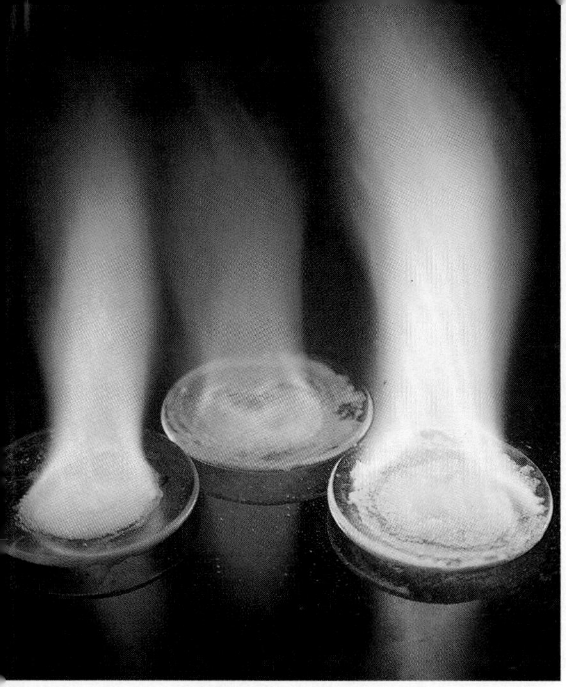

(C.D. Winters)

Contents Overview

(C.D. Winters)

Preface

Approach and Scope

The second edition of *World of Chemistry* continues in the tradition of the first edition in providing a text for nonscience majors that presents chemistry in its broad cultural, social, and economic context. Since the text covers topics that concern students' everyday lives, updating the chapters to include the most recent information on subjects such as energy, new advances in materials, and environmental issues has been an important part of the revision. The comments of reviewers who have used the first edition have been a valuable resource. Mary E. Castellion, who joined the writing team as contributing editor, played a significant role in planning, writing, and editing the second edition. The end result has culminated in a major revision that both preserves and expands the fundamental approach of the first edition—to teach chemical principles within the framework of real-world applications.

To the beginning student there may be a mystery in chemistry, but to leave the workings of the chemist as a mystery suggests that the liberally educated person must be dependent upon the chemist for those chemical decisions that affect society as a whole. *World of Chemistry* is based on the belief that the liberal arts student can see and appreciate the chain of events leading from chemical fact to chemical theory and the ingenious manipulation of materials based on the chemical theories. Thoughtful students will then see that the intellectual struggles in chemistry are closely akin to their own personal intellectual pursuits and will feel that each educated individual should and can have a say in how the applications of chemical knowledge are to affect the human experience.

The topics covered in this book have been selected based on what we and others who have used the first edition have observed to be student interests:

1. Feeling the satisfaction of understanding the cause of natural phenomena.
2. Understanding the scientific bases for making the important personal choices demanded for the use of chemicals and chemical products.
3. Participating on a rational basis in the societal choices that will affect the quality of human life.

4. Helping to preserve and restore the quality of the environment along with a sensible approach to the recycling of natural resources.
5. Developing an insight into the perplexing problem of chemical dependency.
6. Sensing the balance involved in population control, the chemical control of disease, and the ability of the world to produce food.
7. Choosing personal habits in exercise programs and in nutritional selections that are compatible with healthful living.
8. Going places in vehicles that reflect the best uses of the materials used in transportation.
9. Using present energy reserves at a sensible rate, as new energy sources are developed for the long term.

All of these paramount interests, as well as many of lesser note, are featured in *World of Chemistry,* Second Edition.

Organization

World of Chemistry utilizes the common sense approach that is too often lost as the chemical community presents itself to the educated public at large. As in the total human experience, there is in chemical studies a fundamental relationship between cause and effect—structure causes function, chemical periodicity, and consequent material properties. We have selected carefully that thread of chemical history that shows chemistry to be the human endeavor it is. With the aid of ongoing communication with college students, as well as the critical reviews from our peers, we feel this text effectively tells the essential chemical story. The philosophical setting for the text is presented in Chapter 1, whetting the appetite of the students to reach a level of understanding that may have previously been thought to belong only to the scientific elite. Chapters 2 through 15, while laying the groundwork for an intellectual consideration of the effect of chemistry on society, are replete with interesting applications to which the liberal arts students readily relate. The remainder of the text addresses problems that generate intense interest from the general public. These issues include synthetic materials that dramatically alter the human environment; the nutritional basis of healthy living; medicines and drugs; pollution and the conservation of natural resources; consumer chemistry; and the agricultural production of food for a hungry world population.

The book is organized to allow its use for either a one- or two-semester course. The approach followed throughout the book is to link chemical principles within the framework of real-world applications. As a result, selection of several earlier chapters for a one-semester course will still provide the student with a sense of the relevancy of the chemical principles taught in later chapters.

After setting the philosophical tone for the book in **Chapter 1,** basic concepts about atoms, compounds, the periodic table, and nuclear changes are presented in **Chapters 2 through 5.** Chemical bonding, states of matter, chemical reactions, acids and bases, and oxidation and reduction are covered in **Chapters 6 through 10.**

The sources of our natural raw materials, and the energy, and consumer products we derive from them, are the focus of **Chapters 11 through 14.** Inorganic chemicals from the air, land, and sea are discussed in Chapter 11. Organic chemistry and its relevance to energy sources and materials is described in Chapters 12 and 14, with Chapter 13 including a discussion of nuclear energy and other alternate energy sources.

The chemical principles of biochemistry are discussed in **Chapter 15,** and the remaining chapters in the book are topic-centered. **Chapters 16 through 19** address biochemistry-based topics. Consumer chemistry, with an emphasis on personal-care products, is covered in Chapter 16. Health-related topics are then covered in a logical order: nutrition, toxic substances, and medicinal chemistry. Next, in **Chapters 20 and 21,** we look at our environment: water quality and abundance, and air quality. Finally, the text closes with a topic of major importance to us all—feeding the world **(Chapter 22).**

Major Changes in This Revision

Chapter 1 now includes a connection to chemistry-related stories in newspapers and can lead students to discussions of science in the news throughout the course. Chapter 2 introduces the macro-micro aspect of the chemical view of matter, and systematically defines essential terms needed in following chapters. A new chapter, Chapter 7, States of Matter and Solutions, has been added in response to reviewer and user comments about the need for more emphasis on these topics. The properties of water have been moved from the chemical reactions chapter to Chapter 7.

Updated topics include recycling (Chapter 11), energy sources (Chapter 13), consumer products (Chapter 16), global warming and ozone layer depletion (Chapter 21), medicines (Chapter 19), nutrition (Chapter 17), environmental issues (Chapters 18, 20, 21), and feeding the world population (Chapter 22). The number of chapters remains the same as in the first edition, but the order has been changed. Nuclear Reactions (Chapter 5) now comes immediately after the Periodic Table (Chapter 4) rather than after Chemical Reactivity (Chapter 8). The treatment of organic chemistry has been rearranged to relate the chemical principles more directly to the applications: Chapter 12 discusses Energy and Hydrocarbons, and Chapter 14, Organic Chemicals and Polymers, emphasizes industrial and consumer uses of organic chemicals. In keeping with the theme of stressing the application of chemical principles to the students' everyday lives, the chapter on consumer chemistry has been moved from the end of the book to Chapter 16. The topical natures of Chapters 16 through 22 allow for their use individually or in a different sequence. During the complete revision of every chapter, we have, in addition to updating, selectively diminished coverage of some topics.

New to the Second Edition

A major focus during the revision has been to incorporate into each chapter guides to what is important and tools that will assist students in learning.

Chapter opening questions highlight the most important aspects of the chapter topic.

Essential terms are defined upon their first use. They are boldfaced and definitions are given in the new Glossary/Index.

Worked examples within chapters demonstrate how to utilize the simple concepts in this course that require problem solving. Each Example is accompanied by one or more Exercises that are answered in Appendix F, thereby helping students to check their understanding. The Examples and Exercises, together with the Self Tests featured in the first edition, help students gain confidence about smaller segments of material before they try to answer questions and problems at the end of the chapters.

Optional mathematical problem solving sections appear at the ends of certain chapters (unit conversions in Chapter 2, gas law arithmetic in Chapter 7, simple stoichiometry in Chapter 8, pH in Chapter 9, and nutritional calories in Chapter 17). This feature is provided in response to what we have learned from users of the first edition. There is a wide variation in the amount of mathematical problem solving that different instructors choose to incorporate in this course. By placing these sections at the ends of chapters, they are easily skipped by those who wish to do so.

End of chapter questions and problems are all new and have greatly increased in number. They are divided into two groups, to again meet the varying needs of instructors who do or do not include mathematical problem solving: **Questions for Review and Thought** include some questions that review chapter material and others that require thought about the meaning and application of chemical concepts. **Problems** are usually mathematical and appear only in appropriate chapters.

The scope of our emphasis on the relationship of chemistry to everyday life has been broadened with three new types of boxed features:

Frontiers in the World of Chemistry describe new developments related to chapter topics; developments that frequently attract attention in the media.

Science and Society essays focus on science-related societal issues of the type that often raise difficult questions and do not have clearcut answers.

Discovery Experiments are activities that students can carry out on their own. Some are experiments using materials readily available at home or on campus, and others lead the students to examine consumer products, their personal environments, or their lifestyles.

In addition, new **Personal Side** boxes highlight the achievements of recent contributors to the science of chemistry, and **World of Chemistry** boxes feature interviews from *The World of Chemistry Video Series.*

For today's students, visual impact is approaching equal importance with the written word. With the assistance of the team of creative professionals assembled by Saunders College Publishing, the second edition of *World of Chemistry* has a greatly enhanced art and photo program.

Macro-micro visualizations throughout the book, drawn by our outstanding scientific illustrator, George Kelvin, are designed to help students with what is always difficult for them—relating what they can see to the chemist's micro view of materials.

Photos, chosen for their visual appeal, also relate people and everyday circumstances to the chemistry under discussion, in addition to illustrating chemical concepts and the properties of materials.

Connections to the World of Chemistry videotapes have been expanded by introducing marginal quotations from interviews with scientists, new still photos (identified by the program so that the tie-in can be discussed), and some new *World of Chemistry* boxes.

Through the services of the **Harcourt Brace Custom Publishing Group,** portions of *World of Chemistry,* 2e, can be packaged according to individual needs. Instructors who wish to augment *World of Chemistry,* 2e, with their own material or to make selected chapters available in courses with a different focus than that of the textbook as a whole should contact their local Saunders College Publishing sales representative.

Accompanying Materials to *World of Chemistry,* Second Edition

Printed Materials for Instructor and Student

Student Study Guide with Selected Solutions, by Walt Volland of Bellevue Community College, Washington, provides section-by-section list of main topics, listing of objectives, key terms, detailed solutions to all the end-of-chapter odd-numbered Questions for Review and Thought as well as selected Problems, suggested readings on related topics, and ''Bridging the Gap''—a set of intriguing activities designed to connect chapter topics to everyday life.

Instructor's Resource Manual with *The World of Chemistry Video Series Correlation Material* includes three sections:

1. *Notes to the Instructor* includes detailed solutions to the end-of-chapter Problems that do not appear in the text and Student Study Guide with Selected Solutions.
2. *Printed Test Bank* of multiple-choice questions and problems.
3. *Correlation Guide I,* by Cheryl Dembe of Diablo Valley College, California, provides a detailed synopsis of each of the 26 videos from *The World of Chemistry Video Series* with a special section referencing corresponding textbook material.

ExaMaster™ Computerized Test Bank is the software version of the printed test bank. Instructors can create thousands of questions in a multiple-choice format. A command reformats the multiple-choice question into a short-answer question. Adding or modifying existing problems, as well as incorporating graphics, can be done. ExaMaster has gradebook capabilities for recording and graphing students' grades.

Overhead Transparencies set includes 100 full-color acetates with labels enlarged for easy viewing.

Laboratory Manual to accompany World of Chemistry, 2e by John Blackburn, John Craig, Paul Langford, and Melvin Joesten includes 45 experiments that include pre- and post-lab questions, safety considerations, as well as a new, introductory discussion of risk versus benefit.

Instructor's Manual to accompany the Lab Manual provides listing of equipment and reagents needed for each experiment, as well as suggested demonstrations and safety considerations for pre-lab instruction.

Multimedia Materials

The World of Chemistry Video Package. The World of Chemistry, Second Edition, is presented as either a stand-alone course in chemistry for nonscience majors or as an integral part of a comprehensive telecourse package including a series of 26 thirty-minute video programs, with an accompanying telecourse study guide, telecourse faculty manual, and a telecourse laboratory manual. Sponsored by the Annenberg/CPB Project and corporate sponsors, *The World of Chemistry* video series was developed by the late Dr. Isidore Adler of the University of Maryland and Dr. Nava Ben-Zvi of Hebrew University of Jerusalem. The video programs feature Nobel laureate and Priestley medalist Roald Hoffmann and provide a comprehensive survey of the field of chemistry and its impact on modern society. The series was jointly produced by the University of Maryland and The Educational Film Center. For information on ordering the videocassettes, call 1-800-LEARNER.

The World of Chemistry: Selected Demonstrations and Animations I and II are two videodiscs produced by JCE:SOFTWARE. For information on these videodiscs contact JCE:SOFTWARE, John W. Moore and Jon L. Holmes, Department of Chemistry, University of Wisconsin-Madison, 1101 Madison Avenue, Madison, WI 53706.

The Saunders Interactive General Chemistry CD-ROM (Designed exclusively for Saunders College Publishing by Archipelago Productions and authored by Jack Kotz of SUNY-Oneonta and Bill Vining of Hartwick College).

Based on our best-selling general chemistry text *Chemistry & Chemical Reactivity,* 3e, and including the text in its entirety, the CD-ROM serves as a useful multimedia companion to *World of Chemistry,* 2e, by providing imaginative and innovative approaches to learning chemistry.

With the CD-ROM, students navigate through original animation and graphics, interactive tools, and pop-up definitions, as well as over 100 video clips and chemical experiments enhanced by sound effects and narration. A student workbook accompanies the CD-ROM, and important technical information and suggestions for accessing some of the CD-ROM's unique teaching features are available to professors.

CalTech Chemistry Animation Project (CAP) is a set of five video units covering the following chemical topics with unmatched quality and clarity: Atomic Orbitals, Valence Shell Electron Pair Repulsion Theory, Crystals and Unit Cells, Molecular Orbitals in Diatomic Molecules, and Periodic Trends.

With an emphasis on general chemistry, the *Chemistry in Perspective Videodisc* includes over 110 minutes of motion footage, including molecular model animations, chemical reaction videos, animated principles of chemistry, and excerpts from *The World of Chemistry* video series that correlate with the *The*

World of Chemistry boxed essays found in the textbook, as well as 2000 still images from Saunders College Publishing 1996 chemistry textbooks.

With the emphasis on organic and biological chemistry, the *Chemistry of Life Videodisc* contains over 100 molecular model animations, chemical reaction videos, and approximately 2500 still images from a variety of Saunders College Publishing chemistry textbooks.

LectureActive™ Software and *Barcode Manuals* for our videodiscs enable instructors to customize their lectures with our chemistry videodiscs. Available for both IBM and Macintosh.

Reviewers

We are deeply grateful to all the reviewers who have contributed to the improvement of the manuscript and teaching aids for the second edition. We would especially like to thank Walt Volland, Bellevue Community College, Washington, who provided a thorough review of the edited manuscript, contributed chapter questions, and provided solutions to the chapter questions and problems. We also appreciate the thoughtful comments of all our other reviewers:

Steve Albrecht *Oregon State University*
Gretchen Anderson *Indiana University at South Bend*
Frank Brown *Tallahassee Community College*
William Church *East Carolina University*
Jack Cummins *Metropolitan State College*
Karen Eichstadt *Ohio University*
Seth Elsheimer *University of Central Florida*
Joel Goldberg *University of Vermont*
Marie Hankins *University of Southern Indiana*
Lenore Koczon *Northern State University*
Robert Ludt *Virginia Military Institute*
William McMahan *Mississippi State University*
Robert Metzger *San Diego State University*
Marion Rhodes *University of Massachusetts*
Rand Rodewald *Minot State*
Jimmy Rogers *Eastfield College*
Lawrence Snyder *SUNY-Albany*
Louis Trefonas *University of Central Florida*
Jerry Wilson *California State University, Sacramento*
Sid Young *University of South Alabama*

Reviewers of the first edition included:

Robert C. Belloli *California State University at Fullerton*
Rudolph S. Bottei *University of Notre Dame*
Gilbert Castellan *University of Maryland*
Jerry A. Driscoll *University of Utah*
John J. Fortman *Wright State University*
J. Leland Hollenberg *University of Redlands*

Keith Kennedy *St. Cloud State University*
Robert E. Miller *Keene State College*
J. W. Robinson *Louisiana State University*
Tamar Y. Susskind *Oakland Community College*
Gary D. White *Cleveland State University*
Sheldon S. York *University of Denver*

Acknowledgments

We appreciate the help of the entire staff at Saunders College Publishing; they know how to get the job done! Beth Rosato, Developmental Editor, facilitated the flow of information and ideas, prompted the necessary decisions, pushed for meeting deadlines, and worked for excellence in every facet of editorial control. We enjoyed working again with Maureen Iannuzzi, Project Editor, who was patient and helpful throughout the process of turning our manuscript into a book. Her equanimity in dealing with authors who care not only about the words on the page but also about how the page looks was greatly appreciated. Caroline McGowan, Art Director for this edition, managed and orchestrated with calm and humor a major revision to the art program. Charlene Squibb, Senior Production Manager, kept everyone's schedule in time with the publication deadline. We give our special thanks to John Vondeling, Publisher, for his continued support and confidence in us.

Many of the drawings in this book were done by George Kelvin, an outstanding science illustrator. His drawings not only illustrate the principles of chemistry, but also are truly works of art. We also acknowledge Sue Howard, Photo Editor, for her efforts in locating so many beautiful photographs to illustrate the applications of chemical principles and Charles Winters for his creative and excellent photographs.

We initiated the use of *Chemistry You Can Do* experiments that students can do on their own in *The Chemical World: Concepts and Applications* by J. C. Kotz, M. D. Joesten, J. L. Wood, and J. W. Moore, 1994, Saunders College Publishing. We wish to thank John Moore, University of Wisconsin—Madison, for contributing many of the *Chemistry You Can Do* experiments and class testing each of them prior to their publication. Since the *Chemistry You Can Do* experiments have been well received by both students and faculty, we decided to add experiments for students to the Second Edition of *World of Chemistry*. Many of these experiments, called *Discovery Experiments,* previously appeared as *Chemistry You Can Do* experiments and were adapted from activities published by the Institute for Chemical Education as *Fun with Chemistry, Volumes I and II.* These were originally collected for ICE by Mickey and Jerry Sarquis of Miami University, Ohio.

Given the broad scope of topics in our book, we sometimes must rely on the assistance of experts. For this edition we would like to thank Dr. Keith Brown, Director of Hair Color Research, Clairol, Inc., who reviewed sections on hair chemistry. Also, we wish to acknowledge the support we received from the efforts of the authors of two textbooks that we view as outstanding: Lori A. Smolin and Mary B. Grosvenor, *Nutrition: Science and Applications,* Saunders College Publishing, 1994; and Peter H. Raven, Linda R. Berg, and George B. Johnson, *Environment,* Saunders College Publishing, 1993.

We also appreciate the help given us by one of our original writing partners, David O. Johnston, who read early drafts of all chapters in the second edition. While words and ideas were in a state of flux, Dave provided thoughtful comments and suggestions. Thanks, Dave.

Although much help has come our way, the responsibility for the contents of the text rests entirely on us.

As in all of our previous works, we dedicate this effort to our families and gratefully acknowledge their support and understanding during the preparation of this manuscript.

MDJ
JLW
Nashville, Tennessee

NOVEMBER 1995

(Tom Hollyman/Photo Researchers)

Contents

(C.D. Winters)

(C.D. Winters)

(C.D. Winters)

CHAPTER 11 Chemicals from the Air, Sea, and Land 311

CHAPTER 12 Energy and Hydrocarbons 340

CHAPTER 13 Alternate Energy Sources 377

(C.D. Winters)

(C.D. Winters)

(C.D. Winters)

(C.D. Winters)

CHAPTER 22 Feeding the World *705*

Index to Boxed Features

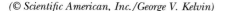

(© *Scientific American, Inc./George V. Kelvin*)

CHAPTER 1

Living in a World of Chemistry

1.1 The World of Chemistry

Here you are about to take a chemistry course. What do you already know about chemistry? Maybe you've seen some chemistry demonstrations that produced explosions or dramatic color changes. Many chemists practicing today were first attracted into the profession by just such demonstrations. They wanted to do chemistry themselves—create dramatic changes in materials and understand why they happen.

■ The real meaning of "chemical" is discussed in Chapter 2.

Maybe "chemistry" means "chemicals" to you. And perhaps you think this word should be used only with the adjective "toxic." That belief wouldn't be surprising, since you have probably heard of "toxic chemical spills" or warnings about "toxics" in the environment. Indeed, some chemicals are toxic—very toxic: the arsenic of mystery stories, the poisonous gases of World War I, the chemicals released by the microorganism that grows in badly canned food and causes severe food poisoning.

Learning about chemistry doesn't, however, mean learning only about harmful substances. Nor does it mean learning only about the wonders of our modern world provided by chemistry, such as the medications that cure once-incurable diseases; the synthetic fabrics that are beautiful, durable, and inexpensive; or the colors on the television screen.

Instead, we suggest you come to the subject with a "What's in it for me?" attitude, not from a selfish viewpoint, but as a citizen of the world of chemistry. The first photographs of planet Earth from the moon provided a forceful reminder to many of us that this planet and the materials on it are finite. The world of chemistry really has but one concern: the materials provided by our planet and what we do with them.

In a global view, some understanding of chemistry will be helpful in dealing with major social issues that lie ahead in the 21st century. The world population is expanding at an ever-increasing rate. The roughly 5.7 billion of the mid-1990s is expected to double in the next 50 years. How can everyone be fed, clothed, and housed? How can we prevent spoiling what planet Earth has provided us? How can we reverse some of the damage done when we knew less

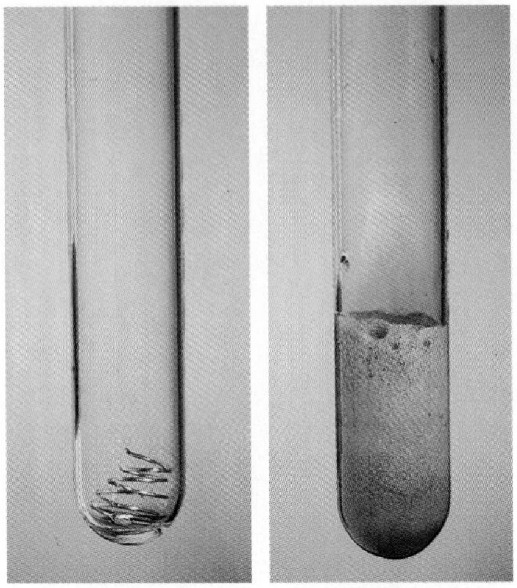

The amazing effect of nitric acid (b) on copper (a). As a young man, Ira Remsen (1846–1927) dropped a copper penny into nitric acid to see what would happen. Remsen, who became an outstanding chemist and teacher, had this reaction: "It resulted in a desire on my part to learn more about that remarkable kind of action. Plainly the only way to learn about it was to see its results, to experiment, to work in a laboratory." *(Larry Cameron)*

■ Population trends and the problem of feeding the world population are discussed in Section 22.1.

about the consequences of our activities? How can we prevent doing new damage to our environment? None of these questions can be fully addressed without some serious applied chemistry.

Knowing something about chemistry adds a new dimension to everyday life, too. If an advertisement proclaims that a product "contains no chemicals," what does that tell you about the producers of that product? Is it worth paying twice the price for something that proclaims itself to be "all natural"? (Sometimes it is and sometimes it isn't.) Do you know that you shouldn't mix household ammonia and chlorine bleach because they react to form a very toxic gas?

Lying ahead in the chapters of this book is a look into the chemical view of the world. It is based on observations and facts. A *scientific fact* results from repeated observations that produce the same result every time. (Water boils at 100°C at sea level—that's a scientific fact.) The chemical world is also based on models, theories, and experiments. Scientists use models and theories to organize knowledge and make predictions. The predictions must then be tested by experiments. If experimental results disagree with the predictions, the reason must be explored. Possibly the theory is wrong.

You will also see that the chemical view of the world requires learning to look at materials in two ways. The first is what direct observation provides; the second is the mental image of what scientists have learned about the submicroscopic world that cannot be observed directly.

Curiosity, skepticism, and attention to small details play an essential role in science. Even though you may not become a professional scientist, it is useful and even fun to bring a scientific approach into everyday life now and then. The author Robert Pirsig has summed it up beautifully in a passage that is about much more than repairing a motorcycle:

> *When you've hit a really tough one, tried everything, racked your brain and nothing works, and you know that this time Nature has really decided to be difficult, you say, "Okay Nature, that's the end of the nice guy," and you crank up the formal scientific method.*
>
> *For this you keep a lab notebook. Everything gets written down, formally, so that you know at all times where you are, where you've been, where you're going and where you want to get. In scientific work and electronics technology, this is necessary because otherwise the problems get so complex you get lost in them and confused and forget what you know and what you don't know and have to give up. In cycle maintenance things are not that involved, but when confusion starts it's a good idea to hold it down by making everything formal and exact. Sometimes just the act of writing down problems straightens out your head as to what they really are. . . .*
>
> *The real purpose of scientific method is to make sure Nature hasn't misled you into thinking you know something you don't actually know. There's not a mechanic or scientist or technician alive who hasn't suffered from that one so much that he's not instinctively on guard. That's the main reason why so much scientific and mechanical information sounds so dull and so cautious. If you get careless or go romanticizing scientific information, giving it a flourish here and there, Nature will soon make a complete fool out of you. It does it often enough anyway even when you don't give it opportunities. One must be extremely careful and rigidly logical when dealing with Nature: one logical slip and an entire scientific edifice comes tumbling down. One false deduction about the machine and you can get hung up indefinitely.*
>
> Robert Pirsig: *Zen and the Art of Motorcycle Maintenance,* p. 100. New York, William Morrow & Co., 1975.

To get on with this introduction to the world of chemistry, we want to offer a glimpse of chemistry-related topics that influence everyday events enough to be newsworthy. Cranking up the scientific method to do this would require gathering data and drawing conclusions from the data. It could be done by surveying a national cross section of television news programs, news magazines, and newspapers. The media should be monitored for several weeks or several months, maybe even a year, and the topics recorded. Ultimately, analyzing enough data would make it possible to list the most commonly covered topics during that period. The list might not represent what was of greatest importance to chemists, but it would certainly show what was deemed of greatest interest to the reading and TV news-viewing public. Almost all of us are part of that group.

Without the means to do the type of data collection and analysis we've just described, we have settled instead on examining three newspapers on randomly chosen Sundays—one a major newspaper from one of the largest cities in the country, the others average newspapers from cities of about 100,000 people. We have chosen a few stories from these papers that illustrate chemistry-related topics of general interest and also open the door to some further insight into the science of chemistry.

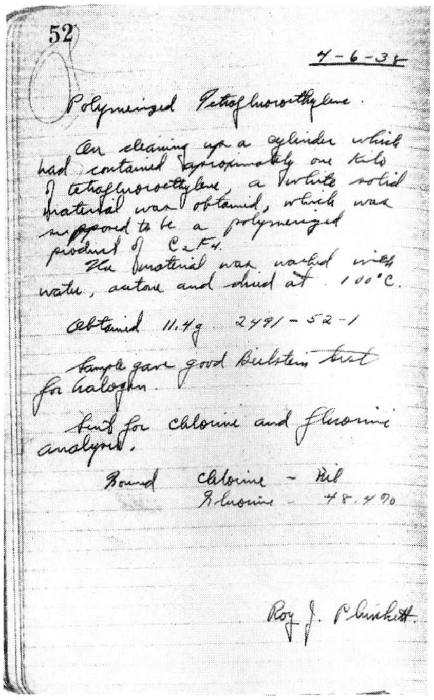
52
7-6-38
Polymerized Tetrafluoroethylene.
On cleaning up a cylinder which had contained approximately one kilo of tetrafluoroethylene a white solid material was obtained, which was supposed to be a polymerized product of C_2F_4.
The material was washed with water, acetone and dried at 100°C.
Obtained 11.4g 2491-52-1
Sample gave good Beilstein test for halogen.
Sent for chlorine and fluorine analysis.
Found Chlorine - Nil
Fluorine - 48.4%
Roy J. Plunkett

Laboratory notebook entry for the discovery of Teflon (described later in this chapter). Recording exactly what is done in a laboratory and what is observed is essential. *(Chemical Heritage Foundation)*

1.2 DNA Fingerprinting, Biochemistry, and Science

First, consider a topic important enough to make the front page of a Sunday newspaper—the beginning of a long feature article about the use of DNA tests in court. DNA, chemical name *deoxyribonucleic acid,* is present in cells throughout our bodies. The reporter has risen to the challenge of explaining the quite-complicated chemistry of how DNA analysis provides the equivalent of a

Science Helps Finger the Suspect
—headline on August 6, 1994

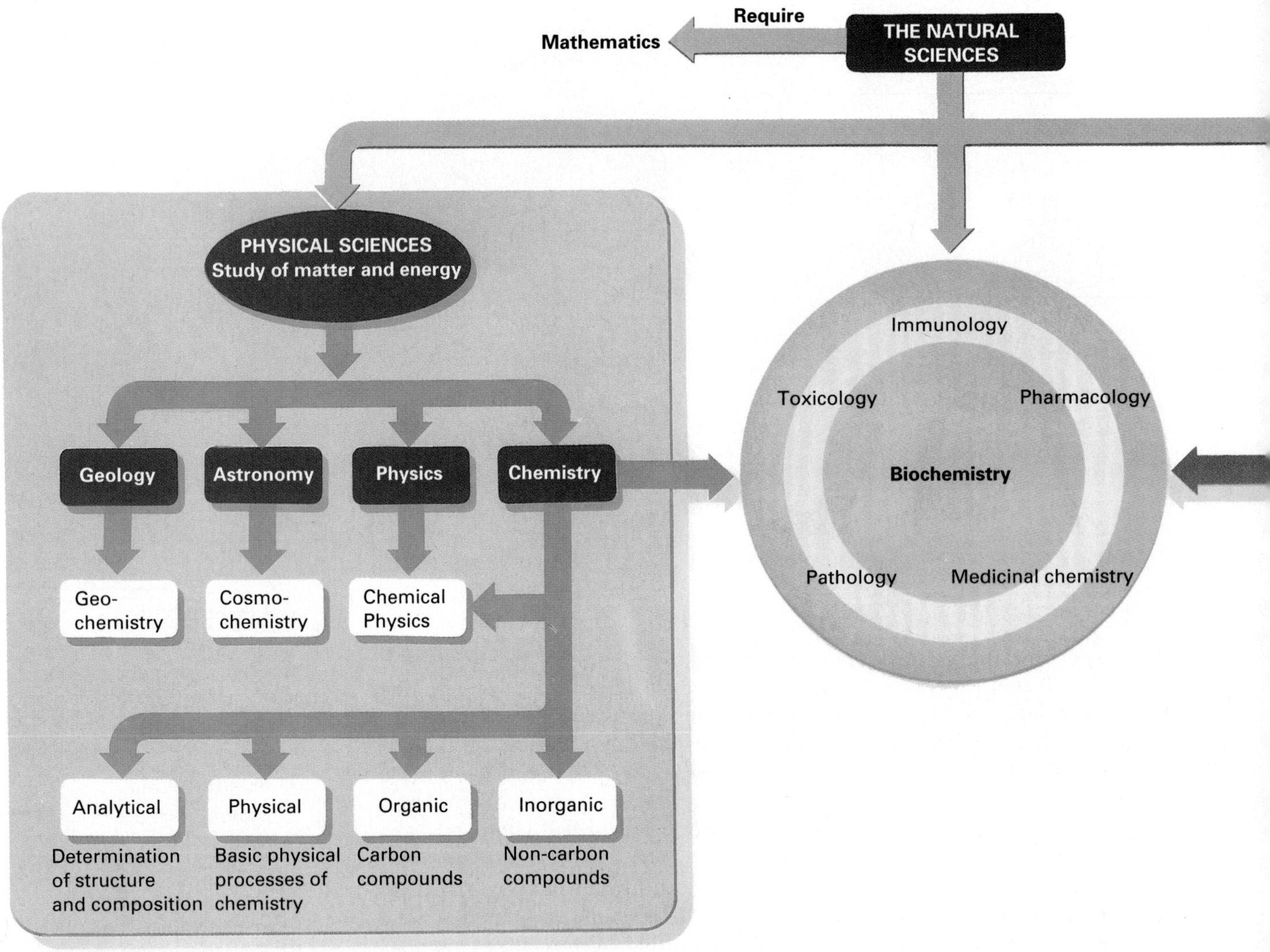

■ The scientific complexities of DNA testing clearly made it to the front page because of the expected use of such testing in the murder trial of a well-known public figure.

■ Chapter 15 is devoted to biochemistry.

fingerprint. Comparing the DNA analysis of a sample of blood or other biological material left behind at the scene of a crime with that of a sample from a suspect can weigh heavily as evidence for guilt or innocence.

DNA analysis as fingerprinting is relatively new. It is still controversial in terms of admissibility in some courts, and legal defense teams have been creative in their efforts to raise doubts about it. (Similar efforts were made to discredit printing of fingers when that was new.) The DNA test is based on sound and fully accepted science, however, and it is being used in thousands of court cases per year. Already, re-examination of evidence predating the test has freed several individuals judged to have been wrongly imprisoned. And the FBI and Scotland Yard report that now one third of those accused of rape are cleared of suspicion by DNA tests before the cases go to court.

The study of DNA is one aspect of *biochemistry,* the natural science that unites chemistry, a physical science, with the biological sciences. **Chemistry** is

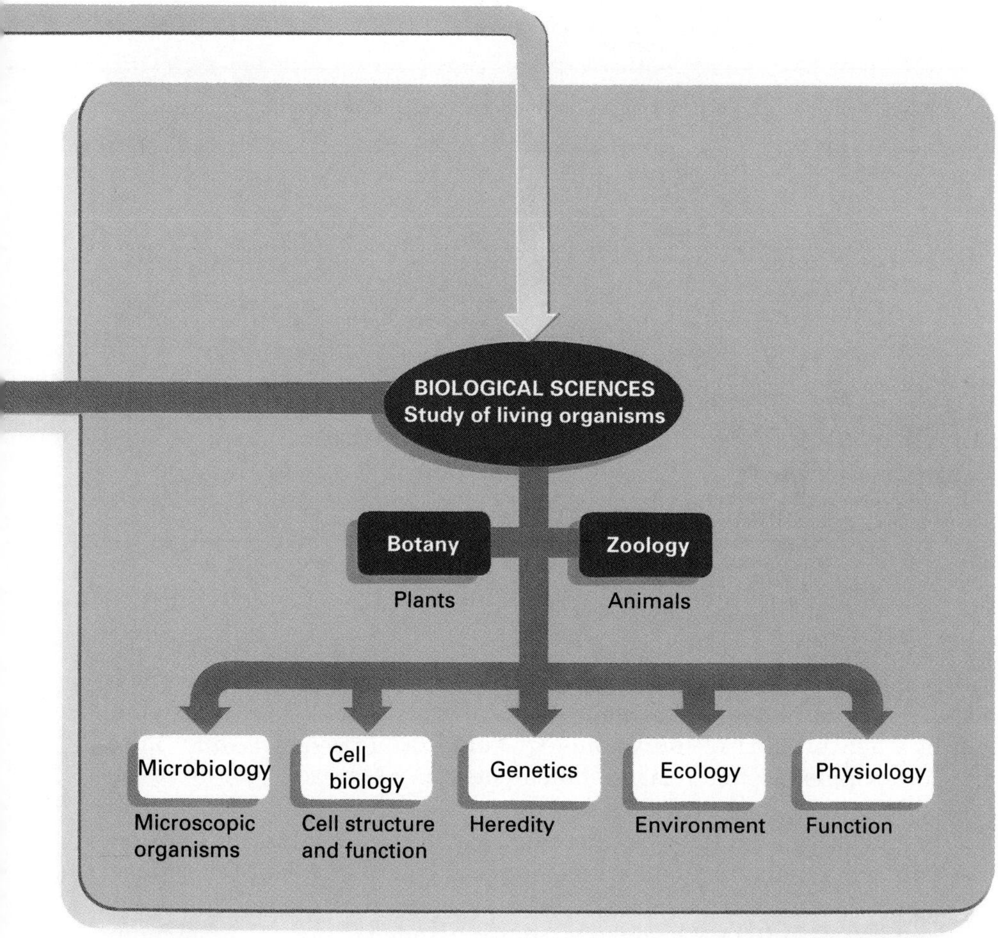

Figure 1.1 The natural sciences, with some of their subdivisions. The number of subdivisions continues to increase as the boundaries between the sciences disappear.

often defined as the study of matter and the changes it can undergo. A moment's thought should convince you that this makes an understanding of chemistry fundamental to an understanding of everything in nature. And, of course, with an understanding of natural matter, chemists are able to modify matter and synthesize new kinds of matter.

Historically, the natural sciences have been associated with observations of nature—our physical and biological environment. A traditional classification of the natural sciences is represented in Figure 1.1, with an emphasis on the relationship of chemistry to other sciences. As each science grows more sophisticated, the boundaries become more blurred, however. The dynamic character of science is illustrated by the emergence of new disciplines. There are now individuals who refer to themselves as biophysicists, bioinorganic chemists, geochemists, chemical physicists, and even molecular paleontologists (they use modern chemical methods to analyze ancient artifacts.)

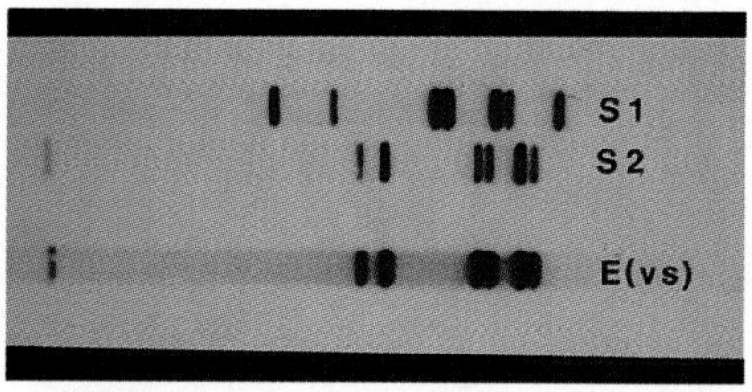

A DNA fingerprint test. The test result shows up as a series of lines that vary with the makeup of an individual's DNA. Comparison shows that the DNA of suspect 2 (S2) matches the evidence blood sample (E(vs)). *(Leonard Lessin/ Peter Arnold, Inc.)*

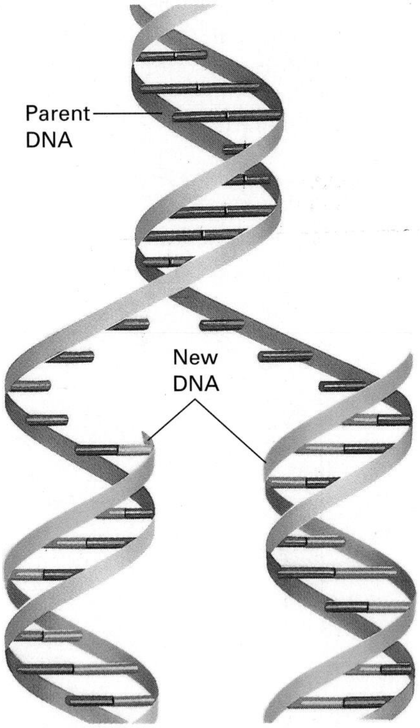

An abstract picture of how DNA unravels and is reproduced. In this way, a parent's DNA is passed on to the offspring. (More information about how this happens is given in Section 15.8.)

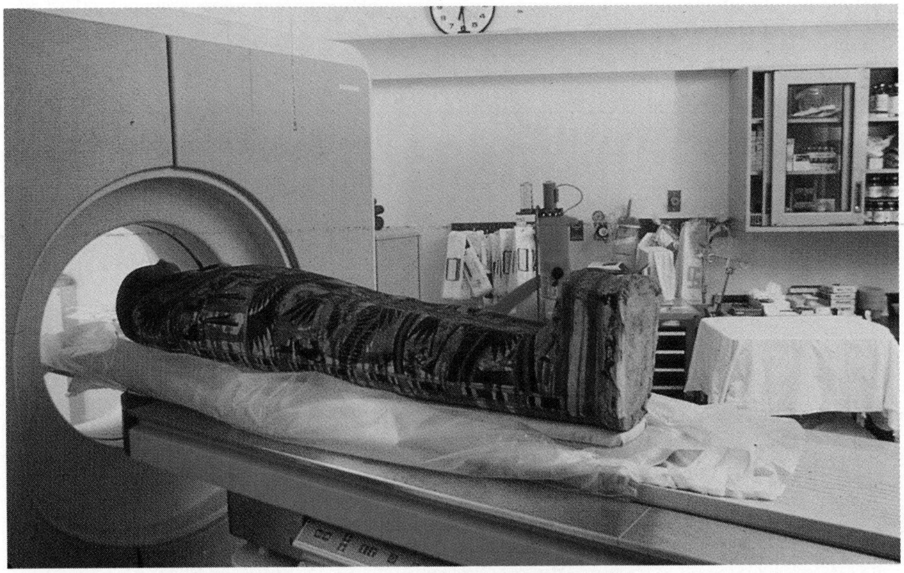

An Egyptian mummy having a CAT scan. Now the archeologists can study the occupant without opening the sarcophagus. *(Alexander Tsiaros/Science Source/Photo Researchers)*

■ Identification of DNA as the basic genetic material in chromosomes was done in 1944 by O. T. Avery, C. M. MacLeod, and M. McCarty. Nine years later James D. Watson and Francis Crick solved the difficult problem of how DNA is constructed (Section 15.8).

The study of DNA is now one of the frontier areas of science. It seems that hardly a week passes without the report of some new development, which might be the unraveling of the cause of an inherited disease or the introduction of a new product of the technology known as *genetic engineering*. Since these developments often relate to human health or behavior, they make the newspapers.

About 50 years ago DNA was identified as the carrier of genetic information from one generation to the next, determining how we will be the same or different from our ancestors. At first, there was a period of intense study of the chemical nature of DNA. The focus has now expanded to how DNA functions in living cells and how it might be manipulated to obtain desired results, such as the cure of inherited diseases or the manufacture of life-saving drugs by genetically engineered bacteria. DNA testing—determination of the exact composition of an individual's DNA—is an analytical tool. Those who use it along with many other kinds of *analytical* chemistry in the investigation of crimes are *forensic* chemists.

Having a Cool Car Is Becoming More Costly
—headline on Sunday, May 15, 1994

1.3 Air Conditioning, the Ozone Hole, and Technology

Next, we consider a story about something that directly affects more of us than DNA testing—the need to convert automobile air-conditioning systems to the use of a new refrigerant. The background of this need is rooted in some interesting history of chemistry. Refrigeration as we know it has been in use since the late 1800s. The process requires a fluid that absorbs heat as it evaporates, releases heat when it condenses, and can be continuously cycled through evaporation and condensation without breaking down.

In the early 1900s, the fluids used as refrigerants were mostly flammable or toxic. One day, Charles Kettering, then director of research at General Motors, passed along a challenge to Thomas Midgley, a young engineer: "The refrigeration industry needs a new refrigerant if they ever expect to get anywhere." Midgley and a colleague vigorously dug into the problem by first hunting for candidates in the extensive tables of data kept at hand in every chemistry laboratory. They combined this information with a survey of systematic variations in the properties of the chemical elements. The result, in excerpts from Midgley's own account, was incredibly successful:

> *Seemingly no one previously had considered it possible that fluorine might be nontoxic in some of its compounds. This possibility had certainly been disregarded by the refrigeration engineers. If the problem were to be solved by the use of a single compound [rather than a mixture] then that compound would certainly contain fluorine. . . .*

They settled on beginning to experiment with a new chemical that they would have to make in the laboratory. One of the necessary ingredients was a fluorine chemical that was scarce, so they quickly purchased the available supply in five small bottles. One of the bottles was used to make the first sample of the new chemical.

> *A guinea pig was placed under a bell jar with it and, much to the surprise of the physician in charge, didn't suddenly gasp and die. In fact, it wasn't even irritated. Our predictions were fulfilled.*

A second sample made from the starting material in a different bottle did, however, kill the guinea pig. Here was a problem to be solved. Some investigation showed that only the first bottle contained pure material. It was a contaminant that had killed the guinea pig, not Midgley's newly prepared chemical.

> *Of five bottles . . . one had really contained good material. We had chosen that one by accident for our first trial. Had we chosen any one of the other four, the animal would have died as expected by everyone else in the world except ourselves. I believe we would have given up what would then have seemed like a "bum hunch."*
>
> *And the moral of this last little story is simply this: You must be lucky as well as have good associates and assistants to succeed in this world of applied chemistry.*

The outcome of this experiment was the widespread presence of refrigerators in our homes. Food spoilage diminished, and after World War II the food processing industry expanded to the production of frozen foods.

So far, this is a success story—the development of chlorofluorocarbons, often referred to as CFCs, as much less hazardous refrigerants. Through the 1950s and 1960s, their use expanded as all kinds of new consumer products were brought to market. The properties of CFCs made them ideal for propellants in aerosol cans and for blowing the tiny holes into materials such as the polyurethane foam used in pillows and furniture cushions. The problem being created did not surface until the 1970s because no one was aware of the fate of these very stable compounds as they were released into the environment.

■ To a chemist, a stable substance is one that is not easily changed into something else. Water is a stable substance.

By the 1980s it had become clear that CFCs drift unchanged into the stratosphere, where they interact with ozone and destroy it. The result is the "ozone hole" and the potential for damage from the resulting increase in solar radiation that reaches the Earth's surface. Herein lies the reason the cost of automobile air-conditioner repair is newsworthy. In 1987 nations that

■ The chemistry of the ozone hole and other aspects of the atmosphere are discussed in Chapter 21.

produce CFCs drafted a plan of action that required industrial production of CFCs to cease by 1995. Our news story explains that the old CFC refrigerant, R-12, and the new one, R-134a, are not compatible. A car designed to use R-12, which includes almost all those made before 1992, cannot use R-134a unless the system is modified, which can be costly. Economic decisions lie ahead for anyone owning an older car with a leaky air-conditioning system.

It is helpful in reading about refrigerants and similar topics to recognize the distinctions among basic science, applied science, and technology. **Basic science,** or basic research, is the pursuit of knowledge about the universe with no short-term practical objectives for application in mind. The biochemists who struggled for years to understand exactly how DNA functions within cells were doing basic science.

The World of Chemistry

Program 1, *The World of Chemistry*

There's still so much to learn. My goodness, the sort of things it takes me years to do, nature does in a matter of seconds. The secrets there still have to be unraveled. Dr. Bertram Fraser-Reid, Duke University

Applied science has the well-defined, short-term goal of solving a specific problem. The search for a better refrigerant by Midgley and his colleagues is an excellent example of applied science: they had a clear-cut, practical goal. To reach the goal, as is done in either basic or applied science, they utilized the recorded observations of earlier studies to make a prediction and then performed experiments to test their prediction.

Technology, also an application of scientific knowledge, is a bit more difficult to define. In essence, it is the sum of the way we apply science in the context of our society, our economic system, and our industry. The first refrigerators and auto air-conditioners designed to use CFCs were the products of a new technology. The rapidly expanding number of ways to manipulate DNA

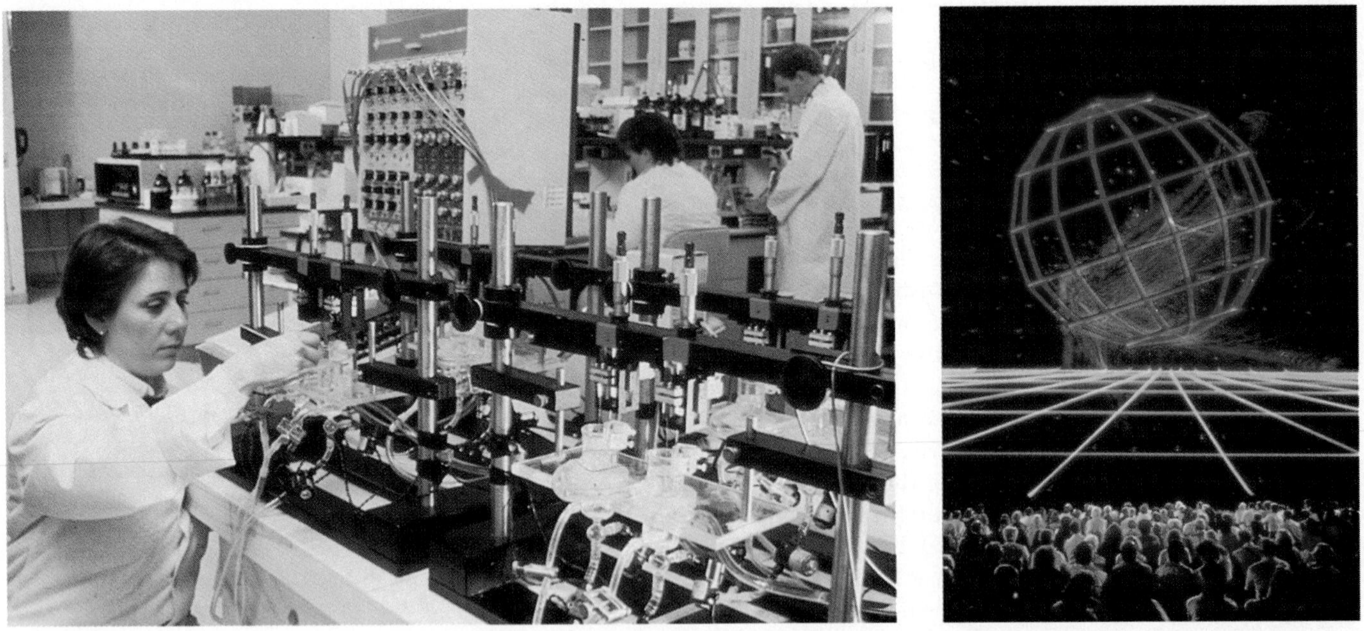

Science and Technology (a) A basic research laboratory in a pharmaceutical company. (b) People enjoying a laser light show set to rock music. Technology has brought the laser out of the laboratory and into an IMAX theater. *(a, Hank Morgan/ Rainbow; b, Courtesy of Audio Visual Imagineering Inc., Orlando, FL)*

to make new medicines or other marketable products is referred to as *biotechnology*.

Regardless of the type of scientific discovery, there is a delay between the discovery and its technological application. The incubation intervals for a number of practical applications of various types are given in Table 1.1.

The important point is that technology, like science, is a human activity. Decisions about the uses of technology and priorities for technological developments are made by men and women. How scientific knowledge is used to promote technology depends on those persons who have the authority to make the decisions. Sometimes those persons are all of us—when we go to the polls in a democratic society, we can influence decisions about technology. When we have this chance, it is important to be well enough informed to critically evaluate the societal issues related to the technology.

As the history of CFCs illustrates, science and technology, like social conditions, are constantly evolving. When CFCs were introduced as refrigerants in 1930 they were a great advance for the economy and replaced hazardous materials. In addition, the number of technologically advanced nations and the world population were smaller. People were in the habit of assuming that natural processes would keep the environment healthy, and to a greater extent than today, that was true. Moreover, some of the sophisticated instruments that have since revealed ozone depletion were not available then. Only by staying informed can we be ready to adjust to changing times.

The World of Chemistry

Program 2, *Color*

Now, the way to find new dyes was to get a large number of guys who were competently trained chemists, get them a laboratory, get them the major reagents, set them down and say, Start making things. . . . To make these dyes they had to build up tens of thousands of intermediate compounds. And what you have then is an enormous stable of things that you can use in experiments. And so, as the base broadens, the combinations just become infinite. John K. Smith, science historian, Lehigh University, commenting on the beginnings of the chemical industry in Germany in the late 1800s.

TABLE 1.1 ■ Time Needed to Develop Technology for Some Fruitful Ideas

Innovation	Conception	Realization	Incubation Interval (Years)
Antibiotics	1910	1940	30
Cellophane	1900	1912	12
Cisplatin (anticancer drug)	1964	1972	8
Heart pacemaker	1928	1960	32
Hybrid corn	1908	1933	25
Instant camera	1945	1947	2
Instant coffee	1934	1956	22
Nuclear energy	1919	1945	26
Nylon	1927	1939	12
Photography	1782	1838	56
Radar	1907	1939	32
Recombinant DNA drug synthesis	1972	1982	10
Roll-on deodorant	1948	1955	7
Self-winding wristwatch	1923	1939	16
Videotape recorder	1950	1956	6
Photocopying	1935	1950	15
X-rays in medicine	Dec. 1895	Jan. 1896	0.08
Zipper	1883	1913	30

Roy J. Plunkett, Who Made Accidental Discovery of Teflon

—obituary headline on Sunday, May 15, 1994

■ The use of the term "serendipity" for accidental discoveries was first proposed in 1754 by Horace Walpole after he read a fairy tale titled "The Three Princes of Serendip." Serendip was the ancient name of Ceylon, and the princes, according to Walpole, "were always making discoveries by accident, of things they were not in quest of."

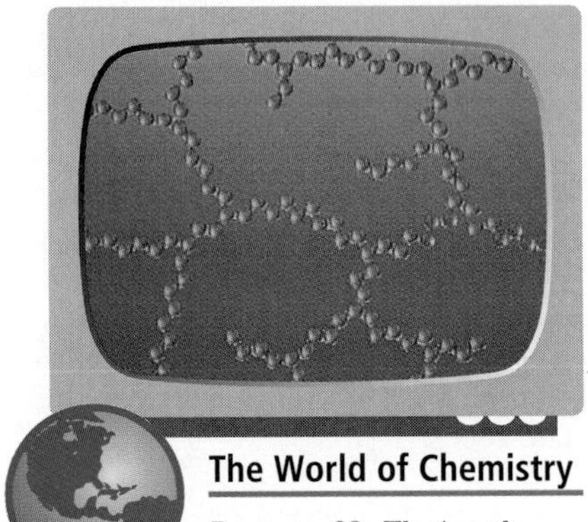

The World of Chemistry

Program 22, *The Age of Polymers*

When a polymer forms, small units are joined together like beads in a necklace.

1.4 Teflon, Scientific Discovery, and Serendipity

By coincidence, the same newspaper that carried the "cool car" story described in Section 1.3 carried a different story also related to the search for a new refrigerant. Sometimes useful discoveries are the result of what we might call accidents or serendipity. Of course, these accidents had to have happened to the right kind of individuals. When an unexpected experimental result turns up, these people look upon it as an opportunity. They want to know more about what happened and why it happened.

Teflon is a material familiar for its use in nonstick cookware. In 1938 Teflon was discovered by accident by Roy J. Plunkett. At the time, Plunkett was making new kinds of CFCs for possible use as refrigerants. To do this required tetrafluoroethylene, a gaseous chemical that was stored in tanks and withdrawn as needed. In his own words, one day the following happened:

> *Soon after the experiment started, my helper called to my attention that the flow of tetrafluoroethylene had stopped. I checked the weight of the cylinder and found that it still contained a sizable quantity of material which I thought to be tetrafluoroethylene. . . . I then removed the valve and was able to pour a white powder from the cylinder. Finally, with the aid of a hacksaw the cylinder was opened and a considerably greater quantity of white powder was obtained.*
>
> *It was obvious immediately to me that the tetrafluoroethylene had polymerized and the white powder was a polymer of tetrafluoroethylene.*

Rather than setting aside the material as useless, Plunkett immediately began to investigate its properties. The white powder turned out to be resistant to heat and to have such a low surface friction that nothing sticks to it. The newspaper story reports that three quarters of the pots and pans sold in the United States are now coated with Teflon or one of its derivatives. On being inducted into the National Inventors' Hall of Fame in 1985, Dr. Plunkett

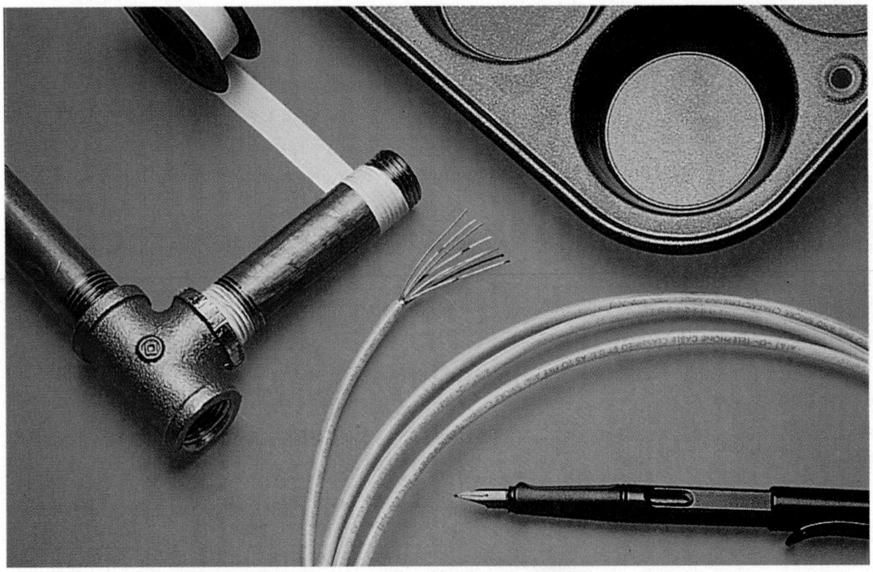

Teflon for the cook, the plumber, the electrician, and the note taker.
(Kip and Pat Peticolas/Fundamental Photographs)

THE WORLD OF CHEMISTRY

Serendipity

Program 2, *Color*

The history of science is replete with examples of serendipity, simply described as a fortuitous and happy discovery or observation from which many important future developments have flowed. There is a special quality to serendipity. It is significant only where the discoverer or observer recognizes through a burst of intuition that there is something that needs further exploration and then proceeds to devote a serious effort to exploit the discovery. Much more often than not, of course, is the requirement of a prepared mind.

A classic example of serendipity is to be found in the story of W. H. Perkin, as told by science historian John K. Smith of Lehigh University.

Little did Perkin know that when he mixed these two substances he would produce a beautiful purple dye.

He was a brilliant young chemist who, while working in his home laboratory in 1856 in an effort to synthesize badly needed quinine, succeeded instead in creating the dye "mauve." Perkin recognized the importance of his accidental discovery and as a consequence of his efforts succeeded in establishing the beginning of the dye industry. The spinoffs were enormous. There are today as a consequence a large variety of materials such as drugs, explosives, fertilizers which play such an important role in the affairs of society. One of the most important consequences, for example, is aspirin, easily one of the most useful drugs in the history of pharmaceutical chemistry.

expressed his pleasure at having invented something that "has been of great personal benefit to people—not just indirectly, but directly to real people whom I know."

1.5 Automobile Tires, Hazardous Waste, and Risk

Some risks are pretty obvious, and there can be no doubt that the next chemical story in our randomly chosen newspapers identifies such a risk. Lying in the midst of a woods is (or was, depending on what has happened since the story was written) the biggest tire dump in New England and possibly in the country. Thirty-three million slowly decomposing tires have been dumped on 14 acres in an inaccessible location near a large population. There is no source of water that could be used to fight a fire. The heat generated as the tires undergo spontaneous chemical transformations, like the heat that sometimes ignites garbage dumps, could ignite the whole pile. According to the news story, a tire fire can burn for months and reach temperatures so high that water would instantly be vaporized.

Neighbors Fear Fire at Massive Tire Dump
—headline on Sunday, August 7, 1994

To imagine the consequences if this pile of tires caught fire, consider what a tire is made of. The essential ingredient is rubber, often not natural rubber from a rubber tree, but a synthetic chemical product. The development of synthetic rubber in the 1940s was a triumph of applied chemistry and technology, accomplished under more pressing circumstances than the development of better refrigerants. Faced with the loss of a supply of natural rubber from Asia during World War II, the United States mounted a major synthetic rubber project. By 1945, U.S. industry was producing one million tons of synthetic rubber a year. Synthetic rubber is similar in composition to natural rubber or petroleum, another plant-derived material, and burns to produce high temperatures, soot, and gases such as carbon dioxide, nitrogen oxides, hydrogen sulfide (which gives rotten eggs their odor), and sulfur dioxide. The black material in a tire is carbon black (like soot), a filler that provides extra strength and stiffness. Petroleum is also added to rubber to smooth the mixture for processing and reduce the cost.

■ The reclamation of petroleum from recycled tires can definitely be done if the economics of collecting the tires and processing them is beneficial.

As you might imagine, the tire dump is the subject of legal action involving the owner of the site, and state and federal environmental protection agencies. There are questions about whose responsibility it is to pay for the cleanup. To make matters even more interesting, the dump is on top of an area that has been designated as a hazardous waste site by the federal government.

What should we do with these tires? *(William McCoy/Rainbow)*

A comprehensive survey of newspapers probably isn't needed to convince you that stories about risks to public health or safety are always newsworthy. Certainly none of us would want to live close to that tire dump, and it is doubtful that any scientist could come up with data to prove that the dump does not present a risk.

Matters are not always so simple, however. How serious a risk is the presence of a pesticide residue on a piece of fruit if that pesticide has the *probability* of causing an estimated one additional cancer death among one million people over the course of a normal 70-year lifetime? There is an equivalent one in one million risk from traveling 10 minutes by bicycle (Table 1.2). Should the government be regulating how often we ride bicycles?

TABLE 1.2 ■ Estimates of Risk: Activities with a Probability of One Additional Death per One Million People Exposed to the Risk

Activity	Cause of Death
Smoking 1.4 cigarettes	Cancer, lung disease
Living 2 months with a cigarette smoker	Cancer, lung disease
Eating 40 tablespoons of peanut butter	Liver cancer caused by the natural carcinogen aflatoxin B
Drinking 40 cans of saccharin-sweetened soda	Cancer
Eating 100 charcoal-broiled steaks	Cancer
Traveling 6 minutes by canoe	Accident
Traveling 10 minutes by bicycle	Accident
Traveling 300 miles by car	Accident
Traveling 1000 miles by jet aircraft	Accident
Drinking Miami tap water for 1 year	Cancer from chloroform
Living 2 months in Denver	Cancer caused by cosmic radiation
Having one chest x-ray in a good hospital	Cancer
Living 5 years at the boundary of a typical nuclear power plant	Cancer

Risk assessment for individuals involves a consideration of the probability of harm and the severity of the hazard. Risk assessment for society as a whole adds to these concerns the number of people who will be affected, the size of the geographical area involved, and how long-lasting the damage might be.

The science of risk assessment is still evolving, but it is clear that the perceptions of risk by the public and as the result of scientific study are often different. For example, statistics show that the risk of injury or death during commercial airplane travel is much lower than that from automobile travel, yet most of us know people who avoid airplane flights out of fear, but who ride in automobiles every day.

What factors influence the public perception of risk? Catastrophic events such as major chemical explosions, oil spills, or nuclear power plant accidents obviously affect public perceptions. People tend to judge involuntary exposure to risk (such as living hear a hazardous waste site) as being worse than voluntary risk (such as smoking cigarettes or riding in automobiles). In other words, people rate risks they can control lower than those they cannot control.

■ Some individuals seem to believe that a sufficient amount of public activism and government regulation can guarantee risk-free living. Do you think this is possible?

No absolute answer can be provided to the question, "How safe is safe enough?" The determination of acceptable levels of risk requires value judgments that are difficult and complex, involving the consideration of scientific, social, economic, and political factors. Over the years a number of laws designed to protect human health and the environment have been enacted. The existence of three types of laws in this area adds to confusion about risk assessment and its meaning.

Risk-based laws are zero-risk laws that allow no balancing of health risks against possible benefits. The Delaney clause of the federal Food, Drug, and Cosmetic Act is such a law. It specifically bans the use of any intentional additive to processed foods that is shown to be a **carcinogen** (a cancer-causing

■ Food additives are discussed in Section 17.9.

A few people have been killed when they tried bungee jumping. Should there be a law against it? *(Reuters/Bettmann)*

Industrial plant in Romania. The black smoke shows that pollutants are being produced from the waste gases being burned. *(Earl Dibble/Photo Researcher)*

■ Clean air is discussed in Chapter 21. Clean water is discussed in Chapter 20.

A 1937 portrayal of chemistry: "Chemical industry, upheld by pure science, sustains the production of man's necessities." This illustration appeared in a book entitled "Man in a Chemical World" by A. C. Morrison, published in 1937. *(From a painting by Leon H. Soderston, 1894–1955).*

agent) in humans or animals, no matter how the study was done. In some cases, it has been argued, a dose for a human equivalent to the harmful dose for a laboratory animal would be more than a human could possibly consume in a given period. The rationale for this law is the nonthreshold theory of carcinogenesis, which assumes that there is no safe level of exposure to any cancer-causing agent.

Balancing laws balance risks against benefits. The Safe Drinking Water Act, the Toxic Substances Control Act, and the Clean Air Act are laws of this type. The federal Environmental Protection Agency (EPA) is required to balance regulatory costs and benefits in its decision-making activities. Risk assessments are used here. Chemicals are regulated or banned when they pose "unreasonable risks" to or have "adverse effects" on human health or the environment.

Technology-based laws impose technological controls to set standards. For example, parts of the Clean Air Act and the Clean Water Act impose pollution controls based on the best economically available technology or the best practical technology. Such laws assume that complete elimination of the discharge of human and industrial wastes into water or air is not feasible. Controls are imposed to reduce exposure, but true balancing is not attempted. The goal is to provide an "ample margin of safety" to protect the public. As technology advances and the cost of the technology decreases, the margin of safety can be adjusted.

Many of our present environmental problems stem from decades of neglect. The Industrial Revolution brought prosperity, and little thought was given to possible harmful effects of the technology that was providing so many visible benefits. Recognizing this, the government and the chemical industry have initiated serious efforts to solve the problems caused by previous lack of foresight and to evaluate the potential problems of new technology.

Risk management requires value judgments that integrate social, economic, and political issues with risk assessment. Risk assessment is the province of scientists, but determination of the acceptability of the risk is a societal issue. It is up to all of us to weigh the benefits against the risks in an intelligent and competent manner.

1.6 What is Your Attitude Toward Chemistry?

Before proceeding with this study of chemistry and its relationship to our society, you might want to examine your prejudices (if any) and attitudes about chemistry, science, and technology. Many nonscientists regard science and its various branches as a mystery and think that they cannot possibly comprehend the basic concepts and their relation to societal issues. Many also have what has been christened "chemophobia" (an unreasonable fear of chemicals) and a feeling of hopelessness about the environment. These attitudes are most likely the result of news stories about the harmful effects of technology. Some of these harmful effects are indeed tragic.

What is needed, however, is a full realization of both the benefits and the harmful effects of science and technology. In the analysis of these pluses and minuses, we need to determine why the harmful effects occurred and whether the risk can be reduced for future generations. This book will give you the basics in chemistry along with some insight on the role of chemistry in both positive and negative aspects of the world today. We hope this knowledge will afford you a healthier and more satisfying life by allowing you to make wise decisions about personal problems and problems of concern to society as a whole.

Cottonwood Lake near Buena Vista, CO. It's up to us to keep such places beautiful.
(Grant Heilman/Grant Heilman Photography)

CHAPTER 2

The Chemical View of Matter

Practically everything we use has been changed from a natural state of little or no utility to one of very different appearance and much greater utility. Some of these changes are mechanical, such as producing the lumber used to build houses. Many others are chemical, such as the baking of clay to make pottery. Exploring, understanding, and managing the processes by which natural materials can be changed are basic to the science of chemistry. These activities require close examination of the composition and structure of matter, which we begin to do in this chapter.

- What are elements and chemical compounds made of?
- What is the difference between a mixture and a pure substance?
- What is the difference between a chemical and a physical process?
- What is the basic theme of chemistry?
- How are symbols for the elements used in formulas and equations to communicate chemical information?
- Why are standards and measurements essential to chemistry?
- What are the metric units most commonly used in chemistry?

Many times a day, you expect things to behave in certain ways and you act on these expectations. If you are cooking spaghetti, you know that you must first boil some water and that after the spaghetti has been in the boiling water for a while it will get soft. If you are pouring gasoline into the fuel tank of your lawnmower, you know that you must protect the gasoline from sparks or it will explode into fire. To cook the spaghetti or safely fill the gas tank, you don't need to think about why these expectations are correct.

Chemistry, however, has taken on the job of understanding why matter behaves as it does. Answering questions such as, "Why does water boil at 100°C?" or "Why does gasoline burn, but water doesn't?" or "Why does spa-

In the chemical view of matter, chemical reactions between molecules in the leaves are made possible by the energy from sunlight. *(Runk/Schoenberger/Grant Heilman Photography)*

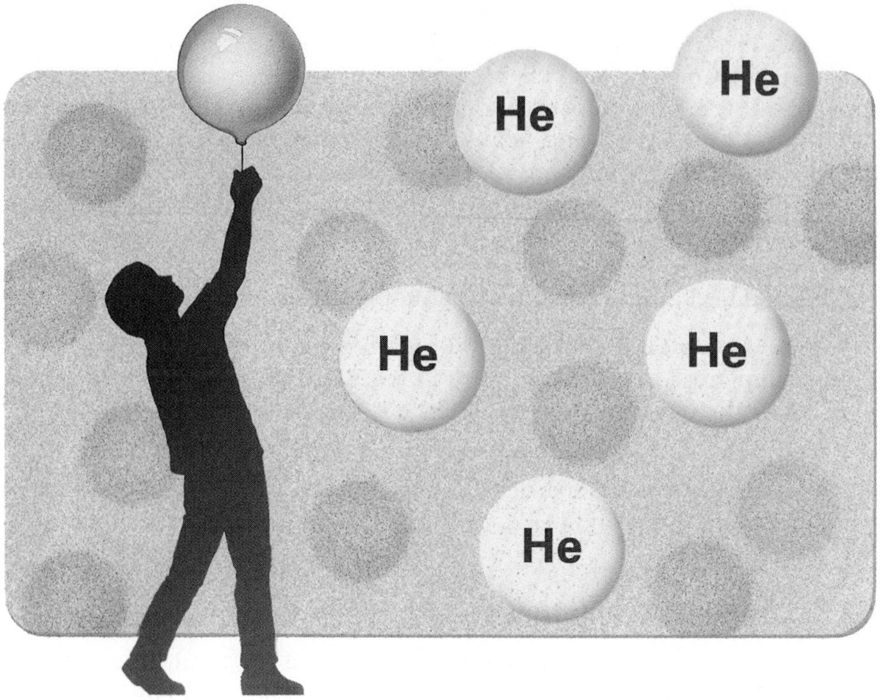

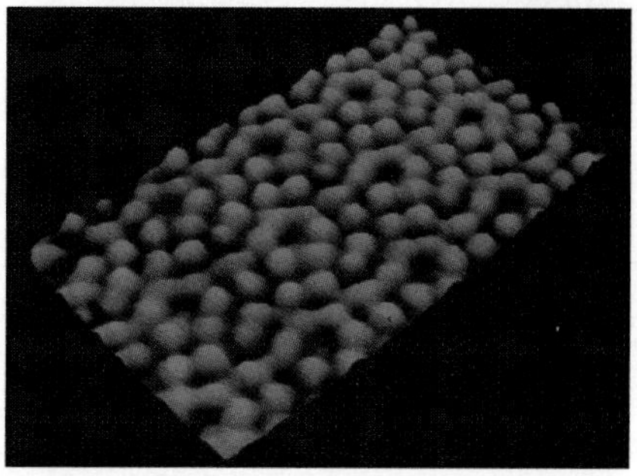

Figure 2.1 Silicon atoms. A scanning tunnelling microscope photo shows the regular pattern of atoms on the surface of a piece of silicon. Pictures like this have been possible since the invention of the STM in 1986. (See *The World of Chemistry* box on the STM in Chapter 3.) *(Science VV/IBMR/Visuals Unlimited)*

ghetti get soft in boiling water?'' requires a different approach than direct observation. We have to look more deeply into the nature of matter than we can actually see, at least under everyday conditions.

Samples of matter large enough to be seen and felt and handled, and thus large enough for ordinary laboratory experiments, are called **macroscopic** samples. In contrast, **microscopic** samples are so small that they have to be viewed with the aid of a microscope. The structure of matter that really interests chemists, however, is at the **submicroscopic** level. Our senses have very limited access into this small world of structure, although in recent years new kinds of instruments are beginning to change this condition (Figure 2.1).

In the chemical model of matter, everything around us, matter of every possible kind, is pictured as collections of very small particles. The picture begins with elements and their atoms—*all matter is composed of atoms.*

■ The historical development of the chemical model of matter is discussed in Chapter 3.

2.1 Elements—The Most Simple Kind of Matter

Most of the materials you handle every day are pretty complex in their structure, and many are mixtures. Only occasionally do you encounter elements not combined with something else. One example is the helium used to fill a birthday balloon. The gas is helium, an element. From the chemical point of view, what does this mean? First, it means that helium cannot be separated or broken down into any other kind of matter that exists independently. Second, it means that the balloon contains the very, very small particles known as helium atoms. Third, it means that helium has a unique and consistent set of properties by which it can be identified.

■ **Solids, liquids,** and **gases** are the three common **states of matter.** The fourth state, plasmas, occurs in flames, stars, and the outer atmosphere of the earth.

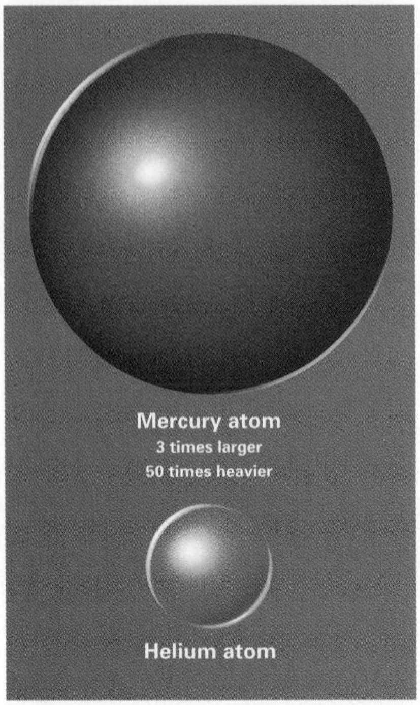

Figure 2.2 The relative sizes of helium and mercury atoms. As you might guess, mercury atoms are not only larger than helium atoms, but are much heavier (about 50 times heavier).

Another element you have seen around your house is mercury. It is used to conduct electricity in switches like those in some thermostats and as the liquid in thermometers. Mercury is certainly very different from helium in its properties. It is a liquid, not a gas, and it conducts electricity, which helium cannot do. What mercury and helium have in common is that they are both elements and each is composed of a single kind of atom. The liquid in the mercury thermometer is made of mercury atoms. Because the properties of helium and mercury are so different, we can safely conclude that helium and mercury atoms are not the same (Figure 2.2).

The chemical model of matter, therefore, starts with the following:

A **pure substance** is something with a uniform and fixed composition at the submicroscopic level. As you will see, pure substances can be recognized by the unchanging nature of their properties.

An **element** is a pure substance composed of only one kind of atom.

An **atom** is the smallest particle of an element, and the atoms of different elements are different.

2.2 Chemical Compounds—Atoms in Combination

If a pure substance is not an element, it is a chemical compound. To take our most common example, what does it mean that water is a chemical compound? You probably know that water is referred to as "h-2-oh" even if you have never studied chemistry. The "h-2-oh" is a way of reading the notation that chemists use to represent water: H_2O. In this kind of notation, H represents the element hydrogen, and O represents the element oxygen.

Chemical compounds are pure substances made of atoms of different elements combined in definite ways. We know that water is a pure substance because it has the same properties no matter where it comes from. We know that water is a chemical compound because passing energy through it in the form of an electric current causes decomposition to the elements hydrogen and oxygen (Figure 2.3). Also, under the right conditions, hydrogen and oxygen combine to form water.

■ The symbolism used to represent elements and compounds is explained further in Section 2.7.

■ Some properties of water: colorless, odorless liquid; melts at 0°C; boils at 100°C; able to dissolve many substances

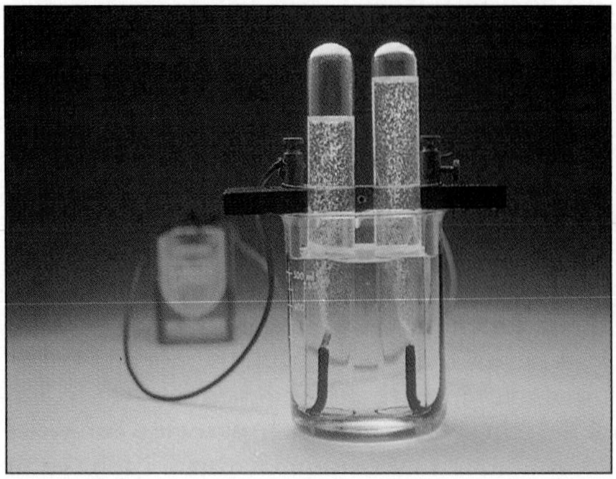

Figure 2.3 Hydrogen, oxygen, and water. Water can be produced by the combination of hydrogen and oxygen. When electrical energy is added, water is decomposed to give back hydrogen and oxygen. *(C.D. Winters)*

Remember that, from the chemical point of view, all matter consists of tiny particles. In water, each particle is composed of two hydrogen atoms (shown by the $_2$ in H_2O) and one oxygen atom (shown by the absence of a subscript in H_2O), combined in what is called a *water molecule.* Experimentally, then, pure substances are classified into two categories: (1) chemical compounds, which can be broken down into simpler pure substances (elements), and (2) elements, which cannot be broken down in this way.

Once elements are combined in compounds, the original, characteristic properties of the elements are replaced by the characteristic properties of the compounds. Consider the difference between table sugar, a white crystalline substance that is soluble in water, and its elements: carbon, which is usually a black powder and is not water soluble; hydrogen, the lightest gas known; and oxygen, the atmospheric gas needed for respiration.

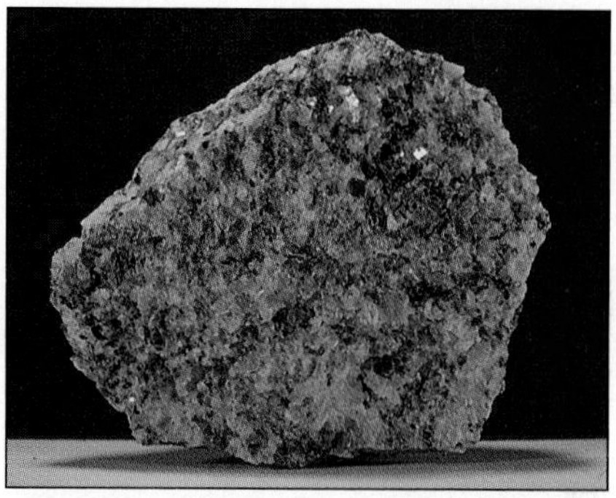

Figure 2.4 A natural heterogeneous mixture. *(Barry L. Runk/Grant Heilman Photography)*

2.3 Mixtures and Pure Substances

Most samples of matter as they occur in nature are mixtures. Some mixtures are obviously **heterogeneous**—their uneven texture is clearly visible, as in the different kinds of crystals in many rocks (Figure 2.4). Other mixtures appear to be **homogeneous,** when actually they are not. For example, the air in your room appears homogeneous until a beam of light enters the room and reveals floating dust particles. Blood appears homogeneous as it gushes from a cut in your finger, but a microscope shows that it is quite a heterogeneous mixture (Figure 2.5).

■ **Homogeneous mixtures** are uniform in composition and **heterogeneous mixtures** are not.

Homogeneous mixtures do exist—we call them **solutions.** No amount of optical magnification will reveal a solution to be heterogeneous, because it is a mixture of particles too small to be seen with ordinary light. In addition, different samples of the *same* homogeneous mixture have the same

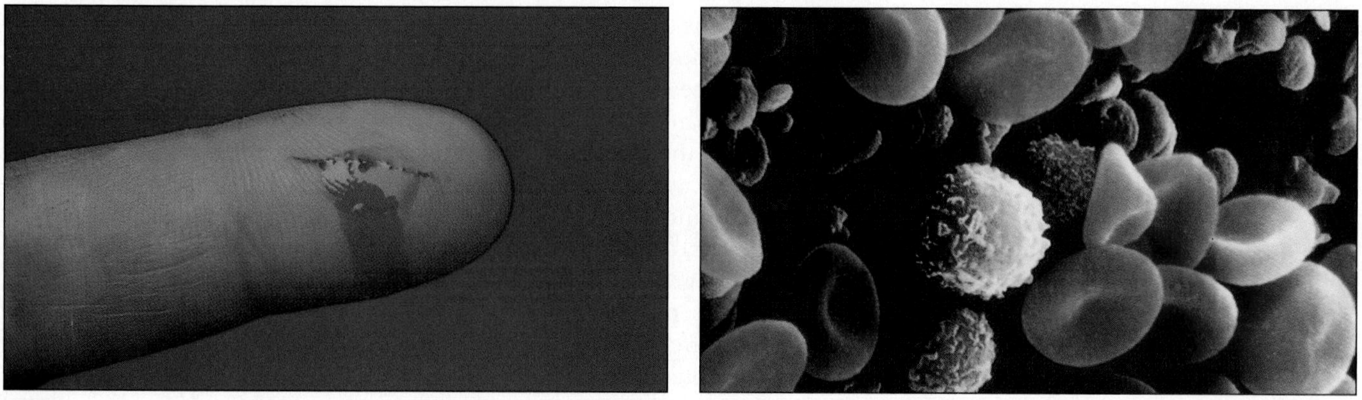

(a) (b)

Figure 2.5 Blood, another heterogeneous mixture. (a) Blood looks homogeneous as it flows from a cut. (b) A microscope shows that it is quite a mixture. Color has been added to this electron microscope picture to show the different kinds of blood cells. The doughnut shaped ones are red blood cells; the fuzzy ones are immune system cells; the small tan ones are platelets. *(a, Tom Pantages; b, Ken Eward/Science Source/Photo Researchers)*

(a) (b)

Figure 2.6 Iron and sulfur. (a) Piles of iron powder and sulfur. (b) Separating iron and sulfur after the piles have been stirred together. Separating elements from each other is not always so easy. *(C.D. Winters)*

composition throughout. Examples of solutions are clean air (mostly the elements nitrogen and oxygen), freshly brewed tea, and some brass alloys, which are homogeneous mixtures of the elements copper and zinc. As these examples show, *solutions can be found in the gaseous, liquid, and solid states.*

■ As science advances, smaller and smaller concentrations of impurities can be detected.

When a mixture is separated into its components, the components are said to be *purified.* Efforts at separation are usually incomplete in a single step, and repetition of the process is necessary to produce a purer substance. Ultimately, the goal is to arrive at substances that cannot be purified further.

Separation of mixtures is done by taking advantage of the different properties of substances in the mixture. For example, many people want to separate certain substances from their tap water before they drink it. One way to do this is based on the forces of attraction between particles at the submicroscopic level. The water is passed through a material that attracts the undesirable substances and holds them back. As another example, consider separation of a mixture of sulfur and finely divided iron, which have particles large enough to photograph (Figure 2.6). Since iron is magnetic and sulfur is not, the iron can be separated from the sulfur by repeatedly stirring the mixture with a magnet. After the bright yellow color is obtained, it is assumed that the sulfur has been purified.

Drawing a conclusion based on one property of the mixture may, of course, be misleading because other methods of purification might change some other properties of the sample. It is safe to call the sulfur a pure substance only when all possible methods of purification fail to change its properties. *The assumption that all pure substances have a set of distinctive properties by which they can be recognized is fundamental to the chemical view of matter.*

2.4 Changes in Matter: Is It Physical or Chemical?

Sulfur is yellow; iron is magnetic; water boils at 100°C. These are different properties of matter, but with something in common. We can observe the color of a substance, pick it up with a magnet, or measure its boiling point without changing its chemical identity. Such properties are classified as **physical properties.** In chemistry, the word *physical* is used to refer to processes that do not change chemical identities. Separating sulfur and iron with a magnet is a *physical* separation. Boiling water is a *physical* change—the steam that forms is still water, H_2O.

■ Some physical properties: color, odor, melting point, boiling point, solubility, hardness, density, state (solid, liquid, or gas)

Whenever possible, the physical properties of substances are pinpointed by making numerical measurements. You could say that acetone, used to remove paint and nail polish, boils at a lower temperature than water and that ethylene glycol, the essential ingredient in many antifreezes, boils at a higher temperature than water. But the measured boiling points give us a better understanding of the differences.

■ Boiling points:

Acetone	56°C
Water	100°C
Ethylene glycol	197°C

You might describe lead as heavier than aluminum—if you pick up same-sized pieces of these metals, lead will definitely feel heavier. The difference is better represented, though, by giving the numerical value of the **density** of each metal, which takes into account the need to compare pieces of equal size, that is, of equal volume. One cubic centimeter of lead weighs 11.3 grams, and one cubic centimeter of aluminum weighs 2.7 grams. Lead is *denser* than aluminum. Stated in the usual way,

■ Don't confuse density and heaviness. A large piece of aluminum could be heavier than a small piece of lead, but the lead is still denser.

Densities (at 20°C):	
Lead	11.3 g/cm^3
Aluminum	2.7 g/cm^3

■ **Density** is mass per unit volume. The customary measurement units (Section 2.8) for density are grams (g) for mass (weight) and cubic centimeters (cm^3) for the volume of solids or milliliters (mL) for the volume of liquids.

By contrast to physical changes such as boiling, there are processes that do result in changes of identity. When gasoline burns, it is converted to a mixture of carbon dioxide, carbon monoxide, and water. Burning in air, a property that gasoline, kerosene, and similar substances have in common, is classified as a **chemical property.**

The word *chemical* is used to describe processes that result in a change in identity. The combination of hydrogen and oxygen to form water is a *chemical change* or **chemical reaction**—a process in which one or more substances (the **reactants,** which can be elements or compounds, or both) are converted to one or more different substances (the **products,** which can also be elements or compounds, or both).

$$\text{Reactants} \longrightarrow \text{products}$$
$$\text{Hydrogen} + \text{oxygen} \longrightarrow \text{water}$$

A chemical reaction produces a new arrangement of atoms. The number and kinds of atoms in the reactants and products remain the same, but the reactants and products are different substances that can be recognized by their

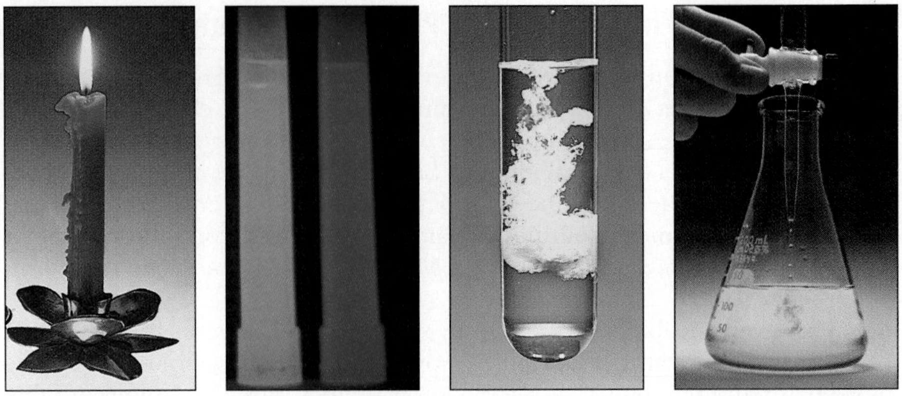

(a) (b) (c) (d)

Figure 2.7 Observable indications of chemical reactions. Chances are that a chemical reaction has occurred when (a) a flame is visible, (b) mixing substances produces a glow, (c) a solid appears when two solutions are mixed, (d) there is a color change when two substances are mixed. *(C. D. Winters)*

different properties. To distinguish between a hydrogen-oxygen mixture and a compound composed of hydrogen and oxygen, we might say that in water, hydrogen and oxygen are *chemically* combined. The meaning is that the hydrogen and oxygen atoms are held together strongly enough to form the individual units we call water molecules. For many compounds, a **molecule** is the smallest unit of the compound; molecules retain the composition of the compound and can have a stable, independent existence.

■ Compounds composed of molecules at the submicroscopic level are referred to as **molecular compounds.**

Some easily observed results of chemical reactions are the rusting of iron, the change of leaf color in the fall, and the formation of carbon dioxide bubbles by an antacid tablet. Often, though not always, the occurrence of a chemical reaction can be detected because there is some observable change (Figure 2.7).

The word *chemical* is often used as a noun. In this sense, every substance in the universe is a *chemical.* You might read of polluted waters that contain chemicals. Used this way, the word really stands for "chemical compound."

DISCOVERY EXPERIMENT

A Heterogeneous Mixture for Breakfast

Breakfast cereal is a heterogeneous mixture and so is instant breakfast drink. Many breakfast foods are fortified with iron, but do you know whether the iron is there as an element or in a compound?

To do this experiment you will need a reasonably strong magnet, a packet of Carnation Instant Breakfast (vanilla works best), a couple of plastic sandwich bags, a pencil, and a rubber band. First, attach the magnet to the pencil with the rubber band and wrap one of the sandwich bags around the magnet and pencil. Next, empty an individual serving packet of the breakfast drink into another sandwich bag. Then insert the plastic-covered magnet into the bag of drink mix and, holding on to the other end of the pencil, stir with the plastic-covered magnet for several minutes. After this, remove the plastic-covered magnet and examine it carefully. It will help to move the plastic around and hold it against a white background. What do you see? What does this tell you about the iron in this instant breakfast drink? Read the label to see what information it provides about the form of iron that is present. (Hint: Iron in compounds is rarely magnetic.)

TABLE 2.1 ■ Some Examples of Conversion of Chemical Potential Energy to Other Forms of Energy

Conversion to—	**Electrical Energy**	**Heat**	**Light**	**Mechanical Energy**
Done by—	Batteries Fuel cells	Combustion Digestion of food Many kinds of chemical reactions	Burning candle, logs in fireplace Luminescence in firefly Luminescence from chemical reaction	Rocket Animal muscles Dynamite explosion

The "chemicals" in polluted water may be unhealthful or harmful. Anyone who has studied even a little chemistry understands that useful and healthful substances—penicillin, table salt, nylon fabrics, laundry detergents, and all natural herbs and spices—are also made of chemical compounds. So are all plants and animals.

Some physical changes and almost all chemical reactions are accompanied by changes in energy. Frequently, the energy is taken up or released in the form of heat. Heat must be added for the physical changes of melting or boiling to occur. Heat is released when the chemical reactions of combustion, or burning, take place. But many chemical reactions, such as those of photosynthesis, require energy from their surroundings in order to take place.

Energy is the ability to cause change or, in the formal terms of physics, to do work. Energy in storage, waiting to be used, is **potential energy.** There is potential energy in gasoline, known as *chemical energy*—it is released as heat when gasoline burns. Chemical energy can also be released in the form of electrical energy, light, or mechanical energy (Table 2.1). Energy in use, rather than in storage, is **kinetic energy,** the energy associated with motion.

The diver has potential energy because of his position above the surface of the water.
As diver falls potential energy is converted to kinetic energy.

EXAMPLE 2.1 *Physical and Chemical Change*

Which of the photos below represent physical changes and which represent chemical changes? Explain each choice.

(a) *(Larry Cameron)*

(b) *(Charles Steele)*

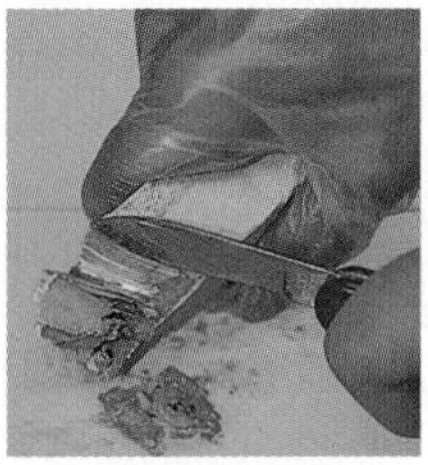

(c) *(C.D. Winters)*

(continued on next page)

SOLUTION

(a) Physical process. Pouring a mixture through a filter will separate a solid from a liquid but will not change the identity of either.
(b) Chemical process. It's a sure thing that when something is burning, its identity is being changed.
(c) Physical process. Changing the size or shape of a substance does not change its identity.

Exercise 2.1
Repeat the question of Example 2.1 for the photos below.

(a) *(C.D. Winters)*

(b) *(C.D. Winters)*

2.5 Classification of Matter

The kinds of matter we have described—elements, compounds, and mixtures—can be classified according to their composition and how they can be separated into other substances, as shown in Figure 2.8.

Heterogeneous samples of matter are all mixtures and can be physically separated into various kinds of homogeneous matter (Figure 2.9). Homogeneous matter can be a pure substance or a mixture. If it is a mixture, it is described as a solution and has the same composition throughout. Different solutions of the same substances, however, can vary in composition, sometimes over a very wide range. A teacup full of water, for example, might dissolve anywhere from a few grains of sugar to more than 1 measuring cup of sugar. The water and sugar can be separated by the physical process of evaporating the water.

If the homogeneous matter is a pure substance, it has a fixed composition and must be either an element or a compound. While elements contain only atoms of that element, compounds contain atoms of different elements combined in distinctive ways. At the submicroscopic level, each sugar molecule contains 12 carbon (C) atoms, 22 hydrogen (H) atoms, and 11 oxygen (O) atoms, represented by $C_{12}H_{22}O_{11}$. At the macroscopic level, every 100 g of table sugar contains 42.1 g of carbon, 6.5 g of hydrogen, and 51.4 g of oxygen. Only chemical reactions could produce pure carbon, hydrogen, and oxygen from sugar.

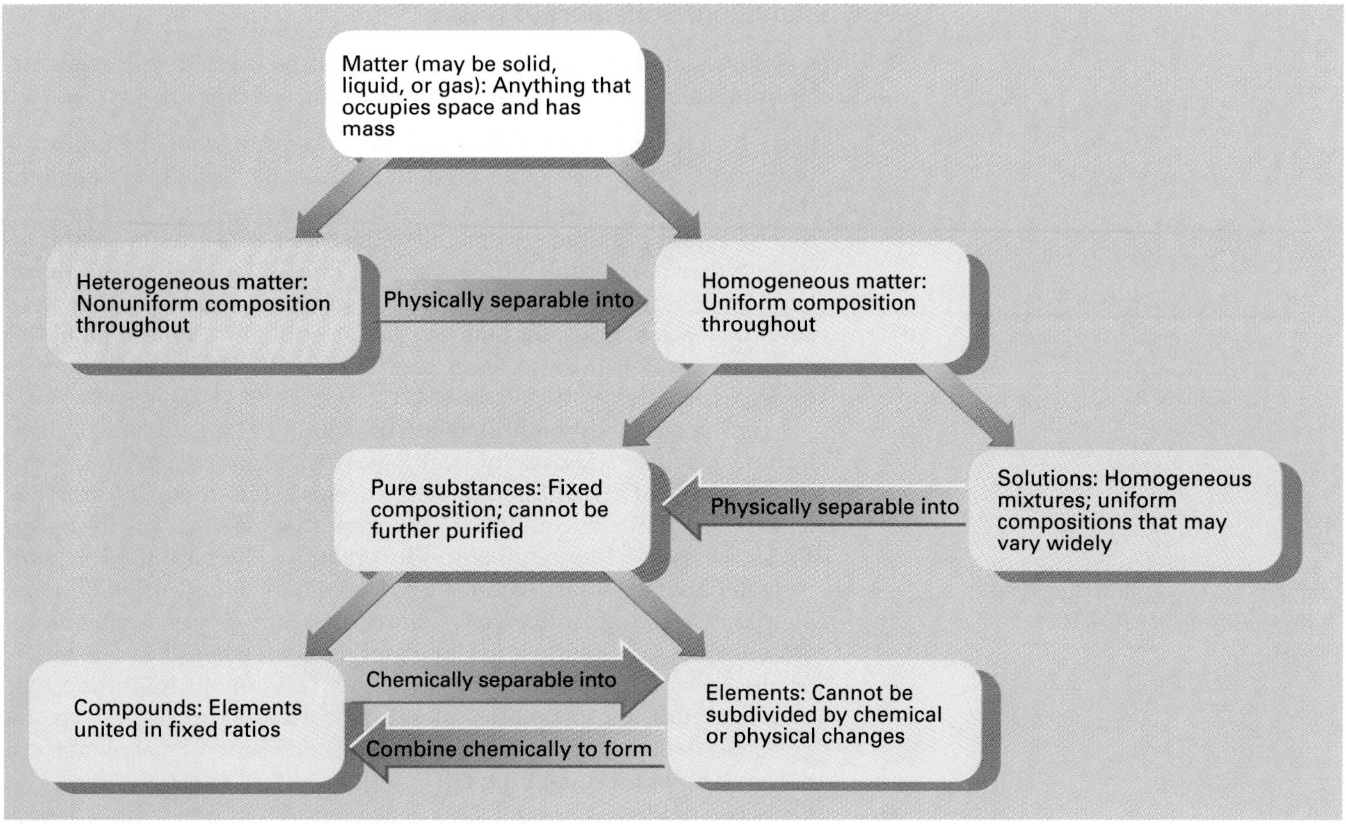

Figure 2.8 A classification of matter. Once elements have combined into compounds, only chemical reactions can separate them.

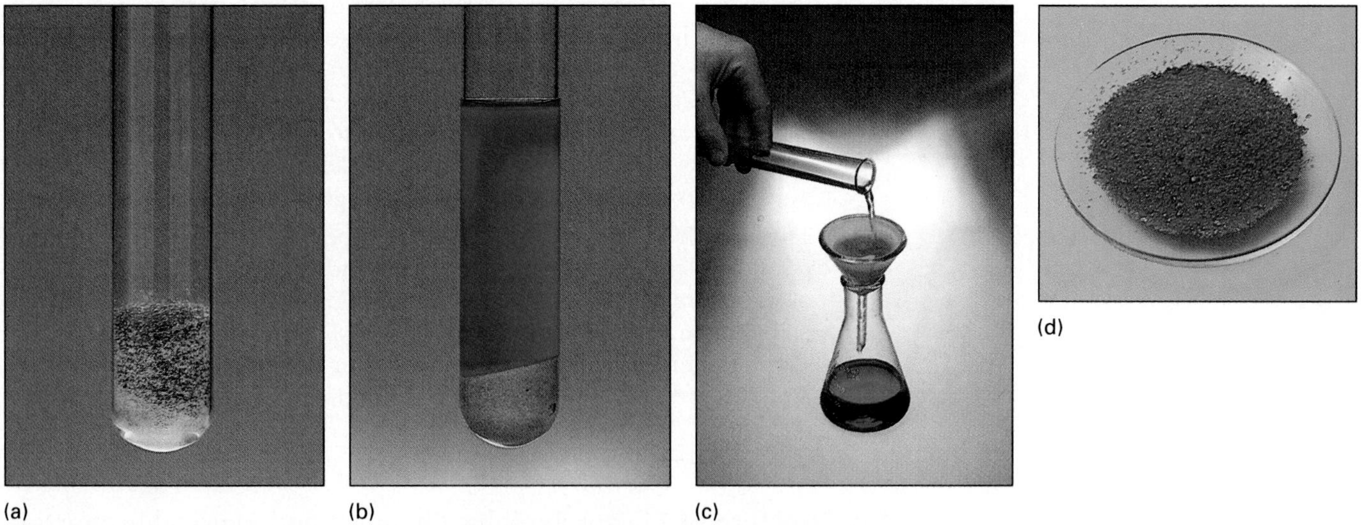

Figure 2.9 A physical separation. A mixture of blue copper sulfate and sand (a) can be separated by (b) dissolving the copper sulfate in water, (c) filtering the sand from the solution, and (d) evaporating the water to leave behind the solid copper sulfate. *(C.D. Winters)*

Why Study Pure Substances?

Perhaps by now you are wondering why we should be interested in elements and compounds and their properties. There are three basic reasons.

1. Only by studying pure matter can we understand how to utilize its properties, design new kinds of matter, and make desirable changes in the nature of everyday life. A long time ago, for example, it was known that swallowing extracts from the willow tree could relieve pain, although there were some unpleasant side effects. Once chemists identified the pain-relieving substance in the mixture of extracts, they were able to create a molecule that was similar enough to relieve pain, but different enough to have fewer side effects. We know this pure substance as aspirin.

    At one time most natural materials could be changed only by physical means. Only a few useful materials, such as pottery and iron, were the products of chemical change. Otherwise, the design of everyday objects was limited to utilizing the properties of natural substances. Today we have evidence all around us that the chemical modification of materials has indeed changed the quality of life. Synthetic fibers, plastics, life-saving drugs, latex paints, new and better fuels, photographic films, and audio and video tapes are but a few of the materials produced by controlled chemical change. It is equally important that understanding the properties of substances allows us to prevent *undesirable* chemical changes, such as the corrosion of metal objects, the spoiling of food, and the effects of hereditary diseases.

■ We shall return to examine the chemistry of many of these modern materials in later chapters.

2. Studying the properties of matter helps us to deal intelligently with our environment, both in everyday life and in the social and political arena. Knowing something about the properties of fertilizers or pesticides helps us decide which ones to use, which ones to avoid, and when their use is or is not necessary. In the context of community life, many decisions are enlightened by some knowledge of the properties of matter. When you pour waste down a sewer drain, what happens to it? Does it matter whether the waste is motor oil or the water from washing your car? Does your town or city really need to spend millions of dollars to upgrade its sewage treatment plant to keep nitrogen out of natural waters or to remove asbestos from the high school auditorium ceiling?
3. The third reason for studying the properties of elements and compounds is simple curiosity. Chemicals and chemical change are a part of nature that is open to investigation. For some people, this is as big a challenge as a high mountain is to a climber. If we hope to understand matter, the first steps are to discover the simplest forms of matter and study their interactions. Curiosity draws many chemists to basic research.

The Structure of Matter Explains Chemical and Physical Properties

An essential part of basic research is investigation into the **structure of matter**—how atoms of the elements are connected in larger units of matter and how

(a) (b) (c)

Figure 2.10 Three ways to look at snow. The beautiful patterns of snowflakes seen under a microscope (b) reflect the submicroscopic, orderly arrangement of water molecules (c). *(a, Grant Heilman; b and c;* The World of Chemistry, *Program 12, "Water")*

these units are arranged in samples of matter large enough to be seen (Figure 2.10). It is at this level of organization that answers to fundamental questions are found.

The properties of a sample of matter are determined by the nature of its parts, just as the abilities of a computer are determined by the parts that have been assembled. If we even hope to understand the nature of matter, it is absolutely necessary that we understand the minute parts and how they are related to each other. Indeed, *the basic theme of this text and of chemistry itself is the relationship between the structure of matter and its properties.* Armed with an understanding of this relationship, chemists are developing an ever-increasing ability to create new substances and predict the properties of these substances.

THE PERSONAL SIDE

Alfred Bernhard Nobel (1833–1896)

(The Granger Collection)

To receive a Nobel Prize is looked upon by most scientists as the highest possible honor, and the accomplishments recognized by the Prize are always among the most significant advances in science. Textbooks usually make note of achievements and individuals worthy of Nobel Prizes as a way of highlighting their importance. It was breakthroughs in the chemistry of explosives that made the Nobel Prizes possible.

Alfred Nobel was a Swedish chemist and engineer, the son of an inventor and industrialist. Together they established a factory to manufacture nitroglycerin, a liquid explosive that is extremely sensitive to light and heat. After two years in operation, there was an explosion at the factory that killed five people, including Alfred Nobel's younger brother. Perhaps motivated by this event, Alfred searched for a way to handle

(continued on next page)

THE PERSONAL SIDE *(continued)*

nitroglycerin more safely. He found the answer and in 1867 patented dynamite. By mixing nitroglycerin with a soft, absorbent, nonflammable natural material known as *kieselguhr,* or ''diatomaceous earth,'' Nobel had stabilized the explosive power of nitroglycerin. His dynamite could be shipped safely and would explode only when this was desirable. He also invented the blasting caps (containing a mercury compound) needed to set off the dynamite explosion.

Nobel's talent as an entrepreneur combined with his many inventions (he held 355 patents) made him a very rich man. He never married and left his fortune to establish the Nobel Prizes, awarded annually to individuals who ''have conferred the greatest benefits on mankind in the fields of physics, chemistry, physiology or medicine, literature and peace.''

■ SELF-TEST 2A

1. Which of the following questions would a chemist study?
   (a) Does a grapefruit contain vitamin E?
   (b) What gives a grapefruit its yellow color?
   (c) How many grapefruits does the average tree produce in a season?
2. If a sample of matter contains only one kind of atom it is a (an) ____________ and ____________ be broken down to simpler substances.
3. The two kinds of pure substances are ____________ and ____________.
4. A solution is a ____________ mixture.
5. Solutions may exist in the solid, liquid, or gaseous states. (a) True, (b) False.
6. If you burn the butter in the frying pan, you have carried out a ____________ change because the butter changed in ____________.
7. Which of the following are chemicals? (a) Aspirin, (b) Baking soda, (c) Rubbing alcohol, (d) Vanilla, (e) Soap.
8. Which of the following mixtures is homogeneous? (a) Sugar and water, (b) Oil and water, (c) Cement, sand, and gravel.
9. Identify by name the reactant(s) shown in the two equations below:
   (a) $\underset{\text{Methane}}{CH_4} + \underset{\text{Oxygen}}{O_2} \longrightarrow \underset{\text{Carbon dioxide}}{CO_2} + \underset{\text{Water}}{H_2O}$
   (b) $\underset{\text{Hydrogen peroxide}}{H_2O_2} \longrightarrow \underset{\text{Water}}{H_2O} + \underset{\text{Oxygen}}{O_2}$
10. Which of the following are physical properties and which are chemical properties?
    (a) Flammability (b) Density
    (c) Souring of milk
11. Potential energy is converted into light energy in combustion. (a) True, (b) False.

THE WORLD OF CHEMISTRY

Models and Modeling

Program 4, *Modeling the Unseen*

The history of science is filled with examples of models—mental concepts that scientists and investigators use to explain the unseen. Many of these models are triumphs of the human mind and represent extraordinary flights of the imagination. Among the classic examples are models of the atom, the kinetic molecular model of gases as composed of molecules rapidly flying about in mostly empty space, and the biochemists' model of how molecules of DNA unwind to form copies in new generations of plants and animals. Models must agree with facts, be modified to agree with facts, or be abandoned.

A remarkable example of the use of models to explain observed phenomena is found in the observations of Jupiter's moon Io by that astonishing spacecraft Voyager. As explained by Dr. Torrence Johnson, director of the Voyager Project,

Prior to the Jupiter visit, the model of Io, based on terrestrial observations, was of a moon similar to our own in age, composition, etc. The only inconsistency was that the surface was too bright. The model proposed at the time attributed its high reflectivity to the possibility that Io was covered by a layer of ice which had long since evaporated and left behind deposits of highly reflecting salt.

The visit of Voyager showed that this model is invalid. Instead, Io proved to be one of the most active volcanic bodies in the solar system, showing at the time of the Voyager encounter in 1978 at least seven active volcanoes. Enormous quantities of sulfur and sulfur dioxide were being spewed into space. Where does the energy for these volcanoes come from? And why isn't the volcanic material molten rock, as it is in volcanoes on Earth? A new model was needed. It pictures Io with a central core of silicate rock surrounded by the sulfur. The conflicting gravitational pull of Jupiter on one side and another moon on the other side provide the energy to keep subsurface sulfur and sulfur dioxide in a molten state.

Dr. Johnson goes on to explain that

When you build a model like this, you shouldn't get to thinking that the model represents reality. It will always change. In fact, we're not doing our job right if it doesn't change. It will always evolve as we get more information and revise our methods of observation based on the models.

Voyager photograph of Io, one of Jupiter's moons. Note the volcanic activity that appears above the surface at the upper left. The cold-moon model of Io had to be changed as a result of these observations.

2.6 The Chemical Elements

Iron, gold, sulfur, tin, and a few other substances that occur in the earth's crust were recognized as elements long before the modern meaning of the term developed. Other naturally occurring elements were identified only after long and laborious efforts to separate them from the compounds in which they are found. Although it isn't always easy, chemists can now separate and identify the elements present in the most complex mixtures and compounds.

The names of the elements are listed inside the front cover of this book. About 18 of these elements are not found anywhere in nature, but have only been produced artificially in laboratories, usually in extremely small quanti-

Three nonmetals: chlorine, a gas; bromine, a liquid; and iodine, a solid. *(Larry Cameron)*

■ The spontaneous decomposition of unstable heavy atoms is responsible for the phenomenon of radioactivity, which is discussed in Chapter 5.

■ By "everyday conditions" chemists refer to usual temperatures and pressures at the earth's surface. Changing these conditions can change the state of a substance, for example, the freezing or boiling of water.

ties. Their syntheses became possible starting in the 1940s. At that time, chemists and physicists had developed the high-energy tools needed to create heavy atoms that are unstable and spontaneously break down to give atoms of previously unknown elements.

The elements range widely in their properties. Under everyday conditions some, including helium, hydrogen, nitrogen, oxygen, and chlorine, are gases. Only two elements—mercury and bromine—are ordinarily liquids. Most of the elements are solids, and of these most are metals. The distinguishing properties of metals are that they conduct electric current and have a shiny appearance. A second major class of elements, the nonmetals, do not conduct electricity and are not lustrous. The familiar nonmetals include all the elemental gases, and carbon, sulfur, phosphorus, and iodine, which are solids.

Symbols for the Elements

Sometimes people are discouraged about understanding chemistry when they encounter chemical formulas and names: CH_3OH, H_2SO_4, polychlorinated biphenyl, nitric oxide. . . . There is nothing unique about chemistry, however, in needing its own vocabulary. Surely, the symbols and language in the margin are equally mysterious to anyone who has never been a musician or played football. Chemistry, just like music or football, needs special symbols and language. Without them communication would be impossible.

Since elements are the building blocks of all matter, symbols for the elements are fundamental to communicating about chemistry. The symbols for the elements are listed next to their names inside the front cover. Some symbols are single letters that are the first letter of the name. Others are two letters from the name, always with the first letter capitalized and the second lower case. *I*odine, for example, is represented by I, and *in*dium is represented by In;

TABLE 2.2 ■ Elements with Symbols Not Based on Their Modern Names

Modern Name	Symbol	Origin of Symbol
Antimony	Sb	*anti + monos:* Greek name, meaning an element not found alone
Copper	Cu	*cuprum:* Latin name, meaning from the island of Cyprus
Gold	Au	*aurum:* Latin name, meaning shining dawn
Iron	Fe	*ferrum:* Latin name for the element
Lead	Pb	*plumbum:* Latin name for the element
Mercury	Hg	*hydrargyrum:* Greek name, meaning liquid silver or quick silver
Potassium	K	*kalium:* Latin name, meaning alkali
Silver	Ag	*argentum:* Latin name for the element
Sodium	Na	*natrium:* Latin name for the element
Tin	Sn	*stannum:* Latin name for the element
Tungsten	W	*wolfram:* Swedish name, meaning devourer of tin (because it interferes with purifying tin)

*n*itrogen is represented by N and *ni*ckel by Ni; *mag*nesium is represented by Mg and *man*ganese by Mn. You can see that it is important to recognize the symbols and know how they differ for different elements.

Eleven metals are represented by symbols not derived from their modern names, but rather from older names in other languages. Most of these metals, listed in Table 2.2, were known in ancient times because they are found free in nature or are easily obtained from their naturally occurring compounds. The names, symbols, and properties of a few of the most common metals are listed in Table 2.3.

■ Symbols for the elements are the alphabet for the language of chemistry.

TABLE 2.3 ■ Some Common Metals

Element	Symbol	Properties
Iron	Fe	Strong, malleable, corrodible
Copper	Cu	Soft, reddish-colored, ductile
Sodium	Na	Soft, light metal, very reactive, low melting point
Silver	Ag	Shiny white metal, relatively unreactive, good conductor of electricity and heat
Gold	Au	Dense yellow metal, very unreactive, ductile, good conductor
Chromium	Cr	Resistant to corrosion, hard, bluish-gray, brittle
Aluminum	Al	Blue-white, soft, highly malleable and ductile, good conductor of heat and electricity, coated by oxide in air
Magnesium	Mg	Silver-white, hard, quite reactive
Lead	Pb	Dark gray, very dense, soft, highly malleable and ductile, poor conductor of heat and electricity
Zinc	Zn	Blue-white, hard, brittle, resistant to corrosion

Of the known elements, only 17 are nonmetals. Most of the common nonmetals have single letter symbols:

■ Consult the alphabetical listing inside the front cover to find the symbol from the name of any element, or vice versa.

H Hydrogen	C Carbon	N Nitrogen	O Oxygen
P Phosphorus	S Sulfur	F Fluorine	I Iodine

Earlier we noted that a sample of gaseous helium (also a nonmetal) contains helium atoms. The symbol He is therefore used to represent a single He atom or a large collection of helium atoms. One of the first milestones in understanding the structure of matter was the discovery that a number of nonmetals are composed not of atoms, but of molecules. A sample of pure hydrogen gas does contain only hydrogen atoms, as expected for an element, but the atoms are joined together in two-atom hydrogen molecules, represented by H_2. The seven nonmetals that exist under everyday conditions as two-atom molecules, known as **diatomic molecules,** are listed in Table 2.4.

2.7 Using Chemical Symbols

Formulas for Chemical Compounds

■ Chemical formulas are the words in the language of chemistry.

We have already used a few **chemical formulas,** combinations of the symbols for the elements that represent the stable combinations of atoms in compounds: H_2 for elemental hydrogen, H_2O for water, and $C_{12}H_{22}O_{11}$ for table sugar. The principle is the same no matter how many atoms are combined. The symbols represent the elements and the **subscripts** (like the $_2$ in H_2O) indicate the relative numbers of atoms of each kind. The formulas and properties of a few simple molecular compounds are given in Table 2.5. Such formulas can represent one molecule or a large sample of the compound.

TABLE 2.4 ■ Elements Composed of Diatomic Molecules*

Element	Symbol	Properties
Hydrogen	H_2	Colorless, odorless, occurs as a very light gas, burns in air
Oxygen	O_2	Colorless, odorless gas, reactive, constituent of air
Nitrogen	N_2	Colorless, odorless gas, rather unreactive
Chlorine	Cl_2	Greenish-yellow gas, very sharp choking odor, poisonous
Fluorine	F_2	Pale yellow, highly reactive gas
Bromine	Br_2	Dark red liquid, vaporizes readily, very corrosive
Iodine	I_2	Dark purple solid, sublimes easily†

*These elements are all nonmetals.

†Solids that *sublime* go directly from the solid state to the gaseous state without melting.

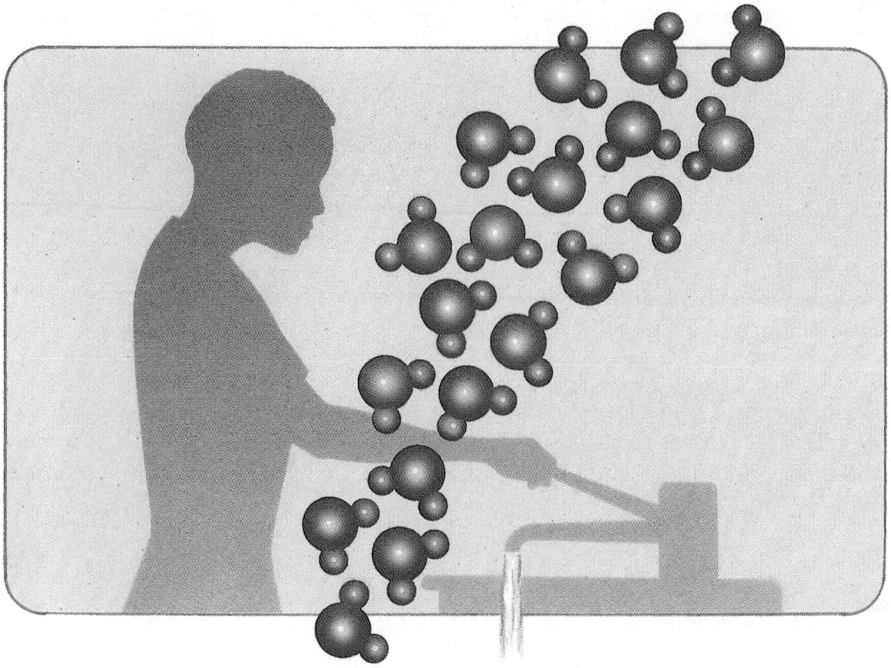

Sometimes a line is drawn between symbols to indicate which atoms are connected in molecules:

$$H—H \qquad H—\overset{\displaystyle H}{\underset{}{N}}—H \qquad H—\overset{\displaystyle H}{\underset{\displaystyle H}{C}}—H$$

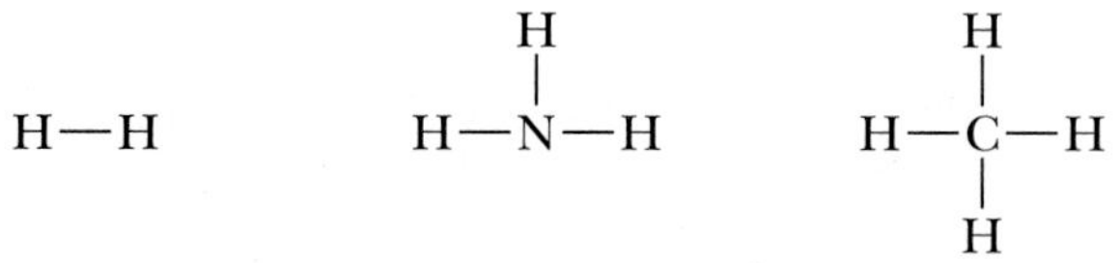

A hydrogen molecule An ammonia molecule A methane molecule

Formulas that show the connections in this way are known as **structural formulas,** whereas formulas that give just one symbol for each element present are called **molecular formulas.** The molecular formula for methane is CH_4.

■ The lines in structural formulas represent chemical bonds, which we'll have much more to say about in Chapter 6.

TABLE 2.5 ■ Some Simple Molecular Compounds

Compound	Formula	Properties
Water	H_2O	Odorless liquid
Carbon monoxide	CO	Odorless, flammable, toxic gas
Carbon dioxide	CO_2	Odorless, nonflammable, suffocating gas
Sulfur dioxide	SO_2	Nonflammable gas; suffocating odor
Ammonia	NH_3	Colorless, nonflammable gas; pungent odor
Methane	CH_4	Odorless, flammable gas
Carbon tetrachloride	CCl_4	Nonflammable, dense liquid

(a)

(b)

(c)

(d)

(e)

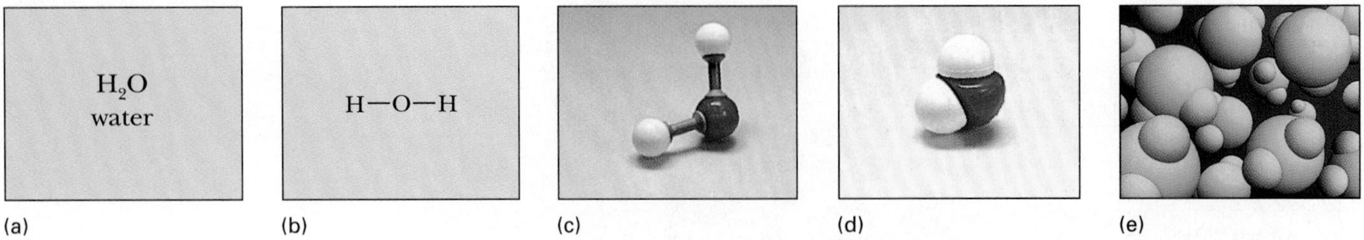

Figure 2.11 The water molecule, as represented in print (a and b), in physical models that can be handled (c and d), or on a computer screen (e). *(e, Ken Eward/ Science Source/Photo Researchers)*

■ The molecular formulas for ethyl alcohol and butane are C_2H_6O and C_4H_{10}. You can see that these formulas give little information about how the atoms are put together.

As molecules get larger, it becomes helpful to write formulas yet another way. The structural formulas for ethyl alcohol, the active ingredient in wine and beer, and butane, the fuel in pocket cigarette lighters, are cumbersome to write:

$$
\text{Structural formulas:}\quad
\underset{\text{Ethyl alcohol}}{
\begin{array}{c}
\;\;\;\;\text{H}\;\;\text{H}\\
\;\;\;\;|\;\;\;\;|\\
\text{H}-\text{C}-\text{C}-\text{OH}\\
\;\;\;\;|\;\;\;\;|\\
\;\;\;\;\text{H}\;\;\text{H}
\end{array}}
\qquad
\underset{\text{Butane}}{
\begin{array}{c}
\;\;\;\;\text{H}\;\;\;\text{H}\;\;\;\text{H}\;\;\;\text{H}\\
\;\;\;\;|\;\;\;\;|\;\;\;\;|\;\;\;\;|\\
\text{H}-\text{C}-\text{C}-\text{C}-\text{C}-\text{H}\\
\;\;\;\;|\;\;\;\;|\;\;\;\;|\;\;\;\;|\\
\;\;\;\;\text{H}\;\;\;\text{H}\;\;\;\text{H}\;\;\;\text{H}
\end{array}}
$$

A compromise is to write formulas on one line, but with the groups of atoms that are connected written together in what are known as **condensed formulas.**

$$
\text{Condensed formulas:}\quad
\underset{\text{Ethyl alcohol}}{CH_3CH_2OH}
\qquad
\underset{\text{Butane}}{CH_3CH_2CH_2CH_3}
$$

■ Organic compounds are so numerous and so important that they are the basis for an entire field of chemistry—**organic chemistry** (Chapters 12 and 14). Most compounds produced by living things, most medicines, and the components of most plastics are organic compounds.

These two compounds and millions of others that contain carbon combined with hydrogen—and often also with nitrogen, phosphorus, or sulfur—are known as **organic compounds.**

Compounds not based on carbon are referred to as **inorganic compounds.** A few that are in everyday use are

	$NaCl$	NH_4NO_3	H_2SO_4	$Mg(OH)_2$
Chemical name:	Sodium chloride	Ammonium nitrate	Sulfuric acid	Magnesium hydroxide
Use:	Table salt	Fertilizer	Battery acid	A laxative

When it is necessary to explore the shapes of molecules beyond what can be shown on a flat piece of paper, chemists resort to physical models or computer-drawn pictures such as those shown in Figure 2.11.

EXAMPLE 2.2 *Chemical Symbols*

Which of the following represent elements and which represent compounds?

(a) KI (b) Co (c) Ag (d) NO (e) Cl_2

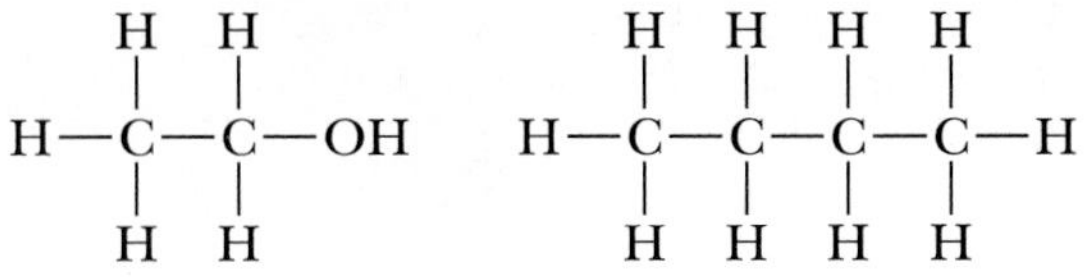

SOLUTION

(a) and (d) represent compounds—the second letter in an element symbol is never a capital letter; (b) and (c) represent the elements cobalt and silver; (e) represents the element chlorine, which exists as diatomic molecules (Table 2.4).

Exercise 2.2

Write the symbols or formulas for (a) lead, (b) phosphorus, (c) a molecule containing one hydrogen atom and one chlorine atom, (d) a molecule containing one aluminum atom and three bromine atoms, (e) elemental fluorine.

EXAMPLE 2.3 *Structural Formulas*

The compound whose structural formula is shown below is used in making heat-resistant glass and, in solution in water, as an eye wash. (a) Write the molecular formula. (b) Give the number of atoms in the molecule. (c) Name the elements present in the molecule.

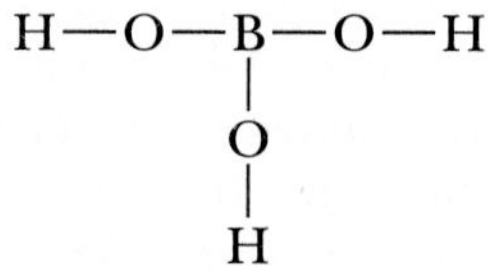

SOLUTION

(a) Using one symbol for each element and using subscripts to show the number of each kind of atom gives H_3BO_3. (b) There are seven atoms in the molecule—three H atoms, one B atom, and three O atoms. (c) The elements are hydrogen (H), boron (B), and oxygen (O). (The compound is boric acid.)

Exercise 2.3

Repeat parts (a), (b), and (c) of Example 2.3 for the structure shown below.

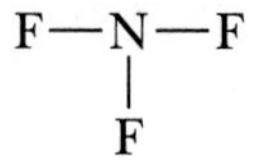

Chemical Equations

To concisely represent chemical reactions, symbols and formulas are arranged in **chemical equations.** For example, in words, a chemical reaction could be expressed as "carbon reacts with oxygen to form carbon monoxide," a reaction that happens whenever something containing carbon burns incom-

■ Chemical equations are the sentences in the language of chemistry.

pletely. Like most solid elements, carbon is represented just by its symbol, C; oxygen must be represented by the molecular formula for its diatomic molecules, O_2; and carbon monoxide molecules, which contain two atoms, one each of carbon and oxygen, are represented by their molecular formula, CO.

Earlier, we pointed out that in chemical reactions atoms are rearranged into different substances, but the number of atoms stays the same. To represent this correctly, chemical equations must be **balanced**—the number of atoms of each kind in the reactants and products must be the same. In this case, one oxygen molecule will combine with two carbon atoms to form two carbon monoxide molecules. All this information is contained in the equation

■ States of matter are shown in equations as (*g*) for a gas, (*ℓ*) for a liquid, and (*s*) for a solid; (*aq*) is used for a substance dissolved in water (an **aqueous solution**).

$$2\,C(s) + O_2(g) \longrightarrow 2\,CO(g)$$

We have also included here the information that carbon is a solid (*s*) and that oxygen and carbon monoxide are gases (*g*). The arrow ($\longrightarrow$) is often read as "yields." The equation, then, states the following:

At the macro level: Carbon, a solid, plus oxygen gas yields carbon monoxide gas.

At the submicroscopic level: Two atoms of carbon plus one diatomic molecule of oxygen yield two molecules of carbon monoxide.

The number written before a formula in an equation, the **coefficient,** gives the relative amount of the substance involved. The subscripts give the composition of the pure substances. Changing the coefficient changes only the amount of the element or compound involved, whereas changing a subscript would change the identity of the reactant or product. For example, 2 CO represents two molecules of carbon monoxide, whereas CO_2 represents a molecule of carbon dioxide, a very different substance formed in the complete combustion of carbon-containing materials:

$$C(s) + O_2(g) \longrightarrow CO_2(g)$$

EXAMPLE 2.4 *Chemical Equations*

Nitrogen dioxide (NO_2) is a red-brown gas often visible in the haze during a period of air pollution over a city. (a) Interpret in words the information given in the equation for formation of nitrogen dioxide from nitrogen monoxide (NO):

■ NO is commonly known as nitric oxide.

$$2\,NO(g) + O_2(g) \longrightarrow 2\,NO_2(g)$$

(b) Prove that the equation is balanced.

SOLUTION

(a) Nitrogen monoxide, a gas, reacts with oxygen gas to give nitrogen dioxide, also a gas. The coefficients show that two molecules of NO react with one molecule of O_2 to give two molecules of NO_2.

(b) *Reactants*: The coefficient 2 in 2 NO shows the presence of two molecules of NO, which means two N atoms and two O atoms. The O_2 has no coefficient, so there is one O_2 molecule, but the subscript shows that the single molecule has two O atoms. Thus, there are two N atoms and four O atoms on the reactants side. *Products*: The coefficient in 2 NO_2 shows two molecules of NO_2, which gives a total of two N atoms (N has no subscript) and four O atoms (two in each of the two molecules, as shown by the subscript on O). The equation is balanced.

Exercise 2.4

Repeat the question of Example 2.4 for the equation for the formation of hydrogen chloride (HCl):

$$H_2(g) + Cl_2(g) \longrightarrow 2\ HCl(g)$$

SELF-TEST 2B

1. Chemists strive to understand the ______________ of matter in order to create useful new materials.
2. Identify the elements with the following properties and write their symbols:
   (a) A soft reddish metal used in electrical wiring
   (b) A greenish-yellow poisonous gas with a sharp odor
3. Name two elements that are solids, two that are liquids, and two that are gases.
4. Write the symbols for the six elements you named in Question 3.
5. Name the elements combined in the compound K_2HPO_4. How many atoms in total are represented in this chemical formula?
6. Write a condensed formula for methyl alcohol, the simplest of the organic family of alcohols, whose structural formula is

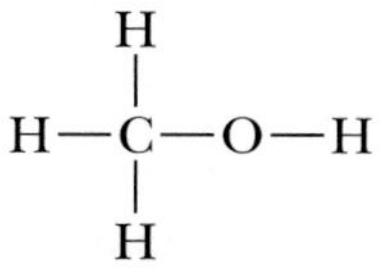

7. Are the following equations balanced?
   (a) $2K(s) + Cl_2(g) \rightarrow KCl(s)$
   (b) $2Mg(s) + O_2(g) \rightarrow 2MgO(s)$
   (c) $SO_3(g) + H_2O(\ell) \rightarrow 2\ H_2SO_4(aq)$
   (d) $NaCl(aq) + AgNO_3(aq) \rightarrow NaNO_3(aq) + AgCl(s)$
8. Consider the chemical equation $CH_4 + 2\ O_2 \rightarrow CO_2 + 2\ H_2O$. Explain what is meant by the symbols:
   (a) O ______________________________
   (b) 2 O_2 ______________________________
   (c) CH_4 ______________________________
   (d) $\rightarrow$ ______________________________

(e) H_2O ____________________

(f) 2 H_2O ____________________

9. A solid element does not conduct electricity. Is it a metal or a nonmetal?
10. Which of the following elements have symbols not based on their modern names? (a) Lead, (b) Carbon, (c) Oxygen, (d) Potassium.

The World of Chemistry

Program 5, *A Matter of State*

It would be inconvenient to measure the components of this circuit board in meters. The millimeter is much more convenient.

2.8 The Quantitative Side of Science

Standards and Measurement

Several times in the preceding sections, we used the numerical results of measurements of the boiling points, masses, or densities of pure substances. These and hundreds of other kinds of measurements are fundamental to chemistry and every other science. The result of a measurement is what we refer to as **quantitative** information—it uses numbers. Weighing yourself is a quantitative experiment. By contrast, there is **qualitative** information that does not deal with numbers. Eating an artichoke if you have never eaten one before is a qualitative experiment.

■ Anyone making scientific measurements must understand the *significance* of the digits when they are using the results of measurements. This becomes especially important in using electronic calculators. If a 3-cm^3 piece of something weighs 4.52 g, is its density 1.5 g/cm^3, or 1.50 g/cm^3, or, as given by a calculator, 1.506666667 g/cm^3? Appendix A explains how questions like this are dealt with.

The result of a measurement is recorded as a number plus a unit. "A boiling point of 52," doesn't mean anything without the unit. Was the temperature measured in Fahrenheit degrees, as is done by the usual household thermometer in the United States? Or in Celsius degrees, as would be done in the rest of the world?

To report the result of a measurement correctly requires (1) reading numbers from the measuring device, (2) choosing how many digits to include in the numerical part of the measurement, and (3) using the units of the measuring device or, in some cases, converting the result of the measurement to other units.

As an example of (1) and (2), take a look at Figure 2.12. The temperature could be reported as 52°C or 53°C, or as 52.7°C if an estimated third digit is included. It should be pretty obvious that it would be faulty to report the temperature from this thermometer as 52.7753°C. Such a measurement would require a much more sensitive thermometer.

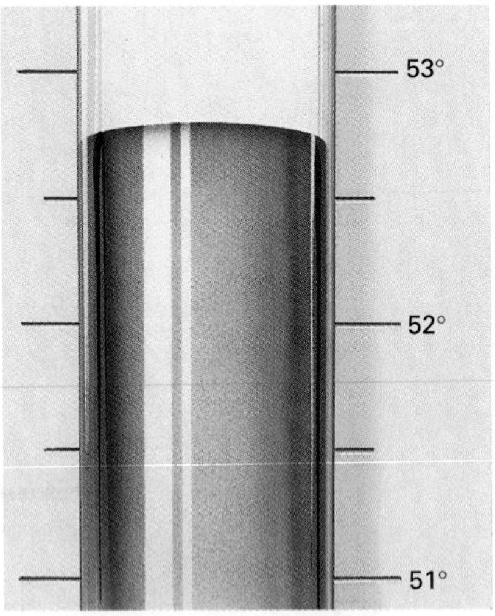

Figure 2.12 What is the temperature?

The establishment of scientific facts and laws is obviously dependent on accurate observations and measurements. Although measurements can be reported as precisely in one system of measurement as another, there has been an effort since the time of the French Revolution in the late 1700s to have all scientists embrace the same simple system. The hope was and is to facilitate communication in science. The metric system, which was born of this effort, has two advantages. First, it is easy to convert from one unit to another, since smaller and larger units for the same physical quantity differ only by multiples of ten. Consequently, to change millimeters to meters, the decimal point need only be moved three places to the left:

1 millimeter = 0.001 meter therefore, 5.0 millimeters = 0.0050 meter

Compare the decimal shift to the problem of changing inches to yards. Knowing that

$$1 \text{ ft} = 12 \text{ inches} \qquad 1 \text{ yard} = 3 \text{ feet}$$

we can calculate that

$$5.0 \cancel{\text{inches}} \times \frac{1 \cancel{\text{ft}}}{12 \cancel{\text{inches}}} \times \frac{1 \text{ yard}}{3 \cancel{\text{ft}}} = 0.14 \text{ yard}$$

The second advantage of the modern metric system is that standards for most fundamental units are defined by reproducible phenomena of nature. For example, the metric unit for time—the second—is now defined in terms of a specific number of cycles of radiation from a radioactive cesium atom, a time period believed never to vary.

The units still in everyday use in the United States have evolved from the English system of measurement, which began with the decrees of various English monarchs. The ''yard'' started out as the length of the waist sash worn by Saxon kings and obviously varied with the girth of the king. A step toward standardization came when King Henry I decreed that the yard should be the distance from the tip of his nose to the end of his thumb. Standards for the English units in use in the United States today are designated by the National Institute of Standards and Technology (NIST).

Since its introduction in 1790 the metric system has been continually modified and improved. The current version, the International System of Units, abbreviated SI (from the French Système International d'Unités), was adopted by the International Bureau of Weights and Measures in 1960. The seven fundamental units listed in Table 2.6 are defined by the SI system; other units are derived from them. Area, for example, is length × length, so the official SI unit for area is the square meter (m^2); volume is length × length × length, so the SI unit for volume is the cubic meter (m^3). Multiples and fractions of the fundamental units are designated by adding prefixes (Table 2.7). Thus, when a length unit smaller than the meter is needed, we can choose the *centi*meter (cm) or the *milli*meter, and when a larger length unit is needed, we can use the *kilo*meter. For mass units, prefixes are added to *gram* to indicate smaller quantities, such as the milligram or microgram.

The World of Chemistry

Program 3, *Measurement: The Foundation of Chemistry*

It doesn't really matter too much what you define a meter to be—you could define it as a very small length like the length of your pen. But unless everyone in the world agrees that that's a meter then you don't have a standard of measurement that's adequate to today's world. Stanley Rasberry, Director, Standard Reference Materials Division, National Institute of Standards and Technology, formerly the National Bureau of Standards.

TABLE 2.6 ■ The Fundamental SI Units

Physical Quantity	Name of Unit	Symbol
Length	Meter	m
Mass	Kilogram	kg
Time	Second	s
Thermodynamic temperature	Kelvin	K
Luminous intensity*	Candela	cd
Electric current	Ampere	A
Amount of a substance†	Mole	mol

*A quantity rarely used in chemistry

†An important quantity in chemistry that is explained in Section 8.2

■ The prefixes in Table 2.7 are represented by the powers of 10 used in scientific, or exponential, notation for writing large and small numbers. For example, $10^3 = 10 \times 10 \times 10 = 1000$. Appendix B reviews this notation.

TABLE 2.7 ■ Common Prefixes for Multiples and Fractions of SI Units

Prefix	Abbreviation	Meaning	Example
mega-	M	10^6 (1 million)	1 megaton = 1×10^6 tons
kilo-	k	10^3 (1 thousand)	1 kilogram (kg) = 1×10^3 g
deci-	d	10^{-1} (1 tenth)	1 decimeter (dm) = 0.1 m
centi-	c	10^{-2} (1 one-hundredth)	1 centimeter (cm) = 0.01 m
milli-	m	10^{-3} (1 one-thousandth)	1 millimeter (mm) = 0.001 m
micro-	μ*	10^{-6} (1 one-millionth)	1 micrometer (μm) = 1×10^{-6} m
nano-	n	10^{-9} (1 one-billionth)	1 nanometer (nm) = 1×10^{-9} m
pico-†	p	10^{-12} (1 one-trillionth)	1 picometer (pm) = 1×10^{-12} m

*This is the Greek letter mu (pronounced "mew").

†This prefix is pronounced "peako."

EXAMPLE 2.5 *Metric Unit Prefixes*

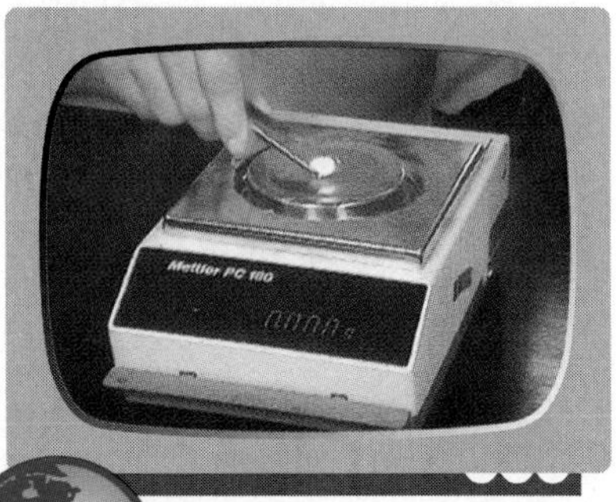

The World of Chemistry

Program 3, *Measurement: The Foundation of Chemistry*

A typical laboratory balance.

How many meters are in 2 km?

SOLUTION

The prefix kilo- (k) means 1000 times, meaning that 1000 m = 1 km. It is hardly worth the trouble to write anything down in this solution. One just thinks 1000 meters for every kilometer as one thinks ten dimes for every one-dollar bill. The best way to write down this problem, or any more complicated problem, however, is to include the units and cancel them out (like numbers in algebra). This way, if you are left with the correct unit, you have probably done the problem correctly.

$$2\ \cancel{\text{km}} \times \frac{1000\ \text{m}}{1\ \cancel{\text{km}}} = 2000\ \text{m}$$

Exercise 2.5

How much in dollars is 20 *mega*bucks?

EXAMPLE 2.6 *Sizes of Metric Units*

■ The symbol μ in the abbreviation for microgram, μg, is the Greek letter mu.

A vitamin pill contains 18 mg (milligrams) of iron and 150 μg (micrograms) of iodine. To compare the quantities, convert both of these values to grams.

SOLUTION

Table 2.7 shows that one *milli*gram equals 0.001 g (10^{-3} g) and one microgram equals 0.000 001 g (10^{-6} g). The conversions can be made by moving decimal points three places to the left for milligrams-to-grams conversion and six places to the left for micrograms to grams. *Or*, the problem can be done as in Example 2.5 by recognizing that the prefixes show that 1 gram = 1×10^3 mg

(1000 mg) and 1 gram = 1×10^9 μg (1,000,000 μg), writing down units, and canceling them. Choose whichever method you find easier to do.

$$18 \text{ mg of iron} = 0.018 \text{ g of iron } or\ 18\ \cancel{\text{mg}} \times \frac{1 \text{ g}}{1000\ \cancel{\text{mg}}} = 0.018 \text{ g}$$

$$150\ \mu\text{g of iodine} = 0.000150 \text{ g of iodine } or\ 150\ \cancel{\mu\text{g}} \times \frac{1 \text{ g}}{1{,}000{,}000\ \cancel{\mu\text{g}}} = 0.000150 \text{ g}$$

Exercise 2.6

Which way would you move the decimal point to convert grams to milligrams or micrograms? In the same vitamin pill described in Example 2.6 there are 0.060 g of vitamin C and 0.000400 g of folic acid. Express these quantities in milligrams and micrograms, respectively.

Units Common in Chemistry and Everyday Use

Because of their convenient sizes, the five units listed in Table 2.8 are commonly used in chemistry and indeed are in everyday use in the rest of the world (see *Science and Society: Coming Attractions—The Metric System,* p. 44). We will use the unit symbols listed in Table 2.8 in this text. The volume unit of liters is preferable to the SI-derived unit of cubic meters, which is much too large (1 cubic meter = 1000 liters). Examples 2.7 and 2.8 illustrate how quantities are converted from one unit to another. See Table 2.9 and Figure 2.13 for some other unit equivalences.

■ We tend to use the words *mass* and *weight* interchangeably, and that's all right because we understand what we mean. Strictly speaking, though, *mass* measures the amount of matter in an object and is the same everywhere in the universe, while *weight* measures the force of gravity on that object and is different where the force of gravity is different (e.g., on the moon).

TABLE 2.8 ■ Some Common Units in Chemistry

Name of Unit	Symbol	Common Equivalent
Meter	m	39.4 inches
Liter	L	1.06 quarts
Gram	g	0.0352 ounce
Degrees Celsius	°C	Water boils at 100°C and freezes at 0°C
Calorie	cal	Energy required to heat 1 g of water 1°C

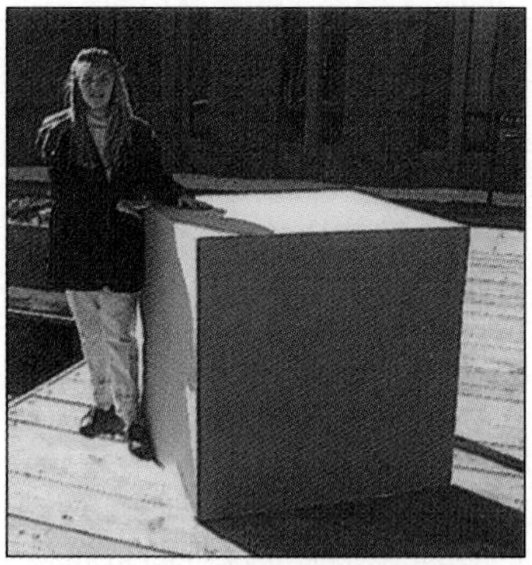

A cubic-meter box. Each edge of the box is 1 meter long. *(C.D. Winters)*

EXAMPLE 2.7 *Unit Conversions*

A typical slice of white bread contains 70 dietary calories, usually represented by Cal or just C. One dietary calorie is equal to 1000 small (scientific) calories (cal): 1 C = 1000 cal. How many small calories are provided by this slice of white bread?

SOLUTION

$$70\ \cancel{\text{Cal}} \times \frac{1000 \text{ cal}}{1\ \cancel{\text{Cal}}} = 70{,}000 \text{ cal}$$

You can see why dietitians like the larger unit; with it, they can use smaller numbers in their notations. When in need of a larger unit for energy, a chemist would use the same size unit, but would call it a *kilo*calorie.

Exercise 2.7

Express in calories and kilocalories the energy provided by a 225-C slice of chocolate cake.

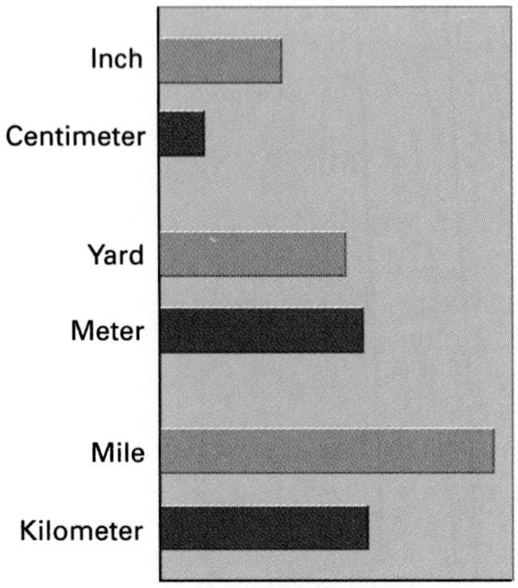

Figure 2.13 Relative sizes of some English and metric length units: 1 inch = 2.54 centimeters; 1 yard = 0.91 meter; 1 mile = 1.61 kilometers.

TABLE 2.9 ■ Ingredients for Perfect Brownies in English and Metric Measurements

Ingredient	English	Metric (SI)
Unsweetened chocolate squares	2 oz	60 g
Butter or margarine	$\frac{1}{2}$ c	112 g
Sugar	1 c	240 g
Eggs	2	2
Vanilla	1 tsp	5 mL
Sifted enriched flour	$\frac{1}{2}$ c	84 g
Chopped walnuts	$\frac{1}{2}$ c	50 g
Oven	325°F	163°C
Pan	8 × 8 × 2 in.	20 × 20 × 5 cm

EXAMPLE 2.8 *Unit Conversions*

(a) How many fluid ounces are in 1.5 L (liter)? How many milliliters? (b) How many liters are equal to 3840 fl oz?

SOLUTION

(a) You might remember that there are 32 fl oz/qt, and Table 2.8 shows that there are 1.06 qt/liter. The first solution is therefore

$$1.5\ \cancel{\text{L}} \times \frac{1.06\ \cancel{\text{qt}}}{1\ \cancel{\text{L}}} \times \frac{32\ \text{fl oz}}{1\ \cancel{\text{qt}}} = 51\ \text{fl oz}$$

Since there are 1000 mL/L,

$$1.5\ \cancel{\text{L}} \times \frac{1000\ \text{mL}}{1\ \cancel{\text{L}}} = 1500\ \text{mL}$$

Notice again how units are included and canceled in solving the problem. This is an excellent way to keep track of what you are doing. If the units of the answer are correct, the calculation set up and the numerical answer are probably correct. (Appendix C provides more discussion of this method.)

Weights and measures may be ranked among the necessaries of life to every individual of human society. They enter into the economical arrangements and daily concerns of every family. They are necessary to every occupation of human industry. . . . The knowledge of them, as in established use, is among the first elements of education. John Quincy Adams in a Report to the Congress in 1821.

(b) Going in the reverse direction requires turning over the conversion factors:

$$3840 \cancel{oz} \times \frac{1 \cancel{qt}}{32 \cancel{fl\ oz}} \times \frac{1\ L}{1.06 \cancel{qt}} = 113\ L$$

■ Any equivalence between units, such as 1 inch = 2.54 cm, provides two conversion factors that can be used to convert between the two units. In this case,

$$\frac{1\ \text{inch}}{2.54\ \text{cm}} \text{ or } \frac{2.54\ \text{cm}}{1\ \text{inch}}$$

Exercise 2.8

What is the volume in liters of a bottle that contains 64 fl oz of grapefruit juice?

EXAMPLE 2.9 *Density*

A large piece of pine wood to be used as a tabletop has a volume of 32,000 cm^3. The density of pine is 0.50 g/cm^3. What is the mass of this piece of pine in grams?

SOLUTION

Density is used in a calculation just like any of the unit conversion factors.

$$32{,}000 \cancel{cm^3} \times \frac{0.50\ g}{1 \cancel{cm^3}} = 16{,}000\ g$$

Exercise 2.9

(a) What is the mass in ounces of the board described above? (b) What is the mass of the board in pounds (16 ounces = 1 pound)?

One-quart and one-liter containers. *(C.D. Winters)*

SCIENCE AND SOCIETY

Coming Attractions—The Metric System

The United States, Liberia, and Myanmar (formerly Burma) have something in common. We are the only countries still weighing and measuring in pounds, feet, and quarts. In all other countries, kilograms, meters, and liters are in everyday use.

Why have we avoided speaking the same measurement language as the rest of the world? It has long been recognized that communication is best served by a common system of weights and measures. Since 1866, by act of Congress, using the metric system has been legal in the United States. And in 1975 the federal government passed the Metric Conversion Act, which was meant to initiate a gradual changeover. But the Act was based on voluntary change, and very little happened.

Among the public in the United States, the fundamental opposition to change has surfaced in concerns about excessive cost. Reflecting this, the Congress in 1994 banned using federal funds for metric highway signs. There is also the usual discomfort with the unfamiliar, attributed by some teachers to widespread math phobia, and the conservatism of Americans for Customary Weight and Measure, whose members believe that what has been customary is what is right.

One individual who experienced the changeover to metric from the English system in Australia in the late 1960s, however, reported that it wasn't long before people threw away their conversion charts. The logic and simplicity of a system based on multiples of 10 overcame initial concerns.

Two ways to convert to the metric system: By retaining the same size containers and relabeling (left) or by adopting metric-size containers (right). (C.D. Winters)

Suppose you go to the supermarket tonight and find that you have to buy kilos of potatoes, milk labeled in liters, and aluminum foil measured in meters or centimeters. Or suppose that you manufacture chocolate bars and will have to design new labels that give their weight in grams? How would you react? Would you be happy that we are finally joining up with the rest of the world? Would you be annoyed at the inconvenience? Would you be uncomfortable about the arithmetic you might have to do? Would you be upset over the cost of new labels? Would the prospect of additional sales outside the United States make conversion more attractive?

The necessity and benefits of using the metric system become more obvious as the movement of people, goods, and information from country to country accelerates. The 1990s may be the decade when avoiding metric units is no longer possible.

We already buy soda in one- and two-liter bottles, with little inconvenience, and most packaged food labels give both metric and English units. Larger numbers of elementary school students, in addition to chemistry students, are learning about grams, meters, and liters. In 1994 the Federal Trade Commission ruled that consumer products such as detergents, toilet paper, and batteries must also be labeled with both types of units. Perhaps public acceptance of metric units will be eased by such exposure. In addition, by 1996 all projects financed by federal highway funds are supposed to be designed in metric units.

So far, the United States has not mandated a general metric changeover for *all* public and private endeavors, and perhaps that will not be needed. According to an estimate from the National Institute of Building Sciences, about 30% of U.S. products had gone metric by 1994. A deadline lies ahead that will undoubtedly motivate metric changeover in any industry that wants to compete globally. As of 1999 the European Union will stop buying nonmetric goods.

■ SELF-TEST 2C

1. The result of a measurement must include both a ____________ and a ____________.
2. The metric system uses ____________ to indicate multiples of ____________.
3. The volume unit derived from the centimeter is a ____________ ____________.

4. The prefix meaning 1000 times bigger is ____________, and the prefix meaning 0.001, or 1000 times smaller, is ____________.
5. What are the units for the answer to the calculation below?

$$0.500\ \text{ft} \times \frac{12\ \text{inches}}{1\ \text{ft}} \times \frac{1\ \text{m}}{39.4\ \text{inches}} \times \frac{1000\ \text{mm}}{1\ \text{m}} = 152\ \_\_\_\_\_\_\_\_$$

6. Fill in the units for these three ways of expressing the volume of a popular large-size bottle for soda: 2 ____________, or 67.6 ____________, or 2000 ____________.
7. Which conversion factor should you use to convert miles to kilometers?
   (a) $\frac{1\ \text{mile}}{1.61\ \text{km}}$, (b) $\frac{1.61\ \text{km}}{1\ \text{mile}}$
8. Which of the following experiments is qualitative and which is quantitative?
   (a) Determination of the distance between two atoms in a molecule.
   (b) Determination of the identity of the metal in a piece of wire.

MATCHING SET

____ 1. Produces a new type of matter	a. Filtration
____ 2. Air	b. Chemical change
____ 3. Unchanged by further purification	c. Element
____ 4. Used to separate a solid from a liquid	d. Properties of pure substance
____ 5. Cannot be reduced to simpler substances	e. Mixture
____ 6. SI system	f. Uses multiples of ten
____ 7. Symbol for iron	g. O_3
____ 8. Volume unit	h. CO
____ 9. Molecule containing three oxygen atoms	i. Fe
____ 10. Carbon monoxide	j. 1 m
____ 11. 100 cm	k. H—S—H
____ 12. SI fundamental unit	l. Liter
____ 13. Deci-	m. One-tenth
____ 14. Centi-	n. Meter
____ 15. Structural formula	o. C_4H_{10}
____ 16. Solution	p. Homogeneous mixture
____ 17. Molecular formula	q. One-hundredth
____ 18. Condensed formula	r. $CH_3CH_2CH_3$

QUESTIONS FOR REVIEW AND THOUGHT

1. Name as many materials as you can that you have used during the past day that were not chemically changed from their natural states.
2. Identify the following as physical or chemical changes. Explain your choices.
   (a) Formation of snowflakes
   (b) Rusting of a piece of iron
   (c) Ripening of fruit
   (d) Fashioning a table leg from a piece of wood
   (e) Fermenting grapes
   (f) Boiling a potato
3. Would it be possible for two pure substances to have exactly the same set of properties? Give reasons for your answer.
4. List three physical properties that can be used to identify pure substances. Give a specific example of each property.
5. Describe in your own words what happens when you flick a BIC lighter.

6. Chemical changes can be both useful and destructive to humanity's purposes. Cite a few examples of each kind of change from your own experience. Also give evidence from observation that each is indeed a chemical change and not a physical change.
7. Classify each of the following as a physical property or a chemical property. Explain your answers.
   (a) Density
   (b) Melting temperature
   (c) Decomposition of a substance into two elements upon heating
   (d) Electrical conductivity
   (e) The failure of a substance to react with sulfur
   (f) The ignition temperature of a piece of paper
8. Write your last name. How many element symbols can you produce from the letters in your name?
9. Classify each of the following as an element, a compound, or a mixture. Justify each answer.
   (a) Mercury
   (b) Milk
   (c) Pure water
   (d) A piece of lumber
   (e) Ink
   (f) Iced tea
   (g) Pure ice
   (h) Carbon
   (i) Antimony
10. Which of the materials listed in Question 9 can be pure substances?
11. Is it possible for the properties of iron to change? What about the properties of steel, which is an alloy and a homogeneous mixture? Explain your answer.
12. You have a mixture of sand (SiO_2) and salt (NaCl). How would you separate these two substances? When you have them as separate substances, how can you prove which is which?
13. Suggest a method for purifying water slightly contaminated with a dissolved solid.
14. Consider the following five elements: nitrogen, sulfur, chlorine, magnesium, and cobalt. Using this text or any other source available at the library, find the major source for these elements and at least one compound that uses the element in combined form.
15. Atrazine is a selective herbicide that has the molecular formula $C_8H_{14}N_5Cl$. This compound is used for season-long weed control in corn, sorghum, and certain other crops. What elements are present in atrazine?
16. Cytoxan, also known as cyclophosphamide, is widely used alone or in combination in the treatment of certain kinds of cancer. It interferes with protein synthesis and in the process kills rapidly replicating cells, particularly malignant ones. Cytoxan has the molecular formula $C_7H_{15}O_2N_2PCl_2$.
   (a) How many atoms are in one molecule of cytoxan?
   (b) What elements are present in cytoxan?
   (c) What is the ratio of hydrogen atoms to nitrogen atoms in cytoxan?
   (d) Would cytoxan be classified as an organic compound?
   (e) Does cytoxan contain any metal atoms?
17. By reading labels, identify a commercial product that contains each of the following compounds:
   (a) Calcium carbonate
   (b) Phosphoric acid
   (c) Water
   (d) Fructose
   (e) Sodium chloride
   (f) Potassium sorbate
   (g) Potassium iodide
   (h) Glycerol
   (i) Aluminum
   (j) Butylated hydroxytoluene (BHT)
18. There are three states of matter—gas, liquid, and solid. Name a material that is a pure substance for each state of matter. Do not use water, oxygen, or salt. Name a material that is a mixture for each state of matter. Do not use air, gasoline, or brass.
19. Given the following sentence, write a chemical equation using chemical symbols that convey the same information: "One nitrogen molecule reacts with three hydrogen molecules to produce two ammonia molecules, each containing one nitrogen and three hydrogen atoms."
20. Is it possible to have a mixture of two elements and also to have a compound of the same two elements? Explain. Can you think of an example?
21. Name four kinds of energy.
22. Describe in words the chemical processes represented by the following equations:
   (a) $2\,Na(s) + Cl_2(g) \rightarrow 2\,NaCl(s)$
   (b) $N_2(g) + 3\,Cl_2(g) \rightarrow 2\,NCl_3$ [named as nitrogen trichloride]
   (c) $CO_2(g) + H_2O(\ell) \rightarrow H_2CO_3(aq)$ [carbonic acid]
   (d) $2\,H_2O_2(aq)$ [hydrogen peroxide] $\rightarrow O_2(g) + 2\,H_2O(\ell)$
23. Prove that each of the equations in Question 22 is balanced.
24. For equations b and d in Question 22, identify the reactant(s) and the product(s).
25. Are the following equations balanced?
   (a) $AgNO_3(aq) + Na_2SO_4(aq) \rightarrow Ag_2SO_4(s) + NaNO_3(aq)$
   (b) $AgNO_3(aq) + HCl(aq) \rightarrow AgCl(s) + HNO_3(aq)$
26. Is the tea in tea bags a pure substance? Use the process of making tea to make an argument for your answer. How would your argument apply to instant tea?
27. Find and list as many pure substances as you can in a kitchen.
28. (a) How many milligrams are there in one gram?
   (b) How many meters are there in one kilometer?
   (c) How many centigrams are there in one gram?

29. What are the most common units in chemistry for mass, length, and volume?
30. Which of the following quantities is a density?
    (a) 9 cal/gram (b) 100 cm/meter
    (c) 1.5 g/mL (d) 454 g/lb
31. A cook wants to pour 1.5 liter of batter into a 2-quart bowl. Will it fit?

PROBLEMS

1. Suppose you have 8 dozen marbles. How many marbles do you own?
2. You have made 10 liters of lemonade to be poured out in 200-mL servings. How many glasses will you need?
3. A common storage capacity for a PC hard drive is 200 Mbytes. Express this number in kilobytes and bytes.
4. A brand of reduced-fat peanut butter advertises 30% less fat than the regular kind. If the serving size is 36 grams and the reduced-fat peanut butter contains 12 grams of fat per serving, how much fat does the regular type of peanut butter contain per serving?
5. A rancher needs one acre of grazing land for 10 cows. How many acres are needed for 55 cows? Solve this problem (and others where appropriate) by including units. In this case, the answer will have the "units" of acres/cow.
6. There are 200 mg of ibuprofen in an Advil tablet. How many grams is this? How many micrograms?
7. A block of wood measures 5 cm by 20 cm by 10 cm. What is the volume of this block of wood?
8. If the block of wood in Problem 7 has a mass of 3000 grams, what is the ratio of mass to volume for this block of wood? What is the name of this quantity?
9. Convert the ratio found in Problem 8 to units of kg per m^3.
10. A medicine dropper delivers 25 drops of water to make 1.0 mL. What is the volume of one drop in cubic centimeters? in millimeters?
11. How many meters are in a 10-km race?
12. If a 1-ounce portion of cereal contains 3.0 g of protein, how many milligrams of protein does it contain?
13. Complete the following:
    (a) 4 cm = ________ m
    (b) 0.043 g = ________ mg
    (c) 15.5 m = ________ mm
    (d) 328 mL = ________ L
    (e) 0.98 kg = ________ g
14. An average African gorilla has a mass of 163 kg. What is this mass in grams?
15. An average adult man has a mass of 70. kg. Convert this mass to milligrams.
16. A paperclip is 3.3 cm long. Express this length in millimeters.
17. A particle of tobacco smoke has a diameter of 0.50 μm. Convert this value to centimeters.
18. The density of gold is 19.3 g/cm^3. What is the mass of 50.0 cm^3 of gold?
19. The density of air at everyday conditions is 0.0012 g/mL. What is the mass of one liter of air?
20. The density of pine wood is 0.50 g/cm^3. What is the volume of a 250-g piece of pine?
21. A piece of metal has a mass of 6.32 g and a volume of 0.8 cm^3. Is this piece most likely to be lead, density 11.4 g/cm^3, or iron, density 7.9 g/cm^3?
22. Which is colder, 0°C or 0°F?

CHAPTER 3

Atoms

As you focus on a dot over an "i," can you visualize the thousands and thousands of individual, very small atoms in the dot? Trying to fathom the minuteness of the atom is as deeply challenging to the human mind as trying to fathom the wholeness of the universe. One lures the mind to unseen smallness, the other to unseen largeness. The nature of atoms is the subject of this chapter.

- What are the milestones in the development of atomic theory?
- What is the experimental evidence for the existence of subatomic particles within atoms?
- What are the three basic subatomic particles of the atom, and where are they found?
- What are isotopes?
- Where are electrons in atoms, and how are they arranged?
- What is the relationship between electron energy levels in atoms of an element and the line emission spectrum of that element?

Why does an element or compound have the properties it has? Why does one element or compound undergo a change that another element or compound will not undergo? Inanimate matter is the way it is because of the nature of its parts. The use of atoms to represent the "parts" dates back to about 400 B.C. when the Greek philosopher Leucippus and his student, Democritus (460–370 B.C.), argued for a limit to the divisibility of matter, which was counter to the prevailing view of Greek philosophers that matter is endlessly divisible. Democritus used the Greek word *atomos,* which literally means "uncuttable," to describe the ultimate particles of matter, particles that could not be divided further.

Beautiful flame colors: they are explained by atomic structure. *(C.D. Winters)*

3.1 The Greek Influence on Atomic Theory

Democritus reasoned that if a piece of matter such as gold were divided into smaller and smaller pieces, one would ultimately arrive at a tiny particle of gold that could not be further divided and still retain the properties of gold. The atoms that Democritus envisioned representing different substances were all made of the same basic material. His atoms differed only in shape and size.

Democritus used his concept of atoms to explain the properties of substances. For example, the high density and softness of lead could be caused by lead atoms packed very closely together like marbles in a box and moved easily one over another. Iron was known to be a less dense metal that is quite hard. Democritus argued that the properties of iron resulted from atoms shaped like corkscrews, atoms that would entangle in a rigid but relatively lightweight structure.

■ Because the writings of Leucippus and Democritus have been destroyed, we know about their ideas only from the recorded opposition to atoms and from a lengthy poem (55 B.C.) by the Roman poet Lucretius.

All atomic theory has been built on the assumption of Leucippus and Democritus: The properties of matter that we can see are explained by the properties and behavior of atoms that we cannot see.

However, the atomic theory developed by Democritus was rejected by Plato (427–347 B.C.) and Aristotle (384–322 B.C.), who persisted in their belief that matter was continuous and could be endlessly subdivided. For many centuries most of those in the mainstream of enlightened thought rejected or ignored the atoms of Democritus, and ideas about atoms drifted in and out of philosophical discussions for about 2200 years without generally affecting how nature was understood. It wasn't until John Dalton (1766–1844) introduced his atomic theory in 1803 that the importance of using atoms to explain properties of matter was recognized.

Material objects are of two kinds, atoms and compounds of atoms. . . . The atoms must be made of imperishable stuff into which everything can be resolved in the end, so that there may be a stock of matter for building the world anew. Lucretius (circa 95–55 B.C.), *De Rerum Naturum*

3.2 John Dalton's Atomic Theory

John Dalton, drawing from his own quantitative experiments and those of earlier scientists, proposed in 1803 that

1. All matter is made up of indivisible and indestructible particles called atoms.
2. All atoms of a given element are identical, both in mass and in properties. Atoms of different elements have different masses and different properties.
3. Compounds are formed when atoms of different elements combine in the ratio of small whole numbers.
4. Elements and compounds are composed of definite arrangements of atoms, and chemical change occurs when the atomic arrays are rearranged.

The Parthenon is contemporary with early ideas about atoms.

John Dalton's atomic theory was accepted because it could be used to explain three scientific laws that were established by Dalton and other scientists of this time. These laws are (1) the law of conservation of matter, (2) the law of definite proportions, and (3) the law of multiple proportions.

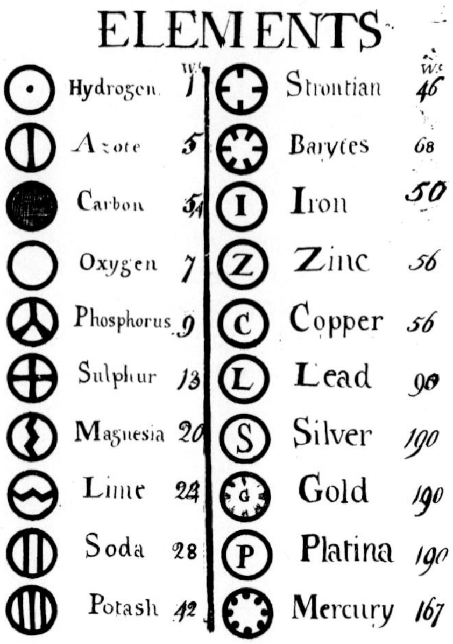

ELEMENTS
Hydrogen 1
Azote 5
Carbon 54
Oxygen 7
Phosphorus 9
Sulphur 13
Magnesia 20
Lime 24
Soda 28
Potash 42
Strontian 46
Barytes 68
Iron 50
Zinc 56
Copper 56
Lead 90
Silver 190
Gold 190
Platina 190
Mercury 167

Diagrams of the symbols and relative weights of atoms used by Dalton in his lecture on atomic theory. *(Stock Montage)*

THE PERSONAL SIDE

John Dalton (1766–1844)

(The Bettmann Archives)

John Dalton, a gentle man and a devout Quaker, gained acclaim because of his work. He made careful measurements, kept detailed records of his research, and expressed them convincingly in his writings. However, he was a very poor speaker and was not well received as a lecturer. When Dalton was 66 years old, some of his admirers sought to present him to King William IV. Dalton resisted because he would not wear the court dress. Since he had a doctor's degree from Oxford University, the scarlet robes of Oxford were deemed suitable, but a Quaker could not wear scarlet. Dalton, being color blind, saw scarlet as gray, so he was presented in scarlet to the court but in gray to himself. This remarkable man was, in fact, the first to describe color blindness. He began teaching in a Quaker school when only 12 years old, discovered a basic law of physics—the law of partial pressure of gases—and helped found the British Association for the Advancement of Science. He kept more than 200,000 notes on meteorology. Despite his accomplishments he shunned glory and maintained he could never find time for marriage.

Law of Conservation of Matter

Some years before Dalton proposed his atomic theory, Antoine Lavoisier (1743–1794) had carried out a series of experiments in which the reactants were carefully weighed before a chemical reaction and the products were carefully weighed afterward. He found no change in mass when a reaction occurred, proposed that this was true for every reaction, and called his proposal the **law of conservation of matter:** *Matter is neither lost nor gained during a chemical reaction.* Others verified his results, and the law became accepted. Points (2) and (4) in Dalton's theory imply the same thing. If each kind of atom has a particular characteristic mass, and if there are exactly the same number of each kind of atom before and after a reaction, the masses before and after must also be the same.

Law of Definite Proportions

Another chemical law known in Dalton's time had been proposed by Joseph Louis Proust (1754–1826) as a result of his analyses of minerals. Proust found that a particular compound, once purified, always contained the same elements in the same ratio by mass. One such study, made by Proust in 1799, involved copper carbonate. Proust discovered that, regardless of how copper carbonate was prepared in the laboratory or how it was isolated from nature, it always contained the same proportions by mass—five parts copper, four parts oxygen, and one part carbon. Careful analyses of this and other compounds led Proust to propose the **law of definite proportions:** *In a compound, the constituent elements are always present in a definite proportion by mass.*

Proust's experimental results for copper carbonate and other compounds are also explained by Dalton's theory—all atoms of copper are alike, all atoms of carbon are alike, and all atoms of oxygen are alike, but the three types of atoms differ from each other and have different masses. Therefore, a chemical compound of these elements will contain a fixed number of copper atoms, carbon atoms, and oxygen atoms.

EXAMPLE 3.1 *Law of Definite Proportions*

If pure water is always 88.81% oxygen and 11.19% hydrogen by mass, how many grams of oxygen and hydrogen are combined to form 20.00 g of water?

SOLUTION

This problem makes use of the law of definite proportions. Since water is always 88.81% (the decimal equivalent, 0.8881) oxygen by mass, the number of grams of oxygen in 20.00 g of water is

$$88.81\% \text{ of } 20.00 \text{ g water or } 0.8881 \times 20.0 \text{ g} = 17.76 \text{ g O}$$

The grams of hydrogen can be found by using the percentage of hydrogen in water:

$$20.00 \text{ g water} \times 0.1119 = 2.24 \text{ g H}$$

To check the calculations, add the masses of H and O; they should equal 20.00.

$$17.76 \text{ g O} + 2.24 \text{ g H} = 20.00 \text{ g } H_2O$$

Exercise 3.1

Ammonia gas contains 82.2% nitrogen and 17.8% hydrogen by mass. (a) How many grams of each element, nitrogen (N) and hydrogen (H), are in 50.0 grams of ammonia? (b) More than 18 billion tons of ammonia are produced each year for use in making fertilizers, plastics, and explosives. How many tons of hydrogen and nitrogen are needed to make 50.0 tons of ammonia?

THE PERSONAL SIDE

Antoine-Laurent Lavoisier (1743–1794)

There are many reasons why Antoine-Laurent Lavoisier has been acclaimed the father of chemistry. He clarified the confusion over the cause of burning. He wrote an important textbook of chemistry, *Elementary Treatise on Chemistry*. He was the first to use systematic names for the elements and a few of their compounds. However, his most notable achievement was to show the importance of accurate weight measurements of chemical changes, and his quantitative experiments on chemical changes led him to propose the law of conservation of matter.

(continued on next page)

THE PERSONAL SIDE (continued)

A drawing by Madame Lavoisier of Lavoisier studying the chemistry of breathing in his laboratory. She is seated at the right, taking notes, as she did for all of his experiments. (The Bettman Archives)

With all of his success, Lavoisier had his problems and disappointments. His highest goal, that of discovering a new element, was never achieved. He lost some of the esteem of his colleagues when he was accused of saying the work of someone else was his own. In 1768 he invested half a million francs in a private firm retained by the French government to collect taxes. He used the earnings (about 100,000 francs a year) to support his research. Although Lavoisier was not actively engaged in tax collecting, he was brought to trial as a "tax-farmer" during the French Revolution. Lavoisier, along with his father-in-law and other tax-farmers, was guillotined on May 8, 1794, just two months before the end of the revolution. On that day, one of Lavoisier's scientific colleagues said, "It took but a moment to cut off that head; perhaps a hundred years will be required to produce another like it."

Law of Multiple Proportions

Dalton's own experimental results from the analysis of different compounds made from the same elements led him to discover that compounds of the same elements often differed in composition but each still obeyed the law of definite proportions. For example, carbon combines with oxygen in a mass ratio of 3.0 to 4.0 to give carbon monoxide, a poisonous gas, and in a mass ratio of 3.0 to 8.0 to give carbon dioxide, a product of respiration and burning fuels. Both compounds contain only carbon and oxygen. Note that for equal amounts of carbon (three parts), the ratio of oxygen in the two compounds is 4 to 8, or 1 to 2. Hence, the formulas of the two compounds are CO and CO_2. These results and similar results for other compounds led Dalton to propose the **law of multiple proportions:** *In the formation of two or more compounds from the same elements, the masses of one element that combine with a fixed mass of a second element are in a ratio of small whole numbers.*

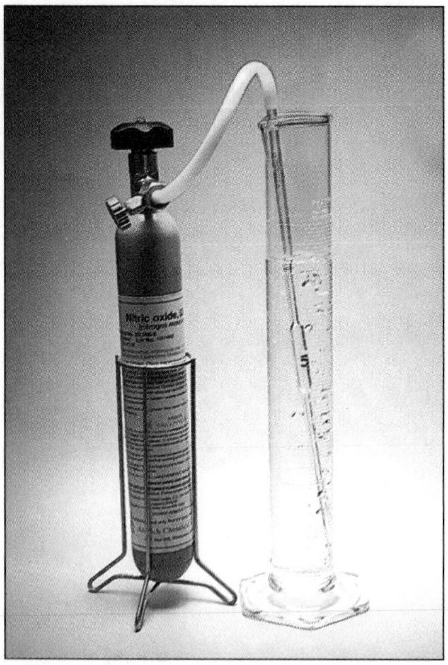

NO and NO_2. The gas NO, nitric oxide, is stored in a tank. It is bubbled through water, where it is evident that the gas is colorless. However, as soon as the bubbles of NO enter the atmosphere, the NO reacts with oxygen to form brown NO_2, nitrogen dioxide. *(C.D. Winters)*

This aspect of Dalton's atomic theory—that it suggested a new law—stimulated Dalton and other scientists to do more research, which resulted in scientific progress. This is an example of the interdependence of theory and experimentation to move science forward. A successful theory not only accounts for existing knowledge but also stimulates the search for new knowledge. Dalton's atomic theory was widely accepted for most of the 19th century.

3.3 Structure of the Atom

Dalton's view of atoms as hard and indivisible yielded to the idea that atoms are themselves composed of smaller particles as a result of several discoveries during the last part of the 19th century and the first decade of the 20th century.

Natural Radioactivity

The discovery of natural radioactivity, a spontaneous process in which some natural materials give off very penetrating radiations, indicated that atoms must have some kind of internal structure. Henri Becquerel (1852–1908) discovered this property in natural uranium and radium ores in 1896 (Section 5.1). In 1898 Marie Sklodowska Curie (1867–1934), a student of Becquerel, and her husband, Pierre, discovered two radioactive elements, radium and polonium. In 1899 Marie Curie suggested that atoms of radioactive substances disintegrate when they emit these unusual rays. She named this phenomenon **radioactivity.** Whether in a compound or uncombined, each radioactive element gives off exactly the same rays; about 25 elements exist only in radioactive forms. Marie Curie's suggestion that atoms disintegrate contradicts Dalton's idea that atoms are indivisible. If atoms can break apart, there must be something smaller than an atom; that is, there must be **subatomic particles.** We now refer to the kinds and arrangements of these particles as **atomic structure.**

Electrical Nature of Matter

Electricity was involved in many of the experiments that led to discoveries of subatomic particles. In fact, electric charge was first observed and recorded by the ancient Egyptians, who noted that amber, when rubbed with wool or silk, attracted small objects. A bolt of lightning, a spark between a comb and hair in dry weather, or a shock upon touching a doorknob results from an electric charge moving from one place to another.

■ Amber is fossilized tree sap, and was prominent in the book and movie Jurassic Park, where the amber contained fossilized blood-sucking insects with dinosaur blood in their stomachs.

Two types of electric charge had been discovered by the time of Benjamin Franklin (1706–1790). He named them positive (+) and negative (−) because they appear as opposites and can neutralize each other. When a glass rod is rubbed vigorously with fur or silk and allowed to touch lightweight balls wrapped with aluminum foil, the balls spring apart immediately (Figure 3.1a). The touching allows the rod and the balls to share the same type of charge (positive), and their movement indicates that *like charges repel.* A pair of aluminum-covered balls can also be charged negatively by touching them with a rod rubbed vigorously with wool, and they also repel each other. If one of these negatively charged balls is brought close to one of the positively charged balls, they are attracted to each other (Figure 3.1b). This movement indicates that *unlike charges attract.*

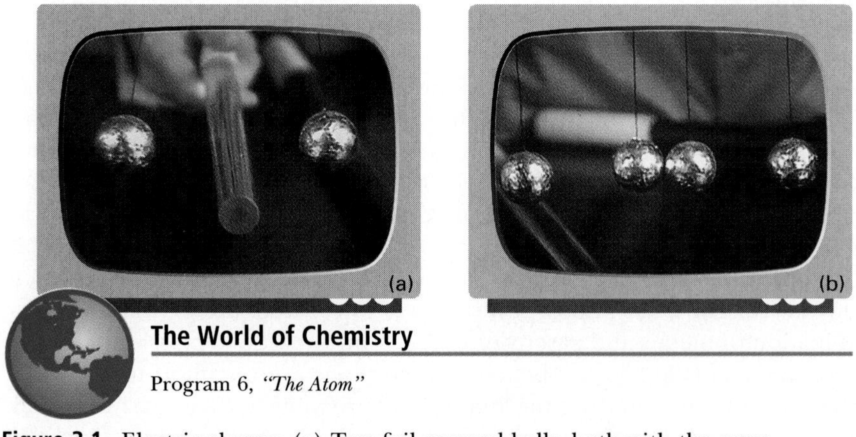

Figure 3.1 Electric charge. (a) Two foil-covered balls, both with the same charge from having touched the positively charged rod, repel each other. (b) Two foil-covered balls with opposite charges attract each other.

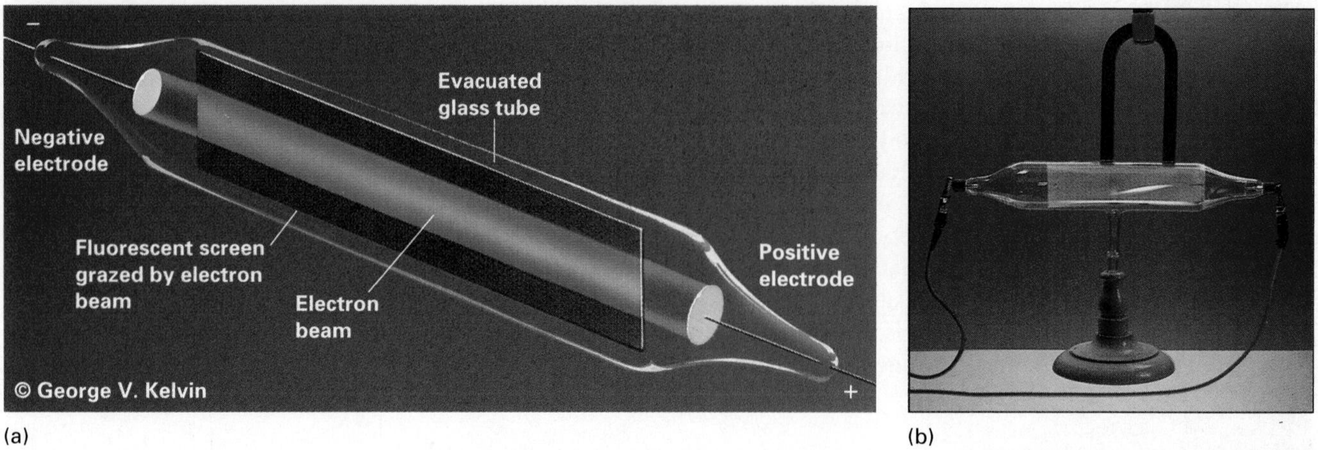

Figure 3.2 Cathode-ray tube. A beam of electrons is produced from the gas in a partially evacuated tube, diagrammed at the left, by the application of a high voltage between the electrodes. The beam becomes visible as it grazes along a fluorescent screen, as shown in the photo at the right. *(b, Richard Megna/Fundamental Photographs)*

The Electron—The First Subatomic Particle Discovered

The first ideas about electrons came from experiments with cathode-ray tubes. The forerunner of neon signs, fluorescent lights, and TV picture tubes, a typical cathode-ray tube is a partially evacuated glass tube with a piece of metal sealed in each end (Figure 3.2). The pieces of metal are called electrodes; the one given a negative charge is the **cathode,** and the one given a positive charge is the **anode.**

■ **Cathode rays** are streams of the negatively charged particles called electrons.

■ A fluorescent material absorbs high-energy radiation and emits visible light.

■ The charge and mass of an electron were determined by a combination of experiments by Sir Joseph John Thomson in 1897 and by Robert Andrews Millikan in 1911. Both scientists were awarded Nobel Prizes, Thomson in 1906 and Millikan in 1923.

When a sufficiently high voltage is applied to the electrodes, a beam of cathode rays flows from the cathode to the anode. Cathode rays travel in straight lines, cast sharp shadows, cause gases and fluorescent materials to glow, can heat metal objects red hot, can be deflected by a magnetic field, and are attracted toward positively charged plates. When cathode rays strike a fluorescent screen, light is given off in a series of tiny flashes. These properties all led to the conclusion that cathode rays are a stream of negatively charged particles. The cathode-ray particles became known as **electrons.**

By using a specially designed cathode-ray tube (Figure 3.3), Thomson applied electric and magnetic fields to the rays. Then, by applying basic laws of electricity and magnetism to the results, he determined the charge-to-mass ratio of the electron. He was able to measure neither the absolute charge nor the absolute mass of the electron, but he established the ratio between the two numbers and made it possible to calculate either one if the other could ever be measured.

An important part of Thomson's experimentation was his use of 20 different metals for cathodes and of several gases to conduct the discharge. Every combination of metals and gases yielded the same charge-to-mass ratio for the cathode rays. This led to the belief that electrons are common to all of the metals and gases used in the experiments and probably to all atoms in general. Thus, it appeared that the electron is one of the atomic building blocks of which atoms are made.

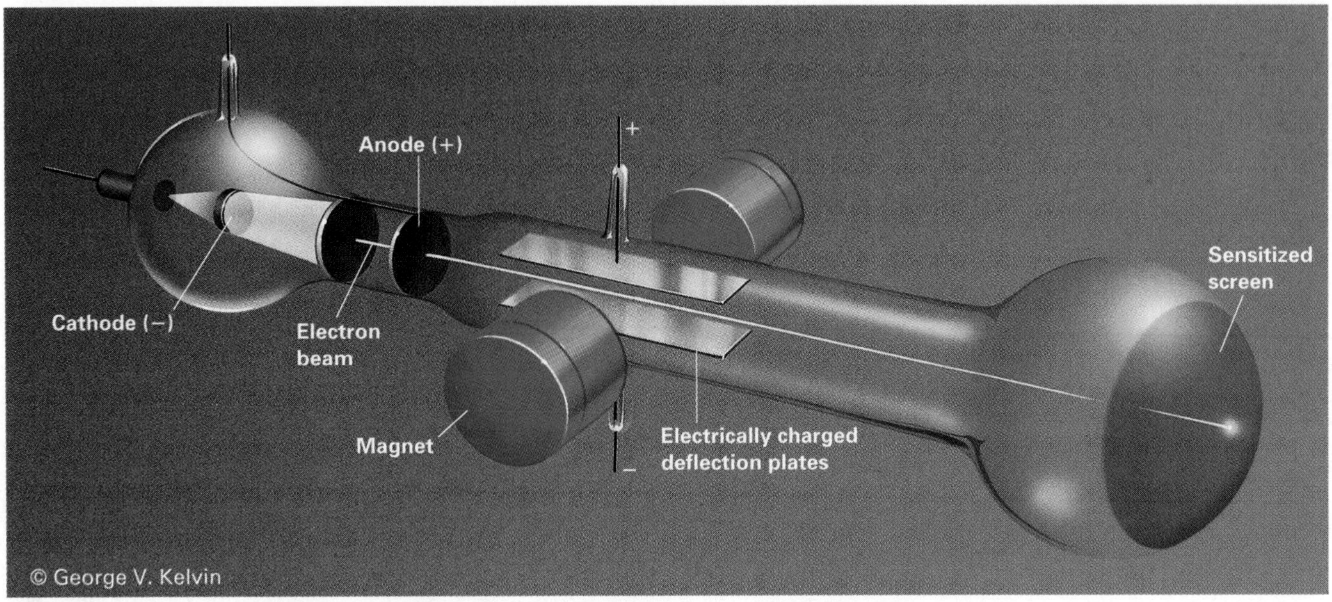

Figure 3.3 A schematic diagram of the J. J. Thomson apparatus. Sir J. J. Thomson determined the mass-to-charge ratio of the electron by making quantitative measurements of the effect of electric and magnetic fields on the path of electrons in the tube.

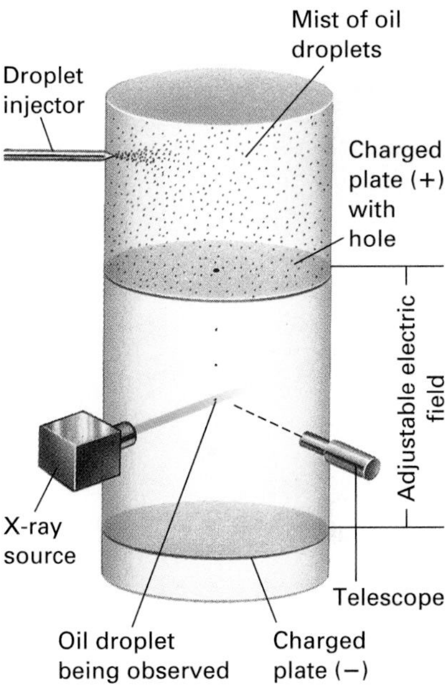

Figure 3.4 Millikan oil-drop experiment. Tiny oil droplets fall through the hole in the upper plate and settle slowly through the air. X rays cause air molecules to give up electrons to the oil droplets, which become negatively charged. From the known mass of the droplets and the applied voltage at which the charged droplets were held stationary, Millikan could calculate the charges on the droplets.

THE PERSONAL SIDE

Joseph John Thomson (1856–1940)

Sir Joseph Thomson was a scientist chiefly working in mathematics until he was elected Cavendish Professor at Cambridge University in 1884. He was not skilled at experimental techniques, but his ability to suggest experiments and interpret their results led to the discovery of the electron. In 1897 Thomson wrote in the *Philosophical Magazine,* "We have in the cathode rays a new state, a state in which the subdivision of matter is carried very much further than in the ordinary gaseous state—this matter being the substance from which the chemical elements are built up." Thomson won the Nobel Prize in Physics in 1906 and was knighted in 1908. Seven of his research assistants later won Nobel Prizes for their own research work. Thomson is buried in Westminster Abbey near the grave of Sir Isaac Newton.

Sir J. J. Thomson, at the left, being shown a vacuum tube in the United States at the Western Electric Laboratory in 1923, 17 years after he won the Nobel Prize in Physics for his studies using cathode ray tubes. (Bell Telephone Laboratories. Courtesy AIP/Emilio Segré Visual Archives)

It remained for Robert Andrews Millikan (1868–1953) to measure the charge on an electron. A simplified drawing of his apparatus is shown in Figure 3.4. When the electric charge on the plates was increased enough to balance the effect of gravity, a droplet could be suspended motionless. At this point, the gravitational force equaled the electric force. Measurements made in the motionless state, when inserted into equations for the forces acting on the droplet, enabled Millikan to calculate the charge carried by the droplet.

Millikan found different amounts of negative charge on different drops, but the charge measured each time was always a whole-number multiple of the same smaller charge. That smaller charge was 1.60×10^{-19} C. (The coulomb, abbreviated C, is a charge unit.) Millikan assumed this to be the fundamental charge, the charge on an electron. From this the mass of the electron could be calculated by using the charge-mass ratio determined by Thomson. The currently accepted mass of the electron is 9.109389×10^{-28} g, and the currently accepted value of the electron's charge is $-1.60217733 \times 10^{-19}$ C. This much charge is represented as -1, the charge on one electron.

■ Because atoms are electrically neutral, discovery of electrons meant there must be a positively charged particle waiting to be discovered.

Protons

■ Canal rays are streams of positive ions derived from gases present in the discharge tube.

■ An **ion** is an atom that has a charge. If it has lost electrons, the charge is positive (+). If it has gained electrons, the charge is negative (−).

The first experimental evidence of a fundamental positive particle came from the study of so-called canal rays (Figure 3.5), which were observed in a special type of cathode-ray tube with a perforated cathode. When high voltage is applied to the tube, cathode rays can be observed passing between the electrodes as in any cathode-ray tube. On the other side of the perforated cathode, a different kind of ray is observed. Since these rays are attracted to a negatively charged plate, they must be composed of positively charged particles. Each gas used in the tube gives a different charge-to-mass ratio for the positively charged particles (unlike the cathode rays, which are the same no matter what the gas is). When the tube contains hydrogen, the largest charge-to-mass ratio is obtained, suggesting that hydrogen provides the positive particles with the smallest mass. The emissions from hydrogen were considered to be the fundamental positively charged particles of atomic structure and are called **protons** (from a Greek word meaning "the primary one").

The mass of a proton is known from experiments to be 1.672623×10^{-24} g, about 1800 times the mass of an electron. The charge on the proton ($+1.60217733 \times 10^{-19}$ C) is equal in size, but opposite in sign, to the charge on the electron; the charge on the proton is designated as $+1$.

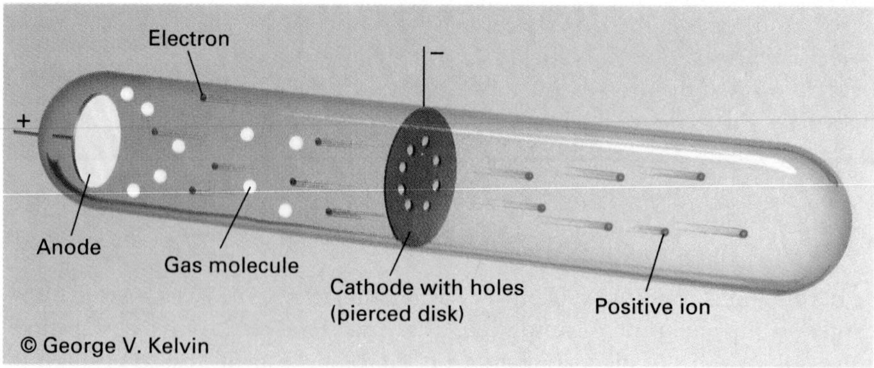

Figure 3.5 Canal ray tube. Positive ions, produced when electrons collide with gas molecules in a cathode ray tube, are attracted to the cathode. If the cathode is perforated, the positive ions pass through the holes and form positive rays (canal rays). The positive rays are deflected by electric and magnetic fields, but much less so for a given value of the field than electrons because positive ions are much heavier than electrons.

Neutrons

Because atoms normally have no charge, there must be equal numbers of protons and electrons in an atom, since protons and electrons have the same quantity of charge, just different signs. Most atoms, however, have masses greater than would be predicted from the sum of the masses of their protons and electrons, indicating that some other particles must be present in the atom. Since this third type of particle has no charge, the usual methods of detecting particles could not be used. Nonetheless, in 1932, many years after the discovery of the proton, James Chadwick (1891–1974) devised a clever experiment that produced neutral particles from a radioactive element and then detected them by having them knock protons, a detectable species, out of paraffin. It is now known that these particles, called **neutrons,** have no electric charge and a mass of $1.6749286 \times 10^{-24}$ g, nearly the same as the mass of a proton.

The Nucleus of the Atom

Ernest Rutherford (1871–1937), a student of J. J. Thomson, began studying the radiation emitted from radioactive elements soon after experiments by the Curies and others had shown that three types of radiation are spontaneously emitted by radioactive elements (Section 5.1). These rays, referred to as alpha (α), beta (β), and gamma (γ) rays, behave differently when passed between electrically charged plates, as shown in Figure 3.6. Alpha and beta rays are deflected, while gamma rays pass straight through undeflected. This implies that alpha and beta rays are electrically charged particles, since particles with a charge would be attracted or repelled by the charged plates. Even though an alpha particle has an electrical charge (+2) twice as large as that of a beta particle (−1), alpha particles are deflected less; hence, alpha particles must be heavier than beta particles. Careful studies by Rutherford showed that **alpha particles** are helium atoms that have lost two electrons (He^{2+}). **Beta particles** were shown to be negatively charged particles identical with the electron. **Gamma rays** have no detectable charge or mass—they behave like light rays.

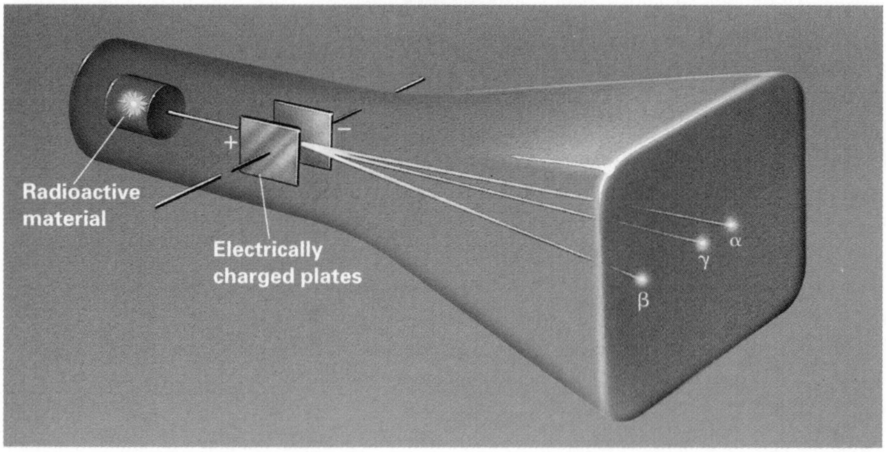

Figure 3.6 Separation of alpha, beta, and gamma rays by an electric field. Alpha rays are deflected toward the negative plate; beta rays are attracted toward the positive plate; and gamma rays are not deflected. Additional studies showed that alpha particles are high energy helium-4 nuclei, beta particles are high energy electrons, and gamma rays are high energy electromagnetic radiation.

Rutherford's experiments with alpha particles led him to consider using them in experiments on the structure of the atom. In 1909 he suggested to two of his co-workers that they bombard a piece of gold foil with alpha particles. Hans Geiger (1882–1945), a German physicist, and Ernest Marsden (1889–1970), an undergraduate student, set up the apparatus diagrammed in Figure 3.7 and observed what happened when alpha particles hit the thin gold foil. Most passed straight through, but Geiger and Marsden were amazed to find that a few alpha particles were deflected through large angles, and some came almost straight back! Rutherford later described this unexpected result by saying, "It was about as incredible as if you had fired a 15-inch [artillery] shell at a piece of paper and it came back and hit you."

What allowed most of the alpha particles to pass through the gold foil in a rather straight path? According to Rutherford's interpretation, the atom is mostly *empty space* and therefore offers little resistance to the alpha particles (Figure 3.8).

What caused a few alpha particles to be deflected? According to Rutherford's model of the atom, all of the positive charge and most of the mass of the atom must be concentrated in a very small volume at the center of the atom. He named this part of the atom containing most of the mass of the atom and all of the positive charge the **nucleus.** When an alpha particle passes near the nucleus, the positive charge of the nucleus repels the positive charge of the alpha particle; the path of the smaller alpha particle is consequently deflected. The closer an alpha particle comes to a target nucleus, the more it is deflected. Those alpha particles that meet a nucleus head-on bounce back toward the source as a result of the strong positive-positive repulsion, since the alpha particles do not have enough energy to penetrate the nucleus.

Rutherford's calculations, based on the observed deflections, indicate that the nucleus is a very small part of an atom. The diameter of an atom is about 100,000 times greater than the diameter of its nucleus.

Truly, Rutherford's model of the atom was one of the most dramatic interpretations of experimental evidence to come out of this period of significant discoveries.

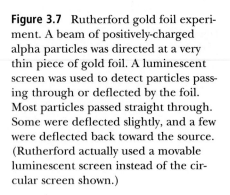

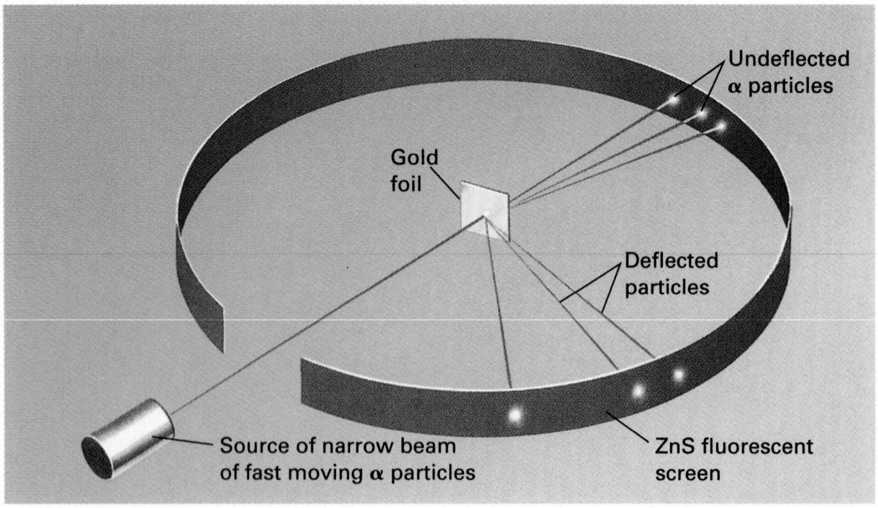

Figure 3.7 Rutherford gold foil experiment. A beam of positively-charged alpha particles was directed at a very thin piece of gold foil. A luminescent screen was used to detect particles passing through or deflected by the foil. Most particles passed straight through. Some were deflected slightly, and a few were deflected back toward the source. (Rutherford actually used a movable luminescent screen instead of the circular screen shown.)

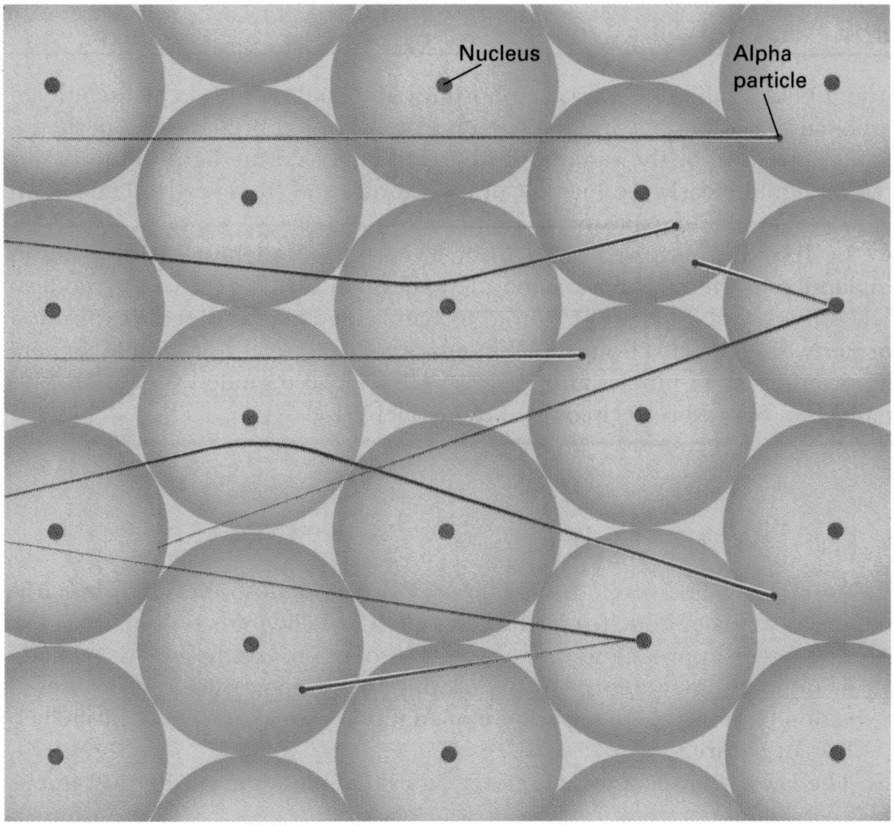

Figure 3.8 Rutherford's interpretation of the gold-foil experiment. Each circle represents an atom and the dots represent their nuclei. The alpha particle pathways show how those that collide directly with a nucleus return toward their source. The gold foil was about 1000 atoms thick.

THE PERSONAL SIDE

Ernest Rutherford (1871–1937)

Lord Ernest Rutherford was born in New Zealand in 1871 but went to Cambridge University in England to pursue his Ph.D. in physics in 1895. His original interest was in a phenomenon that we now call radio waves, and he apparently hoped to make his fortune in the field, largely so he could marry his fiancée back in New Zealand. However, his professor at Cambridge, J. J. Thomson, convinced him to work on the newly discovered phenomenon of radioactivity. Rutherford discovered alpha and beta radiation while at Cambridge. In 1899 he moved to McGill University in Canada where he did further experiments to prove that alpha radiation is actually composed of helium nuclei and that beta radiation consists of electrons. For this work he received the Nobel Prize in Chemistry in 1908.

In 1903 Rutherford and his young wife visited Pierre and Marie Curie in Paris, on the very day that Madame Curie received her doctorate in physics. That evening during a party in the garden of the Curies' home, Pierre Curie

(continued on next page)

Ernest Rutherford (right) with a colleague in the Cavendish Laboratory at Cambridge University. The sign over Rutherford's head in this photo, taken in 1935, is said to have been directed at the booming voice of Rutherford himself. (C. E. Wynn-Williams. Courtesy AIP/Emilio Segrè Visual Archives)

THE PERSONAL SIDE *(continued)*

brought out a tube coated with a phosphor and containing a large quantity of radioactive radium in solution. The phosphor glowed brilliantly from the radiation given off by the radium. Rutherford later said the light was so bright that he could clearly see Pierre Curie's hands were "in a very inflamed and painful state due to exposure to radium rays."

In 1907 Rutherford moved from Canada to Manchester University in England, and there he performed the experiments that gave us the modern view of the atom. In 1919 he moved to Cambridge and assumed the position formerly held by J. J. Thomson. Not only was Rutherford responsible for very important work in physics and chemistry, but he also guided the work of no fewer than ten future recipients of the Nobel Prize.

SELF-TEST 3A

1. Atoms were first proposed (a) by early British philosophers, (b) by early Greek philosophers, (c) in the early 1900s, (d) by John Dalton.
2. The Greek approach to the "discovery" of atoms can best be described as (a) experimentation, (b) philosophy (use of logic), (c) direct observation of atoms, (d) consistent explanation of well-known, established laws of nature.
3. The law of conservation of matter states that matter is neither lost nor ____________ in a ____________ reaction.
4. The composition of sulfur dioxide is 32 parts by mass S and 32 parts by mass O. What is the percent sulfur in SO_2?
5. The law of multiple proportions explains the existence of two compounds of nitrogen and oxygen such as ____________ and ____________.
6. Hydrogen peroxide, H_2O_2, is always 94.12% O. What is the percent H in hydrogen peroxide?
7. According to Dalton's atomic theory, a compound has a definite percentage by mass of each element because (a) all atoms of a given element weigh ____________, and (b) all molecules of a given compound contain a definite number and kind of ____________.
8. If the law of definite composition is true, will the percent mass of silver, Ag, in silver sulfide be the same for all lumps or pieces of Ag_2S? (a) Yes, (b) No. Explain your answer.
9. Natural gas is essentially methane, CH_4. Will methane produced from Texas gas fields have the same composition as methane produced by gas fields in China? (a) Yes, (b) No. Explain your answer.
10. Which of the following pairs of compounds illustrate the concept of multiple proportions at work?

NO and NO_2	CO and CO_2	$CaCl_2$ and NaCl
CH_4 and CO_2	H_2O and H_2O_2	$FeCl_2$ and $FeCl_3$

11. What is the charge on the proton? (a) −1, (b) +1, (c) 3.
12. A beta particle is a high energy (a) neutron, (b) electron, (c) proton, (d) helium nucleus.

13. Ernest Rutherford proposed the modern nuclear model of the atom. (a) True, (b) False.
14. Most of the mass of an atom is concentrated in its (a) nucleus, (b) electrons, (c) protons.
15. The mass of the proton is ____________ times the mass of the electron.

3.4 Modern View of the Atom

Early experiments on the structure of the atom clearly showed that the three primary constituents of atoms are electrons, protons, and neutrons. The nucleus or core of the atom is made up of protons with positive electrical charge and neutrons with no charge. The electrons, with a negative electrical charge, are found in the space around the nucleus (Figure 3.9). For an atom, which has no net electrical charge, *the number of negatively charged electrons around the nucleus equals the number of positively charged protons in the nucleus.*

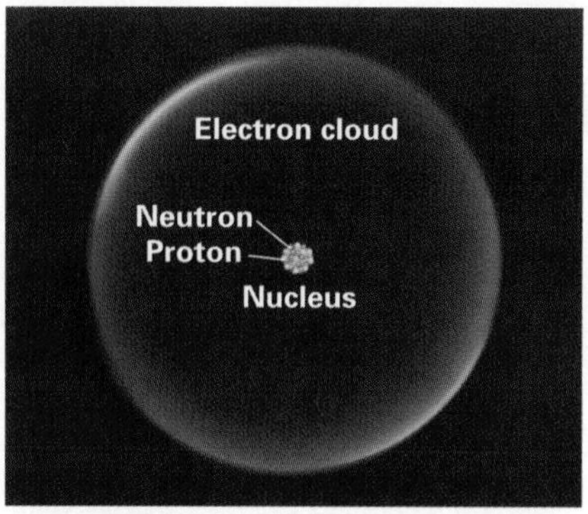

Figure 3.9 Model of atom. All atoms consist of one or more protons (positively charged) and usually at least as many neutrons (no charge) packed into an extremely small nucleus. Electrons (negatively charged) are arranged in a cloud around the nucleus.

Atoms are extremely small, far too small to be seen with even the most powerful optical microscopes. Since 1914, scientists have used **X-ray diffraction,** the study of patterns in the intensity of X rays that passed through single crystals of substances, to determine the locations and dimensions of atoms in a substance. The diameters of atoms range from 1×10^{-8} cm to 5×10^{-8} cm. For example, the diameter of a carbon atom is 1.5×10^{-8} cm. To visualize how small this is, take a sharp pencil and draw a line 3 cm long (________________). Since graphite is made up of carbon atoms, the 3-cm line of carbon contains two hundred million (200,000,000) carbon atoms from end to end and about a million atoms across (the width of the line). If this weren't hard enough to imagine, remember that Rutherford's experiments provided evidence that the diameter of the nucleus is 100,000 times smaller than the diameter of the atom. For example, if an atom were scaled upward in size so that the nucleus were the size of a small marble, the atom would be the size of the Houston Astrodome, and most of the space in between would be empty. Because the nucleus carries most of the mass of the atom in such a small volume, a matchbox full of nuclear material would weigh more than 2.5 billion tons! The interior of a collapsed star is made up of nuclear material that is estimated to be nearly this dense.

In recent years scientists have been able to obtain direct images of atoms (Figure 3.10) using the scanning tunneling microscope (STM) and the atomic force microscope (ATM).

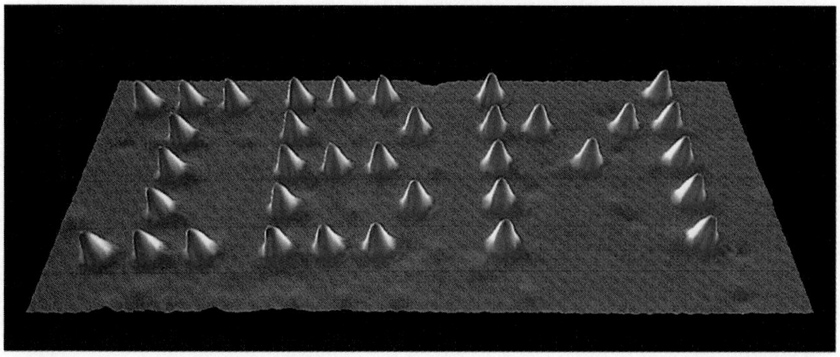

Figure 3.10 IBM spelled out in xenon (Xe) atoms. A few years after invention of the scanning tunnelling microscope, scientists discovered that not only could they take pictures of atoms on a surface, but they could push them around with the tip of the microscope. They created this picture to demonstrate what they could do. The "IBM" is 660 billionths of an inch, or 16.8 nanometers. *(IBM Corporation Research Division, Almaden Research Center)*

THE WORLD OF CHEMISTRY

The Scanning Tunneling Microscope

Program 6, *The Atom*

Chemists and physicists had more than ample evidence of the existence of atoms before the invention of the scanning tunneling microscope (STM). They were able to "see" atoms through a large variety of phenomena, but to say that they saw atoms had a special meaning. What they were seeing by such techniques as X-ray diffraction was a manifestation of many atoms and a composite picture created by the scattering of X rays from many planes of a crystal.

Yet chemists and physicists have always dreamed of being able to see individual atoms directly, that is, of being able to produce images with a direct correspondence to the atom's actual position in the sample.

These dreams began to be realized in the 1950s. An early and spectacular effort was reported by Erwin Mueller using a field ion microscope that he invented, which made it possible to image individual atoms on a crystal's surface. The even more remarkable STM not only makes it possible to see individual atoms and how they are arranged on a surface, but also permits the study of atom migration and atomic dislocations on surfaces. The development of the STM is considered an event of such magnitude that its developers, Gerd Binnig and Heinrich Rohrer of IBM's Zurich Research Laboratory in Switzerland, received the Nobel Prize in physics in 1986. The STM is an astonishing device because of its inherent simplicity. It consists of a tungsten needle, hardly more than a single atom wide at the end. When this needle is lowered to within a few atoms' thickness of the surface to be imaged and a small voltage is applied, electrons tunnel; that is, they pass from the tungsten atom into the electron clouds of the atoms on the surface and produce a measurable current. By adjusting the up-down position of the tungsten needle as it moves across the surface, a constant tunneling current is maintained. As this takes place, however, the positions of the atoms are actually measured, giving a picture of the atomic landscape.

The potential for studying materials using the STM is great, particularly in the area of catalysts. Recently, STM studies have also seen actual amino acid molecules on the surfaces of crystals. Amino acids are the basic building blocks of all living matter.

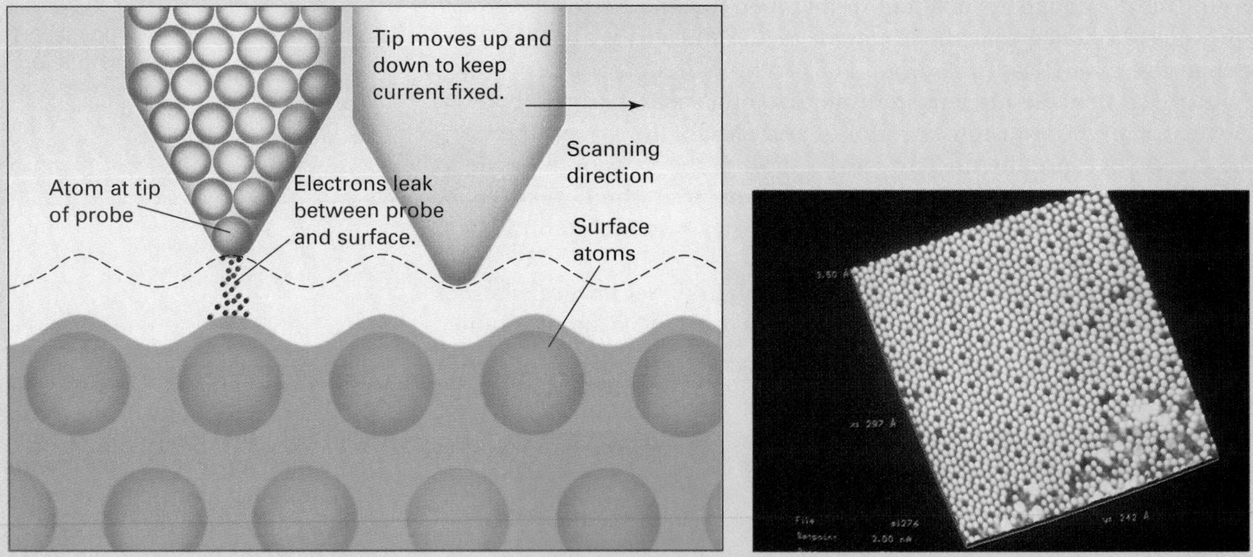

(a) (b)

Scanning Tunneling Microscope (STM). (a) When an electric current passes through a tungsten needle with a narrow tip (atoms's width) into the atoms on the surface of the sample being examined, the electron flow between the tip and the surface changes in relation to the electron clouds around the atoms. By adjusting the position of the needle to maintain a constant current, the positions of the atoms are measured. (b) STM image of atoms of silicon (b, John Ozcomert/Michael Trenary)

EXAMPLE 3.2 *Size of Atoms*

The diameter of a U.S. penny is 1.9 cm. How many carbon atoms would fit side by side along this diameter? The diameter of a carbon atom is 1.5×10^{-8} cm.

SOLUTION

The number of carbon atoms is calculated by dividing 1.9 cm by the diameter of one carbon atom.

$$\left(\frac{1.9\text{ cm}}{1\text{ penny}}\right)\left(\frac{1\text{ carbon atom}}{1.5 \times 10^{-8}\text{ cm}}\right) = \frac{1.3 \times 10^{8}\text{ carbon atoms}}{1\text{ penny}},$$

or 130,000,000 carbon atoms

Exercise 3.2

The diameter of a gold atom is 2.9×10^{-8} cm. How many gold atoms would fit side by side along a gold coin with a diameter of 5.0 cm?

The World of Chemistry

Program 6, *The Atom*

Well the first time I saw atoms, it was about two or three o'clock in the morning. Of course, I was doing most of my work at night because the instrument itself is extremely sensitive to vibration, so most of our experiments are done at night or on weekends. And it was about two or three o'clock in the morning, and I was staring at this image, and I started to notice this regular pattern appearing, which I knew had to be the positions of the atoms, and it just got better and better, and I was just elated. Eventually, tears came to my eyes because I was so happy, having worked on this day and night for so long. Dr. Robert Hamers, research scientist at IBM

Atomic Number—Each Element Has a Number

The **atomic number** of an element indicates the number of protons in the nucleus of the atom. All atoms of the same element have the same number of protons in the nucleus. In the periodic table (inside back cover) the atomic number for each element is given above the element's symbol. Beginning with the atomic number 1 for hydrogen, there is a different atomic number for each element. Phosphorus, for example, has an atomic number of 15, so the nucleus of a phosphorus atom contains 15 protons.

In a neutral atom the number of protons is equal to the number of electrons, so the atomic number also gives the number of electrons outside the nucleus in an atom of an element.

■ The great significance of the periodic table to our understanding of the elements is discussed in Chapter 4.

Mass Number

The **mass number** of a particular atom is the total number of neutrons and protons present in the nucleus of an atom. Since the atomic number gives the number of protons in the nucleus, the difference between the mass number and the atomic number equals the number of neutrons in the nucleus.

A notation frequently used for showing the mass number and atomic number of an atom places subscripts and superscripts to the left of the symbol:

Mass number ⟶ ${}^{19}_{9}\text{F}$ ⟵ Symbol of element
Atomic number ⟶

The subscript giving the atomic number is optional because the element symbol tells you what the atomic number must be. For example, the fluorine atom described above would have the notation ${}^{19}_{9}\text{F}$ or just ${}^{19}\text{F}$. For an atom of fluorine, ${}^{19}_{9}F$, the number of protons is 9, the number of electrons is also 9, and the number of neutrons is 19 − 9 = 10.

EXAMPLE 3.3 *Atomic Composition*

How many protons, neutrons, and electrons are in an atom of gold (Au) with a mass number of 197?

SOLUTION

The atomic number of an element gives the number of protons and electrons. If the element is known, its atomic number can be obtained from a periodic table (or the alphabetical list inside the front cover). The atomic number of gold is 79. Gold has 79 protons and 79 electrons. The number of neutrons is obtained by subtracting the atomic number from the mass number.

$$197 - 79 = 118 \text{ neutrons}$$

Exercise 3.3

How many protons, neutrons, and electrons are there in a neutral ${}^{59}_{28}Ni$ atom?

Isotopes

A natural sample of most elements, when analyzed with a special type of canal-ray tube in an instrument called a mass spectrometer (Figure 3.11), is found to

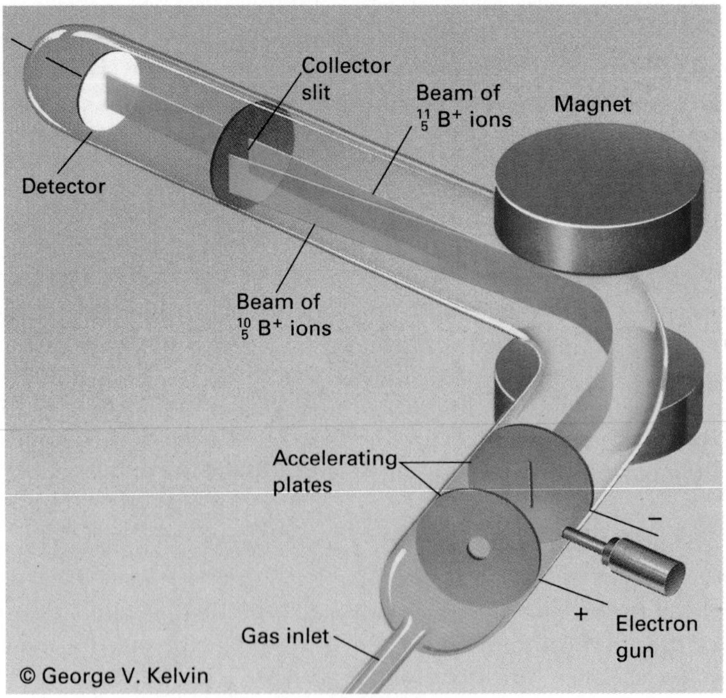

Figure 3.11 Mass spectrometer. A gas sample is injected at the gas inlet. An electron beam ionizes the gas sample, in this case a sample of boron, by knocking electrons from the neutral atoms or molecules. Charged plates are arranged to accelerate positive ions toward the first slit and into the rest of the apparatus. Positive ions that pass the first slit move into a magnetic field perpendicular to their path. In this field, the path of each ion curves as determined by its charge-to-mass ratio. The separation of ions of two boron isotopes is shown. A collector plate, behind the second slit, detects charged particles passing through the second slit. The relative magnitudes of the electrical signals are a measure of the numbers of the different kinds of positive ions.

be composed of atoms with different mass numbers. Atoms of the same element having different mass numbers are called **isotopes** of that element.

The element neon is a good example to consider (Figure 3.12). A natural sample of neon gas is found to be a mixture of three isotopes of neon:

$$^{20}_{10}\text{Ne} \quad ^{21}_{10}\text{Ne} \quad ^{22}_{10}\text{Ne}$$

The fundamental difference between isotopes is the different number of neutrons per atom. All atoms of neon have 10 electrons and 10 protons; about 90% of the atoms have 10 neutrons, some have 11, and others have 12. Because they have different numbers of neutrons, they must have different masses. Note that all the isotopes have the same atomic number. They are all neon.

Over 100 elements are known, yet more than 1000 isotopes have been identified, many of them produced artificially (see Section 5.6). Some elements have many isotopes; tin, for example, has 10 natural isotopes. Hydrogen has three isotopes, and two of them are the only known isotopes generally referred to by different names: $^{1}_{1}$H is commonly called hydrogen, $^{2}_{1}$H is called deuterium, and $^{3}_{1}$H is called tritium. Tritium is radioactive.

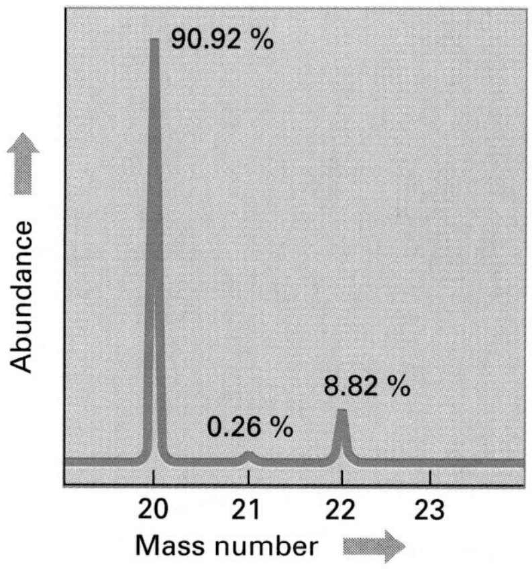

Figure 3.12 Mass spectrum of neon (+1 ions only). The principal peak corresponds to the most abundant isotope, neon-20. Percent relative abundance is shown.

■ To represent isotopes with words rather than symbols, the mass number is added to the name; for example, neon-20, neon-21, and neon-22.

EXAMPLE 3.4 *Isotopes*

Carbon has seven known isotopes. Three of these have 6, 7, and 8 neutrons, respectively. Write the complete chemical notation for these isotopes that gives mass number, atomic number, and symbol.

SOLUTION

The atomic number of carbon is 6. The mass number for the three isotopes is 6 + the number of neutrons, which equal 12, 13, and 14, respectively.

$$^{12}_{6}C \qquad ^{13}_{6}C \qquad ^{14}_{6}C$$

Exercise 3.4

Silver has two isotopes, one with 60 neutrons and the other with 62 neutrons. Give the complete chemical notation for these isotopes.

Atomic Masses and Atomic Weights

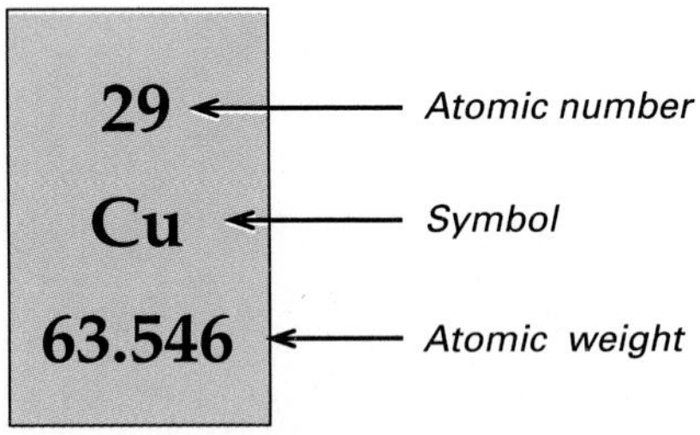

The periodic table entry for copper. (The complete periodic table and its significance are discussed in Chapter 4.)

Although Dalton knew nothing about subatomic particles, he proposed that atoms of different elements have different masses. He and other scientists of his time carried out experiments that verified this. For example, they found that 100 g of water contains 11.1 g of hydrogen and 88.9 g of oxygen. Dalton incorrectly assumed the formula of water was HO, which led him to propose an atomic mass of 8 for an oxygen atom and 1 for a hydrogen atom. After other scientists had determined that the correct formula for water is H_2O, oxygen was assigned an atomic mass of 16.

The fact that an oxygen atom is about 16 times heavier than a hydrogen atom does not tell us the mass of either atom. These are relative masses in the same way that a grapefruit may weigh twice as much as an orange. This information gives neither the mass of the grapefruit nor that of the orange. However, if a specific number is assigned as the mass of any particular atom, this fixes the numbers assigned to the masses of all other atoms. The present atomic mass scale, adopted by scientists worldwide in 1961, is based on assigning the mass of a particular isotope of the carbon atom, the carbon-12 isotope, as exactly 12 **atomic mass units (amu).**

■ The definition of "amu" is 1/12 the mass of the carbon-12 isotope.

■ Chemists usually use the term "atomic weight" of an element rather than "atomic mass" when they are referring to a naturally occurring sample of an element. Although the quantity is more properly called a mass than a weight, the term "atomic weight" is so commonly used that it has become accepted. Atomic weights are established by exacting experiments. The same values are used by all scientists because the relative abundances of the elements are essentially the same everywhere on our planet.

The atomic masses given in the periodic table are average masses, which take into account the relative abundances of the different isotopes as found in nature. This average is often referred to as the **atomic weight** of the element. For example, boron has two naturally occurring isotopes, $^{10}_{5}B$ and $^{11}_{5}B$, with natural percent abundances of 19.91% and 80.09%. The masses in amu are 10.0129 and 11.0093, respectively. The atomic weight listed in the periodic table is the average mass of a natural sample of atoms, expressed in atomic mass units. The percent abundance gives the necessary information needed to calculate the atomic weight. The natural percent abundances of 19.91% and 80.09% mean that in a sample of 10,000 atoms of boron found in nature, 1991 atoms would be the $^{10}_{5}B$ isotope and 8009 atoms would be the $^{11}_{5}B$ isotope. The average atomic weight (the value found in the periodic table) is calculated by multiplying the percent abundance times the mass of the isotope and adding up the resulting masses.

$$\text{atomic weight} = (19.91\%)(10.0129) + (80.09\%)(11.0093)$$

Changing the percent abundance to their decimal equivalents, 0.1991 and 0.8009, respectively, gives

$$\text{atomic weight} = (0.1991)(10.0129) + (0.8009)(11.0093) = 10.81 \text{ amu}$$

This is the value listed for the atomic weight of boron in the periodic table (inside back cover).

SELF-TEST 3B

1. If an atom has an atomic number of 10, then it has ________ protons and ________ electrons. If its mass number is 21, then it has ________ neutrons.
2. In the symbol $^{80}_{35}Br$, the number 35 is the ________, and the number 80 is the ________.
3. Isotopes of an element are atoms that have nuclei with the same number of ________ but different numbers of ________.
4. The negatively charged particles in an atom are ________; the positively charged particles are ________; the neutral particles are ________.
5. In a neutral atom there are equal numbers of ________ and ________.
6. The number of protons per atom is called the ________ number of the element.
7. An atom of arsenic, $^{75}_{33}As$, has ________ electrons, ________ protons, and ________ neutrons.
8. The diameter of the nucleus is ________ smaller than the diameter of the atom.
9. Draw the nuclei of $^{1}_{1}H$, $^{2}_{1}H$, and $^{3}_{1}H$, representing protons and neutrons with circles.
10. The nuclei of all helium atoms contain exactly 2 protons. (a) True, (b) False.
11. All of the isotopes of an element have the same relative abundance. (a) True, (b) False.

3.5 Where Are the Electrons in Atoms?

The Bohr theory of the atom, proposed in 1913 by Niels Bohr for the hydrogen atom, is still a useful model for representing the relative positions and energies of electrons in an atom. Bohr's theory was based on experimental measurements of the hydrogen **emission line spectrum.** What is a line spectrum and how was this information used by Bohr?

Continuous and Line Spectra

We are familiar with the spectrum of colors that make up visible light. The spectrum of white light is a display of separated colors. This type of spectrum is referred to as a **continuous spectrum** and is obtained by passing sunlight or light from an incandescent light bulb through a glass prism. When we see a rainbow, we are looking at a continuous visible spectrum formed when raindrops act as prisms and disperse the sunlight. The different colors of light correspond to different wavelengths. Red light has longer wavelengths than

The World of Chemistry

Program 2, *Color*

Spectrum of white light produced by refraction in a glass prism. The different colors blend into one another smoothly.

(a)

(b)

Figure 3.13 Neon. (a) Partially evacuated tube that contains neon gas gives a reddish-orange glow when high voltage is applied. (b) Line emission spectrum of neon is obtained when light from a neon source passes through a prism. *(a, Grant Heilman, Runk/Schoenberger)*

does blue light, but in the continuous spectrum the colors merge from one to another with no break in the spectrum. There are literally millions of colors. The spectrum of white light is a combination of all the colors of different wavelengths.

If a high voltage is applied to an element in the gas phase in a cathode-ray tube (Section 3.3), the atoms absorb energy and are said to be "excited." The excited atoms emit light. An example of this is a neon advertising sign, in which excited neon atoms emit orange-red light (Figure 3.13a). When light from such a source passes through a prism, a different type of spectrum is obtained, one that is not continuous but has characteristic lines at specific wavelengths (Figure 3.13b). This type of spectrum is called a line emission spectrum (Figure 3.14).

Several hundred years ago, chemists discovered that compounds of some elements give characteristic colors when heated in a flame, and they used flame tests to identify these elements. Potassium salts, for example, give a fleeting lavender flame; strontium salts, a brilliant red flame; and barium salts, a green flame. Fireworks utilize these and other salts for their colors. Every element has a unique line spectrum (though not all in the visible range), and

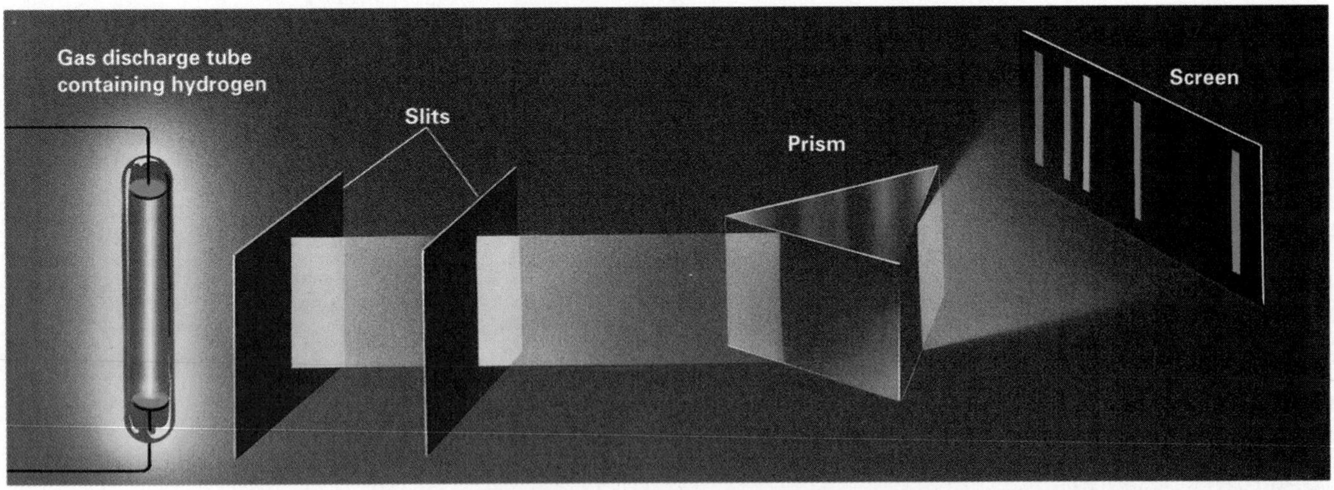

Figure 3.14 Line-emission spectrum apparatus. A line-emission spectrum of excited hydrogen atoms is obtained by passing the emitted light from a hydrogen discharge tube through a prism to separate the light into its component wavelengths. A photographic plate or other instrument detects the separate wavelengths as individual lines.

DISCOVERY EXPERIMENT

Characteristic Flame Colors

Fireworks displays are the result of excited electrons in metal ions emitting energy when they return to lower energy states. If you have a fireplace, you can observe the different colors that metal ions give to flames. Take two small pieces of kindling. Soak one piece in a solution of table salt (NaCl) for about 10 minutes. Soak the other piece in a solution of calcium chloride ($CaCl_2$) for about 10 minutes. (Calcium chloride is available at farm supply stores and some hardware stores. It is used to remove ice from sidewalks.) Observe the color given off when each piece of kindling is placed on the fire in a fireplace. What colors do you see? Explain your observations.

the characteristic lines can be used in chemical analysis to identify an element and even to determine how much of it is present. The brightness of the spectrum can be used to measure the amount of each element.

Visible light is only a small portion of the electromagnetic spectrum (Figure 3.15). Ultraviolet radiation, the type that leads to sunburn and some forms of skin cancer, has wavelengths shorter than those of visible light; X rays and gamma rays (the latter emitted from radioactive atoms) have even shorter wavelengths. Infrared radiation, the type that is sensed as heat from a fire, has longer wavelengths than visible light. Longer still are the wavelengths of the types of radiation in a microwave oven and in television and radio transmissions. Although the examples of line spectra shown here are only for the visible region, excited atoms of elements also emit characteristic wavelengths in other regions of the electromagnetic spectrum, as demonstrated by the experiments described in the next section.

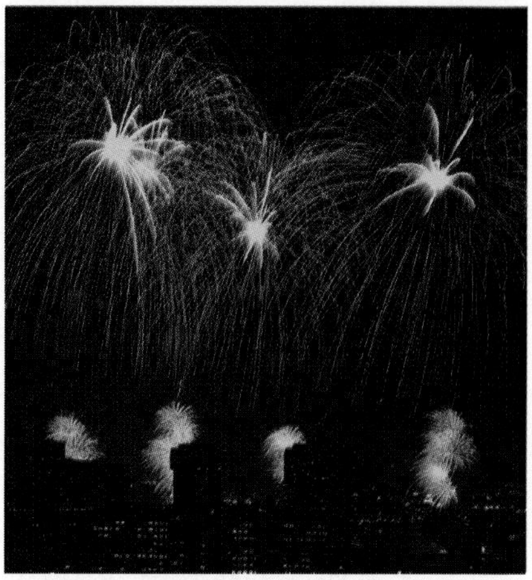

Aerial fireworks, with red, blue, and yellow produced by salts of strontium, copper, and sodium, respectively. *(Richard Megna, Fundamental Photographs)*

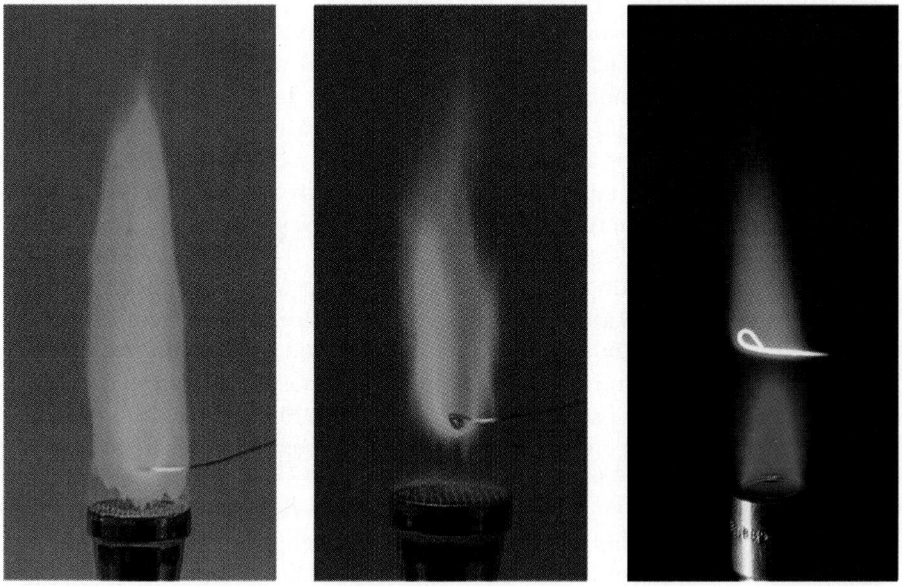

(a) (b) (c)

Flame tests. Many alkaline and alkaline-earth salts give characteristic colors when heated in a flame. For example, (a) strontium salts give a brilliant red flame, and (b) barium salts give a green flame, and (c) potassium gives a lavender flame. *(a and b, C.D. Winters; C, VV/Rich Treptow/Visuals Unlimited)*

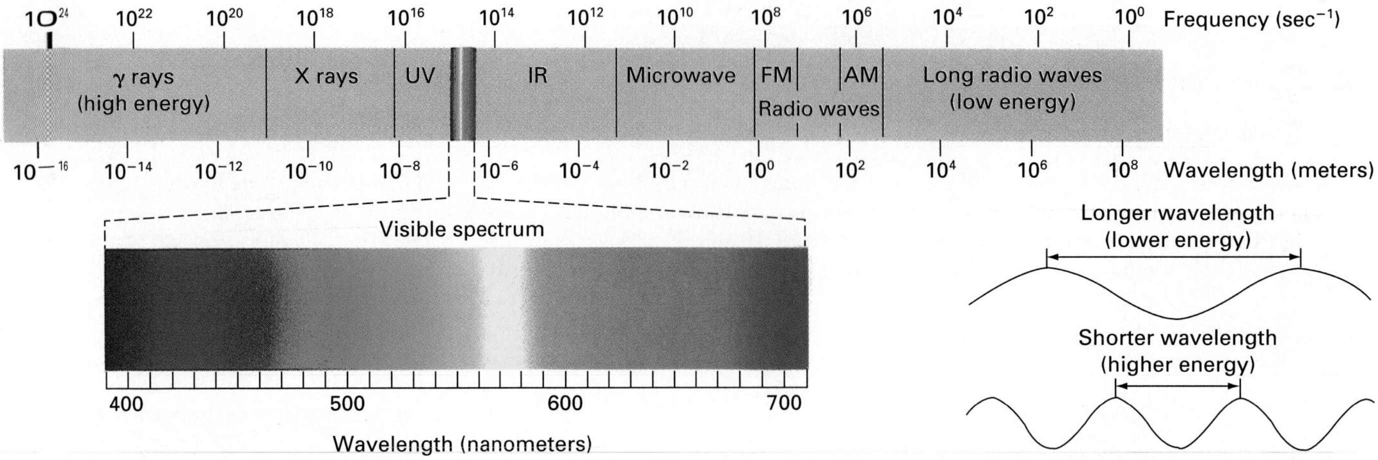

Figure 3.15 The electromagnetic spectrum. Visible light (enlarged section) is but a small part of the entire spectrum. The energy of electromagnetic radiation increases from the radio-wave end to the gamma-ray end. The frequency of electromagnetic radiation is related to the wavelength by $\nu\lambda = c$ where ν = frequency, λ = wavelength, and c = speed of light, 3.00×10^8 meters/second. The higher the frequency, the lower the wavelength, the larger the energy.

Bohr Model of the Atom

■ Bohr assumed that atoms can exist only in certain energy states.

■ In 1900 Max Planck proposed that energy is not continuous but comes in discrete "bundles" or "packets" called **quanta.** Something that can have only certain values, with none in between, is referred to as *quantized.*

■ Bohr used the term "orbits," but the modern equivalents of his orbits are called shells or energy levels.

■ Ultimately, all scientific theories and models must stand up to the kind of experimental test represented in Table 3.1 for the Bohr model of the atom. Although the Bohr model must be modified to account for properties of atoms other than hydrogen, the existence of distinct energy levels for electrons remains part of the modern atomic model.

In 1913 Niels Bohr introduced his model of the hydrogen atom. He proposed that the single electron of the hydrogen atom could occupy only certain energy levels. He referred to these energy levels as orbits and represented the energy difference between any two adjacent orbits as a single **quantum** of energy. When the hydrogen electron absorbs a quantum of energy (for example, in a gas discharge tube), it moves to a higher energy level. When this electron returns to the lower, more stable energy level, the quantum of energy is emitted as a specific wavelength of light (Figure 3.14).

In Bohr's model, each allowed orbit is assigned an integer, n, known as the principal quantum number. The values of n for the orbits range from 1 to infinity. The radii of the circular orbits increase as n increases. The orbit of lowest energy, with $n = 1$, is closest to the nucleus, and the electron of the hydrogen atom is normally in this energy level. Any atom with its electrons in their normal, lowest energy levels is said to be in the **ground state.** Energy must be supplied to move the electron farther away from the nucleus because the positive nucleus and the negative electron attract each other. When the electron of a hydrogen atom occupies an orbit with n greater than 1, the atom has more energy than in its ground state and is said to be in an **excited state.** The excited state is an unstable state, and the extra energy is emitted when the electron returns to the ground state. According to Bohr, the light forming the lines in the bright-line emission spectrum of hydrogen comes from electrons moving toward the nucleus after having first been excited to orbits farther from the nucleus (Figure 3.16). Since the orbits have only certain energies, the emitted light has only certain wavelengths.

With brilliant imagination, Bohr applied a little algebra and some classic mathematical equations of physics to his tiny solar-system model of the hydro-

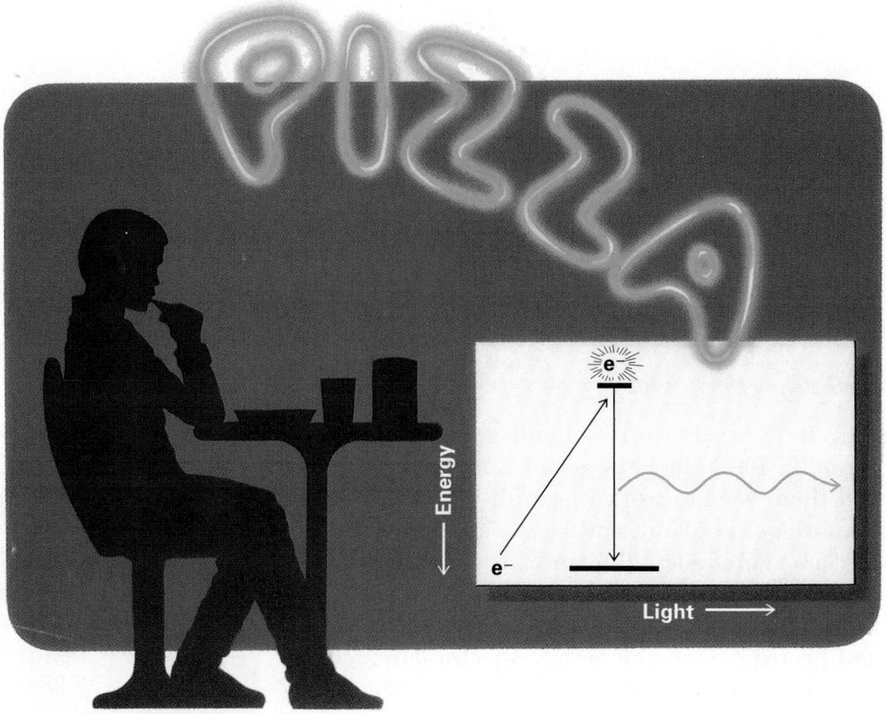

gen atom. Bohr was able to calculate the wavelengths of the lines in the hydrogen spectrum. By 1900 scientists had measured the wavelengths of lines for hydrogen in the ultraviolet, visible, and infrared regions. In Table 3.1, Bohr's calculated values are compared with the measured values. Note the close agreement between them. Niels Bohr had tied the unseen (the interior of the atom) with the seen (the observable lines in the hydrogen spectrum)—a fantastic achievement. The concept of quantum number and energy level is valid for all atoms and molecules.

The Bohr model was accepted almost immediately after its presentation, and Bohr was awarded the Nobel Prize in physics in 1922 for his contribution

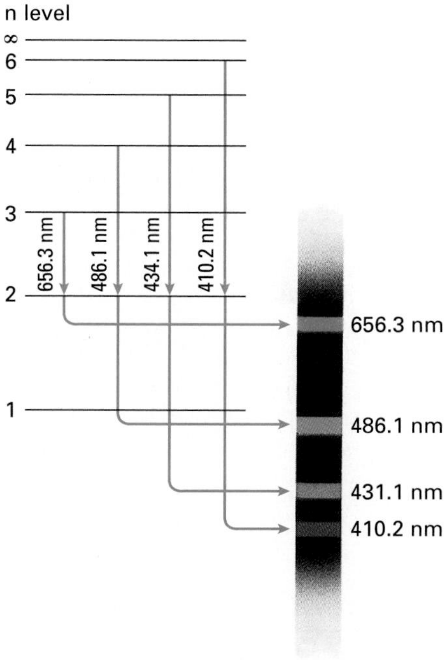

Figure 3.16 Some of the electronic transitions that can occur in an excited H atom. The lines in the visible region result from transitions from levels with values of n greater than 2 to $n = 2$.

TABLE 3.1 ■ Agreement Between Bohr's Calculations and the Lines of the Hydrogen Spectrum*

Changes in Energy Levels	Wavelength Predicted by Bohr's Theory (nm)	Wavelength Determined from Laboratory Measurement (nm)	Spectral Region
$2 \rightarrow 1$	121.6	121.7	Ultraviolet
$3 \rightarrow 1$	102.6	102.6	Ultraviolet
$4 \rightarrow 1$	97.28	97.32	Ultraviolet
$3 \rightarrow 2$	656.6	656.7	Visible red
$4 \rightarrow 2$	486.5	486.1	Visible blue-green
$5 \rightarrow 2$	434.3	434.1	Visible blue
$4 \rightarrow 3$	1876	1876	Infrared

*These lines are typical; other lines could be cited as well, with equally good agreement between theory and experiment. The unit of wavelength is the nanometer (nm), 10^{-9} m.

to the understanding of the hydrogen atom. However, his model gave only approximate agreement with line spectra of atoms having more than one electron. Later models of the atom have been more successful by considering electrons as having both particle and wave characteristics (discussed later in this chapter).

THE PERSONAL SIDE

Niels Bohr (1885–1962)

(Niels Bohr Institute. Courtesy AIP Emilio Segré Visual Archives)

Niels Bohr was born in Copenhagen, Denmark. He earned a Ph.D. in physics in Copenhagen in 1911 and then went to work first with J. J. Thomson in Cambridge, England, and later with Ernest Rutherford in Manchester, England. It was there that he began to develop the ideas that a few years later led to the publication of his theory of atomic structure and his explanation of atomic spectra. He received the Nobel Prize in 1922 for this work. After being with Rutherford for a very short time, Bohr returned to Copenhagen, where he eventually became the director of the Institute of Theoretical Physics.

Bohr was still in Denmark when Hitler's army suddenly invaded the country in 1940. In 1943, to avoid imprisonment, he escaped to Sweden. There he helped to arrange the escape of nearly every Danish Jew from Hitler's gas chambers. He was later flown to England in a tiny plane, in which he passed into a coma and nearly died from lack of oxygen.

He went on to the United States, where until 1945 he worked with other physicists on the atomic bomb development at Los Alamos, New Mexico. His insistence upon sharing the secret of the atomic bomb with other allies, in order to have international control over nuclear energy, so angered Winston Churchill that he had to be restrained from ordering Bohr's arrest. Bohr worked hard and long on behalf of the development and use of atomic energy for peaceful purposes. For his efforts, he was awarded the first Atoms for Peace Prize in 1957. He died in Copenhagen on November 18, 1962.

Atom Building Using the Bohr Model

Level	$2n^2$	Maximum Number
Level 1	$2(1)^2$	2
Level 2	$2(2)^2$	8
Level 3	$2(3)^2$	18
Level 4	$2(4)^2$	32
Level 5	$2(5)^2$	50
Level 6	$2(6)^2$	72

Recall that the atomic number is the number of electrons (or protons) per atom of an element. Imagine building atoms by adding one electron to the appropriate energy level as another proton is added to the nucleus. As part of his theory, Bohr proposed that only a fixed number of electrons could be accommodated in any one orbit, and he calculated that this number was given by the formula $2n^2$, where n equals the number of the orbit, or energy level. For the lowest energy level (first orbit), n equals 1, and the maximum number of electrons allowed is $2(1)^2$, or 2. For the second energy level, the maximum number of electrons is $2(2)^2$, or 8. Using $2n^2$, the maximum number of electrons allowed for levels 3, 4, and 5 is 18, 32, and 50, respectively. A general

TABLE 3.2 ■ Electron Arrangements of the First 20 Elements*

Element	Atomic Number	Number of Electrons in Each Energy Level			
		1st	*2nd*	*3rd*	*4th*
Hydrogen (H)	1	1 e			
Helium (He)	2	2 e			
Lithium (Li)	3	2 e	1 e		
Beryllium (Be)	4	2 e	2 e		
Boron (B)	5	2 e	3 e		
Carbon (C)	6	2 e	4 e		
Nitrogen (N)	7	2 e	5 e		
Oxygen (O)	8	2 e	6 e		
Fluorine (F)	9	2 e	7 e		
Neon (Ne)	10	2 e	8 e		
Sodium (Na)	11	2 e	8 e	1 e	
Magnesium (Mg)	12	2 e	8 e	2 e	
Aluminum (Al)	13	2 e	8 e	3 e	
Silicon (Si)	14	2 e	8 e	4 e	
Phosphorus (P)	15	2 e	8 e	5 e	
Sulfur (S)	16	2 e	8 e	6 e	
Chlorine (Cl)	17	2 e	8 e	7 e	
Argon (Ar)	18	2 e	8 e	8 e	
Potassium (K)	19	2 e	8 e	8 e	1 e
Calcium (Ca)	20	2 e	8 e	8 e	2 e

*Valence electrons are shown in color.

overriding rule to the preceding numbers is that the outside energy level can have no more than eight electrons for a stable atom.

You might like to follow along in Table 3.2 as the building-up process is described. Hydrogen (H), with atomic number 1, has one electron. In its ground state, this electron is in the first energy level. The two electrons of helium (He) are in its first energy level, since this level can have a maximum of two electrons. For all atoms of other elements, two electrons are in the first energy level, and the other electrons are placed into higher-numbered energy levels. In atomic-number order, lithium (Li, atomic number 3) through neon (Ne, atomic number 10) have two electrons in the first energy level (which fill it), and into the second energy level are placed one, two, three, and so on to eight electrons (for Ne). Eight electrons fill the second energy level.

Sodium (Na), with 11 electrons, has the first two energy levels filled with two and eight electrons, respectively, and has one electron in the third energy level. Each succeeding element in atomic-number order, magnesium (Mg) through argon (Ar), adds one more electron to the third energy level of its atoms.

■ The reasons for the limit of eight outermost electrons in an atom will become clearer in the study of the periodic table.

At Ar, the rule about a maximum of eight electrons in the outside energy level comes into play. When 19 electrons are present, as in an atom of potassium (K), the first energy level has two electrons, the second energy level has eight electrons, and the third energy level could have the other nine electrons (maximum of 18 electrons) if it were not the outside energy level. So to acommodate 19 electrons, there are two choices: 2-8-9 or 2-8-8-1. The first choice violates the requirement of having no more than eight electrons in the outside energy level. The second is the proper choice. Calcium (Ca), with 20 electrons per atom, has an electronic arrangement of 2-8-8-2.

Beginning with scandium (Sc, atomic number 21) and continuing through zinc (Zn, atomic number 30), ten electrons are added to the third energy level to complete its maximum of 18. Zinc has the electronic arrangement 2-8-18-2.

Electrons in the highest occupied energy level listed for the elements in Table 3.2 are at the greatest stable distance from the nucleus. These are the most important electrons in the study of chemistry because they are the ones that interact when atoms react with each other. This important observation was first proposed by G. N. Lewis who, independent of Bohr, conceived of the idea that electrons in atoms might be arranged in concentric shells, with the nucleus at the center. He proposed that each shell could hold a characteristic number of electrons, and only those electrons in the outermost shell were involved when one atom combined with another by forming ions or molecules. These outermost electrons came to be known as **valence electrons.**

For example, look at phosphorus (P). The Bohr arrangement of electrons is 2-8-5. This means that the stable state of the phosphorus atom has electrons in three energy levels. The one closest to the nucleus has two electrons; the second energy level has eight electrons; and the highest energy level has five electrons. The energy level with five electrons is farthest from the nucleus, so P has *five valence electrons,* and these are the electrons that are most available for interactions with valence electrons of other atoms in chemical reactions.

EXAMPLE 3.5 *Electron Arrangement*

Give the Bohr arrangement of electrons for an atom of aluminum (Al). How many valence electrons does Al have?

SOLUTION

The atomic number of Al is 13, so there are 13 electrons. The maximum number of electrons per energy level is $2n^2$. The number of electrons that can be accommodated in the first energy level is $2(1)^2 = 2$, and the number of electrons that can be accommodated in the second level is $2(2)^2 = 8$. This leaves three electrons for the third level. Therefore, the Bohr arrangement of electrons for an atom of Al is 2-8-3. Since the number of valence electrons is equal to the number of electrons in the highest energy level of the ground state, Al has three valence electrons.

Exercise 3.5

Draw the Bohr arrangement of electrons for an atom of S and give its number of valence electrons.

The Wave Theory of the Atom

The Bohr model failed when applied to elements other than hydrogen because it could not account exactly for the line spectra of atoms with more than one electron. After Bohr's work, a more modern, highly sophisticated mathematical theory of the atom was developed. In this theory, electrons are treated as having both a particle and a wave nature. The locations of the electrons are treated as **probabilities,** without seeking to locate the exact spot for an electron at a given time. This approach suggested that the Bohr theory describing the electrons with fixed orbits sought more precision than nature would allow.

■ Electrons are described by both particle and wave theories.

A Frenchman, Louis de Broglie, was the first to suggest (in 1924) that electrons and other small particles should have wave properties. In this respect, he said, electrons should behave like light, a suggestion that scientists of the time found hard to accept. However, in a few years experiments verified de Broglie's hypothesis. The **electron microscopes** found in many research laboratories today are built and operated on our understanding of the wave nature of the electron.

It should not be surprising to find that matter can be treated by both wave and particle theories (the duality of matter), since its convertible counterpart—light—has been treated successfully by both theories for a long time. Keep in mind that we do not really know if matter or light is a wave or a particle. However, because there are limits on what we can visualize in our physical world, in talking about something like subatomic behavior we are forced to use physical models based on known behavior, rather than more sophisticated models that would describe some type of intermediate behavior with which we are unfamiliar in our macroscopic world.

Electron microscope.

The wave theory of the atom was developed in the 1920s, principally by Erwin Schrödinger (1887–1961). The most fundamental aspects of the theory are the mathematical wave equations used to describe the electrons in atoms. Solutions to the equations are called wave functions, or **orbitals.** Calculations involving the wave equations are complicated and time-consuming, but we do not need to do the elaborate calculations in order to use the results.

The orbitals of the wave theory are actually subdivisions of the Bohr energy levels that predict a volume of space where the electrons are. Bohr's energy levels are expanded to include these sublevels called orbitals. Only the **probability** of finding an electron in a given volume of space around the nucleus can be calculated from the orbital resulting from the Schrödinger equation. In order to portray the probabilities of finding an electron, usually

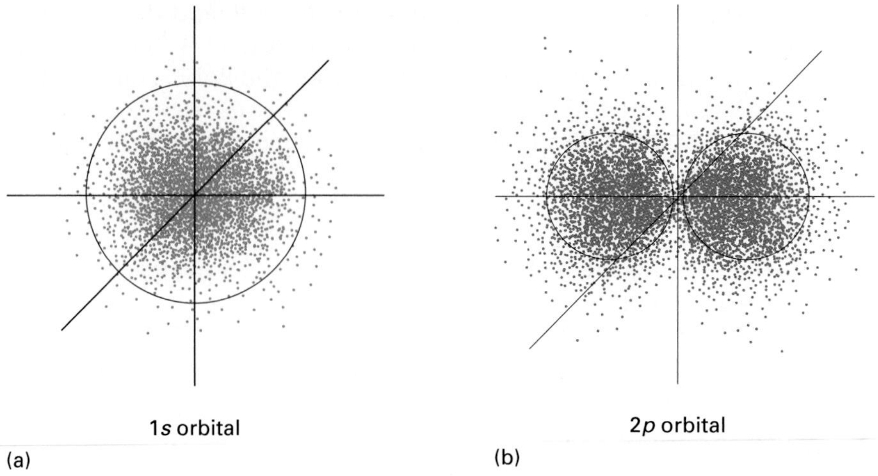

Figure 3.17 (a) The 1*s* orbital. The nucleus is located at the center. This probability map of electron positions is called an electron cloud. The circle encloses 90% of the dots and the resulting spherical shape is used to represent the *s* orbital. (b) A 2*p* orbital. The probability map is shown with the outer line enclosing 90% of the dots to give a dumbbell shape.

■ Orbitals or sublevels are labeled *s, p, d,* and *f.* These letters were derived from terms in spectroscopy (sharp, principal, diffuse, and fundamental, respectively) and emphasize again that atomic theory developed very closely with atomic spectra.

the surface of a region in space (similar to the surface of a balloon) is plotted that will enclose the volume where the electron will be expected to be found 90% of the time (Figure 3.17).

Figure 3.17 illustrates two of the orbital shapes that result from probability plots—the *s* orbital has a spherical shape and the *p* orbital has a dumbbell shape. Imagine the dots as representing places where the electron has been during a given time period. Enclosing about 90% of the dots in these "electron cloud" representations gives the characteristic shapes commonly used for orbitals.

The energy level diagram for orbitals is given in Figure 3.18. The number indicates the Bohr energy level (n = 1, 2, 3,. . . .) and the letter indicates the energy sublevel or orbital (*s, p, d, f*). Each line represents an orbital that can hold two electrons. Table 3.3 lists the electron configurations for the first 20 elements. The superscripts are the number of electrons in the orbitals. For example, neon has the electron configuration $1s^22s^22p^6$. Figures 3.19(a, b) illustrate the individual orbitals occupied by the electrons of neon in the ground state, and Figure 3.19(c) shows a "fuzzy" view of the neon atom that results from superimposing all the orbitals.

Compare the electron arrangements in Table 3.3 with the electron arrangements given in Table 3.2. Notice that the totals for each Bohr energy level are the same, and the number of valence electrons are the same. The wave mechanical model adds energy sublevels (orbitals) and emphasizes the importance of representing electron locations as "electron clouds" whose shape depends upon the orbital.

What, then, is an atom really like? The atomic concepts have changed over a long period of time. We have Dalton's concept of an atom as a hard sphere similar to a small billiard ball. We have Bohr's concept of the atom as a small three-dimensional solar system with a nucleus and electrons in paths called energy levels, or orbits. In the modern theory, we have more detail in that energy levels are now sub-levels called orbitals, and we are given approximate spaces where electrons exert their greatest influence in an atom. Why present all three theories? First, an understanding of the simpler Dalton and Bohr

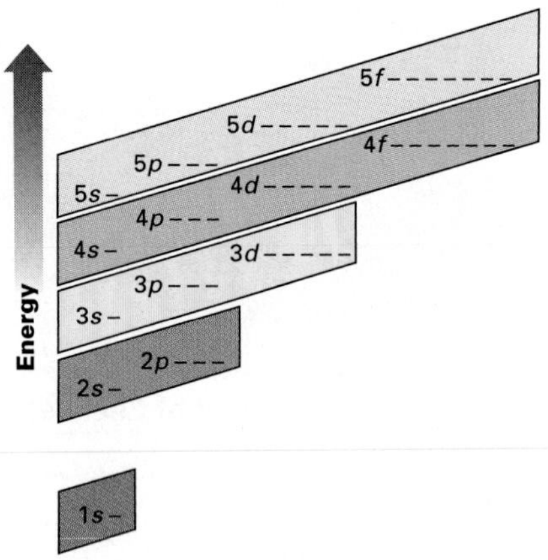

Figure 3.18 The order of orbital filling. The orbitals for a given Bohr energy level (indicated by the number) are the same color. The orbitals fill in order of increasing energy (1*s*, 2*s*, 2*p*, 3*s*, 3*p*, 4*s*, 3*d*, 4*p*, 5*s*, 4*d*, 5*p*, . . .).

TABLE 3.3 ■ Electron Configurations of the First 20 Elements*

Element	Atomic Number		Element	Atomic Number	
Hydrogen (H)	1	$1s^1$	Sodium (Na)	11	$1s^22s^22p^6\ 3s^1$
Helium (He)	2	$1s^2$	Magnesium (Mg)	12	$1s^22s^22p^6\ 3s^2$
Lithium (Li)	3	$1s^2\ 2s^1$	Aluminum (Al)	13	$1s^22s^22p^6\ 3s^23p^1$
Beryllium (Be)	4	$1s^2\ 2s^2$	Silicon (Si)	14	$1s^22s^22p^6\ 3s^23p^2$
Boron (B)	5	$1s^2\ 2s^22p^1$	Phosphorus (P)	15	$1s^22s^22p^6\ 3s^23p^3$
Carbon (C)	6	$1s^2\ 2s^22p^2$	Sulfur (S)	16	$1s^22s^22p^6\ 3s^23p^4$
Nitrogen (N)	7	$1s^2\ 2s^22p^3$	Chlorine (Cl)	17	$1s^22s^22p^6\ 3s^23p^5$
Oxygen (O)	8	$1s^2\ 2s^22p^4$	Argon (Ar)	18	$1s^22s^22p^6\ 3s^23p^6$
Fluorine (F)	9	$1s^2\ 2s^22p^5$	Potassium (K)	19	$1s^22s^22p^63s^23p^6\ 4s^1$
Neon (Ne)	10	$1s^2\ 2s^22p^6$	Calcium (Ca)	20	$1s^22s^22p^63s^23p^6\ 4s^2$

*Valence electrons are shown in color.

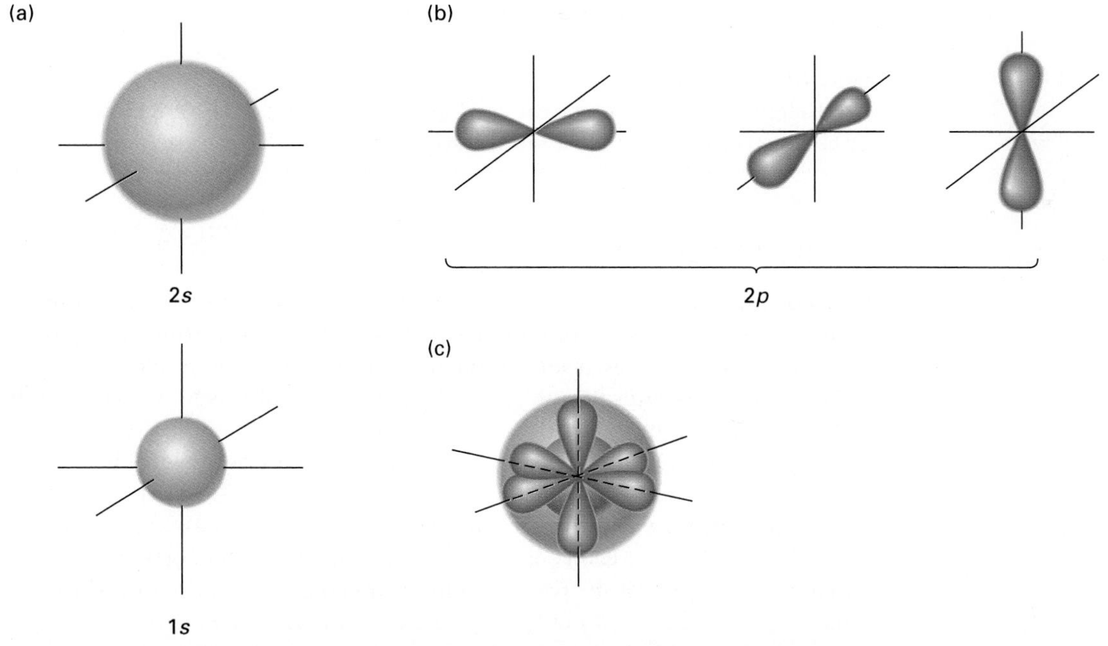

Figure 3.19 Orbitals used for the ten electrons in the neon atom, which has the electron configuration $1s^22s^22p^6$. (a) $1s$ and $2s$ orbitals each have two electrons. (b) Each $2p$ orbital has two electrons. (c) The fuzzy view of a neon atom results from superimposing all the orbitals used by the electrons in neon. The nucleus is located at the center.

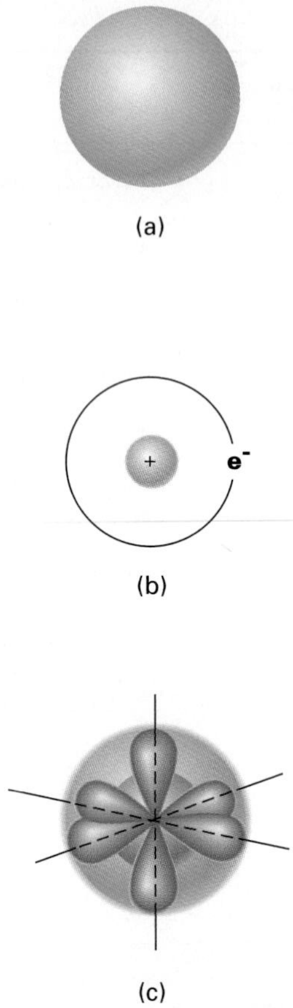

Comparison of models of the atom. (a) Dalton model. (b) Bohr model. (c) Wave mechanics model.

theories helps us to understand the more complicated, more detailed modern theory of the atom. Second, all three theories help us to understand the phenomena we observe. We simply use whatever detail is necessary to explain what we see. For example, the simpler Dalton concept adequately explains many properties of the gaseous, liquid, and solid states. Most bonding between atoms of the light elements can be explained by application of the orbits of Bohr. The shapes of molecules and the arrangement of atoms with respect to each other can best be explained by the orbital representations of the modern theory. In the explanations given in this text, we shall follow the principle that simplest is best.

EXAMPLE 3.6 *Electron Configuration*

What element has the electron configuration $1s^22s^22p^63s^23p^2$? How many valence electrons does this element have? What is the Bohr electron arrangement?

SOLUTION

The total number of electrons is 14, obtained by adding the superscript numbers. Silicon has an atomic number of 14. The number of valence electrons is obtained by adding up the total number in the highest energy level. The highest energy level of silicon that contains electrons in the ground state is level 3, which has 4 electrons so there are 4 valence electrons. The Bohr arrangement is the total for each energy level or 2,8,4.

Exercise 3.6
What element has the electron configuration $1s^22s^22p^63s^23p^5$? How many valence electrons does this element have?

■ SELF-TEST 3C

1. According to Bohr's theory, light of characteristic wavelength is emitted as an electron drops from an energy level (closer to/farther from) the nucleus to an energy level (closer to/farther from) the nucleus.
2. Which has less energy? (a) Blue light, (b) Red light, (c) Ultraviolet light, (d) Infrared light.
3. The maximum number of the electrons in $n = 3$ energy level is ______________.
4. The ground-state Bohr representation for electrons in an atom of K is ______________. K has ______________ valence electron(s).
5. The ground-state Bohr representation of electrons in an atom of Cl is ______________. Cl has ______________ valence electron(s).
6. Which color of light has the shortest wavelength? (a) Blue, (b) Green, (c) Orange, (d) Red.
7. All colors of light travel at the same speed. (a) True, (b) False.
8. The location of electrons in an atom is described as an electron cloud in the ______________ model of the atom.

MATCHING SET

____ 1. Mass number	a.	Attract
____ 2. Unlike electric charges	b.	Equal to number of protons in nucleus
____ 3. $2n^2$	c.	Cathode-ray particle
____ 4. Nucleus	d.	Neutrons plus protons
____ 5. Electron	e.	Discovered the nucleus
____ 6. ^{22}Ne and ^{20}Ne	f.	Proton and an electron
____ 7. Atomic number	g.	Developed a theory for representing electrons in energy levels
____ 8. Particles in an H atom	h.	Isotopes
____ 9. Neutron	i.	Repel
____ 10. Rutherford	j.	Maximum number of electrons in an energy level
____ 11. Bohr	k.	Uncharged elementary particle
	l.	Equal to the number of neutrons in the nucleus
	m.	Central part of an atom consisting of protons and neutrons

QUESTIONS FOR REVIEW AND THOUGHT

1. What is the law of conservation of matter? Give an example of the law in action.
2. State the law of definite proportions. Give an example to illustrate what it means.
3. When the Greeks described matter as being endlessly divisible what did they mean?
4. What kinds of evidence did Dalton have for atoms that the early Greeks (Democritus, Leucippus) did not have?
5. How does Dalton's atomic theory explain
   (a) the law of conservation of matter?
   (b) the law of constant composition?
   (c) the law of multiple proportions?
6. Consider two compounds formed with phosphorus and oxygen, P_4O_6 and P_4O_{10}. How does the law of multiple proportions explain these results?
7. State the law of multiple proportions and explain how it relates to SO_2, 

, and SO_3, 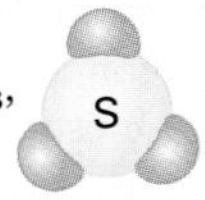

.
8. Describe in detail Rutherford's gold-foil experiment under the following headings:
   (a) Experimental setup
   (b) Observations
   (c) Interpretations
9. What are cathode rays?
10. What is meant by the term subatomic particles? Give two examples.
11. What is a ground state for an atom? What is an excited state for an atom?
12. What is a practical application of cathode-ray tubes?
13. Give short definitions for the following terms:
    (a) Atomic number
    (b) Mass number
    (c) Atomic weight
    (d) Isotope
    (e) Natural abundance
    (f) Atomic mass unit
14. Give short definitions for the following terms:
    (a) Ion
    (b) Visible light
    (c) Wavelength
    (d) Emission spectrum
    (e) Emission spectrum line
15. Are any of the following an exception to the Law of Conservation of Matter? Explain.
    (a) The loss of water during the evaporation of a puddle of water
    (b) The burning of a piece of wood so that only a small amount of ash remains
16. If electrons are a part of all matter, why are we not electrically shocked continually by the abundance of electrons about and in us?
17. There are more than 1000 kinds of atoms, each with a different weight, yet there are only 109 elements. How does one explain this in terms of subatomic particles?
18. What do all the atoms of an element have in common?
19. Which of the following pairs are isotopes? Explain your answers.
    (a) ^{50}Ti and ^{50}V

(b) ^{12}C and ^{14}C
(c) ^{40}Ar and ^{40}K

20. A common isotope of Li has a mass of 7. The atomic number of Li is 3. How can this information be used to determine the number of protons and neutrons in the nucleus?
21. The element iodine (I) occurs naturally as a single isotope of atomic mass 127; its atomic number is 53. How many protons and how many neutrons does it have in its nucleus?
22. Which pairs of atoms are isotopes?

	A	B	C	D
Mass number	53	53	52	54
Atomic number	25	24	24	25

23. The following table contains information about five different atoms. Complete the table:

	Number of Protons	Number of Neutrons	Number of Electrons	Atomic Number	Mass Number
(a)	32	____	____	____	73
(b)	____	14	14	____	____
(c)	____	____	____	28	59
(d)	48	64	____	____	____
(e)	____	115	____	77	____

24. Identify the elements in Question 23.
25. What number is most important in identifying an atom?
26. The atomic mass listed in the periodic table for magnesium is 24.305 amu. Someone said that there wasn't a single magnesium atom on the entire earth with a mass of 24.305 amu. Is this statement correct? Why or why not?
27. Complete the following table.

Isotope	Atomic No.	Mass No.	No. of Protons	No. of Neutrons	No. of Electrons
Bromine-81	____	81	____	____	____
Boron-11	5	____	____	____	____
^{35}Cl	17	____	____	____	____
^{52}Cr	____	52	____	____	____
Ni-60	____	____	____	____	____
Sr-90	____	____	____	____	____
Lead-206	____	____	____	____	____

28. Distinguish between a continuous spectrum and a bright-line spectrum under the two headings:
(a) General appearance
(b) Source
29. What is constant about a compound? (a) The weight of a sample of the compound, (b) The weight of one of the elements in samples of the compound, (c) The ratio by weight of the elements in the compound.
30. If you found the number of wheels received by an assembly plant to be twice the number of motors, what type of vehicle would you assume to be assembled there? Of what chemical law does this remind you?
31. In recent years we have found that pure substances, such as some plastics, do vary in composition and that some elements can be decomposed (nuclear fission). What does this say to you about concepts and progress in science?
32. Why is it impossible to produce a positive charge without producing a negative charge at the same time?
33. Krypton is the name of Superman's home planet and also that of an element. Look up the element krypton and list its symbol, atomic number, atomic weight, and electron arrangement.
34. Explain in your own words why alpha particles are deflected in one direction in an electric field, whereas beta particles are deflected in the opposite direction.
35. For each of the following isotopes determine the number of protons, neutrons, and electrons.
(a) $^{24}_{12}Mg$ (b) $^{56}_{26}Fe$
(c) $^{115}_{49}In$ (d) $^{127}_{53}I$
(e) $^{107}_{47}Ag$ (f) $^{222}_{86}Rn$
36. Write out the placement of electrons in their ground-state energy levels according to the Bohr theory for atoms having 6, 10, 13, and 20 electrons.
37. Write the wave-mechanical electron configuration ($1s2s2p$ . . .) for the atoms in Question 36.
38. Write the Bohr electron notation for atoms of the elements sodium through argon.
39. How many valence electrons do atoms of each of the elements in Question 38 contain?
40. Write the Bohr electron notation for atoms of the elements beryllium through neon.
41. Write the wave-mechanical electron configurations for atoms of the elements beryllium through neon.
42. How many valence electrons do atoms of each of the elements in Question 41 contain?
43. Name five regions of the electromagnetic spectrum that you use most every day. Arrange these regions in order of increasing wavelength.
44. The visible region of the electromagnetic spectrum ranges from 400 nm to 700 nm. Energy varies inversely with wavelength. Consider the colors of a rainbow, and rank these colors in order of increasing energy.

45. What change in energy levels will give the smallest energy increase for the electron in the hydrogen energy diagram shown at right?
46. The visible emission spectrum for hydrogen includes red, 656.7 nm, for the jump between levels 2 and 3. Would the light for the jump between levels 2 and 4 have longer wavelength? Explain.
47. The spectral lines from the emission spectrum of helium were first observed during a solar eclipse in 1868 when the main body of the Sun was hidden by the Moon. Helium had not yet been observed on Earth. What made it possible to decide that the light seen in the solar eclipse didn't come from another element?

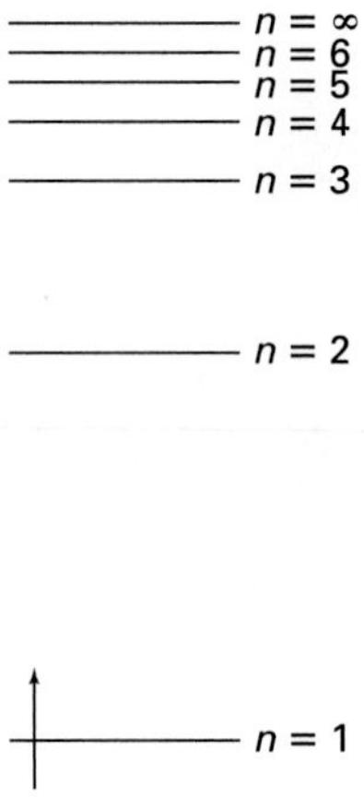

PROBLEMS

1. If butane, the fuel in a cigarette lighter, contains 82.6% carbon and 17.4% hydrogen by mass, how many grams of carbon and hydrogen atoms are combined to make 60.0 grams of butane?
2. Suppose Millikan had determined the following charges on his oil drops:

1.33×10^{-19} C $\quad 2.66 \times 10^{-19}$ C $\quad 3.33 \times 10^{-19}$ C

4.66×10^{-19} C $\quad 7.92 \times 10^{-19}$ C

What do you think his value for the electron's charge would have been?
3. In the game of marbles a large, fancy marble used for shooting is called a taw. Suppose you have a taw that is 1 inch in diameter (2.54 cm). You line up 1000 of these taws in a row. How long is the row in inches? In meters?
4. Let's now reduce the diameter of the taw to 0.20 inch. How long is the row in inches? In meters?
5. Now make the diameter of each taw 1.0×10^{-8} cm. A little-bitsy taw is what you have. In order to have a line of these taws that you could measure, what variable in Problem 3 would you have to change?
6. How many of these little-bitsy taws would you have to use to make a line 1.0 inch long?
7. If 5.00 grams of calcium carbonate (Tums) contains 2.00 g of calcium atoms, 0.600 g of carbon atoms, and 2.40 g of oxygen atoms, what are the mass percentages of each element in this compound?
8. The diameter of an aluminum atom is 286 pm. If aluminum atoms could be lined up in a row, how many atoms would it take to make a line 1 dm long?
9. If a copper atom has a diameter of 376 pm, how many copper atoms would it take to make a line 1 cm long?
10. Lithium has two isotopes, one with a mass of 6.015 that is 7.42% abundant and another with a mass of 7.016 that is 92.58% abundant. Find the atomic weight of lithium.
11. What are the wavelengths for the light waves illustrated below? Which one would carry more energy?

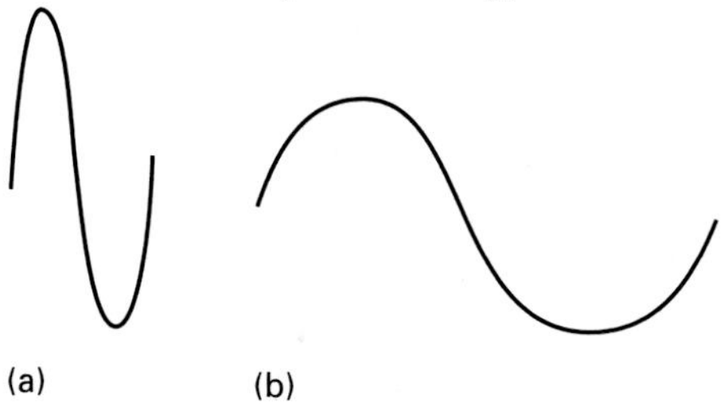

12. What is the smallest energy level jump for an electron in the n = 1 level in a hydrogen atom?
13. What is the wavelength for a microwave illustrated below? Microwaves have wavelengths in the range 0.1 cm to 10. cm. (Actually measure wave with a ruler.)

CHAPTER 4

The Periodic Table

When a natural science is young, classification is a major activity. Information-gathering focuses on questions such as, How many kinds of butterflies are there? Which stars are similar to our Sun? Are there different kinds of metals with properties in common?

Sometimes the early classifications hold up as a science matures and more information is gathered. Sometimes they don't. The periodic table, which groups the chemical elements according to their properties, is an amazing example of a classification that has stood the test of time. Since its origin in 1869, adjustments have been made, but the fundamental principles have remained unchanged. Furthermore, as new elements have been discovered, where to place them has become obvious. Even more amazingly, although the table was devised before atomic structure was understood, the arrangement of elements in the periodic table has made more and more sense as the roles of protons, neutrons, and electrons in atoms have become clear. In particular, the influence of the location and number of electrons in atoms on the properties of elements has become one of the essential ideas of chemistry.

- How was the periodic table developed?
- What is the concept of valence electrons, and why is it useful?
- Why do elements in the same group in the periodic table have similar chemical properties?
- How is the size of atoms of an element related to the element's position in the periodic table?
- How is the activity of an element related to its position in the periodic table?

Salt being spread to dry. The periodic table shows why sodium and chlorine form this compound so easily. *(Luiz C. Marigo/Peter Arnold, Inc.)*

So far, in exploring the chemical view of matter, you have seen that everything is made of atoms. Atoms of different elements combined in specific ratios are present in chemical compounds, and all matter consists of elements, or compounds, or mixtures of elements or compounds. You have also been introduced to the subatomic composition of atoms—the electrons, protons, and neutrons. The next part of the story, which begins in this chapter, is to see how atomic structure and the properties of elements and compounds are related to each other.

You have probably already noticed a periodic table in your classroom. Why is such a table on the wall in almost every chemistry classroom and laboratory? The periodic table is the single most important classification system in chemistry because it summarizes, correlates, and predicts a wealth of chemical information. Chemists consult it every day during every possible kind of work. It can be simply a reminder of the symbols and names of the elements, of which elements have similar properties, and of where each element lies on the continuum of atomic numbers. It can also be an inspiration in the search for new compounds or mixtures that will fulfill a specific need.

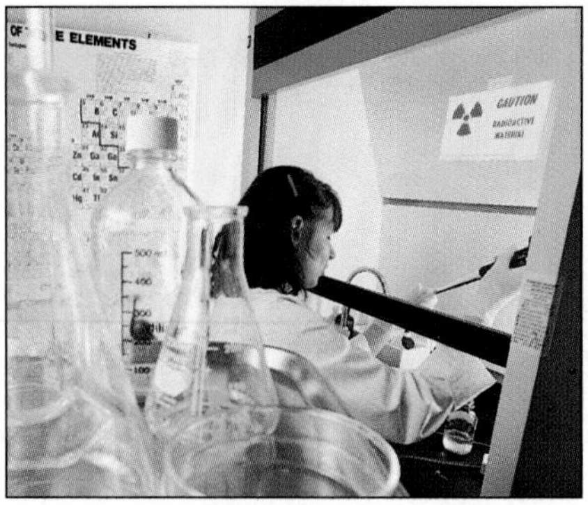

The periodic table is on the wall in most chemistry laboratories. *(Robert Holmgren/Peter Arnold, Inc.)*

Memorizing the periodic table is no more necessary than memorizing the map of your home state. But in both cases, it's very helpful to have a general idea of the major features. This chapter describes the topography of the periodic table.

4.1 Development of the Periodic Table

On the evening of February 17, 1869, at the University of St. Petersburg in Russia, a 35-year-old professor of general chemistry, Dmitri Ivanovich Mendeleev (1834–1907) was writing a chapter of his soon-to-be-famous textbook on chemistry. He had the properties of each element written on cards, with a separate card for each element. While he was shuffling the cards trying to gather his thoughts before writing his manuscript, Mendeleev realized that if the elements were arranged in the order of their atomic weights, there was a trend in properties that repeated itself several times!

Thus the periodic law and table were born, although only 63 elements had been discovered by 1869 (for example, the noble gases were not discovered until after 1893), and the clarifying concept of the atomic number was not known until 1913. Mendeleev's idea and textbook achieved great success, and he rose to a position of prestige and fame as he continued to teach at St. Petersburg.

At the same time that Mendeleev was developing his periodic table in 1869, Lothar Meyer (1830–1895), a German physician and professor of chemistry at the University of Tübingen, prepared a table very similar to Mendeleev's. Apparently, the men were unaware of each other's work, yet both had left gaps for undiscovered elements. Meyer's table was based primarily on the repeatable trends in physical properties. Mendeleev's table, which was based on chemical properties, received more recognition because Mendeleev used it to predict properties of undiscovered elements. However, Mendeleev acknowledged Meyer's claim to independent discovery, and in 1882 the British Davy Medal was jointly awarded to both men.

By 1871 Mendeleev published a more elaborate periodic table (Figure 4.1). This version was the forerunner of the modern table currently seen in classrooms and textbooks. The genius and daring of Mendeleev are

Row	Group I — R_2O	Group II — RO	Group III — R_2O_3	Group IV RH_4 RO_2	Group V RH_3 R_2O_5	Group VI RH_2 RO_3	Group VII RH R_2O_7	Group VIII — RO_4
1	H = 1							
2	Li = 7	Be = 9.4	B = 11	C = 12	N = 14	O = 16	F = 19	
3	Na = 23	Mg = 24	Al = 27.3	Si = 28	P = 31	S = 32	Cl = 35.5	
4	K = 39	Ca = 40	_ = 44	Ti = 48	V = 51	Cr = 52	Mn = 55	Fe = 56, Co = 59, Ni = 59, Cu = 63
5	(Cu = 63)	Zn = 65	_ = 68	_ = 72	As = 75	Se = 78	Br = 80	
6	Rb = 85	Sr = 87	?Yt = 88	Zr = 90	Nb = 94	Mo = 96	_ = 100	Ru = 104, Rh = 104, Pd = 106, Ag = 108
7	(Ag = 108)	Cd = 112	In = 113	Sn = 118	Sb = 122	Te = 125	I = 127	
8	Cs = 133	Ba = 137	?Di = 138	?Ce = 140				
9								
10			?Er = 178	?La = 180	Ta = 182	W = 184		Os = 195, Ir = 197, Pt = 198, Au = 199
11	(Au = 199)	Hg = 200	Tl = 204	Pb = 207	Bi = 208			
12				Th = 231		U = 240		

Figure 4.1 Mendeleev periodic table. This English translation of Mendeleev's periodic table was published in 1871. The formulas for simple oxides, chlorides, and hydrides are shown under each group heading. R represents the element in each group. Mendeleev predicted several elements that were unknown by leaving empty spaces where the elements would fall based on their predicted properties.

(a)

(b)

Dimitri Mendeleev (a) and Lothar Meyer (b) who independently developed the periodic table by recognizing that the properties of elements repeat in a periodic fashion. *(a, Novosti/SPL/Photo Researchers; b, Meiman Photo Studio Courtesy AIP Emilio Segré Visual Archives)*

demonstrated by the empty spaces he left in the table for the not-yet-discovered elements. The spaces were necessary to retain the rationale of ordered arrangement based on periodic recurrence of the properties. For example, the order of atomic weights for elements known at the time was copper (Cu), zinc (Zn), and then arsenic (As), with atomic weights of 63, 65, and 75, respectively. If arsenic had been placed next to zinc, arsenic would have fallen under aluminum (Al). But arsenic forms compounds similar to those formed by phosphorus (P) and antimony (Sb), not aluminum. Mendeleev reasoned that two as yet undiscovered elements existed and moved arsenic over two spaces to the position below phosphorus in Group V, as shown in Figure 4.1. The two missing elements were soon discovered: gallium (Ga) in 1875 and germanium (Ge) in 1886. In later years the gaps in this 1871 periodic table were filled as the predicted elements were discovered.

Mendeleev aided the discovery of the new elements by predicting their properties with remarkable accuracy, and he even suggested the geographical regions in which minerals containing the elements could be found. The properties of a missing element were predicted by consideration of the properties of its neighboring elements in the table. An example of Mendeleev's prediction of the properties of an undiscovered element is shown in Table 4.1, which compares his predictions for the element he called ''*eka-silicon*'' and that we know as germanium. The term *eka* comes from Sanskrit and means ''one''; thus, ''*eka-silicon*'' means ''one place away from silicon.'' He also predicted the properties of eka-boron (scandium) and eka-aluminum (gallium).

TABLE 4.1 ■ Some of Mendeleev's Predicted Properties of Eka-silicon and the Corresponding Observed Properties of Germanium

Property	Eka-silicon (Es) (Mendeleev's Predictions Made in 1871)	Observed Properties of Germanium (Ge) (Discovered in 1886)
Atomic weight	72	72.6
Color	Gray	Gray
Density of element (g/cm^3)	5.5	5.36
Melting point (°C)	High	947
Formula of oxide	EsO_2	GeO_2
Density of oxide (g/cm^3)	4.7	4.228
Formula of chloride	$EsCl_4$	$GeCl_4$
Density of chloride	1.9	1.844
Boiling point of chloride (°C)	Under 100	84

The empty spaces in the table and Mendeleev's predictions of the properties of missing elements stimulated a flurry of prospecting for elements in the 1870s and 1880s. As a result, in addition to gallium and germanium, scandium (Sc), samarium (Sm), holmium (Ho), and thulium (Tm) were discovered in 1879, gadolinium (Gd) in 1880, neodymium (Nd) and praseodymium (Pr) in 1885, and dysprosium (Dy) in 1886. Many of these elements are not common even today, yet they are important as ingredients in catalysts and color television screens.

Mendeleev found that a few elements did not fit under other elements with similar chemical properties when arranged according to increasing atomic weight. To make the elements fit, Mendeleev predicted that several atomic weights were incorrect. For example, he placed tellurium (Te) in the same group as sulfur (S), and iodine (I) in the same group as chlorine (Cl) because of chemical similarities, even though this inverted the atomic weight order. However, as can be seen from the modern periodic table in Figure 4.2, tellurium does have a larger atomic weight than iodine (127.6 versus 126.9). The point is that the *atomic weight is not the property that governs periodicity.* This was discovered in 1913 by H.G.J. Moseley (1888–1915), a young scientist working with Ernest Rutherford. Moseley bombarded many different metals with electrons in a cathode-ray tube and observed the X rays emitted by the metals. He found that the wavelengths of X rays emitted by a particular element are related in a precise way to the atomic number of that element. He quickly realized that other atomic properties may be similarly related to atomic number and not, as Mendeleev had believed, to atomic weight. Indeed, if the elements are arranged in order of increasing atomic number, the questions about the Mendeleev table are answered.

■ Notice that Mendeleev's table, shown in Figure 4.1, does not have the transition and inner transition elements in separate parts of the table, as in the modern periodic table in Figure 4.2.

■ Other reversed pairs of atomic weights in the modern periodic table are uranium (U) before neptunium (Np), argon (Ar) before potassium (K), cobalt (Co) before nickel (Ni), and thorium (Th) before protactinium (Pa).

Building on the work of Mendeleev and Meyer and others, and using the concept of the atomic number, we are now able to state the modern **periodic law:** *When elements are arranged in the order of their atomic numbers, their chemical*

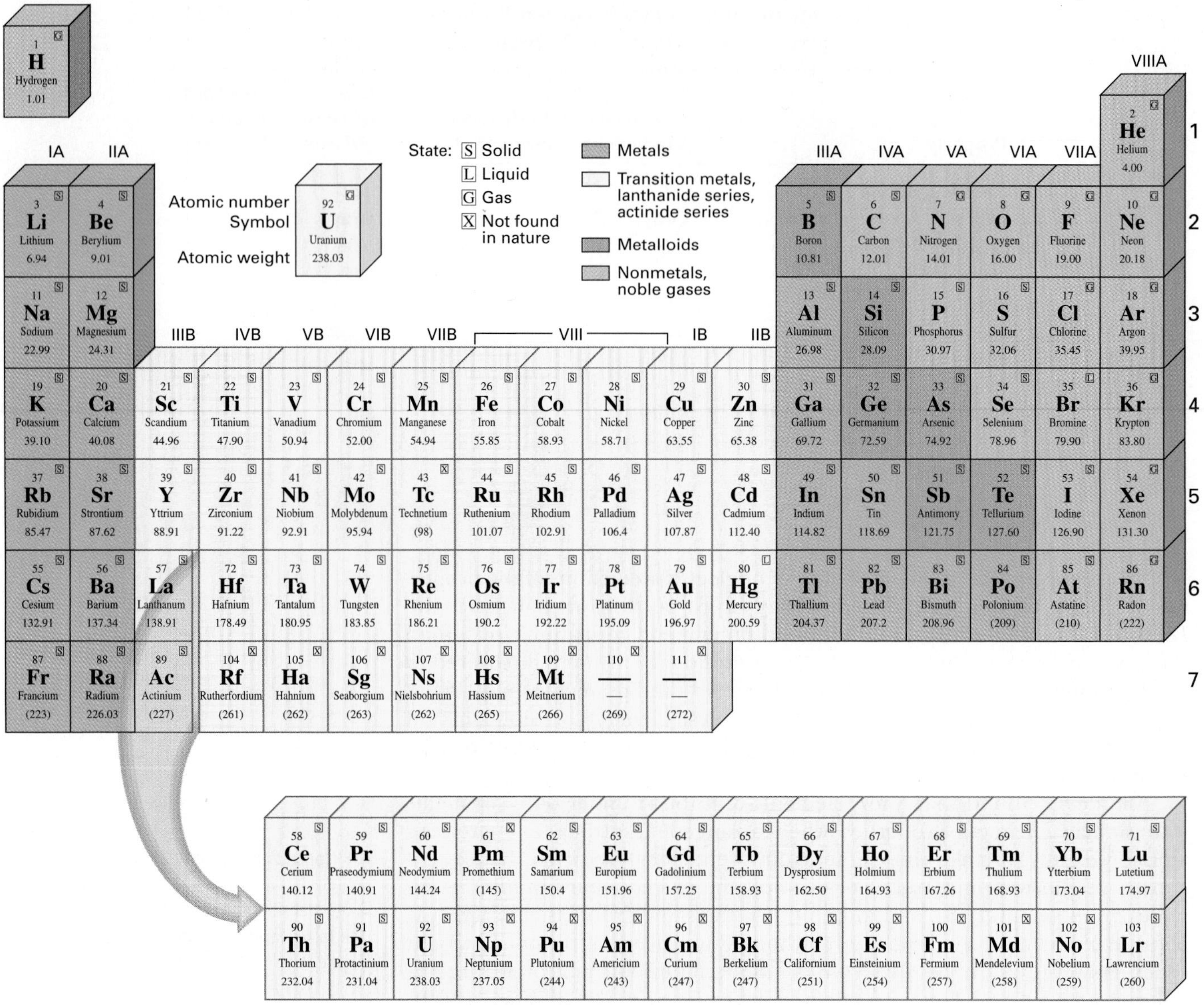

Figure 4.2 Modern periodic table of the elements.

and physical properties show repeatable, or periodic, trends. Other familiar periodic phenomena include the average daily temperature, which is periodic with time in a temperate climate. A shingle roof has the same pattern over and over and is, therefore, periodic.

So, to build up a periodic table according to the periodic law, the elements are lined up in a horizontal row in the order of their atomic numbers. At an element with similar properties to one already in the row, a new row is started. The columns then contain elements with similar properties. Some chemical and physical properties of the first 20 elements are summarized in Table 4.2. Do you see any trends and similarities among the elements in Table 4.2? For

TABLE 4.2 ■ Some Properties of the First 20 Elements

Element	Atomic Number	Description	Compound Formation* With Cl (or Na)	With O (or Mg)
Hydrogen (H)	1	Colorless gas; reactive	HCl	H_2O
Helium (He)	2	Colorless gas; unreactive	None	None
Lithium (Li)	3	Soft metal; low density; very reactive	LiCl	Li_2O
Beryllium (Be)	4	Harder metal than Li; low density; less reactive than Li	$BeCl_2$	BeO
Boron (B)	5	Both metallic and nonmetallic; very hard; not very reactive	BCl_3	B_2O_3
Carbon (C)	6	Brittle nonmetal; unreactive at room temperature	CCl_4	CO_2
Nitrogen (N)	7	Colorless gas; nonmetallic; not very reactive	NCl_3	N_2O_5
Oxygen (O)	8	Colorless gas; nonmetallic; reactive	Na_2O, Cl_2O	MgO
Fluorine (F)	9	Greenish-yellow gas; nonmetallic; extremely reactive	NaF, ClF	MgF_2, OF_2
Neon (Ne)	10	Colorless gas; unreactive	None	None
Sodium (Na)	11	Soft metal; low density; very reactive	NaCl	Na_2O
Magnesium (Mg)	12	Harder metal than Na; low density; less reactive than Na	$MgCl_2$	MgO
Aluminum (Al)	13	Metal as hard as Mg; less reactive than Mg	$AlCl_3$	Al_2O_3
Silicon (Si)	14	Brittle nonmetal; not very reactive	$SiCl_4$	SiO_2
Phosphorus (P)	15	Nonmetal; low melting point; white solid; reactive	PCl_3	P_2O_5
Sulfur (S)	16	Yellow solid; nonmetallic; low melting point; moderately reactive	Na_2S, SCl_2	MgS
Chlorine (Cl)	17	Green gas; nonmetallic; extremely reactive	NaCl	$MgCl_2$, Cl_2O
Argon (Ar)	18	Colorless gas; unreactive	None	None
Potassium (K)	19	Soft metal; low density; very reactive	KCl	K_2O
Calcium (Ca)	20	Harder metal than K; low density; less reactive than K	$CaCl_2$	CaO

*The chemical formulas shown are lowest ratios. The molecular formula for $AlCl_3$ is Al_2Cl_6, and that for P_2O_5 is P_4O_{10}.

example, lithium (Li) is a soft metal with low density that is very reactive. It combines with chlorine gas to form lithium chloride with the formula LiCl. The other elements in Table 4.2 that have properties similar to those of lithium are sodium (Na) and potassium (K). According to the periodic law, lithium, sodium, and potassium should be in the same group, and they are. Look for similarities among other elements listed in Table 4.2 and check your grouping with that shown in the periodic table in Figure 4.2.

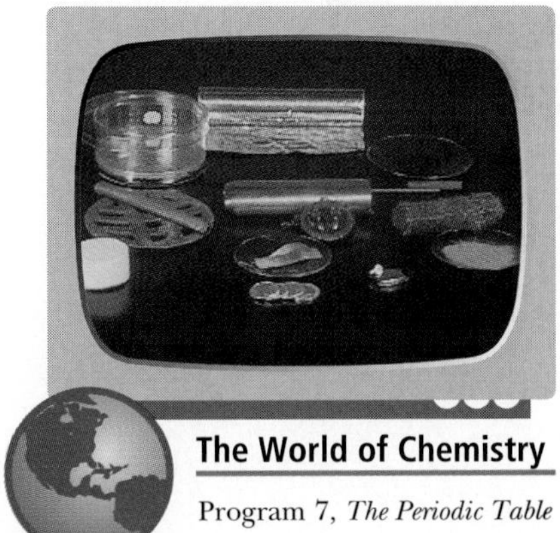

The World of Chemistry

Program 7, *The Periodic Table*

An assortment of pure elements.

4.2 The Modern Periodic Table

Note the following features in the periodic table in Figure 4.2. The vertical columns that list elements with similar chemical and physical properties are called **groups.** The periodic table commonly used in the United States has groups numbered I through VIII, with each Roman numeral followed by a letter A or B. The A groups are the **representative** or **main-group** elements.

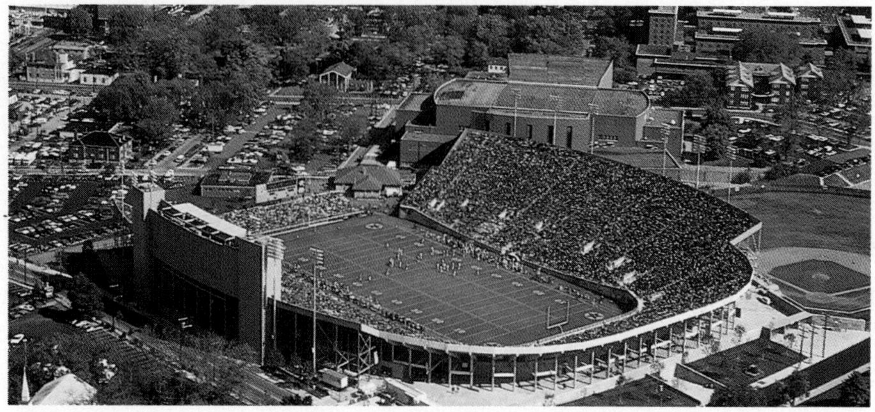

Figure 4.3 Analogy of periods in periodic table to rows in a football stadium. Larger periods in the periodic table as atoms of elements get larger are similar to longer rows and more seats per row in a stadium as the rows are further from the playing field. *(Courtesy of Department of Athletics, Vanderbilt University)*

■ Chemists from all over the world belong to the International Union of Pure and Applied Chemistry (IUPAC). IUPAC has recommended that groups be labeled 1 through 18 consecutively from left to right. The periodic table on the inside back cover includes both 1 through 18 and A and B group labels for comparison.

The B groups are the **transition elements** that link the two areas of representative elements. The **inner transition elements** are the **lanthanide series** and the **actinide series.** They are placed at the bottom of the periodic chart because the similarity of properties within the two series would require their placement between lanthanum and hafnium (lanthanide series) and between actinium and rutherfordium (actinide series).

The horizontal rows are called **periods.** These periods or rows are related to energy levels for electrons in atoms (Figure 3.18). The length of a row is linked to the maximum number of electrons, $2n^2$, that can fit into an energy level. The periods are not equal in size because the maximum number of electrons per energy level increases as the distance of the energy level from the nucleus increases. Periods one through seven have 2, 8, 8, 18, 18, 32, and 23 (incomplete) elements, respectively. Larger periods as the atoms of elements get larger are similar to longer rows and more seats per row in a stadium as you proceed from the field to higher in the stands (Figure 4.3).

Transition elements begin in period four with the expansion of the third energy level from 8 to 18 elements by the addition of 10 electrons one by one to scandium through zinc. As a result, main group elements that come at the end of period four have 18 electrons in the third energy level. For example, the electron arrangement of gallium is 2-8-18-3. Period five is similar to period four with the expansion of the fourth energy level from 8 to 18 and the addition of 10 electrons one by one to atoms of yttrium (Y) through cadmium (Cd). In period six an additional 14 electrons are used for the lanthanide series, and this brings the fourth energy level up to 32 before the expansion of the fifth energy level by 10 electrons in the transition elements of period six. Main-group elements that come at the end of period six have 32 electrons in the fourth energy level and 18 electrons in the fifth energy level. For example, the electron arrangement of lead (Pb) is 2-8-18-32-18-4.

Eighty-seven of the elements are **metals** and are found in Groups IA, IIA, parts of Groups IIIA to VIA (red in Figure 4.2), and the B groups (yellow).

Characteristic physical properties of metals include malleability (ability to be beaten into thin sheets such as aluminum foil), ductility (ability to be stretched or drawn into wire such as copper), and good conduction of heat and electricity. As you shall see from the discussion of chemical bonding, the chemical properties of metals are based on their tendency to give up electrons to form positive ions when they react with nonmetals.

Seventeen elements are **nonmetals** (in green), and except for hydrogen they are found in the upper right-hand corner of the periodic table. Hydrogen is shown in Group IA because its atoms have one electron. However, hydrogen is a nonmetal and probably should be in a group by itself, although you may see H in both Group IA and Group VIIA in some periodic tables. Hydrogen forms compounds with formulas similar to those of the Group IA elements, but with vastly different properties. For example, compare NaCl, table salt, with HCl, a strong acid, or compare Na_2O, an active metal oxide with H_2O, which is, of course, water. Hydrogen also forms compounds similar to those of the Group VIIA elements: NaCl and NaH (sodium hydride); $CaBr_2$ and CaH_2 (calcium hydride).

■ Hydrogen—the element without a home on the periodic table

The physical and chemical properties of nonmetals are opposite those of metals. For example, nonmetals are **insulators;** that is, they are extremely poor conductors of heat and electricity. Chemically, nonmetals have a tendency to

SCIENCE AND SOCIETY

Hydrogen: Fuel of the Future?

What will we use for energy sources when fossil fuels are depleted? Hydrogen gas, H_2, has long been touted as "the fuel of the future," but there are currently two major barriers to extensive use of hydrogen as an alternative to fossil fuels. The first is the need for an inexpensive method of making hydrogen that avoids the use of fossil fuels, which are the major starting materials for the production of H_2 at the present time. Hydrogen gas can be produced by electrolysis of water, but the electrical energy needed is too expensive to yield hydrogen at a reasonable cost. However, solar cells may soon be cheap enough to provide the electrical energy needed for this purpose. For example, the HYSOLAR project, carried out jointly by Saudi Arabia and Germany, uses solar energy to power an electrolysis plant that produces hydrogen. (See I. Dostrovsky: *Scientific American,* December 1991, pp. 102–107.)

The second barrier to using hydrogen gas as a fuel is a means of convenient storage. The space program has demonstrated that H_2 can be stored relatively easily and safely as a liquid, even though cold temperatures and high pressures are required. This might be appropriate for large-scale industrial applications where H_2 could be piped to the plant and burned to heat water to steam, which in turn could generate electricity. Use of H_2 as a fuel for vehicles, however, requires a more convenient form of storage, such as an interstitial hydride. Interstitial metallic hydrides are metals that adsorb hydrogen atoms into the holes or interstices of their metallic structure. Heating the interstitial hydride releases hydrogen gas.

A prototype car run by hydrogen combustion in a slightly modified combustion engine. The hydrogen is stored in a solid hydride, where it fits into spaces between other atoms. Heat from the engine exhaust is used to release hydrogen gas from the hydride. (Reuters/Bettmann)

■ Metals and nonmetals usually form ionic compounds like sodium chloride, NaCl. Nonmetals with other nonmetals usually form covalent compounds like carbon dioxide, CO_2.

gain electrons to form negative ions when they react with metals. Nonmetals also react with other nonmetals to form molecules.

Elements that border the staircase in Figure 4.2 between metals and nonmetals are six **metalloids** (in blue). Their properties are intermediate between those of metals and nonmetals. For example, silicon (Si), germanium (Ge), and arsenic (As) are **semiconductors** and are the elements that form the basic components of computer chips. Semiconductors conduct electricity less than metals such as silver and copper but more than insulators such as sulfur. The six **noble gases** in Group VIIIA have little tendency to undergo chemical reactions. The classifications of metals, nonmetals, and metalloids will enable you to predict the kind of compounds formed between elements.

EXAMPLE 4.1 *Periodic Table*

For the elements with atomic numbers 17, 33, and 82, give the names and symbols and identify the elements as metals, metalloids, or nonmetals.

SOLUTION

Chlorine, Cl, is the element with atomic number 17. It is in Group VIIA. Chlorine and all of the other elements in Group VIIA are nonmetals.

Arsenic, As, is the element with atomic number 33. It is in Group VA. Since it lies along the line between metals and nonmetals, arsenic is a metalloid.

Lead, Pb, is the element with atomic number 82. It is in Group IVA. Like other elements at the bottom of Groups IIIA to VIA, lead is a metal.

Exercise 4.1

List the main groups (the A groups) in the periodic table that (a) consist entirely of metals, (b) consist entirely of nonmetals, and (c) include metalloids. Identify the numbers of valence electrons in atoms from groups listed under (a), (b), and (c).

4.3 The Periodic Table and Chemical Behavior

■ See Section 3.5 for a discussion of valence electrons.

■ The "reactivity" of an element or a compound is its tendency to undergo chemical reactions. Some chemical pollutants, such as DDT, have low reactivity and therefore remain in the environment unchanged for a long time. An element such as potassium that on exposure reacts immediately with water or oxygen in the air is highly reactive.

Why do elements in the same group in the periodic table have similar chemical behavior? Why do metals and nonmetals have different properties? G.N. Lewis was seeking answers to these questions during his development of the concept of valence electrons. He assumed that each noble gas atom had a completely filled outermost shell, which he regarded as a stable configuration because of the lack of reactivity of noble gases. He also assumed that the reactivity of other elements was influenced by their numbers of valence electrons.

Lewis used the element's symbol to represent the atomic nucleus together with all but the outermost shell of electrons; he called this the kernel of the atom. The valence electrons, which he represented by dots, are then placed around the symbol one at a time until they are used up or until all four sides are occupied; any remaining electron dots are paired with the ones already

TABLE 4.3 ■ Lewis Dot Symbols for Atoms

IA	IIA	IIIA	IVA	VA	VIA	VIIA	VIIIA
H·							He:
Li·	·Be·	·B·	·C·	·N·	:O·	:F·	:Ne:
Na·	·Mg·	·Al·	·Si·	·P·	:S·	:Cl·	:Ar:
K·	·Ca·						

there. Lewis dot symbols for atoms of the first 20 elements are shown in Table 4.3. Notice that all atoms of elements in a given A group have the same number of valence electrons and that the number of valence electrons equals the group number. All atoms of Group IA elements have one valence electron; Group IIA atoms have two valence electrons, and so forth. The importance of valence electrons in the study of chemistry cannot be overemphasized. It is the identical number of valence electrons that primarily account for the similar properties of elements in the same group. *The chemical view of matter is primarily concerned with what valence electrons are doing in the course of chemical reactions.*

Lewis dot symbols will be used extensively in Chapter 6 in the discussions of bonding. For example, atoms of metals lose valence electrons to form positive ions:

Group IA	Na· $\longrightarrow$ Na^{+} + e^{-}
Group IIA	·Mg· $\longrightarrow$ Mg^{2+} + 2 e^{-}

and nonmetals gain valence electrons to form negative ions:

Group VA	·N· + 3 e^{-} $\longrightarrow$:N:$^{3-}$
Group VIA	:O· + 2 e^{-} $\longrightarrow$:O:$^{2-}$

EXAMPLE 4.2 *Lewis Dot Symbols*

Draw the Lewis dot symbols for barium and selenium.

SOLUTION

The Lewis dot symbol for an element is the element symbol with a number of dots around the symbol equal to the number of valence electrons. The number of valence electrons is the number of outer electrons and is identical with the group number of the element for A group elements. In the case of barium (Ba) and selenium (Se), both members of A groups, the number of valence electrons is the same as the group number. Barium is in Group IIA, so atoms of barium have two valence electrons. Selenium is in Group VIA, so atoms of selenium have six electrons. Hence the Lewis dot symbols are

·Ba·   :Se·

Exercise 4.2
Draw the Lewis dot symbols for rubidium (Rb) and bromine (Br).

SELF-TEST 4A

1. Give the Lewis dot symbol for boron, an element in Group IIIA.
2. How many valence electrons are in an atom of chlorine, Cl, in Group VIIA?
3. How many valence electrons are in each of the following atoms?
   (a) Sodium (Na), Group IA (b) Calcium (Ca), Group IIA
   (c) Boron (B), Group IIIA (d) Aluminum (Al), Group IIIA
   (e) Neon (Ne), Group VIIIA (f) Iodine (I), Group VIIA
4. The periodic table organized by Mendeleev placed elements in order of increasing atomic mass. (a) True, (b) False.
5. The modern periodic table has the elements placed in order of increasing atomic mass. (a) True, (b) False.
6. Metals typically have low numbers of valence electrons. (a) True, (b) False.
7. Nonmetals are typically found in Groups VA through VIIIA. (a) True, (b) False.
8. Which of the following is true about metals?
   (a) They are found in Group IA and IIA.
   (b) They are good conductors of heat.
   (c) They are good insulators.
   (d) They usually react to form positive ions.
9. Group IIA includes Be, Mg, Ca, Sr, Ba, Ra. What is the predicted formula for the compound formed between Sr and Cl, if Be and Ba form $BeCl_2$ and $BaCl_2$?
10. Beryllium, Be, has the electron arrangement 2-2, and magnesium has the electron arrangement 2-8-2. What do these electron arrangements have in common?
11. Which of the following atoms has the fewest valence electrons? F, O, S, Be.
12. Which of the following atoms has the greater number of valence electrons?
    (a) Lithium, Li or oxygen, O (b) Calcium, Ca or sodium, Na
    (c) Oxygen, O or fluorine, F (d) Boron, B or nitrogen, N
    (e) Sulfur, S or arsenic, As (f) Neon, Ne or carbon, C
13. Group IA elements are all metals except for hydrogen. (a) True, (b) False.
14. Isotopes have different numbers of valence electrons. (a) True, (b) False.
15. Which of the following pairs of atoms will be more reactive?
    (a) Lithium, Li or cesium, Cs (b) Fluorine, F or bromine, Br
    (c) Beryllium, Be or calcium, Ca (d) Oxygen, O or sulfur, S
    (e) Lithium, Li or sodium, Na (f) Neon, Ne or xenon, Xe

4.4 Periodic Trends in Atomic Properties

■ Metallic character means having properties of metals.

From left to right across each period, metallic character gives way to nonmetallic character (Figure 4.2). The elements with the most metallic character

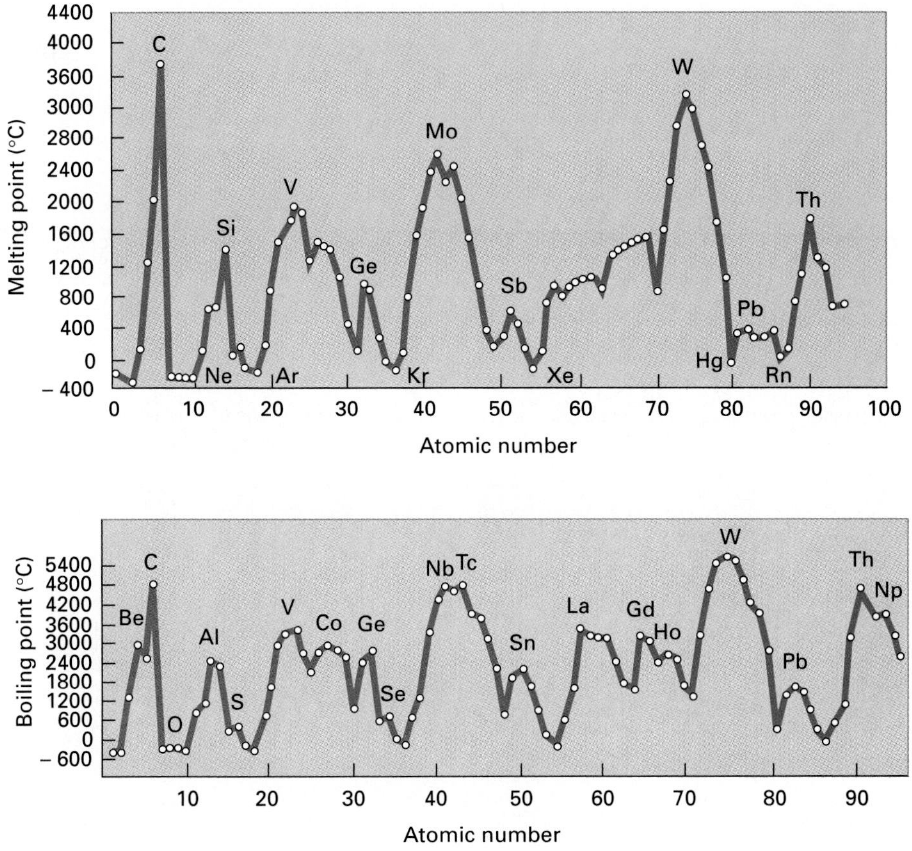

Figure 4.4 Periodic trends. The periodic nature of the melting points and boiling points of the elements are evident when (a) melting points and (b) boiling points are plotted versus atomic number.

are at the lower left part of the periodic table near cesium (Cs). The elements with the most nonmetallic character are at the upper right portion of the periodic table near fluorine. The heavy line on the periodic table that begins at boron and moves down like a staircase to astatine (At) roughly separates the metals and the nonmetals.

Notice the periodic patterns of melting points of the elements when plotted against atomic number (Figure 4.4a). The boiling points follow a similar trend when plotted against atomic number (Figure 4.4b). The trends are not smooth, but a general periodic pattern is obvious. In which groups of the periodic table are the elements with the lowest melting and boiling points? Which groups have the highest melting and boiling points?

Atomic radii also show periodicity with atomic number (Figure 4.5). Why do atoms get larger from the top to the bottom of a group? Do you suppose it has something to do with more layers making a larger onion? Yes, by analogy the larger atoms simply have more energy levels inhabited by electrons than do the smaller atoms.

Atomic radii decrease across a period from left to right (Figure 4.5). You may see a paradox in adding electrons and getting smaller atoms, but protons also are being added. The greater nuclear charge pulls electrons in the same energy levels closer to the nucleus and causes contraction of the atom.

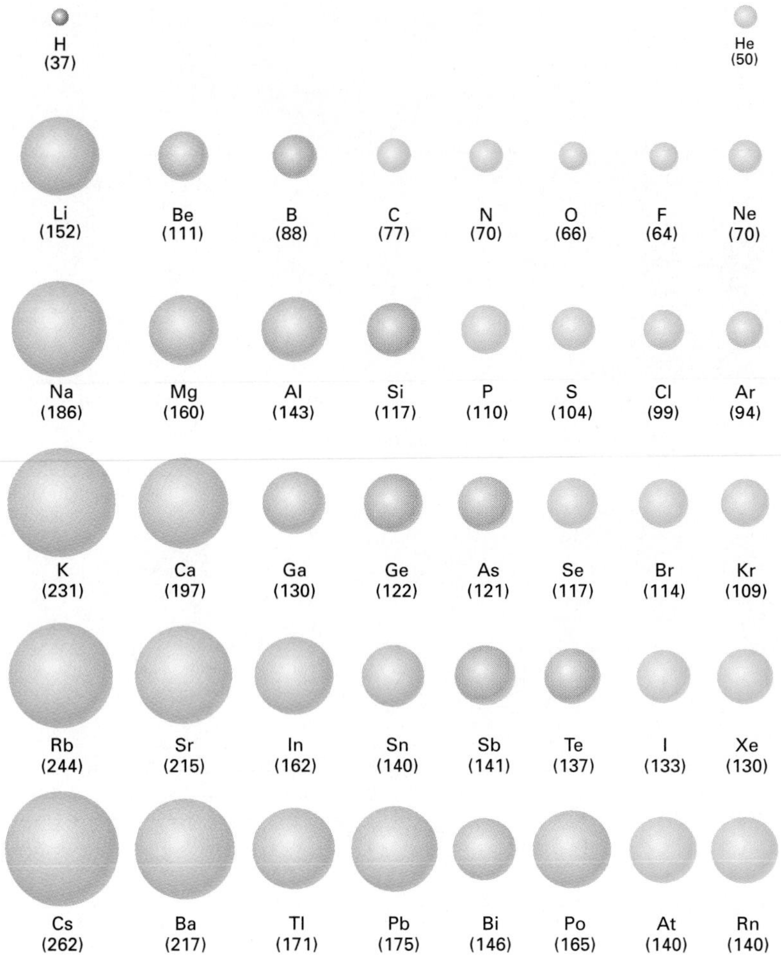

Figure 4.5 Atomic radii of the A group elements (picometers, pm). Atoms increase in size down a group and, in general, decrease in size across a period in the periodic table.

We can also use the trends in size of atomic radii to predict trends in reactivity. The larger the atom, the easier it is to remove the valence electrons because the attractive forces between protons in the nucleus and valence electrons decrease with increasing size of the atom.

Metal atoms lose valence electrons to form positive ions. The larger the metal atom, the greater the tendency to lose valence electrons and the more reactive the metal. Therefore, we would predict that the most reactive metal in Figure 4.5 is cesium (Cs), the metal with the largest radius. The result of *increase* in reactivity of metals down a given group in the periodic table is dramatically illustrated by lithium, sodium, and potassium, the first three metals in Group IA. Their atoms increase in size in this order down the group. Each element reacts with water—lithium quietly and smoothly, sodium more vigorously, and potassium much more quickly. The reactions of both sodium and potassium give off enough heat to ignite the hydrogen gas produced by the reaction, but as shown in Figure 4.6, potassium reacts with explosive violence. For elements at the bottom of Group IA, just exposure to moist air produces a vigorous explosion.

■ A general equation for the reaction of water and a metal from Group IA, using M to represent any of the IA metals, is

$$2\ M(s) + 2\ H_2O(\ell) \longrightarrow 2\ MOH(aq) + H_2(g)$$

Nonmetal atoms gain electrons from metals to form negative ions. The smaller the nonmetal atom, the higher the reactivity of the nonmetal. For example, fluorine atoms are the smallest of Group VIIA elements, and fluorine is the most reactive nonmetal. It reacts with all other elements except three noble gases—helium, neon, and argon. The reaction of Group VIIA elements with hydrogen illustrates how the reactivity of nonmetals *decreases* down the group. Fluorine reacts explosively with hydrogen, but the reaction with hydrogen is less violent for chlorine and is very slow for iodine.

■ A general equation for the reaction of hydrogen with a nonmetal from Group VIIA, using X to represent any of the VIIA nonmetals, is

$$H_2(g) + X_2(g) \longrightarrow 2\ HX(g)$$

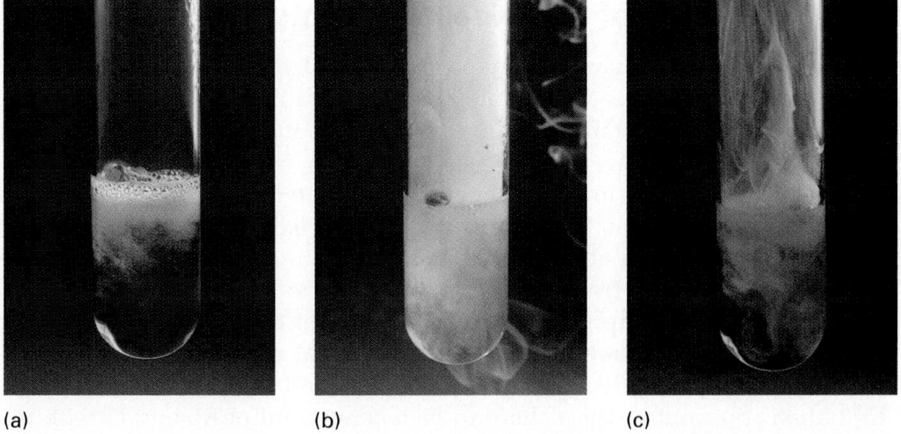

Figure 4.6 Reaction of alkali metals with water: (a) lithium, (b) sodium, (c) potassium. *(C.D. Winters)*

Why are there repeatable patterns of properties across the periods in the periodic table? Again, it is because there is a repeatable pattern in atomic structure, and properties depend upon atomic structure. Each period begins with one valence electron for atoms of the elements in Group IA. Each period builds up to eight valence electrons, and the period ends. This pattern repeats across periods two through six. As more elements are made by nuclear accelerators (Section 5.7), period seven may be completed someday. When this happens, the periodic table and atomic theory predict the last element in period seven (the final member of the noble gases) will be element number 118. Atoms of element 118 will have eight valence electrons.

EXAMPLE 4.3 *Atomic Radii and Reactivity*

Which element in each pair is more reactive: (a) O or S, (b) Be or Ca, or (c) P or As?

SOLUTION

(a) Oxygen and sulfur are Group VIA nonmetals. Generally, the smaller the atomic radius of a nonmetal, the more reactive the nonmetal is. Since oxygen atoms are smaller than sulfur atoms, oxygen should be more reactive than sulfur. (b) Beryllium (Be) and calcium (Ca) are Group IIA metals. Reactivity of metals in a given group increases down the group as the atomic radius increases. Therefore, calcium should be more reactive than beryllium. (c) Phosphorus and arsenic are Group VA nonmetals so phosphorus, which has a smaller atomic radius, is predicted to be more reactive than arsenic.

Exercise 4.3A

Which element in each pair has the larger atoms: (a) Ca or Ba, (b) S or Se, (c) Si or S, or (d) Ga or Br?

Exercise 4.3B

Which element in each pair is more reactive: (a) Mg or Sr, (b) Cl or Br, or (c) Rb or Cs?

4.5 Properties of Main-Group Elements

Elements in a group have similar properties, but not the same properties. Some properties, as already illustrated for atomic radii and reactivity, increase or decrease in a predictable fashion from top to bottom of a periodic group.

Other properties are similar but do not vary in a regular pattern. For example, the densities (in grams per cubic centimeter, g/cm^3) of the metals of Group IA, lithium through cesium, are 0.53, 0.97, 0.86, 1.53, and 1.87 g/cm^3, respectively. Prediction of densities is not exact, but overall there is a consistent increase in density from the top of the group to the bottom.

Elements in a group generally react with other elements to form similar compounds, a fact accounted for by their identical numbers of valence electrons. For example, if the formula for the compound composed of Li and Cl is LiCl, then you would expect there to be a compound of Rb and Cl with the formula RbCl and a compound of Cs and Cl with the formula of CsCl. Likewise, if the formula Na_2O is known, then a compound with the formula Na_2S predictably exists, since oxygen and sulfur are in the same group. This ability to predict formulas from the periodic table has limitations. For example, Na, K, Rb, and Cs all form superoxides (formula MO_2) as well as oxides (M_2O), but no superoxide with Li is known. However, these limitations do not prohibit use of the periodic table for predicting formulas. In general, elements in the same group of the periodic table form some of the same types of compounds.

Glass. Common window glass is essentially a mixture of sodium and calcium oxides with silicon dioxide (SiO_2), which is sand. Alkali metals are very reactive, but their oxides melt with sand to give stable glasses. Pyrex glass, formed by adding boric oxide to molten glass, has a low thermal expansion and is used for laboratory glassware. Blue glass is formed by adding cobalt compounds to molten glass. *(Larry Cameron)*

Alkali Metals

The Group IA elements (Li, Na, K, Rb, Cs, and Fr) are called the **alkali metals.** The name *alkali* derives from an old word meaning "ashes of burned plants." All the alkali metals are soft enough to be cut with a knife. None are found in nature as free elements, since all combine rapidly and completely with virtually all the nonmetals and, as illustrated in Section 4.4, with water. For example, the reaction of sodium with chlorine to form sodium chloride is a very violent reaction (Figure 4.7).

$$2\ Na(s) + Cl_2(g) \longrightarrow 2\ NaCl(s)$$

Because the alkali metals react readily with the oxygen and water vapor in the atmosphere, they must be stored under an inert liquid, such as mineral oil. Francium, the last member of Group IA, is found only in trace amounts in nature, and all of its 21 isotopes are naturally radioactive.

Figure 4.7 Reaction of sodium with chlorine. Like all alkali metals, sodium reacts vigorously with chlorine and other halogens. After a piece of sodium is placed in the flask with chlorine gas, the sodium reacts with the chlorine to give salt (sodium chloride, NaCl). The reaction evolves energy in the form of heat and light. *(C.D. Winters)*

THE WORLD OF CHEMISTRY

Making Glass Stronger

Program 7, *The Periodic Table*

That glass was once so rare it was prized more highly than jewels or gold is hard to realize. Today, household items made of glass are common and inexpensive, which is not surprising since the basic ingredient of glass is sand.

Glassmaking requires that the sand, which is mainly silicon dioxide (SiO_2, also known as "silica"), is mixed with other substances; heated to a temperature high enough to melt the mixture; and then molded, blown, or otherwise formed into the shape it will retain when cooled. The properties of glass, such as its color, melting point, sparkle, and hardness, are determined by the other substances in the mixture, most of which are metal oxides. Everyday glass, for example, contains oxides of two alkali metals, sodium and calcium, plus smaller amounts of magnesium, aluminum, and boron oxides. Chemical laboratory glassware and oven glassware are made of Pyrex, the borosilicate glass developed at Corning Glass Works, which maintains an active glass research program.

Pouring an experimental glass in the laboratory at Corning Glass Works. (James L. Amos/Peter Arnold, Inc.)

As Dr. Gerry Fine, a glass chemist at Corning explains, the periodic table is a key to choosing ingredients to modify the properties of glass.

I cannot imagine working without the periodic table, because scientists are interested in looking for systematic relationships. . . . If we take ordinary glass that contains sodium [ions] and substitute potassium [ions] in the surface—an element that behaves basically in the same way but is slightly larger—we can enhance the strength of that glass and make strong glass. After ordinary window glass has been dipped into molten potassium, *the potassium literally stuffs the surface of that glass,*

The result is a glass that will not shatter when struck by a steel ball bearing dropped from a height of 20 feet.

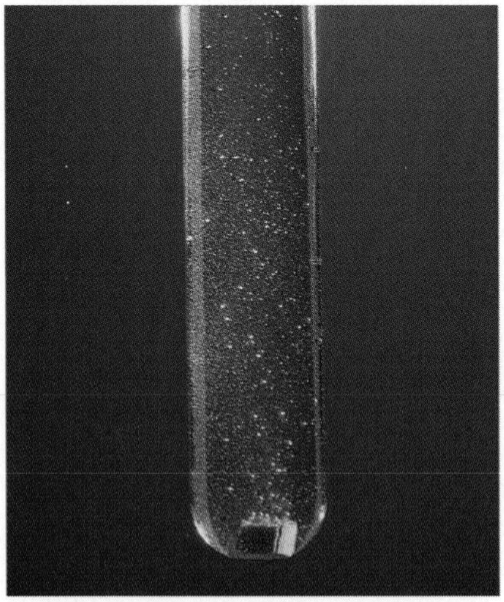

Reaction of calcium with water. It is easy to see by comparison with Figure 4.6 that calcium is less reactive than the alkali metals. *(C.D. Winters)*

Alkaline Earth Metals

The Group IIA elements (Be, Mg, Ca, Sr, Ba, and Ra) are called the **alkaline earth metals.** Compared with the alkali metals, the alkaline earth metals are harder, are more dense, and melt at higher temperatures. In the Middle Ages, an *earth* was any solid substance that did not melt and was not changed by fire into some other substance. Under these conditions, many of the alkaline earth compounds change into oxides having high melting temperatures (in excess of 1900°C) and the same general white appearances of the original compounds. Because these early investigators could not attain temperatures high enough to melt the oxides, the name "earth" was applied and has stuck to this day.

The atomic radii of the alkaline earth metals are smaller than those of the adjacent alkali metals. Since the valence electrons of the alkaline earths are held more tightly, they are less reactive than their alkali metal neighbors. All alkaline earth metals react with oxygen to form an oxide MO, where M is the alkaline earth. The trend of increasing reactivity within the group is illustrated by the behavior of the elements toward water. Beryllium does not react with water or steam, even when heated. Although magnesium does not react with liquid water, it does react with steam to form magnesium oxide and hydrogen.

$$Mg(s) + H_2O(g) \longrightarrow MgO(s) + H_2(g)$$

Calcium and the elements below it react readily with water at room temperature:

$$Ca(s) + 2\ H_2O(\ell) \longrightarrow Ca(OH)_2 + H_2$$

In the presence of oxygen (O_2), magnesium metal is protected from many chemicals by a thin surface of water-insoluble magnesium oxide (MgO). As a result, magnesium can be incorporated into lightweight structural alloys even though it is a very reactive metal. Several hundred thousand tons of magnesium are produced annually, and most of it is used in lightweight alloys for the manufacture of aircraft and automotive parts.

■ An alloy is a mixture of two or more metals.

The heavier alkaline earth metals (Ca, Sr, and Ba) are even more reactive toward nonmetals than magnesium and must be stored in such a way as to protect them from oxidation by oxygen and water vapor in the air.

Important Nonmetals and Metalloids of Groups IIIA to VIA

The elements of Groups IIIA to VIA are a bridge between the metals of Groups IA and IIA and the nonmetals of Groups VIIA and VIIIA. Groups IIIA to VIA include all the metalloids, all the nonmetals other than the halogens (Group VIIA), the noble gases (Group VIIIA), and, near the bottom of each group, additional metals.

Boron (B, Group IIIA) is a metalloid that is very low in relative abundance. Its common minerals are found in concentrated deposits, especially in California (Figure 4.8). Borax has been used for centuries as a low-melting flux in metallurgy because of the ability of molten borax to dissolve other metal oxides, thus cleaning the surfaces to be joined and permitting good metal-to-metal contact.

■ A flux is added to a metal or mineral to lower its melting point and protect it from forming oxides.

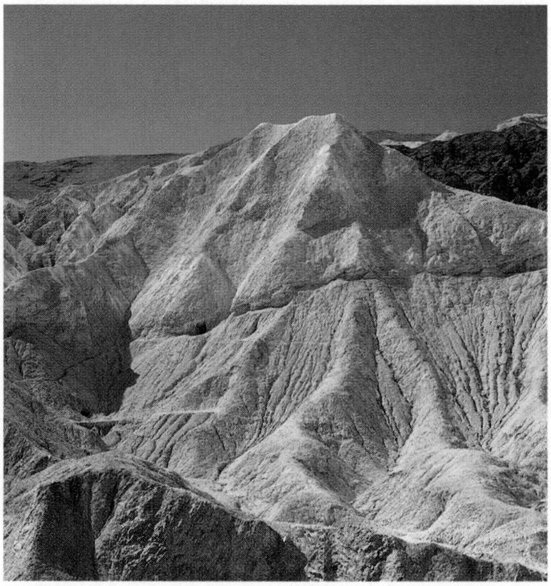

Figure 4.8 A borax mine. The mine tunnels and roads are visible in this borax formation at Furnace Creek in the Death Valley National Monument in California. *(W. Kleck/Terraphotographics/BPS)*

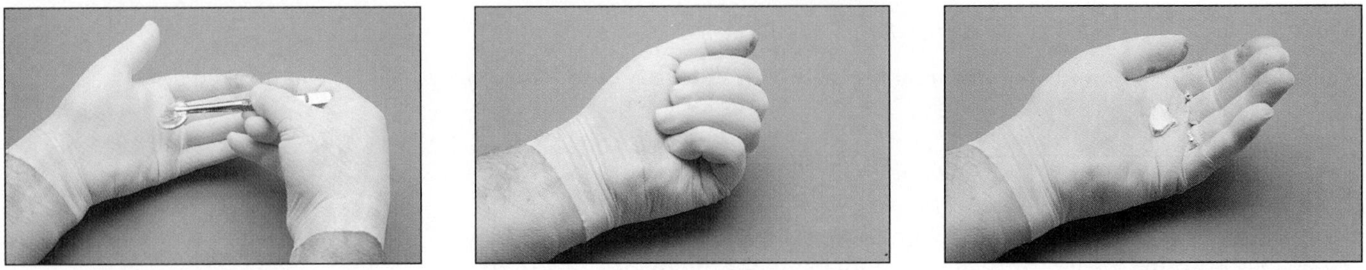

(a) (b) (c)

Figure 4.9 Metallic gallium melts at 30°C, which is less than body temperature (37°C). *(Larry Cameron)*

The largest use of borax and of boric oxide (B_2O_3) is in the manufacture of borosilicate glass. The presence of boric oxide gives the glass a higher softening temperature, a better resistance to attack by acids, and makes it expand less on heating.

The production of aluminum (Al, Group IIIA) has grown rapidly because aluminum has thousands of uses as a structural material and in packaging. However, pure aluminum is rarely used, since it is soft and weak. What is more, it loses strength rapidly above 300°C. To strengthen the metal and improve its properties, aluminum is alloyed with small amounts of other metals. A typical alloy, for example, may contain about 4% copper with smaller amounts of silicon, magnesium, and manganese; a large passenger plane today may use more than 50 tons of this alloy. A softer, more corrosion-resistant alloy for window frames, furniture, highway signs, and cooking utensils, however, may include only manganese.

■ The production, uses, and recycling of aluminum are discussed in Section 11.4.

Gallium (Ga, Group IIIA) is truly a remarkable element. It has the greatest liquid range of all known elements; it can melt in your hand (mp = 30°C) (Figure 4.9), but it does not boil until the temperature reaches 2403°C. Like water (Section 7.4), gallium is one of the few known materials that expands upon freezing. The greatest use for gallium is in the semiconductor gallium arsenide, GaAs. However, new arsenic-free semiconductors are being sought because of the toxicity of arsenic.

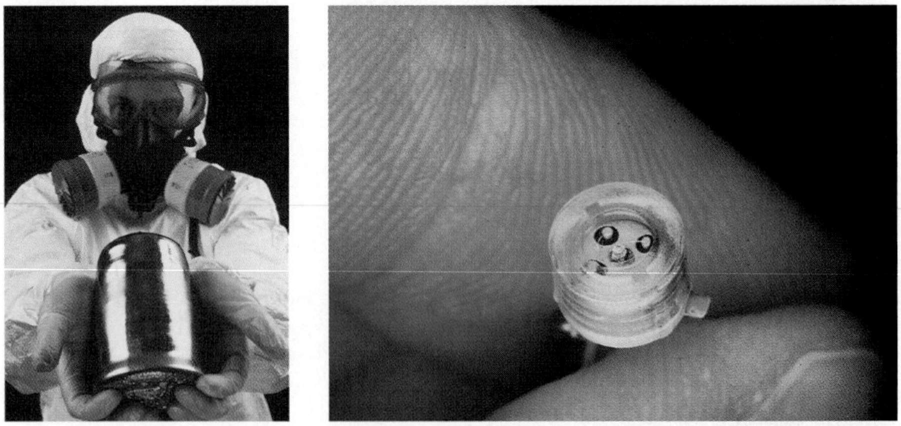

(a) (b)

Gallium arsenide. (a) A metallurgist with very pure gallium arsenide to be used in making computer chips. (b) A gallium arsenide light-emitting diode with the glass cover removed. *(a, Hank Morgan/Rainbow; b, Mike McNamee/SPL/Photo Researchers)*

Carbon (C, Group IVA), which provides the structural basis for all organic chemistry (Chapters 12 and 14), is found in the fossil fuels and in all living matter (Chapter 15). The largest natural sources of carbon, however, are the carbonate minerals limestone and dolomite. Carbon occurs as carbon dioxide (CO_2) in the atmosphere, where it is a byproduct of natural and industrial combustion. More than 30 million tons of carbon dioxide are also manufactured annually in the United States for commercial applications. Roughly half of the CO_2 is used as a refrigerant in the solid form (sublimes at $-78°C$) known by the trade name ''Dry Ice.'' Approximately 25% of the carbon dioxide manufactured is used to carbonate beverages. More than 400 bottles of carbonated beverages are produced per person in the United States per year.

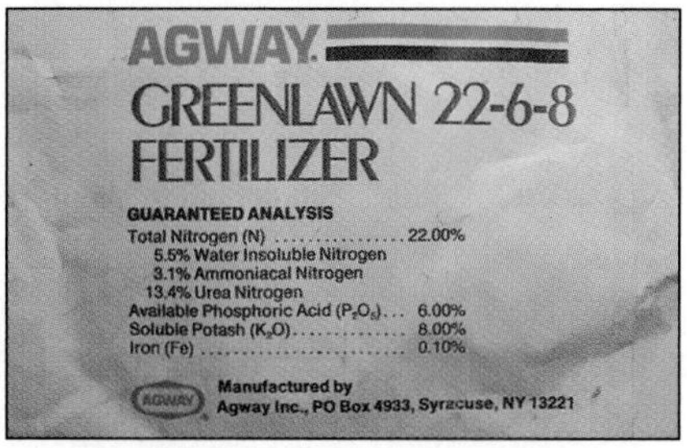

Fertilizer label, showing nitrogen, phosphorus, and potassium content. The composition of a fertilizer is always shown in this order: percent total nitrogen; percent phosphorus as the oxide (P_2O_5); percent potassium as the oxide (K_2O) (see Section 22.4). *(C.D. Winters)*

Nitrogen (N, Group VA), phosphorus (P, Group VA), oxygen (O, Group VIA), sulfur (S, Group VIA) and their compounds have many important commercial uses. Thousands of tons of nitrogen and four of its compounds—ammonia, nitric acid, urea, and ammonium nitrate—are produced in the United States each year for use as fertilizers, explosives, and in the preparation of plastics. A major use of phosphorus is in the preparation of phosphoric acid (Section 9.1), which is essential to the production of fertilizers, food phosphate additives, and detergents.

Oxygen forms compounds with all of the elements except the noble gases helium, neon, argon, and possibly krypton. Oxygen ranks third of all chemicals produced annually in the United States. Both oxygen and nitrogen, which ranks second, are obtained by distillation of liquid air. More than 50% of the manufactured oxygen is used in the production of steel (Section 11.4). Liquid oxygen is also used to oxidize fuels in rocket propulsion. Pure oxygen is also used for medical patients with respiratory problems and for breathing systems in high-altitude aircraft to prevent pilot black outs.

The largest use of sulfur is in the production of sulfuric acid, the number one chemical in the United States. The cost of sulfuric acid, about 1 cent per pound, has not changed much in 300 years, a tribute to improving technology in its production from natural sources and pollution wastes. More than 80 billion pounds are produced and sold each year in the United States alone. More than 70% of sulfuric acid is used in the manufacture of fertilizers, with the rest being used in the petroleum industry, car batteries, the production of steel, and the manufacture of organic dyes, plastics, drugs, and many other products.

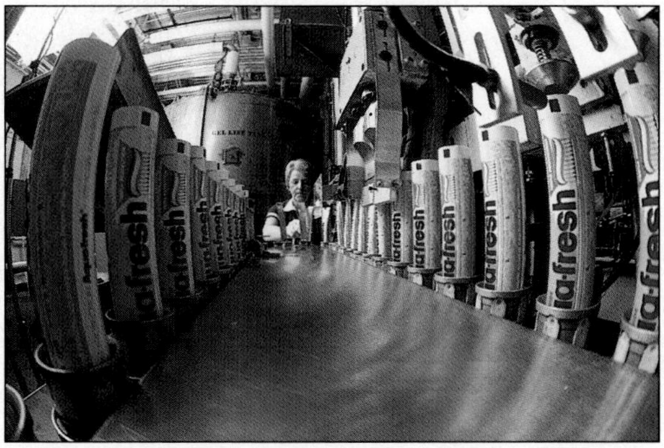

Many toothpastes contain phosphate salts. *(Dan McCoy/Rainbow)*

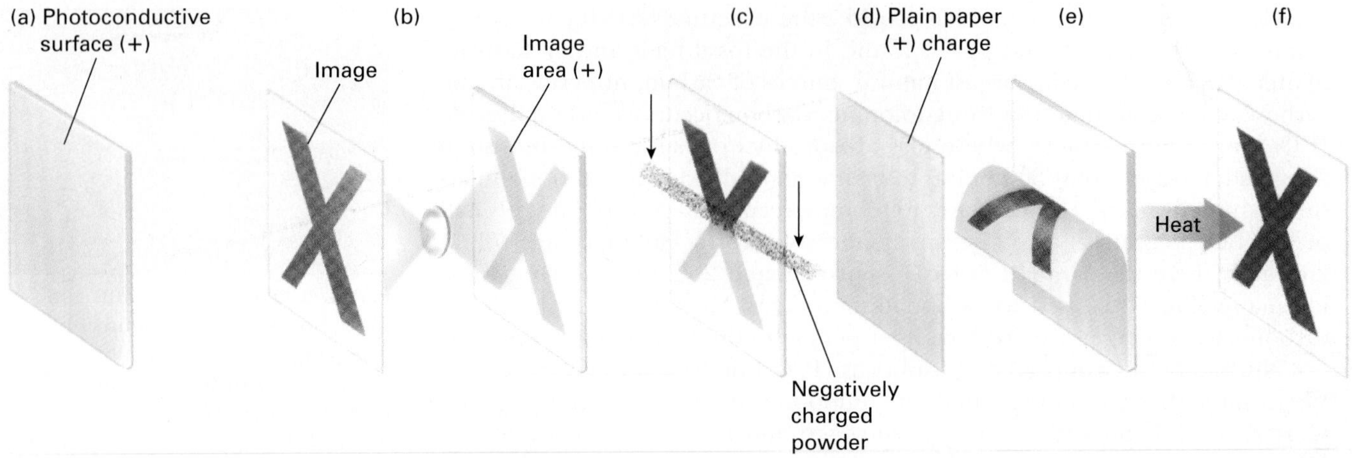

Figure 4.10 Basic Xerography. (a) A photoconductive surface is given a positive electrical charge (+). (b) The image of a document is exposed on the surface. The light energy causes the charge to drain away from the surface in all but the image area, which remains unexposed and charged. (c) Negatively charged carbon powder is cascaded over the surface. It adheres electrostatically to the positively charged image area, making a visible image. (d) A piece of plain paper is placed over the surface and given a positive charge. (e) The negatively charged powder image on the surface is electrostatically attracted to the positively charged paper. (f) The powder image is fused to the paper by heat.

Although selenium is rare, it has a wide range of uses. The use you are most familiar with is in "xerography." At the heart of most photocopying machines is a photoreceptor belt coated with a film of selenium. Light coming from the imaging lens selectively discharges a static electric charge in the selenium film, and the black "toner" sticks only to the areas that remain charged (Figure 4.10). A copy is made when the toner is transferred to a sheet of plain paper.

The Halogens

■ **Salts** (Section 6.1) are compounds that have in common properties like those of table salt. Most salts are solids with high melting points and are composed of crystals that may be white, like table salt crystals, or colored.

The **halogens**—fluorine (F), chlorine (Cl), bromine (Br), and iodine (I)—are Group VIIA elements. In the elemental state, each of these elements exists as diatomic molecules (X_2). Fluorine (F_2) and chlorine (Cl_2) are gases at room temperature, whereas bromine (Br_2) is a liquid, and iodine (I_2) is a solid. All isotopes of astatine (At) are naturally radioactive and disintegrate quickly. If you could accumulate enough astatine, it would be a solid at room temperature.

The name *halogen* comes from a Greek word and means "salt-producing." The best known salt containing a halogen is sodium chloride (NaCl), table salt. But there are many other halogen salts, including calcium fluoride (CaF_2), a natural source of fluorine; potassium iodide (KI), an additive to table salt that prevents goiter; and silver bromide (AgBr), the active photosensitive component of photographic film.

The Noble Gases

■ Aliases: noble gases, inert gases, rare gases

The **noble gases**—helium (He), neon (Ne), argon (Ar), krypton (Kr), xenon (Xe), and radon (Rn)—are all colorless gases composed of single atoms at

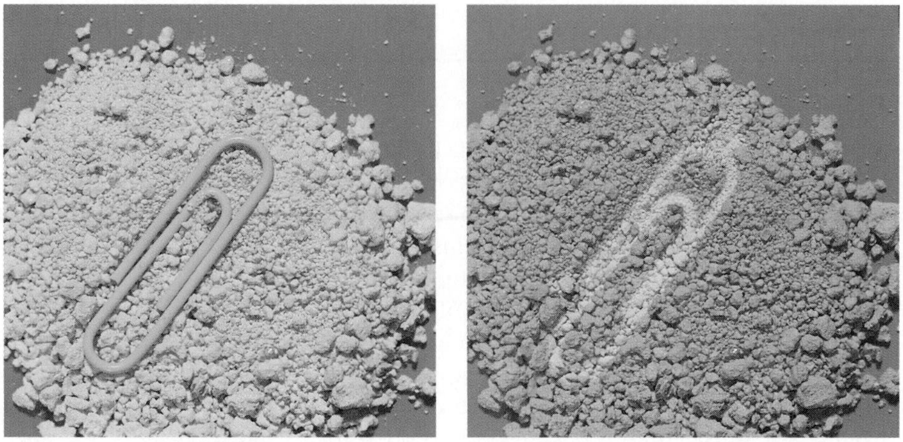

(a) (b)

Silver chloride crystals (a) exposed to light (b). Like silver bromide, silver chloride darkens on exposure and is used in photographic chemicals. *(©1995 Richard Megna/ Fundamental Photographs)*

room temperature. They are referred to as noble because they generally lack chemical reactivity. None of the noble gases were known when Mendeleev proposed his periodic table in 1869, so his table (Figure 4.1) does not include them. Sir William Ramsay discovered all the noble gases except radon. He proposed the names of the noble gases he discovered. The derivations of some of the names of these elements are consistent with their inactivity: argon (from the Greek *argon,* meaning "inactive"), xenon (from the Greek *xenon,* meaning "stranger"). Helium (Greek *helios,* meaning "the sun") was discovered by analysis of the sun's light; later it was found on Earth. Neon (Greek *neos,* meaning "new") is the common gas that glows in the tubes of "neon" lights. Neon glows red when excited in a discharge lamp, and argon glows blue (Figure 4.11). Other gases and painted tubes are used to give different colors. Radon is naturally radioactive.

■ The name "radon" comes from "radium" (Latin *radius,* meaning "ray"). At first, radon was called niton (Latin *nitens,* meaning "shining"). The connection between radon, cancer, and basements is discussed in Section 5.9.

■ The existence of helium was predicted in 1868 from studies of the spectrum of sunlight. The isolation of helium from uranium ores is possible because of the disintegration of radioactive uranium to give alpha particles (^4_2He).

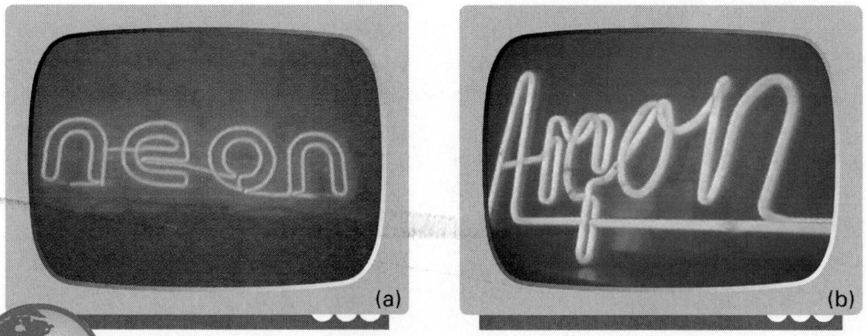

The World of Chemistry

Program 8, *Chemical Bonds*

Figure 4.11 Neon and argon signs. Neon glows orange-red and argon glows blue in what are essentially sign-shaped cathode-ray discharge tubes.

THE PERSONAL SIDE

William Ramsay (1852–1916)

(Chemical Heritage Foundation)

Sir William Ramsay discovered argon, helium, krypton, neon, and xenon during the period 1894 through 1898. These include all but radon, and Ramsay is the only individual to be involved in the discovery of essentially an entire group of elements. The first hint of the existence of noble gases came from experiments done by Lord Rayleigh in 1892 on the density of nitrogen. When Rayleigh removed all the other known gases from air, the remaining nitrogen was denser than nitrogen prepared from ammonia. After learning of Rayleigh's results, Ramsay began studying nitrogen samples from air to determine the cause of the discrepancy in the density of nitrogen from two sources. He and Rayleigh began discussing the results of Ramsay's experiments, and they concluded that the nitrogen from air must contain an unknown component. They were able to isolate small quantities of the unknown gas, and from additional experiments they determined that the gas was a new element. Rayleigh and Ramsay announced their discovery of a new element in 1894, which they called argon.

Later in 1894, Ramsay isolated helium when he heated uranium ores. Then in 1898, Ramsay and his co-workers isolated neon, krypton, and xenon from air. The discoveries of these elements and their lack of chemical reactivity led to placement of the entire group after the halogens.

Sir William Ramsay received the Noble Prize in Chemistry in 1904 for his discovery of the noble gases and his determination of their place in the periodic table.

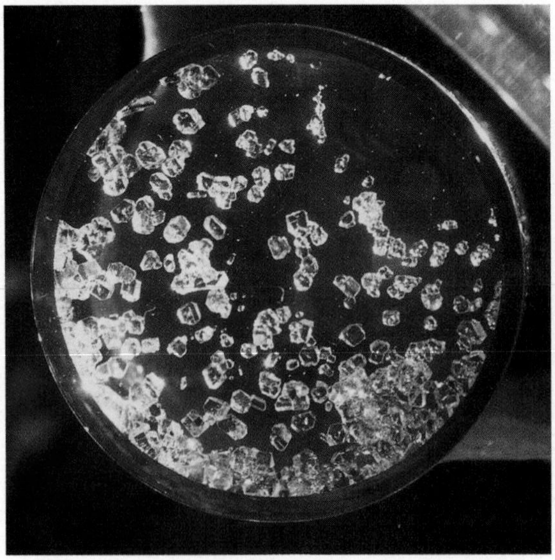

Crystals of xenon tetrafluoride (XeF_4) formed by direct reaction of xenon with fluorine at 400 °C. That this compound could exist as stable crystals came as quite a surprise to most chemists. *(Argonne National Laboratory)*

Until 1962, it was thought that all of the noble gases had absolutely no chemical reactivity. On some older periodic tables, the noble gas column was called "inert gases." Many reasons were presented to explain why the noble gases were inactive and why they never would react. Beginning in 1962, however, the situation began to change. Neil Bartlett prepared the first noble gas compound, $XePtF_6$. His discovery was followed quickly by the work of scientists at Argonne National Laboratory, who made some 30 compounds involving the heavier members of the noble gases combined with fluorine or oxygen. Some of the first prepared compounds were KrF_2, KrF_4, XeF_2, XeF_4, XeF_6, XeO_3, XeO_4, and RnF_4. No compounds with helium, neon, or argon have yet been reported.

(a)

(b)

(c)

A desirable application of the properties of silver (a), platinum (b), and gold (c). *(a and c, Tiffany & Co.; b, Platinum Guild International USA)*

EXAMPLE 4.4 *Group Properties*

By referring to Table 4.2, identify the periodic table main group in which an element (X) with the following properties most likely falls: a metal that forms the compounds XCl_2 and XO.

SOLUTION

Beryllium, magnesium, and calcium, which are from Group IIA, form compounds with chlorine and oxygen with the general formulas XCl_2 and XO, and all elements from Group IIA are metals. Assuming the unknown element is from a main group, it must also be a Group IIA element.

Exercise 4.4

By referring to Table 4.2, identify the periodic table main group in which an element (X) with the following properties most likely falls: a nonmetal that forms the compounds XH_3 and XCl_3.

4.6 Properties of Transition Group Elements

In the main-group elements, chemical similarities occur principally within a group, and the chemistry changes markedly across a given period as the number of valence electrons changes. In contrast, the transition elements that fill the fourth, fifth, and sixth periods in the center of the periodic table have similar chemistry within a given period as well as within a group. Some, such as iron (Fe), are abundant in nature and very important commercially. Others, including silver (Ag), gold (Au), and platinum (Pt), are much less abundant but also less reactive; they can be found in nature as the pure element and are coveted for their beauty.

Two rows at the very bottom of the table encompass the inner transition elements, which are subdivided further into the lanthanides and the actinides. These are much less abundant and less important commercially than many of the transition metals, though some lanthanide compounds are used in color television picture tubes.

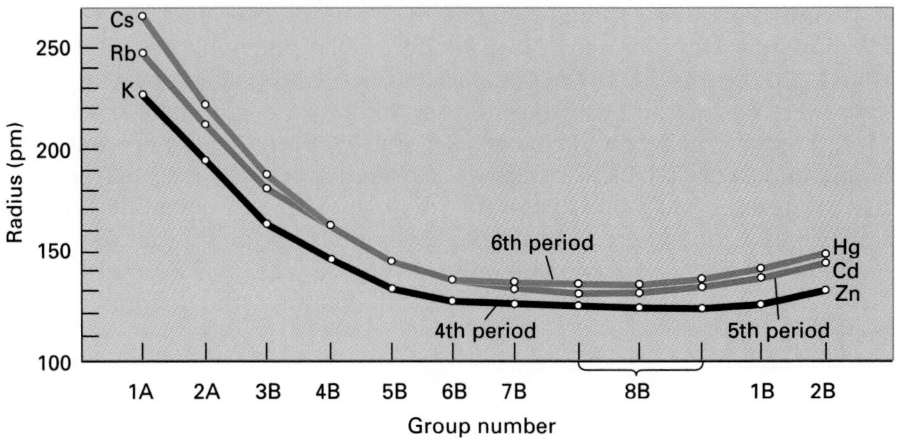

Figure 4.12 Trends in atomic radii of transition metals. The atomic radius decrease of transition metals across a period is more gradual than for main-group elements, and there is little change in size within a group of transition metals.

All of the transition and inner transition elements are metals, and their chemical similarities can be explained by their similarities in atomic radii (Figure 4.12). The atomic radius decreases in size across a period from left to right as is observed for main-group elements (Figure 4.5). However, the change is much more gradual for the transition metals, and there is a size minimum near the center of each row. Another difference is seen in the trends within a group. Except for IIIB (the scandium, Sc, group), there is no change in size in going from period 5 to period 6, unlike the increase in size observed going down a group in the main-group elements. The reason for this is the existence of the 14 elements of the lanthanide series that fall between lanthanum (La) and hafnium (Hf). The gradual decrease in size of atomic radii across the lanthanide series is sufficient to cause the elements that follow the lanthanides to be nearly the same size as the elements above them in the fifth period. As a result, the chemical properties of the second and third members of transition metal groups are very similar, but different from the

DISCOVERY EXPERIMENT

Preparing a Pure Sample of an Element

The following items are needed to do this experiment: two glasses or plastic cups that will each hold about 250 mL (milliliters) of liquid; about 100 mL of vinegar; soap; an iron nail or paper clip; a piece of steel wool, a Brillo pad, sandpaper, or a nail file; about 40 to 50 cm of thread; table salt; a magnifying glass (optional); 15 to 20 dull pennies (shiny pennies won't work).

Wash the piece of iron with soap, dry it, and clean the surface further with steel wool or sandpaper until the surface is shiny. Tie one end of the thread around one end of the piece of iron.

Place the pennies in one cup (A) and pour in enough vinegar to cover them. Sprinkle on a little salt, swirl the liquid around so that it touches all the pennies, and observe what happens. When nothing more seems to be happening, pour the liquid into the second cup (B), leaving the pennies in the first cup (A).

Suspend the piece of iron from the thread so that it is half submerged in the liquid in the second cup (B).

Observe the piece of iron over a period of 10 minutes or so, and then use the thread to pull it out of the liquid. Observe it carefully, using a magnifying glass if you have one. Compare the part that was submerged with the part that remained above the surface of the liquid.

1. What did you observe happening to the pennies?
2. What did you observe happening to the piece of iron?

first member. For example, nickel readily reacts with oxygen to form nickel oxide, but platinum and palladium are called noble metals because they do not form oxides readily and are found in nature as the free elements.

SELF-TEST 4B

1. What are the formulas for chlorides of Group IA elements?
2. What are the formulas for the highest oxides of Group IVA?
3. Group IIA, the alkaline earth metals, react with oxygen, O_2, to form an oxide with the formula MO. (a) True, (b) False.
4. Many Group IA and IIA metals must be stored under mineral oil to keep them from reacting with atmospheric oxygen and water vapor. (a) True, (b) False.
5. Which of the elements in Group IA is the most reactive? Li, Na, K, Rb, Cs
6. Carbonated beverages are produced by dissolving carbon dioxide, CO_2, in water. (a) True, (b) False.
7. Compounds of nitrogen, phosphorus, and oxygen are used in commercial food additives and fertilizers. (a) True, (b) False.
8. Elements in Group IA all have one valence electron. (a) True, (b) False.
9. Which of the following noble gases is the most reactive? He, Ne, Ar, Kr, Xe
10. Sulfuric acid, H_2SO_4, is a major chemical and the acid inside automobile lead storage batteries. (a) True, (b) False.

MATCHING SET

____ 1. Periodic	a. Electron arrangement 2-8-2
____ 2. Larger atoms	b. Outermost occupied shell
____ 3. Two valence electrons	c. Chromium (Cr)
____ 4. A noble gas	d. Repeated pattern
____ 5. A transition metal	e. At the bottom of a group
____ 6. A main-group metal	f. Praseodymium (Pr)
____ 7. Decrease in atomic radius	g. Eight valence electrons
____ 8. A halogen	h. Electron arrangement 2-8-1
____ 9. An inner transition element	i. Seven valence electrons
____ 10. Valence electrons	j. Across a period
	k. Boron

QUESTIONS FOR REVIEW AND THOUGHT

1. State the periodic law.
2. Give definitions for the following terms:
   (a) Group (b) Period
   (c) Chemical properties (d) Transition element
   (e) Inner transition element (f) Representative element
3. Give the definitions for "metals" and "metallic properties."
4. Give a definition for "nonmetals."
5. Give a definition for "metalloids."
6. How did the discovery of the periodic law lead to the discovery of elements?
7. Find as many examples as you can of elements that would be incorrectly placed if arranged according to Mendeleev's periodic law, which was based on atomic weight rather than atomic number.
8. What is the significance of the empty spaces in Mendeleev's periodic table?
9. Elements can be classified as metals, nonmetals, and metalloids. Give two examples of each category.
10. Describe the relative locations of metals, nonmetals, and metalloids in the periodic table.
11. How do metals differ from nonmetals?

12. Identify each of the following elements as either a metal, nonmetal, or metalloid.
    (a) Nitrogen (b) Arsenic
    (c) Argon (d) Calcium
    (e) Uranium
13. Consider the periodic table:
    (a) How many periods are there?
    (b) How many representative groups or families are there?
    (c) How many groups consist of all metals?
    (d) How many groups consist of all nonmetals?
    (e) Is there a period that consists of all metals?
14. Why are Group VIIIA elements called noble gases?
15. How are the elements in a group related to each other?
16. What is meant by the terms "kernel," "outermost energy level," and "valence electrons"?
17. What is meant by the terms "atomic radius" and "reactivity"?
18. Write the symbols for the family of elements whose atoms have three valence electrons.
19. What do the electron structures of alkali metals have in common?
20. Pick the electron arrangements below that represent elements in the same chemical family:
    (a) 2,1, (b) 2,6, (c) 2,4, (d) 2,8,6, (e) 2,8,8, (f) 2,8,8,2, (g) 2,8,8,1, (h) 2,8,9,2.
21. Give the number of valence electrons for each of the following:
    (a) Ba (b) Al
    (c) P (d) Se
    (e) Br (f) K
22. Give the symbol for an element that has
    (a) 3 valence electrons (b) 4 valence electrons
    (c) 7 valence electrons (d) 1 valence electron
23. Draw the Lewis dot symbols for Be, Cl, K, As, and Kr.
24. From their position in the periodic table, predict which will be more metallic:
    (a) Be or B (b) Be or Ca
    (c) Ca or K (d) As or Ge
    (e) As or Bi
25. Which atom in the following pairs is more metallic?
    (a) Li or F (b) Li or Cs
    (c) Be or Ba (d) C or Pb
    (e) B or Al (f) Na or Ar
26. What general electron arrangement is conducive to chemical inactivity?
27. How many valence electrons does the last element in each period have?
28. Use the information in the periodic table to supply the following:
    (a) the nuclear charge on cadmium (Cd)
    (b) the atomic number of As
    (c) the atomic mass (or mass number) of an isotope of Br having 46 neutrons
    (d) the number of electrons in an atom of Ba
    (e) the number of protons in an isotope of Zn
    (f) the number of protons and neutrons in an isotope of Sr, atomic mass (or mass number) of 88
    (g) an element forming compounds similar to those of Ga
29. Complete the following table:

Atomic Number	Name of Element	Number of Valence Electrons	Period	Metal or Nonmetal
6	______	______	______	______
12	______	______	______	______
17	______	______	______	______
37	______	______	______	______
42	______	______	______	______
54	______	______	______	______

30. Sodium reacts violently with water and forms hydrogen in the process. Magnesium will react with water only when the water is very hot. Copper does not react with water. Suppose you find a bottle containing a lump of metal in a liquid and a label that says "Cesium (Cs)." Based on your knowledge of the periodic table, what danger is there, if any, in disposing of the metal by throwing it into a barrel of water?
31. Write the symbols of the halogen family in the order of increasing size of their atoms.
32. Why does Cs have larger atoms than Li?
33. Write the names and symbols of the alkaline earth elements.
34. How does the atomic radius for a metal atom relate to the reactivity of the metal?
35. How does the atomic radius for a nonmetal atom relate to reactivity?
36. Which atom in the following pairs is more reactive?
    (a) Li or Rb (b) Mg or Ba
    (c) Na or Ar (d) Ne or O
    (e) He or Xe (f) Br or F
37. Take these periodic trends—atomic size, metallic tendency—and indicate with specific examples how these trends vary down a family and across a period.
38. For each pair of elements, state which is larger:
    (a) K or Br (b) Li or K
    (c) O or Na
39. Why do trends exist in the periodic table?
40. Describe the variation in atomic size (a) across a period, (b) down a group.
41. Rank the following atoms by size, with the largest on the left and the smallest on the right: K, S, Al, P, Cl
42. Rank the following atoms by size, with the largest on the left and the smallest on the right: P, Sb, N, As, Bi
43. Find a family in which there is a total change from nonmetallic to metallic behavior. Show with specific information how you know that this change is taking place.

44. Using the following descriptive words or phrases, choose an element that matches. Find a common compound of this element that is available in the grocery store, your apartment, your dorm room, or at a discount store.

alkali element, halogen, semiconductor,
alkaline earth element, nonmetal gas

45. What are some uses for the noble gases? Be specific.
46. Generally how does the density of elements in Group IIA vary from top to bottom?
47. Until the early 1960s, the group VIIIA (or group 18) elements were called inert gases. Why is this name no longer appropriate?
48. The elements at the bottom of Groups IA, IIA, VIA, VIIA, and VIIIA are all radioactive, so less is known about their physical and chemical properties. Using group and period trends described in this chapter, predict which of these five elements—Fr, Ra, Po, At, and Rn—would
   (a) be the most metallic.
   (b) be the most nonmetallic.
   (c) have the largest atomic radius.
   (d) be the most unreactive.
   (e) react most readily with water.
49. Consider the following oxides of the elements in period 3: Na_2O, MgO, Al_2O_3, SiO_2, P_4O_{10}, and SO_3. What trend do you see in the ratio of oxygen atoms to element atoms in each compound? Based on this trend, what would you predict for the formula of the oxide between chlorine and oxygen?
50. If element 36 is a noble gas, in what groups would you expect elements 35 and 37 to occur?
51. Oxygen and sulfur are very different elements, in that one is a colorless gas and the other a yellow crystalline solid. Why, then, are they both in Group VIA?
52. Suppose the popular press reports the discovery of a large deposit of pure sodium in northern Canada. What is your reaction as an informed citizen?
53. There are more nonmetallic elements than metallic elements. (a) True, (b) False.
54. Write the symbol for an alkali metal, a lanthanide, an alkaline earth, a halogen, an actinide, and a transition metal.
55. Below are some selected properties of Li and K. Before looking up the number, estimate values for the corresponding properties of Na.

	Lithium	**Sodium**	**Potassium**
Atomic weight	6.9	—	39.1
Density (g/cm³)	0.53	—	0.86
Melting point (°C)	180	—	63.4
Boiling point (°C)	1330	—	757

56. Give the names and symbols for two elements most like selenium (Se), atomic number 34.
57. If element 119 is ever produced, what will be its position in the periodic table?
58. Here is a list of descriptive words or phrases: metallic, nonmetallic, hard, soft, gas, liquid, solid, less dense than air, semimetallic, transition element, main-group element, inert, very reactive toward water, colored, radioactive, will form negative ions
   For each listed element, match as many of the above descriptive words or phrases as you can.
   (a) Neon (b) Potassium
   (c) Lead (d) Mercury
   (e) Silicon (f) Iron
   (g) Sulfur (h) Radon
   (i) Magnesium (j) Chlorine

What would these alchemists have thought of artificial transmutation? Entitled **Lo Stradano: L'Alchimista,** *this painting was done in 1570 by Joanes Stratensis (1523–1605), also known as Jan vander Straet.* (Scala/Art Resource, NY)

CHAPTER 5

Nuclear Changes

Radioactivity. With what do you associate that word? Hazardous waste? Nuclear power plants? Medical diagnosis? Cancer risks? Nuclear weapons? Cancer cures? All of these are appropriate associations with the kind of change that happens only in atomic nuclei. Nuclear changes are very different from ordinary chemical reactions, as you will see. Nuclear changes are usually accompanied by the emission of radiation and can, in some cases, be accompanied by the release of large amounts of energy. In this chapter you will learn something about how and when nuclear changes occur. Later, in Chapter 13, the production of energy from nuclear changes is included in a discussion of alternate sources of energy.

In this chapter you will see that nuclear changes are pictured in much the same way as chemical changes: reactants going to products. Equations can be written for nuclear changes, and although there are similarities between nuclear and chemical changes, nuclear changes are different in some rather significant ways. Certainly the discovery of nuclear changes has changed our lives, some might say for the worse, but you or one of your friends may be alive today because of an application of what is known about how radioactive isotopes undergo change.

- What are the characteristics of nuclear changes?
- Why do some atoms undergo spontaneous nuclear decay?
- How can an atom of one element be transformed into an atom of another element?
- Why are some radioactive isotopes more dangerous than others?
- What are some of the harmful effects of nuclear radiation?
- What are some of the useful applications of radioactive isotopes?

The nuclear age is often considered to have started either in the late 1800s and early 1900s with the discovery of radioactive elements or in 1945 with the first explosion of an atomic bomb. Actually, radioactivity has been a part of the universe since the beginning. It is true that early atomic theory said nothing about radioactivity, but that was because radiation cannot be detected directly by our five senses. It took the maturing of the sciences—with such diverse discoveries as how to produce a vacuum, photographic film, fluorescent materials, electricity, and magnetic fields—to lead to the knowledge that some atoms disintegrate spontaneously and, in the process, produce radiation.

5.1 The Discovery of Radioactivity

In February 1896, Henri Becquerel was experimenting in France with the relation between the recently discovered X rays and the phosphorescence of certain minerals. X rays had been found to penetrate black paper and expose photographic plates. Becquerel had already discovered that phosphorescing uranium minerals also exposed the plates. During several cloudy days some samples were left waiting in a drawer. Out of curiosity, Becquerel developed the plates. He had quite a surprise—the plates were exposed. Obviously, the emission of radiation that penetrated the black paper had nothing to do with phosphorescence but was a property of the mineral. In fact, all uranium compounds, and even the metal itself, exposed photographic plates. Becquerel also discovered that uranium (U) emitted radiation that was capable of causing air molecules to ionize (that is, to lose electrons and become positively charged particles). He showed this by bringing uranium samples close to charged gold leaves in a vacuum jar (Figure 5.1), with the result that the negative charges were neutralized.

■ Phosphorescent minerals re-emit light after they have been exposed to light.

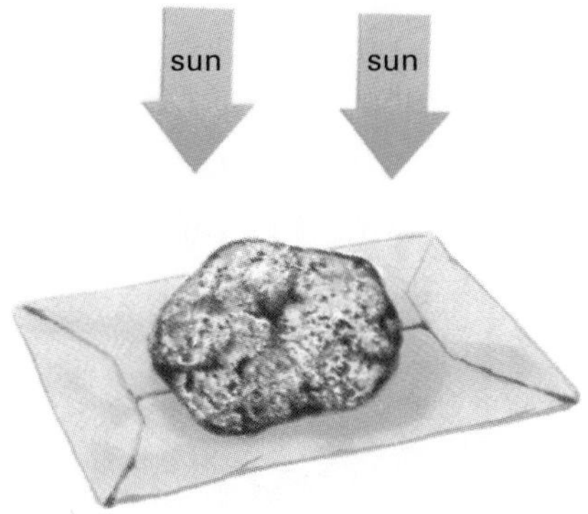

Becquerel's experiment. The goal was to find out if phosphorescence, known to occur when the mineral was exposed to sunlight, included X rays. The presence of X rays would be shown by exposure of the photographic plate.

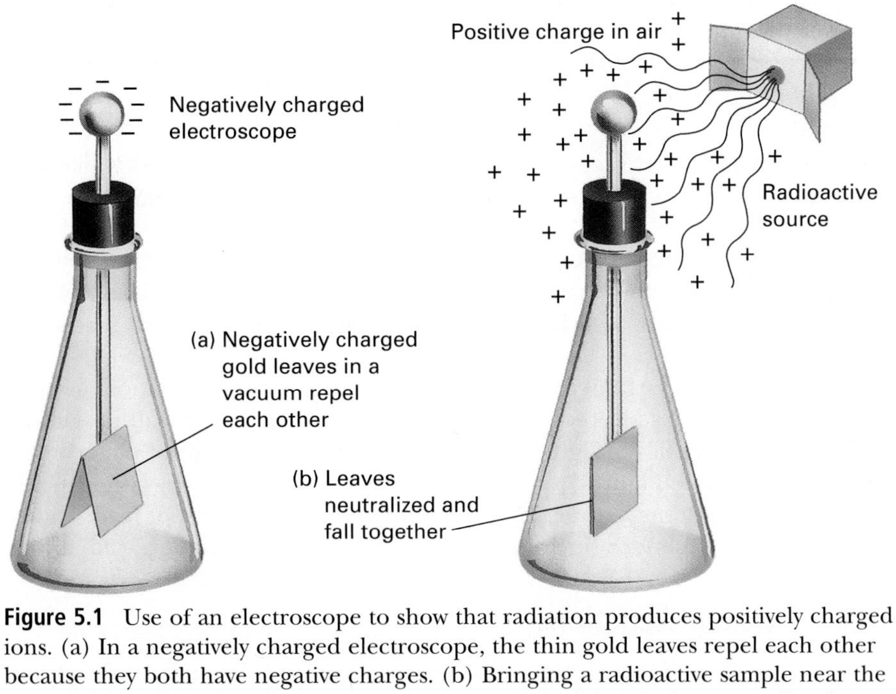

Figure 5.1 Use of an electroscope to show that radiation produces positively charged ions. (a) In a negatively charged electroscope, the thin gold leaves repel each other because they both have negative charges. (b) Bringing a radioactive sample near the negatively charged electroscope creates positive ions in the air that neutralize the electroscope charge, as shown by the leaves falling together.

THE PERSONAL SIDE

The Curies

Pierre and Marie Curie with their daughter, Irene. Irene grew up to continue the study of radioactivity with her husband, Frédéric Joliot. Together Irene and Frédéric won a Nobel Prize in 1935 for production of the first artificial radioactive isotope. (Stock Montage)

Soon after Becquerel's discovery of uranium's radioactivity, Marie Sklodowska Curie (1867–1934), also working in France, studied the radioactivity of thorium (Th) and began to search systematically for new radioactive elements. She showed that the radioactivity of uranium was an atomic property—that is, its radioactivity was proportional to the amount of the element present and was not related to any particular compound. Her experiments indicated that other radioactive elements were probably also present in certain uranium samples. With painstaking technique, she and her husband, Pierre Curie (1859–1906), separated the element radium (Ra) from uranium ore and found that it is more than one million times more radioactive than uranium. In 1903 Marie and Pierre Curie shared the Nobel Prize in physics with Henri Becquerel for their discovery of radioactivity. After Pierre died, Marie continued her research and discovered polonium (Po), which she named after her native Poland. In 1911 she became the first person to win a second Nobel Prize, this one for the discoveries of radium and polonium. In 1921 Marie Curie came to the United States where she was given 1 gram of pure radium, purchased with donations from American women interested in her work.

■ As you saw in Section 3.3, identification of alpha rays as helium nuclei (^4_2He) and beta rays as electrons played a significant role in our understanding of the structure of the atom.

Before long it was recognized that the radiation from elements like uranium and radium consisted of the three types known as alpha, beta, and gamma rays.

In 1899 Ernest Rutherford found that alpha rays could be stopped by thin pieces of paper and had a range of only about 2.5 cm to 8.5 cm in air before being absorbed, while beta rays were capable of penetrating far greater distances in air.

In 1900 Paul Villard identified the third form of natural radiation, gamma (γ) rays. These he discovered were not streams of particles, but rather had the general characteristics of light or X rays. Gamma rays, a form of electromagnetic radiation (see Figure 3.15), are extremely penetrating; they are capable of passing through more than 22 cm of steel and about 2.5 cm of lead. Figure 5.2 compares the penetrating ability of the three forms of natural radiation.

5.2 Nuclear Reactions

After discovery of the natural radioactivity of uranium, thorium, and radium, many other elements were found to have radioactive isotopes. All the elements heavier than bismuth (Bi, atomic number 83) and a few lighter than bismuth

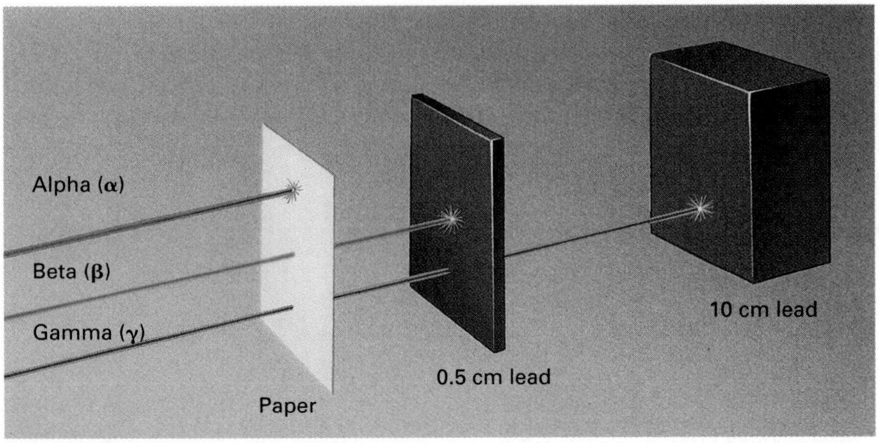

Figure 5.2 The relative penetrating abilities of alpha, beta, and gamma rays. The heavy, highly charged alpha particles are stopped by a piece of paper (or the skin). The lighter, less highly charged beta particles penetrate paper, but are stopped by a 0.5-cm sheet of lead. Because gamma rays have no charge and no mass, they are the most penetrating.

have natural radioactivity. While studying radium, Rutherford found that besides emitting alpha particles, radium was also producing radioactive radon gas (Rn). This led Rutherford and one of his students, Frederick Soddy, in 1902 to propose the revolutionary theory that *radioactivity is the result of a natural change of an isotope of one element into an isotope of a different element.* Such a change is a **nuclear reaction,** a process in which an unstable nucleus emits radiation and is converted into a more stable nucleus of a different element. Thus, a nuclear reaction, sometimes referred to as a **transmutation,** results in a change in atomic number and often a change in mass number as well. When nuclear reactions are compared with chemical reactions, we see some differences and some similarities.

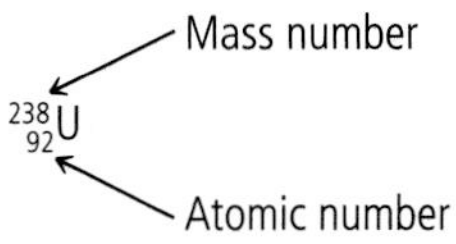

Nuclear Reactions	**Chemical Reactions**
	Differences
New elements are often formed.	New elements are never formed.
Particles in the nucleus are involved.	Only electrons are involved.
Relatively large amounts of energy are involved.	Small amounts of energy are involved.
Conditions such as temperature and pressure do not influence the rate of reaction.	Conditions such as temperature and pressure greatly influence the rate of reaction.
	Similarities
The number of nucleons remains the same.	The number of atoms remains the same.
The sum of the atomic numbers (positive charge) remains the same.	The total number of electrons (negative charge) remains the same.

In chemical reactions atoms or ions are rearranged, but they are never created or destroyed. The number and kinds of atoms always remain the same.

In nuclear reactions the total number of nuclear particles, called **nucleons** (protons plus neutrons), remains the same, but the identities of atoms can change. Just as with chemical equations, nuclear equations reflect the fact that matter is conserved. As a result, the sum of the mass numbers of reacting nuclei must equal the sum of the mass numbers of the product nuclei. There must also be nuclear charge balance—the sum of the atomic numbers of the products must equal the sum of the atomic numbers of the reactants.

Equations for Nuclear Reactions

■ When 226 g of radium-226 has completely decayed, 222 g of radon-222 and 4 g of helium will have been formed.

■ Radium-226 is another way of writing $^{226}_{88}Ra$.

Consider the equation for the nuclear reaction that Rutherford studied:

$$\underset{\text{Radium-226}}{^{226}_{88}Ra} \longrightarrow \underset{\text{Alpha particle}}{^{4}_{2}He} + \underset{\text{Radon-222}}{^{222}_{86}Rn}$$

The mass number on the left equals the sum of the mass numbers on the right. Similarly, the atomic number on the left equals the sum of the atomic numbers on the right.

mass number: 226 = 4 + 222

atomic number: 88 = 2 + 86

Nuclear equations usually do not show any charges on symbols for the nuclei involved. For example, the alpha particle ($^{4}_{2}He$) that is ejected from a decaying radioactive nucleus is really a doubly charged positive particle (refer to Section 3-3, where the attraction of alpha particles to opposite charges is discussed).

The isotope of uranium with atomic mass 238 is an alpha-emitter. When the $^{238}_{92}U$ nucleus gives off an alpha particle, made up of two protons and two neutrons, four units of atomic mass and two units of atomic charge are lost. The resulting nucleus has a mass of 234 and a nuclear charge of 90, showing that it is an isotope of thorium, which has 90 protons in its nucleus and an atomic number of 90.

$$\underset{\text{Uranium-226}}{^{238}_{92}U} \longrightarrow \underset{\text{Alpha particle}}{^{4}_{2}He} + \underset{\text{Thorium-234}}{^{234}_{90}Th}$$

mass number: 238	=	4	+	234
atomic number: 92	=	2	+	90

Some unstable nuclei are beta-emitters. For example, uranium-235 emits a beta particle, which is an electron ($^{0}_{-1}e$).

$$^{235}_{92}U \longrightarrow {}^{0}_{-1}e + {}^{235}_{93}Np$$

mass number: 235	=	0	+	235
atomic number: 92	=	−1	+	93

Knowing that some nuclei emit beta particles leads to a basic question: How can a nucleus containing protons and neutrons emit a beta particle, which is an electron? It has been established that an electron and a proton can combine outside the nucleus to form a neutron. Therefore, the reverse process is proposed to occur inside the nucleus. When a beta particle is emitted from a decaying nucleus, a neutron decomposes, giving up an electron and changing itself into a proton. The ejected electron is the beta particle. The resulting proton remains in the nucleus and increases the atomic number by one.

$$\text{Beta particle production:} \quad \underset{\text{Neutron}}{{}^{1}_{0}\text{n}} \longrightarrow \underset{\text{Electron}}{{}^{0}_{-1}\text{e}} + \underset{\text{Proton}}{{}^{1}_{1}\text{H}}$$

Gamma radiation (γ) may or may not be given off simultaneously with alpha or beta rays, depending on the particular nuclear reaction involved. Gamma rays are emitted when the product nucleus must lose some additional energy to become stable. Being electromagnetic radiation, gamma rays have no charge and essentially no mass. The emission of a gamma ray, therefore, cannot alone account for a transmutation event.

EXAMPLE 5.1 ***Writing an Equation for a Beta Emission***

Write an equation for beta emission from a lead-210 nucleus.

SOLUTION

First, write a partial equation that includes what is known: lead-210 is a reactant, and a beta particle is a product. Then, to aid in determining the mass and atomic number changes, set up a table like those used earlier. The atomic number of lead is 82.

	$^{210}_{82}Pb \longrightarrow$	$^{0}_{-1}e$	$+ ?$
mass number:	$210 \longrightarrow$	0	$+ ?$
atomic number:	$82 \longrightarrow$	-1	$+ ?$

The sum of the mass numbers of the products must equal 210, the mass number of the decaying lead isotope. Since the mass of the beta particle is essentially zero, the mass number of the product nucleus must be 210. The sum of the atomic numbers of the products must also equal the atomic number of lead, 82. So the atomic number of the product must be 83 [$83 + (-1) = 82$], which is the atomic number of bismuth (Bi). The product nucleus is $^{210}_{83}Bi$.

Exercise 5.1
Write an equation showing the emission of a beta particle from $^{234}_{91}Pa$.

EXAMPLE 5.2 *Writing an Equation for an Alpha Emission*

Write an equation for alpha emission from a polonium-218 isotope.

SOLUTION

First, write the partial equation and set up a table of mass and atomic number changes under it. The atomic number of polonium is 84.

	$^{218}_{84}Po \longrightarrow$	$^{4}_{2}He$	$+ ?$
mass number:	$218 \longrightarrow$	4	$+ ?$
atomic number:	$84 \longrightarrow$	2	$+ ?$

The mass number of the product must be 214 since its mass number plus that of the alpha particle must equal 218, the mass number of the decaying polonium-218 isotope. The atomic number of the product must be 82 since its atomic number plus that of the alpha particle must equal 84, the atomic number of the decaying isotope. Since the element with an atomic number of 82 is lead, the product is $^{214}_{82}Pb$.

Exercise 5.2A
Write an equation showing the emission of an alpha particle by an isotope of neptunium, $^{237}_{93}Np$.

Exercise 5.2B

Thorium-230 decays by emitting an alpha particle. Its decay product also decays, emitting an alpha particle. Write two equations showing these nuclear reactions.

5.3 The Stability of Atomic Nuclei

Why are some nuclei unstable and radioactive, while others are stable and not radioactive? The fact that there are strong repulsions among all those protons packed inside the nucleus of an atom has something to do with nuclear stability. In addition, the relative numbers of neutrons, which are not charged, play some role. Figure 5.3 shows a plot of the number of protons versus the number of neutrons in the known isotopes from hydrogen ($Z = 1$) to bismuth ($Z = 83$). The nonradioactive (stable) isotopes (black dots) are far fewer in number than the radioactive (unstable) isotopes (red dots).

■ Z is the symbol for atomic number.

The stability of nuclei is apparently dependent on the relative numbers of protons and neutrons. The nucleus of the simplest atom, hydrogen, contains only a proton. Its two isotopes, deuterium (^2_1H) and tritium (^3_1H), contain one and two neutrons, respectively. Looking at Figure 5.3 you can see that the band of stable nuclei, the black dots, curves upward toward the neutron axis,

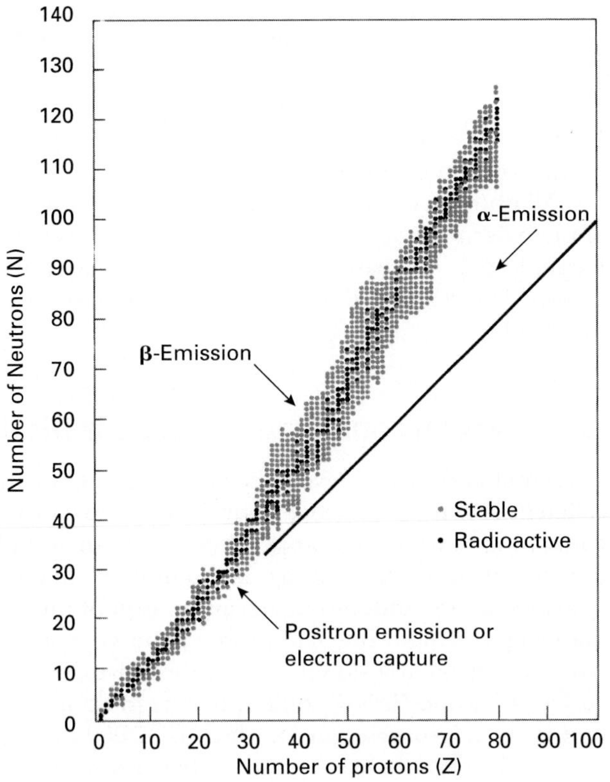

Figure 5.3 The band of nuclear stability, as shown by the black dots for stable isotopes. When radioactive nuclei (red dots) decay, their neutron-to-proton ratio moves closer to the band. Isotopes above the band undergo β decay (the number of protons increases), isotopes below the band undergo positron emission (the number of protons decreases), and those with more than 83 protons undergo α decay (the number of protons decreases by 2 and the number of neutrons decreases by 2). *(Redrawn from Oxtoby, Nachtrieb, Freeman:* Chemistry: Science of Change)

■ Perhaps it is because there are so many unstable (radioactive) isotopes that people sometimes think of "radioactive" when the word "isotope" is mentioned.

■ Beta particles result from the conversion of neutrons into protons and electrons.

$$^{1}_{0}n \rightarrow ^{1}_{1}H + ^{0}_{-1}e$$

■ The positron is sometimes called the "anti-electron." The positron is one of a group of "antimatter" particles known to exist. An electron will react with a positron to annihilate each other and produce two high-energy photons. As will be seen later in this chapter, positron-emitting isotopes have use in medical diagnosis. (See Section 5.10.)

showing that in stable nuclei, the number of neutrons is equal to or greater than the number of protons. From hydrogen to bismuth, except for $^{1}_{1}H$ and $^{3}_{2}He$, *the mass numbers of stable isotopes are always twice as large as the atomic number or even larger.* It appears that the larger numbers of protons in the nuclei of heavier atoms require extra neutrons to gain stability. Any unstable isotope (a red dot in Figure 5.3) will decay in such a way that its decay product falls closer to the stable band (the black dots).

Beta emission occurs in isotopes that have *too many neutrons* to be stable. These isotopes appear as red dots above the stable band in Figure 5.3. When beta decay occurs, the proton resulting from the conversion of a neutron into a proton and an electron increases the atomic number while lowering the number of neutrons, as was illustrated in Example 5.1, and the new isotope moves toward the stable region.

Those isotopes with *too few neutrons* (red dots below the band of stability) decay as well, but in a manner that increases the number of neutrons relative to the number of protons. One way this can happen is by emission of a type of subatomic particle discovered in 1932, a **positron**—a positively charged electron, $^{0}_{+1}e$. For example, the decay of nitrogen-13, an isotope with too few neutrons, is by positron emission.

$$^{13}_{7}N \longrightarrow ^{0}_{+1}e + ^{13}_{6}C$$

The positron results from the decay of a proton.

$$\underset{\text{Proton}}{^{1}_{1}H} \longrightarrow \underset{\text{Neutron}}{^{1}_{0}n} + \underset{\text{Positron}}{^{0}_{+1}e}$$

Because the positron, like the electron, has a mass number of zero, the mass number of the product nucleus is the same as that of the starting nucleus.

All isotopes of the elements beyond bismuth ($Z = 83$) are unstable. Most of them decay by ejecting an alpha particle. This kind of decay, as illustrated in Example 5.2, decreases the mass number by four and the atomic number by two. The types of radioactive decay we have discussed are summarized in Table 5.1.

5.4 Activity and Rates of Nuclear Disintegrations

The number of radioactive nuclei that disintegrate in a sample per unit time is called its **activity** (this is the "activity" in radioactivity). The activity of a sample containing radioactive isotopes depends on the number of nuclei present and the rate at which they decay. If a sample of matter is "highly radioactive," many atoms are undergoing decay per unit of time. A small number of nuclei decaying at a rapid rate can produce the same activity as a larger number of atoms decaying at a slower rate. Radioactive disintegrations are measured in **curies** (Ci); one Ci is 37 billion disintegrations per second. A more suitable unit is the **microcurie** (μCi), which is 37,000 disintegrations per second. Another unit used to measure radioactive disintegrations is the **becquerel** (Bq), where 1 Bq = 1 disintegration per second.

TABLE 5.1 ■ Changes in Atomic Number and Mass Number Accompanying Radioactive Decay

Type of Decay	Symbol	Charge	Mass	Change in Atomic Number	Change in Mass Number
beta	${}_{-1}^{0}e$	−1	0	+1	none
positron	${}_{+1}^{0}e$	+1	0	−1	none
alpha	${}_{2}^{4}He$	+2	4	−2	−4
gamma	${}_{0}^{0}\gamma$	0	0	none	none

To illustrate the differences in rates of decay of radioactive nuclei, consider first how cobalt-60 is used in medicine to treat malignancies in the human body. When cobalt-60 decays, it produces beta particles as well as gamma rays.

$$ {}_{27}^{60}Co \longrightarrow {}_{28}^{60}Ni + {}_{-1}^{0}e + {}_{0}^{0}\gamma $$

Although the cobalt-60 isotope is radioactive, it is stable enough so that only half of a sample will decay in 5.27 years. A cobalt-60 sample is installed in a well-shielded apparatus that emits a focused beam of gamma rays. Because the half-life of the cobalt-60 radioisotope is fairly long, the sample does not have to be replaced very often. By rotating the radiation source around the patient, the physician can concentrate the rays in the cancerous region being treated, while limiting the radiation somewhat to other parts of the body (Figure 5.4).

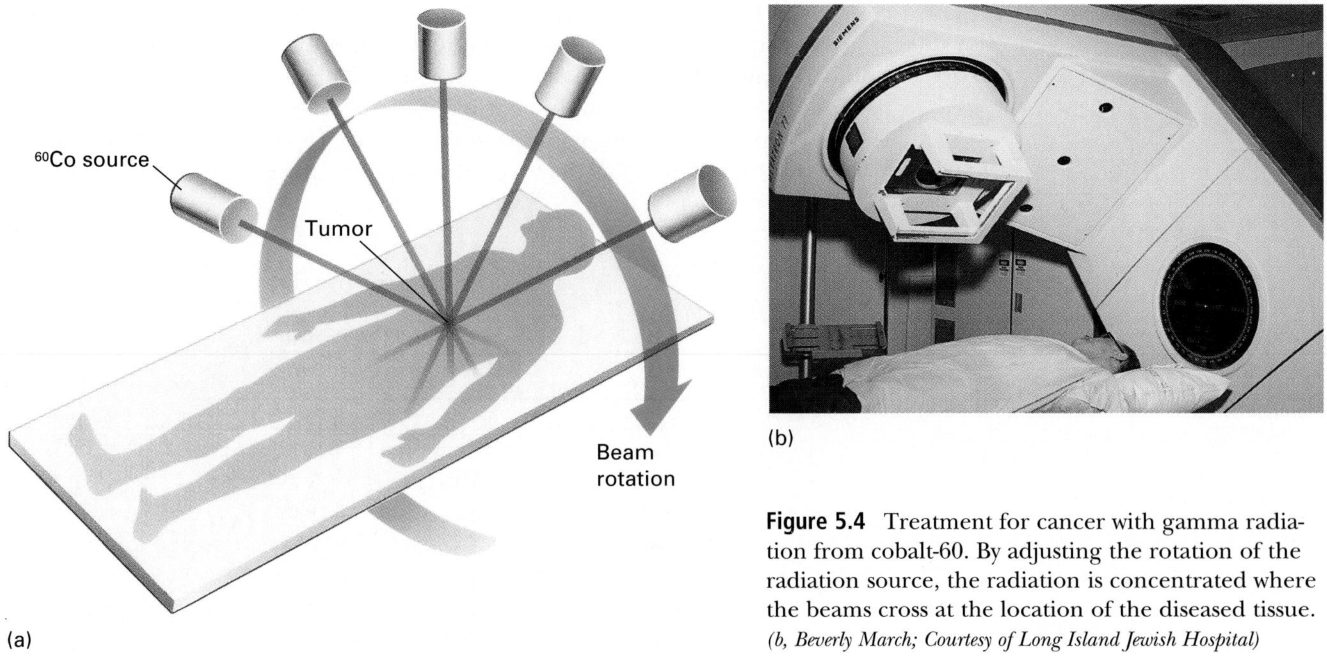

Figure 5.4 Treatment for cancer with gamma radiation from cobalt-60. By adjusting the rotation of the radiation source, the radiation is concentrated where the beams cross at the location of the diseased tissue. *(b, Beverly March; Courtesy of Long Island Jewish Hospital)*

Copper-64, in the form of copper acetate, is also used in medicine, but as a diagnostic tool rather than for treatment. Also a gamma-emitter, copper-64 decays much more rapidly than cobalt-60; half the copper-64 atoms in a sample will decay in 12.9 hours. When injected into a patient's blood, the copper compound is carried to the brain and concentrates in any tumorous region that is present. A camera that detects gamma rays is used to locate the tumor. The radiologist knows that the copper-64 nuclei used in diagnosis will have decayed into safer, more stable isotopes in a matter of a few hours or days because of the rapid rate of decay of these nuclei.

■ Further examples of the use of radioisotopes in medicine are given in Section 5.10.

Half-Life

The rate of decay of any radioactive isotope can be represented by its characteristic **half-life,** the period required for one half of the radioactive material originally present to undergo radioactive decay. Short half-lives are the results of high rates of decay, while long half-lives are the results of low rates of decay.

The half-life of an isotope is independent of the amount of radioactive material present and is essentially independent of temperature and the chemical form in which the radioactive atoms are present. Table 5.2 gives the half-lives of some radioactive isotopes. The 12.9-hour half-life of $^{64}_{29}Cu$, for example, means that one half of the original amount of copper-64 atoms will remain after 12.9 hours. In another 12.9 hours, half of the original half, $(\frac{1}{2})^2 = (\frac{1}{4})$, will remain. This process continues indefinitely until virtually all of the copper-64 isotopes have decayed. Figure 5.5 illustrates graphically how the concept of half-life works for a radioactive isotope. No matter what its half-life, the fraction of a radioactive isotope remaining will be $\frac{1}{2}$ after one half-life, $\frac{1}{4}$ after two half-lives, $\frac{1}{8}$ after three half-lives, and so on.

■ Mathematically, the fraction of a radioactive isotope remaining after n half-lives is $(\frac{1}{2})^n$. The fraction after two half-lives is $(\frac{1}{2})^2 = \frac{1}{4}$; after three half-lives it is $(\frac{1}{2})^3 = \frac{1}{8}$, and so on.

Focusing attention on the decay products also helps in understanding the concept of half-life. For example, if a sample contains one million copper-64 atoms at some beginning time, 12.9 hours later only 500,000 copper-64 atoms would remain. However, there would be 500,000 zinc-64 atoms present that had not been there 12.9 hours earlier. After 25.8 hours (two half-lives) only 250,000 copper-64 atoms would remain, and there would be 750,000 zinc-64 atoms resulting from the decay of the copper atoms. After many half-lives

TABLE 5.2 ■ Half-Lives of Some Radioactive Isotopes

Decay Process	Half-Life
$^{238}_{92}U \longrightarrow {}^{234}_{90}Th + {}^{4}_{2}He$	4.51×10^9 years
$^{3}_{1}H \longrightarrow {}^{3}_{2}He + {}^{0}_{-1}e$	12.3 years
$^{14}_{6}C \longrightarrow {}^{14}_{7}N + {}^{0}_{-1}e$	5730 years
$^{131}_{53}I \longrightarrow {}^{131}_{54}Xe + {}^{0}_{-1}e$	8.05 days
$^{64}_{29}Cu \longrightarrow {}^{64}_{30}Zn + {}^{0}_{-1}e$	12.9 hr
$^{69}_{30}Zn \longrightarrow {}^{69}_{31}Ga + {}^{0}_{-1}e$	55 min

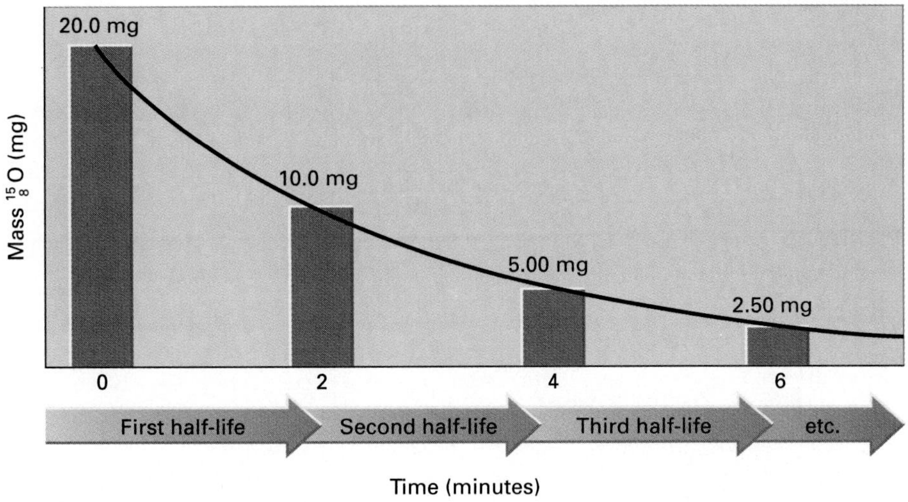

Figure 5.5 Radioactive decay of 20 mg of oxygen-15, which has a half-life of 2.0 minutes. The plotted data are given in the table below. After each half-life period, the quantity present at the beginning of the period is reduced by 1/2.

Number of Half-Lives	Fraction of Initial Quantity Remaining	Quantity Remaining (mg)
0	1	20.0 (initial)
1	1/2	10.0
2	1/4	5.00
3	1/8	2.50
4	1/16	1.25
5	1/32	0.625

almost all of the copper atoms will have decayed, and there will be almost one million zinc-64 atoms, which are stable and do not undergo decay.

EXAMPLE 5.3 *Calculations Using Half-Life*

Tritium ($^{3}_{1}$H), a radioactive isotope of hydrogen, has a half-life of 12.3 years.

$$^{3}_{1}\text{H} \longrightarrow {}^{0}_{-1}\text{e} + {}^{3}_{2}\text{He}$$

If a sample contains 1.5 mg of tritium, how many milligrams will remain after 49.2 years?

SOLUTION

First, find the number of half-lives in the time given. Since the half-life is 12.3 years, the number of half-lives is

$$49.2 \text{ years} \times \frac{1 \text{ half-life}}{12.3 \text{ years}} = 4.00 \text{ half-lives}$$

Four half-lives means that the initial quantity of tritium will have decreased four times by $\frac{1}{2}$, or $(\frac{1}{2})^4$. Written another way,

$$1.5 \text{ mg} \times \tfrac{1}{2} \times \tfrac{1}{2} \times \tfrac{1}{2} \times \tfrac{1}{2} = 1.5 \times (\tfrac{1}{2})^4 = 1.5 \times \tfrac{1}{16} = 0.094 \text{ mg}$$

After 49.2 years, only 0.094 mg of the original 1.5 mg of tritium will remain.

Exercise 5.3

The half-life of copper-67, a beta emitter, is 61 hours. If a sample originally contains 21.0 mg of copper-67, how many milligrams of copper-67 will remain after 244 hours?

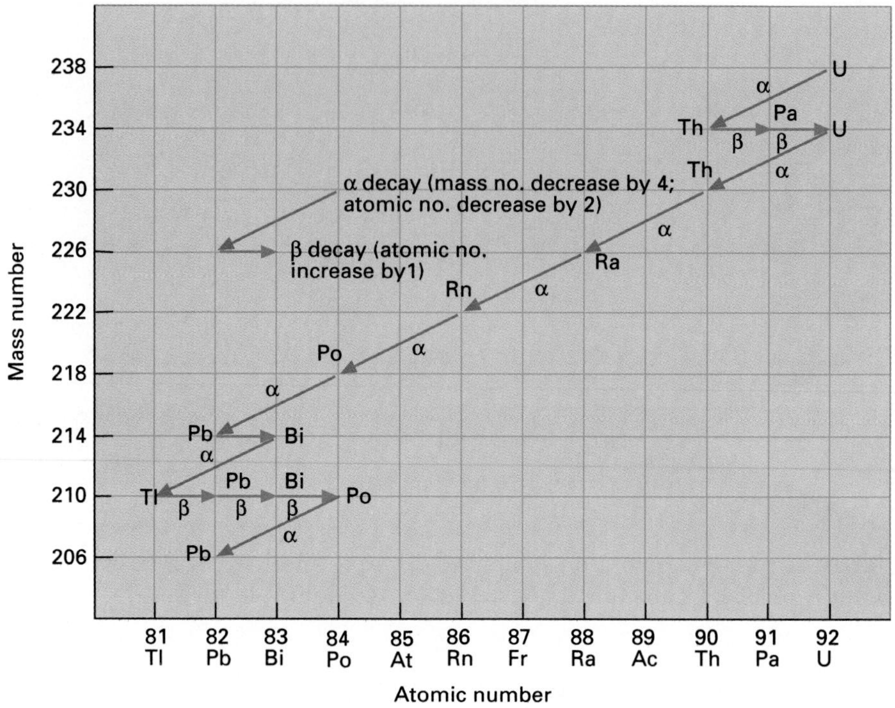

Figure 5.6 The uranium-238 decay series. Radium (Ra) and polonium (Po), the two elements discovered by Marie Curie, are part of this series. Radon (Rn), the radioactive gas of environmental concern, is generated as shown here wherever rocks contain uranium. Lead-206 is not radioactive.

Natural Radioactive Decay Series

Some half-lives are extremely long, and others are extremely short. The longest half-lives are of the order of 10^{10} years, and the shortest are approximately 10^{-10} seconds. The half-life for the uranium-238 alpha decay is 4.5 *billion* years. The half-life of thorium-234, on the other hand, is 24 days. As one would expect, relatively large amounts of uranium-238 can be found in certain rocks and mineral deposits, whereas only trace amounts of thorium-234 are present.

The radioactive decay of thorium-234 into protactinium-234 is the second step in a natural series of nuclear decays that starts with uranium-238. After 14 decays this **uranium series** ends with a stable, nonradioactive isotope of lead, lead-206 (Figure 5.6). Table 5.3 gives the half-lives of the isotopes in the uranium series. You would not expect to find much of the short-lived isotopes in a sample of rock, and indeed you have to look carefully for them, but they are there. The longer-lived isotopes such as uranium-234 and thorium-230 are readily detected. Two other similar natural decay series exist, each of which starts out with a different isotope and proceeds through a different set of radioactive decay products. The thorium series begins with thorium-232 (a different isotope from the two Th isotopes that occur in the uranium series) and ends with stable lead-208. The actinium series begins with uranium-235 and ends with lead-207. Most of the naturally occurring radioactive isotopes are members of one of these three decay series.

TABLE 5.3 ■ Half-Lives of the Naturally Occurring Radioactive Elements in the Uranium-238 ($^{238}_{92}U$) Series

Isotope	Type of Disintegration	Half-Life
^{238}U	alpha	4.5 billion years
^{234}Th	beta	24.1 days
^{234}Pa	beta	1.18 min
^{234}U	alpha	250,000 years
^{230}Th	alpha	80,000 years
^{226}Ra	alpha	1620 years
^{222}Rn	alpha	3.82 days
^{218}Po	alpha, beta	3.05 min
^{214}Pb	beta	26.8 min
^{214}Bi	alpha, beta	19.7 min
^{210}Tl	beta	1.32 min
^{210}Pb	beta	22 years
^{210}Bi	beta	5 days
^{210}Po	alpha	138 days
^{206}Pb	stable	

5.5 Atomic Dating

The concept of radioactive half-life was almost immediately recognized as a useful tool for measuring the age of radioactive materials. As early as 1905, Ernest Rutherford suggested during a lecture at Yale University that one indication of the age of objects would be the measurement of helium gas formed from alpha particles produced by radioactive decay.

■ Helium gas results when an alpha particle gains two electrons from its environment.

$$\underset{\text{Alpha particle}}{^{4}_{2}He^{2+}} + 2\,e^- \longrightarrow \underset{\text{Helium atom}}{^{4}_{2}He}$$

At about the same time, it was suggested that the $^{238}_{92}U/^{206}_{82}Pb$ ratio could be measured to date rock samples. The half-lives of all the decay reactions in the uranium series are relatively short (see Table 5.3); except for the uranium-238 → lead-206 decay. Because it occurs at the slowest rate, the decay of uranium-238 becomes a *rate-determining step* in the whole process. The overall decay series can be summarized as

$$^{238}_{92}U \longrightarrow ^{206}_{82}Pb$$

for every atom of uranium-238 that decays, an atom of lead-206 will be formed. By using the ratio of the atomic masses of the atoms involved, we can calculate that 0.50 g of uranium-238 would become 0.43 g of lead-206.

$$\frac{206 \text{ amu Pb}}{238.07 \text{ amu U}} \times 0.50 \text{ g U} = 0.43 \text{ g Pb}$$

■ No matter how much of a radioactive substance is present at the beginning, only half of it remains at the end of one half-life.

■ The rate-determining step is the slowest step in a multistep process.

Now, if the ratio of 0.50 g of uranium-238 to 0.43 g of lead-206 is found in an ore of uranium, it would follow that the rock is 4.5 billion years old, the half-life of uranium-238. Ratios like this are actually found in some samples of rock, such as granite, from locations around the world.

Carbon-14 Dating

Cosmic rays, which enter the earth's atmosphere from outer space, are composed of many kinds of very high-energy particles. The particles undergo nuclear reactions with stable nuclei in the upper atmosphere to produce other nuclei and slow-moving neutrons. These neutrons can then react with nitrogen-14 ($^{14}_{7}N$) nuclei present in nitrogen molecules in the upper atmosphere to produce carbon-14 ($^{14}_{6}C$), a radioactive isotope of carbon that has a half-life of 5730 years.

$$^{1}_{0}n + ^{14}_{7}N \longrightarrow ^{14}_{6}C + ^{1}_{1}H$$

The $^{14}_{6}C$ decays by beta emission.

$$^{14}_{6}C \longrightarrow ^{0}_{-1}e + ^{14}_{7}N$$

■ There is about one atom of ^{14}C for every 10^{12} atoms of ordinary carbon.

Although only about 7.5 kg of $^{14}_{6}C$ is produced in the entire atmosphere per year, this tiny amount is eventually incorporated into carbon dioxide (CO_2), which mixes with the ordinary CO_2 in the atmosphere and in turn is incorporated into the structure of all living matter through photosynthesis and natural food chains. It has been observed that the beta activity of carbon-14 in living plants and the atmosphere is about 14 disintegrations per minute per gram of carbon present. Upon the death of the organism, the intake of food ceases, and the natural level of radioactive carbon present within it begins to decrease at the rate of 50% every 5730 years.

To understand how carbon-14 dating works, consider how the age of a sample of cotton cloth might be determined. As a cotton plant grows, it incorporates carbon from atmospheric CO_2. This CO_2 contains a natural abundance of carbon-14, and that carbon, along with atoms of carbon-12 and carbon-13, is incorporated into the cellulose molecules in the cotton fiber (see Section 15.5). In effect, during the short growing period of the cotton plant the carbon-14 in the plant reflects the same abundance of carbon-14 that is found in the atmosphere. When the cotton is harvested, the radioactive carbon atoms are no longer replaced in new growth, and the disintegrations due to carbon-14 gradually diminish from the background carbon-14 disintegrations found in the atmosphere and in living things. If an older sample of cotton cloth has *half as many disintegrations* due to carbon-14 as another, newer sample, it is safe to assume that the older sample is about 5730 years, or one half-life, *older*. Many interesting ancient articles have been dated using the carbon-14 technique.

■ SELF-TEST 5A

1. The first evidence for radioactivity occurred when photographic films were exposed when placed near samples of uranium. (a) True, (b) False.

2. Which of the following scientists discovered gamma rays? (a) Becquerel, (b) Marie Curie, (c) Villard.
3. Two kinds of nuclear decay that lead to a reduction in atomic number are ____________ and ____________.
4. The most penetrating nuclear radiation is ____________ radiation.
5. The mass number of a nucleus is unchanged after beta particle emission. (a) True, (b) False.
6. The mass number of a uranium-238 nucleus will decrease by 2 units after alpha emission. (a) True, (b) False.
7. When a $^{216}_{84}Po$ nucleus emits an alpha particle, the nuclear species that results is ____________.
8. Cosmic ray action leads to conversion of nitrogen-14 in the atmosphere to radioactive carbon-14. (a) True, (b) False.
9. Nuclei with atomic numbers greater than 83 are all radioactive. (a) True, (b) False.
10. The half-life of ^{210}Bi is 5 days. If the initial activity was $2\mu Ci$, or 74,000 dps, the activity after 15 days is expected to be ____________.

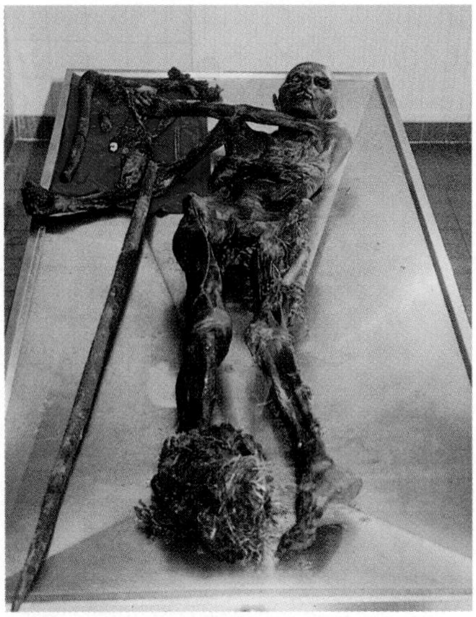

The Iceman. In 1991 the Iceman was found in a glacier near the border between Italy and Austria. Remarkably, his body had been dehydrated by cold winds before he was buried by the ice and thus was not chemically decomposed. Radiocarbon dating has shown the Iceman to be about 5300 years old. His possessions, also perfectly preserved, are of intense interest to archaeologists, and include an unusual pure copper (rather than bronze) ax. *(Sygma)*

5.6 Artificial Nuclear Reactions

After it was realized that the nuclei of some of the heavier isotopes were unstable, scientists wondered if nuclear reactions could be initiated by producing new unstable nuclei. To achieve this, one nucleus could be made to approach another with sufficient kinetic energy to overcome the repulsions of their positive charges so they would collide. One of these nuclei would have to bombard the other nucleus at very high speed. When a collision occurs, one could postulate an unstable **compound nucleus** that would emit particles, energy, or both in seeking stability.

In 1919, Rutherford was successful in producing the first artificial nuclear change by bombarding nitrogen (N_2) with alpha particles. All of the results of the experiment could be explained if one assumed the nuclear reaction to be

$$^{14}_{7}N + ^{4}_{2}He \longrightarrow [^{18}_{9}F] \longrightarrow ^{17}_{8}O + ^{1}_{1}H$$

where $^{18}_{9}F$ is the unstable compound nucleus. Both of the product nuclei are stable. Rutherford had observed an **artificial transmutation,** the conversion of one element into another during a laboratory experiment. Following Rutherford's original experiment, there was considerable interest in discovering new nuclear reactions. Many isotopes were subjected to beams of high-energy particles.

■ Alchemists in ancient times tried in vain to transmute one element into another—mainly lead into gold. Nevertheless, they would have been extremely excited at the news of Rutherford's accomplishment.

■ Natural fluorine consists exclusively of one isotope, $^{19}_{9}F$.

In 1934 Irene Curie Joliot, daughter of Marie and Pierre Curie, and her husband Frédéric Joliot bombarded aluminum (Al) with alpha particles and observed neutrons and a positron. The Joliots discovered that when the alpha particles striking the Al were stopped, the neutrons stopped, but the positron emissions continued. They reasoned that the reactions taking place were

$$^{27}_{13}Al + ^{4}_{2}He \longrightarrow ^{30}_{15}P + ^{1}_{0}n$$

$$^{30}_{15}P \longrightarrow ^{30}_{14}Si + \underset{\text{Positron}}{^{0}_{+1}e}$$

The second reaction continued because the phosphorus-30 was decaying more slowly than it was being produced. Phosphorus-30 was the *first radioactive isotope to be produced artificially*. Today, more than 1000 other radioactive nuclides have been produced.

A fundamental question arises about why the alpha particles were scattered by the gold foil in Rutherford's gold-foil experiment (Figures 3.7 and 3.8), and yet the same alpha source can produce a nuclear change with a smaller atom such as $^{9}_{4}Be$ (Figure 5.7). The answer lies in the fact that the charge on the gold (Au) nucleus is + 79, whereas the charge on the beryllium (Be) nucleus is only + 4. Most of the alpha particles emitted from natural radioactive decay do not have enough energy to penetrate a heavy, positively charged nucleus such as that of gold. Therefore, if artificial nuclear reactions for the heavier elements are to be studied, the kinetic energy of the subatomic projectile particles must be increased.

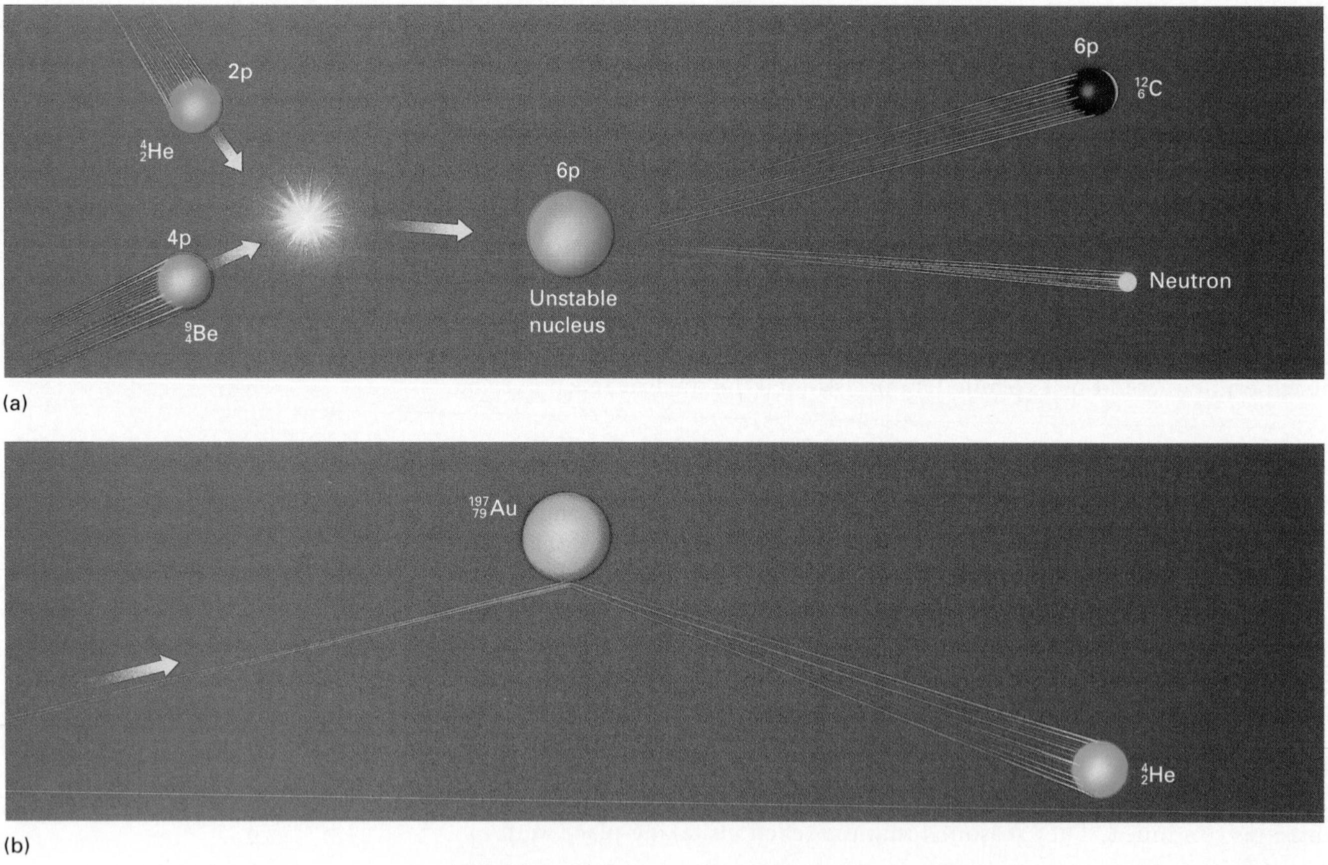

Figure 5.7 Alpha-particle bombardment. (a) The alpha particle has sufficient energy to overcome repulsions from the positively charged beryllium nucleus. A nuclear reaction occurs, producing a carbon (C) atom and a neutron. (b) The alpha particle is not energetic enough to penetrate the gold nucleus and is instead deflected.

Numerous devices have been built to accelerate nuclear particles for use as projectiles in bombardment reactions. One of the most important of these is the cyclotron, a machine that uses alternating charged regions in an intense magnetic field to accelerate small charged particles produced from a radioactive source (Figure 5.8). Hospitals and medical research facilities often have cyclotrons at their disposal for making short-lived isotopes for treatment and diagnosis.

Figure 5.8 A hospital cyclotron used to produce short-lived radioisotopes for medical treatment or diagnosis. Beams of high-energy particles from the cyclotron travel down the tubes to bombard target samples placed in the yellow boxes in the foreground. A shortage of approved facilities for disposal of the radioactive waste threatens to curtail medical use of radioisotopes. *(Adam Hart-Davis/SPL/Photo Researchers)*

5.7 Transuranium Elements

The heaviest known element before 1940 was uranium. The invention of the cyclotron and other devices to produce high-energy particles made it possible to carry out bombardment of heavier nuclei. Thus **transuranium elements,** which have atomic numbers greater than 92, were prepared.

In 1940, at the University of California, E. M. McMillan and P. H. Abelson prepared element 93, the synthetic element neptunium (Np). The experiment involved directing a stream of high-energy deuterons ($^{2}_{1}H$) onto a target of uranium-238. The initial reaction was the conversion of uranium-238 to uranium-239.

$$^{238}_{92}U + ^{2}_{1}H \longrightarrow ^{239}_{92}U + ^{1}_{1}H$$

Uranium-239 has a half-life of 23.5 min and decays spontaneously to the element neptunium by the emission of beta particles.

$$^{239}_{92}U \longrightarrow ^{239}_{93}Np + ^{0}_{-1}e$$

Neptunium is also unstable, with a half-life of 2.33 days; it converts into a second new element, plutonium (Pu).

$$^{239}_{93}Np \longrightarrow ^{239}_{94}Pu + ^{0}_{-1}e$$

Plutonium-239, like neptunium-239, is radioactive, but it has a half-life of 24,100 years. Because of the relative values of the half-lives, very little neptunium could be accumulated, but the plutonium could be obtained in larger quantities. The $^{239}_{94}Pu$ is important because it is a fissionable material and can be used to make bombs (see Section 13.3). The names of neptunium and plutonium were taken from the mythological names Neptune and Pluto in the same atomic number sequence as the order of the planets Uranus (uranium), Neptune, and Pluto out from the sun.

■ No one knows for certain how much plutonium has been made—probably in excess of several million kilograms. Most of this amount has been used to make nuclear warheads by the United States, the former Soviet Union, and the other nuclear nations including Britain, France, China, Israel, and India.

■ Plutonium is one of the most toxic substances known.

Although Neptune and Pluto are the last of the known planets in the solar system, their namesakes are not the last in the list of elements. The rush of bombardment experiments that followed the synthesis of plutonium produced additional elements: americium (Am), curium (Cm), berkelium (Bk), californium (Cf), einsteinium (Es), fermium (Fm), mendelevium (Md), nobelium (No), lawrencium (Lr), and elements 104 through 111. Up to element 101, mendelevium, all the elements can be made by bombarding the target nuclei with small particles such as $^{4}_{2}He$ or $^{1}_{0}n$. Beyond element 101, special techniques using heavier particles are required. For example, lawrencium is made by bombarding californium-252 with boron nuclei. In this reaction five neutrons are produced.

$$^{252}_{98}\mathrm{Cf} + {}^{10}_{5}\mathrm{B} \longrightarrow {}^{257}_{103}\mathrm{Lr} + 5\,{}^{1}_{0}\mathrm{n}$$

The other transuranium elements, as well as the reactions employed to produce them, are shown in Table 5.4.

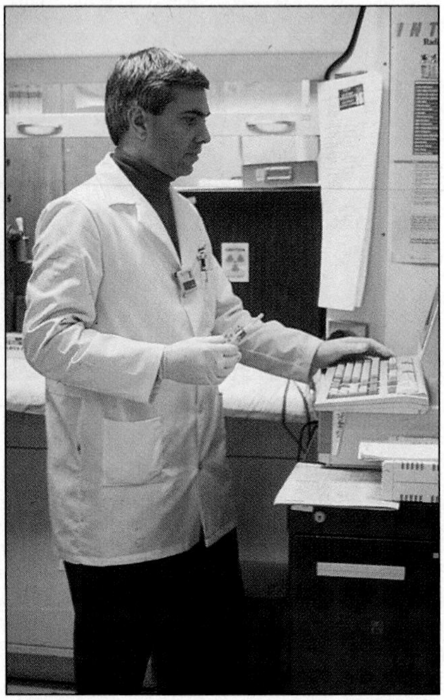

A worker in nuclear medicine. Such individuals wear badges that monitor their radiation exposure. *(Tom Pantages)*

5.8 Radiation Effects

We are bombarded constantly by nuclear radiation from a number of sources. This radiation includes cosmic rays, medical X rays, radioactive fallout from countries that do nuclear testing, and widespread, naturally occurring radioisotopes. All types of nuclear radiation can disrupt normal cell processes in living organisms, and the potential for serious radiation damage to humans is well documented. These include the biological effects of the atomic bomb explosions at Hiroshima and Nagasaki, Japan, and of the Chernobyl nuclear power plant accident in the former Soviet Union (see Section 13.4).

Three principal factors determine how hazardous a radioactive substance will be: (1) the number of disintegrations per second, or activity (measured in curies, microcuries, or becquerels), (2) the type or energy of the radiation

TABLE 5.4 ■ Nuclear Reactions Used to Produce Some Transuranium Elements

Element	Atomic Number	Reaction
Neptunium (Np)	93	$^{238}_{92}\mathrm{U} + {}^{1}_{0}\mathrm{n} \longrightarrow {}^{239}_{93}\mathrm{Np} + {}^{0}_{-1}\mathrm{e}$
Plutonium (Pu)	94	$^{238}_{92}\mathrm{U} + {}^{2}_{1}\mathrm{H} \longrightarrow {}^{238}_{93}\mathrm{Np} + 2\,{}^{1}_{0}\mathrm{n}$
		$^{238}_{93}\mathrm{Np} \longrightarrow {}^{238}_{94}\mathrm{Pu} + {}^{0}_{-1}\mathrm{e}$
Americium (Am)	95	$^{239}_{94}\mathrm{Pu} + {}^{1}_{0}\mathrm{n} \longrightarrow {}^{240}_{95}\mathrm{Am} + {}^{0}_{-1}\mathrm{e}$
Curium (Cm)	96	$^{239}_{94}\mathrm{Pu} + {}^{4}_{2}\mathrm{He} \longrightarrow {}^{242}_{96}\mathrm{Cm} + {}^{1}_{0}\mathrm{n}$
Berkelium (Bk)	97	$^{241}_{95}\mathrm{Am} + {}^{4}_{2}\mathrm{He} \longrightarrow {}^{243}_{97}\mathrm{Bk} + 2\,{}^{1}_{0}\mathrm{n}$
Californium (Cf)	98	$^{242}_{96}\mathrm{Cm} + {}^{4}_{2}\mathrm{He} \longrightarrow {}^{245}_{98}\mathrm{Cf} + {}^{1}_{0}\mathrm{n}$
Einsteinium (Es)	99	$^{238}_{92}\mathrm{U} + 15\,{}^{1}_{0}\mathrm{n} \longrightarrow {}^{253}_{99}\mathrm{Es} + 7\,{}^{0}_{-1}\mathrm{e}$
Fermium (Fm)	100	$^{238}_{92}\mathrm{U} + 17\,{}^{1}_{0}\mathrm{n} \longrightarrow {}^{255}_{100}\mathrm{Fm} + 8\,{}^{0}_{-1}\mathrm{e}$
Mendelevium (Md)	101	$^{253}_{99}\mathrm{Es} + {}^{4}_{2}\mathrm{He} \longrightarrow {}^{256}_{101}\mathrm{Md} + {}^{1}_{0}\mathrm{n}$
Nobelium (No)	102	$^{246}_{96}\mathrm{Cm} + {}^{12}_{6}\mathrm{C} \longrightarrow {}^{254}_{102}\mathrm{No} + 4\,{}^{1}_{0}\mathrm{n}$
Lawrencium (Lr)	103	$^{252}_{98}\mathrm{Cf} + {}^{10}_{5}\mathrm{B} \longrightarrow {}^{257}_{103}\mathrm{Lr} + 5\,{}^{1}_{0}\mathrm{n}$
Rutherfordium (Rf)	104	$^{242}_{94}\mathrm{Pu} + {}^{22}_{10}\mathrm{Ne} \longrightarrow {}^{260}_{104}\mathrm{Rf} + 4\,{}^{1}_{0}\mathrm{n}$
Hahnium (Ha)	105	$^{249}_{98}\mathrm{Cf} + {}^{15}_{7}\mathrm{N} \longrightarrow {}^{260}_{105}\mathrm{Ha} + 4\,{}^{1}_{0}\mathrm{n}$
Seaborgium (Sg)	106	$^{249}_{98}\mathrm{Cf} + {}^{18}_{8}\mathrm{O} \longrightarrow {}^{263}_{106}\mathrm{Sg} + 4\,{}^{1}_{0}\mathrm{n}$
Nielsbohrium (Ns)	107	$^{209}_{83}\mathrm{Bi} + {}^{54}_{24}\mathrm{Cr} \longrightarrow {}^{262}_{107}\mathrm{Ns} + {}^{1}_{0}\mathrm{n}$
Hassium (Hs)	108	$^{208}_{82}\mathrm{Pb} + {}^{58}_{26}\mathrm{Fe} \longrightarrow {}^{265}_{108}\mathrm{Hs} + {}^{1}_{0}\mathrm{n}$
Meitnerium (Mt)	109	$^{209}_{83}\mathrm{Bi} + {}^{58}_{26}\mathrm{Fe} \longrightarrow {}^{266}_{109}\mathrm{Mt} + {}^{1}_{0}\mathrm{n}$
[First made in 1994]	110	$^{62}_{28}\mathrm{Ni} + {}^{208}_{82}\mathrm{Pb} \longrightarrow {}^{269}_{110}\mathrm{A} + {}^{1}_{0}\mathrm{n}$
[First made in 1994]	111	$^{209}_{83}\mathrm{Bi} + {}^{64}_{28}\mathrm{Ni} \longrightarrow {}^{272}_{111}\mathrm{A} + {}^{1}_{0}\mathrm{n}$

THE WORLD OF CHEMISTRY

A Revision to the Periodic Table

Program 7, *The Periodic Table*

Among the most significant contributions to the modern periodic table is that made by Nobel Laureate Glenn Seaborg. Among other things, he demonstrated the importance of maintaining the courage of one's convictions.

Thanks to his insights, it is now very well established that the transuranium elements (atomic numbers greater than 92), a number of which he either discovered or helped to discover during the Manhattan Project, are members of the actinide series. Actinides are the elements following actinium and belonging in a row usually placed at the bottom of the periodic table.

Until Seaborg offered his version of the periodic table, chemists were convinced that Th, Pa, and U belonged in the main body of the table, Th under Hf, Pa under Ta, and U under W. When Seaborg proposed that Th was the beginning of the actinides and that the transuranium elements belonged as a group under the rare earths, some prominent and famous inorganic chemists, many of them Seaborg's friends, tried to discourage him from publishing his idea in the open literature. One very prominent inorganic chemist felt that Seaborg would ruin his scientific reputation. Nevertheless Seaborg, strongly convinced, persisted and, as a result, properly placed this most important class of elements where they are today. Based on Seaborg's expansion of the periodic table, it was possible to predict accurately the properties of many of the as yet undiscovered transuranium elements. Subsequent preparation in atomic accelerators of these elements proved him right, and it was fitting that he was awarded the Nobel Prize in 1951 for his outstanding work.

Glenn T. Seaborg (1912–) began his college education as a literature major, but changed his major to science in his junior year. He has been co-discoverer of 10 of the transuranium elements, which are known only from laboratory synthesis.

produced (alpha, beta, or gamma), and (3) whether the radioactive isotope is of such a chemical nature that it can be incorporated into a food chain or otherwise easily enter a living organism. A large dose of radioactive iodine-131, for example, will easily enter the body and accumulate in the thyroid gland.

Radiation is measured using several units. The **röntgen** (R) is used to measure the intensity of X rays or gamma rays. A single dental X ray represents about 1 R. The **rem** (meaning *r*öntgen *e*quivalent *m*an) is used to quantify the biological effects of radiation in general. One rem is *a dose of any radiation that has the same effect of 1 R.* Since one rem is a large amount of radiation, the **millirem** or **mrem** is commonly used. One mrem = 10^{-3} rem.

■ One röntgen is the quantity of X-ray or gamma-ray radiation delivered to 0.001293 g of air, such that the ions produced in the air carry 3.34×10^{-10} coulomb of charge.

Biological tissue is easily harmed by radiation. A flow of high-energy particles may cause destruction of a vital enzyme, hormone, or chromosome needed for the life of a cell. In general, those cells that divide most rapidly are most easily harmed by radiation (Figure 5.9), for example, cells in the bone marrow, hair follicles, and gastrointestinal tract, as well as white cells in blood.

Whole-body radiation effects are divided into **somatic effects,** which are confined to the population exposed, and **genetic effects,** which are passed on to subsequent generations.

Often somatic effects are delayed. Perhaps the best studied of the delayed effects are the incidences of cancer from exposure to radiation. It has been

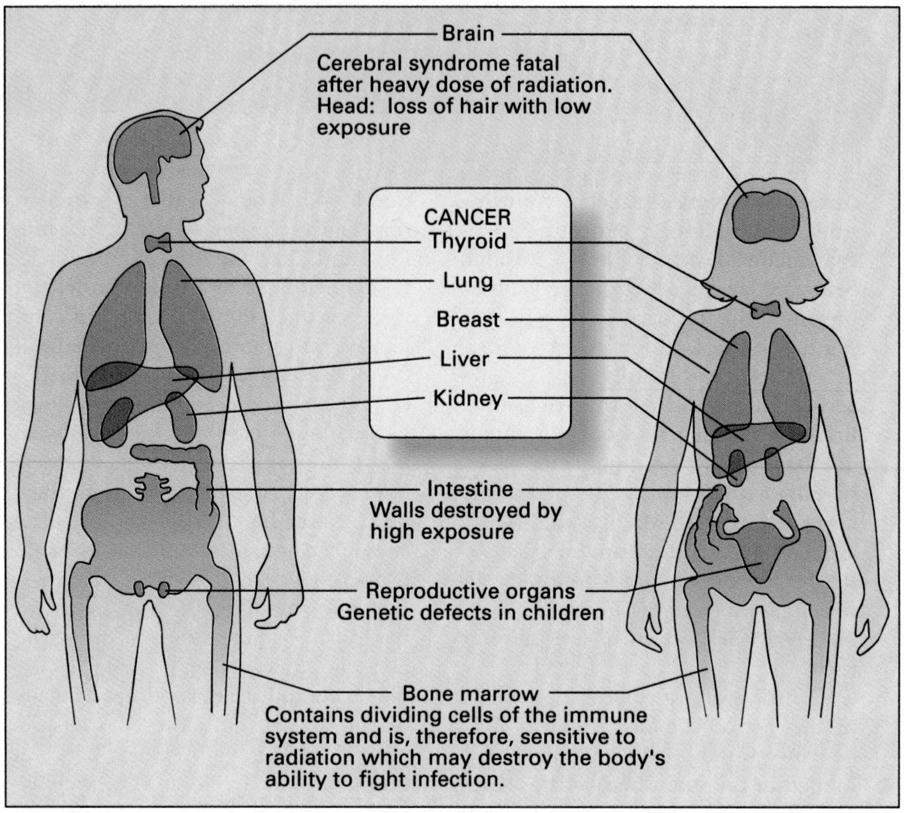

Figure 5.9 How radiation affects various regions of the body.

estimated that 11% of all leukemia cases and about 10% of all forms of cancer are attributable to background radiation. Certainly, individuals who are exposed to higher-than-normal levels of radiation over considerable time increase their chances of cancer. The alteration of normal cells to cancerous cells caused by radiation is undoubtedly a series of changes, since in almost all cases the onset of cancer lags behind the exposure to radiation by an induction period of 5 to 20 years.

The genetic effects of radiation are the result of radiation damage to the germ cells of the testes (sperm) or the ovaries (egg cells). Ionization caused by radiation passing through a germ cell may break a DNA strand or cause it to be altered in some other way. When this damaged DNA is replicated (i.e., when the DNA structure is copied during cell division; see Section 15.8), the result may be the transmission of a new message to successive generations, a **mutation.** Every type of laboratory animal on which radiation damage experiments have been performed has responded with an increased incidence of mutation. Therefore, the necessity of protecting people of childbearing age from radiation should be apparent. Theoretically at least, one photon or one high-energy particle can ionize a chromosomal DNA structure and produce a genetic effect that will be carried to the next generation.

SCIENCE AND SOCIETY

Higgs Will Have to Wait

One of the largest scientific experiments ever planned, called the Superconducting Super Collider (SSC), was begun in 1984 with plans to build a gigantic underground ring of superconducting magnets to contain counter-circulating beams of fast-moving protons. In 1988 Waxachachie, Texas, just south of Dallas, was chosen for the site, and work was begun on the SSC, which was to be 53 miles in diameter. The final cost was initially projected to be just over $4.5 billion. After five years of construction, and with final construction costs projected to run in excess of $11 billion, Congress, in October 1993, defeated the project by a vote of 282 to 143. Citing reasons such as failure to contain costs, high projected operating expenses (more than $1 billion per year), and a lack of clearcut objectives, Congress was in no mood to continue the massive project.

When members of Congress and the public were told what the SSC was going to do, it sounded simple enough. On command, the proton beams, traveling at near the speed of light, were to collide with each other (hence, the "collider" part of the name) and produce showers of subatomic particles that would ultimately lead to an understanding of how the universe began, among other benefits. Some medical uses of the particle beams were also planned, but critics of the SSC claimed Congress was initially oversold on the medical uses.

The SSC was to allow scientists to discover evidence for even more particles, including the Higgs particle, which is sought to explain mass. In the end, Congress, believing that it expressed the interests of the taxpayers, was unconvinced that the discoveries were worth the costs and killed the SSC. Scientists have not given up, however. They now have plans for the Next Linear Collider (NLC), capable of giving the particles much more energy.

5.9 Radon and Other Sources of Background Radiation

You are at risk whenever you are exposed to ionizing radiation. Some of these risks can be minimized by exercising appropriate safety measures, by worldwide agreements prohibiting nuclear weapons testing, and so on. Figure 5.10 illustrates, however, that the bulk of our exposure to radiation is from natural sources. When these are coupled with desirable exposures from medical applications, scarcely 3% of the exposures are truly avoidable to society as a whole. To illustrate how unavoidable most exposures to radiation are,

■ Even the altitude above sea level where you live is related to your annual dose of background radiation—the higher the altitude, the greater the cosmic radiation.

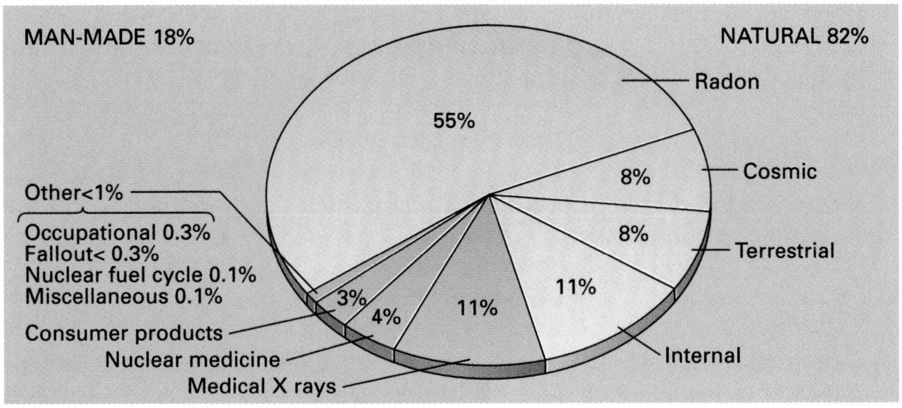

Figure 5.10 Average radiation exposure in the United States.

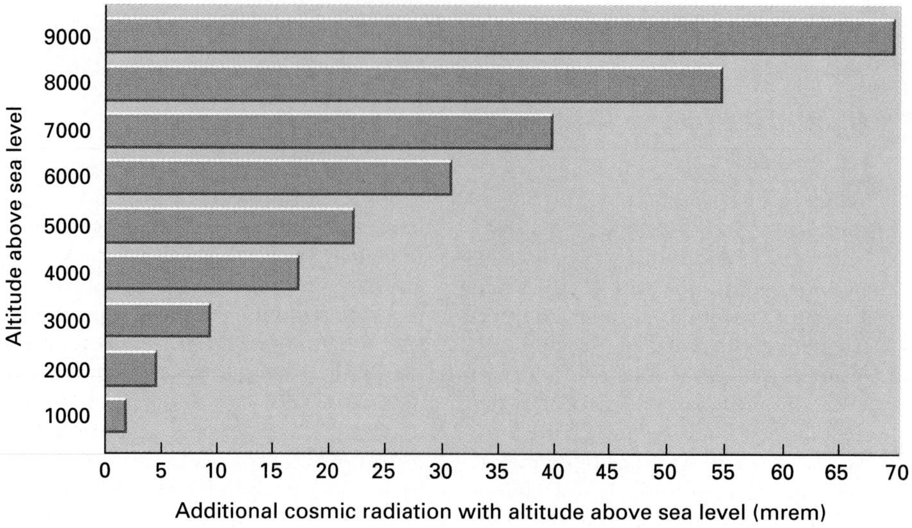

consider the element potassium, an element essential to human life that helps regulate water balance in our bodies and is involved in nerve transmissions. A 60-kg (132-lb) person has about 200 g of potassium. It turns out that just over 0.01% of all potassium atoms are potassium-40 atoms, which are radioactive, with a half-life of 1.25×10^9 years. This means the 60-kg person has roughly 20 mg of radioactive potassium, which disintegrates at a rate determined by the half-life of that isotope. These disintegrations contribute to the background radiation we all receive. In addition, the beta particle from the decay of a potassium-40 disintegration in your neighbor might pass into you to contribute to your background radiation! Normal background radiation from all sources is 2 or 3 disintegrations per second.

Another element that contributes to our background radiation, and hence to our risks of radiation-caused damage, is radon. Radon (Rn) is the heaviest member of the noble gas family of elements (He, Ne, Ar, Kr, Xe, Rn). Radon-222, the most common isotope of radon, is radioactive, with a half-life of 3.82 days. It is a product of the uranium decay series (Figure 5.7) and results from the alpha decay of radium-226:

$$^{226}_{88}\text{Ra} \longrightarrow {}^{222}_{86}\text{Rn} + {}^{4}_{2}\text{He}$$

When radon decays, it produces alpha particles and another short-lived radioisotope, polonium-218:

$$^{222}_{86}\text{Rn} \longrightarrow {}^{218}_{84}\text{Po} + {}^{4}_{2}\text{He}$$

Polonium-218 (half-life 3.05 min) also decays, producing an alpha particle and lead-214 (half-life 26.8 min):

$$^{218}_{84}\text{Po} \longrightarrow {}^{214}_{82}\text{Pb} + {}^{4}_{2}\text{He}$$

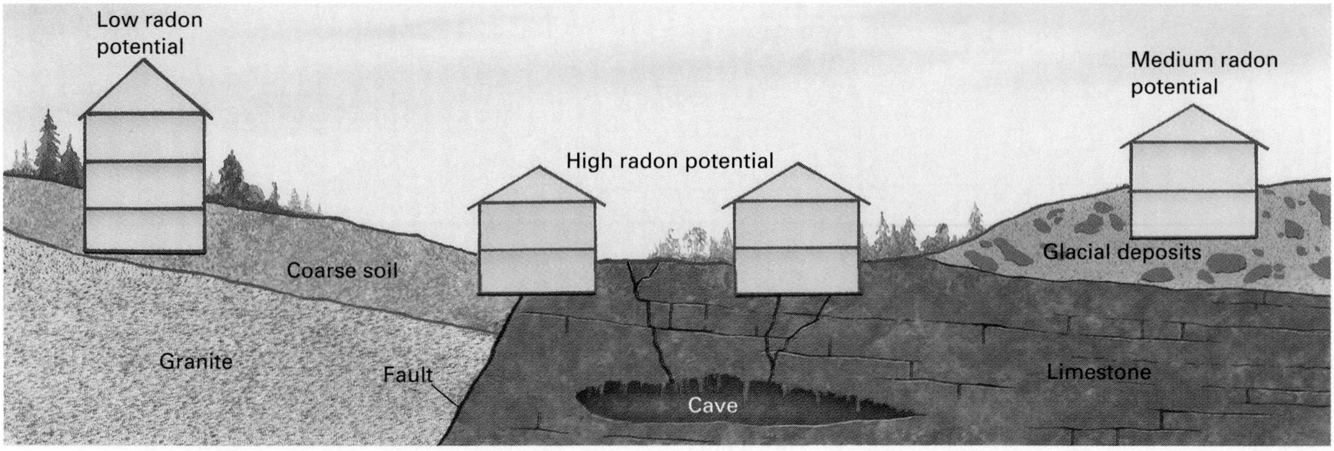

Relation between geology and radon potential in homes.

Radon exists as a gas, and all rocks contain *some* uranium, although in most the amount is small (one to three parts per million). Therefore, radon is constantly being formed in amounts that vary with the type of rock or soil. How much of this radon escapes into the outside air or a building on the surface depends on the porosity and moisture content of the soil and how finely divided it is (escape is easier from a small particle than from a large one). The overall geology of the site is also important.

Because radon is chemically unreactive, radon atoms in the air we breathe are inhaled and exhaled without any chemical change, although some may dissolve in lung fluids. If a radon atom happens to decay within the lungs, however, the nongaseous and radioactive "radon daughters" can remain inside the lungs, where they will continue to decay. Those up to lead-210 (see Table 5.3) have short half-lives and include several alpha emitters. Because alpha particles can travel up to 0.7 mm, the approximate thickness of the epithelial cells of the lung, they can damage delicate lung tissue and create a higher-than-normal risk of lung cancer.

■ "Daughters" is a term used to describe radioactive decay products. Radon daughters come from the decay of radon.

Miners in deep mines are exposed to far more than average radon levels, and as early as 1950 government agencies began monitoring radon exposures and incidences of lung cancer. Today, it is well known that radon exposure increases one's chance of developing lung cancer. If you smoke, the chances are even greater, since there seems to be a synergistic effect between smoking and radon levels in causing lung cancer. As always, it is wise to keep in mind that all such data are based on statistical studies and risk assessment models. A number of studies of radon risk are still underway. Also as always, there is disagreement among scientists and among legislators about the level of exposure at which the risk becomes serious enough to warrant remedial action.

When buildings are built over soil containing the heavy radioactive elements that decay to radium-226, some of the radon gas produced seeps through minute fissures in the soil or rock and migrates into the air in these buildings. The building literally funnels the radon through it by "chimney effects" from such sources as clothes dryers, fireplaces, furnaces, and warm air rising and leaving through openings near the roof. There are various remedies

■ DNA is the large biochemical molecule that controls heredity and the production of cells in living organisms. The role of DNA in biochemistry is discussed in Section 15.8.

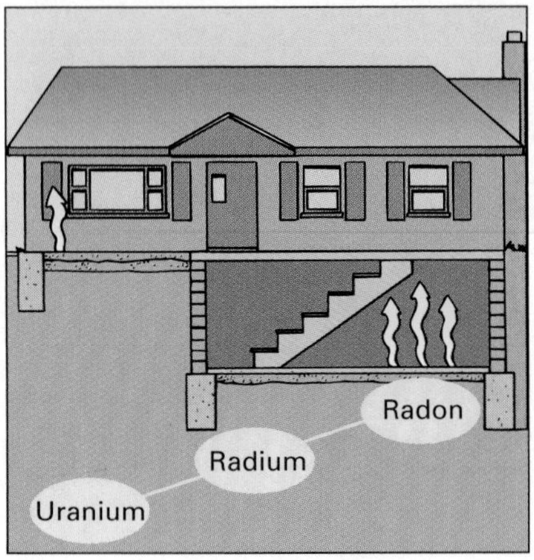

Common route of radon entry to homes—seepage through the foundation.

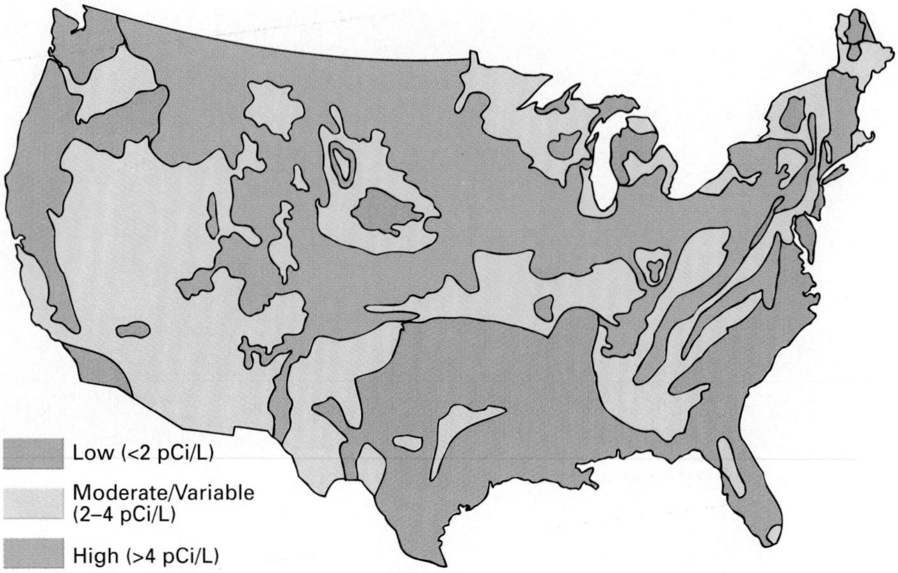

Figure 5.11 Radon potential according to geologic areas. This map was developed by the U.S. Geological Survey by evaluating indoor radon measurements, geology, aerial measurements of radioactivity, soil types, and foundation construction. The map is a statistical evaluation of *potential* indoor radon levels and reflects available data. Within any area on the map, there will be variations, and the EPA recommends testing regardless of geographic location.

for buildings with a high concentration of radon in their air. If a building is built on a foundation over a crushed stone ballast, holes are drilled into the ballast. Then a suction pump is attached to produce a negative pressure. For buildings not built on concrete foundations, increasing ventilation both in the basement and inside the building is effective in removing radon. Of course, this means energy losses, but these might be acceptable when the alternatives of the radiation damage from the radon daughters are considered.

It has been estimated that perhaps eight million homes in the United States are affected by radon contamination at levels higher than 4 pCi per liter of air (4 pCi/L), which is the "action level" the U.S. Environmental Protection Agency has set. Radon has been detected in homes in almost every state. Some areas have a higher potential for having homes with high radon concentrations than others (see Figure 5.11).

This EPA action level is important for two reasons. First, levels of radon higher than 4 pCi/L should be reduced because levels above this are judged to lead to unacceptable risks of lung cancer. Second, it is difficult to lower the level of radon in most contaminated homes below 4 pCi/L. This last point is an acceptance of the fact that radiation exposure will always be with us.

■ If the radon radiation level is above 4 pCi/L, EPA requires that some form of remediation action be taken.

■ 4 pCi/L is a little less than 1.5 disintegrations every 10 s.

■ Outdoor radon concentrations are approximately 0.1 to 0.15 pCi/L worldwide.

■ SELF-TEST 5B

1. In the following reaction, what is the compound nucleus?.

$$^{7}_{3}\text{Li} + ^{1}_{1}\text{H} \rightarrow [\_\_\_\_\_\_\_\_\_\_] \rightarrow ^{7}_{4}\text{Be} + ^{1}_{0}\text{n}$$

2. When unstable uranium-239 goes through beta decay the product nucleus is ______________.
3. Somatic effects of radiation are immediate and rarely delayed. (a) True, (b) False.
4. The first transuranium element is ______________.
5. Normal background radiation from all sources produces two to three disintegrations per second. (a) True, (b) False.
6. The term "rem" stands for ______________ ______________ ______________.
7. Bombardment experiments to make elements 104 through 111 require using projectiles bigger than alpha particles or neutrons. (a) True, (b) False.
8. Radiation damage to germ cells (sperm or egg cells) may alter DNA strands and produce an error in the message carried by the DNA. (a) True, (b) False.
9. Radon gas comes from the decay of what naturally occurring radioactive element?
10. Cosmic rays are more penetrating than alpha particles and are more hazardous. (a) True, (b) False.
11. The EPA action level for radon gas found in homes is ______________.

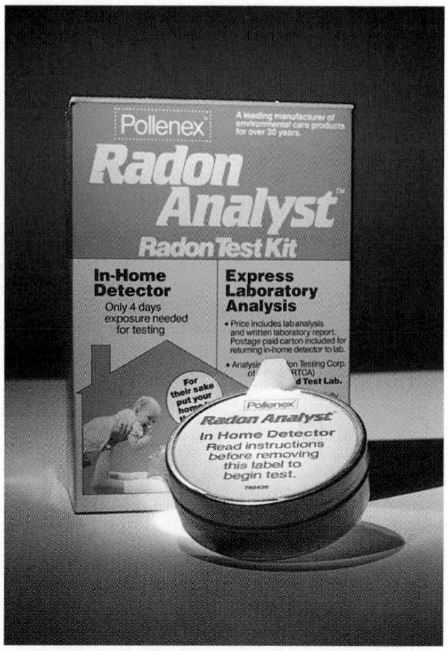

A commercially available kit for testing for radon in the home. *(C.D. Winters)*

5.10 Useful Applications of Radioactivity

Food Irradiation

The damaging aspects of nuclear radiation must always be kept in mind, especially when the possibilities of accidental or unintended exposures are great. However, the radiation from radioisotopes can be put to beneficial use. Consider the important application of killing harmful pests that would destroy our food during storage. In some parts of the world, stored-food spoilage may claim up to 50% of the food crop. In the United States, refrigeration, canning, and chemical additives lower this figure considerably. Still, there are problems with food spoilage. Food protection costs amount to a sizable fraction of the final cost of food.

Foods may be irradiated to retard the growth of organisms such as bacteria, molds, and yeasts. Food irradiation for this purpose using gamma rays from sources such as cobalt-60 or cesium-137 is common in European countries, Canada, Mexico, and the United States. Such irradiation prolongs shelf life under refrigeration in much the same way that heat pasteurization protects milk. Normally, chicken has a three-day refrigerated shelf life. After irradiation, chicken may have a three-week refrigerated shelf life.

Foods irradiated at sufficient levels will keep indefinitely when sealed in plastic or aluminum-foil packages (Figure 5.12). In 1994 NASA was granted permission to use irradiated frozen beefsteak on space missions. The U.S. Food and Drug Administration (FDA) granted NASA's request because there were no known problems associated with beef irradiated this way and because there would be limited consumption of the product by the crew over a short period of time.

Figure 5.12 Cartons of food ready to undergo irradiation. *(Hank Morgan/Rainbow)*

DISCOVERY EXPERIMENT

Your Annual Dose

The Committee on Biological Effects of Ionizing Radiation of the National Academy of Sciences issued a report in 1980 that contained a survey for an individual to evaluate his or her exposure to ionizing radiation. The following table is adapted from this report. By adding up your exposure, you can compare your annual dose to the United States annual average dose of 180 to 200 mrem.

Adapted from A. R. Hinrichs: *Energy,* pp. 335–336. Philadelphia, Saunders College Publishing, 1992.

	Common Sources of Radiation	Your Annual Dose (mrem)
Where You Live	**Location:** Cosmic radiation at sea level	26
	For your elevation (in feet), add this number of mrem	______
	Elevation *mrem* — 1000: 2; 2000: 5; 3000: 9; 4000: 15; 5000: 21; 6000: 29; 7000: 40; 8000: 53; 9000: 70	
	Ground: U.S. average	26
	House construction: For stone, concrete, or masonry building, add 7	______
What You Eat, Drink, and Breathe	**Food, water, air:** U.S. Average	24
	Weapons test fallout	4
How You Live	**X ray and radiopharmaceutical diagnosis:**	
	Number of chest X rays ______ × 10	______
	Number of lower gastrointestinal tract X rays ______ × 500	______
	Number of radiopharmaceutical examinations ______ × 300	______
	(Average dose to total U.S. population = 92 mrem)	
	Jet plane travel: For each 2500 miles add 1 mrem	______
	TV viewing: Number of hours per day ______ × 0.15	______
How Close You Live to a Nuclear Plant	**At site boundary:** average number of hours per day ______ × 0.2	______
	One mile away: average number of hours per day ______ × 0.02	______
	Five miles away: average number of hours per day ______ × 0.002	______
	More than 5 miles away: none	______
	Note: Maximum allowable dose determined by "as low as reasonably achievable" (ALARA) criteria established by the U.S. Nuclear Regulatory Commission. Experience shows that your actual dose is substantially less than these limits.	
	Your total annual dose in mrem	______

Elevation	*mrem*	*Elevation*	*mrem*	*Elevation*	*mrem*
1000	2	4000	15	7000	40
2000	5	5000	21	8000	53
3000	9	6000	29	9000	70

Compare your annual dose to the U.S. annual average of 180 mrem.
One mrem per year is equal to increasing your diet by 4%, or taking a 5-day vacation in the Sierra Nevada (CA) mountains.

Based on the "BEIR Report III"—National Academy of Sciences, Committee on Biological Effects of Ionizing Radiation: "The Effects on Populations of Exposure to Low Levels of Ionizing Radiation," National Academy of Sciences, Washington, DC, 1980.

TABLE 5.5 ■ Irradiated Foodstuffs Approved by U.S. FDA

Food	Purpose
Uncooked pork	Control of *Trichinella spiralis*
All fresh foods	Growth and maturation inhibition
All foods	Disinfestation of anthropod pests
Dry enzyme preparations	Microbial disinfection
Dried herbs and spices	Microbial disinfection
Uncooked poultry	Control of food-borne pathogens

Source: United States Food and Drug Administration regulations, April 1993

International symbol for use on irradiated food.

At present, the FDA has approved irradiation of many classes of foods (Table 5.5). It is important to note that in no case is there any chance of irradiated food becoming radioactive. The ongoing debate about the safety of irradiated foods revolves around the possibly harmful nature of would-be "radiolytic products"—products of chemical changes in food caused by the high energy of the radiation. For example, could irradiation of food produce a chemical capable of causing genetic damage? To date, no evidence has been found for harmful radiolytic products, but further animal feeding studies are under way. Meanwhile, based on extensive review of current data, the FDA-approved radiation limits have been conservatively set at relatively low levels.

Recent findings regarding the potentially harmful health effects of several common agricultural fumigants have indicated that irradiation of fruits and vegetables could be an effective alternative to some chemical fumigants. The agricultural products may be picked, packed, and readied for shipment. After that, the entire shipping container can be passed through a building containing a strong source of radiation. This type of sterilization offers greater worker safety because it lessens the chances of exposure to harmful chemicals, and it protects the environment because it lessens the chances of contamination of water supplies with these toxic chemicals. Furthermore, irradiation of poultry at currently approved levels could greatly reduce the periodic outbreaks of food poisoning due to the *Salmonella* organisms whose presence in poultry is otherwise very difficult to eliminate. Widespread poultry irradiation could be accomplished with current technology, but awaits evidence of public acceptance.

Materials Testing

Cobalt-60, a gamma-ray emitter, has also proved quite useful in the testing of metal castings in industry. Contained within an aluminum thimble, the cobalt radioisotope is placed inside a casting, and a piece of photographic film is positioned on the outside of the object. The gamma rays penetrate the metal part and make observable any structural flaws in the metal by exposing the photographic film. The intensity of the gamma rays passing through the flawed portion of the casting is different from the intensity passing through the rest of the metal. After it is developed, the photographic film reveals the presence of any flaws. Aviation safety has been increased by the use of radiation detection of flaws and weaknesses in structural members of aircraft. Radiation is also frequently used to monitor the thickness of plastic films and plastic coatings.

Radioactive Tracers

Because radioisotopes act chemically in a manner almost identical to that of the nonradioactive isotopes of the same element, and because the location of a radioisotope can always be found by a radiation detector, radioisotopes can be used as **tracers.** For example, plants' roots are known to take up phosphorus (P) from the soil. By supplying plants the radioactive phosphorus-32 isotope, a beta emitter, the uptake of phosphorus can be followed and the speed of the uptake measured under various conditions. By using this technique, plant biologists can search for hybrid strains of plants that absorb phosphorus quickly and work to develop faster-maturing crops, better yields per acre, and more food or fiber at less expense.

By tagging a pesticide with a short-lived radioisotope, applying it to a test field, and following its movement, information can be gathered about its tendency to accumulate in the soil, to be taken up by the plant, and to accumulate in runoff surface water. This is done with a high degree of accuracy by counting the disintegrations of the radioactive tracer.

Medical Imaging

Radioisotopes are used in **nuclear medicine** in two distinctly different ways, diagnosis and therapy. In the diagnosis of internal disorders, physicians need information on the locations of the disorders. An appropriate radioisotope is introduced into the patient's body, either alone or combined with some other chemical, and it accumulates at the site of the disorder. There, acting like a homing device, the radioisotope disintegrates and emits its characteristic radiation, which can be detected. Modern medical diagnostic instruments not only determine where the radioisotope is located in the patient's body but also construct an image of the area.

Four of the most common diagnostic radioisotopes are listed in Table 5.6. Each produces gamma radiation, which in low doses is less harmful to the tissue than ionizing radiations such as beta or alpha rays. By the use of special carriers, these radioisotopes can be made to accumulate in specific areas of the body. For example, the pyrophosphate ion, $P_4O_7^{4-}$, a simple polyatomic ion, can bond to the technetium-99m radioisotope (the *m* denotes "metastable," meaning the isotope decays to a more stable isotope with the same mass number), and together they accumulate in the skeletal structure where abnormal

■ The half-life of ${}^{99m}_{43}Tc$, 6 hours, makes it an ideal radioisotope for medical purposes. The more stable isotope ${}^{99}_{43}Tc$ has a half-life of 2.12×10^5 years. This means it has such a low activity that it will be eliminated from the patient's body before many disintegrations can occur.

TABLE 5.6 ■ Diagnostic Radioisotopes

Radioisotope	Name	Half-Life (Hours)	Uses
^{99m}Tc*	Technetium-99m	6	As TcO_4^- to the thyroid, brain, and kidneys
^{201}Tl	Thallium-201	21.5	To the heart
^{123}I	Iodine-123	13.2	To the thyroid
^{67}Ga	Gallium-67	78.3	To various tumors and abscesses

* The technetium-99m isotope is the one most commonly used for diagnostic purposes. The *m* stands for "metastable," a term explained in the text.

bone metabolism is taking place. Such investigations often pinpoint bone tumors. When the technetium-99m radioisotope decays it loses energy by emitting a gamma ray:

$$^{99m}_{43}\text{Tc} \longrightarrow {}^{99}_{43}\text{Tc} + \underset{\text{Gamma ray}}{{}^{0}_{0}\gamma}$$

The imaging method is based on the emission of gamma rays from the target organ. The gamma rays strike photosensitive sodium iodide in a gamma-ray camera. The signal is then processed by a computer and fed to a video display for construction of the image on the screen (Figure 5.13).

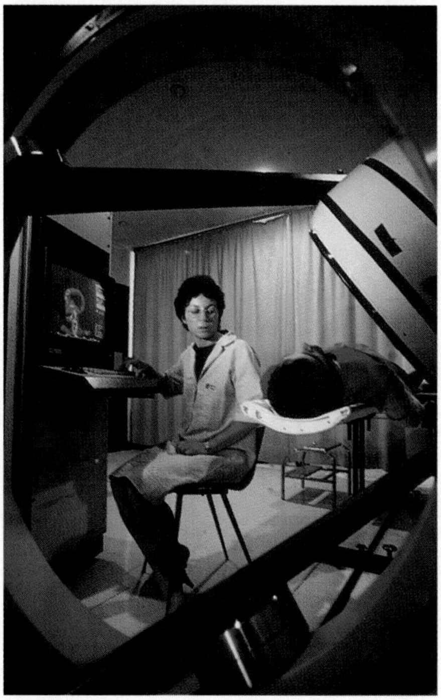

(a)

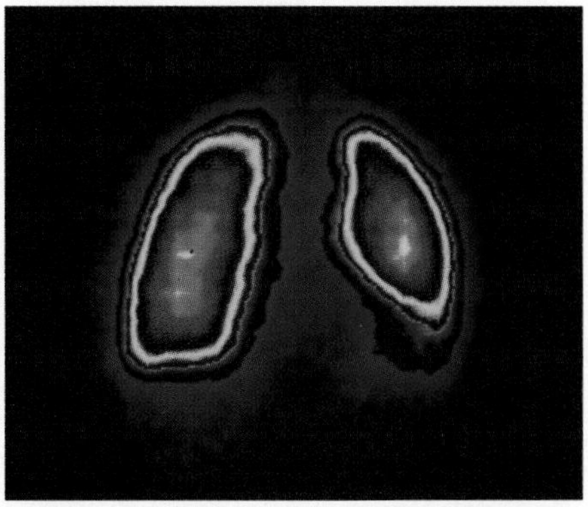

(b)

Figure 5.13 Radioisotopes in medical diagnosis. (a) The technician is monitoring the image of a patient with bone cancer. A gamma-emitting radioisotope concentrates more strongly in cancerous bone, showing as red in the false-color image. (b) A gamma-ray scan of healthy lungs, produced by technicium-99m. *(a, Philippe Plailly/SPL/Photo Researchers); b, Jean-Perrin/CNRI/Science Photo Library/Photo Researchers)*

SCIENCE AND SOCIETY

Naming New Elements Isn't Easy

Transuranium elements, those produced by bombarding heavy elements with atoms of lighter elements, are generally named by allowing the discovery team to pick a name that they think is appropriate. That's how elements 93 through 100 were named—the research group at the Lawrence Berkeley Laboratory in California chose names to honor America, the state of California, the city of Berkeley, and eminent deceased scientists who had made significant contributions to the understanding of the atom. After element 100 was discovered and named (after Enrico Fermi, who built the first nuclear reactor), an international group of scientists called the Transfermium Working Group (TWG) was established to deal with naming problems that were arising as different laboratories were discovering new elements. In the case of elements 104 and 105, both the Berkeley laboratory and one in Russia, at Dubna, claimed first discovery.

In 1992 the TWG agreed that both laboratories would share credit for the discoveries of these elements, but did not agree on the names for them. The Russians have wanted to call element 104 "kurchatovium," after Igor Kurchatov, who was instrumental in making the first Soviet nuclear bomb. For 105, the Russians proposed the name "nielsbohrium," after Niels Bohr. Since 1970, however, the Americans have been calling element 104 "rutherfordium" and element 105 "hahnium," after Ernest Rutherford and Otto Hahn, a German radiochemist. These names have become commonly used in most western nations. A similar controversy also exists concerning element 102, which the Swedish laboratory that claimed discovery named "nobelium," while the Russians have called it "joliotium," after Frédéric Joliot-Curie, a French physicist.

In the summer of 1994, the Berkeley laboratory proposed to name element 106 "seaborgium," after Glen Seaborg. In August, 1994, before the name seaborgium (Sg) could become commonly accepted, a group of scientists at the International Union of Pure and Applied Chemistry (IUPAC), an international governing body that claims jurisdiction in such matters, decided to *rename* many of the transfermium elements, addressing many of the grievances and controversies that have existed. A careful look at the table below indicates that the IUPAC group was attempting to please all of the groups involved, with the possible exception of the Americans who named element 106. In their defense for not using Seaborg's name, the IUPAC stated that it was unwise and unprecedented to name an element for a living scientist.

Which names will end up being used? Probably what is going to happen is that the names proposed by the discovering groups will be commonly used. This will be especially true in laboratories, textbooks (including this one), and scientific writings in their respective countries. In trying to please everyone, the IUPAC naming committee actually pleased almost no one.

Atomic Number	Name Proposed by Discovering Group	IUPAC Name
102	Nobelium (Swedish) Joliotium (Russian)	Nobelium
103	Lawrencium (American)	Lawrencium
104	Rutherfordium (American) Kurchatovium (Russian)	Dubnium
105	Hahnium (American) Nielsbohrium (Russian)	Joliotium
106	Seaborgium (American)	Rutherfordium
107	Nielsbohrium (German)	Bohrium
108	Hassium (German)	Hahnium
109	Meitnerium (German)	Meitnerium

Note: Hassium is named after a state in Germany and Meitnerium is named after Lise Meitner, who, with Otto Hahn, co-discovered nuclear fission.

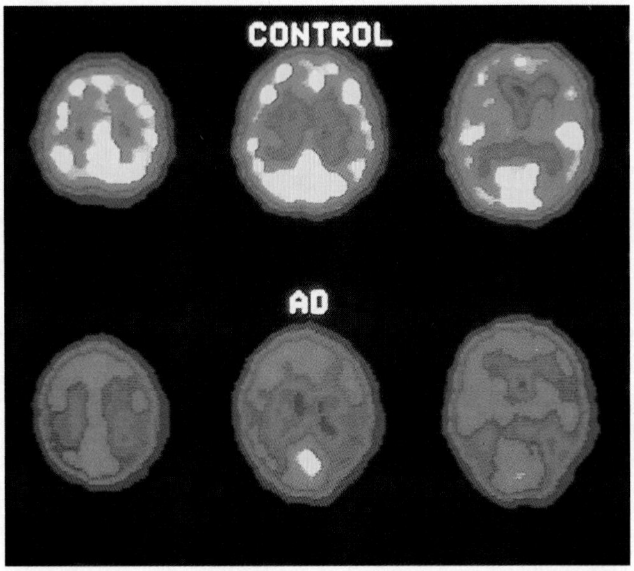

Figure 5.14 Study of Alzheimer's disease with PET scans. The concentration in the brain of radioisotope-labeled glucose in a healthy individual (top) and a patient with Alzheimer's disease (bottom). *(Dr. Mony DeLeon/Peter Arnold, Inc.)*

Positron emission tomography (PET) is a form of nuclear imaging that uses positron-emitters, such as carbon-11, fluorine-18, nitrogen-13, or oxygen-15. All of these radioisotopes are neutron deficient and have short half-lives and therefore must be prepared using a cyclotron immediately before use.

The positrons travel less than a few millimeters before they encounter an electron and undergo antimatter-matter annihilation. The annihilation event produces two gamma rays,

$$\underset{\text{Positron}}{{}^{0}_{+1}\text{e}} + \underset{\text{Electron}}{{}^{0}_{-1}\text{e}} \longrightarrow \underset{\text{Gamma ray}}{2\,{}^{0}_{0}\gamma}$$

which radiate in opposite directions and are detected by two detectors located 180° apart in the PET scanner. By detecting several million annihilation gamma rays within a circular slice about the subject over approximately 10 minutes, the region of tissue containing the radioisotope can be imaged using computer signal-averaging techniques (Figure 5.14).

■ Mass is converted into energy as the particles cease to exist.

■ SELF-TEST 5C

1. Name three uses of radioactive isotopes.
2. Which radioisotope is used for examining metal castings? (a) ^{60}Co, (b) ^{32}P, (c) ^{67}Ga, (d) ^{99m}Tc.
3. Name a radioisotope that might be useful as a tracer in agricultural research.
4. The process of concentrating a radioisotope at a particular site of the body in order to locate and measure the extent of a disorder is called ______________.

5. In the symbol for the radioisotope technetium-99m, the *m* stands for (a) middle, (b) mathematical, (c) metastable.
6. With its half-life of approximately 6 hours, how much technetium-99m would remain 18 hours after injection into a patient?
   (a) one eighth of the original dose
   (b) one half of the original dose
   (c) one sixth of the original dose
   (d) one fourth of the original dose
7. If two radioisotopes were available for diagnosis, worked equally well, and each decayed by giving off gamma rays, but one had a half-life of 13 hours and the other had a half-life of 6 hours, which one would you recommend?
8. How does living in a masonry or stone building alter a person's annual dose of nuclear radiation?
9. What is the U.S. average annual dose of nuclear radiation in mrem? (a) 180–200 mrem, (b) 20–50 mrem, (c) 2–12 mrem.

MATCHING SET

____ 1. Somatic effect	a. Intake stops when organism dies
____ 2. 1 microcurie	b. Radiation effect on general population
____ 3. ${}^{14}_{6}C$	c. First suggested by Rutherford
____ 4. ${}^{0}_{+1}e$	d. Discovered the positron particle
____ 5. Genetic effect	e. Radiation damage to DNA
____ 6. ${}^{238}_{92}U$-dating	f. Tissue easily damaged by radiation
____ 7. Element 111	g. Radioisotope used in medical diagnosis
____ 8. Half-life	h. Time required for half of the nuclei to disintegrate
____ 9. ${}^{218}_{84}Po$	i. 37,000 disintegrations per second
____ 10. Cyclotron	j. Used to detect defects in metal castings
____ 11. Joliots	k. Opposite of the electron
____ 12. 4 pCi/L	l. Latest synthetic element
____ 13. Bone marrow	m. A machine used to make radioisotopes
____ 14. ${}^{99m}Tc$	n. EPA action level for radon
____ 15. ${}^{60}Co$	o. Alpha decay product of ${}^{222}_{86}Rn$
	p. Half-life of 0.5 seconds

QUESTIONS FOR REVIEW AND THOUGHT

1. Which of the following three types of radiation is the most penetrating? (a) Alpha rays, (b) Beta rays, (c) Gamma rays.
2. Give the type and approximate amount of material required to stop each of the following:
   (a) Alpha particles (b) Beta particles
   (c) Gamma rays
3. Which radioisotope sample is more hazardous, one gram of ${}^{238}U$ with a half-life of 4.5 billion years or one gram of ${}^{222}_{86}Rn$ with a half-life of 3. days? Explain your choice.
4. Tell how genetic radiation effects differ from somatic radiation effects.
5. What is the nuclear reaction for the collision of a beta particle with a positron?
6. What makes Rn-222 a health hazard? What health problem results from Rn-222 exposure?
7. Why must the sum of the mass numbers of the products of a nuclear reaction always equal the sum of the mass numbers of the reactants?
8. If the alpha particles have the same low kinetic energy,

which nuclear reaction is more likely to occur? Explain your reasoning. (a) $^{4}_{2}He + ^{27}_{13}Al \longrightarrow ^{30}_{15}P + ^{1}_{0}n$, (b) $^{4}_{2}He + ^{239}_{94}Pu \longrightarrow ^{242}_{96}Cm + ^{1}_{0}n$.

9. What kinds of projectile particles were used for the transmutations to form elements up to atomic number 101?
10. Nobel Laureate Glenn Seaborg proposed the present-day form of the periodic table. What change did he propose for the positions of the rows?
11. What are transuranium elements? How do they differ from the other elements in the periodic table?
12. What is meant by the term "uranium series"?
13. What is meant by the term "actinide series"?
14. What are the two major applications of radioisotopes in nuclear medicine?
15. What is the reason for irradiating food with gamma rays? What is the effect of the radiation?
16. What are the electrical charges and relative masses for the following?
    (a) Beta particles (b) Alpha particles
    (c) Gamma rays (d) Positrons
    (e) Neutrons
17. Tell how cyclotrons are used to produce artificial radioisotopes.
18. Tell why it is necessary to use high energy accelerators to produce the transuranium radioisotopes from smaller nuclei.
19. Describe what effect gamma rays from cobalt-60 have on cancer cells and why this makes them useful in radiation therapy for cancer.
20. Why does airplane travel increase a person's annual dose of ionizing radiation?
21. Which source of ionizing radiation contributes more to a person's annual dose, weapons test fallout or natural radiation in food, water, and air?
22. Would you expect a salesperson working 200 days a year in a television electronics store with 10 televisions constantly operating to have an annual dose higher than the average of 180 mrem? Explain your answer.
23. *E. coli* contaminated food in fast food establishments have caused poisonings and deaths in recent years. Contaminated meat was often the source. What effect would food irradiation have on the *E. coli* residues on meat?
24. Hahnium-260 (atomic number 105) was produced by the transmutation of californium-249 (atomic number 98) by bombardment with high energy $^{15}_{7}N$ projectiles. What other particles resulted from the collisions?
25. What was the significance of exposed photographic films found near samples of uranium during the first experiments with radioactive materials?
26. What was the contribution made by each of the following scientists in the study of radioactivity?
    (a) Becquerel (b) Marie Curie
    (c) Rutherford
27. What kinds of nuclear chemistry observations and experiments occurred around each of the following dates?
    (a) 1899 (b) 1934
    (c) 1945 (d) 1896
    (e) 1946
28. Describe how the penetrating power of beta rays compares with the penetrating power of gamma rays.
29. What effect does the emission of a beta particle have on the mass number of a nucleus?

PROBLEMS

1. Balance the following nuclear equations. Give symbols, nuclear charges, and mass numbers.
   (a) $^{64}_{29}Cu \rightarrow$ ________ $+ ^{0}_{-1}e$
   (b) $^{69}_{30}Zn \rightarrow$ ________ $+ ^{69}_{31}Ga$
   (c) $^{131}_{53}I \rightarrow ^{131}_{54}Xe +$ ________
2. Balance the following beta decay equations. Give symbols, nuclear charges, and mass numbers.
   (a) $^{14}_{6}C \rightarrow$ ________ $+ ^{0}_{-1}e$
   (b) $^{210}_{82}Pb \rightarrow$ ________ $+ ^{0}_{-1}e$
3. Balance the following alpha decay equations. Give symbols, nuclear charges, and mass numbers.
   (a) $^{222}_{86}Rn \rightarrow ^{4}_{2}He +$ ________
   (b) $^{225}_{90}Th \rightarrow ^{4}_{2}He +$ ________
4. How many atoms of tritium will remain after four half-lives if there were initially 300,000 atoms?
5. A sample had an initial activity of 40,000 bq (1 bq = 1 disintegration per second). What is the expected activity after three half-lives?
6. A medical radioisotope presently has an activity of 8000 bq. It sat waiting to be used for four half-lives. What was the original activity?
7. A patient was given a dose of technetium-99m with an initial activity of $10\mu C$ (370,000 bq). The half-life is 6 hours. What is the expected activity after 48 hours?
8. A corn sample from a sealed tomb gives a carbon-14 count of 4 ± 1 disintegrations per minute per gram of carbon. How many years ago was the corn harvested?
9. What is the annual dose of radiation for a U.S. resident living at sea level in San Diego, who lives in a wood house, eats an average diet, has had two chest X rays, has never traveled by airplane, watches television 3 hours per day, and lives 100 miles away from a nuclear power plant? Which source is the greatest contributor to this person's annual dose?
10. What is the annual dose of radiation for an airline attendant who is a U.S. resident living at 5000 feet in Denver, who lives in a concrete apartment house, eats an average diet, has had one chest X ray, has traveled 2,500,000 miles by airplane, watches television 1 hour per day, and lives one mile away from a nuclear power plant 12 hours a day? Which source is the greatest contributor to this person's annual dose?

11. An unknown radioisotope has an activity of 1600 bq. The radiation can be stopped by five sheets of paper. What is the most likely type of emitted radiation?
12. Normal carbon-14 activity is approximately 15.3 disintegrations per minute per gram of carbon present. How many disintegrations occur per minute in a 110-pound (55-kg) human body? (Assume carbon is 0.4% by mass of the human body.)
13. Radioactive decay of ^{238}U can yield ^{234}Th and an alpha particle. An initial sample contained 200,000 atoms of ^{238}U. How many alpha particles will be produced from this sample after one half-life of 4.5 billion years?

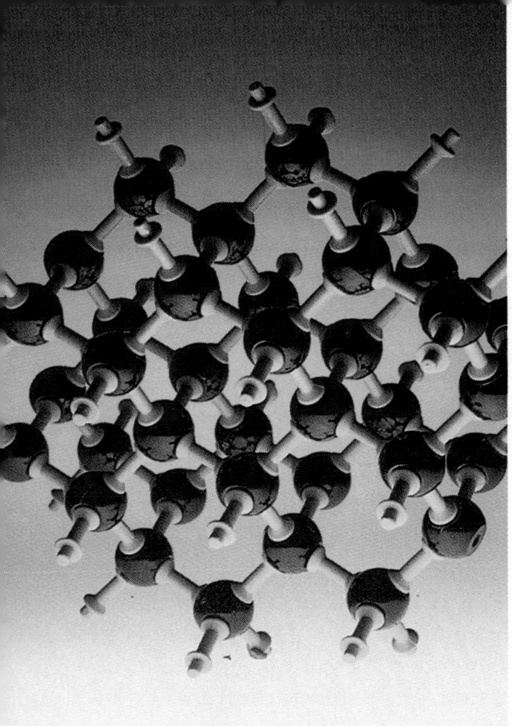

CHAPTER 6

Chemical Bonds: The Ultimate Glue

Atoms of elements are only rarely found by themselves in nature, but atoms of the 90 naturally occurring elements are the basic building blocks of all matter. What makes atoms stick together? Valence electrons form the glue, but how? Atoms form a few different types of bonds in combining with other atoms. In this chapter we're going to describe two of them—ionic bonds and covalent bonds. Transfer of valence electrons from an atom of a metal to an atom of a nonmetal produces ionic bonds. Sharing of electrons between atoms of nonmetals produces covalent bonds. These simple ideas about valence electrons are the basis for understanding the bonding in two major classes of chemical compounds, ionic compounds and molecular compounds.

- How are ionic bonds formed?
- How can the periodic table be used to predict the formulas of ionic compounds?
- How are covalent bonds formed?
- How can Lewis structures be used to predict the number of bonds between atoms in a molecule?
- Why are some covalent bonds polar?
- How can Lewis structures be used to predict the geometry of molecules?
- What are polar molecules?

The concept of valence electrons developed by G. N. Lewis is useful for developing an understanding of how atoms of different elements interact and why elements in the same group have similar properties (Section 4.3). Lewis assumed that each noble gas atom had a completely filled outermost shell, which he regarded as a stable configuration because of the lack of reactivity of noble gases. Since all noble gases (except He) have eight valence electrons, the observation came to be known as the **octet rule:** *When atoms of elements react,*

Structure of a diamond. The regular arrangement of bonded carbon atoms gives diamond its properties of crystallinity, hardness, and brilliance. *(C.D. Winters)*

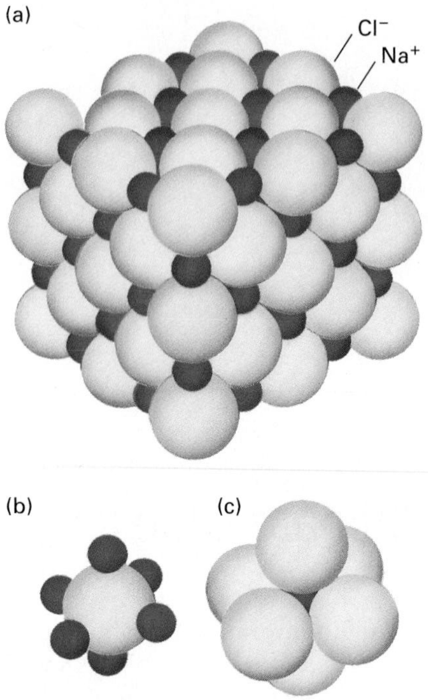

Figure 6.1 Structure of sodium chloride. (a) Model of the three-dimensional sodium chloride crystalline lattice. (b) Each Cl^- ion is surrounded by 6 Na^+ ions. (c) Each Na^+ ion is surrounded by 6 Cl^- ions.

they tend to lose, gain, or share electrons to achieve the same electron arrangement as the noble gas nearest them in the periodic table. Metals can achieve a noble gas electron arrangement by giving up electrons, and nonmetals can achieve a noble gas electron arrangement by adding or sharing electrons. This is the basis for our discussion of the two major classes of bonding—ionic bonding and covalent bonding.

6.1 Ionic Bonds

Sodium chloride is always the best starting point for discussing ionic bonding and ionic compounds. We are all familiar with some of its properties because it's in every kitchen and on every dining table. Let's review what you know so far about sodium chloride. It is a white crystalline solid and is representative of a class of compounds known as "salts." It is composed of two of the most reactive elements—the metal sodium (Na) and the nonmetal chlorine (Cl)—and the reaction between them is very energetic (Figure 4.7). Sodium is from Group IA and has a single valence electron, while chlorine is from Group VIIA and therefore has seven valence electrons:

$$\mathrm{Na}\cdot \qquad \cdot\ddot{\underset{\cdot\cdot}{\mathrm{Cl}}}\colon$$

Applying the octet rule to sodium and chlorine atoms shows how they can form a compound. If sodium loses one electron, it will have the same outer electron arrangement as neon, which precedes it by one atomic number in the periodic table. Since the atom now has one less electron than it has protons, it acquires a single positive charge and is converted to what we call a *sodium ion,* Na^+. If chlorine gains one electron, it has the same outer electron arrangement as argon, which immediately follows it in the periodic table. In this case, the neutral atom has been converted to a *chloride ion,* Cl^-, which has a single negative charge because it has one more electron than it does protons.

The reaction of sodium with chlorine to form sodium chloride is therefore fundamentally the transfer of an electron from a metal atom to a nonmetal atom.

■ To represent the reaction of sodium with chlorine as it actually occurs requires writing Cl_2 for elemental chlorine:

$$2\ \mathrm{Na} + \mathrm{Cl_2} \longrightarrow 2\ \mathrm{NaCl}$$

$$\mathrm{Na}\cdot + \cdot\ddot{\underset{\cdot\cdot}{\mathrm{Cl}}}\colon \longrightarrow \mathrm{Na^+} + \colon\ddot{\underset{\cdot\cdot}{\mathrm{Cl}}}\colon^-$$

The strong electrostatic attraction between the positive and negative ions is known as the **ionic bond,** and compounds that are held together by ionic bonds are known as **ionic compounds.** Because chemical compounds are overall electrically neutral, sodium chloride must be composed of one sodium ion for every chloride ion. To show this composition, the formula of the compound is written as NaCl. (Note that the ionic charges are not indicated in the formula.) In ionic compounds the simplest ratio of oppositely charged ions that gives an electrically neutral unit is represented in the formula and is called a **formula unit.** The formula unit for sodium chloride is NaCl, or one sodium ion and one chloride ion.

Ionic crystals are made up of large numbers of formula units to form a regular three-dimensional crystalline lattice. A model of the sodium chloride crystalline lattice is shown in Figure 6.1. Note that each Na^+ ion has six Cl^- ions around it. Similarly, each Cl^- ion has six Na^+ ions around it. In this way,

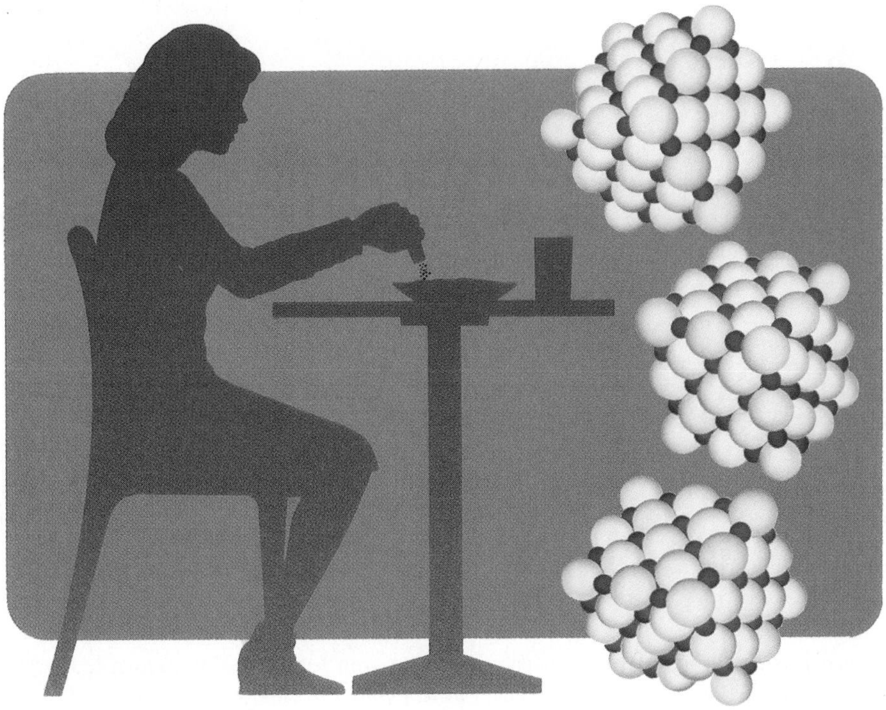

the one-to-one ratio of the singly charged ions is preserved. There is no *unique molecule* in ionic structures; no particular ion is attached exclusively to another ion, but each ion is attracted to all the oppositely charged ions surrounding it.

When atoms become ions, properties are drastically altered. For example, a collection of Br_2 molecules is red, but bromide ions (Br^-) contribute no color to a crystal of a compound such as NaBr. A chunk of sodium atoms (Figure 6.2, left) is soft, metallic, and violently reactive with water, but Na^+ ions are stable in water. A large collection of Cl_2 (Figure 6.2, center) molecules constitutes a greenish yellow poisonous gas, but chloride ions (Cl^-) produce no color in compounds and are not poisonous. We have everyday evidence of this because we use NaCl (Figure 6.2, right) to season food. When atoms become ions, atoms obviously change their nature.

The strong electrostatic forces acting between oppositely charged ions in a crystalline lattice depend on both the size and charge of the ions. The higher the charges on the ions, the stronger the attraction. If two ions have the same charge, the smaller ion has a more concentrated charge and can get closer to another ion to form a stronger ionic bond.

Ion sizes are different from parent atom sizes. Positive ions are smaller than the atoms from which they are made (Figure 6.3); negative ions are larger than their atoms (Figure 6.4). This can be understood by reviewing what happens when sodium atoms become sodium ions and chlorine atoms become chloride ions. When the sodium atom gives up its valence electron to form a sodium ion, the remaining ten electrons are more strongly attracted to the nucleus because of the charge imbalance (-10 from 10 electrons and $+11$ from 11 protons in the nucleus), so the sodium ion is smaller than the sodium atom. The same phenomenon is observed for all metal ions, and the

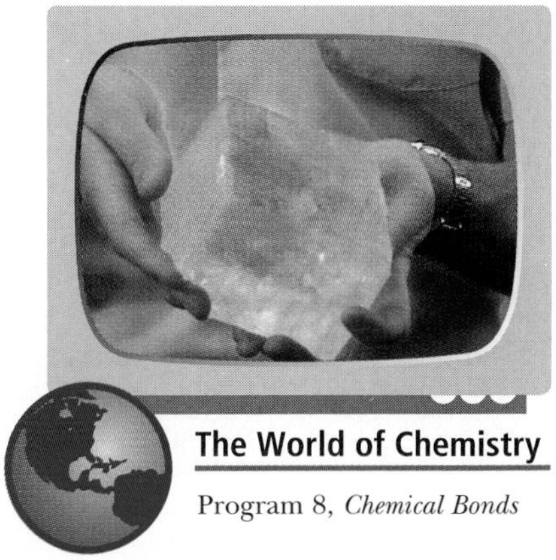

The World of Chemistry

Program 8, *Chemical Bonds*

If you were to take a magnifying glass and look at the salt crystals from your salt shaker on your dinner table at home, you'd see that each of those little crystals looks exactly like this crystal. Dr. Jeffrey Post, crystallographer, Smithsonian Institution

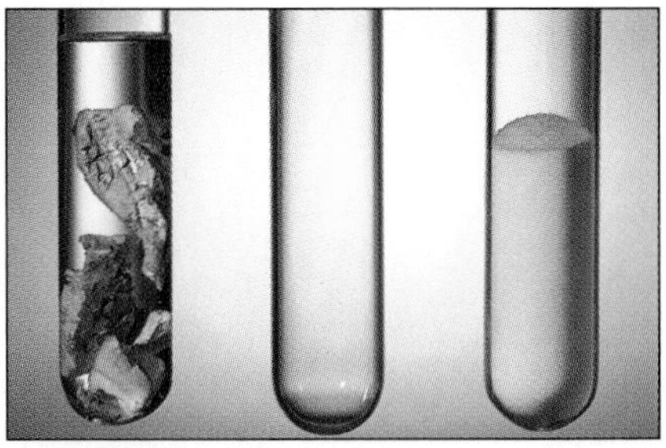

Figure 6.2 Sodium, chlorine, and sodium chloride. *(Larry Cameron)*

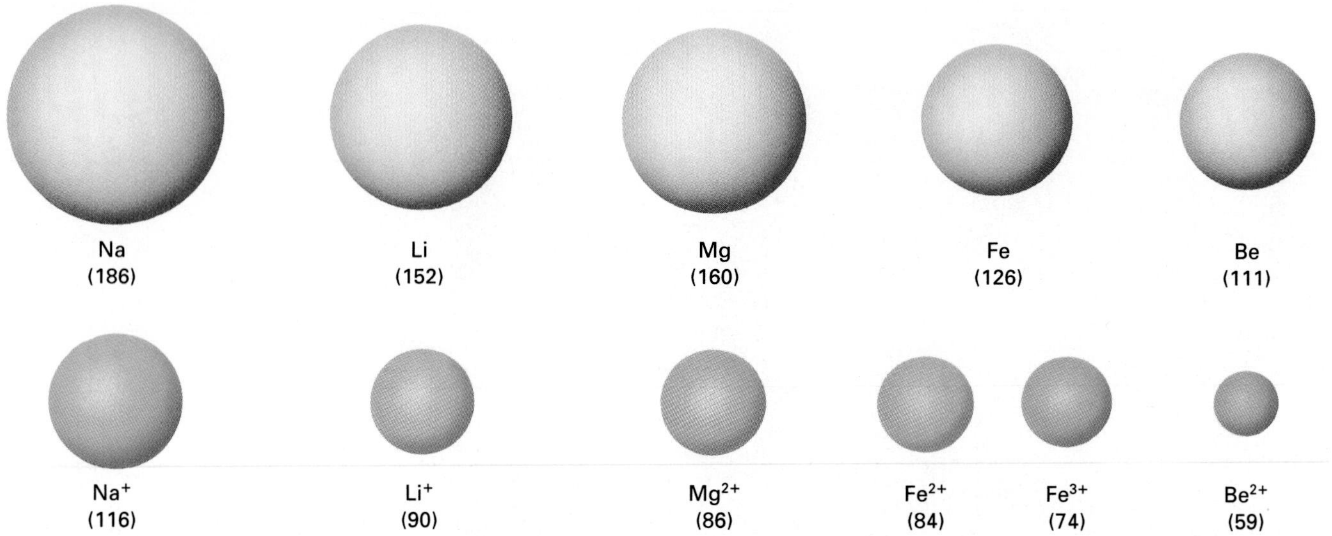

Figure 6.3 Comparison of sizes of metal atoms and their ions (in picometers, pm). Positive ions are smaller than their parent atoms. pm = 10^{-12} m.

more highly charged the positive ion, the smaller it is. For example, Fe^{3+} is smaller than Fe^{2+}. To summarize, *positive ions are always smaller than the atoms from which they are formed.*

The opposite effect occurs when negative ions are formed. Using the chlorine atom as an example, the addition of an electron to form the negative chloride ion adds an electron to the outer level without increasing the positive charge on the nucleus. This charge imbalance increases the size of the electron cloud and causes the negative chloride ion to be larger than the chlorine atom (Figure 6.4). Again this is a general effect: *Negative ions are larger than the atoms from which they are formed.*

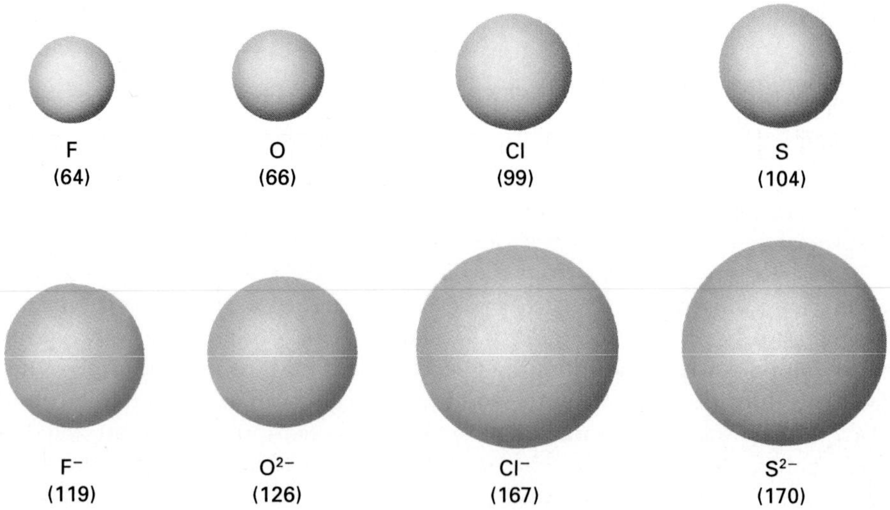

Figure 6.4 Comparison of sizes of nonmetal atoms and their ions (in picometers, pm). Negative ions are larger than their parent atoms.

6.2 Ionic Compounds

Predicting Formulas

Lewis dot symbols can be used along with the octet rule to predict formulas for ionic compounds. For example, consider the reaction of calcium with oxygen. Calcium is a Group IIA metal and has two valence electrons. According to the octet rule, it will give up two electrons to form an ion with a + 2 charge, Ca^{2+}, that has the same configuration as the nearest noble gas (Ar, 2-8-8). Oxygen is a Group VIA element whose atoms have six valence electrons and must gain two electrons to have the same configuration as the nearest noble gas (Ne, 2-8). The oxide ion will have a − 2 charge, and the formation of the compound is essentially the transfer of two electrons:

$$\cdot\text{Ca}\cdot + \cdot\ddot{\text{O}}: \longrightarrow \text{Ca}^{2+} + :\ddot{\underset{\cdot\cdot}{\text{O}}}:^{2-}$$

The formula of the ionic compound formed by this reaction is CaO, and it is called calcium oxide.

In another example, the formula for the ionic compound formed by the reaction of magnesium with nitrogen can be predicted as follows: Mg from Group IIA loses two electrons to form Mg^{2+}, and N from Group VA gains three electrons to form N^{3-}. Since compounds are electrically neutral, once the correct ions for the metal and nonmetal are predicted by the octet rule, the formula of the compound can be predicted by having the appropriate number of each ion that makes the total number of positive charges equal to the total number of negative charges. Hence, three Mg^{2+} ions (+ 6) are needed to balance two N^{3-} ions (− 6), and the formula is Mg_3N_2.

$$3\cdot\text{Mg}\cdot + 2\cdot\ddot{\text{N}}\cdot \longrightarrow 3\ \text{Mg}^{2+} + 2\ :\ddot{\underset{\cdot\cdot}{\text{N}}}:^{3-}$$

Generally, metals in Groups IA, IIA, and IIIA react with nonmetals in Groups VA, VIA, and VIIA to form ionic compounds. The formulas of thousands of ionic compounds can be predicted by using the periodic table and the octet rule to determine the ions formed by elements from these groups. This procedure can be summarized as follows:

1. Form a positive ion from a metal in an A group by removing the number of electrons equal to the group number.
2. Form a negative ion from a nonmetal in an A group by adding the number of electrons to the group number that gives a total of eight.
3. Write the formula unit for the ionic compound that gives the simplest ratio needed to produce an electrically neutral unit.

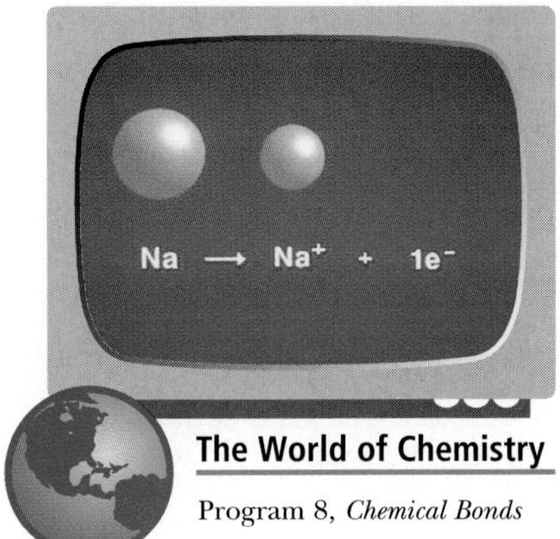

The World of Chemistry

Program 8, *Chemical Bonds*

Loss of an electron to form a sodium cation.

The World of Chemistry

Program 8, *Chemical Bonds*

Gain of an electron to form a chloride anion.

EXAMPLE 6.1 *Predicting Formulas of Ionic Compounds*

What is the formula of aluminum oxide, the ionic compound formed by the combination of aluminum and oxygen?

SOLUTION

1. Aluminum (Al) is in Group IIIA, so remove three electrons from an Al atom to give the Al^{3+} ion.
2. Oxygen (O) is in Group VIA, so add two electrons to give eight; the resulting ion is O^{2-}.
3. To make this formula electrically neutral: Al^{3+} ion is +3, and two ions would give a total charge of +6; O^{2-} ion is −2, and three ions would give a total charge of −6; +6 and −6 = 0; the formula is Al_2O_3.

Exercise 6.1
What is the formula of calcium fluoride, the ionic compound formed by the combination of calcium, Ca, and fluorine, F?

■ Positive ions are called **cations.** Negative ions are called **anions**.

Our examples of formula predictions have come from A group metals because these metals are more predictable in ion formation. Many of the metals in B groups can give up different numbers of electrons to form more than one positive ion and are less predictable using the octet rule. For example, iron forms both Fe^{2+} and Fe^{3+} ions. The stable ions formed by some A group metals and nonmetals are given in Figure 6.5 along with some B group transition metals. Notice in Figure 6.5 that A group metals give up electrons according to the octet rule and A group nonmetals add electrons according to the octet rule.

EXAMPLE 6.2 *Predicting Formulas for Transition Metal Compounds*

What are the formulas for the possible chloride compounds with the iron ions, Fe^{2+} and Fe^{3+}?

SOLUTION

Since iron forms both Fe^{2+} and Fe^{3+} ions, two compounds are possible. Since chlorine is in Group VIIA, the chloride ion is formed by adding one electron to the chlorine atom to give Cl^-. The compound with Fe^{2+} will require two Cl^- to give an electrically neutral compound with the formula $FeCl_2$. The compound with Fe^{3+} will require three Cl^- to give an electrically neutral compound with the formula $FeCl_3$.

Fe^{2+} chloride (green) and Fe^{3+} chloride (mustard yellow) salts. The differently charged iron ions give different colors to these salts. *(Richard Megna/Fundamental Photographs)*

Exercise 6.2
What are the formulas of the oxide compounds of Co^{2+} and Co^{3+}?

Naming Binary Ionic Compounds

Starting once again with table salt, we can use the formula and chemical name of this compound, NaCl, sodium chloride, to help remember the rules for naming compounds. The positive ion is named first, followed by the negative

IA	IIA	IIIB	IVB	VB	VIB	VIIB	VIIIB	VIIIB	VIIIB	IB	IIB	IIIA	IVA	VA	VIA	VIIA	VIIIA
H^+																H^-	
Li^+													C^{4-}	N^{3-}	O^{2-}	F^-	
Na^+	Mg^{2+}											Al^{3+}		P^{3-}	S^{2-}	Cl^-	
K^+	Ca^{2+}		Ti^{2+}		Cr^{2+} Cr^{3+}	Mn^{2+}	Fe^{2+} Fe^{3+}	Co^{2+} Co^{3+}	Ni^{2+}	Cu^+ Cu^{2+}	Zn^{2+}				Se^{2-}	Br^-	
Rb^+	Sr^{2+}									Ag^+	Cd^{2+}		Sn^{2+}		Te^{2-}	I^-	
Cs^+	Ba^{2+}										Hg_2^{2+} Hg^{2+}		Pb^{2+}	Bi^{3+}			

Figure 6.5 Common ions. Metals usually form positive ions with a charge given by the group number in the case of the main-group metals (blue). For transition metals (red), the positive charge is variable, and other ions in addition to those illustrated are possible. Nonmetals (yellow) generally form negative ions with a charge equal to 8 minus the group number.

ion. The element name is used for the positive ion. If the compound is made up of only one metal and one nonmetal (a **binary compound**), the name of the negative ion ends in "ide." Examples we have already used besides sodium chloride include calcium oxide, CaO; aluminum oxide, Al_2O_3; and magnesium nitride, Mg_3N_2. (Note that "binary" means only two elements are present, but the number of atoms in the formula can be more than two.) Many transition metals can form ions with different charges. In these cases, roman numerals are used with the names to indicate the charge. For example, iron(II) ion refers to Fe^{2+}, and iron(III) ion means Fe^{3+}. The name of $FeCl_2$ (Example 6.2) is iron(II) chloride, and the name of $FeCl_3$ is iron(III) chloride.

EXAMPLE 6.3 *Naming Binary Ionic Compounds*

Name the following ionic compounds: (a) K_2S, (b) $BaBr_2$, (c) Li_2O, (d) Fe_2O_3

SOLUTION

All of the compounds are binary compounds—made up of ions of two elements. The positive ion is named first, followed by the negative ion. The positive ion is named as the element. The negative ion is named by adding "–ide" to the stem of the name of the element. The correct names of the first three are (a) potassium sulfide, (b) barium bromide, and (c) lithium oxide. Compounds of iron require the use of a roman numeral after "iron" to identify its charge. The charge on the iron ion in Fe_2O_3 is determined by calculating the total negative charge (-6 from 3 O^{2-}) and dividing by 2 (because there are two iron ions). This gives a charge of $+3$, so the correct name of Fe_2O_3 is iron(III) oxide.

Exercise 6.3

Name the following compounds: (a) RbCl, (b) Ga_2O_3, (c) $BaBr_2$, (d) Fe_3N_2

EXAMPLE 6.4 *Writing Formulas for Binary Ionic Compounds*

Write formulas for the following compounds: (a) cesium bromide, (b) cobalt(III) chloride, (c) barium oxide

SOLUTION

(a) The correct formula is CsBr since Cs (Group IA) forms Cs^+ and Br (Group VIIA) forms Br^-. (b) The roman numeral III indicates the Co^{3+} ion, and Cl (Group VIIA) forms Cl^-, so the correct formula is $CoCl_3$. (c) The correct formula is BaO since Ba (Group IIA) forms Ba^{2+} and O (Group VIA) forms O^{2-}.

Exercise 6.4

What is the formula of (a) cobalt(II) sulfide, (b) beryllium fluoride, and (c) potassium iodide?

Ionic Compounds with Polyatomic Ions

■ Most common polyatomic ions are negatively charged; ammonium ion, NH_4^+, is the major exception.

Two or more elements can also combine to form a polyatomic ion, a chemically distinct species with an electrical charge. Communication in the world of chemistry and understanding many applications require one to know the names, formulas, and charges of the common polyatomic ions listed in Table 6.1.

TABLE 6.1 ■ Names and Composition of Some Common Polyatomic Ions

Cation (Positive Ion)			
NH_4^+	Ammonium ion		
Anions (Negative Ions)			
OH^-	Hydroxide ion	CO_3^{2-}	Carbonate ion
$CH_3CO_2^-$	Acetate ion	HCO_3^-	Hydrogen carbonate ion (or bicarbonate ion)
NO_2^-	Nitrite ion		
NO_3^-	Nitrate ion	PO_4^{3-}	Phosphate ion
SO_3^{2-}	Sulfite ion	HPO_4^{2-}	Hydrogen phosphate ion
HSO_3^-	Hydrogen sulfite ion	$H_2PO_4^-$	Dihydrogen phosphate ion
SO_4^{2-}	Sulfate ion	ClO^-	Hypochlorite ion
HSO_4^-	Hydrogen sulfate ion (or bisulfate ion)	ClO_3^-	Chlorate ion
		ClO_4^-	Perchlorate ion
CN^-	Cyanide ion		

TABLE 6.2 ■ Some Commercially Important Ionic Compounds with Polyatomic Ions

Formula	Name (Common Name)	Uses
NH_4NO_3	Ammonium nitrate	Fertilizers and explosives
$NaNO_2$	Sodium nitrite	Food preservative; metal treatment
KNO_3	Potassium nitrate	Gunpowder and matches
$NaOH$	Sodium hydroxide (lye)	Extract Al from ore; prepare soaps, detergents, rayon; pulp and paper industry
KOH	Potassium hydroxide (caustic potash)	Soaps and detergents; fertilizer manufacture; K_2CO_3 manufacture
$Mg(OH)_2$	Magnesium hydroxide	Milk of magnesia
Na_2CO_3	Sodium carbonate (washing soda, soda ash)	Water softening; detergents and cleansers; pulp and paper industry; glass and ceramics
K_2CO_3	Potassium carbonate (potash)	Glass and ceramics; dyes and pigments
$NaHCO_3$	Sodium bicarbonate (baking soda)	Household use; food industry; fire extinguisher
Na_3PO_4	Sodium phosphate	Food additive
$CaHPO_4$	Calcium hydrogen phosphate	Fertilizer
$Ca(H_2PO_4)_2$	Calcium dihydrogen phosphate	Fertilizer
$CaSO_4$	Calcium sulfate	Gypsum, drywall (wallboard)
$Al_2(SO_4)_3$	Aluminum sulfate	Water purification

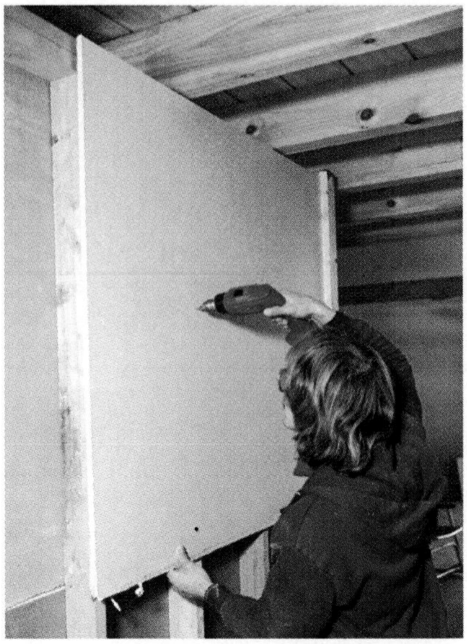

Installation of gypsum wall board. Uses of gypsum, $CaSO_4 \cdot 2\ H_2O$, include the production of wall board, portland cement, plaster of Paris, and building plasters. There is evidence that the interiors of some of the great pyramids in Egypt were coated with gypsum plaster. *(C.D. Winters)*

The bonding between the atoms within polyatomic ions is covalent, but the group of atoms has either more or fewer electrons than protons and therefore has an overall charge. Compounds that contain polyatomic ions are ionic, and their formulas are written by the same procedure as described for binary ionic compounds. The only difference is that the polyatomic ion is enclosed in parentheses when the subscript is larger than one. For example, the formula of aluminum nitrate is $Al(NO_3)_3$. The compounds are also named in the same manner as binary ionic compounds, with the name of the positive ion followed by the name of the negative ion. Examples of some important ionic compounds with polyatomic ions are given in Table 6.2.

EXAMPLE 6.5 *Writing Formulas for Compounds of Polyatomic Ions*

Write the formulas of (a) magnesium sulfate, (b) calcium hydrogen sulfite.

SOLUTION

(a) The sulfate ion is SO_4^{2-} and the charge on the magnesium ion is $+2$, so the

formula is $MgSO_4$. No parentheses are needed around the sulfate ion because only one SO_4^{2-} is present.
(b) The hydrogen sulfite ion is HSO_3^- and the charge on the calcium ion is +2, so the formula is $Ca(HSO_3)_2$.

Exercise 6.5
Write the formulas of (a) magnesium carbonate, (b) sodium dihydrogen phosphate.

SELF-TEST 6A

1. The attractive forces between positive and negative ions in a crystal lattice are called ________ bonds.
2. Positive ions are formed from neutral atoms by (a) losing electrons (b) gaining electrons.
3. What charge is expected when the following atoms combine in ionic compounds?
   (a) Lithium, Li (b) Aluminum, Al
   (c) Sulfur, S (d) Bromine, Br
4. Which of the following atoms form positive ions? (a) Potassium, K, (b) Bromine, Br, (c) Nitrogen, N, (d) Oxygen, O.
5. Negative ions are formed from neutral atoms by (a) losing electrons (b) gaining electrons.
6. Positive ions are (smaller/larger) than their parent atoms.
7. Negative ions are (smaller/larger) than their parent atoms.
8. In each of the following pairs which has the smaller diameter?
   (a) Li or Li^+ (b) Ca or Ca^{2+}
   (c) Br or Br^- (d) Cs^+ or Al^{3+}
9. When nutritionists refer to the importance of low salt intake, they are referring to the compound with the formula ________ and the name ________.
10. Match the following names and formulas.

Potassium bromide	Cl_2O
Magnesium oxide	N_2O_4
Dinitrogen tetroxide	SO_3
Sulfur trioxide	KBr
Dichlorine monoxide	SF_6
Sulfur hexafluoride	MgO

6.3 Covalent Bonds

What holds together the atoms in molecules of carbon monoxide (CO), methane (CH_4), water (H_2O), quartz (SiO_2), ammonia (NH_3), carbon tetrachloride (CCl_4), and millions of other compounds in which all the elements are nonmetals? G. N. Lewis proposed that the bonds that hold atoms together in molecules consist of one or more pairs of electrons *shared* between the bonded atoms. The attraction of positively charged nuclei for electrons between them pulls the nuclei together. In many molecular compounds, atoms

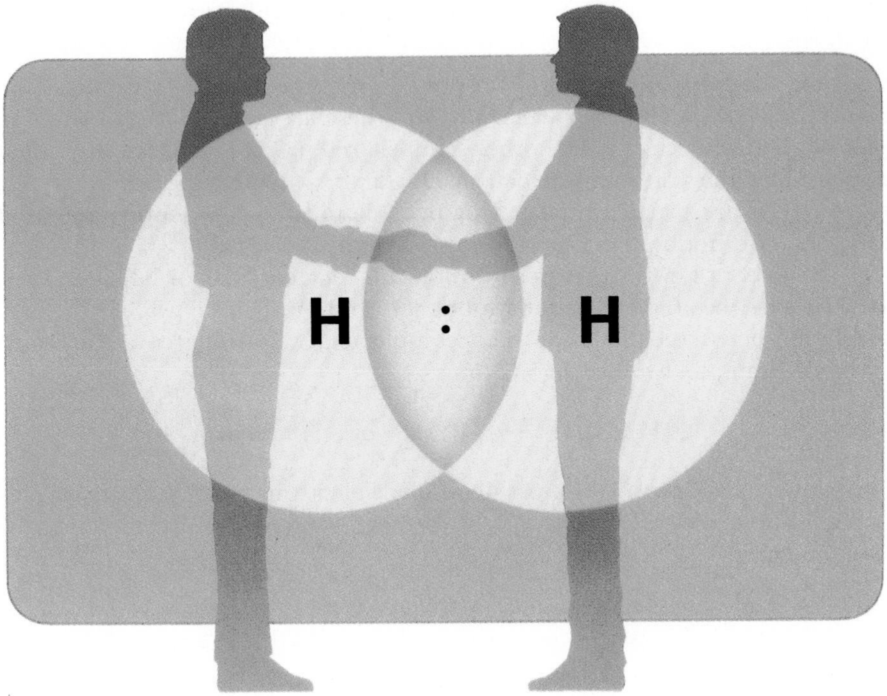

of nonmetals achieve noble gas electron arrangements (octet rule) by sharing electrons. The bond formed between two atoms that share electrons is called a **covalent bond.**

Single Covalent Bonds

A hydrogen atom has one electron. If it can share its electron with another atom that has an unpaired valence electron, a stable pairing of the two electrons can be achieved, and the H atom can then have the electron structure of helium, a noble gas. This arrangement can be achieved by two H atoms sharing their single electrons. The shared electrons are attracted by the positive nuclei of both atoms.

■ A single covalent bond is formed when two atoms share a single pair of electrons.

Lewis dot symbols are used for the elements combining to form a molecule, and the resulting electron dot representation of the valence electrons in the molecule is called the **Lewis structure.** For example, the Lewis structure for H_2 shows two electrons (two dots) shared between two hydrogen nuclei (two H·)

■ Ionic compounds are almost all solids, but molecular compounds can be gases, liquids, or solids.

$$H\cdot + H\cdot \longrightarrow \underset{\text{Lewis structure}}{H\!:\!H \quad \text{or} \quad H\!-\!H}$$

Since each fluorine atom has one unpaired electron ($\cdot\ddot{\underset{\cdot\cdot}{F}}\!:$), two fluorine atoms also can share an electron each to form a single covalent bond and an F_2 molecule. Each fluorine atom needs one electron to complete its outer shell. Shared electrons are counted toward the completion of the shells of both atoms.

$$2\,\cdot\ddot{\underset{\cdot\cdot}{\mathrm{F}}}: \longrightarrow :\ddot{\underset{\cdot\cdot}{\mathrm{F}}}:\ddot{\underset{\cdot\cdot}{\mathrm{F}}}: \quad \text{or} \quad :\ddot{\underset{\cdot\cdot}{\mathrm{F}}}-\ddot{\underset{\cdot\cdot}{\mathrm{F}}}:$$

Only the shared pair of electrons represented between the two symbols (the two Fs) are bonding valence electrons, and these are referred to as a **bonding pair** of valence electrons. The other six unshared pairs of electrons are called **nonbonding pairs** of valence electrons. In Lewis structures the bonding pairs of electrons are usually indicated by lines connecting the atoms they hold together, and nonbonding pairs are usually indicated by dots.

■ Nonbonding pairs of electrons are also called lone pairs.

What about Lewis structures for molecules such as H_2O or NH_3? Oxygen (Group VIA) has the Lewis dot symbol $\cdot\ddot{\underset{\cdot}{\mathrm{O}}}:$ and must share two electrons to satisfy the octet rule. This can be accomplished by forming covalent bonds with, for example, two hydrogen atoms.

$$2\,\mathrm{H}\cdot + \cdot\ddot{\underset{\cdot}{\mathrm{O}}}: \longrightarrow \mathrm{H}:\ddot{\underset{\cdot\cdot}{\mathrm{O}}}:\mathrm{H} \quad \text{or} \quad \mathrm{H}-\ddot{\underset{\cdot\cdot}{\mathrm{O}}}-\mathrm{H}$$

Nitrogen ($\cdot\ddot{\underset{\cdot}{\mathrm{N}}}\cdot$, Group VA) must share three electrons to achieve a noble gas configuration, which can be done by forming covalent bonds with three hydrogen atoms.

$$3\,\mathrm{H}\cdot + \cdot\ddot{\underset{\cdot}{\mathrm{N}}}\cdot \longrightarrow \mathrm{H}:\underset{\underset{\mathrm{H}}{\cdot\cdot}}{\ddot{\mathrm{N}}}:\mathrm{H} \quad \text{or} \quad \mathrm{H}-\underset{\underset{\mathrm{H}}{|}}{\ddot{\mathrm{N}}}-\mathrm{H}$$

THE PERSONAL SIDE

Gilbert Newton Lewis (1875–1946)

(Photo by Frances Simon Courtesy AIP Emilio Segre Visual Archives)

G.N. Lewis was born in Massachusetts but raised in Nebraska. After earning his B.A. and Ph.D. degrees at Harvard University, he began his academic career. In 1912 he was appointed Chairman of the Chemistry Department at the University of California, Berkeley, and he remained there for the rest of his life. Lewis felt that a chemistry department should both teach and advance fundamental chemistry, and he was not only a productive researcher but also a teacher who profoundly affected his students. He developed his concepts about valence electrons and the stability of the noble gas electron configuration (octet rule) to explain periodic trends to students in his introductory chemistry course. Although he was teaching these concepts as early as 1902, he didn't publish them until 1916. Lewis also made major contributions to other areas of chemistry such as thermodynamics, isotope studies, and acid–base theory.

Single Bonds in Hydrocarbons

■ All organic compounds can be pictured as derived from hydrocarbons by the addition of various other kinds of atoms or groups of atoms.

The largest class of organic compounds is the **hydrocarbons,** compounds containing only carbon and hydrogen. Hydrocarbons that contain only C—C and C—H single bonds are called **alkanes.** Methane, CH_4, the simplest alkane, is the main component of natural gas. Is the formula CH_4 in agreement with what we would predict using Lewis dot symbols, the octet rule, and the Lewis structure? Carbon is in Group IVA, so a carbon atom has four valence electrons and the Lewis dot symbol is $\cdot \dot{\underset{\cdot}{C}} \cdot$. To satisfy the octet rule, a carbon atom needs to gain a share of an additional four electrons. In this case, four hydrogen atoms with one valence electron each are needed, so the Lewis structure predicted for methane is

$$\begin{array}{c} \text{H} \\ \text{H} \ddot{\text{C}} \text{H} \\ \ddot{\text{H}} \end{array} \quad \text{or} \quad \begin{array}{c} \text{H} \\ | \\ \text{H}-\text{C}-\text{H} \\ | \\ \text{H} \end{array}$$

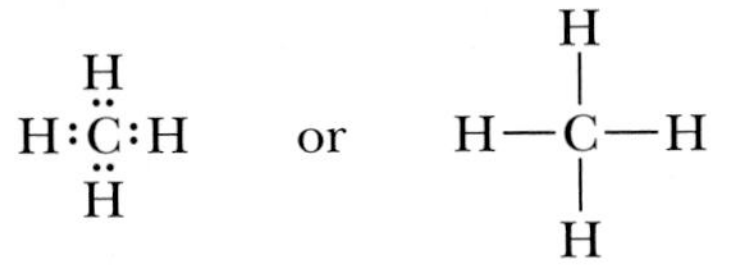

Each line represents a pair of shared valence electrons.

An alkane always has four single bonds around each carbon atom. These may be C—C or C—H bonds. For example, the alkane molecule with three carbon atoms, propane, has the following Lewis structure:

$$\begin{array}{ccc} \text{H} & \text{H} & \text{H} \\ | & | & | \\ \text{H}-\text{C}-\text{C}-\text{C}-\text{H} \\ | & | & | \\ \text{H} & \text{H} & \text{H} \end{array}$$

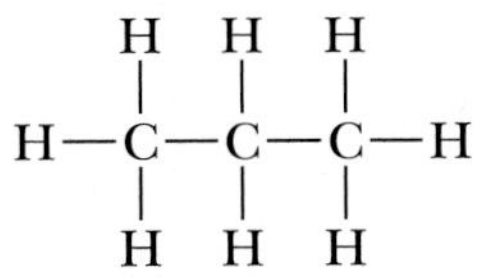

Propane is one of the principal components of bottled gas. Alkanes are often referred to as **saturated hydrocarbons** because they have the highest possible ratio of hydrogen to carbon atoms bonded in a molecule. The general formula for saturated hydrocarbons is C_nH_{2n+2} where n is the number of carbon atoms.

Multiple Covalent Bonds

An atom with fewer than seven valence electrons can form covalent bonds in two ways. The atom may share a single electron with each of several other atoms that can each contribute a single electron. This leads to the single covalent bonds described in the previous sections. The atom can also share two (or three) pairs of electrons with a single other atom. In this case there will be two (or three) bonds between these two atoms.

When two shared pairs of electrons join the same two atoms, the bond is called a **double bond.** For example, carbon dioxide has two carbon-oxygen double bonds, represented by putting a pair of lines between the atom symbols, C=O. This can be predicted by using the octet rule and the Lewis dot symbols.

$$2 \cdot \ddot{\text{O}} \colon + \cdot \dot{\underset{\cdot}{C}} \cdot \longrightarrow \colon \ddot{\text{O}} \colon\colon \text{C} \colon\colon \ddot{\text{O}} \colon \quad \text{or} \quad \colon \ddot{\text{O}} = \text{C} = \ddot{\text{O}} \colon$$

Each O atom needs two electrons, and C needs four electrons to satisfy the octet rule. To accomplish this, the carbon atom shares two electrons with each of the two oxygen atoms, forming two double bonds. Each double bond consists of two pairs of electrons and counts as part of the octet of each bonded atom. The result is that carbon is surrounded by four pairs of bonding electrons, and each oxygen is surrounded by two pairs of bonding electrons and two pairs of nonbonding electrons.

The World of Chemistry

Program 8, *Chemical Bonds*

Formation of the strong triple bond in N_2 helps to account for the energy released by many explosives.

■ The most common double or triple bonds involve carbon, nitrogen, oxygen, or sulfur atoms.

FRONTIERS IN THE WORLD OF CHEMISTRY

Fullerenes

Structure of fullerene, C_{60}. The soccer ball is a model of the C_{60} structure. The surface of C_{60} is made up of five-membered rings (black rings on soccer ball) and six-membered rings (white rings on soccer ball). Seams of the soccer ball represent covalent bonds between carbon atoms at the intersection of each seam with other seams. (C.D. Winters)

Geodesic domes at Elmira College, Elmira, N.Y. (Grant Heilman)

The tendency for carbon to form single bonds to itself is well illustrated in the different forms of carbon, known as allotropic forms. Until the mid-1980s, only two allotropic forms of carbon were known—graphite and diamond. However, in 1985 Richard Smalley at Rice University and Harry Kroto of the University of Sussex, England, and their co-workers detected another form of carbon in the soot formed from laser vaporization of graphite. They proposed that the new form of carbon was a C_{60} molecule in the shape of an icosahedron. The C_{60} molecule resembles a hollow soccer ball; the surface is made up of five-membered rings linked to six-membered rings (like those in graphite). The discoverers named the new allotrope "buckminsterfullerene" (or simply "buckyball") after the innovative American philosopher and engineer R. Buckminster Fuller, who popularized the icosahedral shape by using it in his patented geodesic dome.

Recent research indicates that C_{60} is the first of a family of "fullerenes" that contain an even number of carbon atoms arranged in closed, hollow cages. Others that have been discovered include C_{70}, C_{76}, C_{90}, C_{94}, and even giant fullerenes such as C_{240}, C_{540}, and C_{960}. The original buckyball, C_{60}, though, is the one found in greatest abundance in soot. Some possible uses include lightweight batteries, new lubricants, antitumor therapy for cancer patients (by enclosing a radioactive atom within the cage), and microscopic ball-bearings.

When two atoms share three pairs of bonding electrons, the result is a **triple bond.** In the N_2 molecule, each nitrogen atom needs to share six valence electrons (or three pairs) with the other to satisfy the octet rule.

$$2\cdot\ddot{\underset{\cdot}{N}}\cdot \longrightarrow :N:::N: \quad \text{or} \quad :N{\equiv}N:$$

Notice that each N atom has eight electrons around it, three bonding pairs and one nonbonding pair.

Single, double, and triple bonds differ in length and strength. Triple bonds are shorter than double bonds, which in turn are shorter than single bonds. Bond energies normally increase with decreasing bond length. **Bond energy** is the amount of energy required to break a mole of the bonds. Some typical bond lengths and energies are listed in Table 6.3.

TABLE 6.3 ■ **Some Bond Lengths and Bond Energies**

Bond type	C—C	C=C	C≡C	N—N	N=N	N≡N
Bond length (nm)	0.154	0.134	0.120	0.140	0.124	0.109
Bond energy (kcal/mol)*	83	146	200	40	100	225

*kcal/mol (kilocalories per mole) = thousands of calories necessary to break 6.02×10^{23} bonds.

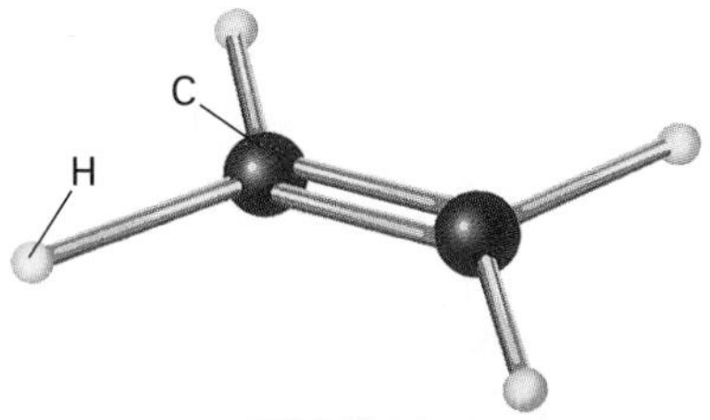

Figure 6.6 Structure of ethylene molecule.

Multiple Covalent Bonds in Hydrocarbons

Ethylene (Figure 6.6) contains a double bond between the carbon atoms and single bonds between the hydrogen atoms and the carbon atoms. Ethylene is the first member of the **alkene** series of hydrocarbons, compounds that have one or more C=C bonds, that is, carbon–carbon double bonds.

■ The official name of ethylene is ethene.

THE WORLD OF CHEMISTRY

Nitrogen Fixation

Program 8, *Chemical Bonds*

There is nitrogen in all living things. Muscles, hair, and DNA all contain nitrogen bonded to other elements. But 80% of the atmosphere is nitrogen molecules held together by strong triple bonds. How do living things get the form of nitrogen they need? Lightning helps. The electrical flash in the sky has enough energy to break apart nitrogen molecules, which then react with oxygen in the air, eventually forming nitric acid. The natural acid dissolves in rain and falls to earth as a dilute solution. There it is absorbed and metabolized by plants.

Some plants, though, convert molecular nitrogen in a different way. Soybeans and other legumes, such as peas and peanuts, host a unique bacterium in their roots. This bacterium converts the nitrogen molecule into a nitrogen compound, ammonia, which the plant can then use to make amino acids.

Exactly how the bacterium works is the subject of vigorous research.

Soybean roots. The nodules are the location of nitrogen fixation.

Don Keister, of the U.S. Department of Agriculture, claims, "This is one of the very unique enzymes in all of nature, because it is the only solution that nature has evolved for biologically reducing nitrogen."

The soybean and the bacterium have a symbiotic relationship. The plant houses and feeds the bacterium and, in turn, it receives the nitrogen it needs. But not all plants can host these nitrogen fixers. They have to rely on rain and natural fertilizers, as well as expensive manufactured fertilizers, such as ammonium nitrate.

Keister goes on to point out, "We are currently using something like 300 million barrels of oil per year in the United States alone to produce nitrogen fertilizers. We forget sometimes that we're going to need to double the food supply over the next 20 years." He further asks the unanswered questions, "Where is that energy going to come from? Where is the fertilizer going to come from?"

For feeding the world, there are two basic options: We can either produce more fertilizer at greater cost and some risk to the environment, or we can create new varieties of nitrogen-fixing plants. Both options are being pursued worldwide.

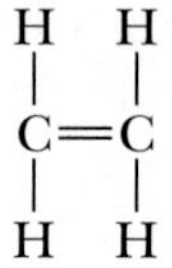

■ Polyethylene is the most widely used polymer. Examples of plastics made from polyethylene include milk bottles, sandwich bags, garbage bags, toys, and molded objects.

More than 48 billion pounds of ethylene are produced annually in the United States. About half is used in the manufacture of polyethylene plastics (Chapter 14). The structural formula of ethylene illustrates why alkenes are said to be **unsaturated hydrocarbons;** they contain fewer hydrogen atoms than the corresponding alkanes and react with hydrogen to form alkanes.

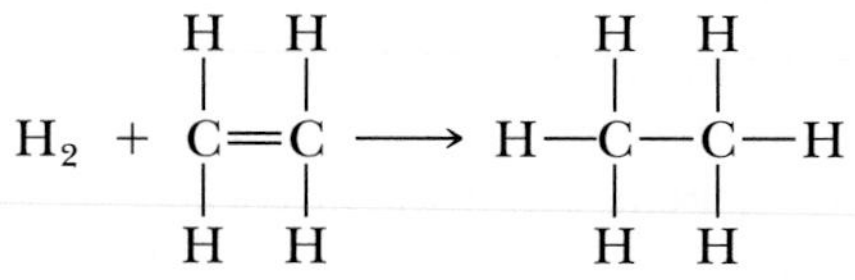

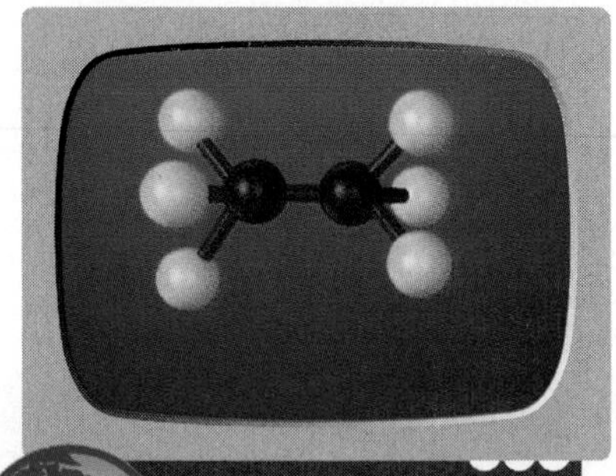

The World of Chemistry

Program 9, *Molecular Architecture*

Molecular model of ethane.

Acetylene, the gas that produces a flame hot enough to cut steel when it is mixed with oxygen and burned, has a carbon-carbon triple bond, $C\equiv C$.

$$H-C\equiv C-H$$

EXAMPLE 6.6 *Bonds in the Propylene Molecule*

The structural formula for propylene, which ranks seventh in annual U.S. production for its use in the manufacture of polypropylene plastics, is

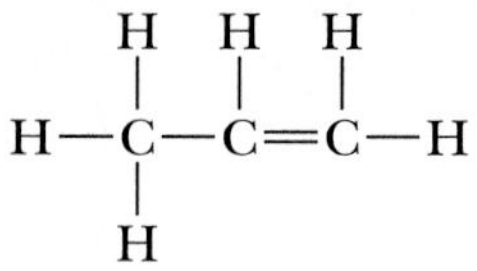

Give the total number of (a) valence electrons, (b) single bonds, (c) double bonds, (d) bonding pairs.

SOLUTION

(a) Three carbon atoms have 4 valence electrons each, to give 12 valence electrons; six hydrogen atoms have one valence electron each, to give 6 valence electrons; the total for the ethylene molecule is 18.
(b) Seven single bonds account for 14 of the 18 valence electrons.
(c) One carbon-carbon double bond accounts for the other 4 valence electrons.
(d) There are nine bonding pairs and no nonbonding pairs in the propylene molecule.

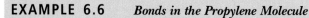

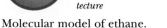

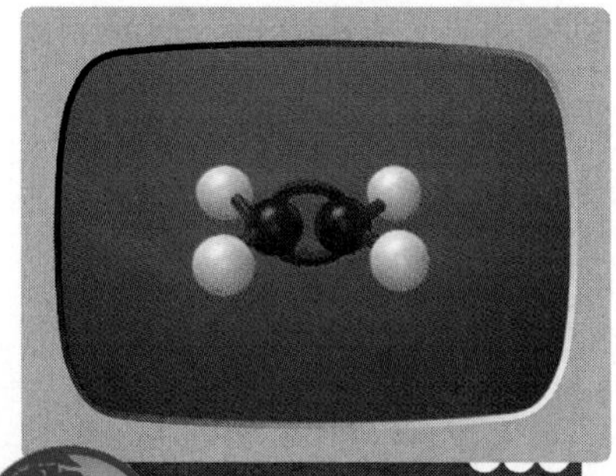

The World of Chemistry

Program 9, *Molecular Architecture*

Molecular model of ethylene.

Exercise 6.6

Draw the structural formula for a compound that has 3 C atoms and enough H atoms to contain one triple bond, 5 single bonds, and 16 valence electrons.

Oxyacetylene torch. A mixture of acetylene and oxygen is called oxyacetylene and is hot enough to cut steel. *(Joseph Nettis/Photo Researchers)*

EXAMPLE 6.7 *Bonds in the Ethyl Acetate Molecule*

The structural formula for ethyl acetate, the organic solvent used in fingernail polish remover, is

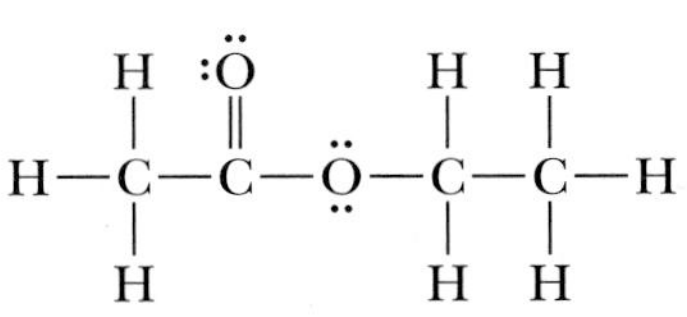

Give the total number of (a) valence electrons, (b) single bonds, (c) double bonds, (d) bonding pairs of electrons, and (e) nonbonding pairs of electrons.

SOLUTION

(a) Four carbon atoms have four valence electrons each, for a total of 16; two oxygen atoms have six electrons each, for a total of 12; eight hydrogen atoms have one valence electron each, for a total of 8; the total for the ethyl acetate molecule is 36.
(b) Twelve single bonds account for 24 valence electrons, or 12 bonding pairs.
(c) One C═O double bond accounts for 4 valence electrons, or 2 bonding pairs.
(d) There are 14 bonding pairs; 8 are C—H single bonds; 2 are single C—C bonds; 2 are single C—O bonds; and 2 are in the C═O double bond.
(e) There are 4 nonbonding pairs, 2 on the oxygen in the C═O and 2 on the oxygen in C—O—C. As a check, 14 bonding pairs plus 4 nonbonding pairs equals 18 pairs of electrons, or 36 valence electrons, which agrees with the number in the ethyl acetate molecule.

Exercise 6.7

Acetic acid, CH_3COOH, the organic acid in vinegar, has the following structural formula:

$$
\begin{array}{ccc}
H & & \ddot{\ddot{O}} \\
| & & \| \\
H-C- & C- & \ddot{\ddot{O}}-H \\
| & & \\
H & &
\end{array}
$$

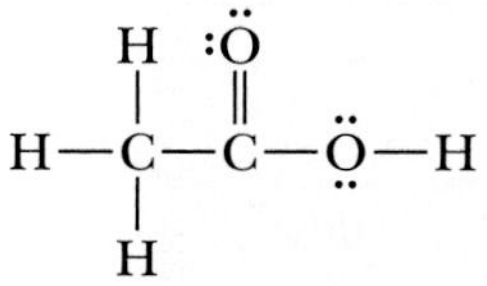

Give the total number of (a) valence electrons, (b) single bonds, (c) double bonds, (d) bonding pairs of electrons, and (e) nonbonding pairs of electrons.

Names for Binary Molecular Compounds

TABLE 6.4 ■ Prefixes for Number of Atoms in a Compound

Number of Atoms	Prefix	Number of Atoms	Prefix
1	mono-	6	hexa-
2	di-	7	hepta-
3	tri-	8	octa-
4	tetra-	9	nona-
5	penta-	10	deca-

Hydrogen forms binary compounds with all the nonmetals (except the noble gases). For compounds of oxygen, sulfur, and the halogens, the H atom is generally written first in the formula and is named first using the element name. The other nonmetal is named as if it were a negative ion. For example, HF is hydrogen fluoride and H_2S is hydrogen sulfide.

When there is more than one possible combination of two elements, the number of atoms of a given type in the compound is designated with a prefix such as *mono-, di-, tri-, tetra-*, and so on. Table 6.4 lists the prefixes for up to ten atoms, and some common compounds and their names are given in Table 6.5.

Many molecular compounds were discovered years ago and have names so common they continue to be used. Examples include water (H_2O), ammonia (NH_3), nitric oxide (NO), and nitrous oxide (N_2O).

TABLE 6.5 ■ Common Molecular Compounds

Compound	Name	Use
CO	Carbon monoxide	Preparation of methanol and other organic chemicals
CO_2	Carbon dioxide	Carbonated beverages, fire extinguisher, inert atmosphere, Dry Ice
NO	Nitrogen monoxide (nitric oxide)	Preparation of nitric acid
NO_2	Nitrogen dioxide	Preparation of nitric acid
N_2O	Dinitrogen oxide (nitrous oxide)	Spray can propellant, anesthetic
SO_2	Sulfur dioxide	Preparation of sulfuric acid, food preservative, metal refining
SO_3	Sulfur trioxide	Preparation of sulfuric acid
CCl_4	Carbon tetrachloride	Solvent
SF_6	Sulfur hexafluoride	Insulator in electric transformers
P_4O_{10}	Tetraphosphorus decaoxide	Preparation of phosphoric acid

EXAMPLE 6.8 *Naming Molecular Compounds*

What is the name of N_2O_4?

SOLUTION

The prefix for 2 N atoms is "di-" and the prefix for 4 O atoms is "tetra-," so the name of N_2O_4 is dinitrogen tetraoxide.

Exercise 6.8
What is the name of P_4S_3?

EXAMPLE 6.9 *Writing Formulas of Molecular Compounds*

Write the formula of silicon tetrachloride.

SOLUTION

The name indicates one molecule is a combination of one silicon atom and four chlorine atoms, so the formula is $SiCl_4$.

Exercise 6.9
Write the formula of dinitrogen tetrafluoride.

6.4 Guidelines for Drawing Lewis Structures

The previous sections have illustrated the importance of Lewis structures for representing the sharing of electrons in covalent molecules. It is important to practice drawing Lewis structures to be able to make predictions about the types of bonds formed.

1. To predict the arrangement of atoms within the molecule, use the following rules:
   a. **H** is always an end or *terminal* atom. It is connected to only one other atom, never to two or more.
   b. Halogens and oxygen are often terminal atoms.
   c. In binary compounds the central atom has the lowest subscript and is usually listed first.
2. Given the formula, calculate the total number of valence electrons for the molecule. For ions, also add the ion charge for a negative ion or subtract the ion charge for a positive ion.
3. Write the skeletal structure, and place one pair of electrons between each pair of bonded atoms. (A line can be used for bonding pairs.)
4. Place valence electrons about terminal atoms so that each (except hydrogen) has an octet. If the central atom is not yet surrounded by four electron pairs, convert one or more terminal-atom nonbonding pairs to bonding pairs. In this way the pairs are still associated with the

terminal atom, and they are also a part of the central atom. Not all elements form multiple bonds. You can use the general rule that *only C, N, O, and S form a multiple bond to another atom of the same element or with another atom of this group of four elements.* That is, there can be bonds such as C=C, C=N, S=O, and so on.

EXAMPLE 6.10 *Drawing Lewis Structures*

Draw Lewis structures for (a) PH_3 (b) SCl_2 (c) CO

SOLUTION

(a) The total number of valence electrons is 5 from P plus 3 (1 each from 3 H) to give 8. P is the central atom, and placing 3 bonding pairs of electrons and 1 nonbonding pair of electrons on P gives the following Lewis structure:

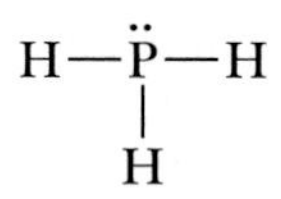

To check the structure: The total number of valence electrons is 8, in agreement with the total available, and P has an octet.
(b) The total number of electrons is 20 (6 from S and 14 from 2 Cl). S is the central atom, and placing 2 bonding pairs and 2 nonbonding pairs around S leaves 3 nonbonding pairs for each Cl.

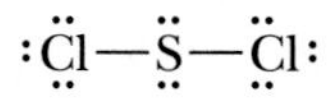

To check the structure: All the atoms have an octet, and the total number of electrons is 20.
(c) The total number of electrons is 10 (4 from C and 6 from O). Starting with one bonding pair between C and O and placing the remaining 8 electrons as 4 nonbonding pairs around the C and O, as shown below, doesn't satisfy the octet rule.

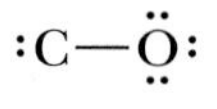 (not correct)

Satisfying the octet rule will require moving two of the nonbonding pairs of electrons on the O to bonding electrons between the C and the O to give a triple bond.

:C:::O: or :C≡O:

Exercise 6.10
Draw Lewis structures for (a) N_2O (N is the central atom) (b) CCl_4 (c) H_2S

SELF-TEST 6B

1. The formula of phosphorus pentachloride is ______________.
2. The name of SO_2 is ______________.

3. The formula of sulfur trioxide is ______________________________.
4. The name of HBr is ______________________________.
5. The formula of dichlorine monoxide is ______________________________.
6. The formula of sulfur dichloride is ______________________________.
7. The name of $SiCl_4$ is ______________________________.
8. How many electrons are shared in (a) a double bond? (b) a triple bond?
9. How many nonbonding pairs of electrons are in each of the following molecules?
   (a) H_2 (b) H_2O
   (c) CH_4 (d) CO
   (e) NH_3
10. Which is a stronger bond? (a) C—C, (b) C=C, (c) C≡C.
11. Which is a shorter bond? (a) C—C, (b) C=C, (c) C≡C.

6.5 Exceptions to the Octet Rule

Although all the Lewis structures shown in Section 6.4 obey the octet rule, there are some exceptions.

Fewer than Eight Valence Electrons

Several compounds have fewer than four electron pairs in the valence shell of the central atom. Hydrogen, of course, can accommodate at most two valence electrons, so it shares only two electrons with another atom. In BeH_2 there are only four valence electrons around Be, which is from Group IIA and has only two valence electrons of its own. There are only six valence electrons around boron in BF_3, since boron is from Group IIIA and has only three valence electrons:

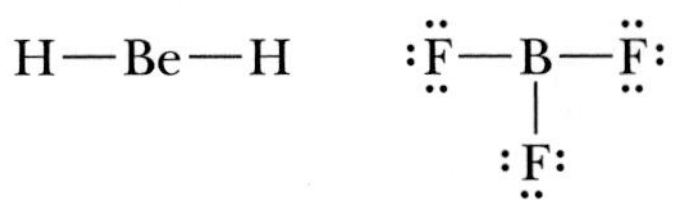

Expanded Valence

Elements of the third or higher periods can be surrounded by more than four valence pairs in certain compounds and are said to display **expanded valence.** The number of bonds formed depends on a balance between the ability of the nucleus to attract electrons and the repulsion between the pairs. An example is SF_6, a gas used as an insulator, in which S has six bonding pairs, or 12 bonding valence electrons, not eight.

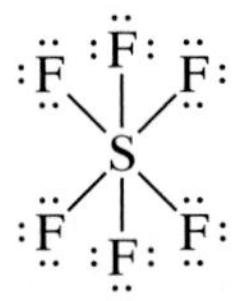

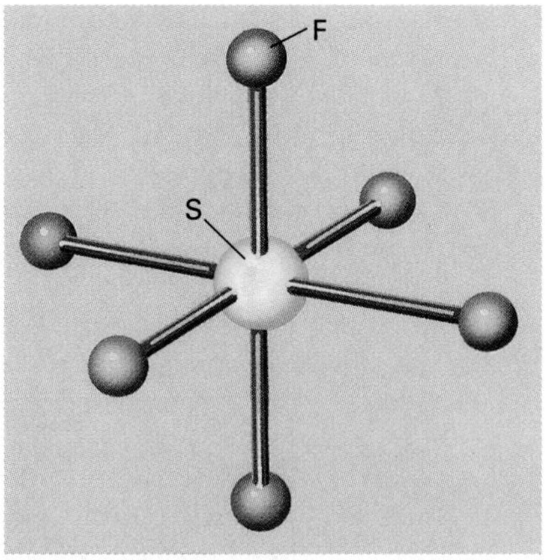

Model of the SF_6 molecule.

Odd-Electron Compounds

A few compounds contain an odd number of electrons and thus cannot obey the octet rule. The molecule NO, commonly known as nitric oxide, has five valence electrons from N and six from O, for a total of 11 valence electrons. The most plausible Lewis structure is

$$:\dot{\mathrm{N}}=\ddot{\mathrm{O}}:$$

Nitric oxide, a colorless gas, is produced from the reaction of nitrogen and oxygen at high temperatures.

$$N_2 + O_2 + \text{heat} \longrightarrow 2\ NO$$

Since the operation of internal-combustion engines in vehicles mixes air with the fossil fuel (gasoline, diesel fuel, natural gas) and then ignites it, the heat produced is sufficient to cause small amounts of the nitrogen and oxygen in the air to form NO. Even though the amount emitted from any one vehicle is very small (less than one part per trillion), the thousands of vehicles operating in an urban area make NO one of the primary air pollutants (Section 21.4). Nitric oxide reacts quickly with oxygen to give nitrogen dioxide, another odd-electron molecule with 17 valence electrons.

■ 1 part per trillion means 1 NO molecule per 1,000,000,000,000 air molecules.

$$2\ NO + O_2 \longrightarrow 2\ NO_2$$

Nitrogen dioxide is responsible for the brown haze in heavily polluted urban air.

FRONTIERS IN THE WORLD OF CHEMISTRY

NO News Is Good News

Until the late 1980s nitric oxide was primarily known for its contribution to environmental pollution. But since then, NO has been found to occur naturally in at least a dozen cell types in various parts of the body. In 1992 NO was named Molecule of the Year by *Science* magazine in recognition of hundreds of research papers that reported on NO activities in the brain, arteries, immune system, liver, pancreas, peripheral nerves, and lungs. Researchers reported that NO is essential to activities ranging from regulation of digestion and blood pressure to antimicrobial defense. Even the cellular basis for learning and memory appears to be connected to the presence of NO.

Cells rely on various forms of an unusual enzyme called nitric oxide synthase (NOS) to prepare NO for a specific cellular function. For example, in blood vessels, NO prepared by endothelial cells on the inside of the vessel wall migrates to nearby muscle cells and relaxes them. This is actually the mechanism of action for the century-old use of nitroglycerin (see Section 19.10) in the treatment of angina pectoris associated with cardiovascular disease. Sufferers of angina pectoris have severe chest pains that result from the constriction of blood vessels of the heart. The pain is relieved a few minutes after taking a nitroglycerin tablet because the drug dilates the blood vessels throughout the body.

As with other biological processes, several illnesses have been linked with improper levels of NO. Septic shock, a leading cause of death in intensive-care units of hospitals, has been shown to be related to high levels of NO in the body. Septic shock is initially caused by a bacterial infection. The blood pressure drops drastically when the body responds by producing high levels of NO. The low blood pressure causes failure of a number of organ systems, such as the liver, the kidney, and the heart. Nitric oxide inhibitors have been used successfully with several patients to raise their blood pressure to a safe level.

Brain damage to stroke patients is apparently caused by excess release of NO, which destroys neighboring neurons. Tests on mice and cultured nerve cells have shown that NO inhibitors can reduce brain damage. The death of neurons in the brains of stroke patients and of those suffering from Parkinson's disease and Alzheimer's dementia may be the result of high levels of NO being produced by improper activation of the NOS enzyme.

References S. H. Snyder and D. S. Bredt: Biological roles of nitric oxide. *Scientific American,* pp. 68–77, May 1992; J. R. Lancaster, Jr., Nitric oxide in cells. *American Scientist,* Vol. 80, pp. 248–259, 1992.

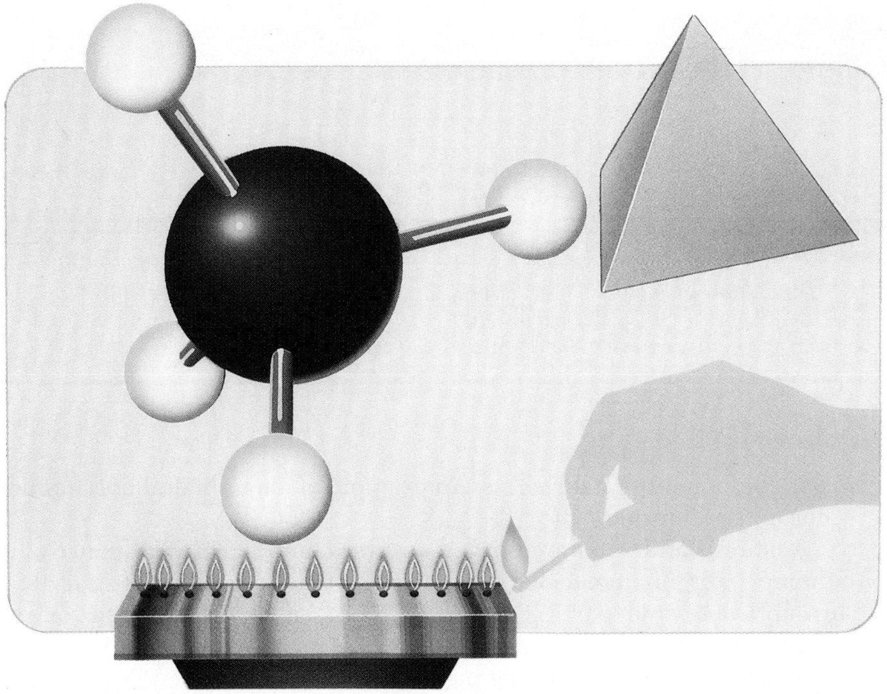

6.6 Predicting Shapes of Molecules and Polyatomic Ions

A simple, reliable method for predicting the shapes of molecules and polyatomic ions is the **valence-shell electron-pair repulsion model (VSEPR),** which is based on the idea that valence electron pairs repel each other. In fact, this model assumes that valence electron pairs around a central atom behave like a group of electrically charged balloons that are connected to a central point. If similarly charged, the balloons would tend to be as far apart as possible.

The arrangement of the balloons in Figure 6.7*a* illustrates that the greatest distance between two electron pairs connected to a central atom will occur when the pairs are 180° apart. This gives a linear geometry (shape) for a molecule with a central atom connected to two other atoms. A triangular planar geometry is favored for three electron pairs (Figure 6.7*b*). Since the noble-gas configuration of eight valence electrons gives a stable configuration for many common nonmetals, the most common geometry is tetrahedral, with four electron pairs at the corners of a tetrahedron (Figure 6.7*c*). An example of a molecule with four bonding pairs of electrons is methane (CH_4), the

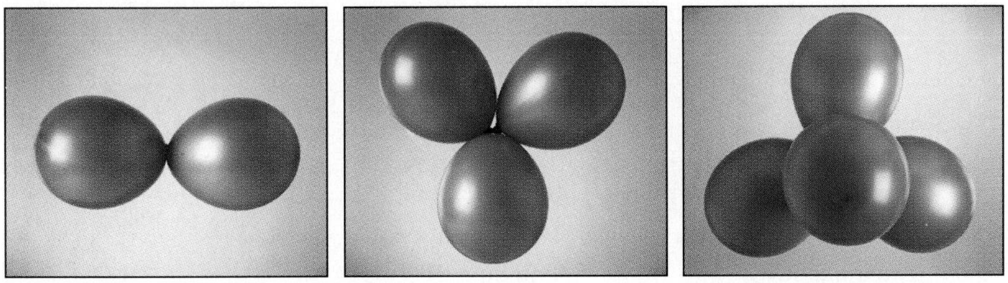

(a) (b) (c)

Figure 6.7 Balloon models of VSEPR geometries predicted for two, three, and four electron pairs. *(Kristen Brochmann/Fundamental Photographs)*

DISCOVERY EXPERIMENT

Using Balloons as Models

Blow up balloons and tie the ends together as shown in Figure 6.7 for combinations of two, three, and four balloons. These make excellent models for visualizing the linear, triangular planar, and tetrahedral geometries assumed by two, three, and four electron pairs, respectively. After the balloon models are assembled, write answers to the following questions:

1. Identify the apexes for each shape (the farthest point of each balloon from the central point). Sketch the figure you would get if you connected the apexes with lines.
2. Identify the faces (the flat surfaces bounded by the lines between apexes) in each sketch, and count how many apexes and faces there are for each shape.

major component of natural gas. An example of a tetrahedral polyatomic ion is boron tetrafluoride, $[BF_4]^-$.

Ammonia and water are two important examples of the influence of nonbonding pairs on molecular geometry. In both cases the central atom is surrounded by four pairs of electrons. Nitrogen in ammonia has one nonbonding pair and three bonding pairs, whereas oxygen in water has two nonbonding pairs and two bonding pairs.

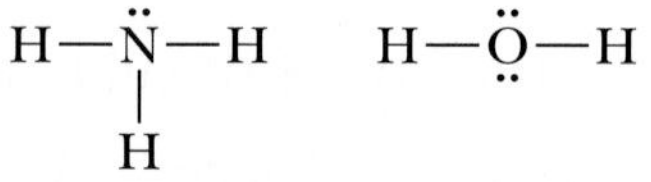

Figure 6.8 shows the tetrahedral representation of four electron pairs in CH_4, NH_3, and H_2O. The bonding angles in ammonia and water are slightly smaller than the normal tetrahedral angle of 109.5°. The VSEPR model attributes this to the larger volume occupied by nonbonding pairs compared with bonding pairs. The bonded electrons are attracted to the positive nuclei and drawn into the space between the atoms. This creates a smaller volume for the shared pair. The nonbonded pairs are attracted to only one atom and their volume is slightly greater. The increased volume spreads the nonbonding pairs farther apart and squeezes the bonding pairs closer together. Hence, the

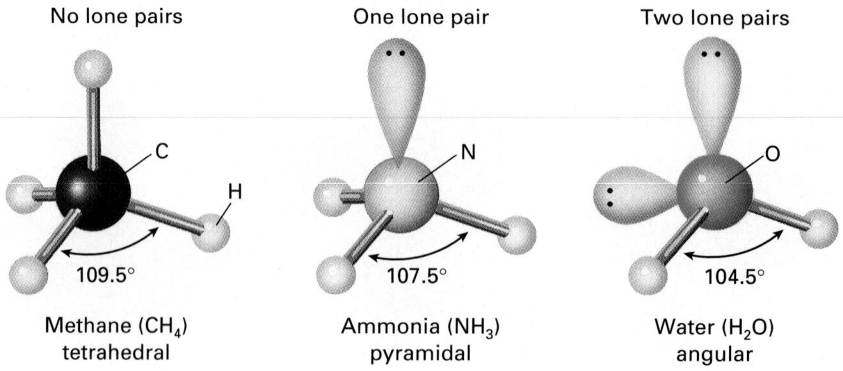

Figure 6.8 Examples of electron-pair geometries and molecular shapes for molecules and ions with four electron pairs around the central atom.

repulsions between nonbonding pairs are larger than those between bonding pairs.

Regarding the shapes of molecules with nonbonding electrons, only the shape formed by the atoms in the molecule is stated, since X ray structure studies locate atoms, not electron pairs. You can visualize the shape of the molecule by simply ignoring the nonbonding pairs. For example, NH_3 is pyramidal and H_2O is bent. To avoid confusion, the term **electron-pair geometry** is used when the geometry around a central atom of both bonding pairs and nonbonding pairs is being considered, and the term **molecular geometry** is used to describe the spatial arrangement of atoms in a molecule or polyatomic ion.

The VSEPR model can also be used to predict the shapes of molecules or ions that contain double or triple bonds by treating the double or triple bonds as a single bonding pair. Why? Electron pairs involved in a multiple bond are all shared between the same two nuclei and therefore occupy the same region as a single bond between the same two nuclei. Therefore, multiple bonds count as a single bond for purposes of predicting geometry. For example, the C atom in CO_2 has no nonbonding pairs and is linked to two O by two double bonds.

:Ö=C=Ö:

Since the double bond counts as one for the purpose of predicting the geometry, the structure of CO_2 is linear.

Another example is formaldehyde, which is predicted to have a triangular planar shape based on three bonding pairs around the central atom. (The two bonding C=O pairs are counted as one bonding pair for determining the geometry of the molecule.) This is the correct shape of the formaldehyde molecule.

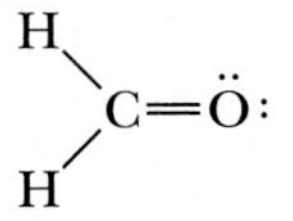

The VSEPR model can also be used to predict the shapes of polyatomic ions. For example, the Lewis structure of NO_2^- is

$[:\ddot{O}=\ddot{N}-\ddot{\underset{..}{O}}:]^-$

Two bonding pairs and one nonbonding pair suggest a triangular planar geometry for the electron pairs and a bent geometry for the bonded atoms in the ion.

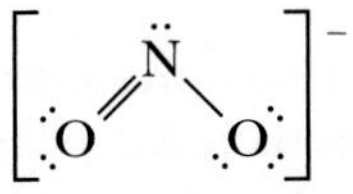

In summary, the VSEPR model can be used to predict shapes of molecules and polyatomic ions (Table 6.6) by following a few simple rules.

TABLE 6.6 ■ Examples of Molecular Geometries Predicted by VSEPR Model

Formula (X = electron pairs)	Number of Nonbonding Pairs on Central Atom	Example	Geometry of Molecule or Ion
AX_2	None	CO_2, $BeCl_2$	Linear
AX_3	None	BCl_3, CO_3^{2-}, SO_3	Triangular planar
	One	NO_2^-, O_3, $SnCl_2$	Bent
AX_4	None	SiH_4, BF_4^-, NH_4^+	Tetrahedral
	One	PF_3, ClO_3^-, NH_3	Triangular pyramidal
	Two	H_2O, NH_2^-	Bent

1. Draw the Lewis structure.
2. Determine the total number of single bonding pairs (including multiple bonds counted as single bonding pairs) and nonbonding pairs attached to the central atom.
3. Pick the appropriate electron-pair geometry, and then choose the shape that matches the total number of single bonding pairs and nonbonding pairs.
4. Predict the bond angles, remembering that nonbonding pairs occupy more volume than bonding pairs.

EXAMPLE 6.11 *VSEPR*

Use the VSEPR model to predict the molecular shape of $SiCl_4$, OF_2, and $[NH_4]^+$.

SOLUTION

Follow the guidelines for drawing Lewis structures. Draw the Lewis structure of each molecule to determine the number of bonding pairs and nonbonding pairs on the central atom.

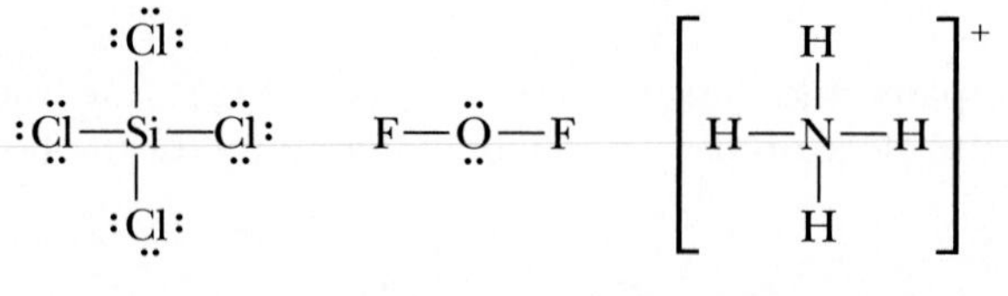

$SiCl_4$ has four bonding pairs and no nonbonding pairs on the central atom, so the molecular shape would be predicted to be tetrahedral. In OF_2 the O has two bonding pairs and two nonbonding pairs. This gives a tetrahedral electron-pair geometry, but the question asks for the molecular shape (the shape described by the atoms in the molecule). The molecular geometry is bent, the same as H_2O. The predicted shape of $[NH_4^+]$ is tetrahedral based on the presence of four bonding pairs around the central atom.

Exercise 6.11
Use the VSEPR model to predict the molecular shape of ClO_3^- and NH_2^-.

6.7 Polar and Nonpolar Bonding

In a molecule like H_2 or F_2, where both atoms are alike, there is equal sharing of the electron pair. Where two unlike atoms are bonded, however, the sharing of the electron pair is unequal and results in a shift of electric charge toward one partner. The more nonmetallic an element is, the more that element attracts electrons.

The attraction an atom has for the electrons in a chemical bond can be expressed on a quantitative basis and is called **electronegativity.** Nonmetallic character increases across and up the periodic table toward F, which has the largest electronegativity of all the elements. In 1932 Linus Pauling first proposed the concept of electronegativity. The currently accepted values for electronegativities are shown in Figure 6.9. (Electronegativity values are relative numbers, with an arbitrary value of 4.0 assigned to fluorine.) The electronegativities generally increase along a diagonal line drawn from 0.7 for cesium (Cs) to 4.0 for F. The values for other elements are between these two extremes.

When two atoms are bonded covalently and the electronegativities of the two atoms are the same, there is an equal sharing of the bonding electrons, and the bond is a **nonpolar** covalent bond. The bonds in H_2, F_2, and NCl_3 (N and Cl have the same electronegativity, 3.0) are nonpolar.

Two atoms with different electronegativities bonded covalently form a **polar** covalent bond. The bonds in HF, NO, SO_2, H_2O, CCl_4, and BeF_2 are polar. In a molecule of HF, for example, the bonding pair of electrons is drawn more towards the highly electronegative fluorine atom and away from

H 2.1						
Li 1.0	Be 1.5	B 2.0	C 2.5	N 3.0	O 3.5	F 4.0
Na 0.9	Mg 1.2	Al 1.5	Si 1.8	P 2.1	S 2.5	Cl 3.0
K 0.8	Ca 1.0	Ga 1.8	Ge 1.8	As 2.0	Se 2.4	Br 2.8
Rb 0.8	Sr 0.9	In 1.7	Sn 1.8	Sb 1.9	Te 2.1	I 2.5
Cs 0.7	Ba 0.9	Tl 1.8	Pb 1.9	Bi 1.9	Po 2.0	At 2.2

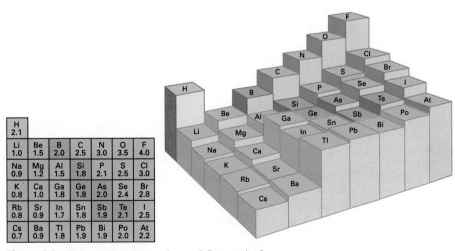

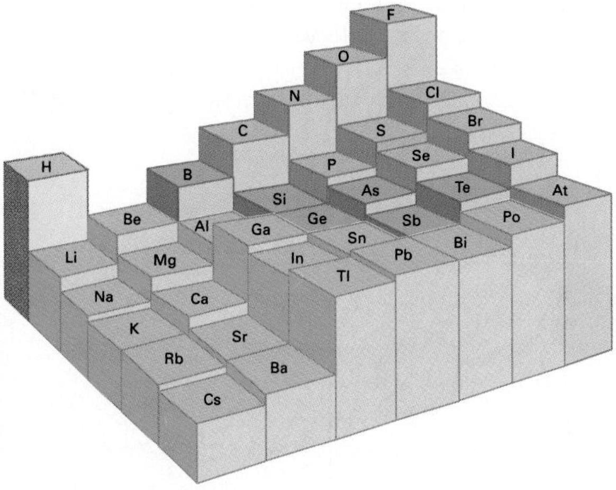

Figure 6.9 Electronegativity values of Group A elements.

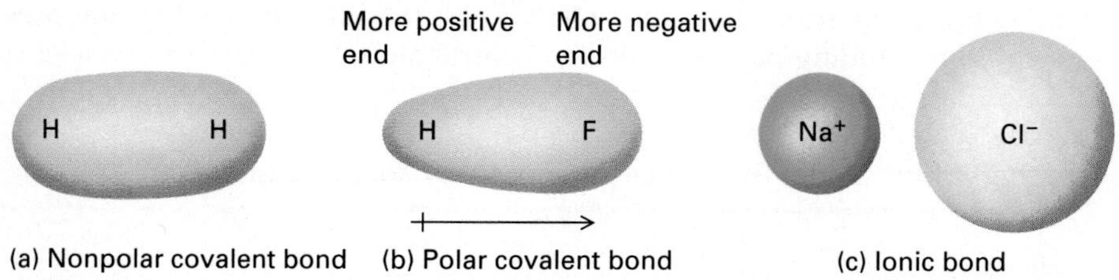

Figure 6.10 Nonpolar, polar, and ionic bonds. (a) In a nonpolar molecule such as H_2, the valence electron density is equally shared by both atoms. (b) In a polar molecule like HF, the valence electron density is shifted toward the more electronegative fluorine atom. An arrow is used to show the direction of molecule polarity, with the arrow head pointing toward the negative end of the molecule and the plus sign at the positive end of the molecule. (c) In ionic compounds such as NaCl, the valence electron(s) of the metal is transferred completely to the nonmetal to give ions.

the less electronegative hydrogen atom (Figure 6.10). The unequal sharing of electrons makes the fluorine end of the molecule more negative than the hydrogen end.

Polar bonds fall between the extremes of pure covalent and ionic bonds. In a pure covalent bond, there is no charge separation; that is, the negative charge of the electrons is evenly distributed over the bond. In ionic bonds there is complete separation of the charges, and in polar bonds the separation falls somewhere in between.

If a molecule has polar bonds, it may or may not be a polar molecule; it all depends on the three-dimensional shape of the molecule. If the electron shifts within the molecule balance out (are symmetrical), the molecule is a **nonpolar molecule.** If the electron shifts within the molecule do not balance out (are asymmetrical), the molecule is a **polar molecule** (H_2O and $CHCl_3$ in Figure 6.11).

Figure 6.11 Polar and nonpolar molecules. Polar bonds may or may not result in polar molecules. The polar bonds in BeF_2 and CCl_4 are arranged about the center atom in such a way as to cancel out the polar effect. In contrast, the polar bonds in H_2O and $CHCl_3$ molecules do not cancel as a result of the molecular shape but combine to give a polar molecule.

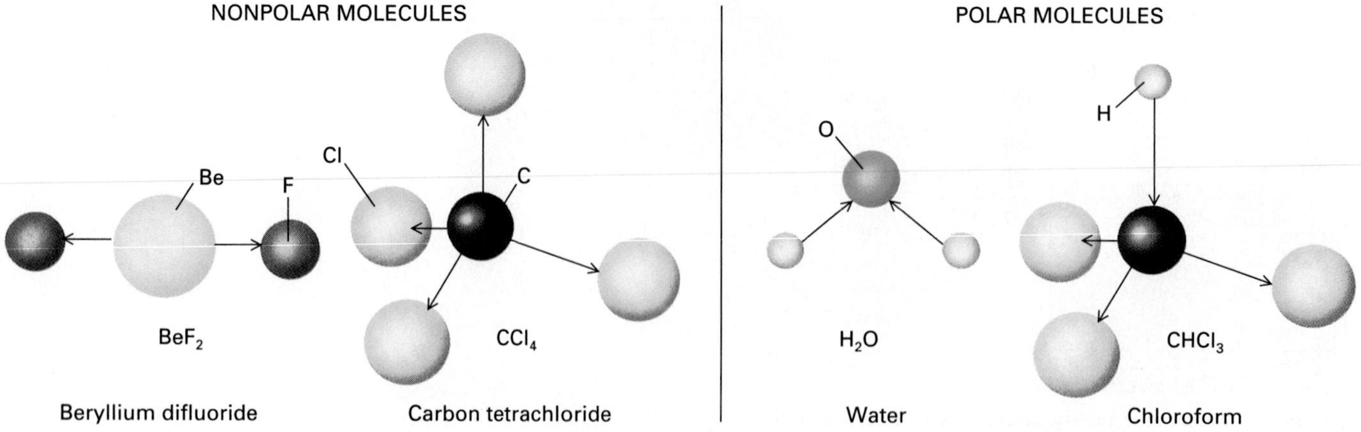

For example, carbon dioxide is a nonpolar molecule that has polar bonds. In Section 6.6 the use of the Lewis structure and VSEPR to predict the linear shape of the CO_2 molecule was described.

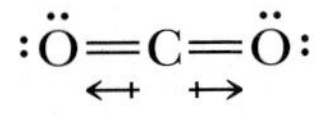

Arrows (⟷) can be added to the polar bonds to see whether the partial charges cancel. Since the electronegativity of oxygen is greater than that of carbon, the negative end of the arrow points toward the oxygen. Carbon dioxide is a nonpolar molecule because the partial charges on the oxygen atoms counteract each other.

The water molecule is an example of a polar molecule. The shape of the water molecule is bent, so the partial charges of the polar bonds do not counteract each other.

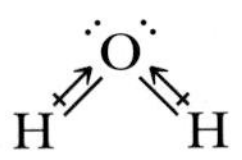

6.8 Properties of Molecular and Ionic Compounds Compared

Whether a substance is made up of polar or nonpolar molecules can have a great effect on the chemical and physical properties of the substance. For example, a rough rule of thumb is that **like dissolves like:** polar or ionic substances dissolve in polar liquids; nonpolar substances dissolve in nonpolar liquids. (See Section 7.5.) Therefore, if rubbing alcohol (isopropyl alcohol) dissolves in polar water, rubbing alcohol must be a polar molecule or have polar groups (such as —OH groups) in its structure that attract polar water molecules. Likewise, if gasoline will not dissolve in polar water, gasoline is nonpolar and will dissolve in nonpolar carbon tetrachloride (CCl_4). Water is the best solvent for ionic compounds because of the highly polar nature of water molecules (further explained in Section 7.3), although in some cases the forces between ions are so strong that not even water will dissolve an ionic compound.

The general properties of ionic and molecular compounds are summarized in Table 6.7. All these properties can be interpreted using the chemical view of matter. Ionic compounds form hard, brittle, crystalline solids with high melting points. Their crystalline nature is explained by the regular arrangement of positive and negative ions needed to place each ion in contact with those of the opposite charge. Their hardness is accounted for by the strong electrostatic forces of attraction that hold the ions in place. The differences in melting and boiling points (high for ionic compounds and low for molecular compounds) are largely accounted for by the greater strength of the electrostatic attraction between ions than between molecules. Melting requires adding enough energy to overcome these forces.

Most highly polar molecules are still more weakly attracted to each other than positive or negative ions. Whether a molecular compound is a gas, liquid, or solid is a function of the strength of the attractions between molecules.

TABLE 6.7 ■ Properties of Ionic and Molecular Compounds

Ionic Compounds	Molecular Compounds
Many are formed by combination of reactive metals with reactive nonmetals	Many are formed by combination of nonmetals with other nonmetals or with less reactive metals
Crystalline solids	Gases, liquids, and solids
Hard and brittle	Solids are brittle and weak, or soft and waxy
High melting points	Low melting points
High boiling points (700°C to 3500°C)	Low boiling points (−250°C to 600°C)
Good conductors of electricity when molten; poor conductors of heat and electricity when solid	Poor conductors of heat and electricity
Many are soluble in water	Many are insoluble in water but soluble in organic solvents
Examples: NaCl, CaF_2	Examples: CH_4, CO_2, NH_3, $CH_3CH_2CH_3$

■ The forces of attraction between molecules are discussed in Section 7.3.

Only when there are huge molecules, with attractions between many polar regions in the molecule, as in polymers (Section 14.5), do we find very hard and strong molecular substances.

Another major difference between ionic and molecular compounds is in their ability to conduct electricity, an ability that depends on the presence of mobile carriers of positive or negative charge. A solid ionic compound cannot conduct electricity because the ions are held in fixed positions. When melted or dissolved in water, the situation is different—the ions are free to move and current can flow, as demonstrated in Figure 6.12. Positive ions move toward the negative electrode and negative ions move toward the positive electrode. A compound that conducts electricity under these conditions is referred to as an

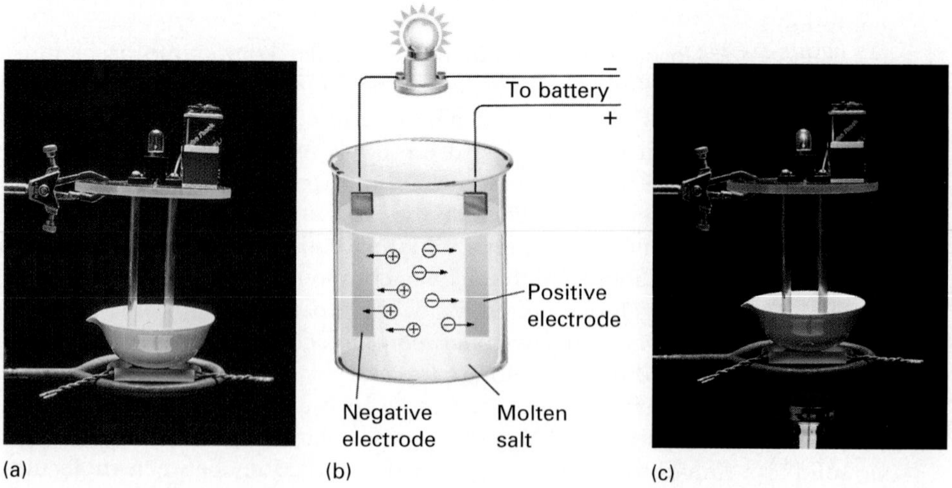

Figure 6.12 Conductivity. (a) Solid salts do not conduct electricity. (b) Molten salts do conduct electricity. Ions in a melted salt are free to move and migrate to the electrodes dipping into the melt. (c) The lighted bulb shows that the electric circuit is complete. *(c, C.D. Winters)*

electrolyte. Ionic compounds, to whatever extent they dissolve, are electrolytes in water solution because the ions separate from the crystal and can move about in the solution. Molecules, whether in pure compounds or dissolved in water or any other liquid, are not sufficiently charged to carry current and are referred to as **nonelectrolytes**.

THE PERSONAL SIDE

Linus Pauling (1901–1994)

Linus Pauling with a colleague at the Linus Pauling Institute of Science and Medicine in Palo Alto, California. Photo taken in 1985. (*UPI/Bettmann*)

Linus Pauling was born in Portland, Oregon in 1901, where he grew up as the son of a druggist. He earned a B.Sc. degree in chemical engineering from Oregon State College in 1922 and completed his Ph.D. in chemistry at the California Institute of Technology in 1925. Before joining the California Institute of Technology as a faculty member, he traveled to Europe where he worked briefly with Erwin Schrödinger and Niels Bohr (see Section 3.5). In chemistry Pauling was best known for his work on chemical bonding. Pauling, along with R. B. Corey, proposed the helical and sheetlike structures for proteins. For his bonding theories and his work with proteins, Pauling was awarded the Nobel Prize in Chemistry in 1954. Shortly after World War II, Pauling and his wife began a crusade to limit nuclear weapons, a crusade that came to fruition in the limited test ban treaty of 1963. For this effort, Pauling was awarded the 1963 Nobel Peace Prize. Never before had any person received two unshared Nobel Prizes.

■ SELF-TEST 6C

1. (a) An example of a molecule with covalent bonds in which the electrons are equally shared between the atoms is ________; (b) one where electrons are unequally shared is ________.
2. ________ is the most electronegative of all elements.
3. Electronegativity of elements (increases/decreases) down a group.
4. What are the bond angles in CH_4?
5. What are the expected bond angles in SO_3?
6. What is the AX classification for the central atom in ammonia, NH_3?
7. Which is a more polar bond?
   (a) H—H or H—F
   (b) C—H or C—O
   (c) H—N or C—N
8. Which is a polar molecule? (a) H_2O, (b) H_2, (c) O_2, (d) CCl_4.
9. Which is a nonpolar molecule with polar covalent bonds? (a) H_2O, (b) H_2, (c) O_2, (d) CCl_4.

10. (Bonding/Nonbonding) pairs of electrons repel more than (bonding/nonbonding) pairs of electrons.
11. According to VSEPR, the water molecule, with ________ bonding pairs and ________ nonbonding pairs of electrons, would be predicted to have a ( ) linear ( ) bent shape.
12. Which of the following molecules will have a bent shape? (a) CO_2, (b) SO_2, (c) NH_3, (d) SO_3, (e) H_2CO.
13. Which of the following molecules will have a bond angle of 120 degrees? (a) CO_2, (b) SO_2, (c) NH_3, (d) SO_3, (e) H_2CO.

MATCHING SET

____ 1. Ionic bonds	a. An electrically neutral arrangement of covalently bonded atoms
____ 2. Covalent bonds	b. Positive ions attracted to negative ions
____ 3. Single covalent bond	c. Smaller than parent atom
____ 4. Double covalent bond	d. Larger than parent atom
____ 5. Triple covalent bond	e. Polyatomic ion
____ 6. Metal ion	f. Bond with four shared electrons
____ 7. Nonmetal ion	g. Bond with two shared electrons
____ 8. NaCl	h. Covalent molecule
____ 9. Molecule	i. Ionic compound
____ 10. NH_3	j. Bond with six shared electrons
____ 11. Phosphate ion	k. Shared electrons

QUESTIONS FOR REVIEW AND THOUGHT

1. Give definitions for the following terms:
   (a) Cation (b) Anion
   (c) Octet rule (d) Formula unit
2. Give definitions for the following terms:
   (a) Nonmetal (b) Metal
3. Give definitions for the following terms:
   (a) Shared pair (b) Double bond
   (c) Triple bond (d) Unshared pair
   (e) Single bond (f) Multiple bond
4. Give definitions for the following terms:
   (a) Covalent bond (b) Polyatomic ion
   (c) Ionic bond (d) Binary compound
5. Give definitions for the following terms:
   (a) Nonpolar bond (b) Polar bond
   (c) Electronegativity
6. What does the term "like dissolves like" mean?
7. Describe what each of the following terms means:
   (a) Hydrocarbon
   (b) Saturated hydrocarbon
   (c) Unsaturated hydrocarbon
   (d) Fullerene
   (e) Alkenes
8. Is Ca^{3+} a possible ion under normal chemical conditions? Why or why not?
9. Predict the ions that would be formed by
   (a) Br (b) Al
   (c) Na (d) Ba
   (e) Ca (f) Ga
   (g) I (h) S
   (i) All Group IA metals
   (j) All group VIIA nonmetals
10. An ion has 12 protons, 13 neutrons, and 10 electrons. What is its charge? Consult the periodic table and write the symbol of the ion.
11. An ion has 16 protons, 16 neutrons, and 18 electrons. What is its charge? Consult the periodic table and write the symbol of the ion.
12. Why are ionic compounds neutral even though they contain charged particles?
13. Which is larger, (a) a cation or the neutral atom from which it forms? (b) an anion or the neutral atom from which it forms? Explain your answers.
14. Write the formula and name of the ionic compounds formed from atoms of each of the following pairs of elements:
   (a) Al and I (b) Sr and Cl
   (c) Ca and N (d) K and S

(e) Al and S (f) Li and N

15. What holds ionic solids together?
16. Using Figure 6.5 as your guide, take each metal ion in period 3 and combine it with each of the nonmetal ions from the same period. Write the resulting formulas.
17. Consult the *Handbook of Chemistry and Physics* in your library and find each of the compounds that you made in Question 16. Are they solids? List each melting point. Can you find any sequence to these data? Try all sodium compounds or all oxygen compounds.
18. Go to your local grocery store and see if you can find at least ten different products that have an ionic compound as a component. Don't use NaCl; it is too common. Try to find as many different ionic compounds in use as you can.
19. For each of the A groups of the periodic table, give the number of bonds an element is expected to form if it obeys the octet rule.
20. Name the following compounds:
    (a) $CaSO_4$ (b) Na_3PO_4
    (c) $NaHCO_3$ (d) K_2HPO_4
    (e) $NaNO_2$ (f) $Cu(NO_3)_2$
21. The formula for terbium phosphate is $Tb_3(PO_4)_4$. Based upon this information, write the formula for the sulfate of terbium. Would you expect this compound to be ionic? Why?
22. Write correct formulas for the ionic compounds you expect to be formed when the following pairs of elements react:
    (a) Li and Te (b) Mg and Br
    (c) Ga and S
23. Write formulas for the following compounds:
    (a) aluminum sulfate
    (b) calcium dihydrogen phosphate
    (c) potassium phosphate
    (d) ammonium nitrate
    (e) sodium carbonate
    (f) calcium hydrogen phosphate
24. Explain the difference between an ionic bond and a covalent bond.
25. Predict the general kind of chemical behavior (i.e., loss, gain, or sharing of electrons) you would expect from atoms with the following electron arrangements:
    (a) 2-8-1 (b) 2-7
    (c) 2-4 (d) 2-8-8-2
26. Predict the type of bond formed between each of the following pairs of elements:
    (a) sodium and sulfur
    (b) nitrogen and bromine
    (c) calcium and oxygen
    (d) phosphorus and iodine
    (e) carbon and oxygen
27. Many elements are known to form compounds with hydrogen. Letting E be an element in any group, the following table represents the possible formulas of such compounds.

Group:	**IA**	**IIA**	**IIIA**	**IVA**	**VA**	**VIA**	**VIIA**
	EH	EH_2	EH_3	EH_4	EH_3	H_2E	HE

Following the pattern in the table, write the formulas for the hydrogen compounds of
    (a) Na (b) Mg
    (c) Ga (d) Ge
    (e) As (f) Cl
28. Complete the following table by writing the predicted formulas for each pair of elements:

NaF	F	O	Cl	S	Br	Se
Na						
K						
B						
Al						
Ga						
C						
Si						

29. Name the following compounds:
    (a) NO (b) SO_3
    (c) N_2O (d) NO_2
30. Name the following compounds:
    (a) SF_6 (b) PF_5
    (c) NO_2 (d) SO_2
31. How many valence electrons in each of the following molecules?
    (a) Hydrogen peroxide, H_2O_2

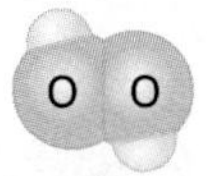

    (b) Ammonia, NH_3

    (c) Hydrogen sulfide, H_2S

32. Draw Lewis structures for the following:
    (a) CO (b) SeF_4
    (c) C_2H_4 (d) H_2S
    (e) C_2H_2 (f) C_2H_6
    (g) OH^- (h) NF_3
33. Draw Lewis structures for the following:
    (a) N_2H_4 (b) CH_3OH

(c) BCl_3

(d) PH_3

(e) ClO_3^-

(f) SO_3^{2-}

(g) NH_4^+

(h) N_2

34. What is the valence-shell electron-pair repulsion (VSEPR) model? What is the physical basis of the model?

35. What is the difference between the electron-pair geometry and the molecular geometry of a molecule? Use the water molecule as an example in your discussion.

36. Designate the electron-pair geometry for two, three, and four electron pairs around the central atom.

37. Draw the Lewis structure for each of the following molecules or ions. Use the VSEPR model to predict the electron-pair geometry and the molecular geometry.

(a) NH_2Cl

(b) HOF

(c) CS_2

(d) $SnCl_3^-$

(e) CBr_4

(f) BF_4^-

38. Draw Lewis structures and use the VSEPR model to predict the molecular shape of the following:

(a) SiF_4

(b) OCl_2

(c) CO_2

(d) BF_3

(e) PF_3

39. Draw the Lewis structure and predict the molecular shape of hydrogen cyanide, a poisonous gas.

40. Describe the trends in electronegativity in the periodic table.

41. Which is the most polar bond in the following molecules?

(a) Chloroethane, CH_3CH_2Cl

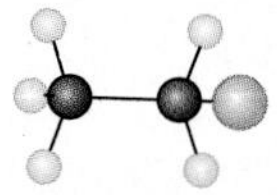

(b) Freon 12, CCl_2F_2

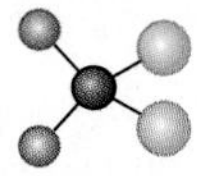

42. Which of the following molecules is polar and which is nonpolar? Explain.

(a) Acetone, CH_3COCH_3, a common solvent

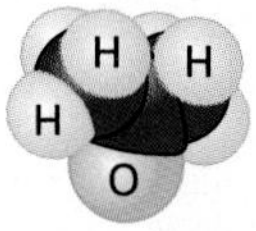

(b) Diethyl ether, $(CH_3CH_2)_2O$, an anesthetic

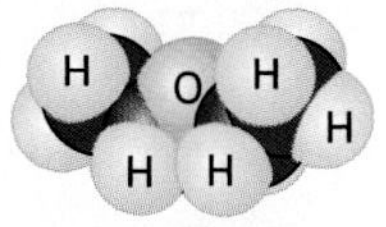

(c) Butane, $CH_3CH_2CH_2CH_3$, a common fuel

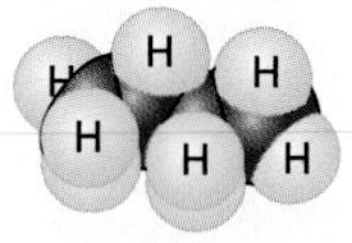

(d) Ammonia, NH_3

43. $BeCl_2$ is a compound known to contain polar Be—Cl bonds, yet the $BeCl_2$ molecule is not polar. Explain.

44. Summarize the differences between ionic, polar covalent, and nonpolar covalent bonding.

45. Which of the following compounds has the most polar bonds? (a) H—F, (b) H—Cl, (c) H—Br.

46. Which of the following molecules is (are) not polar? For each polar molecule, which is the negative and which is the positive end of the molecule? (a) CO, (b) GeH_4, (c) BCl_3, (d) HF.

47. Explain the 106.5° H—N—H bond angles of NH_3, the ammonia molecule.

48. The structural formula for ethanol, the alcohol in alcoholic beverages, is

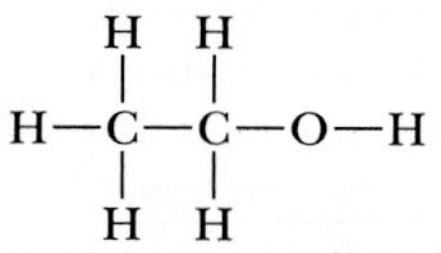

$$\begin{array}{c} \quad H \quad H \\ \quad | \quad\;\; | \\ H-C-C-O-H \\ \quad | \quad\;\; | \\ \quad H \quad H \end{array}$$

Give the total number of (a) valence electrons, (b) single bonds, (c) and bonding pairs of electrons. How many extra pairs of electrons are left? What are these called, and where should they be placed in the structural formula?

49. Acetone is a very widely used industrial solvent. It is also used as fingernail polish remover. Acetone has this structural formula:

$$\begin{array}{c} \quad H \quad O \quad H \\ \quad | \quad\;\; || \quad\;\; | \\ H-C-C-C-H \\ \quad | \quad\quad\;\; | \\ \quad H \quad\quad H \end{array}$$

Give the total number of (a) valence electrons, (b) single bonds, (c) double bonds, (d) bonding pairs of electrons, and (e) nonbonding pairs of electrons. Draw the structural formula with the nonbonding pairs included.

50. The molecule below is urea, a compound used in plastics and fertilizers.

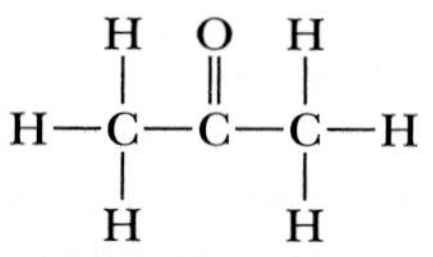

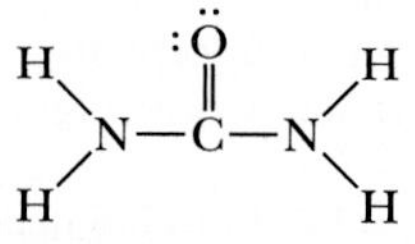

$$\begin{array}{c} \quad\quad\;\; :\overset{..}{O} \\ H \quad\quad || \quad\quad H \\ \;\backslash \quad\quad | \quad\quad / \\ \;\; N-C-N \\ / \quad\quad\quad\quad \backslash \\ H \quad\quad\quad\quad\; H \end{array}$$

Give the total number of (a) single bonds, (b) double bonds, (c) bonding pairs of electrons, and (d) nonbonding pairs of electrons.

51. Try the following in order to understand molecular shapes. Get a package of soft gumdrops, toothpicks, and a protractor.

(a) Place a toothpick in a gumdrop. Use your protractor to measure an angle of 120°. Place your second toothpick at this position relative to the first toothpick. Now measure a second angle of 120°. Place your third toothpick at this position. Make sure all three toothpicks are in the same plane. This

represents three bonding pairs on the surface of a sphere (your gumdrop) and is referred to as trigonal planar.

b. Place a toothpick in a gumdrop. Use your protractor to measure an angle of 109.5°. Place a second toothpick at this position. Measure another angle of 109.5°. If you place the third toothpick in the same plane as the other two you will not have room for the fourth toothpick. We would like to have four toothpicks in the gumdrop, all an equal distance from each other and at angles of 109.5°. See if you can accomplish this task. Four bond pairs on the surface of a sphere with the conditions previously described represent a tetrahedral geometry.

52. Choose five ionic and five covalent compounds. Using the table of properties for each class of compounds and the *Handbook of Chemistry and Physics,* show how each compound that you have chosen will meet these criteria for classification.

PROBLEMS

1. If a cation has a radius of 85 pm, what is its diameter? If an anion has a radius of 145 pm, what is its diameter?
2. Consider the ions from Problem 1. Imagine that they are quarters and pennies. Which one is which? Now arrange the quarters and pennies in a row as quarter-penny-quarter-penny-quarter-penny-quarter. How long is this row?
3. Let's put another row of ions into this arrangement. Directly under each quarter place a penny. The penny needs to be touching the quarter. Will quarters now fit into the vacant spaces under the pennies? What you have made is the start of a lattice. This is a very important structural feature of ionic compounds. As you can observe, the ion size plays a major part in the construction of a lattice.

CHAPTER 7

States of Matter and Solutions

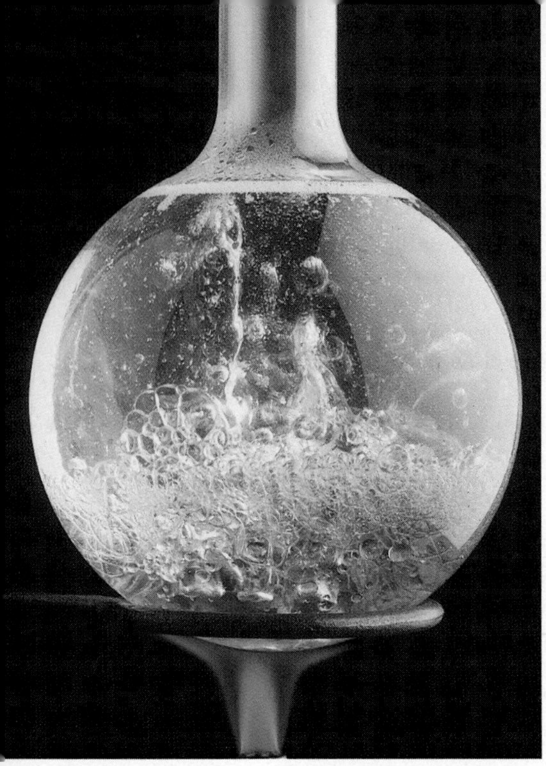

A boiling solution. *Boiling is one of the changes of state described in this chapter.* *(Richard Megna/Fundamental Photographs)*

Why study the states of matter? For one thing, to understand what happens to atoms, molecules, and ions when energy is added to or removed from a sample of matter. It is the interactions among these particles that determine whether a substance exists as a gas, a liquid, or a solid at a particular temperature. Water is our best example. When many water molecules are in rigid positions at a low temperature, we call the sample ice—water in the *solid state.* When the water molecules are much farther apart, we call the sample water vapor or steam—water in the *gaseous state.* Between ice and water vapor or steam is the *liquid state* of water, the most common compound on the face of this planet, since more than 70% of the earth's surface is covered by liquid water.

- How are the states of matter related to one another?
- How is energy related to changes in the states of matter?
- What theoretical model explains the behavior of the states of matter?
- What factors cause gases to condense to form liquids and liquids to solidify?
- What factors determine solubility?
- Why do metals conduct electricity while nonmetals do not?
- What are superconducting solids, and what promise do they hold for the future?

The state of matter or, simply, the *physical state* of a sample of matter depends on the temperature, and the state determines many of the properties of the sample. You might be harmed if you were hit with 200 g of solid water (ice), but only annoyed by 200 g of liquid water squirted on you from a garden hose. On the other hand, 200 g of hot steam might also harm you. In these three examples, the quantity of water is the same but the physical states are different, and because of that the properties are different.

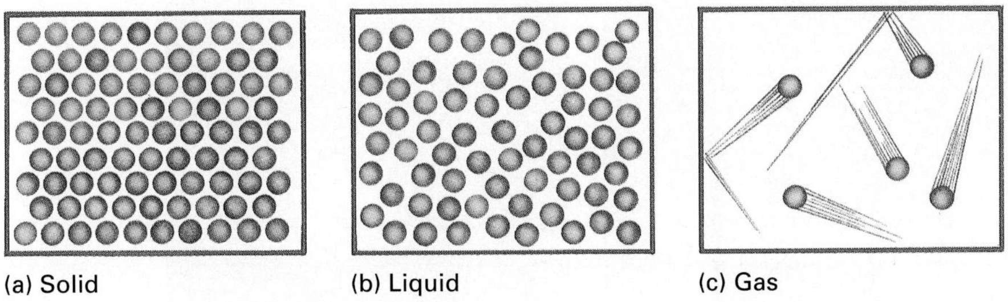

Figure 7.1 The three states of matter. The particles represented by the circles can be atoms, molecules, or ions (in liquids and solids). (a) In a solid, the particles or ions are in fixed positions and can only vibrate in place. (b) In a liquid, the particles move about at random, alone or in clusters. (c) In a gas, particles are very far apart and move rapidly in straight lines.

Many substances can exist in all three states, that is, solid, liquid, and gas. Water is a perfect example of this. Other substances, because they decompose on heating, can exist only as solid and liquid and still fewer can exist only as solids. Have you ever, for example, seen sugar boil? Have you seen it melt? Perhaps, but sugar, like many compounds, partially decomposes as it melts, to produce a tasty caramel-flavored mixture. Such decomposition occurs when substances are not stable to heat. Instead of melting, these substances undergo chemical reactions that break bonds and form new ones. The sample of matter is no longer the same kind of matter.

The properties associated with the gaseous, liquid, and solid states are examined in this chapter. First, it is important to understand the model of behavior at the atomic level that is used to explain these properties.

7.1 The Kinetic Molecular Theory: How and Why Does Matter Change State?

A theory that helps us interpret the physical properties of solids, liquids, and gases is the **kinetic molecular theory** of matter. According to this theory, the atoms, ions, or molecules in every kind of matter are in constant motion. In a solid these particles are packed closely together in a regular array, as shown in Figure 7.1*a*. The particles vibrate back and forth about their average positions, but seldom does one squeeze past its immediate neighbors to come into contact with a new set of particles. Because the particles are packed so tightly and in such a regular arrangement, a solid is rigid, its volume is fixed, and the volume of a given mass is small.

■ Particles, as referred to in the kinetic molecular theory, are atoms, ions, or molecules.

Sometimes the external shape of a solid reflects the internal arrangement of its particles, like the cubic sodium chloride crystals (see Figure 7.12). This relation between the observable structure of the solid and the internal orderly arrangement of its particles is one reason scientists have long been fascinated by the shapes of crystals.

The kinetic molecular theory of matter can also be used to interpret the properties of liquids and gases, as shown in Figure 7.1*b* and *c*. Liquids and gases have the property of being **fluid**—that is, they flow—because their atoms, ions, or molecules are not so strongly attracted to each other as they are in solids. Not being confined to specific locations, the particles in a liquid can move past one another. For most substances, the particles are a little farther apart in a liquid than in the corresponding solid, so that the volume occupied by a given mass of the liquid is a little larger than the volume occupied by the same mass of the solid. This means the liquid is less dense than the solid, and

(a) (b)

Figure 7.2 Water (H_2O) and benzene (C_6H_6). On the left, ice floats on water. On the right, solid benzene (melting point 5.5°C) sinks in liquid benzene. *(C.D. Winters)*

the solid form of a sample of matter sinks in its liquid (Figure 7.2). There is a rather important exception to this rule: solid water floats on liquid water. The importance of this property of water is discussed later.

According to the kinetic molecular theory, no particle in a liquid goes very far without bumping into one of its neighbors (somewhat like people in a crowd exiting a stadium). This means the particles in a liquid collide with each other constantly. By contrast, in a gas the atoms or molecules are very far from one another and, although they are moving quite rapidly, they spend very little time in close contact with one another. (In air at room temperature, for example, the average molecule is going faster than 1000 miles per hour and moves about 400 times its own diameter before striking another molecule.) Since the particles in a gas are in constant motion, they move about to fill any container they are in; hence, a gas has no fixed shape or volume.

■ We don't ordinarily encounter gases composed of ions because they exist only at high temperatures. Such gases, called *plasmas*, result when a gas is subjected to a high-energy source. A lightning strike, for example, creates a plasma consisting of ionized nitrogen and oxygen molecules. These ions quickly re-form more stable molecules. In the laboratory, a plasma can be maintained indefinitely if sufficient energy is continually supplied.

In addition, the kinetic molecular theory states that the higher the temperature, the faster the particles move. A solid melts when its temperature is raised to the point at which the particles vibrate fast enough and far enough to get away from the attraction of their neighbors and move out of their regularly spaced positions. As the temperature goes higher, the particles move even faster, until finally they escape their neighbors and become independent; the substance becomes a gas.

Compressed air in use. *(Tom Pantages)*

7.2 Gases and How We Use Them

Gases surround us in our atmosphere. We breath a mixture of nitrogen, oxygen, and other gases. Every breath we take carries with it the oxygen gas we need to burn the foods we eat. When we exhale, the carbon dioxide gas that is produced by the food-burning processes in our cells leaves our bodies. This is one of the ways we rid ourselves of waste materials.

Compressed air also has many uses. As children, perhaps we played on toys containing compressed air, and balloons are a favorite of all ages. SCUBA

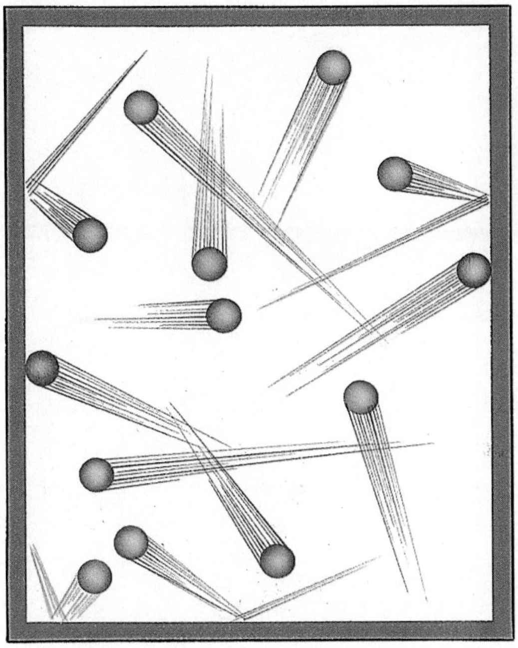

Figure 7.3 Gas pressure. The pressure is caused by gas molecules bombarding the walls. Common pressure units are pounds per square inch (psi), atmospheres (atm), and millimeters of mercury (mmHg).

(self-contained underwater breathing apparatus) diving gear uses compressed air, and we rely on compressed air in automobile and bicycle tires to help cushion some of the bumps on the road.

Interestingly, all gases possess a set of common properties. At constant temperature, all gases expand when the surrounding pressure decreases and contract when the pressure increases. At constant pressure, all gases expand with increasing temperature and contract with decreasing temperature. And all gases have the ability to mix in any proportion with other gases. These properties are explained by the kinetic molecular theory: The gas particles are far apart, move fast, and have very little chance to interact. Their molecular properties, such as size, number of electrons, and shape, have virtually no effect on the properties of the gas as a whole.

A sample of gas confined inside a container exerts a **pressure,** which, according to the kinetic molecular theory, is caused by the individual particles of the gas sample striking the walls of the container (Figure 7.3). The earth's atmosphere, a mixture of gases, exhibits a pressure that is dependent on the altitude relative to sea level and temperature.

Another general property of gases is **compressibility.** All gases can be compressed by applying a pressure on a confined sample (Figure 7.4). This property of all gases, known as *Boyle's law* after Robert Boyle who discovered it in 1661, is explained by the great distances between gas particles—applying pressure only confines the particles in a smaller space. Of course, a gas will always expand if the pressure is reduced. If a sample of gas is released into deep space, where the pressure is effectively zero, the molecules would begin randomly moving away from one another and eventually would become attracted to some nearby star system.

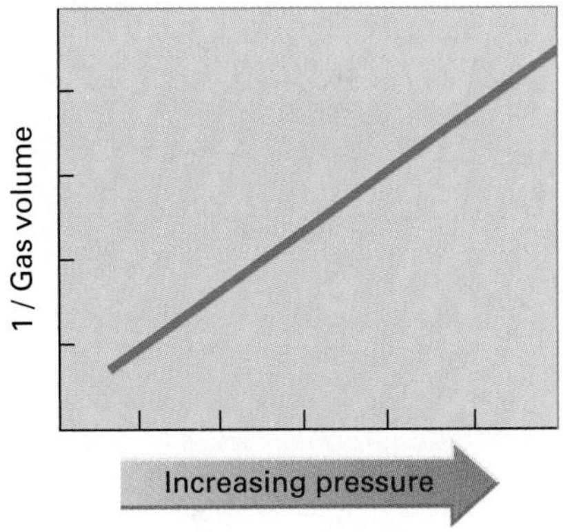

Figure 7.4 Pressure and volume relationship for a gas (at constant temperature). As the pressure increases, the volume decreases. To get a straight line, the pressure is plotted versus 1/gas volume (an example of an *inversely proportional* relationship).

■ The quantitative application of Boyle's and Charles' laws is illustrated in Section 7.8.

Firing jets to adjust course of the Space Shuttle Columbia. The gas released will continue to expand into the universe until the molecules are trapped in the gravity field of our Sun, some other star, or some other planet. *(NASA)*

Yet another general property of all gases is that of expansion on heating and contraction on cooling, as illustrated in Figure 7.5. This behavior of gases is called *Charles' law* after Jacques Charles who discovered it in 1787. The plots in Figure 7.5 represent the changes in volume of different-size samples of hydrogen and oxygen when they were cooled in separate containers. You can see that the gases occupy smaller volumes as the temperature drops. This behavior is in agreement with the kinetic molecular theory: As the temperature of a gas decreases, the gas particles move more slowly, collide with the container walls with less force and less frequency, and thus exert less pressure.

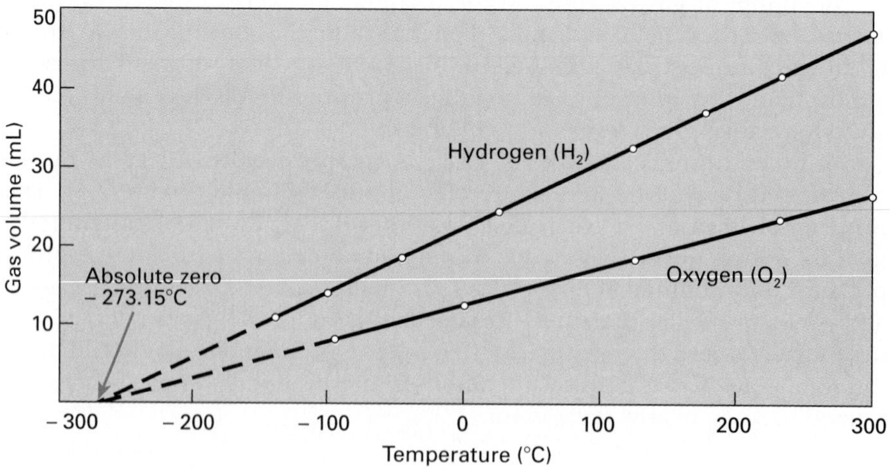

Figure 7.5 Temperature and volume relationship for a gas (at constant pressure). As the temperature increases, the pressure increases, so that plotting the temperature versus the volume gives a straight line (a *directly proportional* relationship).

Interestingly, the two plots of volume versus temperature shown in Figure 7.5 intersect at a temperature of −273.15°C, where the volume appears to be zero. The plot for any gas would reach zero volume at this temperature, although the volume does not actually become zero—all gases liquefy before reaching this temperature.

In 1848 William Thompson, known as Lord Kelvin, proposed that it would be convenient to have a temperature scale on which the zero point is −273.15°C, the temperature known as **absolute zero.** This temperature is now called *0 kelvin* and is the basis of the Kelvin temperature scale.

■ To convert a temperature from the Celsius scale to the Kelvin scale requires adding 273.15. For example, a common room temperature of 23.5°C is 23.5° + 273.15 = 296.7 kelvins (written 296.7 K). To go the other way, from a Kelvin temperature to a Celsius temperature, requires subtracting 273.15. The kelvin is the SI unit for temperature.

Perhaps one of the more interesting properties of all gases is that of **miscibility**—the ability to mix in all proportions with other gases. The miscibility of gases is explained by the great distances between gas molecules. In effect, "there is always room for some more molecules." John Dalton first recorded this property of gases; when two or more gases are mixed, the total pressure they exert on a container is the sum of the pressures each gas would exert if it were the only gas in the container. For example, the earth's atmosphere—a mixture of nitrogen, oxygen, water vapor, argon, carbon dioxide, and other gases—can easily allow any other gas to mix with it, and the total barometric pressure of the atmosphere is actually the sum of the pressures of the component gases. If the pressures of the four main components of the atmosphere are written as P_{O_2}, P_{N_2}, P_{H_2O}, P_{Ar}, and P_{CO_2}, then the atmospheric pressure, P_{atm}, is

$$P_{atm} = P_{O_2} + P_{N_2} + P_{H_2O} + P_{Ar} + P_{CO_2}$$

The kinetic molecular theory explains the miscibility of gases as being due to the great distances between the particles in the gas sample.

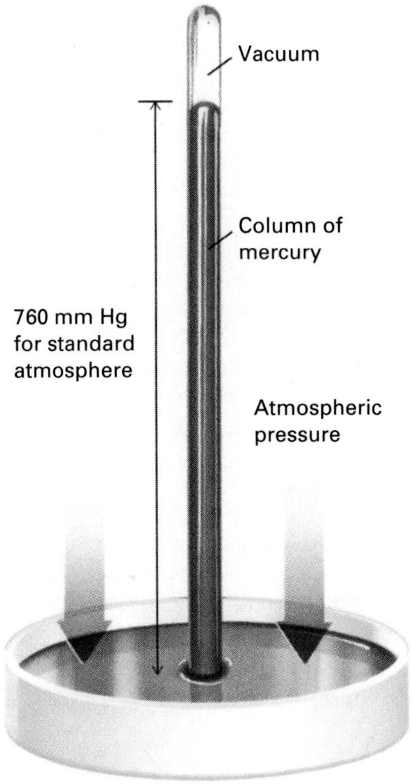

A Toricellian barometer, an inverted column of mercury standing in a pool of mercury. The pressure of the mercury column is balanced by the atmospheric pressure on the mercury in the dish. The height of the mercury column, a measure of pressure, is reported in inches or millimeters of mercury. A pressure of 1 atmosphere (atm) supports a mercury column 760 mm high. The barometer was invented by Evangelista Torricelli in 1643.

THE PERSONAL SIDE

Jacques Alexandre Cesar Charles (1746–1823)

Jacques Charles, a French chemist, was most famous in his lifetime for his experiments in ballooning. After an earlier flight in June 1783 by the Montgolfier brothers using a large spherical balloon made of linen and filled with hot air, Charles led a group to fill a balloon with hydrogen for its maiden voyage in August of 1783. Since linen would easily allow the smaller hydrogen molecules to escape, Charles made the balloon of silk coated with a rubber solution. Inflating the balloon to its final diameter took several days and required nearly 500 pounds of acid and 1000 pounds of iron to generate the hydrogen gas. A huge crowd watched the ascent on August 27, 1783. The balloon stayed aloft for almost 45 minutes and traveled about 15 miles. When it landed in a nearby village, the people were so terrified that they tore it to shreds.

An early engraving of Jacques Charles and his balloon adventure. (Stock Montage)

SCIENCE AND SOCIETY

Limiting "Air Toxics"

The earth's atmosphere is a gigantic mixing bowl for gaseous substances. Besides the nitrogen, oxygen, and other gases naturally present, numerous molecules of potentially harmful substances called *air toxics,* as defined by the Clean Air Act Amendments of 1990, can be found in air samples. These "foreign" molecules are especially common near cities and industrial plants, but natural processes such as volcanoes and decaying animal and vegetable matter can also account for them. Many of these substances find their way into the air because they are gases themselves at room temperature or because they are the vapor from low-boiling liquids (volatile liquids). Although the natural sources of these air toxics cannot be easily controlled, governmental agencies have begun an active program of limiting them at their other major sources—industrial sites that release them into the atmosphere during normal business activity or as the result of accidents.

Virtually all types of businesses, large or small, are affected by the air toxics release rules. These include, but are not limited to, auto manufacturing and repair, bakeries, distilleries, dry cleaners, furniture manufacturing and repair shops, gasoline service stations, hospitals, and print shops. Smaller

Some Air Toxics Listed By the Clean Air Act Amendments of 1990

Compound	Uses
Chlorine, Cl_2	Water treatment, manufacturing other chemicals
Chloroform, CH_3Cl	Solvent in manufacturing other chemicals
Ethylene dibromide, $H_2C{=}CBr_2$	Fumigant for certain agricultural crops
Formaldehyde, H_2CO	Making plywood, adhesives
Hexane, C_6H_{14}	Solvent for making other chemicals
Styrene, $C_6H_5CH{=}CH_2$	Making plastics
Vinyl chloride, $CH_2{=}CHCl$	Making plastics

■ Individuals living near a large commercial bakery in Connecticut expressed great sorrow that soon, as a result of regulations of the Clean Air Act, the smell of baking bread will no longer fill the air in their neighborhood.

To illustrate how gases mix with one another, consider what happens when a person wearing a strong perfume enters a room. The smell of the perfume is noticed immediately by those close by and eventually by everyone in the room. The perfume contains **volatile,** meaning easily vaporized, compounds that gradually mix with the other gases in the room's atmosphere. Even if there were no apparent movement of the air in the room, the smell of perfume would eventually reach everywhere in the room. This mixing of two or more gases due to random molecular motion is called **gaseous diffusion.** Given time, the molecules of one component in a gas mixture will thoroughly and completely mix with all the other components to form a homogeneous mixture.

7.3 Intermolecular Forces

■ Methane, CH_4, is the major component of natural gas.

Although atoms within molecules are held together by strong chemical bonds (see Chapter 6), the attractive forces between two separate molecules are much weaker. For example, only about 1% of the energy required to break one of the C—H bonds in a methane (CH_4) molecule is required to pull two methane molecules away from one another. These small forces between mole-

businesses are allowed to emit up to 50 tons per year of any regulated air toxic and up to 75 tons per year of all the regulated compounds taken together. Larger businesses are allowed to emit only 10 tons per year of any one air toxic or 25 tons annually of a combination of the compounds.

Some small business owners, for example, local dry cleaners and bakeries, are complaining bitterly about the cost of complying with the law. Some even predict that they will be forced to close. Do you think exceptions should be made? If so, how would the decisions be made?

Towards cleaner air. The gas pump shown here is equipped with a hose and nozzle designed to minimize escape of gasoline vapors. *(Craig Newbauer/Peter Arnold, Inc.)*

cules are called **intermolecular forces.** Even though these attractions are small, they give samples of matter exceedingly important properties. For example, intermolecular attractions are responsible for the fact that all gases condense to form liquids and solids. Both pressure and temperature play a role in the liquefaction of a gas. When the pressure is high, the volume decreases and the molecules of a gas are close together. This condition allows the effect of attractive forces to be appreciable. When the temperature of a gas is lowered, the average kinetic energy of the molecules is lowered until the molecules are not energetic enough to overcome one another's attractions. At some combination of temperature and pressure, enough gas molecules stick together to form small droplets of liquid. This phenomenon is called **condensation** and will occur at different temperatures and pressures for different gases. As a sample of condensed gas is cooled further, the liquid eventually solidifies as intermolecular forces of attraction confine the molecules to specific positions.

For polar molecules (Section 6.7), intermolecular forces act between the positive end of one polar molecule and the negative end of an adjacent molecule. Molecules of SO_2, for example, are polar, with the partially negative regions of one molecule (represented by δ^-) being attracted to the partially positive region (represented by δ^+) of an adjacent molecule.

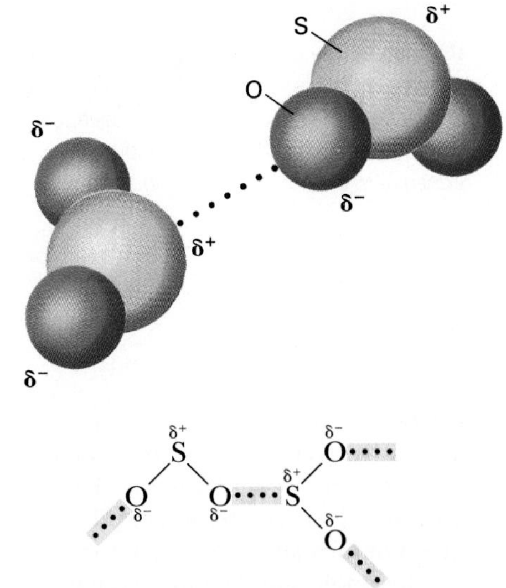

■ Polar sulfur dioxide molecules attracted to one another (oxygen atoms are more electronegative than sulfur atoms)

■ For a sample of water vapor being cooled at atmospheric pressure from some high temperature like 110°C, condensation will occur at 100°C.

Even nonpolar molecules can have momentary unequal distribution of their electrons, which results in weak intermolecular forces. For example, why would nitrogen, composed of nonpolar molecules, liquefy? Or why would carbon dioxide, also composed of nonpolar molecules, form a solid? When the molecules get close enough, one molecule will cause an uneven distribution of charge in its neighbor. The two molecules become momentarily polar and are attracted to one another (Figure 7.6). These induced attractive forces are weak but become more pronounced when the molecules are larger and contain more electrons.

Hydrogen Bonding

An especially strong intermolecular force called **hydrogen bonding** acts between molecules in which a hydrogen atom is covalently bonded to a highly electronegative atom with nonbonding electron pairs. The highly electronegative atom may be fluorine, oxygen, or nitrogen. The hydrogen bond is the attraction between such a hydrogen atom with a partial positive charge on one molecule (positive because it is attached to a very electronegative atom) and a small, very electronegative atom (F, O, or N) of another molecule which has a partial negative charge (see Section 6.7). The greater the electronegativity of the atom connected to H, the greater the partial positive charge on the H and, hence, the stronger the hydrogen bond between it and a partially negative atom on another molecule. Hydrogen bonds are typically shown as dotted lines between the atoms, and partial charges are shown as δ^+ and δ^-.

$$\overset{\delta^+}{H}-\overset{\delta^-}{F} \cdots\cdots \overset{\delta^+}{H}-\overset{\delta^-}{F} \cdots\cdots \overset{\delta^+}{H}-\overset{\delta^-}{F}$$

↑ Hydrogen bond (the dotted line) ↑ Covalent bond (the H—F line)

■ It is possible for a hydrogen bond to exist within a molecule when an H atom, bound to an electronegative element like nitrogen, oxygen, or fluorine, is attracted to another electronegative atom in the same molecule.

Water provides the most common example of hydrogen bonding. Hydrogen compounds of oxygen's neighbors and family members in the periodic table are all gases at room temperature. But water is a liquid at room temperature, and this indicates a strong degree of intermolecular attraction. Figure 7.7 illustrates that the boiling point of H_2O is about 200°C higher than would be predicted if hydrogen bonding were not present.

Since each hydrogen atom can form a hydrogen bond to an oxygen atom in another water molecule, and since each oxygen atom has two nonbonding electron pairs, each water molecule can form a maximum of four hydrogen

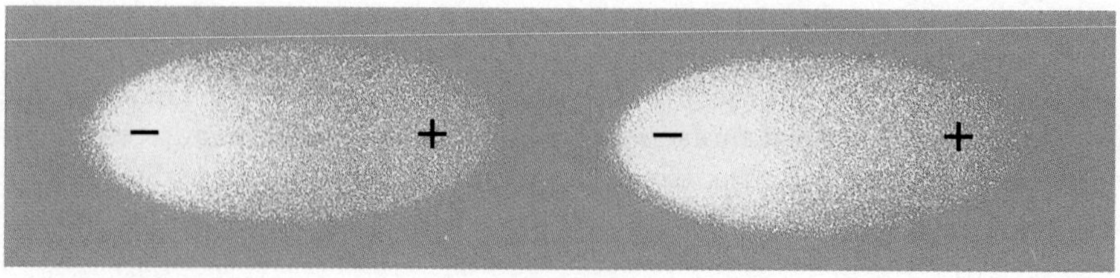

Figure 7.6 Momentary attraction of molecules. A shift in electrons to the left in one molecule creates a slightly positive region on the other end, which in turn attracts electrons from another molecule. At a low enough temperature, this effect pulls together even the small, nonpolar molecules of nitrogen (N_2).

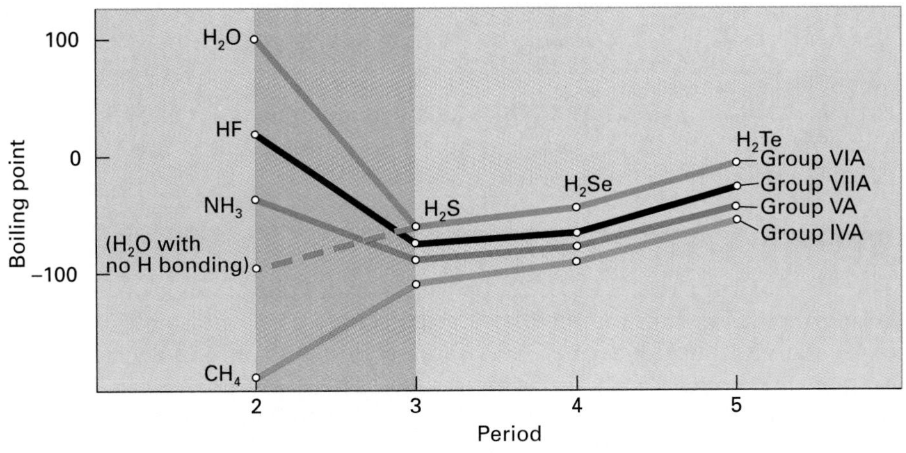

Figure 7.7 Boiling points of simple hydrogen-containing compounds. Lines connect molecules in which hydrogen combines with atoms from the same periodic table group. As shown for water, the point at which the compounds would boil if there were no hydrogen bonding is found by following the straight parts of the lines on the right down to the left.

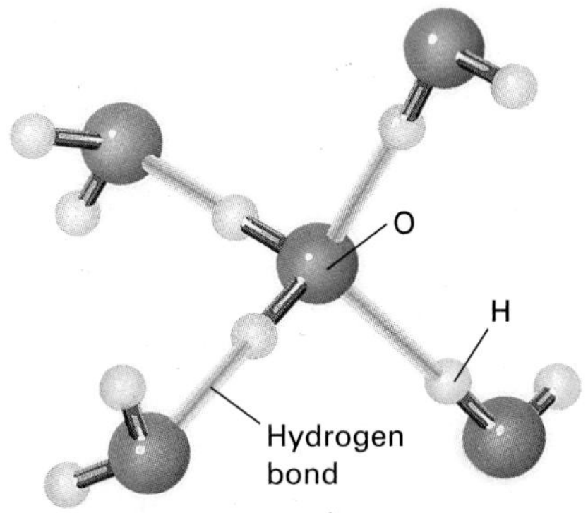

Figure 7.8 The four hydrogen bonds between one water molecule and its neighbors.

bonds to four other water molecules (Figure 7.8). The result is a tetrahedral cluster of water molecules around the central water molecule. In ice, hydrogen bonding is more extensive than in liquid water, and the resulting open structure is less dense than that of liquid water.

Although hydrogen bonds are much weaker than ordinary covalent bonds, they play a key role in the chemistry of life. Later chapters in this text will discuss hydrogen bonding in connection with the properties of a number of substances such as alcohols, proteins, carbohydrates, and DNA.

THE WORLD OF CHEMISTRY

Making Snow the Artificial Way

Program 12, *Water*

Nature doesn't always cooperate by sending snow when it's needed. Instead, snow usually comes when it's not wanted or least expected. On ski slopes good snow cover is a must. When nature doesn't provide, machines and chemistry must step in. Making snow on a cold day is relatively easy. Water freezes at temperatures below 0°C, so blowing small water droplets into cold air will get the job done, although some care must be taken that the water forms snowflakes and not larger ice pellets that easily become treacherous for skiers. What happens when the temperature doesn't cooperate? Snow-

Snow-making at ski resort. (Tom Pantages)

Max, a freeze-dried protein produced by bacteria, can be used to make water freeze at temperatures *above* 0°C. This is possible because proteins contain sites that can hydrogen-bond with water molecules. (Hydrogen bonding with water is the main reason your hair, a protein, becomes softer and more managable when wet.) The snow-making, bacteria-produced proteins are especially capable of forming hydrogen bonds with water molecules, which then become as extensively hydrogen bonded as they would if the temperature were at 0°C or below. The result is a good artificial snow, produced at temperatures above freezing. The next time you are enjoying skiing on artificial snow, you can thank some lowly bacteria that made hydrogen bonding possible that day.

EXAMPLE 7.1 *Intermolecular Forces*

Explain why 1-propanol, $CH_3CH_2CH_2OH$, has a boiling point of 97.2°C, while a compound with the same number of atoms, ethyl methyl ether, $CH_3CH_2OCH_3$, has a boiling point of 7.4°C.

SOLUTION

The alcohol contains —OH groups whose hydrogen atoms can hydrogen bond to —OH groups on neighboring molecules. The ether molecules have oxygen atoms, but unlike the oxygen atoms in the alcohol molecules, these oxygens are bonded only to carbon atoms. There are no hydrogen atoms bonded directly to the oxygen atom in the ether molecule that can become partially charged and form hydrogen bonds with neighbor molecules. As a result, the ether has a much lower boiling point than the alcohol.

Exercise 7.1
Which compound would have the higher boiling point, CH_3CH_2OH or CH_3OCH_3? Explain why.

■ SELF-TEST 7A

1. Arrange the states of matter (liquid, gas, solid) by increasing order of the particles.
2. According to the kinetic molecular theory, as temperature increases, molecular motion (decreases/increases).
3. Name the two states of matter in which the particles are very close to each other, on average.
4. Which two states of matter can be described as fluid?
5. Which state of matter is compressible? Which two states of matter are noncompressible?
6. As the pressure on a sample of gas is increased, the ____________ decreases at constant temperature.
7. As the temperature of a sample of gas decreases, the volume ____________ at constant pressure.
8. All gases mix with one another in all proportions. (a) True, (b) False.
9. Which kind of intermolecular force is stronger—hydrogen bonding or attractions caused by shifting electrons in molecules?
10. Two electronegative elements associated with hydrogen bonding are fluorine and nitrogen. The other element is ____________.

7.4 Water and Other Pure Liquids

There would be no life as we know it on Earth without water and its unique properties. Certainly there are other media on Earth and in the universe wherein much chemistry occurs. However, on Earth the chemistry in water solutions and the chemistry of water dominate. Water plays an important role as a reactant, a product, or a solvent in most of the chemical reactions in our environment.

Ice melting under pressure. *(Yoav Levy/ Phototake)*

Some Properties of Water

1. ***Water is a liquid at room temperature as a direct consequence of hydrogen bonding between adjacent water molecules.*** Pure water is a liquid between 0° and 100°C.
2. ***The density of solid water (ice) is less than that of liquid water.*** Put another way, water expands when it freezes. If ice were a normal solid, it would be denser than liquid water, and lakes would freeze from the bottom up. This would have disastrous consequences for marine life, which could not survive in areas with winter seasons. The application of pressure causes ice to melt. This is a consequence of the structure of ice. The pressure causes ice to change to a form with a smaller volume, and since liquid water occupies a smaller volume than does ice, the ice converts to a liquid.
3. ***Water has a relatively high heat capacity.*** Water can absorb large quantities of heat without large changes in temperature, because the added heat can break hydrogen bonds instead of increasing the temperature.

DISCOVERY EXPERIMENT

Melting Ice with Pressure

You can do this experiment in a kitchen sink. Obtain a piece of thin, strong, single-stranded wire about 50 cm long, two weights of about 1 kg each (two small milk jugs filled with water will work nicely), and a piece of ice about 25 cm by 3 cm by 3 cm. The ice can be in the form of either a cylinder or bar and can be made by pouring some water into a mold made of several thicknesses of aluminum foil and placing it into a freezer. A short piece of plastic pipe, sealed at one end, or a paper milk carton will also work fine as a mold.

Fasten a weight to each end of the wire. Then support the bar of ice in the sink in such a way that the wire can lie across the ice with the weights hanging on either side. Allow sufficient distance for the weights to travel downward. Observe what happens every minute or so. Is the wire applying pressure to the ice? Why does pressure cause the ice to melt? What happens as the wire moves through the ice?

■ **Heat capacity** is defined as the amount of heat required to raise the temperature of a sample of matter of a given size by 1°C. The heat capacity of water is 1 calorie per gram.

■ **Heat of vaporization** is defined as the heat required to vaporize a given quantity of liquid at its boiling point. The heat of vaporization of water at 100°C is 540 cal per gram, or 9.72 kcal per mole.

■ Benzene, chloroform, ethyl alcohol, and octane—all organic compounds that are liquids at room temperature—have surface tensions about one-third as strong as that of water.

For comparison, the heat capacity of water is about ten times that of copper or iron. Water's heat capacity accounts for the moderating influence of lakes and oceans on the climate. Huge bodies of water absorb heat from the Sun and release the heat at night or in cooler seasons. The Earth would have extreme temperature variations if it were not for this property of water. By contrast, the temperatures on the surface of the Moon and the planet Mercury vary by hundreds of degrees through the light and dark cycles.

4. ***Water has a high heat of vaporization.*** For a liquid to vaporize, heat is required. The **heat of vaporization** of a liquid is a measure of the intermolecular attractions holding the molecules together in the liquid. Water has one of the highest heats of vaporization—for 1 g of water at 100°C it is 540 cal. A consequence of this high heat of vaporization is the cooling effect that occurs when water evaporates from moist skin. Evaporating water molecules take with them a considerable amount of energy, which was needed to overcome the attractions between the leaving molecules and those remaining behind.

5. ***Water has a high surface tension.*** Unlike gases, liquids have surface properties, and these are extremely important in the overall behavior of many liquids. Molecules beneath the surface of the liquid are completely surrounded by other molecules and experience forces in all directions due to intermolecular attractions. By contrast, molecules at the surface are attracted only by molecules below or beside them (Figure 7.9). This unevenness of attractive forces at the surface of the liquid causes the surface to contract, making it act like a skin. The energy required to overcome the "toughness" of this liquid skin is called the **surface tension,** and it is higher for liquids that have strong intermolecular attractions. Water's surface tension is high compared with those of most other liquids because of the extensive hydrogen bonding that holds water molecules to each other.

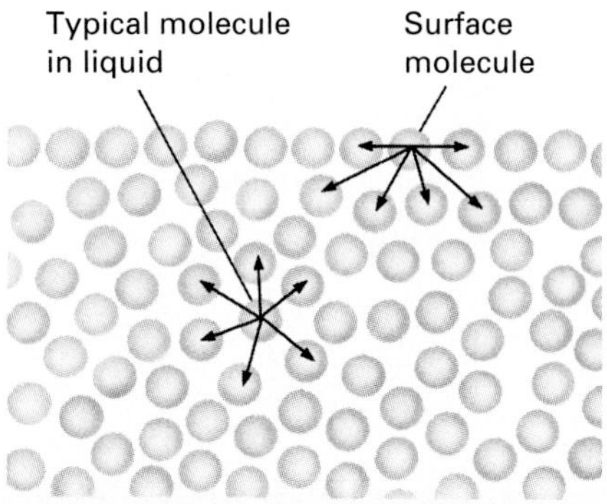

(a)

Figure 7.9 Surface tension. (a) The surface is strengthened by intermolecular forces attracting surface molecules. (b) The water strider, a lightweight insect that does not provide enough force per unit area to break through the surface tension. Note that the strider does not walk on the sharp ends of its "toes." (c) With care, a paper clip can be placed so it won't sink in the water. (d) On a dirty car, this wouldn't happen. The dirt would overcome the surface tension of the water droplets. *(b: Runk/Schoenberger/Grant Heilman Photography; c: C.D. Winters; d: Richard Megna/Fundamental Photographs)*

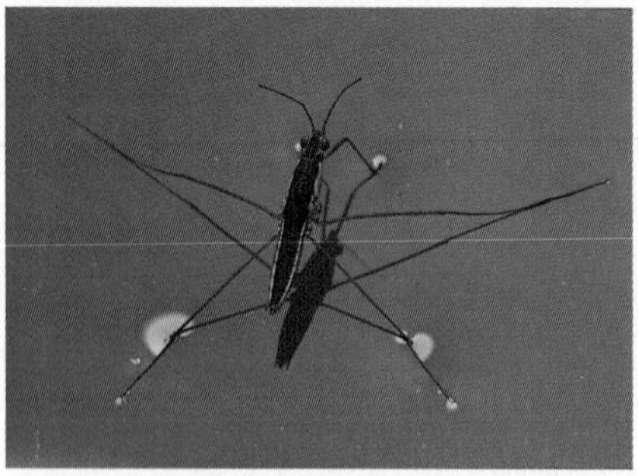

(b)

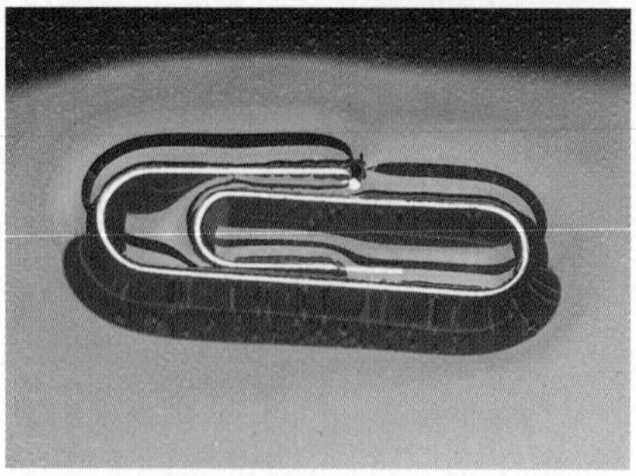

(c)

(d)

6. ***Water is an excellent solvent, often referred to as the universal solvent.*** Because it is such a good solvent, water from natural sources is not pure water but a solution of substances dissolved by contact with water. (See Section 20.3.)

Some More About Properties of Pure Liquids

In a liquid the molecules are close enough together that, unlike a gas, a liquid is only slightly compressible. However, the molecules remain mobile enough that the liquid flows. Because they are difficult to compress and their molecules are moving in all directions, confined liquids can transmit applied pressure equally in all directions. This property is used in the *hydraulic fluids* that operate automotive brakes and airplane wing surfaces, tail flaps, and rudders.

Every liquid has a **vapor pressure,** which is the pressure in the gaseous state of those molecules that have escaped from the liquid at a given temperature. As you would expect, the vapor pressure of a liquid increases with increasing temperature because more molecules escape the liquid (vaporize) and enter the gaseous state. The higher the temperature, the greater the volatility, because a larger fraction of the molecules have sufficient energy to overcome the attractive forces at the liquid's surface. Our everyday experiences such as heating water on a stove or spilling a liquid on a hot pavement in the summer tell us that raising the temperature of the liquid makes evaporation take place more readily. Conversely, at lower temperatures, the volatility of a liquid will be lower. The same amount of water spilled on the pavement on a winter day will remain there much longer. Figure 7.10 shows how the vapor pressures increase for three common liquids as the temperature is increased. The three compounds have different volatilities due to the different strengths of their intermolecular attractions, but in all cases vapor pressure increases with increasing temperature. This is true for all liquids.

The temperature at which the equilibrium vapor pressure equals the atmospheric pressure is the **boiling point** of the liquid. If the atmospheric

Gypsum sand at White Sands National Monument in New Mexico. The sand has formed as the result of weathering, erosion, and the water solubility of gypsum ($CaSO_4 \cdot 2H_2O$). *(Jon Mark Stewart/Biological Photo Service)*

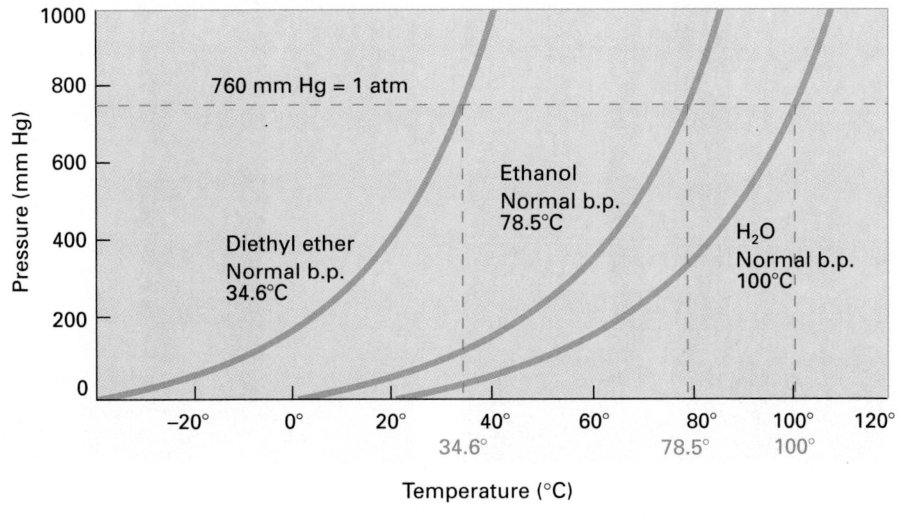

Figure 7.10 Vapor pressures. The curves show how boiling points change with pressure. The normal boiling point is where the curves reach 760 mm Hg. The higher its vapor pressure at a given temperature, the more volatile the liquid.

pressure is 1 atm, the boiling temperature is designated the **normal boiling point.** When the atmospheric pressure is low, the vapor pressure required for boiling is lowered and hence the boiling temperature is lower than the normal boiling point. For example, in Salt Lake City, Utah, where the average barometric pressure is about 650 mm Hg, water boils at about 95°C. It takes longer to hard-boil an egg in Salt Lake City than it does at sea level, because the water at the higher elevation is boiling at a lower temperature.

7.5 Solutions

Solutions, as explained in Section 2.3, are homogeneous mixtures that can be in the gaseous, liquid, or solid states. Most commonly, however, we encounter liquid solutions. How many liquid solutions are familiar to you? How about sugar or salt dissolved in water, or oil paints dissolved in turpentine, or grease dissolved in gasoline? In each of these solutions, the substance present in the greater amount, the liquid, is the **solvent,** and the substance dissolved in the liquid, the one present in a smaller amount, is the **solute.** For example, in a glass of tea, water is the solvent, and sugar, lemon juice, and the components extracted from the tea leaves that impart taste and color are solutes.

■ As noted in Section 6.8, a common rule of thumb is that "like dissolves like."

A substance's **solubility** is defined as the quantity of solute that will dissolve in a given amount of solvent at a given temperature. Solubility is determined by strength of the forces of attraction between solvent molecules, between solute atoms, molecules, or ions, and between solute and solvent. The forces acting between the solvent and solute particles must be greater than those within the solute. In other words, the solute and solvent must like each other more than they like themselves. If the forces between a liquid solvent and a solute are strong enough, then a solution will form (Figure 7.11).

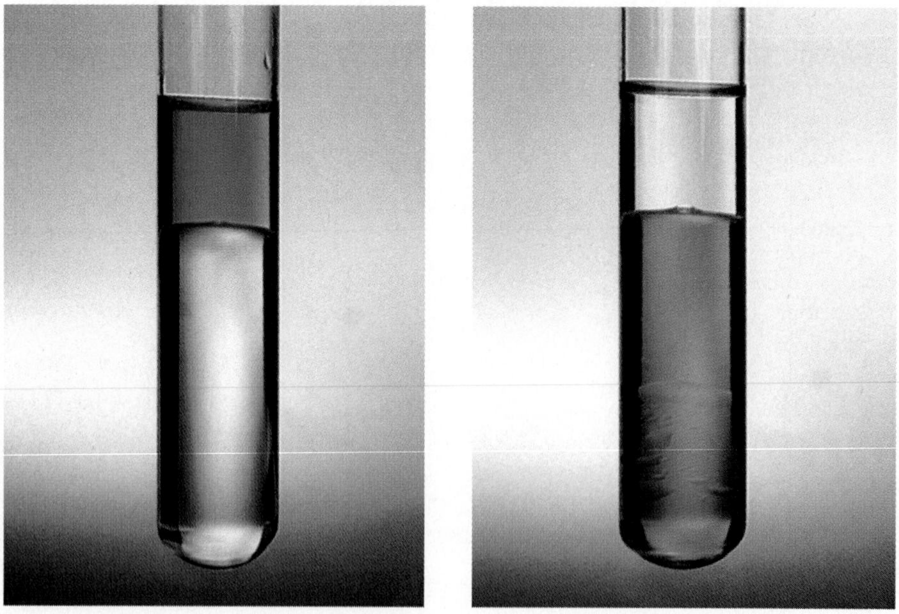

(a) (b)

Figure 7.11 Solubilities. (a) Polar water, with a bit of nonpolar iodine (I_2) dissolved in it, floats on top of nonpolar carbon tetrachloride (CCl_4), with which it is immiscible. (b) Nonpolar iodine is much more soluble in nonpolar carbon tetrachloride than in water. Therefore, shaking the mixture in (a) causes the iodine molecules to migrate into the carbon tetrachloride, where they produce a purple color.

Although some solutes are so much like the solvent they are dissolving in that they dissolve in all proportions (a property called **miscibility,** like that of gases), there are limits to the solubility of most solutes in a given solvent. When a quantity of solvent has dissolved all the solute it can, the solution is said to be **saturated.** If more solute can be dissolved in the solution, the solution is said to be **unsaturated.** Our everyday experiences illustrate this kind of behavior in solutions. For example, if a spoonful of sugar is added to a glass of iced tea, it quickly dissolves with a little stirring which hastens the mixing process. The resulting solution is unsaturated (it can dissolve more sugar), and if you like your iced tea a little sweeter, you just add another spoonful of sugar and stir. However, if sugar is continually added to the solution, a point is reached when no more appears to dissolve and the added sugar simply sinks to the bottom of the glass. At this point the solution has become saturated in the dissolved sugar.

■ Air, a mixture of gases, is a solution. Sterling silver, a mixture of silver and copper metals, is also a solution.

If almost none of the solute dissolves in a solvent, then it is said to be **insoluble** in that solvent. Metals, except those that react with water, are insoluble in water. Oily substances are also insoluble in water, as is water in oily substances. The low solubility of oily substances in water is caused by a lack of attraction between the highly polar water molecules and the nonpolar molecules of an oily substance. The water solubility of a nonpolar molecule can be increased if a polar part can be added. Conversely, the water solubility of a molecule decreases as the nonpolar portion of the molecule increases. Table 7.1 shows the water solubility of some alcohols. The simpler alcohols are very water soluble because the —OH group is polar and forms strong hydrogen bonds with water. As the hydrocarbon chain lengthens, the influence of the —OH group decreases, and with it, the water solubility of the molecule.

TABLE 7.1 ■ Solubilities of Some Alcohols in Water

Name	Formula	Solubility in Water ($g/100\ g\ H_2O$ at 20°C)*
Methanol	CH_3OH	Miscible
Ethanol	CH_3CH_2OH	Miscible
1-propanol	$CH_3CH_2CH_2OH$	Miscible
1-butanol	$CH_3CH_2CH_2CH_2OH$	7.9
1-pentanol	$CH_3CH_2CH_2CH_2CH_2OH$	2.7
1-hexanol	$CH_3CH_2CH_2CH_2CH_2CH_2OH$	0.6
1-heptanol	$CH_3CH_2CH_2CH_2CH_2CH_2CH_2OH$	0.09

*Grams of solute per 100 grams of solvent.

Solubility of Oxygen in Water at Various Temperatures

Temperature (°C)	Solubility of O_2 ($g\ O_2/L\ H_2O$)
0	0.0141
10	0.0109
20	0.0092
25	0.0083
30	0.0077
35	0.0070
40	0.0065

These data are for water in contact with air at 760 mm mercury pressure.

Solutions are important not only because they are so commonplace, but also because so many chemical reactions take place in them (see Chapters 8 and 9). Particles in the solution (whether they are atoms, molecules, or ions) become mobile and move about randomly in the same way as the molecules of the liquid. In this way the particles come into contact with one another, and if a chemical reaction is possible, it will take place upon collision of the reacting particles.

The effect of temperature on the solubility of solutes in various solvents is difficult to predict. Experience with everyday solutions such as sugar in water would lead us to predict that increasing the temperature of the solvent will cause more solute to dissolve. Although this is true for sugar and most other nonionic solutes, it is not true for all ionic compounds. In fact, the solubility of table salt in water is about the same at all temperatures.

Gases dissolve in liquids to an extent dependent on the similarity of the gas molecules and the solvent molecules. Polar gas molecules dissolve to a

A rainbow trout retreating to a cool, well-oxygenated spot in the stream. *(Hervé Berthoule-Scott/JACANA/Photo Researchers)*

FRONTIERS IN THE WORLD OF CHEMISTRY

Supercritical Fluids

Can you imagine a fluid that has "super" dissolving properties, or one capable of penetrating almost any solid—acting like both a gas and a liquid? Such fluids, called supercritical fluids, exist and have uses as diverse as extracting caffeine from coffee beans and toxic chemicals from solid wastes. At appropriate conditions of temperature and pressure, all gases and many liquids can exist as supercritical fluids. Carbon dioxide is a supercritical fluid above 31°C and 73 atm pressure. The temperature of 31°C is called the *critical temperature,* and 73 atm is the *critical pressure.* At temperatures above the critical temperature it is impossible to liquefy carbon dioxide, regardless of how much pressure is applied. So the pressure that must be applied to a sample of carbon dioxide at 31°C to liquefy it is 73 atm. If the pressure is less than 73 atm, carbon dioxide would be a gas; above 73 atm, it would be a liquid. If the temperature of the gas is above 31°C *and* the pressure is above 73 atm, however, the gas cannot be liquefied—it is a supercritical fluid. If a CO_2 fire extinguisher is cool (below 31°C), you can hear the liquid slosh about when you rock the cylinder back and forth. On a warm day (above 31°C) you will not hear the sloshing liquid. The CO_2 is in the supercritical fluid form.

Supercritical fluids are excellent solvents. The principle of "like dissolves like" still holds, but sometimes the supercritical solvent will even dissolve unlike molecules. The nonpolar CO_2 molecules can dissolve molecules such as caffeine

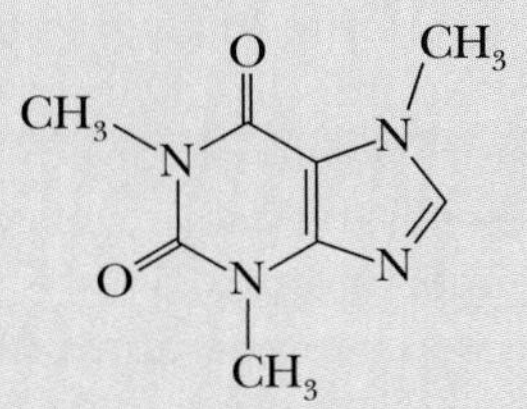

Caffeine

from coffee beans in a supercritical extractor that operates much like a gigantic percolator. Because the supercritical CO_2 can penetrate the intact coffee beans much better than ordinary liquid solvents, the extraction process is more efficient than older methods that used chlorinated hydrocarbons as solvents. Decaffeinating coffee with CO_2 leaves no foreign solvent residues in the coffee beans. Food processors have even tried supercritical CO_2 to extract fats from certain foods such as dairy products and potato chips.

Supercritical water is even more promising as an extraction fluid, partly because of water's already excellent solvent properties. Compared with CO_2, water has the disadvantage of a much higher critical temperature and pressure. One of the most useful potential applications for supercritical water is in the separation of hazardous components from industrial wastes. Even nonpolar organic compounds like polychlorinated biphenyls (PCBs),

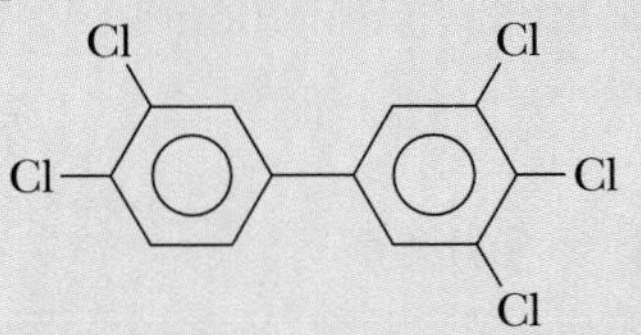

3,4,3′,4′,5′-Pentachlorobiphenyl—
a typical polychlorinated biphenyl (PCB)

notorious hazardous wastes (Section 20.4), readily dissolve in supercritical water. When oxygen is added to a mixture of an organic compound in supercritical water, a reaction takes place that converts the organic compound to carbon dioxide, water, and other small molecules, most of which have much less harmful potential than the original organic compound. In just a matter of minutes, more than 99.9% of all the organic molecules in a sample are destroyed. By regulating the pressure slightly, the products of the reaction can be separated from the supercritical water. Some waste streams that can be treated in this manner are sewage sludge and pulp paper wastes, two of the largest-volume wastes produced. The destruction of hazardous wastes by supercritical water offers advantages over incineration because the operating temperatures are lower and the reaction products can be more carefully controlled. Incinerators generally put most of their reaction products into the atmosphere.

Compound	Critical Temp (°C)	Critical Pressure (atm)
Sulfur dioxide	157.4	78.6
Ammonia	132.5	43.5
Propane	96.8	42
Carbon dioxide	31	73
Water	374.1	218.3

greater extent in polar solvents than nonpolar molecules do. Pressure also affects gas solubility. At higher pressure more gas dissolves in a given volume of liquid than at lower pressure. When the pressure is lowered, gas will be evolved from a gas-in-liquid solution. The behavior of a carbonated beverage when the cap is removed is a common illustration of this principle.

Temperature has a greater effect on gas solubility in liquids than it has on solids or liquids dissolving in liquids. Without exception, lower temperatures cause more gas to dissolve in a given volume of liquid. Higher temperatures cause less gas to dissolve. As every fisherman knows, fish prefer deeper water in the summer months. This is because more oxygen dissolves in colder water, and the water temperature is generally cooler at greater depths.

SELF-TEST 7B

1. Most solids will not float in their liquids. (a) True, (b) False.
2. The density of solid water is less than that of liquid water. (a) True, (b) False.
3. You have 100 g of copper and 100 g of water. Which will increase in temperature more when 100 cal of heat is added to each?
4. The attraction of neighbor molecules to those on the surface of a liquid give rise to the property called ________ ________.
5. A liquid that vaporizes easily is described as ________.
6. The vapor pressure of a liquid will (increase/decrease) with increasing temperature.
7. The temperature at which the vapor pressure of a liquid equals the atmospheric pressure is called the ________ ________ of the liquid.
8. Where would a boiled egg cook more quickly, Miami or Denver?
9. In a solution, the substance dissolved is called the ________, and the substance doing the dissolving is called the ________.
10. Carbon tetrachloride is a nonpolar compound. In which solvent would you expect it to dissolve? (a) Water (a polar solvent), (b) Corn oil (a nonpolar solvent).
11. More of a gas will dissolve in warm water than in cold water. (a) True; (b) False.
12. Increasing the pressure of a gas over a liquid (increases/decreases) its solubility in the liquid.

7.6 Solids

In contrast to gases and liquids, in which molecules are in continual random motion, the movement of atoms, molecules, or ions in solids is restricted to vibration and sometimes rotation around an average position. This leads to an orderly array of particles and to properties very different from those of liquids or gases.

Because the particles that make up a solid are very close together, solids are very difficult to compress. In this respect solids are like liquids, the parti-

The World of Chemistry

Program 5, *A Matter of State*

Crystals from the Smithsonian Museum Collection.

cles of which are also very close together. Unlike liquids, however, solids are rigid, so they cannot transmit pressure in all directions. Solids have definite shapes, occupy fixed volumes, and have varying degrees of hardness. Hardness depends on the kinds of bonds that hold the particles of the solid together. Talc is one of the softest solids known; diamond is one of the hardest. Talc is used as a lubricant and in talcum powder, which we apply to our skin to absorb moisture. At the atomic level, talc consists of layered sheets that contain silicon, magnesium, and oxygen atoms. The attractive forces between the sheets are very weak, so one sheet can slide along another and be removed easily from the rest. In diamond, on the other hand, each carbon atom is strongly bonded to four neighbors, and each of those neighbors is strongly bonded to three more carbon atoms, and so on throughout the solid. Because of the number and strength of the bonds holding each carbon atom to its neighbors, diamond is so hard that it can scratch or cut almost any other solid. The cutting and abrasive uses of diamonds are far more important commercially than their gemstone uses.

■ The bonding in diamond is shown on the first page of Chapter 6.

When atoms, molecules, or ions are packed closely together in a solid, the result is often strikingly beautiful because of its symmetry. Studying crystals gives us some insight about how the crystals are assembled on the atomic level. The extremely ordered nature of crystals, including their faces and edges and the angles between their faces, can often be observed with the naked eye. In sodium chloride, for example (Figure 6.1), the cations and anions are arranged in a three-dimensional cubic pattern. Small grains of salt also show a cubic shape that can be traced back to this atomic level cubic pattern. In fact, crystals of the same substance, when prepared in the same way, always have the same shape, regardless of whether the crystal is large or small (Figure 7.12). When crystalline solids shatter, they usually produce small crystals that look like the larger crystals they came from. Of course, some solids, such as wood, glass, and plastics, contain molecules in less regular arrangements. We can see a manifestation of this when we observe how crystalline solids compare with noncrystalline solids.

The World of Chemistry

Program 5, *A Matter of State*

I think the most fascinating thing about crystals, at least from the standpoint of a scientist who wants to study matter, is the unique insight that a crystal gives you about the nature of matter itself, into the way that chemical elements behave toward each other. Dan Appleman, geologist, Smithsonian Institution

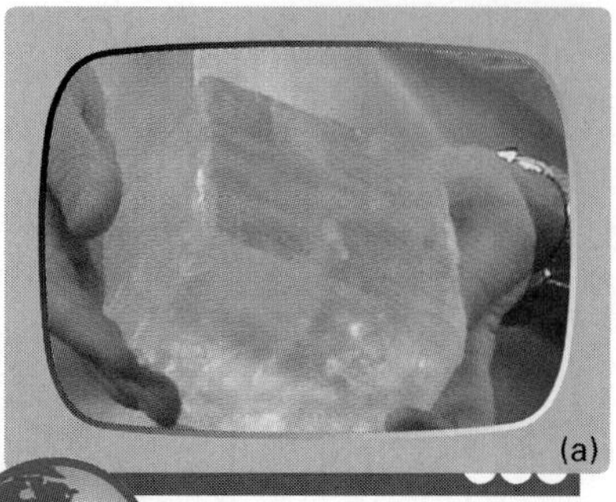

The World of Chemistry

Program 8, *Chemical Bonds*

Figure 7.12 Sodium chloride. (a) A large chunk of sodium chloride. (b) Sodium chloride crystal growing out of sodium chloride solution in water.

(b) *C.D. Winters*

Properties of Solids

When a solid is heated to a temperature at which molecular motions are violent enough to partially overcome the interparticle forces, the orderliness of the solid's structure collapses, and the solid melts. The temperature at which melting occurs is the **melting point** of the solid. Melting requires energy to overcome the attractions between the particles in a solid lattice. In the reverse of melting, **solidification** or **crystallization,** energy is evolved. Melting points of solids depend on the kinds of forces holding the particles together in the solid, since it is these forces that must be overcome for a solid to become a liquid. Table 7.2 gives some melting points for several types of solids.

Molecules can also escape directly from the solid to enter the gaseous state by a process known as **sublimation.** A common substance that sublimes at

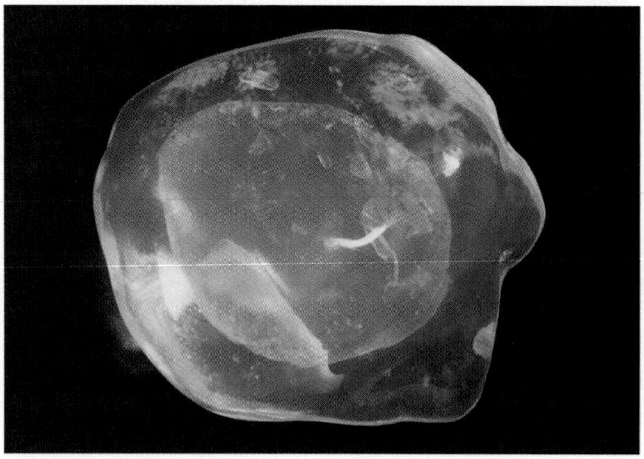

The gemstone opal. Opal is one of the few naturally occurring mineral substances that is an *amorphous* solid, that is, like glass, it is not crystalline. The composition of opal is mostly water combined with silicon dioxide. *(Karl Hartmann/Sachs/Phototake)*

TABLE 7.2 ■ Melting Points of Some Solids

Solid	Melting Point (°C)	Type of Intermolecular Forces
Molecular Solids: Nonpolar Molecules		
H_2	−259	Depends on number of electrons.
O_2	−248	
N_2	−220	
F_2	−103	
Cl_2	−7.2	
Molecular Solids: Polar Molecules		
HCl	−114	Polar ends of molecules attract oppositely charged atoms on adjacent molecules. Hydrogen-bonded molecules (highlighted) are more strongly bound to one another, resulting in higher-than-normal boiling points.
HBr	−87	
H_2S	−86	
HF	−83.1	
NH_3	−77.7	
SO_2	−72.7	
HI	−51	
H_2O	0	
Ionic Solids		
NaI	662	Ionic attractions between oppositely charged particles. Very high melting points occur for solids containing very small ions with multiple charges—such as Mg^{2+}, Al^{3+}, and O^{2-}.
NaBr	747	
$CaCl_2$	772	
NaCl	800	
Al_2O_3	>2200	
MgO	2800	

normal atmospheric pressure is solid carbon dioxide, which has a vapor pressure of − 78°C at 1 atm. Solid CO_2 is known by its trade name, Dry Ice, because it is cold like ice (although much colder) and produces no liquid residue because it does not melt. Ice can sublime or melt. Have you ever noticed that ice and snow slowly disappear even if the temperature never gets above freezing? The reason is that ice sublimes readily in dry air (Figure 7.13). Given

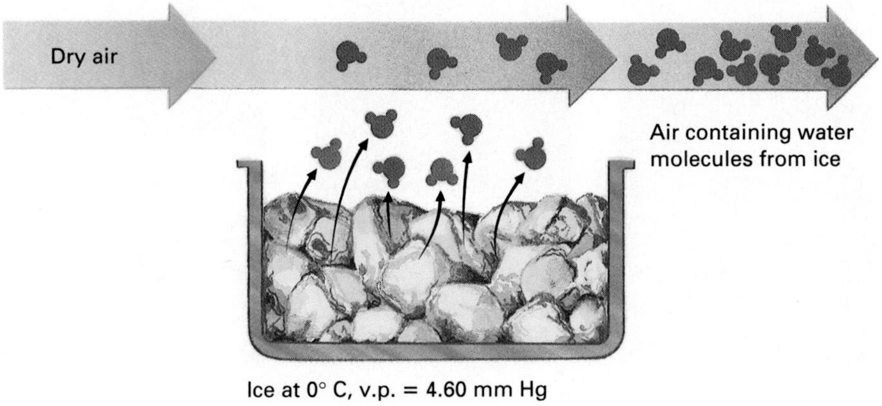

Figure 7.13 Why ice cubes shrink in the freezer—sublimation.

enough air passing over it, a sample of ice will sublime away, even at temperatures well below its melting point, leaving no trace behind. This is what happens in a frost-free refrigerator. A current of dry air periodically blows across any ice formed in the freezer compartment, taking away water vapor (and hence the ice) without having to warm the freezer compartment to melt the ice.

7.7 Metals, Semiconductors, and Superconductors

Metals have some properties totally unlike those of other substances. Except for mercury, which is a liquid at room temperature, and gallium, which is a liquid at slightly above room temperature, all metals are solids. Some remain solids even at very high temperatures. Tungsten has a melting point of 3410°C. There are several properties common to the metals:

- ***High electrical conductivity.*** Metal wires easily carry electrical currents. The electrons in metals are quite mobile.
- ***High thermal conductivity.*** Some metals are much better conductors of heat than others. Try a sterling silver spoon in a cup of hot coffee or tea. Compare its thermal, or heat, conductivity with that of a stainless steel spoon.
- ***Ductility and malleability.*** Most metals can be drawn into wire (ductility) or hammered into thin sheets (malleability); gold is the most malleable metal. Extremely thin sheets of gold are used for decoration.
- ***Luster.*** Polished metal surfaces reflect light; most metals have a silvery white color because they reflect all wavelengths of light equally well.
- ***Insolubility in water and other common solvents.*** No metal dissolves in water, but a few, such as the Group IA and IIA metals, react with water to form hydrogen gas and solutions of metal hydroxides (see Figure 4.6a).

Any theory of the bonding of metal atoms must be consistent with these properties. Structural investigations of metals have led to the conclusion that solid metals are composed of regular arrays, or *lattices,* of metal ions in which

(a)

(b)

(c)

Putting the properties of metals to use. (a) Installation of electrically conducting wires. (b) A flask coated, like a mirror, with silver. (c) Very thin and stable gold leaf being applied to the cupola of a church. *(a,b: C.D. Winters; c: Paul Silverman/Fundamental Photographs)*

the bonding electrons are loosely held. Figure 7.14 illustrates one model for metallic bonding in which the regular array, or lattice, of positively charged metal ions is embedded in a "sea" of mobile electrons. These mobile valence electrons are delocalized over the entire metal crystal, and the freedom of these electrons to move throughout the solid is responsible for the properties associated with metals. In contrast to metals, the valence electrons in nonmetals are fixed in bonds between like atoms. This means that nonmetals are nonconductors of electricity.

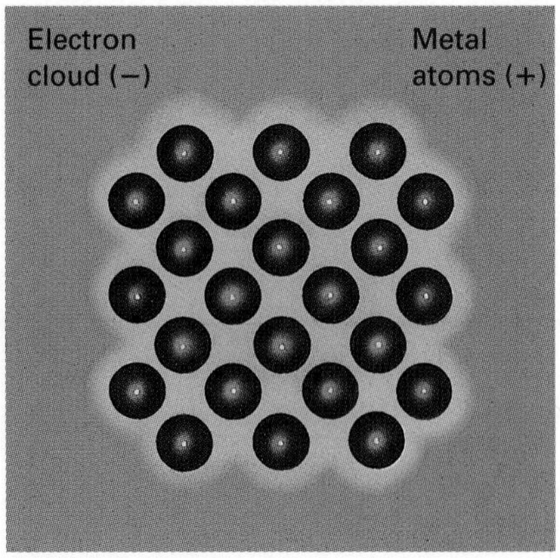

Figure 7.14 "Electron sea" model of bonding in metals. The positively charged metal atom nuclei are surrounded by a "sea" of negatively charged electrons.

Semiconductors—The Basis of our Modern World

You should not be surprised that somewhere between the excellent electrical conductivity of most metals and the nonconductivity of nonmetals, there are some elements that are **semiconductors.** This means they will conduct electricity under certain conditions. Silicon, when it is in a highly purified state, is a semiconductor. Silicon acts like a nonmetal and fails to conduct a current until a certain voltage is applied; then it begins to conduct moderately well. This behavior interested electrical engineers, who recognized that silicon might act like a "gate" for electron flow in electrical circuits. This electron gatelike activity was not realized until a process known as **doping** was discovered. One common dopant is boron, a Group IIIA element just to the upper left of silicon in the periodic table. When boron is added to pure silicon, the boron atoms, with one fewer valence electron than silicon, become *positive holes* in the lattice arrangement of silicon atoms. The presence of these positive holes (which can move about in the solid just like electrons, but in the opposite direction) makes the doped silicon somewhat more conductive—in effect, it becomes a better electron gate. Silicon doped with an element that creates positive holes is called a *p-type* semiconductor. Another dopant is arsenic, from Group VA, which contains one *more* valence electron than silicon. Silicon doped with arsenic is called an *n-type* semiconductor because there are extra negative electrons present in the solid (Figure 7.15). These, of course,

■ The elements shown in (blue) in the periodic table inside the back cover are referred to as the *metalloids* (Section 4.2)

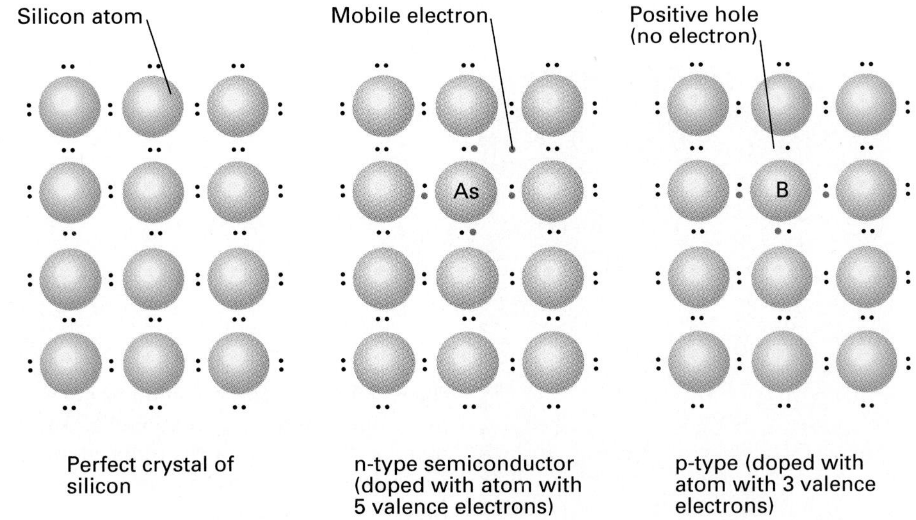

Figure 7.15 Doping of silicon in semiconducting devices. Adding atoms with five valence electrons (e.g., arsenic) introduces extra electrons that can move through the crystal. Adding atoms with three valence electrons introduces holes that can also move through the crystal.

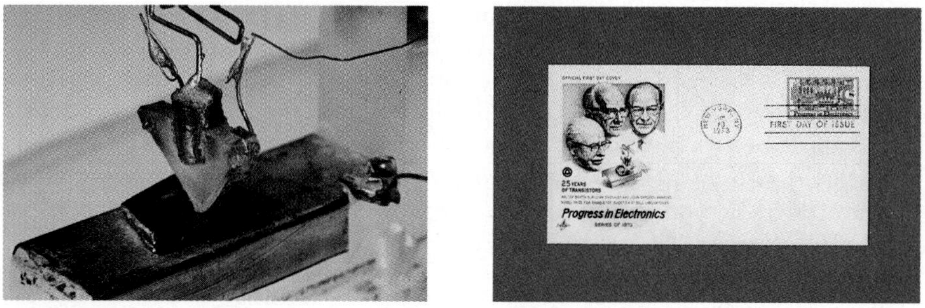

(a) (b)

Figure 7.16 The transistor and its inventors. (a) The first transistor, constructed in 1947 at Bell Laboratories. Electrical contact is made at a single point and the signal is amplified as it passes through a solid semiconductor; modern junction transistors amplify in a similar manner. (b) Envelope and stamp commemorating 25 years of the transistor, with portraits of its inventors Walter Brattain, William Shockley, and John Bardeen. *(Courtesy of AT&T Archives)*

enhance the conductivity of the doped silicon. Through careful control of the amount of dopant, the conductivity of the silicon can be adjusted to a fine degree.

In 1947 a device consisting of a layer of p-type silicon sandwiched between two n-type layers was constructed by John Bardeen, Walter Brattain, and William Shockley at the Bell Laboratories. This device, called the **transistor,** has revolutionized our world (Figure 7.16). Because the transistor can control electron flow in circuits with such accuracy, yet is so small and requires so little power to operate, it is now possible to design electronic circuits to fit into extremely small volumes. Such things as TV cameras as small as a pea, radios small enough to strap to the back of an ant, and other amazing devices can be made using transistors based on doped silicon. Of course, more mundane things such as automatic cameras, microwave ovens, fax machines, and cellular phones also owe their existence to the transistor. The central processing unit

■ Bardeen, Brattain, and Shockley shared the Nobel Prize in Physics in 1956 for the discovery of the transistor. The importance of the transistor was recognized as soon as it was discovered. Although it was first demonstrated at the Bell Laboratories in December of 1947, it wasn't announced until July 1, 1948, after patent applications had been filed. John Bardeen was awarded a second Nobel Prize in Physics in 1972, along with J. R. Schrieffer and Leon N. Cooper, for his work on the theory of superconductivity.

Figure 7.17 A modern computer microprocessor with built-in multimedia capability (a "RISC chip" for a reduced-instruction-set computer). *(Courtesy of Hewlett-Packard Company)*

(CPU) of computers consists of millions of transistors and other circuit elements fabricated on wafers of pure silicon (Figure 7.17). These devices are called integrated circuits.

Can Anything Conduct Better than a Metal?

When metals are heated their electrical conductivity decreases. Lower conductivity at higher temperatures can be explained if the movement of the valence electrons is considered to be limited by rapidly vibrating atoms in the metal lattice. The kinetic molecular theory says that higher temperatures mean more motion, and this applies even when the atoms are in fixed positions, as they are in a metal. When the metal atoms are relatively stationary, as they are when the metal is cool, the electrons can move through the lattice much like a person moving through a room filled with a large number of other people quietly chatting with one another. When the temperature is elevated, the metal atoms begin to vibrate wildly, and the electrons have more trouble getting through the lattice, in much the same way that a person would have trouble moving through the room if all the other occupants suddenly became agitated.

From this picture of electrical conductivity of metals, you might assume that if a sufficiently low temperature were reached, conductivity might be quite high (almost zero resistance, in other words). In fact, the conductivity of a pure metal crystal does approach infinity (zero resistance) as absolute zero is approached. In many metals, however, a more interesting thing happens before absolute zero is reached. At a certain low temperature, the conductivity suddenly increases, as though absolute zero had already been reached. At this temperature, called the *superconducting transition temperature,* the metal becomes a **superconductor** of electricity. The superconductor offers no resistance whatever to electrical flow. This phenomenon means that electric motors made of superconducting wires would be 100% efficient, and electrical transmission lines could be made 100% efficient. It is resistance to electron flow that causes energy loss in motors, transmission lines, and other electrical devices. There is no good theory to explain the phenomenon of superconductivity, but the hindrance of electron flow by vibrating atoms in the metal lattice has been replaced by some kind of cooperative action that allows electron movement. Table 7.3 lists some of the metals that have superconducting transition temperatures. Not all metals display superconducting properties.

TABLE 7.3 ■ Superconducting Transition Temperatures of Some Metals

Metal	Transition Temperature (K)
Aluminum	1.183
Gallium	1.087
Lanthanum	4.8
Lead	7.23
Niobium	9.17

The relatively low transition temperatures of the metals shown in Table 7.3 mean that it would be impractical to make superconducting motors or transmission lines from them. Shortly after the superconductivity of metals was discovered, certain **alloys** (mixtures of metals) were prepared that had much higher transition temperatures than the metals themselves. Niobium alloys showed the most promise, but they still had to be cooled to below 23 K (− 250°C) to exhibit superconductivity. To maintain such a low temperature would require liquid helium, which costs about $4 per liter—an expensive proposition for all but the most exotic applications.

■ The boiling point of helium is 4.35 K (− 268.8°C)

In January 1986, K. Alex Müller and J. Georg Bednorz, scientists at an IBM laboratory in Switzerland, discovered that a barium-lanthanum-copper oxide became superconducting at 35 K. This discovery rocked the scientific world and provoked a flurry of activity that quickly resulted in a substance that became superconducting at 90 K. This compound, $LaBa_2Cu_3O_x$ (where x represents a varying number of oxygen atoms), can superconduct at temperatures above the boiling point of nitrogen (77 K). At less than 6 cents per liter, liquid nitrogen is a much cheaper refrigerant than liquid helium (Figure 7.18).

■ Müller and Bednorz shared the 1987 Nobel Prize in physics for their discovery of superconducting $LaBa_2Cu_3O_x$.

Why the great excitement over the potentialities of superconductivity? Superconducting materials are being used to build more powerful electromagnets, such as those used in nuclear particle accelerators (Section 5.6) and in magnetic resonance imaging (MRI) machines, which are used in medical diagnosis. One of the main factors affecting the efficiency of MRI machines is the heating of the electromagnet due to electrical resistance. Many scientists are saying that the discovery of high-temperature superconductors may prove to be more important than the discovery of the transistor because of its potential effect on electrical and electronic technology. For example, the use of superconducting materials for transmission of electric power could save as much as 30% of the energy now lost because of the resistance of the wire. Superchips for computers could be up to 1000 times faster than existing conventional silicon chips. Electromagnets could be both more powerful and smaller, which could hasten the development of a practical nuclear fusion reactor (see Section 13.3). Since there is a magnetic field around the superconducting material, it is conceivable that cars and trains could be moved along magnetic tracks so that no parts touch, thereby eliminating friction.

Program 26, *Futures*

Figure 7.18 Superconductor. A pellet of a superconducting alloy suspended in air above magnets.

SELF-TEST 7C

1. All particle motion ceases in a solid at room temperature. (a) True, (b) False.
2. When the crystal lattice of a solid breaks up because of added heat, the process is called __________. The temperature at which this process takes place is called the __________ __________.
3. When a solid passes directly from the solid state to the vapor state, the process is called __________.
4. Consider the process named in Question 3. Ice can do this. (a) True, (b) False.
5. Metals are soluble in water. (a) True, (b) False.
6. When the conductivity of a metal suddenly becomes almost infinitely large, the metal is called a __________.
7. Arsenic, a Group 5A element, can be used to dope silicon, a Group 4A element. This doping will produce (n-type/p-type) doped silicon.
8. Boron, a Group 3A element, will produce (n-type/p-type) doped silicon.
9. Doped silicon is used to make (a) magnets, (b) transistors, (c) electrical transmission wires.
10. Which would offer more promise, a compound whose resistance dropped to zero at or just above the boiling point of nitrogen, or one whose resistance dropped to zero at or just above the boiling point of helium?

7.8 Gas Law Arithmetic

An essential aspect of chemistry or any science is learning about nature by making measurements—the quantitative side of chemistry that was introduced in Section 2.8. The consistent behavior of gases under changing conditions is nicely subject to simple quantitative examples. For those who wish to explore this aspect of chemistry further, the following sections expand on the earlier qualitative discussion of the gas laws.

Boyle's Law

Boyle's law can be stated in the following way: *the volume of a fixed amount of gas at a given temperature varies inversely with the applied pressure.* This behavior of gases can be stated mathematically as

$$P \propto \frac{1}{V}$$

The "$\propto$" symbol means "proportional to," and the $1/V$ means that P is inversely proportional to V. The relation can also be written as

$$\text{pressure} \times \text{volume} = \text{a constant} \qquad \text{or} \qquad P \times V = k$$

This equation tells us that the product of the pressure and the volume is a constant. The fact that the temperature and amount of gas are held constant is not shown in the equation, but the value of k is dependent on both the temperature and the amount of gas.

Now, what can be done with this equation, which is an expression of Boyle's law? We can work some simple problems involving changes in pressure and volume.

EXAMPLE 7.2 *Boyle's Law*

A balloon is filled with air to a volume of 1055 milliliters (1055 mL). The pressure at the time of filling the balloon is 755.0 mm Hg. The next day, the temperature is the same but the pressure of the air surrounding the balloon is 739.0 mm Hg (a low-pressure front has moved into the area). Assuming that no air has leaked out of the balloon, what volume will the balloon occupy?

SOLUTION

There are two sets of pressure-volume conditions, although one part is missing—the new volume. Because no air has leaked out of the balloon and the temperature has stayed the same, the constant, k, in the equation $P \times V = k$ has remained the same. If we call the first pressure P_1 (755.0 mm Hg) and the first volume V_1 (1055 mL), and the second pressure P_2 (739.0 mm Hg) and the second volume V_2, then the following is true:

$$P_1 \times V_1 = k \qquad \text{and} \qquad P_2 \times V_2 = k$$

Since the value of k is the same for both equations,

$$P_1 \times V_1 = P_2 \times V_2$$

We can now substitute what is known into the equation.

$$(755.0 \text{ mm Hg})\,(1055 \text{ mL}) = (739.0 \text{ mm Hg}) \times V_2$$

Solving for V_2 we get

$$V_2 = \frac{(755.0 \text{ mm Hg})\,(1055 \text{ mL})}{(739.0 \text{ mm Hg})} = 1078 \text{ mL}$$

The volume of the gas sample has increased, as you would expect from the statement of Boyle's law. If the temperature and quantity of gas are held constant, the Boyle's law relationship can be used to calculate any one of the four variables when three of them are known.

■ A variable is a quantity whose value can change (i.e., not a constant).

Exercise 7.2A

At a pressure of 1.00 atm and some temperature that remains constant, a sample of gas occupies 400. L. What will be the volume of the gas if the pressure is decreased to 0.750 atm?

Charles' Law

Charles' law can be stated in the following way: *the volume of a fixed amount of gas at a constant pressure is directly proportional to the absolute temperature.* If the temperature is expressed in kelvin units, the relation can be written as

$$V \propto T \qquad \text{or} \qquad V = kT$$

The constant k takes into account the amount of gas and the pressure. Notice that the temperature must be expressed on the Kelvin scale when working problems related to gases. If the temperature is given in degrees Celsius, then 273.15 must be added. As in the Boyle's law example shown earlier, if the volume, V_1, and a temperature, T_1, of a sample of gas are known, then the volume, V_2, and temperature, T_2, at the same pressure is given by

$$\frac{V_1}{T_1} = \frac{V_2}{T_2} \qquad \text{or} \qquad V_1T_2 = V_2T_1$$

Remember, when using Charles' law in calculations involving gases, the temperature must always be expressed in kelvins.

EXAMPLE 7.3 *Charles' Law*

Let's use our same 1055-mL sample of gas in the balloon from the previous example. This time the pressure remains constant, and the temperature will change. The temperature was 23°C when the balloon was filled and its volume measured. The balloon was placed overnight in a refrigerator at a temperature of 5°C. What will be the new volume of the gas in the balloon?

SOLUTION

The very first thing to do is convert the temperatures to kelvins.

$$T_1 = 23 + 273 = 296 \text{ K} \qquad T_2 = 5 + 273 = 278 \text{ K}$$

Solving for the new volume of the gas,

$$V_2 = \frac{V_1T_2}{T_1} = \frac{(1055 \text{ mL})(278 \text{ K})}{296 \text{ K}} = 991 \text{ mL}$$

The volume of the gas at the new temperature has decreased, as you would expect from the statement of Charles' law.

Exercise 7.3A

If a sample of gas occupies 765 mL at 31°C, what volume will it occupy at 75°C, assuming the pressure remains constant?

Using Boyle's and Charles' Laws Together

Boyle's and Charles' laws can be used together when both pressure and temperature change for a sample of gas. This is usually the case when dealing with real gas samples. The equations for the two laws can be combined to give

$$\frac{P_1V_1}{T_1} = \frac{P_2V_2}{T_2}$$

This expression allows you to calculate any one of the six variables if the other five are known.

EXAMPLE 7.4 *Boyle's and Charles' Laws Combined*

Again, using the 1055-mL balloon filled with air, the temperature and the pressure are changed by taking the balloon up the side of a mountain. The pressure in the lab where the balloon was filled and where its original volume was measured was 755 mm Hg, and the temperature was 23°C. On the mountain the pressure is 684 mm Hg, and the temperature is a chilly 9°C. What will be the volume of the balloon?

SOLUTION

The gas in the balloon is going to behave according to both Charles' and Boyle's laws. One will tend to counteract the other, with the temperature decrease causing the gas to occupy a smaller volume and the pressure decrease causing the gas to occupy a larger volume. First, convert the temperatures to kelvins.

$$T_1 = 23 + 273 = 296 \text{ K} \qquad P_1 = 755 \text{ mm Hg} \qquad V_1 = 1055 \text{ mL}$$
$$T_2 = 9 + 273 = 282 \text{ K} \qquad P_2 = 684 \text{ mm Hg} \qquad V_2 = ?$$

Solving the equation for V_2, we get

$$V_2 = \frac{P_1V_1T_2}{P_2T_1} = \frac{(755 \text{ mm Hg})(1055 \text{ mL})(282 \text{ K})}{(684 \text{ mm Hg})(296 \text{ K})} = 1110 \text{ mL}$$

Exercise 7.4A

A sample of gas occupies 2125 mL at a temperature of 17°C and a pressure of 714 mm Hg. What is the temperature (in Celsius degrees) if the sample of gas occupies 2333 mL at a pressure of 745 mm Hg? (*Hint:* Remember to deal properly with the temperature conversions or your answer will be way off.)

SELF-TEST 7D

1. The "∝" symbol means ____________ ____________.
2. Boyle's law states that the volume of a fixed amount of gas varies (directly/inversely) with the applied pressure.
3. If a sample of a gas occupies 500 mL at some temperature and pressure, and the pressure is doubled, what volume will the gas occupy?

4. If a sample of a gas occupies 1.0 L at 25°C and the pressure is halved, what volume will the gas sample occupy assuming that the temperature doesn't change?
5. Charles' law states that the volume of a fixed amount of a gas at a constant pressure is (directly/inversely) proportional to the absolute temperature.
6. If the temperature of a sample of a gas at some pressure is increased from 200°C to 400°C while the pressure remains constant, the volume of the gas doubles. (a) True, (b) False.
7. Convert 23°C to kelvins.
8. Convert 298.15 kelvins to °C.
9. A sample of air occupying 1.75 L is cooled from 1255 K to 325 K while the pressure is held constant. The new volume occupied by this gas is ______________.

MATCHING SET

____ 1. Discovered the first high-temperature superconductor
____ 2. Pressure-volume gas behavior
____ 3. Volume-temperature gas behavior
____ 4. Water-soluble alcohol
____ 5. Water-insoluble alcohol
____ 6. Mixable in all proportions
____ 7. Two liquids mixable in all proportions
____ 8. Vapor pressure equals atmospheric pressure
____ 9. Insoluble in water and most common solvents
____ 10. Water's high boiling point
____ 11. Solvent that will dissolve most salts
____ 12. Causes more gas to dissolve in a liquid
____ 13. Causes less gas to dissolve in a liquid
____ 14. Number of neighbors around a water molecule in ice
____ 15. Going from solid to gas
____ 16. Going from gas to liquid
____ 17. Dopant for p-type silicon
____ 18. Dopant for n-type silicon
____ 19. Causes a metal to conduct more
____ 20. Desired temperature for superconductors to operate
____ 21. Number of molecules a single HF molecule can form hydrogen bonds with

a. Hydrogen bonding
b. Any two gases
c. Metal
d. Boiling point
e. Arsenic
f. Lowering the temperature
g. Increasing the temperature
h. Increasing the pressure
i. Condensation
j. Two
k. Boyle's law
l. CH_3OH
m. Sublimation
n. Boron
o. Room temperature
p. Müller and Bednorz
q. Charles' law
r. $CH_3CH_2CH_2CH_2CH_2CH_2CH_2OH$
s. Four
t. Ethyl alcohol and water
u. Water

QUESTIONS FOR REVIEW AND THOUGHT

1. Describe the following states of matter:
   (a) Gaseous state (b) Liquid state
   (c) Solid state
2. Define the following terms:
   (a) Compressibility (b) Absolute zero
   (c) Miscibility (d) Condensation

(cont.)

(e) Heat of vaporization (f) Surface tension
(g) Vapor pressure (h) Solute
(i) Solvent

3. In which state of matter are the particles the greatest average distance apart? (a) Liquid, (b) Solid, (c) Gas.
4. In which state of matter are the particles in fixed positions? (a) Liquid, (b) Solid, (c) Gas.
5. Which state of matter can be described as "particles close together and in constant, random motion"? (a) Liquid, (b) Solid, (c) Gas.
6. Which of the following statements about gases is inconsistent with the kinetic molecular theory of matter? Explain.
   (a) Gases exert more pressure at higher temperatures when the volume is held constant.
   (b) Gases will condense at sufficiently low temperatures.
   (c) Gases will occupy a larger volume at lower pressures when the temperature remains constant.
7. Explain why the pressure of the atmosphere decreases with increasing altitude.
8. What happens to the pressure in an automobile tire in cold weather? Explain.
9. A gas exerts a pressure P_1 inside a container of some volume at a certain temperature. Another gas, in an identical container at the same temperature, exerts a pressure P_2. If the two gases are mixed in the first container, the total pressure, due to both gases, is $P_1 + P_2$. Explain this observation.
10. Explain why molecules of a perfume can be detected a few feet away from the person wearing the perfume.
11. Draw a structure showing four water molecules bonded to a central water molecule by means of hydrogen bonding. Indicate all of the hydrogen bonds by drawing arrows to them.
12. Name two properties of water that are unusual because of the presence of hydrogen bonding between adjacent water molecules.
13. Which of the following compounds would you expect to exhibit hydrogen bonding? Explain your answer.
   (a) CH_3OH, (b) NH_3, (c) SO_2, (d) CO_2, (e) CH_4, (f) HF, (g) CH_3OCH_3.
14. Whenever a liquid evaporates, heat is required. Use this statement to explain why you get chilled when you come out of a swimming pool on a windy day.
15. Explain why doing a "belly flop" in the swimming pool can be painful. In your explanation include one of the properties of water.
16. Explain why "like dissolves like."
17. Why are gases compressible while liquids and solids are not?
18. Explain why the vapor pressure of a liquid increases with increasing temperature.
19. What is the normal boiling temperature of a liquid?
20. Based on Figure 7.7, approximately what would be the boiling point of ammonia if there were no hydrogen bonding between the molecules?
21. What causes surface tension in liquids? Name a compound that has a high surface tension.
22. How would you cause the boiling point of a liquid to be higher than its normal boiling point? How would you make the boiling point be lower?
23. Based on Figure 7.10, approximately what would be the boiling point of water at 200 mmHg pressure?
24. Explain why methyl alcohol, CH_3OH, is soluble in water. In your explanation, consider the structures of both the alcohol molecule and the water molecule including their similarities and differences.
25. Would you expect the molecule shown below to be soluble in water? Explain your answer.

$$CH_3CH_2CH_2CH_2CH_2CH_2CH_2CH_2—OH$$

26. Why is a gas more soluble in a solvent when a higher pressure is applied? Give an example of where you see this behavior of gases dissolving in liquids.
27. Thermal pollution is a name given to industrial discharges of warm waters that contain sufficient heat to warm the water they are discharged into. Why is thermal pollution harmful to fish?
28. What is the temperature called at which the particles making up a solid begin moving from their fixed positions?
29. What is the process called where molecules in the solid escape directly into the gaseous state?
30. Consider the process described in Question 29. Name two compounds that undergo this process.
31. Which is less dense, carbon tetrachloride or water? Refer to Figure 7.11.
32. Name a metal that is a liquid at room temperature.
33. Name one important property of all metals that depends on the easy movement of electrons within the crystal lattice.
34. What is the name of the element used to fabricate most transistors and integrated circuits, such as those used in computers?
35. What is the difference between a p-type and an n-type semiconductor?
36. Why does a metal conduct electricity better when it is cold than when it is hot?
37. Explain how a frost-free refrigerator defrosts itself.
38. What is (a) a superconductor? (b) superconductivity?
39. What is the main goal of present-day superconductivity research?
40. What is an alloy? Name two alloys and give a use for each.

PROBLEMS

1. A balloon is filled with air to a volume of 2090 mL. The pressure at the time of filling is 765 mmHg. The next day, the temperature is the same but the air surrounding the balloon has dropped to 725 mmHg. If no air has leaked from the balloon, what is its new volume?
2. In a lab experiment, 765 mL of a gas is held in a container at a pressure of 0.95 atm. Holding the temperature constant, the pressure is decreased to 0.82 atm. What volume will the gas occupy at this new pressure?
3. A 45-L sample of gas taken from the upper atmosphere at a pressure of 6.25 mmHg is compressed to a new volume of 745 mL while the temperature stays constant. What will be the new pressure of the gas?
4. A cylinder containing 44 L of helium at a high pressure is used to fill balloons. Finally, you reach a point where no more balloons can be filled. What volume of helium gas is inside the cylinder at that time?
5. A sample of oxygen gas occupies 40.0 mL at 0.355 atm. If the temperature remains constant, what volume will the oxygen gas occupy if the pressure decreases to 0.125 atm?
6. A 12-L sample of air is in a container at a pressure of 1.2 atm. What will the new volume be if the pressure on the air sample is increased to 2.4 atm while the temperature is held constant?
7. Convert the following temperatures:
   (a) 22°C to K (b) 300°C to K
   (c) 27.2°C to K (d) −78°C to K
   (e) 29 K to °C (f) 631 K to °C
   (g) 125 K to °C (h) 325 K to °C
8. A tire is filled with air to a pressure of 2.8 atm while the temperature is a chilly −9°C. The next day, the temperature rises to 28°C. The volume of the tire doesn't change. What is the new pressure?
9. A certain gas sample occupies 105 mL at 25°C. If the pressure is held constant and the temperature is increased, the volume occupied by the gas increases to 175 mL. What is the final temperature in degrees C and in kelvins?
10. A sample of nitrogen gas occupies 121 mL at 100°C. If the pressure does not change, what volume will the gas occupy if the temperature increases to 150°C?
11. What would the Celsius temperature be if the sample of gas from Problem 10 occupies 242 mL?
12. If a sample of gas occupies 890 mL at 35°C, what volume will it occupy at 110°C if the pressure remains constant?
13. A 350-mL sample of a gas is held under a pressure of 356 mmHg at a temperature of 23°C. The pressure is increased to 412 mmHg and the temperature is decreased to 21°C. What is the new gas volume?
14. A 325-mL sample of a gas exerts a pressure of 2.25 atm at a temperature of 21°C. What volume will the gas occupy if the temperature is increased to 35°C and the pressure is increased to 4.31 atm?
15. A weather balloon is filled with helium to a volume of 1.05×10^5 L on the ground, where the pressure is 745 mmHg and the temperature is 21°C. When the balloon ascends to a height of about 2 miles, the pressure is only 602 mmHg and the temperature is −33°C. What is the volume of the balloon at this altitude and air temperature?
16. A sample of a gas occupies 375 mL at 1 atm pressure and 25°C. If the temperature is increased to 755°C, what final pressure would be required to keep the volume unchanged?

CHAPTER 8

Chemical Reactivity: Chemicals in Action

Biologists, physicians, chemists, psychologists, and sometimes even sociologists and economics professors depend on experiments to test their theories. Chemists have one advantage in such experiments. Under *identical conditions,* pure chemicals always react with each other in the same way. Sometimes it is difficult, but with effort identical conditions can be achieved. Those who study living things or social interactions can never be sure. Are two plants, two laboratory rats, or two social groups ever identical?

In this chapter the conditions that influence the outcome of chemical reactions are explored.

- What is the meaning of a balanced equation?
- Why are the mole and the molar mass essential concepts?
- How is molar mass calculated and used?
- In what three ways can reaction rates be influenced?
- What is happening in a chemical reaction that has come to equilibrium?
- What two kinds of changes influence the favorability of chemical reactions?
- What are the first and second laws of thermodynamics?

When journalists investigate a story for page one in the newspaper, they want to answer six questions:

Who? What? When? Where? How? Why?

To fully investigate a chemical reaction requires answering a similar list of questions:

What? How much? How fast? How far? Why?

Some scientists spend a lifetime seeking answers to these questions about a single complex reaction. Others devote themselves to answering one of these questions about many chemical reactions.

A candle burning in pure oxygen. *(Richard Megna/Fundamental Photographs)*

In Chapter 2 we introduced chemical reactions and the information needed to answer the "What?" question. What are the reactants and products? Hydrogen and oxygen, the reactants, combine to form water, the product.

$$\underset{\text{Hydrogen}}{2\ H_2(g)} + \underset{\text{Oxygen}}{O_2(g)} \longrightarrow \underset{\text{Water}}{2\ H_2O(\ell)}$$

In the reaction that causes an automobile airbag to inflate, the reactant is sodium azide, and the initial products are sodium and nitrogen gas (which fills the airbag):

$$\underset{\text{Sodium azide}}{2\ NaN_3(s)} \longrightarrow \underset{\text{Sodium}}{2Na(s)} + \underset{\text{Nitrogen}}{3\ N_2(g)}$$

Now we're going to pick up the story of chemical reactions and pursue the meaning of the other questions listed in the introduction to this chapter.

8.1 Balanced Chemical Equations and What They Tell Us

Chemical equations are the best way we have to represent what happens at the submicroscopic level that we cannot see. An equation will not be faithful to reality if the chemical formulas are wrong or if the equation is not balanced.

In balancing chemical equations, we are applying the law of conservation of matter—the atoms in the reactants must all be there in the products. To be sure an equation is balanced requires counting up the atoms of each kind in the reactants and products (Section 2.7). Doing this, of course, requires knowing the identity and correct formulas of the reactants and products. An equation can *never* be balanced by changing the subscript in a chemical formula. This changes the identity of the compound. Only coefficients can be changed to achieve balance.

■ The law of conservation of matter: *Matter is neither lost nor gained in chemical reactions* (Section 3.2). The only known exception to this law is in nuclear reactions, which occur only with radioactive isotopes or under the special conditions of bombardment reactions (Chapter 5). The conservation law (so far) has always been reliable for chemical changes other than nuclear reactions.

The products of the complete burning, or combustion, of any hydrocarbon are always carbon dioxide and water. So, for the burning of methane (CH_4), the major ingredient in natural gas, the *unbalanced* equation is

$$\underset{\text{1 C atom}}{CH_4(g)} + O_2(g) \longrightarrow \underset{\text{1 C atom}}{CO_2(g)} + H_2O(g)$$

The C atoms are balanced here—one C on each side of the arrow. But the O and H atoms are not balanced. Often, it is easiest to first balance atoms that appear in only one formula on each side. Balancing H requires a coefficient of 2 in front of H_2O.

$$\underset{\text{4 H atoms}}{CH_4(g)} + O_2(g) \longrightarrow CO_2(g) + \underset{\text{4 H atoms}}{2\ H_2O(g)}$$

Now the O atoms must be balanced. With four of them on the right, providing two O_2 molecules on the left finishes the job (Figure 8.1).

$$CH_4(g) + \underset{\text{4 O atoms}}{2\ O_2(g)} \longrightarrow \underset{\text{4 O atoms}}{CO_2(g) + 2\ H_2O(g)}$$

$$CH_4 \quad + \quad 2O_2 \quad \longrightarrow \quad CO_2 \quad + \quad 2H_2O$$

Reactant atoms Product atoms

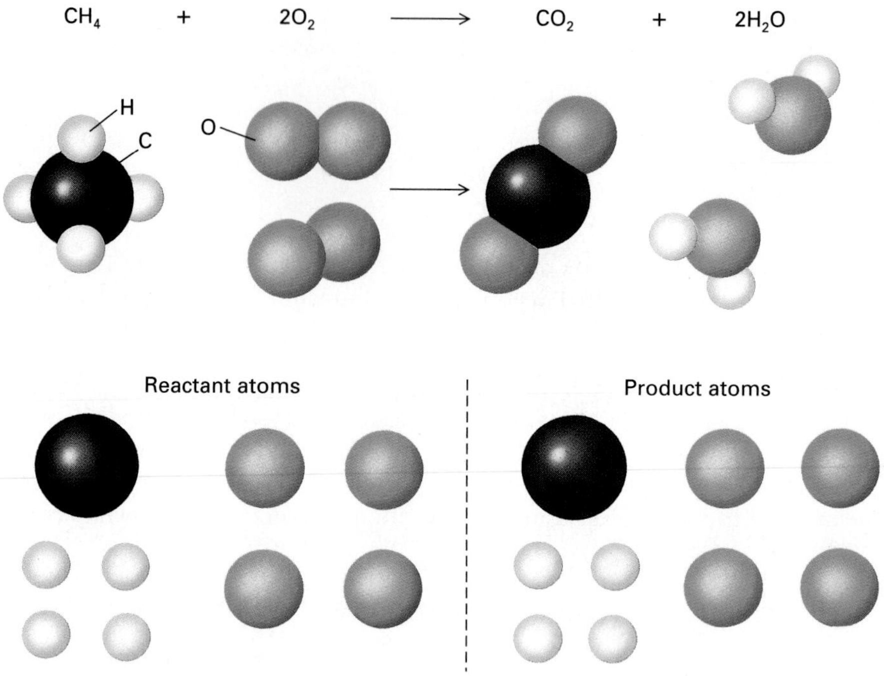

Figure 8.1 Conservation of mass. The numbers of reactant and product atoms are always the same, no matter how complicated the reaction.

Note in this equation the difference between *subscripts* (e.g., the $_4$ in CH_4), which relate to the need for correct formulas, and **coefficients** (e.g., the 2 with O_2 as a reactant), which relate to the need for a balanced equation.

EXAMPLE 8.1 *Equation Balancing*

Balance the equation below for the reaction of hydrofluoric acid [$HF(aq)$] with glass (which can be represented as calcium silicate, $CaSiO_3$). Decorative glass is etched using this reaction.

$$CaSiO_3(s) + HF(aq) \longrightarrow CaF_2(s) + SiF_4(g) + H_2O(\ell)$$

SOLUTION

The Ca and Si atoms are balanced. To balance the three O atoms on the left requires three H_2O molecules on the right. There must then be six H atoms on the left. Putting in both of these coefficients gives

$$CaSiO_3(s) + \boxed{6}\ HF(aq) \longrightarrow CaF_2(s) + SiF_4(g) + \boxed{3}\ H_2O(\ell)$$

There are now six F atoms on each side of the equation, and it is fully balanced.

On the left: 1 Ca 1 Si 6 H 6 F 3 O *On the right:* 1 Ca 1 Si 6 H 6 F 3 O

■ The states of the reactants and products are indicated by
(*g*) for a gas
(*s*) for a solid
(*ℓ*) for a liquid
(*aq*) for something dissolved in water

Exercise 8.1A

Balance the following equation for the preparation of aluminum chloride, which is an ingredient in some antiperspirants.

$$Al(s) + Cl_2(g) \longrightarrow AlCl_3(s)$$

Exercise 8.1B

Are the equations below balanced?

(a) $CaO(s) + H_2O(\ell) \rightarrow Ca(OH)_2$

(b) $SiO_2(s) + C(s) \rightarrow Si(s) + CO(g)$

EXAMPLE 8.2 *Equation Balancing Again*

Balance the following equation for the preparation of silver chloride, an insoluble solid that is sensitive to light and is used in photography.

$$AgNO_3(aq) + CaCl_2(aq) \longrightarrow AgCl(s) + Ca(NO_3)_2(aq)$$

SOLUTION

Note that the NO_3^- ion is in parentheses in $Ca(NO_3)_2$. In balancing equations, such ions can be treated as units. With one NO_3^- ion on the left and two on the right, they can be balanced by adding a coefficient 2 for $AgNO_3$:

$$2\ AgNO_3(aq) + CaCl_2(aq) \longrightarrow AgCl(s) + Ca(NO_3)_2(aq)$$

Now there are two Ag's on the left but one on the right. Balancing Ag's gives

$$2\ AgNO_3(aq) + CaCl_2(aq) \longrightarrow 2\ AgCl(s) + Ca(NO_3)_2(aq)$$

Checking shows that the equation is now balanced:

On the left: 2 Ag 2 NO_3 1 Ca 2 Cl *On the right:* 2 Ag 2 NO_3 1 Ca 2 Cl

Exercise 8.2

Balance the following equation:

$$Ba(NO_3)_2(aq) + Na_2SO_4(aq) \longrightarrow BaSO_4(s) + NaNO_3(aq)$$

8.2 The Mighty Mole and the "How Much?" Question

Moles and Molar Masses

Eventually, anyone who wants to carry out a chemical reaction must figure out how much of the reactants must be combined to make the desired amount of product. Somehow a connection must be made between atoms, molecules,

■ You might want to look back at the description of atomic weights in Section 3.4.

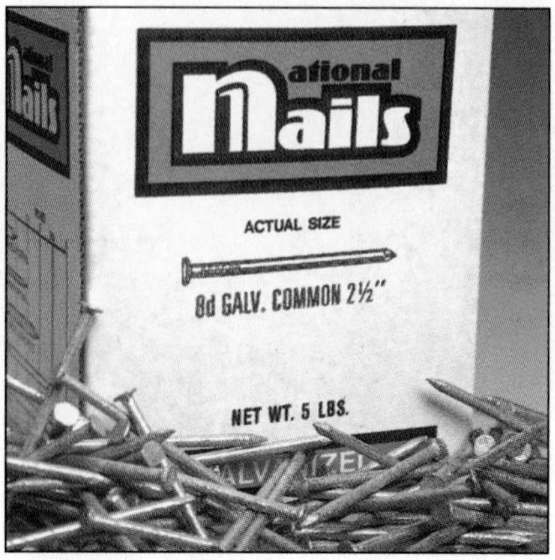

Counting by weighing. It is much easier to buy 5 lb of nails than to count them out. *(C.D. Winters)*

■ To get a more precise dollar value for your quarters, you would have to subtract the weight of the empty jar from the weight of the full jar to get the weight of the quarters alone. You would also need to find the weights on a balance that gives more digits.

■ Before 1982, pennies weighed 3 g and were made of 95% copper and 5% zinc. By 1982, the copper in a penny was worth more than 1 cent, so the composition was changed to 97.6% zinc with 2.4% copper as a thin coating.

■ We don't need all the digits for atomic weights given in the periodic table inside the back cover, so we will round them off.

■ The mole as a unit is symbolized by mol, e.g., 2.5 mol.

and ions at the submicroscopic scale and weighable amounts of chemicals. Here is where balanced equations are essential. The relative atomic weight scale and balanced chemical equations together make it possible to answer "how much?" questions.

To understand the situation, consider the equation for hydrogen burning in chlorine to form hydrogen chloride:

$$H_2(g) + Cl_2(g) \longrightarrow 2\ HCl(g)$$

The equation shows that one molecule of hydrogen and one molecule of chlorine combine to form two molecules of hydrogen chloride. Is this information of any help in figuring out, say, how much hydrogen and chlorine would be needed to make 100 g of hydrogen chloride? The essential problem is that molecules are very small, so small that it is impossible to count them one by one.

The solution to the problem is counting by weighing. To demonstrate how this is done, we can use something you are more familiar with—quarters and pennies. Suppose you have a big jar full of quarters and you want to quickly estimate what it is worth. First, you find that one quarter weighs 5.7 g. Then, you find that the filled jar weighs 3300 g. Some simple arithmetic shows that your jar contains roughly 580 quarters.

$$3300 \text{ g of quarters} \times \frac{1 \text{ quarter}}{5.7 \text{ g}} = 580 \text{ quarters}$$

The jar contains about \$145 (\$0.25 × 580). Now suppose you get to wondering how much 580 pennies would weigh. The weight of one newly minted penny is 2.5 g, so

$$580 \text{ pennies} \times \frac{2.5 \text{ grams}}{1 \text{ penny}} = 1450 \text{ grams}$$

A relative weight scale can be assigned to coins in the same way it has been done for atoms (Figure 8.2). If the newly minted penny is assigned a weight of 1, then on the relative-weight-of-coins scale, a quarter has a weight of 2.3 (5.7 g/2.5 g = 2.3). With this information, you can always get equal numbers of pennies and quarters by weighing them in a weight ratio of 1 to 2.3—it could be 10 g of pennies and 23 g of quarters, or 150 g of pennies and 345 g of quarters, or 1450 g of pennies and 3335 g of quarters, or 1 ton of pennies and 2.3 tons of quarters, or . . .

The atomic weights given for each element in the periodic table are also relative weights. The weight of one average neon atom is 20. amu (atomic mass units), and the weight of one calcium atom is 40. amu, both relative to an atomic weight of exactly 12 amu for carbon-12. Translating these numbers to masses big enough to measure means that 20. g of neon, 12 g of carbon-12, and 40. g of calcium all contain the same number of atoms. This type of relationship is at the heart of the quantitative use of chemical equations.

The chemists' counting unit is named the **mole,** and it is defined as equal to the number of atoms in exactly 12 g of carbon-12. The mole is used in the same way as a dozen. In the grocery store you may need one dozen apples. In the laboratory, you may need 1 mol of carbon. You can count out 12 apples, but to get 1 mol of carbon, you have to weigh out 12 g of carbon.

The actual number of individual atoms, molecules, or ions in one mole is known from experiments. Just as a dozen apples is 12 apples, a mole of atoms is about 602,000,000,000,000,000,000,000, or 6.02×10^{23} atoms. How big is this number? A mole of sand would cover a city the size of Los Angeles to a depth of 600 meters. The number of atoms, molecules, ions or anything else in a mole, 6.02×10^{23}, is known as **Avogadro's number.**

■ If you ever see a car with a bumper sticker that reads, **"Everyone has Avogadro's number,"** you can assume that the car probably belongs to a chemist.

The mass in grams of one mole of a substance is referred to as its **molar mass.** The molar mass of an element (except for those that exist as diatomic gases, e.g., H_2) is the mass in grams numerically equal to the atomic weight of the element (Figure 8.3). The molar mass of helium, the lightest noble gas, is 4.0 g. What about a heavy element like lead? Checking the periodic table shows that the molar mass of lead is 207 g.

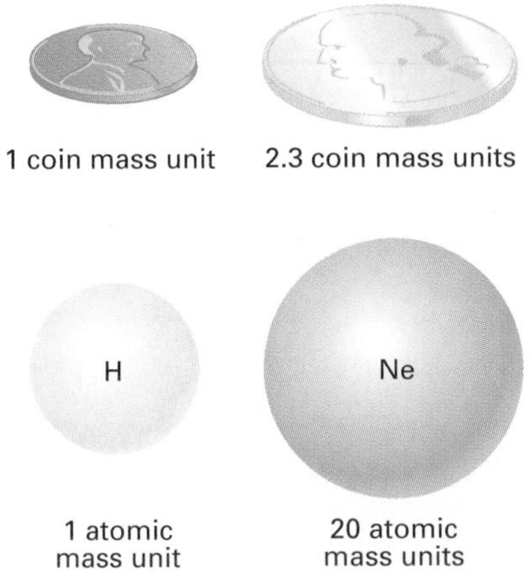

Figure 8.2 Relative weight scales for use in counting by weighing.

Since the mole is a counting unit, every balanced chemical equation can be interpreted in terms of moles—moles of atoms, moles of molecules, moles of ions, or anything else. For example,

$$\underset{1\ H_2\ \text{molecule}}{H_2(g)} + \underset{1\ Cl_2\ \text{molecule}}{Cl_2(g)} \longrightarrow \underset{2\ HCl\ \text{molecules}}{2HCl(g)}$$

also means

$$\underset{1\ \text{mol of}\ H_2}{H_2(g)} + \underset{1\ \text{mol of}\ Cl_2}{Cl_2(g)} \longrightarrow \underset{2\ \text{mol of HCl}}{2HCl(g)}$$

Figure 8.3 One-mole quantities of some elements. The cylinders *(left to right)* hold mercury (201 g), lead (207 g), and copper (64 g). The two Erlenmeyer flasks hold sulfur (*left,* 32 g) and magnesium (*right,* 24 g). All rest on one mole of aluminum in the form of foil (27 g). *(C. D. Winters)*

Figure 8.4 One-mole quantities of some ionic compounds. The crystalline form and variety of colors are common properties of ionic compounds. Clockwise from the upper left, the compounds shown are green nickel(II) chloride ($NiCl_2 \cdot 6H_2O$), orange potassium dichromate ($K_2Cr_2O_7$), red cobalt(II) chloride ($CoCl_2 \cdot 6H_2O$), white sodium chloride (NaCl), and blue copper(II) sulfate ($CuSO_4 \cdot 5H_2O$). *(C. D. Winters)*

■ Sometimes the term "molecular weight" is used instead of molar mass. Both terms refer to the relative mass of a substance according to the atomic weight scale. Strictly speaking, a molecular weight is for one molecule and would be in atomic mass units. We need use only molar masses.

Now, what is needed to interpret this equation in terms of masses of the reactants and products, which happen to be molecular substances? No problem. The molar mass of any pure substance is found by adding up the molar masses of the atoms shown in its formula (Figure 8.4). One mole of hydrogen molecules (H_2) has a molar mass of 2×1.0 g/mol H = 2.0 g/mol H_2. One mole of chlorine molecules has a molar mass of 2×35.5 g/mol Cl = 71.0 g/mol Cl_2. How about hydrogen chloride, HCl? It has a molar mass of 36.5 g.

H molar mass	1.0 g
Cl molar mass	35.5 g
	36.5 g = molar mass of HCl

Masses of Reactants and Products

The mole provides us with the means to answer the "How much?" question about any chemical reaction. If all the reactants are converted to product, 2.0 g of H_2 (1 mol) should react with 71 g of Cl_2 (1 mol) to produce 73 g of HCl (2 mol).

$$H_2(g) + Cl_2(g) \longrightarrow 2\,HCl(g)$$

$H_2(g)$	$Cl_2(g)$	$2\,HCl(g)$
1 H_2 molecule	1 Cl_2 molecule	2 HCl molecules
1 mol of H_2	1 mol of Cl_2	2 mol of HCl
2.0 g of H_2	71 g of Cl_2	2 mol × 36.5 g/mol = 73 g of HCl

■ Rarely can 100% of reactants actually be converted to products. One goal of industrial preparation of chemicals is to come as close to 100% as is possible and practical.

Have you noticed that these masses adhere to the law of conservation of matter? The 2.0 g + 71 g of reactants produce 73 g of product. At the end of the chapter, there are some more examples of molar arithmetic that demonstrate how versatile and important the mole concept is.

THE PERSONAL SIDE

Amadeo Avogadro (1776–1856)

Amadeo Avogadro (The World of Chemistry, *Program 11, "The Mole"*)

Today, Avogadro's name is most often associated with "his" number. The mole and the modern definition of the number of particles in it, however, were not directly his invention. His fame lies in a single simple statement, made in 1811: ". . . *the number of integral molecules in gases is always the same for equal volumes.*" This concept opened the door to understanding atomic weight and the formulas for chemical compounds.

Avogadro was a quiet man who was totally devoted to his studies. He was born into a prominent Italian family of lawyers and, as was surely expected of him, completed his training as a lawyer and entered government service. At age 24 he turned his back on his heritage and devoted the rest of his life to science. He became a professor of physics and mathematics and was content to pursue his studies alone, never attending scientific meetings nor seeking out colleagues with whom to exchange ideas. He published very little, but the intensity of his studies is shown in the 75 volumes of handwritten notes left behind when he died at age 80.

Perhaps because of Avogadro's private nature, the true value of his realization about gas volumes was not recognized until after his death, when another Italian chemist (Stanislao Cannizzarro) brought it to the attention of the scientific community. By comparing the masses of equal volumes of gases at identical temperature and pressure, the relative weights of the molecules could be compared. And by examining the volumes of gases that combined in reactions, it was possible to deduce the formulas. Before this, it was not recognized that a hydrogen molecule consisted of two hydrogen atoms rather than one, or that water molecules consist of two hydrogen atoms and one oxygen atom, rather than just one hydrogen atom and one oxygen atom.

Copper face found in a tomb in Peru. As time passes, green copper sulfide and black copper oxide form on a copper surface as the result of reaction with hydrogen sulfide and oxygen in the atmosphere. *(Heinz Plenge/Peter Arnold, Inc.)*

EXAMPLE 8.3 *Information about Masses from a Chemical Equation*

Interpret the equation for a reaction that produces copper from copper ore in terms of moles, molar masses, and masses of reactants and products.

$$\underset{\text{Copper(I) sulfide}}{Cu_2S(s)} + \underset{\text{Copper(I) oxide}}{2\,Cu_2O(s)} \xrightarrow{\text{Heat}} \underset{\text{Copper}}{6\,Cu(s)} + \underset{\text{Sulfur dioxide}}{SO_2(g)}$$

SOLUTION

The equation in terms of moles shows by the coefficients that 1 mol of copper(I) sulfide reacts with 2 mol of copper(I) oxide to produce 6 mol of metallic copper and 1 mol of sulfur dioxide. The molar masses of the reactants and products are found from the molar masses of the elements combined:

Cu_2S

$$\begin{aligned} 2 \times 64\text{ g Cu/mol} &= 128\text{ g/mol} \\ 32\text{ g S/mol} &= 32\text{ g/mol} \\ \hline & 160\text{ g } Cu_2S\text{/mol} \end{aligned}$$

SO_2

$$\begin{aligned} 32\text{ g S/mol} &= 32\text{ g/mol} \\ 2 \times 16\text{ g O/mol} &= 32\text{ g/mol} \\ \hline & 64\text{ g } SO_2\text{/mol} \end{aligned}$$

Cu_2O

$$\begin{aligned} 2 \times 64\text{ g Cu/mol} &= 128\text{ g/mol} \\ 16\text{ g O/mol} &= 16\text{ g/mol} \\ \hline & 144\text{ g } Cu_2O\text{/mol} \end{aligned}$$

The equation gives the following information about reactants and products:

$Cu_2S(s)$ +	$2\ Cu_2O(s)$	$\longrightarrow$	$6\ Cu(s)$	+ $SO_2(g)$
1 mol Cu_2S	2 mol Cu_2O		6 mol Cu	1 mol SO_2
160 g Cu_2S	2×144 g Cu_2O/mol = 288 g Cu_2O		6×64 g Cu/mol = 384 g Cu	64 g SO_2

Exercise 8.3

Interpret the equation for making methanol in terms of moles, molar masses, and masses of reactants and products.

$$CO(g) + 2H_2(g) \longrightarrow CH_3OH(\ell)$$

The World of Chemistry

Program 11, *"The Mole"*

So using the mole concept, we're able—in the laboratory, in the real world of chemistry—to measure out numbers of molecules that we need for our chemical reactions. John Massingill, research and development chemist, Dow Chemical, Freeport, Texas.

■ SELF-TEST 8A

1. Balancing chemical equations is an application of the ________.
2. The ________ is used by chemists the same way the ________ is used by an egg farmer.
3. Balance the following equations:
   (a) ____ $Si(s)$ + ____ $Cl_2(g) \rightarrow SiCl_4(g)$
   (b) ____ $Al(s)$ + ____ $O_2(g) \rightarrow$ ____ $Al_2O_3(s)$
   (c) ____ $(NH_4)_2CO_3(aq)$ + ____ $Cu(NO_3)_2(aq) \rightarrow$ ____ $CuCO_3(s)$ + ____ $NH_4NO_3(aq)$
4. The molar mass of iron is the mass in grams numerically the same as the ________.
5. In photosynthesis, carbon dioxide combines with water to form oxygen and the simple sugar glucose ($C_6H_{12}O_6$).
   (a) Balance the equation

   ____ $CO_2(g)$ + ____ $H_2O(\ell) \longrightarrow$ ____ $C_6H_{12}O_6(aq)$ + ____ $O_2(g)$

   (b) How many molecules of CO_2 are needed to produce one molecule of glucose?
   (c) How many moles of CO_2 are needed to produce 1 mol of glucose?

(d) What is the molar mass of glucose?
(e) What mass in grams of CO_2 is needed to make 1 mol of glucose?
6. Which is the molar mass of nitrogen gas (N_2)? (a) 7 g, (b) 14 g, (c) 28 g.

8.3 Rates and Reaction Pathways: The "How Fast?" Question

Reaction Pathways

The chemical reactions you may have seen as demonstrations are usually fast. The color change, bubbles of gas, or explosion happens right away as visible proof that a reaction has occurred. Many reactions are also naturally slow, however. At everyday conditions of temperature and pressure, for example, the conversion of carbon monoxide to carbon dioxide is slow:

$$2CO(g) + O_2(g) \longrightarrow 2CO_2(g)$$

It is sometimes unfortunate that this reaction isn't fast. Breathing too high a concentration of carbon monoxide is fatal.

Out of curiosity and also for practical reasons, it is interesting to discover what makes reactions fast or slow. Ideally, to do this a chemist would like to watch the pathway of each atom from its position in the reactants to its position in the products.

What is the connection between the rate of a process and its pathway? Suppose there are several hundred books in a storeroom on the first floor and that you have been hired to move them to the new third-floor library. Depending on the conditions, there are a variety of possible pathways. One pathway might require the following steps: (1) put ten books (the maximum number you can lift) into a carton, (2) carry the carton up one flight of stairs to the

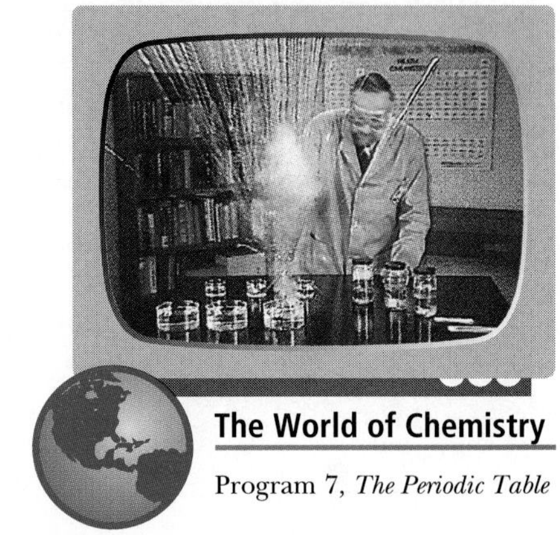

The World of Chemistry

Program 7, *The Periodic Table*

Potassium plus water, $2\ K(s) + 2\ H_2O(\ell) \rightarrow 2\ KOH(aq) + H_2(g)$. A naturally fast reaction.

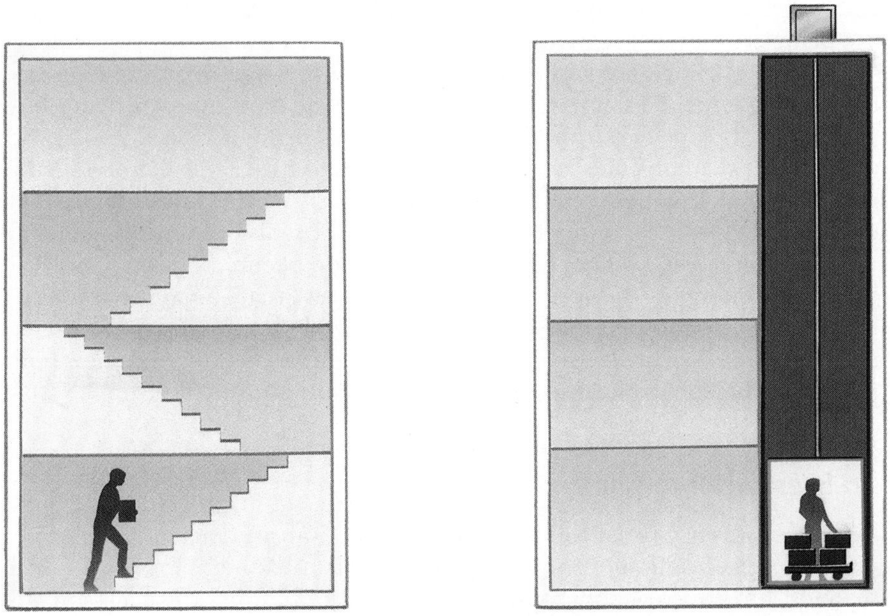

Which pathway has a higher rate of books moved per hour?

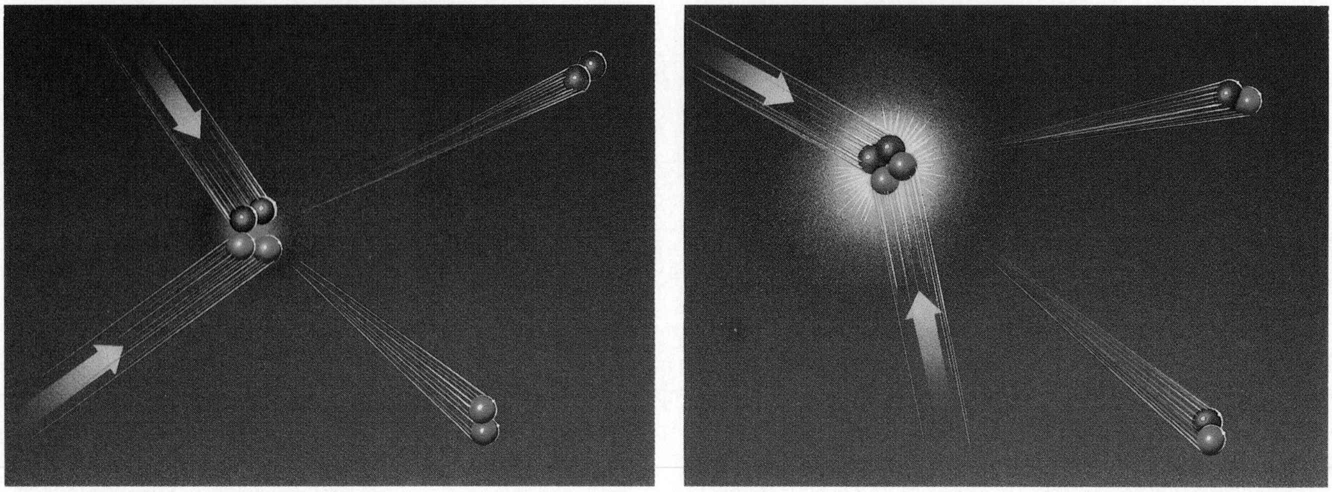

(a) Unsuccessful collision (b) Successful collision

Figure 8.5 Collisions and chemical reactions. Some collisions are successful and some are not. The difference is determined by the kinetic energy of the particles.

second floor, (3) rest a bit, (4) carry the carton up another flight of stairs to the third floor, (5) empty the carton, and (6) return for another load. Another pathway might involve a different series of steps: (1) fill four cartons with books, (2) pile the cartons onto a dolly, (3) push the dolly onto the elevator, (4) ride to the third floor, (5) push the dolly off the elevator, (6) empty the four cartons, (7) return for another load. The second pathway would probably be faster than the first. To compare them quantitatively, the rate for each pathway could be measured in books moved per hour.

The details of chemical reaction pathways can be extremely complex, and they are very hard to study. With the aid of computers, sophisticated electronics, and clever new techniques, however, advances in this area of observation are being announced regularly.

Using your imagination instead, what might be seen in a simple one-step chemical reaction between two different gases? The molecules are flying about at random, as expected from the kinetic molecular theory (Section 7.1). Now and then, two molecules head for a collision. As they get closer together, repulsion builds up between their negatively charged electrons. If their kinetic energies are not great enough to overcome this repulsion, the molecules just veer away from each other. No chemical reaction occurs (Figure 8.5*a*). But if the kinetic energy of the approaching molecules is great enough to drive them together in spite of the repulsion, then they collide. At this instant a chemical reaction might occur. If some of the original bonds break so that new bonds can form, the collision is successful and yields the product (Figure 8.5*b*).

An Energy Hill to Climb

■ In addition to sufficient energy, successful collisions also require that the reacting parts of molecules or ions physically connect with each other.

Reaction rates are usually expressed as the amount of reactant converted to product in a specific unit of time. For fast reactions the time unit might be seconds; for slow reactions it might be days. The number of successful collisions per second, minute, hour, or day determines the reaction rate.

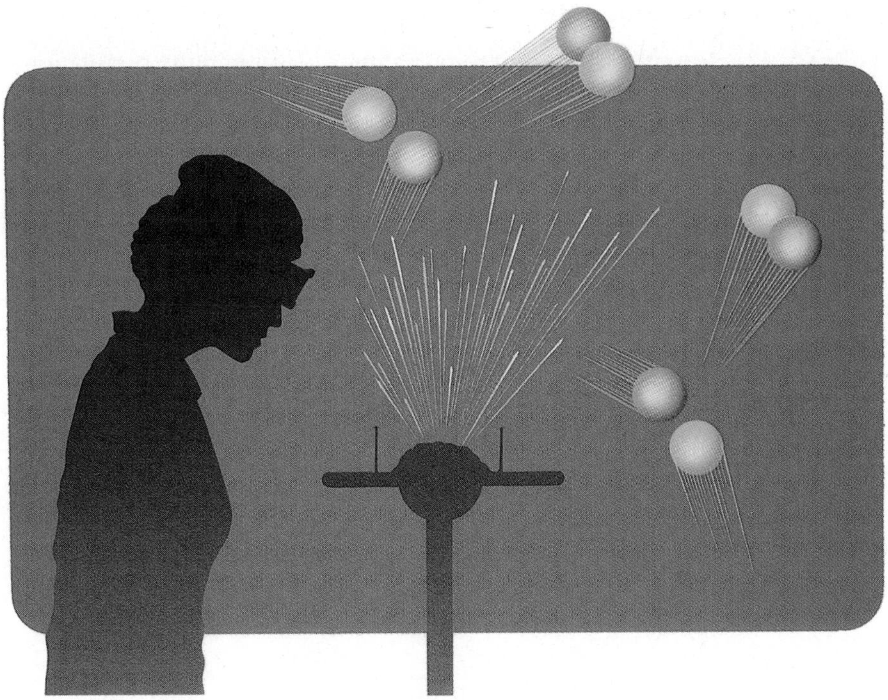

For each reaction, there is a distinctive quantity of energy needed for successful collisions, known as the **activation energy.** A high activation energy is like a steep mountain that must be climbed to reach the valley on the other side. In a given collection of reactant particles, only a small number have enough energy to get over a very high activation energy hill in a given time, so the reaction is slow. In the opposite condition, many reactant particles can get over a low activation energy hill, so the reaction is fast (Figure 8.6).

Some reactions do not take place at all unless enough energy is supplied to get things started by pushing a few reactants over the activation energy barrier. After that, as in the explosion of a hydrogen-oxygen mixture or the combustion of a fuel, the reaction provides enough energy to keep itself going.

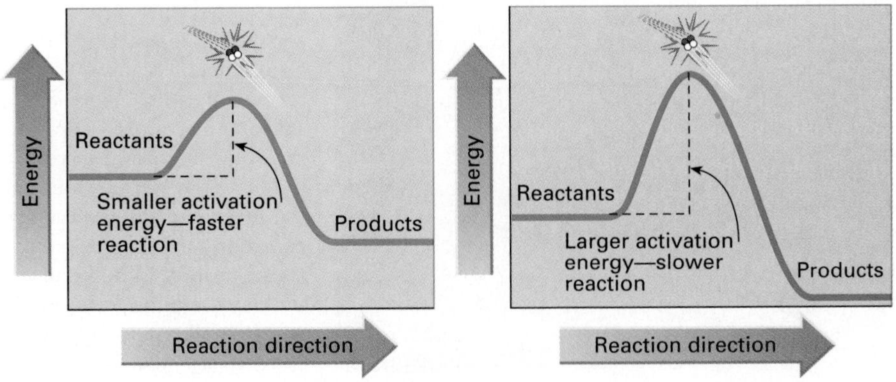

Figure 8.6 Energy pathways of chemical reactions. The height of the activation energy hill determines the rate of the reaction.

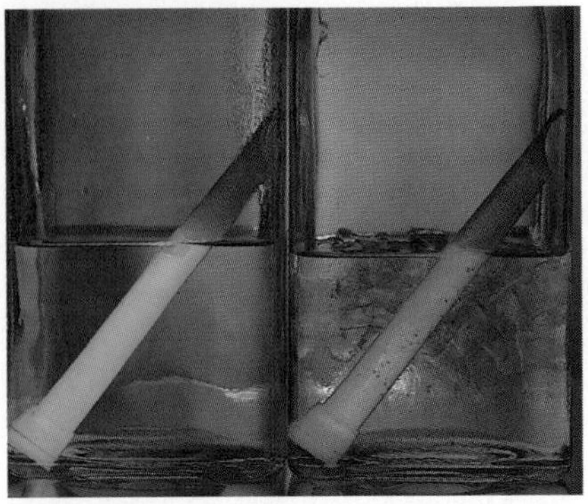

Figure 8.7 Temperature and reaction rate. When a light stick is bent, an inner glass tube breaks, allowing reactants to mix. The result is a chemical reaction that releases energy as light rather than heat. The reaction is faster in hot water, causing the light to glow more brightly (on the left) than in ice water (on the right). *(Richard Megna/Fundamental Photographs)*

Controlling Reaction Rates

To control the rate of a reaction requires either increasing the population of reactants with enough energy to get over the activation energy barrier or lowering the barrier. Three strategies are available: (1) adjust the temperature, (2) adjust the concentration, or (3) add a catalyst.

1. ***Effect of temperature on reaction rate.*** At higher temperatures, molecules (on average) move faster, so more of them have enough energy to get over the activation energy hill and react (Figure 8.7). We make use of this principle in cooking by raising the temperature to speed up roasting a piece of meat and in preserving foods by lowering the temperature to slow down the reactions that spoil the food.
2. ***Effect of concentration on reaction rate.*** The quantity of a substance in a given quantity of a mixture is its **concentration.** For example, a solution might have a concentration of 5 g of sodium chloride in 1 liter of water (5 g/L, or 5 grams per liter). A solution of 10 g of sodium chloride in 1 liter of water (10 g/L) has a higher concentration.

   Increasing the concentration of reactants increases reaction rates. The more molecules, the more frequent the collisions that have enough energy to be successful. There is, for example, a dramatic increase in the rate of combustion in pure oxygen compared with that in air, which is about 20% oxygen. Smoking tobacco in air is dangerous to your internal health, but smoking tobacco in pure oxygen would pose a more immediate risk. Great care must be taken to avoid smoking, flames, or sparks in hospitals or anywhere that pure oxygen is in use.
3. ***Effect of a catalyst on reaction rate.*** Sometimes, a reaction speeds up dramatically when a substance other than the reactants is added to the

An alligator crossing a highway. Alligators are cold-blooded animals—the rate of their metabolism depends on the temperature. Not too long ago, this alligator crawled onto a major highway during an evening when there was a sharp temperature drop. The resulting slow-down in reaction rate put him to sleep in the middle of the road, where considerate police directed traffic around him. Finally, at midday the temperature rose and he walked off without any coaxing. *(AP/Wide World Photos)*

DISCOVERY EXPERIMENT

Enzymes—Biological Catalysts

Raw potatoes contain an enzyme called *catalase,* which converts hydrogen peroxide to water and oxygen. You can demonstrate this by the following experiment:

Purchase a small bottle of hydrogen peroxide at a pharmacy. The peroxide is usually sold as a 3% solution in water. Pour about 50 mL of the peroxide solution into a clear glass or plastic cup. Add a small slice of fresh potato to the cup. (Since potato is less dense than water, the potato will float.)

Almost immediately you will see bubbles of oxygen gas on the potato slice. Does the rate of evolution of oxygen change with time? If so, how does it change? If you cool the hydrogen peroxide solution in a refrigerator and then do the experiment, is there a perceptible change in the initial rate of oxygen evolution? Is there a difference between the time at which oxygen evolution begins for warm and for cold hydrogen peroxide?

mixture. You may have a bottle containing a dilute solution of hydrogen peroxide on your bathroom shelf—hydrogen peroxide is often used as a disinfectant. When stored in a brown or opaque bottle it decomposes very slowly to water and oxygen ($2\,H_2O_2 \rightarrow 2\,H_2O + O_2$). Have you noticed that when you put hydrogen peroxide on a cut, it bubbles vigorously? There is a substance in blood (known as *catalase*) that speeds up the decomposition (Figure 8.8).

Such substances that increase the rate of a chemical reaction without being changed themselves are known as **catalysts.** In the presence of a catalyst, an alternate pathway with a *lower* activation energy is made available (Figure 8.9). More collisions are successful because less energy is required for success. What makes catalysts so practical is that many times they can be recovered after the reaction is over and used again and again. Many industrial processes rely on rare and expensive metals as catalysts, so their recovery is an economic necessity.

Figure 8.8 Catalyzed decomposition of hydrogen peroxide ($2\,H_2O_2[aq] \rightarrow 2\,H_2O[\ell] + O_2[g]$) The liver is well supplied with an enzyme for this reaction. *(Larry Cameron)*

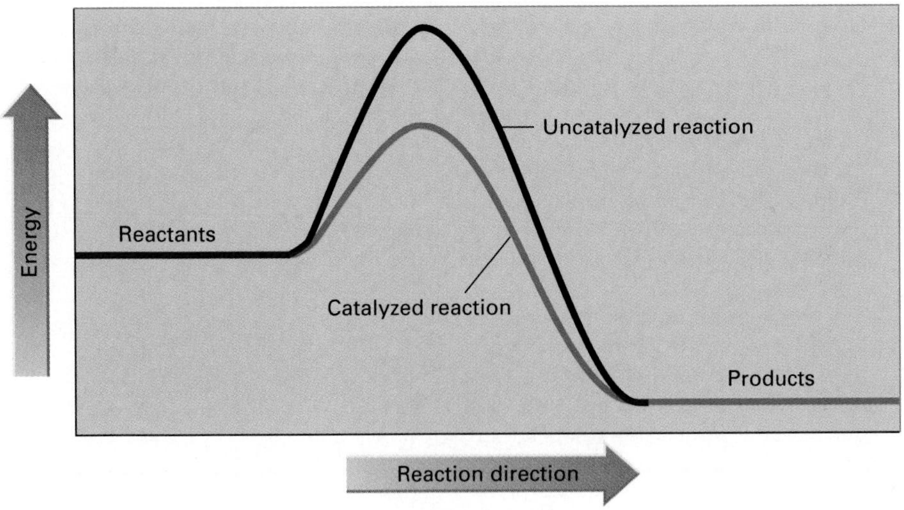

Figure 8.9 Effect of a catalyst. All a catalyst does is lower the activation energy. Because more reactants have enough energy to collide successfully, the reaction speeds up.

The World of Chemistry

Program 13, *The Driving Forces*

Food—it's the study of messy chemistry. In a food which has so many different organic compounds and inorganic compounds, there are lots and lots of different reactions that could have caused spoilage. What we can do is narrow them down to several classes. One is reactions that are enzyme-catalyzed.
Theodore Labuza, Food Chemist

Living things are even more dependent on catalysts than the chemical industry. Engineers can manipulate temperature and concentrations in an industrial process in order to control reaction rates. Our bodies can't. If body temperature varies far from 37°C or if the concentrations of chemicals in body fluids vary much from the normal, we are in serious trouble. Catalysis is the major strategy available for controlling biochemical reactions. The amazing constancy of our internal chemistry is maintained by biological catalysts known as *enzymes* (Section 15.4). There is a good reason why blood and the liver are well supplied with a fast-acting catalyst for the decomposition of hydrogen peroxide. Because it is highly reactive, hydrogen peroxide must be destroyed before it has the chance to damage essential substances in its surroundings.

8.4 Chemical Equilibrium and the "How Far?" Question

In a closed container partly filled with water (Figure 8.10), the air over the liquid soon becomes mixed with water vapor (Section 7.4). Once the air holds all the water vapor it can, some of the water vapor condenses. Eventually, the evaporation and condensation of the water establish a **dynamic equilibrium**—a state of balance between exactly opposite changes occurring at the same rate. To indicate this equilibrium in a chemical equation, a double arrow is placed between the symbols for water in the liquid and vapor states:

$$H_2O(\ell) \rightleftharpoons H_2O(g)$$

FRONTIERS IN THE WORLD OF CHEMISTRY

Atom Economy

While many people are now advocating conservation by limiting what goes into municipal garbage, a chemist has begun a campaign for conservation at the atomic and molecular level. Suppose a useful chemical is manufactured by a reaction that we can represent by $A + B \rightarrow C + D$. Suppose further that C is the useful reaction product and D is a by-product that is essentially "garbage" because it has no use or value.

Barry M. Trost of Stanford University, Professor of Humanities and Sciences, is calling on chemists to stop producing such chemical waste. He has coined the term "atom economy" to describe a conservationist approach to chemical synthesis. It means devising reactions that produce *only* the desired product or perhaps only the desired product plus water.

Like recycling glass, plastic, and aluminum containers, conservation in chemical synthesis requires a new philosophy based on the growing need to protect our resources and environment. Often, if the route to a desirable synthetic chemical is financially economical, little concern is given to the by-products, even if they contain a significant proportion of the atoms in the reactants.

Accomplishing atom economy requires new kinds of chemical reactions. In his own research, Dr. Trost focuses on catalysis by transition metals. How some catalysts work is understood, but in many cases, even for catalysts in successful industrial processes, what happens at the atomic and molecular level is still a mystery. Such a state of affairs presents an exciting challenge to a scientist like Dr. Trost. "We haven't even begun to explore a large fraction of the periodic table. . . . it is going to take very long to appreciate some of the things one will be able to do. . . . There is tremendous untapped power . . . in synthetic chemistry."

Reference Chemical and Engineering News, June 19, 1995; Science, Vol 254, p 1471, 1991.

Figure 8.10 A system at equilibrium. Water has reached equilibrium with its vapor in this Ecosphere. So also have the food and waste products of the inhabitants—a carefully balanced community of plants, shrimp, and a hundred or so kinds of microorganisms. *(Photo courtesy of Ecosphere Associates, Ltd. of Tucson, AZ, who are responsible for developing this first successful sealed community.)*

Chemical reactions establish the same kind of equilibrium. In **chemical equilibrium,** a chemical reaction and its reverse are occurring at equal rates. Theoretically, all chemical reactions are **reversible**—able to take place in either direction and therefore to come to equilibrium. If and how chemical equilibrium is actually established depends on a number of factors.

Heating limestone in masonry pits to produce lime for mortar was one of the first manufacturing processes carried out by settlers in the U.S. colonies. The CO_2 gas escapes from an open pit and all of the calcium carbonate in the limestone can be converted to lime.

$$\underset{\text{Calcium carbonate (limestone)}}{CaCO_3(s)} \xrightarrow{\text{Heat}} \underset{\text{Calcium oxide (lime)}}{CaO(s)} + \underset{\text{Carbon dioxide}}{CO_2(g)}$$

This reaction is reversible and establishes equilibrium *if* the gas cannot escape. When some dry limestone is sealed in a closed container and heated, the decomposition of calcium carbonate begins. As soon as CO_2 accumulates in the container, the reverse reaction starts to occur. Once the concentration of CO_2 reaches a specific point, the system has reached equilibrium.

$$CaCO_3(s) \rightleftharpoons CaO(s) + CO_2(g)$$

If there aren't any changes in the reaction conditions (the pressure and the temperature), the forward and reverse reactions continue to take place at the same rates and the concentration of CO_2 in the container remains the same.

THE WORLD OF CHEMISTRY

A Catalyst in Action

Program 14, "*Molecules in Action*"

Because of their remarkable chemical properties, catalysts are used throughout the industry. Dr. Norman Hochgraf, vice president of Exxon Chemical, shares with us his viewpoint:

Within the petroleum industry, where most of the feedstocks for chemicals come from, catalysts are used to produce motor gasoline, heating oil, and feedstocks for chemicals. And within the chemical industry, catalysts are used not only to purify those feedstocks, but they're also used to polymerize the feedstocks, to make plastics, to make rubbers, and to make synthetic fibers. This industry not only wouldn't be profitable without catalysts, but in a sense, it wouldn't exist. Almost everything we do is the result of catalytic activity, which has been carefully designed, carefully selected, and built into our commercial operations.

Let's look at Eastman Kodak's group of plants in Kingsport, Tennessee. Synthetic products—fabrics, plastics, films, and aspirin—are all made there. These products are all made using acetic anhydride, which is also manufactured at the same location. Yet without the rare South African metal, rhodium, there would be no plant at all. Eastman Kodak used to make acetic anhydride from oil. With the rise in oil prices in the early 1970s, the company began looking for alternative inexpensive materials. Coal was the obvious choice. First it is gasified, producing hydrogen and carbon monoxide. At a later stage, carbon monoxide is reacted with methyl acetate to produce acetic anhydride. This crucial step requires rhodium as a catalyst.

If rhodium is so expensive, how can it be profitable to use the catalyst in such a large-scale reaction? The answer lies in the fact that catalysts are not used up in reactions. Each catalyst molecule may react with thousands and thousands of molecules of reactant. So this catalyst need be present only in tiny amounts. In addition, as the mixture is drawn off, the catalyst is carefully separated from the acetic anhydride and recycled back into the reactor. Very little rhodium is lost.

Without the development of a viable catalyst system for producing acetic anhydride from methyl acetate and carbon monoxide, the chemicals-from-coal complex at Kingsport, Tennessee, would not have been built.

Note that chemical equilibrium is dynamic—the CO_2 concentration doesn't change, but CO_2 is constantly being produced and reacting to form $CaCO_3$.

If the container is opened briefly to let some of the CO_2 out and then sealed again, the forward reaction will outpace the reverse reaction, and the CO_2 concentration will increase until equilibrium is again established. This type of change illustrates a very important principle that applies to all systems at equilibrium: *If a stress is applied to a system at equilibrium, the system will adjust to relieve the stress.* Known as **Le Chatelier's principle** for the French scientist, Henri Le Chatelier, who first stated it in 1884, this principle means that whatever the disruption, the reaction will shift in the direction that re-establishes equilibrium. For a chemical reaction, the stresses might be adding or taking away a reactant or product, changing the temperature or, in some cases, changing the pressure.

By utilizing LeChatelier's principle, the extent and favored direction of a reaction can be controlled. A visible demonstration of the principle at work is in the equilibrium between a pink and a blue cobalt ion. When equilibrium is reached with a high concentration of $[Co(H_2O)_6]^{2+}$, the solution is pink (Figure 8.11*a*). Adding a large amount of chloride ion drives the reaction to the right, so that more product forms, and the solution turns blue (Figure 8.11*b,c*).

$$\text{Cl}^- \longrightarrow$$

$$\underset{\text{Pink}}{[\text{Co}(\text{H}_2\text{O})_6]^{2+}(aq)} + 4\,\text{Cl}^-(aq) \rightleftharpoons \underset{\text{Blue}}{[\text{CoCl}_4]^{2-}(aq)} + 6\,\text{H}_2\text{O}(aq)$$

Figure 8.11 Cobalt ion equilibrium. (a) A solution containing a high concentration of $[Co(H_2O)_6^{2+}$ ions and a small concentration of $[CoCl_4]^{2-}$ ions. (b) As Cl^- ion is added, the blue $[CoCl_4]^{2-}$ ion forms and, because it is denser, sinks to the bottom of the flask. (c) With continued addition of Cl^-, the equilibrium is shifted so far towards $CoCl_4$ that no pink is visible. (This reaction is demonstrated in *World of Chemistry,* Program 14, "Molecules in Action.") *(C.D. Winters)*

Adding a large amount of water drives the reaction back to the left, and the solution becomes pink:

$$\xleftarrow{\quad} H_2O$$

$$\underset{\text{Pink}}{[Co(H_2O)_6]^{2+}(aq)} + 4\ Cl^-(aq) \rightleftharpoons \underset{\text{Blue}}{[CoCl_4]^{2-}(aq)} + 6\ H_2O(aq)$$

As another example, consider a reaction that takes place entirely in the gaseous state, the synthesis of ammonia.

$$N_2(g) + 3\ H_2(g) \rightleftharpoons 2\ NH_3(g)$$

There are more reactant gas molecules than product gas molecules, which means that changing the pressure will stress the reaction. If the pressure is increased, the equilibrium will shift in the direction that decreases the pressure. To do this, the number of gas molecules must be decreased, which for ammonia synthesis means the forward reaction will be favored, and more ammonia will form.

Studies of many, many chemical reactions show that every reaction has its own characteristic equilibrium condition. Understanding what this is provides an answer to the "How far?" question for a reaction. Some reactions go in the forward direction only a little before reaching equilibrium. Such a reaction is described as not proceeding very far towards *completion*—the conversion of all of at least one reactant to products. The formation of ions in water solution by acetic acid is a good example of a reaction that reaches equilibrium with low concentrations of products and high concentrations of unchanged reactants:

■ By a combination of experiments and calculations, chemists can find the amounts of reactants and products present at equilibrium for a given reaction.

■ Acids are an important class of chemical compounds, and Chapter 9 is devoted to their properties and applications.

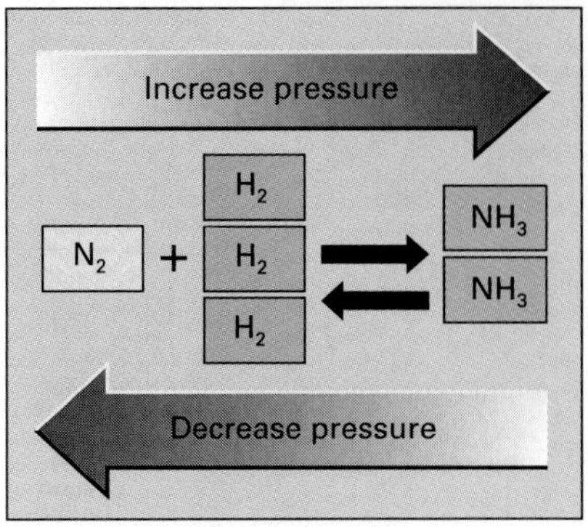

(a)

(b)

Ammonia synthesis. (a) In a demonstration of Le Chatelier's principle, pressure shifts equilibrium of reaction with different amounts of gaseous reactants and products. Higher pressure shifts the reaction toward smaller amounts, and therefore smaller volumes, of gas. (b) An ammonia plant in a rural setting. To increase the amounts of ammonia produced, the synthesis is done under pressure. Some of the ammonia produced here will go directly into the fields as fertilizer. *(John Colwell/Grant Heilman)*

■ The extent to which a reaction proceeds before reaching equilibrium varies with the temperature. Unlike reaction rate, it does not vary with the *concentration* of reactants.

$$\underset{\text{Acetic acid}}{CH_3COOH\ (aq)} + \underset{\text{Water}}{H_2O(\ell)} \rightleftharpoons \underset{\text{Acetate ion}}{CH_3COO^-(aq)} + \underset{\text{Hydronium ion}}{H_3O^+(aq)}$$

Other reactions, like the combination of hydrogen (H_2) and chlorine (Cl_2) to give hydrogen chloride (HCl), reach equilibrium with most of the reactants converted to products.

■ SELF-TEST 8B

1. The rate of a chemical reaction might be expressed as the amount of ____________ converted to ____________ per second.
2. If colliding molecules do not have enough ____________ they cannot react with each other.
3. A reaction can be speeded up by increasing either the ____________ of a reactant or the ____________.
4. A substance that speeds up a reaction without being a reactant or product is called a ____________.
5. The ____________ is like a hill that must be climbed to get a reaction going.
6. Are the amounts of reactants and products always equal to each other at equilibrium?
7. Are the amounts of reactants and products present at equilibrium always the same for the same reaction under all conditions?
8. At equilibrium, are the rates of the forward and reverse reactions equal?
9. Changing the concentration of a reactant in order to change the equilibrium concentrations of the products is an application of ____________.

Rust on the George Washington Bridge. Rusting is slow, but favorable and inevitable. *(C.D. Winters)*

8.5 The Driving Forces and the "Why?" Question

Why does one reaction produce more product at equilibrium than another? Why do some chemicals react when they are mixed, while others have absolutely no tendency to react unless the conditions are changed? In general, **why** is one chemical reaction favorable and another not favorable?

To examine the driving forces that account for such differences requires looking at two kinds of change—changes in energy as heat and changes in the amount of order that accompanies chemical reactions. We experience such changes every day. Turn on the gas stove, and energy is released as heat. Let the refrigerator absorb heat from a mixture of cream and flavorings to produce ice cream. Create major disorder on your desk while rushing to finish a project. Invest the energy needed to restore order to your desk.

Energy Change as a Driving Force

First, it is important to note that how fast a chemical reaction proceeds and the amount of energy associated with it are in no way connected. Our earlier analogy about moving books illustrates this principle. No matter what pathway is chosen and no matter how long the move takes, the total change in potential energy of the books is the same. In terms of a reaction, the difference between the energy stored in the reactants and products is the same, no matter how high or low the activation energy barrier.

■ The potential energy change of books moved up two stories depends only on their mass and the vertical distance they were moved.

The natural direction of chemical reactions is much like the natural direction of more familiar changes in everyday life. We all know that water going over a waterfall is favorable. Likewise, water going back up the waterfall is not favorable. To move water from the bottom of the waterfall back to the top would require energy and work—electricity could drive a mechanical pump that could move the water.

Water held back behind a dam has stored energy—potential energy that can be converted to kinetic energy if the gates in the dam are opened. Like the water behind a dam, chemical compounds have potential energy; it is stored in chemical bonds. In many reactions, this energy is released as heat. The reaction of potassium permanganate and glycerine (an organic compound that can burn) is a good example. When the reactants are mixed, the mixture

■ A **favorable reaction** has a natural tendency to happen, like water running downhill. An **unfavorable reaction** (the reverse of a favorable reaction) can only be made to occur by the expenditure of energy.

A demonstration of the natural tendency towards lower energy. *(Hank Levine/Visuals Unlimited)*

The World of Chemistry

Program 13, *The Driving Forces*

Figure 8.12 A favorable exothermic reaction. The favorable driving forces combine to make the reaction of potassium permanganate and glycerine dramatically exothermic.

bursts into flame (Figure 8.12). Any reaction that releases heat is described as **exothermic,** and most exothermic reactions, once they get going, are favorable—no input of energy is required to keep them going.

The opposite condition is found in an **endothermic** reaction, one that requires energy to take place. Most endothermic reactions, such as the decomposition of limestone in an open pit, don't happen at all without a continuous supply of energy.

So far, you've seen evidence for one answer to the "Why?" question. Like water going over a waterfall, chemical reactions are favorable when they release energy and the products thus have less potential energy than the reactants. Usually the energy is released as heat, although there are chemical reactions that generate light and, under the right conditions, electric current. A few reactions, however, are favorable but endothermic—they absorb heat from their surroundings and keep going without any outside influence. Here is evidence that there must be another driving force for change.

Entropy Change as a Driving Force

Perhaps you have seen a demonstration of the endothermic but favorable reaction of barium hydroxide and ammonium thiocyanate. This reaction absorbs so much heat from its surroundings that it can freeze water in contact with the reaction flask (Figure 8.13).

An oil fire in Kuwait. In an act of environmental sabotage, the oil wells were set on fire in 1991 at the end of the Gulf War that drove Iraqi invaders out of Kuwait. It took months to extinguish the fires, providing a dramatic and devastating demonstration of the favorable, exothermic nature of the chemical reactions once the activation energy had been provided. *(AP/Wide World Photos)*

(a) (b)

Figure 8.13 A favorable but endothermic reaction. The reaction of barium hydroxide and ammonium thiocyanate is one of the uncommon examples of a favorable reaction that absorbs heat from its surroundings. *(C.D. Winters)*

$$Ba(OH)_2 \cdot 8\,H_2O(s) + 2\,NH_4SCN(s) \longrightarrow Ba(SCN)_2(aq) + 2\,NH_3(aq) + 10\,H_2O(\ell)$$

The reactants are both crystalline solids and, as such, their components are held together in a repetitive, orderly arrangement. Look at the products—there is liquid water, gaseous ammonia that mostly dissolves in the water, and an ionic compound ($Ba(SCN)_2$) that dissolves in the water to give separate ions (Ba^{2+} and 2 SCN^-). What a mixture! And this is the clue to the driving force. The system has moved from a highly ordered state to a highly disordered state.

■ Chemical reactions, like more familiar changes, follow the path of least resistance until they reach a lower energy condition.

The melting of ice is another change that naturally proceeds from order to disorder. You should have no trouble identifying this tendency on a personal scale. It is effortless to create a random mixture of possessions in your room, and it certainly takes energy to put them back in order.

■ Some ionic compounds incorporate water molecules in their crystals. To indicate this, the formula is written with a dot: $Ba(OH)_2 \cdot 8\,H_2O(s)$ represents a crystalline solid composed of 2 OH^- ions and 8 H_2O molecules for every Ba^{2+} ion.

Physical scientists have given the name **entropy** to the disorder of matter, and the entropy of substances can be measured and expressed quantitatively.

High and low entropy conditions. Resisting the natural tendency to disorder is harder for some than others. *(C.D. Winters)*

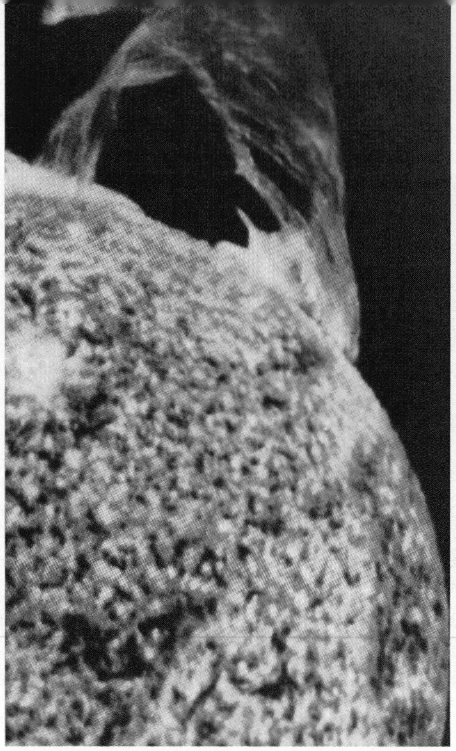

(a)

(b)

Favorable (a) and unfavorable (b) processes in nature. The burning of the sun provides energy for photosynthesis. All unfavorable processes must be driven by favorable processes in such a way that the sum of the coupled processes is favorable. *(a, Courtesy of NASA; b, Beverly March)*

■ For some reactions, changing the temperature can change the favorability. For example, carbon and water will not react at all at room temperature, but at high temperatures this reaction is the basis for an industrial process for making methanol and other organic compounds (Section 12.6).

Gases have a higher entropy than liquids. Liquids have a higher entropy than crystalline solids. Large molecules often have higher entropy than small ones because their atoms can rotate around the bonds in many different ways. The more molecules or the more different kinds of molecules in a mixture, the higher the entropy of the mixture.

Entropy is the second factor that determines the answer to the "Why?" question. Reactions are favorable when they result in a *decrease* in energy and an *increase* in disorder. When one of these changes is favorable but the other is not, the greater effect controls the favorability of the reaction.

What happens when a process is unfavorable because it requires energy and creates order? Such a process is not forbidden by nature—it can be made to happen when energy is supplied from some other process. Ordering processes must be driven by favorable, disordering processes. The result is always that the net disorder of the universe is increased when an unfavorable process is driven by a favorable one.

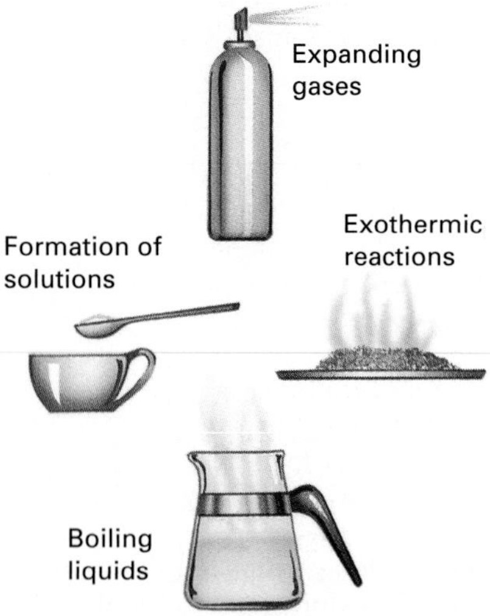

Natural processes that increase entropy.

The First Law

The first and second laws of thermodynamics summarize the universal conditions for changes in energy and entropy (Table 8.1). Because they have such broad application and meaning, not just in science, they are often referred to familiarly as "the first law" and "the second law." Here is a formal statement of the **first law of thermodynamics,** sometimes known as the **law of conservation of energy:**

Energy can be converted from one form to another but cannot be destroyed or created.

TABLE 8.1 ■ Some Statements of the First and Second Laws of Thermodynamics

The First Law
The energy of the universe is constant.
Energy can be converted from one form to another, but cannot be destroyed or created.
You can't get something for nothing.
There's no such thing as a free lunch.
The Second Law
The total entropy of the universe is constantly increasing.
Entropy is time's arrow.
The state of maximum entropy is the most stable state for an isolated system.
Energy is conserved in quantity but not in quality.
Every system that is left to itself will, on the average, change toward a condition of maximum probability.
Things are getting more screwed up every day.
You can't break even.

When gasoline burns in an auto engine, all the energy released could be accounted for if the resulting mechanical energy, the friction of moving parts, the energy that leaves the car in the exhaust, the energy converted to electrical and then chemical potential energy in the battery, and the increase in temperature of the engine and everything surrounding it could be measured.

■ **Thermodynamics** is the movement of energy.

THE WORLD OF CHEMISTRY

Energy, Entropy, and Industrial Design

Program 13, *"The Driving Forces"*

Whether a reaction will go or not depends on the balance of energy and entropy of the reactant and the product. And we can trade off one of these against the other.

In industry, reactions have to work. Both energy and entropy effects determine that. If a reaction does work—if new molecules can be created—then the industrial design is given to the engineers, and they focus on energy and materials.

Let's look at an example at the Union Carbide plant in West Virginia. Chemicals are made for 500 different products: detergents, adhesives, plastic wraps and car seats, paints and waxes. Probe the panoply of pipes and towers and you find more than a flow of materials resulting from the scores of chemical reactions. There's a flow of energy, and the engineers must conserve this valuable commodity. Usually a reaction is exothermic, giving off heat. Plant designers want to reuse this heat to drive other reactions, to minimize the waste of energy in the whole plant. So through some of the pipes, they'll transport steam, and out in the plant, steam is piped from point to point, reaction to reaction. The basic raw materials coming into the plant are coal or petroleum, both high in energy. First, ethane gas is produced, and then, by selective addition of oxygen to ethane, we get a variety of industrial chemicals. The plant is constructed so that each product in the chain of reactions has a successively lower level of energy.

Directing chemicals and heat from one reaction to another requires the maze of pipes visible in chemical manufacturing plants.

Looked at another way, the first law means that *the total quantity of energy in the universe is constant.* The Sun and the energy stored in chemicals on the Earth are what we have to use—that is all! Creation of new energy is not possible. All we can do is change it from one form to another. The first law shows up in conversation whenever someone says, "Oh well, you can't get something for nothing."

■ The implications of the first law for the energy needs of society are explored in Chapters 11 and 12.

The famous insight of Albert Einstein in 1900 when he recognized that matter and energy are interconvertible created an extension of the first law:

The total amount of matter and energy in the universe is constant.

■ The production of energy by nuclear fission and fusion takes advantage of Einstein's extension of the law (Section 13.3).

The Second Law

Even more ways have been found to express the **second law of thermodynamics** than the first (Table 8.1). Each statement reflects the observation that although the energy of the universe is constant, once energy is converted to entropy it is never again available for useful purposes.

In chemistry textbooks, the usual statement of the second law is

The total entropy of the universe is constantly increasing.

The universal truth of this law may be hard to accept at first. The formation of the stars and planets, the formation of continents and oceans, the formation of crystalline mineral deposits, and the growth of plants and animals are all ordering processes. Life itself is a constant struggle against entropy. The essential connection is that every ordering process in one small corner of the universe creates disorder somewhere else in the universe. As we are eating, breathing, and producing new biomolecules, we are emitting disordered waste products and contributing disorder to the atmosphere by giving off heat.

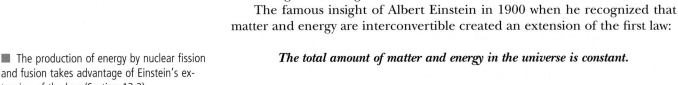

Visualizing body heat. Thermogram photos show heat in reds and yellow, illustrating that the body heats up with exercise. *(Dan McCoy/Rainbow)*

In fact, *every* time energy is generated and used to do work, some of the energy is converted to heat. Consider the release of energy by burning coal, petroleum, or wood. The principal products of combustion, carbon dioxide and water, will not burn and release more energy. The energy from the reactants is dispersed as heat into the random molecular motion of the surroundings, where it is not available to do more work. In the burning process, matter and energy are conserved. But the products and the energy converted to heat are much less useful than the reactants and their stored energy. Observations of this type are the basis of one of the alternative statements of the second law:

Energy is conserved in quantity but not quality.

The implications of the second law are wide-ranging. Our economy is based on extracting raw materials from our surroundings, using energy to process them and, in marketplace terms, "adding value" to the raw materials. The second law reminds us that energy is lost at each manufacturing step.

Recycling metals, paper, and plastics is on the rise as awareness of our dwindling natural resources increases. Recycling, of course, counteracts the natural direction of increased entropy and can be done only with the expenditure of energy. Here, too, some of the energy needed at each step is lost to entropy. A different, more direct approach to conserving resources is to increase the useful lifetime of common household materials and objects. Another is to diminish excessive use of packaging materials, even those that are recyclable.

The second law has been called upon by scholars in exploring economics, politics, history, and religion. Do you think the problems plaguing big cities are founded in the second law? The large population density in a city certainly requires an intense expenditure of energy. Is the accompanying, inevitable increase in the entropy of the surroundings the fundamental cause of the garbage disposal problems, decaying roads and bridges, and possibly even the decaying social structure in our big cities? Some individuals believe this to be the case.

Plastic on its way to recycling. Are there ways to cut down the volume? *(Bernd Wittich/Visuals Unlimited)*

■ SELF-TEST 8C

1. In an exothermic reaction, heat is ______________.
2. In an endothermic reaction, heat is ______________.
3. Some favorable chemical reactions start up as soon as the reactants are mixed, but others do not start till energy is added. (a) True, (b) False.
4. The products of an exothermic reaction store less ______________ than the reactants.
5. Which of the following is a system with higher entropy? (a) A new deck of cards as it comes from the box, (b) A deck of cards after the cards are shuffled.
6. When water freezes, its entropy ______________.
7. The two driving forces for favorable chemical change are a decrease in ______________ and an increase in ______________.
8. "You can't get something for nothing," is one way of stating the ______.
9. "You can't ever break even," is one way of stating the ______________.

8.6 More About the Mole and Chemical Reactions

How to interpret chemical equations in terms of moles and masses of reactants and products was demonstrated in Section 8.2. This section further illustrates the usefulness of the mole in answering the "How much?" question for chemical reactions.

Consider a reaction of interest in freeing metals from their ores. Aluminum reacts with many metal oxides with the release of very large quantities of heat, which is useful where very high temperatures are needed. With chromium oxide, the reaction temperature is so high that both products are molten and can be separated because they have different densities. The reaction can be used to produce chromium.

$$2\,Al(s) + Cr_2O_3(s) \longrightarrow 2Cr(\ell) + Al_2O_3(\ell)$$

What mass of chromium can be made for every 100. g of chromium oxide used?

Such "How much?" questions are answered by using the molar amounts shown by the coefficients. The molar mass of the oxide is 152 g. Using this exactly like a unit conversion factor (Section 2.8) gives the number of moles equivalent to 100. g of the oxide.

Step 1: Finding moles equivalent to known mass:

$$100.\ \text{g}\ \cancel{Cr_2O_3} \times \frac{1\ \text{mol}\ Cr_2O_3}{152\ \text{g}\ \cancel{Cr_2O_3}} = 0.658\ \text{mol}\ Cr_2O_3$$

The coefficients in the balanced equation are 1 (understood) for Cr_2O_3 and 2 for Cr, showing that every 1 mol of the oxide produces 2 mol of chromium. The coefficients are combined in a **mole ratio,** 1 mol Cr_2O_3/2 mol Cr. Then, the mole ratio is used as a conversion factor to calculate how much chromium (in moles) can be made from a known amount of the oxide (in moles).

Step 2: Using coefficients to find moles of a product or reactant:

$$0.658\ \cancel{\text{mol}\ Cr_2O_3} \times \frac{2\ \text{mol Cr}}{1\ \cancel{\text{mol}\ Cr_2O_3}} = 1.32\ \text{mol Cr}$$

Now, to find the mass of chromium in grams requires using the molar mass of chromium, as shown below:

Step 3: Finding mass equivalent to known number of moles:

■ The numerical values of the answers to these calculations are governed by the rules discussed in Appendix A.

$$1.32\ \cancel{\text{mol Cr}} \times \frac{52\ \text{g Cr}}{1\ \cancel{\text{mol Cr}}} = 69\ \text{g Cr}$$

By this reaction, 69 g of chromium can be made from every 100. g of chromium oxide.

EXAMPLE 8.4 *Moles Equivalent to a Known Mass*

What is the equivalent in moles of 3.7 g of water, roughly the amount in a teaspoonful?

SOLUTION

The molar mass of water, H_2O, is 18 g. Therefore,

$$3.7\ \text{g}\ \cancel{H_2O} \times \frac{1\ \text{mol}\ H_2O}{18\ \text{g}\ \cancel{H_2O}} = 0.21\ \text{mol}$$

Exercise 8.4

What is the equivalent in moles of 150 g of table salt, NaCl?

EXAMPLE 8.5 *Mass Equivalent to a Known Number of Moles*

What mass is equivalent to 0.500 mol of strontium nitrate, $Sr(NO_3)_2$, which gives the red color to fireworks?

SOLUTION

It is important to remember in finding molar masses to take into account any parentheses in the formula. In this case, there are two nitrate ions, which means there are two N atoms and six O atoms.

	Moles of Atoms		Mass per Mole of Atoms		
Sr	1	×	88 g	=	88 g
N	2	×	14 g	=	28 g
O	6	×	16 g	=	96 g
			Molar mass of $Sr(NO_3)_2$	=	212 g

$$0.500\ \cancel{\text{mol Sr(NO}_3)_2} \times \frac{212\ \text{g Sr(NO}_3)_2}{1\ \cancel{\text{mol Sr(NO}_3)_2}} = 106\ \text{g Sr(NO}_3)_2$$

Exercise 8.5

What is the mass equivalent to 50. mol of barium nitrate, $Ba(NO_3)_2$, which gives a green color to fireworks?

EXAMPLE 8.6 *Moles of a Product*

How many moles of tin can be made by the reaction of 0.5 mol of tin(IV) oxide, SnO_2, with carbon?

$$SnO_2(s) + 2\ C(s) \longrightarrow Sn(s) + 2\ CO(g)$$

SOLUTION

Since both SnO_2 and Sn have coefficients of 1, 0.5 mol of Sn can be made from 0.5 mol of SnO_2.

$$0.5\ \cancel{\text{mol SnO}_2} \times \frac{1\ \text{mol Sn}}{1\ \cancel{\text{mol SnO}_2}} = 0.5\ \text{mol Sn}$$

Exercise 8.6

How many moles of chromium(III) oxide, Cr_2O_3, can be made by the combination of 2 mol of chromium metal with oxygen?

$$4\ Cr(s) + 3\ O_2(g) \longrightarrow 2\ Cr_2O_3(s)$$

EXAMPLE 8.7 *Mass of a Reactant*

What mass of carbon is needed to react completely with 2500 g of SnO_2 by the reaction in Example 8.7?

SOLUTION

The steps needed here are (1) convert grams of SnO_2 to moles of SnO_2, (2) use the mole ratio to find the number of moles of carbon, and (3) find the mass of carbon from its molar mass.

(1) The molar mass of SnO_2 is 151 g (119 g Sn + (2 × 16) g O).

$$2500\ \cancel{g\ SnO_2} \times \frac{1\ mol\ SnO_2}{151\ \cancel{g\ SnO_2}} = 17\ mol\ SnO_2$$

(2)

$$17\ \cancel{mol\ SnO_2} \times \frac{2\ mol\ C}{1\ \cancel{mol\ SnO_2}} = 34\ mol\ C$$

(3)

$$34\ \cancel{mol\ C} \times \frac{12\ g\ C}{1\ \cancel{mol\ C}} = 410\ g\ C$$

Exercise 8.7
What mass of chromium is needed to make 550 g of Cr_2O_3 by the reaction described in Exercise 8.6?

SELF-TEST 8D

1. The equation below shows that hydrogen reacts with oxygen in the mole ratio of ______ mol H_2/ ______ mol O_2 and that water is formed from oxygen in the mole ratio ______ mol H_2O/ ______ mol O_2.

$$2\ H_2(g) + O_2(g) \longrightarrow 2\ H_2O(\ell)$$

2. The molar mass of $Ba(OH)_2$ is ______________.
3. The mass of 0.3 mol of $Ba(OH)_2$ is ______________.
4. A mass of 14 g of $Ba(OH)_2$ is equivalent to ______________ mol.
5. How many moles of water would be made by the combination of 250 g of hydrogen with enough oxygen for a complete reaction?

MATCHING SET

____	1. Number of atoms shown by the formula of $(NH_4)_3PO_4$	a. Speeds up a reaction
____	2. Mass of one mole of H_2O_2	b. 48 g
____	3. Adding a catalyst	c. Same number of atoms on both sides of the arrow
____	4. Cannot be destroyed	d. 34 g

____ 5. Mass of one mole of titanium
____ 6. Balanced chemical equation
____ 7. Lowering the temperature
____ 8. Forward and reverse reactions proceeding at equal rates
____ 9. Drives a reaction to completion
____ 10. Decreases chemical potential energy
____ 11. Number of moles of N_2 needed to produce 30 mol of ammonia,
$N_2 + 3\ H_2 \rightarrow 2\ NH_3$
____ 12. A mole ratio

e. Slows down a reaction
f. Dynamic equilibrium
g. Formation of gas that escapes
h. 20
i. 15
j. An exothermic reaction
k. Energy
l. 3 mol H_2/2 mol NH_3

QUESTIONS FOR REVIEW AND THOUGHT

1. Look at the following balanced chemical equation for burning ethanol.

$$CH_3CH_2OH(g) + 3\ O_2(g) \longrightarrow 2\ CO_2(g) + 3\ H_2O(g)$$

   (a) How many hydrogen atoms are on the products side of the equation? How many are on the reactants side of the equation?
   (b) How many oxygen atoms are on the products side of the equation? How many are on the reactants side of the equation?
   (c) What are the balancing *coefficients* in the reaction?
   (d) Explain, in your own words, how this balanced equation obeys the law of conservation of matter.
2. Equations can only be balanced by adjusting the coefficients of the reactants and products, while the subscripts within the formulas of the reactants and products cannot be changed and still keep the original sense of the equation. Explain why this is so.
3. Balance the following chemical equations.
   (a) $Al + Cl_2 \rightarrow AlCl_3$
   (b) $Mg + N_2 \rightarrow Mg_3N_2$
   (c) $NO + O_2 \rightarrow NO_2$
   (d) $SO_2 + O_2 \rightarrow SO_3$
   (e) $H_2 + N_2 \rightarrow NH_3$
4. Balance the following chemical equations.
   (a) $CH_3OH(\ell) + O_2(g) \rightarrow CO_2(g) + H_2O(g)$
   (b) $C_3H_8(\ell) + O_2(g) \rightarrow CO_2(g) + H_2O(g)$
   (c) $C_6H_6(\ell) + O_2(g) \rightarrow CO_2(g) + H_2O(g)$
   (d) $C_3H_8O(\ell) + O_2(g) \rightarrow CO_2(g) + H_2O(g)$
5. Balance the following equations.
   (a) $Ba(s) + H_2O(\ell) \rightarrow Ba(OH)_2(aq) + H_2(g)$
   (b) $Fe(s) + H_2O(\ell) \rightarrow Fe_3O_4(s) + H_2(g)$
   (c) $Na(s) + H_2O(\ell) \rightarrow NaOH(aq) + H_2(g)$
   (d) $Li(s) + H_2O(\ell) \rightarrow LiOH(aq) + H_2(g)$
6. Balance the following equations.
   (a) $HBr(aq) + KOH(aq) \rightarrow KBr(aq) + H_2O(\ell)$
   (b) $H_2S(g) + NaOH(aq) \rightarrow Na_2S(aq) + H_2O(\ell)$
   (c) $HNO_3(aq) + Ca(OH)_2(aq) \rightarrow Ca(NO_3)_2(aq) + H_2O(\ell)$
   (d) $HCl(aq) + Al(OH)_3(s) \rightarrow AlCl_3(aq) + H_2O(\ell)$
7. Balance the following equations.
   (a) $Sn(s) + HBr(aq) \rightarrow SnBr_2(aq) + H_2(g)$
   (b) $Mg(s) + HCl(aq) \rightarrow MgCl_2(aq) + H_2(g)$
   (c) $Ca(s) + H_2O(\ell) \rightarrow Ca(OH)_2(aq) + H_2(g)$
   (d) $Zn(s) + HNO_3(aq) \rightarrow Zn(NO_3)_2(aq) + H_2(g)$
   (e) $Cs(s) + H_2O(\ell) \rightarrow CsOH(aq) + H_2(g)$
8. Give two interpretations of the number "2" in front of HCl in the balanced equation below.

$$H_2(g) + Cl_2(g) \longrightarrow 2\ HCl(g)$$

9. Identify the atom used as the basis of the atomic weight scale.
10. What do 35.5 g of chlorine and 12.011 g of carbon have in common?
11. What do 103.5 g of lead (Pb) and 6.006 g of carbon have in common?
12. Look up the unit called the *gross.* How many pairs of jeans are in 4 gross of jeans? In what way is the gross unit like Avogadro's number?
13. What is the numerical value of Avogadro's number?
14. What is meant by the term *molar mass?*
15. What is the definition for the mole?
16. How many carbon atoms are in
   (a) 1 mol of CO_2?
   (b) 1 mol of pentane, C_5H_{12}?
   (c) $\frac{1}{2}$ mol of methane, CH_4?
   (d) 5 mol of methane?
   (e) 0.01 mol of butane, C_4H_{10}?
17. How many molecules are in
   (a) 1 mol of CO_2?
   (b) 0.5 mol of H_2?
   (c) 10 mol of butane?
   (d) 32 mol of O_2?
   (e) 0.001 mol of helium?
18. Interpret the equation for the incomplete combustion of octane in terms of moles of reactants and products, and their molar masses.

$$2\ C_8H_{18}(\ell) + 25\ O_2(g) \longrightarrow 16\ CO_2(g) + 18\ H_2O(\ell)$$

19. Give at least one reason why some chemical reactions are fast while others are slow. Use your own analogy to explain the reason.
20. What influence does temperature usually have on the rate of a chemical reaction? Explain why this is so.
21. What effect does freezing food have on reaction rates? Why is freezing used for preservation of food, tissue samples, and biological samples?
22. Explain the term *activation energy* as it applies to chemical reactions.
23. How does the magnitude of the activation energy influence the rate of a chemical reaction?
24. What is the effect of a catalyst on the activation energy of a chemical reaction?
25. Explain why hydrogen and oxygen can remain mixed at room temperature without any noticeable reaction, yet they will combine explosively to form water if the mixture is ignited with a tiny spark.
26. Describe how each of the following changes will affect the rate of a reaction.
    (a) Increase in temperature
    (b) Increase in concentration
    (c) Increase in surface area
    (d) Introduction of a catalyst
27. Paper burns fairly rapidly in air. How would you expect paper to burn in pure oxygen? Explain.
28. What is the general name for biological catalysts?
29. What is meant by the term *reversible* when describing chemical reactions?
30. What is meant by the term *dynamic equilibrium?*
31. If you heat limestone in a closed vessel, which way would you expect the equilibrium to shift if
    (a) you added more CO_2?
    (b) you allowed some of the CO_2 to escape from the vessel?

$$CaCO_3(s) \rightleftharpoons CaO(s) + CO_2(g)$$

32. Give a statement of Le Chatelier's principle.
33. What is meant when it is said that the reaction shifts in favor of the products of the reaction?
34. What is meant when it is said that the reaction shifts in favor of the reactants?
35. The reaction between HCl and NaOH is described as going to completion. Explain what this means in terms of how much of the reactants remain unreacted.
36. The reaction between N_2 and H_2 to produce ammonia (NH_3) is described as an equilibrium reaction:

$$N_2(g) + 3\ H_2(g) \rightleftharpoons 2\ NH_3(g)$$

After the reaction reaches equilibrium, what is present in the reaction mixture?

37. Define the following terms.
    (a) Potential energy
    (b) Exothermic
    (c) Endothermic
    (d) Entropy
    (e) Favorable reaction
38. State the first law of thermodynamics. What does it mean?
39. State the second law of thermodynamics. What does it mean?
40. Does entropy increase or decrease in the following reactions? Explain your answer.
    (a) Nitrogen reacting with oxygen to form nitrogen dioxide:

$$N_2(g) + 2\ O_2(g) \longrightarrow 2\ NO_2(g)$$

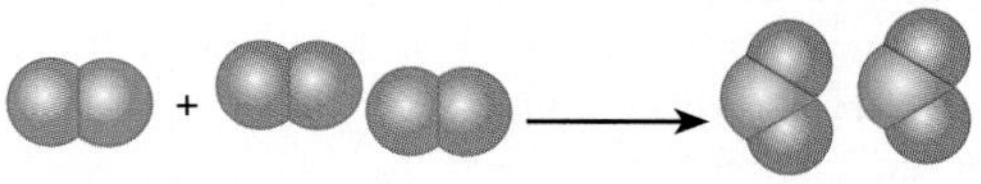

    (b) Acetylene reacting with oxygen to form carbon dioxide and water:

$$2\ HC{\equiv}CH(g) + 5\ O_2(g) \longrightarrow 4\ CO_2(g) + 2\ H_2O(g)$$

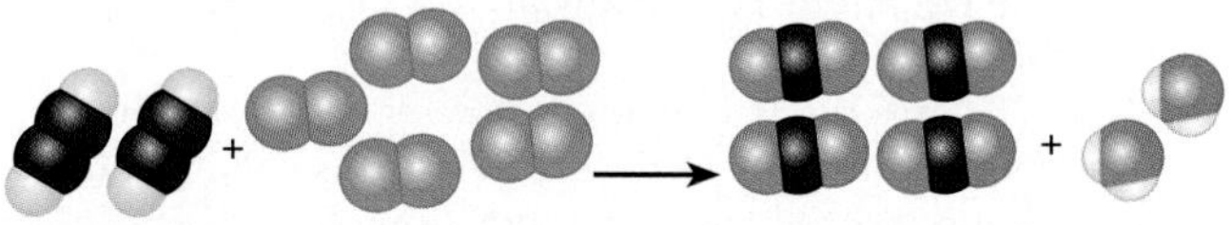

41. Burning methane in air gives off heat. Is this reaction endothermic or exothermic?
42. Give three examples of how your everyday world tends to become disordered. Are these favorable processes? Explain.
43. Photosynthesis involves taking the small molecules CO_2 and H_2O and making larger, more complex molecules like glucose ($C_6H_{12}O_6$). Is this a favorable process? Explain. What is the source of energy for photosynthesis?
44. What does photosynthesis have in common with building a house? Draw as many similarities as you can including ones between the building materials, the finished products, and the energy involved.
45. Does recycling of waste contradict the second law, which predicts the continuous increase in entropy?
46. What is the difference between *quality* of energy and *quantity* of energy?

PROBLEMS

1. Calculate the molar mass of the following.
   (a) H_2O (b) I_2
   (c) KOH (d) NH_3
   (e) CO_2 (f) CO
2. Calculate the molar mass of the following.
   (a) $C_6H_{12}O_6$ (b) H_2SO_4
   (c) Na_2HPO_4 (d) $Ca(NO_3)_2$
   (e) $C_{57}H_{104}O_6$
3. Which will have the larger mass? (a) 1 mol of Cu, (b) 1 mol of Pb, (c) 1 mol of Na.
4. Which will have the larger mass? (a) 0.5 mol of CO_2, (b) 2 mol of Li, (c) 12 mol of H_2.
5. How many moles are equivalent to each of the following:
   (a) 58 g of lighter fuel, butane, C_4H_{10}
   (b) 0.63 g of antacid, magnesium hydroxide, $Mg(OH)_2$
   (c) 230. grams of sodium metal, Na
   (d) 45 milligrams of laughing gas, N_2O
6. How many grams are equivalent to each of the following:
   (a) 4.5 mol of ammonia, NH_3
   (b) 0.0023 mol of hydrogen peroxide, H_2O_2
   (c) 0.45 mol of baking soda, $NaHCO_3$
   (d) 9 mol of neon
7. Consider the balanced equation below.

$$H_2(g) + Br_2(\ell) \longrightarrow 2\,HBr(g)$$

   If 159.8 g of Br_2 reacts with 2.0 g of H_2, how many grams of HBr will be formed?
8. Based on the balanced equation

$$Cu(s) + Cl_2(g) \longrightarrow CuCl_2(s)$$

   (a) How many moles of copper will be required to react with 0.5 mol of chlorine?
   (b) How many moles of $CuCl_2$ can be formed from 1.5 mol of copper?
9. Balance the equation for the fermentation of glucose and then answer the following questions:

$$C_6H_{12}O_6(aq) \longrightarrow CH_3CH_2OH(aq) + CO_2$$

   (a) In this reaction, how many moles of ethanol (CH_3CH_2OH) can be prepared from 6 mol of glucose?
   (b) If, during the fermentation process, 10.5 mol of carbon dioxide are found to be produced, how many moles of ethanol will have been produced?
10. According to the balanced equation

$$C(s) + O_2(g) \longrightarrow CO_2(g)$$

   (a) How many mol of CO_2 can be prepared from 1 mol of carbon?
   (b) How many mol of CO_2 can be prepared from 0.64 mol of carbon?
   (c) What mass in grams of CO_2 can be prepared from 6.0 g of carbon?
   (d) What mass in grams of CO_2 can be prepared from 1201 g of carbon?
11. Balance the equation below and then answer the following questions.

$$Sn(s) + O_2(g) \longrightarrow SnO(s)$$

   (a) What mass in grams of SnO can be prepared from 11.9 g of Sn?
   (b) What mass in grams of O_2 will be required to completely react with 5.4 g of Sn?
   (c) What mass in grams of SnO can be produced when 5.4 g of Sn react?
12. Consider the balanced equation below.

$$C(s) + H_2O(g) \longrightarrow CO(g) + H_2(g)$$

   (a) How many mol of hydrogen can be produced from 120. g of carbon?
   (b) How many mol of carbon monoxide can be produced from 24 g of carbon?
   (c) What mass in grams of carbon monoxide can be produced from 2.4 mol of carbon?

CHAPTER 9

Acids and Bases—Chemical Opposites

The chemistry of acids and bases influences everything in and around us—our personal health, the health of our environment, and the health of the economy. Even a small deviation in the delicate balance of acid–base reactions in our body can be life-threatening. Damage to trees and marine life from acid rain is an example of environmental damage caused by improper balance of acid–base chemistry. The health of the economy can be predicted by sales of sulfuric acid, the number-one chemical produced in the United States. Other acids and bases of commercial importance include muriatic acid (hydrochloric acid), lime (calcium oxide), and lye (sodium hydroxide).

- What are the chemical properties of acids and bases?
- What is the difference between a weak acid and a strong acid or between a weak base and a strong base?
- What happens chemically when neutralization occurs?
- What does the term "pH" mean?
- What causes acid indigestion, and how do antacids cure it?
- Why are sulfuric acid, hydrochloric acid, lime, and lye of commercial importance?
- What is acid rain?
- What are buffers, and why are buffers in the body so important to our health?
- How are acid and base concentrations calculated?

Measuring the acidity of soil. The proper acid-base balance is essential to the health of these plants. *(© Leonard Lessin/Peter Arnold, Inc.)*

Everyone should know a few practical things about acids and bases, which we usually encounter as solutions in water. They can be harmful. The harm can range from the stinging sensation when you accidentally squirt lemon juice (citric acid) into your eye, to the severe and persistent burns of the skin

Figure 9.1 Natural foods that contain organic acids. Citrus fruits aren't the only acidic fruits.

that result if you spill battery acid (sulfuric acid) or lye (sodium hydroxide, the active ingredient in Drano and some oven cleaners) on your skin and do not flush it immediately with lots of water. Acidic and basic solutions can also eat holes in clothing, very quickly if they are strong solutions. The first lesson, then, is that strongly acidic or basic substances must be handled with care.

■ The hazards of acids and bases are discussed in detail in Section 18.2.

The potential for harm from strong acids and bases, however, does not mean all acids and bases are to be avoided. The dilute solutions of acids and bases that are in everyday use would be sorely missed—orange juice, vinegar, soda pop, household ammonia, and most of our soaps and detergents, to name just a few.

It turns out that all acidic and basic water solutions, whatever their sources or applications might be, have some chemical properties in common. Acids and bases are closely related classes of chemical compounds that are highly reactive, both with other substances and with each other.

9.1 Acids

What is an acid? The word "acid" comes from the Latin *acidus,* meaning "sour" or "tart," since in water solutions, acids have a sour or tart taste (Figure 9.1). Lemons, grapefruit, and limes taste sour because they contain citric acid and ascorbic acid (vitamin C). A common acidic substance known since antiquity is vinegar, the sour liquid produced when the fermentation of apple cider, grape juice, or other plant juices proceeds beyond the formation of alcohol. The acid of vinegar is acetic acid.

■ You should never taste anything in a laboratory.

What are some other characteristic properties of acids? Aqueous solutions of acids react with active metals such as zinc and magnesium to liberate hydrogen,

$$Zn(s) + H_2SO_4(aq) \longrightarrow ZnSO_4(aq) + H_2(g)$$

react with most bases to produce salts and water,

$$HCl(aq) + NaOH(aq) \longrightarrow NaCl(aq) + H_2O(\ell)$$

and change the color of litmus (a vegetable dye) from blue to red.

Since the most common kinds of acid–base reactions take place in aqueous solution, the earliest definition of acids, by Svanté Arrhenius in 1887, stated that *an **acid** ionizes in aqueous solution to produce hydrogen ions (which are protons), H^+, and anions.* Later studies of aqueous solutions provided evidence for the combination of the small, positively charged hydrogen ion with a water molecule to form a hydrated proton, $H^+(H_2O)$ or H_3O^+, which is called the **hydronium ion:**

■ H^+ is actually a proton, since removal of the single electron of a hydrogen atom leaves a nucleus containing one proton. Although the hydrated hydrogen ion is represented as $H^+(H_2O)$ or H_3O^+, studies have shown that several water molecules are hydrogen-bonded to the hydrogen ion in water. A more correct representation is $H^+(H_2O)_n$, where n is between 4 and 5 in dilute solutions at room temperature.

$$H^+ \ + \ \overset{..}{\underset{|}{\overset{}{:\underset{H}{O}}}}-H \longrightarrow \left[H-\overset{..}{\underset{|}{\underset{H}{O}}}-H \right]^+$$

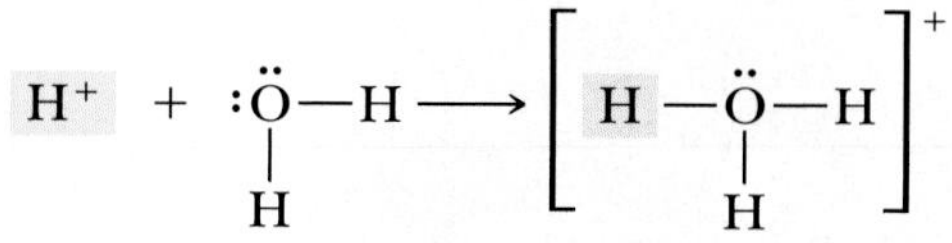

Hydronium ion

The hydronium ion is responsible for the properties by which early chemists recognized substances as acids (Table 9.1).

Hydrogen chloride, a gas, is an acid because it reacts with water to form hydronium ions and chloride ions.

$$HCl(g) + H_2O(\ell) \longrightarrow H_3O^+(aq) + Cl^-(aq)$$

A water solution of HCl is referred to as hydrochloric acid and can be represented in equations as $HCl(aq)$.

Acidic Oxides

As a general rule, nonmetal oxides are **acidic oxides**—they react with water to produce acids. For example, the slight acidity of natural rainwater is caused by the reaction of the carbon dioxide in the air with water to give carbonic acid:

TABLE 9.1 ■ **Properties of Acids and Bases**

Acids	Bases
Sour taste	Bitter taste
React with active metals to give hydrogen	Slippery feeling
Change colors of indicators, e.g., litmus turns from blue to red	Change colors of indicators, e.g., litmus turns from red to blue
Produce CO_2 when added to limestone	
Neutralize bases	Neutralize acids
Some Acidic Substances	**Some Basic Substances**
Vinegar	Household ammonia
Tomatoes	Baking soda
Citrus fruits	Soap
Carbonated beverages	Detergents
Black coffee	Milk of magnesia
Gastric fluid	Oven cleaners
Vitamin C	Lye
Aspirin	Drain cleaners

THE PERSONAL SIDE

Svanté August Arrhenius (1859–1927)

(Special Collections, Van Pelt Library, University of Pennsylvania)

Arrhenius, a Swedish chemist, proposed a revolutionary idea in his 1884 doctoral thesis that was not accepted by his professors and resulted in the lowest possible passing grade for his thesis. He proposed that electrolytes such as sodium chloride dissociate into ions in water. This was a revolutionary idea because the electron had not yet been discovered, and chemists had no basis for understanding how sodium and chlorine atoms could become charged ions. Arrhenius' ideas gained support after Thomson's experiments on the electron during the 1890s (Section 3.3) showed that atoms contain charged particles. In 1903 Arrhenius received the Nobel Prize in Chemistry for the same thesis that had nearly been rejected 19 years earlier.

Arrhenius also proposed the concept of the greenhouse effect—the process in which carbon dioxide in the atmosphere traps heat (Section 21.11), and he introduced the idea of activation energy in his studies of factors that affect rates of chemical reactions (Section 8.3).

$$CO_2(g) + H_2O(\ell) \longrightarrow \underset{\text{Carbonic acid}}{H_2CO_3(aq)}$$

Acidic oxides are used in the manufacture of sulfuric acid, nitric acid, and phosphoric acid.

The many uses of ***sulfuric acid*** make it the number-one chemical of commerce (Section 4.5). The primary raw material for making sulfuric acid is sulfur or sulfur dioxide. Sulfur is found in underground deposits in the United States and is brought to the surface by the **Frasch process** (Figure 9.2). Sulfur is converted to sulfuric acid in four steps, called the contact process. In the first step the sulfur is burned in air to give mostly sulfur dioxide:

$$S(s) + O_2(g) \longrightarrow SO_2(g)$$

The gaseous SO_2 is then converted to SO_3 by passing SO_2 over a hot, catalytically active surface, such as platinum or vanadium pentoxide:

$$2\ SO_2(g) + O_2(g) \xrightarrow{\text{Catalyst}} 2\ SO_3(g)$$

Although SO_3 can be converted directly into H_2SO_4 by passing SO_3 into water, the enormous amount of heat released in the reaction causes the formation of a stable fog of H_2SO_4. This is avoided by passing the SO_3 into H_2SO_4

$$SO_3(g) + H_2SO_4(aq) \longrightarrow H_2S_2O_7(\ell)$$

and then diluting the $H_2S_2O_7$ with water:

$$H_2S_2O_7(\ell) + H_2O(\ell) \longrightarrow 2\ H_2SO_4(aq)$$

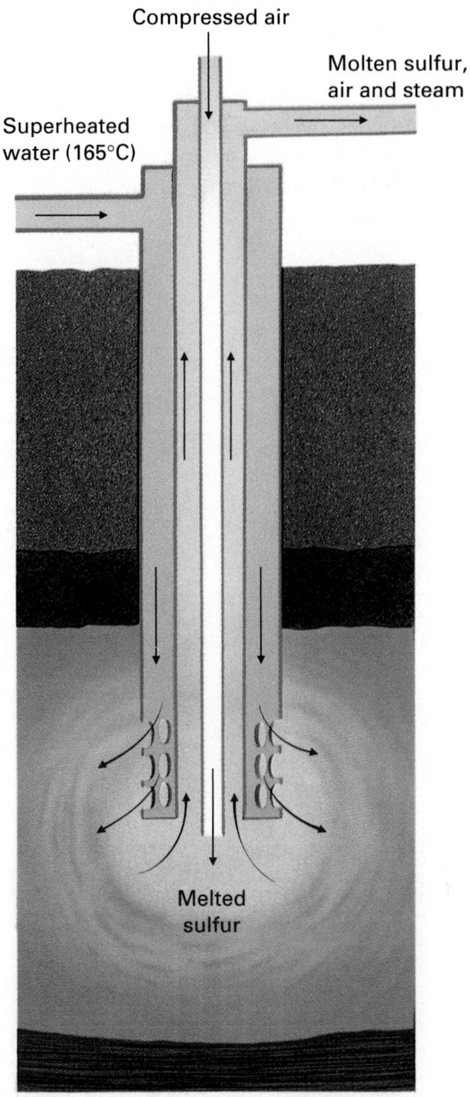

Figure 9.2 The Frasch process for mining sulfur. Superheated steam is injected along with compressed air into a sulfur-bearing stratum underground. A molten sulfur froth pours out of the inner tube. Most of the sulfur mined is converted to sulfuric acid.

■ The top 25 chemicals of commerce are listed inside the front cover of this book. Table 14.1 is a list of the top 50 industrial chemicals.

■ Compounds with formulas that include water molecules, such as $CaSO_4 \cdot 2\ H_2O$, are called **hydrates.** The **anhydrous** form (the form without water), in this case $CaSO_4$, can be obtained by heating the hydrate.

More than half of the sulfuric acid manufactured each year is used to produce phosphate fertilizers. Phosphorus occurs in nature as calcium phosphate in apatite minerals with the general formula $Ca_5(PO_4)_3X$, where X may be F^-, Cl^-, or OH^-. The most common form is with $X = F^-$, a mineral called fluoroapatite. The reaction of sulfuric acid with fluoroapatite is used to produce superphosphate fertilizer, a mixture of calcium dihydrogen phosphate and gypsum:

$$\underset{\text{Fluoroapatite}}{2\ Ca_5(PO_4)_3F(s)} + 7\ H_2SO_4(aq) + 17\ H_2O(\ell) \longrightarrow \underset{\text{Calcium dihydrogen phosphate}}{3\ Ca(H_2PO_4)_2 \cdot H_2O(s)} + \underset{\text{Gypsum}}{7\ CaSO_4 \cdot 2\ H_2O(s)} + 2\ HF(g)$$

■ The only difference in the equations for producing phosphoric acid and calcium dihydrogen phosphate is the ratio of reactants as indicated by the different coefficients in each equation.

Phosphoric acid, the number-nine chemical on the list of top 50 U.S. chemicals, is produced in the reaction of sulfuric acid with fluoroapatite.

$$Ca_5(PO_4)_3F(s) + 5\ H_2SO_4(aq) + 10\ H_2O(\ell) \longrightarrow 3\ H_3PO_4(aq) + 5\ CaSO_4 \cdot 2\ H_2O(s) + HF(g)$$

It is also produced by reacting fluoroapatite with sand (SiO_2) and coke (C) in an electric-arc furnace at 2000°C:

$$4\ Ca_5(PO_4)_3F(\ell) + 18\ SiO_2(\ell) + 15\ C(s) \longrightarrow 18\ CaSiO_3(\ell) + 2\ CaF_2(\ell) + 15\ CO_2(g) + 3\ P_4(g)$$

The phosphorus is then condensed from the gas state and purified. Elemental phosphorus is transformed into phosphoric acid by oxidation with air to give P_4O_{10}, which is then hydrated by absorption into hot phosphoric acid containing about 10% water.

$$P_4(s) + 5\ O_2(g) \longrightarrow P_4O_{10}(s)$$
$$P_4O_{10}(s) + 6\ H_2O(\ell) \longrightarrow 4\ H_3PO_4(aq)$$

About 90% of phosphoric acid is used to make fertilizers. Most of the rest is used in the manufacture of detergents and food additives.

The third most important commercial acid is ***nitric acid,*** number 13 in the list of top 50 U.S. chemicals, used in the manufacture of fertilizers, explosives, and plastics. The manufacture of nitric acid starts with the reaction of ammonia, the number 6 chemical, with oxygen, the number-3 chemical, in the presence of a platinum-rhodium catalyst:

$$4\ NH_3(g) + 5\ O_2(g) \xrightarrow{\text{Catalyst}} 4\ NO(g) + 6\ H_2O(g)$$

Some common products that contain phosphates as additives. *(C.D. Winters)*

The NO reacts readily with O_2 from the air to form NO_2,

$$2\ NO(g) + O_2(g) \longrightarrow 2\ NO_2(g)$$

and NO_2 in turn reacts with water to yield nitric acid and NO:

$$3\ NO_2(g) + H_2O(\ell) \longrightarrow 2\ HNO_3(aq) + NO(g)$$

The NO produced in this step is reacted with more O_2 to produce more NO_2.

The burning of fossil fuels produces the sulfur oxides (SO_2 and SO_3) and the nitrogen oxides (NO and NO_2) which are primary air pollutants (Sections 21.4 and 21.7). Acid rain is a mixture of acids formed by these oxides (Section 21.8).

Many cleaning products contain bases. *(C.D. Winters)*

9.2 Bases

Water solutions of bases are slippery or soapy to the touch, change litmus from red to blue, and react with most acids to form a salt and water. Arrhenius defined a **base** as *a substance that ionizes in aqueous solution to produce hydroxide ions, OH^-, and cations.* Some of the most common bases of this type are the hydroxides of the alkali metals (sodium hydroxide and potassium hydroxide, NaOH and KOH) and of the alkaline earth metals (calcium hydroxide and magnesium hydroxide, $Ca(OH)_2$ and $Mg(OH)_2$). Sodium hydroxide, for example, is a water-soluble ionic compound:

$$\text{Na}\,\text{OH}\,(s) \xrightarrow{\text{Water}} \text{Na}^+(aq) + \text{OH}^-\,(aq)$$

The classic properties by which bases are recognized (Table 9.1) are caused by the hydroxide ion. A **basic,** or **alkaline, solution** contains a higher concentration of OH^- ions than H_3O^+ ions, and an **acidic solution** contains a higher concentration of H_3O^+ ions than OH^- ions.

The word ''alkali'' comes from the Arabic word *al-qali,* meaning ''plant ashes.'' Potassium carbonate, commonly known as potash, is found in ashes from wood fires. It dissolves in water to yield a solution that feels slippery, tastes bitter, and reacts with acids. Such properties are characteristic of alkalis.

Basic Oxides

Oxides of metals are called **basic oxides** because they react with water to form bases. Perhaps the best-known example is calcium oxide (CaO) often called

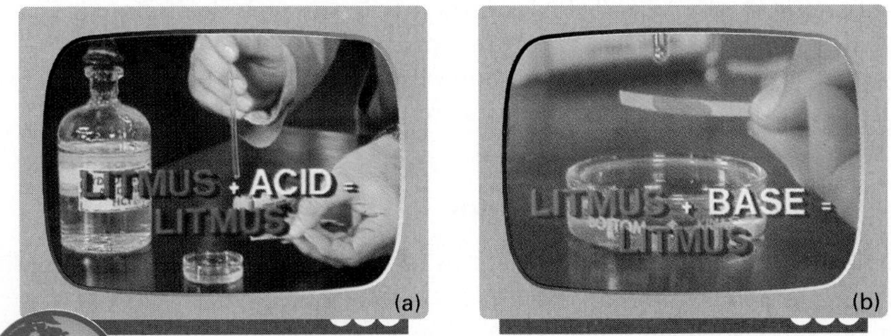

The World of Chemistry

Program 16, *The Proton in Chemistry*

Litmus paper, our oldest acid-base indicator. The color change of organic dyes like litmus indicates whether they have reacted with an acid or a base. An acid turns litmus from blue to pink and a base turns litmus from pink to blue.

The Great Wall of China. Still held together by lime mortar. *(AP/Wide World Photos)*

lime or quicklime. This metal oxide reacts with water to give calcium hydroxide, commonly called slaked lime. Although calcium hydroxide is not very soluble, enough dissolves to give a strongly basic solution:

$$\underset{\text{Lime}}{CaO(s)} + H_2O(\ell) \longrightarrow \underset{\text{Calcium hydroxide}}{Ca(OH)_2(aq)}$$

Lime, an important commercial chemical, ranked fifth in the top 50 chemicals produced in the United States in 1994 with more than 38 billion pounds produced for use in the metals industry, sewage and pollution control, water treatment, agriculture, and the construction industry. Indeed, lime is one of the oldest construction materials. Lime plaster was used in Crete in 1500 B.C.; the Great Wall of China was largely laid with lime mortar; and many structures built with lime mortar by the Romans are still standing.

9.3 Neutralization Reactions

In a reaction of an acid with a base in aqueous solution, the acid supplies hydrogen ions, H^+, the base provides hydroxide ions, OH^-, and the H^+ reacts with the OH^- to form water, H_2O. The second product is a **salt,** which is composed of the positive metal ion from the base and the negative ion from the acid. For example, the product of the reaction of potassium hydroxide (KOH) with hydrochloric acid is potassium chloride (KCl):

$$KOH(aq) + HCl(aq) \longrightarrow KCl(aq) + H_2O(\ell)$$

Since aqueous solutions of KOH, HCl, and KCl form ions when they dissolve in water, the solutions conduct an electric current, and so these solutions are electrolytes (Section 6.8). Water is a nonelectrolyte. Therefore, the preceding equation can be written as

$$K^+(aq) + OH^-(aq) + H^+(aq) + Cl^-(aq) \longrightarrow K^+(aq) + Cl^-(aq) + H_2O(\ell)$$

Ions common to both sides can be canceled to give

$$H^+(aq) + OH^-(aq) \longrightarrow H_2O(\ell)$$

which is referred to as the **net ionic equation.** Reactions that involve ions are often written in this manner as net ionic equations, in which the ions common to both sides ***(spectator ions)*** are canceled. The equation then consists of the formulas for ions that are reactants and products plus the complete formulas for any reactants or products that are gases, liquids, or solids. This type of equation emphasizes formation of products that drive reactions of ions in solution toward completion. Gases that escape from the solution, insoluble solids, and molecular compounds that do not ionize have this effect. In the case of the acid–base neutralization reaction, the driving force is the formation of water molecules, which ionize only slightly.

■ Spectator ions are unchanged in the reaction.

Representation of neutralization reactions as the formation of water from hydrogen and hydroxide ions accounted for much of the acid–base chemistry known at the time of Arrhenius. If $H_3O^+(aq)$ is substituted for $H^+(aq)$, the neutralization equation becomes

$$H_3O^+(aq) + OH^-(aq) \longrightarrow 2\ H_2O(\ell)$$

When equivalent amounts of acid and base react so that all of the H_3O^+ and OH^- ions are used up in forming water and a salt that is neither acidic nor

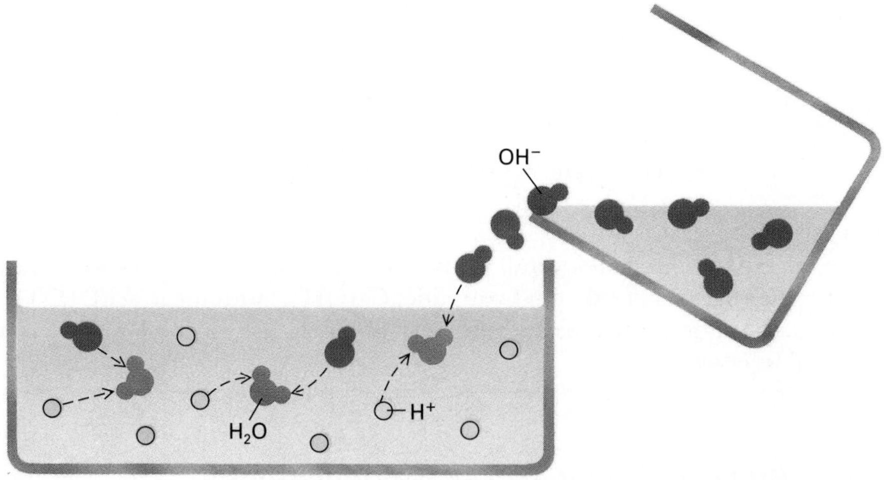

A neutralization reaction. The reaction of equal numbers of OH^- ions and H^+ ions eliminates all acid and base properties.

basic, the result is **neutralization** of all the properties associated with acids and bases. For example, the reaction of a solution containing one mole of HCl with a solution containing one mole of NaOH would eliminate all acid and base properties:

$$HCl(aq) + NaOH(aq) \longrightarrow NaCl(aq) + H_2O(\ell)$$

EXAMPLE 9.1 *Acid–Base Reactions*

Write both the overall equation and the net ionic equation for the neutralization reaction of hydrochloric acid with magnesium hydroxide.

SOLUTION

The overall reaction is

$$2\ HCl(aq) + Mg(OH)_2(aq) \longrightarrow MgCl_2(aq) + 2\ H_2O(\ell)$$

Hydrochloric acid and magnesium hydroxide form ions in aqueous solution, and all soluble salts are also ionized in aqueous solution. However, water is a nonelectrolyte and doesn't ionize. This can be represented as

$$2\ H^+(aq) + 2\ Cl^-(aq) + Mg^{2+}(aq) + 2\ OH^-(aq) \longrightarrow Mg^{2+}(aq) + 2\ Cl^-(aq) + 2\ H_2O(\ell)$$

The ions $Mg^{2+}(aq)$ and $2\ Cl^-(aq)$ cancel out from both sides to give a net ionic equation

$$2\ H^+(aq) + 2\ OH^-(aq) \longrightarrow 2\ H_2O(\ell)$$

dividing through by two gives

$$H^+(aq) + OH^-(aq) \longrightarrow H_2O(\ell)$$

or with the more correct representation of $H^+(aq)$ as $H_3O^+(aq)$ we have

$$H_3O^+(aq) + OH^-(aq) \longrightarrow 2\ H_2O(\ell)$$

Exercise 9.1

Write the equation for the overall acid–base reaction and the net ionic equation for the reaction of calcium hydroxide, $Ca(OH)_2$, with nitric acid, HNO_3. Both compounds form ions in aqueous solution, and calcium nitrate is a water-soluble salt.

9.4 Brønsted-Lowry Acid–Base Definitions

A major problem with Arrhenius' acid–base concept is that certain substances such as ammonia (NH_3) produce basic solutions and react with acids, yet contain no hydroxide ions. In 1923, J. N. Brønsted in Copenhagen, Denmark, and T. M. Lowry in Cambridge, England, independently proposed the acid and base definitions generally used by chemists today: *An **acid** donates hydrogen ions (also called a proton donor); a **base** accepts hydrogen ions (also called a proton acceptor).* These definitions not only explain all the acids and bases listed in Sections 9.1 and 9.2 but also explain the basicity of ammonia and ions such as CO_3^{2-} and S^{2-}. To illustrate these definitions, consider again the reaction between gaseous hydrogen chloride and water to give hydronium ions. Arrhenius defined HCl as an acid because it produced H^+ ions in solution. Brønsted-Lowry defined HCl as an acid because the HCl molecule donates a H^+ ion to the H_2O molecule:

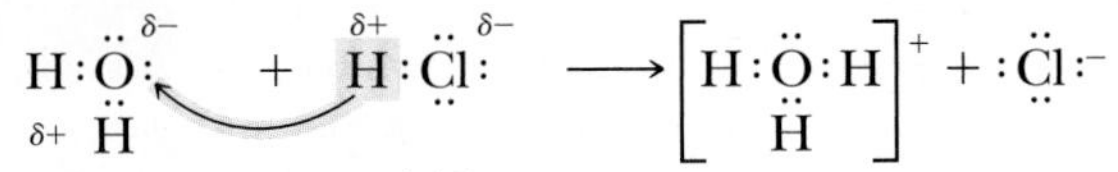

$$H:\overset{..}{\underset{\delta+\ H}{\overset{..}{O}}}{}^{\delta-}\ +\ \overset{\delta+}{H}:\overset{..}{\underset{..}{Cl}}{}^{\delta-}\ \longrightarrow\ \left[H:\overset{..}{\underset{H}{O}}:H\right]^{+}\ +\ :\overset{..}{\underset{..}{Cl}}:^{-}$$

$$\underset{\text{(proton acceptor)}}{\text{Base}}\qquad\qquad\underset{\text{(proton donor)}}{\text{Acid}}$$

Why is ammonia a base? In the reaction

$$\underset{\underset{\text{(base)}}{\text{Ammonia}}}{NH_3(g)}\ +\ \underset{\underset{\text{(acid)}}{\text{Water}}}{H_2O(\ell)}\ \rightleftharpoons\ \underset{\text{Ammonium ion}}{NH_4^+(aq)}\ +\ \underset{\text{Hydroxide ion}}{OH^-(aq)}$$

ammonia is a hydrogen ion acceptor (base) and water is a hydrogen ion donor (acid). Water acts as an acid in this reaction because it donates a H^+ ion to NH_3.

$$H-\underset{\underset{H}{|}}{\overset{\overset{H}{|}}{N}}:\ +\ H-\overset{..}{\underset{H}{O}}:\ \rightleftharpoons\ \left[H-\underset{\underset{H}{|}}{\overset{\overset{H}{|}}{N}}-H\right]^{+}\ +\ \left[:\overset{..}{\underset{..}{O}}-H\right]^{-}$$

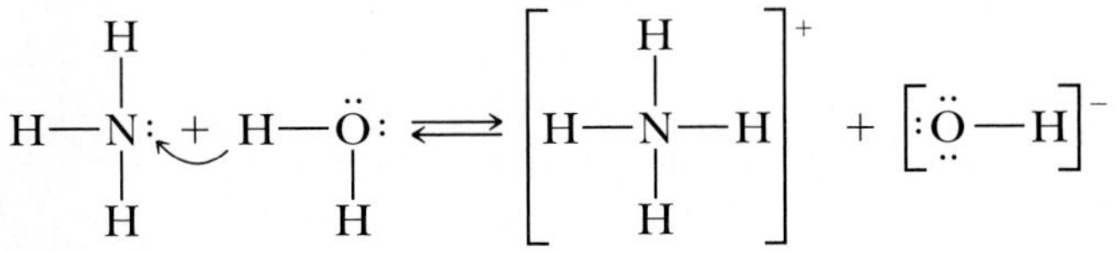

The Brønsted-Lowry definitions also explain why carbonate salts such as sodium carbonate (washing soda) dissolve in water to give basic solutions. How does a carbonate salt produce hydroxide ions in water? Carbonate ion removes a hydrogen ion from a water molecule, which leaves behind a hydroxide ion:

$$\underset{\text{Carbonate ion}}{CO_3^{2-}(aq)}\ +\ H_2O(\ell)\ \rightleftharpoons\ \underset{\text{Bicarbonate ion}}{HCO_3^-(aq)}\ +\ \underset{\text{Hydroxide ion}}{OH^-(aq)}$$

The most important point about the Brønsted-Lowry concept of acids and bases is the involvement of water as a reactant. In the reaction with hydrogen chloride, water reacts as a base. In the reaction with ammonia molecules and carbonate ion, water reacts as an acid. Since all water molecules are the same, water must be able to react in either way, depending on whether it reacts with a base or an acid. A species such as water that can either donate or accept H^+ ions is said to be **amphiprotic**—*it can donate or accept hydrogen ions,* depending on the circumstances.

THE PERSONAL SIDE

Johannes Nicolaus Brønsted (1879–1947)

The son of a civil engineer, Brønsted planned a career in engineering, but became interested in chemistry and switched his major in college. The University of Copenhagen selected Brønsted for a new professorship of chemistry after he earned his doctorate there in 1908. Brønsted published many important papers and books on solubility, the interaction of ions in solution, and thermodynamics. During World War II Brønsted distinguished himself by firmly opposing the Nazi takeover of Denmark. In 1947 he was elected to a seat in the Danish parliament, but he died in December of that year, before he could take his new office.

Johannes N. Brønsted. (Special Collections, Van Pelt Library, University of Pennsylvania)

Thomas M. Lowry (1874–1936)

Lowry was awarded his doctorate from the University of London in 1899. His positions included lecturer in chemistry at the Westminster Training College, head of the Chemical Department in Guy's Hospital Medical School, and professor of chemistry at the University of London. In 1920 he was appointed to a new chair of physical chemistry at Cambridge University. Working independently of Brønsted, Lowry developed many of the same ideas about acids and bases but did not carry them as far. For this reason, a number of textbooks do not acknowledge Lowry's contributions to the development of acid–base chemistry.

Lowry was widely recognized by organic chemists and physical chemists for his extensive studies of the optical rotatory power of optical isomers (Section 15.1).

Thomas M. Lowry. (Special Collections, Van Pelt Library, University of Pennsylvania)

Kinship of Acids and Bases

Whenever an acid donates a hydrogen ion to a base, a new acid and a new base are formed that are called the **conjugate acid** and **conjugate base,** respectively. This is another idea that is central to the Brønsted-Lowry acid–base definitions. For example, in the reaction of nitric acid with water, we can add the labels for the acid and base that are formed in the reaction.

$$\underset{\text{Acid}}{HNO_3(aq)} + \underset{\text{Base}}{H_2O(\ell)} \longrightarrow \underset{\text{Conjugate acid}}{H_3O^+(aq)} + \underset{\text{Conjugate base}}{NO_3^-(aq)}$$

A pair of molecules or ions related to one another by the gain or loss of a single hydrogen ion is called a **conjugate acid–base pair.** In the preceding reaction, H_3O^+ is the conjugate acid of H_2O, and NO_3^- is the conjugate base of HNO_3. *Every Brønsted-Lowry acid has a conjugate base, and every Brønsted-Lowry base has a conjugate acid.*

The conjugate base formula can be derived from the formula of an acid by removing a H^+ ion, that is, by removing H and making the charge of the acid one unit more negative. The conjugate acid formula can be derived from the formula of a base by adding a H^+ ion, that is, by adding H and making the charge of the base one unit more positive. Familiarity with conjugate acid–base pairs is important to understanding the relative strengths of acids and bases.

EXAMPLE 9.2 *Conjugate Acids and Bases*

Label the acid, base, conjugate acid, and conjugate base in the following equation:

$$NH_3(g) + H_2O(\ell) \rightleftharpoons NH_4^+(aq) + OH^-(aq)$$

SOLUTION

Ammonia is a base because it accepts a hydrogen ion from water. Hence, the ammonium ion is the conjugate acid of ammonia. Water is an acid because it is donating a hydrogen ion to ammonia. The hydroxide ion is the conjugate base of water.

$$\underset{\text{Base}}{NH_3(g)} + \underset{\text{Acid}}{H_2O(\ell)} \rightleftharpoons \underset{\text{Conjugate acid}}{NH_4^+(aq)} + \underset{\text{Conjugate base}}{OH^-(aq)}$$

Exercise 9.2

Label the acid, base, conjugate acid, and conjugate base in the following equation:

$$CO_3^{2-}(aq) + H_2O(\ell) \rightleftharpoons HCO_3^-(aq) + OH^-(aq)$$

9.5 The Strengths of Acids and Bases

What is the difference between a strong acid such as hydrochloric acid, sold in hardware stores to clean brick and concrete, and a weak acid such as acetic acid, found in vinegar? The strength of an acid or base is determined by the *extent* of ionization in aqueous solution. The greater the ionization, the stronger the acid or base. Strong acids, such as hydrochloric acid, are 100% ionized in aqueous solution, while weak acids, such as acetic acid, are only about 1% ionized.

■ The strength of an acid or base is unrelated to its concentration in a solution. The point is that *all* of a strong acid or base is ionized, whether its concentration in the solution is large or small.

Experimentally, the extent of ionization can be determined by measuring the electrical conductance of solutions. Solutions of **strong acids** and **strong bases** are **strong electrolytes.**

■ It could be said that the reaction of a strong acid with water goes to completion.

Common strong acids like hydrochloric acid, sulfuric acid, and nitric acid can be assumed to react completely with water to give hydronium ions and anions. For example, in the reaction of hydrogen chloride gas with water, all of the hydrogen chloride molecules are converted to hydronium ions and chloride anions:

$$\underset{\text{Acid}}{HCl(g)} + \underset{\text{Base}}{H_2O(\ell)} \longrightarrow \underset{\text{Conjugate acid}}{H_3O^+(aq)} + \underset{\text{Conjugate base}}{Cl^-(aq)}$$

Knowing the strength of an acid also provides information about the strength of its conjugate base. The conjugate base of a strong acid is weak. If this weren't the case, the conjugate base would combine with the conjugate acid to form the acid, and 100% ionization would not occur. In the equation above, the conjugate base, Cl^-, is a very weak base because all the hydrogen chloride molecules are ionized. The chloride ion does not react at all with water to form a basic solution.

If the electrical conductance of a solution of an acid such as acetic acid is measured in an apparatus like that in Figure 9.3, the light glows dimly, which indicates the solution is a **weak electrolyte.** How can this be explained using the Brønsted-Lowry concept? Acetic acid, like most organic acids, is a **weak acid**—it is only slightly ionized in aqueous solution. An equilibrium is established (Section 8.4):

$$CH_3COOH(aq) + H_2O(\ell) \rightleftharpoons H_3O^+(aq) + CH_3COO^-(aq)$$

The CH_3COOH molecules undergo ionization while H_3O^+ and CH_3COO^- ions simultaneously recombine to give CH_3COOH molecules and water. Since about 1% of the acetic acid molecules are ionized at any given time, aqueous solutions of acetic acid contain mostly acetic acid molecules with a few hydronium ions and acetate anions.

Weak acids react only slightly with water. Acetic acid, the organic acid in vinegar, is a common organic acid. The molecular structure of acetic acid is

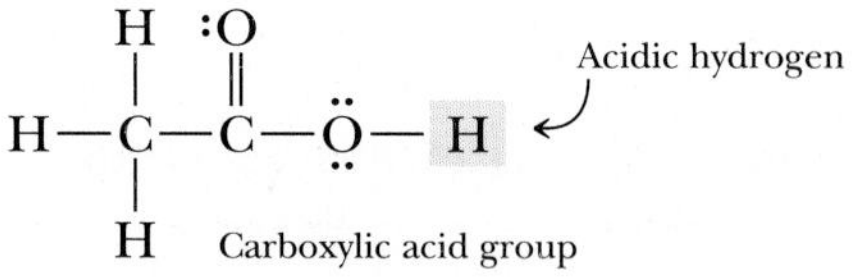

The hydrogen atom bonded to the highly electronegative oxygen atom is the only hydrogen positive enough to be donated to a base in aqueous solution; for that reason, it is designated the acidic hydrogen atom in the structural formula. The other hydrogen atoms in the acetic acid molecule are not acidic because the C—H bonds are nonpolar. The common organic acids all contain the carboxylic acid (—COOH) group.

An acetic acid solution contains relatively few acetate ions because they react so easily with H_3O^+ and reach equilibrium with acetic acid. The acetate ion, like other anions from weak acids, also establishes an equilibrium with water molecules in which OH^- ions are formed:

$$CH_3COO^-(aq) + H_2O(\ell) \rightleftharpoons CH_3COOH(aq) + OH^-(aq)$$

Because of this reaction, a solution of sodium acetate ($NaCH_3CO_2$) is basic.

Figure 9.3 Acetic acid solution is a weak electrical conductor. (See also Figure 6.12.) *(C. D. Winters)*

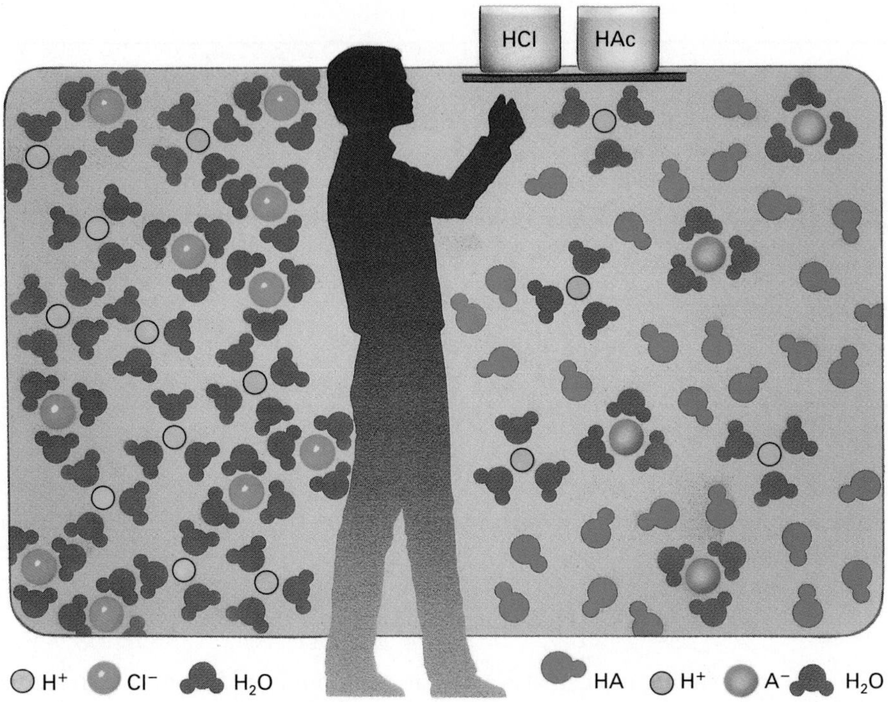

The most common weak base is ammonia. The reaction between ammonia and water is

$$\underset{\text{Base}}{NH_3(g)} + \underset{\text{Acid}}{H_2O(\ell)} \rightleftharpoons NH_4^+(aq) + OH^-(aq)$$

Since aqueous ammonia is a weak electrolyte (Figure 9.5), relatively few ammonium ions and hydroxide ions are produced in the reaction. In fact, the amount of ionization is about the same as for acetic acid. Thus, the ammonium ion is a strong conjugate acid. To summarize: *Strong acids have weak conjugate bases, and weak acids have strong conjugate bases; strong bases have weak conjugate acids and weak bases have strong conjugate acids.* Some common acids are listed in Table 9.2, and Table 9.3 lists common bases (p. 260).

Polyprotic Acids

Several common acids in Table 9.2 have more than one ionizable hydrogen ion. Each successive hydrogen ion in these polyprotic acids ionizes less readily.

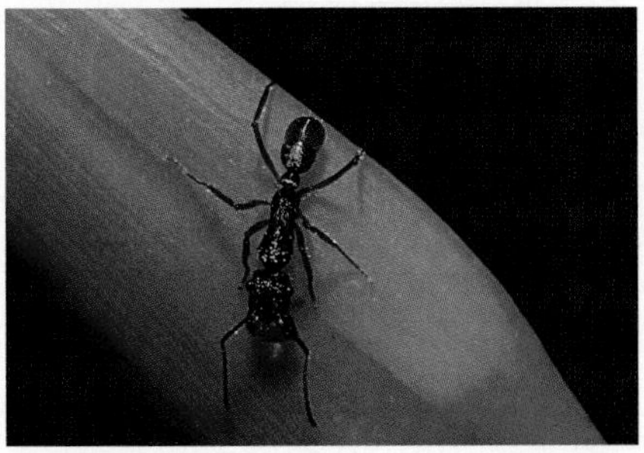

A fire ant from the Amazon Basin on a ginger plant. The sting of these and other ants contain the simplest organic acid, formic acid (HCOOH). In the Amazon, fire ant colonies can be so large that streams become polluted by their formic acid. *(©1990 M & A Doolittle/Rainbow)*

TABLE 9.2 ■ Common Acids

Name	Formula	Strength	Use
Sulfuric acid	H_2SO_4	Strong	Cleaning steel, car batteries, making plastics, dyes, fertilizers
Hydrochloric acid	HCl	Strong	Cleaning metals and brick mortar
Nitric acid	HNO_3	Strong	Making fertilizers, explosives, plastics
Phosphoric acid	H_3PO_4	Moderate	Making fertilizers, detergents, food additives
Acetic acid	CH_3COOH	Weak	Vinegar
Propionic acid	CH_3CH_2COOH	Weak	Swiss cheese
Citric acid	$HOC(COOH)(CH_2COOH)_2$	Weak	Fruit
Carbonic acid	H_2CO_3	Weak	Carbonated beverages
Boric acid	H_3BO_3	Weak	Eye drops, mild antiseptic

For example, sulfuric acid is a strong acid because of the full ionization of the first H^+ ion:

$$H_2SO_4(aq) + H_2O(\ell) \longrightarrow H_3O^+(aq) + HSO_4^-(aq)$$

The hydrogen sulfate ion can also act as an acid, but it is of moderate strength; the ionization is not complete because it is more difficult to remove H^+:

$$HSO_4^-(aq) + H_2O(\ell) \rightleftharpoons H_3O^+(aq) + SO_4^{2-}(aq)$$

Phosphoric acid has three ionizable hydrogen ions. Each step in the stepwise ionization of phosphoric acid occurs to a lesser extent than the one before it. Phosphoric acid is a moderately strong acid because in the first step it ionizes to a greater extent than does a weak acid such as acetic acid.

■ $H_2PO_4^-$ is called the dihydrogen phosphate ion. HPO_4^{2-} is called the monohydrogen phosphate ion.

$$H_3PO_4(aq) + H_2O(\ell) \rightleftharpoons H_3O^+(aq) + H_2PO_4^-(aq)$$

However, $H_2PO_4^-$ is a weaker acid than acetic acid because the second-step ionization is much smaller than the first step. The reason for this is the greater difficulty of removing a positively charged H^+ from the negatively charged $H_2PO_4^-$ than from a neutral H_3PO_4 molecule.

TABLE 9.3 ■ Common Bases

Name	Formula	Strength	Use
Sodium hydroxide	$NaOH$	Strong	Drain cleaner, producing aluminum, rayon, soaps and detergents
Potassium hydroxide	KOH	Strong	Producing soaps, detergents, fertilizers
Calcium hydroxide	$Ca(OH)_2$	Strong	Producing bleaching powder, paper and pulp, softening water
Ammonia	NH_3	Weak	Producing fertilizer, explosives, plastics, insecticides, detergents
Sodium bicarbonate	$NaHCO_3$	Weak	Antacid
Sodium carbonate	Na_2CO_3	Weak	Detergents, glass-making

$$H_2PO_4^-(aq) + H_2O(\ell) \rightleftharpoons H_3O^+(aq) + HPO_4^{2-}(aq)$$

For the same reason, the third step produces about 10,000 times fewer hydronium ions than the second step because HPO_4^{2-} has an even higher negative charge.

$$HPO_4^{2-}(aq) + H_2O(\ell) \rightleftharpoons H_3O^+(aq) + PO_4^{3-}(aq)$$

The anions of phosphoric acid can also accept hydrogen ions and act as bases, as shown in the following reversible reactions:

$$HPO_4^{2-}(aq) + H_3O^+(aq) \rightleftharpoons H_2PO_4^-(aq) + H_2O(\ell)$$
$$H_2PO_4^-(aq) + H_3O^+(aq) \rightleftharpoons H_3PO_4(aq) + H_2O(\ell)$$

■ HPO_4^{2-} and $H_2PO_4^-$ are amphiprotic—they can accept or donate H^+ ions.

Salts such as Na_2HPO_4 and NaH_2PO_4 have important uses as buffers (Section 9.8) because they can act as either acids or bases.

■ SELF-TEST 9A

1. An acid is a hydrogen ion __________, and a base is a hydrogen ion __________.
2. A compound HA reacts with water to form H_2A^+. HA is a(n) (acid/base). If compound HA reacts with water to form A^-, HA is a(n) (acid/base).
3. The conjugate acid of CO_3^{2-} is __________.
4. The conjugate base of $H_2PO_4^-$ is __________.
5. Metal oxides usually react with water to produce __________.
6. Nonmetal oxides usually react with water to produce __________.
7. The acetate ion, CH_3COO^-, is a (weak/strong) conjugate base because acetic acid is a weak acid.
8. The chloride ion, Cl^- is a (weak/strong) conjugate base because hydrochloric acid is a strong acid.
9. Baking soda, $NaHCO_3$, dissolves in water to give a(n) (acidic/basic) solution because HCO_3^- is a (weak/strong) conjugate base.
10. The net ionic equation for the reaction $HCl(aq) + KOH(aq) \rightarrow KCl(aq) + H_2O(\ell)$ is __________.
11. The net ionic equation for the reaction of all strong acids with all strong bases is __________.

9.6 Concentration of Acid and Base Solutions

The acidity or basicity of a solution of an acid or base depends on two factors: the strength of the acid or base and the concentration of the acid or base. The previous section described how strength was related to the extent of ionization. The **concentration of a solution** is the quantity of a solute dissolved in a specific quantity of a solvent or solution. Concentrations of solutions, including those of acids and bases, are often expressed in **molarity,** which is the number of moles of solute per liter of solution. For example, a 1 **molar** (1 **M**) solution contains one mole of solute per liter of solution.

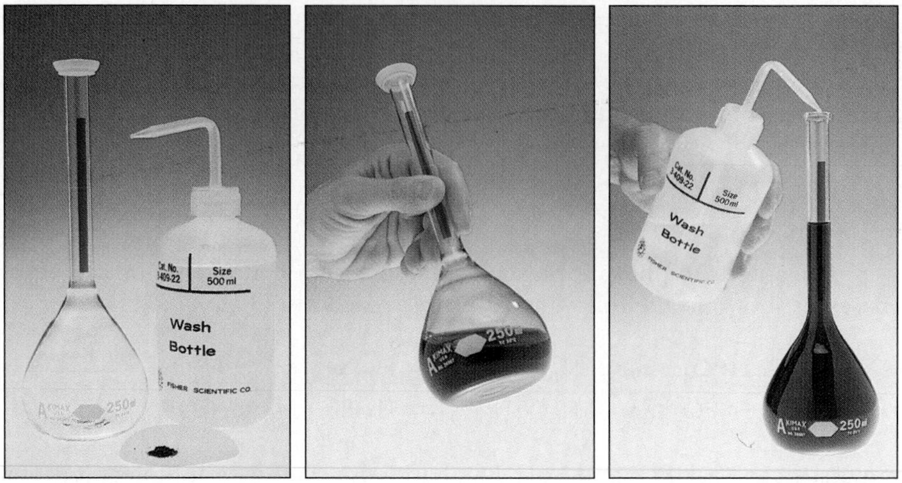

(a) (b) (c)

Figure 9.4 Preparation of an aqueous solution of known concentration. (a) A volumetric flask with a carefully marked volume, distilled (or deionized) water, and a carefully weighed amount of solid are used in the preparation. (b) All of the weighed solid is added to the flask along with some water. The flask is shaken until all of the solid is dissolved. (c) The flask is filled to the mark with distilled (or deionized) water. The flask must be shaken until a homogeneous solution is obtained. *(C.D. Winters)*

■ Molarity = moles of solute/liter of solution. Some other ways of expressing solution concentration are grams of solute/liter of solution, milligrams of solute/100 mL of solution, and grams of solute/liter of solvent.

If the solute is hydrochloric acid, a 1.00 M solution contains one mole, or 36.5 g of HCl per liter of solution. If the solute is sodium hydroxide, a 1.00 M solution contains one mole, or 40.0 g of NaOH per liter of solution. Calculations of molarity are based on determining the number of moles per liter of solution. Usually this requires knowing the molar mass of the solute, and it may require changing other volume units to liters (Figure 9.4).

EXAMPLE 9.3 *Molarity*

How many grams of NaOH must be dissolved in water to make 0.50 L of a 0.20 M solution? Refer to Figure 9.4 and describe how to prepare such a solution.

SOLUTION

To find out how many grams of NaOH are needed requires first knowing how many moles are needed and then finding the grams needed by using molar mass as a conversion factor.

A 0.20 M solution has a concentration of 0.20 mol NaOH/L, but only 0.50 L is needed. Including all units in the calculation and canceling units (as illustrated in Section 8.2) shows that 0.10 mol of NaOH is needed.

$$\text{moles NaOH needed} = 0.50\ \cancel{\text{L}} \times \frac{0.2\ \text{mol}}{1.0\ \cancel{\text{L}}} = 0.10\ \text{mol NaOH}$$

■ See Appendix *A* for a discussion of significant figures.

The molar mass of NaOH (the sum of the molar masses of Na, 23.0 g; O, 16.0 g; and H, 1.0 g) is 40.0 g/mol. Using this as a conversion factor shows that 4.0 g of NaOH must be dissolved in enough water to make 500 mL of solution.

$$0.10\ \cancel{\text{mol NaOH}} \times \frac{40.\ \text{g NaOH}}{1\ \cancel{\text{mol NaOH}}} = 4.0\ \text{g NaOH}$$

To prepare the solution, 4.0 g of NaOH is added to a 500-mL volumetric flask along with some water. The flask is shaken until all of the NaOH has dissolved and then filled carefully to the mark with water.

Exercise 9.3

How many grams of NaOH must be dissolved to make 1.5 liters of a 0.10 M NaOH solution?

TABLE 9.4 ■ Concentrated Acids and Bases

Acids	Molarity	Density*
Acetic acid	17.4	1.05
Hydrochloric acid	12.1	1.19
Nitric acid	15.7	1.42
Sulfuric acid	18.0	1.84
Base		
Ammonia	14.8	0.90

*g/mL solution

Several acids and bases are sold as concentrated solutions. Table 9.4 lists concentrations of some common ones. The most common of these is hydrochloric acid, sold in hardware stores in its concentrated form as *muriatic acid,* which is used to clean metals and cement and to remove excess mortar from bricks.

Since concentrated acids are denser than water, they should always be added to water to avoid spattering from localized heating caused by the exothermic reaction of the water with the acid. Dilute solutions are often prepared from the concentrated acids and bases. A common method of calculation is to use the formula $M_1V_1 = M_2V_2$, where M_1 and V_1 refer to the molarity and volume of the concentrated solution, and M_2 and V_2 refer to the molarity and volume of the dilute solution. The formula $M_1V_1 = M_2V_2$ is derived from the definition of molarity. Since M = moles/liter; moles = M × liters. In a dilution problem, the number of moles is not changing, so (M × liters) for the concentrated solution = (M × liters) for the dilute solution, or $M_1V_1 = M_2V_2$. It doesn't matter which solution is labeled "1" and which is labeled "2"; what is important is that the units of volume are the same in both cases—usually both are expressed in liters.

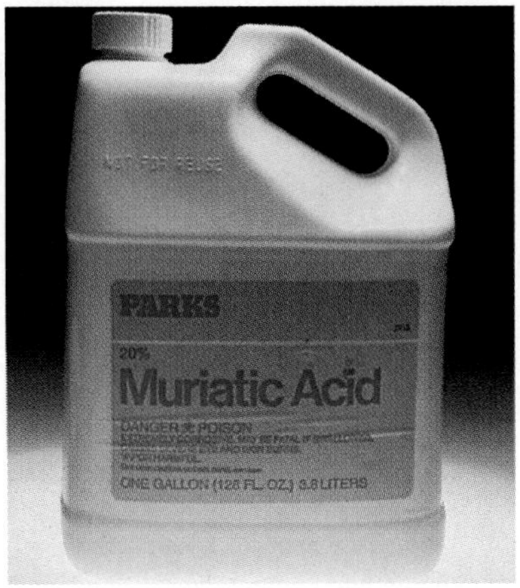

Muriatic acid as sold in a hardware store. Such a 20% hydrochloric acid solution is used (carefully) to clean brick and stone surfaces. *(C.D. Winters)*

EXAMPLE 9.4 *Molarity*

Suppose you need 2.0 liters of 6.0 M sulfuric acid, H_2SO_4. How much concentrated acid, 18.0 M, would be needed to make this solution? Describe how you would make the solution.

SOLUTION

Concentrated sulfuric acid is 18.0 M. Using $M_1V_1 = M_2V_2$, M_1 is 18.0 M, V_1 is unknown, V_2 is 2.0 liters, and M_2 is 6.0 M.

$$18.0\ \text{M} \times V_1 = 6.0\ \text{M} \times 2.0\ \text{L}$$

$$V_1 = \frac{6.0\ \text{M} \times 2.0\ \text{L}}{18.0\ \text{M}} = 0.67\ \text{L}$$

or

$$0.67\ \text{L} \times \frac{1000\ \text{mL}}{1\ \text{L}} = 670\ \text{mL}$$

Lye, the strongly basic chemical used to unclog drains, is sodium hydroxide.
(David Oxtoby)

The solution is prepared by adding 670 mL of concentrated sulfuric acid to enough water to make 2.0 liters of solution.

Exercise 9.4

How much concentrated hydrochloric acid (12.1 M) is needed to prepare 5.0 liters of 3.0 M solution? Describe how you would make this solution.

For strong acids that ionize completely, the concentration of hydronium ions is the same as the concentration of the strong acid. For example, a 1 M solution of hydrochloric acid is 1 M in hydronium ions, and the solution contains one mole of hydronium ions in each liter of solution (along with one mole of chloride ions). Following the same line of reasoning, a 1 M solution of a strong base such as sodium hydroxide is 1 M in hydroxide ions (and 1 M in sodium ions).

The concentration of hydronium ions in solutions of weak acids is not the same as the molarity of the solution because only partial ionization occurs. This is why strong acids such as sulfuric acid and hydrochloric acid are corrosive and must be handled with care, while weak acids, such as vinegar, have everyday uses, such as cleaning glass. For example, the concentration of H_3O^+ is 1.0 M for a 1.0 M solution of hydrochloric acid but only 4.3×10^{-3} M for a 1.0 M solution of acetic acid. Since many common substances have low H_3O^+ concentrations (in the range of 0.1 M to 10^{-14} M), the pH scale was developed to provide a more convenient way of expressing low concentrations of H_3O^+.

9.7 The pH Scale

The Brønsted-Lowry acid–base definitions are based on the amphiprotic properties of water—water is capable of acting as both a proton donor and a proton acceptor, depending on the acidic or basic properties of the dissolved substance. Water can also act as a proton donor and proton acceptor toward itself.

$$H_2O + H_2O \rightleftharpoons H_3O^+ + OH^-$$

Pure water is **neutral** because it contains equal numbers of hydronium, H_3O^+, and hydroxide ions, OH^-. Although pure water ionizes only slightly, as evidenced by its very small electrical conductivity (Figure 9.5a), about one of every 550,000,000 water molecules is ionized at any given time. The actual concentration of hydronium ions and hydroxide ions in pure water at 25°C is 1.0×10^{-7} M.

The product of the molarity of the hydronium ions and hydroxide ions in pure water is $(1.0 \times 10^{-7})(1.0 \times 10^{-7}) = 1.0 \times 10^{-14}$. *The value of 1.0×10^{-14} is important to our study of aqueous solutions of acids and bases because it is a constant that is always the product of the molar concentration of H_3O^+ and OH^-.*

■ See Appendix *B* for a discussion of exponential notation.

■ Brackets are used to represent molar concentrations. For example, $[H_3O^+]$ stands for molar concentration of hydronium ion, and $[H_3O^+] = 0.5$ indicates a solution with 0.5 mol of H_3O^+ per liter.

$$[H_3O^+][OH^-] = 1.0 \times 10^{-14}$$

If acid is added to pure water, the concentration of H_3O^+ will be greater than 1.0×10^{-7}, and the concentration of OH^- will be less than 1.0×10^{-7}. However, the product of the two must equal 1.0×10^{-14}. This relationship is the basis for calculating the concentration of one of the two ions, hydronium or hydroxide, when the other one is known.

EXAMPLE 9.5 *Calculating Hydronium and Hydroxide Ion Concentrations in an Acidic Solution*

What are the $[H_3O^+]$ and $[OH^-]$ in 0.10 M hydrochloric acid?

SOLUTION

The strong acid is 100% ionized, so the H_3O^+ concentration is the same as the acid concentration, 0.10 or 1.0×10^{-1} M. Since $[H_3O^+][OH^-] = 1.0 \times 10^{-14}$ then

$$[1.0 \times 10^{-1}][OH^-] = 1.0 \times 10^{-14}$$

$$[OH^-] = \frac{1.0 \times 10^{-14}}{1.0 \times 10^{-1}} = 1.0 \times 10^{-13}\ M$$

Exercise 9.5
What are the $[H_3O^+]$ and $[OH^-]$ in 0.01 M HCl?

EXAMPLE 9.6 *Calculating Hydronium and Hydroxide Ion Concentrations in a Basic Solution*

What are the $[H_3O^+]$ and $[OH^-]$ in 0.10 M NaOH?

SOLUTION

NaOH is a strong base, so the OH^- concentration is 0.10, or 1.0×10^{-1} M. Since $[H_3O^+][OH^-] = 1.0 \times 10^{-14}$,

$$[H_3O^+][1.0 \times 10^{-1}] = 1.0 \times 10^{-14}, \quad \text{or}$$

$$[H_3O^+] = \frac{1.0 \times 10^{-14}}{1.0 \times 10^{-1}} = 1.0 \times 10^{-13}\ M$$

Exercise 9.6
What are the $[H_3O^+]$ and $[OH^-]$ in 0.025 M NaOH?

The value of the exponent for the H_3O^+ molar concentration in Examples 9.5 and 9.6 goes from -1 in a strong acid solution to -13 in a strong basic solution. The difference in H_3O^+ concentration between 0.1 M HCl and 0.1 M NaOH is one trillion times because each change in the exponent is a power of ten, and the difference between 10^{-1} [H_3O^+] and 10^{-13} [H_3O^+] is 10^{12}, or one trillion. The Danish biochemist S. P. L. Sørensen proposed in 1909 that these exponents be used as a measure of acidity. He devised a scale that would be useful in his work of testing the acidity of Danish beer. Sørensen's scale came to be known as the pH scale, from the French *pouvoir hydrogene,* which means hydrogen power. pH is defined as *the negative logarithm of the hydronium ion concentration.*

$$pH = -\log [H_3O^+]$$

■ With a calculator that includes base-10 logarithms, the pH is easily found from any H_3O^+ concentration by finding its log; the H_3O^+ concentration can be determined from any pH by finding the value of 10^{-pH}.

To find the pH of a solution, write the concentration of the hydronium ion as a power of 10 and use the exponent without the negative sign. For example, a 0.01 M solution of HCl has a hydronium ion concentration of 1×10^{-2}, and the pH is 2.0.

The pH scale includes values between 0 and 14. The pH of pure water at 25°C is 7 because [H_3O^+] is 1.0×10^{-7} M. In acidic solutions, the hydronium ion concentration is greater than 10^{-7} so the exponent of 10 must be *less negative* than -7. For example, if the hydronium ion concentration is 1.0×10^{-5} M, the exponent is -5; the negative of -5 is $+5$, so the pH is 5.0. *Each decrease of one pH unit represents a tenfold increase in acidity, or hydronium ion concentration* (Figure 9.5). For example, a solution with a hydronium ion concentration of 1×10^{-4} M has a pH of 4.0, which is ten times more acidic than a solution with a pH of 5.0.

■ Acidic solution: pH less than 7; [H_3O^+] greater than [OH^-]
Neutral solution: pH 7; [H_3O^+] = [OH^-]
Basic solution: pH larger than 7; [OH^-] greater than [H_3O^+]

A solution with a pH above 7 is basic, with each unit of increase representing a decrease in the hydronium ion concentration by one tenth. In other words, the higher the pH is above 7, the more basic the solution. The relationship between pH and hydronium ion concentration and the pH of some common substances are illustrated in Figure 9.5.

Measuring the pH of Common Substances

Approximate pH values can be determined with a variety of **natural indicators,** which are organic dyes that change color over a fairly short range of pH values.

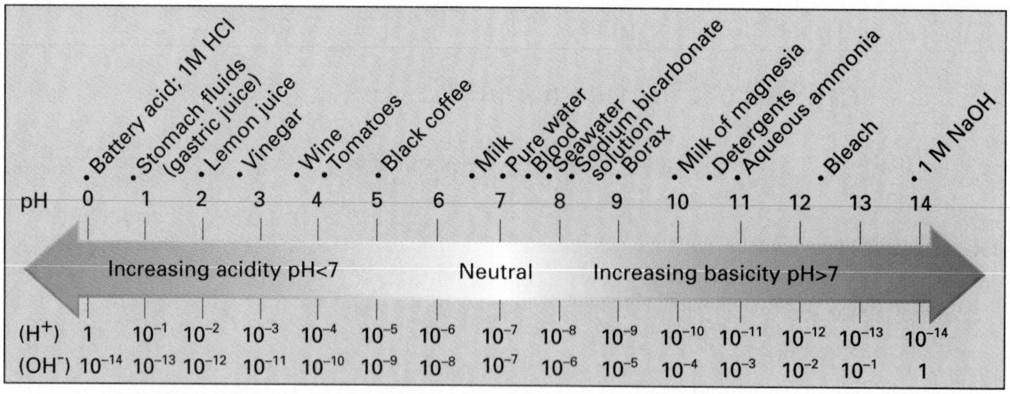

Figure 9.5 Relationship of pH to the concentration of hydrogen ions (H^+ or H_3O^+) and hydroxide ions (OH^-) in water at 25°C. The pH of some common substances are also included in the diagram.

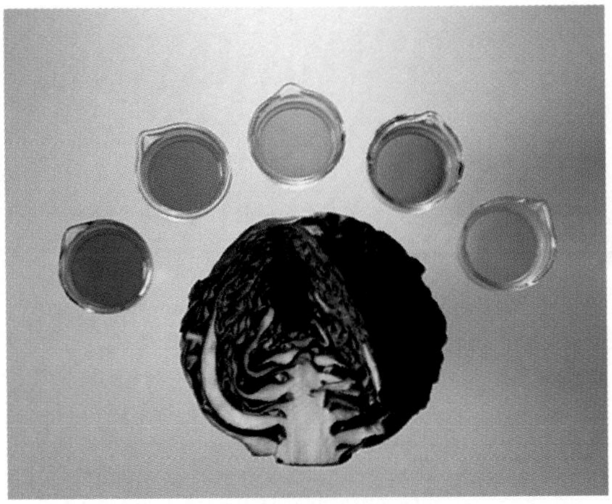

Figure 9.6 Red cabbage boiled in water yields dyes that show a variety of colors over the pH range of 1 through 14. Shown (left to right) is red cabbage juice in solutions of pH 1, 4, 7, 10, and 13. *(Charles Steele)*

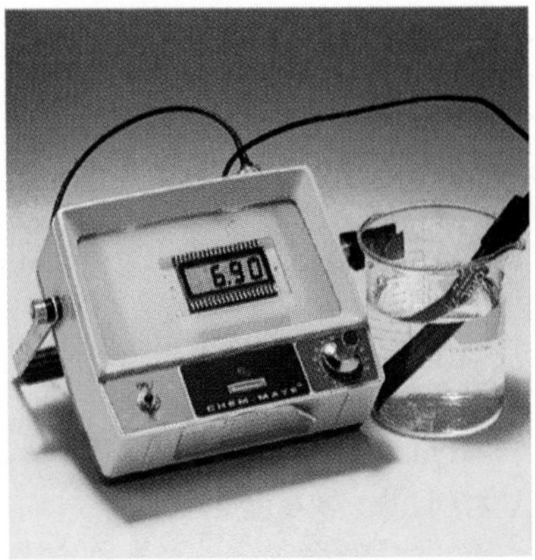

Figure 9.7 A pH meter. *(Leon Lewandowski)*

For example, litmus, a natural indicator known since the time of Arrhenius, is an organic dye extracted from certain lichens. When the pH value changes, litmus undergoes a structural change that results in a color change. Litmus turns red in acidic solutions and blue in basic solutions, but the litmus test provides no additional information about how acidic or how basic the solution is. However, red cabbage juice, another natural indicator, permits visual estimation of the pH within 1 pH unit because it is a mixture of eight different organic dyes (called anthocyanins) that change color at different pH ranges (Figure 9.6). When more accurate values of pH are needed, they can be measured with a pH meter (Figure 9.7).

9.8 Acid–Base Buffers

An acid–base buffer is like a shock absorber—something to absorb a disturbance while retaining the original conditions or structure. The control of pH requires maintaining a steady concentration of H_3O^+ even when sudden "shocks" of acid or base are added. *Buffer solutions contain a base and an acid that can react with added acid or base, respectively, and maintain a pH very close to the original value* (Figure 9.8).

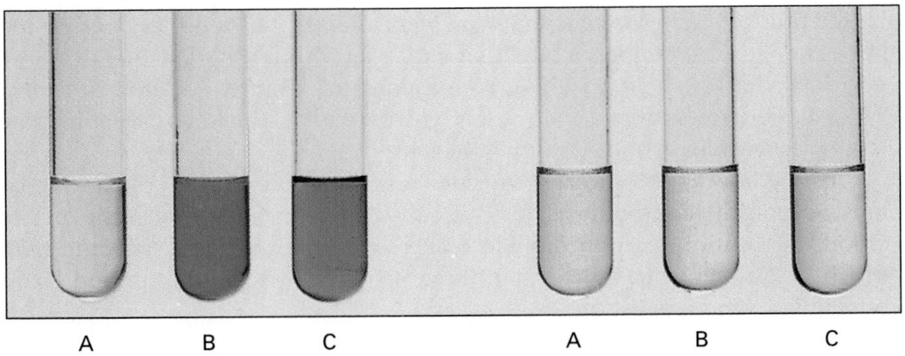

Figure 9.8 Demonstration of buffer action. At the beginning of the experiment, the three test tubes on the left contained water and an indicator that is yellow in neutral solution (pH 7); the three test tubes on the right contained the same indicator and a buffer solution that maintains pH 7. A few drops of strong acid were added to the tubes labeled B. A few drops of strong base were added to the tubes labeled C. On the left, the added acid and base caused a large change in pH, as shown by the strong color changes. On the right, the pH remained essentially unchanged. *(Marna G. Clarke)*

DISCOVERY EXPERIMENT

Natural Indicators

One of the most useful and interesting properties of acids and bases is their ability to change the color of certain vegetable and fruit juices. These include juices obtained from red cabbage, cherries, blueberries, beets, tomatoes, and the skins of radishes, red apples, plums, peaches, pears, turnips, and rhubarb. Red cabbage juice has the widest range of color changes over the entire pH range and is the natural indicator recommended for this experiment. You will need the following:

A few leaves of red cabbage

Food blender or food processor

Coffee filter or food strainer

Three clear glasses

A selection of household chemicals such as baking soda, vinegar, household ammonia, a bar of soap, clear carbonated beverage, milk

Make red cabbage juice by tearing one red cabbage leaf into small pieces and placing these pieces in a blender. Add about 500 mL (two cups) of water to the blender. Turn the blender on for one to two minutes. Filter the mixture through a coffee filter or food strainer and save the juice.

The skins of these fruits and vegetables all contain natural acid-base indicators. (*Dennis Drenner*)

Look at the label on the vinegar bottle. Does it say it contains an acid? Does it say what acid? To see what effect this acid has on cabbage juice, pour about 80 mL (one-third cup) of vinegar into a glass, add about 10 mL (two teaspoonfuls) of red cabbage juice, and stir. What is the color of the mixture? What can you conclude about the effect of an acid on red cabbage juice? Consult Figure 9.6 and estimate the pH of your mixture. (Save this mixture to compare with the result in the next part of the experiment.)

Next examine the effect of soap on the cabbage juice. Shave a bar of soap with a knife to get small pieces, or use a few squirts of liquid soap. Place the soap pieces or liquid soap in a glass containing some warm water and stir. Then add about 10 mL of red cabbage juice and stir. What color is the solution? Does it differ from the color you obtained with vinegar? Soap is a base, so the color you see with cabbage should be different. (Save this mixture for comparison.) Estimate the pH of this mixture. Now test other household chemicals with red cabbage juice. For each substance tested, estimate the pH by comparing the color with Figure 9.6 and with your acidic and basic solutions. Be sure to rinse the glass thoroughly after each test.

An **acid–base buffer** controls the pH of a solution within a narrow range, even when small amounts of acid or base are added. Buffers are very important to many industrial and natural processes. In fact, the control of the pH of your blood is essential to your health. The pH of the blood is 7.40 ± 0.05, and your good health depends on the ability of buffers to maintain the pH of the blood within this narrow range. If the pH falls below 7.35, a condition known as *acidosis* occurs; increasing pH above 7.45 leads to *alkalosis.* Both these conditions can be life-threatening.

How does a buffer solution maintain its pH at a nearly constant value? A buffer must contain an acid that can react with any added base, and at the same time it must contain a base that can react with any added acid. It is also necessary that the acid and base components of a buffer solution not react with each other. Buffers usually are mixtures of a weak acid and its conjugate base, or a weak base and its conjugate acid.

■ The carbonate ion, CO_3^{2-}, is the conjugate base of carbonic acid, H_2CO_3, which is a very weak acid. Thus, sodium carbonate is basic enough to be used as a cleaning agent.

To see how a buffer works, consider human blood. Carbon dioxide is the most important compound (but not the only one) that affects the pH of blood. In solution, carbon dioxide reacts with water to form carbonic acid, which, as a weak acid, ionizes slightly to produce hydronium ion and bicarbonate ion. The reactions are

$$CO_2(aq) + H_2O(\ell) \rightleftharpoons H_2CO_3(aq)$$
$$H_2CO_3(aq) + H_2O(\ell) \rightleftharpoons H_3O^+(aq) + HCO_3^-(aq)$$

Carbonic acid and the bicarbonate ion constitute a buffer since carbonic acid is a weak acid and bicarbonate ion is its conjugate base. If small amounts of a strong base such as NaOH are added to this buffer, carbonic acid will react with the hydroxide ion to give bicarbonate ion:

$$H_2CO_3(aq) + OH^-(aq) \rightleftharpoons HCO_3^-(aq) + H_2O(\ell)$$

If small amounts of a strong acid such as hydrochloric acid are added to this buffer, the bicarbonate ion will react with the hydronium ions to form carbonic acid:

$$HCO_3^-(aq) + H_3O^+(aq) \rightleftharpoons H_2CO_3(aq) + H_2O(\ell)$$

Notice in these reactions of added base or acid that no new substances are produced. The product is always one of the buffer components.

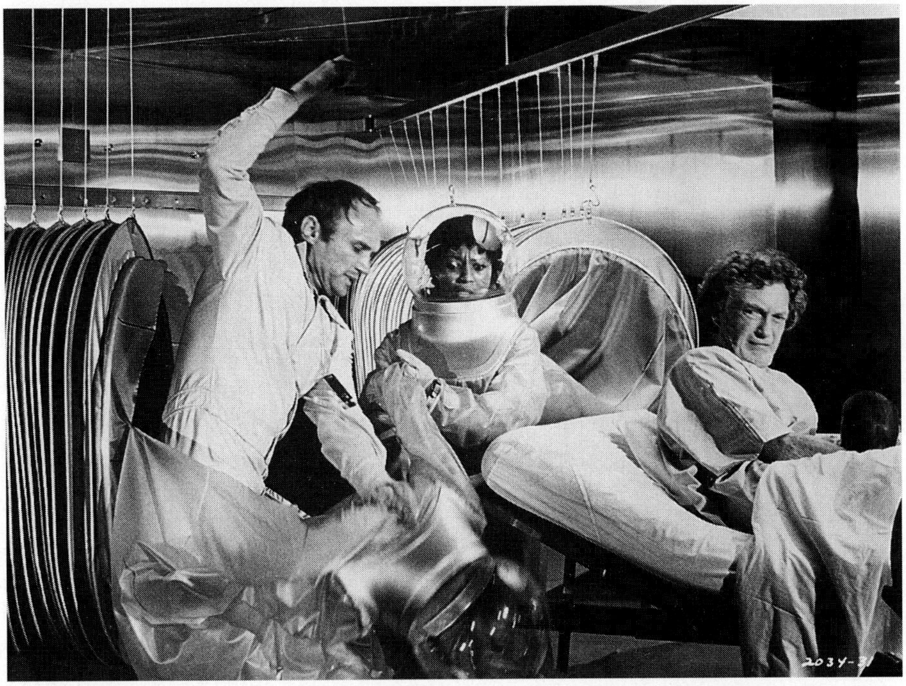

The Andromeda Strain Scientist Dr. Mark Hall is shown quickly getting out of his anti-contamination suit to help in an experiment needed to learn why two people could survive attack by an alien organism. In this Michael Crichton novel made into a movie, the organism infects and kills everyone in town except a screaming baby and the town drunk. The baby was hyperventilating, and her blood was slightly basic due to lower carbon dioxide levels resulting from the rapid breathing. The drunk was drinking Sterno, a solid gel of sodium acetate and ethanol, and taking large doses of aspirin to ease his stomach pain. As a result, his blood was slightly acidic. The alien organism required blood of normal pH. Crichton, who has an M.D. degree, also wrote *Jurassic Park*. *(The Kobal Collection)*

EXAMPLE 9.7 *Buffers*

Another buffer system in blood is one made up of sodium dihydrogen phosphate, NaH_2PO_4, and sodium monohydrogen phosphate, Na_2HPO_4. The buffer reactions are based on the $H_2PO_4^-/HPO_4^{2-}$ acid/conjugate base system. Write the equations for reactions that control the pH when small amounts of hydroxide ion or hydronium ion are added.

SOLUTION

The $H_2PO_4^-$ will neutralize small amounts of added OH^- and produce HPO_4^{2-} according to the following equation:

$$H_2PO_4^-(aq) + OH^-(aq) \rightleftharpoons HPO_4^{2-}(aq) + H_2O(\ell)$$

Small amounts of added H_3O^+ will be neutralized by reaction with HPO_4^{2-} to give $H_2PO_4^-$.

$$HPO_4^{2-}(aq) + H_3O^+(aq) \rightleftharpoons H_2PO_4^-(aq) + H_2O(\ell)$$

Exercise 9.7

For a buffer system that uses sodium carbonate (Na_2CO_3) and sodium bicarbonate ($NaHCO_3$), write equations for reactions that illustrate how the buffer would control the pH when small amounts of acid or base are added.

9.9 Indigestion: Why Reach for an Antacid?

The walls of a human stomach contain thousands of cells that secrete hydrochloric acid, the main purposes of which are to suppress the growth of bacteria and to aid in the digestion of certain foodstuffs. The pH of the stomach ranges from 0.9 to 1.5. Normally the stomach's inner lining is not harmed by the presence of hydrochloric acid with such a low pH, since the mucosa, the inner lining of the stomach, is replaced at the rate of about a half million cells per minute. However, when too much food is eaten, the stomach often responds with an outpouring of acid, which lowers the pH to a point at which the discomfort of "acid indigestion" or "heartburn" is felt. **Antacids** are bases used to neutralize the excess hydrochloric acid in the stomach that causes this discomfort. Many antacids contain bicarbonate or carbonate salts, which are the basic anions of weak acids (Section 9.4). Some common antacids (Table 9.5) are baking soda, milk of magnesia, Alka-Seltzer, Rolaids, Di-Gel, and Tums. How do these antacids or bases work?

■ The bicarbonate ion is the conjugate base of carbonic acid, a weak acid. As a result, the bicarbonate ion can neutralize acids.

■ Recipe for homemade antacid fizz: Mix a teaspoonful of baking soda ($NaHCO_3$) in a glass of orange juice.

■ People with high blood pressure are advised to choose an antacid that doesn't contain sodium ions.

Baking soda (sodium bicarbonate) was used to relieve indigestion before many of the other commercial products became available. The bicarbonate ion reacts with the hydronium ion from hydrochloric acid to form carbonic acid, which decomposes to give carbon dioxide and water.

$$HCO_3^-(aq) + H_3O^+(aq) \longrightarrow H_2CO_3(aq) + H_2O(\ell)$$
$$H_2CO_3(aq) \longrightarrow CO_2(g) + H_2O(\ell)$$

TABLE 9.5 ■ Some Common Antacids

Compound	Formula	Examples of Commercial Products
Magnesium hydroxide	$Mg(OH)_2$	Phillips' Milk of Magnesia
Calcium carbonate	$CaCO_3$	Tums, Titralac
Sodium bicarbonate	$NaHCO_3$	Alka-Seltzer, baking soda
Aluminum hydroxide	$Al(OH)_3$	Amphojel
Aluminum hydroxide and magnesium hydroxide		Maalox, Mylanta, Di-Gel tablets
Aluminum hydroxide, magnesium hydroxide, and magnesium carbonate	$MgCO_3$	Di-Gel liquid
Dihydroxyaluminum sodium carbonate	$NaAl(OH)_2CO_3$	Rolaids
Calcium carbonate and magnesium hydroxide		Sodium-free Rolaids

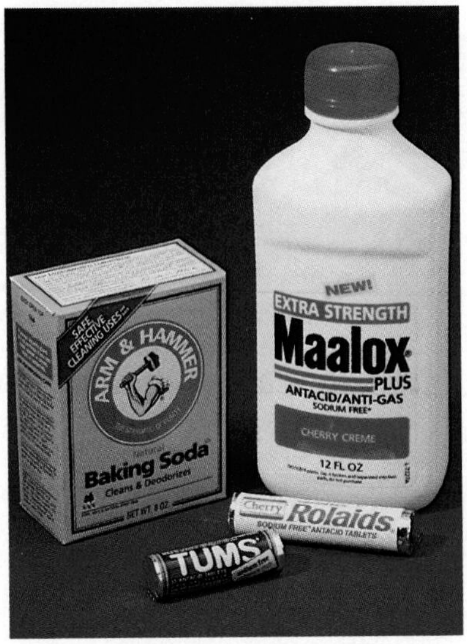

Some common antacids. *(Larry Cameron)*

Alka-Seltzer contains sodium bicarbonate, potassium bicarbonate, citric acid, and aspirin. The fizz of the Alka-Seltzer tablet in water is carbon dioxide gas given off by the reaction of the citric acid with the bicarbonates to give carbonic acid, the same as the reaction above.

Milk of magnesia is a suspension of magnesium hydroxide in water. Magnesium hydroxide acts as an antacid in small doses, but in large doses it is a laxative.

Calcium carbonate is the active ingredient in Tums. The carbonate ion, a base, neutralizes hydronium ion:

$$CO_3^{2-}(aq) + 2\,H_3O^+(aq) \longrightarrow H_2CO_3(aq) + 2\,H_2O(\ell)$$
$$H_2CO_3(aq) \longrightarrow CO_2(g) + H_2O(\ell)$$

Although small amounts of calcium carbonate are safe, regular use can cause constipation. Aluminum hydroxide, the active ingredient in Amphojel, can also cause constipation in large doses. Because calcium carbonate and aluminum hydroxide can cause constipation, antacids such as Maalox and Mylanta contain aluminum hydroxide mixed with magnesium hydroxide to counteract the constipating effects of the former with the laxative action of the latter.

The World of Chemistry

Program 16, *The Proton in Chemistry*

Even the sight and smell of appetizing food is enough to make the stomach produce more acid. Dr. Paul Maton, National Institutes of Health

9.10 Calculations Using pH

The pH scale is designed for hydronium ion concentrations between 1.0×10^0 and 1.0×10^{-14} (1 to 14). Since strong acids and bases ionize completely, concentrations greater than 1 M would fall outside the pH scale. Dilute solutions of weak acids and bases are well within the pH scale since the compounds are only partially ionized. For example, vinegar is about 0.7 M in acetic acid, and if acetic acid were a strong acid, the pH would be close to zero. However, the pH of vinegar is 2.5 since the percentage of acetic acid molecules that are ionized to hydronium ions is very small. For the same

THE WORLD OF CHEMISTRY

Stomach Acidity

Program 16, *The Proton in Chemistry*

The pH of stomach fluids, even in the normal stomach, is 1. The acid is hydrochloric acid. The stomach produces a small amount of acid all the time, but then that amount of acid may be stimulated by food. Even the sight and smell of appetizing food are enough to make the stomach produce more acid.

What actually happens when you eat a meal? Both hydrogen ions and chloride ions, maintaining an electrochemical balance, move through the stomach lining from the surrounding blood plasma. In the stomach the result is a highly acidic medium. That's what it takes to activate certain enzymes for the process of digestion. If the acidity of the stomach becomes excessive, problems can occur—problems that often need antacid solutions.

Dr. Paul Maton of the National Institutes of Health is an expert on stomach acid. He says:

There are a variety of different compounds, basically bases, that can function as antacids. For example, sodium bicarbonate could be used as an antacid, or magnesium hydroxide, or calcium carbonate. All antacids neutralize acid. And if they're given in sufficient amounts, any antacid is as good as another at neutralizing acid.

One hears a lot of talk about hyperacidity—too much acid. In fact, there's very little or no evidence that people with ulcers actually produce more acid than many of the rest of us.

For most of those acid-stomach discomforts then, the tried and tested over-the-counter remedies do their job of neutralization well enough.

reason, household ammonia with a molarity of 2.5 has a pH of about 11 instead of being outside the pH range.

Checking the pH of a lake in the Pocono Mountains. The pH here was 4.4. *(Runk/Schoenberger from Grant Heilman)*

EXAMPLE 9.8 *pH Scale*

Which is more acidic: your stomach's digestive juices, with a pH of 1.4, or your blood, with a pH of 7.4? How much difference is there in the acidity of your digestive juices and your blood?

SOLUTION

Since pH $= -\log [H_3O^+]$, the *lower* the pH, the *larger* the concentration of hydronium ions. A pH of 1.4 is much more acidic than a pH of 7.4. Since the difference between 7.4 and 1.4 is six units, and each unit is a power of ten, a pH of 1.4 is 10^6 or *one million times more acidic* than a pH of 7.4.

Exercise 9.8

Which is more basic: a solution of baking soda, with a pH of 8.4, or milk of magnesia, with a pH of 10.4? How much difference is there in the basicity of these two substances?

EXAMPLE 9.9 *Finding pH from Concentration*

What is the pH of the following solutions?

(a) 0.01 M HCl (b) 0.01 M NaOH

SOLUTION

(a) Since hydrochloric acid is a strong acid, the concentration of hydronium ions is 0.01 M or 1×10^{-2} M. pH is defined as the negative logarithm of the hydronium ion concentration, so the pH is $-(-2) = 2.0$.

(b) Since sodium hydroxide is a strong base, the concentration of hydroxide ions is 0.01 M, or 1×10^{-2} M. The hydronium ion concentration is determined from the expression $[H_3O^+][OH^-] = 1.0 \times 10^{-14}$.

$$[H_3O^+][10^{-2}] = 1.0 \times 10^{-14}$$

$$[H_3O^+] = \frac{1.0 \times 10^{-14}}{1 \times 10^{-2}} = 1 \times 10^{-12}$$

$$pH = -(-12) = 12.0$$

Exercise 9.9
What is the pH of the following solutions?

(a) 0.001 M HCl (b) 0.001 M NaOH

EXAMPLE 9.10 *Finding H_3O^+ Concentration from pH*

The pH of a soft drink is 3.0 because it contains carbonic acid and phosphoric acid. What is the hydronium ion concentration in this soft drink?

SOLUTION

$pH = -\log [H_3O^+]$. This means the exponent of ten will be the negative of the pH value, or -3 in this case. The hydronium ion concentration is 1×10^{-3} M.

Exercise 9.10
Most tomatoes have a pH of around 4. What is the hydronium ion concentration in tomatoes?

SELF-TEST 9B

1. Which is more acidic, a pH of 6 or a pH of 2? What is the magnitude of difference between these two pH values? (a) 4, (b) 4000, (c) 1000, (d) 10,000.
2. High pH means (a) high hydronium ion concentration, (b) low hydronium ion concentration.
3. Low pH means (a) high hydronium ion concentration, (b) low hydronium ion concentration.
4. In the buffer system $H_2PO_4^-/HPO_4^{2-}$, which component neutralizes added base? Which component neutralizes added acid?
5. What is the pH of 1.0×10^{-3} M hydrochloric acid?
6. What is the pH of 1.0×10^{-3} M sodium hydroxide?
7. The pH of household ammonia is 11. What is the hydronium ion concentration? What is the hydroxide ion concentration?

8. The pH of a sodium hydroxide solution is 10. What is the hydroxide ion concentration?
9. A mole of NaOH weighs 40.0 g. How many grams of NaOH are needed to make 5.0 liters of a 2.0 M solution?
10. Concentrated hydrochloric acid is 12 M. How many liters of 6 M hydrochloric acid can be made with 0.5 liter of concentrated hydrochloric acid?
11. What is the $[H_3O^+]$ for a 0.01 M HCl solution?

MATCHING SET

___ 1. Conjugate base of HA^-	a. 7
___ 2. Conjugate acid of A^{2-}	b. Molarity; moles of solute per liter of solution
___ 3. M	c. Hydrogen ion donor
___ 4. pH of pure water	d. Maintains pH
___ 5. Acidic pH	e. Hydrogen ion acceptor
___ 6. Strong acid	f. 10
___ 7. Strong base	g. CH_3COOH, acetic acid
___ 8. Weak conjugate acid	h. Strong conjugate base
___ 9. Acid definition	i. Causes solution to conduct electricity
___ 10. Base definition	j. NaOH
___ 11. Weak acid	k. HA^-
___ 12. Electrolyte	l. H_2SO_4
___ 13. Buffer	m. A^{2-}
___ 14. Basic pH	n. 3

QUESTIONS FOR REVIEW AND THOUGHT

1. What is the Arrhenius definition of an acid?
2. What is the Arrhenius definition of a base?
3. What is a hydronium ion?
4. What do the following terms mean?
   (a) acidic oxide (b) basic oxide
   (c) hydrate
5. What do the following terms mean?
   (a) acidic solution (b) basic solution
   (c) neutral solution
6. What is a neutralization reaction?
7. What is a salt?
8. What is a net ionic equation?
9. Explain the role of spectator ions in reactions. Why can they be deleted from reactions to give net ionic equations?
10. What is a Brønsted-Lowry acid?
11. What is a Brønsted-Lowry base?
12. What are the differences between the Arrhenius theory and the Brønsted-Lowry theory? How are the theories alike?
13. What do the following terms mean?
    (a) Proton donor (b) Proton acceptor
    (c) Amphiprotic (d) Conjugate acid
    (e) Conjugate base (f) Conjugate acid-base pair
14. What do the following terms mean?
    (a) Extent of ionization (b) Strong acid
    (c) Strong base (d) Weak acid
    (e) Weak base
15. Identify four common substances that are acidic.
16. Name four common substances that are basic.
17. Classify the following oxides as acidic or basic.
    (a) $SO_2(g)$ (b) $CaO(s)$
    (c) $MgO(s)$ (d) $CO_2(g)$
    (e) $SO_3(g)$
18. Which of the following oxides are acidic? Which are basic?
    (a) $CO_2(g)$ (b) $Al_2O_3(s)$
    (c) $BaO(s)$ (d) $NO_2(g)$
19. Label each of the following substances as acidic or basic.
    (a) Vinegar (b) Citrus fruits
    (c) Aspirin (d) Lye
    (e) Black coffee (f) Milk of magnesia
    (g) Detergents (h) Household ammonia
    (i) Vitamin C
20. Label each of the following substances as acidic or basic.
    (a) Gastric fluid (b) Tomatoes
    (c) Oven cleaners (d) Soap
    (e) Carbonated beverages (f) Baking soda

21. Indicate which of the following is a property of an acid and which is a property of a base.
(a) Sour taste (b) Bitter taste
(c) Slippery feeling
(d) Change color of red litmus to blue
(e) Change color of blue litmus to red
22. Define the following terms:
(a) Weak electrolyte (b) Strong electrolyte
(c) Nonelectrolyte
23. What do the following terms mean?
(a) Acidic hydrogen (b) Polyprotic acid
(c) Molarity (d) Concentration
24. What is the definition of pH?
25. What is an acid-base indicator? Give three examples and explain how they indicate whether a solution is acidic or basic.
26. What is a buffer?
27. What do the following terms mean?
(a) Acidosis (b) Alkalosis
28. What is an antacid?
29. Write the formula of the product obtained in the following reactions.
(a) $CaO(s) + H_2O(\ell) \rightarrow$ ______
(b) $Na_2O(s) + H_2O(\ell) \rightarrow$ ______
(c) $MgO(s) + H_2O(\ell) \rightarrow$ ______
(d) $Al_2O_3(s) + H_2O(\ell) \rightarrow$ ______
30. Write the balanced equation for the following neutralization reactions.
(a) Acetic acid, $CH_3COOH(aq)$, with potassium hydroxide, $KOH(aq)$
(b) Sulfuric acid, $H_2SO_4(aq)$, with calcium hydroxide, $Ca(OH)_2(aq)$
(c) Sulfuric acid, $H_2SO_4(aq)$, with sodium hydroxide, $NaOH(aq)$
31. Write both the balanced complete and balanced net ionic equations for the following reactions.
(a) $HBr(aq) + Ca(OH)_2(aq) \rightarrow$ ______
(b) $HNO_3(aq) + Al(OH)_3(aq) \rightarrow$ ______
32. What reaction occurs when an antacid like Tums, $CaCO_3(aq)$, reacts with stomach acid, $HCl(aq)$?
33. Which is the stronger acid in each of the following pairs?
(a) $H_2O(\ell)$ or $H_3O^+(aq)$ (b) $H_2S(aq)$ or $HS^-(aq)$
(c) $H_2SO_4(aq)$ or $HSO_4^-(aq)$
(d) $H_3PO_4(aq)$ or $H_2PO_4^-(aq)$
34. Which is the stronger base in each of the following pairs?
(a) $PO_4^{3-}(aq)$ or $H_2PO_4^-(aq)$
(b) $S^{2-}(aq)$ or $HS^-(aq)$
(c) $HCO_3^-(aq)$ or $CO_3^{2-}(aq)$
35. Identify acids, bases, conjugate acids, and conjugate bases in the following reactions.
(a) $HCl(aq) + H_2O(\ell) \rightarrow H_3O^+(aq) + Cl^-(aq)$
(b) $H_2SO_4(aq) + H_2O(\ell) \rightarrow H_3O^+(aq) + HSO_4^-(aq)$
(c) $NH_3(aq) + H_2O(\ell) \rightarrow NH_4^+(aq) + OH^-(aq)$
(d) $CH_3COOH(aq) + H_2O(\ell) \rightarrow H_3O^+(aq) + CH_3COO^-(aq)$
36. Which of the following is amphiprotic? Explain or justify by using reactions.
(a) $H_2O(aq)$ (b) $H_3O^+(aq)$
(c) $OH^-(aq)$ (d) $H_2PO_4^-(aq)$
(e) $HCO_3^-(aq)$
37. Which is more basic: a solution with pH of 2 or a solution with pH of 10? Explain.
38. Which is more acidic: a solution of black coffee with a pH of 5.0 or milk with a pH of 6.5? Explain.
39. Which of the following are acidic and which are basic?
(a) Vinegar, pH = 3.0
(b) Baking soda solution, pH = 8.5
(c) Beer, pH = 4.2
(d) Lye solution, pH = 14
(e) Soft drink, pH = 4.5
(f) Milk of magnesia, pH = 10.4
40. Which is more acidic: gastric juice with a pH of 1 or tomato juice with a pH of 4? What is the difference in acidity of these solutions? (Give answer as a multiple of acidity.)
41. Which is more basic: a soap solution with pH of 10 or a household ammonia solution with pH of 12? What is the difference in basicity of these solutions? (Give answer as a multiple of basicity.)
42. Which of the following pairs of compounds would be useful for making a buffer solution? Explain.
(a) $NaOH(aq)$ and $HCl(aq)$
(b) $NaH_2PO_4(aq)$ and $Na_2HPO_4(aq)$
43. A buffer can be made using $NaHCO_3(aq)$ and $H_2CO_3(aq)$. Explain how this combination resists changes in pH when small amounts of acid or base are added.
44. Two solutions contain 1% acid. Solution A has a pH of 4.6 and solution B has a pH of 1.1. Which solution contains the stronger acid?
45. Moist baking soda is often put on acid burns. Why? Write an equation for the reaction assuming the acid is hydrochloric acid (HCl).
46. Explain how the $H_2PO_4^-/HPO_4^{2-}$ blood buffer keeps the pH of blood constant.

PROBLEMS

1. How many moles of solute are dissolved in each of the following solutions?
(a) 200. mL of 0.15 M HCl
(b) 80. mL of 3.0 M KOH
(c) 0.050 L of 0.25 M HNO_3
(d) 0.80 L of 2.0 M H_2SO_4
2. How many moles of solute are dissolved in one mL of each of the following solutions?
(a) 6.0 M HCl (b) 2.5×10^{-2} M KOH

(c) 12.0 M HCl (d) 18 M H_2SO_4
(e) 14.8 M NH_3 (f) 15.7 M HNO_3

3. How many grams of solute are in each of the following solutions?
   (a) 53 mL of 1.5 M NaOH
   (b) 370 mL of 5.0 M NH_3
   (c) 85 mL of 0.35 M H_2SO_4
   (d) 0.051 L of 3.0 M H_3PO_4
4. How many grams of solute are needed to prepare the following solutions?
   (a) 185 mL of 6.0 M HCl
   (b) 0.017 L of 2.0 M HNO_3
   (c) 0.350 L of 1.1 M $Ca(OH)_2$
   (d) 6.0 L of 0.06 M $Al(OH)_3$
5. How many grams of solute are required to prepare 1.0 L of the following solutions?
   (a) 0.25 M CH_3COOH, acetic acid
   (b) 4.0 M $Al(OH)_3$, aluminum hydroxide
   (c) 2.0×10^{-3} M HCl, hydrochloric acid
   (d) 9.0 M KOH, potassium hydroxide
6. What is the final molarity of each of the following solutions?
   (a) 180 mL of 2.5 M HCl is added to enough water to give 0.5 L of solution.
   (b) 30. mL of 6.0 M NH_3 is added to enough water to give 1.0 L of solution.
   (c) 1.20 L of 12.0 M HCl is added to enough water to give 10.0 L of solution.
   (d) 48 mL of 0.2 M $Al(OH)_3$ is added to enough water to give 500 mL of solution.
7. What is the final molarity of each of the following solutions?
   (a) 75 mL of 3.0 M HBr is added to enough water to give 500 mL of solution.
   (b) 10. mL of 18 M H_2SO_4 is added to enough water to give 1.0 L of solution.
   (c) 700. mL of 8.0 M HNO_3 is added to enough water to give 2.0 L of solution.
   (d) 65 mL of 0.2 M NaOH is added to enough water to give 100. mL of solution.
8. What are the molarities of each of the following solutions?
   (a) 750 mL of solution containing 8.0 g NaOH
   (b) 250 mL of solution containing 36 g $Ca(OH)_2$
   (c) 100. mL of solution containing 9.1 g HCl
   (d) 1.0 L of solution containing 0.02 g $Mg(OH)_2$
9. What are the molarities of each of the following solutions?
   (a) 865 mL of solution containing 65.0 g NaOH
   (b) 35 mL of solution containing 1.5 g H_2SO_4
   (c) 475 mL of solution containing 1.5 g $Ba(OH)_2$
   (d) 18 L of solution containing 12.6 g $Ca(OH)_2$
10. What are the final concentrations for the following solutions if each is diluted to 1.0 L?
    (a) 200. mL of 12 M HCl (b) 25 mL of 3.0 M KOH
    (c) 0.021 L of 18 M H_2SO_4
11. What are the final concentrations for the following solutions if each is diluted to 2.0 L?
    (a) 55 mL of 6.0 M NH_3
    (b) 400. mL of 1.5 M HCl
    (c) 0.628 L of 18.0 M H_2SO_4
    (d) 0.125 L of 0.50 M KOH
12. What is the pH for each of the following solutions?
    (a) 1.0×10^{-2} M HCl (b) 0.001 M HNO_3
    (c) 0.1 M NaOH (d) 0.10 M HBr
13. What is the pH for each of the following solutions?
    (a) 1.0×10^{-3} M NaOH (b) 1.0×10^{-3} M HCl
    (c) 0.01 M KOH (d) Neutral water at 25°C
14. What is the molarity of H_3O^+ for each of the following solutions?
    (a) Solution with a pH of 1.0
    (b) Solution with a pH of 0.0
    (c) Solution with a pH of 5.0
    (d) Solution with a pH of 3.0
15. What is the molarity of OH^- for each of the following solutions?
    (a) Solution with a pH of 8.0
    (b) Solution with a pH of 10.0
    (c) Solution with a pH of 12.0
    (d) Solution with a pH of 14.0

Batteries are all dependent on oxidation-reduction reactions, the subject of this chapter. (Dr. Jeremy Burgess/SPL/Photo Researchers)

CHAPTER 10

Oxidation–Reduction in Chemistry

Equal in importance to the hydrogen ion transfer in acid–base chemistry are the processes called oxidation and reduction. Oxidation, and the reduction that always accompanies it, is a very common kind of chemical reaction. We live in an oxidizing atmosphere. Oxygen is used to derive energy from foods and fuels. Chlorine disinfects drinking water, sewage, and swimming pools by oxidizing harmful bacteria. Controlled oxidation–reduction reactions occur in devices we call batteries, and without them, many of life's conveniences would be impossible. Oxidation and reduction reactions always occur together, so they are often named together as oxidation–reduction. In this chapter we shall look at both the applications of oxidation–reduction, which are of vast importance in our lives, and the theory of oxidation–reduction, which seeks to explain what is going on at the molecular level. In this chapter we shall answer the following questions:

- What is oxidation, what is reduction, and how can you recognize oxidation–reduction reactions?
- How are oxidation–reduction reactions important?
- Why do some chemical reactions produce an electric current and others take place only when an electric current is applied?
- What is corrosion? How is it controlled?
- What are the characteristics of batteries?
- What are fuel cells, and how do they differ from batteries?

10.1 Oxygen—The Element

Oxygen is a very common element. It is 21% by volume of the Earth's atmosphere. This means that for every 1 L of air, 210 mL is oxygen. Oxygen also accounts for 21% of the pressure of the atmosphere at sea level. Table 10.1 gives some of the properties of oxygen.

TABLE 10.1 ■ Some Properties of Oxygen

Formula	O_2
Molar mass	32.00 g
Melting point	−218.4°C
Boiling point	−183.0°C
Description	Colorless and odorless gas
Solubility in water	48.9 mL per liter of liquid water at 0°C

■ Perhaps someday a chemist will discover that helium, neon, or argon also reacts with oxygen. At one time, no one believed that krypton and xenon would react, but they do.

Oxidation got its name from the chemical changes that occur when the element oxygen combines with other elements. In fact, oxygen combines with every element except helium, neon, and argon, which form no compounds with any elements. When oxygen combines with another element, the product is an **oxide.** A few of the many oxides known are listed in Table 10.2. Oxygen gas is very reactive, and for this reason many of the elements found on this planet are in the form of oxides or in compounds in which the element is combined with oxygen in a polyatomic ion. A sizable fraction of all the hydrogen on Earth is found in the oxide we call water. In addition, most of the carbon not found in living things is combined with oxygen as carbonates of various forms. Iron, the most common metal in the Earth's crust, is almost always combined with oxygen. A few of the metals, such as gold, silver, platinum, and copper, do not form oxides readily but can be made to combine with oxygen under certain conditions. Because of their lack of oxide formation when simply exposed to the air, these metals have been used for jewelry and coins since ancient times.

■ Oxygen is the number 3 industrial chemical, with 49.7 billion pounds being produced in the United States in 1994.

Most of the oxygen prepared industrially is used in the making of steel. Tank trucks loaded with liquid oxygen and trucks carrying cylinders of compressed oxygen gas are marked with special *placards* to indicate the hazards of pure oxygen (Figure 10.1).

Figure 10.1 The hazards of oxygen. The U.S. Department of Transportation requires that shipments of elemental oxygen exceeding 1001 lb. carry a warning placard like that shown. The placard indicates that oxygen will support combustion. *(Tom Pantages)*

TABLE 10.2 ■ Oxides of Some of the Elements

Element	Oxide Name	Formula	State at 25°C
Carbon	Carbon monoxide	CO	gas
	Carbon dioxide	CO_2	gas
Hydrogen	Water	H_2O	liquid
	Hydrogen peroxide	H_2O_2	liquid
Nitrogen	Nitrous oxide	N_2O	gas
	Nitric oxide	NO	gas
	Nitrogen dioxide	NO_2	gas
Sulfur	Sulfur dioxide	SO_2	gas
	Sulfur trioxide	SO_3	gas
Boron	Boron oxide	B_2O_3	solid
Aluminum	Aluminum oxide	Al_2O_3	solid
Iron	Iron(II) oxide	FeO	solid
	Iron(III) oxide	Fe_2O_3	solid

10.2 So What Is Oxidation, Anyway?

Oxidation is such a common kind of reaction that three different definitions have evolved: (1) oxidation is reaction with oxygen, in which one or more products of the reaction, the **oxidation products,** have gained oxygen; (2) oxidation is loss of hydrogen by a reactant; and (3) oxidation is loss of valence electrons by an atom, ion, or molecule. Each of these definitions is useful when applied to various types of chemical reactions. It is important to realize that not all chemical reactions involve oxidation–reduction—just those that fit one or more of these three definitions.

■ The three matching definitions of reduction are found in Section 10.4.

Oxidation as Gain of Oxygen

Whenever oxygen combines with another element or compound, oxidation takes place. Iron and aluminum, common metals used in construction, both react readily with oxygen. In such reactions we say the metals have been *oxidized.* The oxidation products are iron oxide and aluminum oxide.

$$4\ \mathrm{Fe}(s) + 3\ \mathrm{O_2}(g) \longrightarrow 2\ \mathrm{Fe_2O_3}(s)$$
$$4\ \mathrm{Al}(s) + 3\ \mathrm{O_2}(g) \longrightarrow 2\ \mathrm{Al_2O_3}(s)$$

The oxide of iron, Fe_2O_3, is red-brown in color. In purified form, Fe_2O_3 is used as a colorant in paints and even in cosmetics. In the presence of moisture, a hydrate of iron oxide forms whenever iron reacts with oxygen. This hydrate, also red-brown in color, is the familiar compound rust (see Section 10.5 for a discussion of corrosion).

Aluminum oxide (also known as corundum) is extremely hard and is used as an abrasive and in certain kinds of cutting tools. Extremely pure forms of aluminum oxide are found in rubies and sapphires. Synthetic varieties of both these gemstones can be prepared by fusing extremely pure aluminum oxide at high temperatures.

Nitrogen, 78% of the atmosphere, readily combines with oxygen during lightning strikes to form nitric oxide (NO), which combines further with oxygen to form nitrogen dioxide (NO_2). Both of these oxides are important in the role they play in air pollution (Section 21.4):

$$\mathrm{N_2}(g) + \mathrm{O_2}(g) \longrightarrow 2\ \mathrm{NO}(g)$$
$$2\ \mathrm{NO}(g) + \mathrm{O_2}(g) \longrightarrow 2\ \mathrm{NO_2}(g)$$

Carbon, in the form of charcoal or even diamond, readily combines with oxygen. Two oxides of carbon can be formed, depending on the availability of oxygen. These are carbon monoxide (CO), a poisonous gas, and carbon dioxide (CO_2). Carbon dioxide is very common. It is a product of respiration and is produced whenever any fuel burns in air. In both of these compounds, the carbon is oxidized. Because there are two oxygen atoms per carbon in CO_2, the carbon is *more oxidized* than in CO.

■ The toxicity of carbon monoxide is discussed in Section 18.3. The role of carbon dioxide in the atmosphere is discussed in Section 21.11.

■ The formation of carbon monoxide when carbon dioxide could be formed is called incomplete combustion. In a limited supply of oxygen, carbon monoxide is the likely product.

$$2\ \mathrm{C}(s) + \mathrm{O_2}(g) \longrightarrow 2\ \mathrm{CO}(g)$$
$$\mathrm{C}(s) + \mathrm{O_2}(g) \longrightarrow \mathrm{CO_2}(g)$$

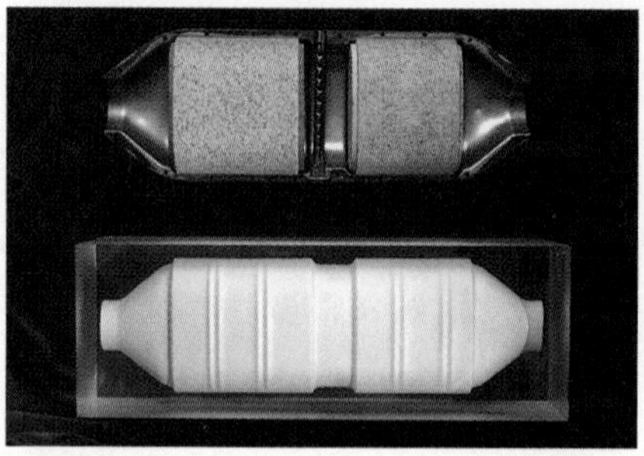

Figure 10.2 An automobile catalytic converter. The converter is an integral part of the exhaust system. It contains two catalysts—one for converting nitric oxide to nitrogen and the other for converting carbon monoxide and hydrocarbons to carbon dioxide. *(© AC/GM/Peter Arnold, Inc.)*

EXAMPLE 10.1 *Oxidation as Oxygen Addition*

Oxygen combines with sulfur to form sulfur dioxide. Write an equation for the reaction.

SOLUTION

The product of oxidation of S is SO_2. So the reaction can be written as

$$S(s) + O_2(g) \longrightarrow SO_2(g)$$

Exercise 10.1A

Calcium, when ignited, combines with oxygen to form calcium oxide, CaO, commonly known as lime. The reaction produces a bright white light. In the early days of indoor theater, the "limelight" used to illuminate actors was produced by this reaction. Write an equation for this oxidation reaction.

Exercise 10.1B

Consider the following pairs of oxides. In each pair, which has the more oxidized second element?

(a) SO_3 or SO_2
(b) FeO or Fe_2O_3
(c) PbO_2 or PbO

■ When a piece of wood combines with oxygen in a fireplace, we call it combustion. The yellow light generated is caused by glowing particles of unburned carbon that may react with additional oxygen unless the particles cool too rapidly. If they do, soot is formed. The same piece of wood may have remained on the forest floor, where bacteria would have aided the oxidation, but at a much slower rate. The oxidation products would have been the same.

■ The catalytic converter also contains a catalyst that can speed the conversion of another unwanted product of automobile exhaust, NO (a reduction, the opposite of oxidation; Section 10.4):

$$2\ NO(g) \xrightarrow{\text{Catalyst}} N_2(g) + O_2(g)$$

When oxygen combines with another element or with a compound, heat is almost always produced. If this energy (as heat) is given off rapidly enough, the oxidation is called **combustion.**

Many oxidation reactions are controlled by the careful addition of fuel or oxygen. These include reactions such as the combustion of fuels in engines, furnaces, fireplaces, and stoves. The modern automobile gasoline engine, for example, uses a computer to regulate the flow of both air (containing O_2, the oxidizing agent) and fuel (Section 21.5). Under perfect conditions, the fuel would burn completely to give CO_2 and H_2O as the exhaust products. In spite of this technology, some CO and unburned fuel does get out of the combustion chambers as part of the exhaust gases. To correct this problem, a catalytic converter (Figure 10.2) is placed in the exhaust system. This device contains the catalyst that allows oxygen to oxidize these unwanted products completely.

$$2\ CO(g) + O_2(g) \xrightarrow{\text{Catalyst}} 2\ CO_2(g)$$

$$2\ C_8H_{18}(g) + 25\ O_2(g) \xrightarrow{\text{Catalyst}} 16\ CO_2(g) + 18\ H_2O(g)$$

Some oxidation reactions, on the other hand, are not easily controlled. These include rusting of things made of iron and steel, forest and house fires, and unwanted explosions. Some of these uncontrolled oxidation reactions may even be life-threatening (Figure 10.3).

Figure 10.3 A brush fire on the horizon at Laguna Niguel, California. *(© Spencer Grant/Photo Researchers)*

EXAMPLE 10.2 *Recognizing Oxidation as Combination with Oxygen*

In which of the following reactions is the element or compound shown in color oxidized?

(a) $H_2(g) + O_2(g) \rightarrow 2\ H_2O(g)$
(b) $2\ HgO(s) \rightarrow 2\ Hg(\ell) + O_2(g)$
(c) $2\ CO(g) + O_2(g) \rightarrow 2\ CO_2(g)$
(d) $2\ Zn(s) + O_2(g) \rightarrow 2\ ZnO(s)$

SOLUTION

(a) The hydrogen atoms form bonds with oxygen atoms; they are oxidized.
(b) The mercury (II) oxide loses oxygen, so it is not oxidized.
(c) The carbon monoxide molecules combine with oxygen; they are oxidized.
(d) Zinc atoms combine with oxygen atoms; they are oxidized.

Exercise 10.2
Which of the following incomplete reactions involve oxidation of the reactant?

(a) $S \rightarrow SO_2$
(b) $2\ H_2O \rightarrow 2\ H_2 + O_2$
(c) $NO \rightarrow NO_2$
(d) $CO_2 \rightarrow CO$

Oxidation as Loss of Hydrogen

A second definition of oxidation is as the loss of hydrogen atoms. Usually this occurs in reactions of organic molecules. The loss of hydrogen is not the cause

■ The acetaldehyde formed by the oxidation of ethanol is further oxidized to CO_2 and H_2O.

of the oxidation but merely one way to *recognize* when oxidation has occurred. For example, ethanol, which has a formula that can be written as CH_3CH_2OH, is oxidized to the compound acetaldehyde, CH_3CHO, in the liver with the aid of enzymes. This is the way ethanol, a depressant, is removed from the blood stream. In this reaction, two hydrogen atoms are removed from each molecule of ethanol:

$$\underset{\text{Ethanol}}{CH_3CH_2OH(aq)} \xrightarrow{\text{In the liver}} \underset{\text{Acetaldehyde}}{CH_3\overset{\overset{\displaystyle O}{\|}}{C}{-}H(aq)}$$

Because the acetaldehyde molecule has fewer hydrogen atoms than the ethanol molecule (hydrogen loss has occurred), it is considered more highly oxidized than ethanol.

EXAMPLE 10.3 *Recognizing Oxidation as Loss of Hydrogen*

The hydrocarbon ethane, C_2H_6, can be converted to ethylene, C_2H_4, by heating the gas over a catalyst of finely divided metal particles:

$$C_2H_6(g) \xrightarrow{\text{Catalyst}} C_2H_4(g) + H_2(g)$$

Is the ethane oxidized?

SOLUTION

Yes. Because the ethane molecule loses hydrogen, it is oxidized.

Exercise 10.3

In the incomplete reactions below, determine which of the reactants are oxidized:

(a) $C_2H_4(g) \rightarrow C_2H_2(g)$
(b) $CO(g) \rightarrow CH_3OH(g)$
(c) $C_2H_4(g) \rightarrow C_2H_6(g)$
(d) $CH_3CH_2OCH_3(g) \rightarrow CH_2{=}CHOCH_3(g)$

Other examples of reactions of this type will be discussed in Chapters 12, 14, and 15.

Oxidation as Loss of Electrons

The third and most general definition of oxidation is as electron loss. An atom or ion is said to be oxidized when it loses electrons. When a neutral atom becomes a positive ion, it has lost electrons and has been oxidized. Sodium is oxidized by bromine to produce sodium bromide, a white, crystalline solid that is an ionic compound composed of equal numbers of Na^+ and Br^- ions (Figure 10.4):

$$2\ Na(s) + Br_2(\ell) \longrightarrow 2\ NaBr(s)$$

The oxidation of sodium atoms in this reaction is the loss of electrons to give sodium ions. An equation for this type of reaction is written using the symbol e^- for the electron:

$$Na \longrightarrow Na^+ + e^-$$

In any reaction in which a metallic element combines with a nonmetallic element to form an ionic compound, the metal has been oxidized because it has lost electrons. To recognize such reactions, it is helpful to remember that compounds containing active metals from the left of the periodic table (e.g., Na, K, Mg, Ca) combined with halogens (F, Cl, Br, I) are ionic compounds (Section 6.1).

In addition to oxygen and bromine, the elements chlorine (Cl_2) and fluorine (F_2) combine with elements and compounds in ways that can be called oxidation. Fluorine is rather exotic in its applications, but Cl_2 is commonly used in oxidation applications such as disinfecting water supplies and in bleaches and cleaning compounds. When chlorine disinfects a sample of water, the Cl_2 molecules oxidize pathogenic organisms by pulling away electrons from molecules within their cellular structures. When this happens, the organism dies. Chlorine bleaches remove stains in clothing by literally destroying through oxidation those organic molecules responsible for the stain.

In summary, oxidation can be defined in three different ways; each definition finds use for a particular type of reaction.

Definition	Used for
Oxidation is the gain of oxygen.	Reactions involving oxygen
Oxidation is the loss of hydrogen atoms.	Reactions involving hydrogen
Oxidation is the loss of electrons.	A variety of reaction types

Any chemical that causes the gain of oxygen, the loss of hydrogen, or the loss of electrons in another reactant is an **oxidizing agent.** Consider the following reaction of carbon with water. Before the reaction, the carbon is not combined with oxygen, but the product of the reaction is oxidized carbon:

$$C(s) + H_2O(g) \xrightarrow{\text{Heat}} H_2(g) + CO(g)$$

This means that water is an oxidizing agent in this reaction. Note that the water molecule contains oxygen. All oxidizing agents that cause a gain of oxygen will contain oxygen.

Identifying oxidizing agents that cause the loss of hydrogen is a bit more challenging. First, the loss of hydrogen by one of the reactants must be recog-

Figure 10.4 Oxidation of sodium by bromine. The reaction between metallic sodium and bromine releases a large amount of energy. The product is sodium bromide, NaBr, a white crystalline ionic compound.

nized, then the reactant causing that loss of hydrogen is identified as the oxidizing agent. In the following reaction, hydrogen sulfide (H_2S) loses hydrogen. The other reactant, fluorine, causes this to take place, so fluorine is the oxidizing agent.

$$H_2S(g) + 3\ F_2(g) \longrightarrow 2\ HF(g) + SF_4(g)$$

Identifying oxidizing agents that cause the loss of electrons is like identifying those that cause loss of hydrogen; first you must recognize that loss of electrons has occurred. If a chemical reaction is written in such a way as to show the charges on ions, this is easy. For example, the oxidation of calcium by bromine can be written as

$$Ca(s) + Br_2(\ell) \longrightarrow Ca^{2+} + 2\ Br^-$$

The calcium ions and bromide ions form $CaBr_2$, a colorless salt. It is obvious that calcium atoms have lost electrons because of the charge on the calcium ions (Ca^{2+}). That oxidation reaction is

$$Ca \longrightarrow Ca^{2+} + 2\ e^-$$

Looking at the equation, you can also see that the bromine atoms in the bromine molecule have gained electrons to form bromide (Br^-) ions.

$$Br_2 + 2\ e^- \longrightarrow 2\ Br^-$$

Since the bromine caused the oxidation of the calcium, bromine is the oxidizing agent in this reaction.

When the same reaction is written without the charges on the ions,

$$Ca(s) + Br_2(\ell) \longrightarrow CaBr_2(s)$$

you might not realize electrons are lost by the calcium and gained by the bromine unless you recall what you learned about the formation of ions and ionic bonds (see Section 6.1).

EXAMPLE 10.4 *Identifying Oxidizing Agents in Chemical Reactions*

Identify the oxidizing agent in the reaction below.

$$2\ S(s) + 3\ O_2(g) \longrightarrow 2\ SO_3(g)$$

SOLUTION

The element sulfur (S) has no oxygen associated with it in its elemental state [written as $S(s)$]. On the products side of the reaction, sulfur is combined with oxygen in SO_3. The element oxygen has caused this, so oxygen is the oxidizing agent.

Exercise 10.4

Identify the oxidizing agent in each of the following chemical equations. Which definition of oxidation applies to each reaction?

(a) $2\ Sn(s) + O_2(g) \rightarrow 2\ SnO(s)$
(b) $2\ H_2(g) + O_2(g) \rightarrow 2\ H_2O(g)$
(c) $NH_3(g) + 3\ F_2(g) \rightarrow 3\ HF(g) + NF_3(g)$
(d) $2\ Na(s) + F_2(g) \rightarrow 2\ Na^+ + 2\ F^-$ [2 NaF(s)]
(e) $2\ K(s) + I_2(s) \rightarrow 2\ KI(s)$

■ SELF-TEST 10A

1. When oxygen combines with an element to form a compound, that compound is called a(n) ____________.
2. When iron oxidizes in the presence of moisture, the product of the reaction is called ____________.
3. Which is more oxidized, (a) CO or (b) CO_2?
4. Which is more oxidized, (a) NO_2 or (b) NO?
5. The products of the complete combustion of methane (CH_4) are ____________ and ____________.
6. The conversion of C_2H_6 to C_2H_4 is an oxidation reaction. (a) True; (b) False.
7. The conversion of Na to Na^+ is an oxidation reaction. (a) True; (b) False.
8. Which is more highly oxidized, (a) ethanol (C_2H_5OH) or (b) acetaldehyde (CH_3CHO)?
9. Which is the oxidized form of sodium, (a) Na^+ ion or (b) Na atom?
10. In the reaction $2\ Li(s) + S(s) \rightarrow Li_2S(s)$, sulfur is the oxidizing agent. (a) True; (b) False.

The reaction of zinc with hydrochloric acid. *(C.D. Winters)*

10.3 Hydrogen—The Element

Unlike oxygen, which is so common on Earth, hydrogen makes up only about 0.9% of the Earth's crust. Hydrogen is, however, very common throughout the universe as a whole—undoubtedly because it is the simplest atom (see Figure 11.1). The huge planet Jupiter has a 1000-km-deep atmosphere that is about 87% hydrogen. Most of the hydrogen on Earth is combined with oxygen in the form of water. Hydrocarbons and the fats, carbohydrates, and proteins present in all living things contain considerable amounts of hydrogen.

■ Below the gaseous hydrogen atmosphere of Jupiter is a sea of liquid hydrogen, and below that is a core consisting mostly of solid hydrogen in a *metallic* state.

Hydrogen, a colorless, odorless, flammable gas, can be made by reacting an acid—a source of hydrogen ions—with most metals. Table 10.3 lists some properties of hydrogen. Hydrochloric acid readily reacts with zinc to form hydrogen gas. This reaction was used in many of the early chemistry laboratories to make hydrogen.

$$Zn(s) + 2\ HCl(aq) \longrightarrow H_2(g) + ZnCl_2(aq)$$

■ Note that zinc is oxidized in this reaction.

THE PERSONAL SIDE

Richard Feynman and the Challenger *Explosion*

Richard Feynman demonstrating experimentally that a cold O-ring of the type used in the Space Shuttle Challenger does not retain its shape and thus would allow gases to escape from the rocket. (Marilyn K. Yee/NYT Pictures)

Richard Feynman (1919–1988), winner of the 1965 Nobel Prize in physics, was one of the most original thinkers of this century. He wrote, "Scientific knowledge is a body of statements of varying degrees of certainty—some most unsure, some nearly sure, but *none* absolutely certain. Now, we scientists are used to this, that it is possible to live and *not* know."

Feynman gave the United States a lesson in how science works when he used a simple experiment to uncover the reason for the disastrous explosion of the space shuttle *Challenger*. On launch day, January 28, 1986, the weather was unusually cold in Florida—the temperature was 29°F. A few moments after launch, the world watched in horror as the shuttle and its rockets exploded in a gigantic fireball, killing all the astronauts aboard.

A commission was appointed to investigate the cause of the explosion. It was Feynman who reasoned that due to the cold temperature, rubber O-rings used to seal joints in the solid-fuel booster rockets had not expanded properly. This failure allowed hot flames from the booster rocket to burn through the hydrogen fuel tank, causing the explosion. To prove his point, Feynman performed a dramatic—but very simple—experiment. During a public hearing Feynman held a sample rubber O-ring like the one in the rocket engine tightly in a clamp and immersed it in a glass of ice water. Everyone watching his experiment could see that the rubber did not spring back to its original shape. The poor low-temperature characteristics of the rubber O-ring had doomed the *Challenger* and its crew.

Commercially, hydrogen is made by the decomposition of water using electricity (see Section 10.8) and by the catalytic decomposition of gaseous hydrocarbons. For example, ethane (C_2H_6) can be dehydrogenated using a catalyst to produce hydrogen and ethylene (C_2H_4), an industrially more useful chemical than ethane (see Section 12.3). Because the reaction between hydrogen and oxygen to produce water is explosive, any reaction that produces hydrogen gas should be considered dangerous.

TABLE 10.3 ■ Some Properties of Hydrogen

Formula	H_2
Molar Mass	2.016 g
Melting Point	−259.23°C
Boiling Point	−252.77°C
Description	Colorless, odorless gas
Density at 25°C	0.0899 g/L

Hydrogen forms hydrides with many of the elements (Table 10.4). Several of these, such as water, the hydride of oxygen, and ammonia, the hydride of nitrogen, are of great importance. The hydrocarbons, which can be considered hydrides of carbon, since they are hydrogen compounds of carbon, are also very important.

Hydrogen has the lowest density of any gas. This property made it useful as a gas for lighter-than-air ships. The German zeppelins in the 1930s were filled

TABLE 10.4 ■ Some Hydrides of the Elements

Element	Compound	Formula
Oxygen	Water	H_2O
	Hydrogen peroxide	H_2O_2
Nitrogen	Ammonia	NH_3
	Hydrazine	N_2H_4
Carbon*	Methane	CH_4
	Acetylene	C_2H_2
	Benzene	C_6H_6
Chlorine	Hydrogen chloride	HCl
Lithium	Lithium hydride	LiH

*See Chapter 12 for other examples of hydrocarbons.

with hydrogen. The flammability of hydrogen made these craft extremely susceptible to disasters, and in 1937 the *Hindenberg* caught fire while landing in New Jersey, resulting in a spectacular fiery crash. In spite of this history, hydrogen was the fuel chosen by NASA for the space shuttle flights. In the shuttle launch rocket, liquid hydrogen reacts with liquid oxygen to produce the enormous energy needed to lift the million-pound vehicle into orbit. In 1986, the shuttle *Challenger* was destroyed when a gasket leaked rocket exhaust gases and ignited the hydrogen fuel during liftoff.

10.4 Reduction—The Opposite of Oxidation

Reduction always accompanies oxidation. The reaction between hydrogen and oxygen to form water is an excellent example of an **oxidation–reduction reaction.** When the mixture of the two gases is sparked, the hydrogen molecules combine with oxygen and are oxidized, but the oxygen combines with hydrogen (gains hydrogen) and is reduced. Thus, in this reaction oxidation and reduction occur together, as they always do. For one reactant to be oxidized, another must be reduced.

$$2\,H_2(g) + O_2(g) \xrightarrow{\text{Spark}} 2\,H_2O(\ell)$$

Reduction is the opposite of oxidation. For each of the three definitions of oxidation, there is a corresponding definition of reduction: (1) Reduction is loss of oxygen by a reactant; (2) reduction is gain of hydrogen by a reactant; (3) reduction is gain of valence electrons by an atom, molecule, or ion.

Reduction as Loss of Oxygen

Hydrogen reacts with copper oxide at high temperatures to produce metallic copper. The copper has lost its oxygen. It has been reduced by hydrogen.

$$CuO(s) + 2\,H_2(g) \longrightarrow Cu(s) + 2\,H_2O(g)$$

Notice how the hydrogen gets oxidized. Hydrogen can reduce other metal oxides as well. Anything that causes the loss of oxygen or any other type of reduction is a **reducing agent.** Hydrogen is an excellent reducing agent.

The making of iron metal from its ores, which are oxides of iron, is one of the most important industrial reactions (Section 11.4). This reaction, which takes place in a blast furnace at temperatures as high as 1500°C, involves the reduction of hematite ($Fe_2O_3 \cdot xH_2O$, same formula as rust) by coke, a form of carbon.

$$Fe_2O_3 \cdot xH_2O + 3\ C(s) \xrightarrow{\text{High temp}} 2\ Fe(\ell) + 3\ CO(g) + xH_2O(g)$$

Looking at the equation for iron oxide reduction in the blast furnace, you can see that the iron loses oxygen (gets reduced). The reducing agent is carbon, which has no oxygen associated with it at the start of the reaction. At the end of the reaction, carbon is combined with oxygen as carbon monoxide. The carbon gets oxidized.

Reduction as Gain of Hydrogen

Hydrogen can react directly with some elements and compounds. It reacts with nitrogen to form ammonia, NH_3, an important agricultural fertilizer. (See Section 11.2 for a discussion of the Haber process for making ammonia.)

$$N_2(g) + 3\ H_2(g) \longrightarrow 2\ NH_3(g)$$

Hydrogen reacts with carbon monoxide to reduce it and form methanol. This is a very useful reaction because methanol is a high-energy liquid fuel and can be used in automobiles, buses, and trucks (Section 12.9):

$$CO(g) + 2\ H_2(g) \xrightarrow{\text{Catalyst}} \underset{\text{Methanol}}{CH_3OH\ (\ell)}$$

EXAMPLE 10.5 *Recognizing Reduction*

Determine which reactant in the following reactions is reduced, and identify the reducing agent.

(a) $H_2(g) + F_2(g) \rightarrow 2\ HF(g)$
(b) $PbO(s) + H_2(g) \rightarrow Pb(s) + H_2O(g)$
(c) $H{-}C{\equiv}N(g) + 2\ H_2(g) \rightarrow CH_3{-}NH_2(g)$
(d) $Na^+(aq) + Cl^-(aq) + Ag^+(aq) + NO_3^-(aq) \rightarrow AgCl(s) + Na^+(aq) + NO_3^-(aq)$

SOLUTION

(a) This reaction involves the gain of hydrogen by fluorine. Fluorine is reduced, and hydrogen is the reducing agent.
(b) Oxygen is lost by the PbO, so PbO is reduced. Hydrogen is the reducing agent.

(c) Hydrogen is added to the HCN molecule, so HCN is reduced. Hydrogen is the reducing agent.
(d) Nothing is reduced (or oxidized) in this reaction. Silver chloride is an insoluble ionic compound.

Exercise 10.5

Indicate whether the reactant is being reduced in these incomplete reactions. Suggest a reducing agent for those reactions in which the reactant is reduced.

(a) $Cu(s) \rightarrow CuO(s)$
(b) $CH_3C{\equiv}N(g) \rightarrow CH_3CH_2{-}NH_2(g)$
(c) $H_2O(g) \rightarrow H_2(g) + O_2(g)$
(d) $SnO(s) \rightarrow Sn(s) + H_2O(g)$

Reduction as Gain of Electrons

Reduction can also be defined as the gain of electrons—the opposite of oxidation, which is the loss of electrons. Consider again the reaction between sodium and bromine. A bromine molecule reacts with two sodium atoms to form NaBr:

$$2\,Na(s) + Br_2(\ell) \longrightarrow 2\,NaBr(s)$$

The ions formed show that the bromine atoms have each gained one electron while each sodium atom has lost one electron. In terms of reduction and oxidation, the bromine atoms have been reduced—they gained electrons. The sodium atoms have been oxidized—they lost electrons. The oxidation and reduction reactions can be written separately.

■ Loss of electrons is oxidation. Gain of electrons is reduction.

■ Oxidation cannot occur without reduction. Electrons lost in oxidation are gained in a matching reduction.

$$2\,Na \longrightarrow 2\,Na^+ + 2\,e^- \quad \text{Oxidation}$$
$$Br_2 + 2\,e^- \longrightarrow 2\,Br^- \quad \text{Reduction}$$
$$2\,Na(s) + Br_2(\ell) \longrightarrow 2\,NaBr(s) \quad \text{The oxidation–reduction reaction}$$

In summary, reduction can be defined in three ways that match the three definitions of oxidation (Section 10.2).

Definition	Used for
Reduction is the loss of oxygen.	Reactions involving oxygen
Reduction is the gain of hydrogen atoms.	Reactions involving hydrogen
Reduction is the gain of electrons.	Variety of reaction types

Many different compounds and elements can promote oxidation or reduction. It is important to remember that oxidation is always accompanied by reduction (and reduction is accompanied by oxidation). That is, if something is oxidized in a chemical reaction, something else gets reduced in that chemical reaction. (This is not unlike what was learned about acid–base reactions in Chapter 9—if something called an acid loses a hydrogen ion, something else,

TABLE 10.5 ■ Some Oxidizing and Reducing Agents

Name	Formula	Uses
	Oxidizing Agents	
Lead dioxide	PbO_2	Automobile batteries
Manganese dioxide	MnO_2	Batteries
Potassium permanganate	$KMnO_4$	Chemical synthesis
Potassium peroxydisulfate	$K_2S_2O_8$	Denture cleaners
Sodium hypochlorite solution	$NaOCl(aq)$	Bleaching cloth, disinfection
Dichlorine oxide	Cl_2O	Water purification, bleaching paper
Oxygen	O_2	Metabolism of foods, burning fuels
Ozone	O_3	Water purification, hazardous chemical destruction
Chlorine	Cl_2	Drinking water and waste-water purification, chemical synthesis
	Reducing Agents	
Hydrogen	H_2	Fuel, chemical synthesis
Sodium borohydride	$NaBH_4$	Chemical synthesis
Sulfur dioxide	SO_2	Chemical synthesis
Carbon	C	Iron production
Hydrocarbons	CH_4 (typical)	Fuels for transportation, heating
Zinc	Zn	Batteries

called a base, must gain that hydrogen ion.) Many molecules and polyatomic ions that contain a large number of oxygen atoms are oxidizing agents; they promote, or cause, oxidation to take place (Table 10.5). Oxidizing agents cannot react unless there is something that can be oxidized. The chemical an oxidizing agent reacts with (and oxidizes) is a reducing agent. So an oxidizing agent requires a reducing agent to react. One without the other means no reaction will take place. As you look at Table 10.5, notice how many common and useful applications of oxidation–reduction reactions there are. This is only a partial listing of many thousands of oxidizing and reducing agents.

Chlorine-containing tablets being added to a swimming pool filter.
(© Yoav Levy/Phototake NYC)

EXAMPLE 10.6 *Identifying Oxidation and Reduction*

Determine what is oxidized and what is reduced in each of the following chemical equations:

(a) $2\ FeS(s) + 3\ O_2(g) \rightarrow 2\ FeO(s) + 2\ SO_2(g)$
(b) $Mg(s) + 4\ HNO_3(aq) \rightarrow Mg(NO_3)_2(aq) + 2\ NO_2(g) + 2\ H_2O(\ell)$
(c) $2\ KBr(aq) + Cl_2(g) \rightarrow 2\ KCl(aq) + Br_2(\ell)$

SOLUTION

(a) S is oxidized because it is combined with oxygen in SO_2. O is reduced because it has added electrons to form O^{2-} in FeO. Fe is not changed; it is +2 in both FeS and FeO.
(b) Mg is oxidized because it loses electrons in going from Mg to Mg^{2+}. N is reduced because it has less O in NO_2 than in NO_3^-.
(c) Br^- is oxidized because 2 Br^- each lose an electron to form $Br_2(\ell)$. Cl_2 is reduced because it gains two electrons to form 2 Cl^-.

Exercise 10.6

Determine what is oxidized and what is reduced in each of the following chemical equations:

(a) $2\,K(s) + 2\,H_2O(\ell) \rightarrow 2\,KOH(aq) + H_2(aq)$
(b) $SnO_2(g) + 2\,C(s) \rightarrow Sn(s) + 2\,CO(g)$
(c) $2\,NaCl(aq) + F_2(g) \rightarrow 2\,NaF(aq) + Cl_2(g)$

SELF-TEST 10B

1. Conversion of carbon monoxide to methanol by the addition of hydrogen is an example of carbon monoxide being (a) oxidized, (b) reduced.
2. When coke (a form of carbon) reacts with iron ore in a blast furnace, the coke is the (a) oxidation product, (b) reducing agent, (c) oxidizing agent.
3. When nitrogen reacts with hydrogen to form ammonia (NH_3), the nitrogen is said to be (a) oxidized, (b) reduced.
4. Which of these two compounds is more reduced than the other: (a) H—C≡N or (b) CH_3—NH_2?
5. An oxidized form of sodium is in the compound NaCl. The reduced form of sodium is ____________.

DISCOVERY EXPERIMENT

Getting Rid of Tarnish

Silverware tarnishes when exposed to air because the silver reacts with the hydrogen sulfide gas in the air to form a thin coating of black silver sulfide, Ag_2S. You can chemically remove the tarnish from silverware and other silver utensils by using a solution of baking soda and some aluminum foil. The chemical cleaning of silver is an electrochemical process in which electrons move from aluminum atoms to silver ions in the tarnish, reducing the silver ions to silver atoms while aluminum atoms are oxidized to aluminum ions. The sodium bicarbonate provides a conductive ionic solution for the flow of electrons and also helps to remove the aluminum oxide coating from the surface of the aluminum foil.

Put 1 to 2 liters of water in a large pan. Add 7 to 8 tablespoons of baking soda. Heat the solution, but do not boil it. Place some aluminum foil in the bottom of the pan, and put the tarnished silverware on the aluminum foil. Make sure the silverware is covered with water. Heat the water almost to boiling. After a few minutes remove the silverware and rinse it in running water.

This method of cleaning silverware is better than using polish, because polish removes the silver sulfide, including the silver it contains; instead, the process described here restores the silver from the tarnish to the surface. If you have aluminum pie pans or aluminum cooking pans, you can use them as both the container and the aluminum source.

6. Magnesium in sea water is in the oxidized form, the Mg^{2+} ion. The reduced form of magnesium is ______________.
7. Which is the reduced form of copper, (a) Cu or (b) Cu^{2+}?
8. A chemical that causes oxidation to occur is a(n) ______________ ______________.
9. A chemical that causes reduction to take place is called a(n) ______________ ______________.

10.5 The Strengths of Oxidizing and Reducing Agents

Not all reducing agents or oxidizing agents have the same ability to cause reduction or oxidation. One reducing agent might be able to reduce a given element or compound, whereas another reducing agent might not be able to do the same. For example, hydrogen can reduce copper oxide to copper metal, but it cannot reduce sodium oxide to sodium metal.

■ The symbol $\not\longrightarrow$ is often used to indicate that a reaction does not occur.

$$CuO(s) + H_2(g) \longrightarrow Cu(s) + H_2O(g)$$
$$NaO(s) + H_2(g) \not\longrightarrow$$

Oxidizing agents display similar differences in ability to bring about oxidation. For example, oxygen can oxidize calcium metal to calcium oxide, but it cannot oxidize silver to its oxide, Ag_2O.

$$2\ Ca(s) + O_2(g) \longrightarrow 2\ CaO(s)$$
$$Ag(s) + O_2(g) \not\longrightarrow$$

We can observe the differences in strengths of reducing agents. A reaction occurs when a piece of metallic zinc is placed in a blue-colored solution containing copper ions (Cu^{2+}). As the reaction proceeds, you can quickly tell that an oxidation–reduction reaction occurs because metallic copper can be seen forming on the surface of the piece of zinc (Figure 10.5). The gradual decrease in the intensity of the blue color of the solution also indicates removal of the Cu^{2+} ions. The gain of electrons by Cu^{2+} ions is reduction.

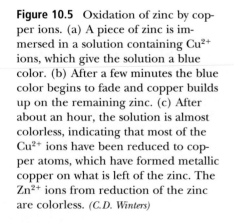

(a) (b) (c)

Figure 10.5 Oxidation of zinc by copper ions. (a) A piece of zinc is immersed in a solution containing Cu^{2+} ions, which give the solution a blue color. (b) After a few minutes the blue color begins to fade and copper builds up on the remaining zinc. (c) After about an hour, the solution is almost colorless, indicating that most of the Cu^{2+} ions have been reduced to copper atoms, which have formed metallic copper on what is left of the zinc. The Zn^{2+} ions from reduction of the zinc are colorless. *(C.D. Winters)*

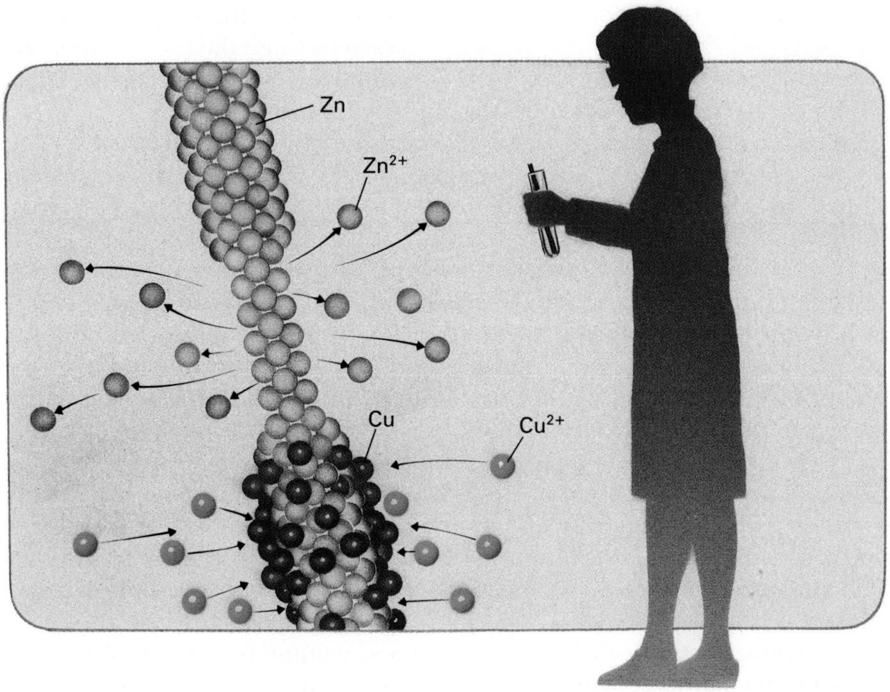

$$Cu^{2+}(aq) + 2\,e^- \longrightarrow Cu(s) \qquad \text{Reduction}$$

Since reduction is always accompanied by oxidation, something must also be oxidized. Observation of the reaction tells us the zinc strip is gradually being consumed in the reaction. The reaction is the oxidation of zinc to form Zn^{2+} ions.

$$Zn(s) \longrightarrow Zn^{2+}(aq) + 2\,e^- \qquad \text{Oxidation}$$

In the terminology discussed earlier, the Cu^{2+} ions are the oxidizing agent, and zinc is the reducing agent.

The oxidation of zinc by copper ions can be thought of as a competition between zinc ions (Zn^{2+}) and copper ions (Cu^{2+}) for the available electrons. If Cu^{2+} ions win the competition and take electrons from zinc atoms, the reaction is written as

$$Zn(s) + Cu^{2+}(aq) \longrightarrow Zn^{2+}(aq) + Cu(s)$$

If Zn^{2+} ions, on the other hand, win the competition for the electrons and take them from copper atoms, the reaction would be the *reverse* of the preceding one, or

$$Cu(s) + Zn^{2+}(aq) \not\longrightarrow Cu^{2+}(aq) + Zn(s)$$

This reaction is unfavorable and *does not* occur. Since the reaction between Cu^{2+} ions and zinc is favorable (that is, it proceeds when the two reactants are mixed), the Cu^{2+} ions must win in the competition for the electrons. Other combinations of metals will compete similarly for electrons.

The tendency of a metal to lose electrons is referred to as its **activity.** Zinc is a more active metal than copper on the basis of the experiment just described. In other words, zinc is a better reducing agent than copper—it loses electrons more easily.

■ Oxidizing agents gain electrons and get reduced, and reducing agents lose electrons and get oxidized.

Experiments of the kind shown in Figure 10.5 with various pairs of metals and other reducing agents yield an **activity series** of the elements, which ranks each oxidizing and reducing agent according to its strength or tendency to gain or lose electrons. Oxidizing agents are shown in the left column of the table, and reducing agents are shown in the right column. The strongest oxidizing agents, those that are most able to take electrons from other chemicals, occupy positions near the top of the left-hand column. The very weakest oxidizing agents are at the bottom of this column. Each entry, or line, in the table is written as the reduction of an oxidizing agent. The product of the gain of electrons is shown in the column to the right. This product is by definition a reducing agent, because the reaction can, under some conditions, go in reverse.

An oxidizing agent shown in the left-hand column of Table 10.6 will take electrons from any reducing agent in the right-hand column below it in the table. For example, look at the first oxidizing agent in Table 10.6, fluorine. Fluorine is able to oxidize any reducing agent in the table. The third reducing agent in Table 10.6 is water. Fluorine can oxidize it. The reaction that occurs is shown by adding the equation for the reduction of fluorine to the equation for the oxidation of water. To do this, the equation in the table with water as the product must be reversed in direction. Since the number of electrons gained

TABLE 10.6 ■ Relative Strengths of Some Oxidizing and Reducing Agents: An Activity Series

Oxidizing Agents		Reducing Agents
$F_2(g) + 2\ e^-$	→	$2\ F^-(aq)$
$Cl_2(g) + 2\ e^-$	→	$2\ Cl^-(aq)$
$O_2(g) + 4\ e^- + 4\ H^+(aq)$	→	$2\ H_2O(\ell)$
$Br_2(\ell) + 2\ e^-$	→	$2\ Br^-(aq)$
$Ag^+(aq) + e^-$	→	$Ag(s)$
$Fe^{3+}(aq) + e^-$	→	$Fe^{2+}(aq)$
$Cu^{2+}(aq) + 2\ e^-$	→	$Cu(s)$
$2\ H^+(aq) + 2\ e^-$	→	$H_2(g)$
$Cd^{2+}(aq) + 2\ e^-$	→	$Cd(s)$
$Fe^{2+}(aq) + 2\ e^-$	→	$Fe(s)$
$Zn^{2+}(aq) + 2\ e^-$	→	$Zn(s)$
$Mg^{2+}(aq) + 2\ e^-$	→	$Mg(s)$
$Na^+(aq) + e^-$	→	$Na(s)$
$Ca^{2+}(aq) + 2\ e^-$	→	$Ca(s)$
$Li^+(aq) + e^-$	→	$Li(s)$

increasing strength of oxidizing agent (↑, left column) — *increasing strength of reducing agent* (↓, right column)

(by fluorine, the oxidizing agent) must equal the number of electrons lost (by water, the reducing agent), the fluorine reaction must be multiplied by 2 to give four electrons gained to match the four electrons lost. Note how in the overall reaction the electrons cancel each other.

$$2\,F_2(g) + \cancel{4e^-} \longrightarrow 4\,F^-(aq) \qquad \text{Reduction}$$
$$2\,H_2O(\ell) \longrightarrow O_2(g) + 4\,H^+(aq) + \cancel{4e^-} \qquad \text{Oxidation}$$
$$2\,F_2(g) + 2\,H_2O(\ell) \longrightarrow O_2(g) + 4\,H^+(aq) + 4\,F^-(aq) \qquad \text{Overall reaction}$$

This reaction could be used to make oxygen from water, but fluorine is far too expensive for this to be a practical source for oxygen. Besides, fluorine is very hazardous to work with because it is such a strong oxidizing agent.

Using the ideas just presented, it should be clear that fluorine could oxidize Ag to Ag^+, Zn to Zn^{2+}, and so on. Can Cu^{2+} oxidize zinc? (You already know the answer to this—see Figure 10.7.) Find Cu^{2+} in the left column of Table 10.6. Now find zinc in the right column of the table. Since zinc is below Cu^{2+}, this means Cu^{2+} will oxidize zinc. This confirmation should help you navigate in this activity table.

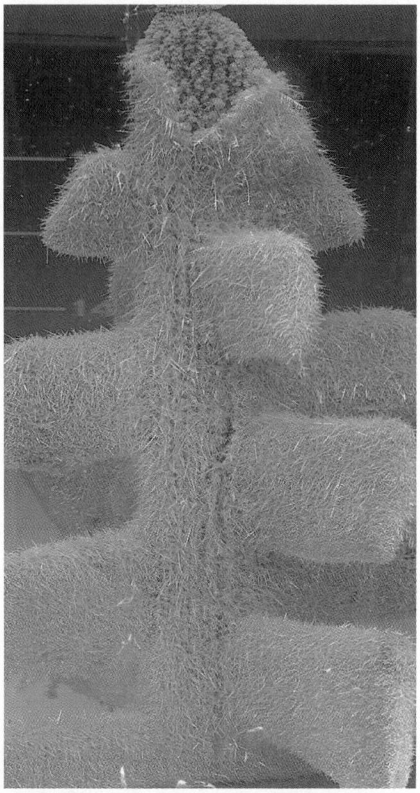

Figure 10.6 The result of placing a copper tree in a solution of silver nitrate ($AgNO_3$), an ionic compound that forms Ag^+ and NO_3^- ions in solution. *(C.D. Winters)*

EXAMPLE 10.7 *Using the Activity Series*

Can Cu^{2+} oxidize an iron nail? What would happen to an iron nail placed in a solution containing Cu^{2+} ions (and no other reactants)? What would happen to a piece of copper metal placed in a solution containing Fe^{3+} (and no other reactants)?

SOLUTION

Look at Table 10.6 and first find Cu^{2+} in the column of oxidizing agents. An iron nail contains Fe atoms, so next look for Fe in the column of reducing agents. Since Fe is lower in the table than Cu^{2+}, Cu^{2+} will be able to oxidize iron. An iron nail will be partly dissolved in a solution of a copper salt containing Cu^{2+} ions, with copper being deposited on the nail that remains. If a piece of copper is immersed in a solution containing Fe^{3+} ions, no reaction would occur.

Exercise 10.7A
Describe two experiments to determine which metal is more active, zinc or iron.

■ The active metals lose electrons more easily; hence, these elements are not found as free metals in nature.

Exercise 10.7B
Look at Figure 10.6 and explain what is happening in terms of oxidation–reduction.

EXAMPLE 10.8 *Strengths of Reducing Agents*

Magnesium metal can reduce Cu^{2+} to Cu. Explain why and write an equation for the reaction.

SOLUTION

The entry in Table 10.6 containing Cu and Cu^{2+} is above the entry containing Mg and its oxidized form, Mg^{2+}. A reducing agent in the table can reduce any oxidizing agent above it in the table. (This is like saying an oxidizing agent can oxidize a reducing agent below it in the table.) Therefore, Mg can reduce Cu^{2+}. The reaction is:

$$Mg(s) + Cu^{2+}(aq) \longrightarrow Mg^{2+}(aq) + Cu(s)$$

Exercise 10.8A

Can magnesium reduce Ag^+ to Ag? Show your reasoning, and write the equation for the reaction if one occurs.

Exercise 10.8B

Can iron reduce magnesium ions (Mg^{2+})? Show your reasoning, and write the equation for the reaction if there is one.

10.6 Corrosion—Unwanted Oxidation–Reduction

Figure 10.7 Corrosion has made this bridge unsafe. *(Tom Pantages)*

In the United States alone, more than $10 billion is lost each year to **corrosion,** a term used for unwanted oxidation of metals. Much of this corrosion is the rusting of iron and steel, although other metals may oxidize as well. The oxidizing agent causing all of this unwanted corrosion is usually oxygen found in the atmosphere. The reason iron is so severely affected by corrosion is that its oxide, rust, does not adhere strongly to the metal's surface once the rust is formed. Because the rust flakes off or is rubbed off easily, the metal surface becomes pitted. The continuing loss of surface iron by rust formation eventually causes structural weakness. Bridges with severely corroded steel reinforcement beams have collapsed under heavy loads (Figure 10.7).

The driving forces behind metal corrosion are the activity of the metal as a reducing agent and the strength of the oxidizing agent. Whenever a strong reducing agent (the metal) and a strong oxidizing agent (such as oxygen) are together, a reaction between the two substances is likely. Factors governing the rates of chemical reaction, such as temperature and concentration (Section 8.3), affect the rate of corrosion as well. Consider the corrosion of an iron spike (Figure 10.8). There are tiny microcrystals composed of loosely bound iron atoms on the surface of the metal that can easily be oxidized.

$$Fe(s) \longrightarrow Fe^{2+}(aq) + 2\ e^-$$

Because iron is a good conductor of electricity, the electrons produced by this oxidation can migrate through the metal to a point where they can reduce something. The fact that iron, like other metals, is a conductor of electricity is important because the corrosion process would come to an abrupt halt as a result of a buildup of excessive negative charge if the electrons were not conducted away. One location on the surface of the iron where electrons can be used is any tiny drop of water containing dissolved oxygen. Here the oxygen gains electrons, forming hydroxide ions. Note oxygen's position relative to iron in Table 10.6.

$$O_2(g) + 2\,H_2O(\ell) + 4\,e^- \longrightarrow 4\,OH^-(aq)$$

The Fe^{2+} ions are further oxidized to Fe^{3+} ions, which react with OH^- to form the hydrated iron oxide known as rust.

$$2\,Fe^{3+}(aq) + 6\,OH^-(aq) \longrightarrow Fe_2O_3\cdot 3H_2O(s)$$

The rate of rusting is enhanced by salts, which dissolve in the water on the surface of the iron and act like tiny salt bridges of an electrochemical cell (discussed in the next section). The hydroxide ions and iron ions migrate more easily in the ionic solutions produced by the presence of the dissolved salts. Automobiles rust more quickly when exposed to road salts in wintry climates. If road salts are used in your driving area, it's a good idea after snowy seasons to wash the undersides of automobiles to remove the accumulated salts.

For rust to form, three reactants are necessary. These are iron, oxygen, and water. Rusting can be prevented by protective coatings such as paint, grease, oil, enamel, or a corrosion-resistant metal such as chromium. Most of these coatings keep out moisture. Some of the metals that are more active than iron form adherent oxide coatings when they corrode. Coating iron with these metals provides corrosion protection. One of these active metals is zinc. Zinc coating of iron and steel is called **galvanizing** and can be done by dipping the iron object into a molten bath of zinc metal. This *hot-dip* galvanizing is a popular corrosion-preventing measure for automobile bodies. When the zinc coating of a galvanized object is exposed to air and water, a thin film of zinc oxide forms that protects the zinc from further oxidation. If the zinc coating should get scratched so that iron is exposed to the air, zinc will quickly reduce any Fe^{2+} ions formed because zinc is more active than iron in giving up electrons.

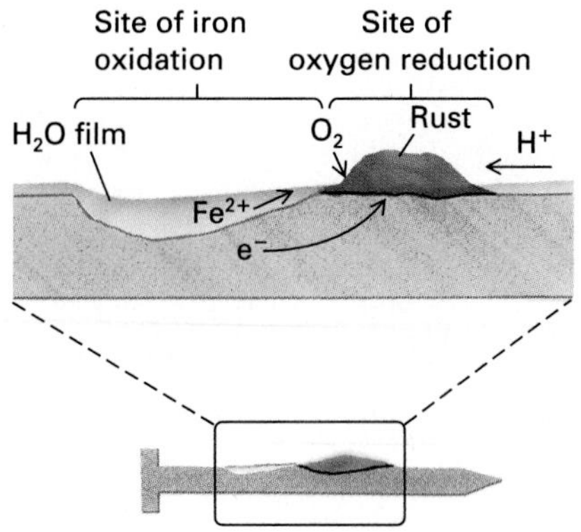

Figure 10.8 Corrosion of iron. The site of iron oxidation may be different than the site of oxygen reduction because iron is a conductor of electricity and electrons can move through it from one site to another.

■ Iron, oxygen, **and** water are required for rust formation.

■ SELF-TEST 10C

1. If zinc loses electrons to Cu^{2+}, then zinc is said to be more ____________ than copper.
2. If zinc loses electrons to Cu^{2+}, then copper will lose electrons to Zn^{2+}. (a) True, (b) False.
3. Look at Table 10.6 (activity series). Will Ca reduce Fe^{2+} ions to Fe?
4. Will Ag^+ oxidize Ca?
5. Will Ag^+ oxidize F^- ion?
6. What three reactants are necessary for rusting?
7. Zinc coating on iron is called ____________.
8. According to the activity series, which is the stronger oxidizing agent, fluorine or oxygen?
9. Which is the stronger reducing agent, iron or lithium?

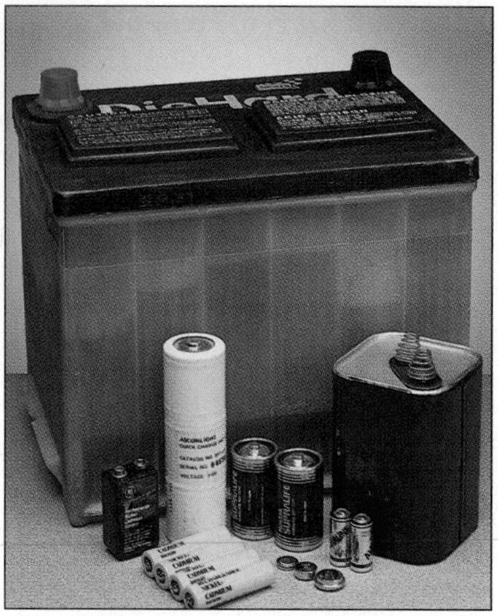

An assortment of batteries. In front of the lead-acid automobile battery are, from left to right, three types of rechargeable nickel-cadmium batteries, three types of non-rechargeable alkaline batteries, and a zinc-graphite dry cell. *(© 1990 Richard Megna/Fundamental Photographs)*

10.7 Batteries

One of the most useful applications of oxidation–reduction reactions is the *production* of electrical energy. A device that produces an electron flow (current) from a chemical reaction is called an **electrochemical cell.** Although a series of such cells is a **battery,** the term "battery" is commonly used even for single cells. Batteries come in all shapes and sizes. Some batteries can be recharged and used again, but some cannot be recharged. Those that can be reused by recharging are generally more economical to operate than those that cannot. A battery can be considered a favorable oxidation–reduction reaction occurring inside a container that has two *electrodes.* At one of these electrodes oxidation takes place, and electrons flow out of the cell (it is marked with a − sign). This electrode is called the **anode.** At the other electrode, reduction takes place as electrons flow into the cell (it is marked with a + sign). This electrode is called the **cathode.** When all the reactants inside

■ *C*athode—electrode where *r*eduction takes place. *A*node—electrode where *o*xidation takes place. As a memory aid, note that "cathode" and "reduction" both start with consonants and that "anode" and "oxidation" both start with vowels.

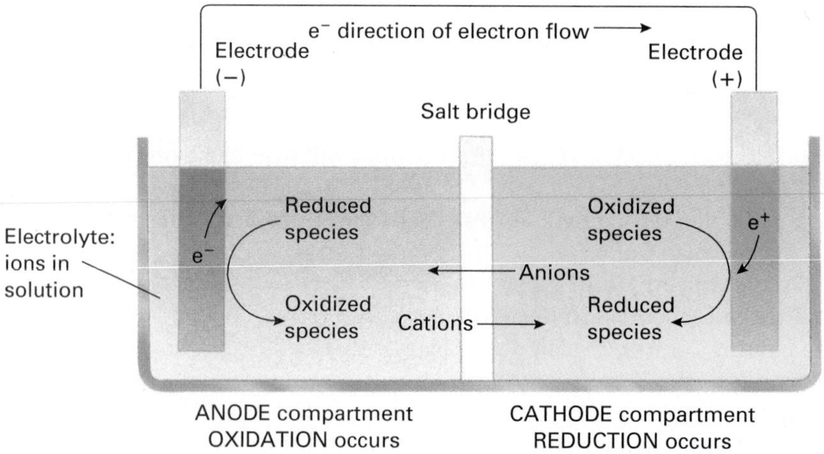

Components of an electrochemical cell. A battery consists of at least one cell containing two compartments. In the anode compartment, oxidation (loss of electrons) occurs and electrons flow out of the cell. In the cathode compartment, electrons flow into the cell and reduction (gain of electrons) occurs. A salt bridge allows ions to flow between the two compartments in order to maintain charge balance. Many batteries consist of several cells joined together.

the battery are used up and it is impossible to convert the reaction products back to their original form, the battery must be discarded. On the other hand, if the reactants can be converted back to their original form, the battery can be used again.

Throw-Away Batteries

Consider the reaction between zinc atoms and copper ions that was discussed previously. If zinc is placed in a solution containing Cu^{2+} ions, electron transfer takes place from the zinc to the copper ions because zinc is a more active metal than copper. If the zinc is separated from the copper solution and the two are connected so that an electric current flows between them, the reaction can proceed, but now the electrons are transferred through the connecting wires. Figure 10.9 shows a device that can be constructed to make use of the oxidation–reduction reaction of Zn with Cu^{2+}. A "salt bridge" carries a current of ions from one electrode chamber to the other. This bridge is necessary because Zn^{2+} ions are produced at the anode (where oxidation is taking place), so negative ions must flow into that chamber to counteract their presence.

The electrons flow from the Zn anode through the connecting wire, light the lamp in the circuit, and then flow into the copper cathode where reduction of Cu^{2+} ions occurs. The important thing to note about this simple battery is that the products of the reaction cannot be converted easily to their original form. For this to happen, the copper deposited on the copper electrode would somehow have to be dissolved back into solution, and the Zn^{2+} ions would have to be converted back into the zinc metal strip. In other words, this reaction is best used in a "throw-away" battery, one that cannot be recharged. Batteries of this type are called **primary** batteries.

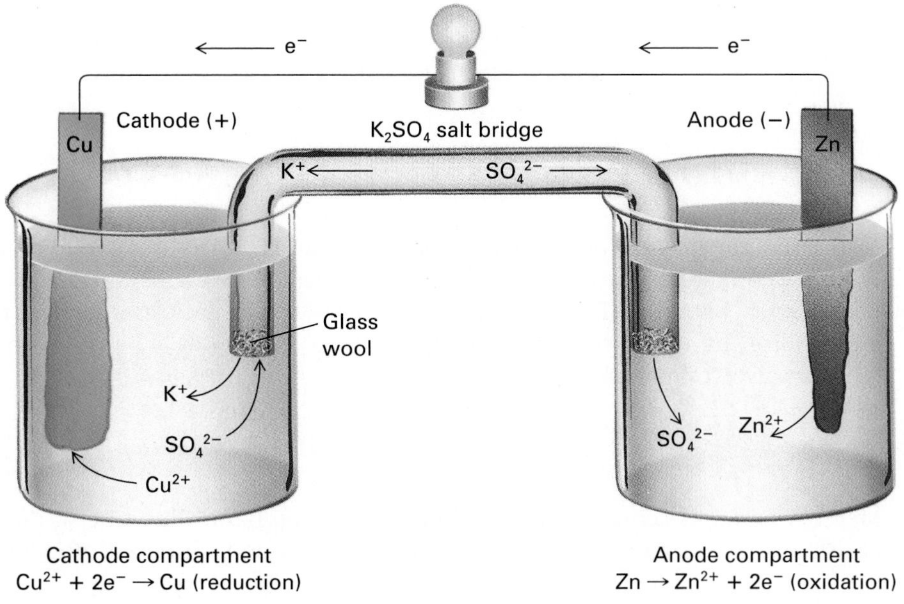

Figure 10.9 A simple electrochemical cell. The cell consists of a copper electrode in a solution containing Cu^{2+} ions (on the left), a zinc electrode in a solution containing Zn^{2+} ions, and a salt bridge that allows ions to flow into and out of the two solutions. When the two metal electrodes are connected by a conducting circuit, electrons flow from the zinc electrode, where zinc is oxidized, to the copper electrode, where copper is reduced. The overall reaction in this cell is $Zn(s) + Cu^{2+}(aq) \rightarrow Zn^{2+}(aq) + Cu(s)$

■ A throw-away battery cannot produce any additional electron flow when the reactants are consumed.

The most common primary battery in use for such applications as powering radios, simple calculators, and flashlights is the "dry cell." The container of the cell is made of zinc. Since this is the electrode that gets oxidized, the structural integrity of the battery decreases as the anode is used up.

$$Zn(s) \longrightarrow Zn^{2+}(aq) + 2\,e^-$$

The zinc is separated from the other chemicals (Figure 10.10) by a porous paper, which acts like the salt bridge discussed earlier. The center electrode of the dry cell is made of graphite, a form of carbon, and is inserted into the surrounding moist mixture of ammonium chloride, zinc chloride, and manganese dioxide. As electrons flow from the cell through a flashlight bulb, for example, the zinc is oxidized. Ammonium ions are reduced at the graphite electrode.

$$2\,NH_4^+(aq) + 2\,e^- \longrightarrow 2\,NH_3(g) + H_2(g)$$

■ The dry cell was invented by Georges Leclanché in France in 1866.

For the same reasons that the zinc-copper battery cannot be reused, the dry cell cannot be reused. The reaction products cannot be converted back to the reactants. Another problem with dry-cell batteries is that the buildup of ammonia and hydrogen gas can cause the voltage of a dry-cell battery to drop off quickly because the ionic flow to the electrode is impeded by bubbles of gas. "Renewable" dry-cell batteries can be rejuvenated over a short period by placing them in electric devices that create resistance heating of the chemicals inside the battery. The gas bubbles formed near the center electrode will migrate away from its surface and allow further reaction of the chemicals. For applications that require continuous heavy-duty use, such as in motorized toys, more robust batteries are needed. The most common heavy-duty batteries are the alkaline batteries.

Alkaline dry cells (alkaline batteries) are similar to ordinary dry cells, except that the oxidizing agent is manganese dioxide (MnO_2). They also contain KOH, which is highly alkaline. The reduction of MnO_2 does not produce any gases at the graphite electrode, and this gives the battery a much longer life.

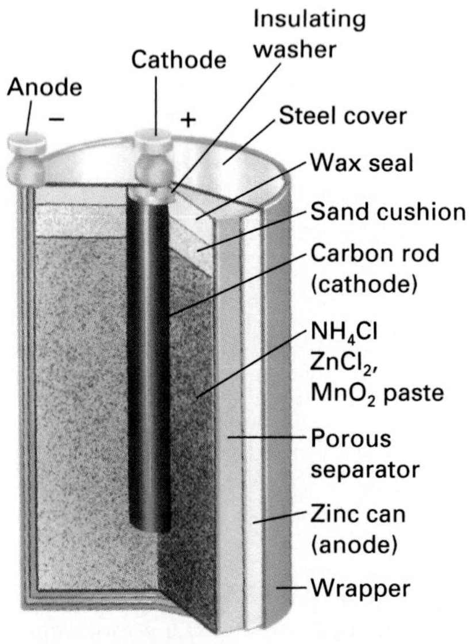

Figure 10.10 A dry-cell battery (a Leclanché cell). Compared to older electrochemical cells, this cell was relatively dry (at least no liquids sloshed around). The zinc container, which is the anode, contains a graphite cathode and the electrolyte, which is a moist paste of manganese dioxide (MnO_2) and ammonium chloride (NH_4Cl). As the anode is used up, this type of battery tends to develop leaks.

$$MnO_2(s) + H_2O(\ell) + e^- \longrightarrow MnO(OH)(s) + OH^-(aq)$$

Notice that the formula of the manganese compound formed in the reduction of MnO_2 contains a hydrogen atom. It is thus a reduced form of MnO_2.

Mercury batteries (Figure 10.11) are similar to alkaline batteries in that both utilize the oxidation of zinc as the source of electrons. The reduction reaction involves mercuric oxide.

$$HgO(s) + H_2O(\ell) + 2\,e^- \longrightarrow Hg(\ell) + 2\,OH^-(aq)$$

Mercury batteries represent a special environmental hazard because of mercury's toxicity. In addition, mercury cells may explode if disposed of in fire

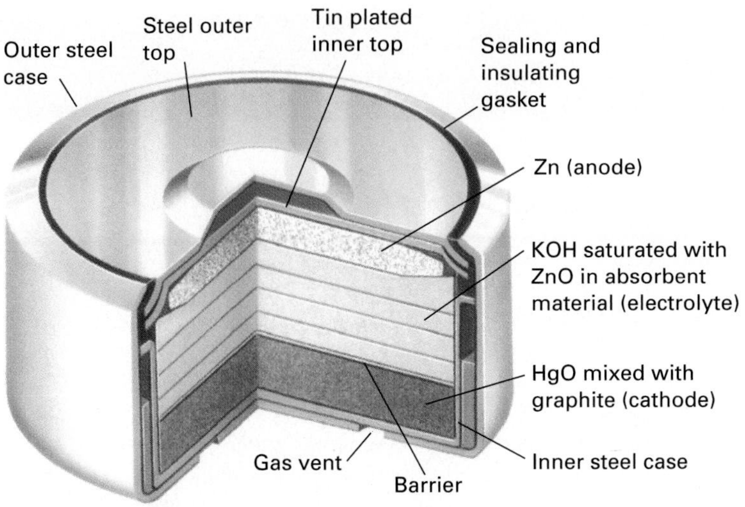

Figure 10.11 A mercury battery. The anode is zinc as in the dry-cell and the oxidizing agent is mercury(II) oxide. The mercury in this battery makes it an environmental hazard. It is also an explosion hazard if heated because it is tightly sealed so that internal pressure can build up.

because the metallic mercury formed by the reduction of HgO will vaporize and rupture the sealed container.

■ Mercury batteries are being phased out by many manufacturers, much to the dismay of some owners of cameras and electronics that require a special-size mercury battery.

Lithium batteries are a variation of the dry cell and like them are not reusable. Lithium batteries are popular because of their light weight and high energy content. Lithium metal is the reducing agent and is a much stronger reducing agent than zinc (refer to Table 10.6). Some lithium batteries use MnO_2 as the oxidizer, as in alkaline batteries, and others, like some pacemaker batteries, use more exotic oxidizers.

■ Lithium metal has the lowest density of any nongaseous element: 0.534 g/mL.

Reusable Batteries

Some batteries are constructed in such a way that the oxidation–reduction reactions at the electrodes can be reversed by the addition of energy, so that the battery is said to be recharged. Batteries of this type are called **secondary** batteries. Under favorable conditions, these secondary batteries may be discharged and recharged many times over.

One of the most widely used secondary batteries is the lead-acid automobile battery (Figure 10.12). As this battery is discharged, metallic lead is oxidized to lead sulfate ($PbSO_4$) at the anode, and lead dioxide (PbO_2) is reduced at the cathode:

■ The lead-acid battery was first presented to the French Academy of Science in 1860 by Gaston Planté.

$$Pb(s) + SO_4^{2-}(aq) \longrightarrow PbSO_4(s) + 2\,e^- \qquad \text{Oxidation}$$

$$PbO_2(s) + 4\,H^+(aq) + SO_4^{2-}(aq) + 2\,e^- \longrightarrow PbSO_4(s) + 2\,H_2O(\ell) \qquad \text{Reduction}$$

The lead-acid battery is reusable because the lead sulfate formed at both electrodes is insoluble and stays on the electrode surface. Then, when the battery needs recharging, the lead sulfate is available for the reverse reaction.

THE WORLD OF CHEMISTRY

The Pacemaker Story

Program 15, *The Busy Electron*

Sometimes an advance in science can come from an unlikely source. Several years ago, the inventor Wilson Greatbatch had an outrageous dream to prolong human life. His story is fascinating:

I quit all my jobs, and with two thousand dollars I went out in the barn in the back of my house and built 50 pacemakers in two years.

I started making the rounds of all the doctors in Buffalo who were working in this field, and I got consistently negative results. The answer I got was, well, these people all die in a year, you can't do much for them, why don't you work on my project, you know.

When I first approached Dr. Shardack with the idea of the pacemaker, he alone thought that it really had a future. He looked at me sort of funny, and he walked up and down the room a couple of times. He said, "you know—if you can do that—you can save a thousand lives a year."

In 1958, a medical team implanted the first heart pacemaker, but for the next few years there was one major problem.

After the first ten years, we were still only getting one or two years out of pacemakers, two years on average, and the failure mechanism was always the battery. It didn't just run down, it failed. The human body is a very hostile environment;it's worse than space; it's worse than the bottom of the sea. You're trying to run things in a warm salt-water environment. The first pacemakers could not be hermetically sealed, and the battery just didn't do the job. Well, after ten years, the battery emerged as the primary mode of failure, and so we started looking around for new power sources. We looked at nuclear sources, we looked at biological sources, of letting the body make its own electricity, we looked at rechargeable batteries, and we looked at improved mercury batteries. And we finally wound up with this lithium battery. It really revolutionized the pacemaker business. The doctors have told me that the introduction of the lithium battery was more significant than the invention of the pacemaker in the first place.

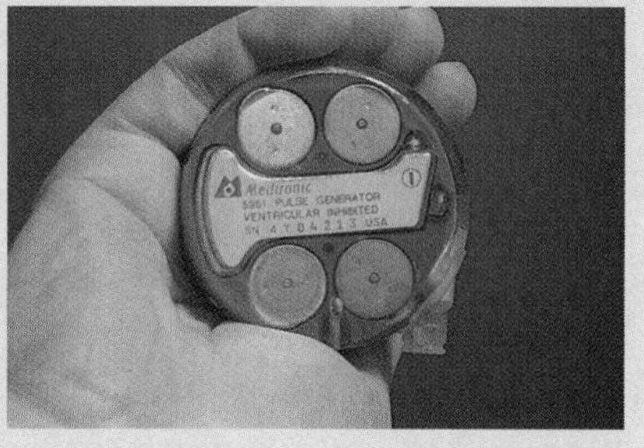

A cardiac pacemaker. (© *Yoav Levy/Phototake NYC*)

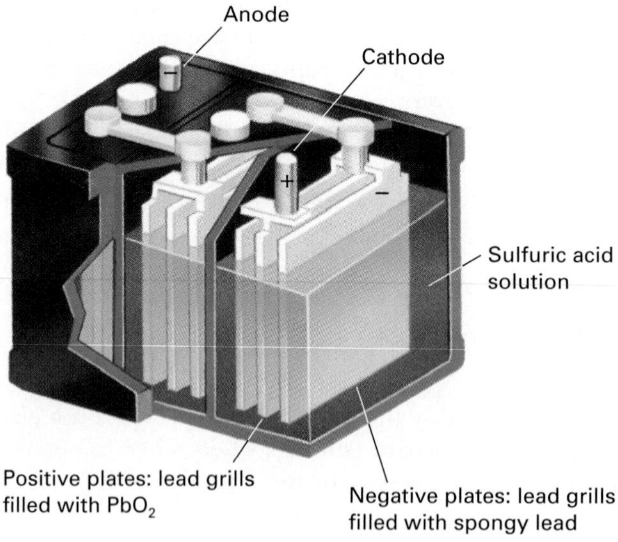

Figure 10.12 A lead-storage battery, the most common secondary battery. Most lead-storage batteries have a useful life of three years or less. The lead in these batteries is toxic and there are stringent environmental requirements regarding their disposal.

Recharging a secondary battery requires reversing the electron flow through the battery, which can be accomplished by a generator or alternator. When recharging occurs, the reactions at the two electrodes are reversed.

Discharge in a battery: chemical energy $\longrightarrow$ electrical energy

Recharging a battery: electrical energy $\longrightarrow$ chemical energy

Normal recharging of an automobile lead-acid battery occurs during driving. The voltage regulator senses the output from the alternator, and when the alternator voltage exceeds that of the battery, electrical energy is added back into the battery and the battery is recharged. During the recharging cycle in most automobile batteries, some water is reduced at the lead electrode, while water is oxidized at the PbO_2 electrode.

$$4\,H_2O(\ell) + 4\,e^- \longrightarrow 2\,H_2(g) + 4\,OH^-(aq) \qquad \text{Reduction of water}$$

$$2\,H_2O(\ell) \longrightarrow O_2(g) + 4\,H^+(aq) + 4\,e^- \qquad \text{Oxidation of water}$$

These reactions produce a mixture of hydrogen and oxygen in the atmosphere in the top of the battery. This represents a hazard because, if this mixture is accidentally sparked, an explosion results. Under normal driving conditions lead-acid batteries do not explode, but internal short circuits can produce explosions in older batteries.

All in all, the lead-acid battery is relatively inexpensive, reliable, simple, and has an adequate life. Its high weight is its major fault. Newer secondary batteries have found use in some applications such as electronics, but none of these newer batteries can perform like the lead-acid battery does for its cost.

Nickel-cadmium batteries (''Ni-Cad'') are another popular secondary battery. Being lightweight and producing a constant voltage until discharge, these batteries are useful in cordless appliances, video camcorders, cellular phones, and other applications.

■ The voltage of a battery is a measure of its potential energy. The numeric value of the voltage depends on the oxidizing and reducing agents in the battery and their concentrations.

In the Ni-Cad battery the oxidation reaction is conversion of cadmium metal into insoluble $Cd(OH)_2$.

$$Cd(s) + 2\,OH^-(aq) \longrightarrow Cd(OH)_2(s) + 2\,e^-$$

The reduction reaction is the reduction of the nickel in NiO(OH):

$$NiO(OH)(s) + H_2O(\ell) + e^- \longrightarrow Ni(OH)_2(s) + OH^-(aq)$$

Note that the reduction product in this reaction, $Ni(OH)_2$, has one more H atom than NiO(OH). This shows it is the reduced form.

Ni-Cad batteries can be recharged because the reaction products are insoluble hydroxides that remain at the electrode surface. The reactions that take place during recharging of a Ni-Cad battery are the reverse of the two reactions above. Normally Ni-Cads are recharged in a device that plugs into an ordinary electrical outlet. Ni-Cad batteries have an annoying problem of having a discharge ''memory''—if their normal use cycle involves short discharge periods followed by a recharge, they ''remember'' that and will not allow long periods of discharge. A user of Ni-Cad batteries should carefully follow the manufacturer's recharging suggestions for maximum battery life.

■ Thomas Edison invented the Ni-Cad battery in 1900.

Like the mercury battery, Ni-Cad batteries have the disadvantage of containing toxic ingredients. Cadmium is a toxic metal, and therefore Ni-Cad batteries should be disposed of properly.

The most promising new reusable battery is the lithium-ion battery. Like the throw-away batteries containing lithium, lithium-ion batteries benefit from the low density of metallic lithium and the strong reducing strength of lithium metal itself (Table 10.6). The electrode where oxidation takes place in a lithium-ion battery is made of lithium metal that has been mixed with a conducting carbon polymer (Section 12.4). The polymer has tiny spaces in its structure that can hold the lithium atoms and lithium ions formed by the oxidation reaction.

$$\text{Li(in polymer)} \longrightarrow \text{Li}^+\text{(in polymer)} + e^- \qquad \text{Oxidation}$$

The electrode where reduction takes place in the lithium-ion battery also contains lithium, but in the lattice of a metal oxide like CoO_2. This oxide lattice, like the carbon polymer electrode, has holes in it that can accommodate Li^+ ions. The reduction reaction is

$$\text{Li}^+(\text{in CoO}_2) + e^- + \text{CoO}_2 \longrightarrow \text{LiCoO}_2 \qquad \text{Reduction}$$

Lithium ion batteries have a very high energy output for their weight, they can be recharged many hundreds of times, and unlike the Ni-Cad batteries, they have no memory effect. Because of these desirable characteristics, lithium-ion batteries are finding applications in cellular telephones, portable computers, and cameras.

10.8 Fuel Cells

Fuel cells, like batteries, have an electrode where oxidation takes place and an electrode where reduction takes place. These are separated by an electrolyte solution. Unlike batteries, which use energy stored in the chemicals in the electrode compartments, fuel cells produce energy from a constant flow of chemicals into them. Most fuel cells convert the energy of oxidation–

Figure 10.13 A fuel-cell powered bus. As a demonstration of the capabilities of fuel cells, this bus is powered by 24 fuel cells. The fuel is compressed hydrogen gas and air, and the cells utilize platinum for catalysis. The bus has a range of 100 miles before refueling and a top speed of 45 mph. *(The Technology Advancement Office of the South Coast Air Quality Management District, 1995)*

reduction reactions of gaseous reactants directly into electricity. The most familiar application of fuel cells has been in the space program on board the Gemini, Apollo, and space shuttle missions, but fuel cell–powered vehicles are beginning to appear (Figure 10.13).

Consider once again the reaction between hydrogen and oxygen to produce water and energy:

$$2\ H_2(g) + O_2(g) \longrightarrow 2\ H_2O(\ell) + \text{energy}$$

As mentioned earlier in this chapter, if a mixture of hydrogen and oxygen is sparked, the energy is released suddenly in the form of a violent explosion. In the presence of a platinum gauze, which acts as a catalyst, these gases will react at room temperature, slowly heating the catalytic surface to incandescence. In a fuel cell (Figure 10.14), the oxidation of hydrogen by oxygen also takes place in a controlled manner, with the electrons lost by the hydrogen molecules flowing out of the fuel cell and back in again at the electrode, where oxygen is reduced. In the spacecraft applications of fuel cells, the water produced in the fuel cell can be purified for drinking purposes.

Because of their light weight and their high efficiencies compared with batteries, fuel cells have proved valuable in the space program. Beginning with Gemini 5, alkaline fuel cells have logged more than 10,000 hours of operation in space. The fuel cells used aboard the space shuttle deliver the same power that batteries weighing ten times as much would provide. On a typical seven-day mission, the shuttle fuel cells consume 1500 lb of hydrogen and generate 190 gal of water suitable for drinking. Other types of fuel cells use air as the

■ Fuel cells are about 60% efficient in converting chemical energy to electricity.

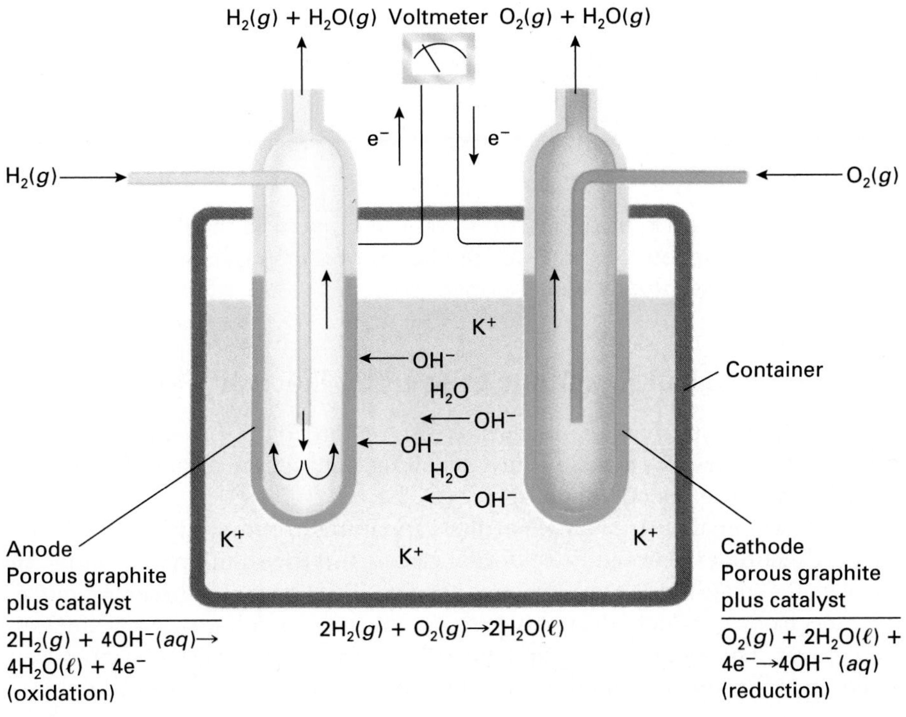

Figure 10.14 A hydrogen-oxygen fuel cell. Hydrogen is oxidized in the anode compartment. Electrons flow out of the cell, provide current to an external circuit, and flow back into the cathode compartment, where oxygen is reduced. The water produced can be purified and used for drinking.

SCIENCE AND SOCIETY

Electric Automobiles—Like Them or Not, Here They Come

Beginning in 1998, 2% of all vehicles sold in the state of California must have zero emissions of air pollutants such as carbon monoxide, oxides of nitrogen, and unburned hydrocarbons. That means these vehicles, about 40,000 by current sales figures, will have to be electric battery–operated. By 2003 the goal is 10% or 200,000 vehicles. That's many more electric vehicles (EVs) in California alone than there were in the entire United States back in the heyday of electric autos in the early 1900s. The early EVs seemed like a good choice compared with the early gasoline-powered cars, whose engines were hard to start and just plain temperamental. By about 1920, however, EVs had all but disappeared from the roads as gasoline engines became more powerful and offered far greater speed and range. By then the electric starter had become standard equipment on almost every car.

That was before air pollution caused by automobiles became almost unbearable. Now California has passed a law requiring that more and more vehicles on its highways be zero-emission vehicles. Achieving California's goals will not be easy for auto manufacturers. They can make EVs, but getting buyers to buy them may not be as easy because many of the shortcomings of the older electric designs are still noticeable today. Many of the current EV designs lack the speed, range, acceleration, and, until recently, style that people want. In the past, driving an EV meant lengthy battery recharges. That problem has been overcome with newer rechargeable battery technology. Chrysler is currently developing a nickel-iron battery, while Ford favors a sodium-sulfur battery. General Motors will use a redesigned lead-acid battery that will last approximately 20,000 miles before it has to be replaced. When all aspects of car maintenance and repair are considered, even a total battery replacement will be less expensive than operating a gasoline-powered car. In the near future, even lithium batteries may see use in electric cars. With all of the new batteries it will be possible to get a quick recharge, good for about 60 to 100 miles, in about 10 minutes. That's about the time it takes to get a complete fill-up with a gasoline-powered car today. Overnight charging at home will become a possibility, with the "fuel" bill being tacked on the home electric bill. And parking lots may offer charging stations as well.

All things considered, electric cars may be good not just for the environment, but will help to limit our dependence on foreign oil as well. An electric vehicle can go approximately 1100 miles per barrel of oil (assuming oil is used to generate the electricity). A gasoline-powered vehicle can go only 670 miles per barrel. There are, however, some disputes arising about just how much pollution EVs will be responsible for when the emissions from electric power plants are considered. If you live in California, or some other state hard-hit by urban air pollution caused by cars, chances are you may have an electric automobile in your future. Would you like to see your state mandate the use of electric cars?

---

oxidizer and less-pure hydrogen or carbon monoxide as the fuel. It is hoped that fuel cells capable of direct air oxidation of cheap gaseous fuels such as natural gas will eventually be developed.

10.9 Chemical Reactions Caused by Electron Flow

■ Electrolysis: electrical energy produces chemical change.
Batteries: chemical change produces electrical energy.

Chemical reactions that are unfavorable can be forced to proceed by the input of energy. **Electrolysis reactions** are oxidation–reduction reactions driven by the application of electrical energy.

The principal parts of an electrolysis apparatus are shown in Figure 10.15. Electrical contact between the external circuit and the solution is obtained by means of *electrodes,* which are often made of graphite or some unreactive metal. The electrode at which electrons enter an electrolysis cell is where reduction takes place. This electrode is called the cathode. While electrolysis reactions can occur in many kinds of solutions, we shall discuss only aqueous solutions here.

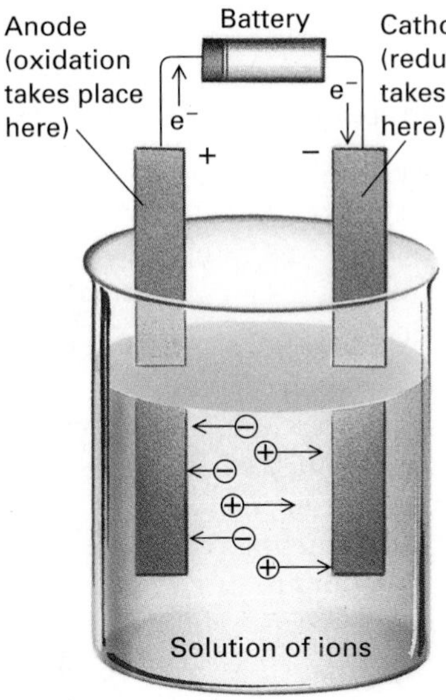

Figure 10.15 The major components of an electrolysis cell. Electrons enter the cell at the cathode, where reduction takes place. Ions flow through the electrolyte solution to maintain charge balance. At the anode, where oxidation takes place, electrons leave the cell.

The flow of electrons through the electrolysis cell is provided by a battery or similar device. Where the electrons flow into the cell, the electrode becomes negatively charged, and the positive ions in solution migrate toward that electrode and reduction takes place. Soon, evidence for a chemical reaction can be seen at the surface of the electrode. Depending on the substances present in the solution, gases may be evolved, metals deposited, or ionic species changed at the electrode. Oxidation takes place at the other electrode.

The electroplating of copper is illustrated in Figure 10.16. Such an electrolysis reaction can be used either to plate an object with a layer of pure copper or to purify an impure sample of copper metal. In either case, copper is transferred from the anode into the solution and eventually to the cathode. If the anode is impure copper to be purified, electrolysis deposits very pure copper on the cathode. This method is used industrially as the final step in conversion of copper ore to pure copper.

If you want to coat an object with copper, you have to render the surface conducting and make the object the cathode in a solution of copper sulfate. The object will become coated with copper as the electrolysis proceeds. The copper coating will grow thicker as the electrolysis is continued. This process is called copper **plating.** If the object to be plated is a metal, it will conduct electricity by itself. But if the object is a nonmetal, its surface can be lightly dusted with graphite powder to render it conducting.

Water can be decomposed by passing an electric current through a solution of salt prepared by dissolving a small amount of NaCl or similar salt. The reduction reaction that produces hydrogen gas is

$$4\,H_2O(\ell) + 4\,e^- \longrightarrow 2\,H_2(g) + 4\,OH^-(aq) \qquad \text{Reduction of water}$$

The oxidation reaction produces oxygen gas.

$$2\,H_2O(\ell) \longrightarrow O_2(g) + 4\,H^+(aq) + 4\,e^- \qquad \text{Oxidation of water}$$

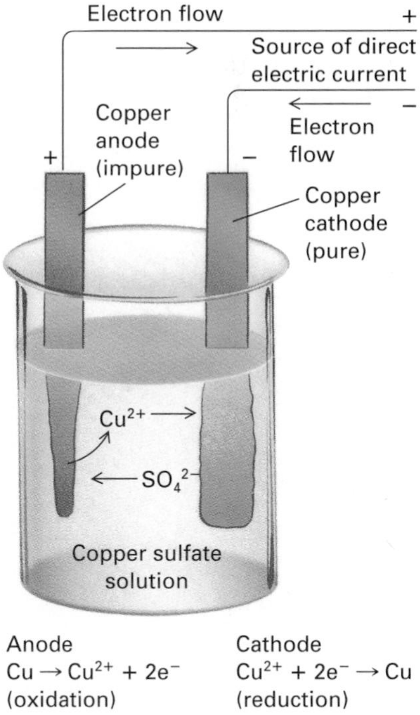

Figure 10.16 Electroplating copper. Copper metal can be electroplated on a cathode from a solution containing Cu^{2+} ions. If the anode consists of impure copper and the cathode is purer copper, then the copper is purified as it moves from the anode to the cathode. Copper is purified industrially in this manner.

Adding these two reactions gives the net reaction.

$$\cancel{4\,H_2O(\ell)} + \cancel{4\,e^-} \longrightarrow 2\,H_2(g) + \cancel{4\,OH^-(aq)}$$
$$2\,H_2O(\ell) \longrightarrow O_2(g) + \cancel{4\,H^+(aq)} + \cancel{4\,e^-}$$
$$\overline{\quad 2\,H_2O(\ell) \longrightarrow 2\,H_2(g) + O_2(g) \quad}$$

Note that the 4 OH^- and 4 H^+ are equal to 4 H_2O. That is why they canceled. This reaction is used commercially to make hydrogen gas.

Many industrially important chemicals are prepared using electrolysis. These include aluminum, chlorine, sodium, magnesium, and fluorine (see Chapter 11). In addition, the electrolysis of water has been suggested as a source for hydrogen. The electricity could be supplied by solar cells (see Section 13.4). The H_2 would then be used as a clean-burning fuel.

SELF-TEST 10D

1. If the reaction products in a battery become mixed with one another while the battery discharges, this means the battery will be a (a) throw-away battery, (b) rechargeable battery.
2. If the reaction products in a battery do not become mixed with one another as the battery discharges, the battery will be a (a) throw-away battery, (b) rechargeable battery.
3. In most "dry-cell" batteries, the ingredient that gets oxidized is (a) iron, (b) zinc, (c) PbO_2.
4. If gases such as ammonia and hydrogen accumulate around one of the electrodes in a battery, the electron flow from the battery will (a) increase, (b) decrease.
5. Which word best describes the chemical reactions at the electrodes of a rechargeable battery? (a) Renewable, (b) Reversible.
6. Which battery has a memory effect? (a) Alkaline, (b) Ni-Cad, (c) Lead-acid, (d) None of these.
7. If a fuel cell uses oxygen to oxidize hydrogen and produce electricity, the chemical product of this fuel cell is ____________.
8. If copper plates out on an electrode, which reaction is taking place? (a) $Cu \rightarrow Cu^{2+} + 2\,e^-$, (b) $Cu^{2+} + 2\,e^- \rightarrow Cu$
9. The products of the electrolysis of water are ____________ and ____________.

MATCHING SET

____	1. Chemicals in automobile lead storage battery	a. CO_2
		b. PbO_2, Pb, and H_2SO_4
____	2. Most active metal in Table 10.6	c. CO
____	3. Reduction of Cu^{2+} ions produces this	d. Hydrogen
		e. Lithium-ion battery
____	4. Oxidizing agent used to purify drinking water	f. Gain of oxygen
		g. Reduction

____ 5. Product of incomplete combustion of carbon
____ 6. Most oxidized form of carbon
____ 7. Oxidation
____ 8. Reduction
____ 9. Gain of hydrogen
____ 10. Oxidized form of sulfur
____ 11. Oxidized form of nitrogen
____ 12. Allows ions to pass from one electrode compartment in a battery to the other.
____ 13. Strongest oxidizing agent
____ 14. Reducing agent found in the typical dry-cell battery.
____ 15. Reduced form of nitrogen
____ 16. Fuel used in some fuel cells
____ 17. Product of ethanol oxidation in the liver
____ 18. Lightweight rechargeable battery
____ 19. Battery used in most automobiles
____ 20. Rechargeable battery with a ''memory''

h. SO_2
i. NO_2
j. Ni-Cad
k. Acetaldehyde
l. Li
m. Loss of oxygen
n. Fluorine
o. Lead-acid battery
p. Chlorine
q. Copper metal
r. Ammonia
s. Salt bridge
t. Zinc

QUESTIONS FOR REVIEW AND THOUGHT

1. Define the following terms:
   (a) Oxidation in terms of electrons
   (b) Reduction in terms of electrons
   (c) Oxidation in terms of oxygen
   (d) Reduction in terms of hydrogen
   (e) Oxidizing agent
   (f) Reducing agent
2. Define the following terms:
   (a) Oxidation product (b) Corrosion
   (c) Anode (d) Cathode
   (e) Primary battery (f) Secondary battery
   (g) Galvanizing
3. Which of the following pairs of compounds or elements is more highly oxidized?
   (a) CO or CO_2 (b) NO_2 or NO
   (c) SO_3 or SO_2 (d) CrO or CrO_3
   (e) CaO or Ca (f) N_2 or NH_3
4. Name two common uses of the oxidizing properties of chlorine (Cl_2).
5. What is the most common oxide of the element hydrogen?
6. Name three ingredients necessary for rust formation.
7. What is the difference between oxidation and combustion?
8. What oxide of carbon is the product of incomplete combustion?
9. Which molecule is more highly oxidized, (a) CH_3CH_2OH or (b) CH_3CHO?
10. Which compound is more oxidized, (a) C_3H_6 or (b) C_3H_8?
11. A potassium atom can form a K^+ ion. When it does, the potassium ion is said to be (a) oxidized or (b) reduced.
12. Beside each of the following reactions, indicate whether oxidation or reduction is occurring to the underlined element.
    (a) $2\ \underline{H}_2(g) + O_2(g) \rightarrow 2\ H_2O(\ell)$
    (b) $2\ \underline{Cu}(s) + O_2(g) \rightarrow 2\ CuO(s)$
    (c) $2\ \underline{Sn}O(s) \rightarrow 2\ Sn(s) + O_2(g)$
    (d) $\underline{Fe}_2O_3(s) + 3\ C(s) \rightarrow 2\ Fe(s) + 3\ CO(g)$
    (e) $\underline{N}_2(g) + H_2(g) \rightarrow NH_3(g)$
    (f) $2\ \underline{S}(s) + 3\ O_2(g) \rightarrow 2\ SO_3(g)$
13. The atmosphere of earth can be described as an oxidizing atmosphere. Explain.
14. The atmosphere of Jupiter contains considerable hydrogen and methane. Would you describe that atmosphere as an oxidizing atmosphere or a reducing atmosphere?
15. Iron ore contains mostly Fe_2O_3. Is this iron in an oxidized or a reduced form?
16. Aluminum ore contains Al_2O_3. Is the aluminum in an oxidized or reduced form?
17. The main source of magnesium is sea water, which contains Mg^{2+} ions. Is this magnesium in an oxidized or reduced form?
18. Name three metals that usually occur free in nature, that is, not combined with oxygen.
19. When zinc metal reacts with an acid, Zn^{2+} ions form. Is the zinc metal oxidized or reduced?

20. In each of the following reactions, tell which substance is oxidized and which is reduced. Then name the oxidizing agent and the reducing agent in each reaction.
   (a) $2\ Al(s) + 3\ Cl_2(g) \rightarrow 2\ AlCl_3(s)$
   (b) $S(s) + O_2(g) \rightarrow SO_2(g)$
   (c) $CuO(s) + H_2(g) \rightarrow Cu(s) + H_2O(g)$
   (d) $C_2H_4(g) + H_2(g) \rightarrow C_2H_6(g)$
   (e) $N_2(g) + 2\ O_2(g) \rightarrow 2\ NO_2(g)$
   (f) $Fe_2O_3(s) + 3\ C(s) \rightarrow 2\ Fe(s) + 3\ CO(g)$
21. Consider the formulas of the following compounds and tell whether you would expect them to be oxidizing agents or reducing agents.
   (a) $NaBH_4$ (b) CH_4
   (c) SO_3 (d) $KMnO_4$
   (e) C_6H_6
22. One of the forms of carbon is diamond. Write formulas for the two oxides that diamond will form.
23. In which compound is carbon more highly oxidized (a) glucose ($C_6H_{12}O_6$) or (b) carbon dioxide (CO_2)? (*Hint:* Determine the ratio of carbon atoms to oxygen atoms in each molecule.)
24. If an electrode is positively charged, it will tend to attract ions of what charge?
25. When water is electrolyzed, what product is produced at the cathode? At the anode?
26. What three chemical substances are necessary for corrosion to occur?
27. Why would rusting of automobiles be less of a problem in Arizona than in Chicago?
28. Would you expect a piece of iron to rust on the surface of the moon? Explain.
29. How is the oxide coating formed when aluminum oxidizes fundamentally different from the oxide coating formed when iron oxidizes?
30. What is the strongest oxidizing agent in Table 10.6? What is the weakest oxidizing agent?
31. What is the strongest reducing agent in Table 10.6? What is the weakest reducing agent?
32. Predict whether the following reactions would be expected to occur.
   (a) $2\ Na + Fe^{2+} \rightarrow 2\ Na^+ + Fe$
   (b) $Ca + F_2 \rightarrow Ca^{2+} + 2\ F^-$
   (c) $2\ H^+ + Cu \rightarrow Cu^{2+} + H_2$
   (d) $Zn^{2+} + Fe \rightarrow Fe^{2+} + Zn$
33. Name three kinds of rechargeable batteries. Do you own anything that uses any of these batteries? If so, name them.
34. Oxidation always occurs at the anode of an electrochemical cell. (a) True; (b) False.
35. Describe a simple battery, naming three essential parts.
36. Describe a simple electrochemical cell, naming three essential parts.
37. A friend shows you his camera that has been ruined because he used an inexpensive dry-cell battery in it and the battery corroded. Recall what you know about the construction of these kinds of batteries and explain what has happened.
38. What is the purpose of the salt bridge found in some simple batteries?
39. Ni-Cad batteries are commonly used in cordless appliances. List two problems with these batteries.
40. Name two environmental dangers of mercury batteries.
41. How is a fuel cell similar to a battery? How is it different?
42. Besides electricity, what do the fuel cells used on the space shuttle produce?
43. A brand new battery has a mass of 249.6 g. After it has been fully discharged its mass is still 249.6 g. Explain this.
44. Think of ways electricity might be distributed to consumers of electric automobiles. Compare all the methods you have thought of with the methods used to distribute hydrocarbon fuels (gasoline). List as many benefits and problems as you can for each.
45. Use the activity series to explain why silver tarnish can be removed by using an aluminum pan filled with a solution of water and baking soda.
46. Draw an apparatus that could be used to make silver-plated jewelry. Label the parts and the solution.
47. Use the appropriate substances from the following list to write an equation for a reaction that could be used in an electrochemical cell: aluminum metal, copper metal, $Al(NO_3)_2(aq)$, $Cu(NO_3)_2(aq)$.

We do some beautiful things with raw materials from the world of chemistry. *(Runk/Schoenberger from Grant Heilman)*

CHAPTER

11

Chemicals from the Air, Sea, and Land

The long view from space has dramatized what we already knew—the crust of the Earth is a very unusual environment, uniquely suited, at least in this solar system, for the production and support of life. Our environment is also quite heterogeneous. Mixtures abound; everywhere we look, the elements and compounds are almost lost in the complicated array of mixtures produced by natural forces acting over very long periods.

Throughout most of history, we had not developed the power to alter our environment significantly. Early everyday objects, such as stone hammers or wooden plows, were only physically changed from the natural material. Then came the chemical reduction of copper from natural minerals, followed by iron, and now the flood of new materials produced each year. We have developed, beyond question, the power to change Earth's natural chemical mixtures in almost any way we choose.

In this chapter, our focus is on inorganic substances—the elements other than carbon and their compounds. We look at the origins of the raw materials for our pots and pans, homes and office buildings, automobiles and airplanes, and a multitude of other manufactured items. Chapter 12 will examine the second major class of raw materials upon which our society depends—the organic compounds derived from fossil fuels.

- What gases are separated from the air, and how is this done?
- What is the Haber process, and why is it of such importance?
- How are chlorine and sodium hydroxide made from salt?
- How are metals extracted from their minerals and ores?
- What are the major issues in recycling metals, and which metals are most recycled?
- What are the general compositions, properties, and methods of production of glass, ceramics, and cement?

Planet Earth, the source of all our raw materials, as seen from the Moon. *(NASA)*

When it comes to the environment, all of the chemistry that you have read about in the preceding chapters applies simultaneously. Acid–base reactions, redox reactions, dissolving and crystallizing solids, evaporation and condensation—anything can happen. The composition of the air, natural waters, and solid land of our environment has been determined by the sum of such processes. So, also, is the fate of all materials that we return to the environment.

Planet Earth, at least at the present time, is the only source we have for the raw materials needed to sustain life and support our lifestyle. We draw gases from the *atmosphere,* water and some elements and compounds from the *hydrosphere,* and metallic and nonmetallic elements, structural materials, and fossil fuels from the Earth's crust. Planet Earth is also the only place we have to put our waste.

11.1 The Whole Earth

■ *Ecology* is the study of the complex interrelationships among living things, the atmosphere, the hydrosphere, and the Earth's solid surface.

This chapter illustrates how elements and compounds are separated from the atmosphere, hydrosphere, and crust of the Earth and put to use. The atmosphere is our major source of nitrogen and oxygen and also our only source for much smaller amounts of the noble gases other than helium. The hydrosphere, which includes salt waters and fresh waters above and below the earth's surface, must supply the water necessary to sustain life. The soluble salts in the oceans are a commercial source of magnesium, bromine, and sodium chloride, which is not only table salt but an essential chemical raw material.

The portion of the solid earth available to us is a very small part of the whole, less than 1% by mass. The deepest mine extends only 3.8 km beneath the surface. Geologists define the Earth's crust as a region between the surface and a depth of about 5 km to 35 km that lies over regions of greater density (Figure 11.1).

Three major types of rocks are found in the earth's crust: *igneous rocks,* formed by solidification of molten rock (e.g., basalt); *sedimentary rocks* (e.g., sandstone; which is cemented sand), formed by deposition of dissolved or suspended substances from oceans and rivers; and *metamorphic rocks* (e.g., marble), formed by the action of heat and pressure on existing rocks. Figure 11.2 gives the average composition of the Earth's crust. The most abundant substances in rocks are silicates, which are composed of silicon, oxygen, and positive metal ions (Section 11.6). The more than 2000 kinds of known **minerals** fall into a few major classes (Table 11.1).

■ **Minerals** are naturally occurring, solid inorganic compounds; they have definite internal structure and chemical composition. Most of the earth's crust consists of *rocks,* which may be single minerals or mixtures of minerals. A natural material that contains a sufficient concentration of an element or compound to economically justify mining it is an **ore.**

Fortunately for the mining industry, the composition of the crust is not uniform. Natural forces have concentrated different minerals in different places. For example, as molten rock gradually cools, the minerals that solidify first (those with the higher melting points) can sink in the remaining liquid and become concentrated. Or minerals can be redistributed according to variations in their solubility in natural waters.

The western United States was once covered by a large, land-locked sea. The water evaporated, leaving huge deposits of sodium carbonate, a soluble salt that is a valuable chemical raw material. While most nations must manufacture sodium carbonate from other chemicals, the United States meets a large proportion of its needs by mining. Other minerals are concentrated elsewhere, of course. Most of the nickel is in New Caledonia and Zimbabwe. Most of the chromium is in Botswana and Turkey. Each country must rely on imports of one kind or another.

Some uses for a few of the less familiar elements are listed in Table 11.2. Methods of purification and uses have been found for even the scarcest of elements.

TABLE 11.1 ■ Major Mineral Groups in the Earth's Crust

Mineral Group	Example	Formula	Uses
Silicates	Quartz	SiO_2	Glass, ceramics, alloys
	Feldspar	$KAlSi_3O_8$	Ceramics
Oxides	Hematite	Fe_2O_3	Iron ore, paint pigment
Carbonates	Calcite	$CaCO_3$	Optical instruments (pure crystals), industrial chemical
Sulfides	Galena	PbS	Lead ore, semiconductors
Sulfates	Gypsum	$CaSO_4 \cdot 2H_2O$	Cement, plaster of Paris, wallboard, paper sizing
Halides	Fluorite	CaF_2	Lasers and electronics (pure crystals), source of fluorine (F_2)

TABLE 11.2 ■ Some Less Familiar Elements

Name (symbol)	Uses
Beryllium (Be)	Watch springs and spark-free tools
Boron (B)	Fibers in metal and ceramic composites
Selenium (Se)	Light meters and photocopying machines
Antimony (Sb)	Solder and mascara
Cesium (Cs)	Photoelectric cells and infrared lamps
Terbium (Tb)	Color TV picture tubes and X ray screens
Osmium (Os)	Fountain-pen points and compass needles
Bismuth (Bi)	Fuses
Americium (Am)	Smoke detectors

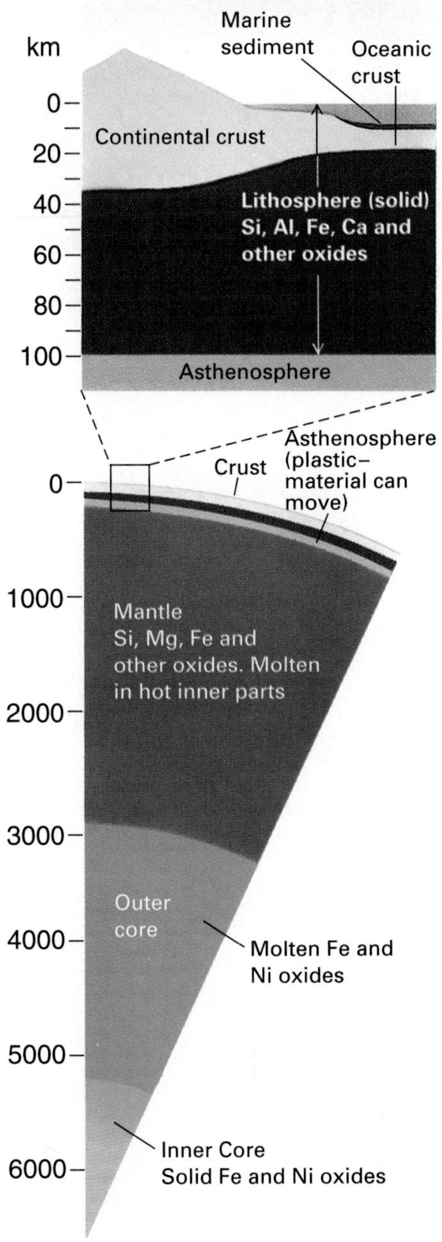

Figure 11.1 A cross section of the Earth. Geologists customarily list the composition of the earth in terms of oxides, as shown here.

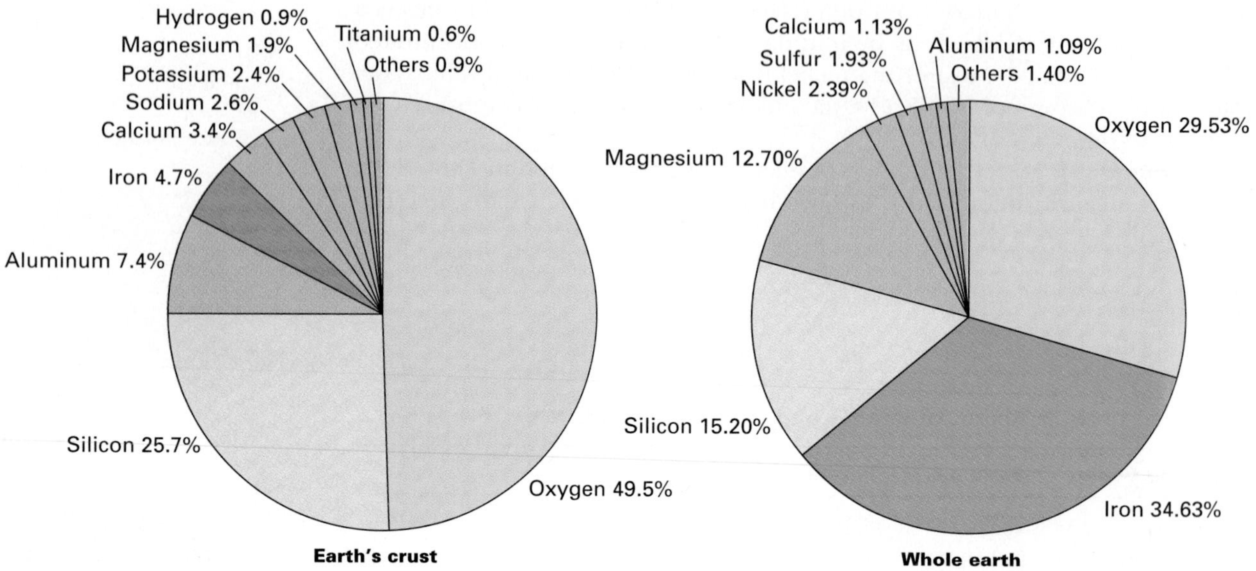

Figure 11.2 Relative abundance (by mass) of elements in the Earth's crust compared to abundance in the whole Earth.

11.2 Gases from the Atmosphere

■ Air, like most gases, heats up when it is compressed and cools down when it expands.

Before the pure elements can be obtained by the process known as ***fractionation of air,*** water vapor and carbon dioxide must be removed. This is usually done by precooling the air to separate ice and frozen carbon dioxide, or by using silica gel to absorb water and lime to absorb carbon dioxide ($CaO + CO_2 \rightarrow CaCO_3$). Afterward, the air is compressed to more than 100 times normal atmospheric pressure, cooled to room temperature, and allowed to expand into a chamber. Overcoming intermolecular forces requires energy (Section 7.3), so as the gas expands, the molecules lose energy, they slow down, and the gas gets cooler. If the compression and expansion are repeated many times and controlled properly, the expanding air cools to the point of liquefaction. Once air has been liquefied, its pure gases can be separated by taking advantage of their different boiling points.

When liquid air is allowed to warm up, nitrogen (bp −196°C) vaporizes first, and the liquid becomes more concentrated in oxygen and argon. Further processing allows separation of high-purity oxygen (bp −183°C) and argon (bp −189°C). The less-abundant noble gases (neon, bp −246°C; xenon, bp −108°C; and krypton, bp −153°C) are also separated from liquid air.

■ Because helium atoms are very light (4 amu), once released they can achieve velocities great enough to escape from earth's gravity. The decay of radioactive elements resupplies helium, creating a roughly constant amount in the atmosphere.

Helium (bp −271°C) is not commercially recovered from air because it is cheaper to isolate it from natural gas, where it is sometimes present in as much as 7% by volume. Not all natural gas, however, is subject to helium removal before it is burned. This has led at times to controversy. Should we recover more helium from natural gas than is currently needed and stockpile it? Once natural gas as a source of helium is depleted, the stockpiles could be tapped, and the much more expensive and difficult process of separating helium from liquid air could be avoided.

Oxygen

Most oxygen produced by the fractionation of liquid air is used in steelmaking. Some is also used in rocket propulsion (to oxidize hydrogen) and in controlled oxidation reactions of other types.

■ A small but vital use of oxygen is in breathing therapy.

Liquid oxygen (known as *LOX*) can be stored and shipped at its boiling temperature of $-183°C$ under atmospheric pressure. Substances this cold are called **cryogens** (from the Greek *kryos,* meaning "icy cold"). Cryogens represent special hazards, since contact produces instantaneous frostbite, and structural materials such as plastics, rubber gaskets, and some metals become brittle and fracture easily at these low temperatures. Because liquid oxygen can accelerate oxidation reactions to the point of explosion, contact between it and substances that can ignite and burn in air must be prevented.

Nitrogen

Since nitrogen gas is so chemically unreactive, it is used as an inert atmosphere for applications such as welding and other high-temperature metallurgical processes. If air is not excluded from these processes, unwanted oxides of the hot metals would form.

■ Nitrogen ions and molecules ranked from the most oxidized (bonded to oxygen) to the most reduced (bonded to hydrogen): NO_3^-, NO_2, NO_2^-, NO, N_2O, N_2, NH_3 and NH_4^+.

Liquid nitrogen is used in medicine *(cryosurgery),* for example, in cooling a localized area of skin prior to removal of a wart or other unsightly or pathogenic tissue. Because of its low temperature and inertness, liquid nitrogen is also widely used in frozen-food preparation and preservation during transit. Trucks or railroad box cars with nitrogen atmospheres present health hazards, since they contain little (if any) oxygen to support life. Workers must either enter such areas with breathing apparatus or first allow fresh air to enter.

Nitrogen is an essential element for plants, but they cannot derive it from gaseous nitrogen. It must first be "fixed"—**nitrogen fixation** is the process of changing atmospheric nitrogen into compounds that dissolve in water and can be absorbed through plant roots. Figure 11.3 represents the cycle in which nitrogen moves from the atmosphere, through its fixation, utilization by plants and animals, and return to the soil.

■ Nitrogen fixation requires breaking the strong triple bond connecting the two atoms in the nitrogen molecule N_2 ($:N{\equiv}N:$), as described in Section 6.3.

The complex interactions among social, economic, and political necessities are well illustrated by the development of the industrial synthesis of ammonia. The need for an industrial process for nitrogen fixation was recognized as early as 1890. Scientists in England noted that the world's future food supply would be determined by the amount of nitrogen compounds available for fertilizers. At the time, the sources of such compounds were limited to sodium nitrate from Chile and rapidly depleting supplies of guano (bird droppings) from Peru.

■ The first commercial ammonia plant in the United States began production in 1921.

While England was interested in nitrogen for fertilizers, Germany was interested in nitrogen for explosives. When World War I broke out in August of 1914, many people thought that a naval blockade of German ports would result in a shortage of nitrogen-based explosives that would force the war to end within a year. What they did not know was that the German chemist Fritz Haber had developed a feasible process for the fixation of nitrogen by direct synthesis of ammonia from its elements.

$$N_2(g) + 3\,H_2(g) \rightleftharpoons 2\,NH_3(g)$$

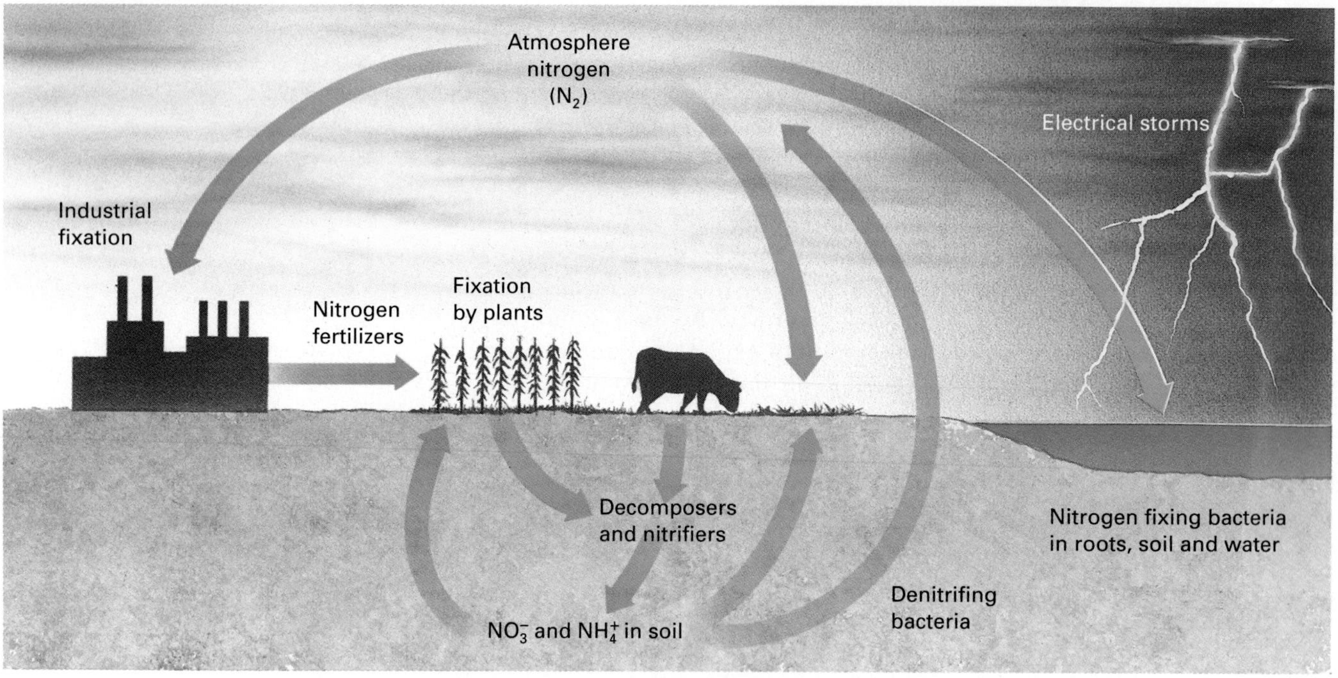

Figure 11.3 The nitrogen cycle in the environment. The chemical conversions that occur in the soil are all carried out by bacteria:

$$\text{organic materials} \xrightarrow{\text{Decomposer \& ammonifying bacteria}} NH_4^+$$

$$NH_4^+ \xrightarrow{\text{Nitrifying bacteria}} NO_3^-$$

$$NO_3^- \xrightarrow{\text{Denitrifying bacteria}} N_2$$

■ The Haber process is based on a delicate balancing act among the reaction energy, the reaction rate, and the reaction equilibrium. The combination of nitrogen with hydrogen is exothermic, but slow. The rate is increased by a higher temperature, but the yield of ammonia decreases with increasing temperature (*because* the reaction is exothermic). Therefore, a catalyst that acts at a moderate temperature is necessary. Also, because four moles of reactant give two moles of product, the extent of the forward reaction can be increased by increasing the pressure (Le Chatelier's principle; Section 8.4).

The first industrial plant based on the **Haber process** (Figure 11.4) for ammonia synthesis had begun operation in Germany in 1911, and by 1914 such plants were being built very rapidly. The availability of ammonia from this process undoubtedly prolonged the war.

Subsequently, the Haber process has provided ammonia for fertilizers on a huge scale and is now largely responsible for our ability to feed a world population of more than 5.7 billion. The Haber process has been so well developed that ammonia is very inexpensive (about $150 per ton) and is usually among the top ten industrial chemicals in terms of quantity produced each year (see list inside the front cover of this book).

Noble Gases

Approximately 250,000 tons of argon, the most abundant noble gas in the air, are recovered each year in the United States. Most of the argon is used to provide inert atmospheres for metallurgical processes. It is also used as a filler gas in incandescent light bulbs in order to prolong the life of the hot filament. Neon is used in many "neon" signs, but argon, krypton, and xenon are also used for this purpose.

THE PERSONAL SIDE

Fritz Haber (1868–1934)

Although Fritz Haber was a fine scientist and the success of his ammonia synthesis made him a rich man, he ultimately had a tragic life. At the start of World War I he joined the German Chemical Warfare Service, where he supervised the use of chlorine as a chemical weapon during the battle of Ypres in France. This first use of a chemical weapon led to further tragic developments in chemical warfare and also to personal tragedy for Haber. His wife pleaded with him to stop his work in this area, and when he refused, she committed suicide. In 1918 he was awarded the Nobel Prize for the ammonia synthesis, but the choice was criticized because of his role in developing chemical warfare. After World War I, Haber continued his academic research in Germany until, shortly before his death in 1934, he left the country to avoid danger because of his Jewish background.

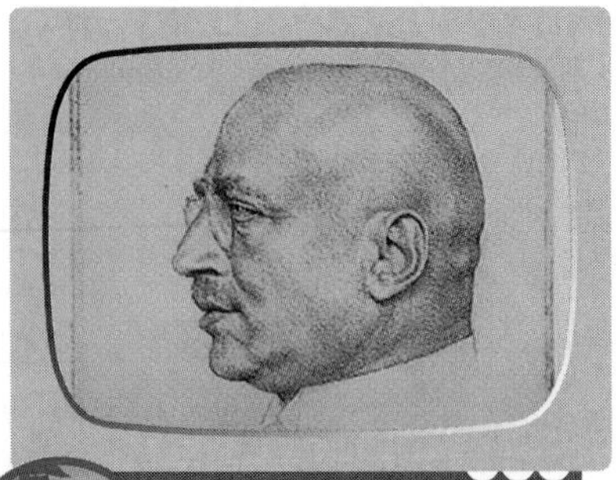

The World of Chemistry

Program 14, *Molecules in Action*

Fritz Haber.

Helium is uniquely valuable because it is the lowest-boiling liquid known (bp −271°C) and is the refrigerant of choice when extremely cold temperatures are required. As the least reactive noble gas, helium has its major use in

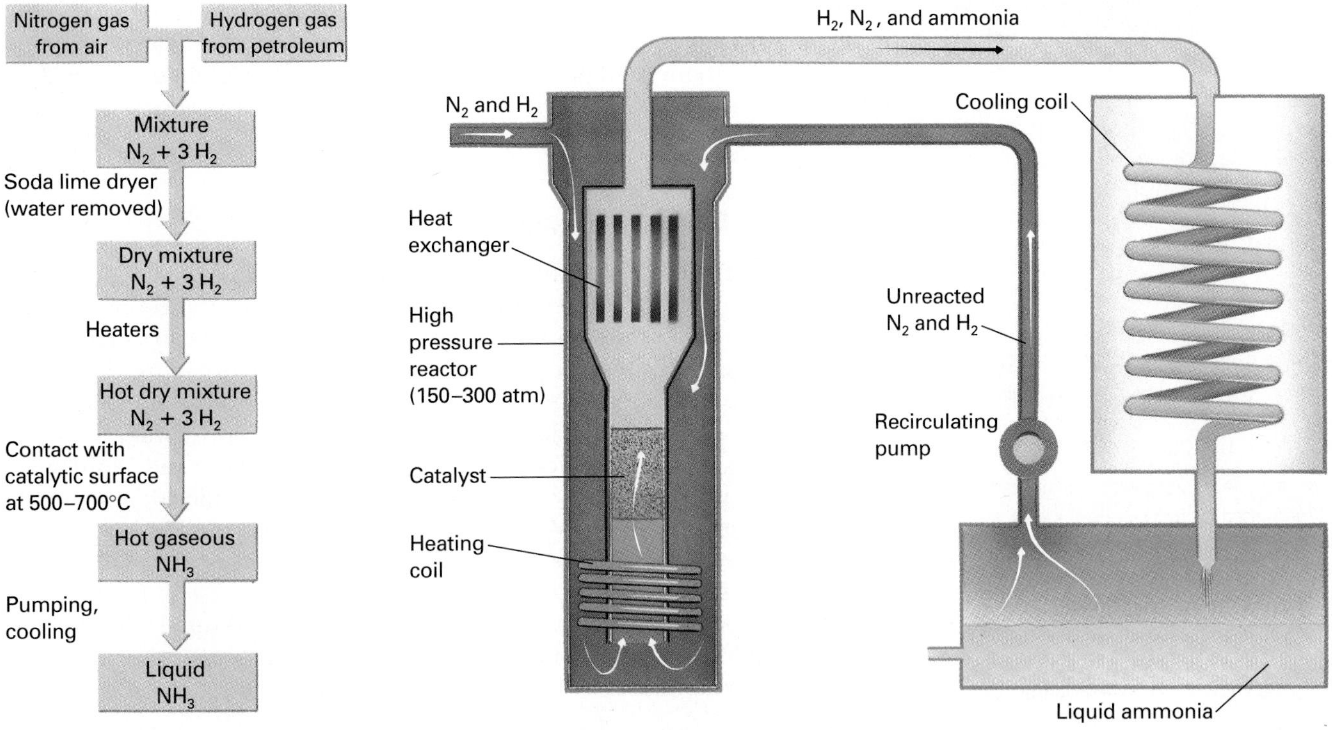

Figure 11.4 The Haber process for ammonia production. A mixture of N_2 and H_2 in the proper proportions for reaction is heated and passed under pressure over the catalyst. The ammonia is collected as a liquid and unreacted gases are recycled.

inert atmospheres, especially in welding. Helium is also unusual in having the lowest solubility in water of any gas. Deep-sea divers take advantage of this by breathing helium-oxygen mixtures instead of air, thus avoiding the dissolution of nitrogen in their blood. Unless a diver returns to atmospheric pressure very gradually, nitrogen comes out of solution as bubbles that block off blood flow in capillaries, causing the painful and potentially fatal condition known as the "bends." Because helium dissolves to a much smaller extent in blood, a diver breathing a helium-oxygen mixture is less likely to suffer decompression difficulties.

11.3 Chemicals from the Hydrosphere

■ Some other uses for sodium chloride: food processing, highway snow melting, animal feed, domestic table salt, and rubber, oil, paper, and textile manufacture

Separation of salt from sea water by evaporation, in San Francisco Bay. *(Christine L. Case/Visuals Unlimited)*

A single mouthful of seawater is enough to convince anyone that it is salty. Indeed, sodium chloride is the major mineral in seawater. But consider that other dissolved minerals are constantly being deposited into the oceans from rivers, undersea volcanoes, and thermal vents. In addition to sodium and chlorine, the major elements present in ions in solution are magnesium, sulfur, calcium, potassium, bromine, carbon, nitrogen, and strontium. Lower concentrations of almost every other element are there. Worldwide, these concentrations could provide huge quantities of economically important metals such as uranium, copper, manganese, and gold.

The lower the concentration of a metal ion in seawater, of course, the higher the cost of isolating the metal is likely to be. Nevertheless, as high-quality mineral deposits on the land are depleted, the economics of mining the sea might become more attractive. Marine organisms may help to solve the problem. For example, one family of such organisms (the *tunicates*) accumulates vanadium to more than 280,000 times its concentration in seawater. Perhaps aquatic farming of such creatures could be put to work extracting metals.

Salt and the Chloralkali Industry

■ Salt has been important throughout history. The Romans gave a *salarium* to those who were "worth their salt"—the origin of the word "salary."

■ For many years, cells with mercury cathodes were used in the chloralkali industry. Mercury is not very reactive or soluble and was thought to be harmless in the environment. Then some individuals who ate fish from mercury-contaminated waters became seriously ill. This event brought to light the fact that aquatic microorganisms convert metallic mercury to a toxic, water-soluble compound that enters the food chain (Section 18.4).

The majority of our salt is obtained from natural waters. In coastal regions salt is separated from seawater by evaporation of the water in large lagoons open to the sun. The natural **brines,** or salty waters, found in wells and lakes such as the Great Salt Lake in Utah are other sources of salt. After isolation, the salt is purified to the degree required for its use. Much of it is destined for the *chloralkali industry,* which produces chlorine and sodium hydroxide.

Chlorine is used to disinfect drinking water and sewage, and in the production of organic chemicals such as pesticides and vinyl chloride, the building block of plastics called PVCs (polyvinyl chlorides, Section 14.5). In 1994 chlorine was tenth on the list of chemicals produced in the United States. Almost all chlorine gas is made by electrolysis of aqueous sodium chloride. The other product of sodium chloride electrolysis, sodium hydroxide, is equally valuable because it is the most commonly used base in industrial processes. The reaction in electrolysis of aqueous NaCl is

$$2\ NaCl(aq) + 2\ H_2O(\ell) \xrightarrow{\text{Electrical energy}} 2\ NaOH(aq) + H_2(g) + Cl_2(g)$$

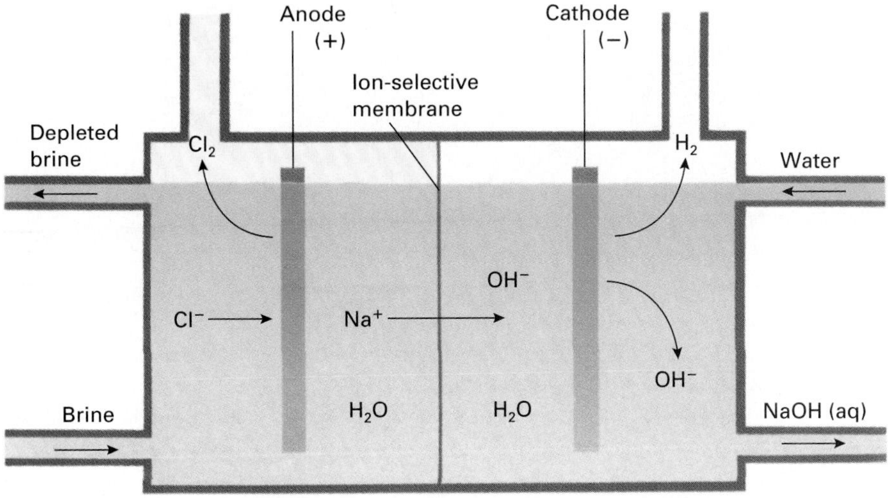

Figure 11.5 A chloralkali cell. Large banks of these cells produce gaseous chlorine and aqueous sodium hydroxide solution, both important industrial chemicals. Because of the need for cheap electricity, chloralkali plants are located near hydroelectric plants, for example, at Niagara Falls.

A complicating factor in designing electrochemical cells for this reaction is that the chlorine and sodium hydroxide, if they remain in contact, react with each other.

A modern electrochemical cell for producing chlorine and sodium hydroxide is illustrated in Figure 11.5. The two electrode compartments are separated by a synthetic membrane that allows only sodium ions to pass through it. The reactions are

■ Electrochemical cells and electrolysis are described in Section 10.9.

$$2\,Cl^-(aq) \longrightarrow Cl_2(g) + 2\,e^- \qquad \text{Oxidation}$$

$$2\,H_2O(\ell) + 2\,e^- \longrightarrow H_2(g) + 2\,OH^-(aq) \qquad \text{Reduction}$$

$$2\,Cl^-(aq) + 2\,H_2O(\ell) \longrightarrow H_2(g) + Cl_2(g) + 2\,OH^-(aq) \qquad \text{Overall reaction}$$

Brine is introduced into the anode compartment, and chloride ion oxidation occurs there. To maintain charge balance within the cell, as Cl^- ions are oxidized, Na^+ ions must pass from the anode to the cathode compartment. Since OH^- ions are produced in the cathode compartment, the product there is aqueous NaOH, with a concentration of 20 to 35% by weight. Most sodium hydroxide is used in industrial processes, although it is also present in some oven, drain, and sewer-pipe cleaners.

Magnesium from the Sea

Magnesium, with a density of 1.74 g/cm^3, is the lightest structural metal in common use. Many alloys designed for light weight and great strength contain magnesium. Most manufactured aluminum objects, for example, contain about 5% magnesium, which is added to improve the mechanical properties and corrosion resistance of the aluminum under alkaline conditions. There are also alloys that have the reverse formulation, that is, more magnesium than aluminum. These alloys are used where a high strength-to-weight ratio is needed and where corrosion resistance is especially important. In the early

■ **Alloy:** a mixture of two or more metals. Some alloys are homogeneous mixtures and others are heterogeneous.

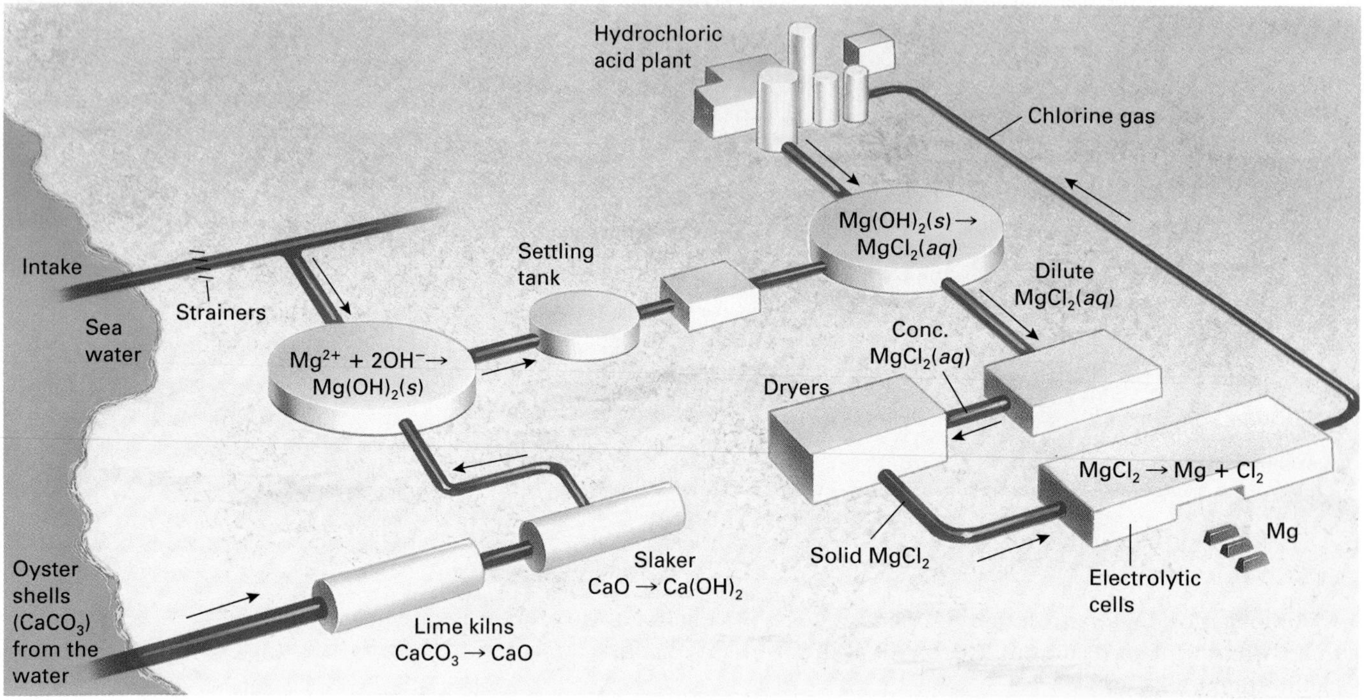

Figure 11.6 Recovery of magnesium from sea water.

1990s American automobiles contained about 4% magnesium by weight. This percentage is expected to increase as manufacturers produce lighter-weight cars in order to meet new federal fuel-economy standards.

Because there are six million tons of magnesium present as Mg^{2+} in every cubic mile of sea water, the sea can furnish an almost limitless amount of this element. The recovery of magnesium from sea water (Figure 11.6) provides an excellent illustration of the stepwise production of a chemical. The process begins with the **precipitation** of the insoluble magnesium hydroxide ($Mg(OH)_2$). The only thing needed for this step is a ready supply of an inexpensive base (a source of OH^- ions), a need fulfilled nicely by seashells, which contain calcium carbonate ($CaCO_3$). Heating the calcium carbonate converts it to lime which then reacts with water to give calcium hydroxide, the base used in the precipitation.

■ When the product of a reaction between ions dissolved in water is not soluble, the reaction is called **precipitation.** The insoluble reaction product *precipitates* out of the solution.

$$\underset{\text{Seashells}}{CaCO_3(s)} \xrightarrow{\text{Heat}} \underset{\text{Lime}}{CaO(s)} + CO_2(g)$$

$$CaO(s) + H_2O(\ell) \longrightarrow Ca(OH)_2(aq)$$

$$Mg^{2+}(aq) + Ca(OH)_2(aq) \longrightarrow Mg(OH)_2(s) + Ca^{2+}(aq)$$

The solid magnesium hydroxide is isolated by filtration and then neutralized by another inexpensive chemical, hydrochloric acid.

$$Mg(OH)_2(s) + 2\ HCl(aq) \longrightarrow MgCl_2(aq) + 2\ H_2O(\ell)$$

When the water is evaporated, solid hydrated magnesium chloride is left. After drying, it is melted (at 708°C) and then electrolyzed in a huge steel pot that serves as the cathode. Graphite bars serve as the anodes. The electrode reactions are

■ Notice how many kinds of chemical reactions are organized into the production of magnesium (Figure 11.6)—decomposition by heat, neutralization, precipitation, and oxidation–reduction by using electrical current (electrolysis).

$$2\,Cl^- \longrightarrow Cl_2(g) + 2\,e^- \qquad \text{Oxidation}$$
$$Mg^{2+} + 2\,e^- \longrightarrow Mg(\ell) \qquad \text{Reduction}$$
$$Mg^{2+} + 2\,Cl^- \longrightarrow Mg(\ell) + Cl_2(g) \qquad \text{Overall reaction}$$

As the molten magnesium forms, it is removed. The chlorine produced in the electrolysis is recycled into the process by reacting it with hydrogen to produce hydrochloric acid.

■ SELF-TEST 11A

1. The most abundant element in the Earth's crust is ____________.
2. The most abundant metal in the Earth's crust is ____________.
3. Identify the source of each of the following substances as (i) the atmosphere, (ii) the hydrosphere, or (iii) the Earth's crust:
   (a) krypton (b) the mineral malachite
   (c) water (d) sand
   (e) xenon (f) marble
4. In fractionation of air, gases are separated according to differences in their ____________ ____________.
5. Which of the following equations represents nitrogen fixation?
   (a) $NH_3(aq) + H_2O(\ell) \rightarrow NH_4^+(aq) + OH^-(aq)$
   (b) $N_2(g) + 3\,H_2(g) \rightarrow 2\,NH_3(g)$
6. Name the properties that make nitrogen and helium useful for (a) protective atmospheres, and (b) cryogens.
7. Which metal is sufficiently concentrated in the oceans so that it is currently extracted commercially: aluminum, copper, magnesium, or iron?

11.4 Metals and Their Ores

The structural materials of our buildings and manufactured objects come from the crust of the earth. Processed materials such as steel, aluminum siding, brick, and glass are made from the minerals that we scrape, dig, and blast out of the Earth's surface. These materials contribute at least $300 billion to the U.S. economy each year.

According to the U.S. Bureau of Mines, the average American in a lifetime will require the quantities of minerals listed in Table 11.3. In this section, we focus on the metals. Some of the chemistry of familiar silicon-based materials —glass, ceramics, cement—is discussed in the next section.

Less active metals such as copper, silver, and gold can be found as free elements. The somewhat more reactive metals are present as sulfides formed early in the Earth's existence (e.g., CuS, PbS, ZnS). Because of their extremely low water solubility, sulfides resist oxidation and reactions with water and

■ The metal activity series is discussed in Section 10.5.

TABLE 11.3 ■ Average Lifetime (73 Years) Supply of New Raw Materials from the Earth's Crust for an American

Element or Mineral	Quantity (pounds)	Major Uses
Stone, sand, gravel	1.3 million	Roads and buildings
Coal	500,000	Generating electricity; iron, steel, and chemicals manufacture
Iron and steel	91,000	Automobiles and ships, structural support
Clays	27,000	Bricks, paper, paint, glass, pottery
Salt	26,000	De-icing, detergents, cooking, chemical manufacturing
Aluminum	3,200	Food and beverage cans, household items, vehicles
Copper	1,500	Electrical motors, wiring
Zinc	840	Brass, galvanized iron, steel
Lead	800	Auto batteries, solder, electronics parts

other ions. The still more reactive metals have been converted over millennia into oxides (e.g., MnO_2, Al_2O_3, TiO_2) and are mined in that form.

The most reactive metals, such as sodium and potassium, are present in nature either as soluble salts in the ocean and mineral springs, in solid deposits of these salts, or in insoluble, stable aluminosilicates such as albite, $NaAlSi_3O_8$, and orthoclase, $KAlSi_3O_8$. Such silicates are found in all parts of the world, but because of their great stability they are not currently used as sources of the metals they contain. These minerals, like the ocean, represent a resource that may have to be tapped when richer and more easily processed ores are depleted.

A selection of beautiful minerals that are also ores is pictured in Figure 11.7. (A mineral is an ore if separating the metal from it is possible and economically practical.) The preparation of metals from minerals requires chemical reduction—the conversion of positive metal ions to free metals. To reduce a metal ion requires a source of electrons, which can be either electrical current (as illustrated in the preceding section for magnesium) or a chemical reducing agent.

Iron

Iron is the fourth most abundant element in the Earth's crust and the second most abundant metal. Our economy depends on iron and its alloys, particularly steel. Most of the world's iron is located in large deposits of iron oxides in Minnesota, Sweden, France, Venezuela, Russia, Australia, and England.

Iron ore is reduced in a blast furnace (Figure 11.8a). The solid material fed into the top of the furnace is a mixture of iron oxide (Fe_2O_3), coke (C), and limestone ($CaCO_3$). A blast of heated air is forced into the furnace near the bottom. The major reactions that occur within the blast furnace result in reduction of iron oxide:

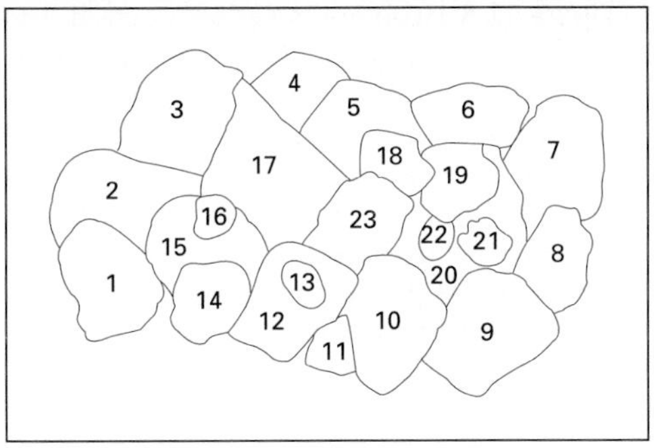

Figure 11.7 A collection of native metals and minerals. The minerals serve as ores for the metals indicated in the key below. *(©1988 Paul Silverman/Fundamental Photographs)*

1. Bornite (iridescent)—COPPER
2. Dolomite (pink)—MAGNESIUM
3. Molybdenite (grey)—MOLYBDENUM
4. Skutterudite (grey)—COBALT, NICKEL
5. Zincite (mottled red)—ZINC
6. Chromite (grey)—CHROMIUM
7. Stibnite (top right, grey)—ANTIMONY
8. Gummite (yellow)—URANIUM
9. Cassiterite (rust, bottom right)—TIN
10. Vanadinite crystal on Goethite (red crystal)—VANADIUM
11. Cinnabar (red)—MERCURY
12. Galena (grey)—LEAD
13. Monazite (white)—RARE EARTHS: Cesium, Lanthium, Neodymium, Thorium
14. Bauxite (gold)—ALUMINUM
15. Strontianite (white, spiny)—STRONTIUM
16. Cobaltite (grey cube)—COBALT
17. Pyrite (gold)—IRON
18. Columbinite (tan, grey stripe)—NIOBIUM, TANTALUM
19. Bismuth (shiny)
20. Rhodochrosite (pink)—MAGNESIUM
21. Rutile (shiny twin crystal)—TITANIUM
22. NATIVE SILVER Filigree on Quartz
23. Pyrolusite (black, powdery)—MANGANESE

$$2\,C(s) + O_2(g) \longrightarrow 2\,CO(g) + \text{heat}$$

$$Fe_2O_3(s) + 3\,CO(g) \longrightarrow 2\,Fe(s) + 3\,CO_2(g) + \text{heat}$$

and conversion of silica present in the ore to molten calcium silicate ($CaSiO_3$), known as **slag.**

$$CaCO_3(s) \xrightarrow{\text{Heat}} CaO(s) + CO_2(g)$$

$$CaO(s) + SiO_2(s) \xrightarrow{\text{Heat}} \underset{\text{Slag}}{CaSiO_3(\ell)}$$

Consequently, as the blast furnace operates, two molten layers collect in the bottom. The lower, denser layer is mostly liquid iron. The upper, less dense layer is the slag. From time to time the furnace is tapped at the bottom, and the molten iron is drawn off. Another outlet somewhat higher in the blast furnace can be opened to remove the liquid slag.

■ The mixture of nonmetallic waste and by-products from the refining of any metal is known as a **slag.**

The iron that comes from the blast furnace, known as ***pig iron,*** contains many impurities (up to 4.5% carbon, 1.7% manganese, 0.3% phosphorus, 0.04% sulfur, and 1% silicon). Iron reacts with the carbon impurity at the

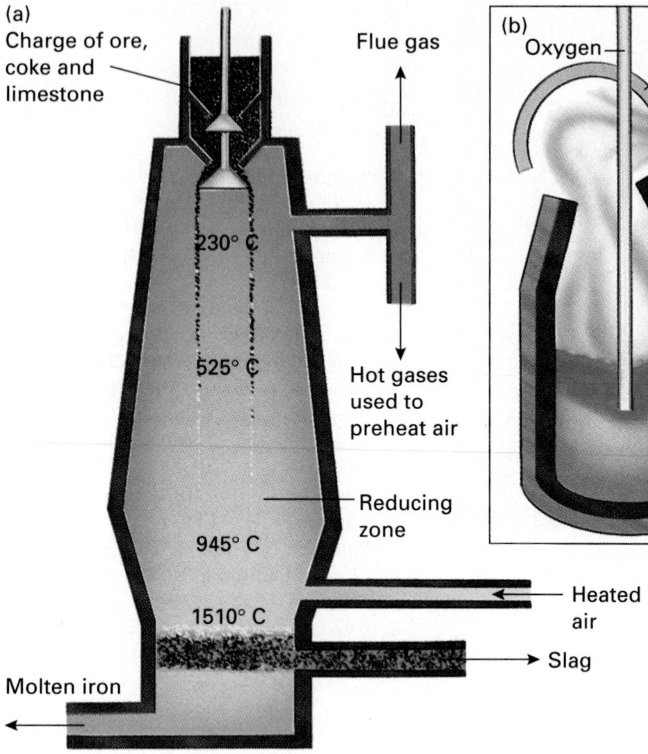

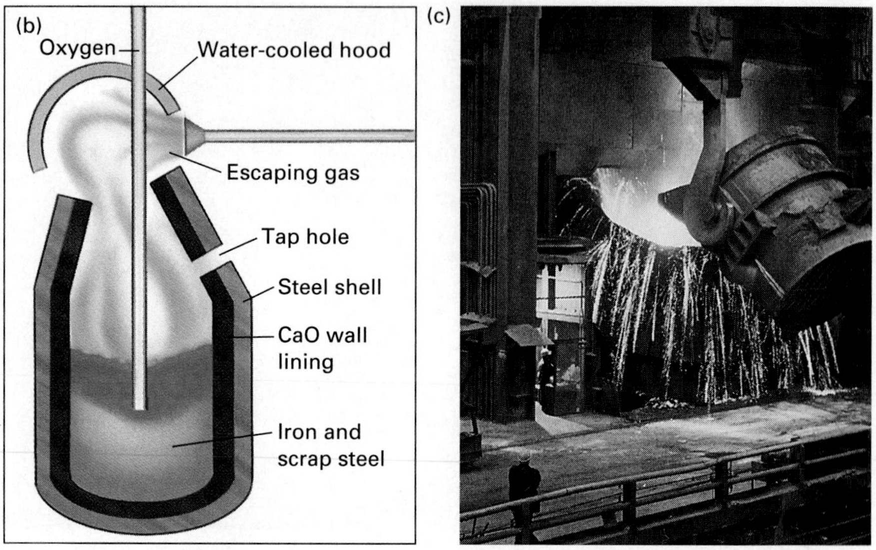

Figure 11.8 Iron and steel production. In the blast furnace (a) the descending materials in the charge are hit by a blast of hot air from coke burning in the heated air. The basic oxygen furnace for steel production (b) is charged with a mixture of molten pig iron from the blast furnace, steel scrap to be recycled, and other metals according to the type of steel being made. After oxygen is blown in for about 20 minutes, the finished steel is poured off through the tap hole, and the furnace is ready for another charge. The photo (c) shows molten iron being poured into a basic oxygen furnace. *(c, Bethlehem Steel)*

temperatures of the blast furnace to form cementite, an iron carbide (Fe_3C), which causes pig iron to be brittle.

$$3\,Fe(s) + C(s) \longrightarrow Fe_3C(s)$$

■ About 75% of the weight of an automobile is iron and steel.

The World of Chemistry

Program 19, *Metals*

We make basic alloys, probably three or four dozen. But there are variations in which we put in a little pinch of this, a little pinch of that. We can make two—three—hundred varieties. I'm still accused of being a witch doctor, so to speak, with the little pinches of this and the little pinches of that. But we do know what we're doing. Gerald Gelazela, Senior Metallurgist, Armco Steel

When molten pig iron is poured into molds of a desired shape (engine blocks, brake drums, transmission housings), it is called **cast iron.** However, pig iron and cast iron contain too much carbon and other impurities for most uses. The structurally stronger material known as **steel** is obtained by removing the phosphorus, sulfur, and silicon impurities and decreasing the carbon content.

Steels

Many kinds of iron alloys are collectively known as *steels.* The most common is *carbon steel,* an alloy of iron with about 1.3% carbon. In order to convert pig iron into carbon steel, the excess carbon is burned out with oxygen.

In the *basic oxygen process* for steel production (Figure 11.8b), pure oxygen is blown into molten iron through a refractory tube, which is pushed below the surface of the iron. A *refractory* is a material that withstands high temperatures

THE WORLD OF CHEMISTRY

Materials from the Earth

Program 18, *The Chemistry of Earth*

In the South African bush veldt, an area the size of New England, interesting and valuable natural chemical separations have been made. Just beneath the surface lie some of the world's richest deposits of platinum, rhodium, and chromium. A mineral is a naturally occurring substance with a characteristic chemical composition. When minerals are concentrated and have economic value, they are called ores. Some places are rich with valuable ores, like South Africa, while others are not, and the fortunes of whole countries can rise and fall based on the wealth found in the natural treasuries.

Since prehistoric times, we have recovered and used a variety of minerals. How did our ancestors obtain the minerals and elements they needed from the earth? One of the techniques used in an ancient tin mine in Cornwall, England, was to build large fires at the base of the rock cliffs. The intense heat cracked the rocks, exposing the tin ore, which was then removed. Today we use dynamite and ammonium nitrate to blast rock away from ore-rich veins. But although we have greatly increased our power to remove useful ores from the earth, we are still limited by effective and economical mining techniques and, to an ever-increasing degree, the availability of exploitable ore deposits.

Ancient tin mines at Cornwall, England.

without melting. At elevated temperatures, the dissolved carbon reacts very rapidly with the oxygen to give gaseous carbon monoxide and carbon dioxide, which escape.

During steelmaking, silicon or transition metals such as chromium, manganese, and nickel can be added to give alloys with specific physical, chemical, and mechanical properties.

■ In carbon steel, the carbon atoms fit in the small spaces (interstices) between the larger iron atoms of the metal. The small structural change of even one part per thousand carbon makes slipping of the iron atoms more difficult. That's why steel is stronger than iron.

The properties of steel can also be adjusted by the temperature and rate of cooling used in its production. If the steel is cooled rapidly by quenching in water or oil, the carbon in the steel remains in the form of cementite (Fe_3C), and the steel will be hard, brittle, and light-colored. Slow cooling favors the formation of crystals of carbon (graphite) rather than cementite. The resulting steel is more ductile (easily drawn into shape).

All of the processes in steelmaking, from the blast furnace to the final heat treatment, use tremendous quantities of energy, mostly in the form of heat. In the production of a ton of steel, approximately one ton of coal or its energy equivalent is consumed.

Steel structures in Chicago. Alexander Calder's steel sculpture, Flamingo, stands in a plaza surrounded by steel-frame office buildings. *(T.J. Florian/Rainbow)*

Aluminum

■ Aluminum usage: 35% of U.S. aluminum production is used in containers and packaging, 20% each in transportation and construction, and the rest in electrical applications, durable consumer goods, machinery and equipment, and other miscellaneous uses.

Aluminum, in the form of Al^{3+} ions, is the third most abundant element in the Earth's crust (7.4%). The quantity and commercial value of aluminum used in the United States each year is exceeded only by that of iron and steel. You probably know aluminum best in beverage cans. Aluminum is also familiar in the kitchen as a food wrap, a use demonstrating its excellent formability, or in a step-ladder, which illustrates its low density (2.70 g/cm^3) and great strength. Just as importantly, aluminum has excellent corrosion resistance provided by the transparent, chemically inert film of aluminum oxide (Al_2O_3) that clings tightly to the metal's surface.

$$4\ \text{Al}(s) + 3\ \text{O}_2(g) \longrightarrow 2\ \text{Al}_2\text{O}_3(s)$$

From the time of its discovery in 1825 until near the turn of the century, aluminum was made by reducing $AlCl_3$ with a more active metal, potassium or sodium,

$$3\ \text{Na}(s) + \text{AlCl}_3(s) \longrightarrow \text{Al}(s) + 3\ \text{NaCl}(s)$$

but only at a very high cost. Even though there was commercial production of aluminum by 1854, aluminum was considered to be a precious metal, as gold and platinum are today, and one of its early uses was for jewelry. In fact, in the 1855 Exposition in Paris, France, some of the first aluminum metal produced was exhibited along with the crown jewels.

Napoleon III saw the possibilities of aluminum for military use, however, and commissioned studies on improving its production. The French had a

THE PERSONAL SIDE

Charles Martin Hall (1864–1914)

While a student at Oberlin College, Charles Martin Hall was inspired by a professor's remark that a fortune awaited the person who could invent a cheap process for producing aluminum. Hall set this as his goal, and shortly after his graduation, he succeeded. In electrolysis experiments in his family woodshed, Hall used batteries and a blacksmith's fire to reduce aluminum oxide dissolved in cryolite. The success depended on the finding that aluminum oxide, which has too high a melting point (above 2000°C) to be reduced in an electrochemical cell, dissolved in cryolite to give a mixture that melts at 960°C. Hall made this great discovery in 1886. Later, he founded the Aluminum Corporation of America (Alcoa) and he did indeed become a millionaire. A memorial to him in the form of an aluminum statue stands on the Oberlin campus.

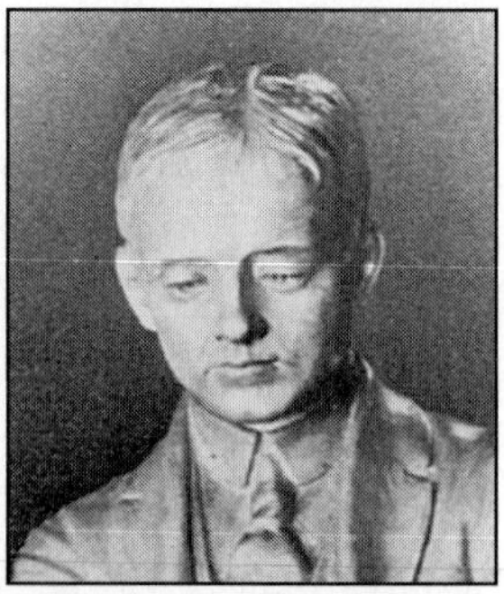

Aluminum statue that stands as a memorial to Charles Martin Hall on the campus of Oberlin College. (Special Collections, Van Pelt Library, University of Pennsylvania)

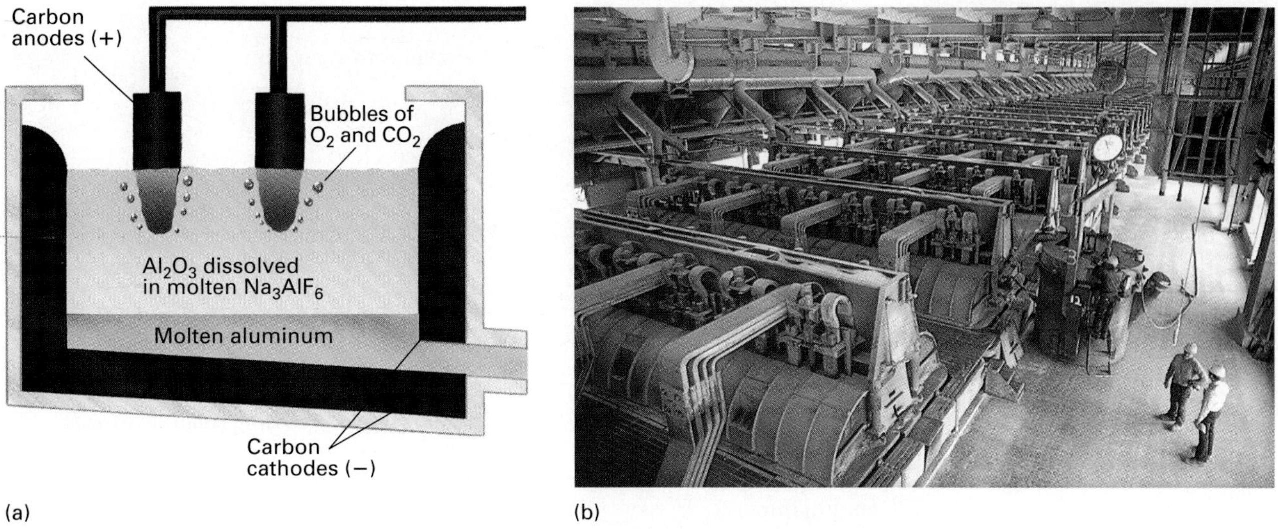

Figure 11.9 Aluminum production by electrolysis. The cathode reaction in the cell (a) is reduction of aluminum ions to aluminum metal. The anode reaction is production of oxygen gas, which reacts with the carbon anodes. (The cell reaction is $2\ Al_2O_3 + 3\ C \rightarrow 4\ Al + 3\ CO_2$.) Molten aluminum is denser than the molten salt mixture in the cell and collects at the bottom. In (b), aluminum is being tapped from one of a bank of cells. Note the scale at the top of the tank for measuring how much aluminum is collected from each cell. *(b, Aluminum Company of America (ALCOA))*

ready source of aluminum-containing ore, bauxite, named for the French town of Les Baux. In 1886 a 23-year-old Frenchman, Paul Héroult, conceived the electrochemical method that is still in use today. In an interesting coincidence, an American, Charles Hall, who was 22 at the time, announced his invention of the identical process in the same year. Hence, the commercial process is now known as the Hall-Héroult process.

■ The top of the Washington Monument is aluminum, made in 1884 by the sodium-reduction method.

In the Hall-Héroult process, purified aluminum oxide from bauxite is dissolved in molten cryolite, Na_3AlF_6. The mixture of cryolite and aluminum oxide is electrolyzed in a cell with carbon anodes and a carbon cell lining that serves as the cathode (Figure 11.9).

Because aluminum is such an active metal, separating it from its oxide requires a lot of energy. This, combined with the energy needed to maintain the molten cryolite bath, makes aluminum production highly energy-intensive. One reason for the success of aluminum recycling is the large saving in energy cost—making aluminum beverage cans from recycled aluminum requires only 5% of the energy used in making the cans from new aluminum, and the process is about 20% cheaper overall.

11.5 Recycling: New Metal for Old

Recycling of municipal solid waste has come to many communities in the United States in recent years. Perhaps you are now putting aluminum beverage cans into a recycling bin. There is a long way to go, however, before we can congratulate ourselves too heartily on major steps to reduce the quantity of

■ Do any of you remember the time when milk and soda pop were sold in glass bottles that were returned, washed, and used again?

How many of these items could have been repaired or manufactured to last longer? *(Courtesy of the Institute of Scrap Recycling Industries, Inc.)*

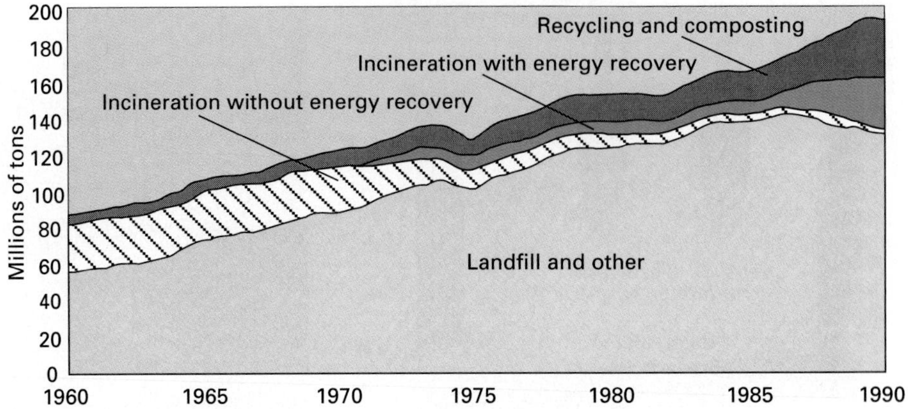

Figure 11.10 Municipal solid waste. The plot shows the rising quantity of waste generated over the years and its fate—recycling or composting of organic material, burning in a municipal incinerator (with or without recapture of heat as energy), or landfill disposal.

material that is simply dumped somewhere. Between 1960 and 1990 the generation of municipal solid waste almost doubled (Figure 11.10), while the United States population increased by only one third. In other words, each of us produced more garbage. The best remedy remains to generate less waste, rather than to expend time, energy, and money in collecting, processing, and recycling it.

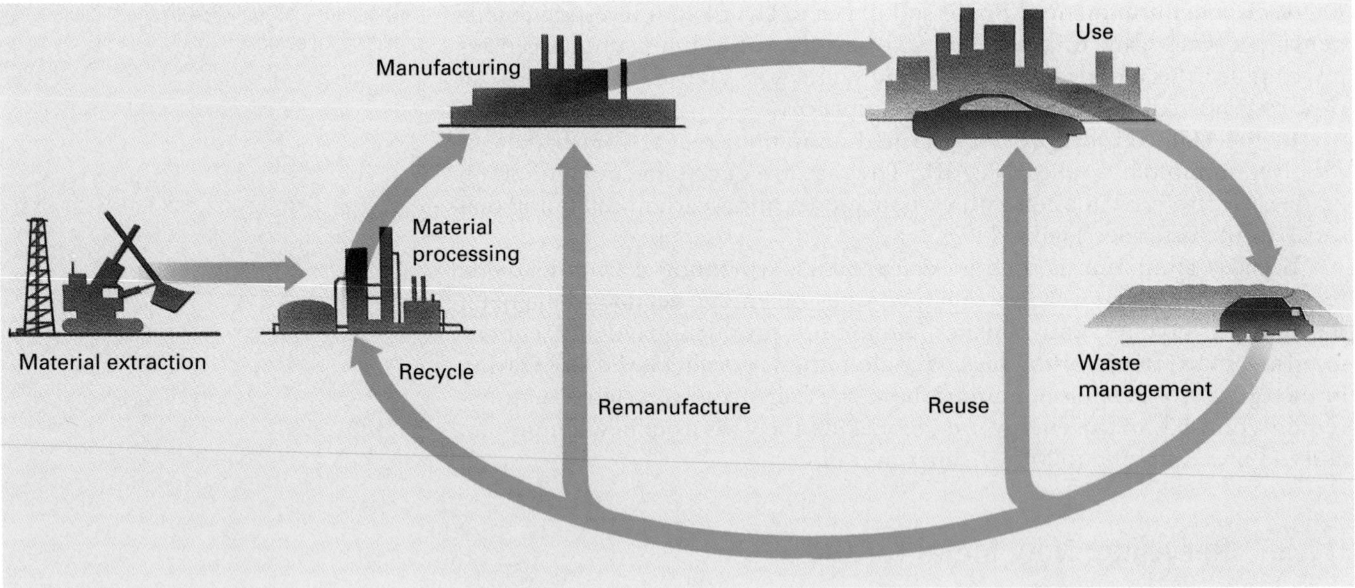

Figure 11.11 Stages in the production, use, and disposal of manufactured products. At each stage energy is used, there is the possibility of air or water pollution, and there is generation of waste. Therefore, there is also the possibility for conservation and recycling at each stage.

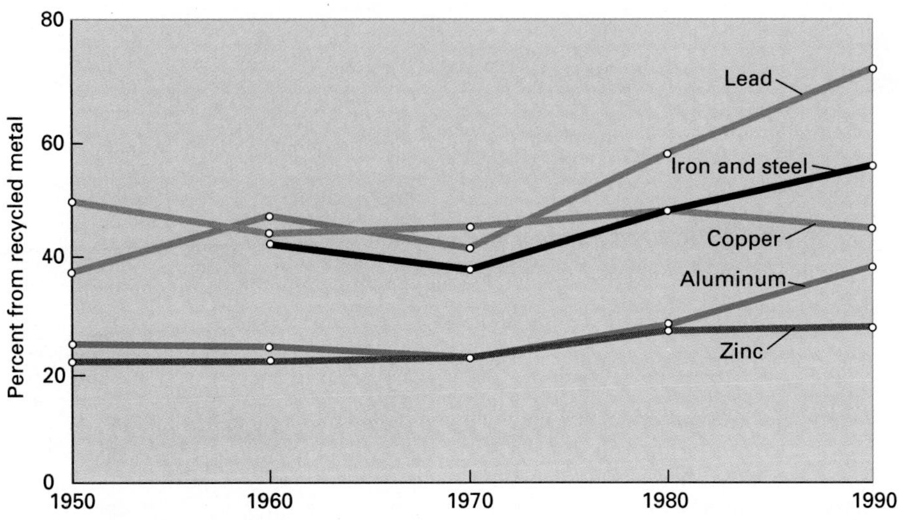

Figure 11.12 The percentage of the annual total consumption of metals from recycled materials. The total of recycled metals includes both production scrap and metal recycled after use in a consumer application. *(Data from U.S. Bureau of Mines)*

To get the whole picture of the amount of waste we generate and its impact on the environment, it's important to keep in mind that municipal waste is only a small fraction of the total. Waste is generated at each stage of a product's life cycle, meaning, of course, that there are opportunities for waste reduction at each stage (Figure 11.11).

■ Other aspects of waste reduction and pollution control are discussed in Chapters 20 and 21.

Metals are not incinerated, but go either to waste dumps or to recycling plants. Many factors determine the extent to which a metal is recycled. The influence of these factors can be seen in the percentage recycled of the five metals produced in largest volume each year (Figure 11.12). According to the U.S. Bureau of Mines, these metals combined account for more than 99% of the quantity and about 92% of the market value of metals recycled.

Factors in Metal Recycling
Market value of the metal
Quantity of the metal in use
Lifespan of products containing the metal
Cost of collection and transport of scrap
Cost of reprocessing the waste metal
Cost of disposal of nonrecycled scrap
Environmental impact of nonrecovery of the metal

Lead is the most recycled metal, for several reasons: it is too toxic to go to landfills (Section 18.4); the major use is in automobile batteries, which have a predictable lifespan; and used batteries are collected at legally designated locations in most states. Iron and steel are second in percentage recycled, most of which is recycled within the industry rather than from consumer products. Iron and steel are used in vastly greater quantities than all other metals combined, resulting in a huge pool of scrap to be dealt with. Not recovering it would be a financial as well as an environmental burden. Also, all steel is potentially recyclable without separation of pure metals from the mixture.

The significant increase in aluminum recycling illustrated in Figure 11.12 has been motivated by the rising cost of the energy needed to make new aluminum, public concern about waste management, and the resulting increase in recycling of beverage cans.

■ Do you think the deposit paid on beverage cans and the resulting motivation for scavengers to collect them has contributed to their greater entry into the recycling stream? It has certainly helped to clean up public areas in some parts of the country.

■ SELF-TEST 11B

1. Which of the following elements can be found as free elements in the Earth's crust: iron, magnesium, gold, krypton, copper, sodium?
2. ____________ are chemical compounds found in the Earth's crust.

FRONTIERS IN THE WORLD OF CHEMISTRY

Green Design

Most products on the market today have been designed either without concern for their impact on the environment or with concern only for their waste-management stage (Figure 11.12). Consumer interest in environmentally friendly products is causing changes. The more desirable approach of considering environmental impact at all stages in a product's life cycle has been christened "green design."

The goals of green design are summarized by the U.S. Office of Technology Assessment as shown in the diagram. Chemistry has a role to play in choices at every stage. Currently, for example, the plastic parts of cars are not recycled; they are shredded along with the metals, but sent to landfills. Applying the green design concept might require formulation by chemists of new plastics with the properties required by automotive parts *and* the ability to be recycled. The plastic parts of cars would then be labeled according to type, and cars would be designed so that the plastic parts could be removed easily at the scrap yard before the metals are shredded for recycling. Economics would still, of course, be a major governing factor in whether this type of change is actually made.

Another approach to green design is providing products that can be remanufactured, that is, restored to new quality and reused, giving them a longer life before they enter the waste stream. Yet another is reducing the mass and volume of material in a product, most noticeably in avoiding overpackaging. The redesign of the compact disc package is an example, although replacement of the larger, more wasteful packages on store shelves has moved slowly.

It is important to understand that decision-making in green design, like that in risk management, must constantly balance positive and negative outcomes. Many people are happy to purchase their own telephones rather than paying a monthly rental, as was the case before the disbanding of the nationwide telephone monopoly. However, most telephones then were designed to be remanufactured numerous times before they were retired. Today, several million phones are discarded each year after use by only one consumer. Many of these phones have broken down and were not designed to be repaired.

In another example, perhaps you have received a notice from your electric utility company urging the use of fluorescent light bulbs that consume less energy than incandescent bulbs—the plus side of their use. Currently some fluorescent bulbs contain enough mercury to be designated as hazardous waste, which is the minus side. But green design is coming to fluorescent bulbs; the amount of mercury in each bulb is gradually being decreased. It's easy to see that "green design" will be a frontier of chemistry for a long time to come.

Cars on their way to metal recycling. Perhaps some day the plastic parts from these cars can be recycled too. (Courtesy of the Institute of Scrap Recycling Industries, Inc.)

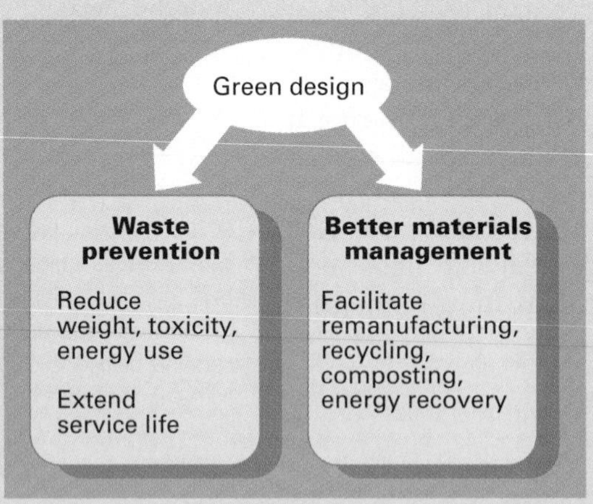

3. ____________, which are usually mixtures, contain a sufficient concentration of a metal to justify mining them.
4. Most metals are found in the Earth's crust as (a) neutral atoms, (b) positive ions, or (c) negative ions.
5. Iron ore is composed of iron ____________ plus impurities.
6. A natural material that is almost pure calcium carbonate and is used in iron production is ____________.
7. When molten in the blast furnace, calcium silicate is part of the ____________.
8. Choose which of the terms below apply to (i) cast iron and which apply to (ii) steel:
   (a) Used in engine blocks
   (b) Used in structural support of buildings
   (c) Brittle
   (d) Made in the basic oxygen process
   (e) Made in a blast furnace
   (f) Can be magnetic
9. For most metals to be prepared from their ores they must be (a) oxidized, (b) reduced.
10. Choose the process by which the metals (i) aluminum and (ii) iron are prepared from their ores: (a) electrolysis, (b) reaction with carbon.

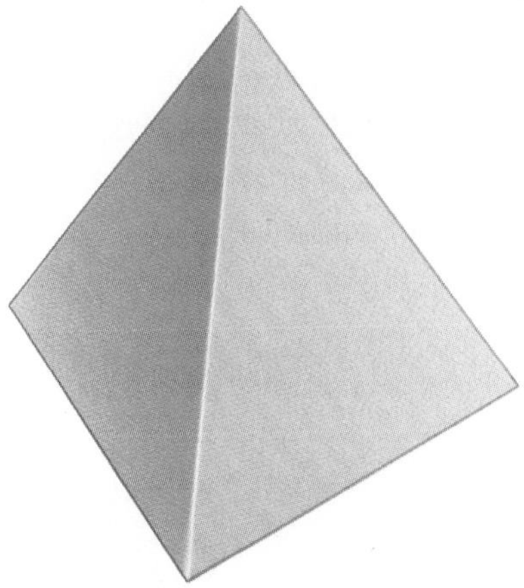

■ A tetrahedron has four sides, each of which is a triangle.

11.6 From Rocks to Glass, Ceramics, and Cement

The earth's crust is largely held together by chemical bonds between silicon and oxygen, in either pure **silica** (SiO_2) or *silicate* minerals in which silicon-oxygen anions are combined with metal cations.

The most common of the many crystalline forms of pure silica is quartz. It is a major component of granite and sandstone and also occurs as pure crystals. The basic structural unit of quartz and most silicates is the tetrahedron. In quartz, every silicon atom is bonded to four oxygen atoms, and every oxygen atom is bonded to two silicon atoms. The result is an infinite array of tetrahedra sharing corners.

Most silicates consist of networks of silicon-oxygen tetrahedra linked together in ways that range from chains, rings, and sheets to three-dimensional networks. Some of the variations are illustrated in Table 11.4.

■ The serpentine form of asbestos is composed of two-dimensional sheets curled over into fibrous tubes.

The simplest network silicates are the *pyroxenes,* which contain extended chains of linked SiO_4 tetrahedra (Figure 11.13). If two such chains are laid side by side, they may link up by sharing oxygen atoms in adjoining chains. The result is an *amphibole.* Because of their double-stranded chain structure, the amphiboles are fibrous materials.

If the linking of silicate chains continues in two dimensions, sheets of SiO_4 tetrahedral units result. Various clays and mica have this sheet-like structure. Clays, which are essential components of soils, are *aluminosilicates*—some Si^{4+} ions are replaced by Al^{3+} ions plus other cations that take up the additional

■ Feldspars make up 60% of the Earth's crust.

positive charge. Feldspar, a component of many rocks and a network silicate, is weathered in the following reaction to form clay:

$$\underset{\text{Feldspar}}{2\ KAlSi_3O_8(s)} + CO_2(g) + 2\ H_2O(\ell) \longrightarrow \underset{\text{Kaolinite (a clay)}}{Al_2(OH)_4Si_2O_5(s)} + 4\ SiO_2(s) + K_2CO_3(aq)$$

Some medications sold in the United States (e.g., Kaopectate) contain highly purified clay that absorbs excess stomach acid and possibly harmful bacteria that cause stomach upset.

Glass

■ Glass and other solid substances (e.g., tar) that lack the internal order and properties of crystals are known as **amorphous** substances.

When silica is melted, some of the bonds are broken and the tetrahedral SiO_4 units move with respect to each other. On cooling, reorganization into the same orderly arrangement in crystalline silica is hard to achieve because of the difficulty the groups experience in moving past one another. Instead, cooling produces a **glass**—a hard, noncrystalline transparent substance with an internal structure like that of a liquid. The random liquid-like molecular arrangement accounts for one of the typical properties of a glass: it breaks irregularly rather than splitting along a plane like a crystal.

■ The n in these formulas represents a large number and is necessary to show that glass contains SiO_3 groups linked together rather than individual SiO_3^{2-} ions.

Common window glass is made by melting a mixture of silica with sodium and calcium carbonates. Bubbles of carbon dioxide are evolved, and the cooled mixture is a glass composed of sodium and calcium silicates:

$$n\ Na_2CO_3(\ell) + n\ SiO_2(s) \xrightarrow{\text{Heat}} Na_{2n}(SiO_3)_n + n\ CO_2(g)$$

$$n\ CaCO_3(\ell) + n\ SiO_2(s) \xrightarrow{\text{Heat}} Ca_n(SiO_3)_n + n\ CO_2(g)$$

White sand is the source of the silica for glass-making. Even the best grade of sand contains a small proportion of iron(III) compounds that give it a brown or yellow color. When this sand is made into glass, the iron is converted to a

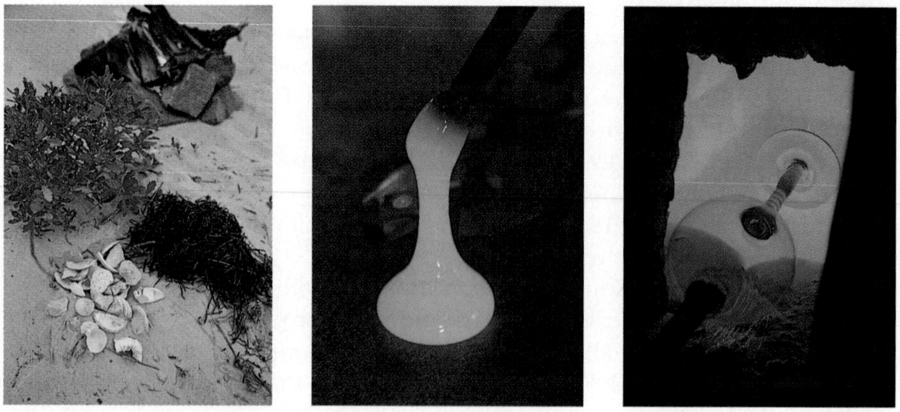

Glass. (a) Natural ingredients for making glass: sand (SiO_2), sea shells for lime ($CaCO_3$), and seaweed for soda ash (Na_2CO_3). (b) Molten glass ready to be shaped. (c) A hand-blown goblet being turned in the annealing oven. *(a, © James L. Amos/Peter Arnold, Inc.; b and c, Tom Pantages)*

(a) (b) (c)

TABLE 11.4 ■ Silicates

Si—O Tetrahedra Present as:	Class Name	Structure
Individual anions	Orthosilicates	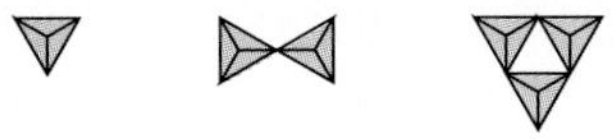
Chains	Pyroxenes (linear chains)	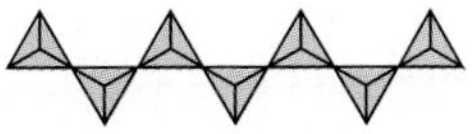
•	Amphiboles (double chains)	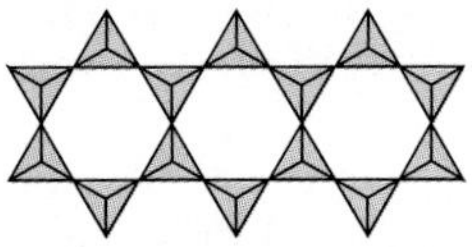
Sheets	Mica, talc, clays	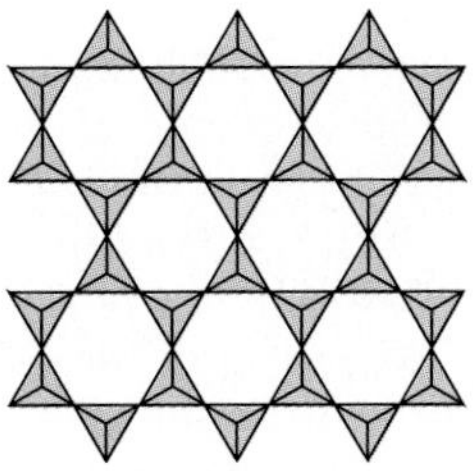
Three-dimensional networks	Silica	—
	Feldspars and zeolites	

(a)

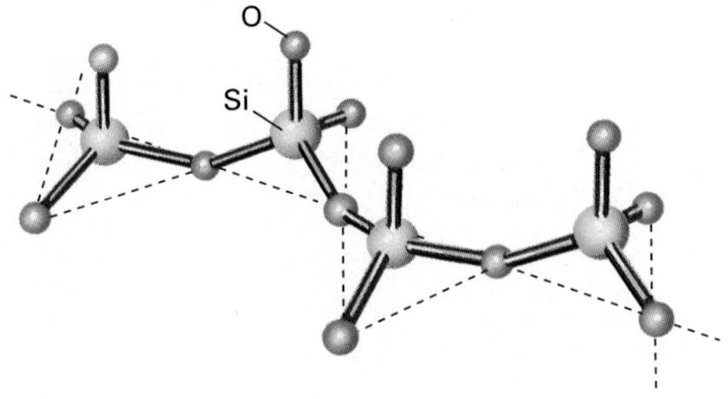

(b)

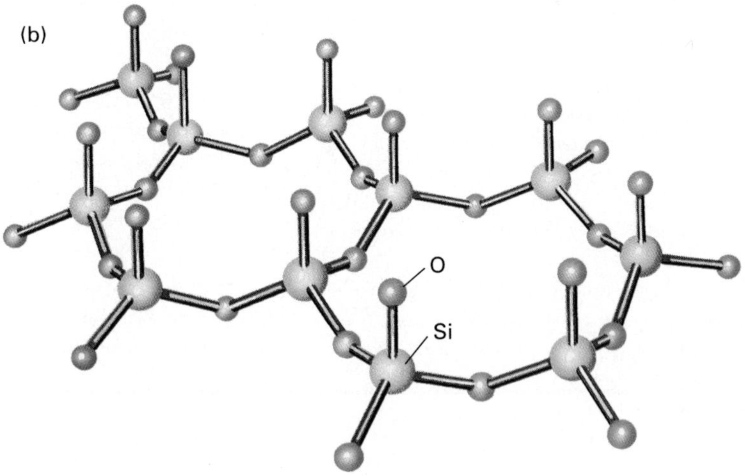

Figure 11.13 Silicate structures. (a) Pyroxene. Tetrahedral SiO_4 units are joined in chains by silicon-oxygen-silicon bonds. (b) Amphibole, which is asbestos. Chains of SiO_4 units are joined side-by-side by silicon-oxygen-silicon bonds.

SCIENCE AND SOCIETY

Asbestos

The term "asbestos" is generally applied to two forms of fibrous natural silicates: the amphiboles and serpentine. The asbestos minerals do not burn, do not rot, and have low thermal and electrical conductivity. They can be woven into fabrics, compressed into mats, and mingled with such binders as rubber and asphalt to produce strong and dimensionally stable composites. These properties have led to widespread use of asbestos in fireproofing materials, brake linings, floor tiles, pipes, and roofing materials.

The serpentine form of asbestos, known as *chrysotile,* is mined chiefly in Canada and the former Soviet Union; more than 90% of the asbestos used in the United States is in this form. The amphibole *crocidolite* is mined in small quantities, mainly in South Africa. The two minerals differ greatly in composition, color, shape, solubility, and persistence in human tissue. Crocidolite is blue, relatively insoluble, and persists in tissue. Its fibers are long, thin, and straight, and they penetrate narrow lung passages. In contrast, chrysotile is white, and it tends to be soluble and disappear in tissue. Its fibers are curly; they ball up like yarn and are more easily rejected by the body. Scientific studies of many types and by groups in many countries have shown that chrysotile asbestos is significantly less of a health hazard than other types. It is important to note that almost all manufactured materials in the United States contain only this form of asbestos.

Long-term occupational exposure to airborne asbestos fibers can, however, lead to a greater-than-average risk of lung cancer and certain other health problems, risks greatly enhanced by cigarette smoking. Understanding of this risk has fostered tight controls during the mining of asbestos and fabrication of asbestos-containing products.

The perceived risk also fostered an assortment of drives in the United States to ban all uses of asbestos and to remove all asbestos-containing products in existing public buildings and homes. One outcome of this activity came in 1989 when the U.S. Court of Appeals struck down the proposed U.S. Environmental Protection Agency (EPA) ban on asbestos, citing failure to provide sufficient evidence and failure to adequately consider alternative measures. EPA bypassed the opportunity to appeal this ruling. Another outcome was a congressionally mandated study of asbestos in buildings, which concluded in 1991 (after seven years of study) that asbestos-containing material in good repair inside buildings creates no higher asbestos fiber concentration in the air inside the buildings than in the air outside the buildings. Subsequently, an EPA advisory, while stressing assessment of asbestos risk in public buildings, stated:

Removal is often not *a school district's or other building owner's best course of action to reduce asbestos exposure. In fact, an improper removal can create a dangerous situation where none previously existed.*

"An Advisory to the Public on Asbestos in Buildings," U.S. Environmental Protection Agency, 1991.

Chrysotile asbestos. (C.D. Winters)

Public fears about asbestos exposure were still in the news in 1993, when school opening in New York City was delayed because of delays in removal. At that time, a group of 17 scientists and physicians from throughout the United States and the United Kingdom, all of them actively involved in asbestos study, stated:

Except under unusual conditions, such as demolition or extensive renovation, we strongly advise against asbestos removal in schools. It is scientifically unsound, economically wasteful, and medically imprudent . . . since it carries potentially unnecessary future risks.

Letter to the Editor, *The New York Times,* December 23, 1993.

Armed with information of this kind, what would you do if your town proposes to spend $10 million of taxpayers money on asbestos removal in your local high school? What questions should be asked of the decision makers?

TABLE 11.5 ■ Substances Used to Color Glass

Substance	Color
Copper(I) oxide	Red, green, blue
Tin(IV) oxide	Opaque
Calcium fluoride	Milky white
Manganese(IV) oxide	Violet
Cobalt(II) oxide	Blue
Finely divided gold	Red, purple, blue
Uranium compounds	Yellow, green
Iron(II) compounds	Green
Iron(III) compounds	Yellow

mixture of light green iron(II) silicates, explaining the green tint of some old bottles. Adding a manganese compound to the melt produces pink manganese silicates, which offset the green of the iron silicates, making the glass appear colorless.

Countless variations in glass composition and properties are possible. Many beautiful colors can be produced by adding the substances listed in Table 11.5. If part of the silica is replaced by boron oxide, the glass has less tendency to crack with changes in temperature. Pyrex, the trademarked glass common in kitchens and laboratories, is a borosilicate glass. The composition and properties of some other types of glasses are listed in Table 11.6.

In the manufacture of glass, proper **annealing** is important. Annealing is the cooling schedule that a glass is put through on its way from a viscous, liquid state to a solid at room temperature. If a glass is cooled too quickly, bonding forces become uneven in local regions as small areas of crystallinity develop. This results in strain that will cause the glass to crack or shatter when subjected to mechanical shocks or sudden temperature changes. High-quality glass, such as that used in optics, must be annealed very carefully. The huge Mt. Palomar, California, observatory mirror was annealed from 500°C to 300°C over a period of nine months!

TABLE 11.6 ■ Some Special Glasses

Special Addition or Composition	Desired Property
Large amounts of PbO with SiO_2 and Na_2CO_3	Brilliance, clarity, suitable for optical structures: crystal or flint glass
SiO_2, B_2O_3, and small amounts of Al_2O_3	Small coefficient of thermal expansion; borosilicate glass: Pyrex, Kimax, and others
One part SiO_2 and four parts PbO	Ability to stop (absorb) large amounts of X rays and gamma rays: lead glass
Large concentrations of CdO	Ability to absorb neutrons
Large concentrations of As_2O_3	Transparency to infrared radiation

DISCOVERY EXPERIMENT

The Importance of Annealing

Glasses are not the only solids whose properties are affected by annealing. Obtain a glass of water, at least four bobby pins, and some kind of tongs or tweezers that you can use to hold the bobby pins in a flame without burning your fingers. Scrape off any plastic coating from the ends of the bobby pins. Hold one bobby pin by its open end and heat the bend red-hot in the flame of a gas stove or other gas burner. Remove the bobby pin from the flame and place it on a nonflammable surface to cool. After it has cooled, try bending the bobby pin by pulling both ends apart; what do you observe?

Take a second bobby pin and also heat its bend until it is red-hot; when you remove it from the flame, plunge it into the glass of water so that it cools very rapidly. Try to bend this bobby pin; what happens?

Heat a third bobby pin red-hot and plunge it into the water, but then gently heat it again until an iridescent blue coating forms on the surface; now slowly cool the bobby pin. What happens this time when you try to bend it?

When you test each bobby pin, compare its response to bending with that of the fourth bobby pin, which serves as a control. What can you conclude about the influence of heat treatment and speed of cooling of a metal on its properties?

Ceramics

What do you know about "ceramics"? Perhaps you associate the term with pottery vases that you see at a crafts show, or bathroom tile, or maybe components of electrical equipment. **Ceramics** are a large and diverse class of materials with the properties of nonmetals. What they have in common is that all are made by baking or firing minerals or other substances, often including silicates and metal oxides.

Ceramic materials have been made since well before the dawn of recorded history. They are generally fashioned from clay or other natural earths at room temperature and then permanently hardened by heat. Silicate ceramics include objects made from clays, such as pottery, bricks, and table china. The three major ingredients of common pottery are clay (from weathering of feldspar as described previously), sand (silica), and feldspar (aluminosilicates). Clays mixed with water form a moldable paste because they consist of many tiny silicate sheets that can easily slide past one another. When the clay-water mixture is heated, the water is driven off, and new Si—O—Si bonds are formed so that the mass of platelets becomes permanently rigid.

A potter at his wheel, with already-fired pieces in the background. *(Blair Seitz/ Photo Researchers)*

A new type of glasslike ceramics with unusual properties has become widely available for home and industrial use. Ordinary glass breaks because once a crack starts, there is nothing to stop the crack from spreading. It was discovered that if glass objects produced in the usual manner are heated until many tiny crystals develop, the resulting material, when cooled, is much more resistant to breaking than normal glass. In molecular terms, the randomness of the glass structure has been partially replaced by the order of a crystalline silicate. The materials produced in this way (an example is Pyroceram) are generally opaque and are used for kitchenware and in other applications in which the material is subjected to stress or high temperatures.

Ceramic materials are attractive for several reasons. The starting materials for making them are readily available and cheap. Ceramics are light-weight in comparison with metals and retain their strength at temperatures above

1000°C, where metal parts tend to fail. They also have electrical, optical, and magnetic properties of value in the computer and electronic industries.

The one severely limiting problem in utilizing ceramics is their brittle nature. Ceramics deform very little before they fail catastrophically, the failure resulting from a weak point in the bonding within the ceramic matrix. However, such weak points are not consistent from object to object, so that the predictability of failure is poor. Since the stress failure of ceramic materials is due to molecular abnormalities resulting from impurities or disorder in the basic atomic arrangements, much attention is now being given to purer starting materials and the control of the processing steps. In addition, ceramic **composite materials,** mixtures of ceramic materials or of ceramic fibers with plastics, can overcome the tendency of plain ceramics to crack.

A concrete chapel in Ronchamp, France. Le Corbusier (1887–1965), a pioneer in the sculptural use of industrial materials, designed this building in the 1950s. *(Paolo Koch/Photo Researchers)*

Cement and Concrete

A **cement** is a material that can bond mineral fragments into solid mass. The most common cement, known as Portland cement, is made by roasting a powdered mixture of calcium carbonate (limestone or chalk), silica (sand), aluminosilicate mineral (kaolin, clay, or shale), and iron oxide at a high temperature in a rotating kiln. As the materials pass through the kiln, they lose water and carbon dioxide and ultimately form "clinker," in which the materials are partially melted together. The clinker is ground to a very fine powder after the addition of a small amount of calcium sulfate (gypsum). A typical composition of a Portland cement, expressed in terms of oxides, is 60% to 67% CaO, 17% to 25% SiO_2, 3% to 8% Al_2O_3, up to 6% Fe_2O_3, and small amounts of magnesium oxide, magnesium sulfate, and potassium and sodium oxides. As in glass, the oxides are not isolated into molecules or ionic crystals, and the submicroscopic structure is quite complex.

■ Portland cement, patented in England in 1824, is named for its similarity to a rock native to the Isle of Portland in England. It is also called *hydraulic cement* because it sets even under water.

Many different reactions occur during the setting of cement. Initially the calcium silicates react with water to give a sticky gel. The gel has very large surface area and is responsible for the strength of concrete. Reactions with carbon dioxide in the air also occur at the surface. After the initial solidification, small, densely interlocked crystals begin to form, a process that continues for a long time and increases the compressive strength of the cement.

■ **Mortar** is a mixture of cement, sand, water, and lime. **Concrete** is a mixture of cement, sand, and aggregate (crushed stone or pebbles).

More than 800 million tons of cement are manufactured each year, most of which is used to make concrete. Concrete, like many other materials containing Si—O bonds, is virtually noncompressible but lacks tensile strength. If concrete is to be used where it will be subject to tension, it must be reinforced with steel.

■ *Tensile strength* refers to the resistance of a material to being stretched.

■ SELF-TEST 11C

1. The two principal nonmetals in glass are ____________ and ____________.
2. Which of the following apply to (i) silica and (ii) silicates?
   (a) SiO_2
   (b) Quartz

(c) Contain metal ions
(d) Clay
(e) Built up from tetrahedral units

3. How does glass differ from silica?
4. What three silicate minerals are used to make common pottery?
5. A distinctive property of ceramics is their ability to withstand high ____________. Another is their tendency to fail in response to ____________.
6. The three principal ingredients of concrete are ____________, ____________, and ____________.

MATCHING SET

___ 1. Helium	a. Reduction
___ 2. SiO_4	b. Lead
___ 3. Steel	c. Magnesium
___ 4. Sodium hydroxide	d. A cryogen
___ 5. Most recycled metal	e. Ore
___ 6. $Fe^{3+} \rightarrow Fe$	f. In mortar and concrete
___ 7. Valuable mixture of minerals	g. Aluminum
___ 8. Ceramics and glass	h. Product of chloralkali industry
___ 9. Liquid nitrogen	i. Element obtained from natural gas
___ 10. Seashells	j. Building block of Earth's crust
___ 11. Production is energy-intensive	k. Used to separate air
___ 12. From the ocean	l. Oxidation
___ 13. Fractionation	m. Source of calcium carbonate
___ 14. Portland cement	n. Alloy
___ 15. $Fe \rightarrow Fe^{3+}$	o. Made of silicates

QUESTIONS FOR REVIEW AND THOUGHT

1. Define the following terms.
   (a) atmosphere (b) hydrosphere
   (c) igneous rock (d) sedimentary rock
2. Define the following terms.
   (a) mineral (b) ore
   (c) fractionation of air (d) LOX
3. Define the following terms.
   (a) cryogen (b) cryosurgery
   (c) nitrogen fixation (d) "the bends"
   (e) brine
4. Define the following terms.
   (a) slag (b) ductile
   (c) brine (d) alloy
5. Define the following terms.
   (a) annealing (b) amorphous
   (c) ceramic (d) cement
   (e) glass
6. Explain what the Haber process is, the raw materials used, its importance to agriculture, and its importance in world history.
7. Why is the Earth's atmosphere richer in nitrogen and oxygen than in helium and hydrogen?
8. Nitrogen boils at −195.8°C and oxygen boils at −183°C. Which one boils at the higher temperature?
9. What is the main source of
   (a) liquid oxygen (b) argon
   (c) helium (d) neon
10. Name two dangers of liquid oxygen. Which one of these hazards is not shared by liquid nitrogen?
11. Why would it be hazardous to enter a refrigerated boxcar in which liquid nitrogen had been used to keep meat frozen?
12. What are the two products of the "chloralkali" process? Name two common uses for each of the products.

13. Name a common metal taken from sea water. Give several uses for this metal.
14. Complete and balance (if needed) the following equations.
    (a) $N_2(g) + H_2(g) \rightarrow NH_3(g)$
    (b) $Al^{3+}(aq) + 3\ e^- \rightarrow$ ________
    (c) $Mg^{2+}(aq) + 2\ OH^-(aq) \rightarrow$ ________
    (d) $Cl^-(aq) \rightarrow Cl_2(g) +$ ________
    (e) $CaCO_3(s) \rightarrow$ ________ $+ CO_2(g)$
    (f) $Mg^{2+}(aq) +$ ________ $\rightarrow Mg(s)$
15. Name two elements that occur in the free metallic state in nature.
16. What are the three ingredients used in a blast furnace to make iron? Which one of these is the reducing agent?
17. Explain why the molten mixture in a blast furnace separates naturally into a layer of slag and a layer of molten iron.
18. What are the differences between pig iron, cast iron, and steel? Mention composition as well as properties and uses.
19. What is the principal impurity in pig iron?
20. What element is used to convert pig iron into steel?
21. Complete and balance (if needed) the following equations.
    (a) $Al(s) + O_2(g) \rightarrow$ ________
    (b) $Na(s) + AlCl_3(s) \rightarrow$ ________ $+ NaCl(s)$
22. What is cryolite and what property does it have that is especially important in the economic production of aluminum from its ore?
23. What is the name of the common ore of aluminum?
24. Aluminum is the most common metal in the Earth's crust. Why is aluminum so expensive to produce?
25. The top of the Washington monument is made of a large casting of aluminum. It was placed there in 1884. Give two reasons why aluminum was used.
26. Discuss the recycling of iron, aluminum, and lead. Explain why each is recycled and describe how they are recycled in your community. Indicate whether you personally have ever recycled any of these metals. Do you intend to recycle any of these metals in the future?
27. Draw the structural unit SiO_4, found in many silicate minerals. What is the name for the shape of this structural unit?
28. Draw a structure consisting of SiO_4 units arranged in a chain with each unit sharing two of its oxygen atoms with neighboring SiO_4 units. What is this silicate structure named?
29. Why is asbestos dangerous?
30. What is the difference between the two types of asbestos, chrysotile and crocidolite? Which poses the lesser health risk?
31. The element silicon occurs in all clays. Name two other elements common in all clays.
32. What element causes glass to have a brown to yellow color? (*Hint:* This is the same element that causes some clays to be red in color.)
33. What is the purpose of annealing glass?
34. Compare ceramics with metals by filling in the following table with a "yes" or a "no" depending on whether the material has that property.

	Metal	**Ceramic**
Hardness	________	________
Strength	________	________
Ductility	________	________
Electrical conductivity	________	________
Brittleness	________	________

35. What is a ceramic composite, and how might it be useful?
36. Write the formula for limestone.
37. What is Portland cement?
38. Cements contain calcium oxide (lime), SiO_2 (sand), and various other metal oxides. Which of the oxides present reacts with CO_2 in the air, which helps bond everything together?
39. How is the structure of a common glass different from the structure of a solid like sodium chloride?
40. When a molten mixture of different materials cools, will all of the substances crystallize at the same time? Silicates and quartz, SiO_2, have the highest melting points. Which will crystallize first, the metallic elements or the silicates?
41. What substance is a mixture of $Na_{2n}(SiO_3)_n$ and $Ca_n(SiO_3)_n$?
42. The extraction of iron from its ore is an oxidation-reduction reaction. The first reaction between iron oxide, $Fe_2O_3(s)$, and carbon monoxide, $CO(g)$, is

$$Fe_2O_3(s) + 3\ CO(g) \longrightarrow 2\ Fe(s) + 3\ CO_2(g) + \text{heat}$$

What is oxidized and what is reduced in the reaction?
43. Aluminum was discovered in 1825. The first method to purify aluminum involved the reaction of aluminum chloride with a more active metal:

$$AlCl_3(s) + 3\ Na(s) \longrightarrow Al(s) + 3\ NaCl(s)$$

What is oxidized and what is reduced?
44. What two materials from the Earth's crust have the highest annual consumption per person in the United States?
45. Explain the differences between concrete, cement, and mortar.
46. To celebrate the opening of an American film *(Aladdin)*, 1.5 million helium-filled balloons were released in mid-1994 in a park in England. It must have been a beautiful sight. A number of questions, however, might be asked about the balloons and their contents. What are some of these questions?
47. Would it make any difference in our ability to extract pure elements if they were uniformly distributed on Earth instead of being concentrated as many of them are?
48. Discuss the pros and cons of asbestos removal in schools and other public buildings.
49. What is the concept behind green design? Name a common product that has been "green-designed."

CHAPTER 12

Energy and Hydrocarbons

The average use of energy per individual is near the highest point in the history of the world. In the United States alone, with only 5% of the world's population, we consume 24% of the daily supply of energy. The combustion of the fossil fuels (coal, petroleum, and natural gas) provides 85% of all the energy used in the world. At current usage rates, world reserves of coal, natural gas, and petroleum are estimated to last 230, 120, and 65 years, respectively.

Hydrocarbons are the principal component of fossil fuels. Natural gas is primarily methane, crude petroleum is a complex mixture of thousands of hydrocarbons, and coal is an even more complex mixture of hydrocarbons. Many of the fuels we use, such as gasoline and jet fuel, are obtained from petroleum.

Fossil fuels are also the major source of hydrocarbons that are used to make thousands of consumer products. This chapter describes the chemistry of hydrocarbons and their importance to our energy needs, and Chapter 14 emphasizes the industrial uses of hydrocarbons. Alcohols and ethers are part of the energy discussion because of the need to improve emissions and reduce pollution.

- What are fuels?
- How do fuels produce energy?
- What are the major classes of hydrocarbons?
- What different types of isomers are possible for hydrocarbons, and why are they important?
- How is petroleum refined?
- How is high-octane gasoline produced?
- What are oxygenated gasolines, and why are they used?
- Why are methanol and ethanol receiving attention as alternate fuels?

Coal, one of our major fossil fuels. *(Grant Heilman/Grant Heilman Photography)*

Why does burning fuels provide energy? Fuels are reduced forms of matter that burn readily in the presence of oxygen, and combustion is an oxidation reaction (Section 10.2) that produces heat. The burning of methane, the principal component of natural gas, can be used to illustrate how the combustion of fuels produces energy.

12.1 Energy from Fuels

The release of energy when methane burns can be represented by the following chemical equation:

$$CH_4(g) + 2\ O_2(g) \longrightarrow CO_2(g) + 2\ H_2O(g) + 192\ \text{kcal}$$

■ Although the SI unit of energy is the joule, the more familiar unit of heat is the calorie. A calorie is the amount of heat required to raise the temperature of 1 g of water 1°C. One calorie equals 4.18 joules (J). 1000 calories = 1 kilocalorie (kcal). For example, you use 140 kcal per hour walking and 80 kcal per hour even when you are asleep.

The balanced equation for the oxidation of a fuel is not only a quantitative expression for mole relationships of products and reactants but also for the energy change of the reaction. The above equation shows that burning one mole of methane produces 192 kcal of energy. Since there are 16 grams in one mole of methane, the amount of energy released per gram of methane is 12 kcal. The heat released in a combustion reaction can be measured quantitatively using a calorimeter (Figure 12.1).

What determines whether energy is released or absorbed in a chemical reaction? For a chemical reaction to occur requires breaking bonds in reactants and forming bonds in products. If the bonds in reactants are weaker than those formed in products, energy will be released. For example, energy is

■ Bond energies were first introduced in Section 6.3.

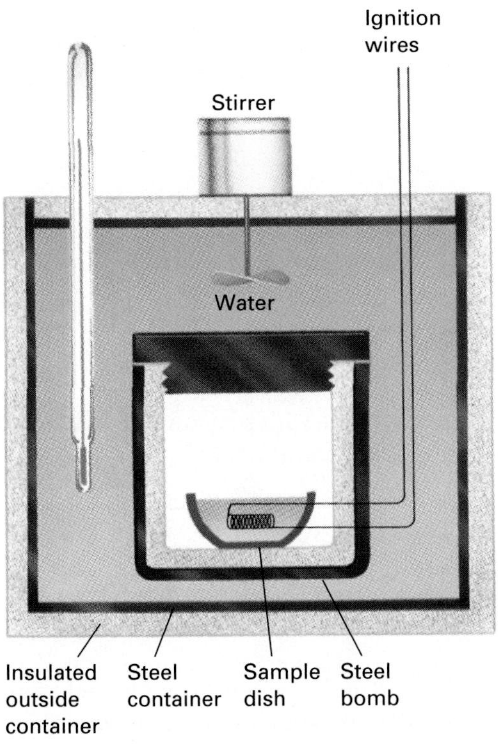

Figure 12.1 A combustion calorimeter. A combustible sample is burned in pure oxygen in a steel "bomb." The heat generated by the reaction flows into the bomb and the water surrounding it, and warms both to the same temperature. By measuring the temperature increase, the heat evolved by the reaction can be determined.

TABLE 12.1 ■ **Average Bond Energies in kcal/mol**

Single Bonds

	H	C	N	O	F	Si	P	S	Cl	Br	I
H	104	99	93	111	135	70	76	81	103	88	71
C		83	70	84	105	69	63	62	79	66	57
N			40	48	65	—	50	—	48	—	—
O				33	44	88	84	—	49	—	48
F					37	129	117	68	61	—	—
Si						42	51	54	86	69	51
P							51	55	79	65	51
S								51	60	51	—
Cl									58	52	50
Br										46	43
I											36

Multiple Bonds

$N=N$	100	$C=C$	146
$N\equiv N$	225	$C\equiv C$	200
$O=O$ (in O_2)	118	$C=O$ (in CO_2)	192
		$C\equiv O$ (in CO)	257

released when methane burns because it takes less energy to break the bonds in the reactants, methane and oxygen, than is produced when the products, carbon dioxide and water, are formed. It is possible to use bond energies to estimate the heat of combustion. For example, the average bond energies for bonds in methane, oxygen, carbon dioxide, and water can be used to calculate the heat of combustion of methane. Table 12.1 lists average bond energies for a number of common bonds.

EXAMPLE 12.1 *Heat of Combustion*

Use bond energies to calculate the heat of combustion of methane, the principal component of natural gas.

SOLUTION

Write the equation for the combustion of methane with structural formulas to determine the type and number of bonds broken and made.

$$
\begin{array}{c}
\text{H} \\
| \\
\text{H}-\text{C}-\text{H} + 2\,\text{O}=\text{O} \longrightarrow \text{O}=\text{C}=\text{O} + 2\,\text{H}-\text{O}-\text{H} \\
| \\
\text{H}
\end{array}
$$

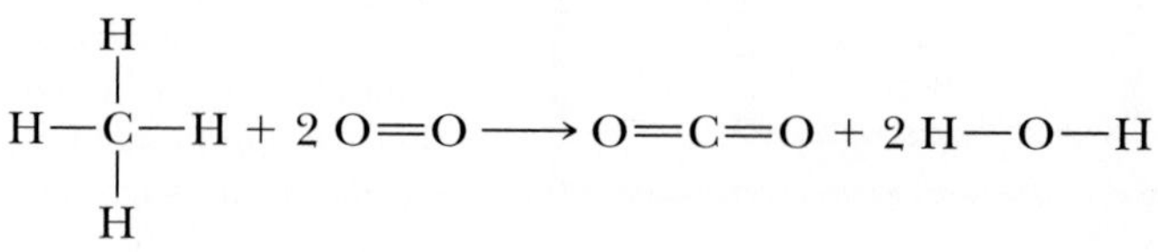

Breaking bonds always requires energy, and making bonds always releases energy. To calculate the energy released by combustion of methane, calculate

the total energy needed to break all the bonds in the reactants and subtract this number from the total energy released when all the bonds in the products are formed.

Breaking bonds of reactants:

The C—H bond energy is 99 kcal per bond, and the O═O bond energy is 118 kcal per bond. Breaking four C—H bonds requires 4 × 99 kcal, and breaking two O═O bonds requires 2 × 118 kcal. The total is 396 kcal + 236 kcal = 632 kcal for breaking the bonds in one mole of methane and two moles of oxygen.

Making bonds of products:

The C═O bond energy is 192 kcal per bond, and the O—H bond energy in water is 111 kcal per bond. Making two C═O bonds releases 2 × 192 kcal, and making four O—H bonds releases 4 × 111 kcal. The total is 384 kcal + 444 kcal, or 828 kcal released when the bonds of the products are formed.

Calculated energy for combustion of methane:

The net release of energy is 828 kcal − 632 kcal = 196 kcal per mole of methane. The experimentally measured value is 192 kcal per mole, so the use of average bond energies provides a reasonable estimate of heat of combustion energies.

■ In chemical reactions, heat can be released, as in the combustion of fuels, or absorbed, as in decomposition. For example, decomposing water into hydrogen and oxygen requires heat. The amount of heat released or absorbed in every case depends on the differences in bond strengths of the reactants and products.

Exercise 12.1

Use bond energies to calculate the heat of combustion of ethane:

$$2\,C_2H_6(g) + 7\,O_2(g) \longrightarrow 4\,CO_2(g) + 6\,H_2O(g)$$

■ Bonding in the alkane, alkene, and alkyne classes of hydrocarbons (compounds of hydrogen and carbon) was discussed in Section 6.3.

TABLE 12.2 ■ Energy Values of Some Fuels

Substance	Heat (kcal/g)
Hydrogen	34.0
Natural gas (mostly methane)	11.7
Gasoline	11.5
Anthracite coal	7.4
Wood (pine)	4.3

The amounts of energy per gram obtained by burning some common fuels are given in Table 12.2. Any hydrocarbon or hydrocarbon mixture can be classified as a fuel, and fossil fuels are complex mixtures of hydrocarbons. Hence, an understanding of hydrocarbon chemistry is central to the study of fossil fuels.

There are four classes of hydrocarbons: the **alkanes,** which contain carbon-carbon single bonds (C—C); the **alkenes,** which contain one or more carbon-carbon double bonds (C=C); the **alkynes,** which contain one or more carbon-carbon triple bonds (C≡C); and the **aromatics,** which consist of benzene, benzene derivatives, and fused benzene rings.

12.2 Alkanes—Backbone of Organic Chemistry

The simplest alkane is methane, CH_4, the principal component of natural gas. Alkanes are **saturated hydrocarbons** (Section 6.3) with the general formula C_xH_{2x+2} (Table 12.3). Notice that all hydrocarbon formulas are traditionally written with the C atom first, followed by the H atom, and that all alkanes have "-ane" as the suffix in their name. When $x = 1$ to 4, the first part of the name is something of historical origin; these are common names that we just have to remember. When $x = 5$ or more, the Greek prefixes (Table 6.3) tell how many carbon atoms are present. For example, the compound with six carbons is called hexane.

■ Alkanes are referred to as saturated hydrocarbons because they contain the highest ratio of hydrogen to carbon possible.

The tetrahedral structure of CH_4 has been discussed (Section 6.6), but it is important to recognize that *every* carbon in an alkane has a tetrahedral environment because all carbon atoms in saturated hydrocarbons have four single bonds. The tetrahedral environment of the carbon atoms in alkanes is difficult to draw in two dimensions. The drawings below depict this for methane and ethane in a perspective drawing that uses solid wedges to represent bonds extending above the page, and dashed lines to represent bonds extending behind the page.

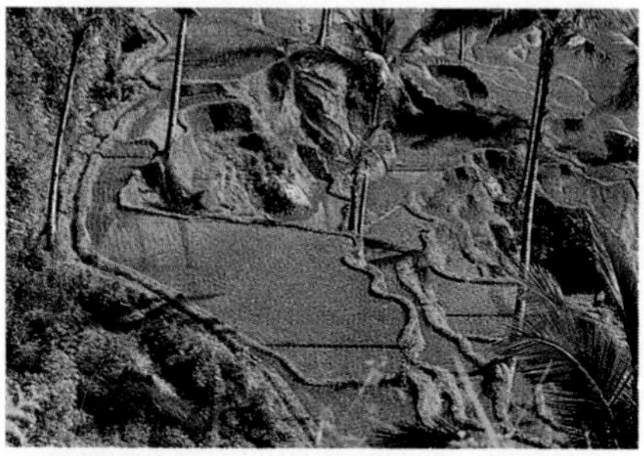

Rice fields in the Phillipines. Decay of organic matter in rice fields is estimated to make up one fifth of all methane emitted each year due to human activities. *(Kjell B. Sandved/Visuals Unlimited)*

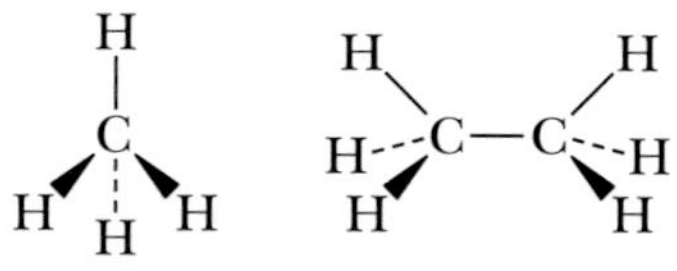

In order to save time and space, chains of carbon atoms are usually represented with straight lines as in Figures 12.2*a*, 12.3*a*, and 12.4*a*. However, keep in mind that these drawings are not an accurate representation of the tetrahedral bond angles, which are 109.5°.

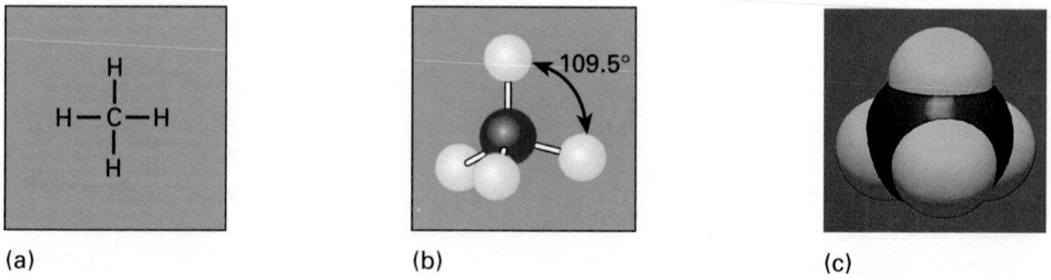

Figure 12.2 The structure of methane, as represented by (a) its structural formula, (b) a ball-and-stick model, and (c) a space-filling model. *(c, C.D. Winters)*

TABLE 12.3 ■ The First Ten Straight-Chain Saturated Hydrocarbons

Name	Formula	Boiling Point, °C*	Structural Formula	Use
Methane	CH_4	−162	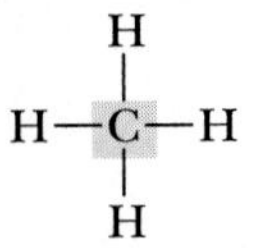	Principal component in natural gas
Ethane	C_2H_6	−88.5	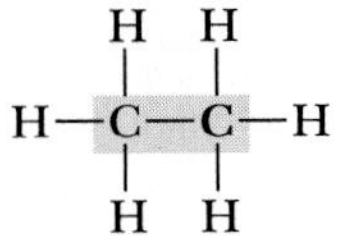	Minor component in natural gas
Propane	C_3H_8	−42	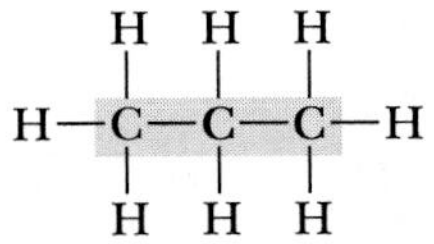	Bottled gas for fuel
Butane	C_4H_{10}	0	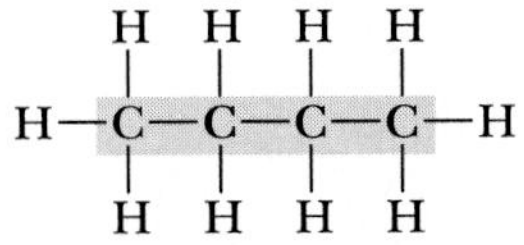	
Pentane	C_5H_{12}	36	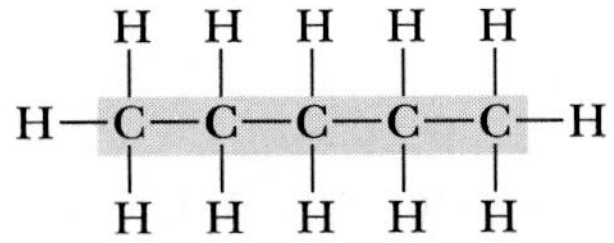	
Hexane	C_6H_{14}	69	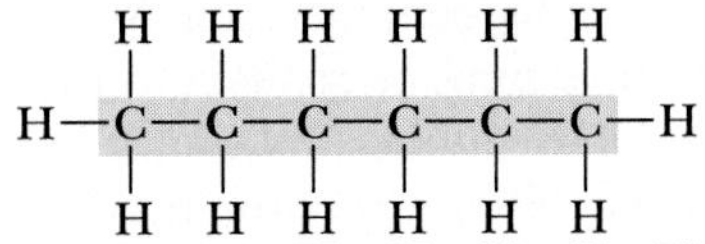	
Heptane	C_7H_{16}	98	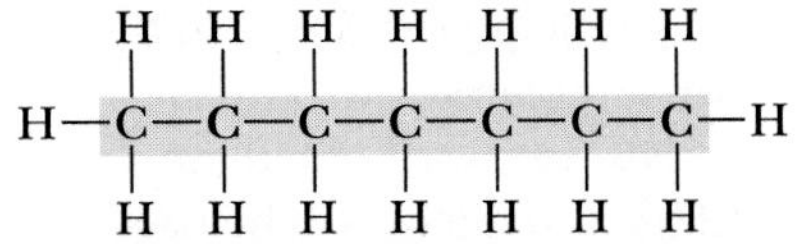	Some of the components of gasoline
Octane	C_8H_{18}	126	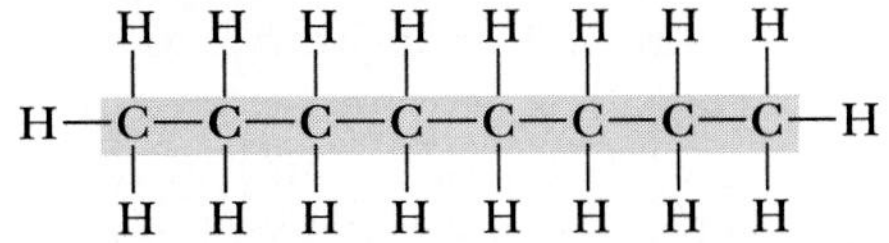	
Nonane	C_9H_{20}	151	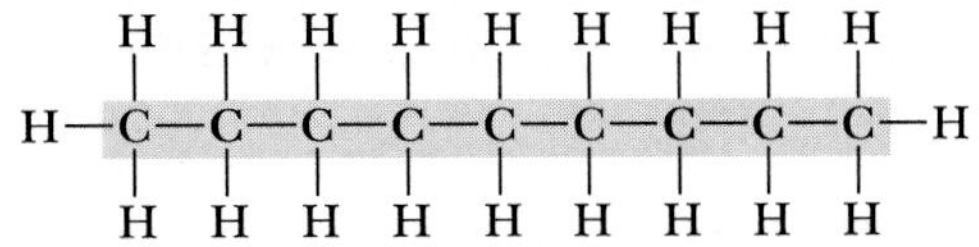	
Decane	$C_{10}H_{22}$	174	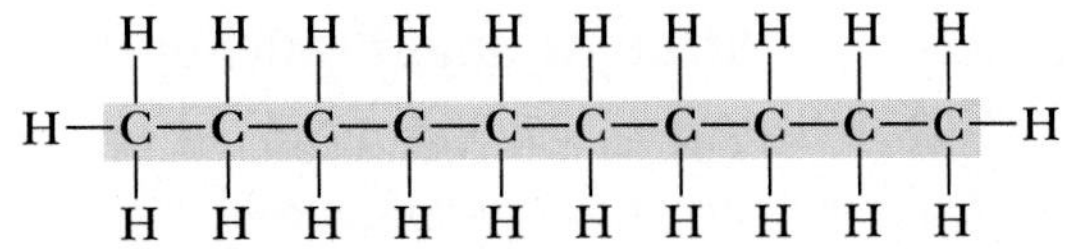	Found in kerosene

*Notice the gradual increase in boiling point as the molecular weight increases. Fractional distillation of petroleum is possible because of these differences (Section 12.7).

Our ability to understand these tetrahedral structures is helped by the use of models. Two types of models are generally used—the "ball-and-stick" model and the "space-filling" model. In a ball-and-stick model, balls represent the atoms, and short pieces of wood or plastic represent bonds. For example,

Figure 12.3 The structure of ethane, as represented by (a) its structural formula, (b) a ball-and-stick model, and (c) a space-filling model. *(c, C.D. Winters)*

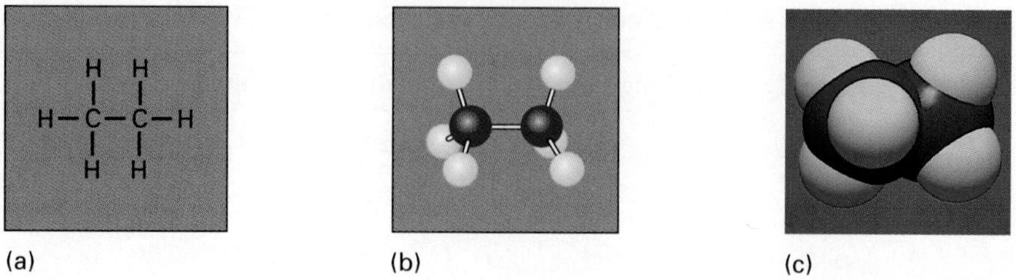

Figure 12.4 The structure of propane, as represented by (a) its structural formula, (b) a ball-and-stick model, and (c) a space-filling model. *(c, C.D. Winters)*

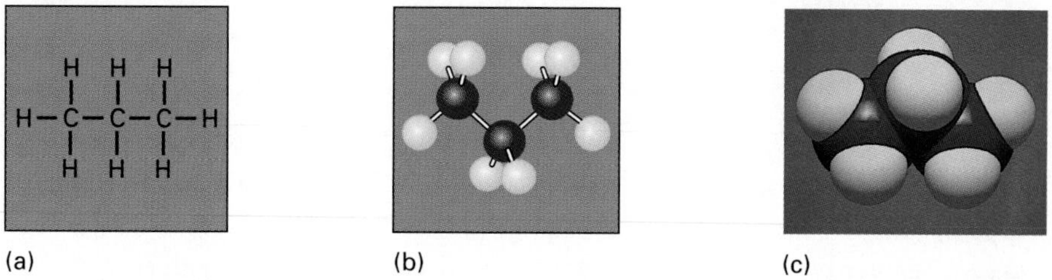

the ball-and-stick model for methane (Figure 12.2*b*) has a black ball representing carbon, with holes at the correct angles connected by sticks to four white balls representing hydrogen atoms. The space-filling model (Figure 12.2*c*) is a more realistic representation because it depicts both the relative sizes of the atoms and their spatial orientation in the molecule. This is done by scaling the pieces in the model according to the experimental values of atoms' sizes. The pieces are held together by links that are not visible when the model is assembled.

Illustrations of ball-and-stick models, such as those shown in Figures 12.2*b*, 12.3*b*, and 12.4*b*, will be used extensively in the discussion of hydrocarbons and hydrocarbon derivatives to help you visualize the molecular geometry of molecules. For example, notice that the three carbon atoms in the ball-and-stick model of propane in Figure 12.4*b* do not lie in a straight line because of the tetrahedral geometry about each carbon atom. This illustrates why straight-chain drawings, such as that for propane in Figure 12.4*a*, are not accurate representations of the tetrahedral H—C—H and C—C—C bond angles.

■ Historically, straight-chain hydrocarbons were referred to as *normal* hydrocarbons, and *n*- was used as a prefix in the name of straight-chain hydrocarbons such as butane (*n*-butane). The current practice is not to use *n*-. If a name is given without indicating that the compound is branched-chain, assume the compound is a straight-chain hydrocarbon.

Straight- and Branched-Chain Isomers of Alkanes

The first three alkanes, CH_4, C_2H_6, and C_3H_8, each have only one possible structural arrangement. However, two structural arrangements are possible for C_4H_{10}—a straight-chain arrangement and a branched-chain arrangement.

Expanded structural formulas:

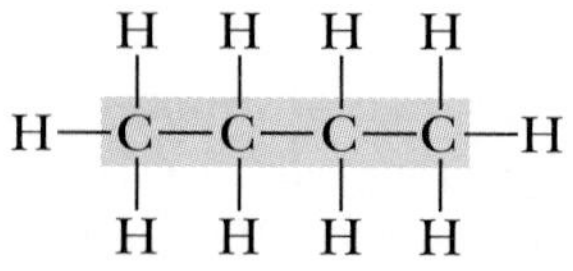

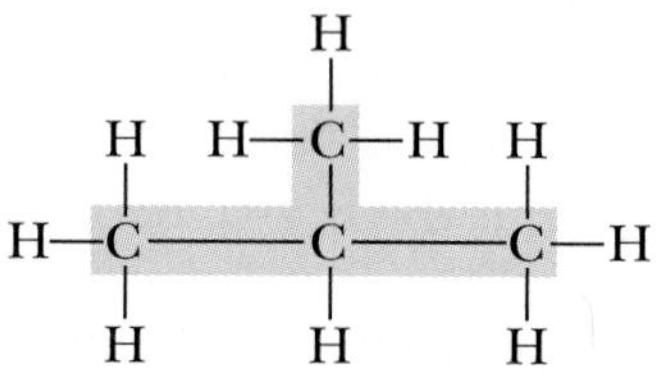

CH_3
Condensed formulas:	$CH_3CH_2CH_2CH_3$ Butane	CH_3CHCH_3 Methylpropane (isobutane)
Melting point	−138.3°C	−160°C
Boiling point (1 atm)	0.5°C	−12°C
Density (at 20°C)	0.579 g/mL	0.557 g/mL

These drawings are examples of the two ways of writing structural formulas—in expanded form and in condensed form (Section 2.7). The expanded form is the same as the Lewis structure. These molecules have different properties even though they have the same number of atoms in the molecule. Ball-and-stick models of the two structures are shown in Figure 12.5.

Two or more compounds with the same molecular formula but different arrangements of atoms are called **isomers.** Isomers differ in one or more physical or chemical properties such as boiling point, color, solubility, reactivity, and density. Several different types of isomerism are possible for organic compounds. **Branched-chain** and **straight-chain** isomers are examples of **structural isomers** that differ in the order in which the atoms are bonded together. Structural isomerism can be compared to the results you might expect from a child building many different structures with the same collection of building blocks and using all the blocks in each structure.

The branched-chain isomer for C_4H_{10}, methylpropane, a common component of bottled gas, has a "methyl" group ($—CH_3$) attached to the central carbon atom. This is the simplest example of the fragments of alkanes known as **alkyl groups.** In this case, removal of H from methane gives a **methyl** group.

$$H-\underset{\underset{H}{|}}{\overset{\overset{H}{|}}{C}}-H \xrightarrow{-H} H-\underset{\underset{H}{|}}{\overset{\overset{H}{|}}{C}}- \quad \text{or} \quad -CH_3 \left(\begin{array}{c}\text{also written}\\\text{as } CH_3-\end{array}\right)$$

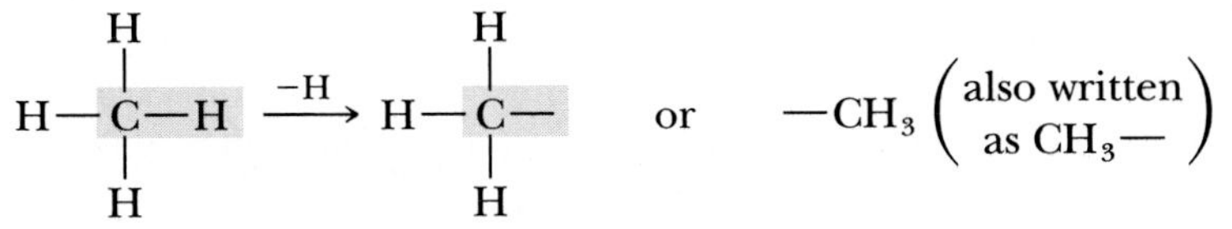

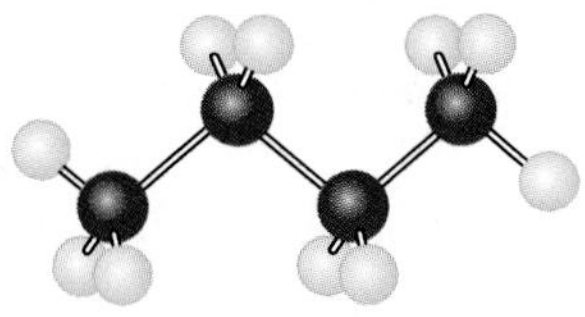

Butane

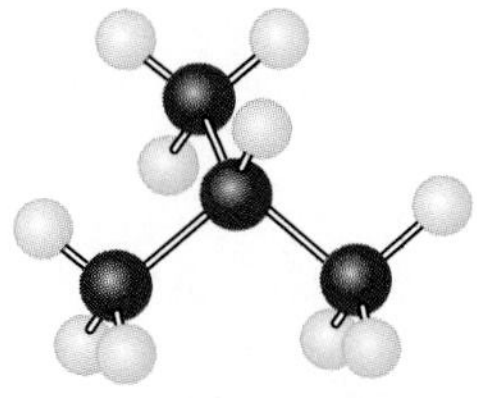

Methylpropane
(isobutane)

Figure 12.5 Ball-and-stick models of butane, a four-carbon straight-chain hydrocarbon, and methylpropane, a four-carbon branched chain hydrocarbon.

Removal of an H from ethane gives an **ethyl** group:

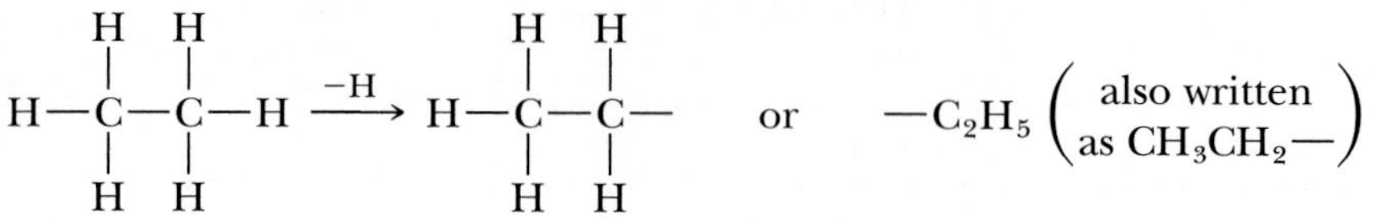

Notice that more than one alkyl group is possible when an H atom is removed from C_3H_8:

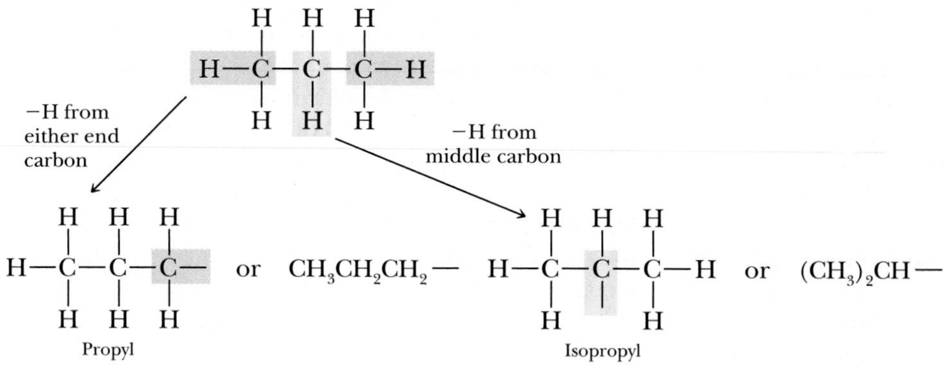

Alkyl groups are named by dropping "-ane" from the parent alkane and adding "-yl." Theoretically, any alkane can be converted to an alkyl group. Some of the more common examples of alkyl groups are given in Table 12.4.

Table 12.5 gives the number of structural isomers predicted for some larger alkane molecules, starting with C_6H_{14}. Every predicted isomer, *and no more,* has been isolated and identified for the C_6, C_7, and C_8 groups. The large number of structural isomers illustrates the complexity and variety organic chemistry can have even for simple hydrocarbons. Although not all of the C_{15} molecules have been isolated, there is reason to believe that with enough time

TABLE 12.4 ■ Some Common Alkyl Groups

Name	Condensed Structural Representation
Methyl	$CH_3—$
Ethyl	$CH_3CH_2—$ or $C_2H_5—$
Propyl	$CH_3CH_2CH_2—$ or $C_3H_7—$
Isopropyl	$CH_3CH—$ $\quad|$ $\quad CH_3$ or $(CH_3)_2CH—$
Butyl	$CH_3CH_2CH_2CH_2—$ or $C_4H_9—$
t-Butyl*	$\quad CH_3$ $\quad|$ $CH_3C—$ $\quad|$ $\quad CH_3$ or $(CH_3)_3C—$

**t* stands for *tertiary,* sometimes abbreviated *tert,* which means that the central C atom is bonded to three other C atoms.

TABLE 12.5 ■ **Number of Predicted Structural Isomers for Some Hydrocarbons***

Formula	Isomers Predicted	Found
C_6H_{14}	5	5
C_7H_{16}	9	9
C_8H_{18}	18	18
$C_{15}H_{32}$	4,347	—
$C_{20}H_{42}$	366,319	—
$C_{30}H_{62}$	4,111,846,763	—
$C_{40}H_{82}$	62,491,178,805,831	—

*R. E. Davies and P. J. Freyd: $C_{167}H_{336}$ is the smallest alkane with more realizable isomers than the observed universe has ''particles.'' *Journal of Chemical Education*, Vol. 66, p. 278, 1989.

Many structural arrangements will be possible. *(Eamonn McNulty/SPL/Photo Researchers)*

and effort, they could be, so structural isomerism certainly helps to explain the vast number of carbon compounds. However, as the number of carbon atoms gets larger than 17, the number of isomers that would be stable enough for isolation is much smaller than predicted. This difference arises because the calculation of predicted isomers does not consider space requirements of atoms within the molecules. Many of the isomers predicted for these larger molecules would require such overcrowding of the atoms that the molecules would not be stable. (See the reference listed in Table 12.5 for further discussion of this point.)

Naming Branched-Chain Alkanes

Many alkanes and other organic compounds have both common names and systematic names. Why are both common and systematic names used? Usually the common name came first and is widely known. Many consumer products are labeled with the common name, and when only a few isomers are possible, the common name adequately identifies the product for the consumer. For example, ''isobutane,'' the common name for methylpropane, is sufficient because there is only one branched-chain isomer possible for C_4H_{10}. However, a system of common names quickly fails when several structural isomers are possible.

You have probably heard of the ''octane rating'' of gasoline. Octane, C_8H_{18}, has 18 possible isomers (Table 12.5). One of these isomers, 2,2,4-trimethylpentane,

$$\underset{1}{CH_3} - \underset{2}{\overset{\overset{\displaystyle CH_3}{|}}{\underset{\underset{\displaystyle CH_3}{|}}{C}}} - \underset{3}{CH_2} - \underset{4}{\overset{\overset{\displaystyle CH_3}{|}}{CH}} - \underset{5}{CH_3}$$

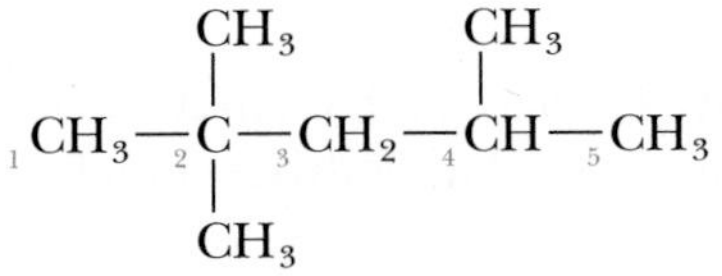

2,2,4-trimethylpentane

■ Rules for naming organic compounds are given in Appendix D.

is used as a standard in assigning "octane ratings" of various gasolines. In this case, a common name such as isooctane would not provide enough information about which isomer was actually being used as the standard. However, the systematic name provides complete information. The "pentane" part, which means a straight five-carbon chain, identifies the longest chain in the molecule. The numbers "2,2,4-" indicate the locations of the three groups attached to the pentane chain, and "tri" is used as a prefix for "methyl" to indicate that all three groups are methyl groups.

EXAMPLE 12.2 *Structural Isomers*

Three isomers are possible for the isomeric pentanes (C_5H_{12}). Draw both expanded and condensed formulas for these isomers.

SOLUTION

A good plan to follow in drawing all possible isomers—no more and no less—is to start with the straight-chain isomer and then remove one methyl group at a time, placing that methyl group on the remaining chain and checking all possibilities before removing another methyl group. In this case, start with the straight-chain five-carbon pentane as the first isomer.

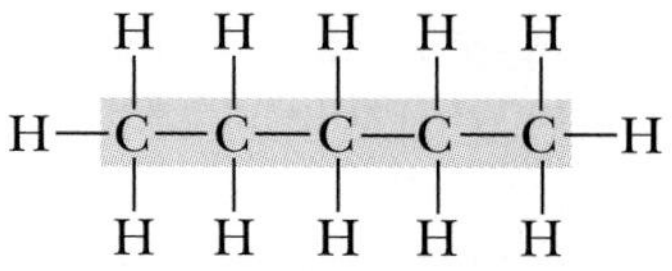

Expanded formula: pentane

$CH_3CH_2CH_2CH_2CH_3$

Condensed formula: pentane

Removing one methyl and placing it on the second carbon gives a second isomer. Convince yourself that this is the only possible one with one methyl attached to a four-carbon chain, since putting the methyl group on the next C gives the same isomer.

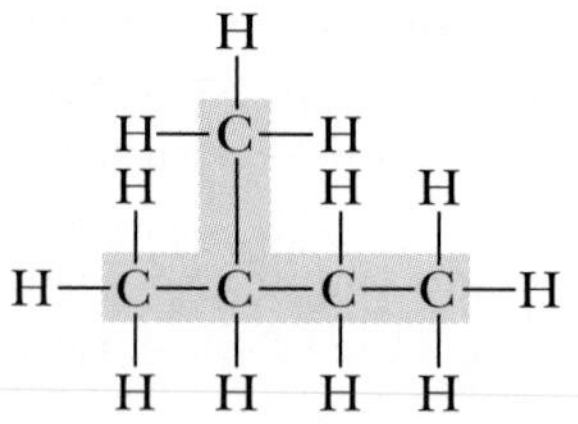

Expanded formula: 2 methylbutane

$CH_3CHCH_2CH_3$ with CH_3 above

Condensed formula: 2-methylbutane

The third possible isomer is obtained by removing a second methyl and placing it on the second C to give 2,2-dimethylpropane.

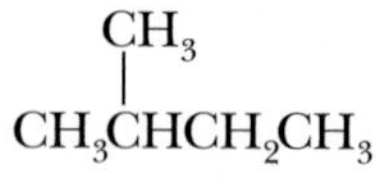

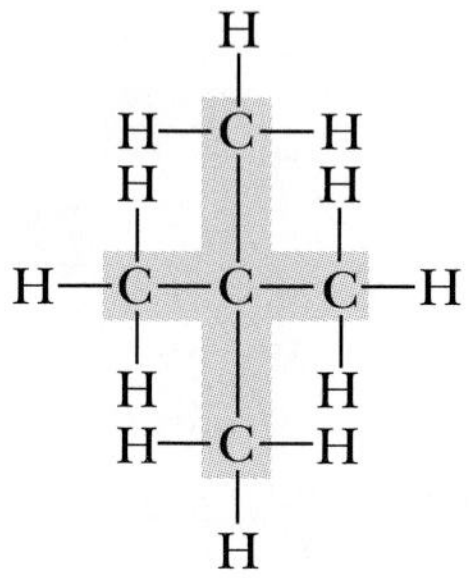

Expanded formula: 2,2-dimethylpropane

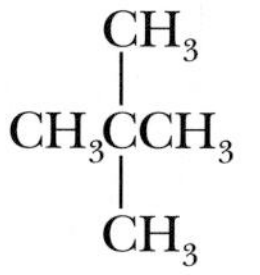

Condensed formula: 2,2-dimethylpropane

Green tomatoes. On their way to market these tomatoes may be ripened by exposure to ethylene gas. *(Jan Halaska/Photo Researchers)*

Exercise 12.2

Draw the condensed structural formulas for the following compounds: (a) 2-methylpentane, (b) 3-methylpentane, (c) 2,2-dimethylbutane, (d) 2,3-dimethylbutane

12.3 Alkenes—Reactive Cousins of Alkanes

Petroleum contains alkenes, and their presence in gasoline raises the octane rating (Section 12.7). Alkenes for use in commercial applications are also obtained from petroleum by a cracking process (Section 12.7). Ethene (ethylene) is the first member of the **alkene** series of hydrocarbons, compounds that have one or more C=C double bonds.

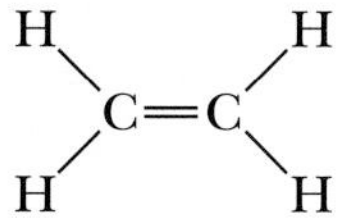

The common name ethylene is used in the present discussion because of its wide use in commercial applications. More ethylene is manufactured each year than any other organic chemical. In 1994 ethylene ranked fourth in production of chemicals in the United States, with the manufacture of more than 24 million tons. Much of the ethylene is used in the production of polyethylene (Section 14.5).

Ethylene is also found in plants, where it is a hormone that controls seedling growth and regulates fruit ripening. The discovery of this property led to the use of ethylene by food processors for ripening fruits and vegetables after harvest.

The "-ene" suffix is used for hydrocarbons with one or more double bonds. The bonding and structural formulas for the first two alkenes, ethene (C_2H_4), and propene (C_3H_6), were illustrated in Section 6.3.

The general formula for alkenes with one double bond is C_nH_{2n}. The second member of the alkene series is propene (propylene). Propylene is manufactured in large quantities for use in the production of polypropylene (Section 14.5). Ball-and-stick models for ethene and propene are shown in Figure 12.6. Unlike alkanes, which undergo few chemical reactions easily (except for combustion), alkenes are quite reactive. The site of the chemical change is usually the double bond. This reactivity is essential to making many kinds of plastics, as discussed in Section 14.5.

Propylene was number 7 in manufactured chemical production in the United States in 1994.

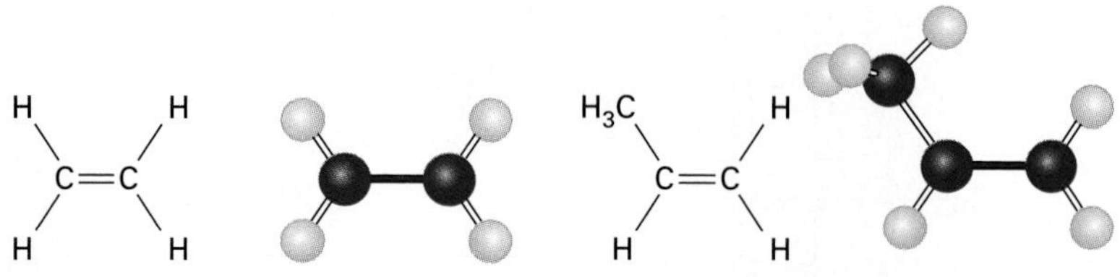

Ethene Propene

Figure 12.6 The two smallest alkenes: ethene, commonly known as ethylene, and propene, commonly known as propylene.

Structural Isomers of Alkenes

In the alkene series, the possibility of locating the double bond between two different carbon atoms adds additional structural isomers. Ethene and propene have only one possible location for the double bond. However, the next alkene in the series, butene, has two possible locations for the double bond.

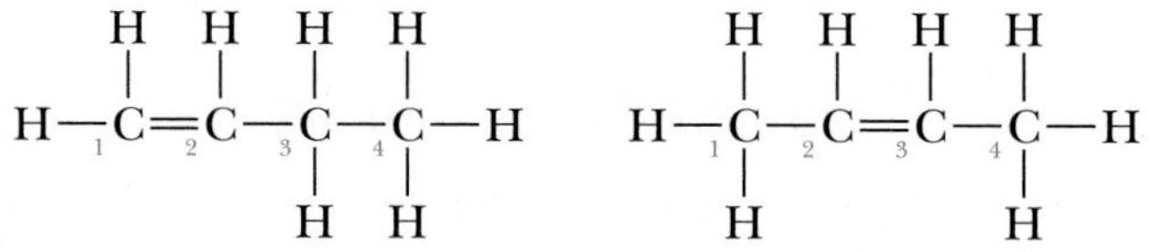

1-butene 2-butene

When groups such as methyl or ethyl are attached to carbon atoms in an alkene, the longest hydrocarbon chain is numbered from the end that will give the double bond the lowest number, and then numbers are assigned to the attached group. For example, in the compound drawn below the longest chain has seven carbons (heptene), the double bond is between C2 and C3 (2-heptene), and the three (tri-) methyl groups are on the third, fourth, and sixth carbons (3,4,6-trimethyl-).

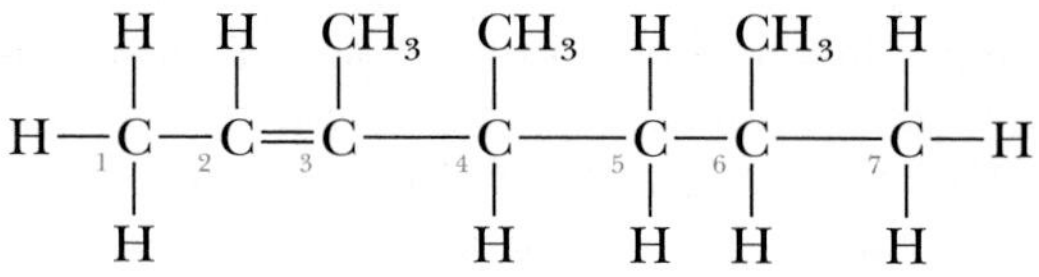

Hence, the name is 3,4,6-trimethyl-2-heptene.

EXAMPLE 12.3 *Drawing Alkenes*

Draw the structure of 2,3-dimethyl-2-pentene.

SOLUTION

First, draw the parent alkene and put the double bond in the correct location. In this case, 2-pentene is the parent.

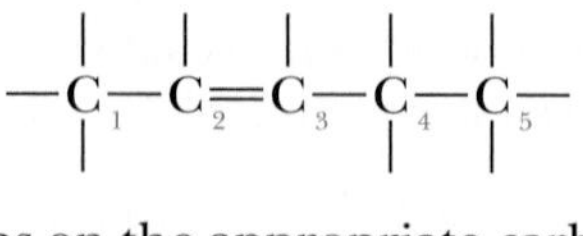

Then place the alkyl groups on the appropriate carbons. In this case there are two methyl groups, one on C2 and one on C3. Remember that numbering the double bond takes precedence. Also check your drawing to make sure you don't have more than four bonds per carbon.

$$H-\underset{\underset{H}{|}}{\overset{\overset{H}{|}}{C}}-\underset{\underset{}{|}}{\overset{\overset{CH_3}{|}}{C}}=\underset{\underset{}{|}}{\overset{\overset{CH_3}{|}}{C}}-\underset{\underset{H}{|}}{\overset{\overset{H}{|}}{C}}-\underset{\underset{H}{|}}{\overset{\overset{H}{|}}{C}}-H$$

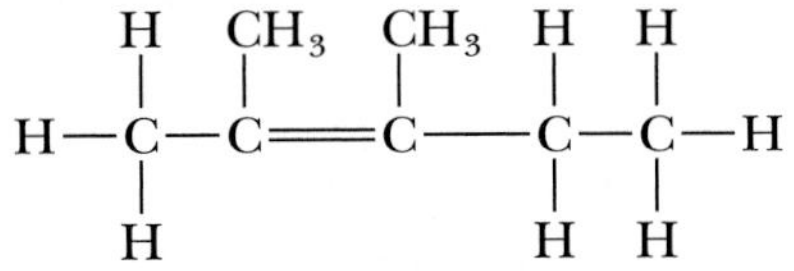

Exercise 12.3
Draw the structure of 3-methyl-1-butene.

Stereoisomerism—*Cis* and *Trans* Isomers in Alkenes

Some alkenes can also have **cis** and **trans isomers,** one of two forms of **stereo-isomerism.** *Here the isomers have the same molecular formulas and the same atom-to-atom bonding sequences, but the atoms differ in their arrangement in space.* The other form of stereoisomerism, **optical isomerism,** is discussed in Section 15.1.

An important difference between alkanes and alkenes is the degree of flexibility of the carbon-carbon bonds in the molecules. Rotation around single carbon-carbon bonds in alkanes occurs readily at room temperature, but the carbon-carbon double bond in alkenes is strong enough to prevent free rotation about the bond. Consider ethene, C_2H_4. Its six atoms lie in the same plane, with bond angles of approximately 120°.

$$\underset{H}{\overset{H}{\diagdown}}C=C\underset{H}{\overset{H}{\diagup}}$$

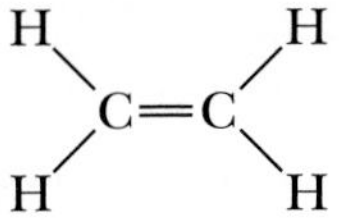

If two methyl groups replace two hydrogen atoms, one on each carbon atom of ethene ($H_2C\!=\!CH_2$), the result is 2-butene, $CH_3CH\!=\!CHCH_3$. Experimental evidence confirms the existence of two compounds with the same set of bonds. The difference in the two compounds is in the location in space of the two methyl groups: the **cis** isomer has two methyl groups on the same side in the plane of the double bond and the **trans** isomer has two methyl groups on opposite sides of the double bond. Note that the properties of the *cis* and *trans* isomers of 2-butene are quite different.

■ If free rotation occurred around a carbon-carbon double bond, these two compounds would be the same.

$$\underset{H_3C}{\overset{H}{\diagdown}}C=C\underset{CH_3}{\overset{H}{\diagup}} \qquad \underset{H_3C}{\overset{H}{\diagdown}}C=C\underset{H}{\overset{CH_3}{\diagup}}$$

	cis-2-butene	*trans*-2-butene
Melting point	−138.9°C	−105.5°C
Boiling point (1 atm)	3.7°C	0.9°C
Density (at 20°C)	0.621 g/mL	0.604 g/mL

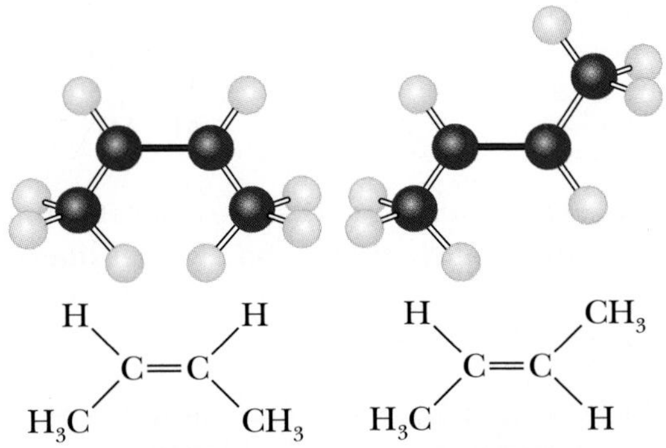

■ Many other *cis* and *trans* isomers are possible. For example, *cis*-1,2-dichloroethene and *trans*-1,2-dichloroethene are possibilities when one hydrogen atom on each carbon atom of ethene is replaced with a chlorine atom.

The third possible isomer, 1-butene (a structural isomer of the *cis* and *trans* isomers), does not have *cis* and *trans* structures, since one carbon atom has two identical groups (H atoms). Its properties are different from the 2-butene isomers.

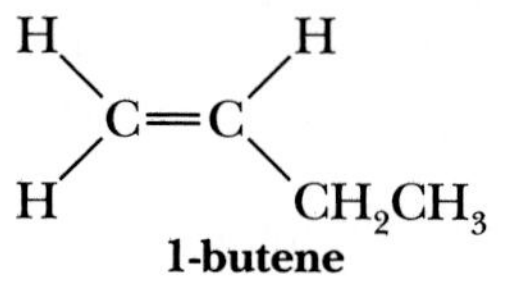

1-butene

Melting point	−185.3°C
Boiling point (1 atm)	−6.3°C
Density (at 20°C)	0.595 g/mL

Cis-trans isomerism in alkenes is possible only when both of the double-bond carbon atoms have two different groups.

12.4 Alkynes

The alkynes have one or more triple bonds (—C≡C—) per molecule and have the general formula C_nH_{2n-2}. The simplest one is ethyne, commonly called acetylene (C_2H_2) (Section 6.3). The naming of alkynes is similar to that of alkenes, with the lowest number possible being used for locating the triple bond.

■ The 180° bond angles around the triple bond make the C—C≡C—C section of the molecule linear.

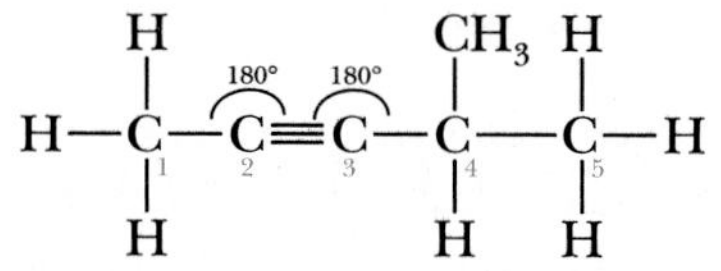

The name of the above compound is 4-methyl-2-pentyne. As with alkenes, changing the location of the multiple bond produces an isomer. For example, 4-methyl-1-pentyne is a different compound than 4-methyl-2-pentyne.

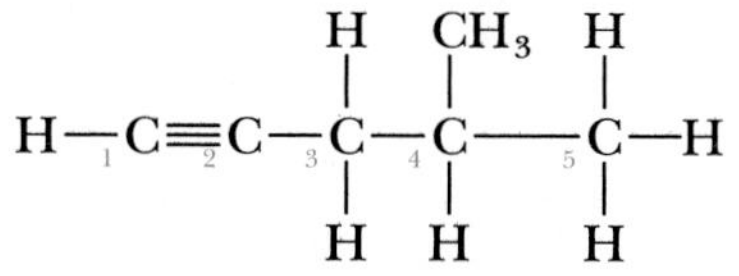

However, *cis* and *trans* isomers are not possible for alkynes because the geometry around the triple-bond carbon atoms is linear.

■ SELF-TEST 12A

1. Which fossil fuel furnishes the most heat energy per gram: (a) coal, (b) petroleum, or (c) natural gas?
2. Which fuel furnishes the most heat energy per gram: (a) natural gas, (b) hydrogen, or (c) coal?
3. All combustions of fossil fuels give off energy. (a) True, (b) False.

FRONTIERS IN THE WORLD OF CHEMISTRY

Organic Metals

Organic compounds are generally good insulators, while metals conduct electricity. However, researchers have been successful in making organic compounds that are conductors. Acetylene can be polymerized in the presence of a catalyst to polyacetylene, a typical plastic that does not conduct electricity.

$$2n\ \text{H}-\text{C}\equiv\text{C}-\text{H} \xrightarrow{\text{Catalyst}}$$

$$\left(\begin{array}{c} \text{H} \quad \text{H} \quad \text{H} \quad \text{H} \\ | \quad\quad | \quad\quad | \quad\quad | \\ \text{C}=\text{C}-\text{C}=\text{C} \end{array}\right)_n$$

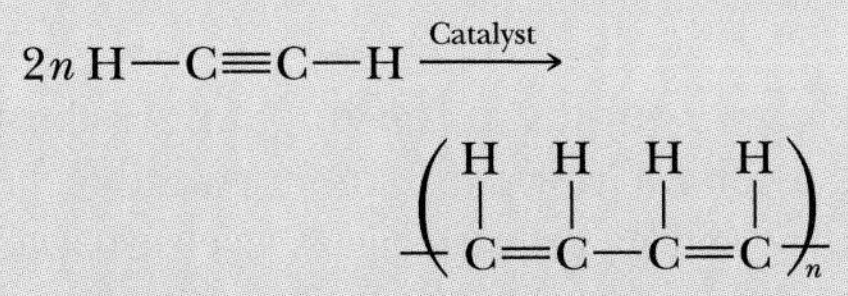

This polymer appears as a black powder in the usual laboratory preparation and received little attention prior to 1970. In that year a Korean university student, having trouble understanding his Japanese instructor, Hidek Shirakawa, prepared the polymer using an excessive amount of the catalyst. The result was a silver film that looked more like a metal than anything else. Furthermore, the polyacetylene film conducted electricity, which was a first for plastic materials.

The conductance of the shiny polyacetylene film can be explained in terms of very long polymer molecules that are lined up in a crystalline structure. Electric charges are more readily passed along the alternating double bonds in the polyacetylene molecules that are ordered in one direction. Evidence of this is the observation that conductance is greater along the aligned chains than perpendicular to the chains. The black-powder form is an insulator because the long chains of polyacetylene are in a jumbled, random fashion, which prevents conductance.

In 1975, at the University of Pennsylvania, Alan MacDiarmid began a systematic study of this new form of polyacetylene. Adding small amounts of iodine (similar to the doping of semiconductors, Section 7.6) during the preparation of polyacetylene increases the electric conductivity of the plastic 10^{12} times, or a trillionfold! This rivals the conductivity of metals. Recall that iodine atoms have an attraction for one additional electron per atom. If an iodine atom removed an electron from a double bond at one end of the polyacetylene molecule, the entire molecule would simply pass negative charge along the conjugated system and into the "positive hole" if an electron were available at the other end of the molecule. This flow of electric charge is electric conduction.

Following polyacetylene, many similar plastics have been conceived theoretically, and several are now being made. For example, a rechargeable battery with electrodes of p-doped and n-doped polyacetylene has been made. Think what this might mean since the one great fault of the electric car has been the weight of the lead electrodes in batteries (Section 10.7).

The World of Chemistry

Program 22, *The Age of Polymers*

A piece of electrically conducting polyacetylene film.

References R.B. Kaner and A.G. MacDiarmid, "Plastics That Conduct Electricity," *Scientific American*, Volume 258, February, 1988, pp. 106–111.
R. Dagani, "Organic Metals," *Chemical and Engineering News*, August 31, 1992, pp. 8–9.

4. Hydrocarbons react with _____ to produce CO_2 and _____.

5. Each carbon in a saturated hydrocarbon has _____ geometry.

6. _____ is the first member of the alkene series.

7. _____ is the first member of the alkyne series.

8. The formula for the ethyl group is _____.

9. Butane and 2-methylpropane are examples of _____ isomers.

10. The number-one organic chemical produced in the United States is _____.

11. The rigidity of the carbon-carbon double bond allows for the possibility of _____ isomers.

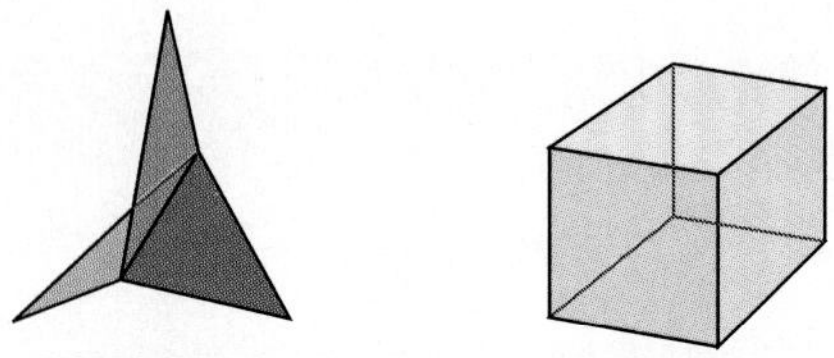

Propellane (C_6H_8) Cubane (C_8H_8)

Two small, strained-ring compounds with appropriate names.

12.5 The Cyclic Hydrocarbons

Hydrocarbons can form rings as well as straight chains and branched chains. Two important classes of cyclic hydrocarbons found in petroleum and coal are the **cycloalkanes** and the **aromatics.**

Cycloalkanes

The simplest cycloalkane is cyclopropane, a highly strained ring compound:

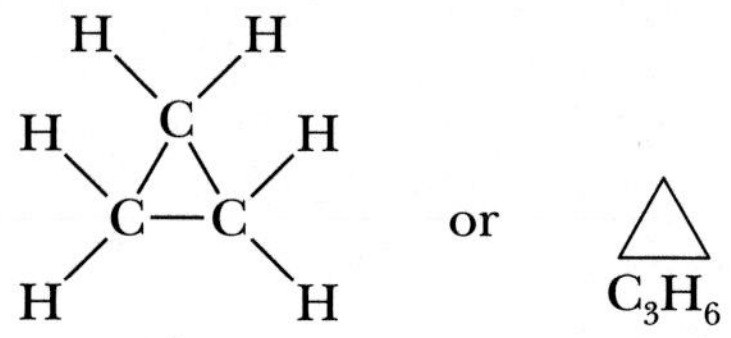

The ring is strained because of the 60-degree angles in the ring; angles larger than 90 degrees show a much greater stability. Cyclopropane, a volatile, flammable gas (b.p. is $-32.7°C$), is a rapidly acting anesthetic. A cyclopropane-oxygen mixture is useful in surgery on babies, small children, and ''bad risk'' patients because of its rapid action and the rapid recovery of the patient. Helium gas is added to the cyclopropane-oxygen mixture to reduce the danger of explosion in the operating room.

The cycloalkanes are commonly represented by polygons in which each corner represents a carbon atom and two hydrogen atoms and the lines represent C—C bonds. The C—H bonds are not shown, but are understood. Other common cycloalkanes include cyclobutane (C_4H_8), cyclopentane (C_5H_{10}), and cyclohexane (C_6H_{12}). These are represented as

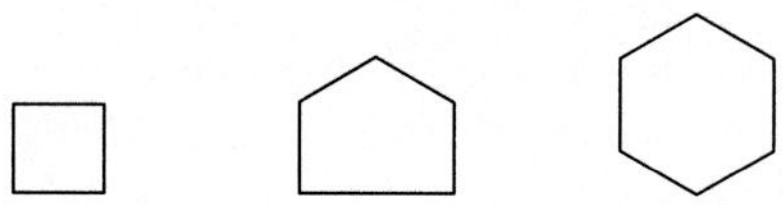

Cyclobutane Cyclopentane Cyclohexane

Cyclohexane exists in two conformations referred to as the ''boat'' and the ''chair'' forms. These are *conformations,* not isomers, because they result from twisting around the C—C single bonds rather than from breaking bonds to put atoms in different positions, such as in *cis* and *trans* 2-butene.

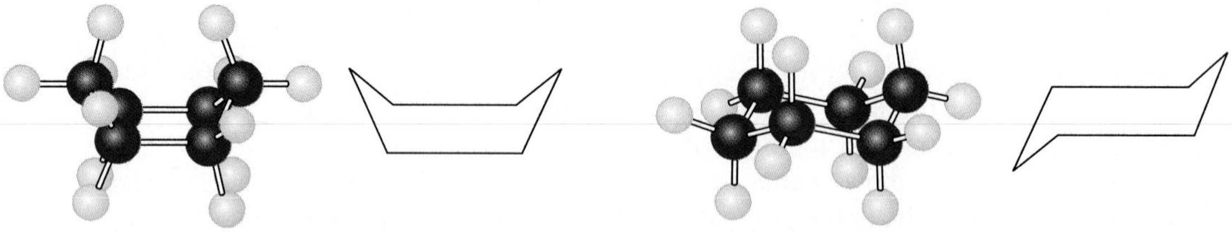

Boat cyclohexane Chair cyclohexane

Since the end groups are farther apart in the chair form, steric (spatial) repulsions between the end groups are less than in the boat form. As a result, the chair form is more stable. The chair form of cyclohexane is the prototype of the six-membered rings found in glucose and other sugars (Section 15.5).

EXAMPLE 12.4 *Cycloalkanes*

The symbol to the right is used to represent cyclohexane. Draw the structural formula that shows all carbon and hydrogen atoms. How many are there of each?

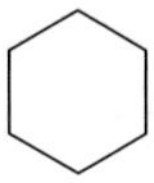

SOLUTION

Each corner represents a CH_2 group, so the structure with all carbon and hydrogen atoms shown is

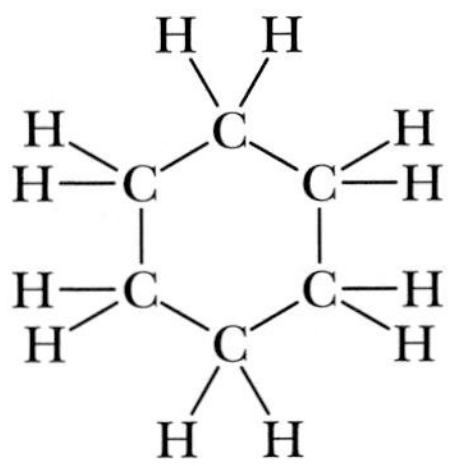

There are six C and 12 H atoms in cyclohexane (C_2H_{12}).

Exercise 12.4

The symbol to the right is used to represent cyclopentane. Draw the structural formula that shows all carbon and hydrogen atoms. How many are there of each?

Aromatic Compounds

Hydrocarbons containing one or more benzene rings (Figure 12.7) are called **aromatic compounds.** The word "aromatic" was derived from "aroma," which describes the rather strong and often pleasant odor of these

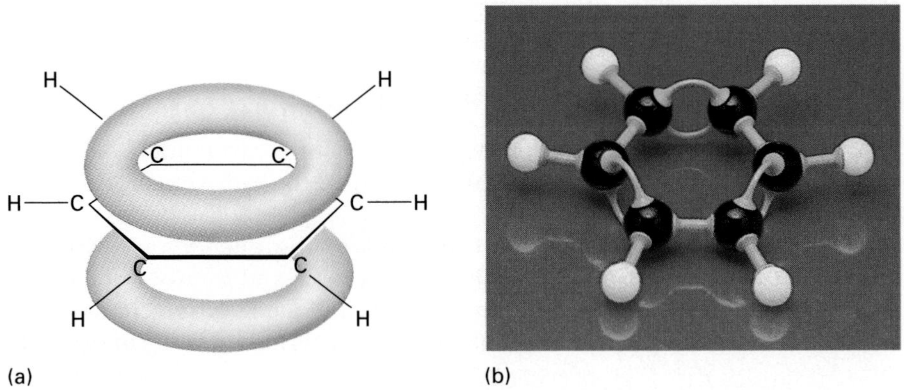

Figure 12.7 Benzene, the smallest aromatic compound. (a) The equal distribution of bonding electrons around the ring can be represented by electron clouds above and below the plane of the ring. (b) Another way to represent the bonding electrons of benzene is as alternating double and single bonds, shown here in a ball-and-stick model. Because all bonds in benzene are the same, (a) is a more correct representation. *(b, C.D. Winters)*

■ Carcinogens are cancer-causing agents. The type of cancer caused may vary from one carcinogen to another. Benzene causes a form of leukemia, and benzopyrene causes skin cancer and lung cancer.

compounds. Benzene and most other aromatic compounds, however, are toxic and often **carcinogenic** (Section 18.8).

The main structural feature, which is responsible for the distinctive chemical properties of the aromatic compounds, is the six-carbon benzene ring. Figure 12.7*a* illustrates *delocalization* of the bonding electrons above and below the plane of the ring. In other words, all of the carbon-carbon bonds are equivalent, and benzene is a planar molecule. Benzene can be represented as

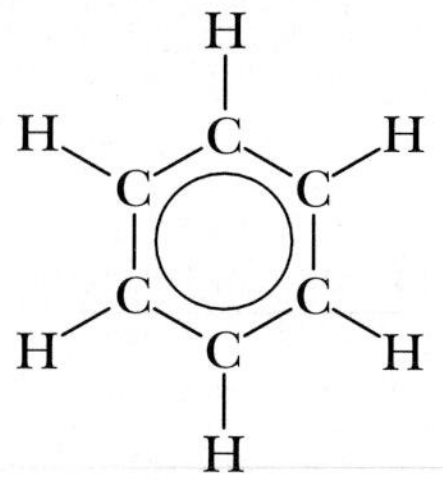

where the circle represents the evenly distributed, delocalized electrons.

When hydrogen and carbon atoms are not shown, benzene is represented by a circle in a hexagon. Each corner in the hexagon represents one carbon atom and one hydrogen atom. Remember that this symbol stands for C_6H_6,

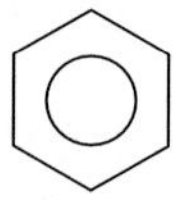

and a hexagon without a circle stands for cyclohexane, C_6H_{12}.

Derivatives of Benzene

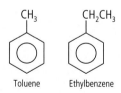

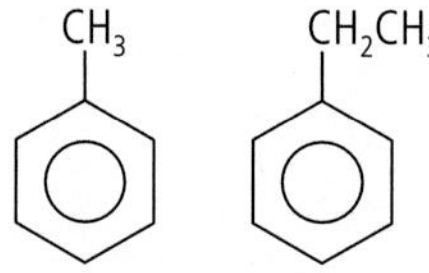

Toluene Ethylbenzene

Benzene and many of its derivatives are on the list of top 50 chemicals (see Table 14.1) because of their use in the manufacture of plastics, detergents, pesticides, drugs, and other organic chemicals. Several important derivatives are monosubstituted benzenes, with one atom or group replacing one of the hydrogen atoms. For example, substitution of a methyl group for one of the hydrogen atoms in benzene gives methylbenzene, usually called toluene, a common solvent, which is number 27 of the top 50 chemicals. Ethylbenzene, which contains an ethyl group substituted for one of the hydrogen atoms of benzene, is number 19, while benzene is number 17.

Structural Isomers of Aromatic Compounds

Since the benzene molecule has a planar structure, structural isomers are possible when two or more groups are substituted for hydrogen atoms on the benzene ring.

Three isomers are possible if two groups are substituted for two hydrogen atoms on the benzene ring. The prefixes *ortho-*, *meta-*, and *para-* or numbers are used to distinguish among the isomers. When the name of the compound is written, usually only the first letter of one of these terms is given. For example,

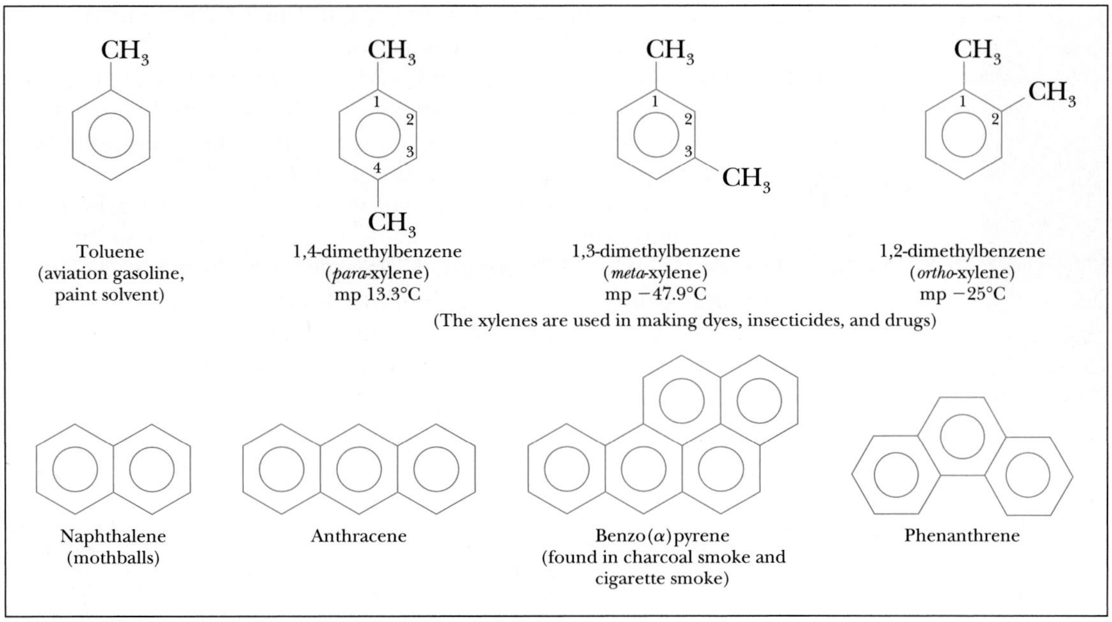

Figure 12.8 Some examples of aromatic hydrocarbons.

when the two groups are methyl groups, the three isomers are commonly known as *o*-xylene, *m*-xylene, and *p*-xylene (Figure 12.8). Substituting two chlorine atoms for two hydrogen atoms gives the three isomers *o*-, *m*-, and *p*-dichlorobenzene.

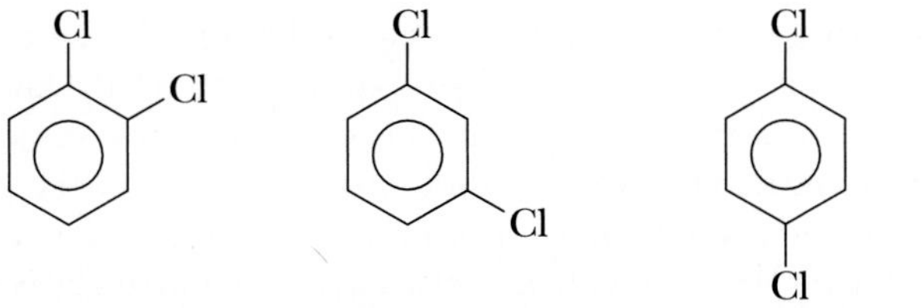

o-dichlorobenzene or 1,2-dichlorobenzene — *m*-dichlorobenzene or 1,3-dichlorobenzene — *p*-dichlorobenzene or 1,4-dichlorobenzene

If more than two groups are attached to the benzene ring, numbers must be used to identify the positions. Consider the following compounds:

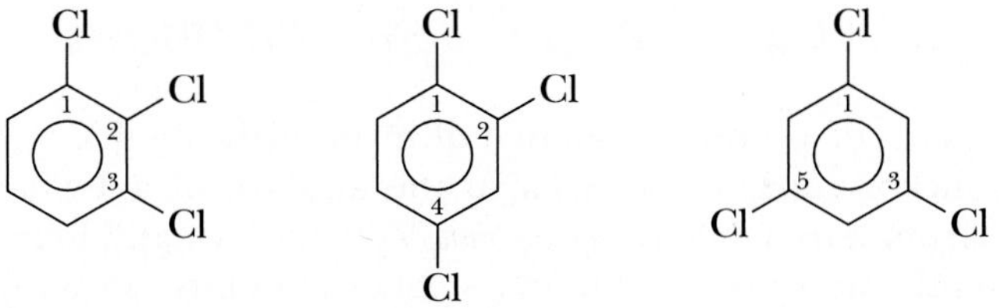

1,2,3-trichlorobenzene — 1,2,4-trichlorobenzene — 1,3,5-trichlorobenzene

One source of aromatic compounds. Both the smoke and the char on the meat contains polycyclic aromatic compounds. *(Hank Morgan/Rainbow)*

There is no other way to attach three atoms of chlorine to a benzene ring, and only three trichlorobenzenes can be prepared.

Another type of aromatic compound has two or more benzene rings sharing ring edges. Examples include naphthalene, anthracene, and benzopyrene (Figure 12.8). Benzene, toluene, and xylenes are important components of gasoline because they raise the octane rating (Section 12.7).

Many organic compounds found in nature are cyclic hydrocarbons that include both aromatic rings and cycloalkane or cycloalkene rings fused together. Steroids (Section 15.6) are good examples. Chemists who isolate organic compounds from plants and develop methods for making them in the laboratory are called natural product chemists. For example, Percy Julian was the first chemist to synthesize hydrocortisone, a steroid, and physostigmine, a compound useful in the treatment of glaucoma.

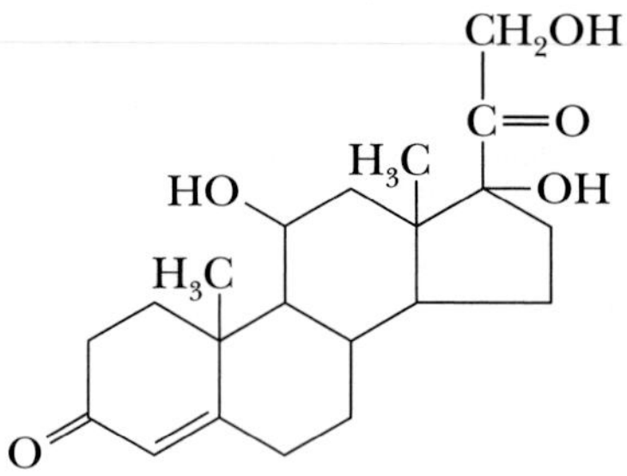

Hydrocortisone

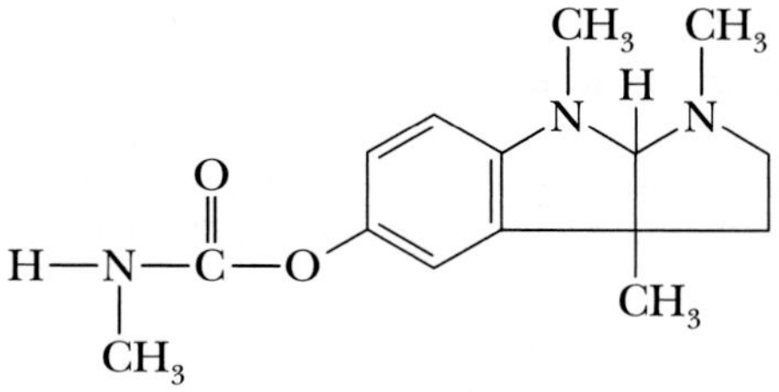

Physostigmine

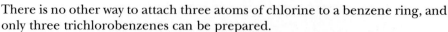

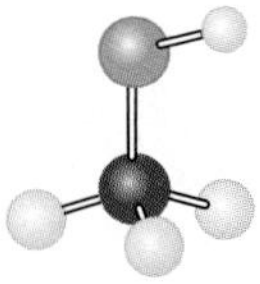

Methanol, CH_3OH.

■ A **functional group** is an atom or group of atoms within a molecule that gives the substance a characteristic chemical behavior. The —OH group is the alcohol functional group. Additional classes of functional groups are discussed in Chapter 14.

■ The International Union of Applied Chemistry (IUPAC) names of alcohols include the name of the hydrocarbon to which the alcohol corresponds and indicate the number of carbon atoms; the suffix "-ol" denotes an alcohol. Common names use the name of the alkyl group (represented as R in ROH) attached to —OH. For example, methyl alcohol, ethyl alcohol, and *tertiary*-butyl alcohol are the common names for methanol, ethanol, and 2-methyl-2-propanol.

12.6 Alcohols—Oxygen Comes on Board

Alcohols

Several alcohols are being used as fuels and fuel additives. These include methanol, ethanol, and 2-methyl-2-propanol, commonly known as *tertiary*-butyl alcohol. Alcohols, one of several classes of organic compounds that have a characteristic **functional group,** contain one or more —OH groups and have the general formula ROH where R is an alkyl group.

Methanol, CH_3OH, also called methyl alcohol, currently is prepared from a mixture of carbon monoxide and hydrogen known as **synthesis gas.** High pressure, high temperature, and a mixture of catalysts are used to increase the yield.

$$\underset{\text{Coal}}{C(s)} + \underset{\text{Steam}}{H_2O(g)} \longrightarrow \underbrace{CO(g) + H_2(g)}_{\text{Synthesis gas}}$$

$$CO(g) + 2\,H_2(g) \xrightarrow[300°C]{ZnO,\ Cr_2O_3} CH_3OH(g)$$

An old method of producing methanol involved heating a hardwood such as beech, hickory, maple, or birch in the absence of air. For this reason, methanol is sometimes called *wood alcohol.* Methanol is highly toxic. Drinking as little as 30 mL can cause death, and smaller amounts (10 to 15 mL) cause blindness.

THE PERSONAL SIDE

Percy Lavon Julian (1899–1975)

(Chemical Heritage Foundation)

Percy Julian's list of achievements reads like that of others who have made it to the top in their professions: a doctorate in chemistry in Vienna in 1931 quickly followed, back home in the United States, by outstanding achievements as a researcher and university professor; 18 years as Director of Research in an industry where he led the way in bringing to market valuable products from soybeans; and the founding of his own research institute, the Julian Laboratories. To grasp the measure of the man, add to this brief outline dozens of scientific publications, over 100 patents granted, numerous academic honors, and positions of responsibility in many civic and humanitarian organizations.

But there were some differences from a successful career path along the way. After completing the eighth grade he had to leave his home in Montgomery, Alabama, for further studies—no more public education was available there for a black man. He enrolled as a "sub-freshman" at DePauw University in Indiana. On his first day, a white student welcomed him with a handshake. Julian later related his reaction: "In the shake of a hand my life was changed, I soon learned to smile and act like I believed they all liked me, whether they wanted to or not."

Early in his career, other challenges had to be met. As a successful businessman, he and his family were the first black residents of an upscale suburb of Chicago. There, on Thanksgiving Day in 1950, his home was attacked by arsonists. The Julian family stayed on to become respected and welcome members of the community.

An organic chemist, Julian built his career around the study of chemicals of plant origin, many of them of medicinal value. The synthesis of a complicated natural molecule is a major goal in such work. Julian was first to achieve synthesis of hydrocortisone, now available in every drugstore because of its value in treating allergic skin reactions. He originated the production of soybean protein and the isolation from soybean oil of compounds from which the first synthetic sex hormone (progesterone) could be made.

Julian's talents were evident early when he and a colleague devised a series of nine chemical reactions that produced a compound identical with natural *physostigmine*. Originally isolated from the Calabar bean from Nigeria, physostigmine had already proven valuable for treating glaucoma by lowering fluid pressure in the eye. Just at the time that Julian was completing this work, one of the most famous organic chemists of the day (Sir Robert Robinson) announced synthesis of the same compound. In a remarkable display of confidence for one so new in the field, Julian wrote in a prominent scientific journal, "We believe the English authors are in error." And indeed they were.

Methanol will probably continue to increase in importance as petroleum and natural gas become too expensive as sources of both energy and chemicals. Although most of the world's methanol currently comes from synthesis gas made from natural gas, coal gasification (described in Section 12.9) will become a more important source of methanol as the natural gas reserves are used up. Since methanol is relatively cheap, its potential as a fuel and as a starting material for the synthesis of other chemicals is receiving more attention.

Ethanol, C_2H_5OH, also called ethyl alcohol or grain alcohol, can be obtained by the fermentation of carbohydrates (starch, sugars).

$$\underset{\text{Glucose}}{C_6H_{12}O_6} \xrightarrow{\text{Yeast}} \underset{\text{Ethanol}}{2\ C_2H_5OH} + 2\ CO_2$$

The yeast contains enzymes that are catalysts for the fermentation process. A mixture of 95% ethanol and 5% water can be recovered from the fermentation products by distillation. Ethanol is the active ingredient of alcoholic beverages. Ethanol is receiving increased attention for use as an alternative fuel and as a fuel additive for oxygenated fuels. At present, most ethanol is used in a blend of 90% gasoline and 10% ethanol (first introduced in the 1970s and known as *gasohol*).

Ethers

Ethers, which contain the R—O—R′ linkage, are formed by the dehydration of alcohols. Methyl-*tertiary*-butyl ether (MTBE) is the most important commercial ether (number 18 in annual production in 1994; see Table 14.1) because of its use in oxygenated and reformulated gasolines.

$$CH_3{-}O{-}\underset{\underset{CH_3}{|}}{\overset{\overset{CH_3}{|}}{C}}{-}CH_3$$

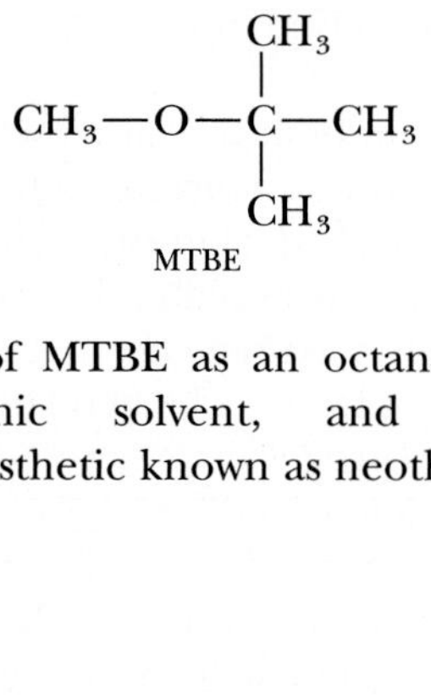

MTBE

Before the development of MTBE as an octane enhancer, diethyl ether, $C_2H_5OC_2H_5$, an organic solvent, and methyl propyl ether, $CH_3OCH_2CH_2CH_3$, an anesthetic known as neothyl, were the most common ethers.

■ SELF-TEST 12B

1. The difference between cyclohexane and benzene is the number of _____ atoms.
2. How many atoms does the symbol ⬡ represent?
3. Synthesis gas is a mixture of _____ and _____.
4. _____ structural isomers are possible for trichlorobenzene.
5. _____ structural isomers are possible for dimethylbenzene.
6. Fermentation of carbohydrates yields (a) ethanol, (b) methanol.
7. Ethers are a class of compounds that contain the _____ linkage.

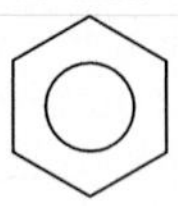

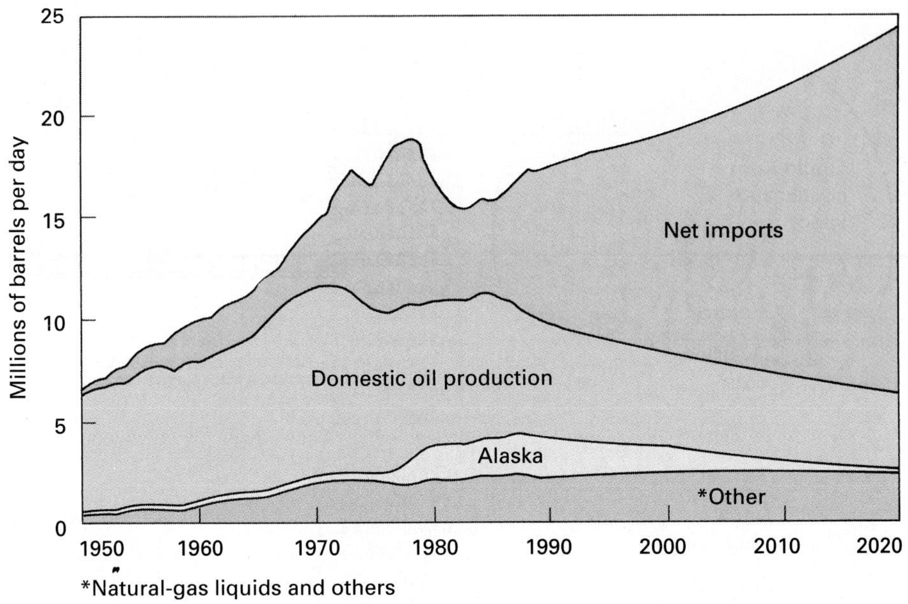

Figure 12.9 United States oil production and oil imports. At present, about 50% of the total oil used is imported. Projections show oil imports will continue to increase.

12.7 Petroleum

Crude petroleum is a complex mixture of thousands of hydrocarbon compounds, and the actual composition of petroleum varies with the location in which it is found. For example, Pennsylvania crude oils are primarily straight-chain hydrocarbons, whereas California crude oil is composed of a larger portion of aromatic hydrocarbons.

How long will petroleum be viable as a source of energy and starting materials for consumer products? At the current rate of use, all of the known petroleum reserves in the world will be consumed by the year 2060. Oil production and oil imports by the United States over the last several years are shown in Figure 12.9. Between 1985 and 1993 the U.S. dependence on imported oil increased from 27 to 44%. With only 4% of global reserves in 1992, the United States produced 11% of the world output of oil. In the United States, the rate of oil used per person each year is 24 barrels. The worldwide average is approximately 4.5 barrels of oil a year for each person. The world use rate is expected to fall to 1.5 barrels per person per year by the year 2030. This dramatic decrease will certainly cause changes in the global energy economy, which will require changes in the way petroleum is used as a major source of hydrocarbons. However, both natural gas and coal are important sources of hydrocarbons, and coal is likely to become more important as petroleum reserves are depleted.

Pumping crude oil. The yield from a well is improved by adding soap-like molecules, which make the oil flow more freely.

■ There are 42 gallons of oil per barrel (Section 13.1).

Petroleum Refining

The refining of petroleum begins with the separation of fractions according to boiling-point ranges by a process called **fractional distillation.** The difference between simple distillation and fractional distillation is the degree of separation achieved for the mixture being distilled. For example, water that contains

■ Fractional distillation is also used to separate liquid air (Section 11.2).

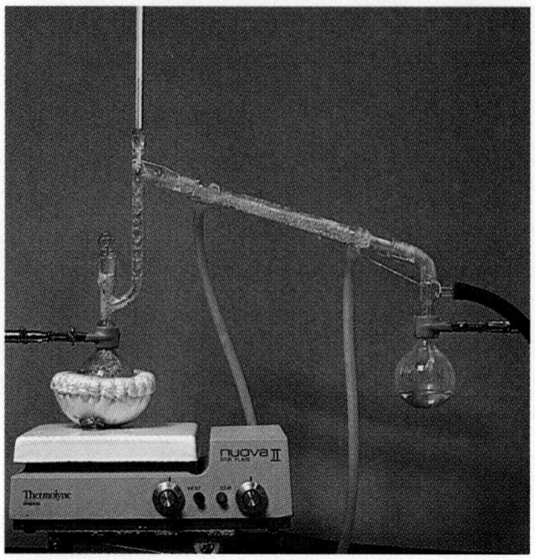

Figure 12.10 Laboratory apparatus for fractional distillation. Vapor from the boiling liquid rises in the vertical column, which has many indentations. As it rises, the vapor is repeatedly condensed and then re-evaporated as it is heated by rising vapors. With each evaporation, the vapor becomes richer in the lowest boiling component of the mixture. The vapor flows down the cooled condenser, and fractions with different boiling points are collected at the end of the condenser. *(J. W. Morgenthaler)*

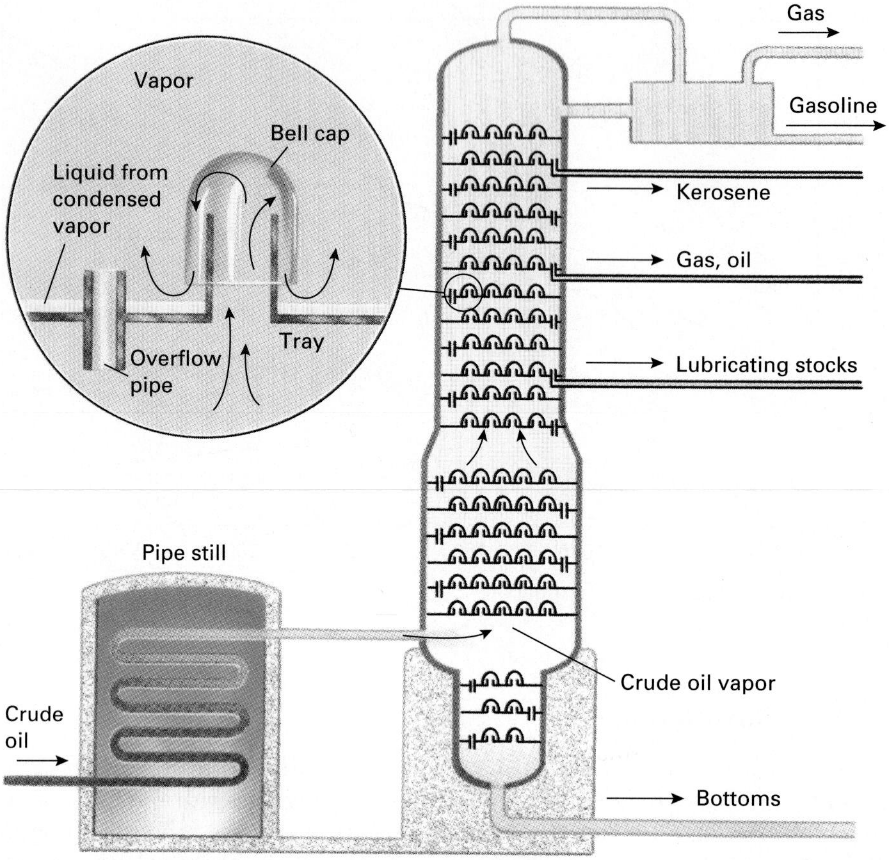

Figure 12.11 Petroleum fractionation. Crude oil is first heated to 400°C in the pipe still. The vapors then enter the fractionation tower. As they rise in the tower, the vapors cool down and condense so that different fractions can be drawn off at different heights. This shows how the rising vapor is repeatedly condensed and collected at the numerous bell caps.

dissolved solids or other liquids can be purified by distillation. The impure solution is heated to boiling; the water vapor is condensed and collected in a separate container (Figure 12.10). Since petroleum contains thousands of hydrocarbons, separation of the pure compounds is not feasible or even necessary. The products obtained from distillation of petroleum are still mixtures of hundreds of hydrocarbons, so they are called **petroleum fractions.**

■ Petroleum fractions are mixtures of hundreds of hydrocarbons with boiling points in a certain range.

Figure 12.11 illustrates a fractional distillation tower used in the petroleum refining process. The crude oil is first heated to about 400°C to produce a hot vapor and liquid mixture that enters the fractionating tower. The vapor rises and condenses at various points along the tower. The lower-boiling petroleum fractions (those that are more volatile) will remain in the vapor stage longer than the higher-boiling fractions. These differences in boiling-point ranges allow the separation of fractions. Some of the gases do not condense and are drawn off the top of the tower, while the unvaporized residual oil is collected at the bottom of the tower. Typical products of the fractional distillation of petroleum are listed in Table 12.6.

TABLE 12.6 ■ Hydrocarbon Fractions from Petroleum

Fraction	Size Range of Molecules	Boiling-Point Range°C	Uses
Gas	C_1-C_4	0–30	Gas fuels
Straight-run gasoline	C_5-C_{12}	30–200	Motor fuel
Kerosene	$C_{12}-C_{16}$	180–300	Jet fuel, diesel oil
Gas-oil	$C_{16}-C_{18}$	over 300	Diesel fuel, cracking stock
Lubricants	$C_{18}-C_{20}$	over 350	Lubricating oil, cracking stock
Paraffin wax	$C_{20}-C_{40}$	Low-melting solids	Candles, wax paper
Asphalt	above C_{40}	Gummy residues	Road asphalt, roofing tar

Octane Rating

The "straight-run" gasoline fraction obtained from the fractional distillation of petroleum is only 55 octane and needs additional refinement because it contains primarily straight-chain hydrocarbons that burn too rapidly to be suitable for use as a fuel in internal-combustion engines. Rapid ignition causes a "knocking" or "pinging" sound in the engine that reduces engine power and may damage the engine.

Isooctane (2,2,4-trimethylpentane) is the standard used to assign octane ratings. The octane rating is an arbitrary scale for rating the relative knocking properties of gasolines, and it is based on the operation of a standard test engine. Heptane knocks considerably and is assigned an octane rating of 0, while 2,2,4-trimethylpentane burns smoothly and is assigned an octane rating of 100. The octane rating of a gasoline is determined by first using the gasoline in a standard engine and recording its knocking properties. The test results are then compared with the behavior of mixtures of heptane and isooctane, and the percentage of isooctane in the mixture with identical knocking properties is called the **octane rating** of the gasoline. Thus, if a gasoline has the same knocking characteristics as a mixture of 13% heptane and 87%

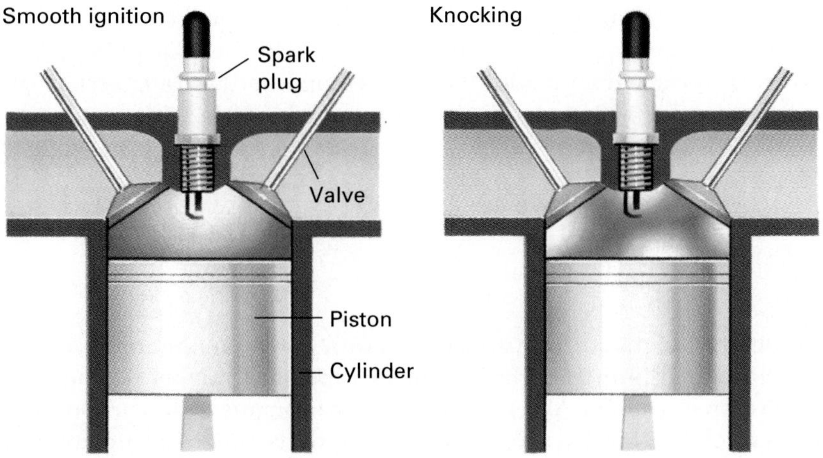

Smooth ignition and knocking.

Typical octane ratings for gasoline available at gas stations. *(C.D. Winters)*

TABLE 12.7 ■ Octane Numbers of Some Hydrocarbons and Gasoline Additives

Name	Octane Number
Octane	−20
Heptane	0
Hexane	25
Pentane	62
1-pentene	91
1-butene	97
2,2,4-trimethylpentane (isooctane)	100
Benzene	106
Ortho-xylene	107
Methanol	107
Ethanol	108
Tertiary-butyl alcohol	113
Methyl *tertiary*-butyl ether (MTBE)	116
Para-xylene	116
Meta-xylene	118
Toluene	118

isooctane, it is assigned an octane rating of 87. This corresponds to regular unleaded gasoline. Other higher grades of gasoline available at gas stations have octane ratings of 89 (regular plus) and 92 (premium).

The octane rating of a gasoline can be increased either by increasing the percentage of branched-chain and aromatic hydrocarbon fractions or by adding octane enhancers (or a combination of both). Since the octane rating scale was established, fuels superior to isooctane have been developed, so the scale has been extended well above 100. Table 12.7 lists octane ratings for some hydrocarbons and octane enhancers.

Catalytic Re-Forming

The **catalytic re-forming process** is used to increase the octane rating of straight-run gasoline by converting straight-chain hydrocarbons to branched-chain hydrocarbons and aromatics. This is accomplished by using certain catalysts, such as finely divided platinum on a support of Al_2O_3.

$$\underset{\substack{\text{Pentane}\\ \text{(octane rating 62)}}}{CH_3CH_2CH_2CH_2CH_3} \xrightarrow{\text{Catalyst}} \underset{\substack{\text{2-Methylbutane}\\ \text{(octane rating 94)}}}{CH_3\overset{\overset{\displaystyle CH_3}{|}}{C}HCH_2CH_3}$$

■ Review the discussion of catalysts in Section 8.3.

In this process, straight-chain hydrocarbons with low octane numbers can be re-formed into their branched-chain isomers, which have higher octane numbers. Catalytic re-forming is also used to produce aromatic hydrocarbons such as benzenes, toluene, and xylenes by using different catalysts and petroleum mixtures. For example, when the vapors of straight-run gasoline, kerosene, and light oil fractions are passed over a copper catalyst at 650°C, a high percentage of the original material is converted into a mixture of aromatic hydrocarbons, from which benzene, toluene, xylenes, and similar compounds may be separated by fractional distillation. This process can be represented by the equation for converting hexane into benzene.

$$\underset{\substack{\text{Hexane} \\ \text{(octane rating 25)}}}{CH_3CH_2CH_2CH_2CH_2CH_3} \xrightarrow{\text{Catalyst}} \underset{\substack{\text{Benzene} \\ \text{(octane rating 106)}}}{C_6H_6} + 4\,H_2$$

The catalytic re-forming process is also a major source of hydrogen gas.

Octane Enhancers

■ Approximately one third of the compounds in refined gasoline are aromatic compounds.

The octane number of a given blend of gasoline can also be increased by adding "antiknock" agents, or octane enhancers. Prior to 1975, the most widely used antiknock agent was tetraethyllead, $(C_2H_5)_4Pb$. The addition of 3 g of $(C_2H_5)_4Pb$ per gallon increases the octane rating by 10 to 15, and before the Environmental Protection Agency (EPA) required reductions in lead content, both regular and premium gasoline contained an average of 3 g of $(C_2H_5)_4Pb$ or tetramethyllead, $(CH_3)_4Pb$, per gallon. However, in the Clean Air Act of 1970 Congress required that 1975-model cars emit no more than 10% of the carbon monoxide and hydrocarbons emitted by 1970 models. The platinum-based catalytic converter chosen to reduce emissions of carbon monoxide and hydrocarbons required lead-free gasolines, since lead deactivates the platinum catalyst by coating its surface. For this reason, new automobiles manufactured since 1975 have been required to use lead-free gasoline to protect the catalytic converter.

■ As little as two tankfuls of leaded gasoline can destroy the activity of a catalytic converter.

Since tetraethyllead can no longer be used, other octane enhancers are being added to gasoline to increase the octane rating. These include toluene, 2-methyl-2-propanol, methyl-*tertiary*-butyl ether (MTBE), methanol, and ethanol. The most popular octane enhancer is MTBE.

■ A side benefit of the removal of lead from gasolines has been a decrease of emissions of this toxic element into the environment (Section 18.4).

EXAMPLE 12.7 *Octane Rating*

Place the following organic compounds in order of decreasing octane rating: pentane, 1-pentene, toluene, 2,2,4-trimethylpentane, heptane, methanol.

SOLUTION

Aromatics and alcohols have the highest octane rating, followed by branched-chain alkanes and alkenes; straight-chain alkanes have the lowest octane rating (see Table 12.7). The decreasing order is toluene, methanol, 2,2,4-trimethylpentane, 1-pentene, pentane, heptane.

Exercise 12.5
Place the following organic compounds in order of decreasing octane rating: 1-butene, ethanol, octane, benzene, methyl-*tertiary*-butyl ether.

Oxygenated and Reformulated Gasolines

The 1990 amendments to the Clean Air Act require cities with excessive levels of ozone and carbon monoxide pollution to use oxygenated and reformulated gasolines to reduce hydrocarbon and toxic compound emissions (Section 21.6). **Oxygenated gasolines** are blends of gasoline with organic compounds that contain oxygen, such as MTBE, methanol, ethanol, and 2-methyl-2-propanol (*tertiary*-butyl alcohol). The oxygenated gasolines can be produced either by blending in additives such as MTBE at the refinery or by adding ethanol or methanol at the distribution terminals.

■ The difference between oxygenated gasoline and reformulated gasoline is in the refining process. Oxygenated gasoline is produced by adding oxygenated organic compounds to refined gasoline. Reformulated gasoline requires changes in the refining process to alter the percentage composition of the different types of hydrocarbons, particularly olefins and aromatics.

Reformulated gasoline is gasoline whose composition has been changed to reduce the percentage of olefins, aromatics, volatile components, and sulfur and to add oxygenated additives such as MTBE. This requires significant changes in the refining process, which makes reformulated gasoline more expensive to produce.

Oxygenated gasolines are required to be used during the four winter months in cities that have serious carbon monoxide pollution (Section 21.9). All gasolines sold in the 41 cities listed in Table 12.8 must contain enough oxygenated organic compounds to provide an average of 2.7% oxygen by weight. Oxygenated gasolines ignite more easily and burn more cleanly, which reduces the need for the fuel-rich operating conditions otherwise required for ignition in winter, and this reduces carbon monoxide emissions. However, oxygenated fuels yield less energy per gram. Use of oxygenated gasolines is estimated to reduce carbon monoxide emissions by 17%.

■ Sulfur in gasoline coats the catalytic converter and reduces its ability to catalyze full combustion of the fuel. This causes an increase in carbon monoxide emissions.

Nine cities with the most serious ozone pollution are required by the 1990 regulations to use reformulated gasolines starting in 1995, and another 87 cities that are not meeting the ozone air-quality standards can choose to use them.

■ The nine cities with the most serious ozone pollution are Baltimore, Chicago, Hartford, Houston, Los Angeles, Milwaukee, New York, Philadelphia, and San Diego.

Catalytic Cracking

Part of the petroleum refinement process involves adjusting the percentage of each hydrocarbon fraction to match commercial demand. For example, the

TABLE 12.8 ■ Cities Using Oxygenated Gasolines During Winter Months

Alaska: Anchorage, Fairbanks	*Maryland:* Baltimore
Arizona: Phoenix, Tucson	*Minnesota-Wisconsin:* Minneapolis-St. Paul
California: Chico, Fresno, Los Angeles-Anaheim-Riverside, Modesto, Sacramento, San Diego, San Francisco-Oakland-San Jose, Stockton	*Montana:* Missolula
Colorado: Colorado Springs, Denver-Boulder, Fort Collins-Loveland	*Nevada:* Las Vegas, Reno
Connecticut: Hartford-New Britain-Middletown	*New Mexico:* Albuquerque
New York: New York metropolitan area, including northern New Jersey; Syracuse	*Ohio:* Cleveland-Akron-Lorain
North Carolina: Greensboro-Winston Salem-High Point, Raleigh-Durham	*Oregon-Washington:* Grant's Pass, Klamath County, Medford, Portland-Vancouver, Seattle-Tacoma, Spokane
	Philadelphia-Trenton-Wilmington (Delaware) metropolitan area
	Texas: El Paso
	Utah: Provo-Orem
	Washington, D.C. metropolitan area

SCIENCE AND SOCIETY

Reformulated Gasoline Regulations

A ruling issued by the Environmental Protection Agency on June 30, 1994, requires that 30% of the oxygenated organic compounds (called oxygenates) used in reformulated gasoline come from renewable sources. Since most gasoline producers have been using MTBE, which is made from methanol, to meet the 1990 Clean Air Act, the new mandate would require more use of ethanol, a renewable resource, and ETBE, which is made from ethanol. Proponents of the mandate argue that the use of ethanol will reduce reliance on oil imports, cut farmers' reliance on federal farm subsidies, and increase the use of renewable resources.

In response to the EPA ruling, the American Petroleum Institute and the National Petroleum Refiners Association filed suit in the U.S. Court of Appeals for the District of Columbia calling for the court to set aside the EPA ruling and institute a stay to prevent the agency from implementing the mandate. The suit accuses EPA of violating the regulatory process by ruling in favor of ethanol producers and farmers who grow the corn used to produce ethanol. The suit also alleges that EPA is ignoring environmental and economic factors related to the use of ethanol as an oxygenate in reformulated fuels. On April 28, 1995 the U.S. Court of Appeals approved the petition by issuing a stay prohibiting the EPA from requiring the use of renewable oxygenates in reformulated gasolines.

Do a literature search on the difference in environmental and economic impact between ethanol-based oxygenates and methanol-based oxygenates.

Do you agree with the U.S. Court of Appeals ruling?

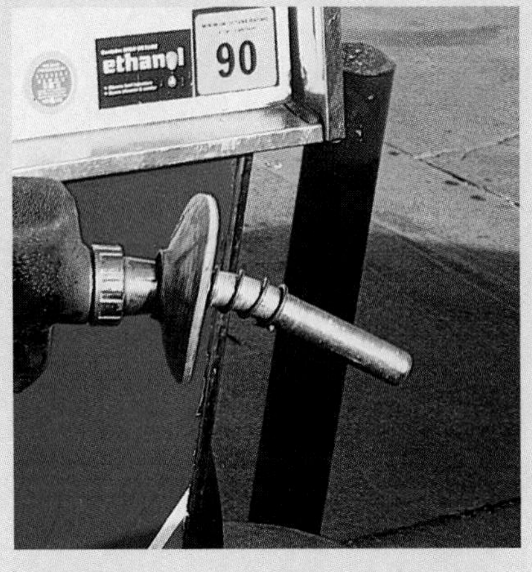

Some gasoline containing ethanol is already available. (Roy D. Farris/Visuals Unlimited)

demand for gasoline is higher than that for kerosene. As a result, chemical reactions convert the larger kerosene-fraction molecules into molecules in the gasoline range in a process called "cracking." The **catalytic cracking process** uses a zeolite catalyst and involves heating saturated hydrocarbons under pressure in the absence of air (Figure 12.12). The hydrocarbons break into shorter-chain hydrocarbons—both alkanes and alkenes, some of which will be in the gasoline range.

■ Cracking breaks larger molecules into smaller ones.

$$\underset{\text{An alkane}}{C_{16}H_{34}} \xrightarrow[\text{Heat}]{\text{Pressure}} \underset{\substack{\text{An alkane} \\ \text{in the gasoline range}}}{C_8H_{18}} + \underset{\text{An alkene}}{C_8H_{16}}$$

Since alkenes have a higher octane rating than alkanes, the catalytic cracking process also increases the octane rating of the mixture. Catalytic cracking is also important for the production of alkenes used as starting materials in the organic chemical industry (see examples in Figure 14.1).

The World of Chemistry

Program 22, *The Age of Polymers*

Figure 12.12 Catalytic cracking unit at a petroleum refinery.

12.8 Natural Gas

Natural gas is a mixture of gases trapped with petroleum in the Earth's crust and is recoverable from oil wells or gas wells where the gases have migrated through the rock. The natural gas found in North America is a mixture of C_1 to C_4 alkanes—methane (60–90%), ethane (5–9%), propane (3–18%), and butane (1–2%)—with a number of other gases, such as CO_2, N_2, H_2S, and

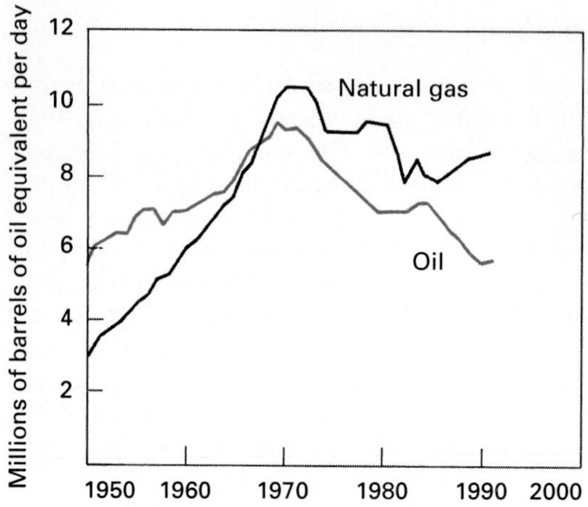

Figure 12.13 Natural gas and oil production in the continental United States.

the noble gases present in varying amounts. In Europe and Japan the natural gas is essentially all methane.

Natural gas is the fastest-growing energy source in the United States, and U.S. production of natural gas supplies 17% more energy than does U.S.-produced oil (Figure 12.13). About half of the homes in the United States are heated by natural gas, followed by electricity (18.5%), fuel oil (14.9%), wood (4.8%), and liquefied gas such as butane and propane (4.6%). Coal and kerosene come in at a low 0.5%, and the percentage of homes using solar heating is even lower. However, the United States has only about 5% of the known world reserves of natural gas, which, at the present rate of use, is enough to last until the year 2050.

Natural gas is also being used as a vehicle fuel, and worldwide there are about 700,000 vehicles powered by natural gas. Although the number of natural-gas-powered vehicles in the United States (10,000) is much smaller than in countries such as Italy (300,000) and New Zealand (100,000), California and several other states are encouraging the use of natural-gas vehicles to help meet new air-quality regulations. Vehicles powered by natural gas emit minimal amounts of carbon monoxide, hydrocarbons, and particulates, and the price of natural gas is about one third that of gasoline. The main disadvantages of natural-gas vehicles include the need for a cylindrical pressurized gas tank and the lack of service stations that sell compressed gas.

Although most natural gas is used as an energy source, it is also an important source of raw materials for the organic chemical industry. (Figure 14.1 shows the uses of the many alkanes produced from natural gas and petroleum.)

12.9 Coal

Coal is a mixture of hydrocarbons with a relatively small amount of sulfur. By way of contrast with petroleum, coal has more fused rings of carbon atoms, and the organic structure of coal is much more complicated.

About 88% of our annual coal production is burned to produce electricity. Only 1% is used for residential and commercial heating. Although the use

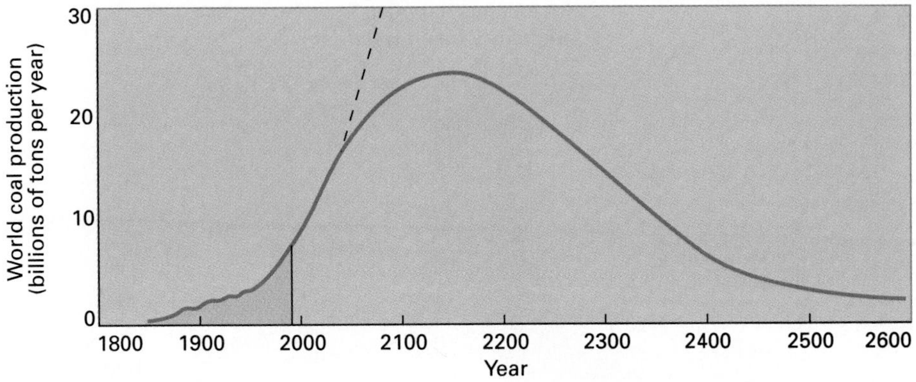

Figure 12.14 Coal resources. The coal mined to date (shaded area) is only a small fraction of the recoverable coal. The rate of increase in coal consumption (dashed line) is 4% per year. It is obvious that such a rise cannot continue.

of coal is on the rise, its use as a heating fuel declined because it is a relatively dirty fuel, bulky to handle, and a major cause of air pollution (because of its sulfur content). The dangers of deep coal mining and the environmental disruption caused by strip mining contributed to the decline in the use of coal.

Given our great dependence on coal for the production of electricity and our smaller but still significant dependence on coal for the production of industrial chemicals, just how much coal do we have and how long is it likely to last? World coal reserves are vast relative to supplies of the other fossil fuels. The known world reserves are estimated to be about 1024 billion tons, of which about 29% is in the United States. How much coal has been used and how long coal is expected to last are summarized in Figure 12.14.

Coal can be converted into a combustible gas (coal gasification) or a liquid fuel (coal liquefaction). In each case environmental problems can be averted, but at additional costs per energy unit obtained from these fuels.

Coal Gasification

When coal is pulverized and treated with superheated steam, a mixture of CO and H_2, synthesis gas, is obtained in a process known as **coal gasification** (Figure 12.15).

$$C + H_2O + 31\ \text{kcal} \longrightarrow CO + H_2$$

■ The formation of synthesis gas is an example of a reaction in which the products have higher bond energies than the reactants, as indicated by the absorption of 31 kcal in the reaction.

Synthesis gas is used both as a fuel and as a starting material for the production of organic chemicals. The heats of combustion of its component gases are

$$2\ CO + O_2 \longrightarrow 2\ CO_2 + 135\ \text{kcal}\ (68\ \text{kcal/mol CO})$$
$$2\ H_2 + O_2 \longrightarrow 2\ H_2O + 116\ \text{kcal}\ (58\ \text{kcal/mol}\ H_2)$$

The sum of these two reactions gives the heat of combustion of synthesis gas (68 kcal + 58 kcal) = 126 kcal/mol. This is less heat per mole than methane provides (192 kcal).

■ The carbon atoms in coal, methane, and other hydrocarbons end up as CO_2 molecules which contribute to global warming (see Section 21.11).

In a newer coal gasification process, methane is the end product. Crushed coal is mixed with an aqueous catalyst; the mixture is dried, and CO and H_2 are

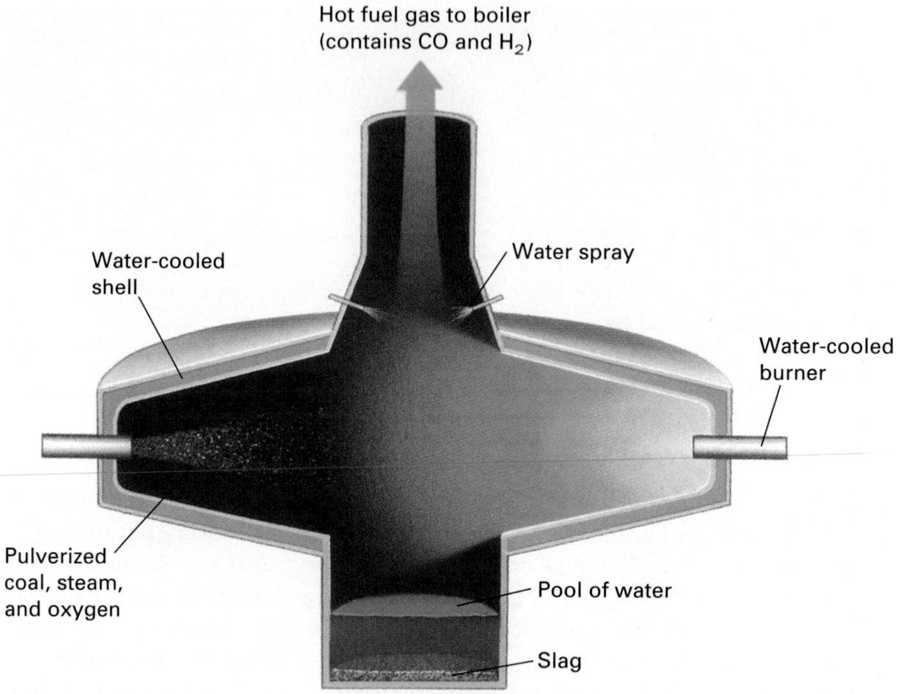

Figure 12.15 A coal gasifier. A relatively cool combustion of powdered coal in a limited supply of oxygen produces a mixture of carbon monoxide and hydrogen, along with other gases. The minerals in the coal collect in the slag.

added. The resulting mixture is then heated to 700°C to produce methane and carbon dioxide. The overall reaction is

$$2\,C + 2\,H_2O + 2\text{ kcal} \longrightarrow CH_4 + CO_2$$

Although the overall reaction is endothermic, the combustion of the methane produced releases 192 kcal/mole, so the process is an energy-efficient way to obtain methane, an environmentally clean fuel.

Coal Liquefaction

Liquid fuels are made from coal by reacting the coal with hydrogen gas under high pressure in the presence of catalysts (hydrogenating the coal). The process produces hydrocarbons like those in petroleum. The resulting crude-oil type of material can be fractionally distilled to give fuel oil, gasoline, and hydrocarbons used in the manufacture of plastics, medicines, and other commodities. About 5.5 barrels of liquid are produced for each ton of coal. At the present time, the cost of a barrel of liquid from coal liquefaction is about double that of a barrel of crude oil. However, as petroleum supplies diminish and the cost of crude oil increases, coal liquefaction will become economically feasible.

12.10 Methanol as a Fuel

Methanol is being considered as a replacement for gasoline, especially in urban areas that have extremely high levels of air pollution caused by motor vehicles. For example, Southern California has been testing methanol-powered cars since 1981. About half the cars use 100% methanol (M100), while the other half are flexible-fueled vehicles (FFVs) that use either M85, a blend of 85% methanol and 15% gasoline, or gasoline. Although methanol fuels M85 and M100 have received more attention than corresponding ethanol fuels E85 and E100, FFVs are being built to test the use of both M85 and E85.

What are the advantages and disadvantages of switching to methanol-powered vehicles? Methanol burns more cleanly than gasoline, and levels of troublesome pollutants such as carbon monoxide, unreacted hydrocarbons, nitrogen oxides, and ozone are reduced. However, there is concern about the higher exhaust emissions of carcinogenic formaldehyde from methanol-powered vehicles. Since the number of methanol-powered vehicles is limited, it is still difficult to assess the extent to which these formaldehyde emissions will contribute to the total aldehyde levels from other sources.

■ Cars at the Indianapolis 500 are powered by methanol.

The technology for methanol-powered vehicles has existed for many years, particularly for racing cars that burn methanol because of its high octane rating of 100. However, methanol only has about one half the energy content of gasoline, which would require fuel tanks to be twice as large to give the same distance per tankful. This is partially compensated for by the fact that methanol costs about half as much as gasoline, so the price per mile would be competitive. Since methanol burns with a colorless flame, something needs to be added (a small amount of gasoline, for example) to methanol so it can be seen when it burns. Another disadvantage is the tendency for methanol to corrode regular steel, so it will be necessary to use stainless steel for the fuel system or have a methanol-resistant coating. Until sufficient numbers of methanol-powered vehicles are on the road, cars equipped to run on *either* methanol or gasoline will be necessary because of the lack of service stations selling methanol. As the problems of distribution and storage are solved, better-engineered methanol-fueled engines will be designed and produced, which will lead to more efficient utilization of methanol as a fuel.

Another option is to use methanol to make gasoline. Mobil Oil Company has developed a methanol-to-gasoline process that is currently not competitive with refined gasoline prices in the United States, but is competitive in those regions of the world, such as New Zealand, where the price of gasoline is much higher. In fact, the production of 92-octane gasoline from methanol is taking place in New Zealand.

Plant in New Zealand that converts natural gas to methanol, which is then converted to gasoline. *(Mobil Corporation)*

$$2\ CH_3OH \xrightarrow[\text{Catalyst}]{} \underset{\text{Dimethyl ether}}{(CH_3)_2O} + H_2O$$

$$2\ (CH_3)_2O \xrightarrow[\text{Catalyst}]{} \underset{\text{Ethylene}}{2\ C_2H_4} + 2\ H_2O$$

$$C_2H_4 \xrightarrow[\text{Catalyst}]{} \underset{\text{Gasoline}}{\text{hydrocarbon mixture in the } C_5\text{–}C_{12} \text{ range}}$$

The New Zealand plant is currently producing 14,000 barrels per day of gasoline with an octane rating of 92 to 94. This is about one third the amount of gasoline used in New Zealand.

■ SELF-TEST 12C

1. The fractions of petroleum are separated by _____.
2. The principal component in natural gas is _____.
3. The octane enhancer used most by gasoline producers at the present time is _____.
4. The _____ process is used to produce branched-chain and aromatic hydrocarbons from straight-chain hydrocarbons.
5. The _____ process is used in refining petroleum to convert molecules in the higher-boiling fractions to molecules in the gasoline fraction.
6. Which of the following hydrocarbons would be expected to have the highest octane rating?
 (a) $CH_3CH_2CH_2CH_2CH_2CH_2CH_3$

 (b) $CH_3CH_2\overset{\overset{\displaystyle CH_3}{\displaystyle |}}{C}HCH_2CH_2CH_3$

 (c) $CH_3-\overset{\overset{\displaystyle CH_3}{\displaystyle |}}{\underset{\underset{\displaystyle CH_3}{\displaystyle |}}{C}}-\overset{\overset{\displaystyle CH_3}{\displaystyle |}}{\underset{\underset{\displaystyle H}{\displaystyle |}}{C}}-CH_3$

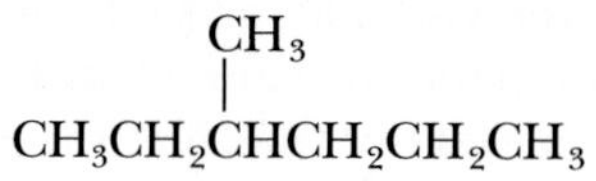

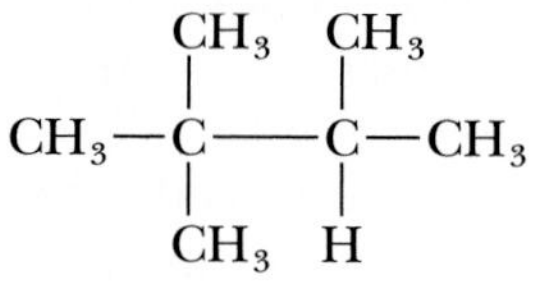

■ MATCHING SET

____ 1. hydrocarbon	a. benzene
____ 2. alkane	b. R—OH
____ 3. alkyl group	c. major component of natural gas
____ 4. alkene	d. contains only C and H
____ 5. alkyne	e. R—O—R'
____ 6. aromatic hydrocarbon	f. C_nH_{2n+2}
____ 7. methane	g. mixture of CO and H_2
____ 8. alcohol	h. contains C=C bond
____ 9. ether	i. hydrocarbon that is missing an H atom
____ 10. synthesis gas	j. contains C≡C bond

■ QUESTIONS FOR REVIEW AND THOUGHT

1. What is the definition of a fossil fuel?
2. What are the three major fossil fuels?
3. What is a hydrocarbon?
4. What is the primary component of natural gas?
5. What is the heat of combustion?
6. Give definitions for the following terms.

 (a) Alkane (b) Alkene
 (c) Alkyne (d) Aromatic
7. Give definitions for the following terms.
 (a) Multiple bond (b) Double bond
 (c) Single bond (d) Triple bond

8. Give definitions for the following terms.
   (a) Isomer
   (b) Straight chain hydrocarbon
   (c) Branched chain hydrocarbon
9. What is an alkyl group? Give the formulas for the methyl and ethyl alkyl groups.
10. Saturated hydrocarbons are so named because they have the maximum amount of hydrogen present for a given amount of carbon. The saturated hydrocarbons have the general formula C_nH_{2n+2}, where n is a whole number. What are the names and formulas of the first four members of this series of compounds?
11. Draw the tetrahedral structure of the methane molecule and label the bond angles.
12. Defend or refute the statement: "The tetrahedral angle is the most common angle found in naturally occurring substances."
13. How does an alkyl group differ from an alkane?
14. Why are alkenes and alkynes considered unsaturated?
15. Draw the structural formulas for the following:
   (a) Isopropyl group (b) *t*-butyl group
   (c) Butyl group
16. What is the structural formula for 1-pentene?
17. Draw the condensed structural formulas for the following:
   (a) 2,2,4-trimethylhexane (b) 3-ethylpentane
   (c) Methylbutane
18. Give the structural formulas for the following:
   (a) 2-methylpentane
   (b) 4,4-dimethyl-5-ethyloctane
   (c) 2-methyl-2-hexene
19. Draw as many different isomers as you can that have the formula C_5H_{12}.
20. Name the isomers in Question 19.
21. Draw the condensed structural formulas of all possible structural isomers for C_6H_{14} and name them.
22. Draw the structures of all possible isomers that are dimethylbenzenes.
23. Draw the structure of 2,3,3-trimethyl-1-pentene.
24. Why does 2-butene have *cis* and *trans* isomers but 1-butene doesn't?
25. Draw the *cis* and *trans* isomers of 1,2-dichloroethene.
26. What unique bond is present in an alkyne hydrocarbon?
27. The symbol □ can be used to represent cyclobutane.

    Draw the structural formula that shows all carbon and hydrogen atoms. How many are there of each?
28. How does the structure of cyclohexane differ from that of benzene?
29. Describe how petroleum is refined, starting with a barrel of crude oil.
30. Explain how fractional distillation is used in the refinement of petroleum.
31. What is "straight-run" gasoline?
32. List three gasoline additives that increase the octane rating of gasoline.
33. What is gasohol?
34. Explain how synthesis gas and methane can be obtained from coal, and write equations that represent these processes.
35. What is meant by the following terms?
   (a) Catalytic re-forming (b) Catalytic cracking
   (c) Octane rating
36. Explain how gasoline can be made from methanol.
37. What types of hydrocarbons have high octane ratings?
38. Methanol is now number 22 in the list of top 50 chemicals produced in the United States. What factors are likely to lead to an increased demand for methanol in the next decade?
39. How is the octane rating of a refined gasoline determined?
40. What are oxygenated gasolines?
41. Why do oxygenated gasolines cause less pollution than regular gasolines?
42. What is the difference between oxygenated gasolines and reformulated gasolines?
43. How is coal gasified?
44. What are the advantages and disadvantages of using methanol as an alternative fuel for vehicles?
45. What do the terms M100, E100, M85, and E85 refer to when describing alternate fuels? What is an FFV?
46. Label each of the following as an alkane, alkene, aromatic, or alkyne.

   (a) methane, CH_4, 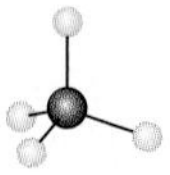

   (b) benzene, C_6H_6, 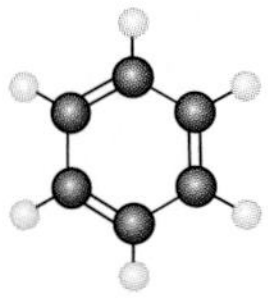

   (c) 1-butene, $CH_2CHCH_2CH_3$, 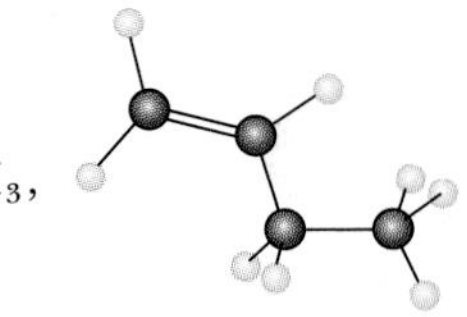

   (d) acetylene, CHCH, 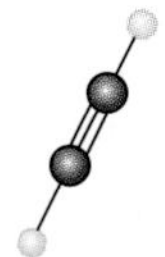

47. The following show the formulas and ball-and-stick models for various compounds. Label each of the following as an alkane, alkene, aromatic, alkyne, alcohol, or ether.

   (a) propane, $CH_3CH_2CH_3$, 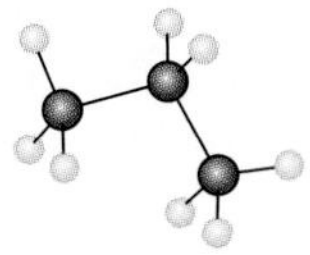

(b) 1-butyne, $CHCCH_2CH_3$,

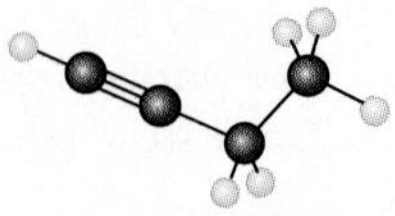

(c) diethyl ether, $CH_3CH_2OCH_2CH_3$,

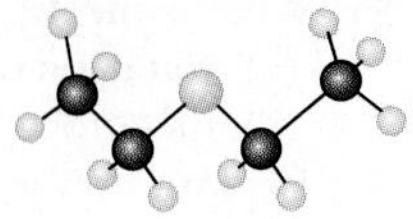

(d) ethanol, CH_3CH_2OH,

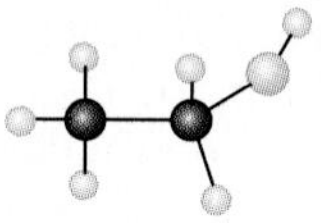

(e) ethylbenzene, $CH_3CH_2C_6H_5$,

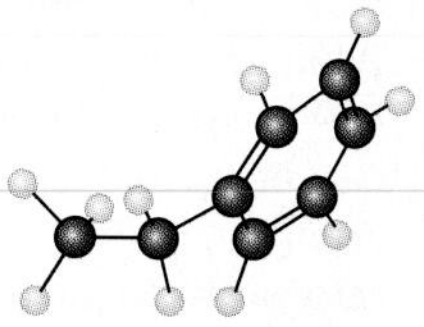

48. The oxygenation of gasolines can be done using ethanol or ETBE. This would make use of grains such as corn to provide the ethanol. Give two advantages and two disadvantages of using agricultural land to produce fuel for machines instead of food for people.
49. The removal of lead-containing additives reduced octane ratings for gasoline. Why is it considered good public policy to prohibit or reduce the lead content of gasoline?
50. Why are methane- and propane-fueled cars considered "clean" even though they still produce carbon dioxide and water? What emission products do other fossil fuels generate that make them dirtier?
51. The octane of gasoline can be enhanced if additional aromatics are blended into the gasoline. What possible disadvantage do aromatics pose if they are added to gasoline?
52. Why do grain producers favor the EPA mandate that gasoline must contain oxygenates and that 30% of these oxygenates must come from renewable sources?

PROBLEMS

1. (a) Write a balanced equation for the combustion of butane (C_4H_{10}), one of the components of natural gas.
(b) Use the equation in (a) and bond energies in Table 12.1 to calculate the heat of combustion per mole of butane.
2. Hydrogen is one of the alternate fuels being considered for vehicles. Using the bond energies in Table 12.1 and the following equation for the combustion of hydrogen, calculate the heat of combustion per gram of hydrogen.

$$2\ H_2(g) + O_2(g) \longrightarrow 2\ H_2O(g)$$

3. (a) Write a balanced equation for the combustion of octane, as a representative equation for burning gasoline, a mixture of hundreds of hydrocarbons.
(b) Use the equation in (a) and bond energies in Table 12.1 to calculate the heat of combustion per mole of octane.
(c) Calculate the heat of combustion per gram of octane from the value calculated in (b) for a mole of octane.
4. How do the energy values you obtained in Problem 2 for hydrogen and in Problem 3 for octane compare with the energy values given in Table 12.2?

CHAPTER 13

Alternate Energy Sources

We are an energy-consuming society, and the rest of the world is rapidly becoming like us. If fossil fuels remain the only source of energy, the entire world will eventually run out of energy. Fortunately, there are alternatives to fossil fuels as sources of energy, and as fossil fuel reserves are used up, these alternatives will become more and more important.

Much of the energy we use is in the form of electricity. We cool and heat buildings and illuminate our homes, workplaces, city streets, and athletic fields with electricity. Soon, many of our vehicles will be powered by electricity as well. The widespread use of electricity is fairly new, having begun in the early 1900s. If new ways are not discovered to make electricity using sources of energy other than fossil fuels, which are needed for other important applications, society will experience some real problems. This chapter explores some of the alternatives to fossil fuels as energy sources. In this chapter we shall address the following questions:

- How is electricity produced and distributed, and how efficient is this process?
- How is energy obtained from the splitting of heavy atoms?
- Why is energy produced when light atoms are fused together?
- How can we call solar energy "free" and yet spend so much money trying to capture it and use it?
- What are some of the ways solar energy can be put to use?
- How can energy be obtained from within the earth?

A question that will be of increasing importance: To what extent can the sun and plants meet the energy demands of an expanding world population? *(Alan Pitcairn/Grant Heilman Photography)*

About 37% of all energy used in the United States in 1993 was devoted to the industrial production of steel, plastics, fertilizers, cars and trucks, and other commodities. Transportation, residential use, and commercial use ac-

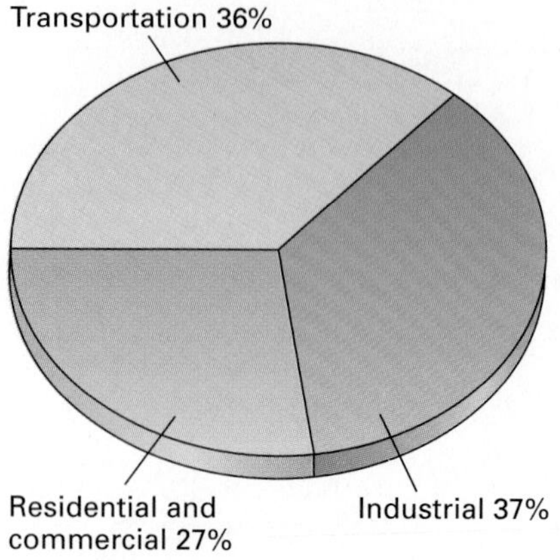

Figure 13.1 Distribution of energy consumed in the United States in 1993.

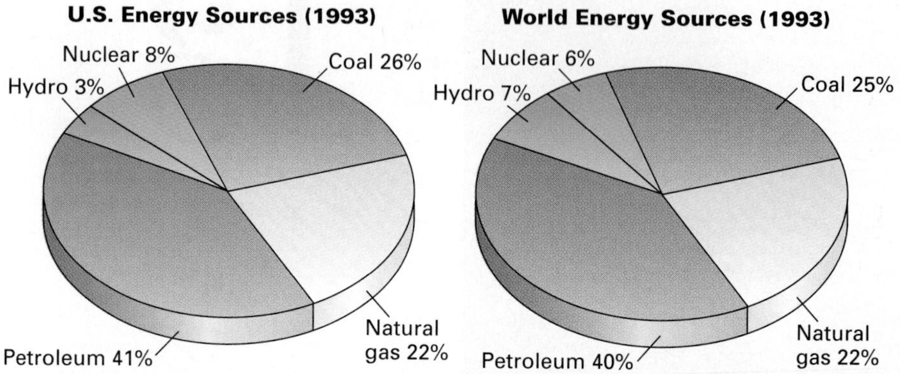

Figure 13.2 Worldwide sources of energy. Note that the nonrenewable fossil fuels (coal, natural gas, and petroleum) account for most of the energy used. The similarities between worldwide usage and that in the United States illustrates parallel technological developments in building hydroelectric facilities and nuclear power plants while remaining primarily dependent on fossil fuels.

count for the remainder (Figure 13.1). The United States uses energy from a variety of sources that are similar to those for the rest of the world (Figure 13.2).

Two important concepts related to energy sources are *reserves* and *renewability*. The reserve of an energy source is the amount that remains to be used. The renewability of an energy source is the extent to which it can be re-created as it is used up. Some energy sources, such as fossil fuels, are nonrenewable but have proven reserves. Other energy sources, such as solar energy and biomass (plants), appear to be renewable for the forseeable future. Complicating matters is the fact that different energy sources have different effects on the environment. Some, like nuclear energy, have especially troublesome problems associated with them. Virtually no source of energy is without some adverse environmental impact.

13.1 Speaking of Energy

Energy comes from a variety of sources. A barrel of petroleum, for example, can be refined into home heating oil, and the heat energy from the furnace burning this oil can warm a home and its occupants. Some of that heat might also be used to heat water for washing, cooking, and other things. Instead of petroleum, natural gas can be used to achieve the same results. Although a different amount of natural gas would be required, the same amount of heat energy could be obtained from either natural gas or petroleum. The same could be said for coal, firewood, or any other energy source. This means that a barrel of petroleum, a cubic foot of natural gas, or a ton of coal can be related to one another by their energy content. For this purpose, *units* of energy are needed. The calorie (cal), the amount of heat necessary to raise the temperature of 1 gram of water 1°C is our most familiar energy unit (Section 2.8). A related unit is the British thermal unit (Btu), the amount of heat necessary to raise the temperature of 1 pound of water by 1°F. This means that 2 Btus of

TABLE 13.1 ■ A Chart of Energy Units*

Cubic Feet of Natural Gas	Barrels of Oil	Tons of Bituminous Coal	British Thermal Units (Btu)	Kilowatt Hours of Electricity	Joules	Kilocalories†
1	0.00018	0.00004	1000	0.293	1.055×10^6	252
1000	0.18	0.04	1×10^6	293	1.055×10^9	0.25×10^6
5556	1	0.22	5.6×10^6	1628	5.9×10^9	1.40×10^6
25,000	4.50	1	25×10^6	7326	26.4×10^9	6.30×10^6
1×10^6	180	40	1×10^9	293,000	1.055×10^{12}	0.25×10^9
3.41×10^6	614	137	3.41×10^9	1×10^6	3.6×10^{12}	0.86×10^9
1×10^9	180,000	40,000	1×10^{12}	293×10^6	1.055×10^{15}	0.25×10^{12}
1×10^{12}	180×10^6	40×10^6	1×10^{15}‡	293×10^9	1.055×10^{18}	0.25×10^{15}

*Based on normal fuel heating values, 10^6 = 1 million; 10^9 = 1 billion; 10^{12} = 1 trillion; 10^{15} = 1 quadrillion (quad)

†1 food calorie = 1000 calories = 1.000 kcal

‡1×10^{15} Btu = 1 quad

heat would raise the temperature of 1 pound of water 2°F, and so on. The SI energy unit is the joule (J). The calorie, Btu, and joule are related to one another:

$$1 \text{ Btu} = 252 \text{ calories (cal)} = 1055 \text{ joules (J)}$$

All kinds of relationships can be worked out using the heat energy equivalents of the various sources of energy. For example, the heat equivalent of one barrel of petroleum can be given in calories or Btus or joules. Table 13.1 is a chart of energy equivalencies in units that will be useful throughout this chapter. Any entry along a horizontal row in Table 13.1 is equivalent to any other entry on that row. For example, 1 barrel of oil (a defined unit) is equivalent to 5.9×10^9 joules, which is also equivalent to 0.22 ton of bituminous coal. Energy is used in the United States and worldwide in such huge quantities that a special unit called the **quad** is defined. One quad is one quadrillion (10^{15}) Btu.

■ Bituminous coal is the most common type of coal and is the kind used in coal-fired electric power plants and in industry to make coke, an impure form of carbon used in steelmaking. Bituminous coal is soft and contains a larger percentage of hydrogen than anthracite coal, which can be almost rock-hard and contains a higher percentage of carbon.

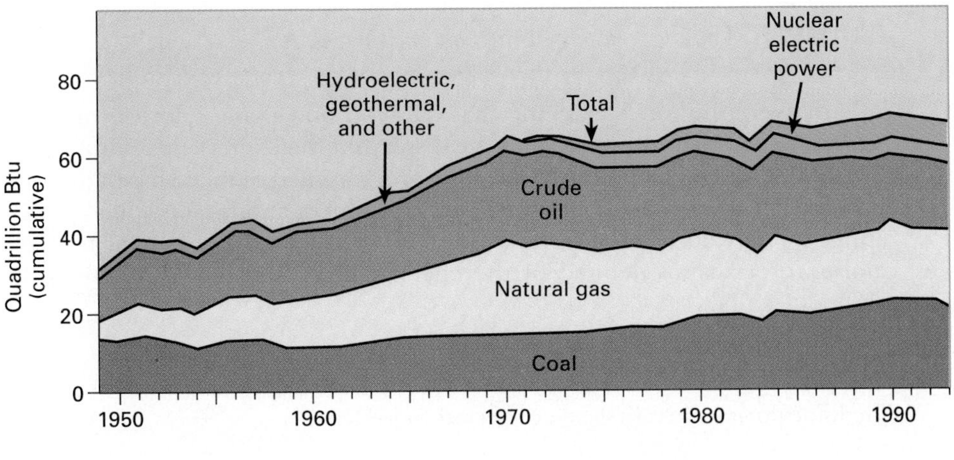

U.S. sources of energy from 1949 to 1993. The regions of the plot, which are additive, show a total of about 60 quads since 1970. About 20 additional quads of energy were produced in the United States from imported fossil fuels.

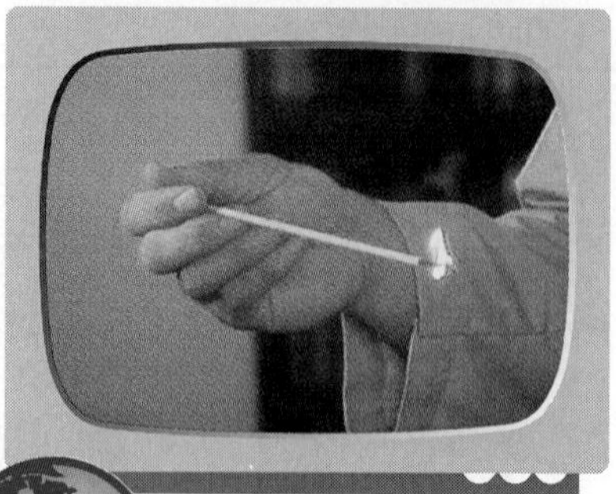

The World of Chemistry

Program 13, *The Driving Forces*

One Btu is approximately equal to the energy released in striking a match.

The United States consumes more than 84 quads of energy annually. From Table 13.1, you can see that 1 quad is equivalent to 1×10^{12} cubic feet of natural gas or 180×10^6 barrels of oil.

EXAMPLE 13.1 *Energy Equivalencies*

A tanker contains 1.5 million barrels of oil. This amount of oil is equivalent to how many tons of bituminous coal?

SOLUTION

First find the heating equivalence between these two sources of energy from Table 13.1: 1 barrel (bbl) of oil = 0.22 ton coal. Next, arrange the equivalence as a ratio, allowing the units to cancel.

$$1.5 \times 10^6 \text{ \sout{bbl oil}} \times \frac{0.22 \text{ ton coal}}{1 \text{ \sout{bbl oil}}} = 3.3 \times 10^5 \text{ ton coal}$$

Exercise 13.1

Calculate the number of barrels of oil required to produce the heat equivalent of one million kilowatt-hours of electricity.

Closely related to energy is **power,** which is the *rate* at which energy is produced, transferred, or used. We have all seen power ratings on light bulbs. One **watt** is 1 J/sec, so a 100-W lamp uses electrical energy at a rate of 100 joules per second. In one hour, the light bulb uses 100 J/s × 3600 s = 3.6×10^5 J of energy. That's about equal to lighting 360,000 matches. (No wonder Thomas Edison worked so hard to invent the electric light bulb!) ''Power'' is not strictly an electrical term. Any use of energy over a period of time can be related to a power rating. A human sitting consumes energy at a rate of about 60 watts. Running, the same person would consume energy at a rate of about 1000 watts.

■ A power rating used for automobile engines is horsepower. One horsepower is 746 watts. To develop maximum horsepower, an engine must burn fuel at a fast rate—many joules each second.

EXAMPLE 13.2 *Energy Unit Conversions and Power*

You are exercising on a machine that tells you how many calories you have ''burned.'' After about 31 minutes the display on the exercise machine shows 133 calories have been used (this is only a programmed estimate based on your age and sex). How many joules have you consumed, and what is your ''power rating''? The calorie used here is the food calorie, which is equal to 1000 calories, or 1 kcal as defined by the chemist. The calorie is also defined as 4.184 joules (J).

SOLUTION

The kilocalories must first be converted to joules:

$$133 \text{ \sout{food cal}} \times \frac{1000 \text{ \sout{cal}}}{1 \text{ \sout{food cal}}} \times \frac{4.184 \text{ J}}{1 \text{ \sout{cal}}} = 5.56 \times 10^5 \text{ J}$$

Power can now be calculated in watts:

$$\text{power} = \frac{5.56 \times 10^5 \text{ J}}{31 \text{ min}} \times \frac{1 \text{ min}}{60 \text{ sec}} = 2.9 \times 10^2 \text{ W (that is, J/sec)}$$

This means you were using energy equivalent to a 300-W light bulb while you were exercising.

Exercise 13.2

For a candy bar label that says it contains 500 (food) calories of energy, calculate the energy in joules. If you ate the candy bar and used all of its energy in a 10-minute jog, what would be your power rating?

13.2 Electricity—Energy Converted from One Form To Another

Electricity is produced when a wire loop rotates in a magnetic field, a phenomenon discovered by Michael Faraday in 1831. In electric-power generating stations, large generators, containing many loops of wire, are rotated by steam turbines or falling water in a hydroelectric facility. The steam can be produced by many means: burning wood, coal, oil, natural gas, or by using heat from a nuclear reactor or even the sun. By its nature, electricity finds a wider variety of uses than any other form of energy.

■ A generator is a device that changes mechanical energy into electrical energy.

■ Lighting accounts for about 20 to 25% of all the electrical usage in the United States.

At one time in this country, electricity was fairly cheap. Today that is no longer the case. The cost of electricity is closely tied to the cost and availability of all of the sources of energy from which electricity is derived. About 36% of the energy consumed in the United States is used in the production of electricity. As convenient as electricity is, its production and distribution are inefficient. Most of the energy lost in the production of electricity is simply lost to the environment as waste heat.

■ According to the second law of thermodynamics, some of the energy generated is lost to heat that cannot be captured for useful work, but simply increases the entropy of the universe.

In 1994 about 27.3 quads of energy were used in the production of electricity. These 27.3 quads yielded only about 8.0 quads of energy in the form of electricity. The 19.3 quads difference was lost in the production and the transmission of electricity. Part of this loss is expected because of the second law of thermodynamics (see Section 8.5), and nothing can be done about it. In spite of the second law, the energy losses in a large electrical power plant are surprisingly large because of other factors. There are additional losses due to friction in the electrical turbines and resistance in the electrical transmission lines. Even an electric light bulb is very inefficient in converting electricity into light. Only 5% to 15% of the electricity consumed by the incandescent bulb is converted to light—the remainder is radiated as heat energy. Fluorescent light bulbs do a somewhat better job, since they are about 20% efficient. Superconducting magnets and transmission lines would provide some remedy for the losses in electricity generation and transmission (Section 7.7).

■ When an electric current flows through a conductor, heating of the conductor occurs as a result of its resistance. This property is used in electric heaters, electric hair dryers, and similar appliances. Even electric transmission lines experience heating, so that some of the electrical energy is lost in transmission. The greater the distance the electric current must travel, the greater the loss. Ideally, an electric power generator would be near the cities where most electricity is used. Environmental and other considerations usually mean this is not possible.

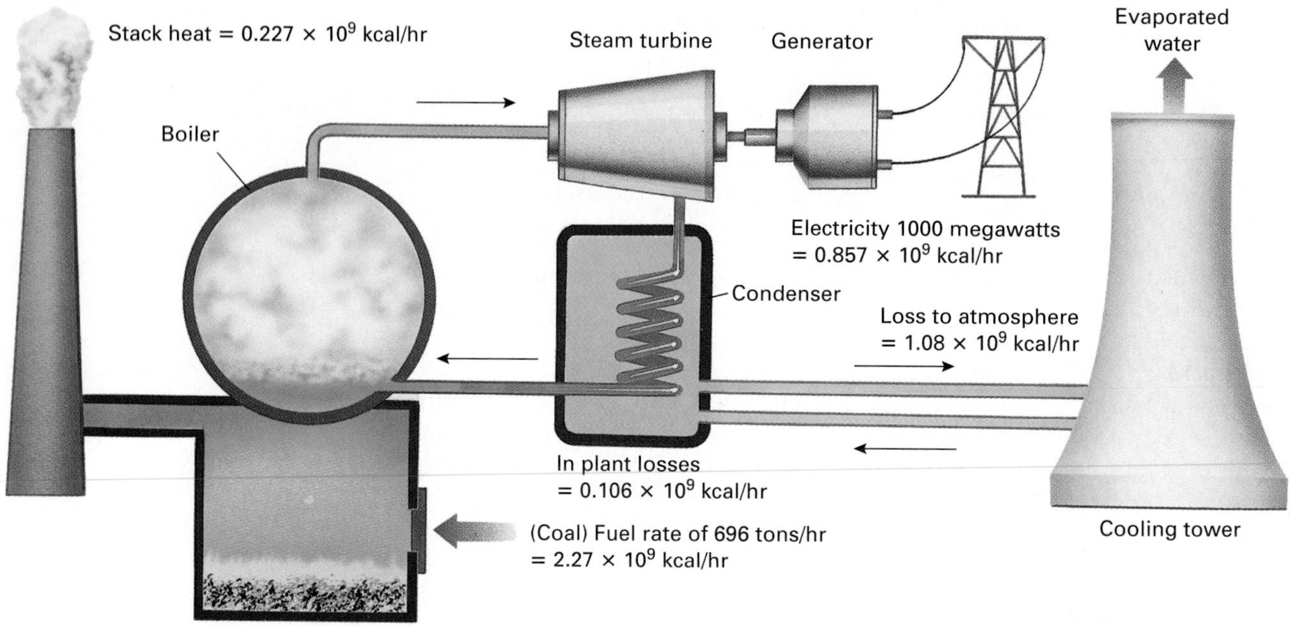

Figure 13.3 Fossil fuel electric power generation. In this 1000-MW coal-burning electric generating plant, 696 tons of coal are burned every hour. Of the heat liberated by the burning coal, less than 38% is converted to electrical energy available for transmission. Note the large amounts of energy lost in the form of heat to the environment.

However, given the current investment in the installed electrical power plants and power transmission lines, it is unlikely that a quick fix such as superconductivity will take place any time soon.

The energy losses in making electricity by burning coal can be seen in the following example. For a 1000-megawatt (MW) coal-burning plant (Figure 13.3), one hour of operation might look like this:

Coal consumed	696 tons producing 2.270 billion kcal
Smokestack heat loss	0.227 billion kcal
Heat loss in plant	0.106 billion kcal
Heat loss in evaporator to cool condenser	1.080 billion kcal
Electrical energy delivered to power lines	0.857 billion kcal
Percentage of energy delivered as electricity before transmission losses	$\frac{0.857 \text{ billion kcal}}{2.27 \text{ billion kcal}} \times 100\% = 37.8\%$

■ One megawatt (MW) = 1 million watts = 1.0×10^6 watts

The efficiency figure for the overall conversion of a fuel into electricity and its distribution to users is about 30%. It is important to note that when you pay for

1000 kcal of heat energy in the form of coal or fuel oil, you receive less than 300 kcal of energy in the form of electricity. Obviously, it requires much less fuel to heat homes with the fuel itself than with electricity made from the fuel. Even for fuels like natural gas, which is one of the simplest to distribute, transmission and distribution costs must be considered.

EXAMPLE 13.3 *Calculating Energy Usage from Your Electric Bill*

Your electric bill for January is \$106.42. Looking closely, you see the bill is for 1578 kilowatt-hours of electricity. Kilowatt-hours are energy units. Use Table 13.1 and calculate the barrels of oil and cubic feet of natural gas that might have been used to produce and distribute this electricity. (In actual practice, most electricity is sold from a ''grid'' with electricity produced at a number of sites using a variety of fuels including coal, fuel oil, natural gas, nuclear, water power, and even wind power.)

SOLUTION

First, look at Table 13.1 and find the entry for 1 barrel of oil. Then move to the right to find the number of equivalent kilowatt-hours (kWh) of electricity. Use this value as your conversion factor. Then enter a conversion factor for the kilowatt-hours consumed for every kilowatt-hour generated (1000 kWh generated/300 kWh consumed). This is the efficiency factor (30%) described in the text. Each of these factors is multiplied by the kilowatt-hours that appeared on your electric bill.

$$\text{oil used} = 1578\ \cancel{\text{kWh consumed}} \times \frac{1000\ \cancel{\text{kWh generated}}}{300\ \cancel{\text{kWh consumed}}} \times \frac{1\ \text{bbl oil}}{1628\ \cancel{\text{kWh}}}$$

$$= 3.23\ \text{bbl}$$

Do the same for the conversion between natural gas and kilowatt-hours, again entering the efficiency factor.

$$\text{natural gas used} = 1578\ \cancel{\text{kWh consumed}} \times \frac{1000\ \cancel{\text{kWh generated}}}{300\ \cancel{\text{kWh consumed}}} \times \frac{1\ \text{ft}^3\ \text{gas}}{0.293\ \cancel{\text{kWh}}}$$

$$= 17{,}900\ \text{ft}^3$$

In effect, much more energy is consumed making and distributing the electricity than the user actually receives. However, the user must still pay for the energy that was used to make the electricity.

Exercise 13.3

A manufacturing company wants to relocate to your community, creating about 50 new jobs. The company will require about 6.5×10^6 kWh of electricity annually. Calculate the tons of bituminous coal, cubic feet of natural gas, and barrels of oil that will be required to produce this electricity and distribute it to the new manufacturing facility. Don't forget to use the efficiency factor of 30%.

DISCOVERY EXPERIMENT

Counting Kilowatts

Check the wattage rating on six electrical appliances, such as a TV set, stereo, fan, iron, hair dryer, and toaster. Estimate the average time in hours that each appliance is operated per day, and calculate the kilowatt-hours of electrical energy consumed by each appliance. Calculate the dollars per kilowatt-hour that your electric company charges from your family's most recent electric bill. Then calculate the cost of operating each appliance per month. If you live in a dorm, your instructor will supply you with the kWh charge in your area. Prepare a chart like the one below. Use a new row for each of the six appliances you choose.

Appliance	Watt Rating	Hours Used per Day	Hours Used per Month	kWh of Electricity Consumed per Month	Electricity Cost = $/kWh	Cost for this Appliance
1.						
2.						
3.						
4.						
5.						
6.						

Adapted from J. D. Wilson: *Home Study Experiments in Practical Physics.* Philadelphia, Saunders College Publishing, 1986.

SELF-TEST 13A

1. The heat energy in a barrel of petroleum is the same as the heat energy in a cubic foot of natural gas. (a) True, (b) False.
2. Approximately how much energy is used annually in the United States? (a) 1 quad, (b) 80 quads, (c) 10^6 quads.
3. A 1000-W light bulb will use how much energy in 1 second? (a) 1000 J, (b) 1000 kcal, (c) 10,000 cal.
4. If 10,000 Btu of energy is used to make electricity and that electricity is distributed to your home, approximately how many Btu of energy will you receive? (a) 30,000 Btu, (b) 1000 Btu, (c) 3000 Btu.
5. The watt is a unit of power. (a) True, (b) False.
6. Electricity production accounted for ____________ percent of all the energy used in the United States in 1993.

13.3 Nuclear Energy

■ Nuclear reactions and radioactivity were discussed in Chapter 5.

Few issues have captured the awe, imagination, and scrutiny of mankind to quite the extent that nuclear energy has in the past five decades. Nuclear energy has been acclaimed, on the one hand, as the source of all our energy needs and accused, on the other hand, of being our eventual destroyer.

In 1938 Otto Hahn, Fritz Strassman, Lise Meitner, and Otto Frisch discovered that $^{235}_{92}U$ is fissionable by neutrons with an accompanying energy release (Figure 13.4). Scientists were quick to make use of the energy of nuclear

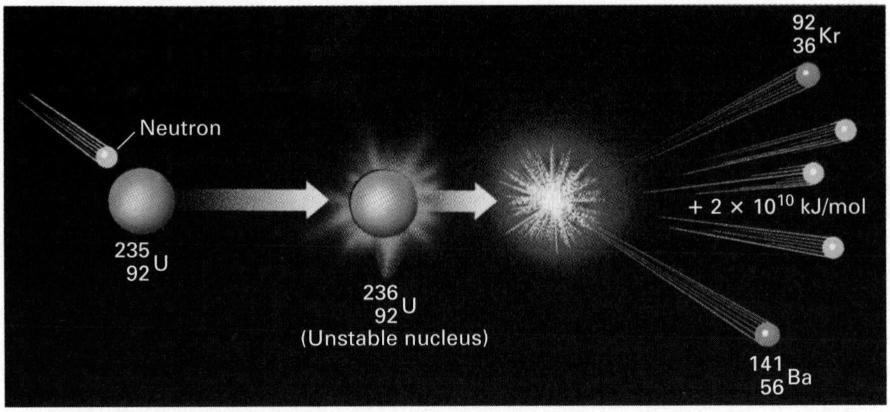

Figure 13.4 Nuclear fission. A slow-moving neutron strikes a uranium-235 atom, which becomes an unstable uranium-236 isotope. This unstable atom then quickly splits into two more-stable atoms of about equal mass. The three neutrons produced here are then available to cause fission in other adjacent uranium-235 atoms.

fission—first by creating the atomic bomb and then with nuclear power plants. In the 1950s it was hoped that nuclear energy would soon relieve the shortage of fossil fuels. To date this has not been accomplished, although the production of nuclear energy has grown in recent years. Uranium, the source of nuclear fuel, is a nonrenewable resource. In the first three decades of the growth of the nuclear power industry, it was thought that uranium sources would be used up quickly. This has not proved to be the case. New deposits of uranium ore have been discovered, while the growth of nuclear energy production has been slower than originally projected.

A vast amount of energy is released when heavy atomic nuclei split—the nuclear **fission** process—and when small atomic nuclei combine to make heavier nuclei—the **fusion** process. Consider the huge difference in the energy released in the combustion of a fossil fuel and in a nuclear fission reaction. When 1 mol (6.02×10^{23} molecules, or 16 g) of methane is burned, 192 kcal of heat are liberated (Section 12.1). Nuclear fission of a mole of uranium-235 releases 25 million times as much energy.

The World of Chemistry

Program 6, *The Atom*

Atomic bomb explosion.

Fission Reactions

Nuclear fission can occur when a neutron ($^{1}_{0}n$) enters a heavy nucleus. Certain nuclei with an odd number of neutrons ($^{235}_{92}U$, $^{233}_{92}U$, $^{239}_{94}Pu$) will undergo fission when struck by slow-moving *thermal* neutrons (neutrons with a kinetic energy about the same as that of a gaseous molecule at ordinary temperatures). The splitting of the heavy nucleus produces two smaller nuclei, two or more neutrons (an average of 2.5 neutrons for $^{235}_{92}U$), and much energy. A mole of uranium-235 atoms, each undergoing fission, will produce a number of different reaction products. One fission reaction that can take place is shown below.

$$^{235}_{92}U + ^{1}_{0}n \longrightarrow \underset{\text{Unstable nucleus}}{^{236}_{92}U} \longrightarrow \underbrace{^{141}_{56}Ba + ^{92}_{36}Kr}_{\text{Fission products}} + 3\ ^{1}_{0}n + 4.6 \times 10^{9}\ \text{kcal}$$

Thus, the fission of a mole of uranium, which has a mass of about 235 grams, produces 25 million times more energy than burning a mole of methane gas.

■ One kilogram of uranium fuel undergoing fission in a nuclear reactor can produce the same amount of energy as the combustion of 3000 tons of coal or 14,000 barrels of oil.

The same unstable $^{236}_{92}U$ atom may undergo fission to produce other products. For example,

$$^{235}_{92}U + ^{1}_{0}n \longrightarrow ^{236}_{92}U \longrightarrow ^{103}_{42}Mo + ^{131}_{50}Sn + 2\ ^{1}_{0}n + \text{energy}$$

Note that the same nucleus may split in more than one way. The lighter nuclei produced by the fission reaction are called **fission products.** These fission products, such as $^{141}_{56}Ba$ and $^{92}_{36}Kr$, can also be unstable and emit beta particles ($^{0}_{-1}e$) and gamma rays (γ), and may have long half-lives (see Section 5.4). Eventually, the decay of these fission products leads to stable isotopes.

The neutrons emitted by the fission of one uranium-235 atom can cause the fission of other uranium-235 atoms. For example, the three neutrons emitted in the uranium fission (Figure 13.4) could produce fission in three more uranium atoms; the nine neutrons emitted by those nuclei could produce nine more fissions; the 27 neutrons from these fissions could produce 81 neutrons; the 81 neutrons could produce 243 additional neutrons; those 243 neutrons could produce 729, and so on. This process is called a **chain reaction** (Figure 13.5), and it occurs at a maximum rate when the uranium sample is large enough for most of the neutrons emitted to be captured by other nuclei before passing out of the sample. A sample of fissionable material of sufficient size to self-sustain a chain reaction is termed the **critical mass.** If a critical mass of fissionable material is suddenly brought together, an explosion will occur

The critical mass for uranium-235 is about 10 kg.

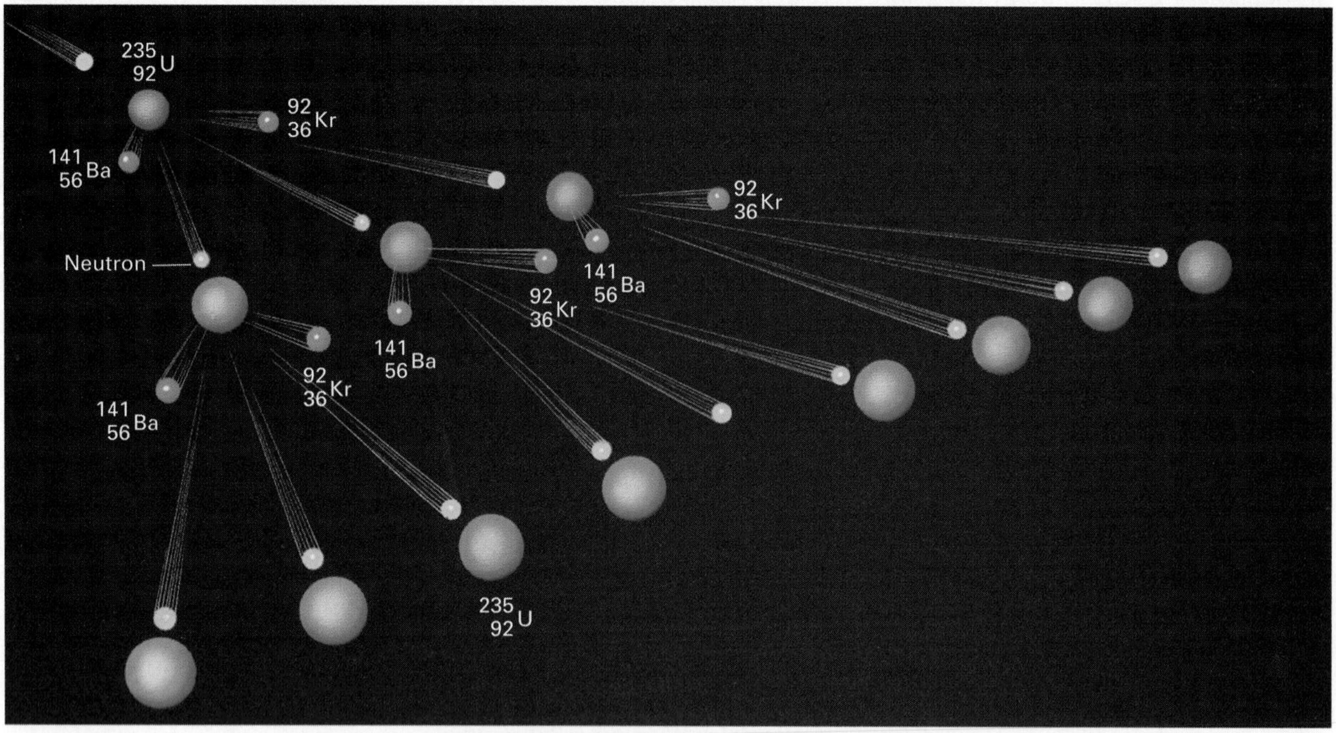

Figure 13.5 A nuclear chain reaction. A single neutron causes the fission of one uranium-235 atom, producing, in this case, three neutrons. Each of these neutrons, in turn, strike nearby uranium-235 atoms, which undergo fission. Their neutrons then cause other uranium-235 atoms to fission, and so on. In a nuclear reactor, the number of fission events is controlled by a moderator (see text), but in a nuclear fission explosion (atomic bomb), the chain reaction causes fission of as many atoms as possible in a short time. The release of all the fission energy in a fraction of a second is the explosive force that causes the devastation of an atomic bomb.

THE PERSONAL SIDE

Lise Meitner (1878–1968)

(AIP Emilio Segré Visual Archives Herzfeld Collection)

Lise Meitner was born in Vienna, Austria, one of seven children. In 1902, while studying science in Vienna, she became fascinated with the accounts of the discovery of radium by Pierre and Marie Curie and decided to pursue a career studying atomic physics. In 1906 she received her doctorate in physics from the University of Vienna. Two years later she moved to Berlin, where she became associated with Otto Hahn and Fritz Strassman, with whom she discovered atomic fission almost 20 years later. In 1918, together with Hahn, she discovered the element protactinium, which decays to form actinium. While working with Hahn and Strassman, Meitner discovered that something highly unusual was taking place when atoms of uranium-235 were struck by neutrons. In 1938, before she could solve this apparent puzzle, she fled to Sweden to escape the repression of Nazi Germany. After she established herself in Sweden, Otto Hahn announced the discovery that the uranium atom was being split into two approximately equal-sized parts by the neutron and sent Meitner his results for her analysis. Based on the data, she calculated the energy released when a uranium atom is split by the neutron and named the phenomenon nuclear fission. She even reported her findings in the British journal *Nature* in 1939. The uranium nucleus, she wrote, would split into a barium atom and a krypton atom under the influence of a neutron, at the same time producing more than 7.6×10^{-12} cal of energy (equal to 4.6×10^{9} kcal per mole of uranium atoms undergoing fission). The enormous energy released by atomic fission was immediately recognized to have military potential, and in the United States the "Manhattan Project" was begun to create the first atomic bomb. In 1945, when she heard that the first atomic bomb had been dropped on Japan, Dr. Meitner said,

> *I must stress that I myself have not in any way worked on the smashing of the atom with the idea of producing death-dealing weapons. You must not blame us scientists for the use to which war technicians have put our discoveries.*

because the energy released during each fission reaction cannot dissipate rapidly enough.

In a nuclear fission bomb the critical mass is kept separated into several smaller subcritical masses until detonation, at which time the masses are driven together by an explosive device (Figure 13.6). During the split second that the chain reaction occurs, the tremendous energy of billions of fission reactions is liberated, and everything in the immediate vicinity is heated to temperatures of 5 to 10 million kelvins and vaporized. The sudden expansion of hot gases literally pushes aside everything nearby and scatters the radioactive fission fragments over a wide area.

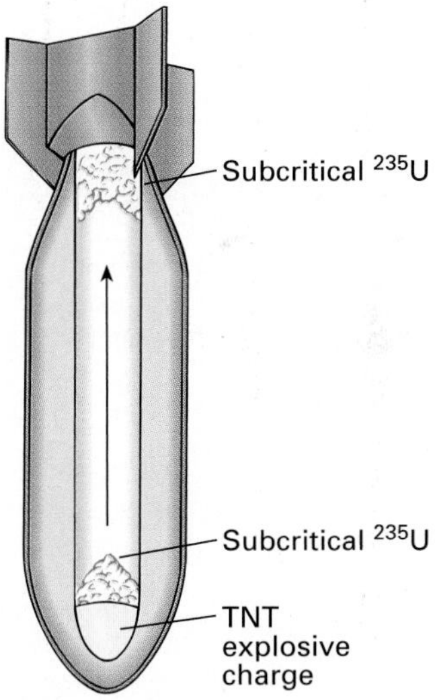

Figure 13.6 Implosion. By carefully shaping a conventional explosive, enough fissionable material can be brought together quickly to produce a massive chain reaction, causing many of the atoms to undergo fission almost simultaneously.

There is no danger of an atomic explosion in the uranium mineral deposits in the earth for two reasons. First, uranium is not found pure in nature—it is found only in compounds, which in turn are mixed with other compounds. Second, only 0.711% of the uranium found in nature is the easily fissionable $^{235}_{92}U$. The other 99.289% is $^{238}_{92}U$, which is not fissionable by thermal neutrons. In order to make nuclear bombs or nuclear fuel for generation of electricity, the naturally occurring uranium must first be *enriched,* a process that increases the relative proportion of $^{235}_{92}U$ atoms in a sample. It is possible to enrich uranium so that the percentage of $^{235}_{92}U$ is between 2 and 5% by making use of slight differences in the ability of the volatile compound uranium hexafluoride, UF_6, to pass through porous barriers.

The Mass-Energy Relationship

What is the source of the tremendous energy of the fission process? It ultimately comes from the conversion of mass into energy, according to Einstein's famous equation, $E = mc^2$, where E is energy that results from the loss of an amount of mass m, and c is the speed of light (186,000 miles/sec, or 3.00×10^8 m/sec). If the masses of the products of the fission of a uranium-235 atom by a neutron are compared with the masses of the reactants, it is found that the products have less mass than the reactants. In the case of the fission reaction

$$^{1}_{0}n + ^{235}_{92}U \longrightarrow ^{93}_{37}Rb + ^{141}_{55}Cs + 2\,^{1}_{0}n + \text{energy}$$

the difference in mass is about 0.000214 g per mol of uranium-235. According to Einstein's equation, this mass difference is equivalent to 4.6×10^9 kcal/mol of $^{235}_{92}U$. The mass "lost" as a result of the fission process is the source of the tremendous energy that is released.

The small nucleus of an atom is crowded with neutrons and protons. The very fact that nuclei hold together indicates that there must be some sort of force binding the particles together. This force is called the nuclear **binding energy** and is directly related to the stability of the nucleus. The binding energy depends on the number of particles in the nucleus. For example, if separate neutrons and protons (collectively, these particles are called **nucleons**) are combined to form any particular nucleus, the resulting nucleus always has less mass than the starting nucleons. This mass difference, converted into energy units according to Einstein's equation, is the binding energy and can be expressed for each atom as its *binding energy per nucleon* by dividing the total binding energy for the atom by the number of nucleons in the nucleus.

■ "Nucleon" is a collective term referring to protons and neutrons.

The mass of one mole of $^{4}_{2}He$ atom calculated from the sum of the masses of the constituent particles is 4.032982 g:

$$\begin{aligned} 2 \text{ mol of protons} \times 1.007826 \text{ g/mol of protons} &= 2.015652 \text{ g} \\ \underline{2 \text{ mol of neutrons} \times 1.008665 \text{ g/mol of neutrons}} &= \underline{2.017330 \text{ g}} \\ \text{total} &= 4.032982 \text{ g, the calculated mass} \\ &\quad\ \text{of one mole of } ^{4}_{2}\text{He atoms} \end{aligned}$$

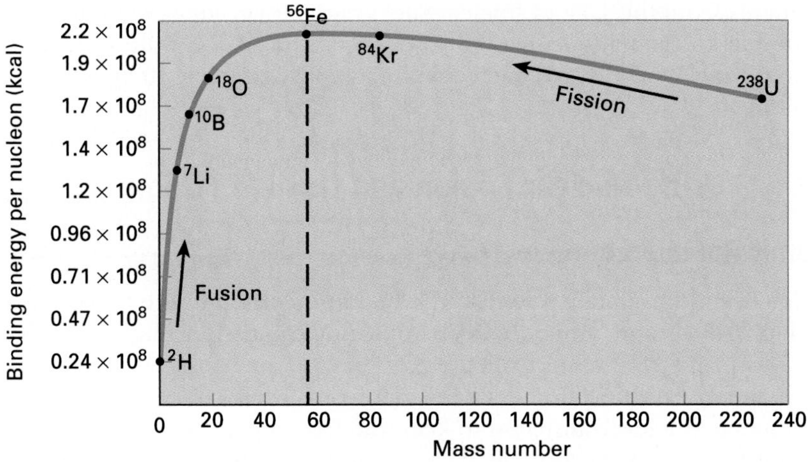

Figure 13.7 Relative stability of nuclei with different mass numbers. Both lighter and heavier isotopes are less stable (less binding energy per nuclear particle) than those isotopes with masses between 50 and about 65 amu. The most stable isotope is that of iron-56. Light isotopes can be fused together to form more stable atoms (nuclear fusion) while heavier isotopes can be split into more stable, lighter atoms (nuclear fission).

Since the measured mass of a mole of $^{4}_{2}He$ atom is 4.002604 g/mol, the difference in mass is 0.030378 g/mol:

$$4.032982 \text{ g/mol} - 4.002604 \text{ g/mol} = 0.030378 \text{ g/mol (the mass difference)}$$

The 0.030378 g of mass lost per mole of helium-4 nuclei created from their nucleons would be released in the form of energy (binding energy) if the $^{4}_{2}He$ atoms were made from separate protons and neutrons. The binding energy is analogous to the earlier concept of bond energy, in that both are a measure of the energy necessary to separate the whole (nucleus or molecule) into its parts. Figure 13.7 shows that light atoms and heavy atoms have lower binding energies per nucleon than atoms with mass numbers between 50 and 65. This means energy can be released when extremely light atoms are combined to make heavier atoms and when extremely heavy atoms are split to make lighter atoms. The greatest nuclear stability (greatest binding energy per nucleon) is at iron-56 ($^{56}_{26}Fe$). This is why iron is the most abundant of the heavier elements in the universe.

Because of their relative stabilities, most fission products fall into the intermediate range of atomic numbers (Figure 13.7). Therefore, when fission occurs and smaller, more stable nuclei result, these nuclei will contain less mass per nuclear particle. In the process, mass must be changed into energy —the tremendous energy released in a nuclear bomb or, under controlled conditions, in a nuclear power plant. It takes only about 1 kg of $^{235}_{92}U$ or $^{239}_{94}Pu$ undergoing fission to produce energy equivalent to that released by 20,000 tons (20 kilotons) of ordinary explosives such as TNT or dynamite. The energy equivalence of matter is further dramatized when it is realized that the atomic fragments from the 1 kg of nuclear fuel weigh 999 g, so only one tenth of 1% of the mass is actually converted to energy. The fission bombs dropped on Japan during World War II each contained approximately 10 kg of fissionable material.

The nuclear power plant at Indian Point, N.Y. *(Dan McCoy/Rainbow)*

13.4 Using Nuclear Fission and Nuclear Fusion

Atomic Reactors and Electricity

Enrico Fermi (an Italian scientist who had immigrated to the United States in the late 1930s) and others believed that nuclear fission might somehow be made to proceed at a controlled rate. They reasoned that if a way could be found to control the number of the neutrons, their concentration could be maintained at a level sufficient to keep the fission process going but not high enough to allow an uncontrolled chain reaction. It would then be possible to drain the heat energy away on a continuing basis to do useful work. In 1942, working at the University of Chicago, Fermi was successful in building the first atomic reactor.

■ Ordinary uranium, which is mostly $^{238}_{92}U$, cannot be used as a fuel in an atomic reactor because of the small concentration of the easily fissionable $^{235}_{92}U$ isotope.

An atomic reactor has several essential components. The reactor fuel must contain significant concentrations of atoms such as $^{235}_{92}U$, $^{233}_{92}U$, or $^{239}_{94}Pu$ that are fissionable by slow-moving neutrons. Typically, reactor fuel will contain uranium in the form of an oxide, U_3O_8, that has been enriched to contain about 3 or 4% $^{235}_{92}U$. A **moderator** is required to slow the speed of the neutrons produced in the reactions without absorbing them. Graphite and water have been used as moderators. A **neutron absorber,** such as cadmium or boron steel, is present in order to provide fine control over the neutron concentration. **Shielding,** to protect the workers from dangerous radiation, is an absolute necessity. Shielding tends to make reactors heavy and bulky installations. Finally, a **heat-transfer fluid** provides a large and even flow of heat energy away from the reaction center. Water is used as a heat transfer fluid in many nuclear reactor designs. In addition to water's high heat capacity, its hydrogen atoms are excellent moderators for neutrons. Conventional technology then allows the heat energy carried by the hot water from the reactor to be used to generate electricity, to power ships, or to operate any device that uses heat energy. A system for the nuclear production of electricity is illustrated in Figure 13.8.

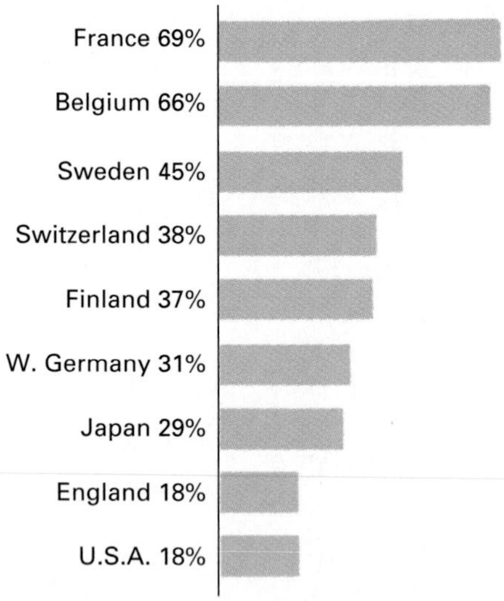

Figure 13.9 The approximate share of electricity generated by nuclear fission in various countries.

Nuclear energy is mostly used for the production of electricity. From 1957, when the first commercial nuclear power plant began operation, until about 1974, more energy was produced in the United States from burning firewood than from nuclear energy. By 1992, 6.6 quads of energy were produced by nuclear fission, and only 2.25 quads were generated by burning firewood. Nuclear energy today produces about 100,000 megawatts of electrical power in 110 power plants throughout the United States. This is about 19% of the electricity used in this country. Some countries produce less energy by nuclear fission than the United States, but in others nuclear energy provides a much larger share of the total energy production (Figure 13.9).

There are several extremely vexing problems associated with nuclear energy, any one of which may have adverse effects on its future as an alternative energy source. One problem, which probably is the greatest in the minds of the public, is the risk of a catastrophic accident at a nuclear power facility. Two

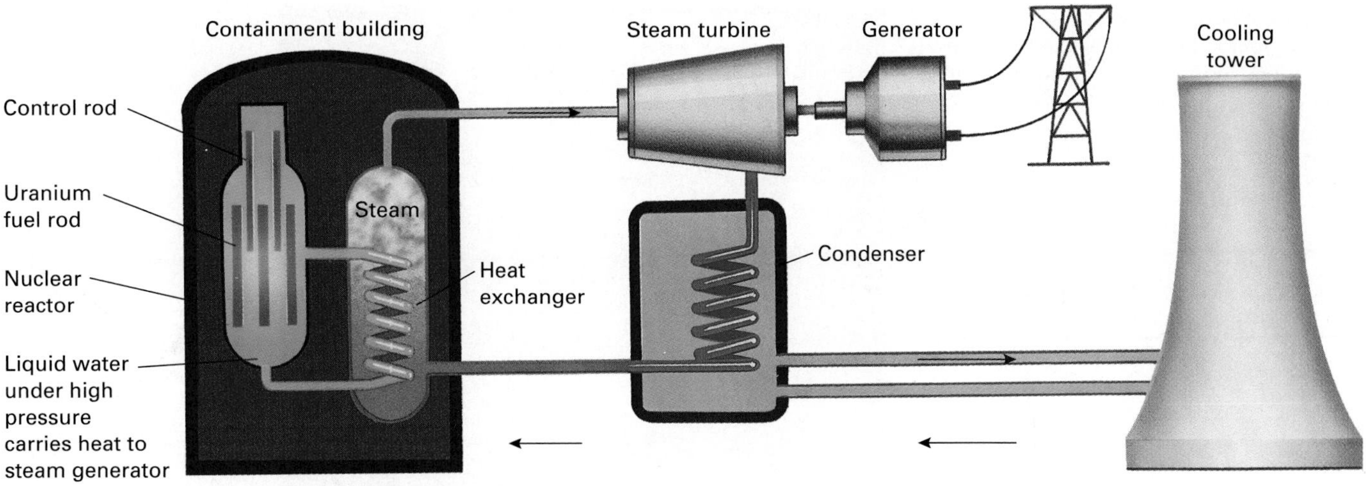

Figure 13.8 A nuclear power plant. In this reactor design (the most commonly used), ordinary water (called light water to differentiate it from D_2O—heavy water—used in some designs) is pressurized and allowed to carry heat energy from the reactor core to a heat exchanger, where steam is generated. This steam passes through a turbine, which generates electricity. Although simple in concept, safety considerations make the design, testing, and operation of a nuclear power plant a complex and costly operation.

accidents, of different degrees of seriousness, are generally known to the public. In late March of 1979, an accident occurred at the Three Mile Island power plant near Harrisburg, Pennsylvania. A water pump failed in the reactor and caused a partial reactor core meltdown. Steam vented inside the reactor vessel, and hydrogen gas was produced by the decomposition of the steam at the very high temperatures. This hydrogen gas, with the oxygen that was also produced, caused a risk of an explosion that would have blown apart the safety containment, releasing fission products. In fact, some radioactive gases were vented into the atmosphere, resulting in an average increased dosage of 2 mrem (Section 5.8) for people near the plant. There were no deaths directly associated with the Three Mile Island accident. Cleanup of the Three Mile Island facility continues today.

Certainly the most catastrophic nuclear accident occurred on April 26, 1986, at the Chernobyl Unit 4 Reactor near Kiev, Ukraine. The accident, a core meltdown, explosion, and fire, killed 31 people, hospitalized 500 others, and exposed many thousands of people to potentially harmful radiation. It has been estimated that the radiation from this accident will cause 17,000 extra cancer deaths in the next 70 years or so. So far, an estimated 5000 people have died from causes attributed to the accident. The Chernobyl reactor, built in 1983, had a design quite different from those used elsewhere in the world. Graphite was used exclusively as the moderator. While this design allows a higher efficiency than reactors with other types of moderators, the graphite can be ignited if sufficiently high temperatures are reached. In addition, the Chernobyl reactor lacked an adequate containment vessel to withstand an explosion in the reactor core.

■ At very high temperatures water will decompose into hydrogen and oxygen.

$$2\ H_2O(g) \xrightarrow{\text{High temp}} 2\ H_2(g) + O_2(g)$$

■ In a core meltdown of a nuclear reactor, the failure of cooling would allow temperatures to rise above the melting point of the metal rods containing the uranium fuel (about 1205°C). In the worst scenario, the resulting mass of highly radioactive molten metal would melt through the steel and concrete of the containment vessel beneath it. Once out of the containment vessel, the radioactivity might contaminate groundwater supplies.

Chernobyl nuclear power plant after the accident. The man at the left in the helicopter is testing the air for radioactivity. Presently, the entire reactor has been encased in concrete. *(The Bettmann Archive)*

■ There are about 10 billion curies of radioactive material within an operating nuclear reactor. (The units used to measure radioactivity are discussed in Section 5.8.)

■ Cesium is in the same chemical family as sodium and potassium and reacts similarly to these elements. Cesium-137 has a long half-life (30 years) and can easily become a part of the food chain.

While engineers were running an unauthorized test on the electrical generator, the Chernobyl reactor suddenly increased its power output, and before neutron absorbers could be lowered into the core, a meltdown occurred. A vast amount of steam was formed which, together with burning blocks of graphite and radioactive fuel, caused the entire reactor roof to blow off. The result was release of a radioactive plume that rose almost 5000 meters into the atmosphere, scattering throughout much of Western Europe an estimated 100 million curies of radioisotopes, including the beta-particle emitters ^{137}Cs (half-life 30 years) and ^{131}I (half-life 8.05 days) (Figure 13.10). About 4.9 million people in Ukraine, Belarus, and Russia were directly affected by the release. Today the reactor is entombed in more than 300,000 tons of concrete, including an underground concrete liner to protect the groundwater. Indications are that the protective concrete may be slowly developing cracks and leaks.

Problems with Nuclear Waste Products

Another problem resulting from nuclear energy production is *unavoidable plutonium production* and its possible diversion to bomb making. A fission reactor makes plutonium because some of the uranium-238 in the fuel rods captures neutrons, first forming $^{239}_{92}U$ (half-life 23.5 minutes), which decays by beta decay to $^{239}_{93}Np$ (half-life 2.35 days), which in turn forms $^{239}_{94}Pu$ (half-life 24,000 years) by beta decay.

$$^{238}_{92}U + ^{1}_{0}n \longrightarrow ^{239}_{92}U + ^{0}_{-1}e \longrightarrow ^{239}_{93}Np + ^{0}_{-1}e \longrightarrow ^{239}_{94}Pu$$

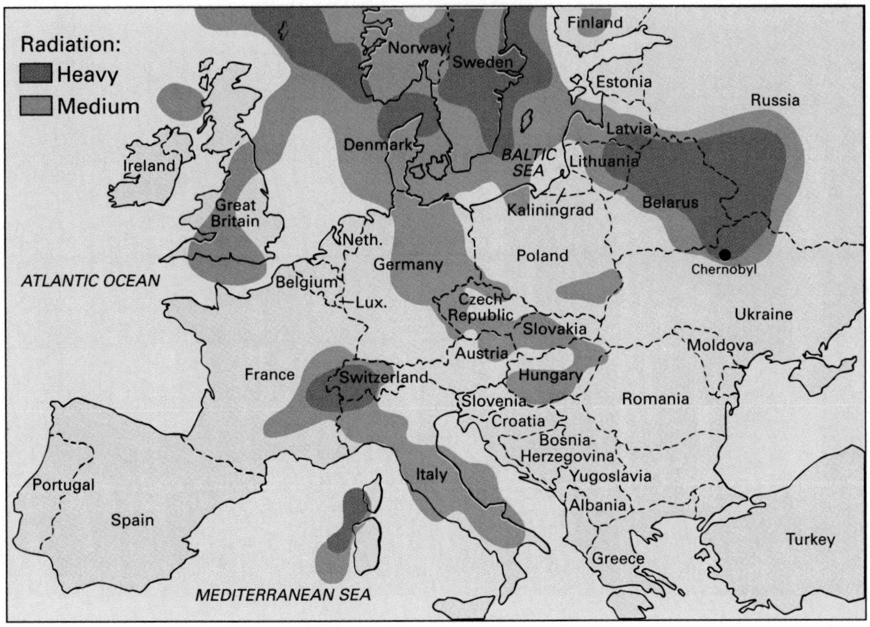

Figure 13.10 Distribution of radiation from the Chernobyl accident, April 26, 1986. For months after the accident, contaminated produce and dairy products were common throughout Eastern and Western Europe. Even cows kept indoors breathed in enough radioisotopes to contaminate their milk.

The short half-lives of the intermediates mean that plutonium-239 quickly accumulates in the fuel rods of the reactor.

There are about 400 nuclear power plants worldwide, and these produce about 70 tons of plutonium annually. Already there is an accumulation of 700 tons of plutonium, and assuming that uranium fuel will be used in the future at its current rate, about 2000 more tons of plutonium will be produced by the time the estimated reserves of uranium fuel are used up.

About once every year a third of the fuel rods in a nuclear reactor are replaced with new ones. In the United States, these fuel rods are currently being stored at the reactor sites while a permanent national storage site is being developed. During the early days of nuclear energy production in this country, nuclear fuel rods were *reprocessed,* allowing unreacted uranium and plutonium to be separated from the fission products and recycled back into new fuel rods. Through the reprocessing of nuclear fuel about 30% of the uranium in any fuel rod assembly could be reused, thus increasing the usefulness of this nonrenewable energy resource. In addition, the plutonium-239 found in the spent fuel rods was also recycled into new reactor fuel. The problem with nuclear fuel reprocessing is that with an additional few simple steps, plutonium of 99% purity can easily be separated from the uranium. This plutonium, mostly the easily fissionable plutonium-239 isotope, can be used to make fission bombs! Although the United States does not currently reprocess nuclear fuel, other countries, such as Great Britain, France, and Japan, do. This has led to widespread concern that plutonium from commercial reactors may someday appear in the hands of renegade governments and terrorists in spite of the very best efforts to prevent such a thing from happening.

■ The diplomatic flap in 1993–1994 between North Korea and the other nations who signed the Nuclear Nonproliferation Treaty was caused by North Korea's unwillingness to allow inspections of its spent nuclear fuel to see if any of the plutonium it contained had been diverted. In 1962 the United States exploded a nuclear bomb made from plutonium derived from a civilian reactor.

But the problems with plutonium get even worse. Nuclear disarmament, begun in earnest in 1993, has brought with it the problem of what to do with

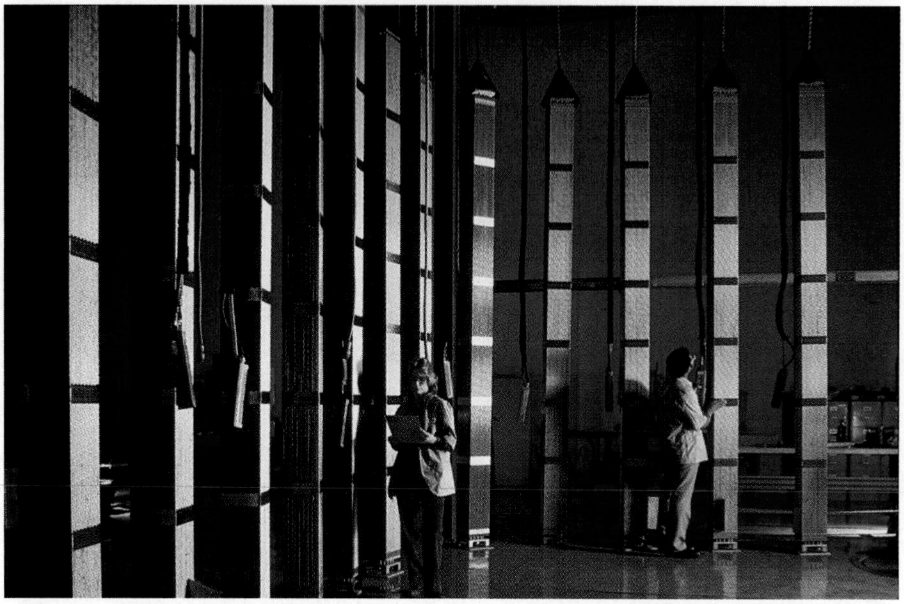

Uranium fuel rod assemblies, ready for a reactor. Each square assembly is a bundle of rods packed with uranium dioxide pellets. *(Courtesy of Westinghouse Electric Corp.)*

the tremendous amount of plutonium that was fabricated into nuclear warhead materials. The United States currently dismantles about 1800 nuclear warheads annually. By 2003, about 50 metric tons of weapons-grade plutonium will have accumulated from dismantled warheads. In addition to the plutonium, an even greater amount—about 400 metric tons—of highly enriched uranium will also have been taken from warheads. Similar amounts of both plutonium and uranium will have been taken from warheads in Russia as well. The enriched uranium from warhead dismantling will probably find uses in civilian nuclear reactor fuel, but currently no large-scale use of fissionable plutonium is planned. Only Japan and Russia are currently showing interest in using plutonium as a reactor fuel. The United States will probably opt to mix weapons-grade plutonium with other highly radioactive fuel waste from reactors and create glass logs that will weigh about 2 metric tons each. These logs would then be placed in geologic repositories. The glass logs would be so highly radioactive that thieves would be unlikely to want to use them as a source of plutonium to make bombs.

One solution to the problem of plutonium accumulation is the so-called **integral fast reactor,** developed by the Argonne Laboratory test facility in Idaho. This reactor uses fast neutrons to cause fission of both uranium *and* plutonium, thus using plutonium as a fuel. A complex recycling process involving electrolysis of the fuel rod metals allows complete recycling of all of the fissionable isotopes until they are entirely used up. The nuclear wastes from the integral fast reactor would be radioactive for a relatively short period of time. After about 200 years, the radioactivity of the reactor wastes would be no greater than the original uranium ore from which the fuel was derived. In spite of these apparent benefits, this reactor will not see commercialization anytime soon because of the current fear within the United States government

TABLE 13.2 ■ Radioactive Wastes from Spent Pressurized Water Reactor Fuel

Fission Products		Actinides	
Nuclide	*Half-Life (years)*	*Nuclide*	*Half-Life (years)*
^{90}Sr	28.8	^{237}Np	2.1×10^6
^{99}Tc	210,000	^{238}Pu	89
^{106}Ru	1.0	^{239}Pu	2.4×10^4
^{125}Sb	2.7	^{240}Pu	6.8×10^3
^{134}Cs	2.1	^{241}Pu	13
^{137}Cs	30	^{242}Pu	3.8×10^5
^{147}Pm	2.6	^{241}Am	458
^{151}Sm	90	^{243}Am	7.6×10^3
^{155}Eu	1.8	^{244}Cm	18.1

Activity (in curies*) after:	10 years	100 years	1000 years
Fission products	300,000	35,000	15
Actinides	10,000	2,200	600

*A curie is the unit used to express the amount of radioactivity contained in a sample. One curie is equal to 3.7×10^{10} decay events per second (irrespective of the type of radiation).

concerning any kind of new reactor design and reuse of plutonium. Interestingly, the integral fast reactor could even use weapons-grade plutonium as a fuel, thus lessening the chances that plutonium from dismantled nuclear warheads will fall into the wrong hands.

What to do with already accumulated *nuclear wastes* is also one of the problems of nuclear energy. All the fission products of uranium are radioactive. Some have long half-lives (Table 13.2). In addition, neutron capture by uranium-238 and other isotopes gives rise to **actinides** (elements with atomic numbers greater than 88), which are both highly radioactive and have extremely long half-lives. Plutonium-239 (half-life 24,000 years) is the best known actinide. Nuclear wastes containing the actinides will be highly dangerous for about 100,000 years or more.

The nuclear wastes from all the United States' nuclear power plants total about 2000 tons per year, and more than 15,000 tons are now being stored. In 1998 construction is scheduled to begin on an underground repository at Yucca Mountain in Nevada (Figure 13.11). The rock formation at Yucca mountain is a dense volcanic ash that will allow any seepage into the groundwater to travel only about a mile every 3400 to 8300 years. Still, it may not be possible to put all of the stored nuclear power plant wastes at Yucca Mountain because of a "not-in-my-backyard" attitude among residents of the region.

■ Military nuclear wastes present yet another problem. Much more military waste exists, and it is more complex than civilian reactor wastes. Much of it is liquid. About 80 million gallons of high-level waste is currently in storage at several sites in the United States. Other countries with nuclear weapons programs have similar waste-disposal problems.

Fusion Reactions

Nuclear fusion produces about the same amounts of energy as fission on a per-mole basis but fewer radioactive byproducts; the products that are radioactive have short half-lives. Hydrogen (^1_1H) has two isotopes, deuterium (^2_1H)

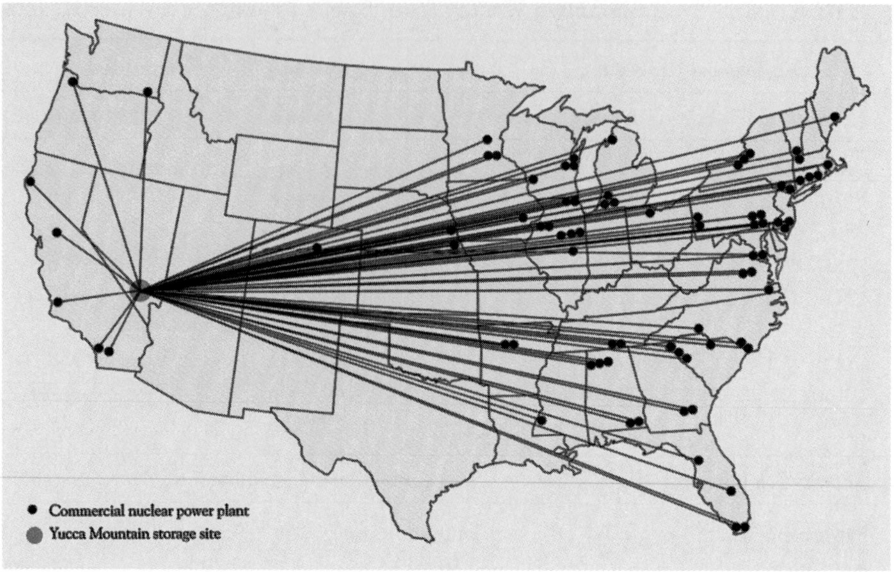

Figure 13.11 Proposed shipments of nuclear reactor wastes to the Yucca Mountain site. Each nuclear power plant (shown as dots on the map) would package and ship its wastes to Yucca Mountain, Nevada, by rail and truck. Residents along established routes have already voiced concerns about the passage of all these loads of nuclear waste through their communities.

and tritium (^3_1H). The nuclei of deuterium and tritium can be fused together at high temperatures to form a helium-4 atom. The energy released is 4.1×10^8 kcal/mol of ^2_1H.

$$^2_1H + ^3_1H \longrightarrow ^4_2He + ^1_0n + 4.1 \times 10^8 \text{ kcal}$$

When very light nuclei, such as those of hydrogen, helium, or lithium, are combined, or *fused,* to form an element of higher atomic number, energy equivalent to the difference between the total mass of the reacting atoms and the smaller total mass of the more stable products is released. This energy, which comes from a decrease in mass, is the source of the energy released by the sun, the stars, and hydrogen bombs. Typical examples of fusion reactions are

$$4\,^1_1H \longrightarrow ^4_2He + 2\,_{+1}^{\;\;0}e + 6.14 \times 10^8 \text{ kcal}$$
$$2\,^2_1H \longrightarrow ^3_2He + ^1_0n + 7.3 \times 10^7 \text{ kcal}$$
$$2\,^2_1H \longrightarrow ^3_1H + ^1_1H + 9.2 \times 10^7 \text{ kcal}$$
$$^3_1H + ^2_1H \longrightarrow ^4_2He + ^1_0n + 4.04 \times 10^8 \text{ kcal}$$

If fusion were to be used as a source of energy here on earth, a suitable source of fusible atoms (fuel) would be needed. Fortunately, the oceans are a potential source of fantastic amounts of deuterium. There are 1.03×10^{22} atoms of deuterium in a *single liter* of sea water. If all the deuterium atoms in a cubic kilometer of sea water were fused to form heavier atoms, the energy released would be equal to that released from burning 1360 billion barrels of crude oil, and this is approximately the total amount of oil originally present on this planet.

SCIENCE AND SOCIETY

How Can the Practical Knowledge of Making Nuclear Bombs Be Controlled?

One of the basic principles of science is the distribution of knowledge in the scientific literature, through meetings of scientists, in government and university publications, and by similar mechanisms. Hardly a year goes by without announcement of a new discovery that radically changes how matter is understood. Other researchers with similar interests quickly attempt to duplicate the experiments leading to the discovery. These experiments often extend the findings of the original discovery into new areas of understanding.

Sometimes, however, new knowledge is suppressed. After its discovery during World War II, the knowledge of how to make nuclear explosive devices was thought to be so potentially dangerous that its details were a closely guarded secret. Yet perhaps even before the United States dropped nuclear bombs on Japan in 1945, the secrets of the process were known by scientists in many nations. In 1949 the Soviet Union exploded its first test nuclear device. Quickly, Britain and France exploded their nuclear bombs and were then joined by China and India. The knowledge of how to enrich uranium-235, make plutonium-239, and fashion these elements into a bomb was spreading—probably as a result of the work of spies, but possibly by scientists sharing bits and pieces of information with one another. In the United States, individuals suspected of passing vital nuclear secrets (including how the critical mass was brought together at the time of detonation) were tried and sometimes executed for treason.

In 1970, 25 years after the explosions of the only nuclear bombs ever to be dropped in war, those nations that by then had acquired the knowledge to make nuclear weapons, as well as most of the other nations of the world, signed the Treaty on the Nonproliferation of Nuclear Weapons. The signers of the treaty agreed "not to manufacture or otherwise acquire nuclear weapons or other nuclear explosive devices." This treaty allowed those countries with nuclear capabilities to keep them, but several countries that have since showed interest in making nuclear explosive devices did not sign the treaty. These countries include India, Pakistan, Israel, Argentina, Brazil, South Africa, and North Korea. By the time the Nonproliferation Treaty was signed, far too many scientists had figured out for themselves virtually all the aspects of how nuclear weapons are made. Only the technological tasks of enriching uranium, separating highly toxic plutonium from nuclear fuel rods, and creating a "trigger" explosive to drive together the critical mass at the time of explosion have remained for any government desiring to make nuclear bombs. Of these challenges, the most common one has become an economic one—that is, finding the money to build facilities to make enough bombs to create an arsenal. One thing that worries leaders of all nations is that someday the knowledge and practical techniques to make nuclear bombs may become so widespread that preventing the spread of nuclear weapons will be an impossibility.

Fusion reactions occur rapidly only when the temperature is of the order of 100 million degrees or more. At these high temperatures atoms do not exist as such; instead, there is a **plasma** consisting of unbound nuclei and electrons. In this plasma nuclei can combine. The first fusion reactions that scientists were able to create artificially were produced in hydrogen bombs, or thermonuclear bombs. In a thermonuclear bomb, the high temperatures needed to initiate fusion are achieved by using the heat of a fission bomb (atomic bomb).

■ The term "thermonuclear" refers to the extreme temperatures required to cause nuclear fusion to take place.

In one type of hydrogen bomb, lithium deuteride (${}^{6}_{3}\text{Li}{}^{2}_{1}\text{H}$, a solid salt) is placed around an ordinary ${}^{235}_{92}\text{U}$ or ${}^{239}_{94}\text{Pu}$ fission bomb. The fission reaction is set off in the usual way. A ${}^{6}_{3}\text{Li}$ nucleus absorbs one of the neutrons produced and splits into tritium and helium.

$$ {}^{6}_{3}\text{Li} + {}^{1}_{0}\text{n} \longrightarrow {}^{3}_{1}\text{H} + {}^{4}_{2}\text{He} $$

The temperature reached by the fission of ${}^{235}_{92}\text{U}$ or ${}^{239}_{94}\text{Pu}$ is sufficiently high to bring about the fusion of tritium and deuterium.

Controlled Nuclear Fusion

Because there is so much available potential fuel in the oceans (as deuterium), controlled fusion seems like a natural candidate as an alternate source of energy. Another attractive feature is the rather limited production of dangerous radioactivity. Most radioisotopes produced by fusion have short half-lives and therefore are a serious hazard for only a short time.

Three critical requirements must be met for controlled fusion to be a source of energy. First, the temperature must be high enough for fusion to occur. The fusion of deuterium ($^{2}_{1}H$) and tritium ($^{3}_{1}H$) requires a temperature of 100 million degrees or more. Second, the plasma must be confined long enough to release a net output of energy. Third, the energy must be recoverable in some usable form.

Fusion reactions are extremely difficult to control, principally because of the difficulty of holding the hot plasma together long enough for the particles to react. A further problem is reaching the very high temperature and maintaining it long enough to sustain the reaction. Yet progress is being made. Using a *magnetic bottle,* in which the plasma is held in place between two magnetic fields, scientists at the Princeton University Plasma Physics Laboratory have achieved record levels of controlled fusion. In November 1994, fusion reactions in a half-and-half mixture of deuterium and tritium generated 10.7 million watts of power. So far, the longest power burst has lasted 0.20 seconds and the highest plasma temperature has been 460 million kelvins. Experiments indicate that alpha particles (He nuclei) from the fusion remain inside the plasma, adding their energy to the plasma and heating it up. If sufficient heating is provided by the alpha particles, the reaction may become self-sustaining because it would need no external source of heat. In this condition, a commercial fusion reactor could produce more energy than it consumes, a condition not yet achieved. Only time will tell whether fusion can be controlled to this extent. If it does happen, abundant and low-cost energy may truly become available to everyone.

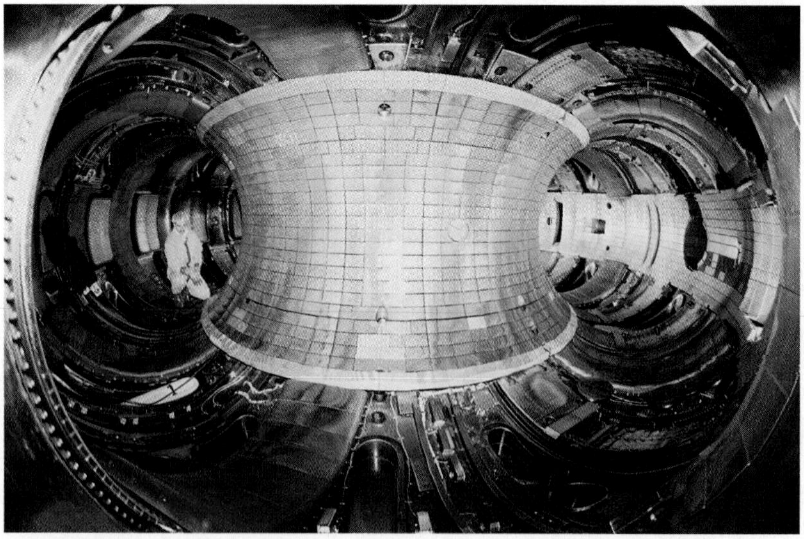

Inside the Tokamak Fusion Test Reactor: the chamber where nuclear fusion takes place. Electric current running through vertical and horizontal coils that surround this donut-shaped space creates a pair of magnetic fields that interact to hold the plasma in place. The chamber is lined with graphite and graphite-composite tiles that can withstand the high temperatures. *(Princeton Plasma Physics Lab)*

■ SELF-TEST 13B

1. Which atom can undergo fission by thermal neutrons? (a) uranium-238, (b) uranium-235, (c) krypton-131.
2. Complete the following nuclear equation:

$$^{1}_{0}\text{n} + ^{235}_{92}\text{U} \longrightarrow [\quad\underline{\hspace{3cm}}\quad] \longrightarrow ^{103}_{42}\text{Mo} + ^{131}_{50}\text{Sn} + 2\,^{1}_{0}\text{n}$$

3. Complete the following nuclear equation:

$$^{1}_{0}\text{n} + ^{239}_{94}\text{Pu} \longrightarrow ^{104}_{42}\text{Mo} + [\quad\underline{\hspace{3cm}}\quad] + 2\,^{1}_{0}\text{n}$$

4. If the masses of the nucleons making up an atom are summed together and compared with the mass of the atom these nucleons can make, the atom will always have a (greater/smaller) mass.
5. What is the most stable element, in terms of binding energy per nucleon? (a) iron, (b) hydrogen, (c) uranium.
6. About what percent of the electricity generated in the United States is from nuclear energy? (a) 100%, (b) 19%, (c) 50%.
7. Plutonium is always produced in nuclear fission reactors. (a) True, (b) False.
8. Which value is closer to the half-life of plutonium? (a) 30 days, (b) 24,000 years, (c) 15.9 minutes.

A garden greenhouse. Solar energy in the form of visible light enters through the transparent roof and warms the plants and soil inside. The warmed objects then reradiate some of this energy in the form of infrared, or thermal, radiation. The roof panels stop most of this radiation, so the inside of the greenhouse warms up. This type of warming is known as the *greenhouse effect.* (The role of the greenhouse effect in global warming is discussed in Section 21.11.) *(R. W. Gerling/Visuals Unlimited)*

13.5 Solar Energy—Almost Free

The Sun is a vast storehouse of energy, and this energy is available to anyone who can collect it. The Sun emits energy as heat, light, X rays, and gamma rays at a rate of about 3.8×10^{26} W (J/s). About 99% of the Sun's energy is in the form of heat and light, with the remainder in the form of high-energy particles and much lower-energy radio waves. Of the 99%, about 9% is ultraviolet radiation, 40% is visible light, and 50% is infrared radiation. The Earth actually receives only a tiny fraction of the Sun's energy—about 1.7×10^{17} W, or 4.5×10^{-8}%, of the total output, and half of that is reflected back into space by clouds, water, and ice. This seemingly small percentage of the Sun's output reaching Earth is still a huge quantity of energy—about 13,000 times all the energy consumed annually on the entire planet.

■ About 46,000 quads of solar energy reach the United States each year, compared with the approximately 84 quads of energy in all forms used annually in the United States.

Solar energy that reaches the Earth's surface undergoes either thermal conversion or photoconversion. In thermal conversion, the infrared portion of the solar spectrum is converted directly into heat. Solar heating causes air to move, creating winds (a direct result of the expansion of gases on heating, Section 7.2), causes water to move in the oceans, and causes water to evaporate and then condense in the form of rain or snow. In photoconversion, the energy in solar radiation brings about chemical change, most notably in the photosynthesis responsible for the growth of most vegetation on Earth.

■ The surface of the Earth is about 250°C warmer than its surroundings in space because of exposure to the Sun.

Air and Water Movement Caused by Solar Heating

Wind, caused by temperature differences in the Earth's atmosphere due to clouds, uneven heating of land and water surfaces, and the earth's rotation,

Figure 13.12 A "wind farm." This windmill farm, operated by U.S. Windpower near Livermore, California, can supply large amounts of electricity when the winds are blowing.

has been used as a source of energy since at least 3000 B.C., when the first sailing ships went to sea. In the United States alone, available wind energy amounts to more than 60 quads annually. If put to use, wind could provide a sizable fraction of the 84 quads of energy used annually. Early windmills pumped water, but modern windmills can produce electricity (Figure 13.12). High unit construction costs and environmental concerns, however, are deterrents to using wind energy.

■ Benjamin Franklin studied the Gulf Stream in 1769.

Energy in the ocean waters is available as waves (caused by winds), currents, and thermal gradients. The Gulf Stream is a warm current that is caused by solar warming and is directly responsible for the moderate climate of Western Europe. A small tributary of the Gulf Stream, called the Florida current, runs offshore of Miami, Florida. It alone contains more water volume than all the rivers of the earth combined.

Differences in temperature are called **thermal gradients.** Because the oceans are constantly bathed in solar radiation, much of which is converted into heat energy, temperature differences occur between the surface and the depths. The total heat energy collected by the oceans each year is estimated to be 1000 quads. As long ago as 1881 it was mentioned that this temperature difference could be a source of energy. The simplest method is to use cool water from below the surface to condense water vapor from evaporation of the warmer surface water. Before it is condensed, the water vapor is used to drive a turbine.

Passive and Active Solar Heating

Solar energy reaching the Earth's surface varies from 0 to 330 Btu per square foot per hour according to the cloudiness, the season, the time of day, and the latitude. It is possible to *concentrate* solar energy by capturing the energy that falls on a large area and putting it to use in a much smaller area. Early solar

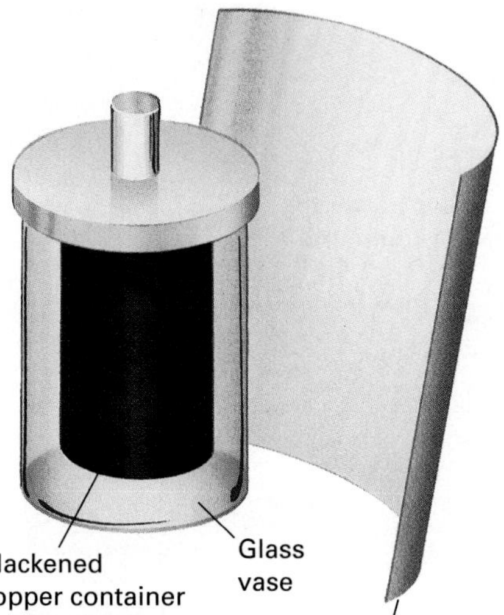

Figure 13.13 A solar cooker from the 1860s. A Frenchman named Mouchot designed this cooker, which he reported could bring 3 L of water to a boil in about 1.5 hours. Mouchot also used steam produced by solar energy to operate his printing press.

cookers used parabolic mirrors to concentrate sunlight on the cooking food (Figure 13.13). The concentration of energy creates much higher temperatures than would be possible if the energy were spread over a larger area (Figure 13.14). The National Renewable Energy Laboratory has had a research program in which solar concentrators achieve the equivalent of more than 21,000 Suns shining on a small area. Under proper conditions, processes such as steelmaking that require large quantities of energy to achieve high temperatures might someday be carried out using solar energy.

As early as the fifth century B.C., buildings were designed and oriented to take advantage of solar heat. Under the right conditions, the roof of a typical-size home (80 m^2) can receive more solar energy in a single day than is needed for heating, cooling, lighting, and appliances. In what is known as **passive solar heating,** the Sun's heat is captured without the use of any pumps or fans to distribute the collected energy. A gardener's greenhouse is a good example of passive solar heating. Some ways to take advantage of passive solar heating of a house are illustrated in Figure 13.15.

In **active solar heating,** energy from the Sun is captured by an endothermic chemical or physical change that can later be reversed to release the stored energy. The reversible formation of a hydrated salt, for example, can be used to store energy for release at night or on a cloudy day. Hydrated sodium sulfate, a solid ($Na_2SO_4 \cdot 10\ H_2O$), can be used in this way. At temperatures above 32.3°C, the hydrated salt absorbs heat and breaks down to form a concentrated salt solution. When the temperature falls below 32.3°C, the salt rehydrates, and the heat is released in this reverse reaction:

$$Na_2SO_4 \cdot 10\ H_2O(s) + \text{heat} \xrightarrow{\text{(above 32.3°C)}} Na_2SO_4(aq) + 10\ H_2O(\ell)$$

$$Na_2SO_4(s) + 10\ H_2O(\ell) \xrightarrow{\text{(below 32.2°C)}} Na_2SO_4 \cdot 10\ H_2O(s) + \text{heat}$$

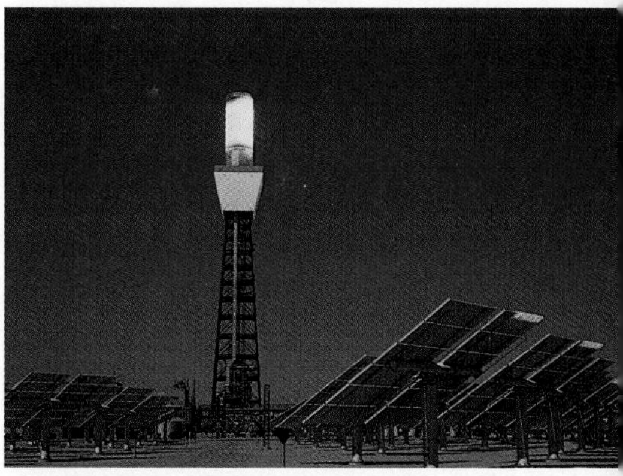

Figure 13.14 Using solar energy to generate electricity. At Solar One, Daggett, California, mirrors that turn with the sun reflect solar energy onto a boiler on the top of a tower. Steam from the boiler is used to drive turbines that generate electricity. *(© William E. Ferguson)*

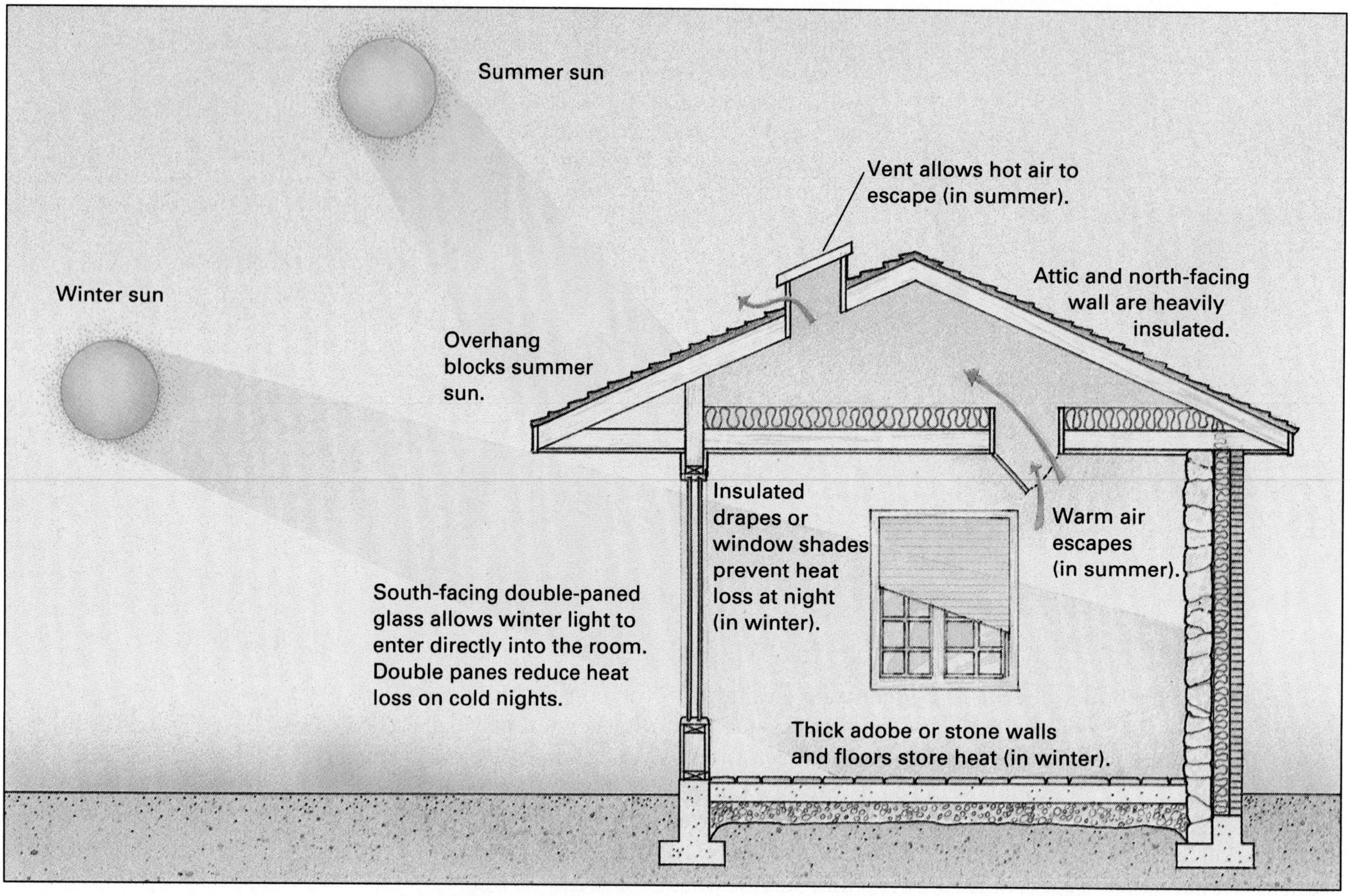

Figure 13.15 Passive solar heating at home. Placement of windows can allow solar energy to enter a room and warm the air as well as the furniture and occupants. The roof overhang allows more solar radiation to enter in the winter than in the summer. Other design features, such as thick, insulating north-facing walls and floors help store heat (in winter).

The heat of this transition is 50 cal/g of hydrated salt. Based on the density of the salt, this is about 9568 Btu/ft^3. A relatively small container filled with this salt could store and then release all the energy needed for heating the living space and water for a typical-size home. One problem with this and all solar heating systems based on a chemical change is that mixing of the reactants can sometimes be incomplete, causing lower efficiency.

Solar Energy to Electricity

Another approach to the direct utilization of solar energy is the *solar cell* or photovoltaic cell, which converts energy from the Sun into electron flow. Solar cells can routinely convert solar energy to electric power at the rate of at least 100 W/m^2 of illuminated surface. Solar cells are now used in calculators,

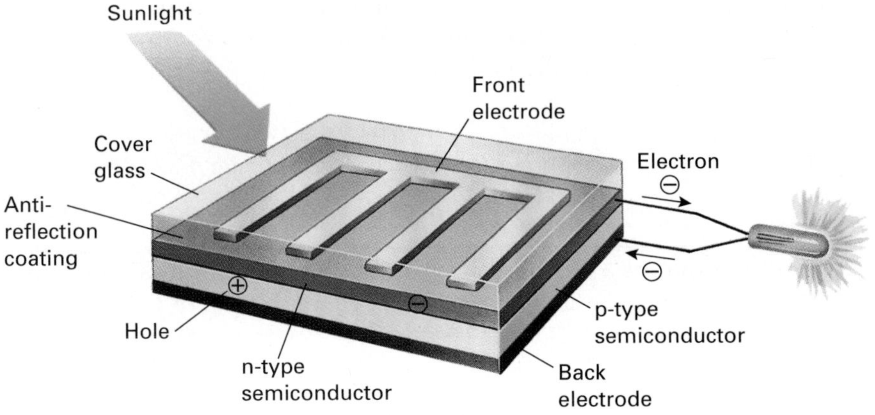

Figure 13.16 A solar cell. Beneath the outer glass is a metal grid that allows as much light as possible to strike the *n*-type semiconductor layer, while serving as the electrode at which electrons leave the cell. The *n*-type semiconductor layer is almost transparent. Beneath it is the *p*-type semiconductor layer and the electrode at which electrons reenter the cell.

watches, space-flight applications, communication satellites, signals for automobiles and trains, and as the source of electricity in undeveloped regions throughout the world, where electrical power grids are virtually nonexistent. It has been estimated that all of the electricity used in the United States could be made by solar cells having only 10% efficiency and covering about 13,000 km^2, which is 0.3% of the land area in the United States.

■ The United States land area is about 9×10^9 km^2.

One type of solar cell consists of two layers of almost pure silicon. The lower, thicker layer contains a trace of boron (B), and the upper, thinner layer has a trace of arsenic (As). As pointed out in Section 7.7, the As-enriched layer is an n-type semiconductor, with mobile electrons, and the B-enriched layer is a p-type semiconductor, with electron deficiencies known as *holes* (see Figure 7.7). There is a strong tendency for the mobile electrons in the n-type layer to pair with the unpaired electrons in the holes in the p-type layer. If the two layers are connected by an external circuit (Figure 13.16) and light of sufficient energy strikes the surface and is absorbed, excited electrons can leave the n-type layer and flow through the external circuit to the p-type layer. As this layer becomes more negative because of added electrons, electrons are repelled internally back into the n-type layer, which is now positive and attracts the electrons. The process can continue indefinitely as long as the cell is exposed to sunlight.

A refrigerator powered by the sun. Photovoltaic cells at the top of the refrigerator generate the electricity needed to keep vaccines cold as they are delivered to remote locations. *(Courtesy of Siemens Solar Industries, Camarillo, CA 93011 USA)*

Solar cells may be the next great technological breakthrough, perhaps comparable to the computer chip and genetic engineering. Experimental solar-powered automobiles are now available and many novel applications of solar cells already exist (such as powering an exhaust fan in an automobile to remove hot air accumulated in a car parked in the sun). The big breakthrough will be in the use of banks of solar cells at a utility power plant to produce huge amounts of electricity. One plant already in operation in California uses solar cells to produce 20 MW of electricity, which is enough to supply the electricity needs of a city of about 600,000 people. In 1993 a consortium of 68 electric utilities began making plans to purchase $500 million worth of solar panels over a six-year period. Since these companies serve more than 40% of the electric customers in the United States, you may be using solar-generated electricity in the near future.

EXAMPLE 13.4 ***Comparing A Solar Cell's Output with the Output of Burning Oil***

Compare the output of a 500 m^2 solar cell that has an output of 100 W/m^2 operating for 10 hours on a sunny day with the energy output of 1 barrel of oil. Refer to Table 13.1.

SOLUTION

$$\text{Solar cell energy output} = 500\ \text{m}^2 \times \frac{100\ \text{W}}{\text{m}^2} \times \frac{1\ \text{kW}}{10^3\ \text{W}} \times 10\ \text{hr} = 50\ \text{kWh}$$

From Table 13.1 you can see that 2.93×10^{-1} kWh = 1.8×10^{-4} bbl oil, so 50 kWh is equal to about 0.03 bbl of oil.

$$\frac{1.8 \times 10^{-4}\ \text{bbl oil}}{2.93 \times 10^{-1}\ \text{kWh}} \times 50\ \text{kWh} = 0.03\ \text{bbl oil}$$

Exercise 13.4

Calculate the energy output in Btu of an array of solar cells covering 100,000 m^2 each with an output of 200 W/m^2 illuminated for 10 hours. Calculate the equivalent energy in tons of bituminous coal, barrels of oil, and kWh of electricity.

13.6 Biofuels

Besides the obvious benefits of photosynthesis in producing foods, photosynthesis annually converts vast quantities of carbon from carbon dioxide into products that can be used as fuels. In fact, about 100 times as much carbon is converted annually to compounds that have potential as fuels as is converted to compounds that are used as food. The oceans contain the largest unharvested crop of products from photoconversion—about 10^{10} metric tons of carbon is converted into compounds annually by photosynthesis in ocean plants. The large amount of carbon fixation occurs in spite of the fact that the efficiency of photosynthesis is very low. Ocean kelp has perhaps the highest utilization of sunlight (2% overall), while some forests have a utilization of only 0.5%.

■ Photosynthesis recycles all of the atmospheric carbon dioxide every 300 years and all of the oxygen every 2000 years.

■ Biomass fuels are derived from living organisms and include wood, corn stalks, and brush.

The most directly useful **biomass fuels** are liquids that can be used in internal-combustion engines. The production of ethanol from corn was described in Section 12.6. Tests are also being carried out on obtaining diesel fuel from microalgae that can grow in arid climates in salty water. Microalgae have the potential to produce the equivalent of 150 to 400 barrels of oil per

FRONTIERS IN THE WORLD OF CHEMISTRY

Solar Electricity for Developing Regions

For about two billion people—one third of the world's population—electricity and the comfort, safety, and entertainment it provides are unknown in their daily lives. They live in countries or regions that are too poor and underdeveloped to provide readily available electricity on a daily basis. The challenge is to provide electricity to these people at a cost they and their governments can afford. Large fossil fuel and nuclear power plants are out of the question because the huge investment would leave the countries with massive debt. Building dams to use the energy of rivers flowing through the land has had undesirable effects by displacing large numbers of people and by flooding fertile land. Transmission and distribution costs of electricity can also be staggering to economies that are hard pressed just to provide basic health care, pure water, and education.

The answer to providing electricity in developing countries seems to be small, affordable solar collection panels connected to storage batteries. Such small solar-powered systems (SSPS) can capture solar energy in the daytime, convert it to electricity, and store it for use at night when the family members have come home from work. Such systems, costing between \$350 and \$750, are already being made available to about 100,000 families, farms, and small businesses in countries such as Sri Lanka, Kenya, the Dominican Republic, and Zimbabwe. The challenges in putting this technology to use are numerous, and many discoveries will have to be made about how solar cells and batteries work and about how to make basic appliances that use electricity more efficiently. In addition to needing more efficient solar collectors, developers need batteries that can take deeper charge and discharge cycles without wearing out. Current systems typically consist of about 36 solar cells, each with a diameter of 10 cm, charging a lead-acid battery like those used in automobiles. When the sunlight is at full intensity, solar units like these can produce enough electricity to operate three fluorescent lamps for three hours in addition to three hours of combined radio listening and television viewing. Electricity could also be used to purify water. Additional research in lighting is helping to better utilize the stored solar electricity. For example, high-efficiency fluorescent lamps use polished aluminum reflectors to lower power requirements. One 9-watt fluorescent bulb of this type can produce as much light as a 50-watt incandescent bulb.

Although there are many opportunities for chemistry and applied chemical technology in improving the lives of people in developing countries, putting the applications into practice will probably depend on economic factors. It may cost \$1 trillion to provide simple solar electrical systems to the developing world's population.

A community in Brazil where solar-powered electricity has been installed. Lead-acid batteries store electricity for use after sunset. (National Renewable Energy Laboratory (NREL))

Reference Williams, N., Jacobson, K., and Burris, H. "Sunshine for light in the night," *Nature,* vol., 362, 22 April: 1993, p. 691.

acre. The compounds produced by these plants are fatty acid esters (see Section 15.6.), which can be converted into a liquid that has characteristics identical to automotive or truck diesel fuel. Similar fuels are already being produced in Europe and Japan from soybean oil. The main problem with biodiesel fuels is cost. At present, diesel fuel from petroleum costs only 50 to 60 cents per gallon to produce, while biodiesel costs about \$2.50 per gallon. Current research with microalgae is directed toward *genetically engineering* the plants so that yields of the biodiesel fuel will increase dramatically.

■ Genetic engineering is discussed in Section 15.9.

13.7 Geothermal Energy

The temperature of the molten core of the Earth is about 4000°C (7200°F), primarily a result of energy from the decay of radioactive atoms trapped there since the Earth was formed. The mantle of the Earth has a temperature of about 1000°C, and there are regions on the Earth where the mantle comes within 2 to 3 km of the surface. These ''hot spots'' are the source of usable geothermal energy and are located roughly along the junctions of the **tectonic plates** that make up the Earth's crust.

Water trapped near one of these hot regions in the Earth's crust is under extreme pressure and can become very hot, reaching temperatures of 370°C. When this hot water reaches the surface, it converts to steam. Places where this occurs are called ''fumaroles.'' A geothermal well can tap this hot water to drive electric turbines. The city of Budapest has been partly heated by geothermal steam since Roman times. Today, greenhouses there are heated with steam all year long to produce flowers and vegetables. Most of the homes in Iceland's capital, Reykjavik, are heated with geothermal steam.

In the United States, about 1400 MW of geothermal electric generating capacity is currently in place. North of San Francisco, Pacific Gas and Electric has a field of geothermal wells that produce enough electricity to supply all the needs of a city the size of San Francisco and Oakland combined. Several other countries have installed electric generating systems using geothermal energy.

In spite of being ''almost free'' energy, geothermal energy is not without its problems. Along with steam often comes hydrogen sulfide (H_2S), a foul-smelling and toxic gas. In addition, the superheated water is such a good solvent that it often carries large amounts of dissolved minerals. These can harm aquatic life if allowed to run into surrounding streams and lakes.

■ SELF-TEST 13C

1. Solar energy is the result of nuclear fusion reactions. (a) True, (b) False.
2. Approximately what percentage of the solar energy reaching Earth is reflected back into space? (a) 50%, (b) 10%, (c) Almost 0%.
3. Which process is not photoconversion of solar energy? (a) Photosynthesis, (b) A photovoltaic cell, (c) Wind energy.
4. Name a structure that uses passive solar heating.
5. When a chemical stores solar heat during the day and releases that heat at night, it is called (passive/active) solar heating.
6. Solar voltaic devices (solar cells) could someday be used in place of fossil fuel–burning power plants. (a) True, (b) False.
7. There is far more carbon fixed annually by photosynthesis than is needed for food by all of the Earth's animals. (a) True, (b) False.
8. For each of the following sources of energy, name one associated environmental problem.
   (a) Solar (b) Wind
   (c) Biomass (d) Geothermal
9. The source of geothermal energy is (a) energy remaining from when the Earth was formed, (b) decay of radioactive elements in the core of the Earth, (c) solar energy that is trapped by the Earth.

■ MATCHING SET

____ 1. Thermal neutrons
____ 2. Amount of energy that can raise the temperature of 1 pound of water by 1°F
____ 3. joule/s
____ 4. 10^{15} Btu
____ 5. Efficient kind of light bulb
____ 6. Most stable atom
____ 7. Amount of uranium-235 that, when split, equals 3000 tons of coal
____ 8. Typical fission products
____ 9. Amount of fissionable material that will self-sustain a chain reaction
____ 10. Percentage of uranium-235 in natural uranium
____ 11. Uranium that contains a greater-than-normal percentage of uranium-235
____ 12. Neutron absorber in a nuclear reactor
____ 13. Reactor core meltdown
____ 14. Dangerous element produced in nuclear reactors
____ 15. Source of 2_1H that may someday be used as a nuclear fuel.
____ 16. Atoms that can be fused together
____ 17. Photosynthesis
____ 18. Solar voltaic cell

a. 3_1H and 2_1H
b. ocean
c. plutonium
d. 0.711%
e. critical mass
f. Btu
g. slow-moving neutrons
h. photoconversion
i. watt
j. quad
k. cadmium
l. silicon
m. enriched
n. fluorescent
o. Chernobyl
p. ^{56}Fe
q. 1 kilogram
r. $^{141}_{56}$Ba and $^{92}_{36}$Kr

■ QUESTIONS FOR REVIEW AND THOUGHT

1. Define the following terms:
 (a) Renewable energy (b) Power
 (c) Quad (d) Thermal neutron
2. Define these terms:
 (a) Fission product (b) Critical mass
 (c) Chain reaction (d) Binding energy
 (e) Photovoltaic
3. Explain why using 100 units of energy to generate electricity produces only about 33 units of useful energy.
4. Name four fuels that can be used to generate electricity.
5. Which fuel generates the greatest amount of energy per mole? (a) Burning gasoline, (b) The fission of uranium-235, (c) The burning of methane (natural gas).
6. Which is more economical, to burn a fuel in your home and generate 1000 Btu of heat, or to burn that fuel in an electric power plant and use that energy in your home?
7. Give two reasons why electricity is a popular source of energy.
8. Describe nuclear fission:

 (a) What is the starting material?
 (b) What is needed to cause fission?
 (c) What are the products of fission?
9. Which will produce more energy, the fission of one mole of uranium-235, or the burning of one mole of gasoline?
10. Why is there no danger of a nuclear explosion in a sample of natural uranium?
11. Explain the enrichment process for uranium.
12. Why is enriched uranium dangerous if it falls into the wrong hands?
13. A helium-4 atom consists of four particles that have significant mass. There are two protons and two neutrons. The total of the masses of these four particles is a larger number than the actual mass of a helium-4 atom. This mass difference can be expressed in terms of energy. What is this energy called?
14. In what city in the United States was the first successful atomic reactor built?
15. What is the purpose of a moderator in a nuclear reactor?

16. About what percentage of the electricity produced in the United States comes from nuclear energy?
17. Name three problems that are associated with nuclear energy from fission.
18. What happened at the Chernobyl Unit 4 reactor in 1986?
19. (a) What does the term "reprocessing of nuclear fuels" mean?
    (b) What danger is associated with it?
20. What is nuclear fusion?
21. Compare nuclear fission and nuclear fusion as sources of energy. Name the fuels, benefits, problems, and current status.
22. Complete the following nuclear equations:
    (a) ${}^{1}_{0}n + {}^{235}_{92}U \rightarrow {}^{142}_{56}Ba + [\quad\quad\quad] + 3\,{}^{1}_{0}n$
    (b) ${}^{1}_{0}n + {}^{235}_{92}U \rightarrow [\quad\quad\quad] + {}^{129}_{50}Sn + 2\,{}^{1}_{0}n$
    (c) ${}^{1}_{0}n + {}^{239}_{94}Pu \rightarrow [\quad\quad\quad] + {}^{123}_{52}Te + 2\,{}^{1}_{0}n$
23. (a) What is the half-life of plutonium-239?
    (b) How does this add to the dangers of this isotope?
24. In what important way are the isotopes uranium-235 and plutonium-239 similar?
25. Describe how the "magnetic bottle" is used to contain a nuclear fusion reaction.
26. About what percentage of solar radiation striking the Earth is reflected back into space?
27. Name the process that is the single largest user of solar energy reaching the Earth.
28. Which number is closest to the amount of solar energy reaching the United States each year? (a) 100 quads, (b) 1000 quads, (c) 46,000 quads.
29. Which number approximates the annual energy usage in the United States? (a) 1 quad, (b) 84 quads, (c) 46,000 quads.
30. Describe (a) passive and (b) active solar heating.
31. Explain how it might be possible to use solar energy to carry out processes that require large amounts of energy and high temperatures, such as making glass or steel.
32. (a) Is it possible to directly convert solar energy into electricity?
    (b) Explain how this might be done.
    (c) List some of the possible benefits.
33. What is the source of geothermal energy?
34. Name one negative environmental consequence of each of the following means of obtaining useful energy.
    (a) Using coal or oil to generate electricity
    (b) Using nuclear fission to generate electricity
    (c) Tapping geothermal energy
    (d) Collecting solar energy with photovoltaic devices
    (e) Harnessing solar energy by using windmills
35. Your classmate tells you that solar and geothermal energy are "free." Can you think of some arguments to refute that statement?
36. Why are biofuels considered a renewable energy source?
37. Describe how genetic engineering may be used to increase biofuel production.
38. What do you think will be the primary source of energy in the years 2005 and 2025?
39. (a) List seven ways that you use energy every day.
    (b) For each way listed, tell if it comes from renewable or nonrenewable sources.
    (c) Give at least one way you could conserve some of that energy.

PROBLEMS

1. Use Table 13.1 and convert the following energy units to other units.
    (a) 8 quads to Btu
    (b) 1×10^9 cubic feet of natural gas to Btu
    (c) 8×10^6 Btu to kcal
    (d) 20 tons of coal to barrels of oil
    (e) 1×10^6 barrels of oil to Btu
2. How much energy does a 1000-MW (megawatt) power plant consume in one second?
3. Firewood is often sold by the cord, a stack roughly 4 feet by 4 feet by 8 feet. One of the highest-energy firewoods is white oak, which produces about 28.8×10^6 Btu per cord. How many barrels of oil is one cord of this firewood equal to?
4. A cord of white oak from Problem 42 costs $125, delivered, while fuel oil costs about $1.22 per gallon, delivered. If a barrel of oil contains 42 gallons, which is more economical as a source of heat, the firewood or the fuel oil?
5. A railroad car can hold 425,000 pounds of coal. How many cubic feet of natural gas is this amount of coal equivalent to?
6. How much power is released if a power plant burns 1 ton (2000 pounds) of coal in one second? Give your answer in watts.
7. Calculate the energy, in joules, derived from an 1800-calorie meal. The calorie units here are food calories.
8. A new office building will have 500 offices as well as meeting spaces, corridors, and other areas. Each room will require about 150 Btu per day to heat during the winter. Ignoring the other heated spaces in the building, how many barrels of oil will be required to heat this building per day
    (a) if an oil-burning furnace is used?
    (b) if electricity is used?

CHAPTER 14

Organic Chemicals and Polymers

Fossil fuels are not only our major source of energy, but they are also the major source of the hydrocarbons that are used to make thousands of consumer products. About 6% of the petroleum refined today is the starting material for the synthesis of organic chemicals of commercial importance. These chemicals are essential to making plastics, synthetic rubber, synthetic fibers, fertilizers, and thousands of other consumer products. For this reason the organic chemical industry is often referred to as the **petrochemical** industry.

A few of the major classes of organic compounds and some of their reactions, especially those used in making polymers, are described in this chapter, with the goal of introducing you to a major segment of chemistry and the chemical industry.

- Why are there so many organic compounds?
- What are the characteristic functional groups?
- How can the list of the top 50 chemicals produced in the United States be used as a guide to the important organic compounds of commerce?
- What are some common addition and condensation polymers?
- Will coal become a major source of chemicals?
- What types of plastics are being recycled?

Plastics! What would we do without them? *(C.D. Winters)*

Carbon compounds hold the key to life on Earth. Consider what the world would be like if all carbon compounds were removed; the result would be much like the barren surface of the moon. If carbon compounds were removed from the human body, there would be nothing left except water and a small residue of minerals. The same would be true for all living things. Carbon compounds are also an integral part of our lifestyle. Fossil fuels, foods, and most drugs are made of carbon compounds. Since we live in an age of plastics

TABLE 14.1 ■ Top 50 Chemicals Produced in the United States in 1994*

1. Sulfuric acid	18. Methyl *tert*-butyl ether	35. Propylene oxide
2. Nitrogen	19. Ethylbenzene	36. Butadiene
3. Oxygen	20. Styrene	37. Carbon black
4. Ethylene	21. Carbon dioxide	38. Potash
5. Lime	22. Methanol	39. Acrylonitrile
6. Ammonia	23. Xylene	40. Vinyl acetate
7. Propylene	24. Terephthalic acid	41. Acetone
8. Sodium hydroxide	25. Formaldehyde	42. Titanium dioxide
9. Phosphoric acid	26. Ethylene oxide	43. Aluminum sulfate
10. Chlorine	27. Toluene	44. Sodium silicate
11. Sodium carbonate	28. Hydrochloric acid	45. Cyclohexane
12. Ethylene dichloride	29. *p*-Xylene	46. Adipic acid
13. Nitric acid	30. Ethylene glycol	47. Caprolactam
14. Ammonium nitrate	31. Cumene	48. Bisphenol A
15. Urea	32. Ammonium sulfate	49. *n*-Butyl alcohol
16. Vinyl chloride	33. Phenol	50. Isopropyl alcohol
17. Benzene	34. Acetic acid	

*Data from *Chemical and Engineering News,* April 10, 1995, p. 17.

and synthetic fibers, our clothes, appliances, and most other consumer goods contain a significant portion of carbon compounds.

Over 11 million of the more than 13 million known compounds are carbon compounds, and a separate branch of chemistry, **organic chemistry,** is devoted to the study of them. Why are there so many organic compounds? The discussion of hydrocarbons and their structural and geometric isomers in Chapter 12 indicates two reasons—(1) the ability of thousands of carbon atoms to be linked in sequence with stable carbon-carbon bonds in a single molecule and (2) the occurrence of isomers. A third reason will be discussed further in this chapter—the variety of functional groups that bond to carbon atoms.

The economic importance of the organic chemical industry can be seen by looking at the list of the top 50 chemicals produced in the United States. Of the top 50 listed in Table 14.1, 31 are organic chemicals.

■ Catalytic cracking was described in Section 12.7.

■ Heating coal at high temperatures in the absence of air produces a mixture of coke, coal tar, and coal gas. The process, called pyrolysis, is represented by

$$\text{coal} \longrightarrow \text{coke} + \text{coal tar} + \text{coal gas}$$

One ton of bituminous (soft) coal yields about 1500 pounds of coke, 8 gallons of coal tar, and 10,000 cubic feet of coal gas. **Coal gas** is a mixture of H_2, CH_4, CO, C_2H_6, NH_3, CO_2, H_2S, and other gases. At one time coal gas was used as a fuel.

14.1 Organic Chemicals

Many of the organic chemicals used in the chemical industry are obtained from fossil fuels. For example, ethylene, propylene, butylene, and acetylene are obtained by catalytic cracking of natural gas or petroleum, and Figure 14.1 summarizes the uses of these as starting materials. Petroleum and coal tar, obtained by heating coal at high temperatures in the absence of air, are the primary sources of aromatic compounds used in the chemical industry (Figure 14.2). Distilling coal tar yields the aromatic compounds listed in Table 14.2.

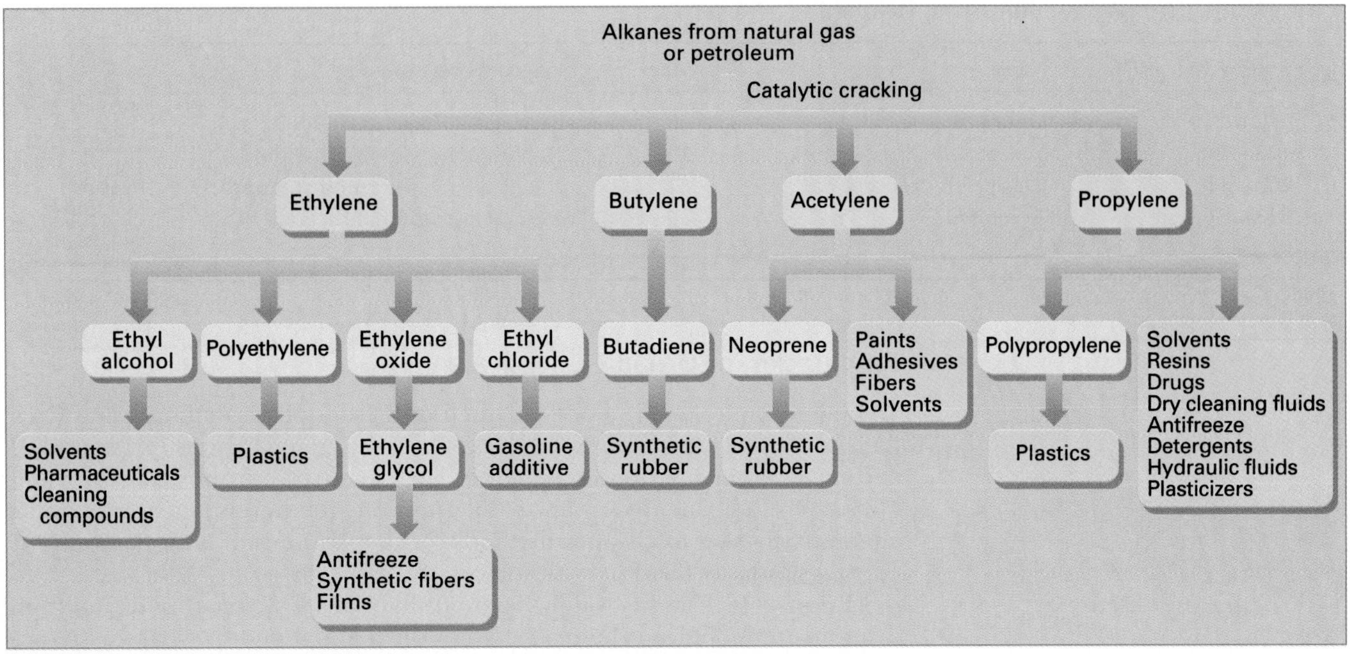

Figure 14.1 Hydrocarbons from petroleum or natural gas as raw materials. Catalytic cracking produces ethylene, butylene, acetylene, and propylene, which are converted into other chemical raw materials and many kinds of consumer products.

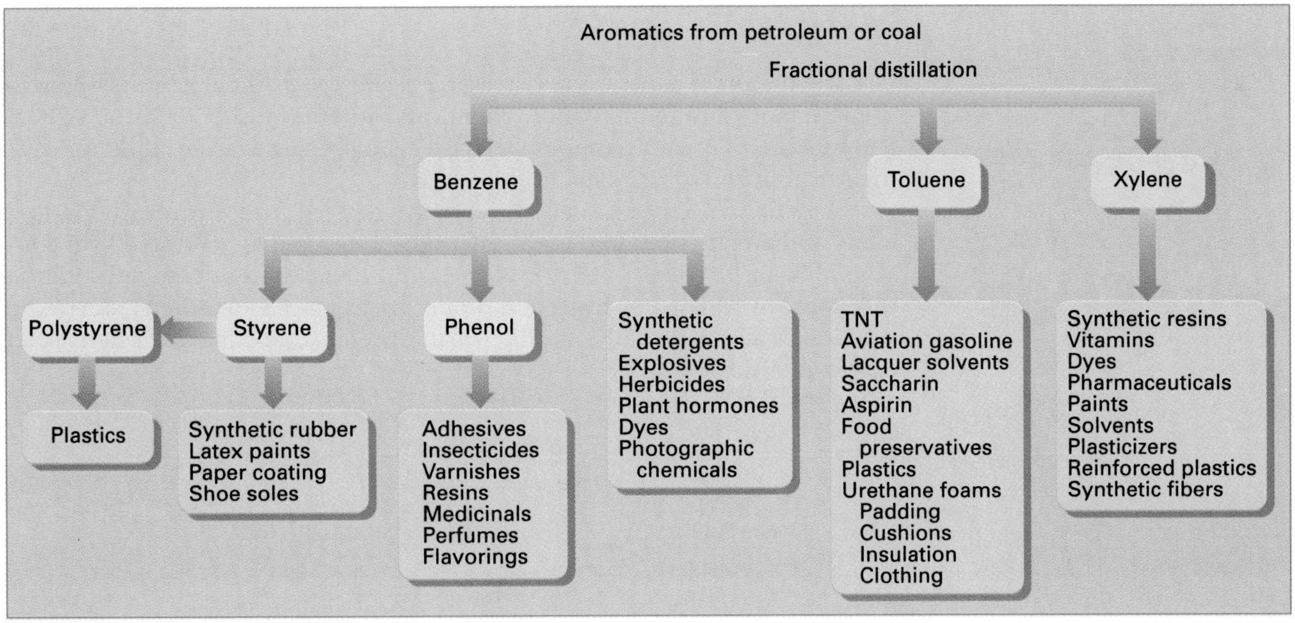

Figure 14.2 Aromatic compounds from petroleum or coal and their uses as raw materials.

TABLE 14.2 ■ Fractions from Distillation of Coal Tar

Boiling Range (°C)	Name	Tar, Mass %	Primary Constituents
below 200	Light Oil	5	Benzene, toluene, xylenes
200–250	Middle Oil (carbolic oil)	17	Naphthalene, phenol, pyridine
250–300	Heavy Oil (creosote oil)	7	Naphthalenes and methylnaphthalenes, cresols
300–350	Green Oil	9	Anthracene, phenanthrene
Residue		62	Pitch or tar

Preparation of Organic Chemicals

Organic chemicals were once obtained only from plants, animals, and fossil fuels, and these are still direct sources of hydrocarbons (Figures 14.1 and 14.2) and many other important chemicals, such as sucrose from sugar cane and ethanol from fermented grain mash. However, the development of organic chemistry has led to cheaper methods for the synthesis of both naturally occurring substances and new substances.

Prior to 1828 it was widely believed that chemical compounds found in living matter could not be made without living matter—a "vital force" was necessary for the synthesis. In 1828 a young German chemist, Friedrich Wöhler, destroyed the vital force myth when he prepared the organic compound urea, a major product in urine, by heating an aqueous solution of ammonium cyanate, an inorganic compound.

■ Urea is an important commercial chemical (number 15) because of its use in fertilizers and as a starting material for making plastics.

$$NH_4^+NCO^- \xrightarrow{\text{Heat}} H_2N-\overset{\overset{\displaystyle O}{\|}}{C}-NH_2$$

Ammonium cyanate　　　　　Urea

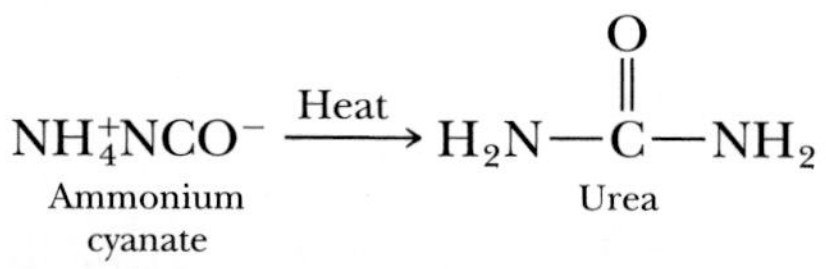

The notion of a mysterious vital force declined as other chemists began to prepare more and more organic chemicals without the aid of a living system, and millions of organic compounds have been prepared in the laboratories of the world since Wöhler's discovery.

Organic molecules can all be viewed as derived from hydrocarbons, and many of them are prepared in this manner. For example, saturated hydrocarbons (alkanes) in the presence of heat and a catalyst undergo **substitution reactions,** in which the hydrogen atoms are replaced by other atoms. The reaction of methane with chlorine in the presence of ultraviolet light is an example of substitution:

$$H-\overset{\overset{\displaystyle H}{|}}{\underset{\underset{\displaystyle H}{|}}{C}}-H + Cl-Cl \xrightarrow[\substack{\text{light or} \\ 120°C}]{\text{Ultraviolet}} H-\overset{\overset{\displaystyle H}{|}}{\underset{\underset{\displaystyle H}{|}}{C}}-Cl + H-Cl$$

Methane　　　　　　　　　　　　Chloromethane

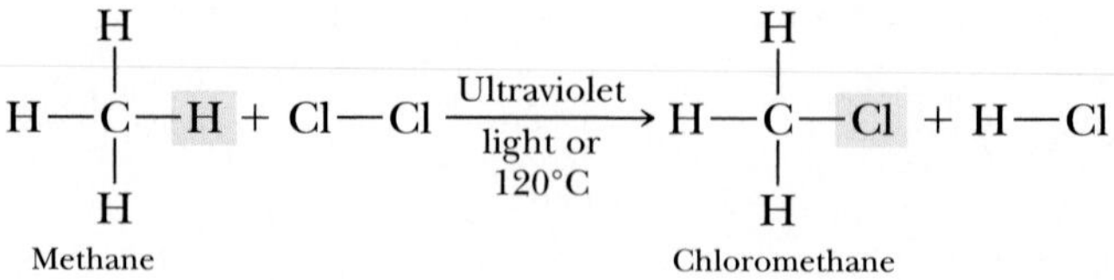

Further reaction can occur to give CH_2Cl_2 (methylene chloride), $CHCl_3$ (chloroform), or CCl_4 (carbon tetrachloride). However, alkanes are generally not very reactive. They do not react with concentrated strong acids, strong bases, or strong oxidizing agents. The only reaction that alkanes undergo readily is combustion, which produces carbon dioxide and water when the reaction is complete.

If alkanes are not very reactive, how can alkanes from petroleum be the starting materials for making organic chemicals with functional groups, such as the alcohols? The answer is that alkanes must first be subjected to catalytic cracking (Section 12.7) to give alkenes and alkynes. Because of the presence of double and triple bonds to which other atoms can be added, alkenes and alkynes are more reactive than alkanes and can be used as starting materials for the preparation of other organic chemicals (Figures 14.1 and 14.2). For example, they undergo **addition reactions,** which occur more quickly than the substitution reactions of alkanes. Compared with substitutions, which require heat and catalysts, the reactions of alkenes and alkynes with chlorine can take place at room temperature in the dark.

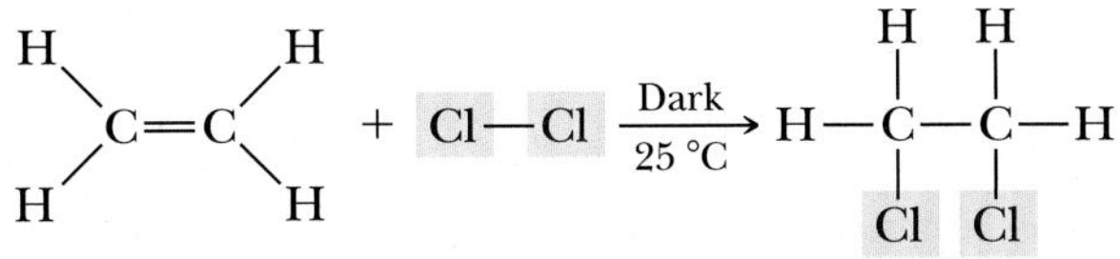

Ethylene (ethene)

1,2-dichloroethane

Other addition reactions include hydrogenation (Section 6.3) and formation of an alcohol by addition of the H and OH of water to the double bond in the presence of an acid.

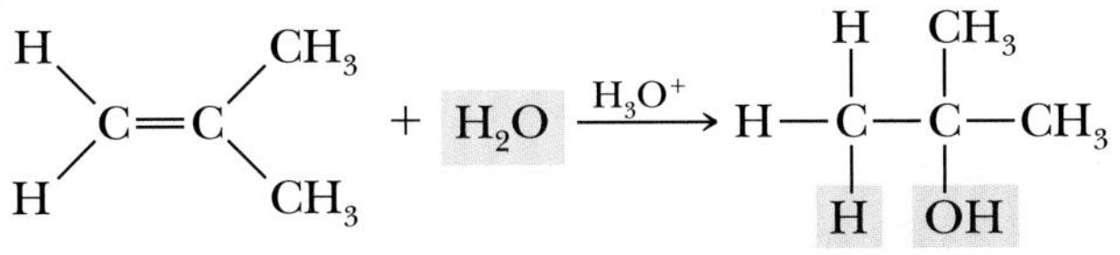

2-methylpropene

2-methyl-2-propanol (*tertiary*-butyl alcohol)

EXAMPLE 14.1 *Addition and Substitution Reactions*

Label the following reactions as addition or substitution reactions:

(a) $CH_3CH{=}CHCH_3 + Cl_2 \rightarrow CH_3CHClCH_2ClCH_3$
(b) $CH_3CH_2CH_3 + Cl_2 \rightarrow CH_3CH_2CH_2Cl + HCl$
(c) $CH_2{=}CH_2 + Br_2 \rightarrow CH_2BrCH_2Br$

SOLUTION

(a) and (c) are addition reactions and (b) is a substitution reaction. C_4H_8 and C_2H_4 in (a) and (c), respectively, are alkenes (general formula C_nH_{2n}). Reagents such as the halogens can easily add across the double bond to give an alkane derivative. Another way to tell whether the reaction is addition or substitution is to note whether another product has been formed because hydrogen atoms have been replaced. You can also tell the difference between addition and substitution reactions by counting the total number of atoms attached to carbon atoms in reactants and products. If the count stays the same, substitution has occurred. If the count is larger in products than in reactants, the reaction is an addition reaction. Notice in (b) that the total number of atoms attached to the carbon atoms has not changed; the reaction is a substitution reaction.

Exercise 14.1

Label the following reactions as addition or substitution reactions:

(a) $CH_3CH_2CH_2CH_3 + Cl_2 \rightarrow CH_3CH_2CH_2CH_2Cl + HCl$
(b) $CH_3CH{=}CH_2 + HBr \rightarrow CH_3CH_2CH_2Br$
(c) $CH_3CH{=}CH_2 + Cl_2 \rightarrow CH_3CHClCH_2Cl$

Functional Groups

The millions of organic compounds include classes of compounds that are obtained by replacing hydrogen atoms of hydrocarbons with atoms or groups of atoms known as **functional groups.** The functional groups for alcohols and ethers were discussed in Section 12.6. These and other important classes of compounds that result from attaching functional groups to a hydrocarbon framework are shown in Table 14.3. The "R" attached to the functional group represents the nonfunctional hydrocarbon framework with one hydrogen atom removed for each functional group added. For example, if the "R" in ROH is the methyl group, the formula of the alcohol is CH_3OH. Refer back to Table 12.4 for some examples of alkyl groups that are represented by "R." The systematic nomenclature for compounds containing functional groups is provided by the International Union of Pure and Applied Chemistry (IUPAC). For example, alcohol names end in "-ol" (methan*ol*), aldehyde names end in "-al" (methan*al*), and carboxylic acid names end in "-oic" (ethan*oic* acid). Further information on the naming of organic compounds is given in Appendix G.

14.2 Alcohols and Their Properties

Alcohols, which contain one or more —OH groups, are a major class of organic compounds. The importance of methanol, ethanol, and 2-methyl-2-propanol as fuels and fuel additives was described in Sections 12.7 and 12.10. Additional uses of these and other commercially important alcohols are listed in Table 14.4. Alcohols are classified according to the number of carbon atoms bonded to the —C—OH carbon as primary (one other C atom), secondary (two other C atoms), and tertiary (three other C atoms).

■ An "R" is used to represent any kind of alkyl group. The use of R, R′, and R″ indicates that all R groups are different.

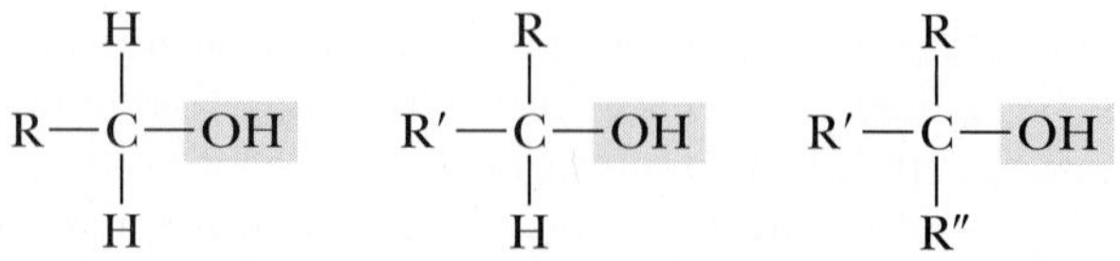

Primary Secondary Tertiary

TABLE 14.3 ■ Classes of Organic Compounds Based on Functional Groups*

General Formulas of Class Members	Class Name	Typical Compound	Compound Name	Common Use of Sample Compound
R—X	Halide	H—$\overset{\displaystyle H}{\underset{\displaystyle Cl}{\overset{\displaystyle \mid}{\underset{\displaystyle \mid}{C}}}}$—Cl	Dichloromethane (methylene chloride)	Solvent
R—OH	Alcohol	CH_3—OH	Methanol (wood alcohol)	Solvent
$R-\overset{\displaystyle O}{\overset{\displaystyle \parallel}{C}}-H$	Aldehyde†	$H-\overset{\displaystyle O}{\overset{\displaystyle \parallel}{C}}-H$	Methanal (formaldehyde)	Preservative
$R-\overset{\displaystyle O}{\overset{\displaystyle \parallel}{C}}-OH$	Carboxylic acid†	$CH_3-\overset{\displaystyle O}{\overset{\displaystyle \parallel}{C}}-OH$	Ethanoic acid (acetic acid)	Vinegar
$R-\overset{\displaystyle O}{\overset{\displaystyle \parallel}{C}}-R'$	Ketone	$CH_3-\overset{\displaystyle O}{\overset{\displaystyle \parallel}{C}}-CH_3$	Propanone (acetone)	Solvent
R—O—R'	Ether	C_2H_5—O—C_2H_5	Diethyl ether (ethyl ether)	Anesthetic
$R-\overset{\displaystyle O}{\overset{\displaystyle \parallel}{C}}-O-R'$	Ester	$CH_3-\overset{\displaystyle O}{\overset{\displaystyle \parallel}{C}}-O-C_2H_5$	Ethyl ethanoate (ethyl acetate)	Solvent in fingernail polish
$R-\overset{\displaystyle H}{\underset{\displaystyle H}{N}}$	Amine‡	$CH_3-\overset{\displaystyle H}{\underset{\displaystyle H}{N}}$	Methylamine	Tanning (foul odor)
$R-\overset{\displaystyle O}{\overset{\displaystyle \parallel}{C}}-\overset{\displaystyle H}{N}-H$	Amide‡	$CH_3-\overset{\displaystyle O}{\overset{\displaystyle \parallel}{C}}-\overset{\displaystyle H}{N}-H$	Acetamide	Plasticizer

*R stands for a hydrocarbon group such as —CH_3 or —C_2H_5. R' could be a different group from R.
†R can be an H atom or a hydrocarbon group.
‡The H atoms in amines and amides can also be replaced by R groups.

Ethanol and 1-propanol are primary alcohols. 2-Propanol, or isopropyl alcohol, is very familiar to us as rubbing alcohol, a 70% solution of 2-propanol sold in drug stores and grocery stores. 2-Propanol is a secondary alcohol and is one of the two structural isomers of an alcohol with three carbon atoms.

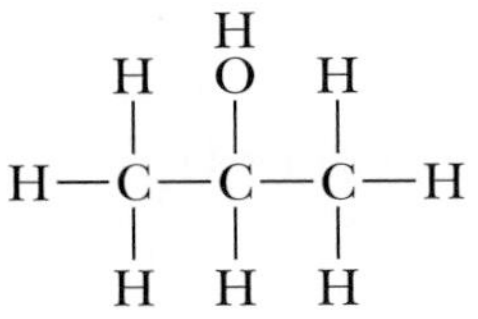

$$H-\overset{\displaystyle H}{\underset{\displaystyle H}{C}}-\overset{\displaystyle \overset{\displaystyle H}{\overset{\displaystyle \mid}{O}}}{\underset{\displaystyle H}{C}}-\overset{\displaystyle H}{\underset{\displaystyle H}{C}}-H$$
2-propanol

$$H-\overset{\displaystyle H}{\underset{\displaystyle H}{C}}-\overset{\displaystyle H}{\underset{\displaystyle H}{C}}-\overset{\displaystyle H}{\underset{\displaystyle H}{C}}-OH$$
1-propanol

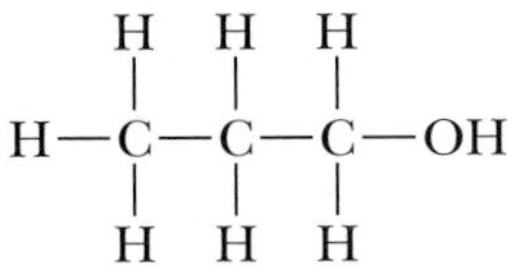

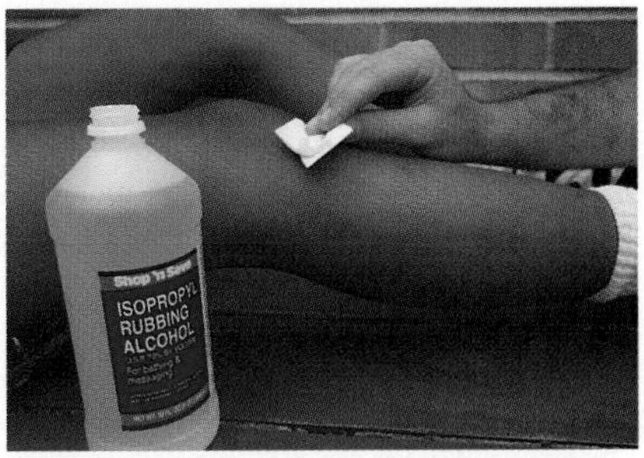

Rubbing alcohol (isopropanol) is used by athletic assistants to clean out cuts and scrapes. *(C.D. Winters)*

TABLE 14.4 ■ Some Important Alcohols

Condensed Formula	Boiling Point (°C)	Systematic Name	Common Name	Use
CH_3OH	65.0	Methanol	Methyl alcohol	Fuel, gasoline additive, making formaldehyde
CH_3CH_2OH	78.5	Ethanol	Ethyl alcohol	Beverages, gasoline additive, solvent
$CH_3CH_2CH_2OH$	97.4	1-Propanol	Propyl alcohol	Industrial solvent
CH_3CHCH_3 \quad OH	82.4	2-Propanol	Isopropyl alcohol	Rubbing alcohol
CH_2CH_2 OH OH	198	1,2-Ethanediol	Ethylene glycol	Antifreeze
$CH_2CH_2CH_2$ OH OH OH	290	1,2,3-Propanetriol	Glycerol (glycerin)	Moisturizer in foods and cosmetics

For an alcohol with four carbon atoms, there are four structural isomers, including 2-methyl-2-propanol (*tertiary*-butyl alcohol), whose use as a gasoline additive was described in Section 12.7.

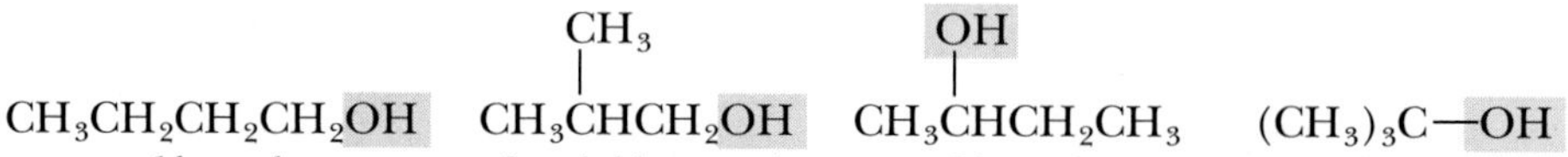

$$CH_3CH_2CH_2CH_2OH \qquad CH_3\overset{\overset{\displaystyle CH_3}{|}}{C}HCH_2OH \qquad CH_3\overset{\overset{\displaystyle OH}{|}}{C}HCH_2CH_3 \qquad (CH_3)_3C\!-\!OH$$

1-butanol \qquad 2-methyl-1-propanol \qquad 2-butanol \qquad 2-methyl-2-propanol
A primary alcohol \quad A primary alcohol \quad A secondary alcohol \quad A tertiary alcohol

Alcohols can serve as the starting substances for the preparation of many other types of organic compounds. Oxidation of alcohols may yield aldehydes, ketones, or carboxylic acids, depending on the alcohol used and the extent of the oxidation. If the starting compound is a primary alcohol, the first oxidation product is an aldehyde, and the second oxidation product is a carboxylic acid. For example, the oxidation of ethanol, a primary alcohol, can be used to make acetaldehyde and acetic acid:

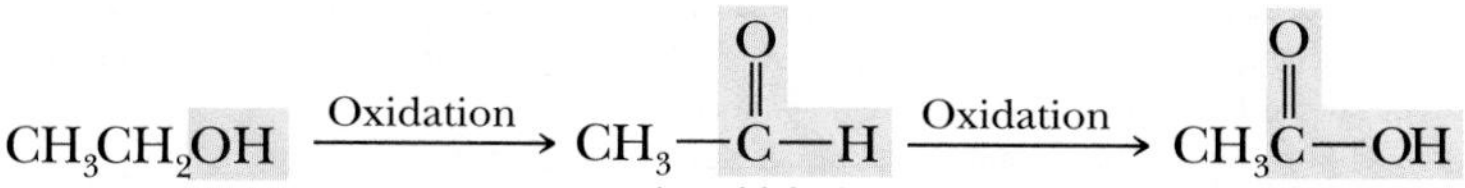

$$CH_3CH_2OH \xrightarrow{\text{Oxidation}} CH_3\!-\!\overset{\overset{\displaystyle O}{\|}}{C}\!-\!H \xrightarrow{\text{Oxidation}} CH_3\overset{\overset{\displaystyle O}{\|}}{C}\!-\!OH$$

Ethanol $\qquad\qquad$ Acetaldehyde $\qquad\qquad$ Acetic acid
A primary alcohol

■ Oxidation of organic compounds is usually the addition of oxygen or the removal of hydrogen, and reduction is usually the removal of oxygen or the addition of hydrogen (see Sections 10.2 and 10.3).

Secondary alcohols are oxidized to give ketones. The oxidation of 2-propanol (isopropyl alcohol) gives acetone, a ketone widely used as an organic solvent.

$$CH_3\overset{\overset{\displaystyle OH}{|}}{\underset{\underset{\displaystyle H}{|}}{C}}CH_3 \xrightarrow{\text{Oxidation}} CH_3\overset{\overset{\displaystyle O}{\|}}{C}CH_3$$

2-propanol $\qquad\qquad$ Acetone

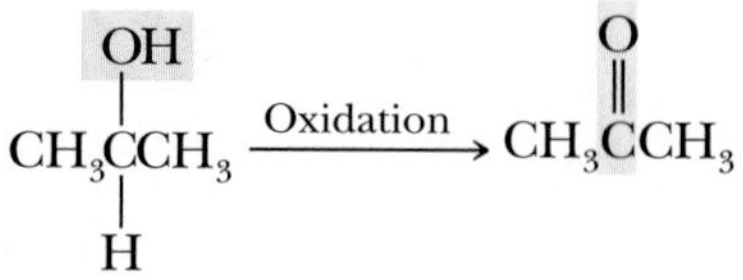

Alcohols can be dehydrated to form alkenes or ethers. Two important dehydration reactions of ethanol illustrate how temperature can be used to determine the product. Sulfuric acid is the dehydrating agent.

$$2\ \underset{\text{Ethanol}}{CH_3CH_2OH} \xrightarrow[H_2SO_4]{140^\circ C} \underset{\text{Diethyl ether}}{CH_3CH_2OCH_2CH_3} + H_2O$$

$$\underset{\text{Ethanol}}{CH_3CH_2OH} \xrightarrow[H_2SO_4]{180^\circ C} \underset{\text{Ethene}}{H-\overset{\overset{H}{|}}{C}=\overset{\overset{H}{|}}{C}-H} + H_2O$$

Hydrogen Bonding in Alcohols

The physical properties of alcohols provide an example of the effects of hydrogen bonding between molecules in liquids, as illustrated by the boiling points in Table 14.4. Hydrogen bonding explains why methanol (molar mass 32 g) is a liquid, whereas propane (molar mass 44 g), which has a higher molar mass but no hydrogen bonding, is a gas at room temperature. The boiling point of methanol is lower than that of water because methanol has only one O—H hydrogen through which it can hydrogen bond. The higher boiling point of ethylene glycol can be attributed to the presence of two —OH groups per molecule. Glycerol, with three —OH groups, has an even higher boiling point.

■ See Section 7.3 for a discussion of hydrogen bonding.

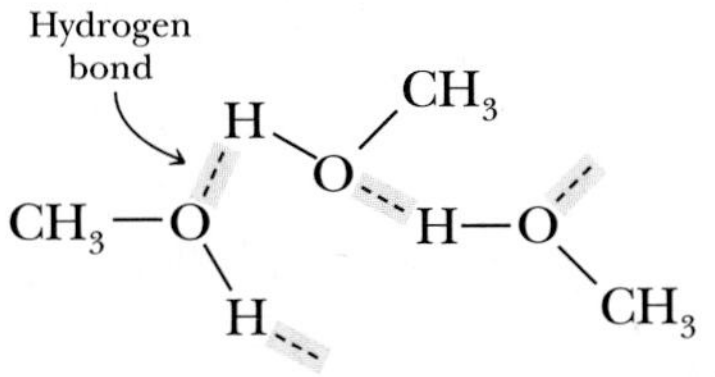

The alcohols listed in Table 14.4 are also very soluble in water because of hydrogen bonding between water molecules and the —OH group in alcohol molecules. However, as the length of the hydrocarbon chain increases, the resulting alcohols are less soluble because the nonpolar hydrocarbon portion has greater influence on the solubility than the hydrogen bonding by the —OH group.

Ethanol

Ethanol is the "alcohol" of alcoholic beverages and is prepared for this purpose by fermentation (Section 12.6) of carbohydrates (starch, sugars) from a wide variety of plant sources (Table 14.5). The growth of yeast is inhibited at alcohol concentrations higher than 12%, and fermentation comes to a stop. Beverages with a higher alcohol concentration are prepared either by distillation or by fortification with alcohol that has been obtained by the distillation of another fermentation product. The maximum concentration of ethanol that can be obtained by distillation of alcohol/water mixtures is 95%. The

Molecular models of the alcohol, aldehyde, and carboxylic acid with two carbon atoms.

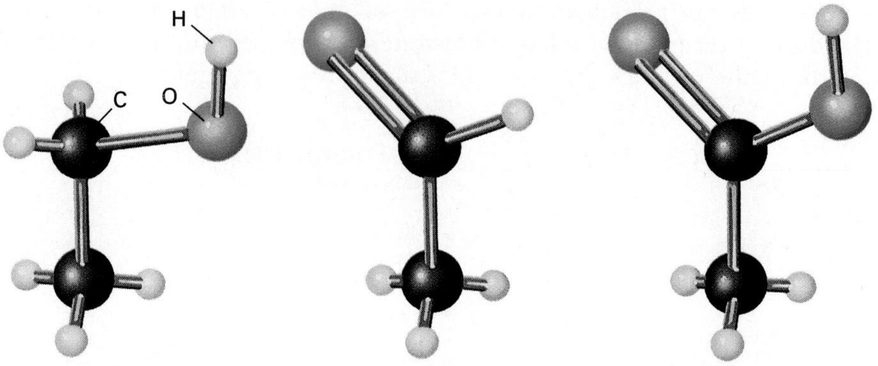

A model of ethanol

(CH_3CH_2OH)

A model of acetaldehyde

$$\overset{O}{\overset{\|}{(CH_3C-H)}}$$

A model of acetic acid

$$\overset{O}{\overset{\|}{(CH_3COH)}}$$

■ Proof = 2 × volume percent

"proof" of an alcoholic beverage is twice the volume percent of ethanol; 80 proof vodka, for example, contains 40% ethanol by volume; 95% ethanol is 190 proof.

Although ethanol is not as toxic as methanol (Section 12.6), 1 pint of pure ethanol, rapidly ingested, would kill most people. Ethanol is a depressant, and the effects of different blood levels of alcohol are shown in Table 14.6. Rapid consumption of two 1-oz "shots" of 90-proof whiskey or of two 12-oz beers can cause one's blood alcohol level to reach 0.05%. Ethanol is quickly absorbed by the blood and metabolized by enzymes produced in the liver. The rate of detoxification is about 1 oz of pure alcohol per hour. Ethanol is oxidized to acetaldehyde, which is further oxidized to acetic acid; eventually, CO_2 and H_2O are produced and eliminated through the lungs and kidneys.

$$H-\overset{\overset{\displaystyle H}{|}}{\underset{\underset{\displaystyle H}{|}}{C}}-\overset{\overset{\displaystyle H}{|}}{\underset{\underset{\displaystyle H}{|}}{C}}-OH \xrightarrow[\text{enzymes}]{\text{Liver}} H-\overset{\overset{\displaystyle H}{|}}{\underset{\underset{\displaystyle H}{|}}{C}}-\overset{\overset{\displaystyle O}{\|}}{C}-H$$

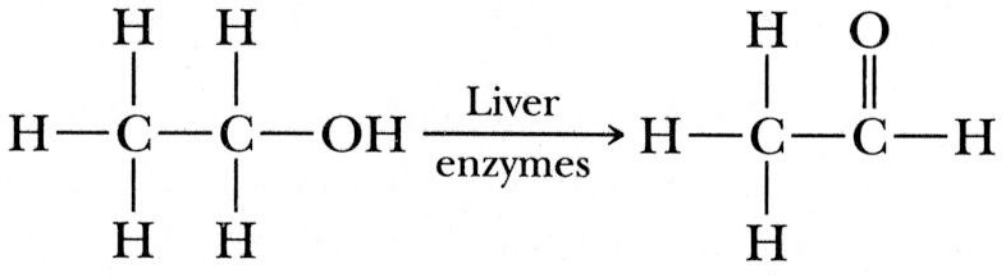

Ethanol Acetaldehyde

TABLE 14.5 ■ Common Alcoholic Beverages

Name	Source of Fermented Carbohydrate	Amount of Ethyl Alcohol	Proof
Beer	Barley, wheat	5%	10
Wine	Grapes or other fruit	12% maximum, unless fortified	20–24
Brandy	Distilled wine	40–45%	80–90
Whiskey	Barley, rye, corn, etc.	45–55%	90–110
Rum	Molasses	~45%	90
Vodka	Potatoes	40–50%	80–100

TABLE 14.6 ■ Effects of Alcohol Blood Level*

(% by Volume)	Effect
0.05—0.15	Lack of coordination, altered judgment
0.15—0.20	Intoxication (slurred speech, altered perception and equilibrium)
0.30—0.40	Unconsciousness, coma
0.50	Possible death

*In many states a person with a blood alcohol level of 0.10% or higher is legally intoxicated.

The federal tax on alcoholic beverages is about $20 per gallon. Since the cost of producing ethanol is only about $1 per gallon, ethanol intended for industrial use must be *denatured* to avoid the beverage tax. **Denatured alcohol** contains small amounts of a toxic substance, such as methanol or gasoline, that cannot be removed easily by chemical or physical means.

Apart from being used in the alcoholic beverage industry, ethanol is used widely in solvents, in the preparation of many other organic compounds, and as a gasoline additive (Section 12.7). For many years industrial ethanol was also made by fermentation. However, in the last several decades, it became cheaper to make the ethanol from petroleum byproducts, specifically by the catalyzed addition of water to ethylene. More than one billion pounds of ethanol are produced each year by this process.

Ethylene Glycol and Glycerol

More than one alcohol group can be present in a single molecule. Ethylene glycol, a di-alcohol (Table 14.4) used in permanent antifreeze and in the synthesis of polymers (Section 14.5), is made in a two-step process, starting with ethylene.

$$CH_2{=}CH_2 + \tfrac{1}{2}O_2 \xrightarrow[300°C]{Ag} H_2C\underset{O}{\diagdown\diagup}CH_2 + H_2O \xrightarrow{Acid} \begin{matrix} H \\ | \\ H-C-OH \\ | \\ H-C-OH \\ | \\ H \end{matrix}$$

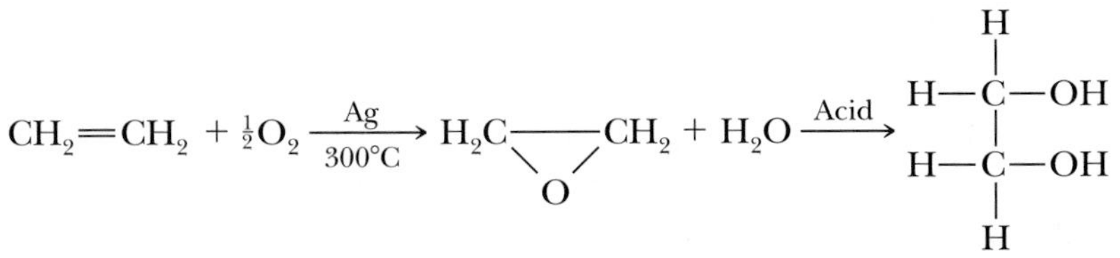

Ethylene Ethylene oxide 1,2-ethanediol Ethylene glycol

A small structure difference makes a big difference in properties. Antifreeze contains ethylene glycol, which is composed of two carbon atoms with an —OH group on each one, and is extremely poisonous. Glycerol, which is composed of three carbon atoms with an —OH group on each one, is non-poisonous, sweet, syrupy, and holds moisture to the skin (a humectant; Section 16.3). It is used in soap and also as a food additive (Section 17.9).
(C.D. Winters)

Glycerol, a tri-alcohol (Table 14.4), is a byproduct from the manufacture of soaps. Because of its moisture-holding properties, glycerol has many uses in foods and tobacco as a digestible and nontoxic humectant (which gathers and holds moisture), and in the manufacture of drugs and cosmetics. It is also used in the manufacture of nitroglycerin and numerous other chemicals. Perhaps the most important compounds of glycerol are its natural esters, which are the fats and oils found in plants and animals (Section 15.6).

EXAMPLE 14.2 *Alcohols*

Label the following alcohols as primary, secondary, or tertiary:
(a) $CH_3CH_2CHOHCH_3$, (b) $CH_3CH_2CH_2CH_2CH_2OH$, (c) $(CH_3)_3COH$

SOLUTION

(a) A secondary alcohol. The alcohol group is attached to an inner carbon that also has one H atom. (b) A primary alcohol. The alcohol group is attached to the end carbon, which also has two H atoms attached. (c) A tertiary alcohol. There are no H atoms attached to the C atom that bonds to the alcohol group.

Exercise 14.2
Label the following alcohols as primary, secondary, or tertiary:
(a) $CH_3CH_2C(CH_3)_2OH$
(b) $CH_3CHOHCH_2CH_2CH_3$
(c) $CH_3CH_2CH_2CH_2CH_2OH$

■ SELF-TEST 14A

1. Identify the functional groups in each of the following molecules:

   (a) $R-OH$ (b) $R-\overset{\overset{\large O}{\|}}{C}-OH$

   (c) $R-\overset{\overset{\large O}{\|}}{C}-H$ (d) $R-\overset{\overset{\large O}{\|}}{C}-R'$
2. Coal tar is the principal source of what major class of hydrocarbons?
3. Ethanol is quickly absorbed by the blood and oxidized to ____________ in the liver.
4. 2-Propanol is commonly known as ____________.
5. The primary component of antifreeze is ____________.
6. Gin that is 84 proof contains what percentage of ethanol?
7. Ethanol intended for industrial use is ____________ by the addition of small amounts of a toxic substance.
8. Glycerol has ____________ alcohol groups.
9. Fats and oils are esters of the alcohol ____________.
10. ____________ is the formula of the alcohol oxidized to formaldehyde.

14.3 Functional Groups with the Carbonyl Group

Aldehydes and Ketones

Aldehydes, ketones, carboxylic acids, and esters all contain the carbonyl ($-\overset{\overset{\large O}{\|}}{C}-$) group. The difference between these functional groups is what is bonded to the carbonyl group. The first column in Table 14.3 shows the general formula for these functional groups.

Formaldehyde, HCHO, the simplest aldehyde, is a gas at room temperature but is often used as a 40% aqueous solution called formalin. Formaldehyde ranks 25th in the top 50 chemicals list (Table 14.1), primarily because of its use in the production of several polymers (Section 14.5). Although formaldehyde is an important starting material for polymers, it presents a number of health hazards because of its toxicity and carcinogenicity. Formaldehyde is also an air pollutant (Section 21.3), being produced in trace amounts in the incomplete combustion of fossil fuels. One of the concerns about methanol as a fuel (Section 12.9) is the potential for increased levels of formaldehyde in urban areas because of the production of formaldehyde from incomplete combustion of methanol.

■ Aldehydes contain the $-\overset{\overset{\displaystyle O}{\|}}{C}-H$ functional group, also represented as —CHO.

Bitter almonds, which contain benzaldehyde, and cinnamon, which contains cinnimaldehyde. *(C.D. Winters)*

Aldehydes with an aromatic ring have pleasant odors, and some are used in food flavorings and perfumes:

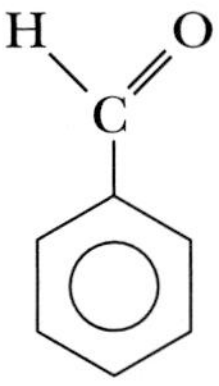

Benzaldehyde
(bitter almonds)

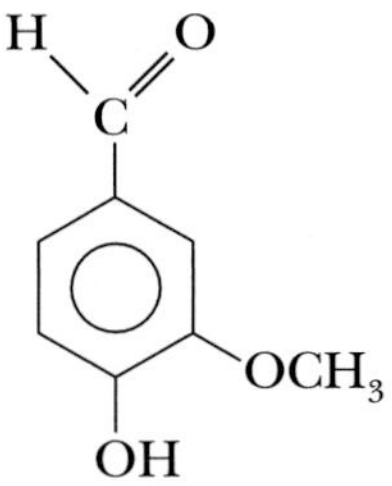

Vanillin
(vanilla bean)

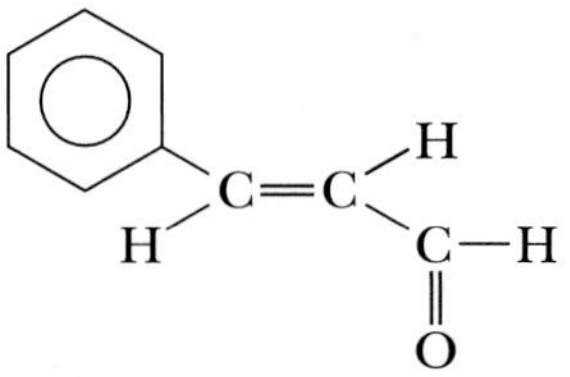

Cinnamaldehyde
(cinnamon)

Acetone is the simplest and the most important ketone. It ranks 41st in the top 50 chemicals list (Table 14.1), primarily because of its wide use as an organic solvent.

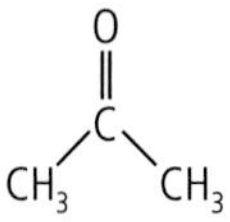

Acetone

Carboxylic Acids

Carboxylic acids contain the —COOH functional group and are prepared by the oxidation of alcohols or aldehydes. These reactions occur quite easily, as evidenced by the souring of wine to form vinegar, which is the oxidation of ethanol to acetic acid in the presence of oxygen from the air.

■ Three ways of representing the carboxylic acid group are —COOH, —CO_2H, and $-\overset{\overset{\displaystyle O}{\|}}{C}-OH$.

Carboxylic acids are polar and readily form hydrogen bonds. This hydrogen bonding results in relatively high boiling points for the acids, even higher than those of alcohols of comparable molar mass. For example, formic acid (46 g/mol) has a boiling point of 101°C, whereas ethanol (46 g/mol) has a boiling point of only 78°C.

All carboxylic acids are weak acids (see Section 9.5) and react with bases to form salts:

$$\underset{\text{Acetic acid}}{CH_3\overset{\overset{\displaystyle O}{\|}}{C}-OH(aq)} + NaOH(aq) \longrightarrow \underset{\text{Sodium acetate}}{CH_3\overset{\overset{\displaystyle O}{\|}}{C}O^-Na^+(aq)} + H_2O(\ell)$$

A number of carboxylic acids are found in nature and have been known for many years. As a result, some of the familiar carboxylic acids are almost always referred to by their common names (Table 14.7).

TABLE 14.7 ■ Some Simple Carboxylic Acids

Structure	Common Name	Systematic Name	Boiling Point (°C)
$H\overset{\overset{\displaystyle O}{\|}}{C}OH$	Formic acid	Methanoic acid	101
$CH_3\overset{\overset{\displaystyle O}{\|}}{C}OH$	Acetic acid	Ethanoic acid	118
$CH_3CH_2\overset{\overset{\displaystyle O}{\|}}{C}OH$	Propionic acid	Propanoic acid	141
$CH_3(CH_2)_2\overset{\overset{\displaystyle O}{\|}}{C}OH$	Butyric acid	Butanoic acid	163
$CH_3(CH_2)_3\overset{\overset{\displaystyle O}{\|}}{C}OH$	Valeric acid	Pentanoic acid	187

The simplest common carboxylic acid is formic acid (Table 14.7), the substance responsible for the sting of an ant bite. Therefore, the name of the acid comes from the Latin word for ant *(formica)*. Acetic acid gives the sour taste to vinegar, and the name comes from the Latin word for vinegar *(acetum)*. Butyric acid gives rancid butter its unpleasant odor, and the name is related to the Latin word for butter *(butyrum)*. The names for caproic acid (C_6), as well as caprylic (C_8) and capric (C_{10}) acids, are all derived from the Latin word for goat *(caper)*, since these acids combine to give goats their characteristic odor. In general, the simple carboxylic acids have unpleasant odors.

Acetic acid is produced in large quantities (number 34 on the top 50 chemicals list) for use in the manufacture of cellulose acetate, a polymer used in the manufacture of photographic film base, synthetic fibers, plastics, and other products (Section 14.5).

The only other acids produced in large quantity are several with two carboxylic acid groups, known as dicarboxylic acids.

Spinach and rhubarb contain oxalic acid. Individuals prone to kidney stones composed of highly insoluble calcium oxalate must limit their intake of foods containing oxalic acid. *(C.D. Winters)*

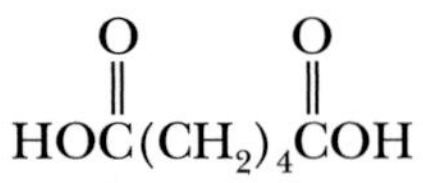

Adipic acid

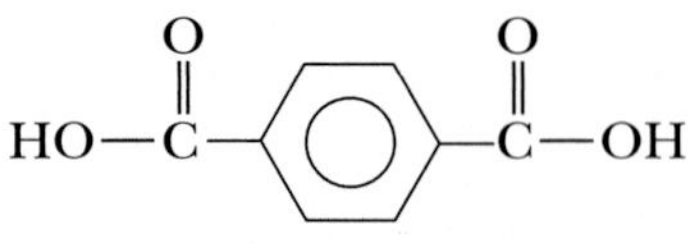

Terephthalic acid

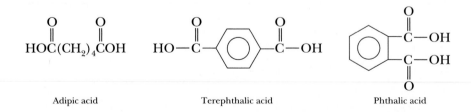

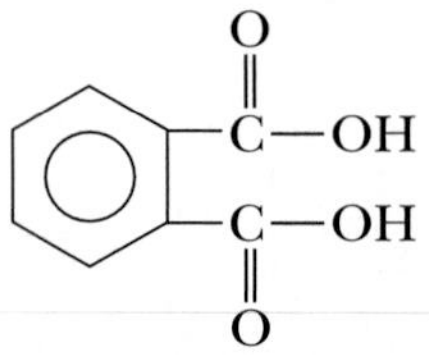

Phthalic acid

These three acids are used to manufacture polymers (Section 14.5). There are other acids, however, whose names are familiar to you (Table 14.8), since they occur in nature.

TABLE 14.8 ■ Some Naturally Occurring Carboxylic Acids

Name	Structure	Natural Source
Citric acid	$HOOC—CH_2—C(OH)(COOH)—CH_2—COOH$ (OH above and COOH below the central C)	Citrus fruits
Lactic acid	$CH_3—CH(OH)—COOH$ (OH below CH)	Sour milk
Malic acid	$HOOC—CH_2—CH(OH)—COOH$ (OH below CH)	Apples
Oleic acid	$CH_3(CH_2)_7—CH{=}CH—(CH_2)_7—COOH$	Vegetable oils
Oxalic acid	$HOOC—COOH$	Rhubarb, spinach, cabbage, tomatoes
Stearic acid	$CH_3(CH_2)_{16}—COOH$	Animal fats
Tartaric acid	$HOOC—CH(OH)—CH(OH)—COOH$ (OH below each CH)	Grape juice, wine

Esters

Carboxylic acids react with alcohols in the presence of strong acids to produce **esters,** which contain the —COOR functional group. In an ester the —OH of the carboxylic acid is replaced by the OR group from the alcohol. For example, when ethanol is mixed with acetic acid in the presence of sulfuric acid, ethyl acetate is formed. This reaction is a dehydration in which sulfuric acid acts as a catalyst and dehydrator.

■ Esters contain the $—\overset{\overset{\Large O}{\|}}{C}—OR$ functional group, also represented as $—CO_2R$ or —COOR.

$$\underset{\text{Acetic acid}}{CH_3\overset{\overset{O}{\|}}{C}—OH} + \underset{\text{Ethanol}}{H—OCH_2CH_3} \xrightarrow{H_2SO_4} \underset{\text{Ethyl acetate}}{CH_3\overset{\overset{O}{\|}}{C}—OCH_2CH_3} + H_2O$$

Names of esters are derived from the names of the alcohol and the acid used to prepare the ester. The alkyl group from the alcohol is named first followed by the name of the acid changed to end in *-ate.* For example, ethyl acetate is the name of the ester prepared from the reaction of ethanol and acetic acid.

Ethyl acetate is a common solvent for lacquers and plastics and is often used as fingernail polish remover.

Unlike the acids from which they are derived, esters often have pleasant odors (Table14.9). The characteristic odors and flavors of many flowers and fruits are caused by the presence of natural esters. For example, the odor and flavor of bananas is primarily caused by the ester 3-methylbutyl acetate (also known as isoamyl acetate).

Food and beverage manufacturers often use mixtures of esters as food additives. The ingredient label of a brand of imitation banana extract reads "water, alcohol (40%), isoamyl acetate and other esters, orange oil and other essential oils, and FD&C Yellow #5." Except for the water, these are all organic compounds.

Some household products containing ethyl acetate, which is an excellent solvent with a pleasant odor.
(C.D. Winters)

TABLE 14.9 ■ Some Acids, Alcohols, and Their Esters

Acid	Alcohol	Ester	Odor of Ester
CH_3COOH Acetic acid	$CH_3\overset{CH_3}{\overset{\vert}{C}}HCH_2CH_2OH$ 3-methyl-1-butanol	$CH_3\overset{O}{\overset{\|}{C}}OCH_2CH_2\overset{CH_3}{\overset{\vert}{C}}HCH_3$ 3-methylbutyl acetate	Banana
$CH_3CH_2CH_2CH_2COOH$ Pentanoic acid	$CH_3\overset{CH_3}{\overset{\vert}{C}}HCH_2CH_2OH$ 3-methyl-1-butanol	$CH_3CH_2CH_2CH_2\overset{O}{\overset{\|}{C}}OCH_2CH_2\overset{CH_3}{\overset{\vert}{C}}HCH_3$ 3-methylbutyl pentanoate	Apple
$CH_3CH_2CH_2COOH$ Butanoic acid	$CH_3CH_2CH_2CH_2OH$ 1-butanol	$CH_3CH_2CH_2\overset{O}{\overset{\|}{C}}OCH_2CH_2CH_2CH_3$ Butyl butanoate	Pineapple
$CH_3CH_2CH_2COOH$ Butanoic acid	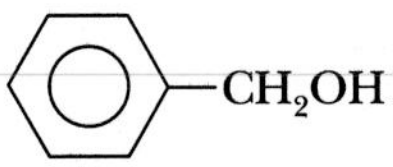 Benzyl alcohol	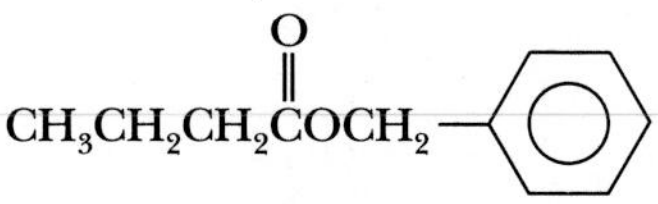 Benzyl butanoate	Rose

As a class, esters are not very reactive. Perhaps their most important reaction is hydrolysis in the presence of a strong base to give the constituents of the ester: the alcohol and a salt of the acid from which the ester was formed.

$$\underset{\text{Ester}}{R\overset{O}{\overset{\|}{C}}-O-R'} + NaOH(aq) \xrightarrow{\text{Heat}} \underset{\text{Carboxylate salt}}{R\overset{O}{\overset{\|}{C}}-O^-Na^+(aq)} + \underset{\text{Alcohol}}{R'OH}$$

$$\underset{\text{Ethyl acetate}}{CH_3\overset{O}{\overset{\|}{C}}-O-CH_2CH_3} + NaOH(aq) \xrightarrow{\text{Heat}} \underset{\text{Sodium acetate}}{CH_3\overset{O}{\overset{\|}{C}}-O^-Na^+(aq)} + \underset{\text{Ethanol}}{CH_3CH_2OH}$$

EXAMPLE 14.3 *Ester Formation*

Write the reaction for the formation of butyl acetate.

SOLUTION

The name of the ester indicates which alcohol and acid are needed to make the ester. The first part, butyl, identifies the alcohol, 1-butanol, and the last part, acetate, identifies the acid, acetic acid. 1-Butanol and acetic acid react in the presence of sulfuric acid catalyst to give butyl acetate.

$$CH_3COOH + CH_3CH_2CH_2CH_2OH \longrightarrow CH_3\overset{O}{\overset{\|}{C}}OCH_2CH_2CH_2CH_3$$

Exercise 14.3

Draw the structural formula of the ester formed in the reaction of (a) acetic acid and methanol, (b) propionic acid and ethylene glycol.

14.4 Organic Chemicals from Coal

Although petroleum is now the source of over 90% of the organic chemicals used to synthesize consumer products, the projected depletion of the U.S. petroleum reserves (Section12.1) is focusing more attention on coal.

An example of the use of coal for the industrial synthesis of organic chemicals is an Eastman Kodak process that produces acetic anhydride used to make cellulose acetate. The first complete "chemicals from coal" plant, built by Eastman Kodak in Kingsport, Tennessee, started production in 1983. Figure 14.3*a* is a schematic drawing of the various components of the plant, which is pictured in Figure 14.3*b*. The basic reactions are to produce synthesis gas from coal, to use the synthesis gas to make methanol, and to use the methanol in the synthesis of acetic anhydride.

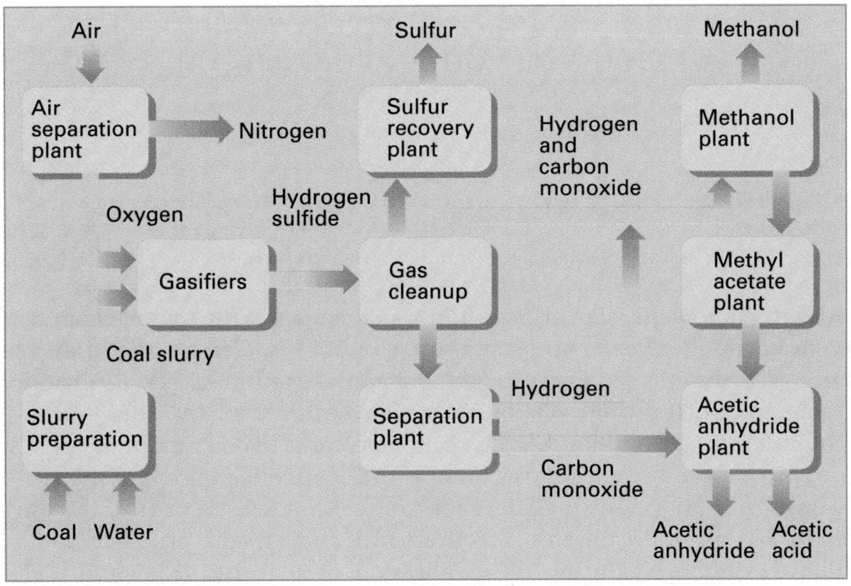

(a)

(b)

Figure 14.3 Chemicals from coal. (a) Flow diagram showing the production of methanol, methyl acetate, and acetic anhydride from coal. Some of the acetic acid produced in the final step is recycled into the preparation of methyl acetate (see equation in text). (b) The Eastman Kodak chemicals-from-coal facility in Kingsport, Tennessee. Numbers in the photograph represent different parts of the plant: 1, coal unloading; 2, coal silos; 3, steam plant; 4, slurry preparation; 5, coal gasification plant; 6, gas cleanup and separation; 7, sulfur recovery plant; 8, gas flare stack; 9, chemical storage; 10, methanol plant; 11, methyl acetate plant; 12, acetic anhydride plant.
(b, Courtesy of Tennessee Eastman)

Within the complex pictured in Figure 14.3*b* are nine separate plants, four related to the gasification of coal (Section 12.9), two for synthesis gas preparation, and three for the synthesis of methanol, methyl acetate, and acetic anhydride. Acetic anhydride is needed to make cellulose acetate, a polymer used in the manufacture of photographic film base, synthetic fibers, plastics, and other products. The main chemical reactions used in the process are

■ Acetic anhydride reacts with water to give acetic acid:

$$\underset{\text{Acetic anhydride}}{CH_3-\overset{\overset{\displaystyle O}{\|}}{C}-O-\overset{\overset{\displaystyle O}{\|}}{C}-CH_3} + H_2O \longrightarrow 2\ CH_3-\overset{\overset{\displaystyle O}{\|}}{C}-OH$$

■ An acid anhydride has the general formula

$$R-\overset{\overset{\displaystyle O}{\|}}{C}-O-\overset{\overset{\displaystyle O}{\|}}{C}-R.$$

(1) $$\underset{\text{Coal}}{C} + \underset{\text{Steam}}{H_2O} \longrightarrow \underset{\text{Synthesis gas}}{CO + H_2}$$

(2) $$CO + 2\ H_2 \longrightarrow \underset{\text{Methanol}}{CH_3OH}$$

(3) $$CH_3OH + \underset{\text{Acetic acid}}{CH_3\overset{\overset{\displaystyle O}{\|}}{C}OH} \longrightarrow \underset{\text{Methyl acetate}}{CH_3\overset{\overset{\displaystyle O}{\|}}{C}OCH_3} + H_2O$$

(4) $$CH_3\overset{\overset{\displaystyle O}{\|}}{C}OCH_3 + CO \longrightarrow \underset{\text{Acetic anhydride}}{CH_3-\overset{\overset{\displaystyle O}{\|}}{C}-O-\overset{\overset{\displaystyle O}{\|}}{C}-CH_3}$$

About 900 tons per day of high-sulfur coal from nearby Appalachian coal mines are ground in water to form a slurry of 55% to 65% by weight of coal in water. The slurry is fed into two gasifiers to make synthesis gas. To produce the same amounts of these chemicals by conventional means would require the equivalent of two million barrels of oil per year. This is an example of how the use of U.S.-produced coal can lower our dependence on imported oil.

The plant design uses the latest control technologies to protect the environment. For example, the sulfur recovery unit converts the hydrogen sulfide gas that was removed during the gasification of coal into free sulfur. This process removes more than 99% of the sulfur from the coal, and this sulfur is then sold to chemical companies for other uses.

■ SELF-TEST 14B

1. The simplest aldehyde is ____________.
2. The organic acid found in vinegar is ____________.
3. Ethyl acetate can be prepared by reacting ____________ with ____________.
4. The name of the ester obtained by reacting methanol with acetic acid is ____________.
5. Citric acid found in citrus fruits has ____________ carboxylic acid groups.
6. The simplest ketone is ____________.
7. The characteristic odors and flavors of many fruits are caused by (a) carboxylic acids, (b) esters, (c) alcohols.

Figure 14.4 Just a few everyday items made from synthetic polymers.
(C.D. Winters)

14.5 Synthetic Organic Polymers

It is impossible for us to get through a day without using a dozen or more synthetic organic **polymers.** The word **polymer** means "many parts" (Greek *poly,* meaning "many," and *meros,* meaning "parts"). Polymers are giant molecules with molar masses ranging from thousands to millions. Our clothes are polymers; our food is packaged in polymers; our appliances and cars contain a number of polymer components (Figure 14.4).

■ Both natural and synthetic polymers are known. Examples of natural polymers include proteins, nucleic acids, starch, cellulose, and rubber. Natural polymers are discussed in Chapter 15.

Approximately 80% of the organic chemical industry is devoted to the production of synthetic polymers. The prominence of synthetic polymers in consumer products is indicated by the fact that about half of the top 50 chemicals produced in the United States are used in the production of plastics, fibers, and rubbers. The average production of synthetic polymers in the United States exceeds 200 pounds per person annually. Many synthetic organic polymers are **plastics** of one sort or another. All plastics are polymers, but not all polymers are plastics. Examples of items often made of plastics include dishes and cups, containers, telephones, plastic bags for packaging and wastes, plastic pipes and fittings, automobile steering wheels and seat covers, and cabinets for appliances, radios, and television sets. In fact, such plastic items, along with textile fibers and synthetic rubbers, are so widely used that they are commonly taken for granted.

■ A plastic is a substance that will flow under heat and pressure and hence is capable of being molded into various shapes. All plastics are polymers, but not all polymers are plastics.

Some of our most useful polymer chemistry has resulted from copying giant molecules in nature. Rayon is remanufactured cellulose (discussed in Section 15.5); synthetic rubber is copied from natural latex rubber. As useful as these polymers may be, however, polymer chemistry is not restricted to nature's models. Polystyrene, nylon, and Dacron are a few examples of synthetic molecules that do not have exact duplicates in nature. We have gone to school on nature and have extended our knowledge to produce polymers that are more useful than natural polymers.

Polymers are made by chemically joining together many small molecules into one giant molecule, or macromolecule. The small molecules used to synthesize polymers are called **monomers.** Synthetic polymers can be classified as **addition polymers,** made by monomer units directly joined together, or **condensation polymers,** made by monomer units combining so that a small molecule, usually water, is split out between them.

■ A *macromolecule* is a molecule with a very high molar mass.

The World of Chemistry

Program 1, *The World of Chemistry*

Most of the chemical industry in the United States today is based on petroleum. You really have a huge network from the raw materials which are primarily petroleum to the refinery, which is primarily the place in which you get starting materials. And those starting materials are then used to make petrochemicals. The petrochemicals are then used to make all kinds of things. Mary Good, Allied Signal Corporation (currently Under Secretary for Technology, U.S. Department of Commerce).

■ An organic peroxide, RO—OR′, produces free radicals, RO·, each with an unpaired electron.

Addition Polymers

Polyethylene

The monomer for addition polymers normally contains one or more double bonds. The simplest monomer of this group is ethylene, $CH_2{=}CH_2$. When ethylene is heated to 100 to 250°C at a pressure of 1000 to 3000 atmospheres in the presence of a catalyst, polymers are formed with molar masses of up to several million. A reaction of ethylene usually begins with breaking of one of the bonds in the carbon-carbon double bond, so an unpaired electron, a reactive site, remains at each end of the molecule. This step, the **initiation** of the polymerization, can be accomplished with initiator chemicals such as organic peroxides that are unstable and easily break apart into **free radicals,** which have unpaired electrons. The free radicals react readily with molecules containing carbon-carbon double bonds to produce new free radicals.

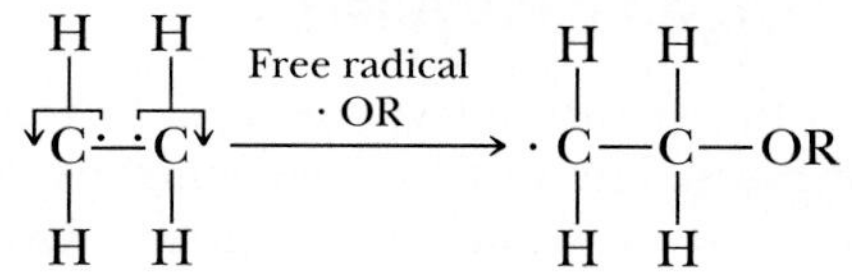

The growth of the polyethylene chain then begins as the unpaired electron bonds to a double bond electron in an unreacted ethylene molecule. This leaves another unpaired electron to bond with yet another ethylene molecule. For example,

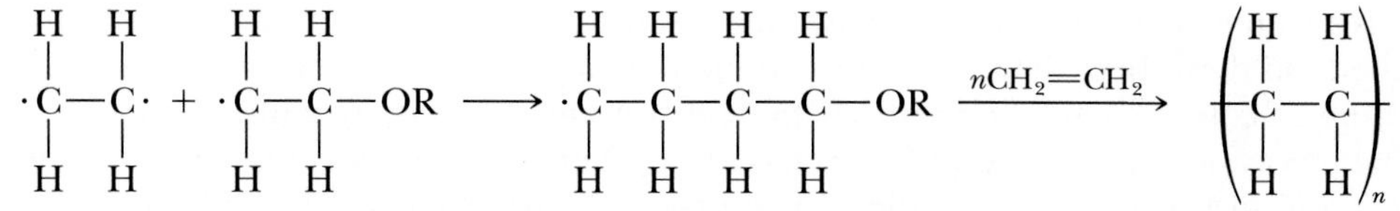

Polyethylene
n ranges from 1000 to 50,000

The World of Chemistry

Program 22, *The Age of Polymers*

Many places where we use metals today, particularly in structural applications, will be taken over by polymers in the future. . . . A prime reason for this is that polymers allow an almost infinite variation in structure, and hence, almost infinite variation in their properties. Whereas we now know almost all there is to know about metals and glasses—obviously not everything, but almost—there are an almost infinite variety of polymers that we can conceive. We can fine tune their structures and fine tune their properties given enough time. Howard E. Simmons, Vice President, Du Pont Company.

In the process, the unsaturated hydrocarbon monomer, ethylene, is changed to a saturated hydrocarbon polymer, **polyethylene.**

Polyethylene is the world's most widely used polymer. About 12 million tons were produced in 1994 in the United States alone. What are some of the reasons for this popularity? The wide range of properties of polyethylene leads to many uses.

Polyethylenes formed under various pressures and catalytic conditions have different molecular structures and hence different physical properties. For example, chromium oxide as a catalyst yields almost exclusively the linear polyethylene shown on page 429—a polymer with no branches on the carbon chain. The zigzag structure represents the shape of the chain more closely because of the tetrahedral arrangement of bonds around each carbon in the saturated polyethylene chain. Long, linear chains of polyethylene can pack closely together and give a material with high density (0.97 g/mL) and high molar mass, referred to as high-density polyethylene (HDPE). This material is hard, tough, and rigid. The plastic milk bottle is a good example of an application of HDPE. Other HDPE containers are shown in Figure 14.5.

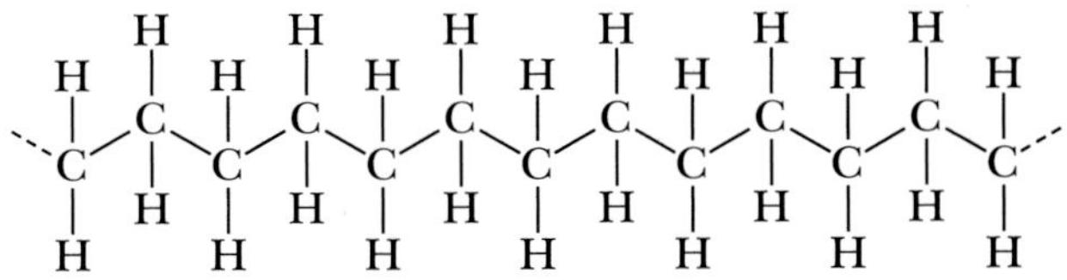

A portion of a polyethylene molecule

or

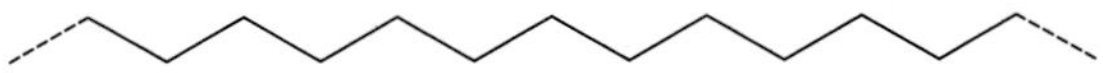

Model of linear polyethylene

Each corner represents a CH_2 group

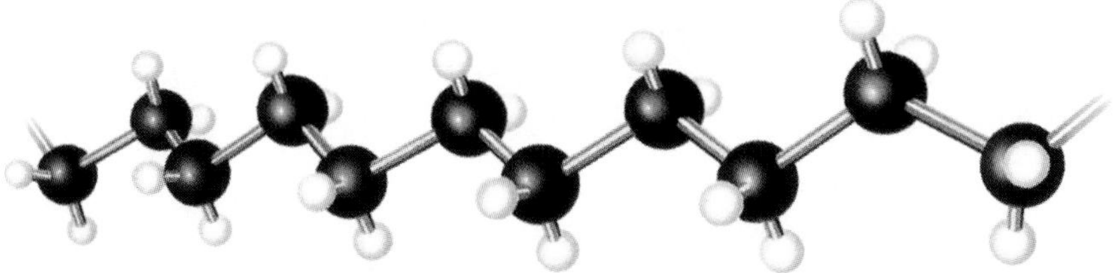

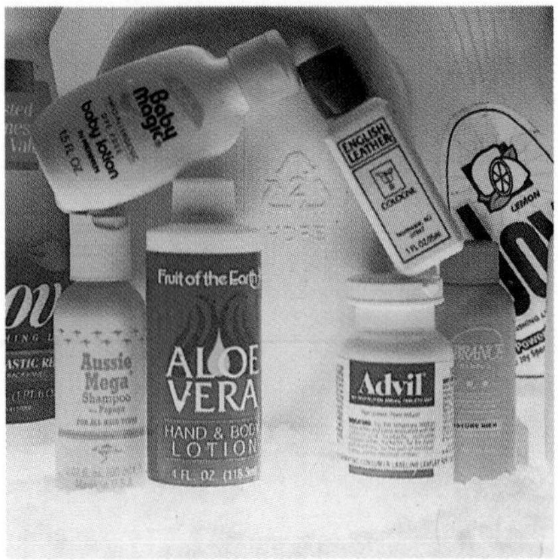

Figure 14.5 Some bottles made of high-density polyethylene. HDPE is one of the most commonly recycled kinds of polymeric materials. *(C.D. Winters)*

If ethylene is heated to 230°C at a pressure of 200 atmospheres, free radicals attack the polyethylene chain at random positions, causing irregular branching (Figure 14.6*a*). Branched chains of polyethylene cannot pack closely together, so the resulting material has a lower density (0.92 g/mL) and is called low-density polyethylene (LDPE). This material is soft and flexible (Figure 14.7). Sandwich bags are made from LDPE. Other conditions can lead to cross-linked polyethylene, in which branches connect long chains to each other (Figure 14.6*b*). If the linear chains of polyethylene are treated in a way that causes cross-links between chains to form cross-linked polyethylene (CLPE), a very tough form of polyethylene is produced. The plastic caps on soft drink bottles are made from CLPE.

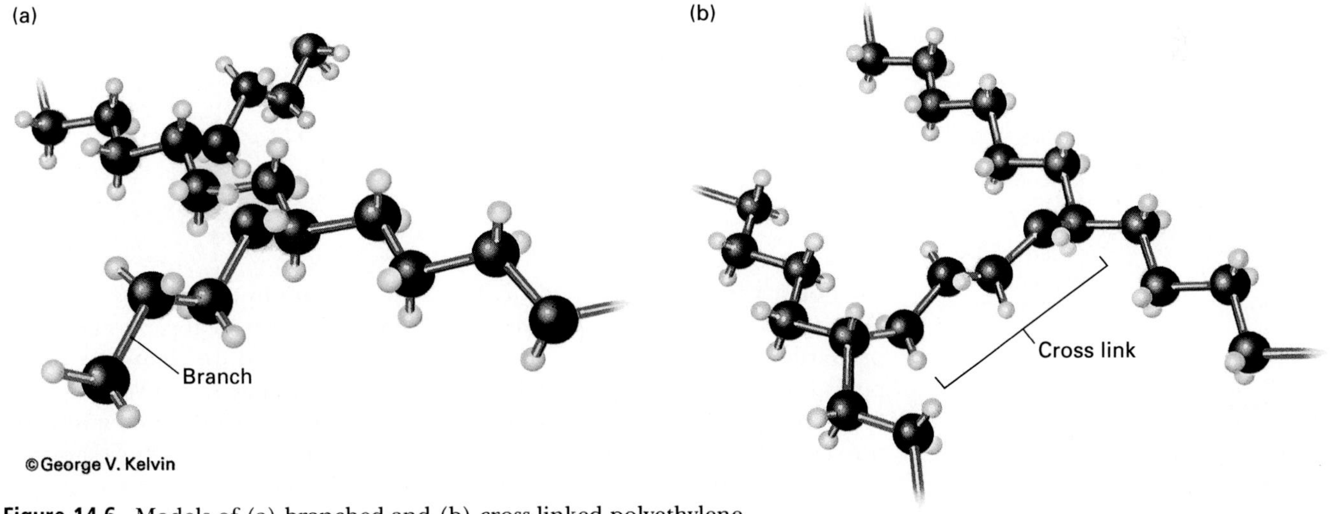

Figure 14.6 Models of (a) branched and (b) cross-linked polyethylene.

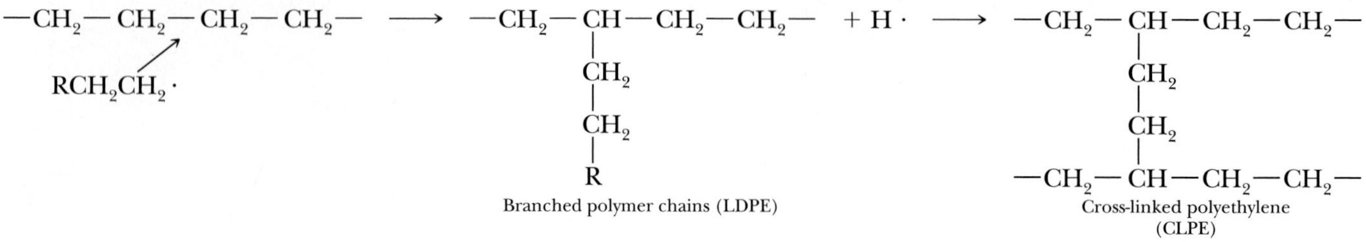

Polymers of Ethylene Derivatives

Many different kinds of addition polymers are made from monomers in which one or more of the hydrogen atoms in ethylene have been replaced with either halogen atoms or a variety of organic groups. If the formation of polyethylene is represented as

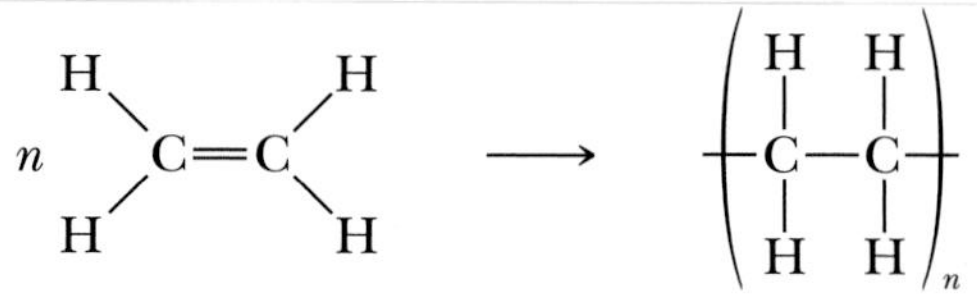

then the general reaction

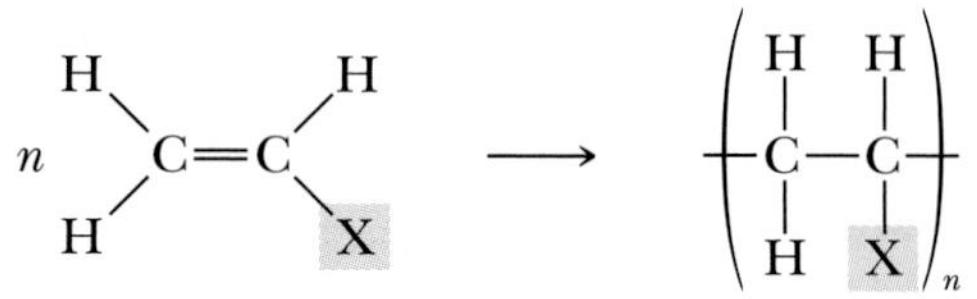

can be used to represent a number of other important addition polymers, where X is Cl, F, or an organic group (Table 14.10).

For example, the monomer for making **polystyrene** is styrene, and n is about 5700.

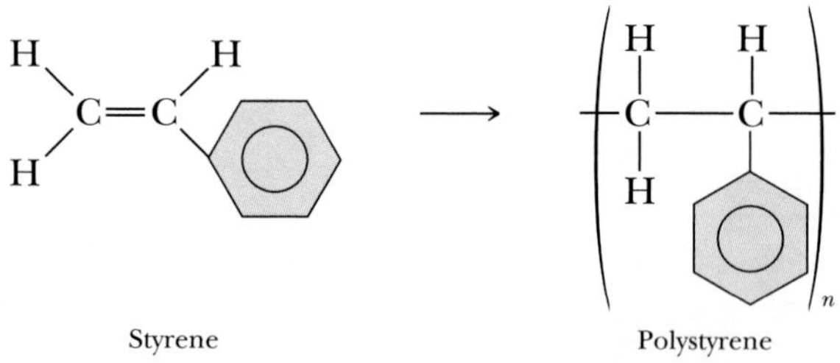

Polystyrene is a clear, hard, colorless solid at room temperature that can be molded easily at 250°C. Styrene is 20th on the list of top 50 chemicals, primarily because of its use in making polystyrene. More than five billion pounds of polystyrene are produced in the United States each year to make food containers, toys, electrical parts, insulating panels, appliance and furniture components, and many other items. The variation in properties shown by polystyrene products is typical of synthetic polymers. For example, a clear polystyrene drinking glass that is brittle and breaks into sharp pieces somewhat

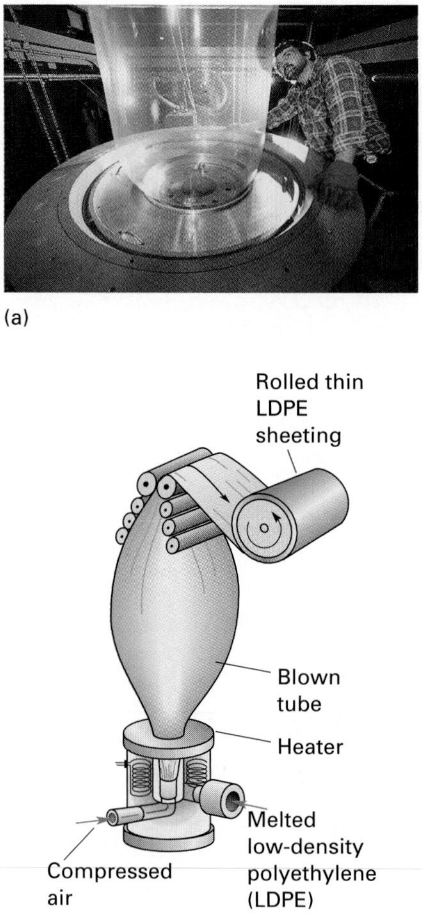

Figure 14.7 Making LDPE film. (a) The blown tube of polyethylene film. (b) Diagram showing how the tube is blown and the film collected in a roll. *(a, Gary Gladstone/The Image Bank)*

TABLE 14.10 ■ Ethylene Derivatives that Undergo Addition Polymerization

Formula	Monomer Common Name	Polymer Name (Trade Names)	Uses	U.S. Polymer Production (Tons/Yr)
$H_2C{=}CH_2$	Ethylene	Polyethylene (Polythene)	Squeeze bottles, bags, films, toys and molded objects, electrical insulation	12 million
$H_2C{=}CH(CH_3)$	Propylene	Polypropylene (Vectra, Herculon)	Bottles, films, indoor-outdoor carpets	5 million
$H_2C{=}CH(Cl)$	Vinyl chloride	Poly(vinyl chloride) (PVC)	Floor tile, raincoats, pipe	5 million
$H_2C{=}CH(CN)$	Acrylonitrile	Polyacrylonitrile (Orlon, Acrilan)	Rugs, fabrics	1 million
$H_2C{=}CH(C_6H_5)$	Styrene	Polystyrene (Styrene, Styrofoam, Styron)	Food and drink coolers, building material insulation	3 million
$H_2C{=}CH(O{-}CO{-}CH_3)$	Vinyl acetate	Poly(vinyl acetate) (PVA)	Latex paint, adhesives, textile coatings	500,000
$H_2C{=}C(CH_3)(CO{-}O{-}CH_3)$	Methyl methacrylate	Poly(methyl methacrylate) (Plexiglas, Lucite)	High-quality transparent objects, latex paints, contact lenses	450,000
$F_2C{=}CF_2$	Tetrafluoroethylene	Polytetrafluoroethylene (Teflon)	Gaskets, insulation, bearings, pan coatings	7000

like glass is much different from the polystyrene coffee cup that is soft and pliable.

A major use of polystyrene is in the production of Styrofoam by "expansion molding." In this process, polystyrene beads are placed in a mold and heated with steam or hot air. The beads, 0.25 to 1.5 mm in diameter, contain 4 to 7% by weight of a low-boiling liquid such as pentane. The steam causes the low-boiling liquid to vaporize, this expands the beads, and as the foamed particles expand, they are molded in the shape of the mold cavity. Styrofoam is used for egg cartons, meat trays, coffee cups, and packing material (Figure 14.8).

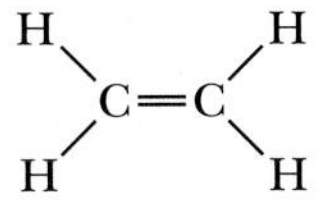

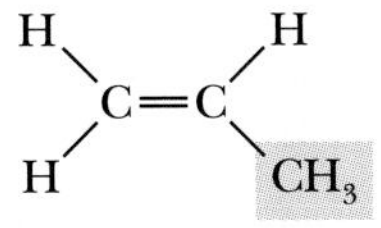

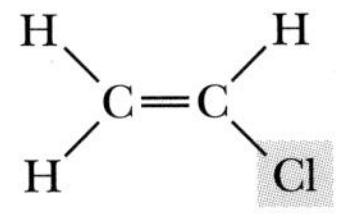

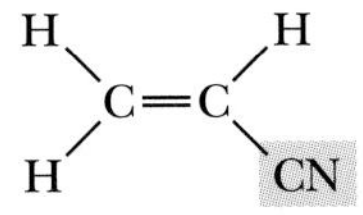

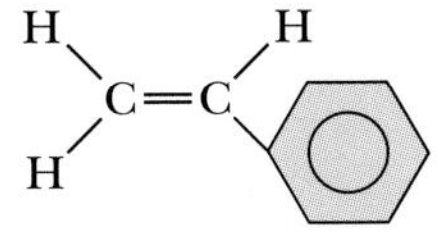

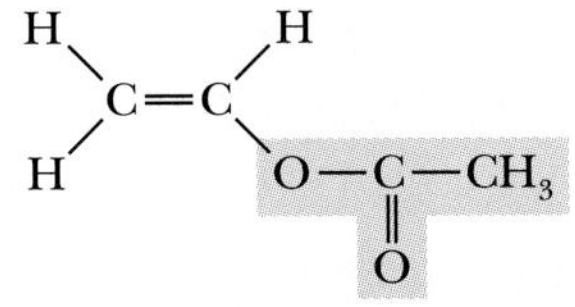

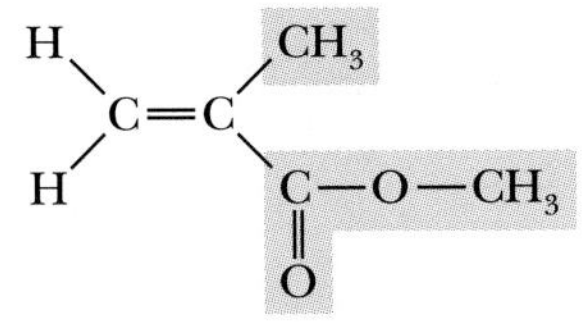

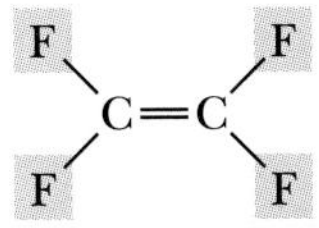

The World of Chemistry

Program 22, *The Age of Polymers*

Large blocks of Styrofoam.

Figure 14.8 Some familiar applications of polystyrene foam, which is an excellent thermal insulator. *(C.D. Winters)*

Polypropylene, used in indoor-outdoor carpeting, bottles, and battery cases, is made from propylene.

$$n \quad \underset{H}{\overset{H}{>}}C=C\underset{CH_3}{\overset{H}{<}} \longrightarrow \left(\begin{array}{cc} H & H \\ | & | \\ -C-C- \\ | & | \\ H & CH_3 \end{array} \right)_n$$

Propylene Polypropylene

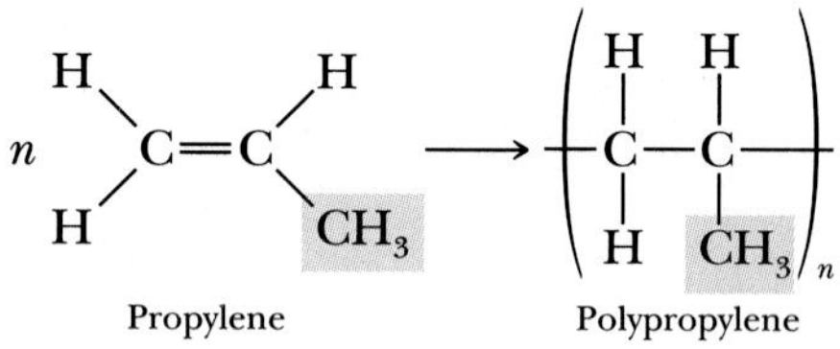

Poly(vinyl chloride) (PVC), used in floor tile, garden hose, plumbing pipes, and trash bags has a chlorine atom substituted for one of the hydrogen atoms in ethylene.

$$n \quad \underset{H}{\overset{H}{>}}C=C\underset{Cl}{\overset{H}{<}} \longrightarrow \left(\begin{array}{cc} H & H \\ | & | \\ -C-C- \\ | & | \\ H & Cl \end{array} \right)_n$$

Vinyl chloride Poly(vinyl chloride)

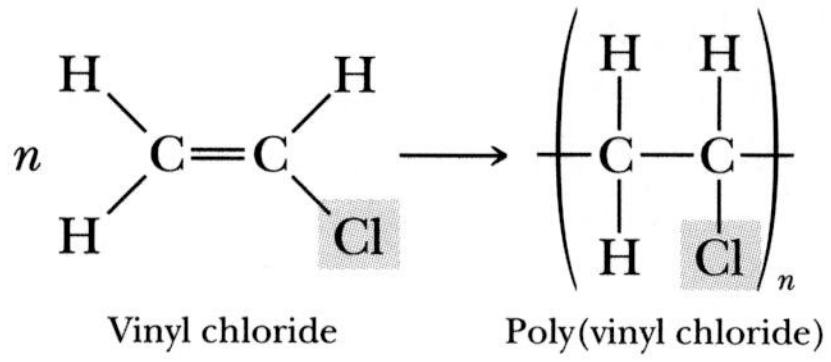

Polyacrylonitrile, the acrylic plastic used to make fibers for rugs and fabrics, is an addition polymer of acrylonitrile.

$$n \quad \underset{H}{\overset{H}{>}}C=C\underset{CN}{\overset{H}{<}} \longrightarrow \left(\begin{array}{cc} H & H \\ | & | \\ -C-C- \\ | & | \\ H & CN \end{array} \right)_n$$

Acrylonitrile Polyacrylonitrile

Although the representation

$$\left(\begin{array}{cc} H & H \\ | & | \\ -C-C- \\ | & | \\ H & H \end{array} \right)_n$$

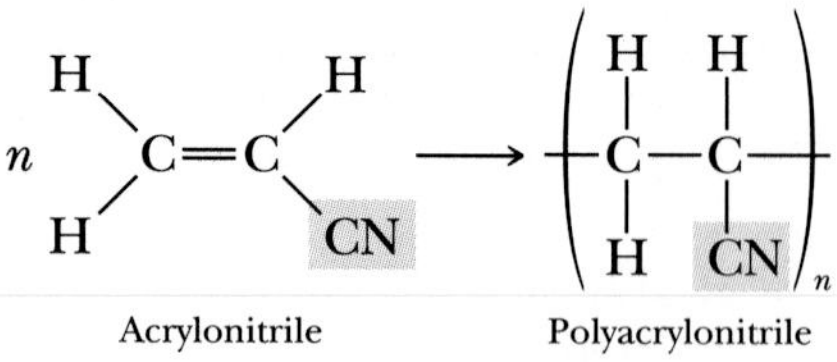

Wigs, among the many items made from acrylic polymers. *(Tom Pantages)*

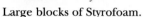

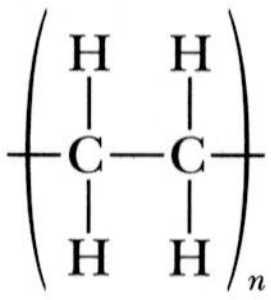

THE WORLD OF CHEMISTRY

Discovery of a Catalyst for Polyacrylonitrile Production

Program 20, *On the Surface*

Oil companies invest considerable sums in equipment and human effort to develop new catalysts to make fuels and chemicals from petroleum. Research and development in this area is a multimillion-dollar gamble. There is no guarantee the money spent will produce anything useful. But if it does, the payoff can be enormous. Just one catalyst breakthrough made more than half a billion dollars for Standard Oil of Ohio, now part of BP America. In the late 1950s, SOHIO researchers came up with a new catalytic process that soon dominated all others in the production of polyacrylonitrile, the polymer used to make textiles, tires, and car bumpers. Oddly enough, SOHIO researchers weren't even trying to produce acrylonitrile at first. They simply wanted to make a metal oxide catalyst to convert waste propane gas from petroleum refining into something more valuable. As Dr. Jeanette Grasselli said,

The theory, the hypothesis at the time, was that we could take the oxygen from the catalyst and insert it into the propane, a relatively unreactive molecule. So this was a tough technical objective. And, in turn, we wanted to generate or take the catalyst back to its original oxidized form by using oxygen from the air.

But the theory didn't hold up. Propane was too stable to react, and the catalyst particles broke down in service. Management gave the research team three more months to show results, so they made some changes. They replaced propane with a more reactive refinery gas, propylene. They made their catalyst from different metal oxides, and they added ammonia to promote, or speed up, the reaction.

To their surprise, ammonia reacted. Rather than just encouraging the reaction to go faster, as a promoter, ammonia reacted and became part of the reaction sequence, and acrylonitrile was made in one step. The researchers had struck pay dirt. Their new catalyst, combined with ammonia, had made a valuable product from a cheap gas. Management quickly saw the value of the new process and wasted no time building a plant to use it. As Dr. Grasselli recalls,

Jeanette Grasselli

In 1960, when our plant came on stream, acrylonitrile was selling for 28 cents a pound. We were making it for 14 cents a pound. And we shut down every other commercial process. Today, 90% of the world's acrylonitrile is manufactured by the SOHIO process.

saves space, keep in mind how large the polymer molecules are. Generally n is 500 to 50,000, and this gives molecules with molar masses ranging from 10,000 to several million. The molecules that make up a given polymer sample are of different lengths and thus are not all of the same molar mass. As a result, only the average molar masses can be determined.

In summary, the numerous variations in substituents, length, branching, and cross-linking make it possible to produce a variety of properties for each type of addition polymer. Chemists and chemical engineers can fine-tune the properties of the polymer to match the desired properties by appropriate selection of monomer and reaction conditions, thus accounting for the widespread and growing use of these giant molecules.

EXAMPLE 14.4 *Addition Polymers*

Draw the structural formula of the repeating unit for the following addition polymers: (a) polypropylene, (b) poly(vinyl acetate), (c) poly(vinyl alcohol)

SOLUTION

The names show that the monomers for these polymers are propylene ($CH_2{=}CHCH_3$), vinyl acetate ($CH_2{=}CHOOCCH_3$), and vinyl alcohol ($CH_2{=}CHOH$). The repeating units in the polymers therefore have the same structures, but without the double bonds.

$$(a) \left(\!\!-\!\!\begin{array}{c} H \quad H \\ | \quad\;\; | \\ C\!-\!C \\ | \quad\;\; | \\ H \quad CH_3 \end{array}\!\!-\!\!\right)_{\!n} \quad (b) \left(\!\!-\!\!\begin{array}{c} H \quad H \\ | \quad\; | \\ C\!-\!C \\ | \quad\; | \\ H \quad O \\ \quad\;\; | \\ \quad O{=}C \\ \quad\quad\; | \\ \quad\quad CH_3 \end{array}\!\!-\!\!\right)_{\!n} \quad (c) \left(\!\!-\!\!\begin{array}{c} H \quad H \\ | \quad\; | \\ C\!-\!C \\ | \quad\; | \\ H \quad OH \end{array}\!\!-\!\!\right)_{\!n}$$

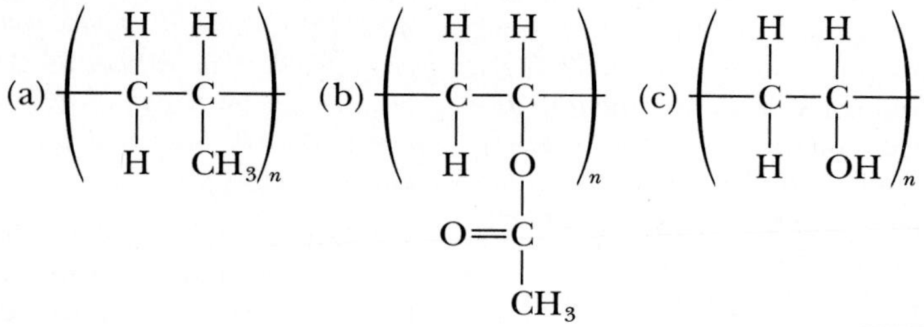

Exercise 14.4

Draw the structural formula of the monomers used to prepare the following polymers: (a) polyethylene, (b) poly(vinyl chloride), (c) polystyrene

Rubber: Natural and Synthetic

Natural rubber, a product of the *Hevea brasiliensis* tree, is a hydrocarbon with the composition C_5H_8. When rubber is decomposed in the absence of oxygen, the monomer isoprene is obtained:

$$CH_2{=}\overset{\displaystyle CH_3}{\overset{\displaystyle |}{C}}{-}CH{=}CH_2 \quad \text{or}$$

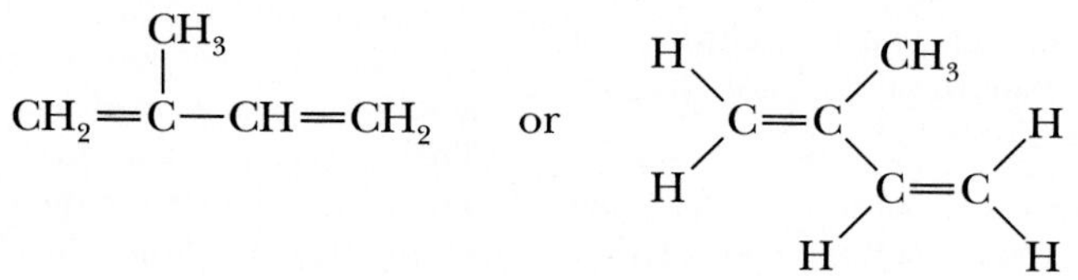

Isoprene (2-methyl-1,3-butadiene)

Natural rubber occurs as *latex* (an emulsion of rubber particles in water), which oozes from rubber trees when they are cut. Precipitation of the rubber particles yields a gummy mass that is not only elastic and water-repellent but also very sticky, especially when warm. In 1839, after five years' work on this material, Charles Goodyear (1800–1860) discovered that heating gum rubber with sulfur produces a material that is no longer sticky but is still elastic, water-repellent, and resilient.

Vulcanized rubber, as the type of rubber Goodyear discovered is now known, contains short chains of sulfur atoms that bond together the polymer chains of the natural rubber and reduce its unsaturation. The sulfur chains help to align the polymer chains, so the material does not undergo a permanent change when stretched but springs back to its original shape and size when the stress is removed (Figure 14.9). Substances that behave this way are called **elastomers.**

In later years chemists searched for ways to make a synthetic rubber so we would not be completely dependent on imported natural rubber during emergencies, such as during the first years of World War II. In the mid-1920s,

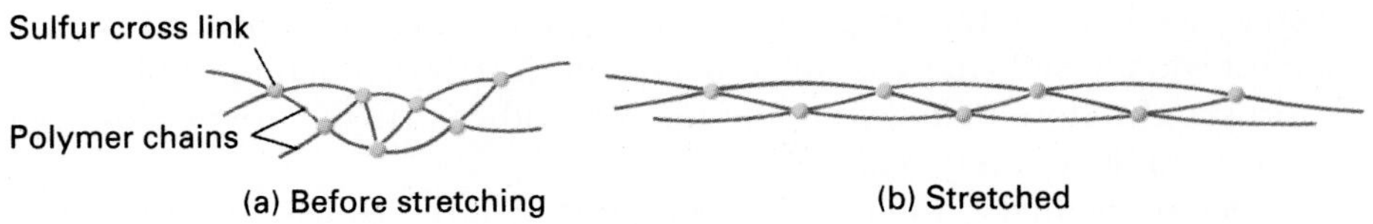

Figure 14.9 Stretching vulcanized rubber. After stretching, the cross links pull it back into the original arrangement.

German chemists polymerized butadiene (obtained from petroleum and structurally similar to isoprene, but without the methyl-group side chain).

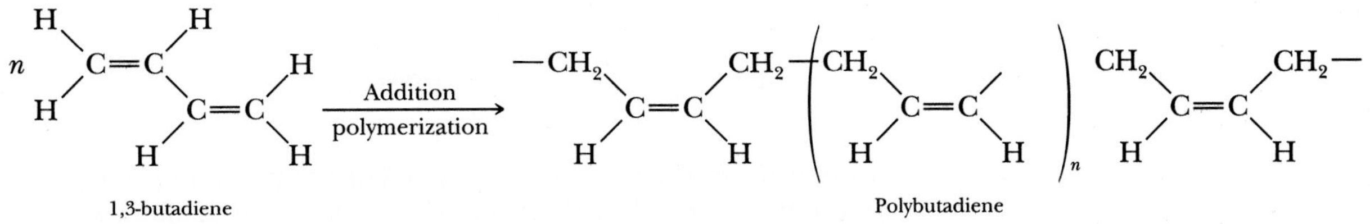

Polybutadiene is used in the production of tires, hoses, and belts.

The behavior of natural rubber (polyisoprene), it was learned later, is due to the specific molecular geometry within the polymer chain. We can write the formula for polyisoprene with the CH_2 groups on opposite sides of the double bond (the *trans* arrangement)

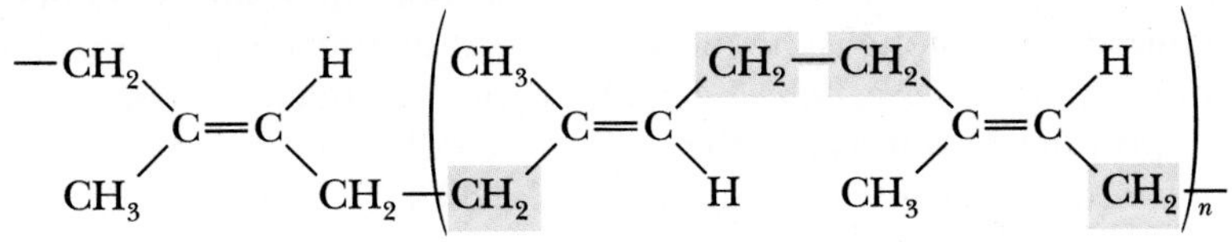

Poly-*trans*-isoprene (the $-CH_2-CH_2-$ groups are *trans*)

or with the CH_2 groups on the same side of the double bond (the *cis* arrangement).

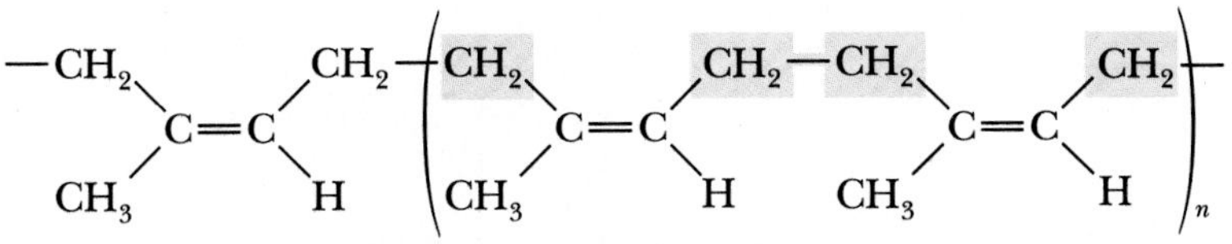

Poly-*cis*-isoprene (the $-CH_2-CH_2-$ groups are *cis*)

Natural rubber is poly-*cis*-isoprene. However, the *trans* material also occurs in nature in the leaves and bark of the sapotacea tree and is known as *gutta-percha.* It is brittle and hard and is used for golf ball covers, electrical insulation, and other applications not requiring the stretching properties of rubber. Without an appropriate catalyst, polymerization of isoprene yields a solid that is like neither rubber nor gutta-percha because it is a random mixture of the *cis* and *trans* geometries. Neither the *trans* polymer nor the randomly arranged material is as good as natural rubber *(cis)* for making automobile tires.

In 1955, chemists at the Goodyear and Firestone companies almost simultaneously discovered how to use **stereoregulation** catalysts to prepare synthetic poly-*cis*-isoprene. This material is structurally identical to natural rubber.

■ Stereoregulation catalysts catalyze reactions that favor the formation of one geometric isomer.

Today, synthetic poly-*cis*-isoprene can be manufactured cheaply and is used almost equally well (there is still an increased cost) when natural rubber is in short supply. More than 2.4 million tons of synthetic rubber are produced in the United States every year.

One of the first synthetic rubbers produced in the United States was **neoprene,** an addition polymer of the monomer 2-chlorobutadiene,

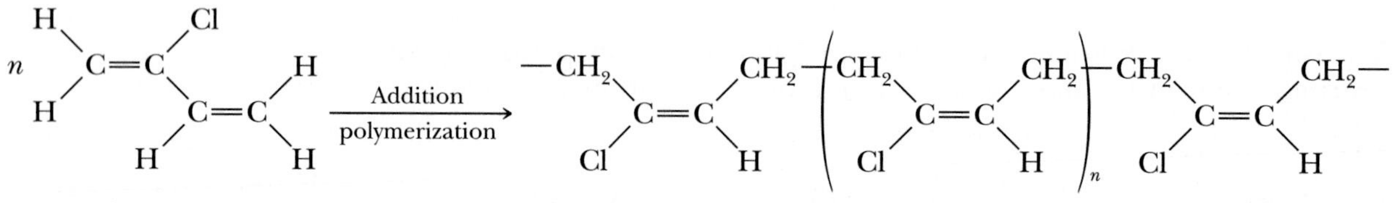

2-chlorobutadiene Neoprene

which has a chlorine atom substituted for the methyl group in isoprene. Neoprene is used in the production of gaskets, garden hoses, and adhesives.

Copolymers

Many commercially important addition polymers are **copolymers,** polymers obtained by polymerizing a mixture of two or more monomers. A copolymer of styrene with butadiene is the most important synthetic rubber produced in the United States. More than 1.4 million tons of styrene-butadiene rubber (SBR) is produced each year in the United States for making tires. A 3 : 1 mole ratio of butadiene to styrene is used to make SBR.

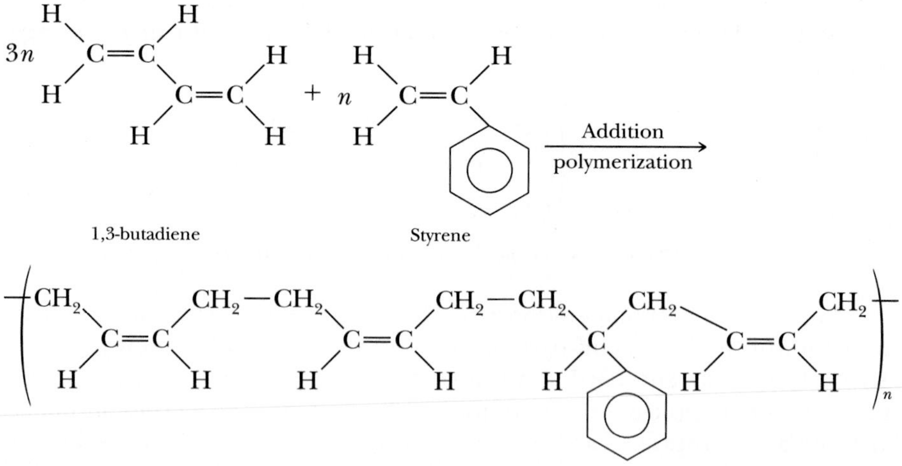

Styrene-butadiene rubber (SBR)

The World of Chemistry

Program 22, *The Age of Polymers*

Tires are made from synthetic rubber polymers.

Other important copolymers are made by polymerizing mixtures of ethylene and propylene or acrylonitrile, butadiene, and styrene. *Saran wrap* is an example of a copolymer of vinyl chloride and 1,1-dichloroethene.

SELF-TEST 14C

1. The individual molecules from which polymers are made are called ______________.
2. The initiation step of an addition polymerization uses ______________ ______________ such as ______________ ______________.
3. Monomers must have a ______________ bond in their structure if they are to participate in an addition reaction.
4. The double bond in ethylene is converted to a ______________ bond during an addition reaction.
5. The monomer in teflon is CF_2CF_2, 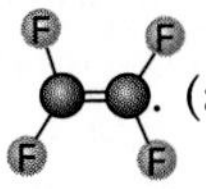

. (a) True, (b) False.
6. PVC is polyvinyl chloride. It is built from the monomer ______________.
7. Natural rubber is poly- ______________ -isoprene.
8. Stereoregulation catalysts are used to prepare synthetic ______________.
9. One of the first synthetic rubbers produced in the United States was ______________.
10. Three different types of polyethylene are LDPE, HDPE, and CLPE.
    (a) LDPE stands for ______________.
    (b) HDPE stands for ______________.
    (c) CLPE stands for ______________.

Condensation Polymers

A chemical reaction in which two molecules react by splitting out or eliminating a small molecule is called a **condensation reaction.** The reaction of alcohols with carboxylic acids to give esters (Section 14.5) is an example of condensation. This important type of chemical reaction does not depend on the presence of a double bond in the reacting molecules. Rather, it requires the presence of two different kinds of functional groups on two different molecules. If each reacting molecule has two functional groups, both of which can react, it is possible for condensation reactions to produce long-chain polymers.

Polyesters

A molecule with two carboxylic acid groups, such as terephthalic acid, and another molecule with two alcohol groups, such as ethylene glycol, can react with each other at both ends.

■ Terephthalic acid is the number 24 commercial chemical.

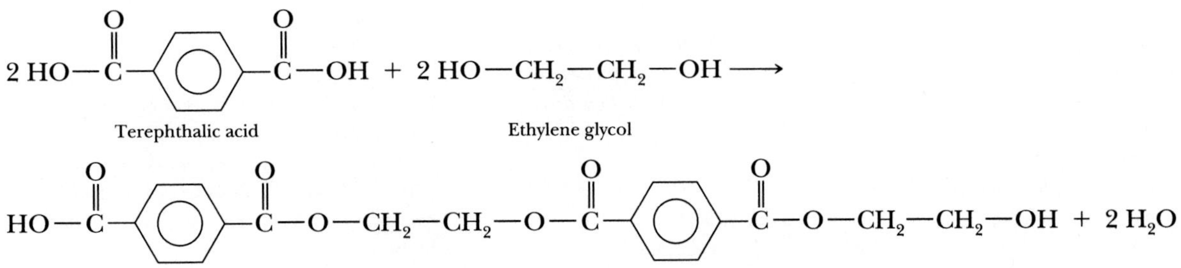

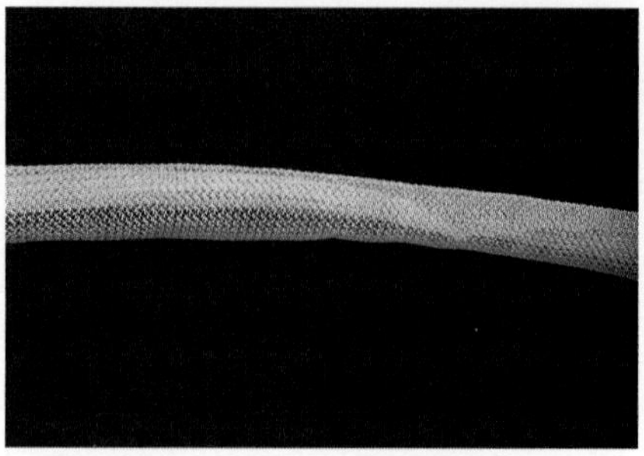

Figure 14.10 Dacron mesh used to replace damaged arteries. *(Dan McCoy/ Rainbow)*

If *n* molecules of acid and alcohol can react in this manner, the process will continue until a large polymer molecule, known as a **polyester,** is produced.

$$\left(\!\! \begin{array}{c} O \\ \| \\ C \end{array} \!\!-\!\! \bigcirc \!\!-\!\! \begin{array}{c} O \\ \| \\ C \end{array} \!\!-\! O \!-\! CH_2 \!-\! CH_2 \!-\! O \!\! \right)_{\!n}$$

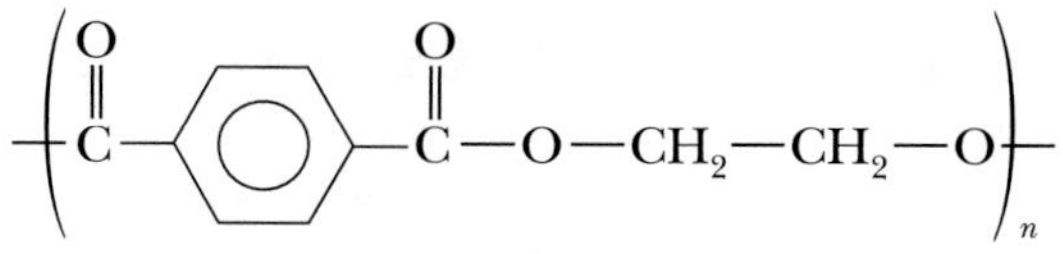

Poly(ethylene terephthalate)

More than 2 million tons of poly(ethylene terephthalate), commonly referred to as PET, are produced in the United States each year for use in beverage bottles, apparel, tire cord, film for photography, food packaging, coatings for microwave and conventional ovens, and home furnishings. Polyester textile fibers are marketed under such names as Dacron and Terylene. Films of the same polyester, when magnetically coated, are used to make audio and video tapes. This film, Mylar, has unusual strength and can be rolled into sheets one-thirtieth the thickness of a human hair.

The inert, nontoxic, nonallergenic, noninflammatory, and non-blood-clotting properties of Dacron polymers make Dacron tubing an excellent substitute for human blood vessels (Figure 14.10) in heart bypass operations. Dacron sheets are used as a skin substitute for burn victims.

THE WORLD OF CHEMISTRY

Inventor of the Poly(ethylene terephthalate) Bottle

Program 22, *The Age of Polymers*

The inventor of the poly(ethylene terephthalate) soft-drink bottle is Nathaniel Wyeth, who comes from the internationally famous family of artists. His brother, Andrew Wyeth, expresses his creativity on canvas, but Nat Wyeth expresses his through chemical engineering. He has an intriguing story:

I got to thinking about the work that Wallace Carothers did for DuPont way back in the days when nylon was born, where he found that, if you took a thread of nylon when it was cold—that is, below the melt point—and stretched it, it would orient itself. That is, the molecules of the polymer would align themselves. This is what you're doing to the molecules when you orient them—you're lining them up so they can give you the most strength. They're all pulling in the direction you want them to pull in.

Wyeth tried this approach to make a poly(ethylene terephthalate) bottle, but the bottles kept splitting. Wyeth estimates that he made 10,000 tries and had 10,000 failures before he made a simple observation.

Well, then I realized what we've got to do now is to align these molecules in the sidewall of the bottle, not only in one direction, but in two directions. So I thought I'd play a trick on this mold, on this problem. I took two pieces of poly(ethylene terephthalate) and turned one of them ninety degrees with the other. So then I had one that would split in this direction and one that would split in that direction. Well, one piece reinforced the other. As soon as I did that, I could blow bottles. That seems almost dirt simple. But as I've often said, quoting Einstein, the biggest part of a problem, and the easiest

Nathaniel Wyeth

way to solving a problem, is to understand it, have the problem in a form you can understand what's going on. And what I was doing here was learning about what was going on. Once I knew, it was simple to solve.

Polyamides (Nylons)

Another useful and important type of condensation reaction is that between a carboxylic acid and a primary **amine,** which is an organic compound containing an $—NH_2$ functional group. Amines can be considered derivatives of ammonia (NH_3), and most of them are weak bases, similar in strength to ammonia. An amine reacts with a carboxylic acid to split out a water molecule and form an **amide,** for example,

$$\underset{\text{Carboxylic acid}}{R—\overset{\overset{O}{\|}}{C}—OH} + \underset{\text{Amine}}{H—\underset{\underset{H}{|}}{N}—R} \xrightarrow{\text{heat}} \underset{\text{Amide}}{R—\overset{\overset{O}{\|}}{C}—\underset{\underset{H}{|}}{N}—R} + H_2O$$

■ Amines are classified as primary, secondary, or tertiary according to whether any of the H atoms in the $—NH_2$ group are substituted by alkyl groups, for example, RNH_2 (primary), R_2NH (secondary), and R_3N (tertiary).

■ Unsubstituted amide group:

$$—\overset{\overset{O}{\|}}{C}—NH_2 \text{ or } —CONH_2$$

Substituted amide group:

$$—CONHR \text{ or } —CONR_2$$

Polymers are produced when diamines react with dicarboxylic acids. Reactions of this type yield a group of polymers that perhaps have had a greater impact on society than any other type. These are the **polyamides,** or nylons.

In 1928, the DuPont Company embarked on a program of basic research headed by Dr. Wallace Carothers (1896–1937), who came to DuPont from the Harvard University faculty. His research interests were high-molecular-weight compounds, such as rubber, proteins, and resins, and the reaction mechanisms that produced these compounds. In February 1935 his research yielded a product known as nylon-66 (Figure 14.11), prepared from adipic acid (a diacid) and hexamethylenediamine (a diamine):

$$\underset{\text{Adipic acid}}{n\,HO—\overset{\overset{O}{\|}}{C}—(CH_2)_4—\overset{\overset{O}{\|}}{C}—OH} + \underset{\text{Hexamethylenediamine}}{n\,H_2N—(CH_2)_6—NH_2} \longrightarrow$$

$$\underset{\text{Nylon-66}}{\left(\underset{\underset{H}{|}}{N}—(CH_2)_6—\underset{\underset{H}{|}}{N}—\overset{\overset{O}{\|}}{C}—(CH_2)_4—\overset{\overset{O}{\|}}{C}\right)_n} + n\,H_2O$$

Figure 14.11 Making nylon. The diacid and the diamine do not mix with each other. The nylon polymer forms at the interface between the two reactants and continues to form there as the nylon "rope" is pulled away.
(C.D. Winters)

■ The name of nylon-66 is based on the number of carbon atoms in the diamine and diacid, respectively, that are used to make the polymer. Since both hexamethylenediamine and adipic acid have six carbon atoms, nylon-66 is the product.

■ In extrusion, a pliable substance is given a shape by being pushed through an opening. Toothpaste is extruded from a toothpaste tube.

■ The amide linkage in nylon is the same linkage found in proteins, where it is called the peptide linkage (Section 15.3).

■ Hair, wool, and silk are examples of nature's version of nylon. However, these natural polymers have only one carbon between each pair of $-\overset{\overset{O}{\|}}{C}-\underset{\underset{H}{|}}{N}-$ units instead of the half dozen or so found in synthetic nylons.

This material could easily be extruded into fibers that were stronger than natural fibers and chemically more inert. The discovery of nylon jolted the American textile industry at almost precisely the right time. Natural fibers were not meeting the needs of 20th-century Americans. Silk was not durable and was very expensive; wool was scratchy; linen crushed easily; and cotton did not have a high-fashion image. All four had to be pressed after cleaning. As women's hemlines rose in the mid-1930s, silk stockings were in great demand, but they were very expensive and short-lived. Nylon changed all that almost overnight. It could be knitted into the sheer hosiery women wanted, and it was much more durable than silk. The first public sale of nylon hose took place in Wilmington, Delaware (the location of DuPont's main office), on October 24, 1939. World War II caused all commercial use of nylon to be abandoned until 1945, as the industry turned to making parachutes and other war materials. Not until 1952 was the nylon industry able to meet the demands of the hosiery industry and to release nylon for other uses.

Figure 14.12 illustrates another facet of the structure of nylon—hydrogen bonding—which explains why nylons make such good fibers. To have good tensile strength, the chains of atoms in a polymer should be able to attract one another, but not so strongly that the plastic cannot be initially extended to form the fibers. Ordinary covalent chemical bonds linking the chains together would be too strong. Hydrogen bonds, with a strength about one-tenth that of an ordinary covalent bond, link the chains in the desired manner. Kevlar, another polyamide, is used to make bulletproof vests (Figure 14.13) and fireproof garments. Kevlar is made from *p*-phenylenediamine and terephthalic acid.

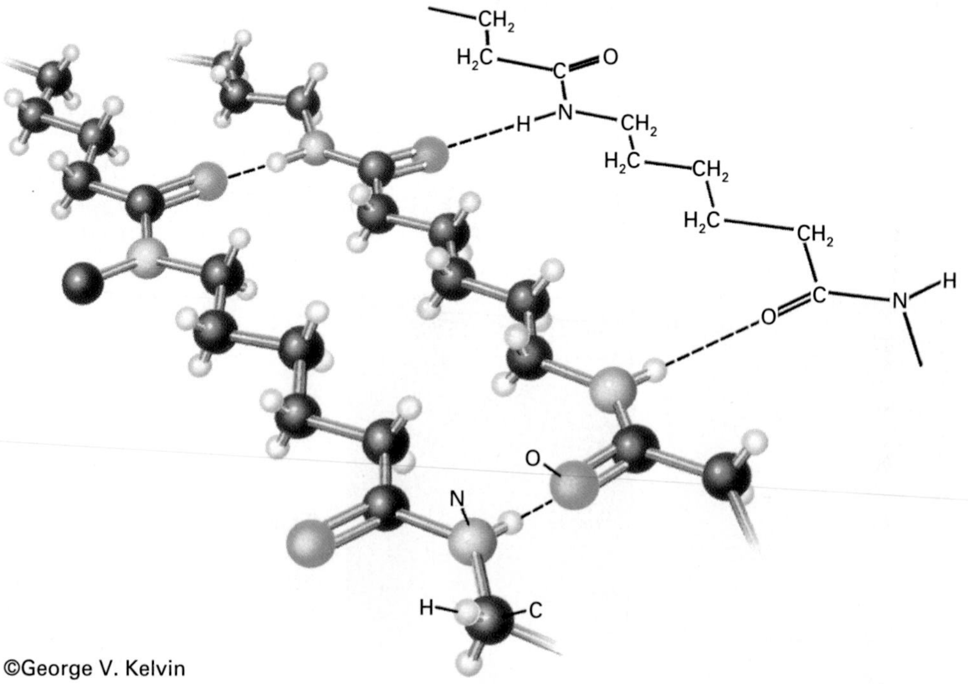

Figure 14.12 Hydrogen bonding in nylon-66.

©George V. Kelvin

THE PERSONAL SIDE

Stephanie Louise Kwolek (1923–)

(DuPont)

Stephanie Kwolek received a Bachelor of Science degree from Carnegie-Mellon University in 1946. Although she wanted to study medicine, she couldn't afford it and decided to take a temporary job at DuPont. She liked her work so well that she stayed for 40 years, retiring in 1986. During her career at DuPont, she received 17 U.S. patents and 86 foreign patents for her work on a variety of polymeric fibers. However, she is best known for her work on the development of Kevlar fiber, which is five times stronger than steel. The patent for Kevlar fiber was issued in 1965, but 15 years passed before it was fully commercialized. Although Kevlar is best known for its use in bulletproof vests, it has a number of other important uses that include brake linings, underwater cables, and high-performance composite materials. In 1994 DuPont featured Stephanie Kwolek in a television commercial about the use of Kevlar in bulletproof vests, which resulted in name recognition for both Kevlar and Kwolek by the American public. However, what isn't widely known are the accomplishments of a woman during a time when women often didn't receive appropriate recognition for their work. She has received many awards, including an honorary Doctor of Science Degree from Worcester Polytechnic Institute in 1981 for her contributions to polymer and fiber chemistry, the American Chemical Society Award for Creative Invention in 1980, and election to the Engineering and Science Hall of Fame in Dayton, Ohio.

Figure 14.13 Vests made of Kevlar have saved many policemen's lives. (Kevlar® is a registered trademark of DuPont for its aramid fiber.)

Polycarbonates

The tough, clear **polycarbonates** constitute another important group of condensation plastics. Compact disks are made of polycarbonate (see chapter-opening photograph). Another type of polycarbonate, commonly called Lexan or Merlon, was first made in Germany in 1953. It is as "clear as glass" and nearly as tough as steel. A 1-inch sheet can stop a .38-caliber bullet fired from 12 feet away. Such unusual properties have resulted in Lexan's use in "bulletproof" windows and as visors in astronauts' space helmets.

The repeating unit of polycarbonates is:

$$\left(\mathrm{-R-O-\overset{\overset{\displaystyle O}{\|}}{C}-O-} \right)_n$$

The name *polycarbonate* comes from the linkage's similarity to an inorganic carbonate ion, CO_3^{2-}.

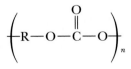

Plastics in action—a dramatic shot of Astronaut Kathryn C. Thornton taken as she worked outside the cargo bay on the orbiting space shuttle Endeavor during a 1992 mission. The shuttle is reflected in her Lexan helmut visor. *(NASA)*

Figure 14.14 Freshly made phenol-formaldehyde polymer (with a bit of pink dye added).

■ Two major categories of plastics are *thermoplastics,* which can be repeatedly softened by heating, molded into shape, and rehardened; and *thermosetting plastics,* which once heated, shaped, and hardened cannot be heated and shaped again. Nylon is a thermoplastic and Bakelite is a thermosetting plastic.

Formaldehyde Polymers

Formaldehyde is number 25 in chemical production in the United States, primarily because of its use in synthesizing a variety of condensation polymers. The first *thermosetting* plastic was the *phenol-formaldehyde* copolymer (Figure 14.14) synthesized by Leo Baekeland in 1909 and produced under the tradename Bakelite. More than 700,000 tons of phenol-formaldehyde resins are produced annually in the United States for use in making plywood adhesive, glass fiber resin, and molding compound for a variety of products such as distributor caps, radios, and buttons.

EXAMPLE 14.6 *Condensation Polymers*

Write the repeating unit of the condensation polymer obtained by combining $HOOCCH_2CH_2COOH$ and $H_2NCH_2CH_2NH_2$

SOLUTION

The dicarboxylic acid reacts with the diamine to split out water molecules and form a polyamide. The repeating unit is

$$\left(\overset{\displaystyle O}{\overset{\|}{C}}CH_2CH_2\overset{\displaystyle O}{\overset{\|}{C}}\underset{\underset{\displaystyle H}{|}}{N}CH_2CH_2\underset{\underset{\displaystyle H}{|}}{N} \right)_n$$

One of the first uses of Bakelite—78 rpm records. This fox trot recording was made in 1923 at Thomas Edison's Laboratories. *(C.D. Winters)*

Exercise 14.6

Draw the structure of the repeating unit of the condensation polymer obtained from reacting terephthalic acid with ethylene glycol.

Silicones

The element silicon, in the same chemical family as carbon, also forms many compounds with numerous Si—Si and Si—H bonds, analogous to C—C and C—H bonds. However, the Si—Si bonds and the Si—OH bonds react with both oxygen and water; hence, there are no useful silicon counterparts to most hydrocarbons. However, silicon does form stable bonds with carbon and especially oxygen, and this fact gives rise to an interesting group of condensation polymers containing silicon, oxygen, carbon, and hydrogen (bonded to carbon).

■ Silane (SiH_4) is structurally like methane (CH_4)—both are tetrahedral.

In 1945 at the General Electric Research Laboratory, E. G. Rochow discovered that a silicon-copper alloy reacts with organic chlorides to produce a whole new class of reactive compounds, the **organosilanes.**

$$\underset{\text{Methyl chloride}}{2\,CH_3Cl} + \underset{\text{Silicon-copper alloy}}{Si(Cu)} \longrightarrow \underset{\text{Dimethyldichlorosilane (a chlorosilane)}}{(CH_3)_2SiCl_2} + Cu$$

The chlorosilanes readily react with water and replace the chlorine atoms with hydroxyl (—OH) groups. The resulting molecule is similar to a dialcohol.

$$(CH_3)_2SiCl_2 + 2\,H_2O \longrightarrow (CH_3)_2Si(OH)_2 + 2\,HCl$$

Two dihydroxysilane molecules undergo a condensation reaction in which a water molecule is split out.

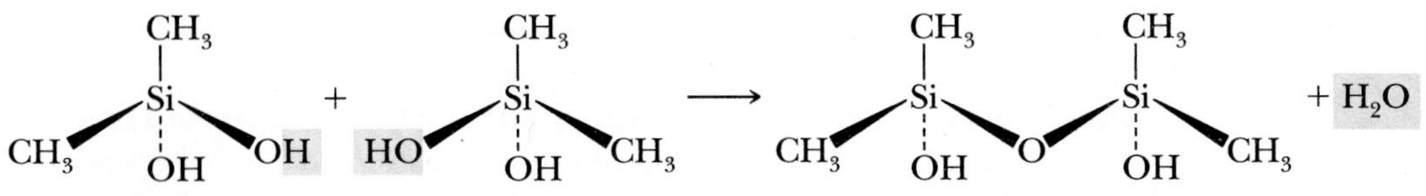

The resulting Si—O—Si linkage is very strong; the same linkage holds together all the natural silicate rocks and minerals. Continuation of this condensation process results in polymer molecules with molecular weights in the millions. Further reaction yields

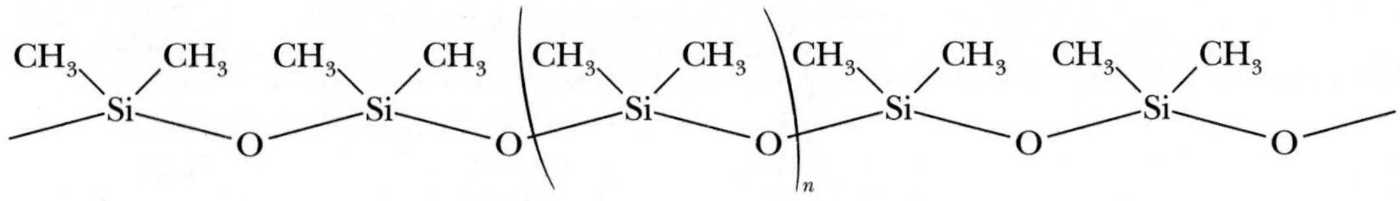

Polymers with different properties result from using different starting silanes. For example, two methyl groups on each silicon atom result in **silicone oils,** which are more stable at high temperatures than hydrocarbon oils and also have less tendency to thicken at low temperatures.

■ Silicon—the element (Si)
Silica—silicon dioxide (SiO_2), quartz
Silicone—polymer with an Si—O—Si backbone and organic groups

Silicone rubbers are very high-molecular-weight chains cross-linked by Si—O—Si bonds. Silicone rubbers that vulcanize at room temperature are commercially available; they contain groups that readily cross-link in the

Figure 14.15 Silicones have a wide variety of uses. *(C.D. Winters)*

presence of atmospheric moisture. The —OH groups are first produced, and then they condense in a cross-linking "cure" similar to the vulcanization of organic rubbers.

More than three million pounds of silicone rubber are produced each year in the United States. The uses include window gaskets, O-rings, insulation, sealants for buildings, space ships, and jet planes, and even some wearing apparel. The first footprints on the moon were made with silicone rubber boots, which readily withstood the extreme surface temperatures.

Silly Putty, a silicone widely distributed as a toy, is intermediate between silicone oils and silicone rubber. It is an interesting material. It has elastic properties on sudden deformation, but its elasticity is quickly overcome by its ability to flow like a liquid when allowed to stand (Figure 14.15).

14.6 New Polymer Materials

Few plastics produced today find end uses without some kind of modification. For example, body panels for the GM Saturn and Corvette automobiles are made of **reinforced plastics,** which contain fibers embedded in a matrix of a polymer. These are often referred to as **composites.** The strongest geometry for a solid is a wire or a fiber, and the use of a polymer matrix prevents the fiber from bending or buckling. As a result, reinforced plastics are stronger than steel. In addition, the composites have a low density—from 1.5 to 2.25 g/cm^3 compared with 3 g/cm^3 for aluminum, 7.9 g/cm^3 for steel, and 2.5 g/cm^3 for concrete. The only structural material with a lower specific gravity is wood, which has an average value of 0.5 g/cm^3. In addition, polymers do not corrode. The low specific gravity, high strength, and high chemical resistance of composites are the basis for their increased use in the automobile, airplane, construction, and sporting goods industries.

Glass fibers currently account for more than 90% of the fibrous material used in reinforced plastics because glass is inexpensive and glass fibers possess high strength, low density, good chemical resistance, and good insulating properties. In principle, any polymer can be used for the matrix material. Polyesters are the number-one polymer matrix at the present time. Glass-

DISCOVERY EXPERIMENT

Homemade Gluep

White school glue, such as Elmer's glue, contains polyvinyl acetate, water, and other ingredients. A type of "Silly Putty" or "gluep" can be made by mixing one cup of a 50/50 glue-water mixture with $\frac{1}{2}$ cup of liquid starch. Roll it together in your hands until it has a putty consistency. Roll it into a ball and let it sit on a flat surface undisturbed. What do you observe? Roll a piece into a ball and drop it on a hard surface. Does it bounce? The gluep can be stored in a zipper-type plastic bag for several weeks. Although gluep does not stick readily to clothes, wall, desks, or carpet, it does leave a water mark on wooden furniture, so be careful where you set it down. Mold will form on the gluep after a few weeks, but the addition of a few drops of Lysol to the gluep will retard formation of mold.

(a)

(b)

(c)

New polymer materials. (a) The flexible fender panel in a Saturn automobile. (b) An array of skis. (c) An F-16 fighter plane. *(a, Tom Pantages; b, Ann Kotowicz/Rainbow; c, Tom Hollyman/Photo Researchers)*

reinforced polyester composites have been used in structural applications such as boat hulls, airplanes, missile casings, and automobile body panels.

Other fibers and polymers have been used, and the trend is toward increased utilization of composites in automobiles and aircraft. For example, a composite of graphite fibers in a polymer matrix is used in the construction of

FRONTIERS IN THE WORLD OF CHEMISTRY

Elephants, Piano Keys, and Polymers

Until recently, no synthetic material has matched ivory in responding to the delicate touch of a concert pianist. Piano keys covered with plastic veneers have been rejected as too slippery and too cool. Eventually, however, a replacement for ivory must be found. In an effort to halt the slaughter of elephants, a global ban on trading in ivory was initiated in 1990 and is reported to be very effective.*

A team of scientists from Rensselaer Polytechnic Institute has patented a new material that may solve the problem. The essential step in their work was analysis of the surface of natural ivory at the microscopic level by a tribologist, an engineer who studies friction and materials that slide against each other. An ivory surface, they found, is covered with ridges, valleys, and tiny pores. When a sweaty finger slides over such a surface, it alternately sticks and slips, creating the feeling that pianists need for better control. The pores make an important contribution by absorbing sweat and oil from the finger.

To duplicate the ridges and valleys, the scientists worked with a finely made cast of a natural ivory surface. Duplicating the pores was challenging. Ultimately, they developed a synthetic ivory made from a mixture of a liquid polyester, a white titanium pigment, and finely powdered poly(ethylene glycol), a water-soluble polymer. When the material is soaked in hot water after it has hardened, the poly(ethylene glycol) is dissolved away, leaving behind pores like those in natural ivory. Their new material has met the ultimate test. Concert pianists have failed to detect a difference between Steinway pianos with keys covered by the new material, known as RPIvory, and those with keys of natural ivory.

Professor Henry Scarton of RPI with a piano key made of RPIvory. (Rensselaer Polytechnic Institute)

**The New York Times*, May 25, 1993, p. C3.

the Lear jet. Graphite/polymer composites are used in a number of sporting goods such as golf-club shafts, tennis racquets, fishing rods, and skis. The F-16 military aircraft was the first to contain graphite/polymer composite material, and the technology has advanced to the point where many aircraft, such as the F-18, use graphite composites for up to 26% of the aircraft's structural weight. This percentage is projected to increase to 40 to 50% in future aircraft.

Although few automobiles contain exterior body panels made of plastics, a number of components are plastic. Examples include bumpers, trim, light lenses, grilles, dashboards, seat covers, steering wheels—enough plastics to account for an average of 250 pounds per car. The increased emphasis on improving fuel efficiency will lead to the use of greater amounts of plastics in the construction of automobiles, both in interior components and exterior body panels.

14.7 Recycling Plastics

Recycling of metals such as aluminum, iron, and lead has been occurring for years (Section 11.5). However, programs for recycling plastics developed much more slowly because of the costs associated with separating different types of plastics and producing usable recycled products from the used plastics.

Disposal of plastics has been the subject of considerable debate in recent years as municipalities face increasing problems in locating sufficient landfill space. The number-one waste is paper products, which make up about 40% of the volume in landfills. (Newspaper alone accounts for 16% of the volume.) Next are plastics, which make up about 20% of the volume in landfills. At the present time, 1% of plastics waste is being recycled, as compared with recycling of 63% of aluminum cans, 20% of paper, and 10% of glass.

Four phases are needed for a successful recycling of any waste material: collection, sorting, reclamation, and end-use. Public enthusiasm for recycling and state laws requiring recycling have resulted in a dramatic increase in the collection of recycled items. About one-third of U.S. households have curbside collection of recyclables. Codes are stamped on plastic containers to help consumers identify and sort their recyclable plastics (Figure 14.16).

A planter made of plastic lumber, which can be worked like wood, but does not rot, splinter, need paint, or get eaten by termites. A mixture of several kinds of recycled plastics is the raw material for this lumber. *(Courtesy of Obex, Inc., Stamford, CT)*

Poly(ethylene terephthalate) (PET), widely used in soft-drink bottles, is the most commonly recycled plastic. The used bottles are available from retailers in states requiring refundable deposits or from curbside pickups. About 40% of the annual production of PET (4 billion bottles) is recycled. Major end-uses for recycled PET include fiberfill for ski jackets and sleeping bags, carpet fibers, and tennis balls. The Coca-Cola Company is using two-liter bottles made of 25% recycled PET.

High-density polyethylene (HDPE) is the second most widely recycled plastic (about 8% of annual production). Milk, juice, and water jugs are the principal source of recycled HDPE. Some products made from recycled HDPE are trash containers, drainage pipe, garbage bags, and fencing.

Although recycling of plastics has shown a dramatic increase in recent years, recycling companies will not be able to increase the percentage of recycled plastics to the 50% goal by the year 2000 without a significant increase in the demand for products made from recycled materials.

Code	Material	Percent of total bottles recycled
1 PETE	Polyethylene terephthalate (PET)*	20–30
2 HDPE	High-density polyethylene	50–60
3 V	Poly(vinyl chloride) (PVC)*	5–10
4 LDPE	Low-density polyethylene	5–10
5 PP	Polypropylene	5–10
6 PS	Polystyrene	5–10
7 OTHER	All other resins and layered multi-material	5–10

*Bottle codes are different from standard industrial identification to avoid confusion with registered trademarks.

Figure 14.16 Plastic container codes, used to identify types of plastic so that containers can be sorted for recycling.

SELF-TEST 14D

1. An example of the formation of a condensation polymer is the reaction of ____________ with ____________ to give ____________ with the elimination of ____________.
2. Nylon is an example of a ____________ polymer.
3. Polyamides are formed when ____________ is split out from the reaction of many organic groups and many amine groups.
4. Polyesters are formed by (a) addition reactions, (b) condensation reactions.
5. Silicone rubber polymers will cross-link if they are allowed to ____________.
6. Successful recycling involves the four phases of collection, sorting, reclamation and end-use. (a) True, (b) False.
7. ____________ fibers currently account for over 90% of the fibrous material used in reinforced plastics.
8. Graphite fibers in a ____________ matrix is a composite used to make skis, fishing rods, and aircraft components.

MATCHING SET

___ 1. Aldehyde
___ 2. Ketone
___ 3. Carboxylic acid
___ 4. Ester
___ 5. Monomer
___ 6. Present primary source of organic chemicals used as raw materials
___ 7. Likely future primary source of organic chemicals used as raw materials
___ 8. Poly-*cis*-isoprene
___ 9. Cross-linking via reaction with sulfur
___ 10. Polyester
___ 11. Nylon

a. Contains —OH
b. Polyamide
c. Vulcanization
d. Contains $R{-}\overset{O}{\overset{\|}{C}}{-}R'$
e. Contains $R{-}\overset{O}{\overset{\|}{C}}{-}H$
f. Contains $R{-}\overset{O}{\overset{\|}{C}}{-}OH$
g. Contains $R{-}\overset{O}{\overset{\|}{C}}{-}O{-}R'$
h. Petroleum
i. Formed from a dialcohol and a diacid
j. Coal
k. Natural rubber
l. Building block for polymer

QUESTIONS FOR REVIEW AND THOUGHT

1. Why are there more than 11 million organic compounds?
2. Use ethene and ethane as examples to illustrate the difference between addition and substitution reactions of hydrocarbons.
3. Identify each of the following molecules as an alcohol, ether, aldehyde, ketone, carboxylic acid, or ester:
   (a) $H\overset{O}{\overset{\|}{C}}CH_2CH_3$
   (b) CH_3CH_2OH
   (c) $CH_3\overset{O}{\overset{\|}{C}}CH_2CH_3$
   (d) $CH_3CH_2CH_2\overset{O}{\overset{\|}{C}}{-}OH$
   (e) $CH_3\overset{O}{\overset{\|}{C}}{-}OCH_2CH_3$
   (f) CH_3OCH_3
   (g) $CH_3\overset{O}{\overset{\|}{C}}H$
4. Write the equation for the formation of an ester from acetic acid (CH_3COOH) and ethanol (C_2H_5OH), and give its name.
5. Draw the structure of
   (a) 2-butanol (b) dimethyl ether
   (c) ethanal (d) trimethylamine
   (e) butanone (f) propanoic acid
   (g) ethanoic acid (h) 1,2-ethanediol
6. What ester is formed from the reaction of methanol and propanoic acid? Draw its structure and give its name.
7. Draw the structures of four alcohols that have the formula $C_4H_{10}O$.
8. Identify the class of each of the following compounds:
   (a) $CH_3CH_2CH_2CH_2COOH$
   (b) $CH_3CH{=}CHCH_2CH_3$
   (c) $C_2H_5OC_2H_5$
   (d) $CH_3CH_2CHOHCH_2CH_3$
   (e) $(CH_3)_3COH$
   (f) $CH_3\overset{O}{\overset{\|}{C}}H$
9. Name the compounds in Question 8.
10. Draw the condensed structure and give the name of
    (a) an amine. (b) an ether.
    (c) a carboxylic acid. (d) a ketone.
    (e) an alcohol. (f) an ester.
    (g) an aldehyde. (h) an amide.
11. What is meant by the following terms?
    (a) Proof rating of an alcohol
    (b) Denatured alcohol
12. What volume percent of ethanol does 90-proof gin contain?
13. Pure ethanol is what proof?
14. What is the difference in the structures of ethanol, ethylene glycol, and glycerol? Give one use of each alcohol.
15. Many naturally occurring carboxylic acids have more than one acid group (Table 14.8). What other functional group is often present?

16. How do the structures of primary, secondary, and tertiary alcohols differ?
17. Give examples of
   (a) two naturally occurring esters and where they are found.
   (b) two naturally occurring carboxylic acids and where they are found.
18. Explain how the liver detoxifies ethanol.
19. Summarize the Eastman Kodak process that produces acetic anhydride from coal.
20. In what ways is a railroad train like polystyrene?
21. What is the origin of the word "polymer"?
22. What is meant by the term "macromolecule"?
23. What do the following terms mean?
   (a) Monomer (b) Polymer
24. What structural features must a molecule have in order to undergo addition polymerization?
25. What do the following terms mean?
   (a) Polyethylene (b) HDPE
   (c) LDPE (d) CLPE
26. What is the repeating unit of natural rubber?
27. What is the difference in the structures of poly-trans-isoprene and poly-cis-isoprene? Sketch the arrangements for the two polymers. How do the two differ in their physical properties?
28. What is the role of sulfur in the vulcanization process?
29. What is the effect of vulcanization on the physical properties of natural rubber?
30. What property does a polymer have when it is extensively cross-linked?
31. Using polyethylene as an example, draw a portion of
   (a) a linear polymer.
   (b) a branched polymer.
   (c) a cross-linked polymer.
32. Explain how polymers could be prepared from each of the following compounds (other substances may be used):
   (a) $CH_3CH{=}CHCH_3$
   (b) $HOOCCH_2CH_2COOH$
   (c) $H_2NCH_2CH_2CH_2CH_2NH_2$
   (d) $(CH_3)_2Si(OH)_2$
33. Draw the structural formula of the monomer used to prepare the following polymers:
   (a) Poly(vinyl chloride) (b) Polystyrene
   (c) Polybutadiene (d) Polypropylene
34. Orlon has the polymeric chain structure shown below:

$$-CH_2-\underset{\displaystyle CN}{\underset{|}{CH}}-CH_2-\underset{\displaystyle CN}{\underset{|}{CH}}-CH_2-\underset{\displaystyle CN}{\underset{|}{CH}}-$$

   What is the monomer from which this structure can be made?
35. Indicate which of the following can undergo addition reaction and which cannot. Explain your choices.
   (a) Styrene, $C_6H_5CH{=}CH_2$, 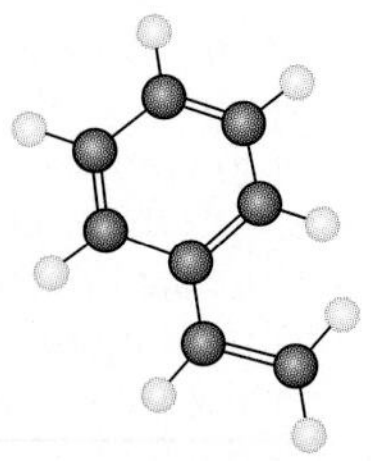
   (b) Propene, $CH_2{=}CHCH_3$, 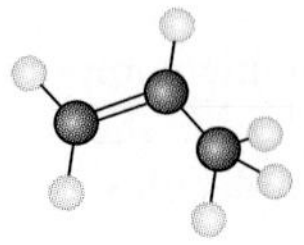
   (c) Ethane, CH_3CH_3, 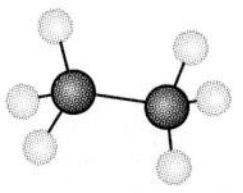
36. What is the monomer unit that yields the kind of addition polymer structure shown below?

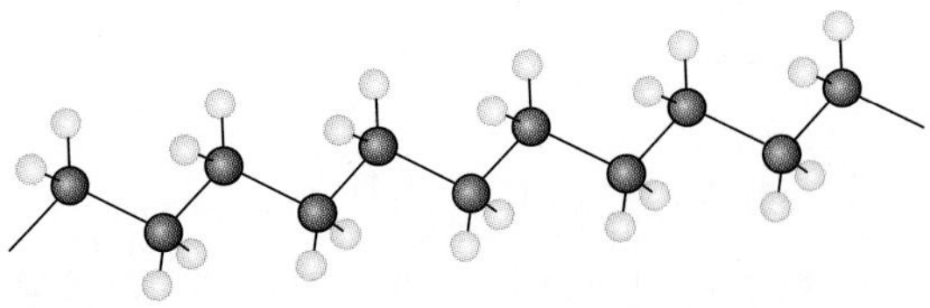

37. Give definitions for the following terms.
   (a) Polyester (b) Polyamide
   (c) Nylon-66 (d) Diacid
   (e) Diamine (f) Peptide linkage
38. What polymer is identified as PET?
39. What is an organosilane?
40. Give definitions for the following terms.
   (a) Silicone oils (b) Silicon rubber
41. What feature do all condensation polymerization reactions have in common?
42. What are the starting materials for nylon-66?
43. Which do you think is the source of most polymers used today, green plants or petroleum? Do you think this will ever change? Explain.
44. Do you think that plastic production will increase in the future? What advantages if any do you see when plastics are used in place of other materials?
45. What are the four phases involved in a successful recycling program? What happens if the public is cooperative in turning in recyclable materials but there isn't an adequate infrastructure to process the collected materials?
46. Germany recently established a recycling program. The public response was so good that the processing plants were unable to handle all of the plastics collected. The excess plastic was offered for export to neighboring countries. What could this export business do to the recycling programs in these other countries? Should this kind of export be allowed?

47. Composite materials are typically reinforced plastics. What properties do the embedded glass fibers or graphite fibers give to composites? Do you forsee any recycling problems with these materials?

48. What was the significance of Friedrich Wöhler's synthesis of urea from inorganic substances? How did his experiment alter the view that a "vital force" was responsible for the production of organic compounds?

49. Discuss what plastics are currently being recycled, and give examples of some products being made from these recycled plastics.

50. What are polymer composite materials? Give two examples that illustrate the importance of polymer composites in the manufacture of consumer products.

51. Many organic compounds have more than one type of functional group. Identify the functional groups in the following naturally occurring molecules.

(a)

$$
\begin{array}{c}
\text{CHO} \\
| \\
\text{H}-\text{C}-\text{OH} \\
| \\
\text{HO}-\text{C}-\text{H} \\
| \\
\text{H}-\text{C}-\text{OH} \\
| \\
\text{H}-\text{C}-\text{OH} \\
| \\
\text{CH}_2\text{OH}
\end{array}
$$

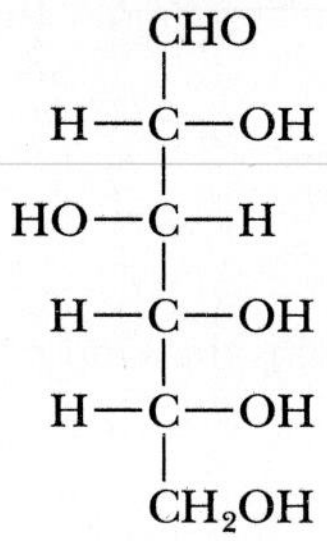

D-glucose

(b)

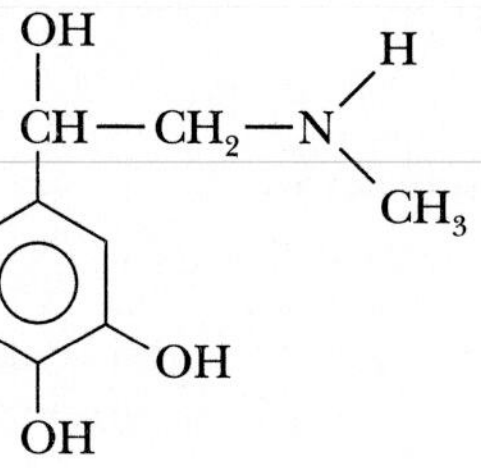

Epinephrine (adrenalin)

(c)

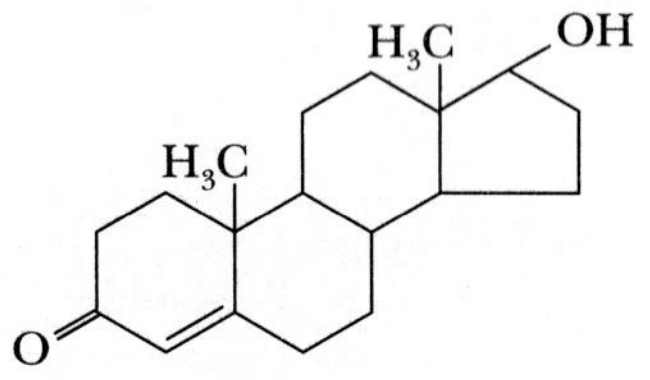

Testosterone

(d)

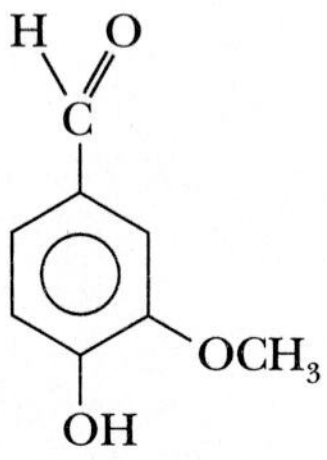

Vanillin
(vanilla bean)

(e)

$$
\begin{array}{c}
\text{CH}_3 \quad \text{O} \\
| \quad \quad \| \\
\text{HO}-\text{C}-\text{C}-\text{OH} \\
| \\
\text{H}
\end{array}
$$

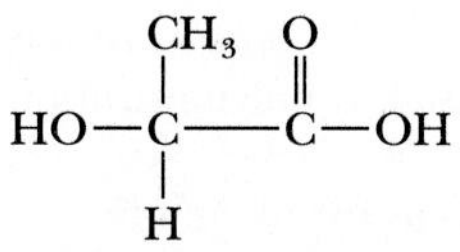

Lactic acid

■ PROBLEMS

1. How many ethylene monomer units, $CH_2=CH_2$, are linked together in a polyethylene molecule with a molar mass of 280,000?

2. How many propylene monomer units (propene,

$CH_2=CHCH_3,$) are linked together in a

polymer molecule with molar mass of 84,000?

3. What is the molar mass of a polymer molecule made up of 1500 chloroprene monomers ($CH_2=CClCH=CH_2,$

)?

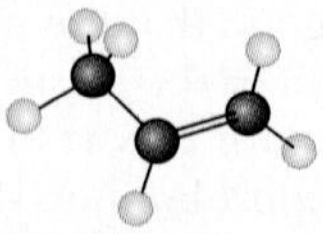

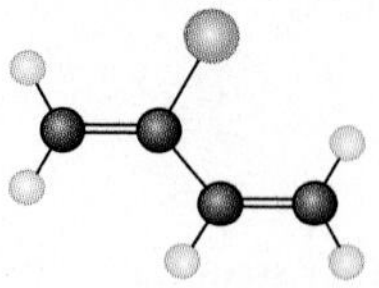

4. What is the polymer formed by the reaction of the following monomers?

(a) Ethylene glycol, $HOCH_2CH_2OH$, 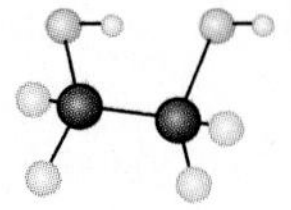

and terephthalic acid, 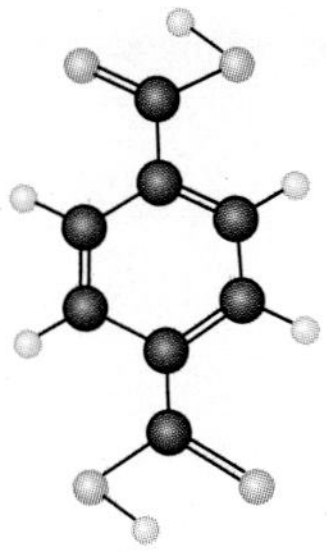

(b) Ethylene, $CH_2{=}CH_2$, 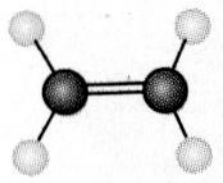

(c) Styrene, $C_6H_5CH{=}CH_2$, 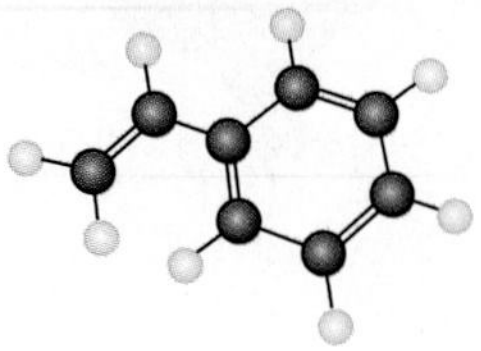

CHAPTER 15

The Chemistry of Life

Cooking, nutrition, personal health, drugs, medicine and dentistry, agriculture, our natural environment—biochemistry is fundamental to them all. Sometimes biochemistry is applied in a practical fashion, for example, when a cook scrambles an egg or someone grabs for an aspirin tablet to quiet a headache. Sometimes it is applied by a practicing professional or research scientist, for example, when an agriculture expert recommends the appropriate pesticide to a farmer or a pharmaceutical chemist designs a molecule to combat disease. In this chapter you will be introduced to the major classes of biochemicals and their functions in the body. Later chapters will relate this material to some of the applied aspects of biochemistry. Questions that will be addressed in this chapter include

- What are optical isomers, and why are they important in biochemical reactions?
- What are amino acids and proteins?
- What are the different types of sugars?
- What is the difference between starch and cellulose?
- What are triglycerides?
- What are the differences among saturated, monounsaturated, and polyunsaturated fatty acids?
- What are steroids?
- What do enzymes do?
- What are DNA and RNA, and how do they relate to the genetic code?
- What is biogenetic engineering?
- What is the human genome project?

Two people with identical DNA, the molecule that controls our biochemistry. *(Gunther/Explorer/Photo Researchers)*

The chemistry of life is referred to as **biochemistry,** and the organic chemicals found in living things are called **biochemicals.** As Wöhler's experiments demonstrated (see Section 14.1), biochemicals are not life-inherent; they are simply part of living systems.

A petrochemical manufacturing plant and the cells in your body are doing the same thing—taking in raw materials, using energy and carefully controlled conditions to perform chemical reactions, and putting out valuable products and waste materials. Many biochemicals are polymers. Starches are condensation polymers of simple sugars; proteins are condensation polymers of amino acids; and nucleic acids are condensation polymers of simple sugars, nitrogenous bases, and phosphoric acid species. All these very large biomolecules and other essential, smaller ones must be assembled from the raw materials in food, our equivalent of petroleum for the industrial plant. At the same time, burning food must provide energy. To carry out these functions simultaneously, living things have evolved an exquisite system for breaking down food molecules and putting them back together again. The overview of biochemistry in this chapter focuses on the molecular structure of the most important kinds of biomolecules. Many of them exhibit "handedness," another type of isomerism, so that's where we begin.

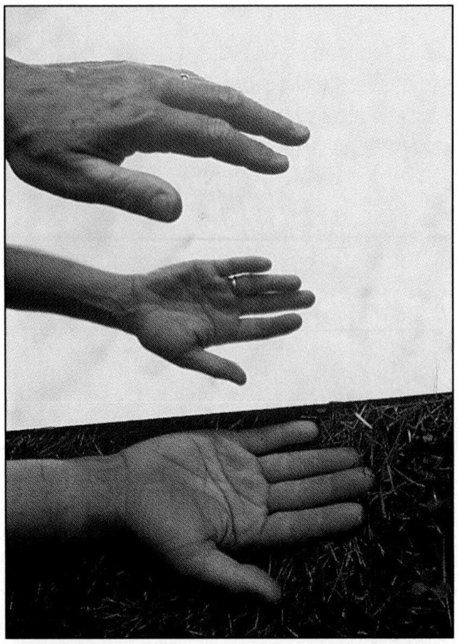

Figure 15.1 Mirror images. The mirror image of your left hand, as shown at the top, looks like your right hand, shown at the bottom. But if you place one hand over the other with palms up, they are not identical. This shows that they are nonsuperimposable mirror images. *(C.D. Winters)*

15.1 Handedness and Optical Isomerism

Are you right-handed or left-handed? Regardless of our preference, we learn at a very early age that a right-handed glove doesn't fit the left hand, and vice versa. Our hands are not identical; they are mirror images of one another and are not superimposable (Figure 15.1). *An object that cannot be superimposed on its mirror image is called **chiral.*** Objects that are superimposable on their mirror images are **achiral.** Stop and think about the extent to which chirality is a part of our everyday life. We've already discussed the chirality of your hands (and feet). Helical seashells are chiral, and most spiral to the right like a right-handed screw. Many vines show a chirality when they wind around trees.

■ Chiral is pronounced "ki-ral" and is derived from the Greek *cheir,* meaning "hand."

What is not as well known is that a large number of the molecules in plants and animals are chiral, and usually only one form of the chiral molecule (left-handed or right-handed) is found in nature. For example, all but one of the 20 naturally occurring amino acids are chiral, and only the left-handed amino acids are found in nature! Most of the natural sugars are right-handed.

A chiral seashell with a righthand spiral. *(C.D. Winters)*

■ Enantiomers have the same set of atoms connected by the same set of chemical bonds, but the atoms have a different orientation in space—just like your left and right hands do.

A chiral molecule and its nonsuperimposable mirror image are called **enantiomers.** Enantiomers are two different molecules, just as your left hand and right hand are different. To have enantiomers a molecular structure must be **asymmetrical** (without symmetry). The simplest case is a tetrahedral carbon atom bonded to four *different* atoms or groups of atoms. Such a carbon atom is asymmetric and is said to be chiral. A molecule that contains a chiral atom is likely to be a chiral molecule.

Some compounds are found in nature in both enantiomeric forms. For example, both forms of lactic acid are found in nature. During the contraction of muscles the body produces only one enantiomer of lactic acid; the other enantiomer is produced when milk sours. Let's look at the structure of lactic acid to see why two different isomers or chiral molecules might be possible. The central carbon atom of lactic acid has four different groups bonded to it: $-CH_3$, $-OH$, $-H$, and $-COOH$.

$$HO \overset{\overset{\displaystyle H}{|}}{\underset{\displaystyle CH_3}{C}} COOH$$

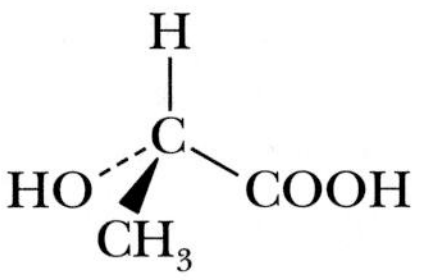

Lactic acid

As a result of the tetrahedral arrangement around the central carbon atom, it is possible to have two different arrangements of the four groups. If a lactic acid molecule is placed so the C—H bond is vertical, as illustrated in Figure 15.2, one possible arrangement of the remaining groups would be that where $-OH$, $-CH_3$, and $-COOH$ are attached in a clockwise manner (isomer I). Alternatively, these groups can be attached in a counterclockwise fashion (isomer II). To see further that the arrangements are different, we place

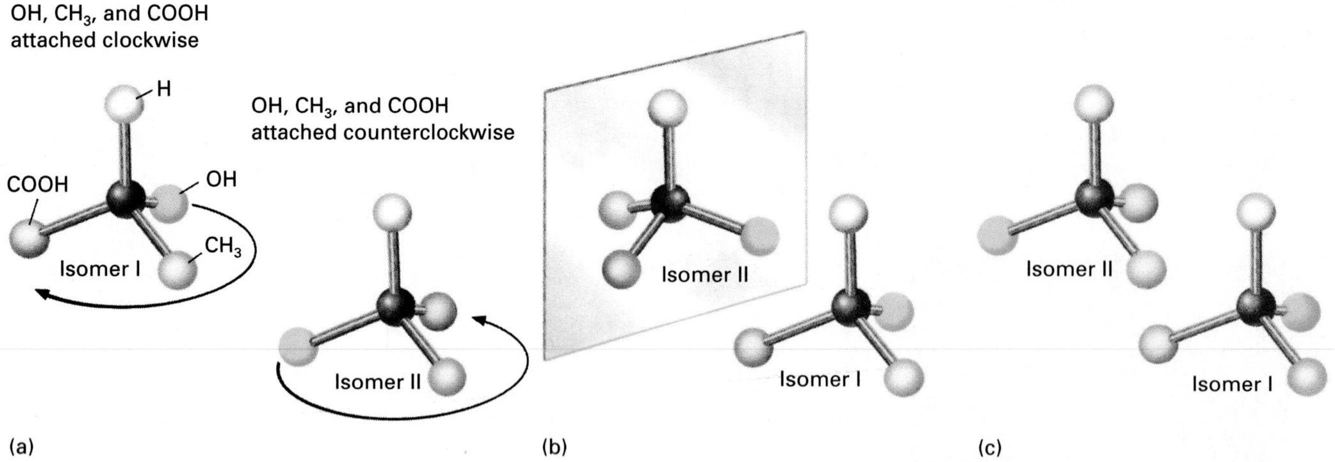

Figure 15.2 The enantiomers of lactic acid. (a) In isomer I, the $-OH$, $-CH_3$, and $-COOH$ groups are attached in clockwise order. In isomer II, the same groups are attached in counterclockwise order. (b) Isomers I and II are mirror images, a result of their having four different groups on the same carbon atom. (c) The two isomers cannot be turned around to make them superimposable.

isomer I in front of a mirror (Figure 15.2*b*). Now you see that isomer II is the mirror image of isomer I. What is important, however, is that these mirror-image molecules *cannot be superimposed* on one another. *These two nonsuperimposable, mirror-image chiral molecules are enantiomers.*

The "handedness" of enantiomers is sometimes represented by D for right-handed (D stands for "dextro," from the Latin *dexter* meaning "right") and L for left-handed (L stands for "levo," from the Latin *laevus* meaning "left"). In the case of lactic acid, the D-form is found in souring milk, and the L-form is found in muscle tissue, where it accumulates during vigorous exercise and can cause cramps.

Enantiomers of a chiral compound have the same melting point, the same boiling point, the same density, and many other identical physical and chemical properties. However, they always differ with respect to one physical property: They rotate a beam of **plane-polarized light** in opposite directions (Figure 15.3). For this reason, chiral molecules are sometimes referred to as **optical isomers** and are said to be ***optically active.*** A **polarimeter,** an instrument based

■ Plane-polarized light consists of electromagnetic waves vibrating in one direction.

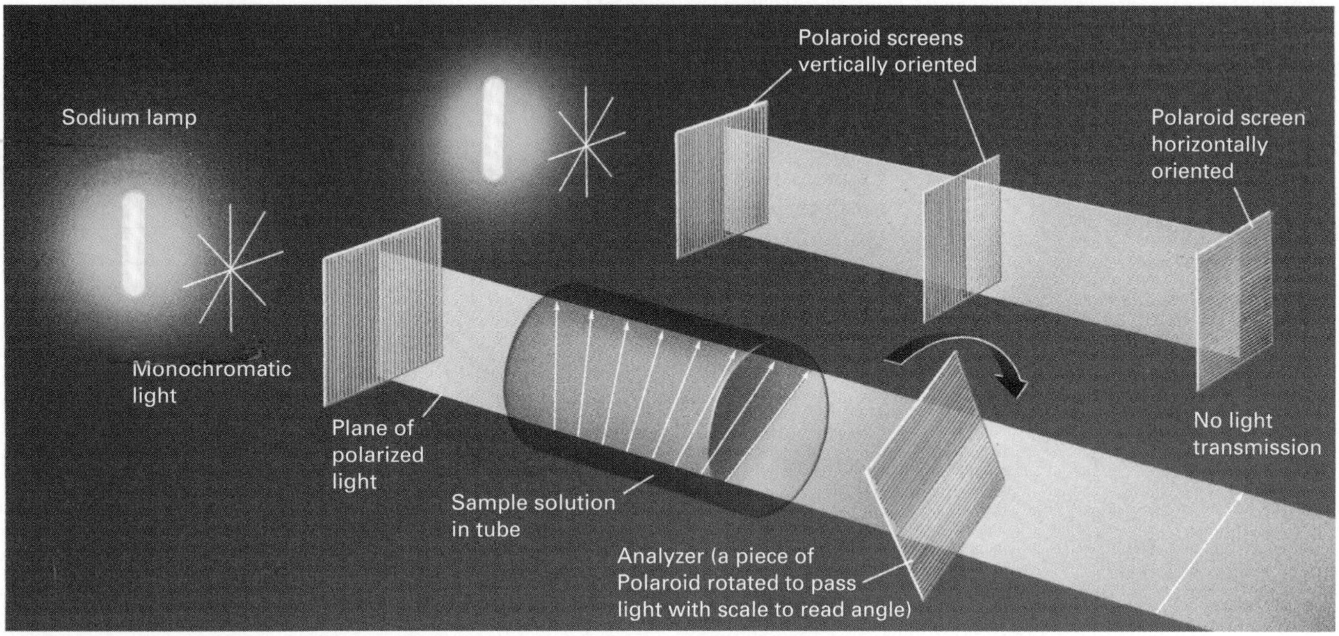

Figure 15.3 Plane-polarized light and its rotation by an optical isomer. (Top) Light of a single wavelength (monochromatic light) passes through a polarizing filter, which allows only light waves vibrating in the same direction to continue on. This plane-polarized light can then pass through a parallel polarizing filter but is blocked by a polarizing filter perpendicular to the first one. (Bottom) A polarimeter for measuring the rotation of light by an optical isomer works on the principle illustrated here. The sample is placed in the path of the plane-polarized light. The extent to which it rotates the plane of the light is determined by rotating a second polarizing filter until the maximum amount of light passes through. The size and direction (counterclockwise or clockwise) of the rotation are distinctive physical properties of the isomer being tested.

■ Although "enantiomers" is now the preferred term for the left-handed and right-handed forms of a chiral molecule, the term "optical isomers" is still used in many references.

on the design shown in Figure 15.3, is used to identify enantiomers, since the direction and angle of rotation are unique properties of each enantiomer.

In 1811 Jean Baptiste Biot (1774–1862) discovered optical activity when he observed that a quartz crystal rotates plane-polarized light. Three years later he found that solutions of sugar also rotate plane-polarized light. Louis Pasteur was the first person to relate this property to the spatial arrangement of atoms in molecules. Pasteur suggested that optical activity is caused by an asymmetric arrangement of atoms in the individual molecule, and the molecules that rotate the plane of polarized light to the right and those that rotate it to the left are mirror images. This was a remarkable deduction, since the tetrahedral nature of carbon was not known at that time, and so he was not able to explain the actual spatial arrangement of atoms. It was not until 1874 that Joseph Achille LeBel (1847–1930) and Jacobus Henricus van't Hoff (1852–1911) independently demonstrated that an arrangement of four different atoms or groups at the corners of a regular tetrahedron would produce two structures, one of which is the mirror image of the other.

THE PERSONAL SIDE

Louis Pasteur (1822–1895)

In 1848, at the age of 25, while Pasteur was experimenting with crystals, he saw something that no one had noticed before—a batch of crystals of optically inactive sodium ammonium tartrate (a salt of tartaric acid, an organic acid) actually was a mixture of two different kinds of crystals, which were mirror images of each other. While looking at the tiny crystals through a hand lens, Pasteur used tweezers to separate the two different kinds of crystals into two piles. When he prepared a solution of the crystals from each pile, he found that the solutions rotated plane-polarized light in opposite directions. His proposal that the molecules making up the crystals were mirror images of each other earned him wide recognition and laid the groundwork for the field of **stereochemistry**—the study of the spatial arrangement of atoms in molecules.

Louis Pasteur in his laboratory, conducting experiments on the fermentation process responsible for the souring of wine and the production of beer. (Jean-Loup Charmet/SPL/Photo Researchers)

Many years later, Pasteur discovered a biological method of separating optical isomers when he observed that a penicillin mold selectively destroyed the right-handed tartaric acid but would not touch the left-handed tartaric acid. This was the first evidence of the handedness of most organic molecules associated with living organisms. Pasteur's continuing studies of microorganisms led to the work for which he is most famous—development of the sterilization method known as *pasteurization* and of vaccines against chicken cholera, anthrax, and rabies.

Enantiomers also differ with respect to their biological properties. They react at different rates in a chiral environment, and many of the molecules in plants and animals are chiral. To understand why this difference in activity might exist, think of the hand-in-glove analogy. Although you can put a right-handed glove on your left hand, it takes longer to put it on and it doesn't fit very well. Since nature has a preference for L-isomers of amino acids and since enzymes, the catalysts for biochemical reactions, are proteins made from L-amino acids, enzymes are chiral molecules. The catalytic activity of enzymes is dependent on their three-dimensional structure, which in turn depends on their L-amino acid sequence. As a result, enzymes have a binding preference for one enantiomer of a reactant.

Even though nature has a preference for one enantiomer, laboratory synthesis of a chiral compound gives a mixture of equal amounts of the enantiomers, which is called a **racemic mixture.** The separation and purification of enantiomers is difficult because of the similarity in their physical properties. Although the first separation of enantiomers by Pasteur was based on his observation of different kinds of crystals in a racemic mixture, this method is rarely an option, since racemic modifications seldom form mixtures of crystals recognizable as mirror images. In fact, even sodium ammonium tartrate does not give two different types of crystals unless it crystallizes at a temperature below 28°C. Thus, a contributing factor to Pasteur's discovery was the cool Parisian climate!

■ Since enantiomers rotate the plane of polarized light an equal amount but in opposite directions, a solution of a racemic mixture will not rotate the plane of polarized light.

The usual method of separating enantiomers is to react them with optically active reagents that have a greater affinity for one enantiomer than the other. Pasteur's discovery of a mold that would selectively destroy one enantiomer of tartaric acid was the first example of this approach.

Nature's preference for one form of optically active amino acids has provoked much discussion and speculation among scientists since Pasteur's discovery of optical activity in 1848. However, there is still no explanation for this preference.

Large organic molecules may have a number of chiral carbon atoms within the same molecule. At each such carbon atom there exists the possibil-

THE WORLD OF CHEMISTRY

Molecular Architecture

Program 9, *Molecular Architecture*

There are various types of isomerism common to organic chemistry. Of these, optical isomers are among the most fascinating. This is because they play such important roles in life processes. In living things, chiral molecules exist in only one form or the other, but not both. How did the selection of one optical isomer over the other occur in nature? This question is addressed by Nobel laureate Christian Anfinsen, who says

How this selection began in nature is anybody's guess. One assumption is that some naturally occurring minerals, for example, might have been involved in binding one form and not the other. In the process a concentration of the form we have now was built up so that when life started, it was stuck with that form. In nature we're stuck pretty much with one isomer. The world has become so evolved that living things are in general composed of one of the two possible mirror images of the basic compounds.

Christian Anfinsen.

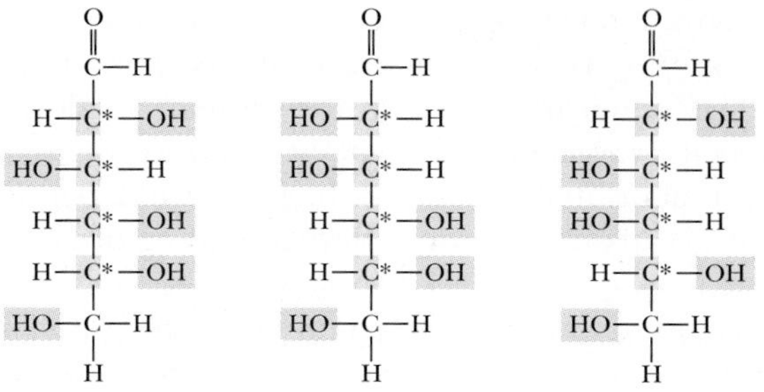

D-glucose D-mannose D-galactose
(C^* = asymmetric carbon atom)

Figure 15.4 Three of the 16 possible isomers for simple sugars with the formula $C_6H_{12}O_6$, which have four asymmetric carbon atoms. D-Glucose is the principal source of chemical energy in human metabolism; D-Mannose is found in plants; and D-galactose is found in milk.

ity of two arrangements of the molecule. The total number of possible molecules, then, increases exponentially with the number of chiral centers. With two asymmetric carbon atoms there are 2^2, or four, possible structures. It should be emphasized that each of the four isomers is composed of the same set of atoms connected by the same set of chemical bonds. Glucose, a simple sugar also known as dextrose or blood sugar, contains four asymmetric carbon atoms per molecule. All of the 16 (2^4) possible isomers are known, but only the three shown in Figure 15.4, D-glucose, D-mannose, and D-galactose, are important (see Section 15.5). Of these, D-glucose is by far the most common and most important.

■ Diabetes and other problems associated with abnormal concentrations of D-glucose in the blood are discussed in Section 15.5.

DISCOVERY EXPERIMENT

Homemade Polarimeter

Optical isomers rotate a plane of polarized light in opposite directions. A device that measures this rotation is called a polarimeter. A polarimeter for demonstration purposes can be made from a 16-oz flat-sided jar (for example, a "Taster's Choice" coffee jar). You will need two polarized filters that have diameters at least as large as the width of the jar. These can be pieces of Polaroid film or polarized lenses from old sunglasses. Tape one piece of polarized filter to the side of the jar. Darken the room and either mount a flashlight or have someone hold it against the taped piece of polarized filter. Rotate the other piece of polarized filter on the opposite face of the jar until a minimal amount of light comes through. Then tape a flat wooden stick such as a tongue depressor to the edge of the untaped polarized filter so that the stick points upward.

Position the flashlight so that its beam of light will shine through the taped piece of polarized filter and turn the flashlight on. Rotate the other piece of polarized filter on the opposite face until a minimum of light gets through. The stick should be pointing upward. Now add water to a level above the taped polarized filter. Rotate the untaped polarized filter until a minimal amount of light gets through. The stick should be pointing upward again because water molecules do not have "handedness" or optical isomers.

Now make enough sucrose solution to fill the jar to the same level as before. The solution needs to be fairly concentrated; 1 gram of sucrose per milliliter of water will work. Empty the water out of the jar and replace it with the sucrose solution. Rotate the untaped polarized filter until a minimal amount of light gets through. What is the position of the wooden stick? Is it to the right or left? The amount of rotation and the direction of rotation are specific physical properties of optical isomers.

References Hambly, G.F. *J. Chem. Educ.*, Vol. 65, p. 623, 1988.
Silversmith, E.F. *J. Chem. Educ.*, Vol. 65, p. 70, 1988.

FRONTIERS IN THE WORLD OF CHEMISTRY

Chiral Drugs

One enantiomer of a chiral drug is usually more active than the other, but 80% of chiral drugs are still sold as racemic mixtures. Only in those cases where one enantiomer is toxic or has harmful side effects is the single-enantiomer drug used. The primary reason for this is the large increase in cost required to isolate an enantiomer from a racemic mixture. Before long, though, many more drugs may be brought to market as the pure optical isomers. You may soon be seeing advertisements for a drug that is *optically pure* or *twice as effective as . . . [the racemic mixture, half of which is inactive]* or a *new and improved* version of a familiar over-the-counter medication. In recent years, techniques for separating enantiomers have improved greatly. In 1992 the U. S. Food and Drug Administration (FDA) released long-awaited guidelines for the marketing of chiral drugs. The decision about whether to sell a chiral drug in the racemic mixture or the enantiomerically pure form has been left to the drug's manufacturer (although the decision is subject to FDA approval). With the regulations finally in place, drug companies will be looking for situations in which production of a single enantiomer can give them a competitive edge. The racemic mixture, however, may remain the best choice in most cases. For example, ibuprofen, the pain reliever contained in Advil, Motrin, and Nuprin, is now sold as a racemic mixture. The left-handed enantiomer of ibuprofen is the active pain reliever; the right-handed isomer is inactive. Since D-ibuprofen is converted to L-ibuprofen in the body, there is probably no therapeutic advantage to the patient to switch from the racemic mixture to the more costly L-ibuprofen.

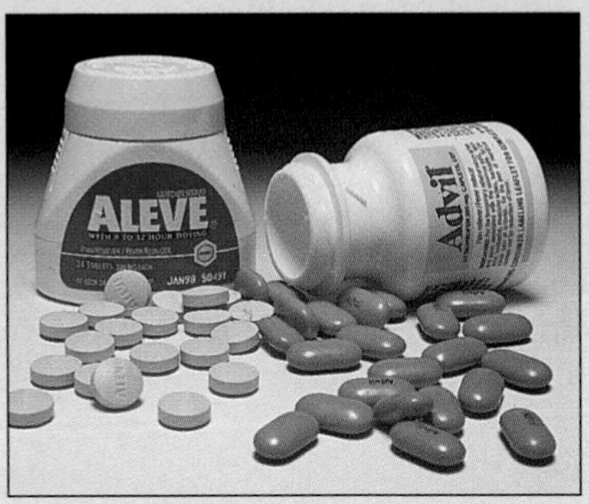

Ibuprofen (e.g., Advil) is sold as a racemic mixture and naproxen (e.g., Aleve) is sold as a single enantiomer. (C.D. Winters)

Naproxen is an example of a chiral drug sold as an enantiomer rather than the racemic mixture. In this case, one enantiomer is a pain-reliever and the other causes liver damage. Originally available only by prescription, naproxen is now available over the counter as a pain-killer and an anti-inflammatory (Aleve).

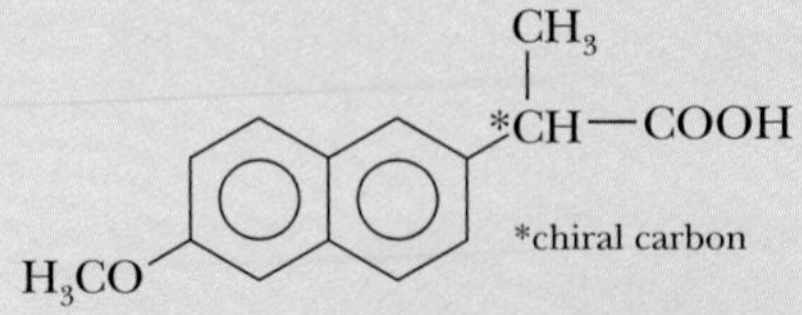

Naproxen

The sale of enantiomeric drugs is already big business. From 1990 to 1993, worldwide sales increased from $18 billion to $35 billion. (More information about classes of drugs is given in Chapter 19.)

World Sales of Enantiomeric Drugs in 1993

Drugs	Sales (in Billions)*
Cardiovascular	$11.3
Antibiotics	10.8
Hormones	4.5
Central nervous system	2.0
Anti-inflammatory	1.5
Anticancer	1.0
Other†	4.5
TOTAL	$35.6

*Wholesale finished dosage forms

†Includes vitamins, amino acids, immunosuppressants, peptide drugs, antiglaucoma drugs, and plant extracts

Reference *Chemical and Engineering News*, September 19, 1994, p. 38.

The chirality of carbohydrates, proteins, and DNA makes the human body highly sensitive to enantiomers. For example, one enantiomer of a drug is usually more active (or more toxic) than the other. A tragic example that called attention to the need for testing both enantiomers occurred in 1963, when horrible birth defects were induced by thalidomide (see Section 18.6). After this was discovered, it was determined that one enantiomer cured morning sickness and the other enantiomer caused birth defects. Often, the difference is one of activity or effectiveness with no difference in toxicity.

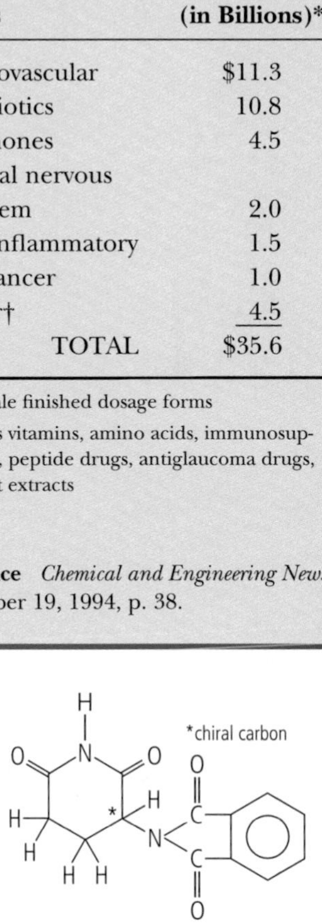

Thalidomide

■ "Metabolism" is a general term for the sum of all the chemical and physical processes in a living organism. When something is "metabolized" it is changed by these processes.

The World of Chemistry

Program 9, *Molecular Architecture*

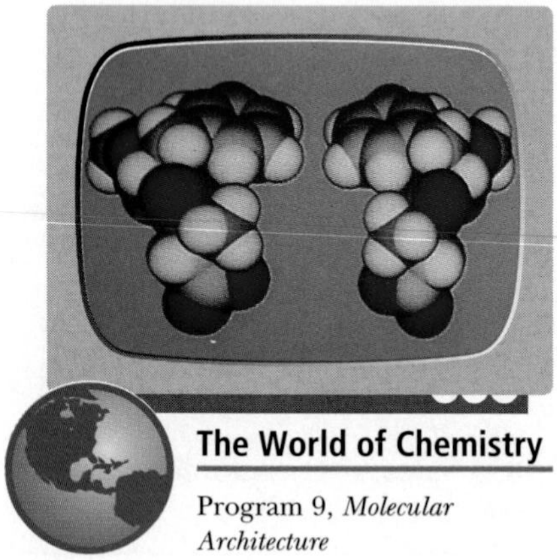

Models of the optical isomers of aspartame.

Aspartame (NutraSweet), widely used as an artificial sweetener, has two enantiomers. However, one enantiomer has a sweet taste, while the other enantiomer is bitter. This indicates that the receptor sites on our taste buds must be chiral, since they respond differently to the "handedness" of aspartame enantiomers! This becomes clearer when looking at the properties of the simple sugars. D-Glucose is sweet and nutritious, whereas L-glucose is tasteless and cannot be metabolized by the body.

EXAMPLE 15.1 *Chiral Molecules*

For each of the following molecules, decide whether the underlined carbon atom is or is not a chiral center. (a) $\underline{C}H_2Cl_2$ (b) $H_2N{-}\underline{C}H(CH_3){-}COOH$ (c) $Cl{-}\underline{C}H(OH){-}CH_2Cl$

SOLUTION

To be a chiral center, an atom must be bonded to four different groups. The underlined carbon atoms in molecules (b) and (c) meet this condition and are chiral centers. The underlined carbon in (a) is bonded to a pair of H atoms and a pair of Cl atoms and is therefore not chiral.

Exercise 15.1

Which of the following molecules is chiral? Draw the enantiomers for any chiral molecule.

(a)
$$\begin{array}{c} OH \\ | \\ C \\ \diagup \quad \diagdown \\ Cl \quad\quad CH_3 \\ | \\ Cl \end{array}$$

(b)
$$\begin{array}{c} OH \\ | \\ C \\ \diagup \quad \diagdown \\ H \quad\quad CH_3 \\ | \\ Cl \end{array}$$

(c)
$$\begin{array}{c} H \\ | \\ C \\ \diagup \quad \diagdown \\ H_2N \quad\quad COOH \\ | \\ H \end{array}$$

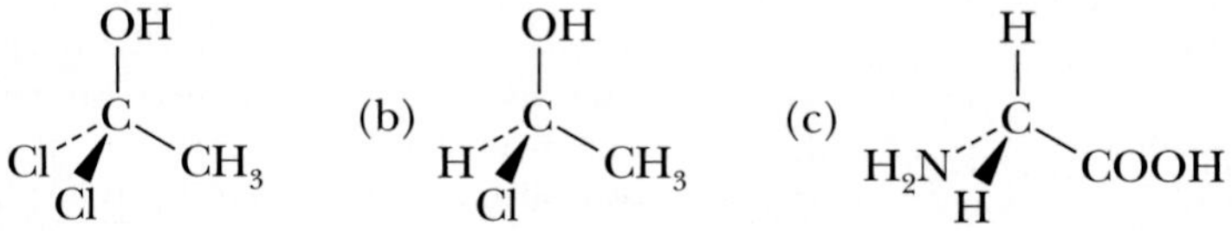

15.2 Amino Acids

All proteins are condensation polymers of **amino acids.** A large number of proteins exist in nature. For example, the human body is estimated to have 100,000 different proteins. What is amazing is that all of these proteins are derived from only 20 different amino acids (Table 15.1). Even more amazing is nature's preference for only the L-enantiomer of these amino acids. All but one of the 20 amino acids found in nature have the general formula

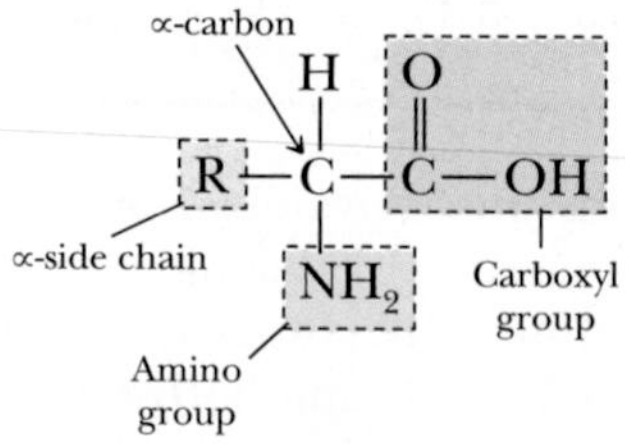

An amino acid

with an amino (—NH₂) group, a carboxylic acid group (—COOH), and an R group attached to the *alpha* carbon, the first carbon next to the —COOH. R is a characteristic group for each amino acid (see Table 15.1), and the α carbon is a chiral carbon atom, with one exception—glycine. R is a hydrogen atom in glycine, the simplest amino acid. Therefore, glycine is achiral and is the only naturally occurring amino acid that does not have enantiomers. The polarity of the R groups in amino acids affects the structure and function of proteins. The amino acids are grouped in Table 15.1 according to whether the R group is nonpolar, polar, acidic, or basic.

Most amino acids dissolve reasonably well in water. In every case, the following equilibrium exists in aqueous solution:

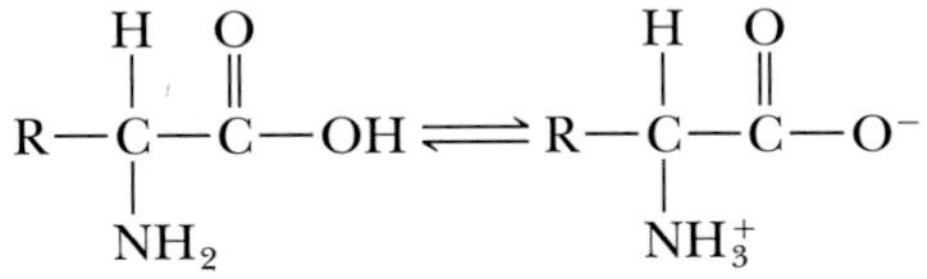

$$
\underset{\substack{|\\ NH_2}}{R-\overset{\overset{\displaystyle H}{|}}{C}-\overset{\overset{\displaystyle O}{\|}}{C}-OH} \rightleftharpoons \underset{\substack{|\\ NH_3^+ \\ \text{Zwitterion}}}{R-\overset{\overset{\displaystyle H}{|}}{C}-\overset{\overset{\displaystyle O}{\|}}{C}-O^-}
$$

Since the amino group is more basic than the carboxylate group, the proton of the —COOH group is donated to the —NH$_2$ group to form the internal salt known as a **zwitterion,** a term taken from the German word *zwitter* meaning "double." In the pure solid state and in aqueous solution near pH 7, amino acids exist almost completely as zwitterions. The importance of the zwitterion structure is that it explains why amino acids have properties typical of salts—high melting points and water solubility.

The **essential amino acids** must be ingested from food; they are indicated by asterisks in Table 15.1. A diet that includes meat, milk, eggs, or cheese provides all the essential amino acids. People who obtain their protein from a diet of grains and vegetables have to make sure all the essential amino acids are included. For example, rice and wheat are deficient in lysine, and corn is deficient in lysine and tryptophan. Peas are deficient in methionine, so a combination of peas and rice would provide the essential amino acids. The other amino acids can be synthesized by the human body.

■ D- and L- are used to identify the two enantiomers of α-amino acids. These symbols represent the spatial orientation of the enantiomer and do not necessarily correlate with the direction of rotation of plane-polarized light.

■ For good nutrition we require all the essential amino acids in our daily diet, but the amount required does not exceed 1.5 g per day for any of them.

EXAMPLE 15.2 *Amino Acids*

Draw the zwitterion structure of alanine by starting with the structure given in Table 15.1.

SOLUTION

The R group in alanine is CH$_3$— (Table 15.1), and in the zwitterion form a proton has been transferred to give —NH$_3^+$ and —COO$^-$ ions.

$$
\underset{\substack{|\\ NH_3^+}}{CH_3-\overset{\overset{\displaystyle H}{|}}{C}-\overset{\overset{\displaystyle O}{\|}}{C}-O^-}
$$

Exercise 15.2

Why do amino acids have the zwitterion structure? Draw the zwitterion structure of valine, beginning with the structure given in Table 15.1.

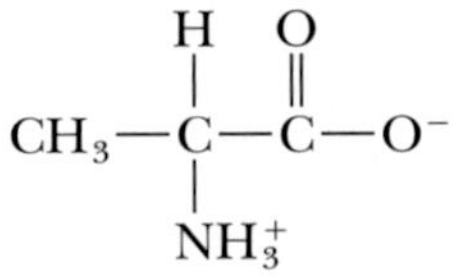

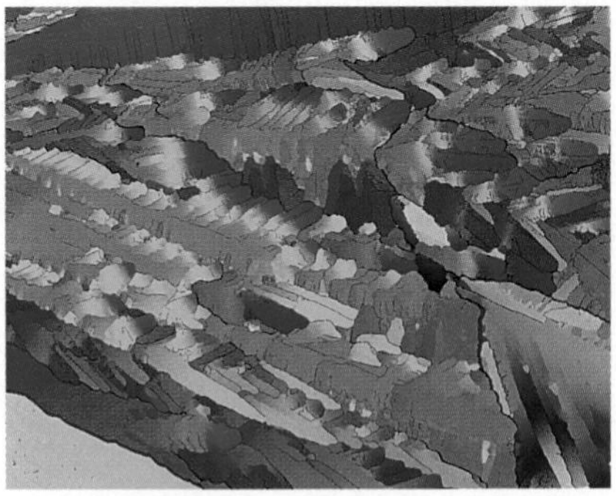

Crystalline phenylalanine, one of the essential amino acids, photographed under polarized light. *(Alfred Pasieka/ SPL/Photo Researchers)*

TABLE 15.1 ■ **Common L-Amino Acids Found in Proteins (The R Group in Each Amino Acid Is Highlighted)**

Amino Acid	Abbreviation	Structure
Nonpolar R Groups		
Glycine	Gly	$H-CH-COOH$ $\quad\quad\mid$ $\quad\quad NH_2$
Alanine	Ala	$CH_3-CH-COOH$ $\quad\quad\quad\mid$ $\quad\quad\quad NH_2$
*Valine	Val	$CH_3-CH-CH-COOH$ $\quad\quad\quad\mid\quad\quad\mid$ $\quad\quad\quad CH_3\quad NH_2$
*Leucine	Leu	$CH_3-CH-CH_2-CH-COOH$ $\quad\quad\quad\mid\quad\quad\quad\quad\mid$ $\quad\quad\quad CH_3\quad\quad\quad NH_2$
*Isoleucine	Ile	$CH_3-CH_2-CH-CH-COOH$ $\quad\quad\quad\quad\quad\mid\quad\mid$ $\quad\quad\quad\quad\quad CH_3\ NH_2$
Proline	Pro	$H_2C\!-\!\!-\!\!-\!CH_2$ $H_2C\quad\ \ CHCOOH$ $\quad\ \ N$ $\quad\ \ \mid$ $\quad\ \ H$
*Phenylalanine	Phe	$\bigcirc\!-CH_2-CH-COOH$ $\quad\quad\quad\quad\quad\mid$ $\quad\quad\quad\quad\quad NH_2$
*Methionine	Met	$CH_3-S-CH_2CH_2-CH-COOH$ $\quad\quad\quad\quad\quad\quad\quad\ \mid$ $\quad\quad\quad\quad\quad\quad\quad\ NH_2$
Polar but Neutral R Groups		
Serine	Ser	$HO-CH_2-CH-COOH$ $\quad\quad\quad\quad\mid$ $\quad\quad\quad\quad NH_2$
*Threonine	Thr	$CH_3-CH-CH-COOH$ $\quad\quad\quad\mid\quad\ \mid$ $\quad\quad\quad OH\ \ NH_2$
Cysteine	Cys	$HS-CH_2-CH-COOH$ $\quad\quad\quad\quad\mid$ $\quad\quad\quad\quad NH_2$

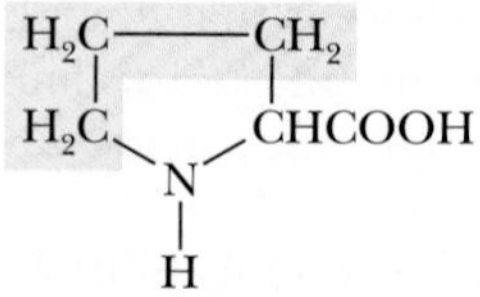

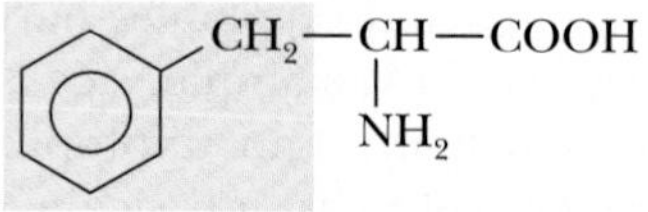

TABLE 15.1 ■ *Continued*

Amino Acid	Abbreviation	Structure
		Polar but Neutral R Groups (cont)
Tyrosine	Tyr	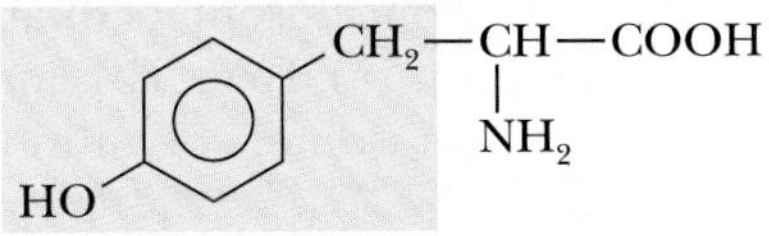
Asparagine	Asn	$H_2N-C(=O)-CH_2-CH(NH_2)-COOH$
Glutamine	Gln	$H_2N-C(=O)-CH_2CH_2-CH(NH_2)-COOH$
*Tryptophan	Trp	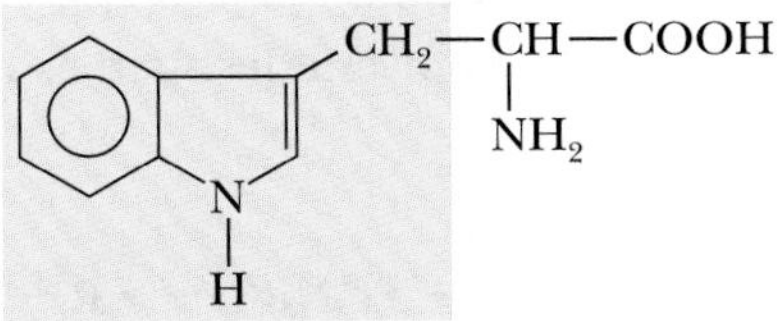
		Polar Acidic R Groups
Glutamic acid	Glu	$HO-C(=O)-CH_2CH_2-CH(NH_2)-COOH$
Aspartic acid	Asp	$HO-C(=O)-CH_2-CH(NH_2)-COOH$
		Polar Basic R Groups
*Lysine	Lys	$H_2N-CH_2CH_2CH_2CH_2-CH(NH_2)-COOH$
†Arginine	Arg	$H_2N-C(=NH)-NH-CH_2CH_2CH_2-CH(NH_2)-COOH$
Histidine	His	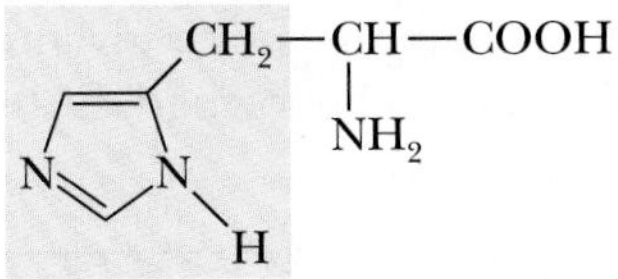

*Essential amino acids that must be part of the human diet. The other amino acids can be synthesized by the body.

†Growing children also require arginine in their diet.

SELF-TEST 15A

1. To have optical isomers in carbon compounds, a carbon atom must have ____________ different groups attached.
2. All amino acids except ____________ have optical isomers.
3. A racemic mixture of optical isomers of alanine will rotate a plane of polarized light to the right. (a) True, (b) False.
4. Only left-handed optical isomers of amino acids are found in your body. (a) True, (b) False.
5. Amino acids that the body cannot synthesize from other molecules are called ____________ amino acids.

15.3 Peptides and Proteins

How are amino acids polymerized to give proteins? The formation of an amide from the reaction of an amine and a carboxylic acid is described in Section 14.5 in the discussion of polyamides such as nylons.

$$R{-}\overset{\overset{\displaystyle O}{\|}}{C}{-}OH + H_2NR' \longrightarrow R{-}\overset{\overset{\displaystyle O}{\|}}{C}{-}\overset{\overset{\displaystyle H}{|}}{N}{-}R' + H_2O$$

Since amino acids have both an amine group and a carboxylic acid group, the $-COOH$ of one amino acid can combine with the $-NH_2$ of a second amino acid.

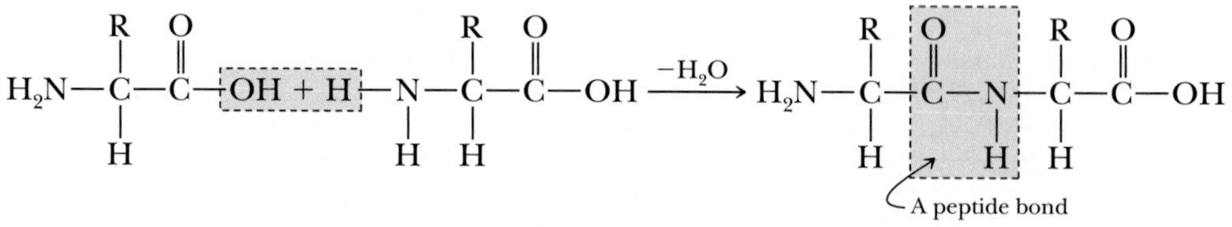

■ Names of peptides are written from left to right starting with the amino- or N-terminal end. The "-ine" ending of all amino acid residues (except the carboxyl or C-terminal end) is changed to "-yl." For example, Gly-Ala-Ser is the tripeptide glycylalanylserine.

In the condensation reaction, one molecule of water is eliminated between the carboxylic acid of one amino acid and the amine group of another. The result is a **peptide** bond (called an amide group in simpler molecules), and the molecule is a **dipeptide.** When two different amino acids are bonded, two different combinations are possible, depending on which amine reacts with which acid group. For example, when glycine and alanine react, both glycylalanine and alanylglycine can be formed. Either end of the dipeptide can then react with another amino acid.

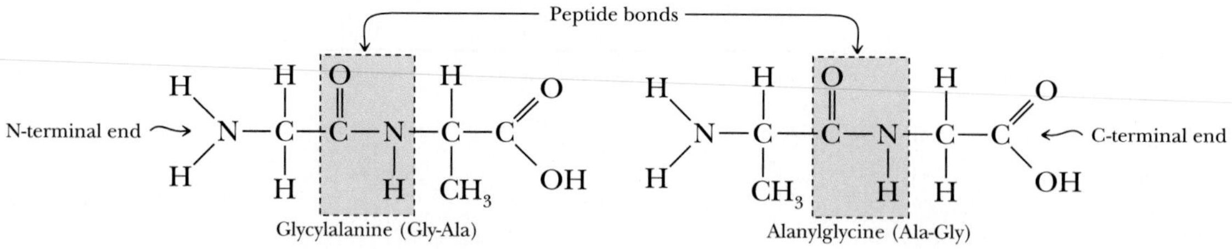

The amino acids that have combined to form the peptide molecule are called **amino acid residues.** Since each dipeptide has a $-COOH$ and an $-NH_2$

group, a tripeptide can be formed from each dipeptide by reaction at either end, and the polymerization process can continue until a large **polypeptide** chain is formed (Figure 15.5).

Proteins are polypeptides containing from fifty to thousands of amino acid residues, and they vary greatly in structure and composition. They can be divided into two classes: simple and conjugated. **Simple proteins** consist only of amino acids. For example, insulin, a hormone that is essential to controlling the concentration of glucose in the blood, is a simple protein that has only 51 amino acid residues in two linked chains (Figure 15.6). In contrast, chymotrypsin, an enzyme that aids in the digestion of proteins in our diet, contains

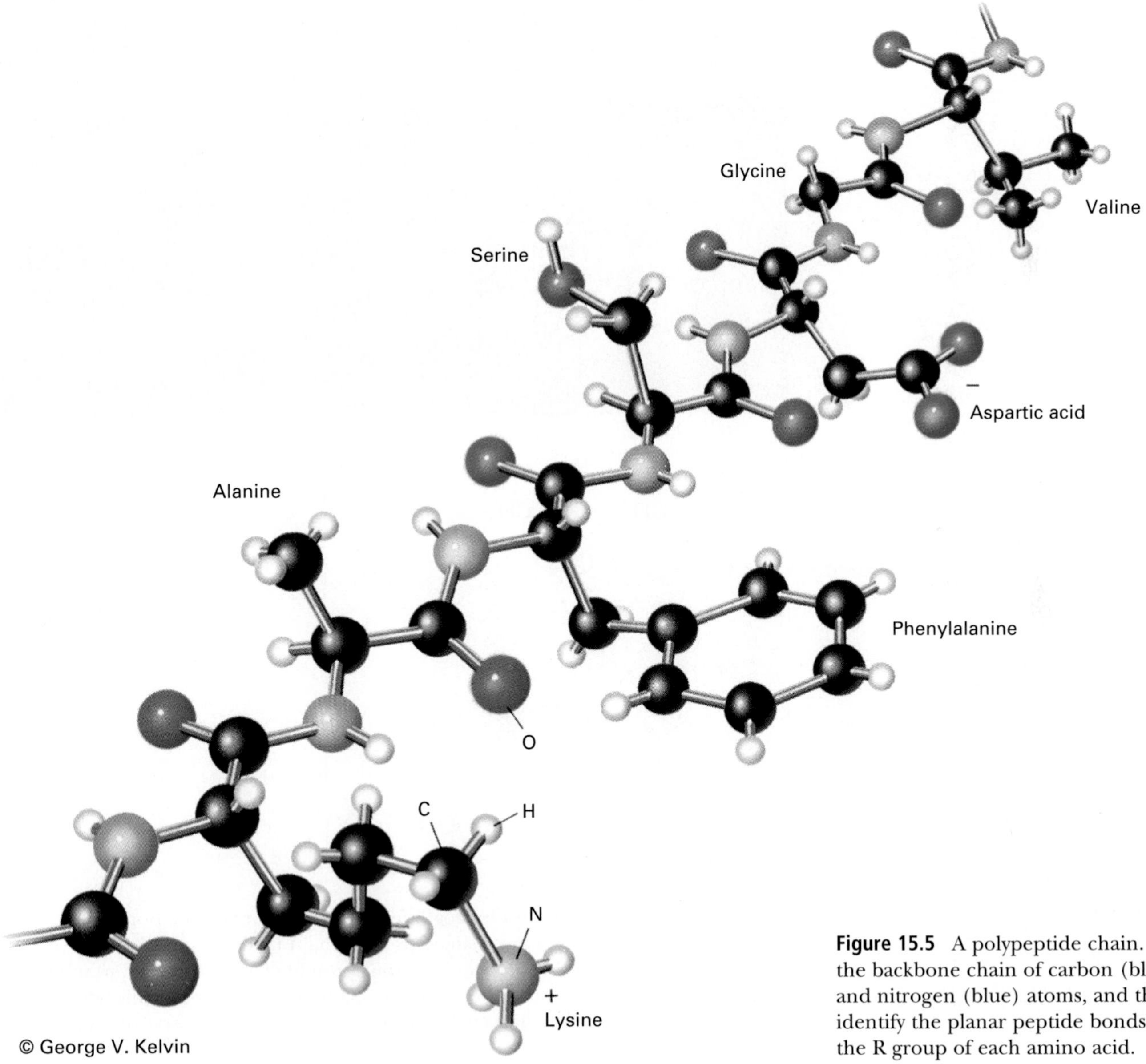

Figure 15.5 A polypeptide chain. Find the backbone chain of carbon (black) and nitrogen (blue) atoms, and then identify the planar peptide bonds and the R group of each amino acid.

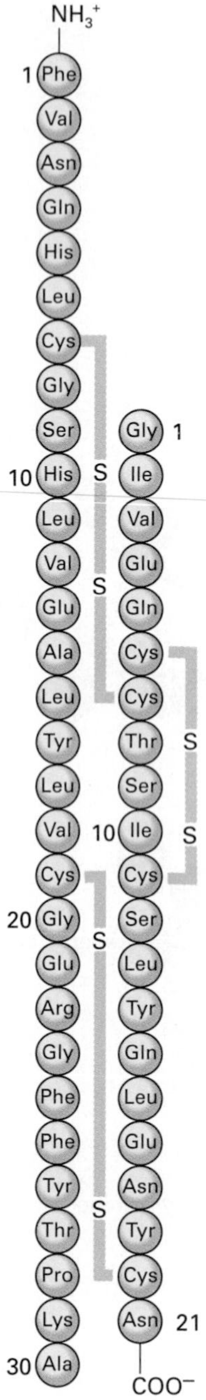

Figure 15.6 The amino acid sequence of insulin, which contains 51 amino acids in two chains linked by disulfide bonds between the side chains of cysteine residues.

245 amino acid residues. Considering all the possible ways that 20 α-amino acids could be put together to form proteins, it is remarkable that so few proteins are in fact known.

Conjugated proteins contain other kinds of groups in addition to amino acids. Examples of conjugated proteins are hemoglobin and myoglobin, which bind oxygen in blood and muscles, respectively. Myoglobin has a single protein chain (Figure 15.7*a*). Human hemoglobin contains four protein chains, two identical ones having 141 amino acid residues and the other two, also identical, having 146 (Figure 15.7*b*). The site at which oxygen connects to

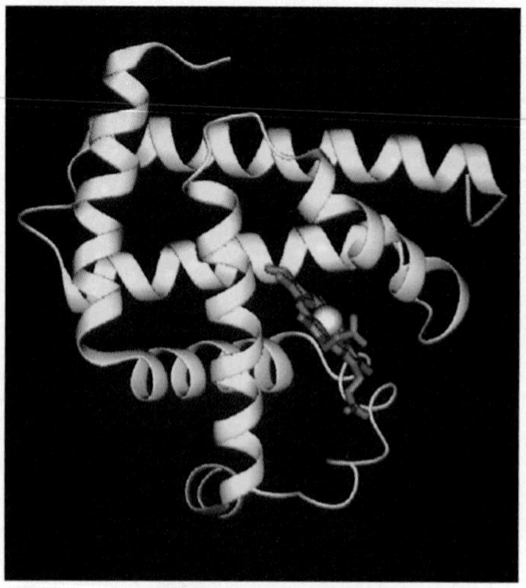
(a)

(b)

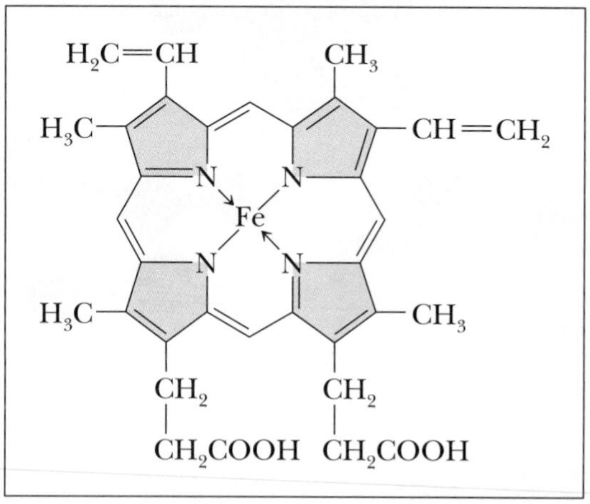

(c)

Figure 15.7 Myoglobin (a) and hemoglobin (b), which contain the non-protein heme group (c). Myoglobin has a single protein chain (yellow) and one heme group (Fe ion shown in white). Hemoglobin has four protein chains of two different kinds (yellow and purple), with one heme group in each protein chain. In this type of representation of proteins, the alpha-helix parts of the protein chain are shown as ribbon-like spirals.

myoglobin and hemoglobin is heme, an organic group that is the non–amino acid part of the molecules (Figure 15.7*c*). The oxygen binds with an iron ion (Fe^{2+}) in the center of the heme group.

To summarize, then, *proteins are polypeptides, which are condensation polymers of amino acids.* They vary greatly in size, and some proteins include non–amino acid groups.

15.4 Protein Structure and Function

The order of the amino acid residues in a peptide or protein molecule is called the **amino acid sequence** of that molecule. As the length of the chain increases, the number of variations in the sequence of amino acids quickly increases. Six tripeptides are possible if three amino acids (for example, glycine, Gly; alanine, Ala; and serine, Ser) are linked in all possible combinations:

Gly-Ala-Ser	Ser-Gly-Ala	Ala-Gly-Ser
Gly-Ser-Ala	Ser-Ala-Gly	Ala-Ser-Gly

If n amino acids are all different, the number of arrangements is $n!$. For four different amino acids, the number of different arrangements is 4!, or $4 \times 3 \times 2 \times 1 = 24$. For five different amino acids, the number of different arrangements is 5!, or 120. If all 20 different naturally occurring amino acids are bonded, the sequences alone make 2.43×10^{18} (2.43 quintillion) uniquely different 20-monomer molecules! *Since proteins can also include more than one molecule of a given amino acid, the possible combinations are essentially infinite.* However, of the many different proteins that could be made from a set of amino acids, a living cell will make only the relatively small, select number it needs.

■ $n!$, called "n factorial," is mathematical shorthand for the result of multiplying n by all the numbers between it and zero.

Many short-chain peptides are important biochemicals. For example, enkephalins and endorphins are referred to as "natural opiates" because they moderate pain in the same manner as opium derivatives. For many years scientists speculated about the action of opiates in the brain and the possible relationship to the human response to pain. Solomon Snyder and co-workers at Johns Hopkins University discovered in 1973 that the brain and spinal cord contain specific bonding or receptor sites that the opiate molecules fit as a key fits into a lock. This enhanced the search for opiate-like natural neurotransmitters. In 1975 John Hughes and Hans Kosterlitz, of the University of Aberdeen, Scotland, isolated two peptides with opiate activity from pig brains. They decided to call these peptides *enkephalins* (from the Greek *en* and *kephale,* meaning "within the head"); specifically, the two pentapeptides they isolated are known as methionine-enkephalin and leucine-enkephalin (Figure 15.8).

■ **Neurotransmitters** are biomolecules that transmit chemical messages along nerve pathways. They act by connecting with other molecules (or parts of molecules) called receptors. The neurotransmitter-receptor interaction plays an important role in the effects on the body of both poisons (as discussed in Section 18.5) and medicines (as discussed in Section 19.5).

A year later, Roger Guillemin and co-workers at the Salk Institute isolated a longer peptide, called *β-endorphin,* from extracts of the pig hypothalamus. β-Endorphin is 50 times more potent than morphine. Since this early work, other enkephalins and endorphins have been isolated. In addition, the relationship of these natural opiates to pain has been studied.

Our bodies synthesize enkephalins and endorphins to moderate pain, and our pain threshold is related to levels of these neuropeptides in our central nervous system. Individuals with a high tolerance for pain produce more neuropeptides and consequently tie up more receptor sites than normal;

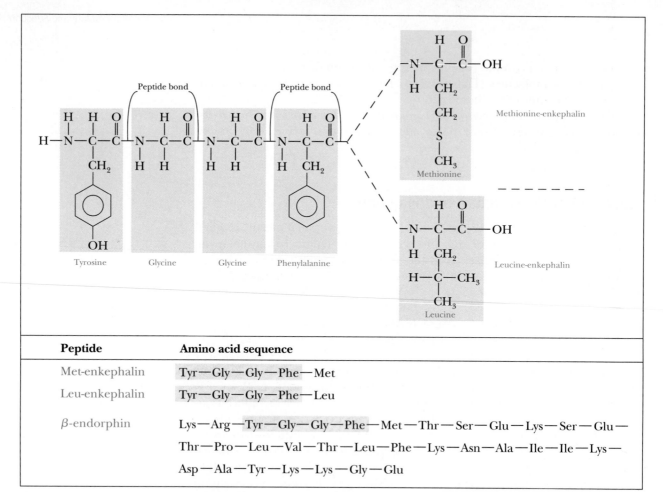

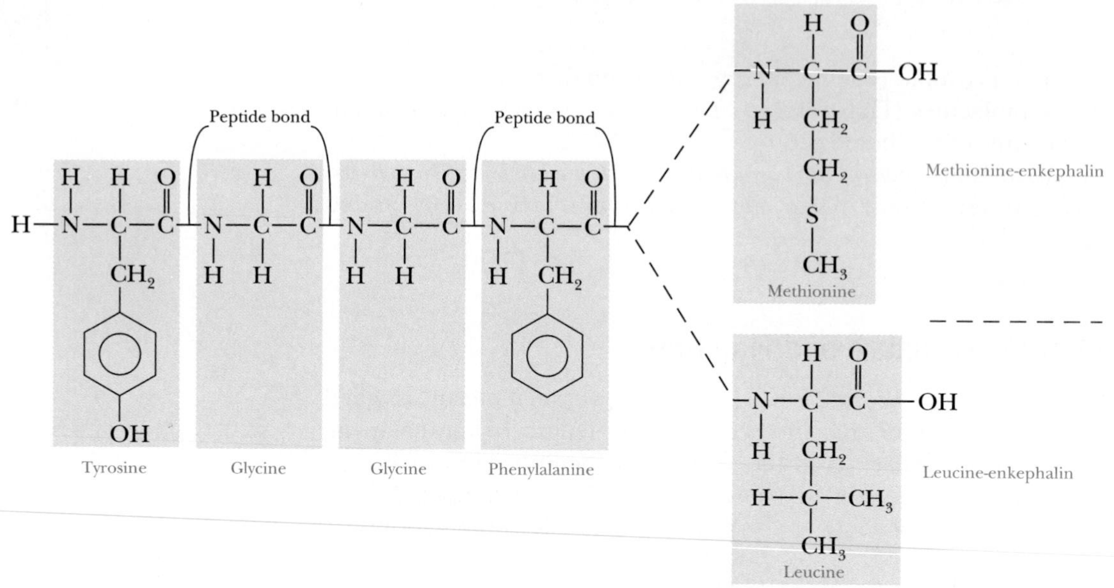

Peptide	Amino acid sequence
Met-enkephalin	Tyr—Gly—Gly—Phe—Met
Leu-enkephalin	Tyr—Gly—Gly—Phe—Leu
β-endorphin	Lys—Arg—Tyr—Gly—Gly—Phe—Met—Thr—Ser—Glu—Lys—Ser—Glu—Thr—Pro—Leu—Val—Thr—Leu—Phe—Lys—Asn—Ala—Ile—Ile—Lys—Asp—Ala—Tyr—Lys—Lys—Gly—Glu

Figure 15.8 Enkephalins and endorphins, polypeptides that are natural opiates. Note that all three polypeptides have one identical amino acid sequence.

hence, they feel less pain. A dose of heroin temporarily bonds to a high percentage of the sites, resulting in little or no pain. Continued use of heroin causes the body to reduce or cease its production of enkephalins and endorphins. If use of the narcotic is stopped, the receptor sites become empty, and withdrawal symptoms occur.

EXAMPLE 15.3 *Peptides*

Use the structures of amino acids in Table 15.1 to draw the structure of the tripeptide represented by Ala-Ser-Gly, and give its name.

SOLUTION

The amino acid sequence in the abbreviated name shows that alanine should be written at the left with a free $H_2N—$ group, glycine should be written at the right with a free $—COOH$ group, and both should be connected to serine by peptide bonds.

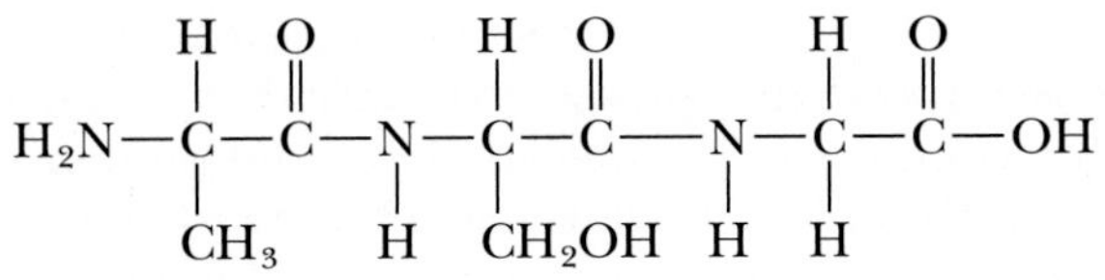

The name is alanylserylglycine.

Exercise 15.3
Use the structures of amino acids in Table 15.1 to draw the structure of the tetrapeptide Cys-Phe-Ser-Ala.

The close relationship between proteins and living organisms was first noted in 1835 by the Dutch chemist G. T. Mulder. He named proteins from the Greek *proteios* (''first''), thinking that proteins are the starting point for a chemical understanding of life. To a great extent he was correct. Proteins are important in a wider variety of ways than other kinds of biomolecules—their functions are summarized in Table 15.2. As *enzymes* they serve as catalysts in

TABLE 15.2 ■ Protein Function

Protein Name	Function
Enzymes	
Amylase	Catalyzes hydrolysis of bonds in starch
Carbonic anhydrase	Catalyzes hydration of carbon dioxide in red blood cells
Transport Proteins	
Hemoglobin	Oxygen transport in the blood
Serum albumin	Fatty acid transport in the blood
Cytochrome c	Transport of electrons
Storage Proteins	
Myoglobin	Storage of oxygen in muscle
Ovalbumin	Protein of egg white—provides amino acids for developing young
Regulatory Proteins	
Insulin	Helps regulate metabolism
Growth hormone	Helps regulate metabolism
Contractile Proteins	
α-Keratin	Allows motion in thick muscle filaments
Collagen	Allows motion in thin muscle filaments
Protective Proteins	
Antibodies	React with foreign particles to destroy them
Fibrinogen and thrombin	Needed for blood clotting

Disc-shaped

Sickle-shaped

Red blood cells (erythrocytes). Normally they are disc-shaped, but with the single amino acid defect of sickle cell anemia, they adopt the pointed, sickle shape. Sickle cell anemia is an inherited (genetic) disease caused by a flaw in DNA.

■ The chemistry of hair care is discussed in Section 16.8.

biological synthesis and degradation reactions. As *hormones* they serve a regulatory role, and as *antibodies* they protect us against disease. They make up the muscle fibers that contract so that we can move. And proteins are the major constituents of cellular and intracellular membranes, skin, hair, muscle, and tendons. Each individual protein has its own group of amino acids arranged in a definite molecular structure that is specific to the function of that protein.

The sequence of amino acids bonded to one another in a protein by peptide bonds is the protein's **primary structure.** Changing the sequence alters the properties of a protein, and just one change may produce a new protein unable to function like the original one. For example, *sickle cell anemia,* a reduction in the ability of hemoglobin to transfer oxygen, is caused by the replacement of only one specific amino acid in two of the four protein chains that make up the hemoglobin molecule (Figure 15.7*b*).

In some entire proteins and in parts of others, the shape of the backbone of the molecule (the chain containing peptide bonds) has a regular, repetitive pattern that is referred to as its **secondary structure.** The two most common secondary structures are the **α-helix** and the **β-pleated sheet.** The alpha-helix is held together by *intramolecular* (within the molecule) hydrogen bonding between backbone amide groups. An N—H group of one amino acid forms a hydrogen bond with the oxygen atom in the third amino acid down the chain (Figure 15.9).

The α-helix is the basic structural unit of the α-keratins in wool, hair, skin, beaks, nails, and claws. Because of the helical protein chains, fibers such as those in human hair are elastic to some extent. Stretching the fibers involves breaking some or all of the relatively weak hydrogen bonds, but not the strong covalent bonds. If a hair fiber is not stretched to the point where the covalent bonds begin to break, the fiber will snap back to its original length. The fiber is elastic because the hydrogen bonds can be re-formed.

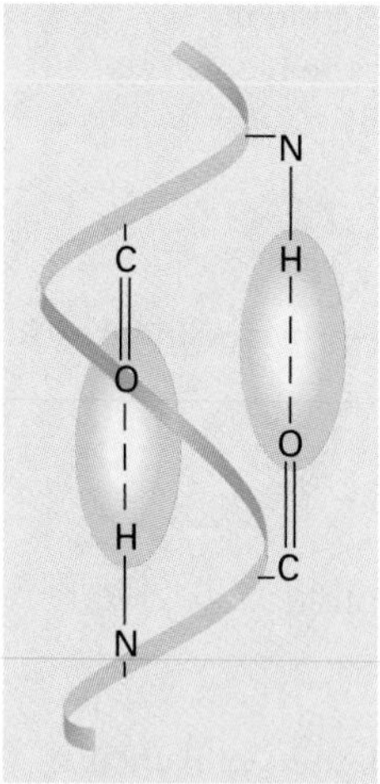

A coiled spring is helical in structure.

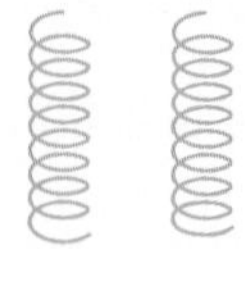

Figure 15.9 Helical protein structure. An illustration of how hydrogen bonding connects a peptide bond nitrogen atom to an oxygen atom in the third amino acid unit down the chain, resulting in a coiled structure.

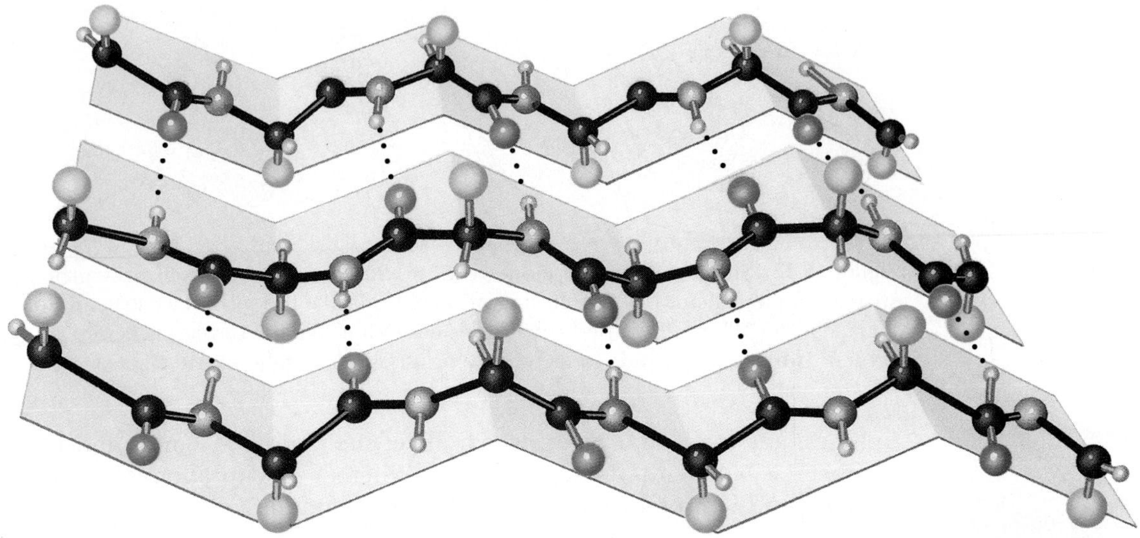

Figure 15.10 β-Pleated sheet protein structure. Hydrogen bonds are shown as dotted lines.

Silk has the β-sheet structure (Figure 15.10) in which several chains of amino acids are joined side-to-side by *intermolecular* (between protein chains) hydrogen bonds. The resulting structure is not elastic, because stretching the fibers would break either covalent bonds or the many hydrogen bonds holding the individual protein strands in the sheet. However, just as you can bend the stack of pages in this book, so too can the stack of protein sheets be bent.

Tertiary structure refers to how a protein molecule is folded. The nature of the R groups of the amino acids in the primary structure determines the tertiary structure. Hydrogen bonding, disulfide bridge bonds (—S—S— bonds formed between adjacent —SH groups in pairs of cysteine residues), and ionic bonds are three types of interactions that affect the folding of the protein molecule. One kind of tertiary structure is found in collagen, a fibrous protein: three amino acid chains twisted into left-handed helices, which in turn are twisted into a right-handed superhelix to form an extremely strong fibril (Figure 15.11). Bundles of fibrils make up the tough collagen. A second kind of tertiary structure is found in globular proteins, in which the polypeptide chain is folded and twisted in a complicated manner (e.g., Figure 5.7*a*). Although these folds may seem random, they form a definite geometric pattern that is the same in all molecules of the same protein.

Quaternary structure is the shape assumed by the entire group of chains in a protein composed of two or more chains. Hemoglobin (Figure 15.7*b*) is a prime example of quaternary structure—its four protein chains, together with the nonprotein heme molecule carried by each, can function only when combined in precisely the right shape.

All the structural features—primary, secondary, tertiary, and quaternary—are critical to the proper functioning of a protein (Figure 15.12). Any physical or chemical process that changes the protein structure and makes it incapable of performing its normal function is called a **denaturation** process. For example, heating an aqueous solution of a protein breaks hydrogen bonds in the secondary and tertiary structure and causes the protein molecule to unfold. Denaturing chemicals include reducing agents, which break disulfide

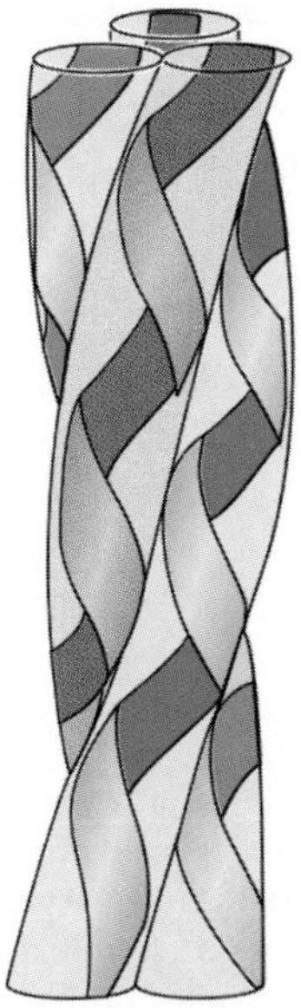

Collagen, a fibrous protein

Figure 15.11 Collagen is composed of three polypeptide chains twisted together into a fiber.

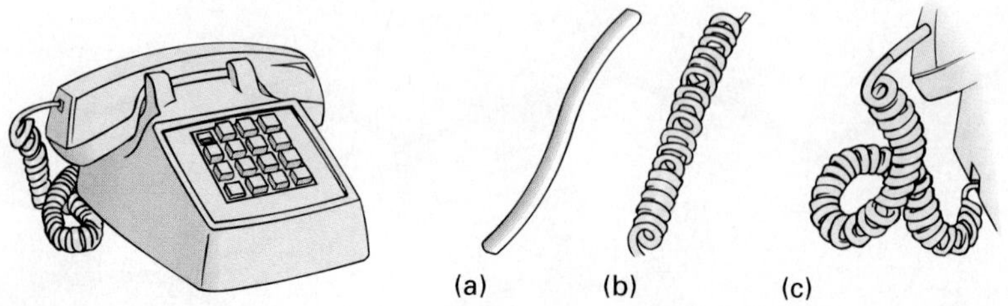

Figure 15.12 Structure of a telephone cord. (a) The straight cord is like the primary structure of a protein. (b) The curling of the cord is like the secondary structure of a protein. (c) When the curled cord is twisted up it is like the tertiary structure of a protein (which may include in its twisted form regions of α-helix and β-sheet).

One way to denature a protein.
(C.D. Winters)

linkages, and acids and bases, which affect the hydrogen bonds and ionic interactions between polypeptide chains. Whether denaturation is reversible depends on the protein and the extent of denaturation.

Enzymes

Enzymes function as catalysts for chemical reactions in living systems. Many enzymes are globular proteins. A typical globular enzyme, like chymotrypsin or lysozyme contains regions where the polypeptide chain forms α-helices, other regions where parallel parts of the chain are organized into β-pleated sheets, and some parts that are just coiled randomly. One outcome of this highly organized, though seemingly random, structure is creation of the region that allows an enzyme to function as a catalyst (the active site). Like all catalysts, enzymes increase the rate of a reaction by lowering the energy of activation (Figure 8.6). Enzymes are very effective catalysts and typically increase reaction rates by anywhere from 10^6 to 10^{16} times.

Many biomolecules are broken down during digestion by hydrolysis reactions, which are essentially the reverse of condensation reactions. In hydroly-

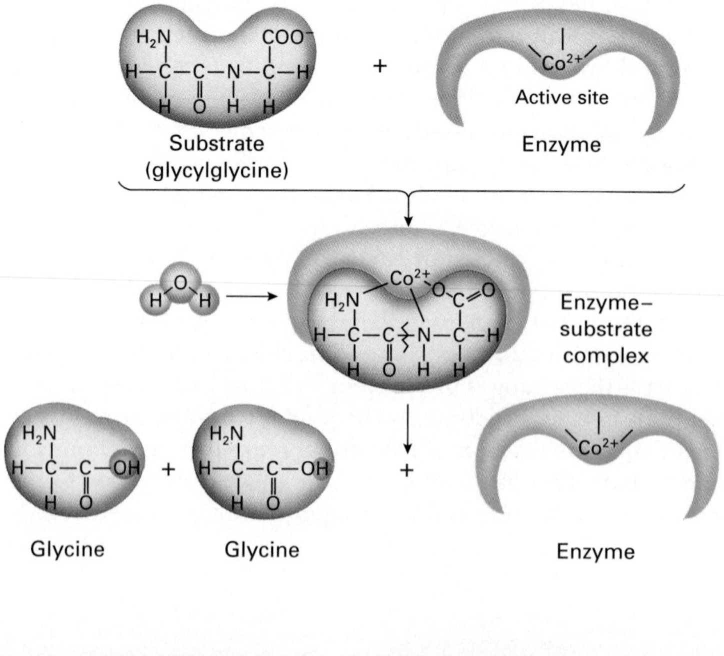

Figure 15.13 An enzyme-substrate reaction. The substrate enters the active site of the enzyme (glycylglycine dipeptidase), where it chemically bonds to a Co^{2+} ion. The bonding of the substrate makes it more susceptible to attack by water. The peptide bond is broken by hydrolysis and the glycine molecules are released by the enzyme, which is then ready to play its catalytic role again.

THE WORLD OF CHEMISTRY

Unraveling the Protein Structure

Program 23, *Proteins: Structure and Function*

One of the key steps in unraveling the mystery of hydrogen bonds in protein structure involved Linus Pauling, a cold, and a Nobel Prize. Early in his career, Pauling worked on the structure of protein molecules. At that time there were several conflicting theories. Pauling and his colleagues thought that the first level of protein structure was a polypeptide chain. Then they asked themselves a fundamental question:

We asked: How is the polypeptide chain folded? We couldn't answer the question, but we said it's probably held together by hydrogen bonds. The conclusion we reached was that there are . . . polypeptide chains in the protein, which . . . are coiled back and forth, and that they are coiled into a very well-defined structure, configuration, with the different parts of the chain held together by hydrogen bonds. In 1937 I spent a good bit of the summer with models for—I assumed that I knew what a polypeptide chain looks like except for the way in which it's folded. And I wanted to fold it to form the hydrogen bonds. I didn't succeed. The fact is, I thought that there was something about proteins that perhaps I didn't know.

Pauling continued to work on this problem, but the solution eluded him. Then one day he had a crucial insight in a completely different and unexpected setting.

I had a cold. I was lying in bed for two or three days, and I read detective stories, light reading, for awhile, and then I got sort of bored with that. So I said to my wife, "Bring me a sheet of paper, and I'm going to—I think I'll work on that problem of how polypeptide chains are folded in proteins." So she brought me a sheet of paper and the slide rule and pencil, and I started working.

Using the knowledge gained from his years of model building, he drew the backbone of a polypeptide chain on a piece of paper. Then it occurred to him to try to fold the paper to see how hydrogen bonds could form along the polypeptide chain. The result was a structure that twisted around like a spring.

Well, I succeeded. It only took a couple of hours of work that day, March of 1948, for me to find the structure, called the alpha helix.

Linus Pauling. Along with R. B. Corey, Pauling proposed the α-helix and β-sheet structures for proteins. For his bonding theories and his work with proteins, Pauling was awarded the Nobel Prize in Chemistry in 1954. For his fight against the dangers of nuclear weapons testing and fallout he received the Nobel Peace Prize in 1962.

sis, a larger molecule is split into smaller molecules with the addition of H— and —OH of water where a bond was broken. The enzyme maltase catalyzes the hydrolysis of the sugar maltose into two molecules of D-glucose. This is the only function of maltase, and no other enzyme can substitute for it. Sucrase, another enzyme, hydrolyzes only sucrose. Some enzymes are less specific. The digestive enzyme trypsin, for example, primarily hydrolyzes peptide bonds in proteins. However, the structure and polarity of trypsin are such that it can also catalyze the hydrolysis of some esters.

The action of an enzyme in hydrolysis is shown in Figure 15.13. The reactant molecule is called the **substrate.** Enzymes and substrates have electrically polar regions, partially charged groupings, or ionic sections that attract and guide the enzyme and substrate together; these regions of chemical activity are the **active sites.** Substrates "sit down" on active sites of enzymes in assembly-line fashion at a remarkably fast rate.

■ The general equation for hydrolysis of an ester is

$$RCOOR + H_2O \longrightarrow RCOOH + ROH$$

DISCOVERY EXPERIMENT

Action of an Enzyme

Casein is the protein substance in milk that forms the curds used to make cheese. The casein also separates from milk when milk sours. Rennin is the enzyme that catalyzes the coagulation of casein. Rennin is available in grocery stores under the tradename Junket and is used in making ice cream, jelly, and cheese.

For this experiment you will need four half-tablets of rennin, about 150 mL ($\frac{5}{8}$ cup) of milk (whole or 2%), a thermometer that reads to at least 80°C (176°F), a saucepan or beaker, vinegar, a clock or watch, five clear plastic cups, paper towels, and ice water.

1. Warm about 60 mL ($\frac{1}{4}$ cup) of milk in a saucepan or beaker to between 40° and 50°C. Do not boil. Pour about 30 mL of the warmed milk into a clear plastic cup. Add half a rennin tablet to the cup, stir for several minutes, and observe the milk forming curds. Use your hand to gather the curds into a ball. Place the curdled mixture on a paper towel, squeeze the remaining liquid from the solid curds, and observe their properties.
2. Heat the milk remaining in the saucepan from Step 1 to 80°C and pour the hot milk into a clear plastic cup. Stir in half a rennin tablet. Observe for about 10 minutes. Did the milk form curds? Allow the milk to cool to 40°C, and stir for a few minutes. Did the milk form curds? Now add an additional half a rennin tablet and stir for a few minutes. Did the milk form curds?
3. Pour about 30 mL of milk into a clear plastic cup and cool with ice water to below 10°C. Add half a rennin tablet, stir for several minutes, and observe. Did the milk form curds?
4. Pour about 30 mL of milk into a clear plastic cup and add 1 tablespoon of vinegar. Stir. Separate the curds from the milk as in Step 1 and compare their properties with those of the curds formed in Step 1.

Explain your observations.

Since the structure of the active site of an enzyme is important, the same factors that cause denaturation will also destroy the activity of the enzyme. For example, most enzymes are effective only over a narrow temperature range and a narrow pH range. Enzymes are denatured irreversibly at high temperatures or at pH values outside their effective range.

Many inherited, or **genetic,** diseases or defects affect how enzymes function. An example is found in **lactose intolerance,** which is an inability to digest lactose, the sugar present in milk from all mammals. Well over half the world's population has this problem to some degree. The pattern of inheritance is noticed largely in those with Asian or African ancestry and to a lesser extent in those of Northern European ancestry. When they are infants, people with lactose intolerance manufacture the enzyme lactase, which is necessary to digest lactose. As they grow older, their bodies stop producing this enzyme, and the ingestion of milk products containing lactose can lead to considerable discomfort in the form of stomachaches and diarrhea. The problem is avoided by eliminating milk products from the diet or by taking enzyme-containing tablets before eating any milk product.

■ Over-the-counter medications containing the lactase enzyme are available without prescription.

■ SELF-TEST 15B

1. The peptide linkage that bonds amino acids together in protein chains has the structure ____________.
2. (a) If we have three different amino acids and can use each one three

times in any given tripeptide, we can make a total of _____ different tripeptides.

 (b) If we can use each amino acid only once, there are still _____ possible different tripeptides.

3. That portion of the enzyme at which the reaction is catalyzed is called the _____.

4. (a) The primary structure of a protein refers to its _____.
 (b) The secondary structure refers to its _____.
 (c) The tertiary structure refers to _____.
 (d) The quaternary structure refers to _____.

5. The best term to describe the general function of enzymes is (a) catalyst, (b) intermediate, (c) oxidant.

6. The activation energy of many biological reactions is decreased if a(an) _____ is present.

7. The peptide bond is the result of an acid-base reaction. (a) True, (b) False.

8. The helical structure of proteins is caused by _____ bonding.

15.5 Carbohydrates

The word "carbohydrate" literally means "hydrate of carbon." Carbohydrates have the general formula $C_x(H_2O)_y$, in which x and y are whole numbers. However, even though the reaction of the carbohydrate sucrose with sulfuric acid produces carbon (Figure 15.14), this does not mean that sucrose is a simple combination of carbon and water. The carbon, hydrogen, and oxygen in carbohydrates are arranged primarily into three organic functional groups:

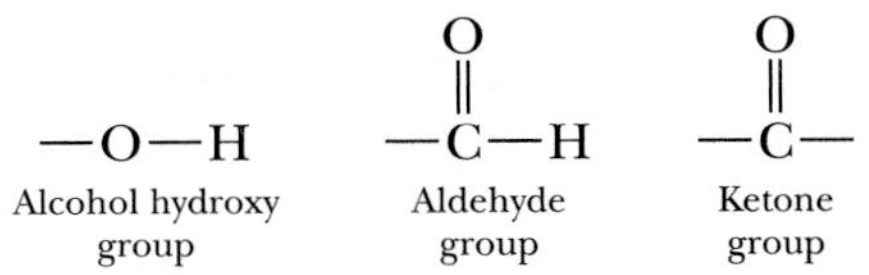

$$\begin{array}{ccc} & O & O \\ & \| & \| \\ -O-H & -C-H & -C- \\ \text{Alcohol hydroxy} & \text{Aldehyde} & \text{Ketone} \\ \text{group} & \text{group} & \text{group} \end{array}$$

Carbohydrates are simple polyhydroxy aldehydes or ketones or larger molecules formed from them by condensation reactions.

Carbohydrates are divided into three groups, depending on how many monomers are combined by condensation polymerization: **monosaccharides** (Latin *saccharum,* "sugar"), **disaccharides,** and **polysaccharides.** Monosaccharides are simple sugars that cannot be broken down into smaller carbohydrate units by acid hydrolysis. In contrast, hydrolysis of a disaccharide yields two monosaccharides (either the same or different), while complete hydrolysis of a polysaccharide produces many monosaccharides (sometimes thousands of them).

Carbohydrates, which make up about half the average human diet, form an essential part of the energy cycle for living things (Section 15.7). Besides energy storage in plants and energy production in animals, carbohydrates serve many other biological purposes. Cellulose is the main structural component of plants. The nucleic acids (Section 15.8) incorporate carbohydrate units in their repeating structure.

Figure 15.14 Reaction of sulfuric acid with sugar. As the sulfuric acid is poured onto the sugar, carbon can be seen forming. *(C. D. Winters)*

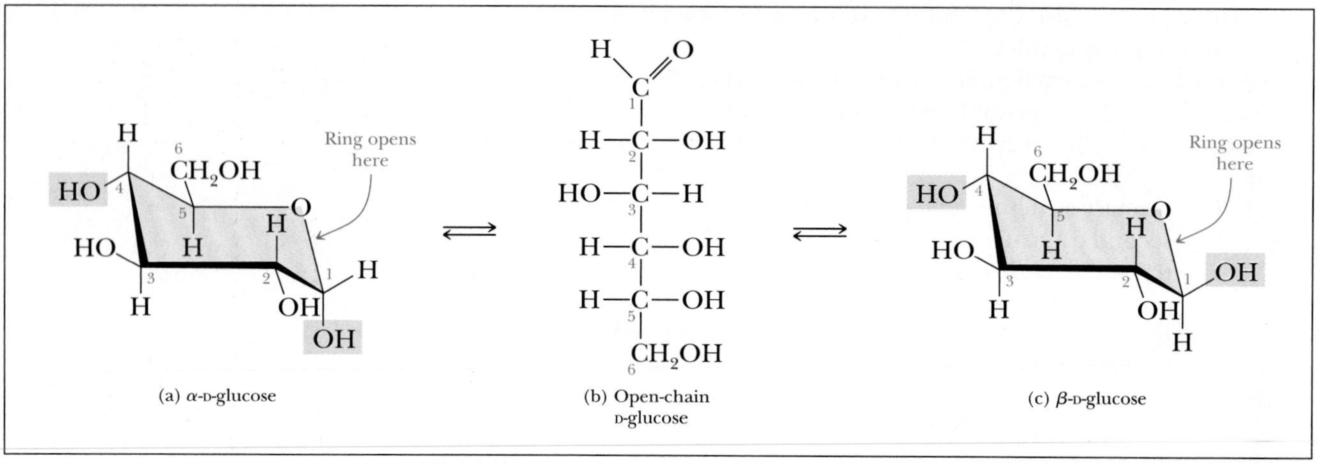

Figure 15.15 Open chain and ring forms of D-glucose. The two rings differ in the orientation of the —H and —OH on carbon 1.

Monosaccharides

The most common simple sugar is D-glucose, or dextrose. As discussed in Section 15.1, glucose is a chiral molecule because it has four asymmetric carbon atoms (Figure 15.4). However, the straight-chain structure shown in Figures 15.4 and 15.15*b* is an oversimplification of the actual structure of D-glucose. An aqueous solution of D-glucose contains all three structures shown in Figure 15.15 in dynamic equilibria involving mostly the two ring forms and less than 1% of the straight-chain form. Because D-fructose, a monosaccharide found in many fruits and table sugar, has a ketone group in its straight-chain form rather than an aldehyde group (Figure 15.16), it forms both a five-membered ring and a six-membered ring.

There is one more structural feature of glucose that will be important in understanding the structural differences in the polysaccharides made from

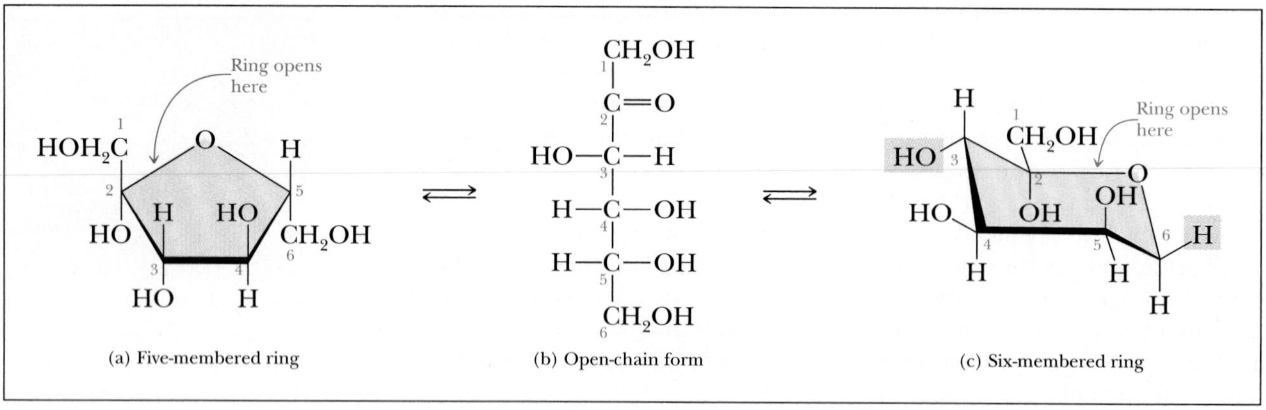

Figure 15.16 Open chain and ring forms of D-fructose.

glucose. Notice the different positions of the —OH group on carbon 1 in the ring forms of glucose in Figure 15.15. The Greek letters α and β are added to the name to indicate this difference.

A pile of sugar beets awaiting collection. *(Nigel Cattlin/Photo Researchers)*

Disaccharides

The three most widely used disaccharides are

Sucrose (from sugar cane or sugar beets), composed of a D-glucose monomer and a D-fructose monomer—table sugar

Maltose (from starch), composed of two D-glucose monomers—used as a sweetener in prepared foods

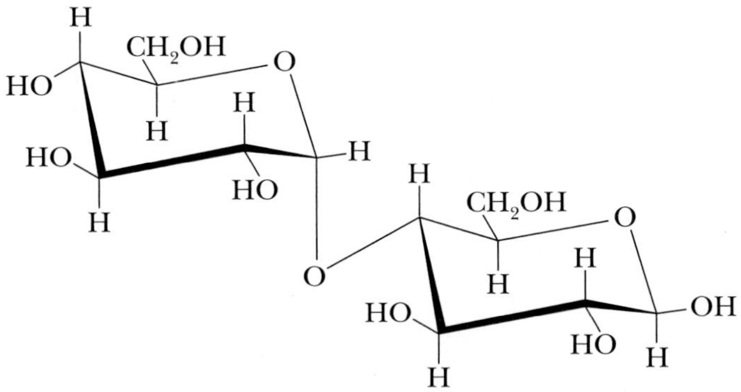

Maltose

Lactose (from milk), composed of a D-glucose monomer and a D-galactose monomer—used in formulating drugs and infant foods, in baking, and in making yeast

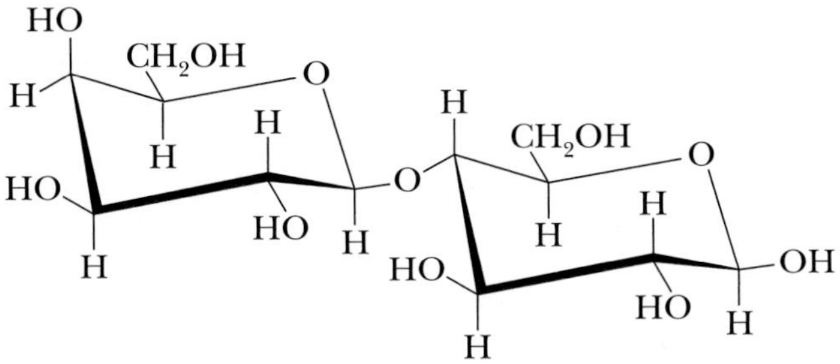

Lactose

■ Disaccharides are condensation products formed from monosaccharides.

The formula for these disaccharides ($C_{12}H_{22}O_{11}$) is not simply the sum of two monosaccharides, $C_6H_{12}O_6 + C_6H_{12}O_6$. A water molecule is eliminated as two monosaccharides are united to form the disaccharide. In the body, enzymes catalyze the breakdown (hydrolysis) of disaccharides to their monosaccharides. Examples of the hydrolysis reactions of disaccharides are given in Figure 15.17.

Sucrose is produced in a high state of purity on an enormous scale—more than 80 million tons per year. About 40% of the world sucrose production comes from sugar beets and 60% from sugar cane. A comparison of the sweetness of common sugars and artificial sweeteners relative to sucrose is given in Table 15.3. Honey, which is a mixture of the monosaccharides glucose and fructose, has been used for centuries as a natural sweetener for foods and is

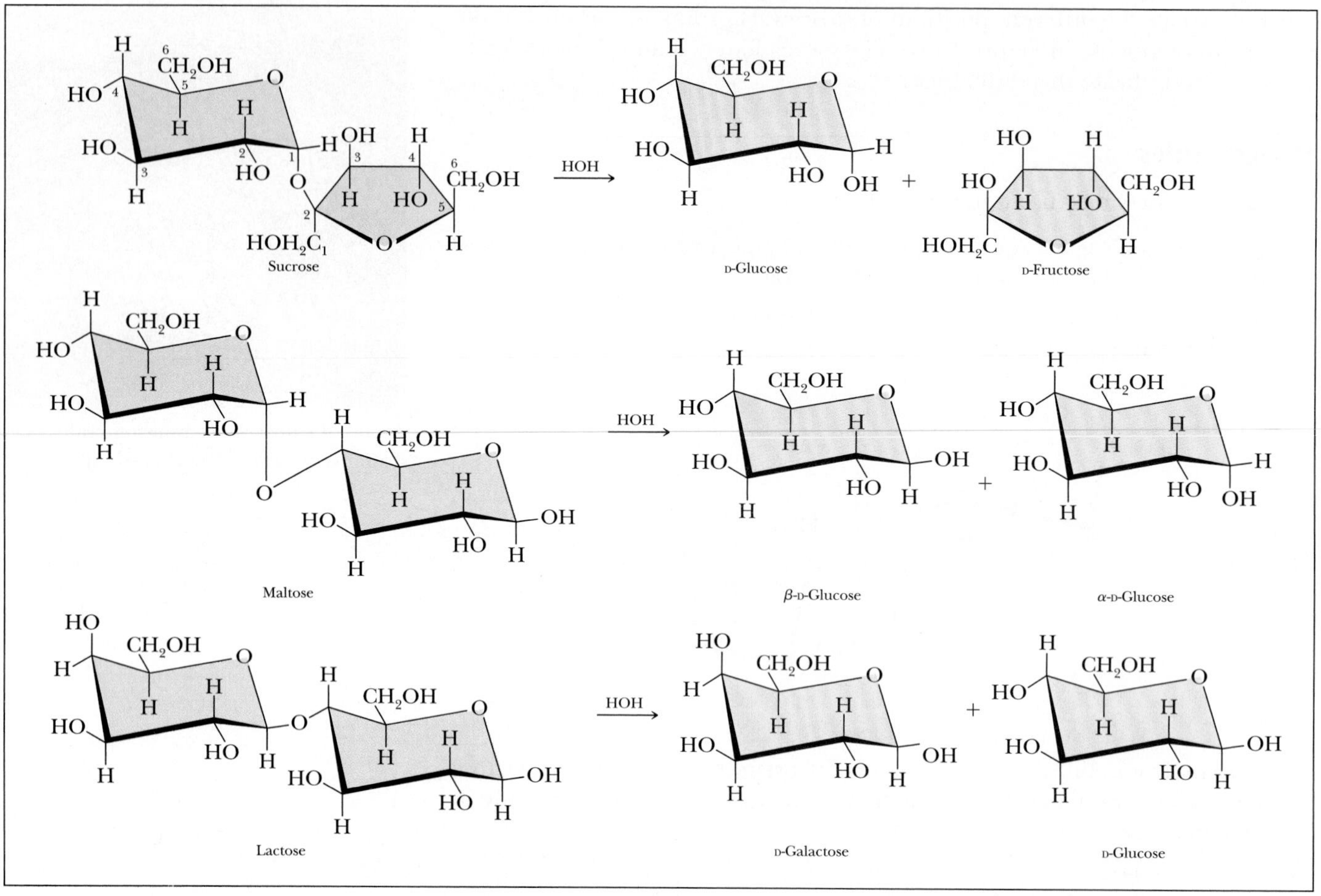

Figure 15.17 Hydrolysis of disaccharides yields two monosaccharides.

sweeter than sucrose, or cane sugar (Table 15.3). To convert cane sugar into glucose and fructose requires treatment with acid or with a natural enzyme called "invertase." The product, known as *invert sugar,* is often used as a sweetener in commercial food products.

Artificial Sweeteners

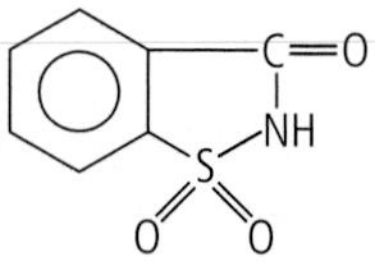

Saccharin

Saccharin was the first common artificial sweetener. Saccharin passes through the body undigested and consequently has no caloric value. It has a somewhat bitter aftertaste that is offset in commercial products by the addition of small amounts of naturally occurring sweeteners. Such products do have a small caloric value because of the natural sweeteners added. High doses of saccharin have been shown to cause cancer in mice, and commercial products containing saccharin are required by law to have a warning label: "This product contains saccharin, which has been determined to cause cancer in laboratory animals."

TABLE 15.3 ■ Sweetness of Common Sugars and Artificial Sweeteners Relative to Sucrose

Substance	Sweetness Relative to Sucrose as 1.0
Lactose	0.16
Galactose	0.32
Maltose	0.33
Glucose	0.74
Sucrose	1.00
Fructose	1.17
Aspartame*	180
Saccharin*	300

*Artificial sweeteners

Aspartame (NutraSweet), which has replaced saccharin as the principal artificial sweetener, is used in more than 3000 products and accounts for 75% of the one-billion-dollar worldwide artificial sweetener market.

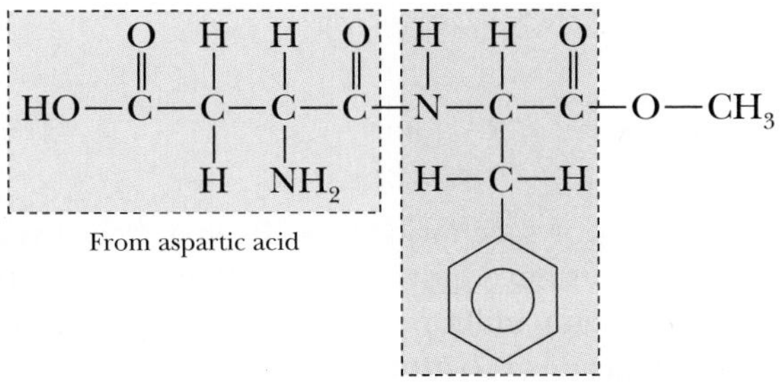

Aspartame

It is a dipeptide derivative made from aspartic acid and the methyl ester of phenylalanine. Aspartame can be digested, and its caloric value is approximately equal to that of proteins. However, since much smaller amounts of aspartame than of table sugar are needed for sweetness, many fewer calories are consumed in the sweetened food. Aspartame is not stable at cooking temperatures, which limits its use as a sugar substitute to cold foods and soft drinks.

Control of Blood Sugar

Serious health problems can be encountered when the glucose level is either too low or too high in the blood. After about ten hours without food, a healthy person has from 80 to 120 mg of glucose per dL of blood. People with low blood sugar, a condition known as **hypoglycemia,** must carefully control the amounts of carbohydrates in their diets in order to maintain normal blood-sugar levels. A low glucose concentration may lead to sluggish or dizzy feelings, and possibly fainting. Eating meals with a very high concentration of carbohydrates can also cause problems for people with hypoglycemia because the body

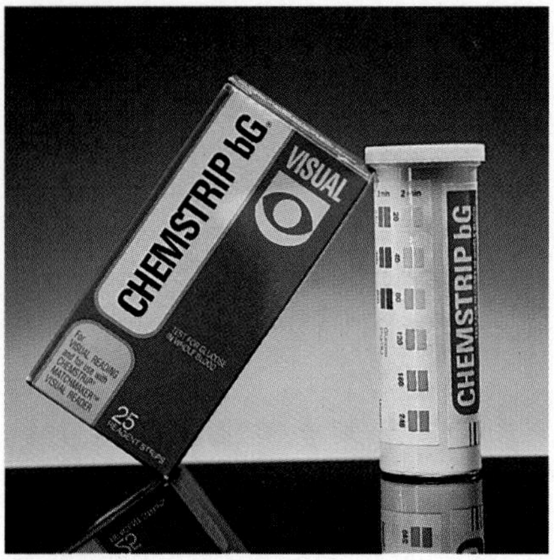

A kit for measuring blood glucose levels. The color of a test strip is matched to the color key on the container. *(C.D. Winters)*

overcorrects with excess insulin, which quickly drives the glucose concentration below the normal level.

Hyperglycemia is the condition of elevated blood-sugar concentrations. Above a concentration of 160 mg of glucose per dL of blood, the kidneys begin to excrete glucose in the urine. If such high blood-sugar concentrations are chronic, the individual has **diabetes mellitus** and, without treatment, is likely to have symptoms of thirst, frequent urination, weakness, low resistance to infection, slowness to heal, and in later stages, blindness and coma. About 5% of Americans have diabetes.

There are two types of diabetes, which have been termed Type I (insulin-dependent, formerly known as juvenile-onset diabetes) and Type II (non–insulin-dependent, formerly known as maturity-onset diabetes). Type I diabetics, about 10% of those who have diabetes, do not produce enough insulin at the *islets of Langerhans* in the pancreas. Insulin, which is necessary for the glucose to move from the blood to the cells, has to be obtained by daily injections; it cannot be taken orally because it is destroyed in the digestive tract. Type II diabetes is usually found in older, obese people. These people generally produce plenty of insulin, but the insulin receptors on their cells do not respond properly and consequently do not move the glucose from the blood into the cells. Type II diabetics can generally control the disease by diet and oral medication.

Polysaccharides

Nature's most abundant polysaccharides are the **starches, glycogen,** and **cellulose.** Some polysaccharides have more than 5000 monosaccharide monomers and molar masses over one million. The monosaccharide most commonly used to build polysaccharides is D-glucose.

Starches and Glycogen

Plant starch is found in protein-covered granules. If these granules are ruptured by heat, they yield a starch that is soluble in hot water, *amylose,* and an insoluble starch, *amylopectin.* Amylose constitutes about 25% of most natural starches.

Structurally, amylose is a straight-chain condensation polymer with an average of about 200 α-D-glucose monomers per molecule. Each monomer is bonded to the next with the loss of a water molecule, just as the two units are bonded in maltose (Figure 15.17). A representative portion of the structure of amylose is shown in Figure 15.18.

A typical amylopectin molecule has about 1000 α-D-glucose monomers arranged into branched chains (Figure 15.19) with a branch every 24 to 30 glucose units. Complete hydrolysis yields D-glucose; partial hydrolysis produces mixtures called **dextrins,** which are used as food additives and in mucilage, paste, and finishes for paper and fabrics.

Glycogen is an energy reservoir in animals, just as starch is in plants. The α-D-glucose chains in glycogen are more highly branched (with a branch every 12 glucose units) than the chains in amylopectin. Glycogen is stored in the liver and muscle tissues and is used for "instant" energy until the process of fat metabolism can take over and serve as the energy source.

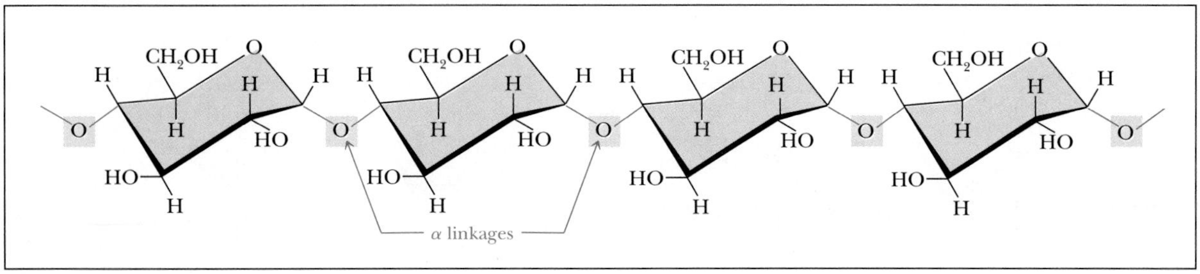

Figure 15.18 The structure of amylose, water-soluble starch. Amylose contains from 60 to 300 α-D-glucose monomers connected by α-linkages between carbons 1 and 4. The —OH groups on carbon atoms 1 and 4 are on the same side of the ring in α-glucose (see Figure 15.15), so all bonds between the α-glucose monomers point in the same direction.

Cellulose

■ Cellulose has 900–6000 glucose units.

Cellulose is the most abundant organic compound on Earth, and its purest natural form is cotton. The woody part of trees, the supporting material in plants and leaves, and the paper we make from them are also mainly cellulose. Like amylose, it is composed of D-glucose units. The difference between the cellulose and amylose lies in the bonding between the D-glucose units; in cellulose all the glucose units are in the β-ring form (Figure 15.20), whereas in amylose they are in the α-ring form. (Review the ring forms in Figure 15.15 and compare the structures in Figures 15.18 and 15.20). This subtle structural difference between starch and cellulose causes their differences in digestibility. Since cellulose is so abundant, it would be advantageous if humans could use it for food. Unfortunately, we cannot digest it, because we lack the necessary enzyme to chew up the β-1,4 bonds (Figure 15.20). On the other hand, termites, a few species of cockroaches, and ruminant mammals such as cows, sheep, goats, and camels do have the proper digestive enzyme.

■ Cotton is about 90% cellulose.

D-glucose can be obtained from cellulose by heating a suspension of the polysaccharide in the presence of a strong acid. At present, wood cannot be hydrolyzed into D-glucose economically enough to satisfy the world's growing need for an adequate food supply.

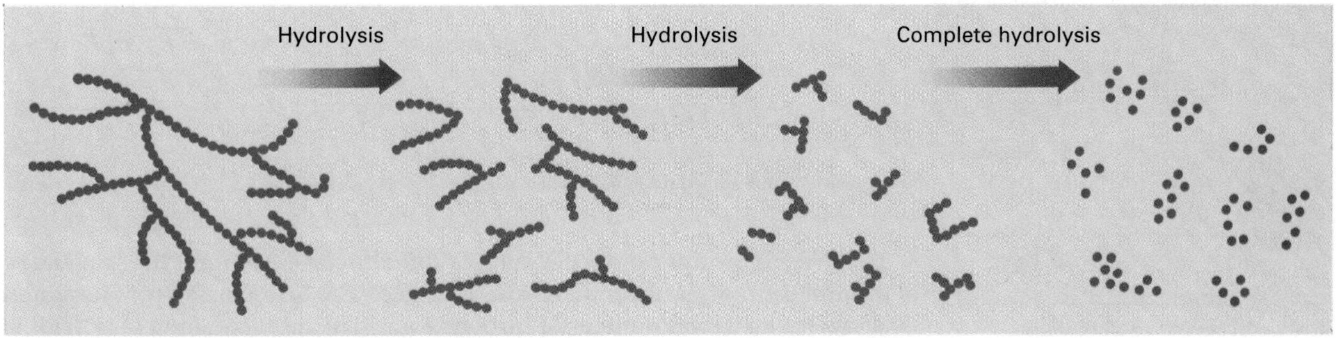

Figure 15.19 Amylopectin (left) is first hydrolyzed to dextrins, then to small polysaccharides, and finally to individual glucose molecules (right).

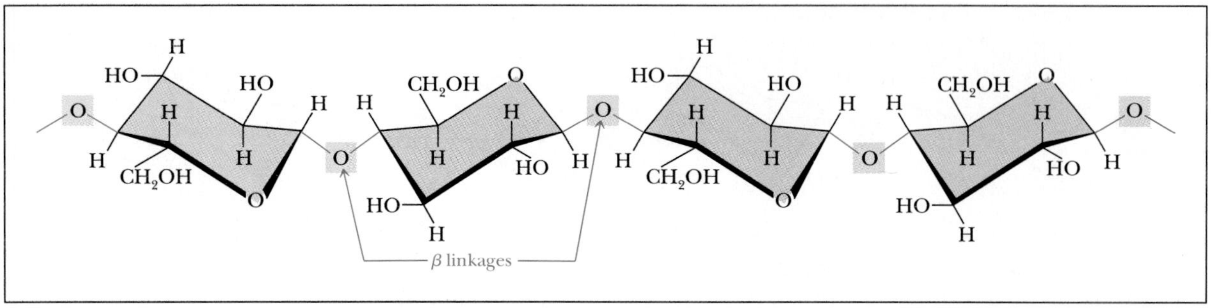

Figure 15.20 The structure of cellulose. Cellulose contains about 280 β-D-glucose monomers connected by β-linkages between carbons 1 and 4. The —OH groups on carbon atoms 1 and 4 in β-glucose are on opposite sides of the ring (see Figure 15.15), so the bonds between the monomers alternate in direction. Note that every other β-glucose monomer is turned over, whereas in amylose all the α-glucose monomers are in the same position.

15.6 Lipids

A lipid is an organic substance found in living systems that is insoluble in water but soluble in organic solvents. Because their classification is based on insolubility in water rather than on a structural feature such as a functional group, lipids vary widely in their structure and, unlike proteins and polysaccharides, are not polymers. Lipids include **fats** and **oils, steroids,** and **waxes.** The predominant lipids are fats and oils, which make up 95% of the lipids in our diet. The other 5% are steroids and several other lipids that are important to cell function.

Fats and Oils

Fats and oils are **triglycerides**—esters of glycerol (glycerin) and fatty acids. The general equation for the formation of a triester of glycerol is

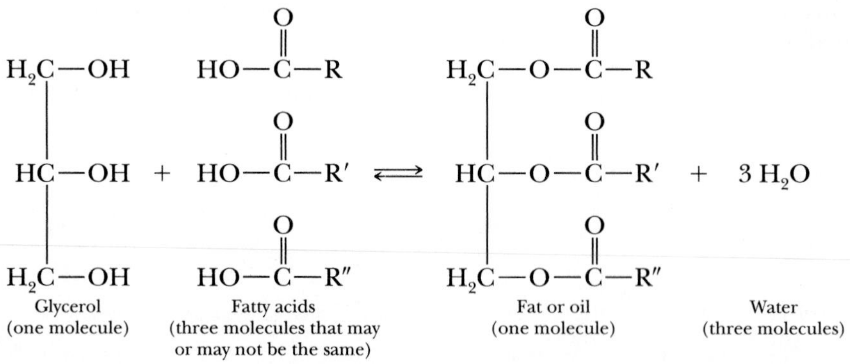

An assortment of vegetable oils.
(C.D. Winters)

The three R groups can be the same or different groups within the same fat or oil, and they can be saturated or unsaturated. The acid portions of fats almost always have an even number of carbon atoms. The most common fatty acids in fats and oils are listed in Table 15.4.

TABLE 15.4 ■ Common Fatty Acids In Fats and Oils

Saturated Acids (All solids at room temperature)		Source
Lauric	$CH_3(CH_2)_{10}COOH$	Coconut oil
Myristic	$CH_3(CH_2)_{12}COOH$	Nutmeg oil
Palmitic	$CH_3(CH_2)_{14}COOH$	Animal and vegetable fats
Stearic	$CH_3(CH_2)_{16}COOH$	Animal and vegetable fats
Unsaturated Acids (All liquids at room temperature)		
Oleic	$CH_3(CH_2)_7CH{=}CH(CH_2)_7COOH$	Animal and vegetable fats
Linoleic*	$CH_3(CH_2)_4CH{=}CHCH_2CH{=}CH(CH_2)_7COOH$	Linseed oil, cottonseed oil
Linolenic	$CH_3CH_2CH{=}CHCH_2CH{=}CHCH_2CH{=}CH(CH_2)_7COOH$	Linseed oil

*An essential fatty acid that must be part of the human diet.

Fatty acids such as oleic acid, which contain only one double bond, are referred to as **monounsaturated acids.** One of the unsaturated acids, linoleic acid, is an **essential fatty acid.** The human body cannot produce this acid, but it is required for the synthesis of the **prostaglandins,** an important group of more than a dozen related compounds. The prostaglandins have potent effects on physiological activities such as blood pressure, relaxation and contraction of smooth muscle, gastric acid secretion, body temperature, food intake, and blood platelet aggregation. Their potential use as drugs is currently under widespread investigation. For example, one of the prostaglandins, PGE_2, lowers blood pressure, but a closely related prostaglandin, $PGF_{2\alpha}$, raises blood pressure. The structures of these two prostaglandins differ only in one functional group (highlighted below).

■ The chemistry of soap, made by saponification of fats, and how soaps get clothes clean is discussed in Section 16.9.

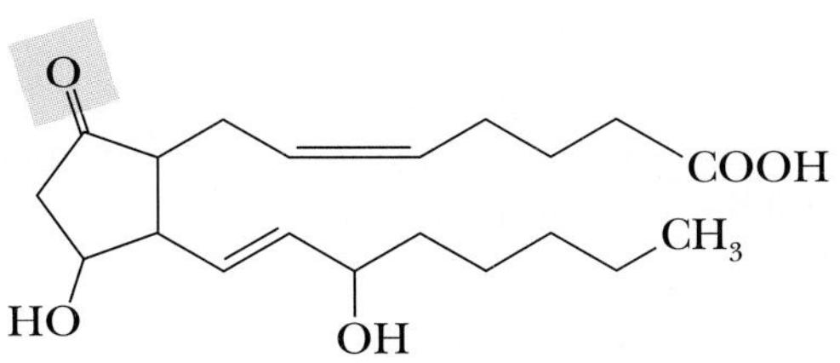

PGE_2

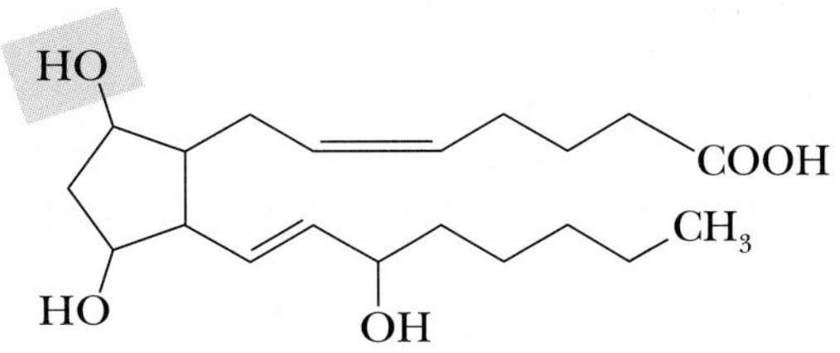

$PGF_{2\alpha}$

Many prostaglandins cause inflammation and fever. The fever-reducing effect of aspirin results from the inhibition of cyclooxygenase, the enzyme that catalyzes the synthesis of prostaglandins.

The term ''fat'' is usually reserved for solid triglycerides (such as butter, lard, and tallow) and ''oil'' for liquid triglycerides (castor, olive, linseed, and tung oils, for example). The R groups in the fatty acid portions of fats are generally saturated, with only C—C single bonds. The R groups in oils are usually either monounsaturated (one C=C double bond) or polyunsaturated (two or more C=C double bonds). Since the C=C bonds interrupt the zigzag pattern of tetrahedral angles with 120° angles, the molecules are irregular in shape and do not pack together efficiently enough to form a solid. Table 15.5 illustrates the percentages of saturated and unsaturated fat found in common dietary oils and fats.

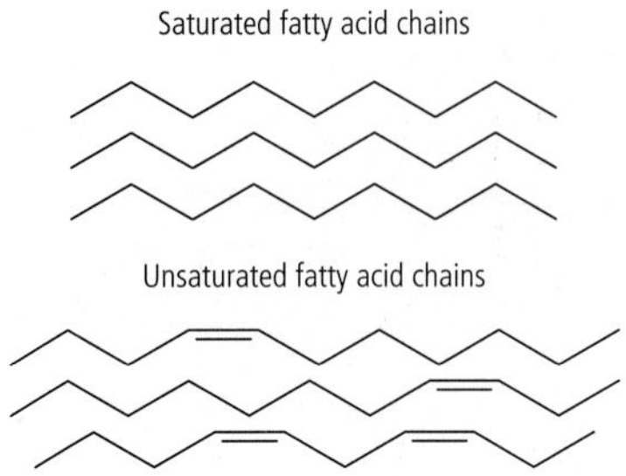

TABLE 15.5 ■ **Amounts of Saturated and Unsaturated Fatty Acids in Fats and Oils**

Dietary Oil / Fat	■ Saturated Fat	■ Polyunsaturated Fat	■ Monounsaturated Fat	
Canola oil	6%	36%	58%	
Safflower oil	9%	78%		13%
Sunflower oil	11%	69%		20%
Corn oil	13%	62%		25%
Olive oil	14%	9%	77%	
Soybean oil	15%	61%		24%
Peanut oil	18%	34%	48%	
Cottonseed oil	27%	54%		19%
Lard	41%	12%	47%	
Palm oil	51%	10%	39%	
Beef tallow	52%	4%	44%	
Butterfat	66%	4%	30%	
Coconut oil	92%		2%	6%

■ The hydrogenation process also forms un-natural *trans* fatty acids from natural *cis* fatty acids (see Science and Society on p. 486).

Hydrogen can be added catalytically to the double bonds of an oil to convert it into a semisolid fat. For example, liquid soybean and other vegetable oils are **hydrogenated** to produce cooking fats and margarine.

$$H_2C-O-\underset{\underset{O}{\|}}{C}-(CH_2)_7CH{=}CH(CH_2)_7CH_3$$
$$HC-O-\underset{\underset{O}{\|}}{C}-(CH_2)_7CH{=}CH(CH_2)_7CH_3 \xrightarrow[200°C]{H_2, Ni}$$
$$H_2C-O-\underset{\underset{O}{\|}}{C}-(CH_2)_7CH{=}CH(CH_2)_7CH_3$$

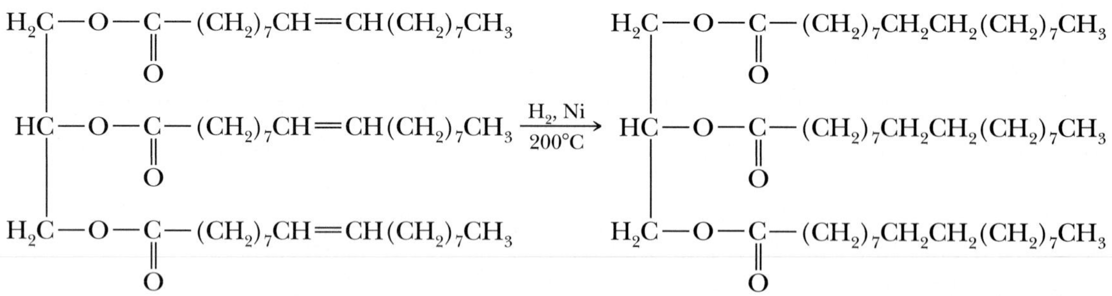

Triolein (a liquid oil)

$$H_2C-O-\underset{\underset{O}{\|}}{C}-(CH_2)_7CH_2CH_2(CH_2)_7CH_3$$
$$HC-O-\underset{\underset{O}{\|}}{C}-(CH_2)_7CH_2CH_2(CH_2)_7CH_3$$
$$H_2C-O-\underset{\underset{O}{\|}}{C}-(CH_2)_7CH_2CH_2(CH_2)_7CH_3$$

Tristearin (a solid fat)

If it is better to consume unsaturated fats instead of saturated ones, why do food companies hydrogenate oils to reduce their unsaturation? There are several answers to this question. First, the double bonds in the fatty acid are a reactive functional group, and oxygen can attack the fat at these bonds. When the oil is oxidized, unpleasant odors and flavors develop. Hydrogenating an oil

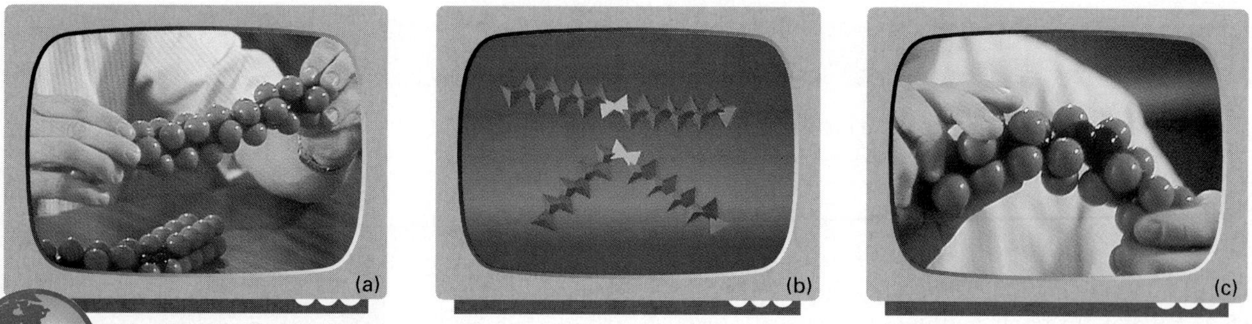

The World of Chemistry

Program 9, *Molecular Architecture*

Saturated fatty acids (a) and *trans*-unsaturated fatty acids (top structure in b) are linear and tend to pack into solid masses. *Cis*-unsaturated fatty acids (c and bottom structure in b) are bent, do not pack as well as straight structures, and tend to be liquid at normal temperatures.

reduces the likelihood that the food will oxidize and become rancid. Second, hydrogenating an oil makes it less liquid. There are many times when a food processor needs a solid fat in a food to improve its texture and consistency. (For example, if liquid vegetable oil were used in a cake icing, the icing would slide off the cake.) Rather than use animal fat, which also contains cholesterol, the manufacturer turns to a hydrogenated or partially hydrogenated oil.

Steroids

Steroids are found in all plants and animals and are derived from the following four-ring structure:

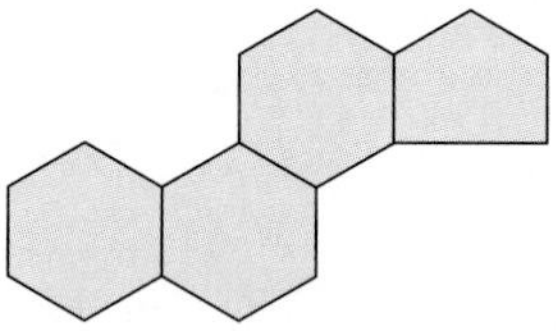

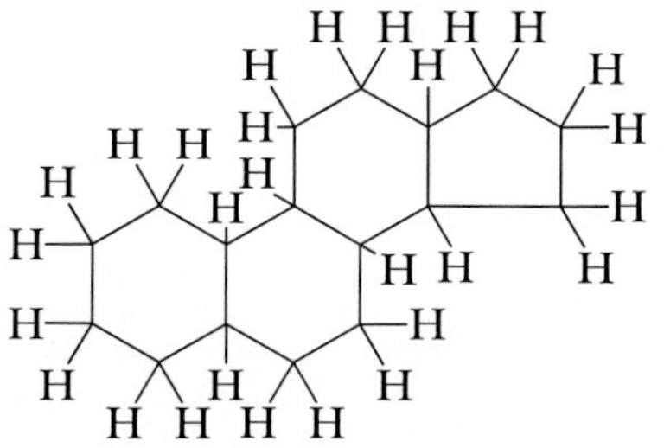

The skeletal four-ring structural drawing on the left is chemical shorthand similar to that described for cyclic hydrocarbons in Section 12.5. There is a carbon at each corner, and the lines represent C—C bonds. Since every carbon atom forms four bonds, additional bonds between carbon atoms and hydrogen atoms are understood to be present whenever the skeletal structure shows fewer than four bonds. The structure on the right shows the hydrogen atoms understood to be present in the four-ring structure shown on the left. Although all the rings in the skeletal drawing are shown as saturated rings, steroids often have one ring that is unsaturated or aromatic. For example,

■ "Aromatic" means a benzene-like structure.

SCIENCE AND SOCIETY

Cis *and* Trans *Fatty Acids and Your Health*

The direct link between diets high in saturated fats and heart disease is well known. Saturated fatty acids are usually found in solid or semisolid fats, while unsaturated fatty acids are usually found in oils. As a result, nutritionists recommend the use of liquid vegetable oils for cooking, and people are generally aware of this. However, what is less well known is that partially hydrogenated oils are a health hazard because of the formation of *trans* fatty acids during the process by which margarine and semisolid cooking fats are made. Hydrogen is added to the C=C double bonds in unsaturated fats (oils) to convert them to a solid or semisolid fat that has better consistency and less chance for spoilage. This hydrogenation process decreases the number of double bonds but also forms unnatural *trans* fatty acids from natural *cis* fatty acids. The *trans* fatty acids are not metabolized in the human system, but they can be stored for the life of the individual because the *trans* fatty acids are "straight" molecular structures and pack together like the saturated fatty acids. By contrast, the *cis* fatty acids are bent and do not pack well (see below).

You can reduce the health risks of *trans* fatty acids by not eating processed vegetable fats. How do you know which products were made with processed vegetable fats? Read the label. For example, a box of cookies or crackers may say on the label "made with 100% pure vegetable shortening . . . (partially hydrogenated soybean oil with hydrogenated cottonseed oil)." Although soybean oil and cottonseed oil are low in saturated fat, the hydrogenation process converts cottonseed oil to a saturated fat, and the partially hydrogenated soybean oil contains *trans* fatty acid.

Some of the many products that contain partially hydrogenated vegetable oils. (C.D. Winters)

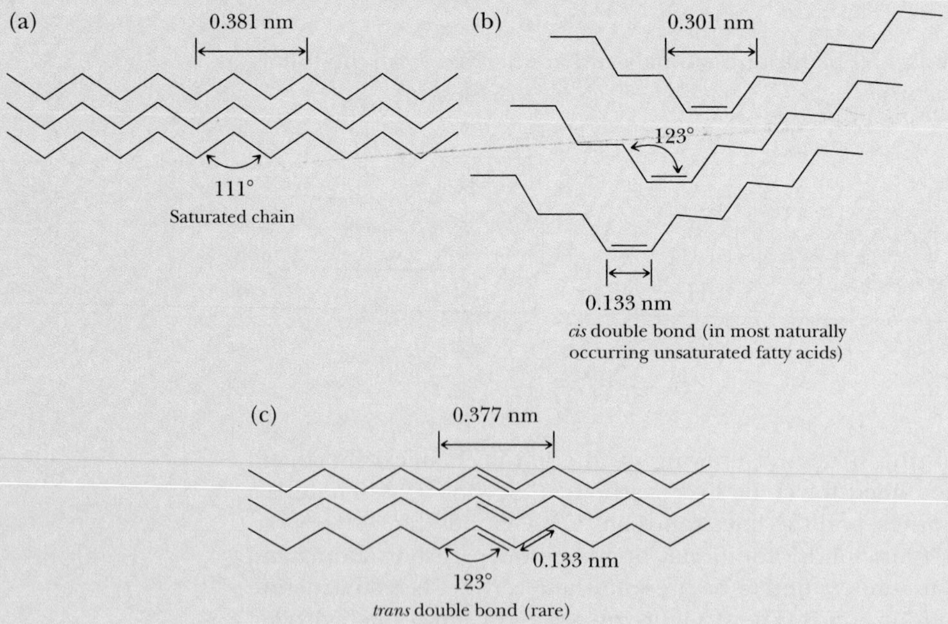

Saturated fatty acids (a) and *trans* fatty acids (c) pack more tightly than *cis* fatty acids (b).

cholesterol has one double bond in the second ring. Note that its structure also includes alkyl groups and an alcohol group. The functional groups replace hydrogen atoms in the skeletal representation.

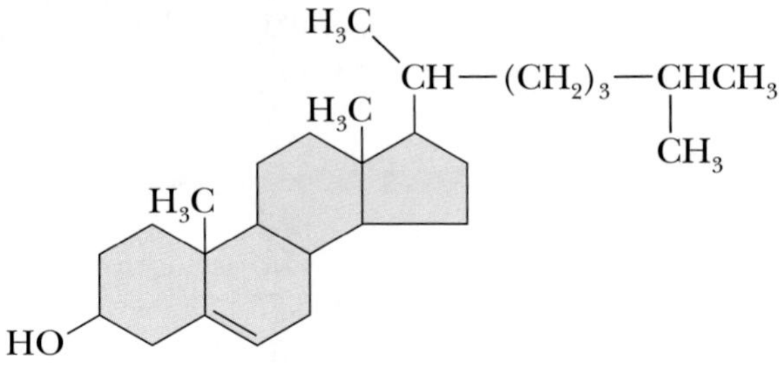

Cholesterol

Cholesterol is the most abundant animal steroid. The human body synthesizes cholesterol and readily absorbs dietary cholesterol through the intestinal wall. An adult human contains about 250 grams of cholesterol. Cholesterol receives a lot of attention because high blood cholesterol levels are associated with heart disease (Section 17.4). Proper amounts of cholesterol are essential to our health, because cholesterol undergoes biochemical alteration to give milligram amounts of many important hormones, such as vitamin D, cortisone, and the sex hormones.

Sex Hormones

Cholesterol is the starting material for the synthesis of steroid sex hormones. One female sex hormone, **progesterone,** differs only slightly in structure from the male hormone, **testosterone.**

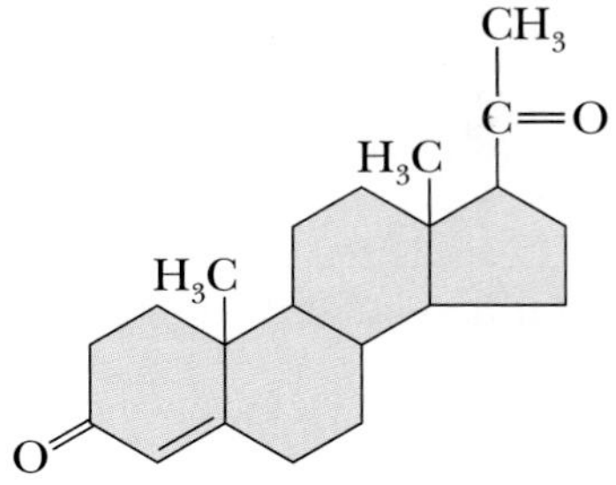

Progesterone

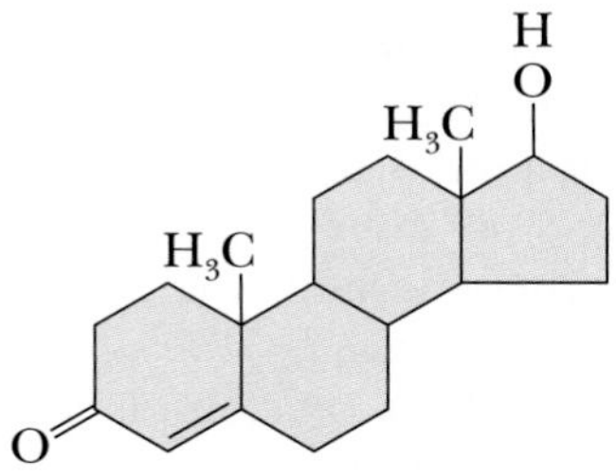

Testosterone

Other female hormones are estradiol and estrone, together called **estrogens.** The estrogens contain an aromatic ring, which differentiates them from the steroids shown above.

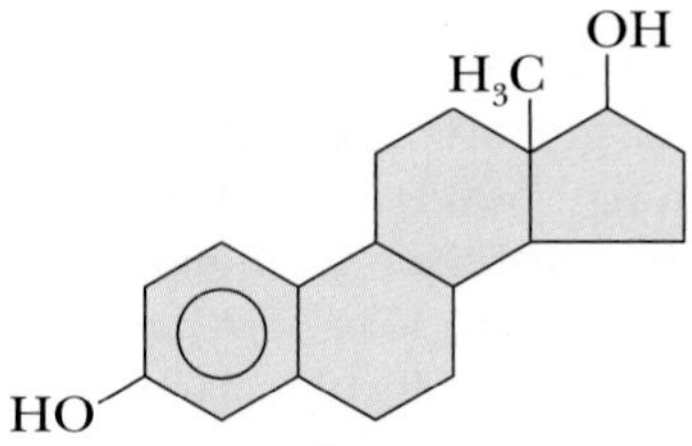

Estradiol

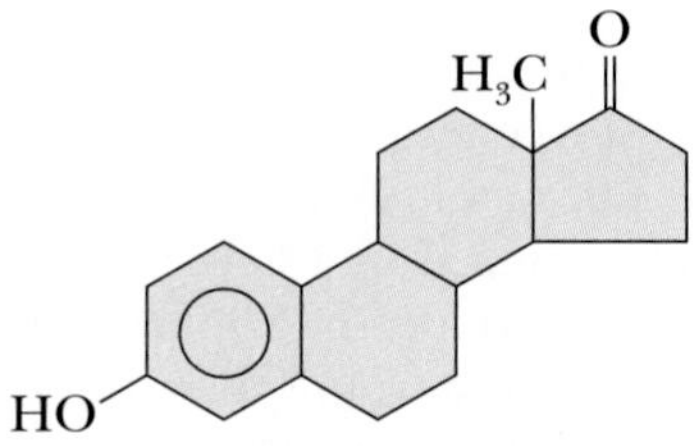

Estrone

The estrogens and progesterone are produced by the ovaries. Estrogens are important to the development of the egg in the ovary, whereas progesterone causes changes in the wall of the uterus and after fertilization prevents release of a new egg from the ovary (ovulation). Birth control drugs use derivatives of estrogens and progesterone to simulate the hormonal process resulting from pregnancy and thereby prevent ovulation (Section 19.4).

Waxes

Waxes are esters formed from long-chain (16 or more carbon atoms) fatty acids and long-chain alcohols. The general formula of a wax is the same as that of a simple ester, RCOOR′, with the qualification that R and R′ are limited to alkyl groups with a large number of carbon atoms. Natural waxes are usually mixtures of several esters. Wax coatings on leaves help to protect the leaves from disease and also help the plant to conserve water. The feathers of birds are also coated with wax. Our ears are protected by wax. Several natural waxes have been used in consumer products. These include carnauba wax (from a Brazilian palm tree), which is used in floor waxes, automobile waxes, and shoe polishes, and lanolin (from lamb's wool), which is used in cosmetics and ointments. Lanolin also contains cholesterol.

■ SELF-TEST 15C

1. The complete hydrolysis of a polysaccharide yields ____________.
2. When a molecule of sucrose is hydrolyzed, the products are one molecule each of the monosaccharides ____________ and ____________.
3. Starch is a condensation polymer built of ____________ monomers.
4. Cellulose is a condensation polymer built of ____________ monomers.
5. ____________ bonding holds polysaccharide chains together, side by side, in cellulose.
6. The sugar referred to as blood sugar, grape sugar, or dextrose is actually the compound ____________.
7. Fats and oils are esters of ____________ and ____________.
8. The structural difference between a saturated fat and an unsaturated fat is ____________.
9. Cholesterol is a (a) steroid, (b) protein, (c) sex hormone, (d) carbohydrate.
10. Cholesterol is essential to our health. (a) True, (b) False.
11. Waxes are esters of ____________ and ____________.

15.7 Energy and Biochemical Systems

Energy for life's processes comes from the sun. During photosynthesis, green plants absorb energy from the sun to make glucose and oxygen from carbon dioxide and water. Glucose is a major energy source for all living organisms. The energy stored in glucose is eventually transferred to the bonds in molecules such as adenosine triphosphate (ATP, Figure 15.21). When living organisms need energy, phosphate bonds in the ATP molecules are broken to give adenosine diphosphate (ADP, Figure 15.22) and energy for other biochemical reactions.

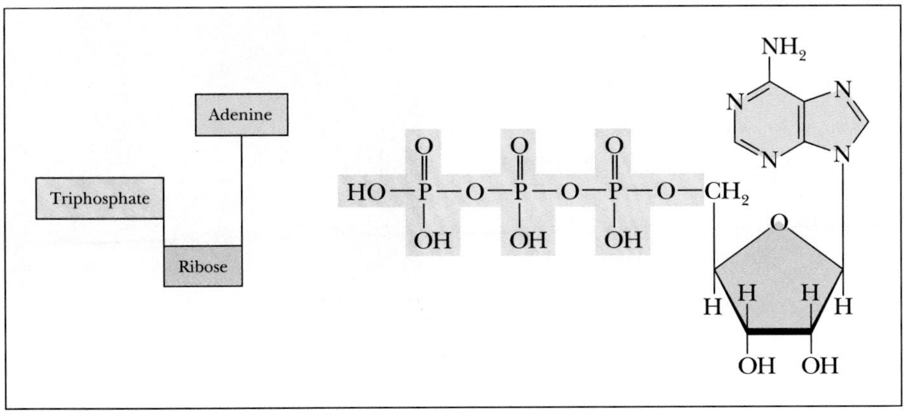

Figure 15.21 Adenosine triphosphate (ATP).

In the complex process of photosynthesis, carbon dioxide is reduced to make sugar, and water is oxidized to oxygen:

$$\underset{\text{Carbon dioxide}}{6\ CO_2} + \underset{\text{Water}}{6\ H_2O} + \underset{\text{Energy (sunlight)}}{688\text{ kcal}} \longrightarrow \underset{\text{Glucose}}{C_6H_{12}O_6} + \underset{\text{Oxygen}}{6\ O_2}$$

The oxygen produced in photosynthesis is the continuing source of all of the oxygen in our atmosphere. We are dependent on the plant life of our planet, and we must live in balance with the oxygen output of that plant life, as well as with the food output of the same plant life. Photosynthesis is absolutely vital to life on Earth.

Photosynthesis is generally divided into a series of **light reactions,** which occur only in the presence of light energy, and a series of **dark reactions,** which do not depend on light energy. The dark reactions feed instead on high-energy compounds (such as ATP) produced by the light reactions.

■ The "dark reactions" also occur in the presence of light but do not require it.

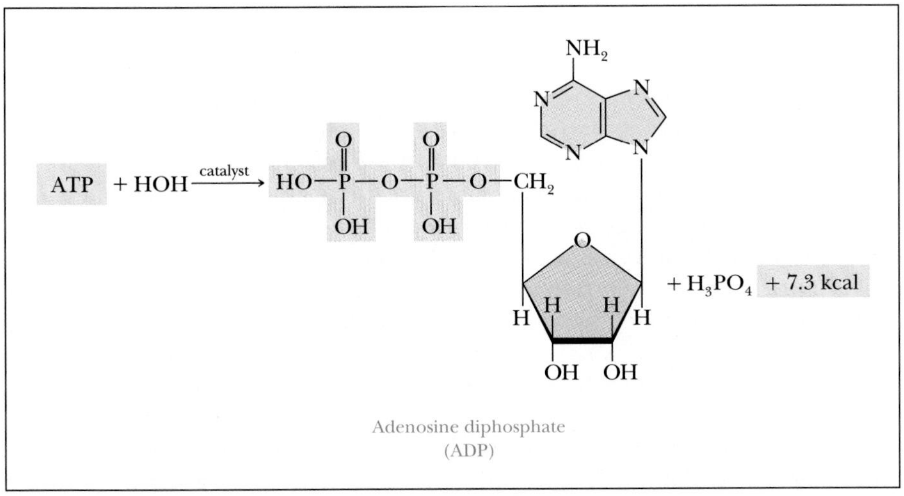

Figure 15.22 Hydrolysis of ATP to give ADP (adenosine diphosphate).

Figure 15.23 (a) Chlorophyll, which is similar in structure to hemoglobin (Figure 15.7*c*). (b) Algae floating on a pond. Chlorophyll is responsible for the green color of algae and other plants. Bubbles of oxygen being formed by photosynthesis can be seen among the algae. *(b, C.D. Winters)*

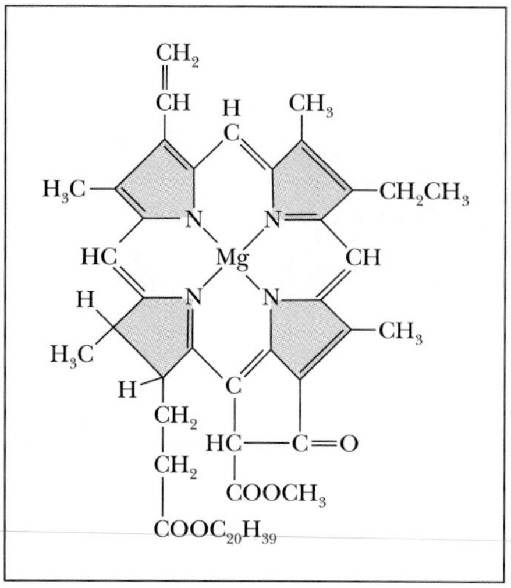

(a)

(b)

Photosynthesis is initiated by light energy. Green plants contain certain pigments that readily absorb light in the visible region of the spectrum. The most important of these is ***chlorophyll*** (Figure 15.23*a*). Note that in chlorophyll Mg^{2+} is bonded to a complex ring structure similar to the one surrounding Fe^{2+} in hemoglobin (Figure 15.7*c*). Chlorophyll is green because it absorbs light in the violet region (about 400 nm) and the red region (about 650 nm) and allows the green light between those wavelengths to be reflected or transmitted.

When chlorophyll absorbs light, electrons are excited to higher energy levels. As these electrons return to the ground state, very efficient subcellular components of the plant cell known as chloroplasts absorb the energy that is released. Through a series of reactions, water is oxidized to oxygen, and energy is stored in the bonds of ATP and a few other energy-storage compounds.

The function of the dark reactions is to convert CO_2 and hydrogen from water into glucose and other carbohydrates, and these reactions are driven by the release of energy stored in ATP by the light reactions. In the presence of a suitable catalyst, ATP undergoes hydrolysis, as shown in Figure 15.22, to give adenosine diphosphate (ADP) with the release of about 7.3 kcal/mol.

A living plant may convert the glucose from photosynthesis to disaccharides, polysaccharides, starches, cellulose, proteins, or oils. The end product depends on the type of plant involved and the complexity of its biochemistry. The various compounds in the plants are then available to be eaten by a plant-eating animal, digested, transported to the cells of the body, and metabolized to provide energy and chemical compounds.

Plants, then, are the primary source of energy for animals and humans. This can be represented by the oxidation of glucose,

$$C_6H_{12}O_6 + 6\ O_2 \longrightarrow 6\ CO_2 + 6\ H_2O + 688\ \text{kcal}$$

which is the reverse of the reaction taking place in photosynthesis. Some of this energy is used immediately, and some is stored through a series of coupled

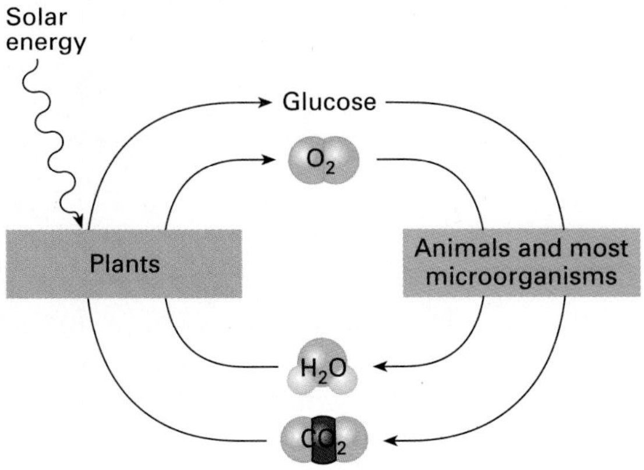

Figure 15.24 The flow of carbon atoms, oxygen atoms, and energy between plants and animals.

reactions involving the conversion of ADP to ATP (the reverse of the reaction illustrated in Figure 15.22). This series of coupled reactions can be represented by the overall equation below, in which the numbers of the reactants are estimations based on current knowledge:

$$C_6H_{12}O_6 + 36\ ADP + 36\ H_3PO_4 + 6\ O_2 \longrightarrow 6\ CO_2 + 36\ ATP + 42\ H_2O$$

Human beings, unlike some animals, eat both plants and animals. Our biochemistry is devoted to converting to energy the complex mixture of fats, carbohydrates, and proteins in our diets. Along the way, we must break these molecules down and use some of the products to synthesize biomolecules according to our own needs. Thus, the carbon and oxygen atoms in our molecules and the energy stored in their bonds are constantly cycling between plants and animals (Figure 15.24). In the next chapter, we'll be looking more closely at the part of this cycle that relates to what we eat.

15.8 Nucleic Acids

The genetic information that makes each organism's offspring look and behave like its parents is encoded in molecules called **nucleic acids.** Together with a set of specialized enzymes that catalyze their synthesis and decomposition, the nucleic acids constitute a remarkable system that accurately copies millions of pieces of data with very few mistakes.

Like polysaccharides and polypeptides, nucleic acids are condensation polymers. Each monomer in these polymers includes one of two simple sugars, one phosphoric acid group, and one of a group of heterocyclic nitrogen compounds that behave chemically as bases. A particular nucleic acid is a **deoxyribonucleic acid (DNA)** if it contains the sugar α-2-deoxy-D-ribose, and it is a **ribonucleic acid (RNA)** if it contains the sugar α-D-ribose.

The five organic bases that play a key role in the mechanism for information storage are ***adenine*** (A) and ***guanine*** (G), ***thymine*** (T), ***cytosine*** (C), and ***uracil*** (U). These bases are mentioned so often in any discussion of nucleic acid chemistry that, in order to save space, they are usually referred to only by the first letter of each name.

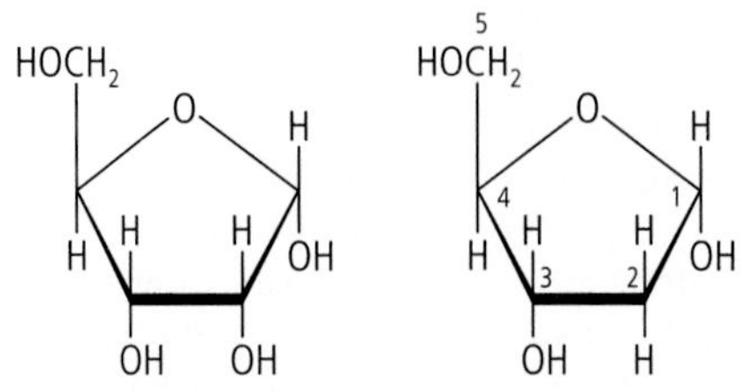

α-D-ribose *α-2-deoxy-D-ribose*

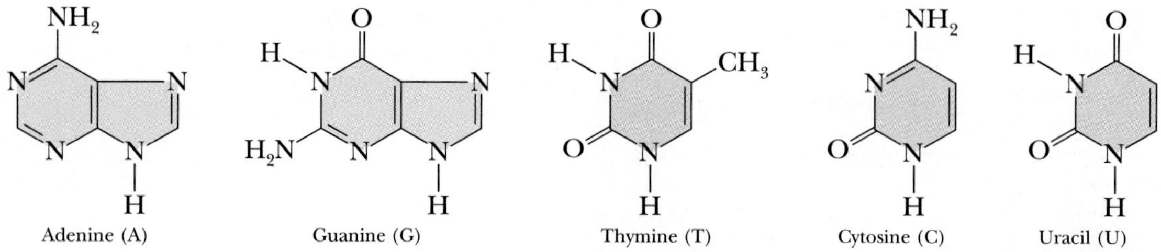

Nucleic acids are found in all living cells, with the exception of the red blood cells of mammals. DNA occurs primarily in the nucleus of the cell, and RNA is found mainly in the cytoplasm, outside the nucleus. There are three major types of RNA, each with its own characteristic size, base composition, and function in protein synthesis (as described later in this section): messenger RNA (mRNA), transfer RNA (tRNA), and ribosomal RNA (rRNA).

The monomers that polymerize to make both DNA and RNA are known as **nucleotides.** They have structures like the one shown in Figure 15.25*a*. Each nucleotide contains a phosphoric acid unit, a ribose sugar unit, and one of the five bases. The nucleotides of DNA and RNA have two structural differences: (1) they contain different sugars (deoxyribose and ribose), and (2) the base uracil occurs only in RNA, whereas the base thymine occurs only in DNA. The other bases—adenine, guanine, and cytosine—are found in both DNA and RNA.

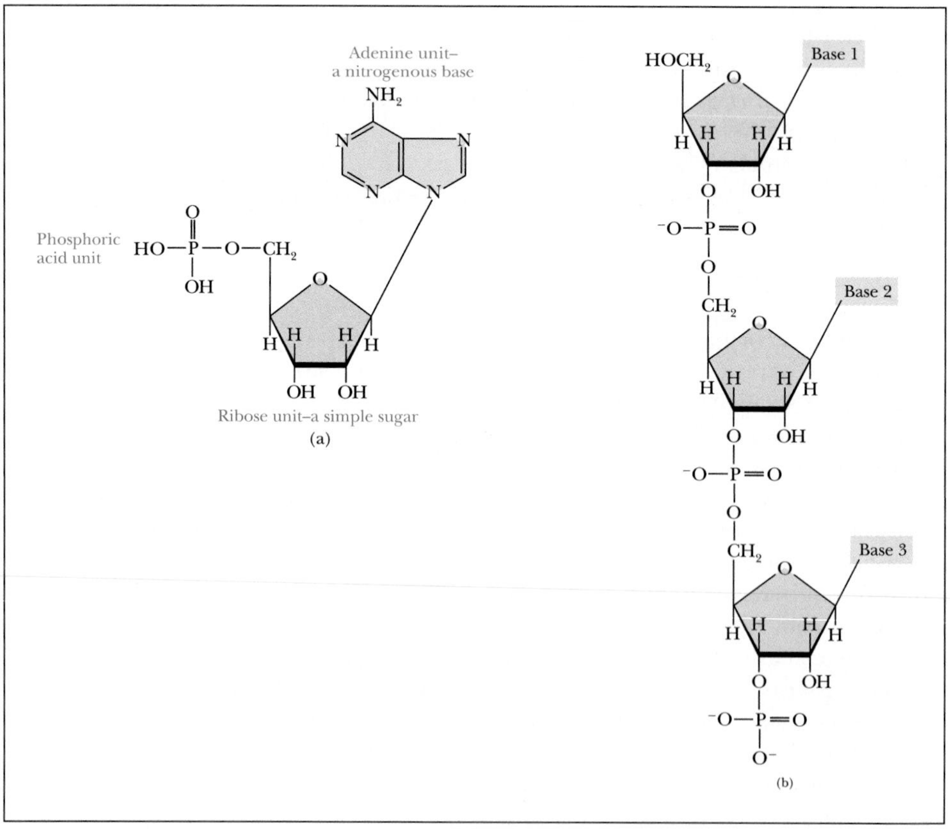

Figure 15.25 (a) A nucleotide. Because its sugar unit is ribose, this is a ribonucleotide. (b) Bonding in a trinucleotide. The sugars could also be deoxyribose units and the bases can be any of the five bases shown in the text. The primary structures of both DNA and RNA are extensions of this structure.

Figure 15.26 James Watson (on the right) and Francis Crick (on the left), shortly after their discovery of the structure of DNA in 1953. The work was carried out in the Cavendish Laboratory at Cambridge, England, where the photograph was taken. Information from studies of the structure of DNA by Maurice Wilkins and Rosalind Franklin contributed to the insight of Watson and Crick. In 1962 the Nobel Prize for Physiology and Medicine was awarded jointly to Watson, Crick, and Wilkins. *(A. Barrington Brown/Photo Researchers)*

When the phosphate group is absent, the remaining fragment (a ribose bonded to one of the five bases) is called a **nucleoside.**

Polynucleotides have molar masses ranging from about 25,000 for tRNA molecules to billions for human DNA. The sequence of nucleotides in the polymer chain (shown by the base sequence) is its primary structure. Polynucleotides are formed by the polymerization of nucleotides to make esters. As an example, Figure 15.25*b* shows three monomers condensed to a trinucleotide.

In 1953, James D. Watson and Francis H. C. Crick (Figure 15.26) proposed a secondary structure for DNA that revolutionized our understanding of heredity and genetic diseases. They proposed that pairs of polynucleotides are arranged in a double helix stabilized by hydrogen bonding between the base groups lying opposite each other in the two chains. The critical point of the Watson-Crick model is that hydrogen bonding can best occur between specific bases. Adenine-thymine (A---T) and guanine-cytosine (G---C) pairs occur exclusively because they are very tightly hydrogen-bonded. The hydrogen bonding between these specific bases, called **complementary hydrogen bonding,** is illustrated in Figure 15.27.

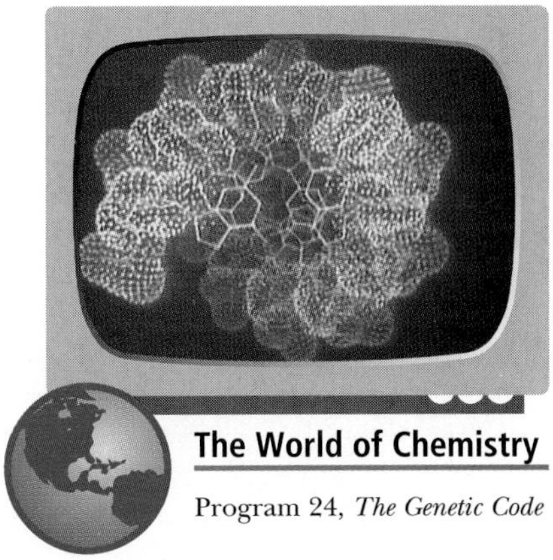

The World of Chemistry

Program 24, *The Genetic Code*

The DNA helix viewed from the top.

The function of polynucleotides is to transcribe hereditary information so that like begets like. The almost infinite variety of primary structures of polynucleotides allows an almost infinite variety of information to be recorded in the molecular structures of the strands of nucleic acids. The different arrangements of just a few different bases give the large variety of structures. In a somewhat similar fashion, the multiple arrangements of just a few language symbols convey the many ideas in this book. The coded information in the polynucleotide controls the inherited characteristics of the next generation as well as most of the continuous life processes of the organism.

■ RNA is generally a single strand of helical polynucleotide.

■ The inherited traits of an organism are controlled by DNA molecules.

Double-stranded DNA forms the 46 human chromosomes, which have specialty heredity areas called **genes.** Genes are segments of DNA that have as few as 1000 or as many as 100,000 base pairs such as those shown in Figure 15.27. Each gene holds the information needed for the synthesis of a single protein. Human DNA (the human **genome**) is estimated to have up to 100,000

■ The total sequence of base pairs in the cells of a plant or animal is called its *genome.*

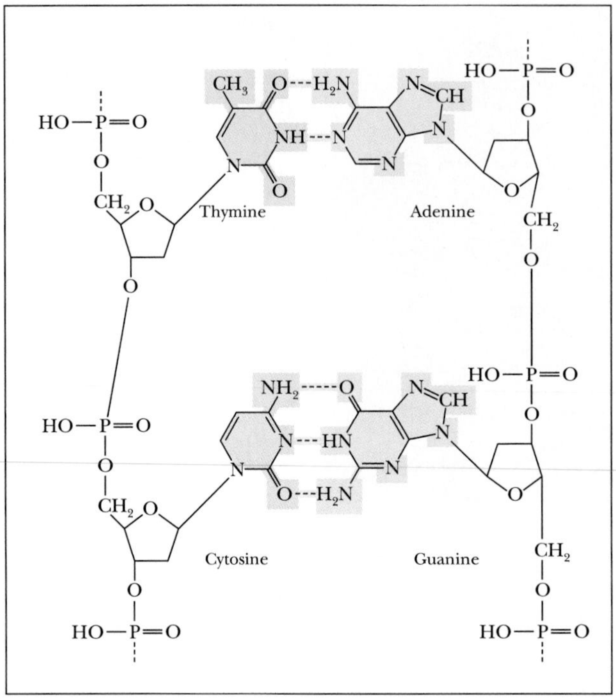

Figure 15.27 Complementary hydrogen bonding in the DNA double helix. Hydrogen bonds in the thymine–adenine (T–A) and cytosine–guanine (C–G) pairs stabilize the double helix.

genes and about three billion pairs of bases (see Section 15.10). However, genes are estimated to make up only 3% of DNA, with each gene sandwiched between ''junk'' or noncoding DNA sequences. There are also short segments that act as switches to signal where the coding sequence begins.

Replication of DNA: Heredity

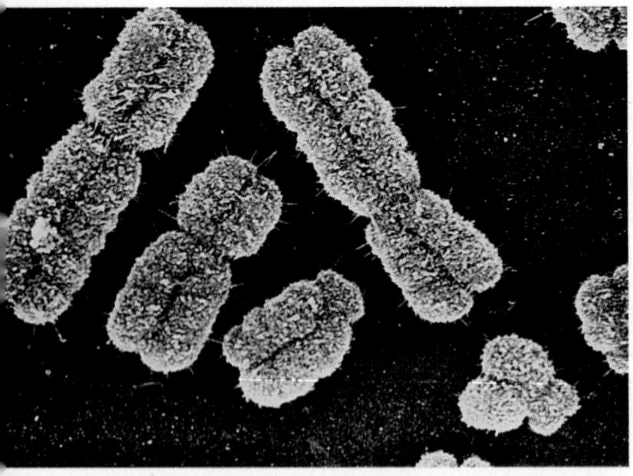

Human chromosomes (magnified about 8000 times). DNA assembles into chromosomes when a cell is about to divide. *(Biophoto Associates/Photo Researchers)*

The transfer of coded information begins with the replication of DNA and continues with natural protein synthesis. Almost all nuclei in an organism's cells contain the same chromosomal composition. This composition remains constant regardless of whether the cell is starving or has an ample supply of food materials. Each organism begins life as a single cell with this same chromosomal composition; in sexual reproduction, half of a chromosome comes from each parent. These well-known biological facts, along with recent discoveries concerning polynucleotide structures, have led scientists to the conclusion that the DNA structure is faithfully copied during normal cell division (mitosis—both strands) and that only half is copied in cell division that produces reproductive cells (meiosis—one strand).

In **replication** the double helix of the DNA structure unwinds, and each half of the structure serves as a template, or pattern, from which the other complementary half can be reproduced from the molecules in the cell environment (Figure 15.28). Replication of DNA occurs in the nucleus of the cell before the cell divides.

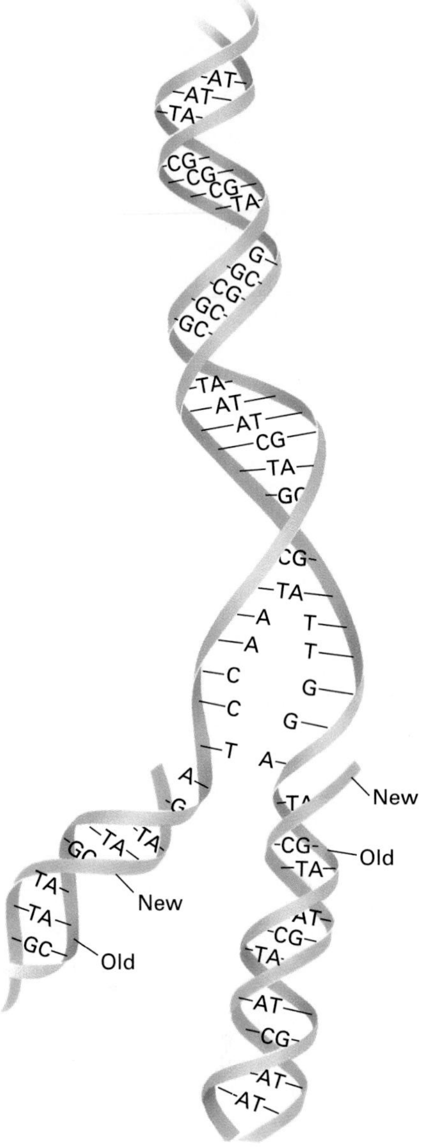

Figure 15.28 Replication of DNA. When the double helix of DNA (blue) unwinds, each half serves as a template on which to assemble nucleotides to form new DNA strands.

Natural Protein Synthesis

The proteins of the body are continually being replaced and resynthesized from the amino acids available to the body. The use of isotopically labeled amino acids (see Section 5.10) has made possible studies of the average lifetimes of amino acids as constituents in proteins—that is, the time it takes the body to replace a protein in a tissue. For a process that must be extremely complex, replacement is very rapid. Only minutes after radioactive amino acids are injected into animals, radioactive protein can be found. Although all the proteins in the body are continually being replaced, the rates of replacement vary. Half of the proteins in the liver and plasma are replaced in six days; the time needed for replacement of muscle proteins is about 180 days, and replacement of protein in other tissues, such as bone collagen, takes even longer.

Recall that each organism has its own kinds of proteins. The number of possible unique arrangements of 20 amino acid units is 2.43×10^{18}, yet proteins characteristic of a given organism can be synthesized by the organism in a matter of a few minutes!

The DNA in the cell nucleus holds the code for protein synthesis, which is carried out in a series of steps summarized in Figure 15.29. First, messenger RNA, like all forms of RNA, is synthesized in the cell nucleus. The sequence of bases in one strand of the chromosomal DNA serves as the template from which a single strand of a messenger ribonucleotide (mRNA) is made in a process known as **transcription.** The bases of the mRNA strand complement those of the DNA strand. Two bases are complementary when each one fits the other and forms one or more hydrogen bonds. Messenger RNA contains only

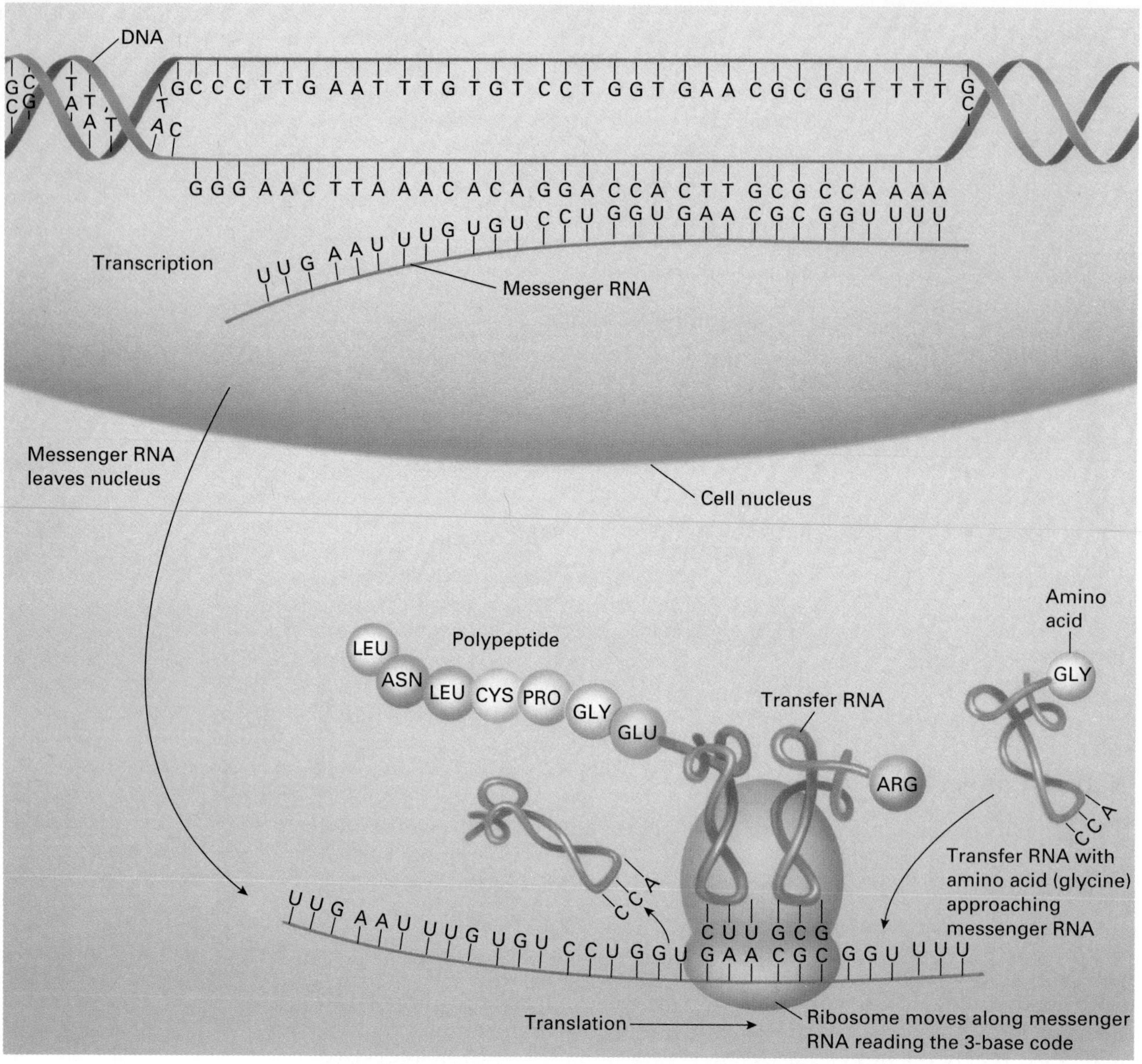

Figure 15.29 Protein synthesis. A section of DNA unwinds and transcription results in production of messenger RNA. Outside the cell nucleus, at a ribosome, the code carried by messenger RNA is translated into the amino acid sequence of a newly synthesized protein. Transfer RNA brings the amino acids into position one at a time.

the four bases: adenine(A), guanine(G), cytosine(C), and uracil(U). DNA contains principally the four bases adenine(A), guanine(G), cytosine(C), and thymine(T). The bases in DNA are transcribed into bases in mRNA as follows:

DNA	mRNA
A	U
G	C
C	G
T	A

TABLE 15.6 ■ Messenger RNA Codes for Amino Acids*

First Letter of Code	Second Letter of Code: U	C	A	G	Third Letter of Code
	U	C	A	G	
U	Phenylalanine	Serine	Tyrosine	Cysteine	U
	Phenylalanine	Serine	Tyrosine	Cysteine	C
	Leucine	Serine	STOP	STOP	A
	Leucine	Serine	STOP	Tryptophan	G
C	Leucine	Proline	Histidine	Arginine	U
	Leucine	Proline	Histidine	Arginine	C
	Leucine	Proline	Glutamine	Arginine	A
	Leucine	Proline	Glutamine	Arginine	G
A	Isoleucine	Threonine	Asparagine	Serine	U
	Isoleucine	Threonine	Asparagine	Serine	C
	Isoleucine	Threonine	Lysine	Arginine	A
	START or methionine	Threonine	Lysine	Arginine	G
G	Valine	Alanine	Aspartic acid	Glycine	U
	Valine	Alanine	Aspartic acid	Glycine	C
	Valine	Alanine	Glutamic acid	Glycine	A
	Valine	Alanine	Glutamic acid	Glycine	G

*In groups of three (called codons), bases of mRNA code the order of amino acids in a polypeptide chain. A, C, G, and U represent adenine, cytosine, guanine, and uracil, respectively. Some amino acids have more than one codon, and hence more than one tRNA can bring the amino acid to mRNA.

This means that, provided the necessary enzymes and energy are present, wherever a DNA has an adenine base(A), the mRNA will transcribe a uracil base(U), and so on.

After transcription, mRNA passes from the nucleus of the cell to a ribosome (which contains the rRNA), where mRNA serves as the template for the sequential ordering of amino acids during protein synthesis by the process known as **translation.** As its name implies, messenger RNA contains the sequence message, in the form of a three-base code (called a **codon**), for ordering amino acids into proteins (Table 15.6). Each of the thousands of different proteins synthesized by cells is coded by a specific mRNA or segment of an mRNA molecule.

■ Protein synthesis is initiated by START codons and terminated by STOP codons (Table 15.6).

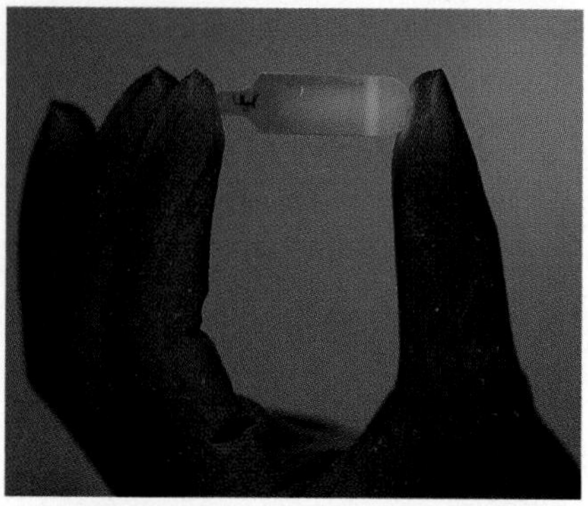

A tube full of DNA. A few dozen years ago, preparing this large a sample of DNA was almost impossible. *(Jean Claude Revy/Phototake)*

Transfer RNAs carry the amino acids to the mRNA one by one. Each of the 20 amino acids found in proteins has at least one corresponding tRNA, and some have multiple tRNAs (Table 15.6). Table 15.6 lists the RNA codes and shows in the first line that, for example, UUU codes for phenylalanine and UCU codes for serine.

At one end of a tRNA molecule is a trinucleotide base sequence (the **anticodon**) that fits a trinucleotide base sequence on mRNA (the codon). At the other end of a tRNA molecule is a specific base sequence of three terminal nucleotides—CCA—with a hydroxyl group on the sugar exposed on the terminal adenine nucleotide group. With the aid of enzymes, this hydroxyl group reacts with a specific amino acid by an esterification reaction.

$$\underset{\text{tRNA}}{(\text{mononucleotides})_{75-90}\text{CCA—OH}} + \underset{\text{Amino acid}}{\text{HO}\overset{\text{O}}{\overset{\|}{\text{C}}}\text{CH(NH}_2\text{)R}} \longrightarrow$$

$$\underset{\text{tRNA-amino acid}}{(\text{mononucleotides})_{75-90}\text{CCA—O}\overset{\text{O}}{\overset{\|}{\text{C}}}\text{CH(NH}_2\text{)R}} + H_2O$$

The tRNA and its amino acid migrate to the ribosome, where the amino acid is used in the synthesis of a protein. The tRNA is then free to migrate back to the cell cytoplasm and repeat the process.

Messenger RNA is used at most only a few times before being depolymerized. Although this may seem to be a terrible waste, it enables the cell to produce different proteins on very short notice. As conditions change, different types of mRNA come from the nucleus, different proteins are made, and the cell responds adequately to a changing environment.

In the schematic illustration in Figure 15.29, which summarizes DNA → RNA → protein, pick a three-base sequence on the bottom DNA strand of the two DNA strands at the top of the figure and follow it through to the bottom, where a tRNA attaches to a three-base codon on mRNA. Assume the DNA strand you have selected serves as the template for the synthesis of the single strand of mRNA shown in the figure. Do you agree with the changes that occur in the letters representing the three-base sequence?

EXAMPLE 15.4 *Nucleic Acids*

If the base sequence in a DNA segment is . . . GCTGTA . . . , what is the base sequence in the complementary mRNA? What is the anticodon order in tRNA?

SOLUTION

The base pairs between DNA and mRNA are G . . . C, A . . . U, and T . . . A. Therefore, the complementary mRNA segment for a DNA segment of GCTGTA is determined by using these allowed base pairs. Start with GCTGTA and place the correct base below it. The resulting mRNA segment is CGACAU. The anticodon order in tRNA is the complement of the mRNA segment. The allowed base pairs are G . . . C and A . . . U, so the anticodon order in tRNA is GCUGUA.

Exercise 15.4

If the sequence of bases along a mRNA strand is . . . UCCGAU . . . , what was the sequence along the DNA template?

15.9 Biogenetic Engineering

The field of biogenetic engineering started after the first successful gene-splicing and gene-cloning experiments produced **recombinant DNA** in the early 1970s. The basic idea is to use the rapidly dividing property of common bacteria, such as *E. coli,* as a microbe factory for producing recombinant DNA molecules that contain the genetic information for the desired product. Bacteria have been produced that can synthesize specific proteins, human growth hormone, and human insulin. To create such bacteria, a gene is removed from the bacterium, part of a gene from a human or other organism (the part that produces human insulin, for example) is spliced in, and the spliced gene is put back into the bacterium. The modified bacterium then serves as a microbe factory to make millions of other insulin-producing bacteria. The process of splicing and recombining genes is referred to as **recombinant DNA technology** or **biogenetic engineering.** Figure 15.30 shows application of this process to production of a disease-free tomato plant.

■ "Recombinant" means capable of genetic recombination, and "recombinant DNA" refers to human control over splitting, splicing, and recombining DNA structures

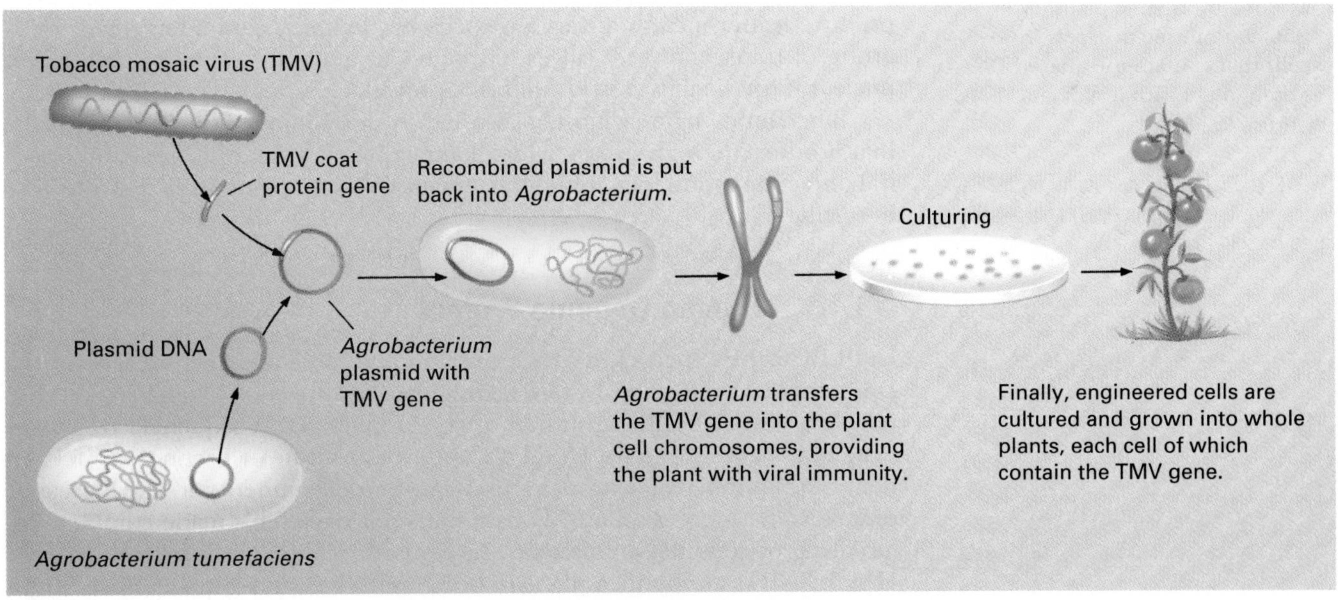

Figure 15.30 Steps necessary to insert a gene from tobacco mosaic virus into a tomato plant in order to make the plant resistant to a viral disease.

A pair of transgenic sheep. These sheep have the human gene for synthesis of α_1-antitrypsin. Persons with an inherited deficiency of this enzyme inhibitor are at high risk for developing emphysema, which affects lung function and causes difficulty in breathing. α_1-Antitrypsin isolated from the sheep's milk was scheduled to begin clinical trials in 1995 for the treatment of emphysema. *(Philippe Plailly/Eurelios/SPL/Photo Researchers)*

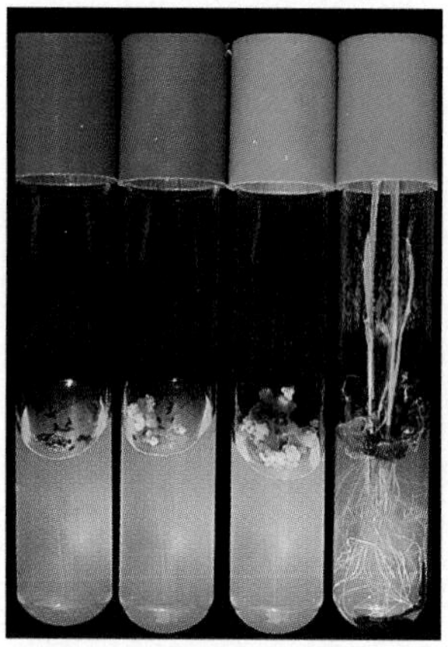

Growth of a transgenic plant. From left to right, you can see progress from a few cells that contain the altered DNA to a young plant with leaves and roots. *(Dan McCoy/Rainbow)*

■ See Section 22.7 for discussion of the Flavr Savr tomato, the first genetically engineered food to be sold to consumers.

One of the earliest benefits of recombinant DNA technology was the biosynthesis of ***human insulin*** in 1978. Millions of people with diabetes depend on the availability of insulin, but many are allergic to animal insulin, which was the only previous source. Biosynthesized human insulin is now being marketed. Biotechnology firms are also producing ***human growth hormone,*** which is used in treating youth dwarfism.

A number of "transgenic animals" have been produced, including goats, rabbits, and mice; these animals are used as drug "pharms" since their new genes cause them to produce marketable quantities of desirable pharmaceuticals. For example, ***tissue-plasminogen-activator (TPA),*** which is used to dissolve blood clots in emergency treatment of heart-attack victims, was originally obtained from a bioreactor of *E. coli,* then from the milk of transgenic mice, and now from the milk of transgenic goats.

Sometimes, transgenic animals are less than ideal. Transgenic pigs did produce leaner meat but they have arthritis, lethargy, and a low sex drive. A group of transgenic beef calves did better in general health, but it is as yet unclear if the quality of beef will be improved.

Biogenetic engineering has also led to developments in growing plants that produce their own insecticide, making plants resistant to viral and bacterial infections, and matching the chemistry of a plant with a protecting herbicide.

15.10 Human Genome Project

Until 1986, the experimental determination of the sequence of base pairs in a DNA strand was laboriously slow, taking place at a speed of about 200,000 base pairs per year. With the invention of an automatic DNA sequencer by a team headed by Professor Leroy Hood at California Institute of Technology in Pasadena, the number of base pairs that can be determined per day increased dramatically (more than 10,000 base pairs per day). This made complete sequencing of genomes a possibility. In January 1989 the National Institutes of Health (NIH) announced plans to determine the complete sequence of the human genome, estimated to contain three billion base pairs. The first five-year phase of the project involves the study of some model genome systems, and research groups are working on sequencing the base pairs in *E. coli,* estimated to contain 4.5 million base pairs; a yeast genome estimated to contain 12.5 million base pairs; and a nematode genome with 100 million base pairs. It is hoped that the techniques and knowledge gained from the study of these model systems along with continued improvement in the number of base pairs that can be sequenced per day will make it possible to sequence the base pairs in human DNA in 15 years. A complete mapping of the human genome will improve our knowledge of the estimated 4000 hereditary diseases and lead to better diagnosis and treatment of these diseases.

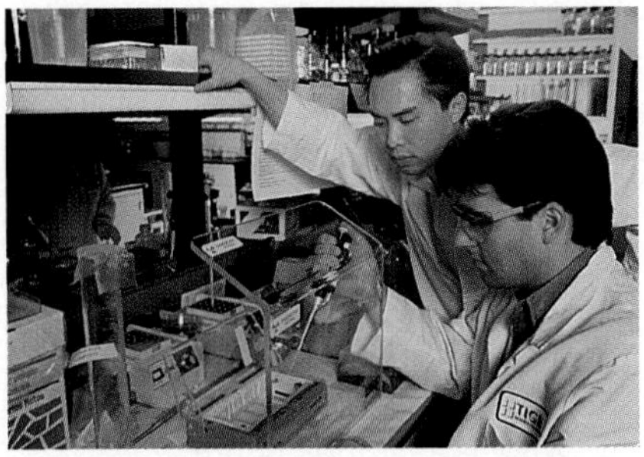

Genome research. The scientist is loading radioactively tagged DNA segments onto a gel for separation according to their composition, one of the steps in mapping the human genome. The photo was taken at The Institute for Genomic Research in Maryland, one of many institutions around the world engaged in such studies. *(Hank Morgan/ Photo Researchers)*

SELF-TEST 15D

1. The reactants in the photosynthesis process are ____________ and ____________; ____________ must also be supplied.
2. Most of the energy obtained by the oxidation of food is used immediately to synthesize the molecule ____________.
3. The hydrolysis of ATP produces ____________ and phosphoric acid. ____________ is also released.
4. The basic code for the synthesis of protein is contained in the ____________ molecule.
5. The sugar in RNA is ____________, whereas the one in DNA is ____________.
6. A nucleotide contains ____________, ____________, and ____________.
7. The secondary structure of DNA is in the shape of a(an) ____________.
8. When DNA replicates itself, each base in the chain is matched to another one via ____________ bonds.
9. Base pairs in DNA are formed between A and ____________, G and ____________, T and ____________, U and ____________.
10. Human DNA is estimated to have ____________ base pairs.

MATCHING SET

____ 1. Energy "cash" in the living cell	a. ADP + energy
____ 2. Natural protein	b. Chlorophylls
____ 3. Product of ATP hydrolysis	c. Structure determined by DNA and RNA
____ 4. Molecules that absorb light energy	d. ATP
____ 5. D-glucose	e. Polymer consisting of α-D-glucose monomers
____ 6. Enzymes	f. Proteins
____ 7. Carbohydrate stored in animals	g. Main sugar present in the blood
____ 8. Starch	h. Polynucleotide
____ 9. Polypeptides	i. Biochemical catalysts
____ 10. DNA	j. Glycogen
____ 11. Fibrous protein	k. Collagen
____ 12. Cellulose	l. Polymer consisting of β-D-glucose monomers

QUESTIONS FOR REVIEW AND THOUGHT

1. What is meant by the term "chiral"?
2. What do the following terms mean?
   (a) L-isomers (b) D-isomers
   (c) Optically active (d) Polarimeter
3. What is a racemic mixture?
4. What is an amino acid? Draw and label a generalized structure for an amino acid.
5. What is a zwitterion? Draw the zwitterion for glycine.
6. What is a peptide bond?
7. What is an essential amino acid?
8. What functional groups are always present in each molecule of an amino acid?
9. What is meant by the following terms?
   (a) Dipeptide (b) Polypeptide
   (c) Amino acid residue (d) Protein
10. What is meant by the term "amino acid sequence"?
11. Define "neuropeptides" and give two examples.
12. What are the meanings of the terms "primary," "secondary," and "tertiary" structures of proteins?
13. Which of the following biochemicals are polymers?
   (a) Starch, (b) Cellulose, (c) Glucose, (d) Fats, (e) Glycylalanylcysteine, (f) Proteins, (g) DNA, (h) RNA.
14. What is meant by the following terms?
   (a) α-helix (b) β-pleated sheet
15. What does it mean to "denature" a protein?
16. What is the relationship between an enzyme and a substrate?
17. What is an active site on an enzyme?
18. What is lactose intolerance?
19. Give definitions for the following terms.
   (a) Carbohydrate (b) Monosaccharide
   (c) Disaccharide (d) Polysaccharide
20. What is an artificial sweetener?
21. What is Type I diabetes? What is Type II diabetes?
22. What are the following substances?
   (a) Starches (b) Glycogen
   (c) Cellulose
23. What is a lipid?
24. What is an unsaturated fatty acid? What is a saturated fatty acid?
25. What is the process of hydrogenation of unsaturated molecules?
26. What are *cis* and *trans* fatty acids? Sketch their structures to show how they differ.
27. Describe or give definitions for the following.
   (a) Photosynthesis (b) Light reactions
   (c) Dark reactions (d) Chlorophyll
28. What three molecular units are found in nucleotides?
29. What are the differences in structure between DNA and RNA?
30. Glutathione is an important tripeptide found in all living tissues. It is also named glutamylcystylglycine. Draw the structure of glutathione. Which enantiomeric forms of the amino acids would you predict are used to synthesize this peptide?
31. (a) How many tetrapeptides are possible if four amino acids are linked in different combinations that contain all four amino acids?
   (b) Write these combinations for tetrapeptides made from glycine, alanine, serine, and cystine. Use three-letter abbreviations for the amino acids. For example, one combination is Gly-Ala-Ser-Cys.
32. Use three-letter abbreviations to write the possible tripeptides that can be formed from phenylalanine, serine, and valine if each tripeptide contains all three amino acids.
33. Name the following tripeptide.

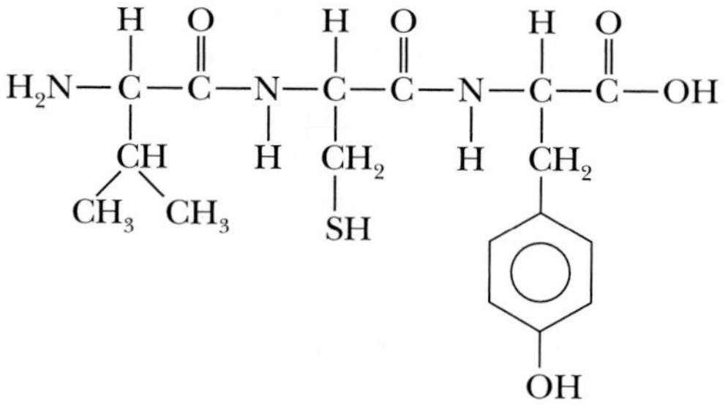

34. Draw the structure of alanylglycylphenylalanine.
35. The human body can metabolize D-glucose but not L-glucose. Explain.
36. What is the purpose of ATP?
37. What are the fundamental components of nucleic acids?
38. Explain the terms "codon" and "anticodon."
39. What is meant by the term "complementary bases"?
40. How many hydrogen bonds are possible between the following complementary base pairs?
   (a) Cytosine and guanine (b) Thymine and adenine
41. What is the Human Genome Project?
42. A segment of a DNA strand has the base sequence . . . GCTGTAACCGAT . . . .
   (a) What is the base sequence in the complementary mRNA?
   (b) What is the anticodon order in tRNA?
   (c) Consult Table 15.6 and give the amino acid sequence in the portion of the peptide being synthesized.
43. A segment of a DNA strand has the base sequence . . . TGTCAGTGGGCCGCT . . . .
   (a) What is the base sequence in the complementary mRNA?

(b) What is the anticodon order in tRNA?
(c) Consult Table 15.6 and give the amino acid sequence in the portion of the peptide being synthesized.

44. Write the overall equation for photosynthesis.
45. (a) Draw the structure of the amino acid in which R = $—CH_2—SH$.
(b) What is its name?
46. What are the basic differences between DNA and RNA structures?
47. What stabilizing forces hold the double helix together in the secondary structure of DNA proposed by Watson and Crick?
48. What is recombinant DNA?
49. Give two examples of successful applications of recombinant DNA technology.
50. (a) What are the three major types of RNA?
(b) What are their functions?
51. What practical problems exist in blending very potent sweeteners into foods?
52. A base sequence in a DNA segment is . . . GTAGC . . . . . What is the base sequence in the complementary mRNA?
53. (a) What are the three bases present in both messenger RNA and DNA?
(b) What is the fourth base found in mRNA?
(c) What is the fourth base found in DNA but not in mRNA?
54. Give the mRNA complementary base sequence that matches the following DNA segment base sequence.

DNA sequence	**mRNA sequence**
G . . .	
A . . .	
C . . .	
A . . .	

55. There are an estimated 4000 genetic diseases. If the origins of these diseases can be identified through the Human Genome Project and additional research, should this information be used to identify high-risk people and should this information be available to health insurers as well as possible employers? Give two reasons for and two reasons against this kind of disclosure.

CHAPTER 16

Consumer Chemistry—Looking Good and Keeping Clean

We routinely spend our money for a variety of chemicals. Some are used to make us look more presentable (cosmetics), while others are used to clean things. To reap the full benefit of the chemicals we purchase, we should know something about their composition, the properties that give the desired effects, and the precautions to be taken in their use. Whether it is the selection of a skin cream, a shampoo, a laundry bleach, a cleaning compound to remove a stain, or a sunscreen to take to the beach, a knowledge of chemistry can help us to make better choices.

- What are the structures of skin and hair, and how can chemicals such as those found in cosmetics affect those structures?
- What are some of the common kinds of cosmetics, what do they contain, and how do they function?
- What are the effects of ultraviolet light on skin, how do sunscreens work, and what are sunscreen protection factors (SPFs)?
- What causes the natural color of hair, and how can its color be changed?
- Why is hair straight or curly, and how is that property changed using chemicals?
- What is dirt, and how do we use chemicals to clean things?
- What are soaps and synthetic detergents, and what kinds of products do we find them in?
- How do stain removers work?

An Egyptian tomb painting, from Thebes. Cosmetics have been with us for a long time. *(Brian Brake/Photo Researchers)*

One thing to keep in mind as you study this chapter is that most consumer products of a given kind are remarkably similar in their composition and function. They do differ, however, in packaging, appearance, and advertising. It is these differences that greatly affect their cost and consumer acceptance. It is probably not a good idea to assume that just because a given product costs a lot or looks good on the shelf that it is a better product. In fact, as you learn more about the ingredients of various products, you may discover some inexpensive alternatives to highly advertised and costly consumer products.

16.1 Cosmetics

The use of chemical preparations that are applied to the skin to cleanse, beautify, disinfect, or alter appearance or smell is older than recorded history. In ancient Egypt red pigments derived from plants were used on the cheeks and lips, and the green mineral malachite was pulverized for use as eye shadow. Such preparations are known as **cosmetics.**

■ Malachite is a copper compound [$Cu_2CO_3(OH)_2$].

■ Approximately $25 billion is spent each year for cosmetics in the United States.

There is at best a fuzzy distinction between cosmetics and drugs. Traditionally, drugs alter body functions, diagnose illness, relieve symptoms, or cure disease, while cosmetics simply improve appearance. In spite of formal definitions for both cosmetics and drugs enacted by Congress in the Food, Drug, and Cosmetic Act, there are gray areas. The U. S. Food and Drug Administration (FDA) is constantly reminding manufacturers of cosmetics to tone down their claims for products that, if they acted as claimed, would make them candidates for the title "miracle drugs." Today there is a high level of governmental control in the introduction of new drugs. Drugs require elaborate safety testing prior to receiving approval by the FDA, but cosmetics do not.

As more has been learned about how chemicals interact with skin and hair, cosmetics are being expanded from their traditional use for aesthetic reasons to scientific applications for skin care, protection from ultraviolet light, and conditioning hair. Consumers are increasingly interested in milder and more environmentally friendly ingredients. Most consumers know that cosmetics are chemicals and that those chemicals have names. Some products, such as Japan's Shiseido brand of cosmetics, have gone so far as to use the molecular formula of one of its ingredients (sodium hyaluronate) right on the product label of a $50 tube of skin cream—an indication of the movement of cosmetic enhancement from an art to a science, or at least a move toward science literacy. Even though molecular formulas are generally not given on the bottles of cosmetics, most do contain a listing of the ingredients, in order of decreasing amounts (Figure 16.1).

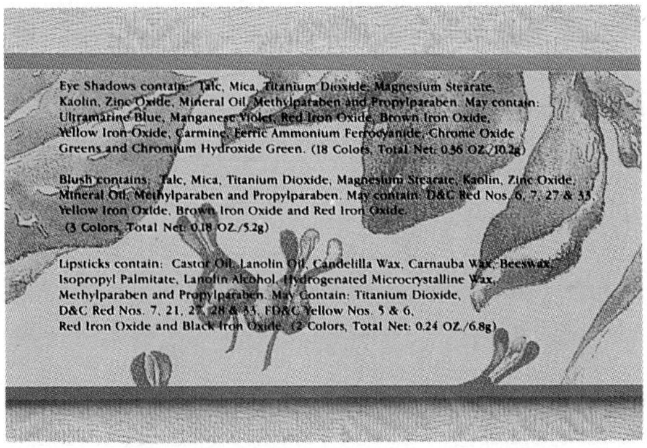

Figure 16.1 The ingredients in several cosmetics. The ingredients are listed in order of decreasing percentage, but the actual percentage values do not have to be given. This could make it too easy for competitors to make copies. As you read this chapter, check back to see if you can identify the purpose of the various ingredients. *(Tom Pantages)*

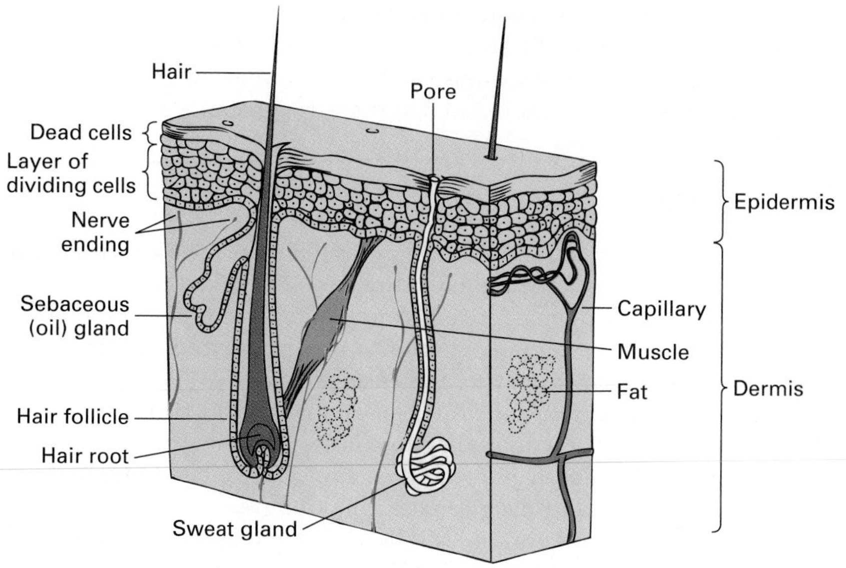

Figure 16.2 Skin structure. Skin is divided into two layers, the epidermis at the surface and the dermis below the surface. The total thickness of skin is about one eighth inch and it covers about 20 sq ft in an adult human.

16.2 The Chemistry of Skin and Hair

The exterior surfaces of the skin and hair are composed of protein (Section 15.3). The toughness of the skin and hair is partly due to bonding between side groups that holds the protein chains in position. As you will see in Section 16.8, these interactions make it possible to curl your hair.

The upper layers of the skin are the epidermis and dermis (Figure 16.2). The skin has several functions made possible by its structure: protection, sensation, excretion, and body temperature control. The exterior of the epidermis is called the *stratum corneum* (corneal layer), and this is where most cosmetic preparations for the skin act. The corneal layer is composed mostly of dead cells; it has a moisture content of about 10% and a pH of about 4. The principal protein of the corneal layer is **keratin** (Section 15.4). Its structure renders it insoluble in, but slightly permeable to, water. New skin cells are continually forming beneath the surface and moving upward while dead surface cells are being shed. Dermatologists have long used "chemical peels" to remove the corneal layer on people with skin that has become aged or that has been weathered by wind and sun. Trichloroacetic acid (CCl_3COOH), a moderately strong acid, is used for chemical peels. When trichloroacetic acid acts on skin it dissolves the older, outer-layer cells, leaving behind only the newer cells. The treatment causes a feeling much like a sunburn. There are a number of skin creams and lotions on the market today that contain alpha-hydroxy acids such as glycolic acid. These acids also dissolve older skin cells, but are much weaker acids and produce a milder skin treatment. Interestingly, washing with ordinary soap also causes old skin cells to be removed. The slipperiness of your skin after it has been wetted by soapy water is an indication that the older, rougher skin cells have been removed, leaving behind younger skin cells.

■ Glycolic acid, an alpha-hydroxy acid used in cosmetics

The dermis is supplied with *sebaceous glands*. They produce an oily secretion, sebum, that protects the skin from excessive moisture loss. To remain healthy, skin must have a moisture content near 10%. Dry skin contains a lower-than-normal amount of sebum, is uncomfortable, and can crack and become infected. An excessively moist skin can also be a problem because it is a very good host for harmful bacteria and fungi. Washing the skin removes the oily sebum that helps retain the right amount of moisture. Routine bathing can cause dry skin for many individuals.

The dermis is also supplied with sweat glands of two types. There are about two million sweat glands of the first type, called *eccrine* glands. They produce a secretion that is mostly water. Both body temperature and emotion stimulate them into action. The cooling effect from evaporation of the water they produce serves to regulate body temperature during exercise, hot weather, or fever. The second type of sweat glands are called *apocrine* glands, and they are located mostly in the armpits and genital area. These sweat glands become active at puberty and are responsive to strong emotions. The sweat produced by both kinds of glands has no odor until it is acted on by bacteria on the surface of the skin. Most people sweat at least a pint of water a day. Under stressful conditions, when it is hot, or during strenuous activity, up to 3 gallons (11.4 L) of sweat per day can pour out of all those sweat glands!

Hair is also composed principally of keratin. An important difference between hair keratin and other proteins is its high content of the amino acid cysteine, which contains the —SH group in its side chain.

■ Fingernails and toenails are composed of "hard" keratin, a very dense type of this protein. The epidermal cells of nails grow from epithelial cells lying under the white crescent at the growing end of the nail. Like hair, the nail tissue beyond the growing cells is dead.

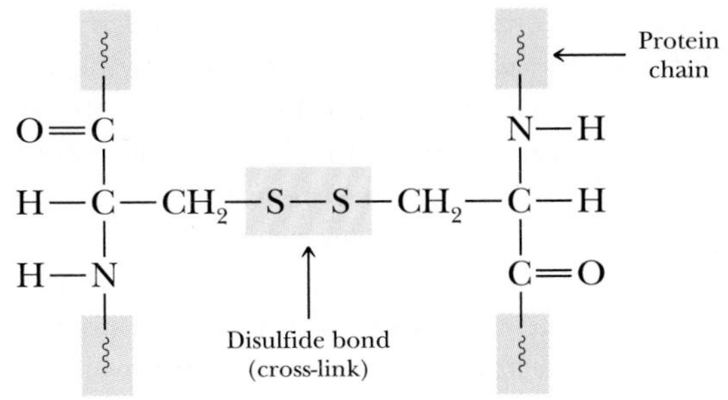

Cysteine plays an important role in the structure of hair by forming sulfide bridges (—S—S—) between protein chains. These protein chains are twisted together into spirals that group together to constitute individual strands of hair.

16.3 Creams and Lotions

If dry skin is treated with an oily substance after washing, the skin will be protected until enough natural oils have been regenerated. The oily substance used can be derived from animal oils, vegetable oils, or even oils from petroleum (called mineral oils). It is not uncommon to see some kind of rare animal oil, such as mink oil, used in a skin preparation, yet the oil from a mink is not any more effective at holding the skin's moisture than a less expensive vegetable or mineral oil. When choosing a skin product containing an oil, perhaps a more important factor than the kind of oil is whether it will be soothing or irritating to the skin.

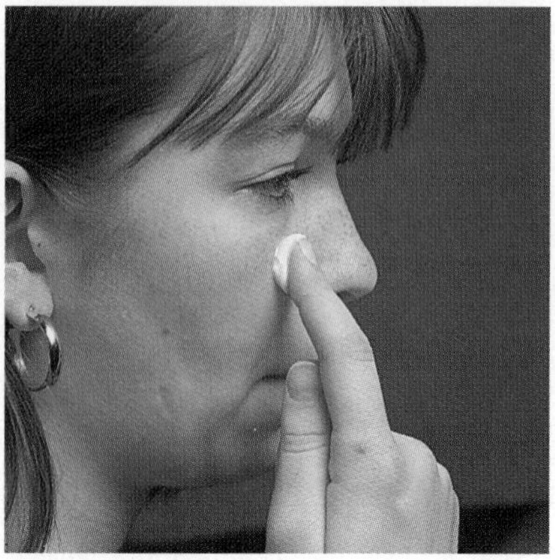

Figure 16.3 Cold cream, which has a thicker consistency than a lotion. *(C.D. Winters)*

Any substance that holds moisture in the skin can be called a **moisturizer.** Since these substances are all oily in nature, they can lubricate the skin (restore the feeling of normal oiliness) and have the noticable effect of softening and soothing the skin. These substances are also called **emollients.** All the oils listed in the first paragraph that help to hold moisture in the skin would be called emollients.

Getting an emollient distributed evenly on your skin is not as easy as it may seem. Rather than just pouring oil over your body after a shower (this would leave your skin too oily!), it would be better to use a mixture containing the emollient. Two kinds of mixtures are commonly used: *creams* and *lotions* are made by mixing an oily component with water and other ingredients in the right proportions to form a stable mixture that can be more like a solid (a cream) or more like a liquid (a lotion) (Figure 16.3). Mixtures like these are called **emulsions.** Emulsions consist of two substances that would normally not mix, such as an oil and water, but that are made to mix by an **emulsifying agent,** a compound whose molecules have a part that is soluble in water and a part that is soluble in oil. With its "dual solubility," the emulsifying agent stabilizes the mixture. Thus, a cream can contain a rather high percentage of an oil and not feel oily or "greasy" because of the presence of the water. When the cream is applied to the skin, some of the water is absorbed by the skin, and the oil (emollient) remains on the surface to hold in the moisture.

■ The name "colloid" comes from the Greek word meaning glue. A close look at a typical glue will show that there are large particles dispersed in water. These particles are large, colloidal-size aggregates of similar molecules.

Emulsions are examples of **colloids,** which are quite common. Fog, foams, foods such as milk, and aerosol sprays are all colloids (see Table 16.1). Colloidal mixtures differ from solutions because the particles (called the *dispersed phase*) distributed in the solvent-like medium (called the *continuous phase*) are much larger than the molecules or ions that are the solutes in true solutions. Two kinds of emulsions can be formed between oil and water: oil droplets of colloid size dispersed in water and water droplets of colloid size dispersed in oil. As you might expect, oil-in-water emulsions have more of the properties of an aqueous solution, while the water-in-oil emulsions have more oil-like properties; for example, they tend to feel more greasy. An oil-in-water emulsion has tiny droplets of an oily or waxy substance dispersed throughout a water medium; homogenized milk is an example. A water-in-oil emulsion has tiny droplets of a water solution dispersed throughout an oil; examples are natural petroleum and butter. An oil-in-water emulsion can be washed off the skin

■ The molecules in the continuous phase of colloidal mixtures can have molar masses as high as several hundred thousand.

TABLE 16.1 ■ Types of Colloids

Continuous Phase	Dispersed Phase	Type	Examples
Gas	Liquid	Aerosol	Fog, clouds, aerosol sprays
Gas	Solid	Aerosol	Smoke, airborne viruses, automotive exhaust
Liquid	Gas	Foam	Shaving cream, whipped cream
Liquid	Liquid	Emulsion	Mayonnaise, milk, face creams
Liquid	Solid	Sol	Milk of magnesia, mud
Solid	Liquid	Gel	Jelly, cheese, butter
Solid	Solid	Solid sol	Milk glass, some gemstones, many alloys such as steel

Note: Most colloids will separate unless stabilized. Food and cosmetic emulsions, foams, and aerosols are usually stabilized with emulsifying agents to give them a long shelf life and consistent properties such as color and texture throughout their life.

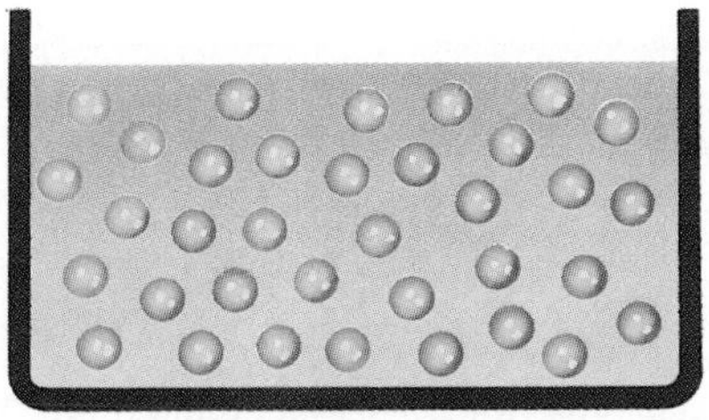

Oil-in-water emulsion

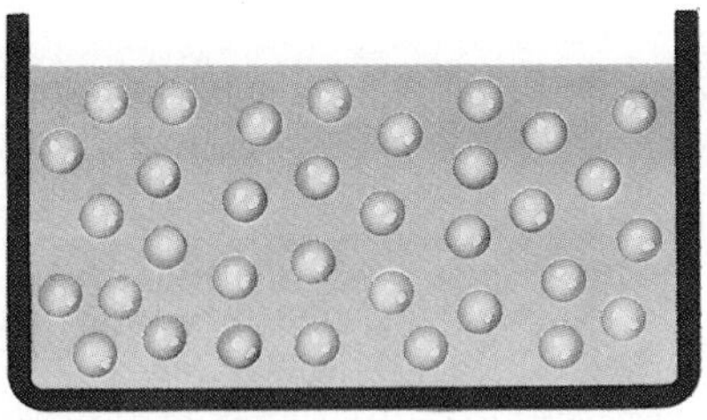

Water-in-oil emulsion

Figure 16.4 Two moisturizing lotions, one an oil-in-water emulsion and the other a water-in-oil emulsion. *(C.D. Winters)*

surface with tap water, whereas a water-in-oil emulsion gives skin a greasy, water-repellent surface that resists being washed off by running water. With careful formulation of the emulsion, a *barrier cream* can be made that will effectively resist aqueous solutions that might contain harmful ingredients. Chemists in the laboratory often apply barrier creams to protect their hands from exposure to toxic chemicals.

■ Spermaceti oil, once obtained from the head of the sperm whale, is chiefly cetyl palmitate, an ester of cetyl alcohol and palmitic acid. Identify the ester group in its formula: $CH_3(CH_2)_{14}COO(CH_2)_{15}CH_3$. Although some spermaceti oil is still derived from whales, it can be readily synthesized from other starting materials.

The ingredient listed first on a cream or lotion label is an indication of the kind of emulsion present. If water is listed first, the emulsion is probably an oil-in-water type. If an oil is listed first, the emulsion is probably the water-in-oil type (Figure 16.4). A lotion might contain the same emollient as a cream but contain a different emulsifier and a different ratio of water to oil. Whether an emollient is distributed on the skin in a cream or a lotion depends more on how the product is perceived by the user than on which kind of product is more effective. All creams and lotions have "shelf lives." Creams and lotions, like all colloids, can tend to "settle out" over a period of time, a property not observed in solutions.

Cold cream, originally formulated by the Greek physician Galen, was an emulsion of rose water in a mixture of almond oil, olive oil, and beeswax. The name "cold" is used because this cream cools the skin as the water evaporates. Subsequently, other ingredients were added to cold creams to get a more stable emulsion. An example of a modified cold cream composition is given in Table 16.2.

TABLE 16.2 ■ Composition of a Cold Cream

Ingredient	Percentage	Purpose
Almond oil	35%	Emollient
Concentrated rose water	35%	Perfume
Beeswax	12%	Emollient
Lanolin	15%	Emulsifier
Spermaceti oil (from whale oil)	8%	Emollient

DISCOVERY EXPERIMENT

An Edible Emulsion

To do this experiment you will need one fresh egg, vinegar or lemon juice, dry mustard, some kind of salad oil (peanut oil is good), and water. You will also need a blender, a food processor, or a bowl and a wire chef's whisk, a glass or cup, and a spoon.

First, to demonstrate that water and oil don't mix, sprinkle a few drops of oil into a glass of water. Stir vigorously, then wait a while. What happened? Now reverse the process—sprinkle a few drops of water into some oil, stir, and wait a while. What happened?

Now for the best part. Gently mix the egg, two tablespoons of vinegar or lemon juice, and $\frac{1}{4}$ teaspoon of dry mustard. Then continuously blend the mixture in the blender, food processor, or bowl while you *slowly* pour in one cup of oil. The product should be the stable emulsion of oil and water that we call mayonnaise.

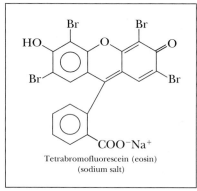

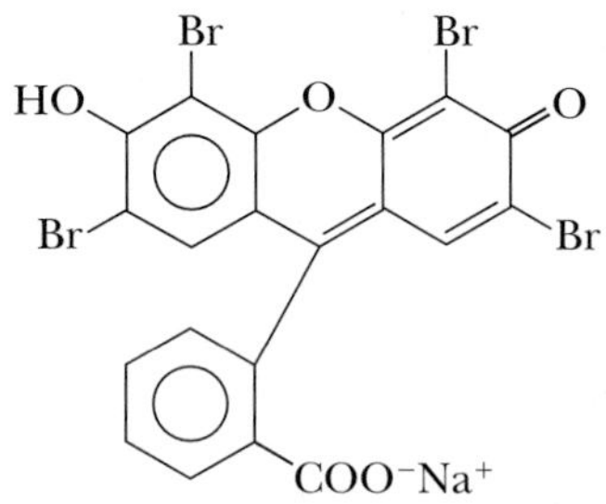

Tetrabromofluorescein (eosin)
(sodium salt)

Figure 16.5 Tetrabromofluorescein, a purple dye used in lipstick. Organic molecules with many double bonds and aromatic groups contain electrons that strongly absorb light of specific colors, leaving behind the colors that we see.

16.4 Lipsticks and Powders

The skin on our lips is covered by a thin corneal layer that has very few sebaceous glands and consequently dries out easily. Moisture is supplied mainly by saliva. In addition to being a beauty aid, lipstick, with or without color, can be helpful under harsh conditions that tend to dry lip tissue.

Lipstick consists of a suspension of coloring agents in a mixture of high-molecular-weight hydrocarbons (waxes) or their derivatives, or both. Consistency of the mixture over a wide temperature range is very important in a product that must have even application, holding power, and resistance to running of the coloring matter at skin temperature. Lipstick is perfumed and flavored to give it a pleasant odor and taste. The color usually comes from a dye such as tetrabromofluorescein (Figure 16.5), which is precipitated with a metal ion such as Fe^{3+}, Ni^{2+}, or Co^{3+}. The metal ion modifies the natural color of the dye and usually produces a more intense color; the metal also keeps the dye from dissolving in the oil medium, thus preventing the color from running. The ingredients in a typical formulation of lipstick are given in Table 16.3.

The purpose of most cosmetic *powders* is to give the skin a smooth appearance and a dry feel. Face powders often contain dyes to impart color or shading to the skin. The principal ingredient in body powder is talc ($Mg_3(OH)_2Si_4O_{10}$), a natural mineral able to absorb both water and oil. The absorptive properties of talc are due partly to its large surface area and partly

TABLE 16.3 ■ A Typical Lipstick Formulation

Ingredient	Percentage	Function
Vegetable or mineral oils or wax	50	Emollient
Lanolin	25	Emulsifier
Carnuba or beeswax	18	Raises melting point; gives hardness
Dye	4–8	Color
Perfume	Small amount	Odor
Flavor	Small amount	Taste

TABLE 16.4 ■ General Formula for Body Powder

Ingredient	Percentage	Purpose
Talc	56	Absorbent
Zinc oxide (ZnO)	20	Astringent
Purified chalk ($CaCO_3$)	10	Absorbent
Zinc stearate*	6	Binder
Dye	trace	Color
Perfume	trace	Odor

*The zinc salt of stearic acid, a fatty acid with 18 carbon atoms, $CH_3(CH_2)_{16}COOH$.

Zinc oxide ointment (with yellow dye added) as sun protection. The zinc oxide (ZnO), which is insoluble in water or oils, is mixed with creams or lotions for use in sun protectants, and also in foundation makeup, rouge, and antiperspirants. *(C.D. Winters)*

to the ionic charges present throughout the structure. In addition to talc, corn starch and calcium carbonate are commonly used. Corn starch is an organic equivalent of talc—it contains numerous binding sites for water molecules and has a large surface area. Calcium carbonate ($CaCO_3$, the same chemical found in marble and limestone) has basic properties and can react with the fatty acids found on skin when bacteria decompose the components of sweat. A binder, such as zinc stearate, is necessary to increase adherence of the powder to the skin. Zinc oxide (ZnO) is often added to skin powders to shrink tissue and reduce sebum flow in oily skin. Compounds that act in this way are called **astringents.** They function by polarizing various charged sites in protein structures and causing a shrinkage of the protein-containing tissue. Zinc compounds such as zinc oxide and zinc stearate can safely be used in cosmetics because they have no harmful effects. A general formulation for body powder is given in Table 16.4.

16.5 Sunscreen Products

■ Stratospheric ozone protects us from harmful uv radiation. See Section 21.10 for a discussion about chemicals that can destroy this ozone layer.

Our world is bathed in ultraviolet (uv) radiation, which is sufficiently energetic to harm living things that are exposed to it. Although the stratospheric ozone layer absorbs most of the extremely harmful uv radiation with wavelengths shorter than 285 nm, plenty of uv rays of longer wavelengths reach the Earth's surface. Two kinds of uv radiation are generally spoken of. The kind called "uvA" has wavelengths of 330 nm or more and is less harmful than the kind called "uvB," which has wavelengths from 285 to 330 nm. Exposure to uv radiation darkens skin because the skin responds to the high energy of the light by increasing the concentration of the natural pigment **melanin.** The melanin molecules absorb some of the ultraviolet energy and covert it to heat, thus protecting the molecular structure of the skin from damage. However, even with melanin's protection, uv light causes a general degradation of the skin and, in extreme cases, can cause skin cancer. The problem is more acute for fair-skinned people, whose skin has smaller amounts of melanin. As everyone knows, exposure to the sun causes skin irritation—a visible reddening (called *erythema*) of the skin is noticeable on light-skinned individuals. Dark-skinned people also experience this reddening, but it is not as noticeable. The **minimum erythemal dose** (MED)—the quantity of solar radiation required to produce a barely observable erythema—depends on the wavelength of radia-

TABLE 16.5 ■ UV Exposure Categories Used by EPA for the UV Index

Exposure Categories	Index Values	Minutes to Burn for "Never Tans"—Most Susceptible	Minutes to Burn for "Rarely Burns"—Least Susceptible
Minimal	0–2	30 min	>120 min
Low	4	15 min	75 min
Moderate	6	10 min	50 min
High	8	7.5 min	35 min
Very high	10	6 min	30 min
Extreme	15	<4 min	20 min

UV rays are only about half as intense three hours before and after the peak. Physical surroundings such as snow, sand, and water reflect more uv and intensify exposure. Latitude and altitude also play a role; exposure increases with proximity to the equator and with altitude. Source: *Science News,* July 23, 1994, p. 61.

tion, with much higher amounts of uvA being required than uvB. The MED, however, is not a measure of long-term damage to the skin because uvA penetrates much deeper into the skin than does uvB.

In June 1994 the National Oceanic and Atmospheric Administration (NOAA) and the Environmental Protection Agency (EPA) began offering a "uv index" as part of the regular weather reports in 58 selected cities in the United States. The index uses satellite measurements as well as ground-based observations, which are fed into a computer model to forecast peak noontime uv levels. The computer model produces a 15-point index that corresponds to six exposure categories set by the EPA (Table 16.5).

If you are going to be exposed to direct sunlight, it is a good idea to protect yourself from as much uv exposure as you can. Besides physical barriers such as long-sleeve shirts and wide-brimmed hats, there are a variety of chemicals that can be applied to the skin that will selectively absorb uv light. These **sunscreens** can be applied as oils, creams, or lotions. The important part of screening against uv radiation is how much of the sunscreen is applied to the skin, how long it stays on the skin, and how effective it is in absorbing the harmful uv rays, especially the uvB type. The sunscreen must function in a manner similar to melanin by absorbing uv light and converting it into heat energy. The first popular sunscreens contained *p*-aminobenzoic acid (PABA) and other chemicals with similar structures as the active ingredients. The absorption spectrum for *p*-aminobenzoic acid is shown in Figure 16.6. This compound, and ones similar to it in structure, absorb an appreciable amount of the radiation in the uvB portion of the spectrum.

Protection from ultraviolet radiation requires a broad-brimmed hat and sunscreen. *(C.D. Winters)*

Today, most sunscreens have sun protection factors listed prominently on the bottle. The **sun protection factor (SPF)** is defined as the ratio of time required to produce a preceptible erythema on a site protected by a specified dose of the sunscreen to the time required for minimal erythema development on unprotected skin. An SPF of 4, then, would provide four times the skin's natural sunburn protection. Looking at Table 16.5 again, it is possible to *estimate* your safe exposure to the sun, given the exposure index and the SPF of the sunscreen you are using. An SPF of 4 would mean that someone who is most susceptible to the sun's rays could remain exposed for about two hours (4×30 minutes) when the exposure index was between 0 and 2. Of course, other factors, such as how well you applied the sunscreen, should be considered as well.

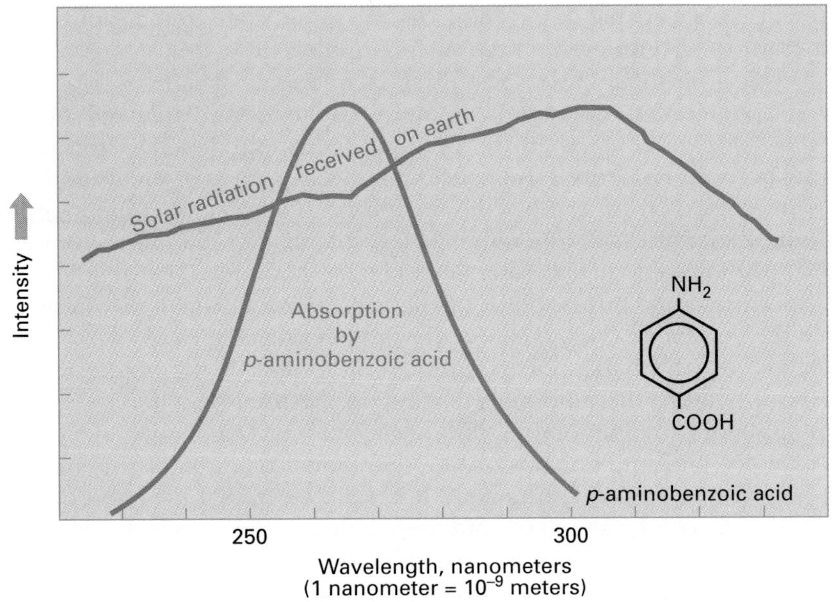

Figure 16.6 Ultraviolet absorption by *p*-aminobenzoic acid (PABA), an example of a sunscreen. The absorption of uv light by PABA is strongest at 265 nm, and the maximum of the deep-burning uv radiation received on Earth is 308 nm.

The World of Chemistry

Program 10, *Signals from Within*

Sunscreen testing.

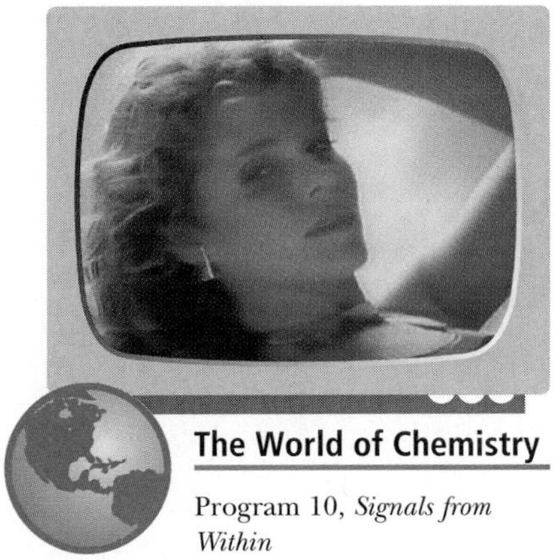

The World of Chemistry

Program 10, *Signals from Within*

Sunbather. The glamour associated with sunbathing and a deep tan is diminishing as the dangers to the skin are recognized.

In general, a mixture of uvA and uvB absorbers is needed if an SPF of 12 or higher is required. The U.S. Food and Drug Administration (FDA), which regulates claims about the effectiveness of sunscreen products in this country, will not allow SPF values above 30 to be advertised. When making a decision about SPFs and exposure times in the sunlight, you should remember that a person's skin color and history of reactions to sun exposure should also be taken into account.

In spite of the dangers of overexposure to ultraviolet light, in a quest for a "perfect" tan, some people will use tanning beds that contain uv lamps that emit both uvA and uvB rays. Skin specialists uniformly warn against tanning lamps because they can produce much higher cumulative exposures to harmful uv rays than does exposure to the sun. But even normal exposure to the sun produces cumulative effects. The first of these is a noticeable toughness and "leatheryness" of the skin, followed quickly by wrinkles and an appearance of skin aging. Precancerous skin lesions such as *actinic keratosis* (also called solar keratosis) and *actinic cheilitis* often appear on the skin and lips of people who have been careless in their exposure to the sun. These lesions can be the first step in the development of skin cancer. When they appear they should be treated promptly because they can develop into any form of skin cancer, including those forms that can spread throughout the body and cause a quick death.

■ Long-term exposure of skin cells to uv light leads to an accumulation of damage with age.

16.6 The Chemistry of Perfumes

Perfumes are used by women and men to make them more appealing to others. For a perfume to function properly, its odorous compounds must be volatile, that is, capable of passing from the liquid state to the vapor state at temperatures encountered on the surface of the skin.

A typical perfume has at least three components of different volatilities. The first component, called the "top note," is the most volatile and therefore the most obvious odor when the perfume is first applied. The top note in natural perfumes is derived from citrus oils, herbs such as rosemary and lavender, and extracts from crushed flower foliage. The second component of a perfume, called the "middle note," is less volatile and is a natural flower oil extract from flowers such as rose, lily, violet, and jasmine. The middle note is more persistent and noticeable as the perfume is worn. The third component, called the "end note," is made of resins or waxy polymers derived from wood, musk, amber, and balsam. The end-note molecules, being higher in molar mass, can bond to the molecules making up the top and middle notes, thus keeping them from evaporating too quickly and making a total effect during the time the perfume remains on the skin. Some researchers think that the end note also acts as a sex attractant. It is well known that the musk odor, for example, is a deer sex attractant, but there is no conclusive evidence that musk or any other odor has this effect on humans. Nevertheless, the concept has made for some interesting ads for various perfumes.

■ Recall that lower molar mass compounds are generally more volatile; that is, they evaporate at lower temperatures.

■ Musk is obtained from a scent gland in the musk deer.

One hundred years ago, perfumes were available only to the very rich. Now they are so common they are included in soaps, detergents, household cleaners, and even bleaches. In addition, inexpensive copies of designer perfumes are now sold in discount stores and boutiques in malls and elsewhere. These copies have resulted from advances in analytical techniques, such as the identification of more than 200 chemicals in jasmine extract, and in organic synthesis (Chapter 14), which made possible the synthesis of the natural compounds and related compounds that sometimes have better qualities than the natural ones. In 1921 the famous Chanel No. 5 perfume became the first fine perfume to employ synthetic chemicals. Since then, essentially all the natural chemicals in the top, middle, and end notes in most perfume formulations have been replaced with synthetics. The natural products are still prepared for those who prefer and can afford them. Although some people claim to be able to tell the difference between natural and synthetic perfumes, in actual practice, few people have the ability to tell the difference.

The World of Chemistry

Program 21, *Carbon*

Fragrance testing during the formulation of a perfume.

What causes a given compound to smell the way it does? Certainly, molecules of different compounds are different in composition, molar mass, and shape. All these differences seem to play a role in the odor a compound imparts to the nose. Experiments have shown that there are a number of primary odors for the human; these include camphorous, musky, floral, pepperminty, ethereal, pungent, and putrid. While most people describe an odor as being "more like" one of these categories, some people will completely miss one or more of these primary odors. (A perfume to allure someone who could not detect certain odors might be a complete waste of money!) Although substances in a given odor class do not always share similarity in molecular and structural formulas, they often do share roughly the same molecular shape and size. They interact with several kinds of receptor sites in the olfactory cells of the nose. When a molecule of the correct size and shape fits into a complementary receptor site, a particular smell impulse is initiated (Fig. 16.7). The molecular structures for the two most active ingredients in jasmine —jasmone and methyl jasmonate—are given in Figure 16.8.

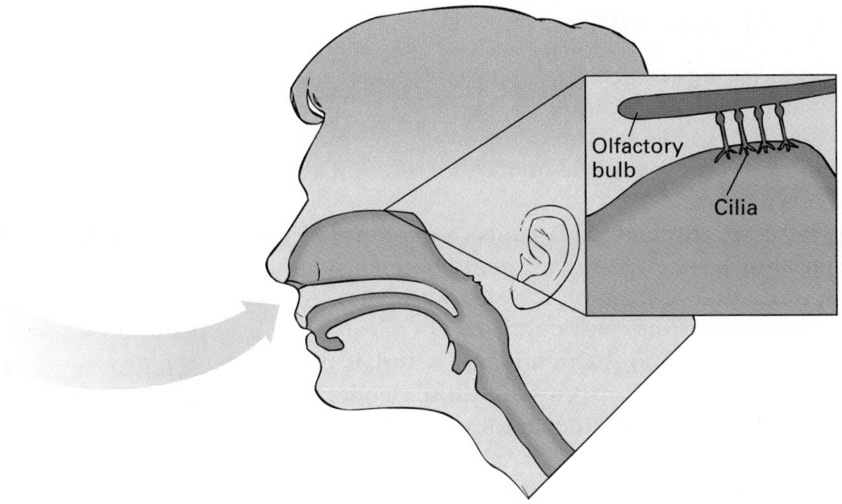

Figure 16.7 Odor receptors. Two small patches of hairlike cilia contain 10 to 20 million receptors for odor-bearing molecules. Binding just a few molecules to receptors initiates transmission of a nerve impulse that registers as an odor.

Typical perfumes are 10 to 25% perfume essence and 75 to 90% alcohol, with a fixative to stabilize the essential oils and keep them in solution. Aftershave lotions and colognes are just diluted perfumes, about one-tenth (or less) as strong as perfumes, with the odors chosen to be pleasing to men, but not smell just like a woman's perfume. Perfumes are mildly bactericidal and antiseptic because of their alcohol content.

(a)

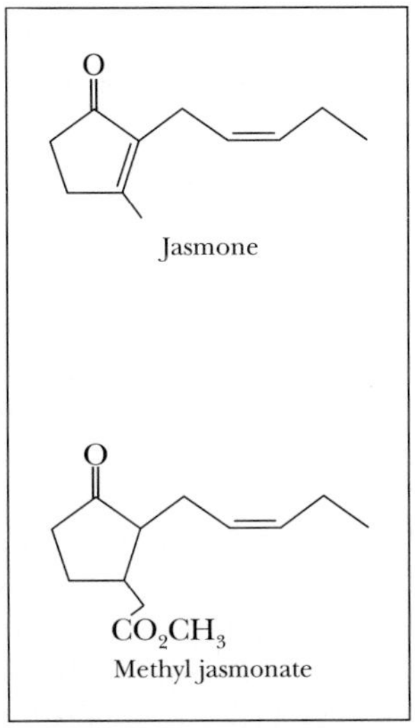

(b)

Figure 16.8 (a) The fragrant jasmine flower. (b) Two of the natural compounds that give the jasmine its beautiful aroma. *(a, Michel Viard/Peter Arnold, Inc.)*

16.7 Deodorants

■ Body odor is promoted by bacterial action.

Body odor results largely from bacterial action on amines and the hydrolysis products of fatty oils (fatty acids, acrolein, and others) that are part of sweat. Sweating is both normal and necessary for the proper functioning of the human body; sweat itself is quite odorless, but the bacterial decomposition products are not.

There are three general kinds of deodorants: those that directly dry up perspiration by acting as astringents (chemicals that close the pores of the skin), those that have an odor to mask the odor of sweat products, and those that remove odorous compounds by chemical reactions. In addition, several deodorant formulations contain antibacterials that inhibit the growth of the bacteria that cause odors. Among those deodorant ingredients that have astringent action are hydrated aluminum sulfate [$Al_2(SO_4)_3 \cdot 6\ H_2O$], hydrated aluminum chloride ($AlCl_3 \cdot 6\ H_2O$), aluminum chlorohydrate [actually aluminum hydroxychloride, $Al_2(OH)_5Cl \cdot 2\ H_2O$], and various alcohols. Compounds that act by masking bad odors include oils from plants and perfumes. Zinc peroxide (ZnO_2) removes odorous compounds by oxidizing the amines and fatty acids found in sweat. One of the most popular antibacterials is trichlosan, which, in concentrations as low as 0.03 ppm, can inhibit the growth of both *Escherichia coli* and staphylococcus bacteria.

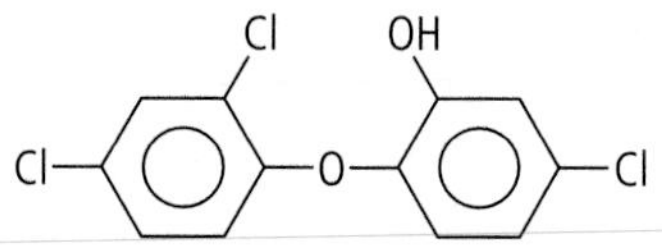

Triclosan (trichlorohydroxydiphenyl ether), an antibacterial

To use deodorants correctly requires that you know something about how they work. For example, astringents take time to act, so if you want to prevent perspiration (''wetness,'' as those TV advertisements describe), you have to apply the compound several hours before the sweating is going to start. If a deodorant is used primarily to mask body odors with perfume, you have to consider how long the perfume will stay on your body. Only trial and error can determine this. Deodorant formulations have recently adopted the ''clear'' marketing concept; anything clear (noncolored) ''has to be better'' than similar products colored red, yellow, tan, pink, or blue. There is no indication that the formulations of colorless products are any more effective than those of colored ones. However, the lack of colored dyes might make the products somewhat less irritating to users with sensitive skin.

Clear consumer products: a deodorant, a glass cleaner, and a skin cleanser. *(C.D. Winters)*

■ SELF-TEST 16A

1. The surface of the skin epidermis is known as the __________ layer.
2. The oily secretion of skin is __________.
3. Keratin is a protein found in __________ and __________.
4. Bridges between protein chains may be __________ linkages or __________ bonds.
5. What is an ideal moisture content for human skin?
6. A skin cream is either an oil-in-water or a water-in-oil __________.
7. Human perspiration is almost odorless. (a) True, (b) False.
8. The major mineral ingredient in face or body powder is __________.
9. The top note is the most volatile odorous component in a perfume. (a) True, (b) False.
10. An example of an active chemical that will absorb ultraviolet light in a sunscreen lotion is __________.
11. Skin darkens because of an increased concentration of the skin pigment __________.

16.8 The Difference Between a Good Hair Day and a Bad Hair Day

The World of Chemistry

Program 23, *Proteins: Structure and Function*

Hair coloring in progress.

The behavior of your hair—how well it remains where you comb it, how it behaves on damp days, whether it curls up after you wash it—depends in part on the chemical bonds that hold the protein fibers together. You can control some of these factors, but some you cannot control. The curl, color, texture, and length of human hair are matters of personal choice that vary considerably from person to person. Means are available to alter these conditions. There is even a drug, Minoxidil, that promotes the growth of hair on skin where none is present. People's interest in hair care is evidenced by the fact that more money is spent in the United States on hair-care products than on any other type of cosmetics and toiletries.

In addition to the sulfide linkages between hair protein structures (discussed in Section 15.4), ionic bonds and hydrogen bonds between protein side chains affect the behavior of hair. In fact, these bonds explain bad hair days. Consider, for example, the interaction between a lysine $—NH_2$ group and a $—COOH$ of glutamic acid on a neighboring protein chain. The acidic $—COOH$ groups lose their protons, forming negatively charged $—COO^-$ groups, while the basic $—NH_2$ groups gain protons to form positively charged $—NH_3^+$ groups. When the $—NH_3^+$ and $—COO^-$ groups on adjacent chains approach each other, an ionic bond that helps hold the two protein chains together is formed.

■ Ionic bonds are described in Section 6.1 and hydrogen bonds are described in Section 7.3.

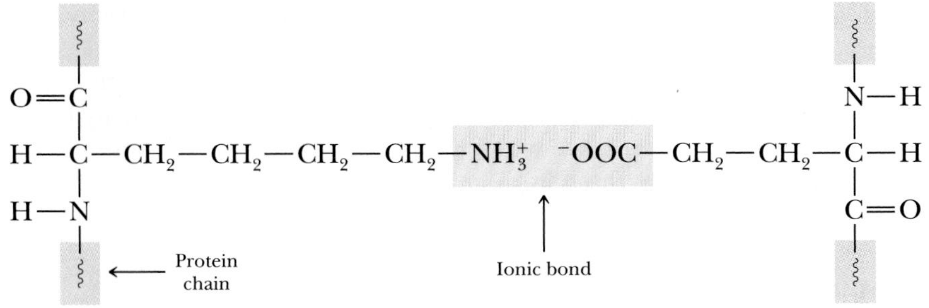

Ionic bond between two protein chains

Or consider how a hydrogen bond might form between side groups.

■ To curl hair using water as the only chemical, wet hair is wrapped on a roller and allowed to dry. The rearranged ionic and hydrogen bonds will hold the curls in place for a while, especially if the humidity doesn't change.

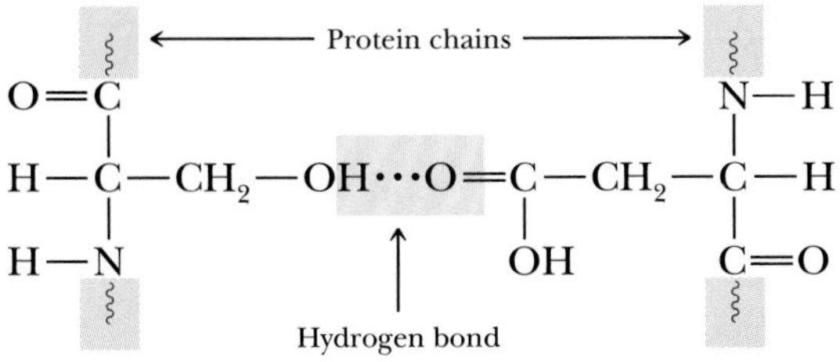

Hydrogen bond between two protein chains

Moisture affects hydrogen bonds between protein chains, and when their number and arrangement changes, hair changes. When hair is wet, it can be

■ The bad smell of permanent wave chemicals is typical of sulfur-containing compounds.

stretched to one and one half times its dry length. On a very moist day, enough bonds will break simultaneously for hair to lose its curl. On a very dry day, enough static electrical charge can build up for individual hairs to repel each other. Different people have bad hair on different days because the bonding patterns in their hair vary. (We're talking about bad hair days here in the practical, not the philosophical, sense.)

As the pH of the hair or of a solution in contact with the hair rises above 4.1 (the pH of normal, dry hair), the $—NH_3^+$ groups begin to lose their extra protons, and the ionic cross-links are broken. The keratin structure swells and becomes soft, and the hair becomes more manageable. This is an important aspect of hair chemistry, since the pH of most shampoos and even that of water is above 4.1.

Making Permanent Waves

For more permanent curling, a more permanent chemical change in the hair structure is needed. The disulfide cross-links between protein chains (see structure in Section 16.2) hold hair in its natural shape. In ''permanent waving,'' these cross-links are broken by a reducing agent (step 1, Figure 16.9), which relaxes the tension. An oxidizing agent generates new cross-links, and the hair retains the shape of the roller, so it appears curly.

The most commonly used hair-waving reducing agent is the ammonium salt of thioglycolic acid ($HSCH_2COO^-NH_4^+$). A typical waving solution contains 5.7% thioglycolic acid, 2.0% ammonia, and 92.3% water. The usual oxidizing agent is hydrogen peroxide (H_2O_2) or a perborate compound that releases hydrogen peroxide in solution ($NaBO_3 \cdot 4\,H_2O$). A typical ''neutralizer'' solution contains this oxidizing agent dissolved in a strongly basic water solution. The strong base also helps to break and re-form hydrogen bonds between adjacent protein molecules. Too-frequent use of these strongly basic solutions can cause the hair to become brittle and lifeless.

Changing Hair Color

Natural hair color for blonds and all shades of brown to black hair is determined by how much melanin the hair contains. Melanin, the same substance that colors skin, is a dark-colored pigment with a complex chemical structure that is not fully understood. Darker hair indicates more melanin is present. The color also varies with the size and shape of the melanin granules. Redheads are blessed with *phaeomelanin,* a red-colored melanin unique to them.

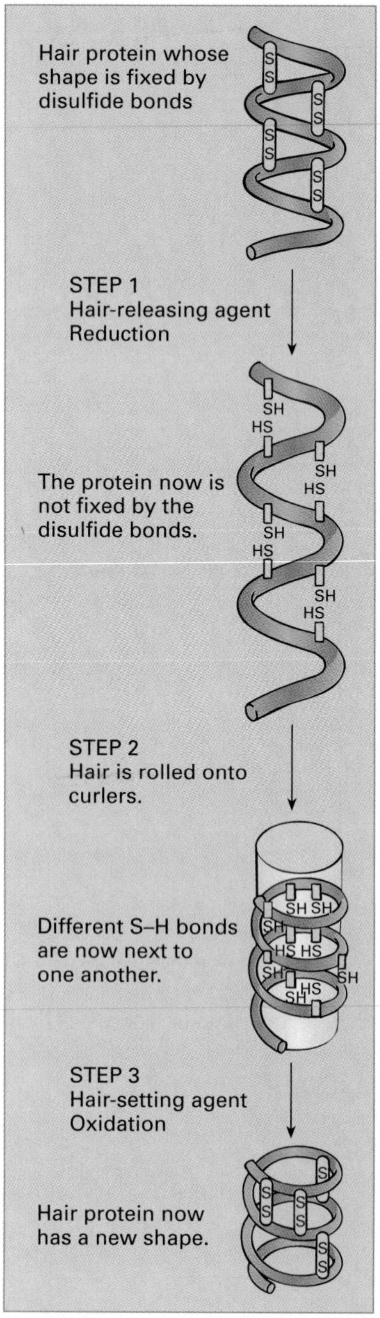
Hair protein whose shape is fixed by disulfide bonds
STEP 1
Hair-releasing agent
Reduction
The protein now is not fixed by the disulfide bonds.
STEP 2
Hair is rolled onto curlers.
Different S–H bonds are now next to one another.
STEP 3
Hair-setting agent
Oxidation
Hair protein now has a new shape.

Figure 16.9 Permanent waving, a chemical oxidation-reduction process. The disulfide bonds in hair protein are broken by a reducing agent (step 1). The protein strands are twisted into a more curly shape (step 2). Then the new shape is fixed by an oxidizing agent that re-forms the disulfide bonds (step 3).

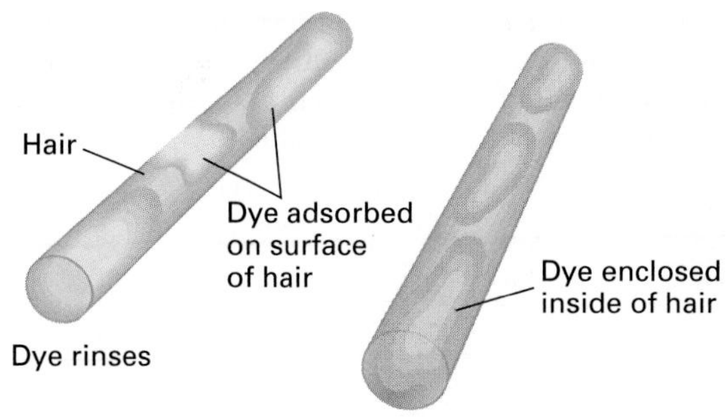

Figure 16.10 Hair dyeing. Rinses are adsorbed onto the surface of the hair. The semipermanent dyes penetrate the hair shaft.

Formulations for dyeing hair vary from temporary coloring (removable by shampoo), to semipermanent (lasts through four or five shampoos), to ''permanent'' (needs to be renewed about once a month). The temporary coloring is usually achieved by means of a water-soluble dye that deposits on the surface of the hair. Semipermanent dyes, which penetrate the hair fibers to a great extent, have low water solubility and are applied mixed with a shampoo or foam that is washed away after a short wait for the dye to diffuse into the hair (Figure 16.10). The more permanent dyes are described as *oxidation dyes.* Most are produced by oxidation of *para*-phenylenediamine, followed by reaction of the product with other simple molecules. The oxidizing agent (often hydrogen peroxide) and other chemicals are mixed just before use on the hair. To get any one of the many desirable hair colors requires a mixture of dyes, making the formulation of hair dyes a challenging business.

■ Some hair dyes have long been suspected of being carcinogenic because they can cause cancers in laboratory animals. Recent tests on large numbers of hair-dye users, however, have found no increases in cancers that can be attributed to the use of these dyes.

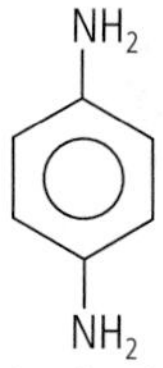

para-phenylenediamine

Hair can be bleached by a more concentrated solution of hydrogen peroxide, which destroys the hair pigments by oxidation. The solutions are made basic with ammonia to enhance the oxidizing power of the peroxide. Treating hair this way does more than just change the color. It may destroy sufficient structure to render the hair brittle and coarse.

Holding Your Hair in Place

Hair sprays are solutions of resins in a volatile solvent. When the solvent evaporates, the hair is coated with a film of sufficient strength to hold the hair in place (Figure 16.11). A suitable resin is a copolymer (Section 14.5) of vinylpyrrolidone and vinyl acetate.

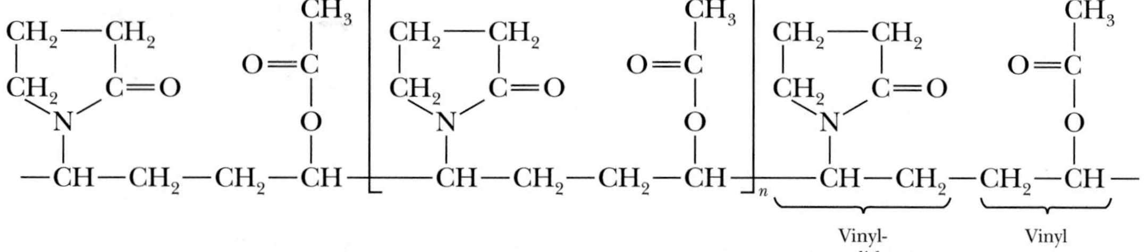

Copolymer of vinylpyrrolidone and vinyl acetate

Figure 16.11 A demonstration of the film left behind by hair spray. The spray contains sticky polymers that partly coat some of the hairs and hold them in place. *(Tom Pantages)*

■ The vapor from hair sprays should not be inhaled. There is the possibility of the chemicals harming the nasal passages and lungs.

In addition to the resin, solvent, and propellant for an aerosol spray (the liquid that gasifies when the valve is opened), a hair spray is likely to have additional ingredients to keep it from flaking off the hair after it is applied and an oil such as silicone oil to impart a sheen to the hair. Silicone oils (Section 14.5) are used because they are odorless and will not discolor on exposure to sunlight.

Removing Unwanted Hair

Unwanted hair can be removed in a number of ways. The purpose of a **depilatory** is to remove hair chemically. Because skin is sensitive to the same kinds of chemicals that attack hair, such preparations should be used with caution, and even then, some damage to the skin is almost unavoidable. Therefore, depilatories should be used only weekly and should never be used on skin that is infected or when a rash is present. Depilatories should not be used with a deodorant that has astringent action. If the sweat pores are closed by the deodorant, the caustic chemicals are retained and can do considerable harm.

The chemicals used as depilatories include sodium sulfide, calcium sulfide, strontium sulfide (all water-soluble sulfides), and calcium thioglycolate [$Ca(HSCH_2COO)_2$], the calcium salt of the compound used to break S—S bonds between protein chains in permanent waving. These active chemicals are usually added to a cream base.

The water-soluble sulfides are all strong bases in water, because of the high affinity of sulfide ions for protons.

$$\underset{\text{Sulfide ion}}{S^{2-}} + H_2O \longrightarrow HS^- + \underset{\text{Hydroxide ion}}{OH^-}$$

■ Hair is also removed by electrical cauterization, commonly called electrolysis. The hair follicle is destroyed by a high-voltage electric spark.

A dilute solution of sodium sulfide will have a pH as high as 13 (strongly basic). This basic solution will break some peptide bonds in the protein chains of hair, and the result will be a mixture of peptides and amino acids that can be washed away in a detergent solution.

■ SELF-TEST 16B

1. Hair protein chains are held together by ____________ and ____________ bonds.

2. Water easily breaks ____________ cross-links found in hair structure, making wet hair more plyable.
3. The dark pigment found in the hair of brunettes is called ____________.
4. Hair sprays are solutions of ____________ in a volatile solvent.
5. A hair dye that remains only on the outer surface of the hair would be considered a (temporary/permanent) hair dye.
6. A substance that can dissolve unwanted hair is called a ____________.
7. A substance capable of dissolving unwanted hair will have a pH of approximately ____________.

Figure 16.12 Matter in the wrong place is a challenge for our soaps and detergents. *(C.D. Winters)*

16.9 Keeping Things Clean—Including Ourselves

Dirt can be defined as matter in the wrong place (Figure 16.12). Tomato soup may be tasty food, but on your shirt it is dirt. To answer the need for keeping things clean, there are available more than 1200 commercial cleaning, or surface-active compounds, called **surfactants** because they can lower the surface tension of water solutions in which they are dissolved. The classic surfactant, soap, dates back in recorded history to the Sumerians in 2500 B.C., in what are now the countries of Iraq and Iran. Soap has always been made from the reaction of a fat with an alkali. The Greek physician Galen referred to this recipe and stated further that soap removed dirt from the body as well as serving as a medicine for cleaning wounds. What is new is that soap can now be made in a very pure state, and many other compounds, both natural and synthetic, have been found to be excellent surfactants.

In strongly basic solutions fats and oils undergo hydrolysis to produce glycerol and salts of fatty acids. Such reactions are called **saponification** reactions, and the sodium or potassium salts of the fatty acids formed are **soaps.**

$$\begin{array}{lclcl}
CH_3{-}(CH_2)_{16}{-}COO{-}CH_2 & & & & HO{-}CH_2 \\
CH_3{-}(CH_2)_{16}{-}COO{-}CH & + \; 3\,NaOH & \longrightarrow & 3\,CH_3{-}(CH_2)_{16}{-}COO^-Na^+ \; + & HO{-}CH \\
CH_3{-}(CH_2)_{16}{-}COO{-}CH_2 & & & & HO{-}CH_2 \\
\text{Tristearin (an animal fat)} & \text{Sodium hydroxide} & & \text{Sodium stearate (a soap)} & \text{Glycerol (glycerin)}
\end{array}$$

Pioneers prepared their soap by boiling animal fat with an alkaline solution obtained from the ashes of hardwood containing potassium carbonate (referred to as "lye"). The resulting lye soap could be "salted out" by adding sodium chloride, because soap is less soluble in a salt solution than in water. The crude soap made this way contained considerable caustic material in addition to the soap molecules, but it did its job of cleaning quite well.

■ Principal fats and oils for soap-making: tallow from beef and mutton, coconut oil, palm oil, olive oil, bone grease, and cottonseed oil

Substances that are water-soluble can readily be removed from the skin or a surface by simply washing with an excess of water. To remove a sticky sugar syrup from your hands, you can dissolve the sugar in water and rinse it away. Many times the material to be removed is oily, and water will merely run over the surface of the oil. Since the skin has natural oils, even substances such as ordinary dirt that are not oily themselves can adhere quite strongly to the skin and to clothing containing these oils. The hydrogen bonding holding water

Making soap the old-fashioned way.
(North Wind Picture Archives)

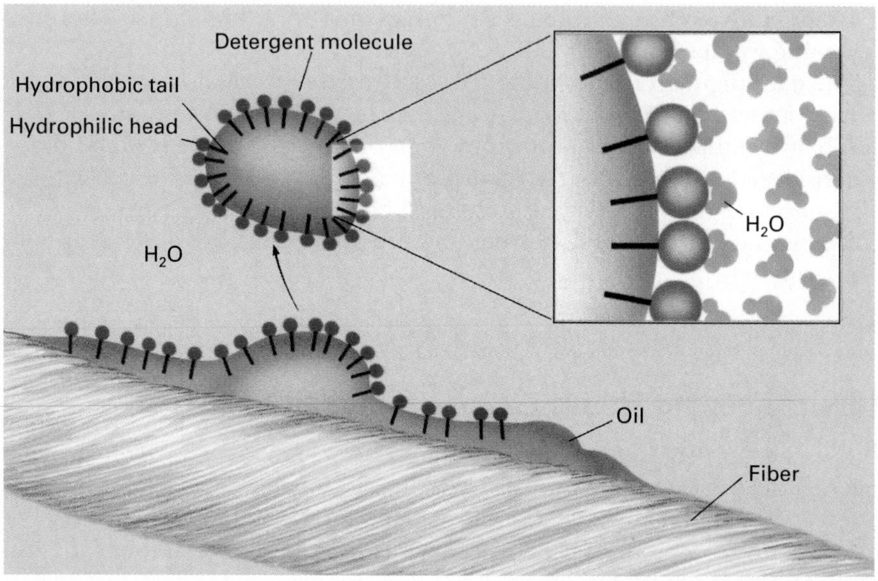

Figure 16.13 How soaps and detergents work. The hydrophilic, salt ends of the molecules interact strongly with water. The hydrophobic, hydrocarbon ends of the molecules avoid water and are drawn to the oily portion of the dirt. As greasy dirt is broken up by agitation, the particles are surrounded and isolated from one another. This prevents them from coming together again and allows them to be carried away in the wash water.

molecules together is too large to allow the oil and water to intermingle (Figure 16.13), so something like soap is needed to loosen the dirt and wash it away.

The cleaning action of soap is explained by its molecular structure. When present in an oil-water system, the stearate anion in the soap moves to the interface between the oil and the water.

$$CH_3CH_2CH_2CH_2CH_2CH_2CH_2CH_2CH_2CH_2CH_2CH_2CH_2CH_2CH_2CH_2CH_2COO^-Na^+$$

Sodium stearate molecule

■ Stearates: sodium—hard soap; potassium—soft soap; ammonium—liquid soap.

■ Soap, water, and oil together form an emulsion, with the soap acting as the emulsifying agent.

■ Floating soaps float because of trapped air.

The hydrocarbon chain, which is a nonpolar organic structure, mixes readily with the nonpolar oil or grease molecules, whereas the highly polar $—COO^-$ group enters the water layer because it becomes hydrated (Figure 16.16*b*). The organic ions then lie across the oil-water interface. The oily material is broken up into small droplets of colloidal size by agitation, with each droplet surrounded by hydrated soap ions. The soap, then, has formed an *emulsion* in which the oil droplets are the dispersed phase and the water solution is the continuous phase. The soap is the emulsifier. The surrounded oil droplets cannot come together again since the exterior of each droplet is covered with $—COO^-Na^+$ groups that interact strongly with the surrounding water. If enough soap and water are available, the dirt trapped in the stabilized oily droplets will be swept away, leaving a clean and water-wet surface.

One undesirable property of soaps is their tendency to form precipitates with Ca^{2+}, Mg^{2+}, and Fe^{2+} ions found in "hard" water. The resulting fatty-acid salts of these doubly positive ions are not as soluble in water as the Na^+ ion salts, so they appear as a "scum" that sticks to laundry and bathtubs, often containing trapped dirt, which makes it appear even worse.

16.10 Synthetic Detergents

Synthetic detergents are derived from organic molecules designed to have even better cleaning action than soaps but less reaction with the doubly positive ions found in hard water. As a consequence, synthetic detergents are often more economical to use and are more effective in hard water than soap. There are many different synthetic detergents on the market. An inventory of cleaning materials in a typical household might include half a dozen or more formulated products designed to be the most suitable for a specific job—cleaning skin, hair, clothes, floors, or the family car.

The molecular structure of a synthetic detergent molecule, like that of a soap, consists of a long oil-soluble (**hydrophobic,** meaning not liking, or fearing water) group and a water-soluble (**hydrophilic,** meaning liking or loving water) group.

$$CH_3CH_2CH_2CH_2CH_2CH_2CH_2CH_2CH_2CH_2CH_2CH_2CH_2CH_2 - \bigcirc - SO_3^-Na^+$$

Oil-soluble part
(hydrophobic)

Water-soluble part
(hydrophilic)

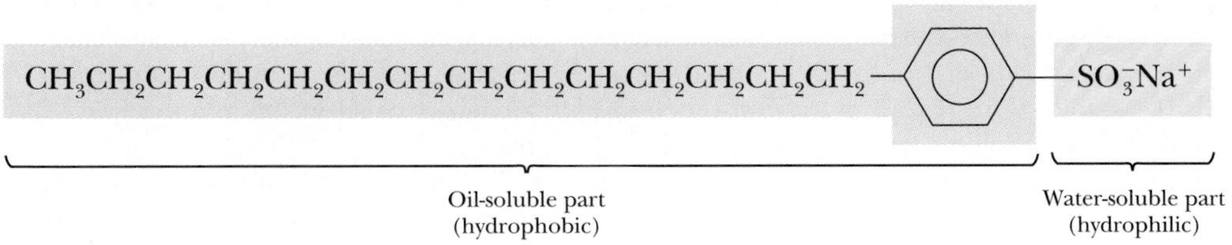

A typical synthetic detergent molecule

Typical hydrophilic groups include negatively charged sulfate ($-OSO_3^-$), sulfonate ($-SO_3^-$), and phosphate ($-OPO_3^{2-}$) groups. Compounds with these groups are called **anionic surfactants.**

Cationic (positively charged) **surfactants** are almost all quaternary ammonium halides (four groups attached to the central nitrogen atom) with the general formula:

$$R_1 - \overset{\overset{\textstyle R_2}{|}}{\underset{\underset{\textstyle R_3}{|}}{N^+}} - R_4 \ \ X^-$$

where one of the R groups is a long hydrocarbon chain and another frequently includes an $-OH$ group. The X^- in the formula represents a halide ion such as chloride (Cl^-) or bromide (Br^-) ion.

Cationic detergents are generally incompatible with anionic detergents. If they are brought together, insoluble products can precipitate from the solution, and the precipitate has none of the desired properties of either detergent. Their reaction with excess anionic detergent, however, makes cationic detergents useful as fabric softeners. When an anionic detergent remains in a fabric after washing, the fabric often loses its softness. Sufficient rinsing would take away the excess detergent, but water-saving rinse cycles in most washing

■ Mayonnaise and peanut butter are also oil-in-water emulsions and will separate without the stabilizing action of surfactants. The surfactant in mayonnaise (at least in homemade mayonnaise) comes from the egg yolk.

■ A detergent with an ammonium end ($-NH_4^+$) would also be cationic.

SCIENCE AND SOCIETY

How Effective Are Disinfectant Substitutes?

In an effort to use more "natural" products, numerous substitutes have been suggested for the germ-killing cationic synthetic detergents called quaternary ammonium compounds. These substitutes include aqueous ammonia (NH_3 in water), vinegar (acetic acid in water), baking soda ($NaHCO_3$), and sodium borate ($Na_2B_2O_7 \cdot 10\ H_2O$, also known as borax). The reason these compounds are often suggested as disinfectant substitutes is probably based more on historical precedence and other factors than on scientific evidence. As early as 1750 sodium borate was reported to inhibit putrefaction, and during the time of the Roman Empire vinegar was used as an antiseptic for abdominal wounds (OUCH!). Ammonia's strong odor has probably led many to believe that it could also be effective in killing germs. But do these compounds really kill harmful bacteria? Consider the results of the experiments described below.

In a series of tests using official methods sanctioned by the Association of Official Analytical Chemists (their tests are also approved by the EPA and other regulatory agencies), stainless steel tube samples, representing food preparation surfaces and utensils, were contaminated with cultures of *Staphylococcus aureus* and *Salmonella choleraesuis,* two potentially lethal but common bacteria. After the samples had been contaminated, each was "disinfected" by placing it in a solution of either a commercial quaternary ammonium disinfectant solution or one of four substitutes. (The substitutes were diluted as suggested by organizations recommending their use as disinfectant substitutes.) Contact with one of the five solutions represented the cleaning process. After contact with one of the solutions, the samples were placed in a growth medium containing neutralizers that reacted with the disinfectants, stopping their action.

If, after being incubated for several days at 35 to 37°C, the samples showed no evidence of bacterial growth, the disinfectant or its substitute would be considered effective. In this "pass-fail" test, contaminating bacteria must be killed on 98% of the samples for the product to be considered a disinfectant. In the test summaries shown below, 0/30 means no bacteria were found on 30 test samples, while 30/30 means bacteria were found on all 30 samples. As the results in the table show, *S. aureus* was not killed by solutions of any of the substitutes. On the other hand, *S. choleraesuis* was killed on seven of the samples using the sodium borate solution (still not enough to make it a disinfectant), but the other solutions were not effective at all. Only in the case of the commercial quaternary ammonium disinfectant were all of the samples free of bacteria (0 out of 30 in each case).

Tests carried out with sponges dipped in the five solutions and wiped on hard surfaces contaminated with bacteria gave similar results—the substitutes did not kill bacteria effectively enough to be considered disinfectants, but the quaternary ammonium surfactant did. In fact, using the substitutes for that purpose might put you at risk of a harmful bacterial infection.

Substance	Dilution	*S. aureus*	*S. choleraesuis*	Disinfection Activity
Sodium borate	1:32	30/30	23/30	No
Vinegar	1:25	30/30	30/30	No
Baking soda	1:25	30/30	30/30	No
Ammonia	1:32	30/30	30/30	No
Quaternary ammonium surfactant	"as is"*	0/30	0/30	Yes

*The quaternary ammonium surfactant tested was an aqueous solution.

Chemical Times and Trends, Chemical Specialties Manufacturing Association, Washington, D.C. October 1994.

machines seldom use enough water to get all of the detergent out. A small amount of a cationic detergent in the rinse water is enough to react with the excess anionic detergent. The result is a uniform surface layer that leaves the fabric feeling softer.

Most cationic detergents are also able to kill bacteria. Quaternary ammonium detergents are particularly good in this regard and find wide use in disinfectants, toilet bowel cleaners, and similar products.

Besides the anionic and cationic detergents, some detergents are *nonionic.* They have an uncharged hydrophilic polar group attached to a large organic group of low polarity, for example,

$$CH_3(CH_2)_{11}COO(CH_2)_2O(CH_2CH_2O)_2CH_2CH_2OH$$

Hydrocarbon chain (hydrophobic) | Ester link (hydrophilic) | Ether link (hydrophilic) | Ether link | Alcohol group (hydrophilic)

A nonionic detergent molecule

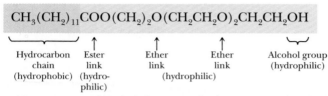

Most of us have a shelf like this. We rely on a wide variety of chemical surfactants to keep things clean. *(C.D. Winters)*

The carbon chain in this molecule is oil-soluble and the rest of the molecule is hydrophilic, the properties needed for the molecule to be a detergent.

The nonionic detergents have several advantages over ionic detergents. Since nonionics contain no ionic groups, they cannot form salts with calcium, magnesium, and iron ions and consequently are totally unaffected by hard water. For the same reason, nonionic detergents do not react with acids and may be used even in relatively strong acid solutions, which makes them useful in toilet bowl cleaners (see Section 16.15).

In general, the nonionic detergents foam less than ionic surface-active agents, a property that is desirable where nonfoaming detergents are required, as in dishwashing. Today, about one third of all detergents sold, including most liquid laundry and dishwashing detergents, are of the nonionic type.

■ In spite of the major impact of various kinds of synthetics on the detergent industry, soap is still the number-one surfactant, holding approximately 39% of the market.

16.11 Shampoos—Detergents that Clean Our Hair

Shampoos are generally more complex formulations than simple soap solutions, with a number of ingredients to satisfy different requirements for maintaining clean and healthy-looking hair. If you use soap to wash your hair, incomplete rinsing will result in a "soap film" on the hair, which makes it appear dull. If soap is used to wash hair in hard water, a very noticeable film can usually be seen and is very difficult to remove. In addition, soaps tend to produce solutions with basic pH values. These harsh conditions are damaging to the hair (recall how depilatories work).

Most shampoos contain anionic detergents. These give the product good foaming characteristics because anionic detergents generally foam more than cationic and nonionic detergents. Nonionic surfactants, such as the products obtained by reacting diethanolamine and lauric acid, are often also present in shampoos. While not such good surfactants as sodium lauryl sulfate and other anionic compounds, the nonionics are useful as thickeners and foam stabilizers. They make a shampoo pour from the bottle more slowly and cause the lather to remain thick for a longer period.

■ $CH_3(CH_2)_{11}OSO_3^-Na^+$
Sodium lauryl sulfate
(an anionic detergent)

$$HN(CH_2CH_2OH)_2 + CH_3(CH_2)_{10}COOH \longrightarrow$$
Diethanolamine Lauric acid

$$CH_3(CH_2)_{10}\overset{\overset{\displaystyle O}{\|}}{-C}-N(CH_2CH_2OH)_2 + H_2O$$
Lauric diethanolamide
(an amide detergent)

■ Lather has nothing to do with cleaning efficiency, but consumers have been taught to expect a good lather whenever a shampoo is used.

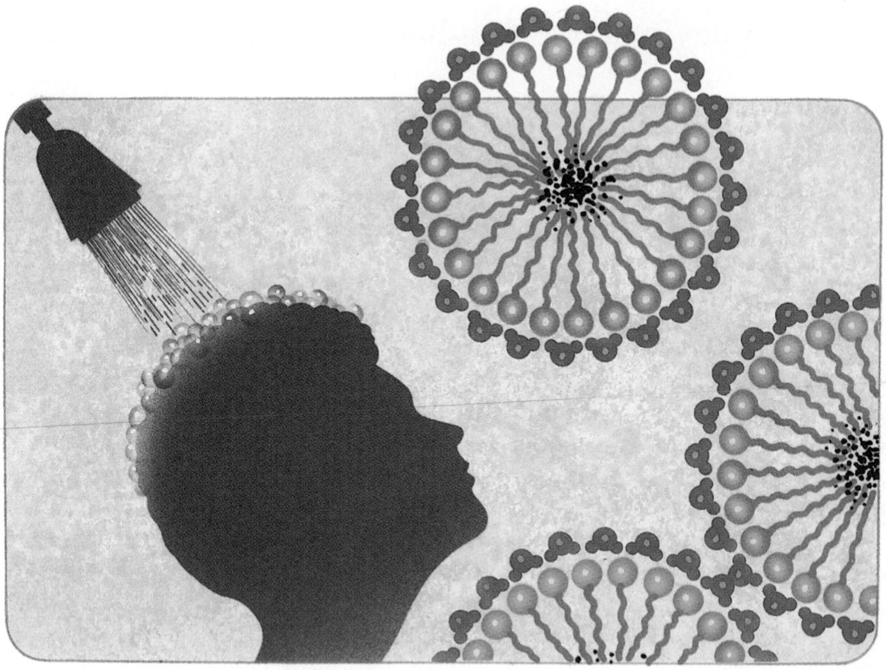

Hair is more manageable, has a better sheen, and has less tendency to attract static charges (causing "fly-away" hair) if all the shampoo is removed after washing. An anionic detergent such as sodium lauryl sulfate can be removed from the hair by using a *rinse,* or conditioner, containing a dilute solution of a cationic detergent, usually a quaternary ammonium compound, which electrically attracts the anions and facilitates their removal. The positive charged end of the quaternary ammonium surfactant also neutralizes negative charges on damaged hair ($—COO^-$ groups from disrupted protein chains), while the alkyl chain attaches to the hair and gives it a smooth feel.

■ Caution should be taken with the cationic rinse because of its possible irritation to the eyes.

An after-shampoo conditioner also attempts to put back into the hair some of the oils that were removed by the detergent. A typical conditioner has a water-alcohol dispersing medium as well as emollients, oils, waxes, resins, and even short amino acid polymers that can adhere to the hair to produce a more pliable and elastic fiber that is not as likely to become dry or be affected by atmospheric conditions. Holding the correct amount of moisture is the key to hair control, because too much water causes the hair to be limp and too little causes the individual hairs to attract static charge. Although many hair preparations make direct or indirect claims that various proteins and other beneficial ingredients can penetrate the hair and "repair" and even strengthen it, there is no scientific evidence for these claims. Protein molecules, for example, are simply too large to pass through the surface of the hair. Only much smaller molecules can do this. In fact, *if* hair preparations did function in this way, they would have to be classed as drugs by the FDA.

■ Lanolin is a waxy substance obtained from wool.

Lanolin and mineral oil (or their substitutes) are often added to shampoos to replace the natural oils in the scalp, thus preventing it from drying out and scaling. The presence of oil additives and stabilizers sometimes gives the shampoo a pearlescent appearance. These ingredients also make the shampoo less foamy, which is popular in European countries.

Mousse has become a very popular medium for the application of chemicals to hair. The French word *mousse* means "froth," foam, lather, or whipped cream (Figure 16.14). A mousse foam can deliver any hair-care chemical with a delightful advantage. It isn't messy, which is very important in coloring and curling hair, and can give pinpoint accuracy with no overspray or runoff.

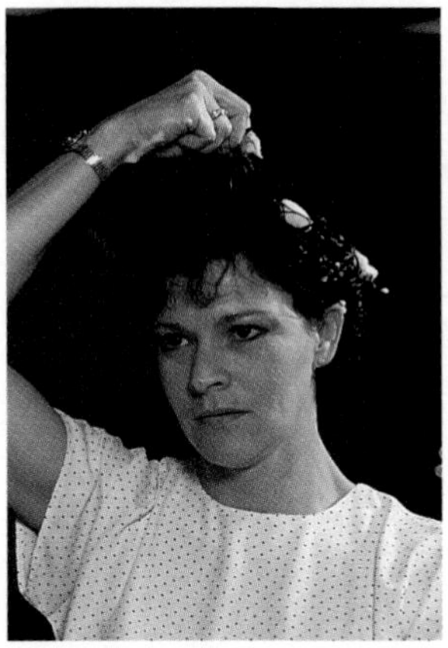

Figure 16.14 A hair-styling mousse in use. *(C.D. Winters)*

16.12 Toothpaste—Detergent Mixtures We Put in Our Mouths

Keeping our teeth clean requires **toothpaste,** a mixture of a detergent and a few **abrasives,** which are hard substances that help remove unwanted materials on the tooth surface. The structure of tooth enamel is essentially that of a stone composed of calcium carbonate ($CaCO_3$) and calcium hydroxy phosphate [apatite—$Ca_{10}(PO_4)_6(OH)_2$]. Despite being the hardest substance in the human body, tooth enamel is readily attacked by acids. Because the decay of some food particles produces acids, it is important to keep our teeth clean.

Within moments after you clean your teeth, a transparent film composed of proteins from saliva begins to coat the teeth and gums. This coating offers a place for food debris to collect and for oral bacteria to multiply. These bacteria convert dextrins, from the breakdown of sugars, into acids. At the same time, a tenacious film composed of these bacteria, food particles, and their breakdown products begins to form. As the film containing a combination of food particles and dead bacteria hardens, it becomes dental **plaque.** If this plaque is not removed regularly and completely from the surface of the teeth and beneath the gumline by brushing and flossing, the generation of acids and other harmful substances continues, eventually destroying the tooth or the bone that holds it in place.

Plaque that is not removed from the teeth becomes calcified from minerals in the saliva. The calcified plaque is known as **tartar.** It is possible to control tartar buildup by using toothpastes containing sodium pyrophosphate ($Na_4P_2O_7$), which interferes with the mineral crystallization that causes tartar buildup. Beneath the gumline, tartar is a special problem because its presence makes it easier for plaque to grow, which irritates gum tissue and allows the gum to become diseased. Only a dentist or oral hygienist can remove tartar from beneath the gumline. By keeping teeth free from plaque and from prolonged contact with the acids produced by plaque bacteria, we can preserve the hard, stonelike enamel of the tooth.

You can't see them, but they're there. Bacterial plaque on the surface of a tooth. *(David Scharf/Peter Arnold, Inc.)*

The abrasive material in toothpaste serves to cut into the surface deposits, and the detergent assists in suspending the particles in the rinse water. Abrasives commonly used in toothpastes include hydrated silica (a form of sand, $SiO_2 \cdot n\,H_2O$), hydrated alumina ($Al_2O_3 \cdot n\,H_2O$), and calcium carbonate ($CaCO_3$). It is difficult to select an abrasive that is hard enough to cut the surface contamination yet not so hard as to cut the tooth enamel. The choice of detergent is easier; any good detergent such as sodium lauryl sulfate will do quite well. Because the necessary ingredients in toothpaste are not very palatable, it is not surprising to see that various flavorings, sweeteners, thickeners, and colors are included to appeal to our senses. The most common flavors in toothpaste are spearmint, peppermint, and cinnamon.

Tooth decay occurs when bacteria eat food particles that remain in tiny fissures and crevices in the teeth and produce acids that attack the tooth enamel. Fortunately, it is possible to modify the crystalline structure of tooth enamel and make it more resistant to decay by the addition of fluoride ion (F^-) to toothpaste. When regularly applied, some of the fluoride ions actually replace the hydroxide ions in the hydroxyapatite structure to form fluoroapatite [$Ca_{10}(PO_4)_6F_2$]. The fluoride ion forms a stronger ionic bond in the crystalline structure because of its high concentration of negative charge, and as a result, the fluoroapatite is harder and less subject to acid attack than the hydroxyapatite. Hence, there is less tooth decay. Fluoride ion is introduced into essentially all of the public water supplies in the United States for this purpose. Concentrations of fluoride ion of 1 part per million (ppm) have proven safe and efficient for reducing tooth decay. About 80% of all toothpaste sold in this country contains fluoride ions in some form. Compounds such as stannous fluoride (SnF_2) and sodium monofluorophosphate (Na_3FPO_3) provide a low level of fluoride ion concentration in toothpastes.

More teeth are now lost as a result of gum disease than from decay. Gum disease results from the lack of proper massage, from irritating deposits below the gum line, from bacterial infection, and from poor nutrition. More attention is being given to toothpastes containing disinfectants such as peroxides in addition to the other ingredients.

16.13 Bleaches and Whiteners

The sun still does a fine job of keeping clothes whiter and brighter. *(C.D. Winters)*

Laundry detergents usually contain bleaching and whitening agents to remove stubborn stains and make whites appear even whiter. Together or separately, these two classes of compounds are used in many commercial laundry and cleaning preparations. **Bleaching agents** are compounds that remove unwanted color from textiles. Most commercial bleaches are oxidizing agents such as sodium hypochlorite and hydrogen peroxide.

In earlier times textiles were bleached by exposure to sunlight and air, with oxygen doing the oxidizing and sunlight providing energy in the form of heat. In 1786 the French chemist Berthollet introduced bleaching with chlorine. Today, this process is carried out with sodium hypochlorite, an oxidizing agent prepared by passing chlorine gas into aqueous sodium hydroxide:

$$2\,NaOH(aq) + Cl_2(g) \longrightarrow \underset{\text{Sodium hypochlorite}}{NaOCl(aq)} + NaCl(aq) + H_2O(\ell)$$

Shortly after Berthollet's discovery of the bleaching power of chlorine, hydrogen peroxide was introduced as a textile bleach. Today, a number of other oxidizing agents based on both chlorine and oxygen are used in bleaching.

Substances exhibit color because they have molecular structures that can absorb various portions of the visible spectrum of light. Many of these light-absorbing structures involve mobile electrons that can be easily excited by light. Chlorophyll, the green pigment found in plants, is just such a compound. One way to decolorize these kinds of substances is to remove or immobilize those electrons in the material activated by visible light. A simple equation for this process is

$$\text{Colored substance (the stain)} - \text{electrons} \longrightarrow \text{white material (without mobile elecrons)}$$

The hypochlorite ion, because it is an oxidizing agent, is capable of removing electrons from many colored materials. In this process, the hypochlorite ion is reduced to chloride and hydroxide ions as it gains electrons.

$$ClO^-(aq) + H_2O(\ell) + 2e^- \longrightarrow Cl^-(aq) + 2\ OH^-(aq)$$

Other bleaches gain electrons in similar ways. The resulting oxidized substance is without those mobile electrons, so it does not absorb light like the unoxidized substance. Usually the loss of the mobile electrons results in a structural change in the oxidized molecules or their complete destruction, forming smaller molecules. The products of the oxidation reaction may be more soluble in the detergent solution, which helps even further in the cleaning process.

THE PERSONAL SIDE

Claude Louis Berthollet (1748–1822)

(North Wind Picture Archives)

Berthollet was born in Italy of French parents and studied medicine in Turin. He became the personal physician to a Parisian noblewoman in 1772 and began studying chemistry there. Chemistry soon became his major interest. The textile industry was beginning to grow in Europe during Berthollet's stay in Paris, and he became curious about how to bleach textiles to make them appear whiter. Berthollet experimented with chlorine gas (which had been discovered by Carl Scheele in Sweden in 1774) and found that chlorine would bleach cotton textiles quite well, but that handling the toxic and corrosive gas limited its usefulness. By passing chlorine through an aqueous solution of potassium hydroxide, Berthollet produced a solution that had all the bleaching characteristics of chlorine but was much easier to handle because it was a liquid. Berthollet marketed his product, named *eau de Javelle* (bleaching water), to laundries around Paris. He also discovered potassium chlorate, and Antoine Lavoisier (who had attained great fame as a chemist by that time) carried out some experiments with that compound, hoping it might replace potassium nitrate in gunpowder. Lavoisier dropped those experiments when a powerful explosion that occurred in his lab killed two of his workers. Berthollet's discovery of bleach solution led to its use as a disinfectant for water supplies and ultimately to the direct use of chlorine itself for that purpose. In the United States, the first application of chlorination to a municipal water supply was in Jersey City, New Jersey, in 1908. Chlorination of municipal water supplies immediately led to dramatic decreases in water-borne diseases such as typhoid fever and cholera.

It is possible to make cleaned objects appear "whiter than white" by the use of **optical brighteners.** These compounds exhibit **fluorescence**; they absorb light of a shorter wavelength and emit light of a longer wavelength

■ Fluorescence: absorption of light at a shorter wavelength and re-emission of it at a longer wavelength.

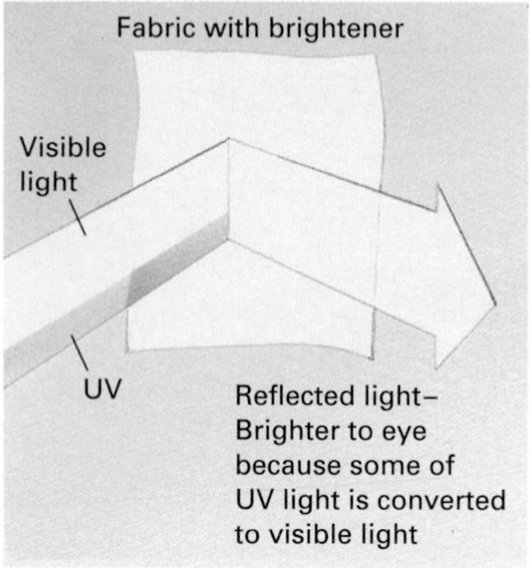
Fabric with brightener
Visible light
UV
Reflected light–Brighter to eye because some of UV light is converted to visible light

Figure 16.15 Optical brighteners are molecules that absorb unseen ultraviolet light and emit the energy as visible light.

Optical brighteners under ultraviolet light. The man's shirt glows because it has been washed in detergents containing optical brighteners. Brighteners are also present in paper and marking pens. White paper is often treated with an optical brightener, making the paper appear whiter and making dark print on its surface somewhat easier to read. *(Richard Megna/Fundamental Photographs)*

(Figure 16.15). When optical brighteners are incorporated into textiles, they make the material appear brighter and whiter, thus giving the appearance of being cleaner than the material might actually be. There are numerous optical brighteners in use. Besides the property of fluorescence when exposed to short-wavelength light, they must adhere to the textile after washing and not exhibit any undesirable properties such as toxicity.

16.14 Stain Removal

Sometimes detergents and bleaches will not remove a spot or stain on a garment. A *stain remover* is required. To a large extent, stain-removal procedures are based on solubility patterns or chemical reactions, usually oxidation. Many stains, such as those from chocolate or other fatty foods, can be removed by treatment with a solvent such as tetrachloroethylene, $Cl_2C{=}CCl_2$. Because of the toxicity of chlorinated solvents such as tetrachloroethylene, this type of stain removal is best left to professional dry cleaners.

Stain removers for the more resistant stains are almost always based on a chemical reaction between the stain and the essential ingredients of the stain remover. Suppose you were a laboratory worker and you stained your white lab coat with an iodine solution. The iodine would appear on the fabric as a dark purple stain that would resist water and many common solvents. An iodine stain remover could be made using a solution of sodium thiosulfate ($Na_2S_2O_3$). The stain-removal reaction is

$$\underset{\text{Purple}}{I_2(s)} + 4\,Na^+(aq) + 2\,S_2O_3^{2-}(aq) \longrightarrow \underbrace{4\,Na^+(aq) + 2\,I^-(aq) + S_4O_6^{2-}(aq)}_{\text{(All soluble in water and colorless)}}$$

Some other examples of stain removers are listed in Table 16.6.

16.15 Corrosive Cleaners

There are some places in the home where really tough cleaning jobs exist (Figure 16.16). For these jobs, cleaners are formulated with extremes in pH, which allow the acidity or alkalinity of the cleaner to quickly attack the un-

TABLE 16.6 ■ Common Stains and Stain Removers

Stain	Stain Remover	Action
Coffee	Sodium hypochlorite (bleach solution)	Oxidation
Lipstick	Isopropyl (rubbing) alcohol	Solubility
Rust and ink	Oxalic acid, methanol, water	Solubility
Airplane cement	Acetone	Solubility
Asphalt	Mineral spirits	Solubility
Blood	Cold water, hydrogen peroxide	Solubility and oxidation
Berry, fruit	Hydrogen peroxide	Oxidation
Grass stain	Sodium hypochlorite solution or rubbing alcohol	Oxidation or solubility
Mustard	Sodium hypochlorite	Oxidation
Antiperspirants	Household ammonia	Acid/base reaction
Perspiration	Household ammonia, hydrogen peroxide	Acid/base reaction, oxidation
Cola drink	Citrus drink (clear)	Solubility
Tobacco	Sodium hypochlorite	Oxidation

wanted dirt, grease, or stain. Toilet-bowl cleaners usually contain hydrochloric acid (HCl solution—also known as muriatic acid), which can dissolve most mineral scale (mostly carbonates) and iron stains. Other acids, such as phosphoric acid and oxalic acid, find use in these products. The pH of these cleaners is usually below 2, and because they contain strong acids they can be quite harmful to skin and eyes on contact. They should be handled with extreme caution, and rubber gloves should be worn when using them. Both phosphoric and oxalic acids are useful in dissolving iron stains because they form colorless or slightly colored, soluble ions that allow the iron to be washed down the drain.

$$\underset{\text{Red stain}}{Fe_2O_3(s)} + \underset{\text{Phosphoric acid}}{6\,H_3PO_4(aq)} \longrightarrow \underset{\text{Colorless ion}}{2\,Fe(PO_4)_3^{6-}(aq)} + 3\,H_2O(\ell) + 12\,H^+(aq)$$

$$\underset{\text{Red stain}}{Fe_2O_3(s)} + \underset{\text{Oxalic acid}}{6\,H_2C_2O_4(aq)} \longrightarrow \underset{\text{Pale green ion}}{2\,Fe(C_2O_4)_3^{3-}(aq)} + 6\,H^+(aq) + 3\,H_2O(\ell)$$

On the other end of the pH scale are drain cleaners and oven cleaners, which have a pH of 12 or higher. These formulations almost always contain the strong alkali (base) sodium hydroxide, NaOH. Drains usually clog as a result of oils, grease, and hair caught on rough edges inside the drainpipes. As the foreign matter builds up it becomes more tightly packed, until the flow of water from the drain practically stops. The only way to get rid of the material clogging the drain is to either dismantle the drain plumbing (often very difficult), use a plumber's "snake," or dissolve the material. Bases like sodium hydroxide are very good at dissolving drain clogs because they can cause rapid hydrolysis of the ester linkages found in oils and greases of animal and vegetable origin (in a manner similar to soap-making, discussed in Section 16.9). Once the ester linkages are broken, smaller, more soluble molecules are formed that can be washed down the drain. Hair and other proteins are also quickly hydrolyzed by sodium hydroxide, so a mat of grease and hair in a clogged drain will quickly succumb to the action of the strong base. Numerous products containing sodium hydroxide are available, including solid NaOH pellets, flakes, and concentrated solutions. The solutions offer the easiest way

Figure 16.16 Corrosive household cleaners. Reading the labels and handling these chemicals with care are essential. *(C.D. Winters)*

SCIENCE AND SOCIETY

Using Environmentally Safe Baking Soda

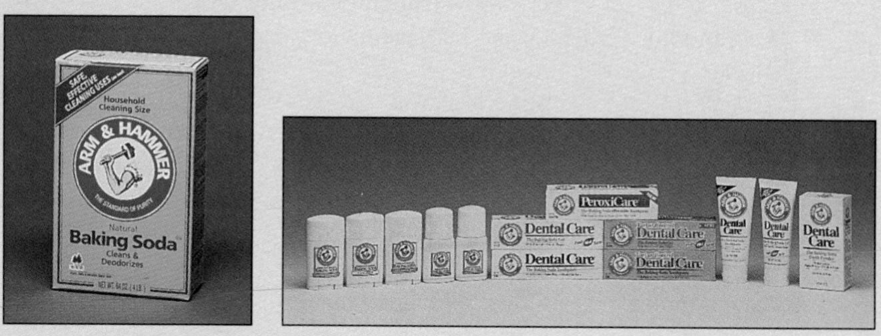

(a) (b)

Baking soda. (a) The old original. (b) Some of the many products on the market that contain baking soda. (Arm & Hammer)

Ordinary baking soda, so named because it releases CO_2 on contact with acids and causes dough to expand during baking, is a remarkably useful chemical. Baking soda is sodium bicarbonate, $NaHCO_3$, the monosodium salt of carbonic acid (H_2CO_3). Because of the presence of the bicarbonate ion, baking soda can react as both an acid and a base in aqueous solutions. The crystals of sodium bicarbonate are moderately hard, and its crystalline structure is porous enough to allow gas molecules to find a large surface area on which they can react with the bicarbonate ion.

These properties make baking soda very useful for a number of applications. For example, baking soda can be used to clean teeth. The abrasive action (not too hard) of the $NaHCO_3$ crystals seems just about right for cleaning away food particles, plaque, and loose tartar. In addition, because it has no taste or odor of its own, baking soda as a toothpaste leaves the mouth clean-feeling and refreshed without an aftertaste. Major manufacturers of toothpastes have gotten the message that consumers are looking for "natural" ingredients and have placed several toothpastes on the market that contain baking soda. Baking soda can also clean dishes (same mechanism as cleaning teeth) and can be used to clean grease from inside an oven. It can also be added to the laundry, where it acts as a fabric softener. Baking soda can be used to put out kitchen fires (it is a major component of several kinds of commercial fire extinguishers), and it was even used to clean the Statue of Liberty during its refurbishing.

Baking soda's odor-absorbing abilities are fascinating. An opened box of baking soda inside a refrigerator that has been made smelly with old food will absorb most of the odors overnight. Baking soda can also be added to carpets to control pet-odor problems, and because it is neither strongly acidic nor basic, it can be used on the most delicate fabrics. It can also be sprinkled in the dirty-clothes hamper to hold down odors until wash time.

Sodium bicarbonate occurs naturally in mineral deposits in several parts of the world, including southwest Wyoming. It can also be made by the reaction of sodium carbonate with carbon dioxide:

$$Na_2CO_3(aq) + H_2O(\ell) + CO_2(g) \xrightarrow{\text{heat}} 2\ NaHCO_3(aq)$$

The sodium carbonate itself is either prepared from a natural mineral or made from other chemicals. Is the sodium bicarbonate in your toothpaste more or less natural depending on how it is made?

Sodium bicarbonate is marketed by numerous companies, including Church & Dwight Company of Princeton, New Jersey, which has sold its Arm & Hammer brand since 1846. Over the past few years, consumers looking for environmentally friendlier and safer products have rediscovered baking soda, but the Arm & Hammer brand got into environmentalism way back in 1888 when it first put wildlife trading cards in boxes of its baking soda.

to apply the drain cleaner, but if they become diluted, their effectiveness is diminished. Some solid formulations also contain pieces of aluminum metal, which reacts with aqueous sodium hydroxide to form hydrogen gas (caution —flammable), which helps agitate the mixture and hastens the unclogging process.

$$2\ Al(s) + 2\ NaOH(aq) + 6\ H_2O(\ell) \longrightarrow 2\ Na^+(aq) + 2\ Al(OH)_4^-(aq) + 3\ H_2(g)$$

Oven cleaners also contain caustics such as sodium hydroxide. Many oven cleaners use aerosol sprays to distribute the cleaner on the inner surface of the oven. Care must be taken when breathing air containing these aerosols because strong bases are quite corrosive to nasal tissue, bronchial tubes, and lungs. Many homemakers advise that a little spilled food inside an oven is not all that bad and that the possibility of injury from exposure to the oven cleaner outweighs the benefits of a clean oven. Besides, if the oven is used often enough, spilled food will probably burn off anyway, or at least char to the point where it can be scaped off without using any chemicals at all. Of course, there are also self-cleaning ovens that use extra energy to bring the oven to a higher-than-normal temperature.

SELF-TEST 16C

1. A fat is a triester of glycerol and ____________ acids.
2. To make soap, a fat is treated with ____________.
3. Is the acid or salt group in a soap molecule more soluble in water or in oil?
4. What makes floating soap float?
5. Which is more likely to precipitate the hard-water ions (Ca^{2+}, Mg^{2+}, Fe^{3+}) as a sticky precipitate: (a) Traditional soaps, (b) Synthetic detergents?
6. "Surfactant" is a short term for ____________ ____________ ____________.
7. Why should a cationic detergent not be mixed with an anionic detergent in a laundry blend?
8. Optical brighteners transform ultraviolet light into ____________ light.
9. What are the two fundamental ingredients in a toothpaste?
10. A compound of what element is added to toothpaste to replace some of the hydroxide ions in apatite?

MATCHING SET

____ 1. Keratin
____ 2. Melanin
____ 3. PABA
____ 4. Moisturizer
____ 5. Apocrine glands
____ 6. Minimal erythemal dose
____ 7. Cysteine
____ 8. Stratum corneum
____ 9. SPF
____ 10. Zinc oxide
____ 11. Top note
____ 12. Depilatory
____ 13. Sebaceous glands
____ 14. Saponification
____ 15. Eccrine glands
____ 16. uvB
____ 17. Trichloroacetic acid

a. Sweat glands that respond to strong emotion
b. Glands in skin that produce a sweat that is mostly water
c. Consists mostly of dead skin
d. Amino acid common in hair
e. More harmful form of uv radiation
f. Dark pigment found in skin and hair
g. More volatile component of a perfume
h. Used in sunscreens
i. A basic solution or cream that can remove unwanted hair
j. Hydrolysis reaction in which a fat or oil produces glycerol and one or more fatty acids
k. Used in skin peels
l. Quantity of solar radiation that produces a barely observable reddening of the skin
m. Protein that makes up skin
n. Glands in skin that produce oils
o. Commonly added to powders to shrink tissues and reduce sebum flow
p. Any substance that holds moisture in the skin
q. A ratio of time required to produce a perceptible erythema on protected skin compared with the time required for production of the same erythema on unprotected skin

QUESTIONS FOR REVIEW AND THOUGHT

1. Describe the two kinds of bonding that hold protein strands together in hair.
2. What is the effect of water on the natural protein bonding in hair?
3. What is the disulfide linkage, and what role does it play in protein structure?
4. Explain what trichloroacetic acid, alpha-hydroxy acids, and soap solution have in common.
5. A certain lotion contains mink oil, while its competitor uses another animal oil. List the factors that might persuade you to purchase one lotion and not the other.
6. A certain cream imparts a somewhat greasy feel to the skin after its application. What kind of emulsion do you suspect this cream to be based on? (a) Water-in-oil, (b) Oil-in-water.
7. A skin cream has its ingredients listed on the back of the jar. Water is listed *after* several other ingredients, some of which have the word "oil" in their names. What kind of emulsion do you think this cream is based on?
8. Name three main ingredients that can be used to make cosmetic powders.
9. Describe the tanning process. Include in your answer the terms "uv radiation," "erythema," and "melanin."
10. What is the "uv index," published by the NOAA?
11. Explain how a chemical sunscreen like PABA works, and compare its action with that of zinc oxide.
12. What is the sunscreen protection factor (SPF)?
13. Compare the volatilities of the top, middle, and end notes of a perfume.
14. What is the solvent in most perfumes?
15. What is the main difference between a perfume and a cologne?
16. Describe the action of the three kinds of deodorants, and give an example of each.
17. Describe how it is possible to change the shape of hair.
18. Name an oxidizing agent that can remove the color from hair.
19. What is the purpose of the residue film left by hairsprays?
20. What acid is used to reduce hair chemically (break the cross-links between protein molecules)?
21. What kind of hair-care product contains para-phenylenediamine?
22. Are the proteins included in some hair conditioners effective? Explain.
23. What is the purpose of aluminum chlorohydrate in deodorants?
24. What advantage does a mousse have over a liquid hair conditioner?
25. What is a danger associated with the use of hairsprays?
26. Describe how a surfactant such as a soap or detergent molecule works.
27. It is possible to mix oil and water and get a stable emulsion using a soap. It is also possible to make a stable oil-in-water emulsion using the yoke of an egg (mayonnaise). Explain how these two emulsifiers work.
28. Explain why synthetic detergents are more effective than soaps in hard water.
29. What are the differences among cationic, anionic, and nonionic detergents?
30. Name two uses for cationic detergents.
31. A shampoo formulation contains sodium lauryl sulfate, an anionic detergent, and an amide detergent, which is nonionic. What is/are the function(s) of the nonionic detergent? What is the function of the anionic detergent?
32. How is a shampoo rinse similar to a fabric softener?
33. Name two kinds of ingredients commonly found in almost all toothpastes, and describe their functions.
34. Name two compounds used as laundry bleaches, and describe, in general terms, how bleaches work.
35. What is an optical brightener, and how does it make a fabric appear cleaner?
36. Consider the following stain-removal compounds: (1) a clear, citrus, carbonated beverage, (2) a chlorinated solvent, (3) water, (4) ethanol, (5) hydrogen peroxide. Which stain remover would you use to remove
    (a) table syrup from your shirt?
    (b) a cola drink stain from the tablecloth?
    (c) a bloodstain from an athletic jersey?
    (d) automotive grease from a pair of jeans?
    (e) chocolate sauce from a sweater?
37. Most toilet-bowl cleaners have a pH in what range? What is the pH range of most drain cleaners?
38. What is the main ingredient of most drain cleaners?
39. Describe the potential hazards of using an aerosol oven cleaner.

Managing good nutrition requires knowing something about the chemistry of food. *(C.D. Winters)*

CHAPTER 17

Nutrition: The Basis of Healthy Living

We are living in a health-conscious society. Newspapers, magazines, and television programs constantly offer us advice on what to eat. The advice ranges from responsible journalism to scare tactics and sensationalism. Fads about what to eat and what not to eat are rapidly translated into marketable products. And advertisements bombard us from every direction with information about food. The advertisements and food stories in the media are bound to attract our attention. After all, many times every day, we make choices about food.

How can we sort out the information so that these choices are good ones? The task is impossible without knowing something about the basic principles of nutrition, which requires starting with the fundamentals of biochemistry presented in Chapter 15. The questions addressed here are the following:

- What are the basic nutrients needed for good health?
- How is our energy generated?
- What are the roles of diet and exercise in control of excessive caloric intake?
- What roles do carbohydrates, fats, and proteins play in our diets and health?
- What are current recommendations for a healthy diet?
- What is the relationship between diet and heart disease?
- What are the categories of vitamins and minerals?
- What are the major functions of vitamins and minerals in our diets?
- What are the types and functions of food additives?

Nutrition is the science that deals with diet and health. The old saying "we are what we eat" is true. The skin that covers us now is not the same skin that covered us seven years ago. The fat beneath our skin is not the same fat that was there just a year ago. Our oldest red blood cells are 120 days old. The entire lining of our digestive tract is renewed every three days. Many chemical reactions are required to replace these tissues, and the energy and raw materials for these reactions are supplied by what we eat.

Nutrition, then, is concerned with the chemical requirements of the body—the nutrients and the chemical energy we get from them. **Nutrients** are chemical substances in foods that provide the energy and raw materials required by biochemical reactions. The five classes of nutrients are carbohydrates, fats, proteins, vitamins, and minerals. In addition, an average adult needs about 2.5 L of water each day.

17.1 Digestion: It's Just Chemical Decomposition

■ **Digestion** is the process of breaking food down into substances small enough to be absorbed into the body from the digestive tract.

Even before you swallow your lunch, chemical reactions begin the process of **digestion.** Like all biochemical processes, digestion is under the control of enzymes (biochemical catalysts, Section 15.4). Amylase, an enzyme in saliva, begins the digestion of the carbohydrates in starch. Carbohydrates are polymers of simple sugars (Section 15.5, and their digestion is the reverse of their formation—bonds are broken to set free the simple sugars. The same is true for digestion of proteins (polymers of amino acids) and the triglycerides in fats and oils (composed of glycerol and fatty acids).

$$\text{Carbohydrates} \xrightarrow{\text{Digestion}} \text{simple sugars (monosaccharides)}$$

$$\text{Proteins} \xrightarrow{\text{Digestion}} \text{amino acids}$$

$$\text{Fats (triglycerides)} \xrightarrow{\text{Digestion}} \text{glycerol} + \text{fatty acids}$$

Digestion begins in the mouth, continues in the stomach, and is completed in the small intestine (Figure 17.1). By the time a meal has been digested, all the nutrient molecules too large to cross cell membranes and enter the bloodstream have been converted into smaller molecules and absorbed.

■ A peanut butter and jelly sandwich on firm, toasted white bread contains approximately 20 g of protein, 86 g of carbohydrate, and 34 g of fat.

Suppose you have a peanut butter and jelly sandwich for lunch. Digestion of starch from the bread begins as you chew and continues in your stomach. It is completed in the small intestine, where sugars from the jelly are also converted to monosaccharides. The indigestible carbohydrates (dietary fiber) from the bread and peanut butter pass unchanged through the small *and* large intestines.

Digestion of protein from the bread and peanut butter starts in your stomach, where large polypeptides are broken down into smaller ones. The job is finished in the small intestine, where the individual amino acids enter the bloodstream. Lipids, mainly fats and oils from the peanut butter, are digested mainly in the small intestine.

Once digestion is finished, what happens to the small molecules that have been produced? They may undergo (1) complete breakdown to produce energy, carbon dioxide, and water, (2) recycling into new biomolecules, or

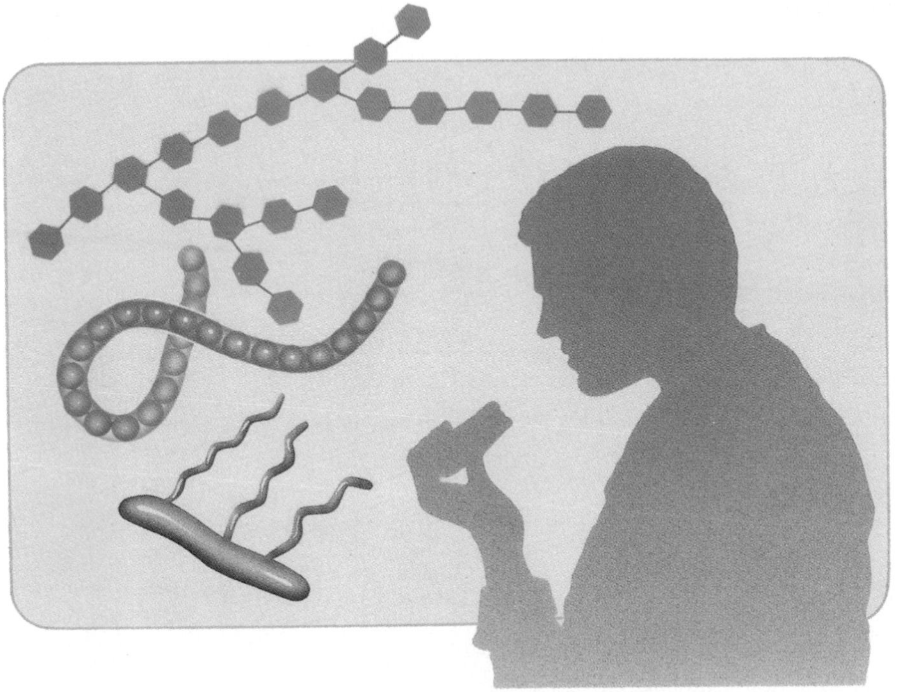

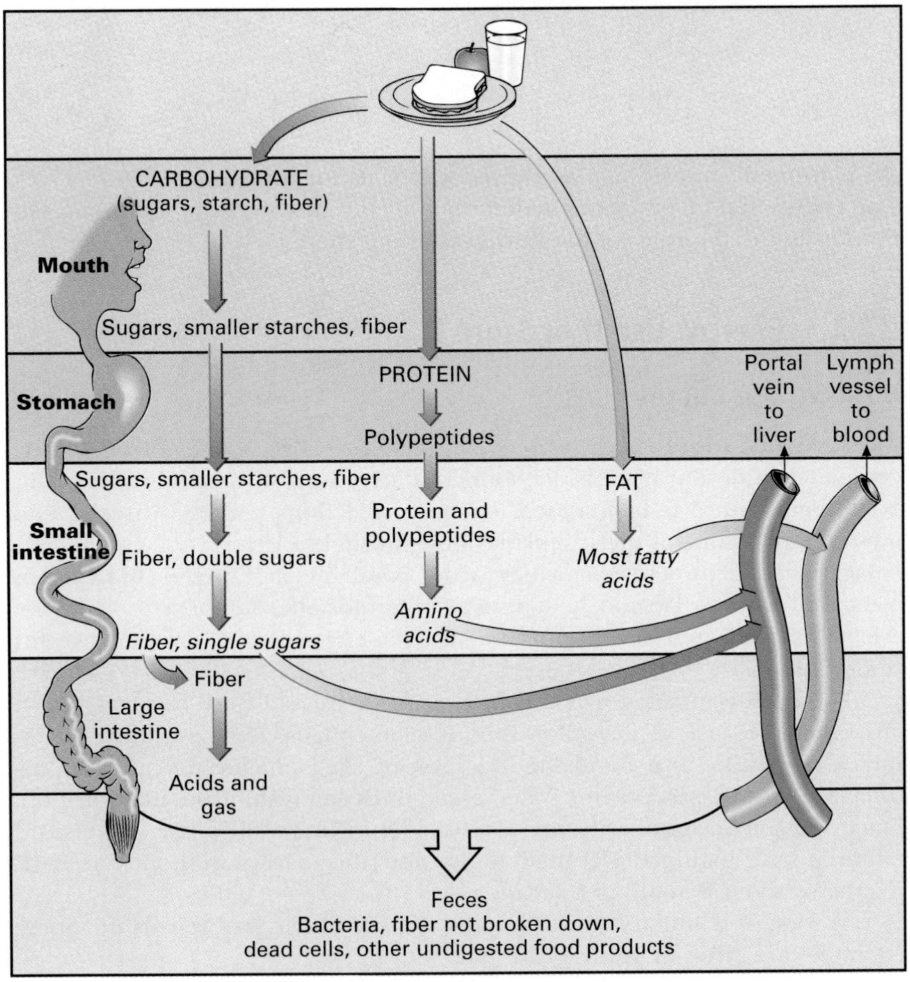

Figure 17.1 Digestion. At each stage, enzymes direct the biochemical reactions of digestion. Eventually all nutrients enter blood via the liver or, for fatty substances, via the lymphatic system.

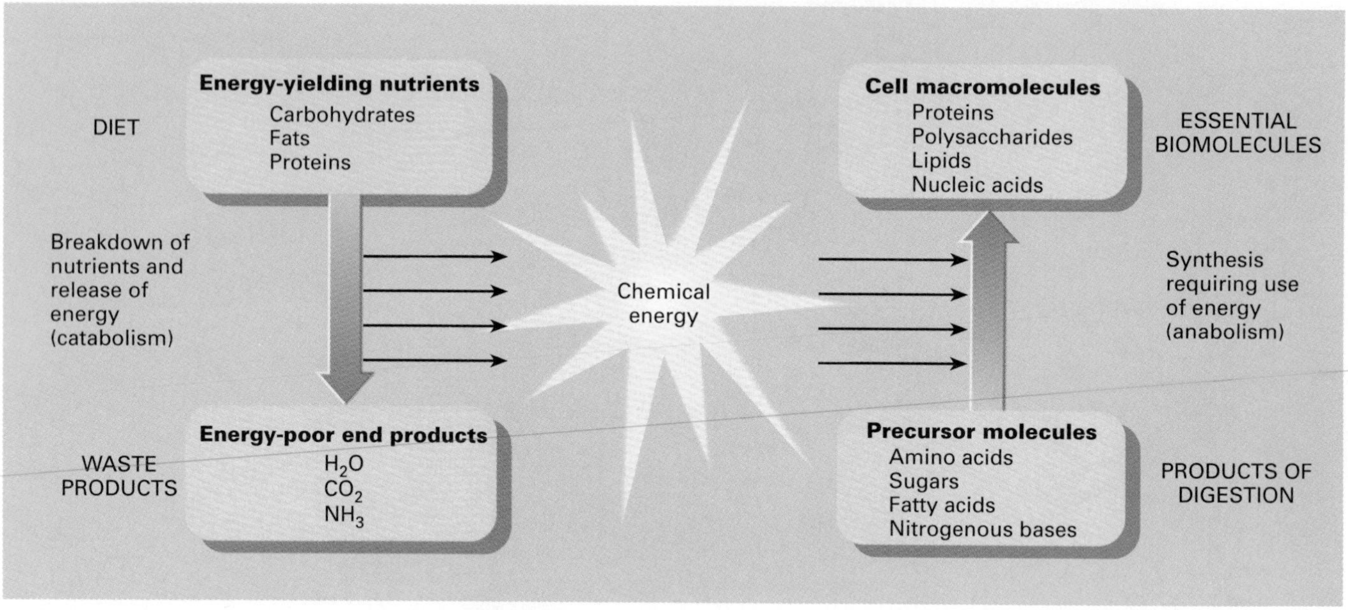

Figure 17.2 The release and use of chemical energy in the breakdown and synthesis of biomolecules.

(3) placement into storage as triglycerides (fat) for future use (Figure 17.2). The energy yield from foods is described in the next section. Following sections examine the roles of the nutrients in our diets.

17.2 Energy: Use It or Store It

How We Use Energy

■ To estimate your basal metabolic rate in kilocalories per day, multiply your weight in pounds by 10.

■ To get a rough estimate of your daily caloric needs, choose your general level of physical activity and multiply your estimated BMR by the factors given in Table 17.2.

A certain amount of energy is needed just to stay alive. While you are sitting completely still, your heart beats, your chest rises and falls as you breathe, your body temperature is maintained, chemical reactions proceed in cells, and messages that control these activities flow through your nervous system. The energy needed for these activities is the **basal metabolic rate (BMR).** It is measured when a person is at rest at a comfortable temperature but not asleep, has not eaten for 12 hours, and has not engaged in vigorous activity for several hours.

The BMR is affected by many factors. An increased BMR can come from anxiety, stress, lack of sleep, low food intake, congestive heart failure, fever, increased heart activity, and the ingestion of drugs, including caffeine, amphetamine, and epinephrine. A decreased BMR can result from malnutrition, inactive tissue due to obesity, and low-functioning adrenal glands. Infants and children have higher BMRs than adults, and after early adulthood the BMR decreases about 2 to 3% per decade.

As soon as voluntary activity begins, the metabolic rate speeds up. Some examples are given in Table 17.1.

TABLE 17.1 ■ Energy Expended in a Variety of Activities

Activity	Energy (kcal/min)	Activity	Energy (kcal/min)
Bicycling, 15 mph	12	Skiing, downhill	8–12
Bicycling, 5 mph	5	Skiing, cross-country	9–17
Bowling, active	7	Sleeping	1.2
Calisthenics	5	Square dancing	7.7
Chopping wood	7.5	Squash and handball	10.0
Driving an automobile	2.8	Standing	1.5
Eating	1.5	Sweeping floors	3.9
Football, touch/tackle	8.8/12	Swimming, pleasure	6
Gardening, weeding	5.6	Tennis, recreational/competitive	7/11
Listening to lecture	1.7	Volleyball, recreational/competitive	3.5/8.0
Pick and shovel work	6.7	Walking uphill, 10% grade	11
Rollerskating, recreational/vigorous	5/15	Walking 3.5 mph on road	5.6
Running, 5 mph (12-min mile)	10	Walking up stairs	10–18
Running, 12 mph (5-min mile)	25	Walking down stairs	7.1
Showering	3.4		

Energy from Foods

Fats in the diet provide more than twice as many calories per gram as do carbohydrates and proteins.

fats—9 kcal/g   carbohydrates—4 kcal/g   proteins—4 kcal/g

For example, if a steak is 49% water, 15% protein, 0% carbohydrate, 36% fat, and 0.7% minerals, then 3.5 oz of steak (about 100 grams) would provide about 384 kcal, or 384 food Cal.

■ 1 food Calorie = 1 kilocalorie

■ Fifteen percent (15%) of 100 g is 0.15 × 100 g

TABLE 17.2 ■ Daily Energy Needs According to Physical Activity

Activity Level	Factor	
Very light		To estimate your daily energy needs, multiply your estimated BMR (10 × your weight in pounds) by the factor listed in the table that best represents your general level of daily activity:
Men	1.3	**Very light:** mostly sitting and standing activities
Women	1.3	**Light:** mostly walking activities
Light		**Moderate:** cycling, tennis, dancing
Men	1.6	**Heavy:** heavy manual digging, climbing, basketball, soccer
Women	1.5	
Moderate		
Men	1.7	
Women	1.6	
Heavy		
Men	2.1	
Women	1.9	

Energy from protein

100 g of steak × 0.15 = 15 g protein

$$15 \text{ g protein} \times \frac{4 \text{ kcal}}{\text{g protein}} = 60 \text{ kcal}$$

Energy from fat

100 g of steak × 0.36 = 36 g fat

$$36 \text{ g fat} \times \frac{9 \text{ kcal}}{\text{g fat}} = 324 \text{ kcal}$$

Total energy 384 kcal

■ Section 17.6 explains how to interpret food labels.

The caloric values of most foods are calculated by this method, and these are the values that are listed in diet books and on food labels. Some representative values are given in Table 17.3.

The human body, like everything else, is subject to the law of conservation of energy. In our case it translates into the following equation:

$$\text{Energy taken in (food)} = \text{energy used} + \text{energy stored (fat)}$$

■ Some additional examples of food and energy arithmetic are given in Section 17.10.

For most people, the secret to dieting is little more than applying this equation. When energy taken in exceeds energy used, the excess enters storage, mostly as triglyceride molecules in fat cells. When energy taken in remains the same, but more energy is used, less is stored as fat. When energy used is more than energy taken in, some must be removed from storage. To be genuinely successful at providing weight loss, a program must integrate management of both diet and exercise.

These two students have different levels of activity and different caloric needs. *(C.D. Winters)*

EXAMPLE 17.1 *Daily Energy Needs*

Julie weighs 110 lb and is a college freshman—she plays tennis or skates several times a week and keeps in shape by running every day. Using Table 17.2, estimate her daily caloric needs.

SOLUTION

Based on her weight, Julie's estimated basal metabolism rate (BMR) is

$$110 \text{ lb} \times 10 \text{ kcal/lb} = 1100 \text{ kcal}$$

From what we know about Julie, her activity level is moderate. Therefore, using the factor of 1.6 from Table 17.2, her estimated daily caloric needs are

$$1100 \text{ kcal} \times 1.6 = 1800 \text{ kcal}$$

Exercise 17.1

Jack weighs 160 lb. As a graduate student he spends most of his time reading in the library or working at the computer. Estimate his daily caloric needs.

TABLE 17.3 ■ The Approximate Percentages of Carbohydrates, Fats, Proteins, and Water in Some Whole Foods as Normally Eaten

Food	Water	Protein	Fat	Carbo-hydrates	kcal/100 g
Vegetables					
Spinach, raw	90.7	3.2	0.3	4.3	26
Collard greens, cooked	89.6	3.6	0.7	5.1	33
Lettuce, Boston, raw	91.1	2.4	0.3	4.6	25
Cabbage, cooked	93.9	1.1	0.2	4.3	20
Potatoes, cooked	75.1	2.6	0.1	21.1	93
Turnips, cooked	93.6	0.8	0.2	4.9	23
Carrots, raw	88.2	1.1	0.2	19.7	42
Squash, summer, raw	94.0	1.1	0.1	4.2	19
Tomatoes, raw	93.5	1.1	0.2	4.7	22
Corn kernels, cooked on cob	74.1	3.3	1.0	21.0	91
Snap beans, cooked	92.4	1.6	0.2	5.4	25
Green peas, cooked	81.5	5.4	0.4	12.1	71
Lima beans, cooked	70.1	7.6	0.5	21.1	111
Red kidney beans, cooked	69.0	7.8	0.5	21.4	118
Soybeans, cooked	73.8	9.8	5.1	10.1	118
Meats and Fish					
Lean beef, broiled	61.6	31.7	5.3	0	183
Beef fat, raw	14.4	5.5	79.9	0	744
Lean lamb chops, broiled	61.3	28.0	8.6	0	197
Lean pork chops, broiled	69.3	17.8	10.5	0	171
Lard, rendered	0	0	100.0	0	902
Calf's liver, cooked	51.4	29.5	13.2	4.0	261
Beef heart, cooked	61.3	31.3	5.7	0.7	188
Brains	78.9	10.4	8.6	0.8	125
Chicken, whole, broiled	71.0	23.8	3.8	0	136
Cod, raw	81.2	17.6	0.3	0	78
Meats and Fish (cont'd)					
Salmon, broiled	63.4	27.0	7.4	0	182
Freshwater perch, raw	79.2	19.5	0.9	0	91
Oysters, raw	84.6	8.4	1.8	3.4	66
Grains and Grain Products					
Wheat grain, hard	13.0	14.0	2.2	69.1	330
Brown rice, dry	12.0	7.5	1.9	77.4	360
Brown rice, cooked	70.3	2.5	0.6	25.5	119
Whole-wheat bread	36.4	10.5	3.0	47.7	243
White bread	35.8	8.7	3.2	50.4	269
Whole-wheat flour	12.0	14.1	2.5	78.0	361
White cake flour	12.0	7.5	0.8	79.4	364
Dairy Products and Eggs					
Milk, whole	87.4	3.5	3.5	4.9	65
Yogurt, whole-milk	89.0	3.4	1.7	5.2	50
Ice cream	62.1	4.0	12.5	20.6	207
Cottage cheese	79.0	17.0	0.3	2.7	86
Cheddar cheese	37.0	25.0	32.2	2.1	398
Eggs	73.7	12.9	11.5	0.9	163
Fruits, Berries, and Nuts					
Apples, raw	84.4	0.2	0.6	14.5	58
Pears, raw	83.2	0.7	0.4	15.3	61
Oranges, raw	86.0	1.0	0.2	12.2	49
Cherries, sweet	80.4	1.3	0.3	17.4	70
Bananas, raw	75.7	1.1	0.2	22.2	85
Blueberries, raw	83.2	0.7	0.5	15.3	62
Red raspberries, raw	84.2	1.2	0.5	13.6	57
Strawberries, raw	89.9	0.7	0.5	8.4	37
Almonds	4.7	18.6	54.2	19.5	598
Pecans	3.4	9.2	71.2	14.6	689
Walnuts	3.5	14.8	64.0	15.8	651

Burning Glucose, the Reverse of Photosynthesis

Glucose is our major fuel molecule. It is the end product of carbohydrate digestion, and if more is needed there are biochemical pathways for making it from stored fat and even protein. These pathways are essential for the brain and red blood cells, which can utilize *only* glucose for their energy needs. In a tightly integrated series of biochemical reactions, the carbon atoms of glucose are converted to carbon dioxide, energy is transferred to ATP, and water is produced from oxygen and hydrogen ions.

■ Oxidation of glucose to produce energy, water, and carbon dioxide is the reverse of photosynthesis (Section 15.7).

$$C_6H_{12}O_6 + 6\ O_2 \longrightarrow 6\ CO_2 + 6\ H_2O + \text{energy}$$

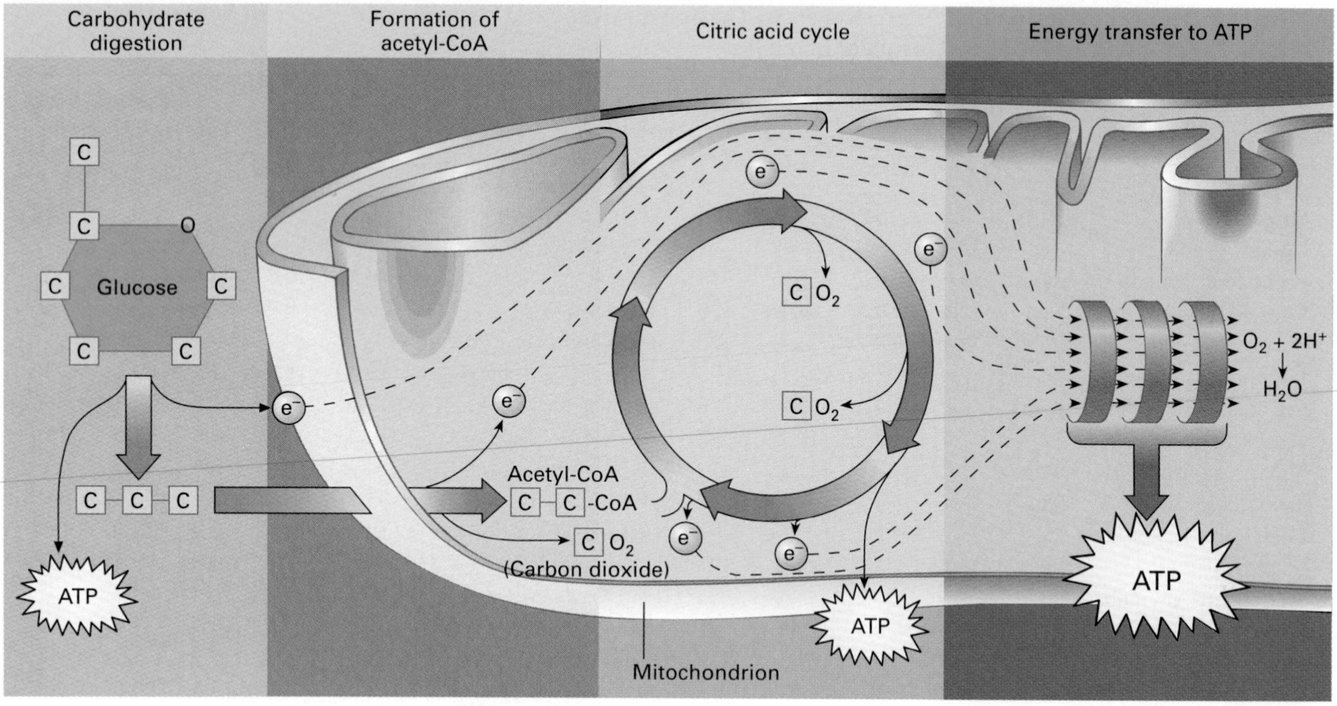

Figure 17.3 Energy from glucose. Glucose, a six-carbon sugar, is the end product of carbohydrate digestion. In a cell, it is first converted to three-carbon molecules and then enters the mitochondrion, the energy-generating "engine" of the cell. Two by two, the carbon atoms are carried into a cycle of eight reactions that convert acetyl groups to carbon dioxide. Simultaneously, the chemical energy of the acetyl groups is transferred to electrons that enter the electron transport chain. In a complex process, ATP, the high-energy molecule that carries energy throughout the body, is then synthesized and water, the other end product of "combustion" of food, is produced.

ATP (adenosine triphosphate), whose structure was given in Figure 15.21, is often called the body's energy currency. It carries energy to be spent wherever the energy is needed. (The coupling of energy-releasing reactions with energy-requiring reactions in biochemistry is a reflection of the need for unfavorable processes to be driven by favorable processes, as described in Section 8.5.) The pathway of carbon atoms from glucose to carbon dioxide while the energy from the chemical bonds of glucose is simultaneously transferred to ATP is summarized in Figure 17.3. Acetyl-coenzyme A (acetyl-CoA) carries carbon atoms two by two into the citric acid cycle. Carbon atoms from the digestion of proteins and lipids can also be transferred to acetyl-CoA, after which their pathway is the same as that of carbon atoms from carbohydrates.

■ Acetyl-coenzyme A is a large biomolecule that carries $CH_3—\overset{\text{O}}{\overset{\|}{\text{C}}}—$ groups, known as *acetyl groups,* from breakdown of food molecules into the energy-generating pathways.

■ SELF-TEST 17A

1. Digestion converts carbohydrates, proteins, and triglycerides into __________, __________, and __________ plus __________, respectively.

2. The quantity of heat required to operate the body at rest is the ______________, which in kilocalories is about ten times your ______________.
3. A food calorie has the same value as ______________ kilocalorie(s).
4. The kilocalories per gram are 4 kcal, 9 kcal, and 4 kcal for ______________, ______________, and ______________, respectively.
5. Our daily caloric needs are determined mainly by our weight and level of physical activity. (a) True, (b) False.
6. Which of the following is the body's major fuel molecule? (a) Triglyceride, (b) Glucose, (c) Maltose.
7. Energy is carried to where it is needed by the molecule abbreviated as ______________.

17.3 Sugar and Polysaccharides, Digestible and Indigestible

The major kinds of digestible carbohydrates in foods are the simple sugars (glucose and fructose), disaccharides (sucrose, maltose, lactose), and polysaccharides (amylose and amylopectin in starch from plants, glycogen from meat) (Table 17.4). The indigestible carbohydrates include cellulose and its derivatives, pectin (the substance that makes jam and jelly gel), and plant gums.

■ Carbohydrate structures are given in Section 15.5.

The indigestible polysaccharides are collectively referred to as **dietary fiber.** All dietary fiber comes from plants. There is insoluble fiber, mainly from the structural cellulose parts of plants, and soluble fiber—the gums and pectins. Barley, legumes, apples, and citrus fruits are foods with a high content of gums and pectins.

■ In 1988–1989, with widespread information about the cholesterol-lowering properties of oat bran, sales of oat bran cereals increased 240%.

It has long been known that dietary fiber prevents constipation. In recent years evidence has been accumulating that dietary fiber has significant other

TABLE 17.4 ■ Carbohydrates in Some Common Foods

Foods Composed of 70% or more Carbohydrates
Sugar (cane, beet, maple, brown)
Honey
Jams, jellies, marmalades
Dry cereals (corn, wheat, oat, bran)
Popcorn (popped)
Cookies (plain)
Foods Composed of 40–70% Carbohydrates
Cakes (plain, no icing)
Bread (white, rye, whole wheat)
Foods with Less than 40% Carbohydrates
Cooked macaroni, spaghetti, noodles, rice (23–30%)
Boiled corn, potatoes, dried beans

Foods high in dietary fiber. *(C.D. Winters)*

■ The results of an evaluation of health claims by the U.S. Food and Drug Administration are given in *Frontiers in the World of Chemistry: Health Claims for Foods* on page 546.

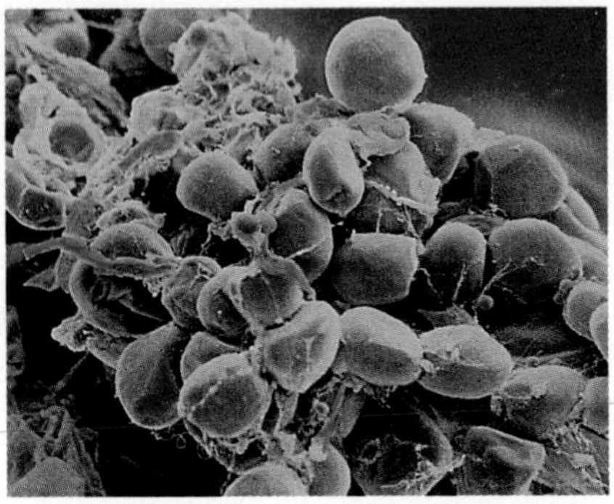

Where the fat goes. The yellow cells are adipocytes, the fat-storing cells of our bodies. Each adipocyte is almost entirely a single droplet of triglycerides. *(Prof. P. Motta/Dept. of Anatomy/University "La Sapienza," Rome/SPL/Photo Researchers)*

■ Almost 7% of new processed foods introduced in 1993 were reduced-fat items.

■ Cholesterol occurs naturally *only* in dairy products, meat, and fish. There is *no cholesterol* in fresh vegetables, fruits, or vegetable oils.

benefits. Adding soluble fiber to the diet decreases blood cholesterol levels, thereby decreasing the risk of heart disease. There is also evidence of a connection between *insufficient* dietary fiber and colorectal cancer. In our increasingly health-conscious society, such information provides a marketing advantage. We have all seen evidence of this in the bran muffins now found on every breakfast counter.

17.4 Lipids, Mostly Fats and Oils

Most of the lipids in the diet, which we usually refer to as "fats," are triglycerides. We get them from meats and fish, vegetables and vegetable oils, and dairy products. They may be solid fats or oils, and they may incorporate saturated or unsaturated fatty acids (Section 15.6). The fat content of some common foods is shown in Figure 17.4.

Fatty tissue is composed mainly of specialized cells, each featuring a large globule of triglycerides. When their energy is needed, the triglycerides in fat cells are hydrolyzed to give glycerol and free fatty acids, which leave the cells and are transported to the liver. There, the fatty acids are broken down to two-carbon molecular fragments that enter the main energy-producing pathway shown earlier (Figure 17.3).

Fat currently makes up about 38% of the average American diet (down from more than 40% in the 1980s). A fat content of 30% or less is strongly recommended by most public health authorities. The fat in today's diet is about 40% saturated, 40% monounsaturated, and 20% polyunsaturated. Lowering the saturated and monounsaturated and raising the polyunsaturated fat content of the diet is also strongly recommended. What is the basis for these recommendations? Heart disease is the number-one cause of death in the United States, and **atherosclerosis,** the buildup of fatty deposits called *plaque* on the inner walls of arteries, reduces the flow of blood to the heart. If a coronary artery is blocked by plaque, a heart attack occurs as a result of the reduced blood flow carrying oxygen to the heart. About 98% of all heart-attack

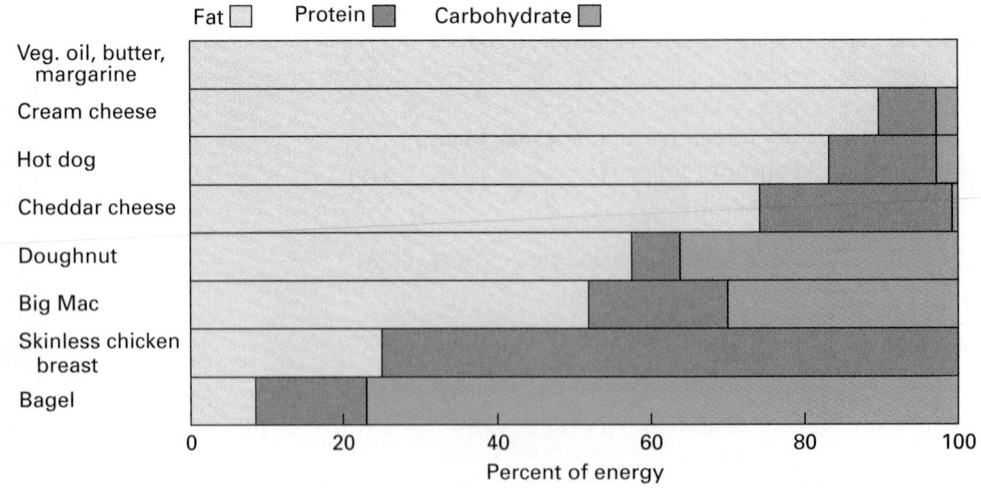

Figure 17.4 The composition of some common foods.

victims have atherosclerosis, and the major components of atherosclerotic plaque are saturated fatty acids and cholesterol.

The relation between blood levels of cholesterol and heart disease is well established. The more saturated fats and cholesterol in the diet, the higher the blood cholesterol is likely to be. Cholesterol, a lipid, has a waxy consistency, so to be transported in the bloodstream it must bond to a more water-soluble substance. Cholesterol combines with proteins to form **lipoproteins,** which are water-soluble because of their many $-NH_3^+$ and $-COO^-$ ions. About 65% of the cholesterol in the blood is carried by low-density lipoproteins **(LDLs),** whereas about 25% is carried by high-density lipoproteins **(HDLs).** (The density difference is caused by the ratios of lipid to protein.)

LDLs are the "bad" cholesterol and HDLs are the "good" cholesterol referred to in discussions of heart disease. LDLs transport cholesterol away from the liver and throughout the body; they are therefore "bad" because they distribute cholesterol to arteries, where it can form the deposits of atherosclerosis. HDLs are "good" because they transport excess cholesterol from body tissues to the liver, where it can be broken down and excreted.

Another approach to altering the content of our diet is to substitute lower-calorie food additives for the real thing. This has been done for years with artificial sweeteners. Now, the time is here for fat substitutes—substances that add a fatty taste and consistency but are not fats or oils.

Emulsified starch is used in Hellmann's light mayonnaise and salad dressing. (For a description of emulsions, see Section 16.3.) This form of starch is not suitable for cooking but can be used in ice cream and yogurt. Emulsified protein is an emulsion of gelatin and water that cuts margarine calories in half and is suitable for baking and light frying.

One new proprietary fat substitute is Simplesse, which contains only 15% of the calories in natural dietary fats. Developed by the makers of NutraSweet (G.D. Searles Company), Simplesse was approved for use in frozen desserts, cheese foods, and other products in 1989, and is most widely available in Baskin–Robbins fat-free ice cream. Simplesse is made from egg white or milk proteins and feels creamy on the tongue. Because heat makes it tough, however, it cannot be used in products that must be cooked.

Some manufacturers are developing natural fats modified to make them healthier. Salatrim (from the laboratories of Nabisco Foods) is a triglyceride in which the acids are stearic acid, a fatty acid that is poorly absorbed by the body, and short-chain acids (e.g., acetic acid) which have fewer calories than fatty acids. Because its ingredients are all present in natural foods, Salatrim may not need extensive testing to receive government approval. Yet another product under study consists of animal fat with the cholesterol removed and replaced by the polyunsaturated oils known to lower blood cholesterol. If neither of these fat substitutes makes it into the marketplace, similar ones will no doubt be developed.

The World of Chemistry

Program 9, *Molecular Architecture*

Solid animal fat.

■ The heart-disease risk/blood cholesterol connection. The values given here are blood levels for persons more than 20 years of age. The values for children and adolescents are about 15% lower.

Risk Level	Total Cholesterol (mg/100 mL)	LDL Cholesterol (mg/100 mL)
Low	200 or lower	130 or lower
Borderline high	200–239	130–159
High	240 or higher	160 or higher

■ In 1991 the National Cholesterol Education Program recommended that children and adolescents whose parents or grandparents had heart disease at 55 years of age or younger undergo cholesterol screening.

■ For a discussion of the fat substitutes see, Designer fats, *Science News,* Vol 145, p. 269 (1993).

17.5 Proteins in the Diet

Meat, fish, eggs, cheese, and beans are high-protein foods. The major function of dietary protein is to provide amino acids for new protein synthesis. Proteins are also the major dietary source of nitrogen for the synthesis of other kinds of nitrogen-containing biomolecules.

FRONTIERS IN THE WORLD OF CHEMISTRY

Health Claims for Foods

Major changes in food labels have been in place since 1994. At that time, the Nutrition Labeling and Education Act of 1990 became effective. To quote the FDA:

The purpose of food label reform is simple: to clear up confusion that has prevailed on supermarket shelves for years, to help consumers choose more healthful diets, and to offer an incentive to food companies to improve the nutritive qualities of their products.

As part of the studies that preceded the new labeling act, the FDA, with the assistance of a large number of experts, evaluated the scientific evidence for a variety of health claims for foods. Seven claims were found sufficiently valid to be allowed. The claims must not be stated on the labels in absolute terms (e.g., the label can't say that eating the food will *cure* a disease), and the foods must contain specified amounts of the ingredients to support the claim. A sample acceptable label statement is

While many factors affect heart disease, diets low in saturated fat and cholesterol may reduce the risk of this disease.

All the allowed diet-health connections are the subject of ongoing scientific research. As seems inevitable, both supporting and conflicting studies are reported from time to time. Meanwhile, the decision to decrease dietary fats and increase dietary fiber, already made independently by many individuals, has the additional backing of the FDA.

Two of the claims relate to ***heart disease.*** The FDA found "strong, convincing, and consistent" evidence that diets low in saturated fat and cholesterol decrease the risk of heart disease by lowering blood cholesterol levels. They also found support for the claim that, in association with a diet low in saturated fat and cholesterol, a diet rich in fiber, and particularly soluble fiber, lowers cholesterol levels and reduces heart disease risk.

Three of the claims relate to ***cancer,*** with the understanding that diet is only one of many factors implicated in an increased risk of cancer (others include smoking, heredity, and environmental factors). The FDA found sufficient evidence for reduction of cancer risk in consumption of (1) fiber-containing fruits, vegetables, and grain products, and (2) fruits and vegetables that are good sources of vitamins A and C. Also, while there is no clearcut evidence for association of a particular kind of fat with a particular kind of cancer, they allow the claim that (3) a diet low in total fat may reduce the risk of some cancers.

The remaining two claims support the relation between increased dietary calcium and decreased ***osteoporosis*** risk (Section 17.8), and the relationship between decreased sodium intake and lowered risk of the problems associated with ***high blood pressure*** (Section 17.8)

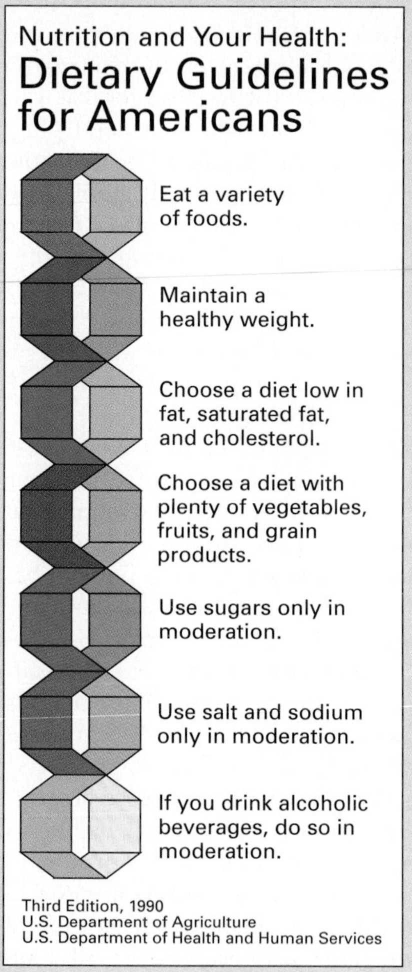
Nutrition and Your Health:
Dietary Guidelines for Americans
Eat a variety of foods.
Maintain a healthy weight.
Choose a diet low in fat, saturated fat, and cholesterol.
Choose a diet with plenty of vegetables, fruits, and grain products.
Use sugars only in moderation.
Use salt and sodium only in moderation.
If you drink alcoholic beverages, do so in moderation.
Third Edition, 1990
U.S. Department of Agriculture
U.S. Department of Health and Human Services

Reference "The New Food Label Summaries," January 6, 1993, Food and Drug Administration, Dept. of Health and Human Services, Washington DC.

Excess amino acids from dietary proteins are not stored in the body. The nitrogen is converted to ammonia and then excreted as urea, while the carbon atoms are cycled to glucose and energy generation or storage as fat. Therefore, it is necessary to eat some protein every day.

Foods high in protein content. *(C.D. Winters)*

A protein-deficient diet is rare in the United States. *Protein-energy malnutrition,* a group of disorders due to various combinations of deficient protein and energy intake, are most common in children in underdeveloped countries.

■ The condition of extreme protein deficiency is called *kwashiorkor.* The name comes from an African dialect and translates as "the evil spirit that infects the first child when the second one is born." Kwashiorkor begins when the earlier child is no longer breastfed and is switched to a carbohydrate-based diet. If nutrition is improved before the condition has progressed too far, health can be restored.

17.6 Our Daily Diet

With implementation of the new food-labeling law in 1994, the most readily available information about the nutritive value of food is on food packages. In addition, some packages are carrying the food pyramid developed by the FDA to graphically represent the latest consensus on a healthy diet.

The pyramid (Figure 17.5) summarizes today's health-conscious approach to eating. In the past, government daily diet recommendations focused on avoiding deficiencies, rather than preventing disease. Reflecting the valid health claims summarized in the *Frontiers* section (p. 546), the emphasis now is on decreasing dietary fat and increasing intake of fruits, vegetables, and grains. The small tip of the pyramid shows fats, oils, and sweets. These foods should be limited because they have "empty" calories—calories accompanied by almost no nutrients. To interpret the pyramid, it's important to be aware of the serving sizes listed in Figure 17.5. A dinner-size serving of spaghetti, for example, counts as two to three servings from the bread group.

An example of the new Nutrition Facts label is given in Figure 17.6. The total Calories and Calories from fat are listed at the top. Next comes a section that lists the weight in grams or milligrams per serving, plus the *% Daily Value,* for the **macronutrients** judged most important to health. It is mandatory that this section include total fat, saturated fat, cholesterol, sodium, total carbohydrate, dietary fiber, sugars and protein, as illustrated. Except for dietary fiber and protein, these are nutrients that should be limited.

■ **Macronutrients** are those nutrients needed by the body in large amounts. **Micronutrients** are those needed by the body in small amounts—mostly the vitamins and minerals.

The *daily reference values* are established as listed at the bottom of the food label and as further explained in Table 17.5. The percent daily values on a label (the *DVs*) are calculated from these values, as represented in a 2000-Calorie daily diet. This many calories is about right for moderately active women, teenage girls, and sedentary men. Teenage boys, active men, and very active or pregnant women have higher caloric requirements. The percentages

■ One reason the FDA settled on listings in terms of 2000 Calories/day is because it is a nice round number. Others must adjust their interpretation of the labels according to their own needs. Labels on large packages, where there is room, include the data for a 2500 Calories/day diet.

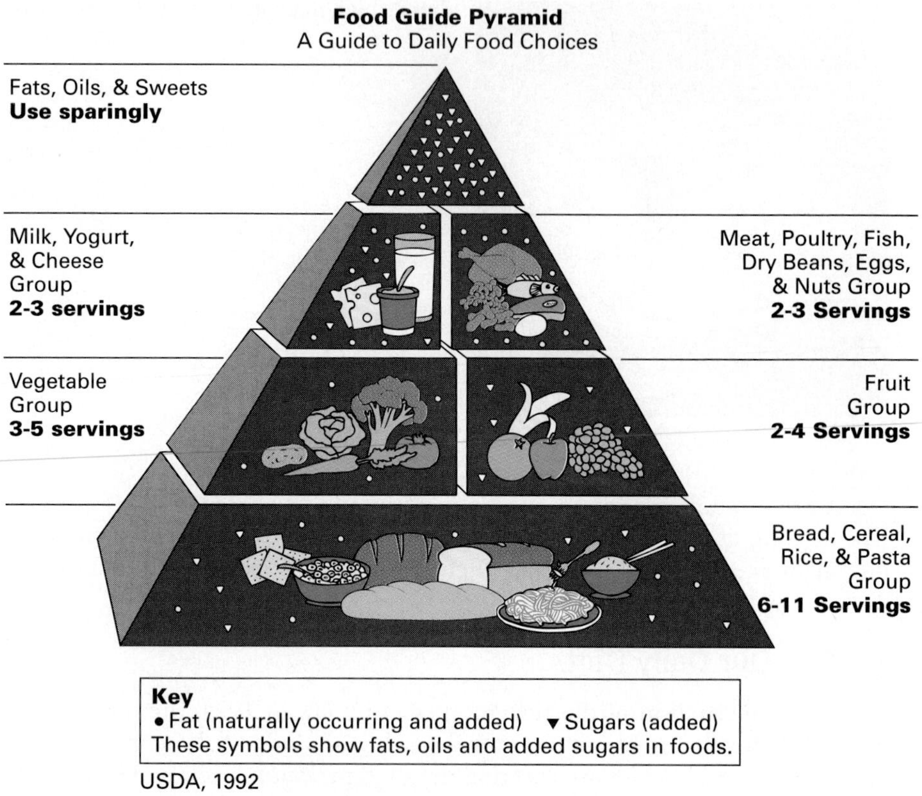

Figure 17.5 The Food Guide Pyramid. The pyramid has been developed by the U.S. Department of Agriculture as a general guide to a healthful diet.

Food Guide Pyramid
A Guide to Daily Food Choices

Fats, Oils, & Sweets
Use sparingly

Milk, Yogurt, & Cheese Group
2-3 servings

Meat, Poultry, Fish, Dry Beans, Eggs, & Nuts Group
2-3 Servings

Vegetable Group
3-5 servings

Fruit Group
2-4 Servings

Bread, Cereal, Rice, & Pasta Group
6-11 Servings

Key
• Fat (naturally occurring and added) ▼ Sugars (added)
These symbols show fats, oils and added sugars in foods.

USDA, 1992

FOOD GROUP SERVING SIZES

Bread, Cereal, Rice, and Pasta
1/2 cup cooked cereal, rice, pasta
1 ounce dry cereal
1 slice bread
2 cookies
1/2 medium doughnut

Vegetables
1/2 cup cooked or raw chopped
 vegetables
1 cup raw leafy vegetables
3/4 cup vegetable juice
10 french fries

Fruit
1 medium apple, banana, or orange
1/2 cup chopped, cooked, or canned
 fruit
3/4 cup fruit juice
1/4 cup dried fruit

Milk, Yogurt, and Cheese
1 cup milk or yogurt
$1\frac{1}{2}$ ounces natural cheese
2 ounces process cheese
2 cups cottage cheese
$1\frac{1}{2}$ cups ice cream
1 cup frozen yogurt

Meat, Poultry, Fish, Dry Beans, Eggs, and Nuts
2–3 ounces cooked lean meat, fish,
 or poultry
2–3 eggs
4–6 tablespoons peanut butter
$1\frac{1}{2}$ cups cooked dry beans
1 cup nuts

Fats, Oils, and Sweets
Butter, mayonnaise, salad dressing,
 cream cheese, sour cream, jam, jelly

can be used to calculate the limits for a higher- (or lower-) calorie diet (for an example, see Section 17.10).

 Beneath the heavy line on every nutrition facts label are mandatory listings for vitamin A, vitamin C, calcium, and iron. Here also, the required listings are

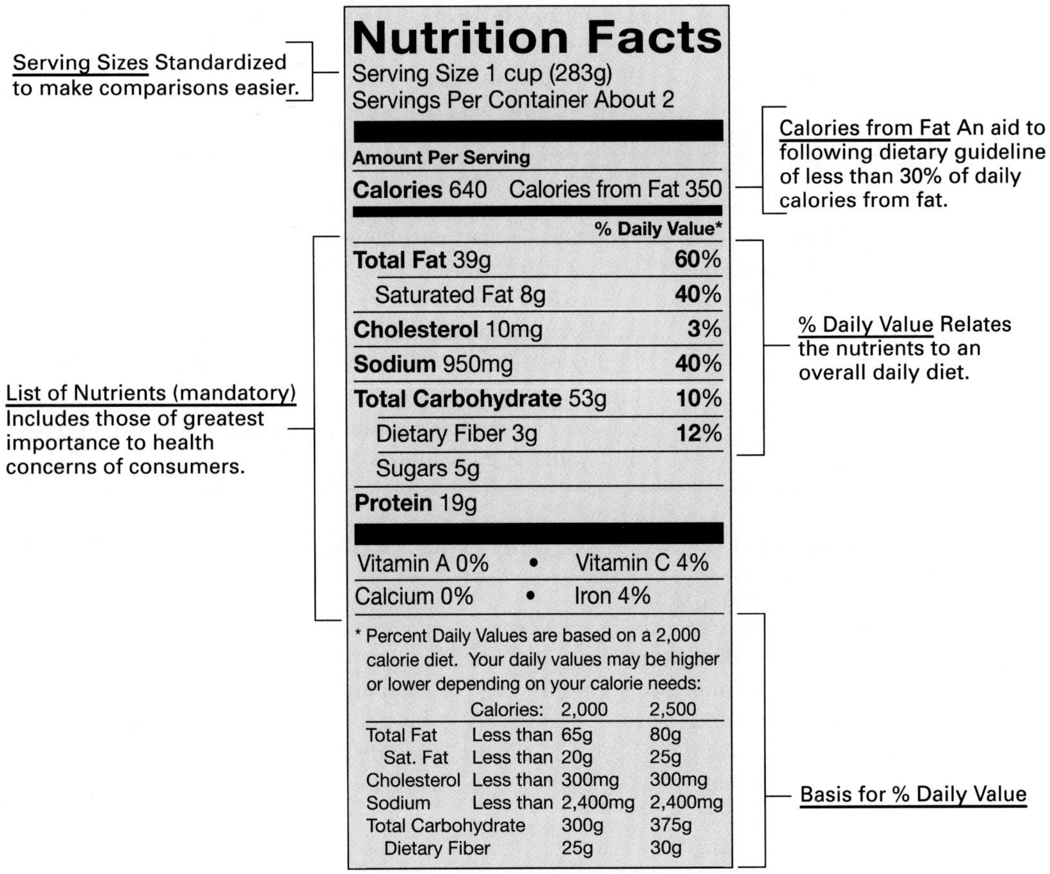
Serving Sizes Standardized to make comparisons easier.
Nutrition Facts
Serving Size 1 cup (283g)
Servings Per Container About 2
Amount Per Serving
Calories 640 Calories from Fat 350
% Daily Value*
Total Fat 39g 60%
Saturated Fat 8g 40%
Cholesterol 10mg 3%
Sodium 950mg 40%
Total Carbohydrate 53g 10%
Dietary Fiber 3g 12%
Sugars 5g
Protein 19g
Vitamin A 0% • Vitamin C 4%
Calcium 0% • Iron 4%
* Percent Daily Values are based on a 2,000 calorie diet. Your daily values may be higher or lower depending on your calorie needs:
Calories: 2,000 2,500
Total Fat Less than 65g 80g
Sat. Fat Less than 20g 25g
Cholesterol Less than 300mg 300mg
Sodium Less than 2,400mg 2,400mg
Total Carbohydrate 300g 375g
Dietary Fiber 25g 30g
Calories from Fat An aid to following dietary guideline of less than 30% of daily calories from fat.
List of Nutrients (mandatory) Includes those of greatest importance to health concerns of consumers.
% Daily Value Relates the nutrients to an overall daily diet.
Basis for % Daily Value

Nutrients that may be listed at the manufacturers' option:

Calories from saturated fat	Soluble fiber
Stearic acid (meat and poultry products only)	Insoluble fiber
	Sugar alcohol (e.g., the sugar substitute sorbitol)
Polyunsaturated fat	Other carbohydrate
Monounsaturated fat	% vitamin A as beta-carotene
Potassium	Other essential vitamins and minerals

Figure 17.6 The Nutrition Facts label. An example of the label required by law since 1994. Variations are allowed for small packages or foods that do not contain significant amounts of certain nutrients (e.g., canned soda need not list vitamins and minerals.)

for those vitamins and minerals deemed to be of greatest public-health concern. The major deficiency diseases of today are anemia, associated with iron deficiency, and osteoporosis, associated with calcium deficiency. The two vitamins in the mandatory list are those associated with decrease of cancer risk.

The percent daily values for vitamins and minerals are based on the quantities listed in Table 17.6, and only these vitamins and minerals can be listed on food labels; others are not allowed.

TABLE 17.5 ■ Basis for the Reference Daily Values on Nutrition Labels

Total Fat	30% of daily calories—an upper limit (65 g/2000 kcal daily diet)
Saturated Fat	10% of daily calories—an upper limit (20 g/2000 kcal)
Cholesterol	300 mg—a daily maximum for all diets
Sodium	2400 mg—a daily maximum for all diets
Total Carbohydrate	60% of daily calories (300 g/2000 kcal)
Dietary Fiber	23 g (based on 11.5 g/1000 kcal)
Sugars	No specified daily value
Protein	Listing the percent daily value is not mandatory, but daily value is 10% of dietary calories from protein (50 g/2000 calories) for adults (excluding pregnant women and nursing mothers) and children over 4.

TABLE 17.6 ■ Reference Daily Values (RDIs)* for Vitamins and Minerals

Nutrient	Amount	Nutrient	Amount
vitamin A	5,000 International units (IU)	folic acid	0.4 mg
vitamin C	60 milligrams (mg)	vitamin B_{12}	6 micrograms (mcg)
thiamin	1.5 mg	phosphorus	1.0 g
riboflavin	1.7 mg	iodine	150 mcg
niacin	20 mg	magnesium	400 mg
calcium	1.0 gram (g)	zinc	15 mg
iron	18 mg	copper	2 mg
vitamin D	400 IU	biotin	0.3 mg
vitamin E	30 IU	pantothenic acid	10 mg
vitamin B_6	2.0 mg		

*The RDIs are renamed but for the time being have the same numerical values as the U.S. RDAs (U.S. Recommended Daily Allowances), which were introduced in 1973 as reference values for voluntary labeling.

■ SELF-TEST 17B

1. Indigestible polysaccharides are known as ______________ ______________.
2. Most of the lipids in our diet are ______________.
3. The ______________ are "good" cholesterol because they ______________.
4. The ______________ are "bad" cholesterol because they ______________.
5. Proteins are the major source of ______________ in the diet.
6. Which nutrient is recommended to be decreased in a healthy diet?
7. Which kinds of foods are recommended to be increased in a healthy diet?
8. Which of these two approaches form the basis of the new Nutrition Facts labels? (a) avoiding nutrient deficiencies, (b) preventing major diseases such as heart disease and cancer
9. A major goal in a healthy diet is to have no more than ______________% calories from fat.

DISCOVERY EXPERIMENT

Your Own Diet

For two or three days, keep a personal food diary. To do this, set up a page with the format illustrated in the table below. In filling out the diary, remember to include oils and fats used in cooking, spreads on bread, dressings, sauces, and soft drinks.

For each day of your diary, convert the food portions you have eaten into the number of servings of the groups on the food pyramid (Figure 17.5). For example, the double hamburger is two servings from the bread group (the bun), two servings from the meat group (the two beef patties), and about half a serving from the vegetable group (the lettuce). Does your daily diet have the recommended numbers of servings of foods from each group? If not, where do you fall short? Where do you overdo?

Using Table 17.3 or a more extensive table of the caloric values of foods, estimate your daily caloric intake. Compare this with your daily caloric needs as estimated by using Table 17.2.

Decide whether you need a plan to modify your diet for either the kinds of food you are eating or the caloric content.

Time	Food	Kind and How Prepared	Amount
7 A.M.	Toast	Whole wheat	1 slice
	Butter	On toast	2 tsp
	Milk	Whole milk	8 oz
	Orange juice	From frozen concentrate	8 oz
12:30 P.M.	Double hamburger	Fast-food restaurant	1

17.7 Vitamins in the Diet

A **vitamin** is an organic compound essential to health that must be supplied in small amounts in the diet. Vitamins provide no energy and are unchanged by digestion. The body does not synthesize vitamins, but needs them to function as coenzymes. There are two kinds of vitamins, fat-soluble and water-soluble.

The fat-soluble vitamins—A, D, E, and K—can be stored in the fatty tissues of the body (especially the liver). They have nonpolar hydrocarbon chains and rings that are compatible with nonpolar oils and fats. Their structures, sources, and functions are listed in Table 17.7. It is important to store enough fat-soluble vitamins, but not too much because excesses are not excreted and can be toxic.

The water-soluble vitamins are the eight B vitamins and vitamin C. Because excesses are excreted rather than stored, it has been assumed that overdosing on water-soluble vitamins has no toxic effects. With the rise in popularity of megadoses of vitamins, however, some toxic reactions have been observed. It does, however, take much greater quantities for water-soluble vitamins to create harmful effects than for the fat-soluble vitamins.

We have our choice of a wide variety of vitamins. *(C.D. Winters)*

Fat-Soluble Vitamins

Vitamin A Vitamin A is essential to vision because it is a component of pigments in the eye. Carrots, long associated with good vision, provide β-carotene, not vitamin A itself. As it passes through the intestinal wall, the β-carotene is converted into vitamin A (also known as *retinol*). Vitamin A also aids in the prevention of infection by barring bacteria from entering and

■ Eat polar bear liver sparingly. Thirty grams contain 450,000 IU of vitamin A; continued ingestion causes peeling of the skin from head to foot.

TABLE 17.7 ■ The Fat-Soluble Vitamins*

Vitamin	Sources	Major Functions	Deficiency Symptoms	Groups at Risk of Deficiency
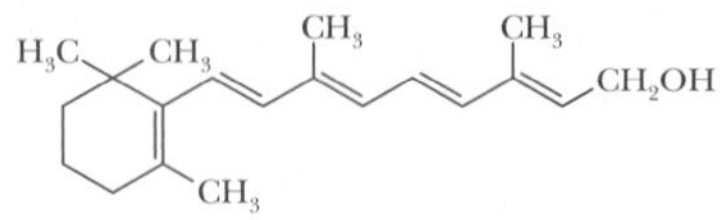 Vitamin A	Liver, carrots, peaches, leafy greens, fortified milk, sweet potatoes, broccoli	Vision, growth, cell differentiation, reproduction, immune function	Night blindness, poor growth, dry skin	Those who live in poverty (particularly children and pregnant women)
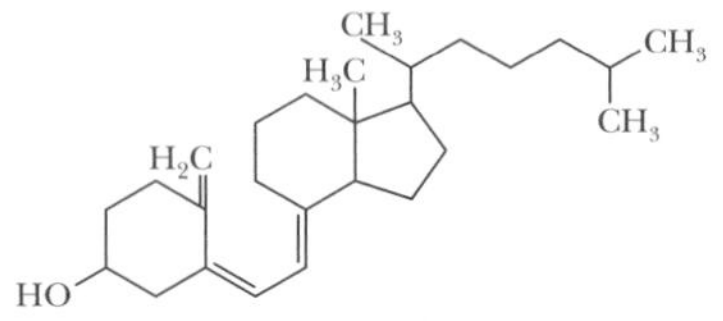 Vitamin D	Egg yolk, margarine, fish oils, tuna, salmon, fortified milk	Absorption of calcium and phosphorus, maintenance of bone	Rickets in children, osteomalacia in adults	Breast-fed infants, children, and elderly
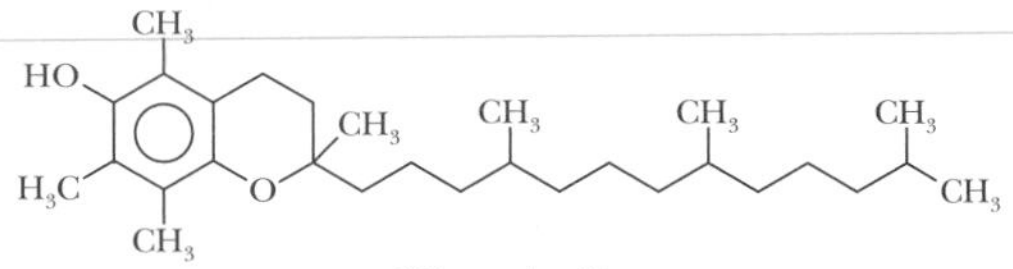 Vitamin E	Corn oil, safflower oil, soybean oil, leafy greens, nuts	Antioxidant, protects cell membranes	Red blood cell breakdown, nerve damage	Those with poor fat absorption
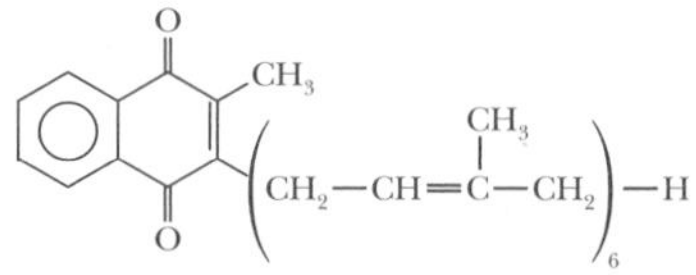 Vitamin K	Beef liver, egg yolk, leafy greens, legumes	Blood clotting	Hemorrhage	People on long-term antibiotics, newborns (especially premature)

*Adapted from L. A. Smolin and M. B. Grosvenor: *Nutrition: Science & Applications*. Philadelphia: Saunders College Publishing, 1994.

passing through cell membranes. A single dose of vitamin A greater than 200 mg (660,000 IU, international units, a unit for vitamin doses used by nutritionists) can cause acute *hypervitaminosis A* in an adult, with symptoms that include vomiting, fatigue, and headache.

Liver and carrots have a high vitamin A content. *(C.D. Winters)*

Vitamin D Vitamin D is produced when ultraviolet light shines on the skin. Its major role is to help the body use calcium, and a deficiency causes rickets in children, the same condition caused by calcium deficiency. Vitamin D supplements are rarely needed, except by those who are almost never exposed to the sun. Both vitamin A and vitamin D are essential to normal growth and development. Overdosage of vitamin D can have serious consequences, however. Calcium deposits can form in the kidney, lungs, or tympanic membrane of the ear (leading to deafness). Infants and small children are especially susceptible to vitamin D toxicity.

Vitamin E Vitamin E is now well established as an **antioxidant.** It is particularly effective in preventing the oxidation of polyunsaturated fatty acids to form peroxides (which contain —O—O— groups). Perhaps this is why vitamin E is always found distributed among fats in nature. The fatty acid peroxides are particularly damaging because they can cause runaway oxidation in the cells. Vitamin E protects the integrity of cell membranes, which are composed mainly of triglycerides. No toxicity has been specifically associated with large doses of vitamin E, although caution in taking megadoses is advised.

■ An **antioxidant** is a substance that prevents oxidation by contributing electrons so that reactive substances are diverted from damaging vital molecules by oxidation reactions. The possible benefits of the antioxidant vitamins are discussed further in the *Frontiers* box at the end of this section.

■ Vitamin E is the only vitamin destroyed by the freezing of food. However, it survives cooking in water.

Vitamin K We need a steady supply of vitamin K because the body uses it up rapidly. Its major and essential role is in regulation of blood clotting, which goes on not only when our skin is cut, but on a daily basis to repair small tears in blood vessels. Because overdoses cause excessive blood clotting and danger of brain damage, vitamin K is the only vitamin for which a prescription is needed for supplements containing the vitamin alone.

Water-Soluble Vitamins

All the water-soluble vitamins (Table 17.8) are **coenzymes** that have polar —OH, —NH_2, and —COOH groups, as illustrated below by vitamin C and several of the B vitamins.

■ A ***coenzyme*** is any nonprotein molecule that cooperates with an enzyme to make the enzyme function possible.

B Vitamins Vitamin B_6, considered the master vitamin, is known to be involved in 60 enzymatic reactions, many in the metabolism and synthesis of proteins. It is also needed for the synthesis of hemoglobin and the white blood cells of the immune system. Consuming more than 60 times the recommended daily quantity of vitamin B_6 causes nerve damage, and in huge doses (2–6 g/day) it can cause paralysis.

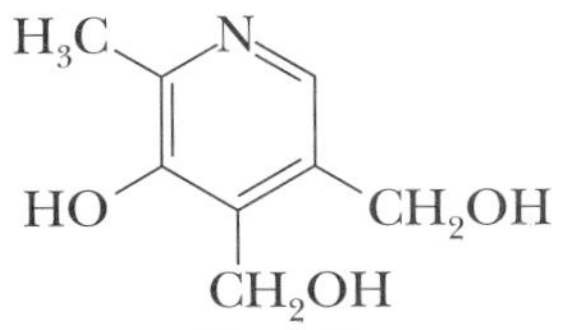
Vitamin B_6

Foods that are high in riboflavin and other B vitamins. *(C.D. Winters)*

Large doses of niacin (50 mg/day or more) can also be toxic, with reactions ranging from a skin rash and nausea to abnormal liver function and changes in heart rhythm. Niacin deficiency is common in countries with a mainly corn diet.

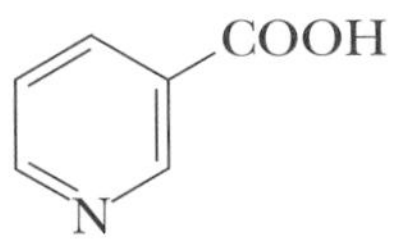
Niacin

■ Once thought to be a single substance, the eight B vitamins are often found together and interact so that a deficiency of one can cause deficiencies of others. The B vitamins are B_6, B_{12}, folic acid, niacin, thiamin, riboflavin, biotin, and pantothenic acid.

When the hulls of wheat, rice, and other grains are removed and the kernels ground to produce flour, a large proportion of riboflavin and certain other vitamins and minerals is lost. To counteract this loss, since the 1940s flour sold in the United States has been *enriched* by adding back some of the

TABLE 17.8 ■ The Water-Soluble Vitamins*

Vitamin	Sources	Major Functions	Deficiency Symptoms	Groups at Risk of Deficiency
Thiamin	Pork, whole and enriched grains, legumes	Coenzyme in nerve function, energy generation	Beriberi	Alcoholics, those in poverty
Riboflavin	Milk, leafy greens, enriched grains	Coenzyme in fat metabolism, energy generation	Inflammation of mouth and tongue	—
Niacin	Enriched grains, peanuts, tuna, chicken, beef	Coenzyme in fat metabolism, energy generation	Pellagra	Those consuming a limited diet high in corn products; alcoholics
Vitamin B_6	Meat, legumes, seeds, leafy greens	Coenzyme in protein metabolism	Headache, nausea, anemia	Women, alcoholics
Folate	Leafy greens, legumes, oranges	Coenzyme in RNA and DNA synthesis	Macrocytic anemia, poor growth	Pregnant women, alcoholics
Vitamin B_{12}	Animal products	Coenzyme in folate metabolism, nerve function	Pernicious anemia, macrocytic anemia, poor nerve function	Vegans,† elderly, those with stomach or intestinal disease
Biotin	Egg yolks	Coenzyme in glucose production and fat synthesis	Dermatitis, depression, anemia	Those consuming large amounts of raw egg whites; alcoholics
Pantothenic acid	Egg, liver, mushrooms	Coenzyme in citric acid cycle, fat metabolism	Nausea, fatigue, headache	Alcoholics
Vitamin C	Citrus fruit, broccoli, strawberries, greens	Collagen synthesis, antioxidant	Scurvy	Alcoholics, elderly men

*Adapted from L. A. Smolin and M. B. Grosvenor: *Nutrition: Science & Applications.* Philadelphia: Saunders College Publishing, 1994.

†Strict vegetarians

lost vitamins and minerals. Customarily, enriched wheat flour contains added riboflavin, thiamin, niacin, and iron. When enriched flour came into use in the United States, the incidence of pellagra, the niacin deficiency disease, almost disappeared.

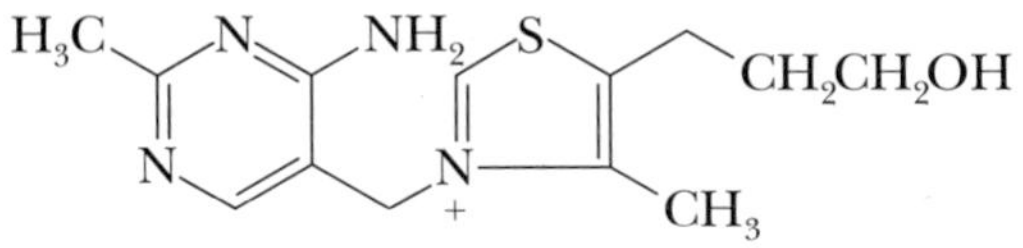

Thiamine

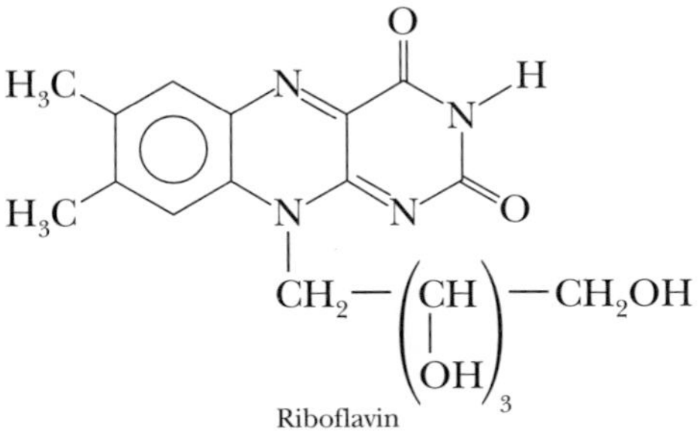

Riboflavin

Vitamin C Vitamin C, *ascorbic acid,* helps destroy invading bacteria; aids the synthesis and activity of interferon, which prevents the entry of viruses into cells; and combats the ill effects of toxic substances, including drugs and pollutants. It also aids in the synthesis of collagen, which is present in cartilage, bone, tendons, and connective tissue and is therefore important in the healing of wounds and for infants and pregnant women.

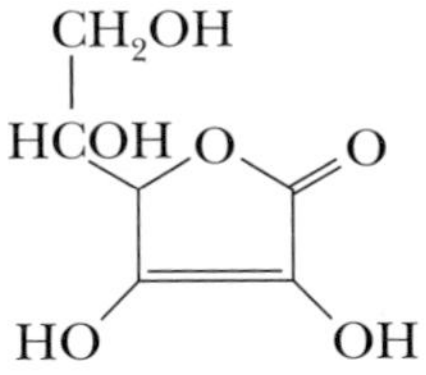

Vitamin C

The role of vitamin C in preventing the common cold has long been debated, and numerous studies have found it to be either effective or ineffective. Currently, there is more evidence in favor of its ability to decrease the severity of cold symptoms than for any ability to prevent colds. It is reported, however, that one third of the U.S. population takes vitamin C supplements to ward off colds. Daily doses of 1 g or more can cause diarrhea, nausea, and abdominal cramps in some people.

■ Vitamin C is a strong enough acid to damage tooth enamel if vitamin capsules are chewed.

■ English sailors have been called ''limeys'' because the British admiralty in 1835 ordered a daily ration of lime juice to prevent scurvy, the vitamin C deficiency disease. The message did not hit home in America, however, where scurvy was common among troops during the Civil War.

17.8 Minerals in the Diet

As nutrients, **minerals** are elements other than carbon, hydrogen, nitrogen, and oxygen needed for good health. Although referred to on labels as ''potassium'' or ''iodine'' and so on, most minerals are present in foods, food supplements, and our bodies as ions. Not only does the human body need minerals, but the minerals must be maintained in balanced amounts, with no

■ One way to ensure ingestion of an ample supply of each mineral nutrient, particularly the trace nutrients, is to eat a variety of whole foodstuffs grown in different places.

FRONTIERS IN THE WORLD OF CHEMISTRY

Antioxidant Vitamins

Who could resist taking a pill that might simultaneously diminish the risk of cancer, heart disease, cataracts, and rheumatoid arthritis, boost the immune system, and slow the effects of aging? Especially if the chemicals in the pill have already been identified as essential to the diet—vitamin A (or β-carotene, which the body converts to vitamin A), vitamin E and vitamin C.

These chemicals are all antioxidants—in the body, this means prevention of oxidation caused by the atoms or molecular fragments known as **free radicals.** A free radical contains an unpaired electron and does not stay around for long without grabbing another electron from a nearby molecule. If this molecule happens to have an important function, then that function will be disrupted. For example, if a free radical connects with the part of a DNA molecule that governs cell division, the result might be abnormal cell division—cancer. According to some theories, aging is the cumulative effect of the action of free radicals running wild.

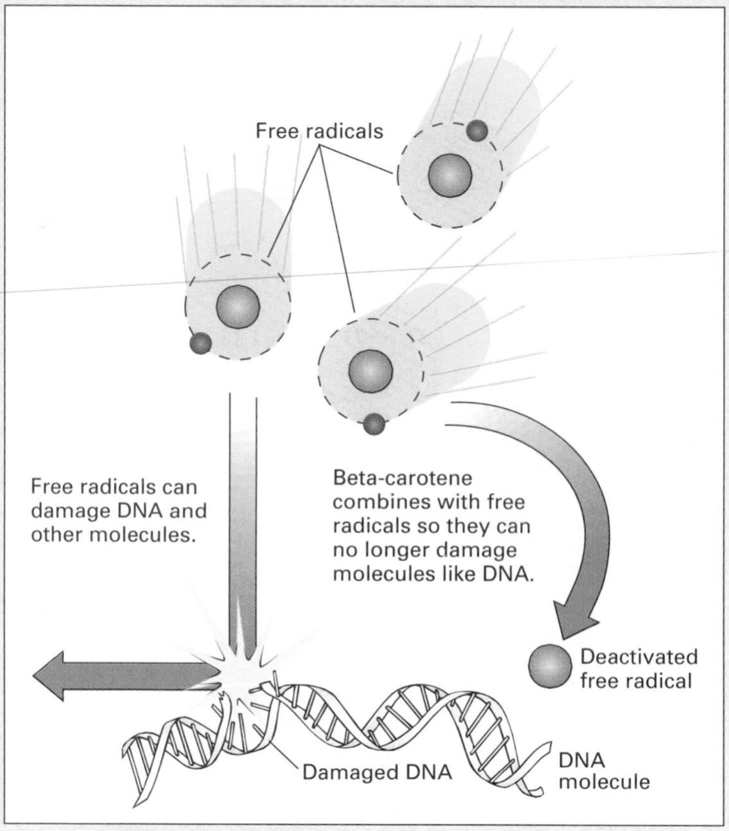

Free radicals
Free radicals can damage DNA and other molecules.
Beta-carotene combines with free radicals so they can no longer damage molecules like DNA.
Deactivated free radical
Damaged DNA
DNA molecule

Beta-carotene as an antioxidant.

Free radicals arise in the body from normal biochemical reactions and are also produced in the presence of toxic substances from cigarette smoke, polluted air, or other sources. Study after study has identified beneficial effects from antioxidant vitamins. To cite just a few of many examples,

- In 128 of 156 studies, high fruit and vegetable consumption were associated with low risk for several kinds of cancer. Fruits and vegetables are the best sources of antioxidant vitamins.
- Of 87,000 people in the Nurses' Health Study, those taking at least 100 IU of vitamin E each day for two years developed heart disease at one-half the rate of those who did not take vitamin E.
- Taking 30 mg of β-carotene every other day reduced the occurrence of heart attacks by 50% among 333 physicians with known heart disease.

By 1993 the evidence from these and other studies was strong enough that *some* nutritionists and physicians began to recommend antioxidant vitamin supplements. It is the nature of medicine and clinical studies, however, that none of the studies so far is considered *definitive.* As a result, the view of antioxidant supplementation remains very cautious. It will continue to be cautious until studies are done that include randomized delivery of vitamin supplements or placebos (look-alike pills with no vitamins) to large numbers of people of clearly documented diet and health status, with long-term follow-up of the health of the participants.

Reference P.J. Skerrett: Vitamighty. *Medical World News,* p. 24, January 1993.

deficiencies and no excesses. Many of the body's minerals are excreted daily and must therefore be replenished each day.

The seven **macronutrient minerals** make up about 4% of body weight. They are calcium, phosphorus, magnesium, sodium, potassium, chlorine, and sulfur, and their sources and functions are listed in Table 17.9. Except for the role of long-term calcium deficiency in osteoporosis, deficiencies of these minerals are rare because of their abundance in a variety of foods.

Sodium, potassium, and ***chloride,*** as ions (Na^+, K^+, Cl^-), are essential to *electrolyte balance* in body fluids. Electrolyte balance, in turn, is essential for fluid balance, acid–base balance, and transmission of nerve impulses. Table salt is the principal source of sodium and chloride ions, and dietary deficiencies are unlikely. When there is extreme fluid loss through vomiting, diarrhea, or traumatic injury, electrolytes must be supplied to restore their concentration in body fluids.

■ Electrolytes are substances that dissolve in water to produce ions that conduct electricity.

■ Potassium chloride (KCl) is sold as a salt (NaCl) substitute.

One of several mechanisms for maintaining normal blood pressure is the adjustment of fluid volume in the body by control of the amounts of sodium

TABLE 17.9 ■ The Macrominerals

Name	Sources	Major Functions	Deficiency Symptoms	Groups at Risk of Deficiency
Sodium (Na^+)	Table salt, processed foods	Major extracellular ion, nerve transmission, regulates fluid balance	Muscle cramps	Those consuming a severely sodium-restricted diet
Potassium (K^+)	Fruits, vegetables, grains	Major intracellular ion, nerve transmission	Irregular heartbeat, fatigue, muscle cramps	Those consuming diets high in processed foods, those taking high blood pressure medication
Chloride (Cl^-)	Table salt	Major extracellular ion	Unlikely	No one
Calcium (Ca^{2+})	Milk, cheese, bony fish, leafy green vegetables	Bone and tooth structure, nerve transmission, muscle contraction, blood clotting	Increased risk of osteoporosis	Postmenopausal women, teenage girls, those with kidney disease
Phosphorus†	Meat, dairy, cereals, and baked goods	Bone and tooth structure, buffers, membranes, ATP, DNA	Bone loss, weakness, lack of appetite	Premature infants, vegans, alcoholics, elderly
Magnesium (Mg^{2+})	Nuts, greens, whole grains	Reactions involving ATP, nerve and muscle function	Nausea, vomiting, weakness	Alcoholics, those with kidney disease
Sulfur‡	Protein foods, preservatives	Part of amino acids and vitamins, glutathione, acid–base balance	None when protein needs are met	No one

*Adapted from L. A. Smolin and M. B. Grosvenor: *Nutrition: Science & Applications.* Philadelphia: Saunders College Publishing, 1994.

†Phosphorus is covalently bonded in organic compounds and is also present in phosphate ions (PO_4^{3-}).

‡Sulfur is covalently bonded in organic compounds and is also present in sulfate ions (SO_4^{2-}).

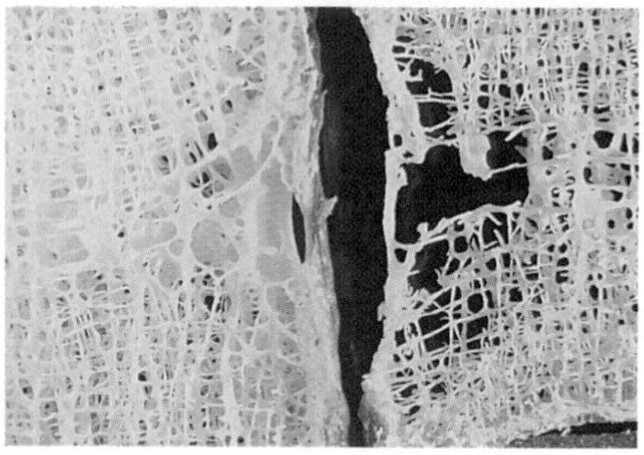

Osteoporosis. The effect of calcium loss on bone is shown by the comparison between a normal bone at the left and a bone afflicted with osteoporosis on the right. *(© Michael Klein/Peter Arnold, Inc.)*

ions and water excreted in the urine. An excess of sodium ions in the diet (possibly also chloride ions) is often associated with high blood pressure. Potassium ions also play a role in blood pressure regulation; some studies suggest that increased potassium ion intake may lower blood pressure.

About 99% of the ***calcium ions*** in the body are in bones and teeth. Together with sodium and potassium ions, calcium ions also participate in transmission of nerve impulses and regulation of the heartbeat. To be absorbed, calcium must be present in solution as Ca^{2+}. Absorption is enhanced in the presence of lactose (milk sugar) and also by a fatty meal, which passes through the intestine more slowly, allowing more time for absorption. A deficiency of vitamin D decreases calcium ion absorption and contributes to the bone deformities that accompany rickets.

■ About ten years ago, an undergraduate chemistry student discovered that many calcium and vitamin pills dissolved hardly at all. Subsequent media reports claimed that undissolved vitamin pills could be observed in large-intestine X rays and septic tanks. The manufacturers of nutrition supplements have since adopted voluntary standards for disintegration and solubility of their products.

A gradual loss of bone mass and density during adulthood is a normal process, and one of every three people older than 65 has some degree of **osteoporosis.** The role of a calcium-deficient diet over a lifetime in hastening osteoporosis is sufficiently well established that a food label health claim on this basis is allowed. Postmenopausal white or Asian women of slight body build with inactive lifestyles are at greatest risk for osteoporosis. After menopause, the body produces less estrogen, which has the ability to slow bone dissolution. A preventive measure against later osteoporosis is consumption of adequate calcium during the adolescent years (ages 12 to 18), when bone is forming. Estrogen replacement for postmenopausal women is effective in slowing bone loss (Section 19.4).

■ By definition, the micronutrient minerals, or trace elements, are those needed in quantities of 100 mg or less per day, or present in the body at less than 0.01% of body weight.

A number of minerals are essential in small amounts **(micronutrient minerals).** Of these, four—iron, copper, zinc, and iodine—are allowed to be listed on Nutrition Facts labels, iron being mandatory and the others optional (see Table 17.7). Others minerals known to be essential are selenium, manganese, fluorine, chromium, and molybdenum. Still other minerals (ten or more) are known to be essential in other mammals and are present and probably essential in humans. The list of trace minerals and our knowledge of their functions is constantly evolving.

Iodide ion has long been known to be essential to thyroid gland function. Hormones produced by the thyroid gland contain iodine and are responsible for growth, development, and maintenance of all body tissues. If there is a deficiency of iodine, the thyroid glands can become extremely enlarged, a condition called goiter. The routine use of iodized salt (0.1% potassium iodide) is the best way to ensure adequate iodine in the diet. More than half the table salt sold in the United States is iodized, and the use of such salt is mandatory in some countries.

Iron deficiency is associated with *anemia*—a decrease in the oxygen-carrying capacity of the blood, as indicated by low red blood cell or heme concentrations. Iron-deficiency anemia is the most common nutritional deficiency disease in both developed and underdeveloped nations. Early symptoms include muscle fatigue and lethargy. The ability to fight off invading bacteria is also diminished. Continued anemia creates defects in the structure and function of skin, fingernails, mouth, and stomach.

Zinc deficiency is known to cause poor growth and development, decreased immune function, and poor wound healing. Because meat is a better source of zinc than plant-based foods, vegetarians are among those at risk of zinc deficiencies (others are the elderly and children with poor diets). Zinc supplements are often touted as able to improve immune function, enhance sexual performance, and increase fertility. It's a good idea to bring a degree of skepticism to such claims for zinc and other minerals. In the case of zinc, adding it to the diet of a person with a mild deficiency might improve wound healing, immune function, and appetite. Currently, however, there is little evidence for such changes in a healthy person with no existing zinc deficiency.

■ Iron supplement pills are a frequent cause of poisoning due to excess intake by children. They should be stored out of reach.

■ A thorough medical evaluation of an individual with anemia is necessary because it can result from a variety of disease conditions as well as from nutritional deficiencies. Vitamin B_{12} and folic acid deficiencies can also cause anemia.

■ The role of fluorine in preventing tooth decay was discussed in Section 16.12.

■ SELF-TEST 17C

1. Vitamins are synthesized by cells of the body. (a) True, (b) False.
2. Vitamins A, D, E, and K are ____________-soluble, while the B-complex vitamins and vitamin C are ____________-soluble.
3. In the body, β-carotene is converted into vitamin ____________.
4. Vitamins A, C, and E, and β-carotene are all ____________.
5. The B-complex vitamins function in the body as parts of ____________.
6. When vitamins are added to flour or other foods, the food is described as ____________.
7. The chemical form of minerals in the diet is mostly as ____________.
8. Sodium and potassium are essential to ____________ and ____________ balance and the transmission of ____________ signals.
9. ____________ is caused by iron deficiency and ____________ is caused by calcium loss.

17.9 Food Additives

Many chemicals with little or no nutritive value are added to commercially processed food. Some of these **food additives** serve to protect the food from being spoiled by oxidation, bacterial attack, or aging. Others add and enhance flavor or color. Still others control pH, prevent caking, stabilize, thicken, emulsify, sweeten, leaven, or tenderize the food.

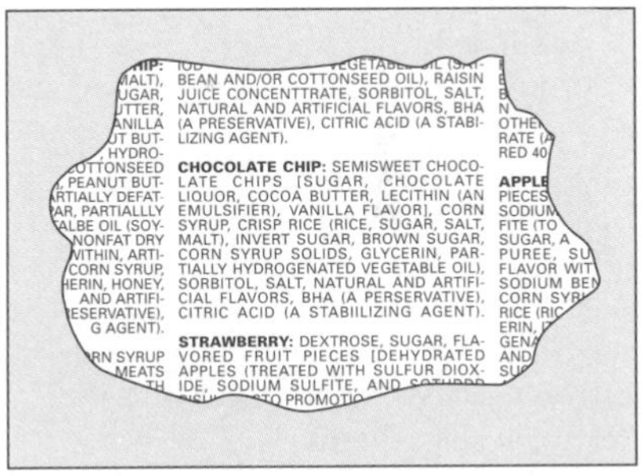

Note the variety of food additives in these cookies.

The GRAS List The Food and Drug Administration lists about 600 chemical substances "generally recognized as safe" (GRAS) for their intended use. Some examples are given in Table 17.10. The GRAS list was published in several installments in 1959 and 1960. It was compiled from the results of a questionnaire asking experts in nutrition, toxicology, and related fields to give their opinions about the safety of various seasonings, artificial flavorings, and

■ Some synthetic substances used as food sweeteners were discussed in Section 15.5.

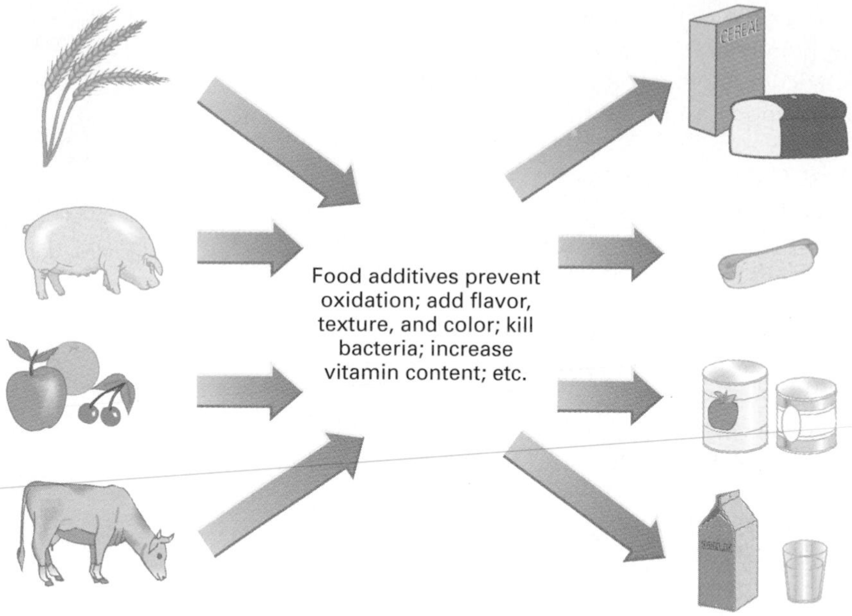
CEREAL
Food additives prevent oxidation; add flavor, texture, and color; kill bacteria; increase vitamin content; etc.

Many foods arrive at the market containing food additives.

TABLE 17.10 ■ A Partial List of Food Additives Generally Recognized as Safe*

Category	Additive
Anticaking Agents	Calcium silicate
	Iron ammonium citrate
Acids, Alkalies, and Buffers	Acetic acid
	Calcium lactate
	Citric acid
	Lactic acid
	Phosphates, Ca^{2+}, Na^{+}
	Potassium acid tartrate
Surface-Active Agents (Emulsifying Agents)	Glycerides: mono- and diglycerides of fatty acids
Preservatives	Sodium benzoate
	Propionates, Ca^{2+}, Na^{+}
	Sorbates, Ca^{2+}, K^{+}, Na^{+}
	Sulfites, Na^{+}, K^{+}
Antioxidants	Ascorbates, Ca^{2+}, Na^{+}
	Butylated hydroxyanisole (BHA)
	Butylated hydroxytoluene (BHT)
	Lecithin
	Sulfur dioxide and sulfites
Flavor Enhancers	Monosodium glutamate (MSG)
	5′-Nucleotides
	Maltol
Sequestrants	Citric acid
	EDTA, Ca^{2+}, Na^{+}
	Pyrophosphate, Na^{+}
	Sorbitol
Stabilizers and Thickeners	Agar-agar
	Algins
	Carrageenan
Flavorings	Amyl butyrate (pearlike)
	Bornyl acetate (piney, camphor)
	Carvone (spearmint)
	Cinnamaldehyde (cinnamon)
	Citral (lemon)
	Ethyl cinnamate (spicy)
	Ethyl formate (rum)
	Ethyl vanillin (vanilla)
	Menthol (peppermint)
	Methyl anthranilate (grape)
	Methyl salicylate (wintergreen)

*For precise and authoritative information on levels of use permitted in specific applications, consult the regulations of the U.S. Food and Drug Administration and the Meat Inspection Division of the U.S. Department of Agriculture.

other substances customarily added to foods. Since its publication, few substances have been added to the GRAS list and some have been removed, mostly due to suspicion that they are cancer-causing agents. *New* food additives—substances not on the original GRAS list—must receive FDA approval for use in foods based on test results submitted by the manufacturers.

Food Preservation Oxidation and microorganisms (bacteria, fungi, and others) are the major enemies in the decomposition of food. Any process that prevents the growth of microorganisms or retards oxidation is generally an effective preservation process. Drying grains, fruits, and meat is one of the oldest preservation techniques. Drying is effective because water is necessary for both the growth of microorganisms and the chemical reactions of oxidation.

There are also chemical additives that can preserve food. Salted meat and fruit in a concentrated sugar solution are protected from microorganisms. The dissolved sodium chloride or sucrose creates a **hypertonic solution** in which water flows by **osmosis** from the microorganism to its environment. Thus, salt and sucrose have the same effect on microorganisms as dryness; both dehydrate them.

A preservative is effective if it prevents multiplication of microorganisms during the shelf life of the product. In general, food is spoiled by toxic substances secreted by the microorganisms. Sterilization by heat or radiation (Section 5.10), or inactivation by freezing, is often undesirable, since it impairs the quality of the food. Chemical agents seldom achieve sterile conditions but can preserve foods for considerable lengths of time. Two common chemical preservatives in packaged foods are sodium benzoate, which is permitted in nonalcoholic beverages and in some fruit juices, fountain syrups, margarines, pickles, relishes, olives, salads, pie fillings, jams, jellies, and preserves, and sodium propionate, which can be used in bread, chocolate products, cheese, pie crust, and fillings.

■ A **hypertonic solution** is more concentrated than solutions in its immediate environment.

■ **Osmosis** is the flow of water from a more dilute solution through a membrane into a more concentrated solution (Section 20.11).

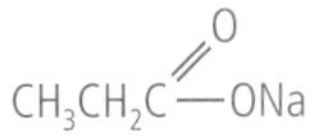

Sodium propionate

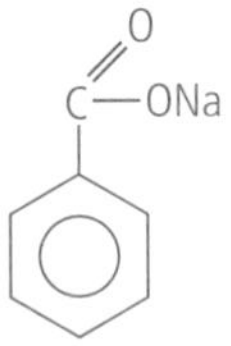

Sodium benzoate

■ A preservative must interfere with microbes but be harmless to the human system—a delicate balance.

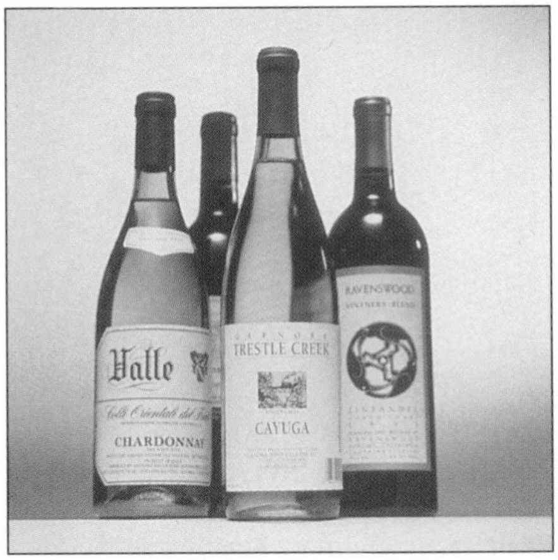

Sulfites as preservatives in wine. Because many individuals are allergic to sulfites, foods that contain more than 10 parts per million of sulfites must list them on the label. The "sulfites" include chemicals such as sulfur dioxide (SO_2), sodium sulfide (Na_2S), and sodium bisulfite (Na_2SO_3). *(C.D. Winters)*

The World of Chemistry

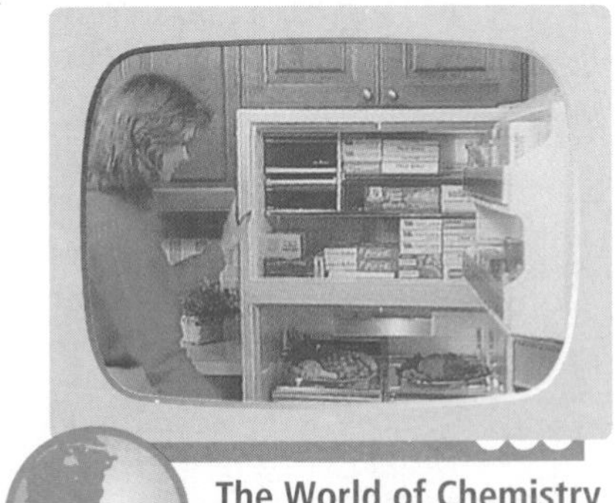

Program 17, *The Precious Envelope*

Food preservation by cooling, which slows down the rate of oxidation and enzyme catalyzed-chemical reactions.

■ Antioxidant additives in foods are serving the same purpose as antioxidant vitamins in the body (Section 17.7).

Antioxidants The direct action of oxygen in the air is the chief cause of the destruction of the fats in food. Oxidation produces a complex mixture of volatile aldehydes, ketones, and acids that causes a rancid odor and taste. Foods kept wrapped, cold, and dry are relatively free of air oxidation. The most common antioxidant food additives are butylated hydroxyanisole (BHA) and butylated hydroxytoluene (BHT), which act by releasing a hydrogen atom from their —OH groups as a free radical (H·).

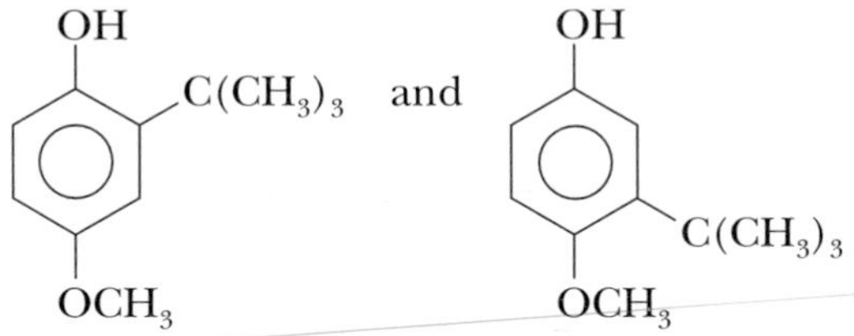

BHA (two isomers)

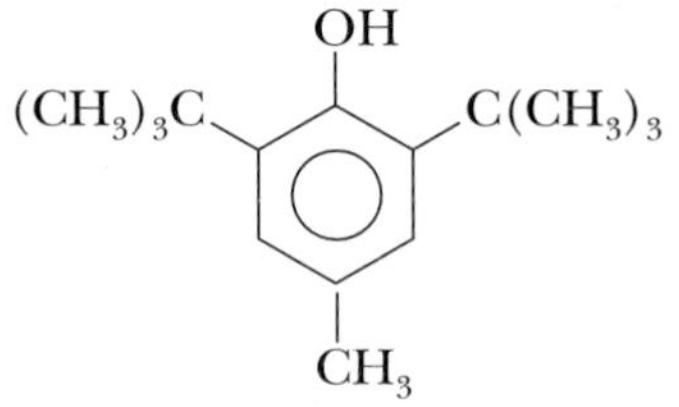

BHT

■ To sequester means "to withdraw from use." The sequestering ability of EDTA accounts for its use in treating heavy-metal poisoning (Section 18.4).

Sequestrants Metals get into food from the soil and from machinery during harvesting and processing. Copper, iron, and nickel, as well as their ions, catalyze the oxidation of fats. The class of food additives known as **sequestrants** react with trace metals in foods, tying them up in complexes so the metals will not catalyze decomposition of the food. With the competitor metal ions tied up (sequestered), antioxidants such as BHA and BHT can accomplish their task much more effectively.

The sodium and calcium salts of EDTA (ethylenediaminetetraacetic acid) are common sequestrants in many kinds of foods and beverages. The structural formula of EDTA bonded to a metal ion is shown in Figure 17.7.

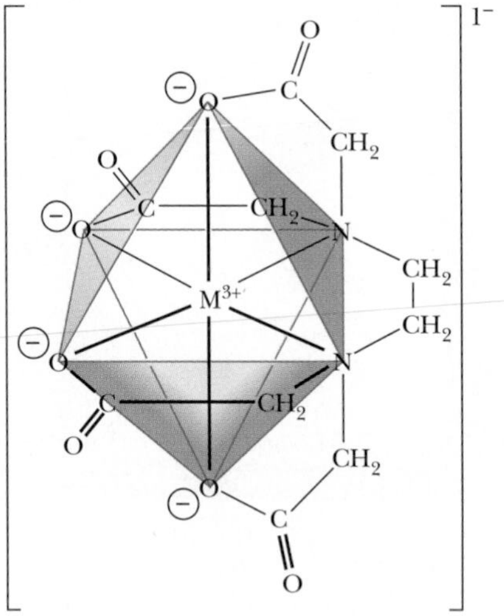

Figure 17.7 EDTA (ethylenediaminetetraacetic acid). The EDTA molecule has the remarkable ability to "sequester," or isolate, a metal ion by forming six bonds to it—two from nitrogen atoms in amino groups and four from oxygen atoms in carboxyl groups.

Food Flavors Much of the sensation of taste in food is from our sense of smell. For example, the flavor of coffee is determined largely by its aroma, which in turn is due to a mixture of more than 500 compounds, mostly volatile oils. Most flavor additives, like many perfume ingredients, originally came from plants. The plants were crushed and the compounds extracted with various solvents such as ethanol or carbon tetrachloride. Sometimes a single compound was extracted; more often, the residue contained a mixture of several compounds. By repeated efforts, relatively pure oils were obtained. Oils of wintergreen, peppermint, orange, lemon, and ginger, among others, are still obtained in this way. These oils, alone or in combination, are then added to foods to produce the desired flavor. Today synthetic preparations of the same flavors are common food additives.

■ Some 1700 natural and synthetic substances are used to flavor foods, making flavors the largest category of food additives.

Flavor Enhancers Flavor enhancers have little or no taste of their own but amplify the flavors of other substances. They exert synergistic and potentiation effects. "Synergism" is the cooperative action of two different substances (or people, or anything) such that the total effect is greater than the sum of the effects of each used alone. "Potentiators" do not have an effect themselves but exaggerate the effects of other chemicals. Some nucleotides (Section 15.8), for example, have no taste but enhance the flavor of meat and the effectiveness of salt. Flavor enhancers were first used in meat and fish but now are also used to intensify flavors or cover unwanted flavors in vegetables, bread, cakes, fruits, and beverages. Three common flavor enhancers are monosodium glutamate (MSG), 5′-nucleotides, and maltol.

■ In some people MSG causes the so-called Chinese restaurant syndrome, an unpleasant reaction characterized by headaches and sweating that usually occurs after an MSG-rich Chinese meal. MSG is a natural constituent of many foods, such as tomatoes, strawberries, and mushrooms. These foods affect some individuals in the same way.

Food Colors Some 30 chemicals are in use as food colors. About half of them are laboratory-synthesized and half are extracted from natural materials. Most food colors are large organic molecules with several double bonds and aromatic rings. Such structures have electrons that can absorb certain wavelengths of light and pass the rest; the wavelengths passed give the substances their characteristic colors. Beta-carotene, the orange-red substance in carrots and a variety of plants (and also an antioxidant), is an example of a natural food color. Because one of the food colors, Yellow No. 5, causes allergic reactions (mainly rashes and sniffles) in susceptible individuals, the FDA requires manufacturers to list Yellow No. 5 on the labels of any food products containing it.

The World of Chemistry

Program 13, *The Driving Forces*

The kinds of reactions that cause food to spoil are not very different than what chemists study in pure chemical solutions. The way I always like to talk about food, it's the study of messy chemistry. And the reason I say that is that, in a food which has so many different organic compounds and inorganic compounds together, there are lots and lots of different reactions that could have caused spoilage. What we can do, however, is narrow them down to several classes. For example, when you bite into an apple, you see it start to brown. That's an enzyme reaction. Dr. Theodore Labuza, Food Chemist

One use of FDA-approved food dyes. *(C.D. Winters)*

■ The acid–base chemistry of buffers was discussed in Section 9.8. Buffers react as a base with an acid or as an acid with a base. The result is to maintain constant pH. (High pH is basic, low pH is acid.)

pH Control in Foods Weak organic acids are added to such foods as cheese, beverages, and dressings to give a mild acidic taste. They often mask undesirable aftertastes, but in some cases, such as in fruit-flavored sodas, the acidic taste is expected. Weak acids and acid salts react with bicarbonate to form CO_2 in the baking process. Buffers are also added to adjust and maintain a desired pH. Potassium acid tartrate, for example, is a buffer because it is a salt of an organic acid that can act as either an acid or a base.

The versatile acidulants also function as preservatives to prevent the growth of microorganisms, as antioxidants to prevent rancidity and browning, as viscosity modifiers in dough, and as melting-point modifiers in such food products as cheese spreads and hard candy.

■ **Hygroscopic** substances absorb moisture from the air. You've probably had trouble getting salt out of the shaker during humid weather.

Anticaking Agents Anticaking agents are added to **hygroscopic** foods (in amounts of 1% or less) to prevent caking in humid weather. Table salt is particularly subjet to caking unless an anticaking agent is present. The additive (magnesium silicate, for example) incorporates water into its structure as water of hydration and does not appear wet as sodium chloride does when it absorbs water physically on the surface of its crystals. As a result, the anticaking agent keeps the surface of sodium chloride crystals dry and prevents crystal surfaces from codissolving and joining together.

SCIENCE AND SOCIETY

How Much Is Too Much? The Delaney Clause

No one wants harmful substances added to their food. Since 1958, the responsibility for testing food additives has been with the manufacturers. In that year the Food, Drug, and Cosmetic Act was amended to transfer responsibility from the FDA, which could not possibly carry out all the work. Now the FDA and the Environmental Protection Agency (EPA) review the results of tests done by others.

The requirement known as the Delaney Clause was introduced into the food additives amendment by Congress. The Delaney Clause prohibits the use in a food of **any** amount of a substance that causes cancer in animals or humans at **any** dose. At first glance, this may seem like a reasonable idea. Many would argue that it is not so reasonable as it seems. As we noted in Chapter 1, the Delaney Clause is a zero-risk law that allows no room for risk-benefit evaluations.

The critical part is the amount. Test animals are often fed diets containing huge quantities of a test substance relative to their diet and body size. But food additives are used in very small amounts. No one can state with certainty that the effect of huge doses on animals will be duplicated in humans who are getting very small doses.

In applying the Delaney Clause to pesticides used on food crops, a further problem arises because of advances in analytical chemistry. At the time of the amendment, pesticide residues could be detected at levels of 100 parts per billion (ppb). Less than that was essentially zero. Now, detection levels have been lowered to parts per *trillion*. For a while, EPA dealt with this problem by granting exemptions for pesticides that left a residue on food so small that it could be judged a negligible health risk.

The exemptions were based on a legal tradition that courts should not apply laws to the extent that the results are pointless *(de minimis)*. The exemption-granting policy was, however, forbidden by an appeals court in 1993. Would you support the necessary change in the law so that the *de minimis* policy could be applied to pesticides or food additives at the discretion of the EPA or the FDA? What questions would you want to ask to make an informed decision on this matter?

Stabilizers and Thickeners Stabilizers and thickeners improve the texture and blends of foods. Like carrageenan, most stabilizers and thickeners are polysaccharides, which have numerous hydroxyl groups as a part of their structure. The hydroxyl groups form hydrogen bonds with water and help to provide a more even blend of the water and oils throughout the food. Stabilizers and thickeners are particularly effective in icings, frozen desserts, salad dressings, whipped cream, confections, and cheeses. In reduced-fat foods, they replace some of the "fatty" texture that would otherwise be missing. The plant gum thickeners (e.g., gum arabic, guar gum, locust bean gum, gum tragacanth) are good sources of soluble dietary fiber (Section 17.3)

Foods containing modified food starch, pectin, and carbohydrate gums as stabilizers and thickeners. *(C.D. Winters)*

17.10 Some Daily Diet Arithmetic

Sometimes doing a little arithmetic helps to interpret food labels or identify your own food needs. A few examples of very practical kinds of diet and food calculations are given here.

EXAMPLE 17.2 *Percent Fat*

The Nutrition Facts labels (Figure 17.7) must list the total Calories in one serving of a food and the total Calories from fat in that serving. They do not list the percent Calories from fat, however. For a person aiming to keep dietary fat to a minimum, it is important to compare the quantity of fat in foods. Which of these cookies has more fat?

Yummie Bites—two cookies per serving

Calories 110 Calories from Fat 45

Chocolate Dandies—two cookies per serving

Calories 210 Calories from Fat 65

SOLUTION

The best way to compare the cookies is to compare the percent calories per serving from fat.

Yummie Bites

$$\frac{45 \text{ Calories from fat}}{110 \text{ Calories}} \times 100\% = 41\% \text{ Calories from fat}$$

Chocolate Dandies

$$\frac{65 \text{ Calories from fat}}{210 \text{ Calories}} \times 100\% = 31\% \text{ Calories from fat}$$

It turns out that Chocolate Dandies have a lower percentage of calories from fat, even though they provide more calories per serving.

Exercise 17.2A

What is the percent fat in each of these soups (from canned soup on the market)? Does either of them exceed the guideline of less than 30% Calories from fat?

Healthy Chicken—Calories 90 Calories from Fat 15

Cream of Mushroom—Calories 140 Calories from Fat 80

Exercise 17.2B

A good way to keep track of Calories from fat is to monitor the total grams of fat in your diet every day. A moderately active 100 lb woman would get 30% of daily Calories from fat by consuming 53 g of fat per day. How many Yummie Bites (2 cookies per serving) can she eat and leave 33 g of fat for other foods eaten that day?

How much of these popular foods should you eat? *(C.D. Winters)*

EXAMPLE 17.3 *Caloric Value of Food and Exercise*

A slice of pepperoni pizza contains 10 g of protein, 20 g of carbohydrate, and 7 g of fat. (a) How many kilocalories does the pizza slice provide? (b) What percent of the total calories is from fat? (c) How long would a person have to run at 5 mph to burn off the calories in the pizza slice (consult Table 17.1)?

SOLUTION

(a) The total caloric value of the pizza is calculated from the quantities of protein, carbohydrate, and fat:

$$10\ \cancel{\text{g protein}} \times \frac{4\ \text{kcal}}{\cancel{\text{g protein}}} = 40\ \text{kcal}$$

$$20\ \cancel{\text{g carbohydrate}} \times \frac{4\ \text{kcal}}{\cancel{\text{g carbohydrate}}} = 80\ \text{kcal}$$

$$7\ \cancel{\text{g fat}} \times \frac{9\ \text{kcal}}{\cancel{\text{g fat}}} = 63\ \text{kcal}$$

$$\overline{183\ \text{kcal}}$$

(b) Percent calories from fat:

$$\frac{63\ \text{kcal}}{183\ \text{kcal}} \times 100 = 34\%$$

(d) Running at 5 mph burns 10 kcal per minute; therefore, to burn off this slice of pizza would require running for

$$183\ \text{kcal} \times \frac{1\ \text{minute}}{10\ \text{kcal burned}} = 18.3\ \text{minutes}$$

Exercise 17.3

A slice of devil's food cake with chocolate frosting contains 3 g of protein, 8 g of fat, and 40 g of carbohydrate. (a) How many calories does the slice of cake provide? (b) What percent of the calories is from fat? (c) How long could a person listen to a lecture fueled by the energy from this piece of cake?

EXAMPLE 17.4 *Daily Diets*

Abdul is a member of the track team. Based on his body weight and intense physical activity, he requires a 4200-kcal daily diet. He wants to maintain the recommended 30% calories from fat, 60% calories from carbohydrate, and 10% calories from protein. How many grams of each nutrient should he consume every day?

SOLUTION

To find the grams of fat, carbohydrate, and protein requires first knowing the number of kilocalories from each.

$$4200 \text{ kcal} \times 0.30 = 1260 \text{ kcal from fat}$$
$$4200 \text{ kcal} \times 0.60 = 2520 \text{ kcal from carbohydrate}$$
$$4200 \text{ kcal} \times 0.10 = 420 \text{ kcal from protein}$$

Then the kilocalories per gram can be used to find the mass in grams of each nutrient that Abdul needs.

$$1260 \text{ kcal from fat} \times \frac{1 \text{ g}}{9 \text{ kcal}} = 140 \text{ g fat}$$
$$2520 \text{ kcal from carbohydrate} \times \frac{1 \text{ g}}{4 \text{ kcal}} = 630 \text{ g carbohydrate}$$
$$420 \text{ kcal from protein} \times \frac{1 \text{ g}}{4 \text{ kcal}} = 105 \text{ g protein}$$

Exercise 17.4

Cerise is trying to lose weight on a 1600-kcal daily diet while maintaining the same percentages of carbohydrates, fats, and proteins as Abdul. How many grams of each should she have each day?

■ SELF-TEST 17D

1. Two of the oldest means of preserving foods are ______________ and adding ______________.
2. GRAS is an acronym for ______________.
3. Antimicrobial preservatives make food sterile. (a) True, (b) False.
4. BHA and BHT are very common food additives because they function as ______________.

5. An important sequestrant goes by the initials ____________.
6. The flavor of a food can usually be traced to a single compound. (a) True, (b) False.
7. Most stabilizers and thickeners, for example, carageenan, are from the class of nutrients known as ____________.
8. If a food yields 230 kcal per serving, of which 100 kcal are from fat, the percent from fat is found from (____________ divided by ____________) × 100%.
9. If you require 2000 kcal per day, then to find how many calories from fat per day will give the maximum of 30%, you would do the following calculation: ____________ × 2000 kcal.

MATCHING SET

A. Match each nutrient with the condition caused by its excess or deficiency.

___	1. Iodine	a. Rickets
___	2. Iron	b. Goiter
___	3. Vitamin C	c. Anemia
___	4. Vitamin A	d. Atherosclerosis
___	5. Saturated fats	e. Kwashiorkor
___	6. Protein	f. Osteoporosis
___	7. Calcium	g. High blood pressure
___	8. Niacin	h. Night blindness
___	9. Sodium	i. Scurvy
___	10. Vitamin D	j. Pellagra

B. Match each type of food additive with its example.

___	1. Antioxidant	a. Trace metals like copper, iron
___	2. Food flavor	b. Sodium EDTA
___	3. Flavor potentiator	c. Monosodium glutamate
___	4. Calorie value of gram of fat	d. Beta-carotene
___	5. Calorie value of gram of protein	e. Refined sugar
___	6. Sequestrant	f. Sodium benzoate
___	7. Source of empty Calories	g. Menthol
___	8. Used for pH control in foods	h. 4 kcal
___	9. Food coloring	i. Butylated hydroxyanisole
___	10. Thickener	j. Tartaric acid
___	11. Preservative	k. 9 kcal
___	12. Catalyze oxidation of fats	l. Carrageenan

QUESTIONS FOR REVIEW AND THOUGHT

1. What is meant by the term "digestion"?
2. What is amylase? To what class of biochemical molecules does it belong?
3. What are the relationships between mass and energy for the following substances?
   (a) Fats (b) Proteins
   (c) Carbohydrates
4. What are the products formed by digestion of the following nutrients?
   (a) Carbohydrates (b) Fats
   (c) Proteins
5. What is meant by the term "BMR," or "basal metabolic rate"?
6. What is meant by the following terms?
   (a) Simple sugars
   (b) Digestible carbohydrates
   (c) Indigestible carbohydrates
   (d) Disaccharides

(e) Polysaccharides
(f) Dietary fiber

7. What are triglycerides?
8. Give definitions for the following terms as applied to fats and fatty acids.
(a) Saturated (b) Monounsaturated
(c) Polyunsaturated (d) Partially hydrogenated
9. What do the following terms mean?
(a) Atherosclerosis (b) Cholesterol
(c) Plaque (d) Lipoproteins
(e) Low-density lipoproteins
(f) High-density lipoproteins
10. What are the following?
(a) FDA (b) FDA food pyramid
11. What is the difference between an empty calorie and a nutritious calorie?
12. Give definitions for the following terms.
(a) Macronutrients (b) Micronutrients
13. What are the major categories of information required on a nutrition facts label?
14. Where do the following substances occur in nature?
(a) Starches (b) Glycogen
(c) Cellulose
15. What are micronutrient minerals? Which one must be listed on nutrition facts labels?
16. What is a vitamin? What is the function of a vitamin?
17. Overdoses of which of the following vitamins should definitely be avoided? Why?
(a) Riboflavin (b) Vitamin C
(c) Vitamin A (d) Vitamin B
(e) Vitamin D
18. What are free radicals? How do antioxidant vitamins and free radicals interact?
19. What is an electrolyte? What is meant when referring to the electrolyte balance of the body?
20. What is the GRAS list?
21. What are the seven most abundant macronutrient minerals in the body?
22. What are the functions of the following food additives?
(a) Salt in salted food (b) Sugar in candied fruit
23. What is the function of each of the following food additives?
(a) Sodium benzoate (b) Sodium propionate
(c) BHA (butylated hydroxyanisole)
(d) BHT (butylated hydroxytoluene)
(e) Citric acid (f) Sodium EDTA
24. Why are sequestrants needed as food additives?
25. What is the function of each of the following food additives?
(a) Monosodium glutamate (MSG)
(b) Yellow No. 5
(c) Acetic acid (d) Calcium silicate
(e) Carrageenan (f) Lecithin
26. What percent fat is currently recommended for today's diet by most public health authorities?
27. Lactose is sometimes referred to as "milk sugar" because it is the principle carbohydrate in milk. Lactose has the structure shown here. Is this a monosaccharide, disaccharide, or polysaccharide? Explain.

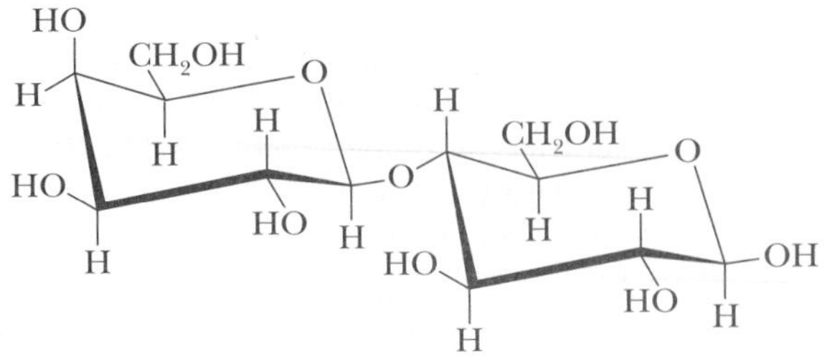

28. Lactose intolerance occurs in people who are unable to hydrolyze milk sugar (lactose). This happens because they do not produce enough of the enzyme lactase. Normally infants and children have adequate amounts of lactase, but many adolescents and adults produce less. Should milk be promoted as a healthy food for the general population? Give one argument for and one against.
29. The body stores excess energy in small globules of fat in specialized cells of adipose tissue. The energy available from a gram of fat is 9 Calories. What would happen to the volume of these cells if the body stored energy by collecting carbohydrates instead of fats? Assume that the density of carbohydrate and fat are the same. (Recall that 1 gram carbohydrate = 4 Calories.)
30. Cholesterol has the following structure. Why is cholesterol insoluble in water? Why is it recommended that people monitor their blood serum cholesterol levels? How many carbon atoms does a cholesterol molecule contain?

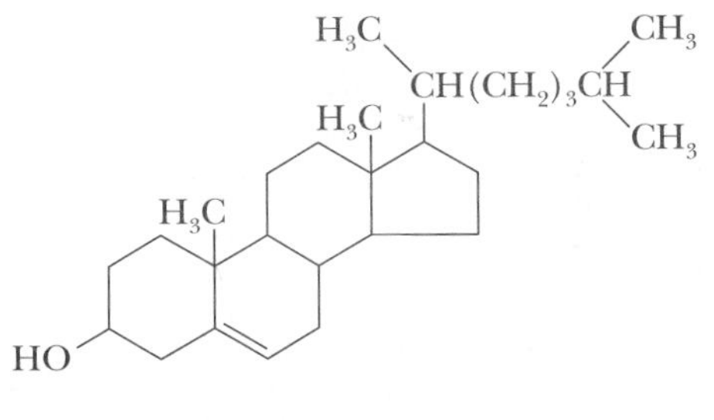

31. Folic acid is found in citrus fruits and leafy green vegetables. The FDA is proposing to have folic acid added to cereals, breads, and pastas. The American Journal of Public Health says such grain fortification would prevent 300–700 birth defects per year. At the level recommended for addition, approximately 3.25 million adults over 50 would receive too much folic acid. What question would you want answered before you would support such a nationwide program?
32. Why is osteoporosis a more significant problem for

women after menopause? What preventive measures can be taken to reduce the chances of experiencing osteoporosis?

33. Should the government embark on a national program to add more calcium to our diets to reduce our chances of calcium deficiency and osteoporosis? Give one reason for such a program and one against.
34. What happens to the human thyroid gland if a person has an iodine deficiency? What is this condition called? What is the link between this problem and iodized salt?
35. When fats and fatty portions of food become rancid what kind of reactions have probably occurred? What food additives are used to minimize these reactions?
36. Which of the following biochemical compounds (i) carries energy to where is it needed, and (ii) carries the carbon atoms from food molecules into the final biochemical energy-generating pathways? (a) Acetyl coenzyme A, (b) Adenosine triphosphate (ATP).
37. Choose a label from a food item and try to identify the purpose of each food additive.
38. What foods have you eaten during the past week that did not contain any food additives?
39. How would you describe to someone else what is meant by a "zero-risk" law?
40. What is the *de minimis* principle in public policy?
41. Get together with a group of two or three others and find out whether the majority would favor repeal of the Delaney Clause.

PROBLEMS

1. What is the basal metabolic rate for a 110-pound woman who has a moderate activity level?
2. What is the basal metabolic rate for a 145-pound man who is sedentary, with very light activity? What would this person's BMR be if he started to exercise regularly so his activity increased to the moderate level?
3. How much time bicycling at 5 mph is needed to consume the 100 calories available from a serving of tomato soup? (*Note:* Energy expended bicycling at 5 mph is 5 Calories/min.)
4. A 100-g serving of ice cream contains 207 Calories. How long would you have to run at 5 mph to consume the energy available from the ice cream?
5. Walking on a road at 3.5 mph consumes 3.5 Calories/minute. How long would you have to walk to consume the energy available from 100 grams of white bread? (*Note:* 100 grams white bread = 269 Calories.)
6. A blood serum cholesterol level of 200 mg/100 mL is supposed to give a low heart disease risk level. How much total cholesterol would be carried by the blood serum of an adult with a blood volume of 12 pints? (1 pint = 473 mL)
7. A blood serum cholesterol level of 240 mg/100 mL is supposed to give a high heart disease risk level. How much total cholesterol would be carried by the blood serum of an adult with a blood volume of 13 pints? (1 pint = 473 mL)
8. Vitamin D is produced by the body when ultraviolet light strikes the skin. The FDA established a RDI of 400 IU for vitamin D. A 240-mL serving of 2% low-fat milk provides 25% of this requirement. How many IU are in this serving? How much milk would a person need to drink to receive a toxic dose of 300,000 IU? (*Note:* This dose would have to be ingested daily for a period of many months.)
9. The quantity of fat in a food is important to people who need to restrict fat intake. Which of the following would be the better low-fat choice, a piece of pecan pie weighing 113 g providing 459 Calories with 180 Calories from fat, or a piece of apple pie weighing 125 g providing 305 Calories with 108 Calories from fat?
10. How many Calories (kcal) are in a fast-food chicken sandwich that contains 27 g of protein, 46 g of carbohydrate, and 34 grams of fat?
11. How many kcal are in a fast-food bacon cheeseburger supreme if it contains 34 g of protein, 44 g of carbohydrate, and 46 g of fat?
12. An exceptionally active female volleyball player who weighs 145 pounds wants to maintain a diet of 30% calories from fat, 60% from carbohydrates, and 10% from protein. How many Calories does the player need per day? How many grams of fat, carbohydrate, and protein does she need?
13. Using the nutrition facts label below, tell what percent of the calories in the product come from fat. What percent of your daily calorie need is provided by a serving if your daily calorie need is 1800 Cal. Would this product be a good food item for a person on a low-sodium diet? Justify your answers.

Nutrition Facts	
Serving Size 1/2 cup (120mL) condensed soup	
Servings Per Container About 6	
Amount Per Serving	
Calories 100	Calories from Fat 20
	% Daily Value*
Total Fat 2g	3%
Saturated Fat 0g	0%
Cholesterol 0mg	0%
Sodium 730mg	30%
Total Carbohydrate 18g	6%
Dietary Fiber 2g	8%
Sugars 10g	
Protein 2g	
Vitamin A 10% •	Vitamin C 30%
Calcium 2% •	Iron 4%

CHAPTER 18

Toxic Substances

Toxic substances can upset the incredibly complex system of chemical reactions occurring in the human body. Sometimes toxic substances cause mere discomfort; sometimes they cause illness, disability, or even death. Often because of where we work or what we eat, we come in contact with toxic substances. The term **poison** usually is limited to substances that are dangerous in small amounts, such as sodium cyanide or the toxin present in spoiled canned food that causes botulism. However, as most of us know, ill effects can be caused by excessive intake of substances normally considered harmless (eating too much candy, for example). The effects of toxic substances can be immediate (**acute toxicity**) or prolonged (**chronic toxicity**). Either kind of effect is undesirable. To protect ourselves from toxic substances, we need to know how to recognize them, both from their chemical names and formulas and from their chemical and physical properties. We need to know how they act on living organisms so we can recognize the symptoms they cause and also how to avoid exposures by understanding where they occur and how they can enter the body.

- How do we measure toxicity?
- How do toxic substances enter our bodies, and what does the body do to rid itself of some of these undesirable substances?
- What are the classes of toxic substances and how do they work?
- What are the mechanisms of toxic substances that affect the nervous system, and how have they helped us learn more about how nerve impulses are transmitted?
- How do toxic substances affect the unborn and possibly affect future generations?
- What substances have been used as chemical warfare agents, and what effects do they have?

Lead compounds that are excellent paint pigments. Because of the toxicity of lead, these pigments are now banned from many kinds of paint. *(C.D. Winters)*

Fortunately for us, in many cases the human body is capable of recognizing harmful chemicals and ridding itself of them. This is essential because we are literally surrounded by substances that could cause harm. During our everyday contacts with the world around us, we breath in air, we drink water and various beverages, and we eat foods, all of which can contain harmful substances. In addition, our skin comes in contact with many things daily that could potentially harm us. In this chapter we shall focus on the chemical mechanisms by which toxic substances work and look at some of the classes of toxic substances.

18.1 The Dose Makes the Poison

■ *Dosis sola facit venenum*—"The dose makes the poison." Paracelsus (1493–1541)

A **dose** of a toxic substance is the amount of the substance that enters the body of the exposed organism. The dose might be lethal (causing death), or it might be sublethal (causing harm, but not death). Whether a dose is lethal or sublethal depends on a number of factors, including the species, age, sex, and general state of health of the organism, and how the dose was administered (slowly over a period of time or all at once). Lethal doses of toxic substances are customarily expressed in milligrams (mg) of substance per kilogram (kg) of body weight of the subject (mg/kg). For example, the cyanide ion (CN^-) is generally fatal to humans in a dose of 1 mg of CN^- per kg of body weight. For a 200-lb (90.7-kg) person, about 0.1 g of cyanide is a lethal dose. Smaller individuals, especially children, because of their lower body weight, can be affected by much smaller doses. Examples of somewhat less toxic substances and the range of lethal doses for humans of average body weights follow:

■ A large enough dose of any compound can result in poisoning.

■ Records are kept on accidental and homicidal poisonings, so *lethal doses* for many substances are known.

Substance	Lethal Dose
Morphine	1–50 mg/kg
Aspirin	50–500 mg/kg
Methanol	500–5000 mg/kg
Ethanol	5000–15,000 mg/kg

TABLE 18.1 ■ LD_{50} Values for Dioxin (2,3,7,8-TCDD)

Species	LD_{50} (mg/kg)
Guinea pig	0.0006
Rat	0.04
Monkey	0.07
Rabbit	0.115
Dog	0.150
Mouse	0.200
Hamster	3.5
Bullfrog	>1.0

A quantitative measure of toxicity is obtained by administering various doses of substances to be tested to laboratory animals (such as rats). The dose found to be lethal in 50% of a large number of test animals under controlled conditions is called the LD_{50} (lethal dose—50%) and is usually reported in milligrams of the substance per kilogram of body weight. Thus, if a statistical analysis of data on a large population of rats showed that a dose of 1 mg/kg was lethal to 50% of the population tested, the LD_{50} for this poison would be 1 mg/kg. Obviously, species differences can produce different LD_{50} values for a given poison. The differences between animal species can be very great. For example, the toxicity of dioxin (2,3,7,8-tetrachloro-*p*-dioxin), a compound produced when chlorinated compounds are incinerated and when chlorine is used to bleach wood pulp, varies over a wide range (Table 18.1). The extremely low LD_{50} for the guinea pig is unusual when compared with the large LD_{50} for other species. Partly because of the low guinea-pig value, some have classed dioxin as one of the most toxic compounds known. Dioxin certainly is toxic to animals, but other data do not rank it as a potent human poison.

■ Because chlorine-containing molecules are found in most plants, forest and grass fires actually produce measurable quantities of dioxin. Dioxin was also found as an impurity in some of the defoliants, such as Agent Orange, used during the Vietnam War. Many veterans of that war feel that this chemical has impaired their health and have sued the U.S. government for damages.

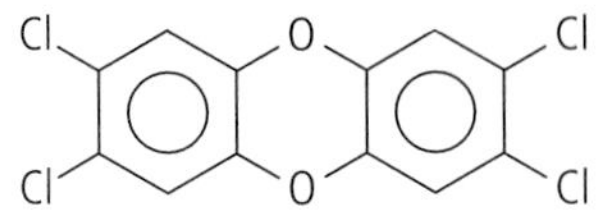

Dioxin
(2,3,7,8-TCDD)

EXAMPLE 18.1 *Calculations Using LD_{50} Values*

When administered orally to the rat, the LD_{50} for parathion, a common insecticide, is 20. mg/kg. For a rat weighing 0.75 kg, what is the dose (in mg) that has a 50% chance of being lethal?

SOLUTION

Multiply the LD_{50} value by the weight of the subject rat to get the dose.

$$\text{Dose} = (20.\ \text{mg/kg}) \times (0.75\ \text{kg}) = 15\ \text{mg}$$

This means that half the rats receiving a dose of 15 mg would be killed or, put another way, that a 0.75-kg rat receiving a 15-mg dose of parathion has a 50% risk of dying.

Exercise 18.1

The LD_{50} of arsenic trioxide is 12 mg/kg when administered orally to rats. What dose, in mg, would have a 50% chance of killing a rather large rat weighing 1.25 kg?

Due to species differences in LD_{50} values, defining risk to human beings based on animal data is difficult. It is, however, generally safe to assume that a chemical with a low LD_{50} value for several species will also be quite toxic to human beings (Table 18.2).

TABLE 18.2 ■ LD_{50} Values for Several Chemicals

Chemical	LD_{50} (mg/kg administered orally to rat)
Aspirin	1750
Ethanol	1000
Morphine	500
Caffeine	200
Heroin	150
Lead	20
Cocaine	17.5
Sodium cyanide	10
Nicotine	2
Strychnine	0.8
Batrachotoxin	0.002*

*From a poisonous frog. LD_{50} in mice.

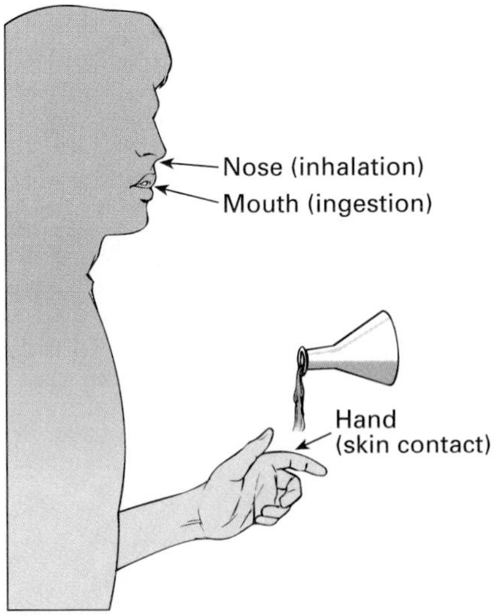

Figure 18.1 Three routes of entry to the body for toxic substances.

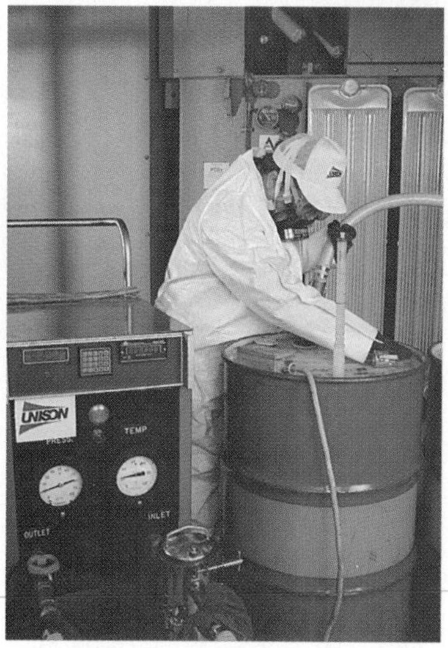

Figure 18.2 Protective gear in the workplace. The worker is pumping toxic oils (PCBs, which are carcinogens) from an electrical transformer in preparation for refilling the transformer with a less harmful substance. *(James Prince/Photo Researchers)*

■ Because the lungs must efficiently transfer oxygen and carbon dioxide between the air and blood, other gases are also efficiently transferred there. The lungs have a very large surface area that is in contact with the blood supply.

Toxic doses can enter an organism by three means: inhalation, ingestion, and skin contact (Figure 18.1). Some substances are gases at room temperature and mix easily with air, so inhalation is their most common route of entry. For substances with acute toxic properties, inhalation is especially dangerous because inhaled air in the lungs immediately comes in contact with the blood. If the substance dissolves in blood or reacts with some component of the blood, an individual can absorb a large dose in a short time.

Ingestion is the common route of entry for many solids and liquids as well as for chemicals that are soluble in water. Skin contact with toxic substances can also be a major route of entry in some environments, such as industrial settings, where workers may routinely handle or otherwise come in contact with chemicals (Figure 18.2).

Toxic substances can be classified according to the way in which they disrupt body chemistry as **corrosive, metabolic, neurotoxic, mutagenic, teratogenic,** and **carcinogenic.** This classification will serve as the basis of our discussion.

18.2 Corrosive Poisons

Toxic substances that actually destroy tissues are **corrosive poisons.** Examples include strong acids and bases (often referred to as alkalis—see Section 16.15) and many oxidizing agents, such as bleaches found in laundry products. Sulfuric acid (H_2SO_4, used in auto batteries) and hydrochloric acid (HCl, also called muriatic acid and used for cleaning purposes) are very dangerous corrosive poisons. So is sodium hydroxide (NaOH), a strong base used around the house to clear clogged drains. Death has resulted from swallowing as little as 1 oz of concentrated (98%) sulfuric acid, and much smaller amounts can cause extensive damage and severe pain.

■ Removal of water is called "dehydration."

Concentrated mineral acids such as sulfuric acid act by first dehydrating cellular structures. The cell dies because its water is removed and its protein is destroyed by the acid-catalyzed hydrolysis of the peptide bonds.

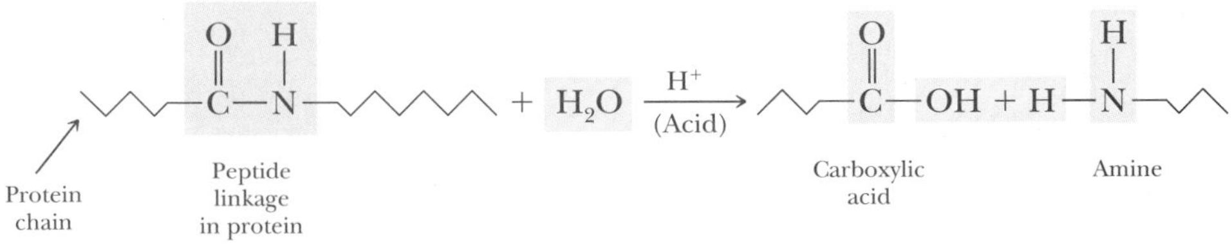

Subsequently, as more bonds are broken, smaller and smaller fragments result, leading to the ultimate disintegration of the tissue. Tissue destroyed by corrosive poisons must be replaced. If the destruction is not too severe, normal growth may replace the lost tissue, although scarring usually results. If external tissue destruction is extensive, skin grafting is necessary. The area of destroyed tissue must be treated much like a severe thermal burn.

Many corrosive poisons have ***warning properties;*** they interact with your senses to let you know you are being exposed. Ammonia, a weakly basic compound, is a good example. Exposure to concentrations of ammonia as low as 0.01% in air can cause violent choking, coughing, gasping for breath, and irritation to the eyes, all of which are valuable warning properties. No one will voluntarily remain in an area of high ammonia concentration.

Acids also warn reasonably well because the nerve endings in the tissue exposed send messages to the brain telling it something is wrong. Not all acids have this effect, however. Hydrofluoric acid (HF, an acid used in the electronics industry) is capable of destroying tissue protein *and* bone, but it does not produce a pain sensation immediately upon exposure. Instead, the exposed skin becomes red, followed by pain when the bone beneath becomes exposed to the acid. This is not good, because by the time the pain is noticed, considerable damage to skin, underlying tissue, and bone has occurred.

■ Toxic substances without warning properties are especially dangerous because harmful exposures can take place before any symptoms are seen.

Sodium hydroxide (also known as caustic soda) is a very strongly alkaline substance that can be just as destructive to tissue as strong acids. The hydroxide ion catalyzes the hydrolysis of peptide bonds and has the same destructive effect on proteins as strong acids. Strong bases lack good warning properties on contact with the skin, although they do impart a slippery feel to the skin as they quickly dissolve its outer layer. If, on the other hand, a strong base contacts your eyes or nose, there will be a strong enough pain sensation to let you know you have been exposed. For example, you might be exposed to a strong base such as sodium hydroxide if you breathe the aerosol produced by a can of oven cleaner. These products have warnings on their labels advising users not to breathe the fumes (Figure 18.3). Both acids and bases, as well as other types of corrosive poisons, continue their action until they are consumed in chemical reactions that neutralize their acidic or basic properties.

Some substances are converted to corrosive poisons by chemical reactions in the body. Phosgene, a deadly gas used during World War I and used today in the chemical industry to make plastics and drugs, acts this way. When inhaled, phosgene is hydrolyzed in the lungs to form hydrochloric acid, which causes *pulmonary edema* (a collection of fluid in the lungs) because it draws water from surrounding tissue. The victim "drowns" because oxygen cannot be absorbed effectively by the flooded and damaged tissues. Phosgene does not have good warning properties. It has a weak odor of new-mown hay, and the hydrolysis reaction producing acid in the lungs is fairly slow, so the victim doesn't experience any effects of the exposure for up to several hours. By then, the pulmonary damage can be quite severe.

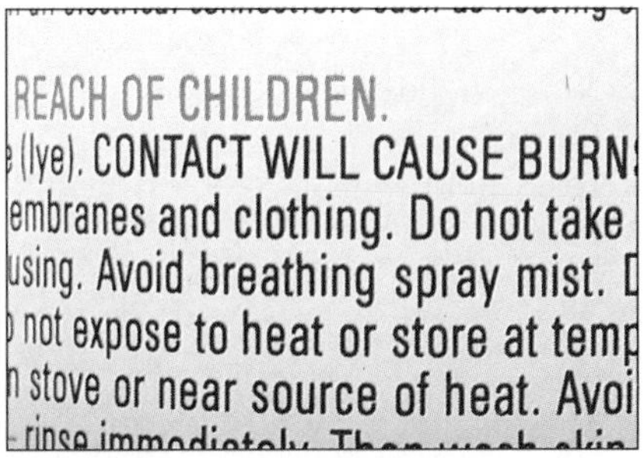

Figure 18.3 Warning label on an aerosol can. *(C.D. Winters)*

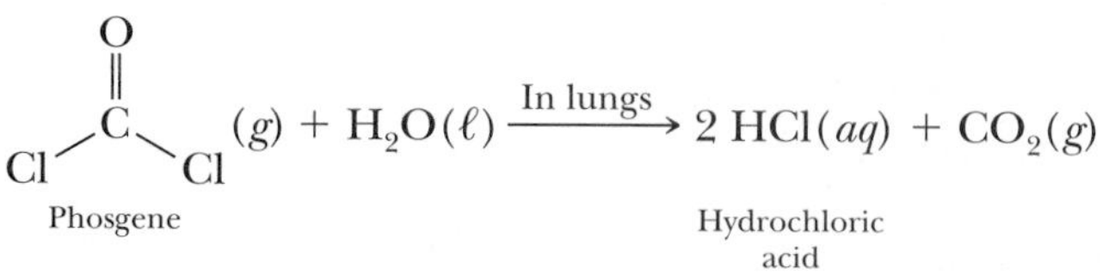

$$\text{ClCOCl}(g) + H_2O(\ell) \xrightarrow{\text{In lungs}} 2\ HCl(aq) + CO_2(g)$$

Some corrosive poisons destroy tissue by *oxidizing* it. This is characteristic of substances such as hydrogen peroxide, ozone, nitrogen dioxide, and the halogens—fluorine, chlorine, bromine, and iodine. These substances can destroy proteins by the oxidation of peptide linkages or the sulfur-sulfur bonds between side chains. When such linkages are destroyed, the protein molecule loses its functionality. Most oxidizing agents will produce enough pain on exposure to sufficiently warn you of exposure. A summary of some common corrosive poisons is presented in Table 18.3.

What do you do if you or someone around you is accidently exposed to a corrosive poison? If the exposure is on the skin or eyes, the affected area must be washed immediately with running water for at least 15 to 30 minutes. A physician (or the emergency medical service) should be called, especially if the exposure is to the eyes. If someone is overcome by the fumes of a corrosive poison, bring the person to fresh air quickly and call for medical help. If someone ingests a corrosive poison, call for medical help immediately. Do not try to induce vomiting, because this action will probably expose more tissue to the corrosive effects of the poison. It is a good idea to protect yourself if you

TABLE 18.3 ■ Some Corrosive Poisons

Substance	Formula	Toxic Action	Possible Contact
Hydrochloric acid	HCl	Acid hydrolysis	Tile and concrete floor cleaner; concentrated acid used to adjust acidity of swimming pools
Sulfuric acid	H_2SO_4	Acid hydrolysis, dehydrates and oxidizes tissue	Auto batteries
Phosgene	$ClCOCl$	Acid hydrolysis	Combustion of chlorine-containing plastics (PVC or Saran)
Sodium hydroxide	$NaOH$	Base hydrolysis	Caustic soda, drain cleaners
Trisodium phosphate	Na_3PO_4	Base hydrolysis	Detergents, household cleaners
Sodium perborate	$NaBO_3 \cdot 4H_2O$	Base hydrolysis, oxidizing agent	Laundry detergents, denture cleaners
Ozone	O_3	Oxidizing agent	Air, electric motors
Nitrogen dioxide	NO_2	Oxidizing agent	Polluted air, automobile exhaust
Iodine	I_2	Oxidizing agent	Antiseptic
Hypochlorite ion	OCl^-	Oxidizing agent	Bleach
Peroxide ion	O_2^{2-}	Oxidizing agent	Bleach, antiseptic
Oxalic acid	$H_2C_2O_4$	Reducing agent, precipitates Ca^{2+}	Bleach, ink eradicator, leather tanning, rhubarb, spinach, tea
Sulfite ion	SO_3^{2-}	Reducing agent	Bleach
Chloramine	NH_2Cl	Oxidizing agent	Produced when household ammonia and chlorinated bleach are mixed
Nitrosyl chloride	$NOCl$	Oxidizing agent	Mixing household ammonia and bleach

DISCOVERY EXPERIMENT

Reading Labels for Information on Toxics

A hardware store or the hardware section of a discount store or supermarket contains numerous commercial products that can cause harm if they are misused or carelessly used. Go to one of these stores and gather information regarding hazards from either the ingredients list or the warnings printed on the label for two products in each of the following categories. For each item, list the product name, its toxic ingredients, the hazards listed, protective measures to take (such as gloves or goggles), and what to do in case of accidental exposure.

1. oven cleaner
2. paint remover
3. brick cleaner
4. flea and tick fogger
5. paint thinner
6. concrete mix
7. fruit tree insect spray
8. flying insect spray
9. liquid drain cleaner
10. weed and grass killer
11. polyurethane floor finish

are going to be working with strong acids, bases, or oxidizing agents around the house or at work. Federal and state regulations help protect workers by calling for minimum protective clothing and adequate ventilation. Around the home, when using products such as paint strippers, drain cleaners, and bathroom cleaners, read the labels and use any protective measures called for, such as gloves, goggles, and protective aprons.

18.3 Carbon Monoxide and Cyanide as Metabolic Poisons

Metabolic poisons are more subtle than the tissue-destroying corrosive poisons. In fact, many of them do their work without actually indicating their presence until it is too late. A metabolic poison can cause illness or death by interfering with a vital biochemical mechanism to such an extent that it ceases to function or is prevented from functioning efficiently.

The human body can usually accommodate small, repeated, or regular doses of many metabolic poisons because detoxification mechanisms exist for them. Over a long period of exposure, however, the effects of subacute doses of these toxic substances can lead to *chronic* effects, which usually result in a lessening of the efficiency of body functions, such as motor skills or cognitive ability (the ability to think). When these functional impairments are not recognized, damage can become cumulative, with serious disabling or even lethal effects. As you read about the metabolic poisons, keep in mind that those that have no warning properties are especially dangerous.

Carbon Monoxide

Carbon monoxide is a colorless, odorless gas at room temperature. This means carbon monoxide has no warning properties. Because it is a gas it mixes

In some occupations, it is difficult to avoid long-term exposure to carbon monoxide. *(© Yoav Levy/Phototake NYC)*

with air, is inhaled, and comes in contact with the blood while in the lungs. The interference of carbon monoxide with oxygen transport in the blood is one of the best-understood kinds of metabolic poisoning. Carbon monoxide, like oxygen, combines with the hemoglobin in red blood cells:

$$O_2(g) + \text{hemoglobin}(aq) \rightleftharpoons \text{oxyhemoglobin}(aq)$$
$$CO(g) + \text{hemoglobin}(aq) \rightleftharpoons \text{carboxyhemoglobin}(aq)$$

Both of these reactions are reversible, as indicated by the double arrow ($\rightleftharpoons$) (see Section 8.4). That is, oxyhemoglobin can give up its oxygen (that's what it does when it transports oxygen to a cell in the body), and carboxyhemoglobin can give up its carbon monoxide molecule, although not as easily. Laboratory tests show that carboxyhemoglobin is 140 times more stable than oxyhemoglobin.

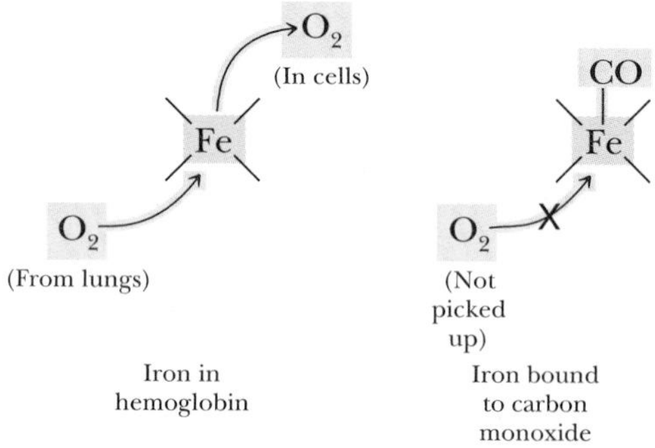

■ "ppm" means parts per million—a measure expressing concentration; 30 ppm CO means 30 mL of CO for every million mL of air. To convert ppm to percent, divide by 10,000. To convert percent to ppm, multiply by 10,000.

■ Air is 21% O_2 by volume; in one million "air molecules" there would be 210,000 O_2 molecules.

■ Carbon monoxide poisoning can occur when kerosene heaters or charcoal burners are used indoors without proper ventilation.

■ Incomplete combustion of carbon; the ratio is 2 C to 1 O_2: $2\ C(s) + O_2(g) \rightarrow 2\ CO(g)$
Complete combustion of carbon; the ratio is 1 C to 1 O_2: $C(s) + O_2(g) \rightarrow CO_2(g)$

Both O_2 and CO bond to the iron atom in the hemoglobin molecule. The greater stability of carboxyhemoglobin occurs because the Fe—CO bond is stronger than the Fe—O_2 bond. Since hemoglobin is so effectively tied up by carbon monoxide, those hemoglobin molecules that contain a CO molecule cannot perform their vital function of transporting oxygen. Breathing air with a concentration of 30 ppm of CO for eight hours is sufficient to cause headache and nausea for most people. Breathing air that is 0.1% (1000 ppm) carbon monoxide for four hours converts approximately 60% of the hemoglobin of an average adult to carboxyhemoglobin (Table 18.3), and death is likely to result unless the carboxyhemogobin molecules can be freed of the attached CO molecules.

Carbon monoxide exposures are quite common. In fact, low exposures are almost impossible to avoid. This means that some of the hemoglobin molecules in your blood are always bound to carbon monoxide. Any organic material that undergoes incomplete combustion will always liberate carbon monoxide. Sources include auto exhausts, smoldering leaves, lighted cigars or cigarettes, and charcoal burners. In the United States alone, combustion sources of all types dump about 200 million tons of carbon monoxide per year into the atmosphere, where it mixes with the other molecules in the air. Carbon monoxide is considered an air pollutant that lowers overall air quality (see Section 21.9).

Since the reactions of both carbon monoxide and oxygen with hemoglobin are reversible, the concentrations of the two gases, in addition to the relative strengths of bonds, affect how much hemoglobin will be combined with either molecule.

■ High O_2 concentrations favor oxyhemoglobin, while high CO concentrations favor carboxyhemoglobin.

$$\text{Carboxyhemoglobin}(aq) + O_2(g) \rightleftharpoons \text{oxyhemoglobin}(aq) + CO(g)$$

As with all chemical equilibria, if the concentration of one of the reactants or one of the products is increased, the equilibrium will shift so that the increased reactant or product with be used and equilibrium is re-established (Le Chatelier's principle, Section 8.4). In air that contains 0.1% or less CO, oxygen molecules outnumber CO by at least 200 to 1. This higher concentration of O_2 molecules helps to counteract the stronger binding between CO and hemoglobin, so the equilibrium favors the formation of oxyhemoglobin. If a person breathes air that has a CO concentration higher than about 0.1%, the equilibrium begins to favor the formation of carboxyhemoglobin. When a victim of carbon monoxide poisoning is exposed to fresh air or, still better, pure oxygen, the equilibrium shifts back in the favor of oxyhemoglobin.

Because the hemoglobin molecules bound to CO molecules can be freed of their CO molecules and thus have the capacity to carry oxygen again, carbon monoxide is not a cumulative poison. The released carbon monoxide is exhaled from the lungs. In spite of this ready reversibility, permanent damage can occur if certain vital cells (e.g., brain cells) are deprived of oxygen for more than a few minutes. Individuals differ in their tolerance of carbon monoxide, but generally those with anemia or an otherwise low reserve of hemoglobin (e.g., children) are more susceptible to its effects. A pregnant woman who smokes can damage her fetus because carbon monoxide from the inhaled tobacco smoke can deprive the fetus of the oxygen it needs during critical developmental stages. Studies have shown that low birth weight is closely related to the mother's smoking habits. Certainly no one is helped by carbon monoxide, and people whose habits (such as smoking) or work environment (such as directing traffic all day long) expose them regularly to carbon monoxide can suffer chronically from its effects.

■ A cumulative poison is one that tends to accumulate in the body because there is no efficient means of getting rid of the poison.

Cyanide

The cyanide ion (CN^-) is the toxic agent in salts such as sodium cyanide (NaCN), which is used industrially in electroplating and metal-ore processing. Since the cyanide ion is a relatively strong base, it reacts easily with many acids (weak and strong) to form volatile hydrogen cyanide (HCN):

$$\underset{\text{Sulfuric acid}}{H_2SO_4(aq)} + \underset{\text{Sodium cyanide}}{NaCN(aq)} \longrightarrow \underset{\text{Hydrogen cyanide}}{HCN(g)} + \underset{\text{Sodium hydrogen sulfate}}{NaHSO_4(aq)}$$

Hydrogen cyanide boils at a relatively low temperature (b.p. 26°C), so it is a gas at temperatures slightly above room temperature. It is often used to fumigate storage bins and the holds of ships because it is toxic to most forms of life and, in gaseous form, can penetrate into tiny openings, even into insect eggs. Hydrogen cyanide has a faint odor of almonds, but this odor is not strong enough to be a good warning property. Because of that, if any odor of cyanide is detected, the concentration may be life-threatening.

SCIENCE AND SOCIETY

Worst-Case Scenarios of Toxic Releases

Toxic chemicals are all around us. As we travel on interstate highways, we sometimes pass trucks loaded with them. A "poison" or "corrosive substance" placard indicates that poisons or corrosive materials are present in the load. Industry uses toxic chemicals for all sorts of purposes, and if you live in certain industrialized communities, your chances of exposure to toxic chemicals are greater than if you live where industry is not as commonplace. Many people do live where toxic chemicals are manufactured, transported, and used, and since accidental releases do happen, the EPA and even local governments regulate industries in an effort to minimize accidents.

One location where the chemical industry is concentrated is in the Kanawha River Valley, near Charleston, West Virginia, an area with about 250,000 people. Here, eight large chemical companies have plants producing or using numerous toxic chemicals. In accordance with the EPA's "community right-to-know" rules, a concerned citizen in 1992 asked industries in the valley to provide her with worst-case scenarios describing just how bad things could be if one or more of the toxic chemicals in use there were to be released in large quantities. In a joint community meeting, held in June 1994, industry representatives presented several scenarios,

from moderate spills to catastrophic accidents that would involve large segments of the community. Industry spokespersons pointed out that their safety measures were even stricter than those called for by the EPA. For example, one plant making methyl isocyanate (MIC), a chemical that was responsible for approximately 4000 deaths after an accidental release in Bophal, India, in 1984, has leak detectors that will sound an alarm when levels reach 1 ppm of MIC in the air. The EPA rule requires reporting leaks starting at 10 ppm. In spite of this ability to detect MIC leaks, citizens questioned why so much of it (about 250,000 pounds) needs to be stored on site. They ar-

Some Chemicals in the Kanawha Valley Subject to Worst-Case Scenarios

Name	Formula	Toxic Action	Warning Properties
Propylene oxide	CH_3CHCHO	Corrosive, narcotic	Poor; faint, sweet odor
Dimethylamine	$(CH_3)_2NH$	Corrosive	Good; fishy odor
Hydrogen chloride	HCl	Corrosive	Good; choking odor
Chlorine	Cl_2	Corrosive, oxidizer	Good; choking odor
Methyl isocyanate	CH_3NCO	Corrosive	Good; choking odor, eye irritation
Sulfur trioxide	SO_3	Corrosive	Good; choking odor
Phosgene	$COCl_2$	Corrosive by slow hydrolysis in lungs	Poor; faint odor-delayed action
Phosphorus trichloride	PCl_3	Corrosive	Good; choking odor of hydrogen chloride

Cyanide occurs naturally in the seeds of cherry, plum, peach, apple, and apricot trees, where it is formed when the compound *amygdalin,* which contains a—CN group, undergoes hydrolysis. Presumably, plants make amygdalin as a defense to protect the seed until it has a chance to germinate.

$$O - C_6H_{10}O_4 \cdot C_6H_{11}O_5 \quad \text{Sugar units}$$

Amygdalin $+ 2 H_2O \longrightarrow HCN + 2 C_6H_{12}O_6 +$ Benzaldehyde

Hydrogen cyanide · Glucose (sugar)

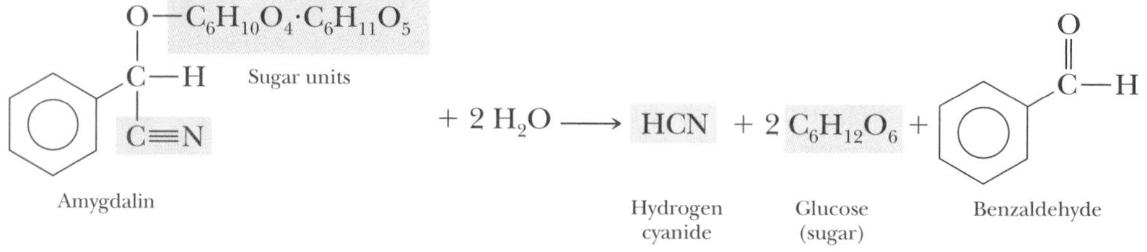

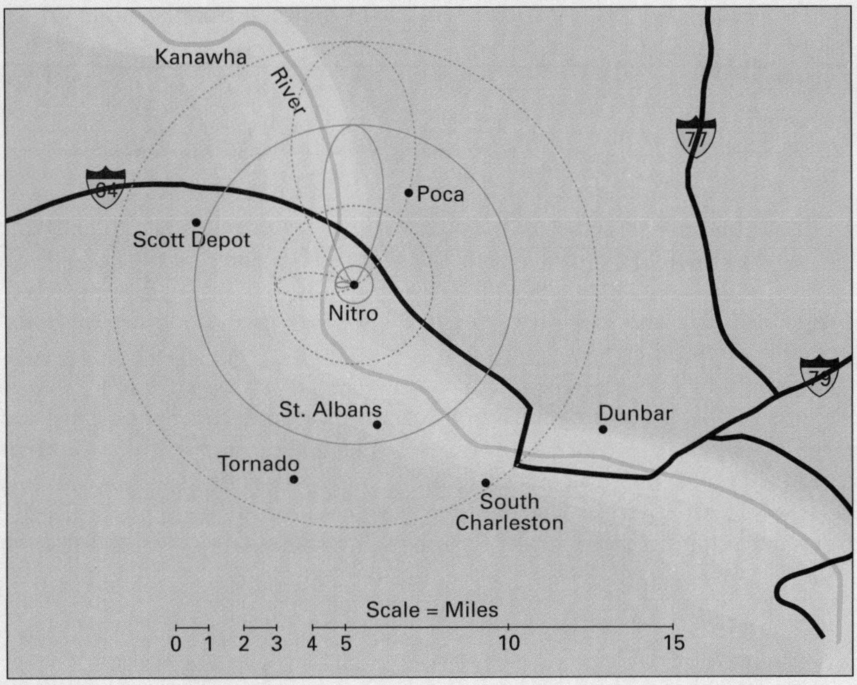

Scenarios for the release of phosphorous trichloride from a chemical plant in Nitro, West Virginia. In this computer-generated map, the ovals represent areas affected when the wind is blowing in that direction. Different sized ovals represent releases of varying magnitudes. Plant personnel and emergency responders use maps like these to plan evacuations and other emergency measures. (Courtesy: FMC Corporation)

gued that perhaps it would be better to prepare the compound as it is needed for other chemical reactions.

Everyone attending the community meeting agreed that the openness of the Kanawha Valley chemical industries went a long way toward showing that they were aware of the potential hazards and had gone to great lengths to prevent accidental exposures to the community. But as Nitro, West Virginia, firechief Steven Hardman, whose fire company is responsible for industrial fires and chemical spills in the valley, said, "If you live in California, you have to worry about earthquakes. If you live in Kansas, you have to worry about tornadoes. If you live in Kanawha Valley, you have to worry about chemicals."

At present, not all communities are as knowledgeable about their chemical exposures as those in Kanawha Valley, West Virginia. Over time, that will probably change as more citizens learn about the community right-to-know rules that the EPA has in place.

While bound in the amygdalin molecule the —CN group is not free to act as a poison, but if a large quantity of apple or peach seeds was hydrolyzed in warm acid (such as the HCl solution in the stomach), sufficient HCN would result to be dangerous. There are a few recorded instances of humans and animals being poisoned by eating large numbers of apple seeds.

The cyanide ion is one of the most rapidly acting poisons when it is in a soluble form that can dissolve in the blood. Small doses of the soluble cyanide salts sodium cyanide (NaCN) and potassium cyanide (KCN) can be lethal and will act in seconds to minutes when taken orally. Gaseous HCN also dissolves quickly in the blood, where it forms CN^- ions. The cyanide ion poisons by the asphyxiation of cells that require oxygen, as does carbon monoxide, but the mechanism of cyanide poisoning is different. Instead of preventing the cells from getting oxygen, cyanide interferes with oxidative enzymes, such as *cytochrome oxidase,* which is found in all cells that use oxygen. Cytochrome oxidase contains an iron ion that is oxidized from Fe^{2+} to Fe^{3+} to provide electrons for the reduction of O_2. The iron regains electrons in other steps in the overall process. If a cyanide ion collides with the cytochrome oxidase molecule at the site of the iron ion, a strong bond forms between the CN^- ion and the iron ion.

$$\underset{\text{Enzyme molecule}}{\text{Cytochrome oxidase(Fe)}(aq)} + \underset{\text{Cyanide ion}}{CN^-(aq)} \longrightarrow \underset{\text{Bound enzyme molecule}}{\text{cytochrome oxidase(Fe)} \cdot CN^-(aq)}$$

■ Cytochrome oxidase is one of several enzymes for the final production of ATP from glucose (see Figure 17.3).

Once the iron ion in the enzyme becomes bound by a cyanide ion, the enzyme is incapable of taking part in the reaction. Plenty of oxygen continues to get to the cells, but the mechanism by which the oxygen is used in the support of life is stopped. Hence the cell dies, and if this occurs fast enough in the vital centers, the victim dies.

The body has a mechanism for ridding itself of cyanide ions, but the reaction is relatively slow. Because the reaction between the cytochrome oxidase molecule and the cyanide ion is reversible, given time, cyanide ions that were bound to enzyme molecules will break away and go into solution. While these ions are in solution the enzyme *rhodanese,* found in almost all cells, converts cyanide ions to relatively harmless thiocyanate ions with the help of thiosulfate ions, which are also present in small concentrations in cells.

$$\underset{\substack{\text{Cyanide}\\\text{ion}}}{CN^-(aq)} + \underset{\substack{\text{Thiosulfate}\\\text{ion}}}{S_2O_3^{2-}(aq)} \xrightarrow{\text{Rhodanese}} \underset{\substack{\text{Thiocyanate}\\\text{ion}}}{SCN^-(aq)} + \underset{\substack{\text{Sulfite}\\\text{ion}}}{SO_3^{2-}(aq)}$$

This mechanism is not as effective in protecting a cyanide-poisoning victim as it might appear, since only a limited amount of thiosulfate is available in the body at a given time. Large doses of thiosulfate can be effective in treating cyanide poisoning if administered in time. Compounds that can counteract the effects of a poison by destroying it or rendering it ineffective are called **antidotes.**

■ SELF-TEST 18A

1. Corrosive poisons such as sulfuric acid destroy tissue by __________ followed by __________ of proteins.
2. Corrosive poisons, such as ozone, nitrogen dioxide, and iodine, destroy tissue by __________ it.
3. Carbon monoxide poisons by forming a strong bond with iron in __________ and thus preventing the transport of __________ from the lungs to the cells throughout the body.
4. CO is a cumulative poison. (a) True, (b) False.
5. The cyanide ion has the formula __________. It poisons by complexing with iron in the enzyme __________, thus preventing the use of __________ in the oxidative processes in the cells.

6. Which of the following are metabolic poisons? (a) CO, (b) NaOH, (c) H_2SO_4, (d) CN^-.
7. Different species of animals, like rats and mice, have ____________ LD_{50} values for the same poison.
8. The three ways a poison can enter the body are ____________, ____________, and ____________.

18.4 Heavy Metals as Metabolic Poisons

Heavy metals are perhaps the most common of all the metabolic poisons. The heavy metals include such frequently encountered elements as lead and mercury, as well as many less common ones such as cadmium, chromium, nickel, and copper. In this group we should also include the infamous poison arsenic, which is really not a metal but is metal-like in many of its properties, including its toxic action. These metals can be acutely toxic, but they are also chronically toxic in very small quantities (a few hundred ppm or less in many cases), and most of them accumulate in the body. In addition, they have no taste or odor so they lack warning properties. It is possible to be poisoned by them and not know it until the symptoms become severe.

■ The heavy metals are considered "heavy" because they have higher atomic weights.

Arsenic

Arsenic, a classic homicidal poison, occurs naturally in small amounts in many foods. Shrimp, for example, contain about 19 ppm arsenic, and corn may contain 0.4 ppm arsenic. The amount of naturally occurring arsenic in foods depends on the surroundings where they were grown and the metabolism of the plant or animal. Some insecticides contain arsenic (Table 18.4), so small amounts of arsenic remain in many grains, fruits, and vegetables. The U.S. Food and Drug Administration (FDA) has set a limit of 0.15 mg of arsenic per pound of food (roughly 330 ppm), and this amount apparently causes no harm. In its ionic forms, arsenic is much more toxic than in its covalently bound compounds. The typical toxic arsenic compounds contain inorganic ions such as arsenate (AsO_4^{3-}) and arsenite (AsO_2^-).

■ The long-running Broadway play *Arsenic and Old Lace* parodied the use of arsenic as a homicidal poison.

■ One of the first drugs ever found effective against syphilis was arsphenamine, which contains covalently bonded arsenic. Today, there are almost no drugs in use that contain arsenic.

Arsenic and heavy metals in general owe their toxicity primarily to their ability to react with sulfhydryl (—SH) groups in enzymes, such as those involved in the production of cellular energy. For example, glutathione (a

TABLE 18.4 ■ Some Arsenic-Containing Insecticides

Name	Formula
Lead arsenate	$Pb_3(AsO_4)_2$
Monosodium methanearsenate	$CH_3—As(=O)(OH)—O^-Na^+$
Copper acetoarsenite (Paris green)	$3\,CuO \cdot 3As_2O_3 \cdot Cu(C_2H_3O_2)_2$

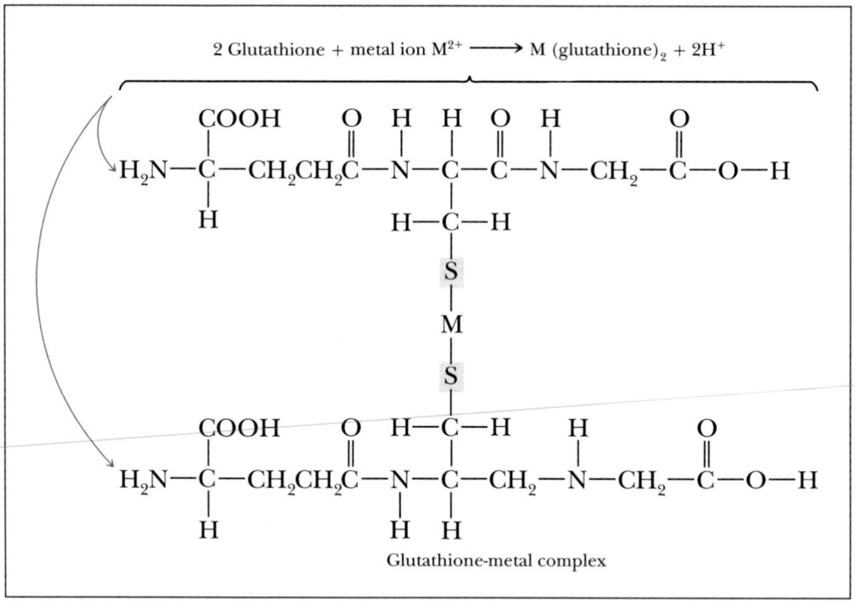

Figure 18.4 Action of a heavy metal poison. As shown in this example for glutathione, which normally aids in transport of amino acids into cells and in protection against damage by oxidizing agents, heavy metals inactivate polypeptides by bonding with their —SH groups.

tripeptide of glutamic acid, cysteine, and glycine) occurs in most tissues and is a coenzyme in, for example, reactions that maintain healthy red blood cells. The reaction of glutathione with metals illustrates their interaction with sulfhydryl groups. The metal replaces the hydrogen on two sulfhydryl groups on adjacent molecules (Figure 18.4), and the strong bond that results effectively prevents the two glutathione molecules from taking part in their normal reactions. Arsenic, like most other heavy metals, tends to accumulate in the body, so it is a cumulative poison.

■ The body can rid itself of many toxic substances if the dose is small enough and sufficient time is allowed.

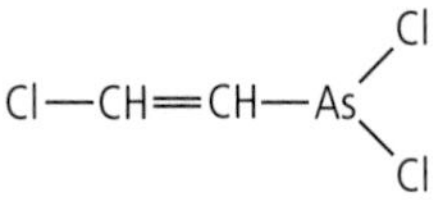

Lewisite
(2-chlorovinyl dichloroasine)

In World War I an arsenic-containing poison gas known as *Lewisite* was developed to be dispersed by shell bursts over concentrated ground troops. The effort to find a compound to counteract Lewisite led to an understanding of how arsenic acts as a poison and subsequently to the development of an antidote. Once it was understood that Lewisite poisoned people by the reaction of arsenic with protein sulfhydryl groups, scientists in England set out to find a suitable compound that contained highly reactive sulfhydryl groups that could compete with sulfhydryl groups in the enzyme for the arsenic and thus render the poison ineffective. Out of this research came a compound now known as British anti-Lewisite (BAL).

The BAL molecule, which bonds to the metal at the two—SH groups, is called a **chelating agent** (Greek *chela,* meaning "claw"), a term applied to any molecule that bonds at two or more sites to the same metal atom or metal ion. A chelating agent encases an atom or ion like a crab or an octopus surrounds a bit of food. Chelating agents are also called "sequestrants." As food additives (Section 17.9), chelating agents perform the same function as heavy metal antidotes. They tie up metal ions so that they cannot participate in other kinds of reactions. With the arsenic or heavy metal ion tied up, the sulfhydryl groups in vital enzymes are freed and can resume their normal functions. BAL is a standard therapeutic item in many hospitals' poison emergency center and is used routinely to treat heavy-metal poisoning. BAL is one of many compounds that can act as chelating agents for metals.

$$CH_2-OH \qquad\qquad CH_2-OH$$
$$| \qquad\qquad\qquad\qquad |$$
$$CH-SH \; + \; M^{2+} \longrightarrow CH-S$$
$$| \qquad\qquad\qquad\qquad | \qquad\;\; M + 2\,H^+$$
$$CH_2-SH \qquad\qquad CH_2-S$$
$$\text{BAL} \qquad \text{Heavy metal} \qquad \text{Chelated metal ion}$$
$$\text{ion}$$

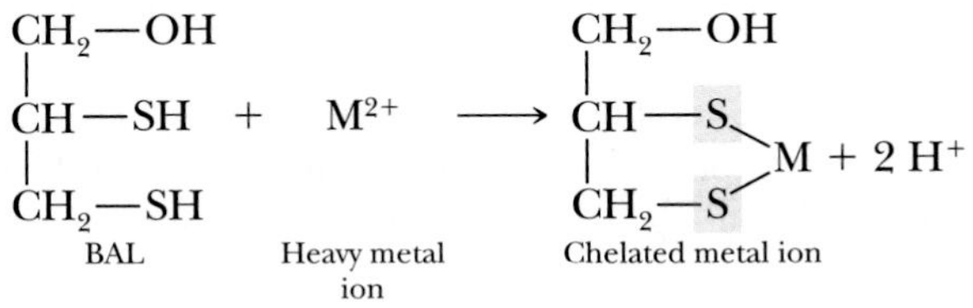

Mercury

Mercury is a fairly common heavy metal. It is a volatile, shiny, metallic-appearing liquid at room temperature. Mercury's volatility (and that of any liquid) depends on the state of subdivision of the mercury droplets. Very fine droplets will volatilize more quickly than a larger pool of mercury because of the increased surface area. Once in the air, mercury atoms can be inhaled, where they can easily be absorbed into the blood. Touching mercury also allows mercury atoms to enter the body. In the body, the metal atoms are oxidized to Hg_2^{2+} [mercury (I) ion] and Hg^{2+} [mercury (II) ion]. Both ionic forms of mercury are quite toxic, and their effects are cumulative. Mercury works like arsenic by bonding with various reactive groups in enzymes and other protein structures.

Mercury is widely used in industry and still can be found in most homes in mercurial thermometers. Mercury's liquid state produces a fascination for some people, especially children, who love to touch it. Others at risk of mercury poisoning include dentists and their assistants (who use mercury in making amalgams for fillings), various medical and scientific laboratory personnel, and some agricultural workers who apply mercury-containing fungicides.

Fluorescent lamps contain small amounts (about 60 mg) of mercury, which helps carry the electric current between the electrodes. The EPA is considering whether to regulate the disposal of used fluorescent lamps in municipal landfills because of this mercury content. In addition, many light switches found in homes and even some children's play shoes contain mercury. The shoes contain a small amount (less than 0.5 g) of mercury in a "step-sensitive" switch which turns a small lamp on and off as the child walks.

■ Mercury thermometers are being replaced by digital thermometers.

■ Amalgam: Any mixture or alloy of metals of which mercury is a constituent. Dental amalgams are silver-mercury alloys. In recent years the amount of mercury used in dentistry has decreased dramatically.

■ A broken mercury thermometer represents a special danger, since the mercury will probably be dispersed into small droplets. Zinc and mercury will readily form an amalgam, which lowers the volatility of mercury. It is possible to pick up small droplets of spilled mercury using a zinc-coated nail. Any remaining mercury bits can be removed by sprinkling them with sulfur, which forms HgS, a nonvolatile salt that can be picked up with a vacuum cleaner.

■ A vivid description of the psychological changes produced by mercury poisoning can be found in the Mad Hatter, a character in Lewis Carroll's *Alice in Wonderland*. The fur felt industry once used mercury(II) nitrate, $Hg(NO_3)_2$, to stiffen the felt used in making hats. Chronic mercury poisoning accounted for the Mad Hatter's odd behavior; it also gave the workers in hat factories symptoms known as "hatter's shakes."

Children's sandal with step-sensitive light switch. Originally such light-up sandals and sneakers contained mercury switches and ran into trouble in states with strict laws against disposal of mercury in landfills. At a high cost to itself, the manufacturer withdrew the sandals and sneakers from the market in some states, offered prepaid envelopes to send discarded shoes to a mercury recycler, and eventually ceased using the mercury switches.
(C.D. Winters)

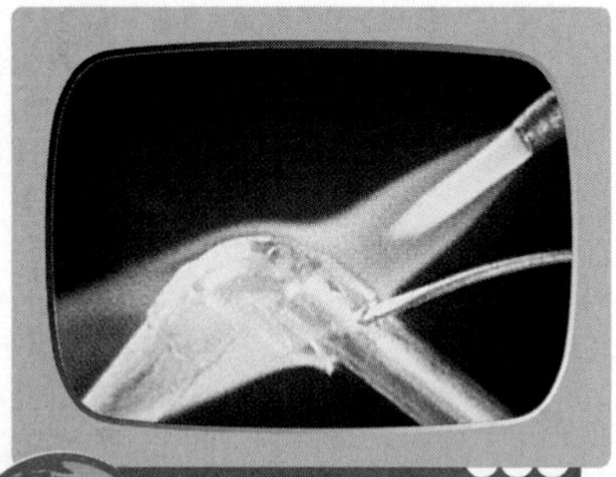

The World of Chemistry

Program 19, *Metals*

Soldering copper water pipes with lead-based solder, a common practice that has been the source of lead in drinking water.

Lead

Lead is another widely encountered poisonous heavy metal. On continuous exposure, lead tends to accumulate in the body, principally in the bones. The body's method of handling lead provides an interesting example of a "metal equilibrium"—the amount of lead entering the body by inhalation, ingestion, and skin absorption is balanced by the amount the body can routinely get rid of in the urine, feces, and perspiration. Lead often occurs in foods (100–300 μg/kg), beverages (20–30 μg/L), public water supplies (100 μg/L, from lead-sealed pipes), and even air. Until the phase-out of lead in automobile fuels in the 1980s, lead in air came primarily from automobile exhausts (up to 2.5 $\mu g/m^3$ of air). Today, because so much lead was deposited from auto exhausts over the years, lead is still found in soil samples and even on city sidewalks and streets. Dry, dusty conditions will usually result in measurable amounts of lead in air, especially in cities. With so many sources and possible doses per day, it is obvious that the body must be able to rid itself of this poison; otherwise everyone would have died long ago of lead poisoning! The average person can excrete about 2 mg (2000 μg) of lead a day through the intestinal tract and kidneys, and fortunately one's daily intake is normally less than this. However, if intake exceeds this amount, accumulation and storage result. In the bones lead acts on the bone marrow. In the soft tissues lead behaves like other heavy-metal poisons, such as mercury and arsenic, reacting with sulfhydryl and other reactive groups in various proteins. Lead, like mercury and arsenic, can also affect the central nervous system.

■ An unusual case of lead poisoning involved a fisherman who chewed on lead "sinkers" while he fished.

Unless they are very insoluble, lead salts are always toxic. But even metallic lead can be absorbed through the skin; cases of lead poisoning have resulted from repeated handling of lead foil, bullets, and other objects made of lead, such as pieces of art or toy soldiers.

Lead compounds have long been used in paints because they are quite dense, which gives the paint a certain "thickness," making it easier to apply. Lead carbonate is a chalk-white solid; lead oxide is red, and lead chromate is bright yellow. Each of these lead compounds, which are pictured on page 571, has been used in many kinds of paints, including those for interior and exterior house painting.

One of the truly tragic unintended consequences of the use of lead in paints is the *lead poisoning* of small children. Even though lead paints have not been used for interior painting in this country during the past 30 years, children are still poisoned by lead from old paint. Health experts estimate that up to 225,000 children become ill from lead poisoning each year, with many experiencing mental retardation or other neurological problems. The reason for this is twofold. Lead-based paints still cover the walls of many older, substandard dwellings. In addition, many children living in such housing are ill-fed and anemic. These children develop a peculiar appetite trait called **pica,** and among the items that satisfy their cravings are pieces of flaking paint, which may contain lead. Lead salts also have a sweet taste, which may contribute to this consumption of lead-based paint.

Children retain a larger fraction of absorbed lead than do adults, and children do not immediately tie up absorbed lead in their bones as adults do. This inability to absorb lead quickly into their bones means the lead stays in a child's blood longer, where it can exert its toxic effects on various organs. Table 18.5 shows the effects of lead in children's blood.

TABLE 18.5 ■ Effects of Lead in Children's Blood

Blood Lead Levels (μg/dL)*	Acute Effects	Chronic Effects
~5	None	Elevated blood pressure
~10	None	Lowered intelligence
15–25	None	Decreased heme synthesis, decreased vitamin D synthesis and calcium metabolism
25–40	None	Impaired central nervous system functions, delayed cognitive development, reduced IQ scores, impaired hearing, reduced hemoglobin formation
40–80	Peripheral nervous system damage	Anemia
>80	Convulsions, coma, possible death	Irreversible mental retardation

*Blood lead levels are measured in μg per deciliter (dL), which is 1/10 of a liter, or 100 mL.

■ 1 μg (microgram) = 10^{-6} g = 1/1,000,000 g
1 mg (milligram) = 10^{-3} g = 1/1000 g

Studies in the mid-1980s by the U.S. Department of Health, Education, and Welfare showed that about 1.5 million children have blood levels above 15 μg/dL (micrograms per deciliter of blood—a deciliter is $\frac{1}{10}$ of a liter, or 100 mL). About 900,000 children have blood levels above 20 μg/dL, and because of this many communities have outreach programs to detect lead poisoning and inform parents (Figure 18.5). In 1991 the Centers for Disease Control (CDC) decreased the **intervention level** for lead in children's blood from 25 μg/dL to 10 μg/dL. Children who are suspected to be at risk of lead poisoning will be treated if their blood lead levels are found to be above the 10 μg/dL intervention level value. For adults in the workplace where lead exposure would be expected, the corresponding acceptable blood lead level is 40 μg/dL.

Figure 18.5 The office of the Lead Poisoning Prevention Program in Chatham County (Savannah), Georgia.

One of the principal sources of lead is drinking water that has contacted lead-containing pipes, joints, and plumbing fixtures. In 1993 the EPA released a list of public water supplies that exceeded its maximum allowable level of 15 ppb (parts per *billion*) lead. Hundreds of cities and towns were on the EPA list, and some had lead levels as high as 484 ppb. Based on these findings, the EPA estimated that as many as one in every six American children under six years of age has a blood lead level above the 10 μg/dL acceptable level.

■ Lead has historically been used in plumbing. The Romans used lead pipes to carry water into their homes.

EXAMPLE 18.2 *Using ppb Concentration Units*

The concentration of lead (Pb) in tap water is found to be 25 ppb. What would be your dose of lead if you drink 2.5 liters of this water?

SOLUTION

The density of water is 1 g/mL near room temperature. This means 1 L of water weighs 1000 g (1 g/mL × 1000 mL/L), so 2.5 L would weigh 2500 g.

Now 25 ppb lead means that for every billion parts (of water) there are 25 parts of lead, or for every billion (10^9) grams of solution there are 25 grams of lead. This information can be written as a factor that will be multiplied by the actual volume of water consumed.

$$25 \text{ ppb} = \frac{25 \text{ g Pb}}{10^9 \text{ g water}}$$

Using the factor given above and canceling units,

$$2500 \text{ g water} \times \left(\frac{25 \text{ g Pb}}{10^9 \text{ g water}}\right) = 63 \times 10^{-6} \text{ g Pb, or } 63\ \mu\text{g Pb}$$

Exercise 18.2

If the concentration of lead in drinking water is 100 ppb, how much lead would be consumed if you drink 1 liter of this water?

An effective chelating agent for removing lead from the human body is ethylenediaminetetraacetic acid, also called EDTA (see Figure 17.7 in Section 17.9). The calcium disodium salt of EDTA $Na_2[Ca(EDTA)]$ is usually used in the treatment of lead poisoning because EDTA by itself would remove too much of the blood serum's calcium.

$$EDTA^{4-}\ (aq) + Ca^{2+}(aq) \longrightarrow Ca(EDTA)^{2-}(aq)$$

TABLE 18.6 ■ Toxicity and Symptoms of Metal Exposure to Humans

Metal	Source of Exposure	Acute Effects	Chronic Effects
Arsenic	Occupational, pesticides, foods, drinking water	Fever, skin lesions, multiorgan effects	Liver injury, cancer
Cadmium	Occupational, food, air	Nausea and vomiting, chemical pneumonia	Kidney damage, pulmonary disease, hypertension, cancer
Chromium	Occupational, food	Respiratory irritation	Irritability, skin reactions, cancer
Copper	Occupational, water	Nausea, vomiting, liver damage	Wilson's disease
Lead	Occupational, food, water, paint, lead objects such as fishing weights	Vomiting, spasms	Neurological effects, anemia, kidney damage
Mercury	Occupational, food, air, pesticides	Bronchitis, bloody diarrhea, kidney damage	Neurological effects, kidney damage
Nickel	Occupational	Headache, nausea, fever, vomiting	Dermatitis, cancer
Zinc	Occupational, food, water, air	Abdominal distress, diarrhea, fever, and chills	Pulmonary effects

Adapted from the *McGraw-Hill Encyclopedia of Science and Technology,* 7th ed., 1992.

In solution, EDTA has a greater tendency to complex with lead (Pb^{2+}) than with calcium (Ca^{2+}). As a result, the calcium is released, and the lead is tied up in the complex:

$$[Ca(EDTA)]^{2-} + Pb^{2+} \longrightarrow [Pb(EDTA)]^{2-} + Ca^{2+}$$

The lead EDTA compound is then excreted in the urine.

Because metals have so many everyday uses, workers and consumers alike are at risk of metal poisoning. Table 18.6 summarizes the acute and chronic effects of several heavy metals. Of the common metals, only aluminum and magnesium show no toxic effects at low doses.

SELF-TEST 18B

1. The heavy metal poison ____________ is present in fluorescent lamps, dental amalgams, and some home light switches.
2. The classic homicidal poison is ____________.
3. Lead is a cumulative poison. (a) True, (b) False.
4. Almost everyone has some lead in their blood. (a) True, (b) False.
5. The CDC intervention level for lead in a child's blood is ____________ μg/dL.
6. BAL is an antidote for ____________. BAL is effective because its sulfhydryl (—SH) groups ____________ arsenic and heavy metals and render them ineffective toward enzymes.
7. Mercury is a cumulative poison. (a) True, (b) False.
8. Name three ways children can be exposed to lead.
9. Two metals that can pass through the skin and cause toxic effects are ____________ and ____________.

18.5 Neurotoxins

Some poisons, known as **neurotoxins,** are known to limit their action to the nervous system. These include poisons such as strychnine (a rat poison used for more than 500 years) and curare (a South American Indian dart poison), as well as the dreaded nerve gases developed for chemical warfare. The exact modes of action of most neurotoxins are not known for certain, but investigations have discovered the action of a few and in the process have taught us more about how the nervous system works.

A nerve impulse is transmitted along a nerve fiber by an electrical impulse carried by the movement of ions. Each nerve fiber leads either to another nerve fiber or to a gland or muscle that will be stimulated by the electrical impulse. Between one nerve fiber and the next is a gap (called a **synapse**). The impulse is carried across the synapse, in a manner illustrated in Figure 18.6, by a messenger molecule known as a *neurotransmitter* that binds to a receptor on the adjacent nerve. Neurotoxins affect the transmission of nerve impulses by interfering with neurotransmitter function in a variety of ways.

One of the best-studied neurotransmitters is acetylcholine, which acts in the brain, the spinal cord, and throughout the body at the connection between nerves and muscles. Very quickly after an acetylcholine molecule

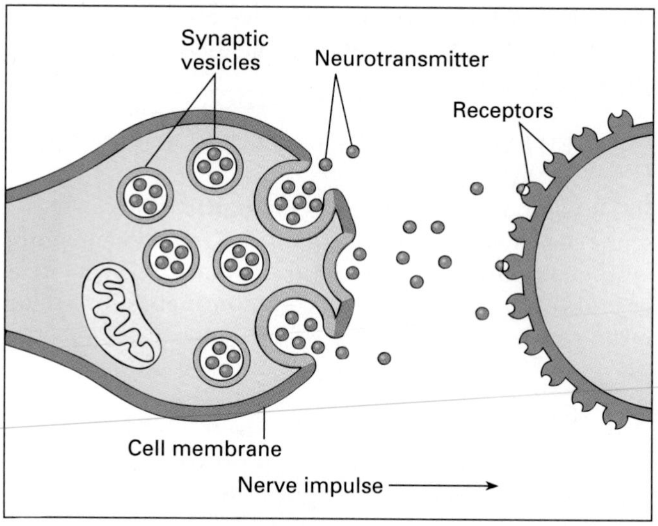

Figure 18.6 Transmission of a nerve impulse between two nerve fibers. The neurotransmitter is stored in *vesicles* until it is needed. When a nerve impulse arrives, the vesicles move to the outer membrane of the nerve cell and join with it so that the neurotransmitter is released. The neurotransmitter crosses the gap and binds to receptors on the surface of the adjacent nerve cell. The receptors in turn initiate chemical changes in that cell that allow the impulse to proceed.

connects with its receptor, it is broken down by the enzyme acetylcholinesterase into acetic acid and choline, freeing the receptor to receive the next impulse.

$$\underset{\text{Acetylcholine}}{CH_3\overset{\overset{O}{\|}}{C}OCH_2CH_2\overset{+}{N}(CH_3)_2{-}CH_3\ OH^-} + H_2O \xrightarrow[\text{cholinesterase}]{\text{Acetyl-}} \underset{\text{Acetic acid}}{CH_3\overset{\overset{O}{\|}}{C}{-}OH} + \underset{\text{Choline}}{HOCH_2CH_2\overset{+}{N}(CH_3)_2{-}CH_3\ OH^-}$$

The choline is taken up by the nerve cell from which it came, where it is converted back into acetylcholine for the next firing of the nerve.

$$\text{Acetic acid} + \text{choline} \longrightarrow \text{acetylcholine} + H_2O$$

One group of neurotoxins, known as **anticholinesterase poisons,** prevent the breakdown of acetylcholine by deactivating cholinesterase. These poisons are usually structurally analogous to acetylcholine, allowing them to bond firmly to the enzyme acetylcholinesterase and preventing it from bonding to acetylcholine. Normal nerve function must await the manufacture of new acetylcholinesterase. In the meantime, the excess acetylcholine overstimulates the nerves leading to glands and muscles, which can cause excessive hormone release, irregular heart rhythms, and even convulsions and death when the doses are high. Two common organic phosphates, parathion and malathion,

TABLE 18.7 ■ Organophosphates Used in Chemical Warfare

Name	Structure	Comment
Sarin	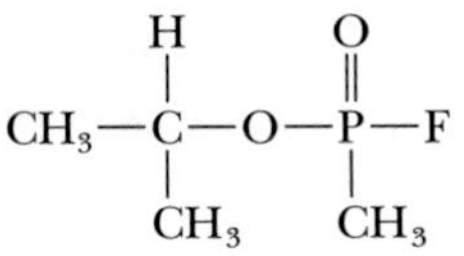	Also known as agent GB; LD_{50}(rat) = 0.55 mg/kg
Soman	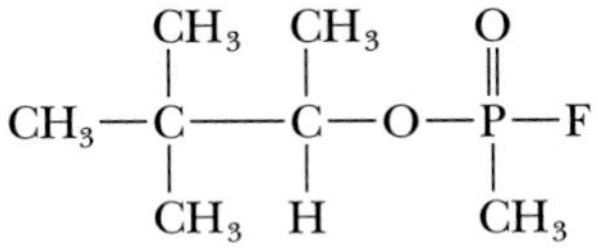	Also known as agent GD; LD_{50}(mice) = 0.78 mg/kg
Tabun	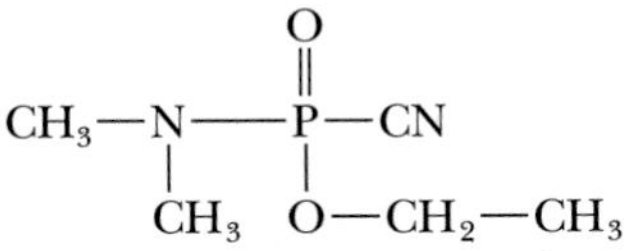	Also known as agent GA; LD_{50}(rat) = 3.7 mg/kg

used as insecticides can act as anticholinesterase poisons, as can the insecticide carbaryl (trade name Sevin).

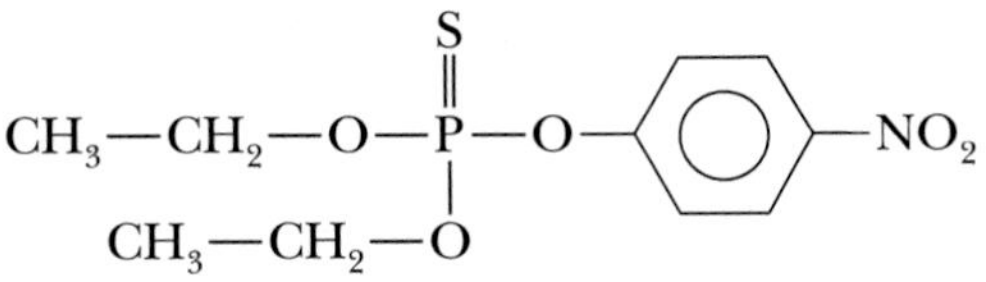

Parathion

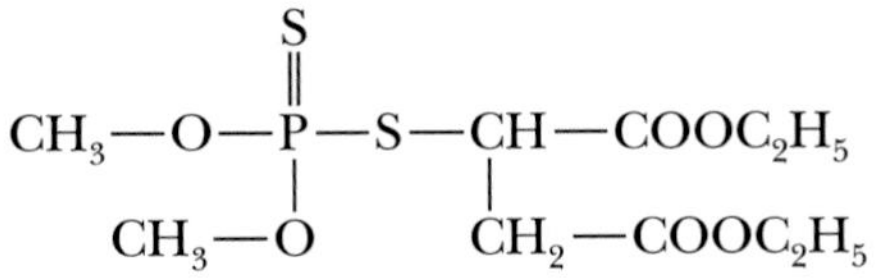

Malathion

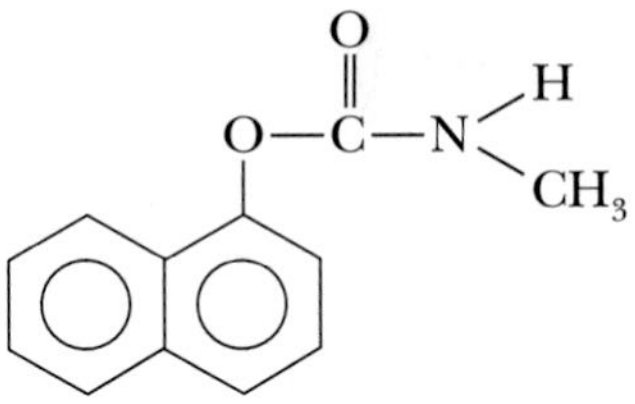

Carbaryl

Minor changes in the molecular structures of the organic phosphate pesticides result in compounds that bind very strongly to acetylcholinesterase. This makes these compounds much more toxic than parathion or malathion. In fact, these organic phosphate compounds have been used as chemical warfare agents (Table 18.7).

In addition to chemicals that can interfere with acetylcholinesterase, there are other classes of compounds that interfere with the normal actions of acetylcholine. Many plants produce nitrogen-containing chemicals called **alkaloids,** some of which can effectively *compete* with acetylcholine at nerve synapses. Because these compounds interfere with the normal functioning of the nervous system, they are also neurotoxins. The alkaloids atropine and curare, used by South American Indians on the tips of their hunting darts, are able to occupy the receptor sites on nerve endings that are normally occupied by the impulse-carrying acetylcholine. When atropine or curare occupies the receptor site, no stimulus is transmitted (Figure 18.7). While anticholinesterase poisons cause a buildup of acetylcholine that results in an irregular heartbeat, a decrease in blood pressure, and excessive saliva, atropine and curare produce the opposite effects: excessive thirst and dryness of the mouth and throat, a rapid heartbeat, and an increase in blood pressure.

SCIENCE AND SOCIETY

Chemical Warfare Agents

Chemical warfare is the use of toxic chemicals to kill and incapacitate enemy troops. The Greeks used choking clouds of sulfur dioxide (SO_2) caused by burning sulfur and pitch during the Peloponnesian War between Sparta and Athens (431–404 B.C.). Modern chemical warfare began in 1915 when the Germans released chlorine gas on Allied troops at Ypres, Belgium, during World War I. After the initial use of chlorine, various other gases were quickly developed and used.

In general, the gases used in World War I caused death if the victim was exposed to high enough doses, but their most significant contribution was dispersing unprotected troops as they ran from the areas of highest concentration.

After World War I, most nations agreed never to use toxic chemicals in warfare—yet development of these agents continued. During World War II, the Germans developed the anticholinesterase poisons tabun and sarin, both organophosphate compounds (Table 18.7). Their discovery led to our present-day organophosphate insecticides. Throughout World War II, war gases were available but were never used.

During the 1980s, chemical agents were used in the Iran-Iraq war against both troops and civilians. Chemical warfare agents are especially devastating against civilians, since they are not only untrained and uninformed about the effects of these chemicals, but are also unprepared to protect themselves. In 1992, during Operation Desert Storm (the Gulf War), there was great concern that chemical agents might be used by Iraq against Coalition troops. In fact, all battle troops and most support personnel were issued complete protective gear against chemical warfare agents including nerve gases and corrosive gases such as chlorine and mustard gases. Although chemical sensors sounded numerous alarms indicating that nerve gases and other agents were in the air during the Gulf War operations, officers in charge concluded that Iraq had not unleashed chemical weapons. Today, most nations, but not including Iraq, have signed the Chemical Weapons Convention, which calls for a systematic destruction of all chemical weapons stocks by 2004.

World War I Chemical Warfare Agents

Type	Example	Mode of Action
Mustard gas	Bis(2-chloroethyl)sulfide [$(C_2H_4Cl)_2S$]	Produces strong blisters on skin, destroys lungs
Choking gas	Chlorine [Cl_2]	Destroys lungs and upper respiratory tract
Blood gas	Arsine [AsH_3]	Destroys vital enzymes
	Hydrogen cyanide [HCN]	Cell death

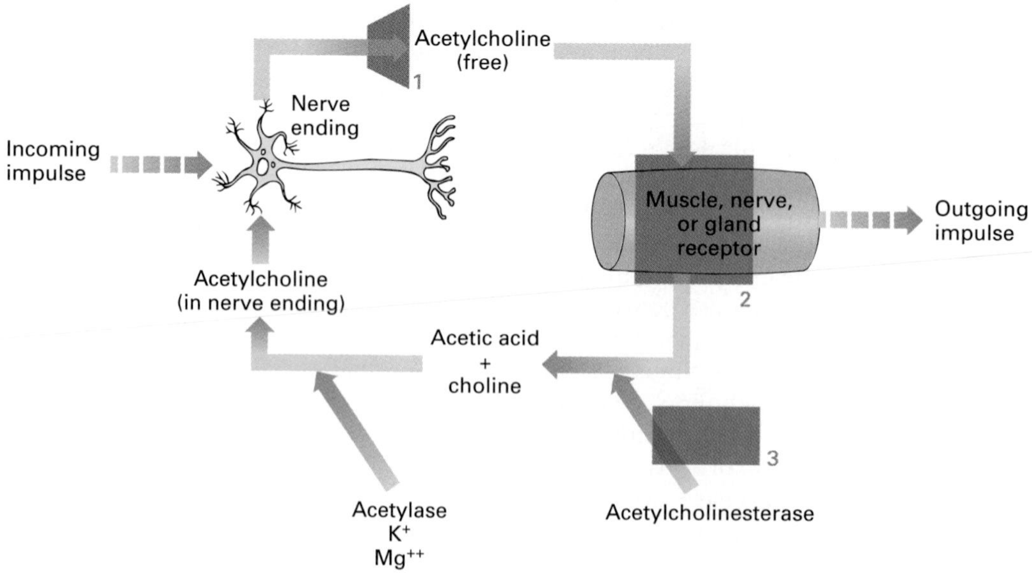

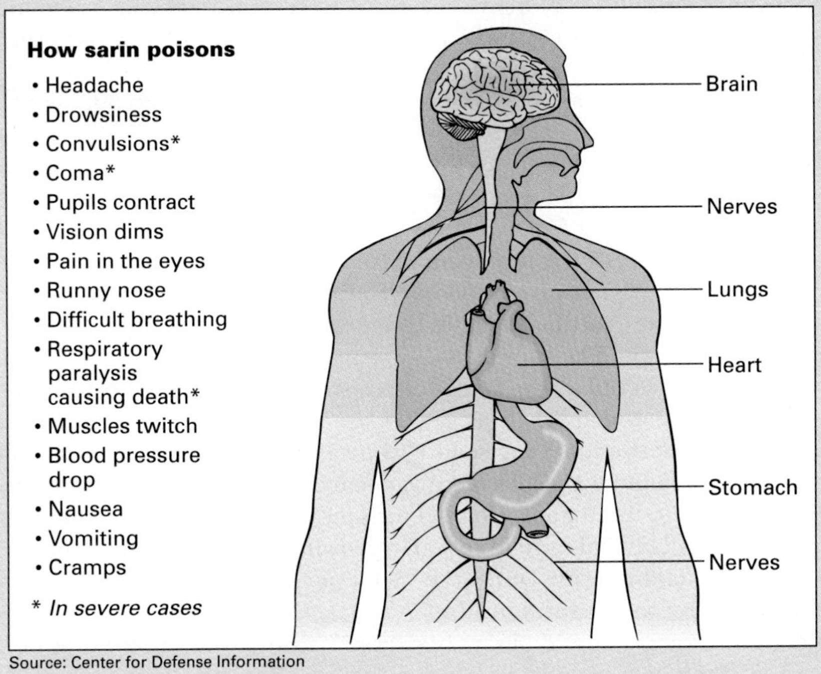

Effects of sarin.

The frightening prospect of chemical warfare agents being used by terrorists against civilians became a reality in March 1995 in Tokyo, Japan. Aerosol clouds of the nerve gas sarin were simultaneously released in several subway cars during rush hour, killing ten passengers and injuring about 5000 others to an extent that required hospitalization. Many of the victims complained of loss of vision, which is a result of the contraction of the pupils as the acetylcholinesterase poison acts on the muscles controling pupil dilation.

The group, a religious cult, found to be responsible for this act was discovered to have a secret laboratory near Tokyo where sarin and similar compounds could be prepared from relatively simple starting materials such as sodium fluoride and phosphoric acid. Although we will probably never know exactly why sarin was used that day in Japan, it was thought that the releases might have been timed to coincide with the arrival of the subway trains at a station where many of Japan's government leaders were arriving for a day's work in Parliament. Unfortunately, of all the agents of mass destruction, chemicals agents such as the nerve gases seem to be the easiest and least expensive to manufacture. So the clandestine manufacture of sarin and other nerve gases becomes yet another worry for law enforcement officers as well as for ordinary citizens everywhere.

Some of these alkaloids are extremely useful in medicine. For example, atropine is used to dilate the pupil of the eye to facilitate examination of its interior. Applied to the skin, atropine sulfate and other atropine salts relieve pain by deactivating sensory nerve endings on the skin. Atropine is also used as an antidote for anticholinesterase poisons, since it acts to counteract the effect of too much acetylcholine. Curare has long been used as a muscle relaxant.

◀ **Figure 18.7** Effect of poisons on nerve impulse transmission by the neurotransmitter, acetylcholine. (1) Botulinus and dinoflagellate toxins inhibit the synthesis or release of acetylcholine from the nerve ending. (2) Curare and atropine block the receptor so that the nerve impulse cannot be transmitted. (3) Normally, the enzyme acetylcholinesterase breaks down acetylcholine at the synapse so that the next impulse can be transmitted. (The choline is then taken up at the nerve ending and acetylcholine is resynthesized.) Anticholinesterase poisons inactivate acetylcholinesterase.

18.6 Teratogens

The effects of chemicals on human reproduction are a frightening aspect of toxicity. Chemical agents that can cause birth defects are called **teratogens.** The root word *terat* comes from the Greek word meaning "monster." In addition to chemical substances, high-energy radiation and some viral agents are known teratogens.

Birth defects occur in 2% to 3% of all births. About 25% of these occur from genetic causes, 5% to 10% are the result of known teratogens, and the remaining 60% or so result from unknown causes.

In the development of the newborn, there are three basic periods during which the fetus is at risk. For a period of about 17 days between conception and implantation of the fertilized egg in the uterine wall, a chemical "insult" will result in cell death. The rapidly multiplying cells often recover, but if a lethal dose is administered, death of the organism occurs, followed by spontaneous abortion or reabsorption.

During the critical embryonic stage (18 to 55 days after fertilization) *organogenesis,* or development of the organs, occurs. At this time the embryo is extremely sensitive to teratogens. Contact with teratogens results in reduction of cell size and number. This is manifested in growth retardation and failure of vital organs to reach maturity. During the fetal period (56 days to term), the fetus is less sensitive to chemical insults.

A pregnant woman should always be advised throughout the term of her pregnancy to limit her exposure to chemicals of unknown toxicity, any of which could be harmful to the developing child. This is especially true during the 18th through 55th days. She should take no drugs or medicines except on the advice of her physician, and she should avoid the use of alcohol and tobacco.

In 1961 worldwide attention was focused on chemically induced birth defects when it was discovered that thalidomide, a tranquilizer and sleeping pill, caused gross deformities (flipperlike or shortened arms, missing arms or legs, and other defects) in children whose mothers used this drug during the first two months of pregnancy. Actually, it was only one enantiomer (see Section 15.1) of the thalidomide molecule that caused the deformities. The use of this drug resulted in more than 4000 surviving malformed babies in West Germany, more than 1000 in Great Britain, and about 20 in the United States. With shattering impact, this incident demonstrated that a compound can appear to be remarkably safe on the basis of animal studies (so safe, in fact, that thalidomide was sold in West Germany without prescription) and yet cause catastrophic effects in humans.

■ The United States was spared large-scale effects of thalidomide by Dr. Francis O. Kelsey of the U.S. Food and Drug Administration. She refused approval of the drug for legal sale in this country.

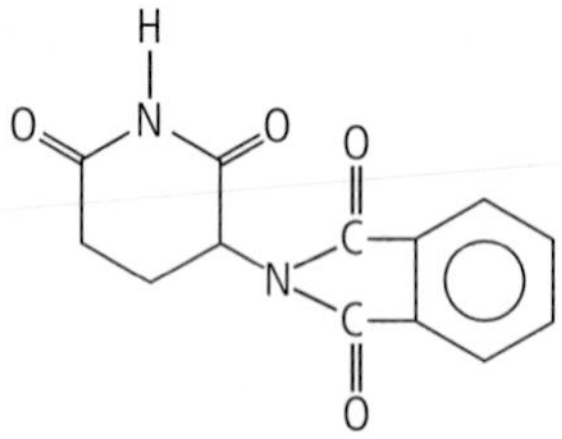

Thalidomide

Recently, thalidomide has found some new uses. Although banned in the United States, thalidomide is still being manufactured in Brazil, where it has been shown to combat the effects of diseases such as leprosy and Behçet's disease, which produces ulcers that eat away the linings of the mouth. Thalidomide has even been used by people with AIDS as a cure for mouth ulcers.

Any chemical substance that can cross the placenta is a potential teratogen, and any activity that introduces harmful chemicals into the mother's blood might prove dangerous for the health and well-being of a fetus. Smoking a cigarette results in higher-than-normal blood levels of such substances as carbon monoxide, hydrogen cyanide, cadmium, nicotine, and benzo(α)py-

TABLE 18.8 ■ Teratogenic Substances

Substance	Species	Effect on Fetus
	Metals	
Arsenic	Mice Hamsters	Increase in males born with eye defects, renal damage
Cadmium	Mice	Abortions
	Rats	Abortions
Cobalt	Chickens	Eye, lower limb defects
Gallium	Hamsters	Spinal defects
Lead	Humans Rats Chickens	Low birth weights, brain damage, stillbirth, early and late deaths
Lithium	Primates	Heart defects
Mercury	Humans	Minamata disease (Japan)
	Mice	Fetal death, cleft palate
	Rats	Brain damage
Thallium	Chickens	Growth retardation, abortions
Zinc	Hamsters	Abortions
	Organic Compounds	
DES (diethylstilbestrol)	Humans	Uterine anomalies
Caffeine (15 cups per day equivalent)	Rats	Skeletal defects, growth retardation
PCBs (polychlorinated biphenyls)	Chickens	Central nervous system and eye defects
	Humans	Growth retardation, stillbirths

rene. Of course, many of these substances are present in polluted air as well. Table 18.8 lists a number of chemical substances known to be teratogenic in humans and laboratory animals.

18.7 Mutagens

Mutagens are chemicals capable of altering the structure of DNA, which composes the genes (and, in turn, the chromosomes) and is responsible for transmitting the traits of parent to offspring (Section 15.8). Although many chemicals are under suspicion because of their mutagenic effects on laboratory animals, it should be emphasized that no one has yet shown conclusively that any chemical induces mutations in human germinal cells. Part of the difficulty of determining the effects of mutagenic chemicals in humans is the extreme rarity of mutation. A specific genetic disorder may occur as infrequently as only once in 10,000 to 100,000 births. Therefore, to obtain meaningful statistical data, a carefully controlled study of the entire population of the United States would be required. In adition, the very long time between generations presents great experimental difficulties, and there is also the problem of tracing a medical disorder to a single specific chemical out of the tens of thousands of chemicals with which we come in contact.

■ Mature sex, or germinal, cells of humans normally have 23 chromosomes; body, or somatic, cells have 23 *pairs* of chromosomes.

TABLE 18.9 ■ Mutagenic Substances as Indicated by Experimental Studies on Plants and Animals

Substance	Experimental Results
Aflatoxin (from mold, *Aspergillus flavus*)	Mutations in bacteria, viruses, fungi, parasitic wasps, human cell cultures, mice
Benzo(α)pyrene (from cigarette and coal smoke)	Mutations in mice
Caffeine	Chromosome changes in bacteria, fungi, onion root tips, fruit flies, human tissue cultures
Captan (a fungicide)	Mutagenic in bacteria and molds; chromosome breaks in rats and human tissue cultures
Chloroprene	Mutagenic in male sex cells; results in spontaneous abortions
Dimethyl sulfate (used extensively in chemical industry to methylate amines, phenols, and other compounds)	Methylates DNA base guanine; potent mutagen in bacteria, viruses, fungi, highter plants, fruit flies
LSD (lysergic acid diethylamide)	Chromosome breaks in somatic cells of rats, mice, hamsters, white blood cells of humans and monkeys
Maleic hydrazide (plant growth inhibitor; trade names Slo-Gro®, MH-30®)	Chromosome breaks in many plants and in cultured mouse cells
Mustard gas (dichlorodiethyl sulfide)	Mutations in fruit flies
Nitrous acid (HNO_2)	Mutations in bacteria, viruses, fungi
Ozone (O_3)	Chromosome breaks in root cells of broadleaf plants
Solvents in glue (glue sniffing) (toluene, acetone, hexane, cyclohexane, ethyl acetate)	4% more human white blood cells showed breaks and abnormalities (6% versus 2% normal)
TEM (triethylenemelamine) (anticancer drug, insect chemosterilants)	Mutagenic in fruit flies, mice

If there is no direct evidence for specific mutagenic effects in human beings, why then is there so much interest in the subject? Because the possibility of a deformed human race is frightening, and the evidence for chemical mutation in plants and lower animals is already established. Many kinds of chemicals are known to alter chromosomes and to produce mutations in rats, worms, bacteria, fruit flies, and other plants and animals. Some of these are listed in Table 18.9.

Experimental work on the chemical basis of the mutagenic effects of nitrous acid (HNO_2) has illustrated much about how chemical mutagens work.

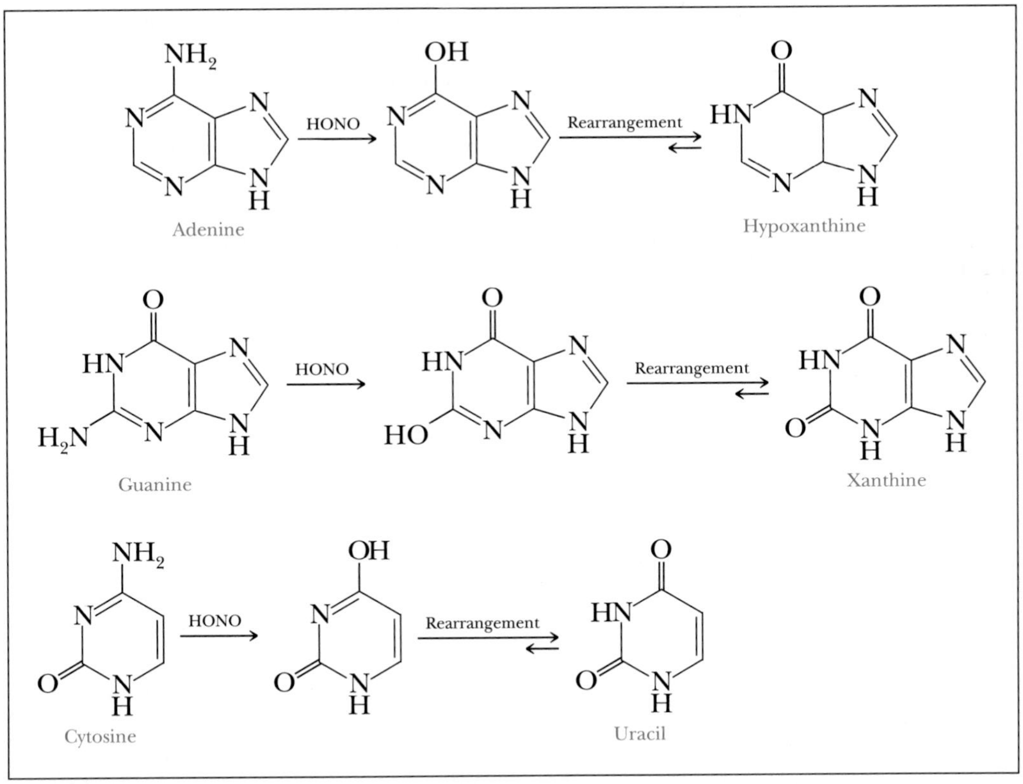

Figure 18.8 The reaction of nitrous acid (HNO_2, or HONO) with the NH_2 groups in the bases of DNA.

Repeated studies have shown that nitrous acid is a potent mutagen in bacteria, viruses, molds, and other organisms. In the early 1950s it was demonstrated experimentally that nitrous acid attacks DNA. Specifically, nitrous acid reacts with the adenine, guanine, and cytosine bases of DNA by removing the amino group of each of these compounds. The eliminated group is replaced by an oxygen atom (Figure 18.8). The changed bases may garble a part of DNA's genetic message, and in the next replication of DNA, the new base may not form a base pair with the proper nucleotide base.

Do all these findings mean that nitrous acid is mutagenic in humans? Not necessarily. We do know that sodium nitrite ($NaNO_2$), the sodium salt of nitrous acid, has been widely used as a preservative, color enhancer, or color fixative in meat and fish products for at least the past 30 years. So human exposures to this mutagen have been occurring throughout several generations. Sodium nitrite is currently used in such foods as frankfurters, bacon, smoked ham, deviled ham, bologna, Vienna sausage, smoked salmon, and smoked shad. Sodium nitrite is converted to nitrous acid by hydrochloric acid in the human stomach:

$$\underset{\text{Sodium nitrite}}{NaNO_2(aq)} + \underset{\text{Hydrochloric acid}}{HCl(aq)} \longrightarrow \underset{\text{Nitrous acid}}{HNO_2(aq)} + \underset{\text{Sodium chloride}}{NaCl(aq)}$$

■ The process of frying bacon produces nitrosamines ($R_2N{-}N{=}O$), which are known carcinogens.

$$\underset{\text{Nitrous acid}}{HNO_2} + \underset{\text{Amine (found in bacon)}}{R_2NH} \longrightarrow \underset{\text{Nitrosamine}}{R_2N{-}N{=}O} + H_2O$$

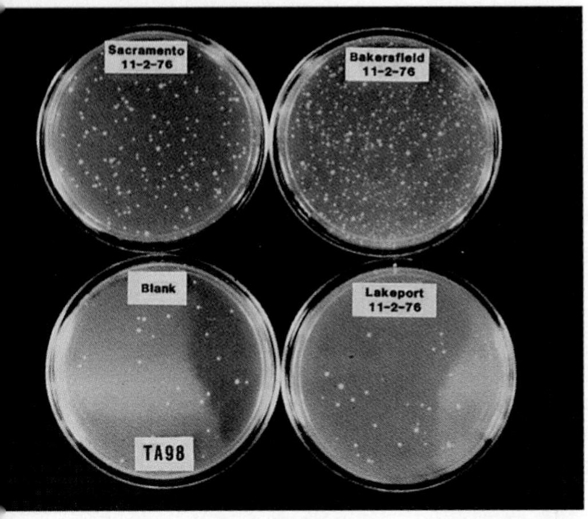

Figure 18.9 The effect of mutagens in the air in various California cities. (© American Chemical Society, from P. Flessel et al.: "Ames Testing for Mutagens and Carcinogens in Air." *J. Chemical Education,* **64,** 391–395, 1987.

The FDA now considers the mutagenic effects of nitrous acid in lower organisms sufficiently ominous to suggest strongly that the use of sodium nitrite in foods be severely curtailed, and a complete ban of this use of sodium nitrite is being considered. Several European countries already restrict the use of sodium nitrite in foods. The concern is that this compound, after being converted in the body to nitrous acid, may cause mutation in somatic cells (and possibly in germinal cells) and thus could possibly produce cancer in the human stomach. Other scientists doubt that nitrous acid is present in germinal cells and, therefore, seriously question whether this compound could be a cause of genetically produced birth defects in humans. The uncertainty of extrapolating results obtained in animal studies to human beings hovers over the mutagenic substances. Meanwhile, sodium nitrate continues to be used as a preservative because it is the most effective agent for preventing the growth of the deadly organism that causes botulism. Since vitamin C has been found to block the harmful effects of nitrites, it is added to many nitrite-containing foods.

In the early 1980s, Professor Bruce Ames and his colleagues at the University of California, Berkeley, developed a simple test that can identify chemicals capable of causing mutations in sensitive strains of bacteria. In this test, about 100 million bacteria unable to synthesize the amino acid histidine are mixed in an agar suspension with a suspected mutagenic chemical. This mixture is then added to a hard agar gel containing salts and glucose, and incubated in a petri dish for several days. If the suspected chemical is a mutagen, some of the histidine-requiring cells mutate and the biosynthesis of histidine resumes. The growth of these bacterial colonies can be seen in the petri dish (Figure 18.9).

The Ames test has utility in identifying not only mutagenic chemicals, but potential carcinogenic chemicals as well, since mutagenic chemicals are often carcinogenic. In studies involving hundreds of chemicals, nearly four out of every five animal carcinogens have been found to be mutagenic in the Ames test.

18.8 Chemical Carcinogens

Every cancer comes from a single cell—one that is a modification of a normal cell. A normal cell functions according to directions stored in its genetic data bank, the DNA. Normally, when a cell divides, each new cell gets its own exact copy of the parent DNA. When anything disrupts this DNA replication process, the genetic code in one of the descendent cells may cause that cell to grow and function differently from a normal cell. **Carcinogens** are chemicals that cause cancer, an abnormal, out-of-control cell growth that manifests itself in at least three ways: (1) The *rate of cell growth* (that is, the rate of cellular multiplication) in cancerous tissue differs from the rate in normal tissue. Cancerous cells may divide more rapidly or more slowly than normal cells. (2) Cancerous cells *spread to other tissues;* they know no bounds. Normal liver cells divide and remain a part of the liver. Cancerous liver cells may leave the liver and be found, for example, in the lung. (3) Most cancer cells show *partial or complete loss of specialized functions.* Although located in the liver, cancerous cells no longer perform the functions of the liver.

TABLE 18.10 ■ Some Industrial Chemicals that are Carcinogenic for Humans

Compound	Formula	Use or Source	Site Affected
		Inorganic Compounds	
Arsenic (and compounds)	As*	Insecticides, alloys	Skin, lungs, liver
Asbestos	$Mg_6(Si_4O_{11})(OH)_6$	Brake linings, insulation	Respiratory tract
Beryllium	Be	Alloy with copper	Bone, lungs
Cadmium	Cd	Metal plating	Kidney, lungs
Chromium	Cr*	Metal plating	Lungs
Nickel	Ni*	Metal plating	Lungs, nasal sinus
		Organic Compounds	
Benzene	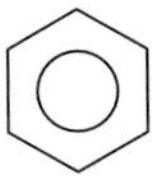	Solvent, chemical intermediate in syntheses	Blood (leukemia)
Acrylonitrile	$CH_2{=}CH(CN)$	Monomer	Colon, lungs
Carbon tetrachloride	CCl_4	Solvent	Liver
Diethylstilbestrol	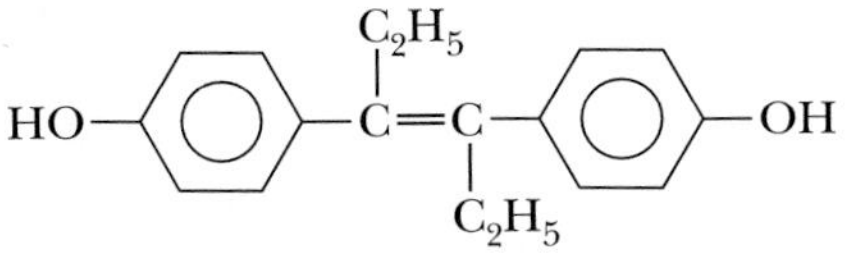	Hormone	Female genital tract
Benzo(α)pyrene	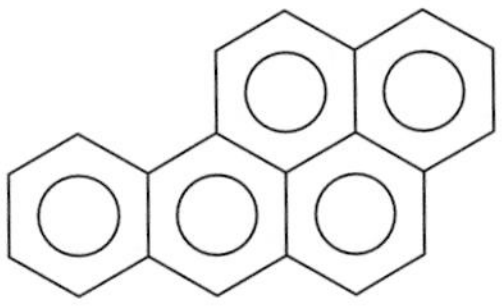	Cigarette and other smoke	Skin, lungs
Benzidine	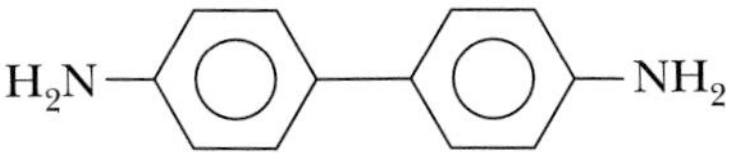	Dye manufacture, rubber compounding	Bladder
Ethylene oxide	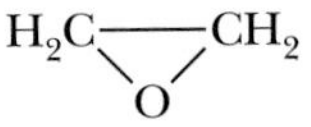	Chemical intermediate used to make ethylene glycol, surfactants	Gastrointestinal tract
Soots, tar, and mineral oils		Roofing tar, chimney soot, oils of hydrocarbon nature	Skin, lungs, bladder
Vinyl chloride	$CH_2{=}CHCl$	Monomer for making PVC	Liver, brain, lungs, lymphatic system

*Certain compounds or oxidation states only.

Carcinogenesis

Attempts to determine the chemical causes of cancer have evolved from early studies in which the disease was linked to a person's occupation. We now know that a person's life style plays a role as well. In 1775, Dr. Percivall Pott, an English physician, first noticed that people employed as chimney sweeps had a higher rate of skin cancer than the general population. It was not until 1933 that benzo(α)pyrene, $C_{20}H_{12}$ (an aromatic hydrocarbon containing five fused carbon rings; see Table 18.10), was isolated from coal dust and shown to be metabolized in the body to produce one or more carcinogens. In 1895 a German physician noted three cases of bladder cancer, not in a random population, but in employees of a factory that manufactured dye intermediates in the Rhine Valley. Rehn attributed these cancers to his patients' occupation. These and other cases confirmed that at times as many as 30 years would pass between the time of the initial employment (beginning of the exposures) and the occurrence of bladder cancer. The principal product of these dye factories was a chemical named aniline. Although aniline was first thought to be the carcinogenic agent, it was later shown that continuous long-term exposure to 2-naphthylamine, one of the dye intermediates, produced bladder cancer in dogs. Since then, other dye intermediates that are structurally similar to 2-naphthylamine have also been shown to be carcinogenic.

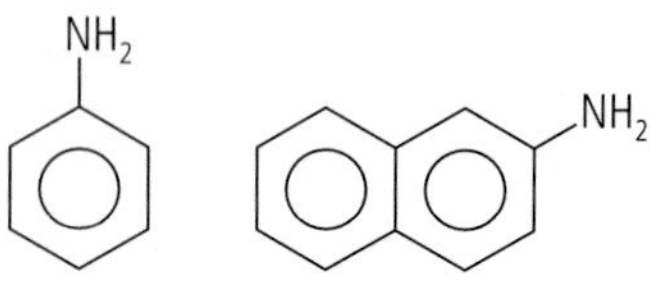

Aniline 2-naphthylamine

Carcinogenesis is often a two-stage process. In the first stage, **initiation,** a chemical, physical, or viral agent alters the cell's DNA. Sometimes a single exposure to some carcinogen causes a rapid onset of a tumor that is composed of rapidly growing, uncontrolled cells, but usually the abnormal cells continue to reproduce in about the same way as normal cells around them. Then a **promotion** occurs. This is the second stage and may occur days, months, or years after the initiation. This promotion may be a physical irritation or exposure to some toxic chemical that is itself not a carcinogen. In either case, the promotion results in the killing of a large number of cells. The destruction of cells is almost always compensated for by a sudden growth of new cells, and the abnormal cells begin to grow in ways the original DNA coding never intended. The cancer has started.

■ When a cancer spreads from one site to another, the process is called *metastasis.*

■ Smoking is thought to play both an initiation and a promotion role in cancer causation.

To illustrate the initiation and promotion aspects of carcinogenesis, consider some experiments performed in 1947 at Oxford University in England. First, very small doses of dimethylbenzanthracene (DMBA), a known carcinogenic component of coal tar, were applied to the skin of a group of mice. These mice were then separated into two groups. In one group the exposed skin was daubed with croton oil, a strongly irritating natural oil. The other group of mice had their skin daubed with croton oil four months later. Almost every mouse in both groups developed a tumor where the DMBA had been applied. In other groups of mice tested, neither croton oil nor DMBA alone produced any tumors, and if croton oil was applied first to the skin, followed by the DMBA, tumors failed to appear. Apparently DMBA had an initiation effect, while croton oil had a promotion effect.

■ Most cancers are concentrated in the epithelial cells. These cells cover the skin and other tissues, make up the glandular organs such as the breasts, and line the lungs and gastrointestinal tract. Cancers in the epithelial cells are known as *carcinomas* and account for about 85% of all cancers. The epithelial cells are those most likely to be exposed to external chemical, physical, and viral agents. *Sarcomas* are cancers of the connective and supportive tissue, *lymphomas* are cancers of the lymph system, and *leukemias* are cancers of the blood.

Chemicals and Cancer

A vast amount of research has verified the carcinogenic behavior of a large number of diverse chemicals. Some of these are listed in Table 18.10. This research has led to a few generalizations concerning the relationship between chemicals and cancer. For example, carcinogenic effects on lower animals are commonly extrapolated to humans. If a chemical causes cancer in a laboratory animal, it is assumed that it will cause cancer in humans also. Many scientists have argued that this extrapolation is probably not correct because an animal is not a human and the way it metabolizes a given chemical, for example, is probably different from the way the human body would metabolize that

■ The mouse has come to be the classic animal for studies of carcinogenicity. Strains of inbred mice and rats have been developed that are genetically uniform and show a standard response to various chemicals.

chemical. Nevertheless, animals are considered the best test subjects for these studies.

One of the biggest problems facing researchers in testing chemicals for carcinogenicity is the *choice of dose.* Some carcinogens are relatively nontoxic in a single large dose but may be quite toxic, often increasingly so, when administered continuously. Thus, much patience, time, and money must be expended in carcinogen studies. The development of a sarcoma (cancer of the connective tissue) in humans, from the activation of the first cell to the clinical manifestation of the cancer, takes from 20 to 30 years. With life expectancy of an average person in the United States now more than 70 years, it is not surprising that the number of deaths due to cancer is increasing. Chemical exposures and other forms of cancer-inducing exposures have sufficient time to manifest themselves into cancers.

■ The Delaney Clause (see *Science and Society* in Chapter 17) is a product of the assumption that anything carcinogenic in an animal is carcinogenic in humans.

■ An abnormal growth is classified as cancerous or malignant when examination shows it is invading neighboring tissue. A growth is benign if it is localized at its original site.

Some compounds cause cancer at the point of contact. Other compounds cause cancer in an area remote from the point of contact. The liver, the site at which most toxic chemicals are removed from the blood, is particularly susceptible to such compounds.

As indicated by the variety of chemicals in Table 18.10, many molecular structures produce cancer, whereas ones closely related to them do not. The 2-naphthylamine mentioned earlier is carcinogenic, but repeated testing gives negative results for 1-naphthylamine. These apparent subtle structural variations make it difficult to be able to predict the carcinogenicity of a chemical by knowing its structure alone. This is why mutagenicity testing in living organisms is essential.

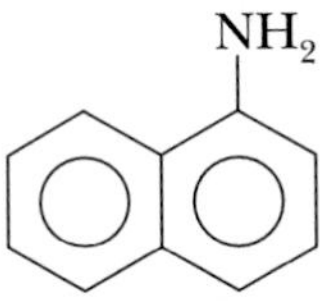

1-naphthylamine
(noncarcinogenic)

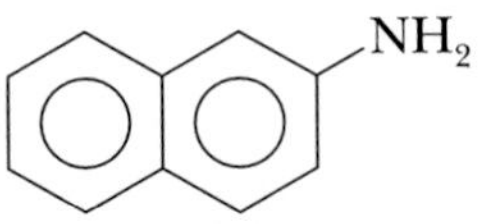

2-napththylamine
(carcinogenic)

Recently, Professor Bruce Ames, whose mutagen test was mentioned earlier, and others have suggested that the way chemicals are tested for carcinogenicity in animals needs to be changed. They have pointed out that animal cancer tests, usually done on rats or mice specially bred for genetic uniformity, are carried out using nearly toxic doses (the maximum tolerated dose, or **MTD**) for the test animal. It has been observed that the MTD over a long period of time can cause chronic *mitogenesis* (cell production). This chronic mitogenesis gives rise to a higher probability of mutagenesis (owing to the higher number of occurrences of cell production) and hence carcinogenesis. In the view of Ames and others, this chronic dosing of chemicals at the MTD is like chronic wounding, which is also known to be a promoter of cancer in animals and a cancer risk factor for humans. The net result is that many more chemicals will appear to be carcinogens because they were tested under the wrong conditions. The classic example of this is the testing of saccharin, an artificial sweetener, on rats. This chemical causes cancer in laboratory rats, but at doses that would correspond to the rats having to consume several gallons of some diet cola or other artificially sweetened soft drink each day of their lives.

Carcinogens in our Diets

In addition to industrial chemicals with which we may come in contact in the workplace and those that may contaminate our atmosphere (see Chapter 21) and our drinking water (see Chapter 20), our everyday diets contain a great variety of natural carcinogens. Some of these chemicals are also mutagens and teratogens. Plants produce these chemicals as defense mechanisms (natural pesticides) and often produce more of them when diseased or stressed. Celery, for example, contains isoimpinellin at a level of 100 μg/100 g. This level increases 100-fold if the celery is diseased. The isoimpinellin in diseased celery is a potent carcinogen when activated by sunlight. Celery pickers often develop skin lesions as a result of their exposure to bruised and diseased celery. Just how many of these lesions develop into skin cancers is unknown.

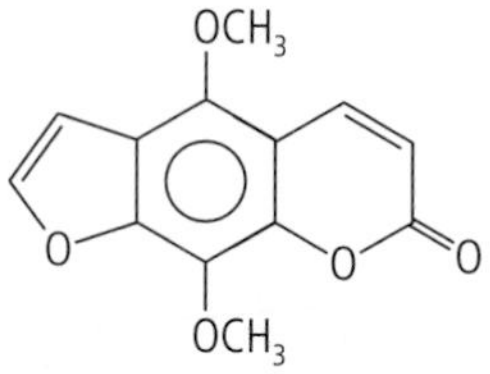

Isoimpinellin

■ It has almost become folk wisdom that "everything causes cancer." That generalization is probably not true, yet we live in a world filled with both natural and synthetic chemicals capable of causing cancer.

Black pepper contains small amounts of *saffrole,* a known carcinogen. Black pepper extracts cause tumors in mice at a variety of sites at dose rates equivalent to 4 mg of dried pepper per day for 3 months. Many people consume more than 140 mg of black pepper per day for life.

Most *hydrazines* that have been tested are carcinogenic. This includes industrial products, such as the dimethylhydrazines used as rocket fuels, and hydrazines produced by plants. The false morel (a sponge mushroom—*Gyromitra esculenta,* considered edible by some) contains eleven hydrazines, three of which are known carcinogens. One of these is *N*-methyl-*N*-formylhydrazine, which is present at concentrations of 50 mg/100 g and which causes lung tumors in mice at very low dietary doses of 20 μg per day. The common mushroom, *Agaricus bisporus,* contains about 300 mg of agaritine per 100 g. Agaritine is a mutagen and is metabolized to a compound that is extremely carcinogenic.

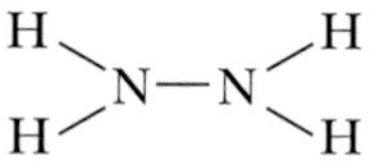

Hydrazine

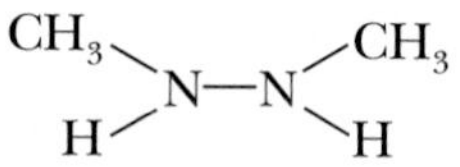

Dimethylhydrazine

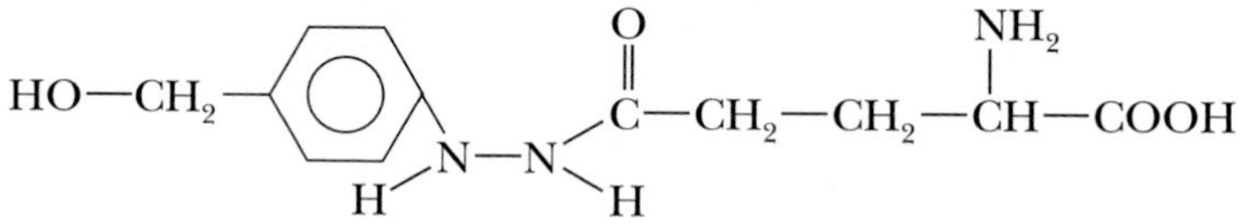

Agaritine

$CH_2{=}CH{-}CH_2N{=}C{=}S$

Allyl isothiocyanate

Allyl isothiocyanate, the main flavor ingredient in oil of mustard and horseradish, has been shown to cause cancers in rats. It also causes chromosome damage in hamster cells at low concentration.

It has been estimated that we consume about 10,000 times more natural carcinogens than man-made pesticides. This means that many of the cancers that may develop as we grow older may have been caused not by the chemical pollutants getting all the publicity, but by the very foods that we have been eating all along. We are not likely to eliminate all carcinogens either from our diets or from our general environment. Table 18.11 contains information on several carcinogens found in foods.

Harvesting celery, which may be hazardous when there are large numbers of bruised or diseased plants. *(Thomas Hovland/Grant Heilman)*

With all the carcinogens in our environment, both natural and synthetic, it is not surprising that cancers are widespread and common. But the disease does not occur with the same frequency in all parts of the world, and not all forms of cancer occur with the same frequency throughout the world. Breast cancer occurs less frequently in Japan than in the United States or Europe and is somehow related to how much dietary fat is consumed. Cancer of the stom-

TABLE 18.11 ■ Some Natural Carcinogens Found in Foods and Beverages

Compound	Formula	Food
Ethanol	CH_3CH_2OH	Alcoholic beverages
Theobromine	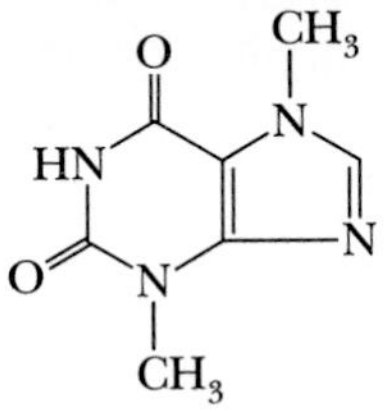	Cocoa-based drinks, tea
Methylglyoxal	$CH_3-C(=O)-C(=O)-H$	Coffee
Prunasin	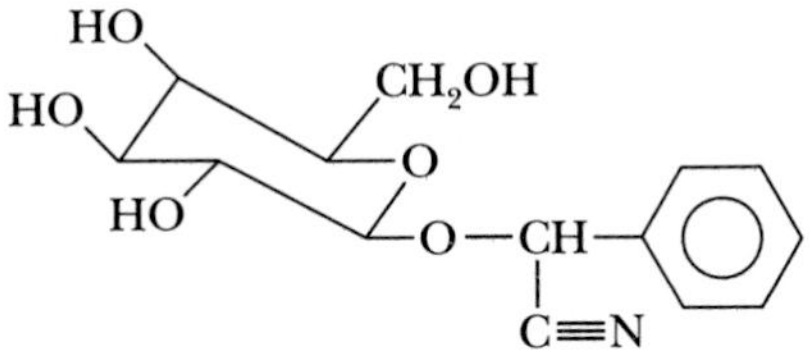	Lima beans
D-Limonene	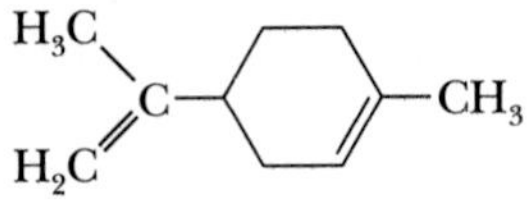	Citrus fruits
Serotonin	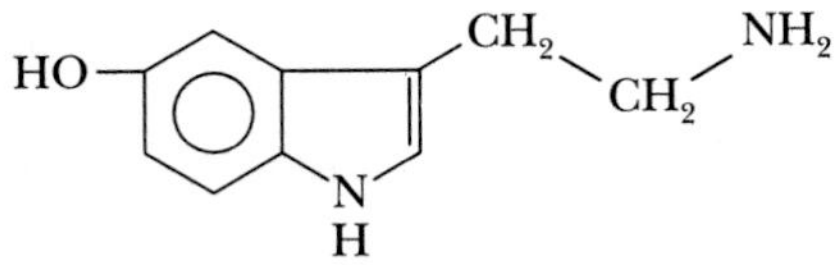	Tomatoes, pineapples, bananas
Aflatoxin B_1	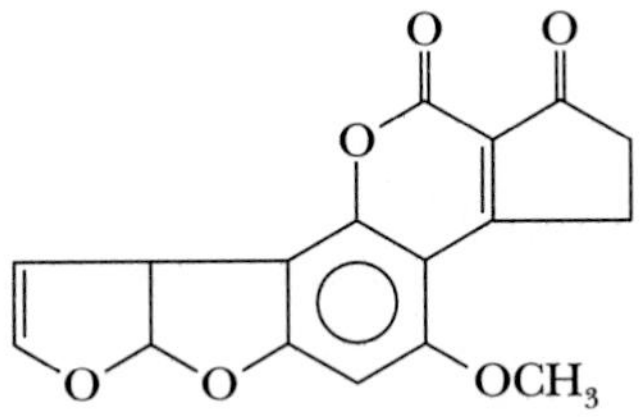	Moldy foods

Some edible varieties of mushrooms.
(C.D. Winters)

ach, especially in men, is more common in Japan than in the United States and is undoubtedly related to the different types of food these two societies eat. Cancer of the liver is not widespread in the Western Hemisphere but accounts for a high proportion of the cancers among the Bantu in Africa and in certain populations in the Far East, again, probably related to foods. The widely publicized incidence of lung cancer is higher in the industrialized world and is increasing at an appreciable rate. Lung cancer among women who smoke is now approaching the same rate as that among men who smoke, for example.

THE WORLD OF CHEMISTRY

Toxic Substances in Perspective

Program 25, *Chemistry and the Environment*

How do scientists perceive the risk from chemicals and the environment? Interestingly, most chemicals, even vitamins, can be toxic at large doses. Most chemicals, in fact, have a threshold dose of toxicity. A dose above the threshold is dangerous; a dose below it is not. There is a great deal of debate about whether potential carcinogens have such thresholds, that is, concentrations below which they won't cause cancer. Thus, given conflicting and imprecise evidence, some scientists talk about carcinogens in terms of risk or probabilities. Risk is relative. For example, some think the cancer risk from hazardous waste is negligible compared with the naturally occurring carcinogens we successfully resist every day.

The originator of a test for determining whether a chemical has the potential to cause cancer is Dr. Bruce Ames, biochemist at the University of California, Berkeley. He claims,

People have been very worried about toxic waste dumps but, in fact, the evidence that they're really causing any harm is really minimal; there's not very much evidence. And the levels of chemicals are very tiny, so we don't really know whether there's no hazard or a little bit of hazard.

The world is full of carcinogens, because half the natural chemicals they've tested have come out as carcinogens. Some plants have toxic chemicals to keep off insects, and we are eating those every time we eat a tomato or potato. Mushrooms have carcinogens, celery has carcinogens, and an apple has formaldehyde in it. So there are an incredible number of carcinogens in nature; we're getting much more of those than man-made chemicals.

Bruce Ames.

Dr. Halina Brown, Professor of Toxicology at Clark University, has stated,

If we accept those risks, why can't we accept small risks from chemical carcinogens in the environment? It's a valid argument. But then there is, of course, the counter argument. The counter argument is, we cannot do much about trace amount of carcinogens that are present in food; why should we add to this burden that we already have by increasing the amount of exposure to carcinogens? But then it boils down to money. Unfortunately, it takes tremendous resources to reduce the levels of exposure in the environment to carcinogens, especially when you get to very low levels. Reducing it by another order of magnitude may take millions of dollars at one hazardous waste site. And the pie is not unlimited. Even those who don't consider hazardous waste dumps a health threat think the money should be spent to clean them up.

Dr. Bruce Ames also states,

I mean, if Congress has put \$10 billion for cleaning up toxic waste dumps, you might as well find the worst ones and clean them up. Now, whether you're getting anything—whether you're gaining much in public health for cleaning them up—is something one could argue about. I think probably very little. But, in any case, you can—you might as well spend the money cleaning up the worst dumps.

These trends in cancer incidences strongly suggest that lifestyle, including the foods we eat and whether we smoke, influences our chances of getting cancer of one form or another. Certainly exposure to known carcinogens, especially in the workplace where exposures can be often repeated, should be avoided, but we probably couldn't eliminate carcinogens from our lives if we tried.

SELF-TEST 18C

1. Substrates that poison the nervous system are called ______________.
2. Most neurotoxins affect chemical reactions that occur in the opening between two nerve cells. These openings are called ______________.

3. The electrical impulse is carried across a synapse by the chemical ____________.
4. Mutagens alter the structures of ____________ or ____________.
5. If a substance is mutagenic in test animals, particularly dogs, it must necessarily be mutagenic in human beings. (a) True, (b) False.
6. The first occupation definitely linked to cancer was ____________.
7. Two dangers associated with smoking are ____________ and ____________.
8. A chemical that can cross the placenta and harm the fetus is called a ____________.

MATCHING SET

____ 1. Metabolic poison	a. Thalidomide
____ 2. Acute toxicity	b. Cyanide ion
____ 3. Corrosive poison	c. Amount of a toxic substance that enters the body of an exposed organism
____ 4. Neurotoxin	d. Immediate toxic effects
____ 5. Mutagen	e. Chemical that causes cancer
____ 6. Carcinogen	f. Volatile, toxic metal that is a liquid at room temperature
____ 7. Mercury	g. Sodium hydroxide (caustic soda)
____ 8. Pica	h. Atropine
____ 9. Dose	i. Alters DNA
____ 10. Chelating agent	j. EDTA
____ 11. Teratogen	k. Associated with lead poisoning in children

QUESTIONS FOR REVIEW AND THOUGHT

1. Define the following terms.
   (a) Dose (b) LD_{50}
   (c) Poison (d) Warning property
   (e) Pulmonary edema
2. What are the six classes of toxic substances?
3. Define the following terms.
   (a) Chelating agent (b) Amalgam
   (c) Pica (d) Metabolic poison
   (e) Antidote
4. Define the following terms.
   (a) Synapse
   (b) Neurotoxin
   (c) Neurotransmitter
   (d) Anticholinesterase poison
   (e) Teratogen
5. Define the following terms.
   (a) Mutagen (b) Carcinogen
   (c) Cancer (d) MTD
6. If a substance has a very low LD_{50} for a rat, what does this say about the toxicity of the substance for a mouse? for a human?
7. Name the three routes of entry of toxic substances into the body. Which one of the three allows toxic substances to enter the bloodstream the quickest?
8. What are the warning properties of a corrosive poison?
9. Which substance will cause pulmonary edema: (a) CO, (b) HCl, or (c) cyanide ion? Explain your answer.
10. Indicate which of the following is a corrosive poison.
    (a) Sodium hydroxide (b) Ozone
    (c) Carbon monoxide (d) Sodium cyanide
11. A paint stripper label shows a warning about skin and eye damage on contact with the product. This means the product probably contains what kind of poison?
12. Name three sources of carbon monoxide in daily life.
13. Describe how carbon monoxide and cyanide act similarly and how they act differently in their toxic action.
14. Briefly describe the properties of carbon monoxide and how these might be related to its warning properties (or lack of them).
15. What molecule in the blood does carbon monoxide bond to?
16. What is meant when carbon monoxide is described as a "reversible poison"?
17. What molecule in the blood does the cyanide ion bond to?

18. What is a natural source of hydrogen cyanide?
19. What is carboxyhemoglobin? How does it differ from hemoglobin?
20. What is a simple treatment for carbon monoxide poisoning?
21. What is the antidote for cyanide poisoning?
22. In general, metabolic poisons like heavy metals can be described as causing harm by reacting with (a) the skin, (b) nerve junctions, or (c) proteins like enzymes. Explain your answer.
23. Two compounds were discussed in this chapter as being used to treat heavy metal exposures by chelating the metal ions. What is chelation? Name the two chelating agents.
24. What is meant by the term "metal equilibrium"?
25. What poison is the substance amygdalin associated with?
26. Indicate which of the following is a metabolic poison: (a) CO, (b) H_2SO_4, (c) HF, (d) Hg.
27. Give definitions for the following.
    (a) Acute poisoning (b) Subacute poisoning
    (c) Chronic poisoning
28. Which heavy metal was used in compound form during World War I?
29. What is BAL? Draw its structure and indicate the groups that bind to heavy metal ions.
30. What is the condition called "pica" and how is it related to lead poisoning?
31. What is the common way of expressing the amount of lead in the human body?
32. What federal agency establishes an "intervention level" in order to protect people exposed to lead? What is the presently accepted intervention level for children?
33. Why is drinking water a principal source of lead?
34. What was thalidomide first prescribed for? What problems arose with its use? What are some uses for this compound today?
35. Describe how acetylcholinesterase works at a nerve synapse.
36. Describe how an anticholinesterase poison works.
37. Name three common pesticides that are anticholinesterase poisons.
38. Name a chemical warfare agent that is a anticholinesterase poison.
39. A new substance is reported to be able to pass through the placenta and into the bloodstream of a developing fetus. What are the dangers of such a substance to the fetus? If the substance is found to cause harm, what type of poison is it?
40. A chemical is found to be capable of causing mutations in certain strains of bacteria. What poison class does this chemical belong to?
41. Describe the Ames test and discuss its importance in testing the hazards of various chemicals.
42. What is a carcinogen? What is the difference between an animal carcinogen and a human carcinogen?
43. Describe the process of initiation in carcinogenesis.
44. What is promotion in the process of carcinogenesis?
45. Discuss the MTD as it applies to testing chemicals for carcinogenic activity in laboratory animals.
46. Name three foods that probably contain carcinogens in potentially harmful amounts.
47. Use the periodic table and identify the groups in which there are elements that are toxic heavy metals.

PROBLEMS

1. Morphine has a lethal dose of approximately 50. mg/kg of body weight for an adult human. How many mg of morphine would be lethal to a 150-pound adult?
2. The LD_{50} for lead is 20 mg/kg in rats. What number of mg would a quarter-pound rat have to eat to have a 50% chance of dying?
3. Sodium benzoate is used as a preservative in soft drinks. The LD_{50} for sodium benzoate in rats is 4 g/kg for oral doses. Assuming the rat data applies to humans, how many grams of sodium benzoate would a 75-kg person have to consume to have a 50% chance of dying from sodium benzoate? If a typical can of soft drink contains 355 g of liquid and there is 0.1% sodium benzoate by weight in the formula, how many cans of this drink would the person have to consume to have a 50% chance of dying from the sodium benzoate?

CHAPTER 19

Chemistry and Medicine

Everyone is interested in developments that influence public health and, most particularly, their own personal health. For even a general understanding of the major public health issues the amount to be learned is large. Knowledge, however, is our personal first line of defense. With chemistry as the framework, this chapter takes a look at the most important classes of diseases and the medications available to treat them. The major questions addressed are the following:

- What are the major fatal diseases in the United States?
- How do antibiotics function?
- What is the cause of AIDS?
- What roles do hormones and neurotransmitters play in body chemistry?
- How do drugs enhance or counteract hormones and neurotransmitter actions?
- What are some important pain-killing and mood-altering drugs?
- What conditions can cold medications attack?
- What are the major chemical weapons against heart disease and cancer, and how do they work?

The average life expectancy for men in the United States rose from 53.6 years in 1920 to 72.1 years in 1990. During this same period, the life expectancy for women rose from 54.6 years to 79.0 years.

What roles do chemistry and chemical technology play in this ongoing increase in life expectancy and the accompanying improvements in the quality of life? Some very important ones. Bacterial diseases have been conquered by antibiotics. Vaccines have vastly decreased the risk from once-epidemic diseases such as polio. And an array of synthetic materials for replacing arteries,

Decorative glassware like this once symbolized apothecary shops, where a pharmacist prepared and sold medicines. *(C.D. Winters)*

hip joints, and other body parts has been developed through laboratory research.

Most importantly, remarkable progress has been made in recent years in understanding the chemical reactions that regulate biological processes. As this understanding grows, the old trial-and-error method of screening chemicals for use as drugs is being replaced. Instead, drug molecules are being designed to have exactly the molecular shape and chemical reactivity needed to counteract disease. Another positive outcome is a growing understanding of what makes for a healthy lifestyle.

This chapter continues the story begun in Chapter 15 with an introduction to biochemistry, related to nutrition in Chapter 17, and extended in Chapter 18 to how we can be poisoned. In this chapter you will see how drugs act by helping the body's own defense mechanisms or blocking disease-related chemical pathways.

19.1 Medicines, Prescription Drugs, and Diseases—The Top Tens

Americans spend more than $34 billion a year on medicines. About $15 billion of this is for **over-the-counter drugs**—those that can be bought in any supermarket or drugstore. The rest is for **prescription drugs,** which cannot be purchased without instructions from a physician. A drug is classified as one or the other by the U.S. Food and Drug Administration (FDA). In general, a substance is available only by prescription if it has potentially dangerous side effects, if it should be used only by people with specific medical problems, or if it treats a condition so serious that a person with that condition should be under a doctor's care.

The top ten prescription drugs, based on the number of prescriptions written in early 1994 (Table 19.1), show an interesting cross-section of medicinal uses. Totalled according to the conditions treated, the largest number of

TABLE 19.1 ■ Ten Most-Prescribed Drugs in the United States in 1994 (First Quarter)*

Trade Name	Generic Name (Drug Class)	Condition Treated
Premarin	Mixture of estrogens (hormones)	Menopausal symptoms
Amoxil	Amoxicillin trihydrate (antibiotic, a penicillin)	Infectious disease
Trimox	Amoxicillin trihydrate (antibiotic, a penicillin)	Infectious disease
Zantac	Ranitidine hydrochloride (acid secretion blocker)	Ulcers†
Synthroid	Levothyroxine sodium (hormone)	Thyroid deficiency
Lanoxin	Digoxin (cardiac glycoside)	Heart disease
Procardia XL	Nifedipine (calcium channel blocker)	Heart disease
Vasotec	Enalapril (enzyme inhibitor)	Hypertension (high blood pressure)
Proventil	Albutarol (bronchial spasm blocker)	Asthma, bronchitis, emphysema
Ceclor	Cefaclor (antibiotic, a cephalosporin)	Infectious disease

*Data for the first quarter of 1994 from *Drug Topics,* Vol. 138, August 22, 1994, p. 46.

†Zantac has since been released as an over-the-counter drug for treating heartburn.

Children in about 1910 at a hospital for the treatment of tuberculosis (Sea Breeze Hospital, Coney Island, NY). At this time, months to years of bed rest in the open air was the major therapy for tuberculosis. *(Brown Brothers)*

prescriptions was for infectious diseases (the antibiotics), and the second-largest number was for heart disease and the related condition of **hypertension**—high blood pressure.

As illustrated in Table 19.1, all drugs have a trade name and a generic name. The **trade name** (brand name) is the name used by the drug manufacturer. For example, the antibiotic amoxicillin is sold under trade names such as Amoxil, Amoxidall, Amoxibiotic, Infectomycin, Moxaline, Trimox, Utimox, and Wymox. Taken together, the total sales of Amoxil and Trimox make this the single most-prescribed drug (in the first quarter of 1994). *Amoxicillin* is a **generic name,** which is a drug's generally accepted common chemical name. Once the patent protection on a drug has expired, it can be manufactured and marketed competitively by many companies and prescribed by its generic name. Often, prescriptions written by the generic name are cheaper to fill.

■ A systematic chemical name for amoxicillin, from which a chemist should be able to write its complete structure, is 6-(*p*-hydroxy-α-aminophenylacetamido)penicillinic acid.

In 1900, five of the ten leading causes of death in the United States were infectious diseases (pneumonia and influenza, tuberculosis, gastrointestinal infections, kidney infections, diphtheria). By 1993, pneumonia was the only one of these diseases that remained in the top ten, causing 4.4% of all deaths of individuals of all ages (Figure 19.1). This dramatic decrease in deaths from infectious diseases can be attributed in large measure to the development of antibiotics.

Meanwhile, however, a new infectious disease caused by the human immunodeficiency virus (HIV) and referred to as "AIDS" (acquired immunodeficiency syndrome) has come onto the scene. With release of the 1993 federal data, HIV infection hit the news by moving ahead of all other causes of death for those in the 25- to 44-years-old age group. Only three years earlier, in 1990, HIV infection placed third as the cause of death in this age group, with accidents and cancer coming first and second. Although vigorous study is underway and numerous leads are being pursued, HIV infection (discussed in Section 19.3) has not yet yielded to any drug.

■ When the causes of death are broken down according to age groups, heart disease is number one for those 65 years and older; cancer is number one for those 45 to 64 years old; HIV infection is number one for those 25 to 44 years old, and accidents are number one for those 15 to 24 years old. In the 15 to 24 years age group, more than 75% of the accidents are motor vehicle accidents.

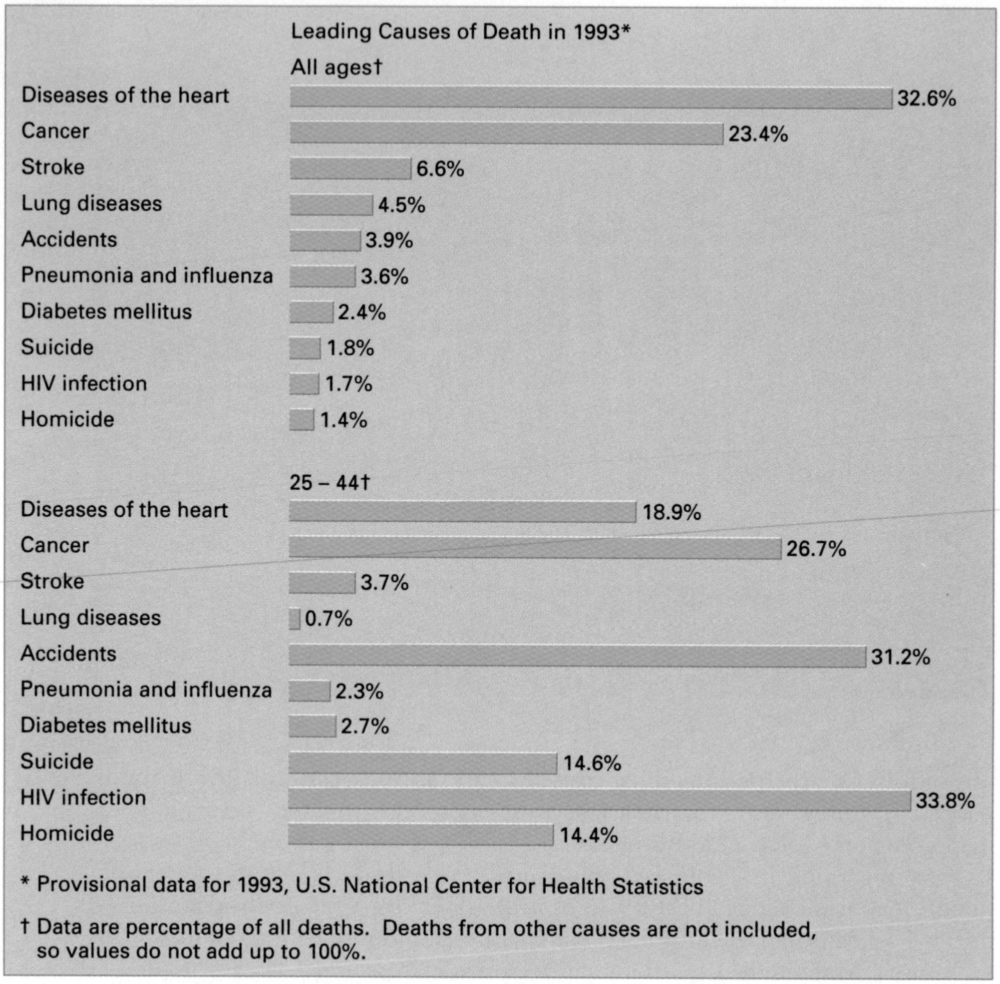

Figure 19.1 Leading causes of death in 1993. With the release of the 1993 data, AIDS moved into first place for those in the 25- to 44-year age range.

■ ***Infectious diseases*** are caused by microorganisms, which include viruses, bacteria, fungi, and other parasites. Ehrlich referred to his chemotherapeutic drugs for infectious diseases as "magic bullets"—they killed the microorganism but not the host.

19.2 Drugs for Infectious Diseases

Modern **chemotherapy**—the treatment of disease with chemical agents—began with the work of Paul Ehrlich (see *The Personal Side* later in this section). In 1904 he concluded that **infectious diseases** could be conquered if chemicals could be found that attack disease-causing microorganisms without harming the host. After observing that dyes used to stain bacteria also killed the

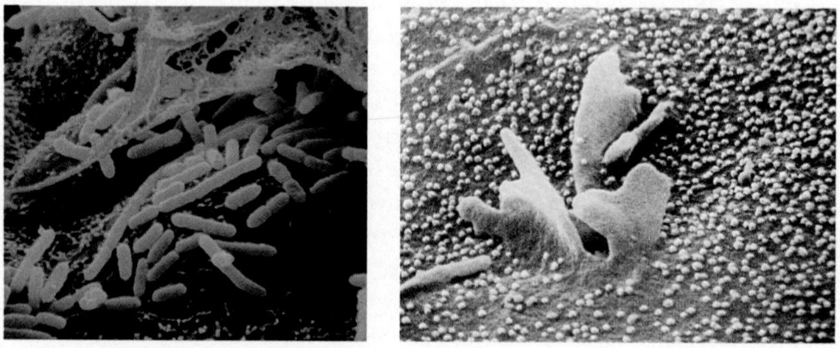

(a) (b)

The enemies: Bacteria and viruses. (a) The blue rods are *Haemophilus influenza* bacteria lying on human nasal tissue. This bacterium causes bronchitis, pneumonia, and a wide range of diseases in children. (b) The dots are *Rubella* viruses emerging from the surface of an infected cell, where the viruses are being reproduced. *Rubella* is the virus that causes German measles. *(a, © Dr. Tony Brain/ SPL/Photo Researchers; b, © NIBSC/SPL/Photo Researchers)*

bacteria, he developed arsenic compounds similar to the dyes. One of these compounds *(arsphenamine)* was the first effective drug for an infectious disease. It revolutionized the treatment of syphilis at the time it was introduced. (Syphilis is now treated with penicillin.)

■ Many drugs in use today are not active themselves but are metabolized to the active agent. Drugs of this type are often referred to as "prodrugs." Sometimes a prodrug is necessary because the drug itself would be broken down by digestion before it reached the site where it could act.

Sulfa Drugs

The sulfa drugs were also developed from a dye. Research showed that the dye Prontosil had antibacterial activity in mice because of its conversion *in vivo* (inside the body) to *sulfanilamide.*

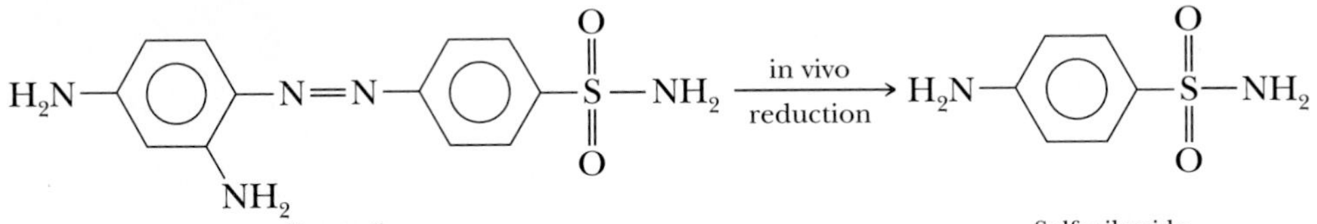

Prontosil (a dye)

Sulfanilamide (first sulfa drug)

Sulfanilamide, introduced in 1908 in Germany, was the first of the sulfa drugs and the first drug for bacterial infections. Sulfanilamide, however, had some harmful side effects. To find effective drugs without these problems, more than 5000 derivatives were synthesized and tested. About 30 of the new sulfonamides, a general term for sulfa drugs, reached the market as useful drugs.

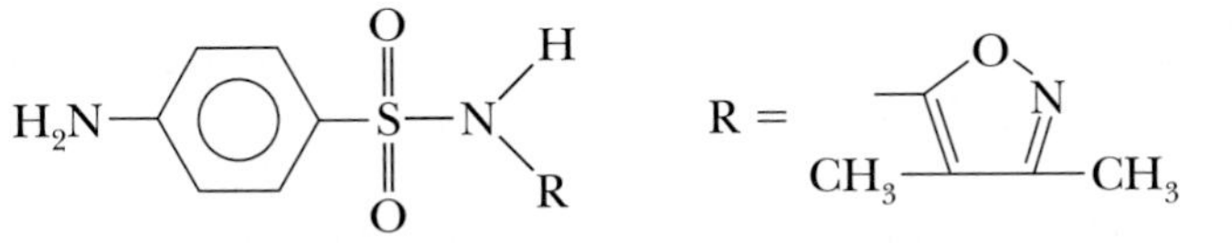

General sulfonamide structure

R group in sulfisoxazole (used to treat urinary tract infection)

■ Sulfa drugs are still important for the treatment of urinary tract infections.

All sulfa drugs act by preventing the synthesis of folic acid, a vitamin essential to the growth of bacteria. The drugs' ability to do this lies in their structural similarity to *para*-aminobenzoic acid, which the body needs to make folic acid (Figure 19.2). Sulfanilamide shuts off production of folic acid, and the bacteria die of vitamin deficiency. Because humans and other higher ani-

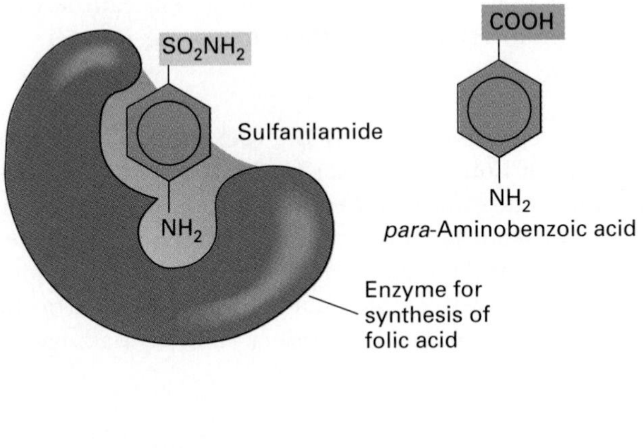

Figure 19.2 Action of a sulfa drug. Sulfanilamide, the first sulfa drug, is shown in the active site of a bacterial enzyme for folic acid synthesis. Because the sulfanilamide structure is so much like that of *para*-aminobenzoic acid, sulfanilamide (and other sulfa drugs) inhibit the synthesis of folic acid.

THE PERSONAL SIDE

Paul Ehrlich (1854–1915)

(SPL/Photo Researchers)

Paul Ehrlich, a German bacteriologist, was a pioneer in the application of chemistry to medicine. While he was a medical student, he became fascinated with chemistry and pursued learning about it on his own. He is an excellent example of a scientist driven by curiosity to do experiments that had incredibly successful practical applications. While investigating the biological properties of the many synthetic dyes available from the German chemical industry, he originated the microscopic study of blood cells. His work with the serum of immunized animals was fundamental to the then-young science of immunology (study of the body's response to and defense against disease-causing invaders). As noted in the text, he created the concept of chemotherapy.

Ehrlich was the first to explain the action of toxic substances and drugs. In what he called his "side-chain" theory, he explained that these substances affect only cells that have matching molecular fragments extending from their surfaces. Today we call these side chains receptors—the parts of cell-surface molecules that extend outside the cell and interact with external messengers and drugs. They are still the object of intensive research.

mals obtain folic acid from foods, they do not need *p*-aminobenzoic acid and are therefore not harmed by sulfa drugs.

Sulfa drugs were the miracle drugs of the late 1930s and early 1940s. During World War II thousands of lives were saved by the use of sulfa drugs to prevent infection in wounds. However, in the 1940s an even more effective group of drugs was discovered—the antibiotics.

Antibiotics

Strictly speaking, an **antibiotic** is a substance produced by a microorganism that inhibits the growth of other microorganisms. Any compound, whether natural or synthetic, that acts in this manner is, however, usually referred to as an "antibiotic." A person falls victim to an infectious disease when invading microorganisms multiply faster than the body's immune system can destroy them. Antibiotics help the immune system by either destroying invaders or preventing their multiplication.

■ *Antimicrobial* is a frequently used, more accurate, classification than *antibiotic*.

Penicillins The penicillins were discovered by Sir Alexander Fleming, a bacteriologist at the University of London. He was working in 1928 with cultures of *Staphylococcus aureus,* a bacterium that causes boils and some other infections. One day he noticed that one culture was contaminated by a blue-green mold. For some distance around the mold growth, the bacterial colonies had been destroyed. Upon further investigation, Fleming found that the broth in which this mold had grown had a similar lethal effect on many **pathogenic**

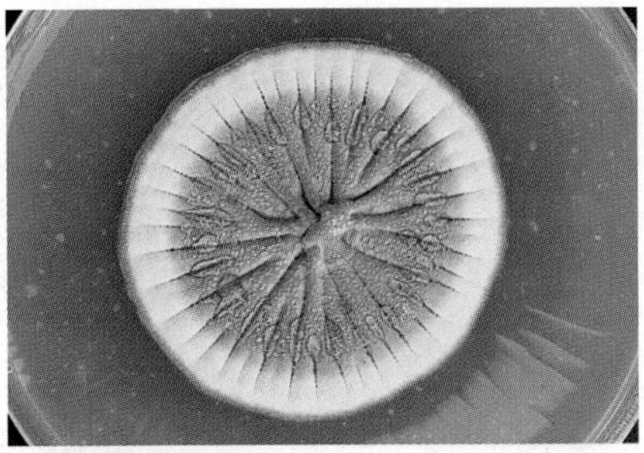

A *Penicillium notatum* culture. *(Andrew McClenaghan/SPL/Photo Researchers)*

(disease-causing) bacteria. The mold was later identified as *Penicillium notatum* (the spores sprout and branch out in pencil shapes, hence the name). Although Fleming showed that the mold contained an antibacterial agent, which he called *penicillin,* he was not able to purify the active substance.

In 1940 the active ingredient, penicillin G, was identified, and by 1943 it was available for clinical use. As World War II drew to a close, penicillin G was saving many lives threatened by pneumonia, bone infections, gonorrhea, gangrene, and other infectious conditions. Since then, numerous penicillins have been developed. Amoxicillin, the most-prescribed drug listed in Table 19.1, is a penicillin.

■ Howard Florey and Ernest Chain, who isolated penicillin, and Fleming, who discovered it, shared the Nobel Prize for medicine and physiology in 1945.

■ Another *Penicillium* strain that proved to be an excellent source of a new antibiotic was discovered on a moldy cantaloupe in a Peoria, Illinois, market.

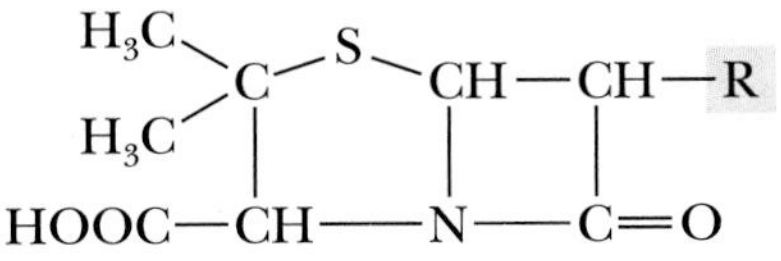

General penicillin structure

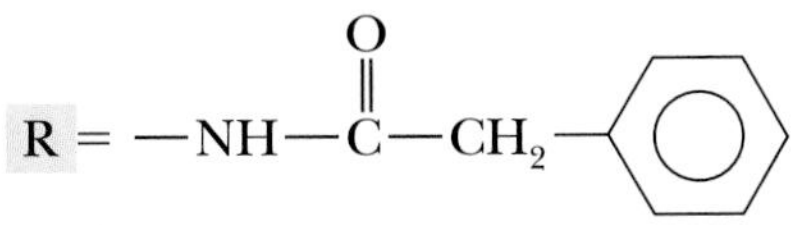

Penicillin G, the first penicillin

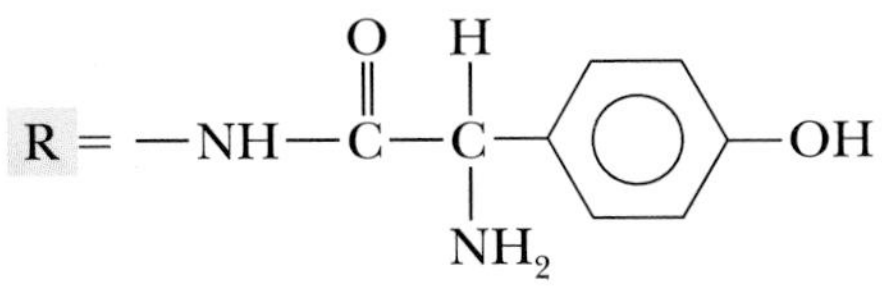

Amoxicillin, a broad-spectrum antibiotic—active against a variety of bacteria

All penicillins kill growing bacteria by preventing normal development of their cell walls. Unlike our cells and those of other mammals, bacteria rely on a rigid cell wall. The rigidity is maintained by cross-linking bonds between peptide chains. Penicillins inhibit the enzyme that forms these cross-links.

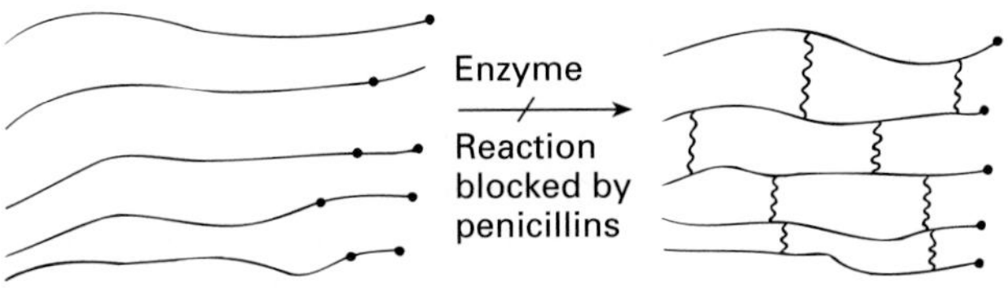

Other Classes of Antibiotics The *cephalosporins* are similar in structure to the penicillins and also act by disrupting cell-wall synthesis. They have become the most widely used antibiotics in hospitals because of their low toxicity and

THE WORLD OF CHEMISTRY

Folk Medicine

Program 21, *Carbon*

There's a long history of folk medicine based on plants, especially in the Near and Far East. Now the question is, Which of these things really works, and how much of it is just a rumor? First, the chemist checks out the plant rumored to be good and analyzes its extracts to see if there is any activity. One such substance is called *fredericamycin.* Fredericamycin is an antibiotic of some interest as a possible antitumor compound. It comes from a soil organism, a bacterium, found in the soil in Frederick, Maryland.

Dr. Kathlyn Parker of Brown University, and other researchers like her, first analyze the naturally occurring substance in the lab. If they find an active component, they look further. As Dr. Parker says,

Then, if you are really interested in drug development, you have to isolate the active component, purify it, determine its structure, and then it gets handed over to the pharmaceutical people who decide how to package it. For some pharmaceuticals, what you really need is to be able to make a large amount of stuff really cheap. One solution is that you would develop methods so that it was so cheap to make something that you could distribute it to people in a way that they could afford it.

Kathlyn Parker.

Tools of the trade for an early American pharmacist. Many medications were made on the spot from plants.

■ Some other antibiotics that act by inhibiting cell-wall synthesis are bacitracin (used only on the skin, not internally) and vancomycin (important for treating bacteria resistant to other antibiotics).

■ Development of antibiotic resistance is something that happens to microorganisms, not to people who take antibiotics.

broad range of antibacterial activity. Cefaclor (see Table 19.1) is currently the most-prescribed cephalosporin.

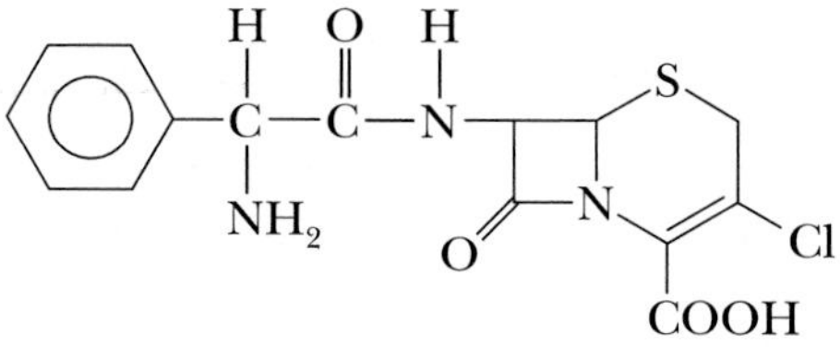

Cefaclor, a cephalosporin

Other kinds of antibiotics attack bacteria in different ways than the penicillins and cephalosporins. Many interfere with the synthesis or functioning of bacterial DNA. The *tetracyclines* (e.g., acromycin and terramycin) and *erythromycin,* for example, inhibit bacterial protein synthesis, and *rifampin* (important in treating tuberculosis) inhibits RNA synthesis from DNA.

A problem of growing seriousness is the evolution of antibiotic-resistant bacteria. Strains of malaria, typhoid fever, gonorrhea, and tuberculosis have emerged that are resistant to antibiotics that were once effective for these diseases. Some bacteria produce enzymes that inactivate the antibiotic; others prevent the antibacterial from entering their cells. A race is on to develop new antibiotics before harmful bacteria emerge that are resistant to all known antibiotics.

19.3 AIDS, A Viral Disease

Acquired immune deficiency syndrome (AIDS, caused by the virus HIV) has become a worldwide epidemic. Because HIV can remain inactive in the body for a time, the rising death rate is partly due to infections acquired many years earlier. Some HIV infections may even have begun before the identification of AIDS as a new disease in 1981. In the United States, the first 100,000 cases were reported over the course of the eight years from 1981 to 1989. About 50,000 new cases are expected to occur each year during the 1990s. The World Health Organization estimates that 1.5 million adults and children in the world now have AIDS, and between eight and ten million adults are infected with HIV.

From a chemical perspective, it is important to understand that very few ways are known to combat viral diseases (other than vaccination). The reason for this is the method of attack used by viruses—they are essentially chemical parasites. They take over the DNA of human cells and put it to work for their own reproduction. It is difficult to attack the cells reproducing the virus without also attacking the host's own cells.

■ Fewer than a dozen drugs are available for attacking viral diseases (as opposed to treating the symptoms).

The HIV is a *retrovirus.* It consists of an outer double lipid layer surrounding a matrix containing proteins, an enzyme called reverse transcriptase, and RNA. The term "retrovirus" is used because the virus enzyme carries out RNA-directed synthesis of DNA rather than the usual DNA-directed synthesis of RNA (see Figure 15.29).

The AIDS retrovirus penetrates the T cells of the immune system (1 in Figure 19.3). Once inside, the virus releases its contents, and the *reverse transcriptase* of the AIDS virus translates the RNA code of the virus into double-stranded DNA (2 in Figure 19.3). The virus DNA enters the T cell's nucleus and is incorporated into the cell's own DNA (3 in Figure 19.3). Then the T cell

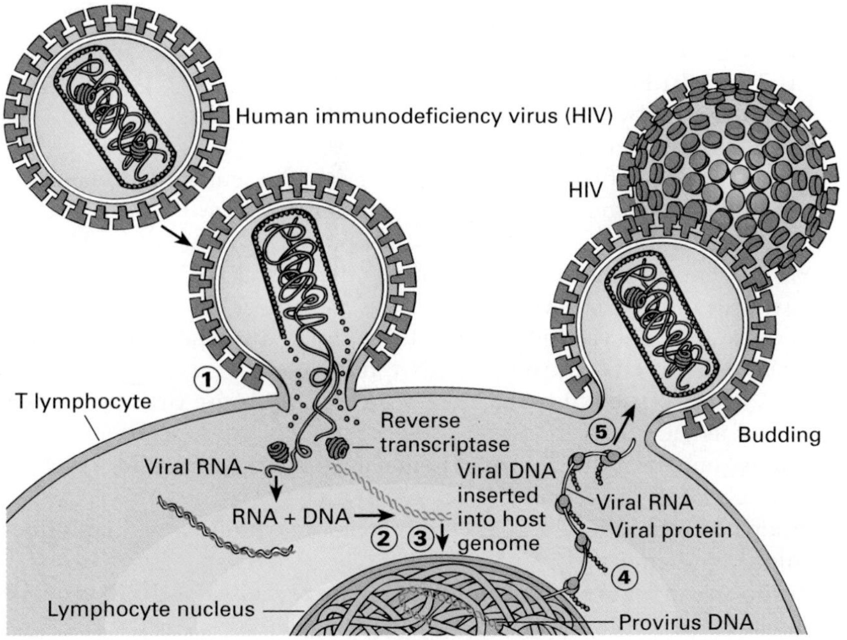

Figure 19.3 Attack of HIV virus on a T lymphocyte. The virus enters the lymphocyte (1) and produces its own DNA (2). The DNA then enters the lymphocyte nucleus (3), where it combines with the lymphocyte's DNA. Then through the cell's normal processing, DNA is transcribed to messenger RNA and, back outside the nucleus, the viral RNA provides the template for the production of proteins (4) that assemble into a new virus particle (5).

makes RNA from viral DNA, the proteins needed for a new virus are made from this RNA (4 in Figure 19.3), and the new virus is released (5 in Figure 19.3). Eventually the T cell swells and dies, releasing more AIDS viruses to attack other T cells. As their T cells are destroyed, individuals are attacked by diseases that are normally defeated by the body's immune system and thus are rare in healthy individuals.

■ AZT differs from the natural nucleoside by having an —N_3 group instead of an —OH group.

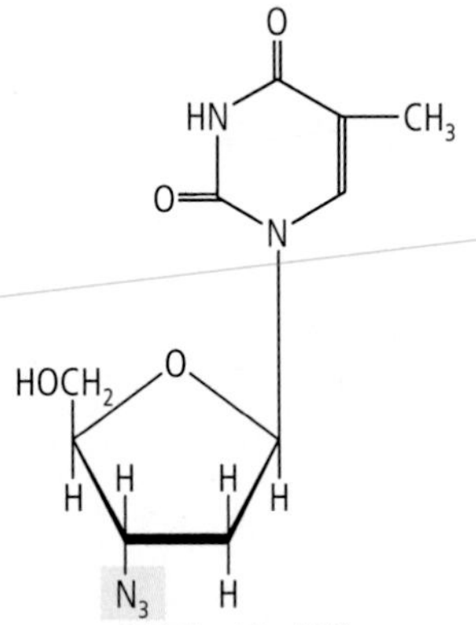

Azidothymidine (AZT)

AZT (*azidothymidine,* also called *zidovudine*) is the most widely used drug for the treatment of AIDS. AZT is a derivative of deoxythymidine, a nucleoside. (Nucleosides are nucleotides without the phosphate group; see Section 15.8). The HIV virus's reverse transcriptase is thought to accept AZT in place of thymidine. After AZT has become a part of the DNA, its structure prevents additional nucleosides from being added onto the DNA chain. Although AZT retards the progression of AIDS, it is not a cure. In addition, there are a number of problems with its use. AZT has toxic side effects that many patients cannot tolerate and the AIDS virus can become resistant to AZT.

Although there are new drugs that offer some hope for the treatment of AIDS, there has been little progress in developing a cure or a vaccine. A cure would require 100% elimination of the HIV virus from the body. Among the problems faced by researchers pursuing this goal are the long period before AIDS develops and the genetic variability of the virus.

■ SELF-TEST 19A

1. The ____________ name applies to the same drug no matter who makes it, but every manufacturer has its own ____________ name for the drug.
2. The three infectious diseases now in the top ten leading causes of death for individuals of all ages in the United States are ____________, ____________, ____________.
3. The microorganisms that cause infectious diseases are referred to as ____________
4. The structure present in all sulfa drugs is $HO—C_6H_5—SO_2—$. (a) True, (b) False.
5. Penicillins act by blocking the enzyme that causes ____________ of peptides in bacterial cell walls.
6. Penicillin G, the first antibiotic, and amoxicillin, currently the most-prescribed antibiotic, differ in structure only by a (an) ____________ group and a (an) ____________ group.
7. Viruses are difficult to combat because they attack from ____________ host cells rather than from ____________ the cells.
8. Because the HIV virus carries out RNA-directed synthesis of DNA rather than DNA-directed synthesis of RNA, it is called a ____________.
9. AZT, which is used to treat ____________, fools the virus by becoming part of its ____________.
10. The treatment of disease with chemical agents is referred to as ____________.
11. Strains of antibiotic ____________ malaria, typhoid fever, and tuberculosis have emerged.
12. According to Figure 19.1, the following three causes of death are most similar in percentage for the two age groups listed: ____________, ____________, ____________.

19.4 Steroid Hormones

In considering hormones, we turn our attention from drugs that fight invading organisms to drugs that make up deficiencies in natural biochemicals or mimic their action.

Hormones are produced by glands and secreted directly into the blood. They serve as chemical messengers, regulating biological processes by interacting with **receptors** sometimes distant from where they are secreted. For example, the adrenal gland just above the kidney releases a group of hormones that act throughout the body to regulate the availability of glucose in the blood. Hormones are chemically diverse but are mostly proteins or steroids. (The male and female sex hormones discussed in Section 15.6 are steroids.)

■ A **receptor** is a molecule or a portion of a molecule, often on the surface of a cell, that interacts with another molecule to cause some change in biochemical activity. Some drugs block receptors and prevent an undesirable change in activity. Other drugs activate a receptor to cause a desirable change in activity.

■ Cortisol, or hydrocortisone, is a steroid hormone. The ring structure common to all steroids is highlighted. Synthetic cortisol is used medically as an anti-inflammatory agent applied to the skin or injected into joints.

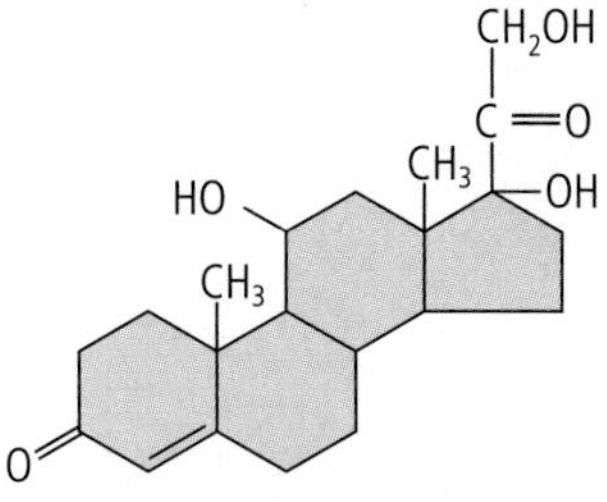

Cortisol

■ Statistics show that the health risks of using the pill are significantly less than the risks associated with pregnancy.

Birth Control Pills

One of the most revolutionary medical developments of the 1950s was the worldwide introduction and use of "the pill." The development of synthetic analogs of the female sex hormones made reliable birth control available to women for the first time. The ongoing search for an equivalent to "the pill" for use by men has not yet been equally successful.

Most oral contraceptive pills used by women today contain a combination of two synthetic hormones. One, most commonly ethynyl estradiol, has estrogen-like activity in regulating the menstrual cycle. The other, most commonly norethindrone, has progesterone-like activity which establishes a state of false pregnancy that prevents ovulation. In a theme that is common in drug development, the successful contraceptives are very similar in molecular structure to the natural molecules with similar activity.

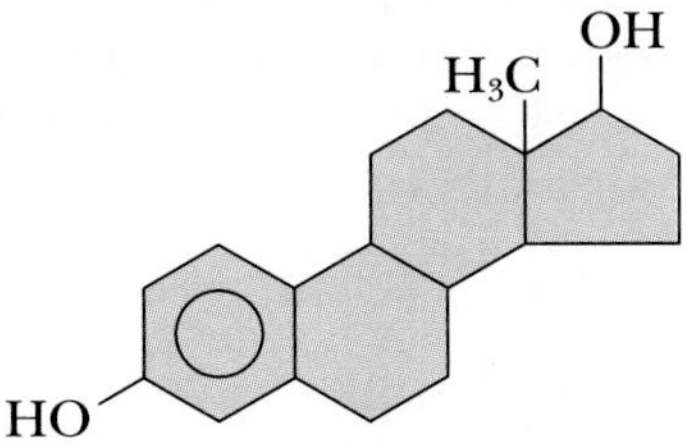

Estradiol
(an estrogen)

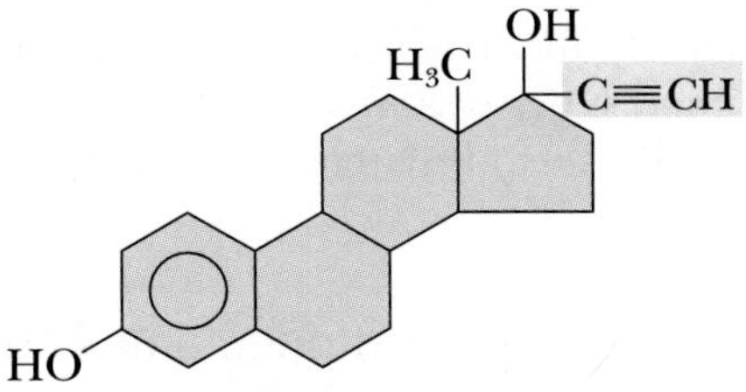

Ethynyl estradiol
(synthetic estrogen)

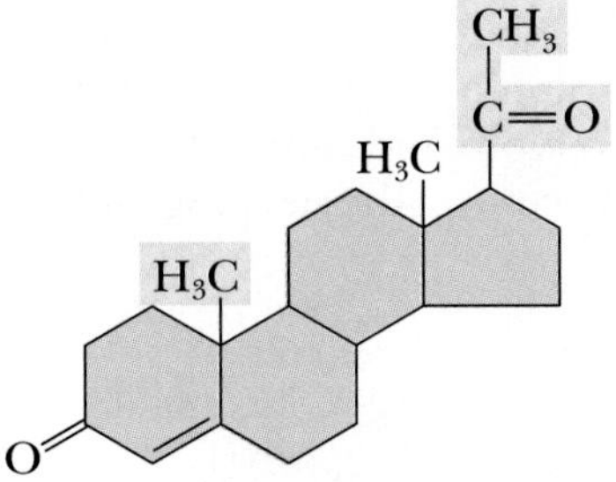

Progesterone

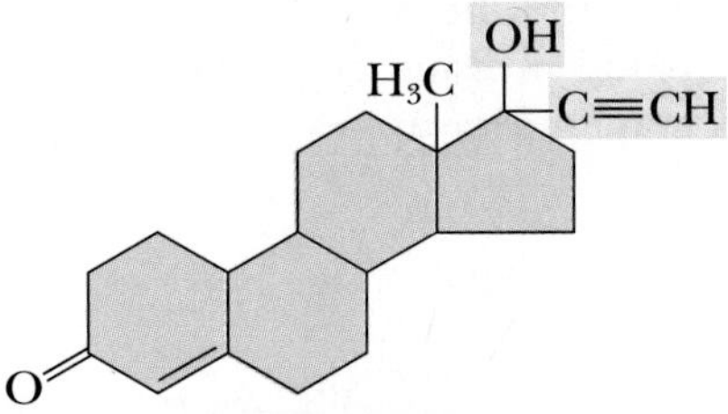

Norethindrone
(synthetic progesterone)

Synthetic steroids with estrogen-like activity are also used medically to replace natural hormones where there is a disease-related deficiency, after a

SCIENCE AND SOCIETY

RU-486, A Drug in Waiting

Mifepristone, also known as RU-486, is a synthetic steroid that blocks the receptors for progesterone. Without the action of progesterone, a pregnancy cannot be established. When RU-486 is taken within 72 hours of unprotected intercourse, it is an effective birth control pill—a "morning after" pill. It has been through extensive clinical testing and is on the market in France, the United Kingdom, and Sweden. Because RU-486 need be taken only when needed, it seems preferable to other birth control pills, which must be taken continuously.

The drug may also be useful in treating recurrent breast cancer, certain brain tumors, and glaucoma. Studies for its use in these conditions are under way.

Whether RU-486 will come to market in the United States is currently an unanswerable question. The difficulty does not lie in any scientific or medical uncertainty. Rather, it is because the drug has a safe and effective use for another purpose—termination of pregnancy at an early stage. This potential use places its availability into the realm of societal interactions, with the usual arguments mounted on both sides of the issue.

hysterectomy, or after menopause. A major benefit is a slowing of the normal bone loss (osteoporosis) that occurs with age. Premarin, one of the most prescribed drugs (Table 19.1), is a female hormone replacement medication.

19.5 Neurotransmitters

Hormones and neurotransmitters have in common their roles as chemical messengers within the body. Neurotransmitters, as explained in Section 18.5, carry nerve impulses from one nerve to the next or to the location where a response to the message will occur. The body has a variety of neurotransmitters, each with its own distinctive molecular receptors and functions. New information about neurotransmitters is being reported frequently and is of great interest because of its usefulness to medicine. Most drugs that affect the brain or the nervous system interact with neurotransmitters or their receptors.

Norepinephrine, Serotonin, and Antidepressive Drugs

Norepinephrine and serotonin are neurotransmitters with receptors throughout the brain. Norepinephrine helps to control the fine coordination of body movement and balance, alertness, and emotion and also affects mood, dreaming, and the sense of satisfaction. Serotonin is involved in temperature and blood pressure regulation, pain perception, and mood. Serotonin and norepinephrine appear to work together to control the sleeping and waking cycle.

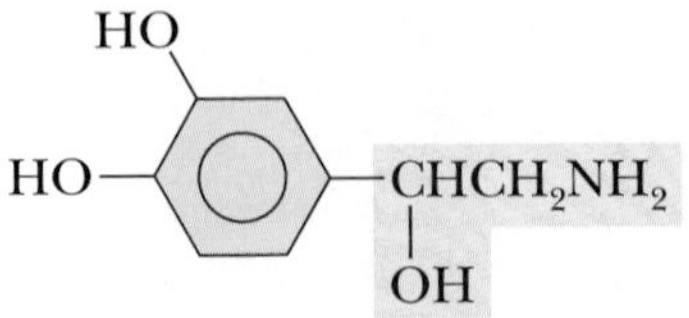

Norepinephrine

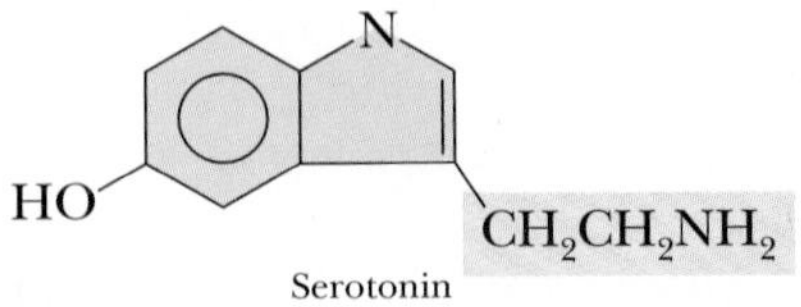

Serotonin

The normal cycle of these two compounds in their action at nerve synapses (Figure 18.6) is (1) release from a neuron and (2) interaction with a receptor, followed by inactivation by either (3) uptake of the neurotransmitter by the same neuron or (4) conversion to an inactive form by an enzyme, a monoamine oxidase. (Both compounds contain one $—NH_2$ group and are thus monoamines). The biochemistry of mental depression is not fully understood, but a deficiency of norepinephrine and serotonin (possibly also dopamine) almost certainly plays a role. Evidence is provided by the manner in which three classes of drugs used to treat depression interfere with this cycle.

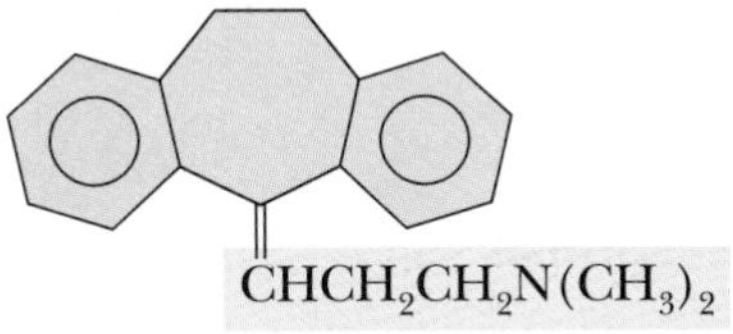

Amitriptyline
(Elavil)

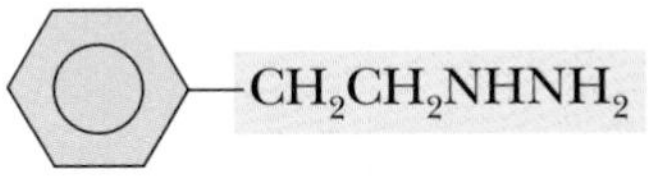

Phenelzine
(Nardil)

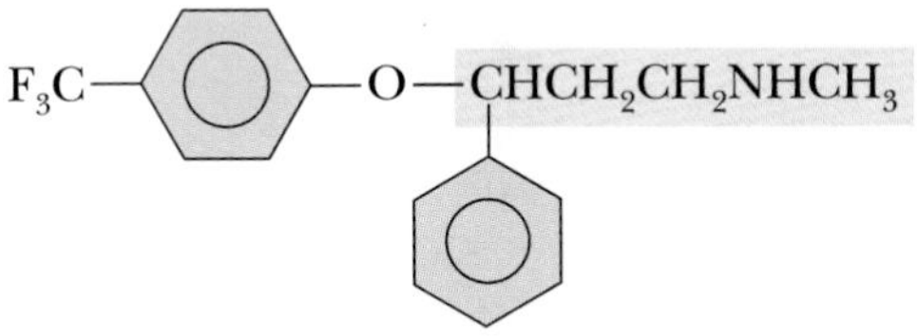

Fluoxetine
(Prozac)

The tricyclic antidepressants (e.g., amitriptyline (Elavil) prevent reuptake (3), which increases the concentration of the neurotransmitters in the synapse. The monoamine oxidase inhibitors prevent inactivation of the neurotransmitters (4), thereby also increasing their concentration. The third type of drug is represented by fluoxetine (Prozac), which acts by selectively preventing the recapture of serotinin by the neuron that released it. It has become the drug of choice for treating serious clinical depression. For each of these drugs we have described the major mechanism of action—increasing the concentration of serotonin at synapses. In each case, there are multiple other modes of action that are not fully understood.

■ It is important to understand that depression as a mental illness is different from depression due to tragic and traumatic real-life events such as death of a loved one or loss of a job.

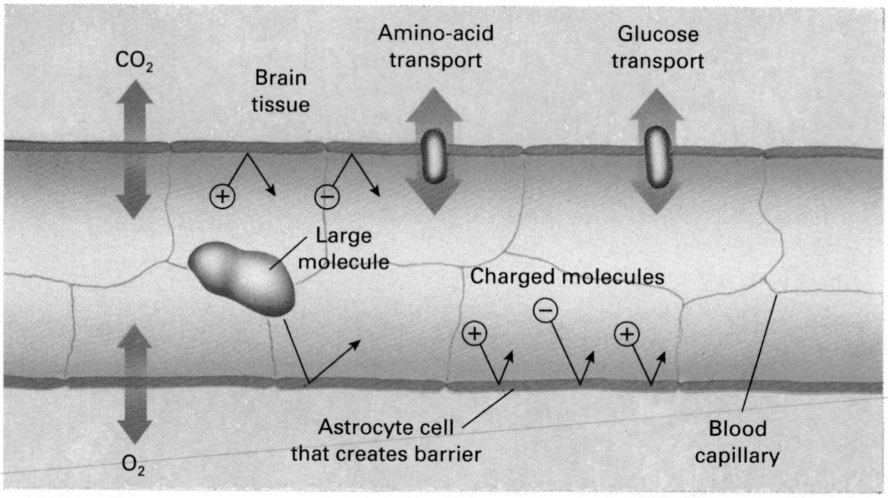

Figure 19.4 Blood–brain barrier. Small openings in brain capillaries keep out large molecules, and the cell membrane can only be crossed by lipid (fat)-soluble molecules. Together, these limitations provide the barrier that protects the brain from toxic compounds, but sometimes present a problem by also keeping out beneficial drugs.

Dopamine

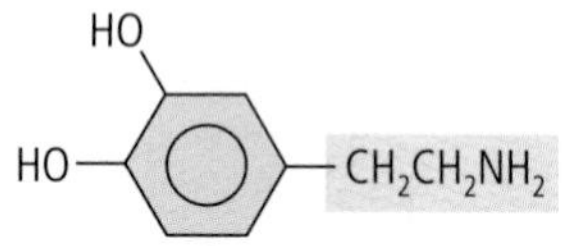

Dopamine
(L-dopa has a —COOH on the C atom next to —NH_2.)

Dopamine is produced in several areas of the brain, where it helps to integrate fine muscular movement as well as to control memory and emotion. An understanding of the brain chemistry of dopamine led to development of an effective treatment for Parkinson's disease. Patients with this disease experience trembling and muscular rigidity, among other symptoms, because of a deficiency of dopamine. Dopamine does not cross the blood-brain barrier (Figure 19.4) and thus cannot be administered as a drug. L-dopa, it was found, could cross the barrier and then be converted to dopamine in the brain. While it does not cure Parkinson's disease, L-dopa can completely alleviate symptoms for several years.

Schizophrenia has been related to an *excess* of dopamine in the brain. All drugs used to treat schizophrenia, none of which were available until the 1950s, act by blocking dopamine receptors.

Epinephrine and the "Fight or Flight" Response

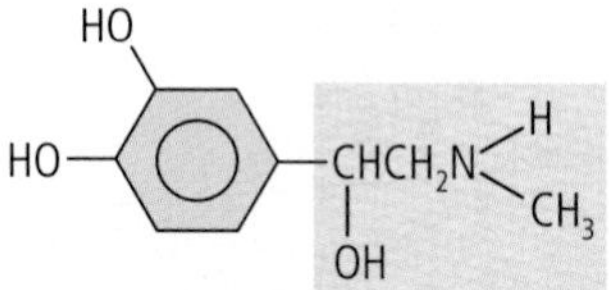

Epinephrine

■ The counteraction of the fight-or-flight reaction to the benefit of heart disease patients who take beta blockers is discussed in Section 19.9.

Epinephrine, or adrenalin, is both a neurotransmitter in the brain and a hormone released from the adrenal gland. Its sudden discharge when we are frightened produces the fight-or-flight response which includes increased blood pressure, dilation of blood vessels, widening of the pupils, and erection of the hair. Because of its widespread and rapid effects, epinephrine has a number of medical uses, notably in crisis situations. It is administered to counteract cardiac arrest (by stimulating heart rate), to elevate dangerously low blood pressure (by dilating blood vessels), to halt acute asthma attacks (by dilating bronchial tubes), and for treating the extreme allergic reaction known as *anaphylactic shock*.

■ SELF-TEST 19B

1. Two main classes of chemical messengers within the body are ____________ and ____________.

2. Chemical messengers act only on cells that contain the appropriate ______.
3. Chemically, most hormones are ______ or ______.
4. Most oral contraceptives contain synthetic derivatives of which two of the following? (a) Progesterone, (b) An androgen, (c) An antihistamine, (d) An estrogen.
5. Patients with Parkinson's disease have a deficiency of the neurotransmitter ______.
6. Depression is counteracted by a moderate excess of the neurotransmitter ______.
7. Our bodies react to a sudden fright with a rapid production of ______.
8. Hormones are produced by ______ and secreted directly into the blood.
9. The chemical structure of synthetic hormones is ______ to the natural hormones.

Seed capsule of an opium poppy. Crude opium is harvested from the sticky liquid that oozes through slits in the capsule. *(Dr. Jeremy Burgess/SPL/Photo Researchers)*

19.6 Painkillers of All Kinds

Drugs that relieve pain are known as **analgesics.** They range from cocaine to morphine to aspirin. Some are illegal drugs, some are prescription drugs, and some are over-the-counter drugs.

Opium and its Relatives

Opium, obtained from the unripened seed pods of opium poppies, contains at least 20 different compounds. Chemically, they are **alkaloids,** organic compounds that contain nitrogen, are bases, and are produced by plants. About 10% of crude opium is ***morphine,*** which is primarily responsible for the effects of opium. Morphine is medically valuable as a strong painkiller able to produce sedation and loss of consciousness.

The term **opioid** is now applied to all compounds with morphine-like activity. As noted in Section 15.4, identification of brain receptors for the plant-derived opioids led to discovery of natural short-chain peptides (endorphins and enkephalins) that activate the same receptors.

■ In an earlier time when few medications were available, Sir William Osler, a famous physician, called morphine "God's own medicine." It was named for the Greek god of dreams, Morpheus, by the German pharmacist who first isolated it from opium in 1803.

Heroin, the diacetate ester of morphine, does not occur in nature but can be synthesized from morphine. As shown in Figure 19.5, their structures differ in only one kind of functional group. Heroin is much more addictive than morphine and for that reason has no legal use in the United States. ***Codeine,*** a methyl ether of morphine, is one of the alkaloids in opium and is used in cough syrup and for relief of moderate pain. Codeine is less addictive than morphine, but its analgesic activity is only about one fifth that of morphine. One of the most effective substitutes for morphine is *meperidine,* first reported in 1931 and now sold as Demerol. It is less addictive than morphine.

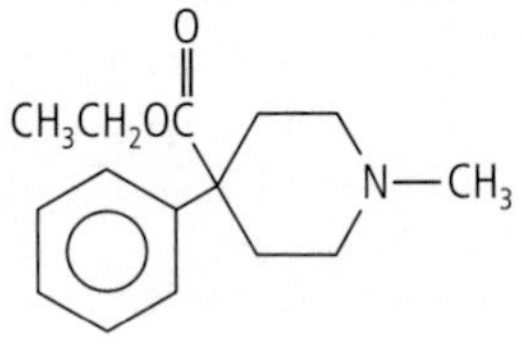

Meperidine

Mild Analgesics

When milder general analgesics are required, few compounds work as well for many people as *aspirin.* Each year about 40 million pounds of aspirin are

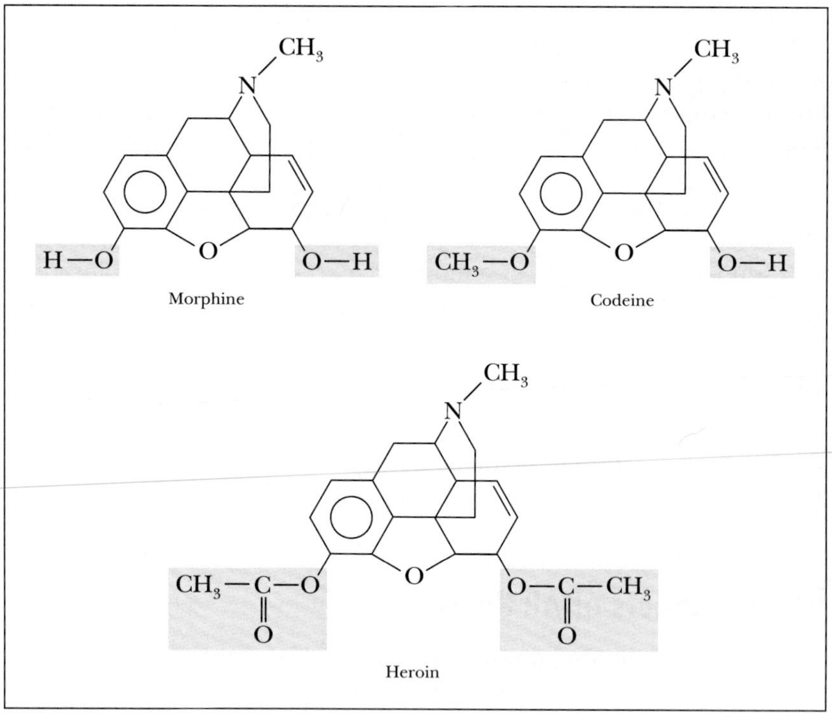

Figure 19.5 Opioids. Morphine and codeine are natural alkaloids in the opium poppy, and heroin is a synthetic derivative with similar activity as a drug, but is dangerously addictive.

manufactured in the United States. Aspirin is also an **antipyretic** (fever reducer) and an **anti-inflammatory** agent, and it has a role in treating heart disease (Section 19.9). Aspirin inhibits cyclooxygenase, the enzyme that catalyzes the reaction of oxygen with polyunsaturated fatty acids to produce prostaglandins (Section 15.6). Excessive prostaglandin production causes fever, pain, and inflammation—just the symptoms aspirin relieves.

Because aspirin is an acid, a dissolving aspirin tablet causes bleeding as it lies against the stomach lining. For most individuals, the blood loss from taking two 5-grain aspirin tablets is between 0.5 mL and 2 mL. Some persons, however, are more susceptible to blood loss. Early aspirin tablets were not particularly fast-dissolving, which enhanced the bleeding problem. Today's aspirin tablets are formulated to disintegrate more quickly.

A bottle of aspirin tablets that has developed the vinegar-like odor of acetic acid should be discarded. Acetic acid is formed as aspirin ages and breaks down.

Willow trees, source of a natural salicylate. Extracts from willow bark were known as pain killers from ancient times. Aspirin, first synthesized in the later 1800s, is a derivative of the natural salicylates, but with milder side effects. *(C.D. Winters)*

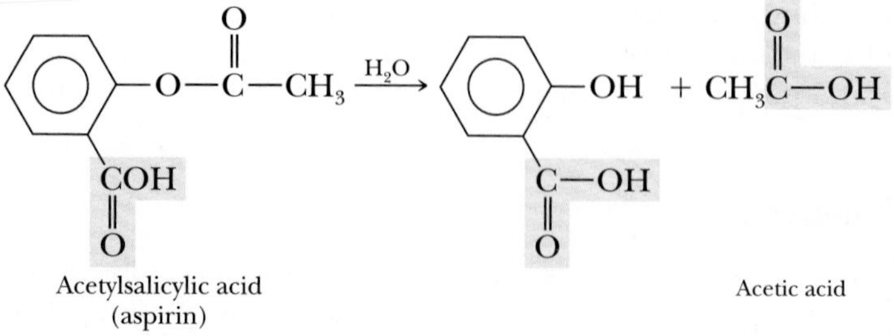

There is little reason to buy anything other than generic aspirin. The active chemical compound in all brands is the same—acetylsalicylic acid. "Buffered" aspirin tablets also contain basic compounds, such as aluminum hydroxide ($Al(OH)_3$) or magnesium carbonate ($MgCO_3$). There is no real "buffering" action (maintenance of constant pH; Section 9.8), but when combined with the basic substances, the aspirin dissolves and is absorbed a bit more quickly. It may also produce less bleeding and stomach irritation.

A greater potential danger of aspirin is its possible link to *Reye's syndrome,* a rare liver disorder. For unclear reasons, Reye's syndrome is sometimes associated with aspirin treatment of children with the flu or chickenpox. Aspirin should not be given to children with fever who may have the flu or chickenpox.

Several over-the-counter aspirin alternatives are now available for pain sufferers. The three principal ones are *acetaminophen (Tylenol), ibuprofen (Advil, Nuprin),* and *naproxen (Aleve).* Because acetaminophen is not an acid, it is the only one that does not cause bleeding in the stomach.

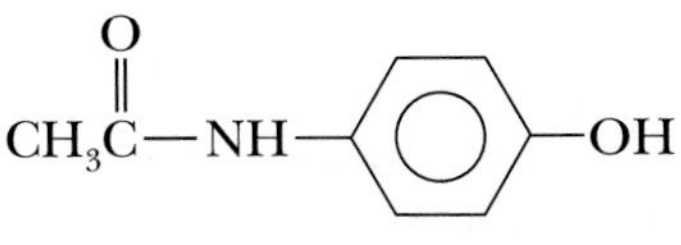

Acetaminophen
(Tylenol)

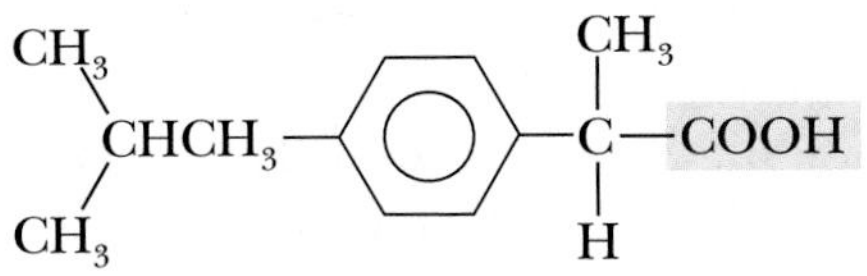

Ibuprofen
(Advil, Motrin)

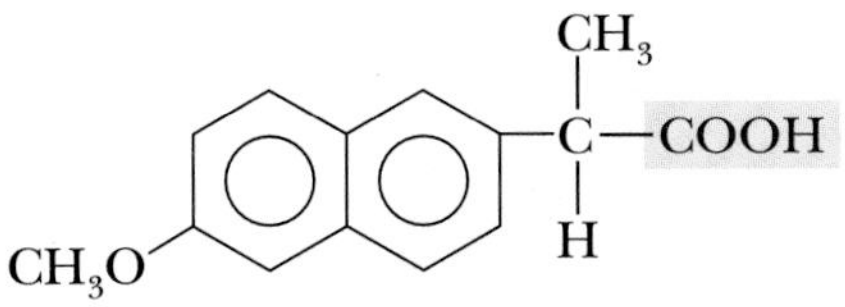

Naproxen
(Aleve)

Acetaminophen is an effective analgesic and antipyretic, but it is not an anti-inflammatory agent. Aspirin, ibuprofen, and naproxen are *nonsteroidal anti-inflammatory drugs (NSAIDs),* as distinguished from anti-inflammatory drugs that are steroids, such as cortisone. Ibuprofen, originally available only by prescription (Motrin), is similar to aspirin in its effectiveness but causes less bleeding. Naproxen was converted from a prescription drug (Anaprox or Naprosen) to an over-the-counter drug in 1994. Its principal advantage is a long period of activity, making twice-a-day administration possible for round-the-clock pain relief.

■ The reason that acetaminophen is the most-used mild analgesic in hospitals is simple —it doesn't cause bleeding, which for hospital patients could be hazardous.

19.7 Mood-Altering Drugs, Legal and Not

Everyone who drinks coffee or alcoholic beverages has personal experience with the mild effects of drugs classified as stimulants (the caffeine in coffee) or depressants (ethyl alcohol). Stimulants and depressants with stronger activity are among the drugs that are often *abused*—used in ways that are socially unacceptable and/or harmful. Although moderate alcohol consumption is an accepted social custom in the United States, alcohol too is an abused drug for many individuals. Other kinds of drugs that are subject to abuse are opioids and hallucinogens.

Although the list of drugs of abuse is wide-ranging, a central theme is their relationship to brain chemistry and the effect of the drug on neurotransmitters or their receptor sites. Before discussing the mood-altering drugs, it's important to understand the legal control of abused drugs. Table 19.2 lists the various classifications of the U.S. Drug Enforcement Administration. Schedule 1 drugs are not legally available in any manner. Schedule 2 drugs can be dispensed only once from a written prescription. A refill requires a new written prescription.

TABLE 19.2 ■ Classification of Drugs

Designation	Description	Examples
Over-the-counter (OTC)	Available to anyone	Antacids, aspirin, cough medicines
Prescription drugs	Available only by prescription	Antibiotics
Unregulated nonmedical drugs	Available in beverages, foods, or tobacco	Ethanol, caffeine, nicotine
Controlled substances*		
Schedule 1	Abused drugs with no medical use	Heroin, LSD, mescaline, marijuana
Schedule 2	Abused drugs that also have medical uses	Morphine, amphetamines, cocaine, codeine
Schedule 3	Prescription drugs that are often abused†	Valium, phenobarbital

*The sale, distribution, and possession of drugs classified as controlled substances are controlled by the Drug Enforcement Administration of the U.S. Department of Justice.

†Prescription can be refilled only up to five times and within six months.

Depressants

The effect of central nervous system depressants depends more on the dose than on the particular drug. The sequence proceeds from **sedation,** or relaxation, to sleep, to general anesthesia, to coma and death. Medically, depressants are used to treat anxiety and insomnia.

The best-known depressants are the *barbiturates.* Variations in chemical structure produce a range of barbiturates from very short-acting to very long-

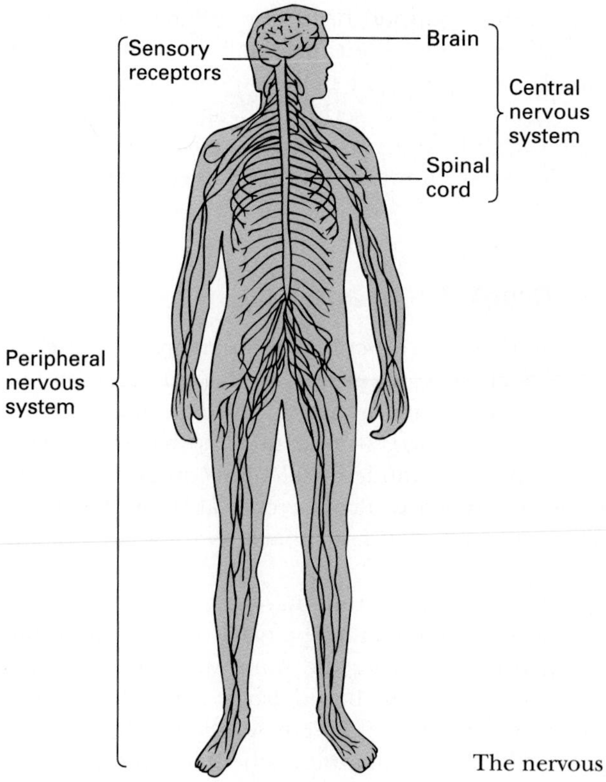

The nervous system.

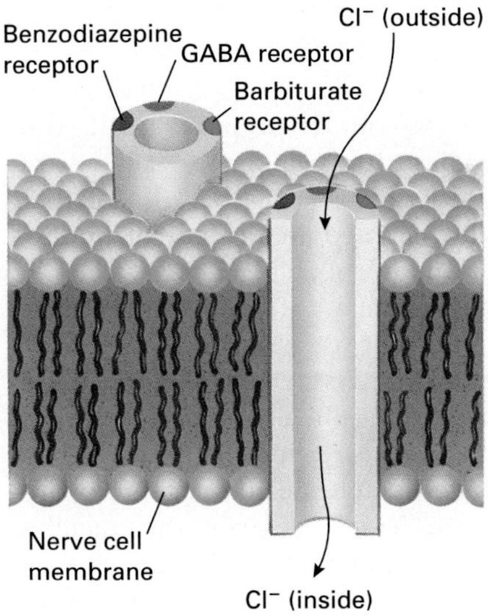

Figure 19.6 Mechanism of depressant action of barbiturates and benzodiazepines. The cylinder represents a large protein complex in the outer membrane of a nerve cell. When open, the channel allows chloride ions to enter the cell and inhibit firing of the nerve. GABA binding to its receptor opens the channel. Benzodiazepine binding prolongs the effect of natural GABA. Barbiturate binding can independently open the channel; the larger the dose, the longer the opening, explaining the greater hazard of barbiturate overdosage.

acting. Barbiturates are potential drugs of abuse, and overdoses cause many deaths. They are especially dangerous when ingested with ethyl alcohol (ethanol), another depressant, because the two together give a synergistic effect.

■ *Synergism* is the working together of two things to produce an effect greater than the sum of the individual effects.

The second major class of depressants is the *benzodiazepines.* The familiar *tranquilizers* Librium (chlordiazepoxide) and Valium (diazepam) are members of this group. Both the barbiturates and the benzodiazepines act at receptors for the neurotransmitter gamma-aminobutyric acid (GABA, $HOOCCH_2CH_2CH_2NH_2$), which normally *inhibits* rather than excites transmission of nerve impulses. Because of differences in their molecular mechanism of action (Figure 19.6), benzodiazepines are safer and less subject to abuse than barbiturates.

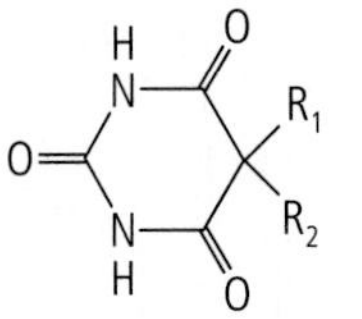

General barbiturate structure

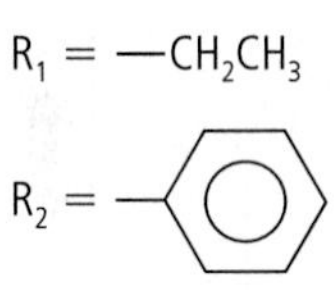

Phenobarbital (Luminal), a long-acting barbiturate used to treat insomnia and seizures

Ethanol (CH_3CH_2OH), like barbiturates, enhances the action of the neurotransmitter GABA. Unlike barbiturates, however, the effects are dose-dependent. At low doses, higher brain functions are affected, producing decreased inhibitions, altered judgment, and impaired control of motion. As dosage increases, reflexes diminish, and consciousness diminishes to the point of coma and death.

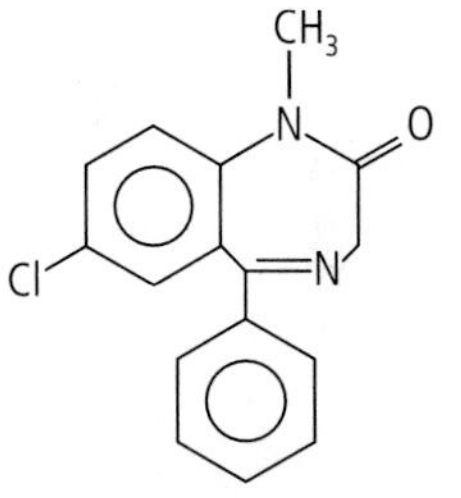

Diazepam (Valium), a benzodiazepam tranquilizer

Stimulants

Stimulants are drugs that excite the central nervous system. They stimulate production of the neurotransmitters norepinephrine, serotonin, and dopamine in the brain, heart (which is stimulated to beat faster), and veins (which become constricted). Note the similarity in the structures of amphetamine and methamphetamine to epinephrine and norepinephrine (pp. 618, 620), which helps to explain their action.

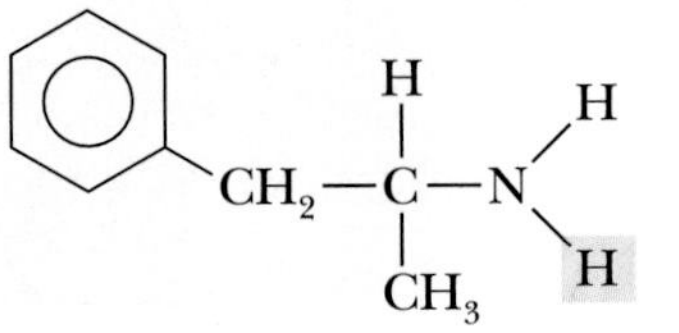

Amphetamine

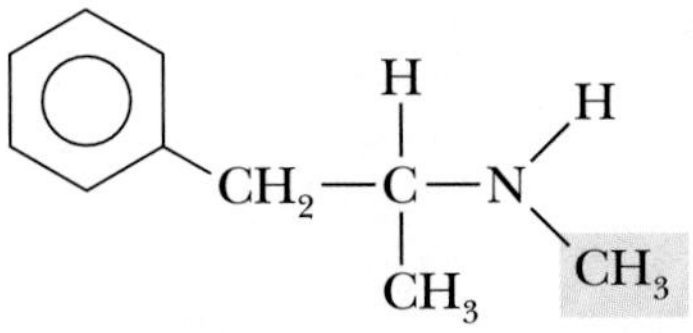

Methamphetamine

Amphetamines were once available in over-the-counter preparations used to stay awake. Now, the only approved ingredient for "stay-awake" pills is caffeine. Amphetamines have become controlled substances because of their great potential for abuse. The only generally accepted medical use of amphetamines is to treat *narcolepsy,* a condition of uncontrollable attacks of sleep.

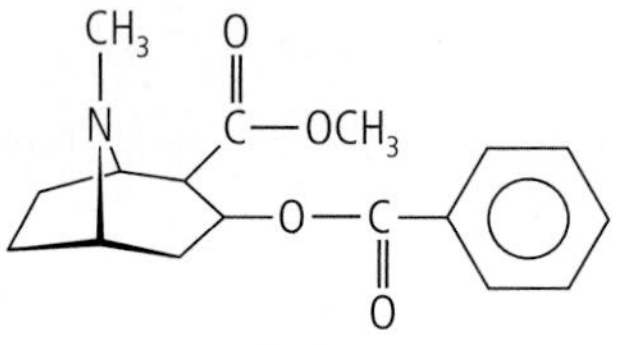

Cocaine

Cocaine, derived from the leaves of the coca plant of South America, is a stimulant and a Schedule 2 drug (used medically as a local anesthetic). By preventing the removal of norepinephrine from nerve endings, it causes uncontrolled firing of the nerves. *Crack* is a form of cocaine obtained by heating a mixture of cocaine and sodium bicarbonate. The reaction is an acid–base reaction since the base, sodium bicarbonate, is neutralizing cocaine hydrochloride, the usual form of cocaine. The term "crack" refers to the crackling sound of the heated mixture during the release of carbon dioxide. The appearance of crack on the illegal drug market has caused an increase in the number of cocaine addicts because crack is much more addictive than cocaine. The "high" lasts less than 10 minutes, creating the need to use crack repeatedly over a short period. Many users become addicted after only one try and there is a high risk of taking a lethal dose.

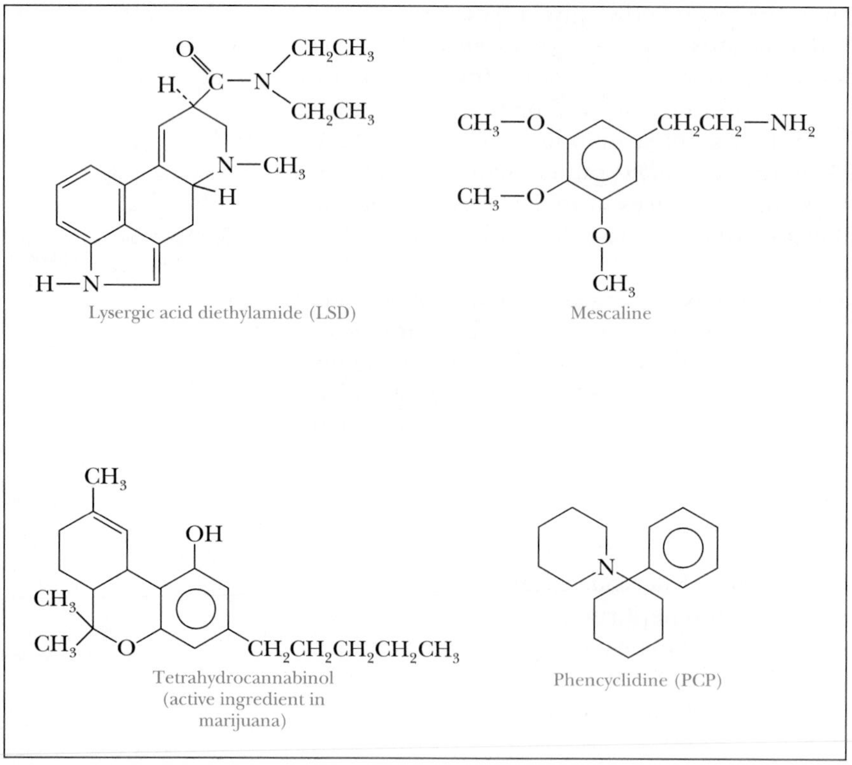

Figure 19.7 Some hallucinogens.

Peyote cactus, source of mescaline. *(© R. Konig/Jacana/Photo Researchers)*

Morning glory, whose seeds contain lysergic acid. *(© Carolina Biological Supply Company, Phototake NYC)*

Marijuana. *(© 1988 Scott Camazine/Photo Researchers)*

Hallucinogens

Hallucinogens are chemicals that cause vivid illusions, fantasies, and hallucinations. Many have been found in plants, including *mescaline,* which comes from the fruit of the peyote cactus, and *lysergic acid diethylamide (LSD),* which is made from lysergic acid derived from either the morning glory or ergot, a fungus that grows on grasses (Figure 19.7). These drugs are not addictive in the same manner as cocaine, but sometimes cause destructive behavior and lingering psychological problems. There are no therapeutic uses of the hallucinogens, which are believed to activate serotonin receptors.

Marijuana is a mild hallucinogen and sedative made from the hemp plant, *Cannabis sativa.* Although the millions of marijuana users regard it as a "safe" drug, this is a controversial conclusion. At high doses, marijuana is moderately addictive and can create paranoia and intense anxiety. Long-term use at moderate doses causes a general state of disinterest in personal achievements. In addition, tetrahydrocannabinol (THC), the active ingredient in marijuana smoke, can cause damage to the lungs, impede brain function, and hamper the immune system. Animal studies also suggest it may produce birth defects in offspring.

PCP (phencyclidine, known as "angel dust") is an especially dangerous drug with a unique pattern of effects. At low doses, its effects resemble those of alcohol. With higher doses, hallucinations set in and behavior can become hostile and self-destructive, promoting psychoses that can last for weeks. Physical effects include seizures, coma, and death from cardiac arrest.

Over-the-counter drugs—we have a great many to choose from. *(C.D. Winters)*

19.8 Colds, Allergies, and Other "Over-the-Counter" Conditions

What is the first thing we do when we feel sick? According to estimates, 75% of all illnesses are treated with products from the drugstore shelves. There are more than 300,000 over-the-counter (OTC) products on the market. But note that there are about 800 active ingredients packaged in all these different combinations. The top three categories of nonprescription medications in terms of sales dollars are used to treat allergies and colds, pain, and gastrointestinal problems.

■ Analgesics are discussed in Section 19.6; antacids are discussed in Section 19.9.

About one person in ten suffers from some form of allergy. Allergic symptoms occur when special cells in the nose and breathing passages release hista-

(a) (b)

Two plants that send many people to the drug store: (a) goldenrod, a major cause of hayfever; (b) poison ivy ("leaflets three, let it be"). *(a, Coco McCoy/Rainbow; b, C.D. Winters)*

■ Some 1993 sales of over-the-counter drugs:
Cough, cold, allergy, and sinus medicines: \$3 billion
Analgesics: \$2.7 billion
Gastrointestinal drugs (laxatives, antacids, antidiarrheals) \$840 million

mine. Histamine is a neurotransmitter that accounts for most of the symptoms of hay fever, bronchial asthma, and other allergies. In a familiar theme in drug design, ***antihistamines*** are medications that block histamine receptors. Chlorpheniramine and compounds with very similar chemical structures (e.g., brompheniramine, diphenhydramine) are ingredients in many OTC preparations for treating hay fever and the "stuffy" noses associated with allergies. A frequent side effect of many antihistamines is drowsiness. In fact, many OTC sleep aids contain antihistamines.

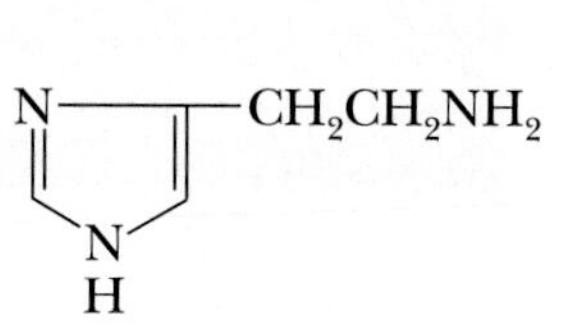

Histamine
(neurotransmitter that causes allergic reaction)

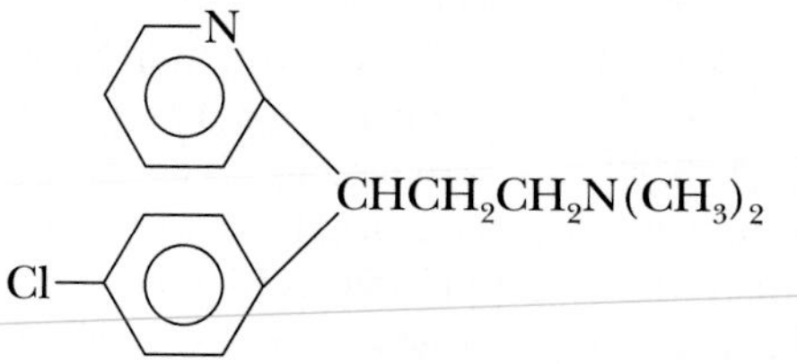

Chlorpheniramine
(antihistamine, e.g., in Chlortrimeton)

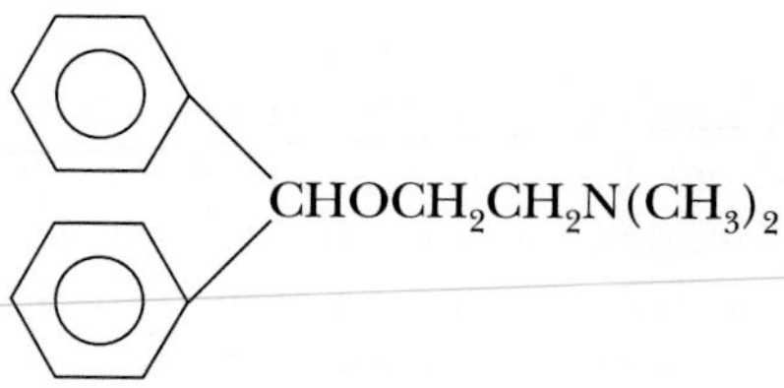

Diphenhydramine
(antihistamine used in sleep aids, e.g., Sleep-Eze)

The common cold is caused by a virus, and like other more serious viral diseases, it cannot truly be cured. The best we can do is treat the symptoms. Most OTC preparations for colds contain a variety of ingredients, usually two or more of those listed in Table 19.3. We have already discussed antihistamines and analgesics. There is some doubt about how useful an antihistamine is in treating a cold.

Decongestants shrink nasal passages to relieve the stuffiness that goes with colds and allergies. The decongestants activate receptors for epinephrine (an ingredient in some cold medications) and similar neurotransmitters. ***Antitussives*** suppress coughing. Opioids all have antitussive activity, and codeine is often used for this purpose. Dextromethorphan (see Table 19.5) is an opioid-related compound that has antitussive activity without the analgesic and other effects of opioids. ***Expectorants*** are meant to stimulate secretions in the respiratory tract so that mucus is dislodged in coughing. Most likely, guaifenesin is the only effective ingredient of this type.

A few guidelines for selecting and using OTC products are recommended by numerous groups concerned with public health:

- Choose single-ingredient products specific to the condition you have.
- Cut down on unnecessary expense by choosing generic products. (The chemical ingredients are the same.)
- *Read* labels and *follow instructions* for dosage.
- *Pay attention* to cautions with respect to drowsiness, or interactions with alcohol or other medications.

Reading the label. It's important when you choose any medication and it's important when you use the product. *(C.D. Winters)*

■ SELF-TEST 19C

1. ____________ is a derivative of morphine that is not found in nature.
2. Codeine is a derivative of morphine that is found in nature and that (does/does not) have a medical use.

TABLE 19.3 ■ Typical Ingredients in a Combination OTC Cold Medication

Type of Ingredient (Purpose)	Common Example	
	Name	*Structure*
Decongestant (shrink nasal tissues)	Pseudoephedrine	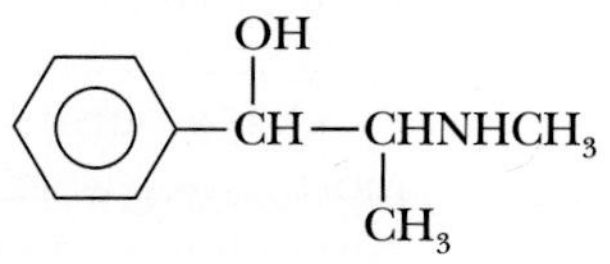
Antitussive (prevent cough)	Dextromethorphan	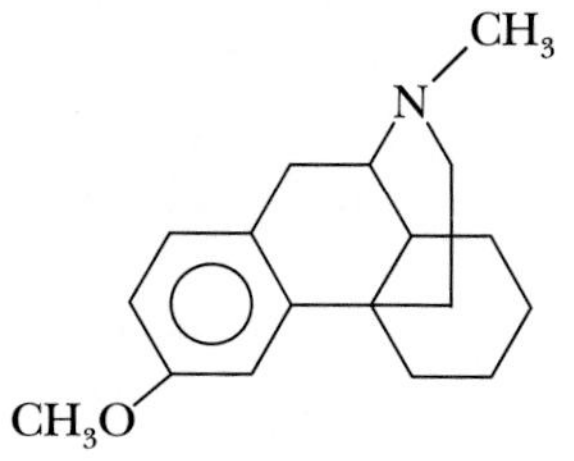
Expectorant (loosen fluids in cough)	Guaifenesin	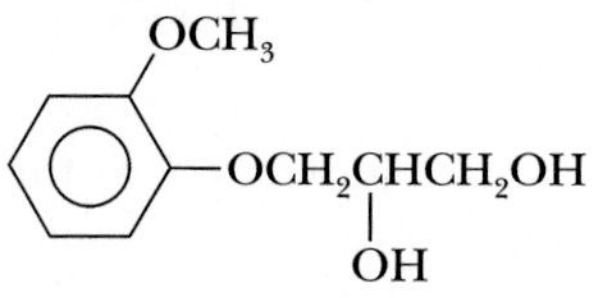
Analgesic (diminish pain, fever)	Acetaminophen	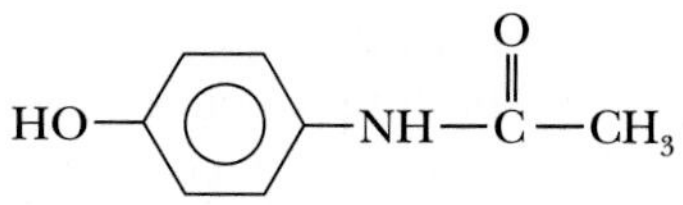
Antihistamine (counteract allergic reaction)	Chlorpheniramine	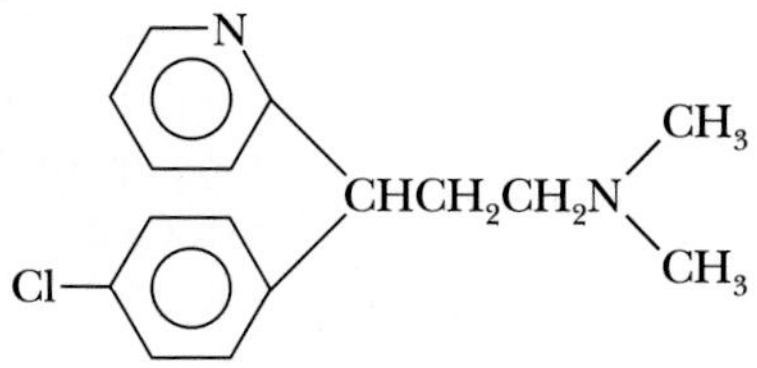

3. Which of these three terms refers to each of the following products: (i) A pain killer, (ii) A medication that combats fever, (iii) A medication that combats the condition that causes muscle pain, swelling, and other symptoms?
   (a) Antipyretic (b) Analgesic
   (c) Anti-inflammatory
4. The chemical name for aspirin is ______________.
5. Acetaminophen differs from ibuprofen and aspirin in two important ways: it does not contain a (an) ______________ functional group,

which means it does not cause intestinal bleeding, and it does not act as an anti- ____________.

6. A Schedule 1 controlled substance is a drug that is ____________ and has no ____________.
7. Barbiturates and benzodiazapines are two major classes of ____________.
8. Crack is a purified form of ____________ and is dangerously even more ____________.
9. Histamine is a neurotransmitter that causes which of the following? (a) Pain, (b) Upset stomach, (c) Allergic reaction.
10. A cold remedy can cure a cold. (a) True, (b) False.
11. Name the five classes of ingredients found in cold medications.
12. The unripened seed pods of ____________ contain at least 20 different alkaloid compounds.
13. Acetaminophen is effective as an analgesic and antipyretic, but is not effective as a (an) ____________.

19.9 Heart Disease and Its Treatment

■ It is important to realize that although death from heart attacks is most common among older people, the condition of atherosclerosis begins many years earlier.

Many new drugs and surgical techniques are being used to decrease the death rate and improve the quality of life for people suffering from heart disease. Known medically as **cardiovascular disease,** heart disease results from any condition that decreases the flow of blood, and consequently oxygen, to the heart or diminishes the ability of the heart to beat regularly and in a normal manner. The most common cause of heart disease is the condition of plaque buildup on artery walls known as atherosclerosis, which was discussed in Section 17.4 in conjunction with the role of diet in plaque formation.

The dietary guidelines developed by the FDA and discussed in Section 17.6 (see Figure 17.5) are largely based on ways to decrease the risk of heart disease, atherosclerosis, and the related condition of high blood pressure, known as hypertension.

Atherosclerotic heart disease results in *angina* (chest pain on exertion) and *ischemia* (partial deprivation of oxygenated blood to the heart). Accumulated damage to the heart muscle by ischemia and heart attacks can lead to *arrhythmias* (abnormal heart rhythms) and heart failure (inability of the heart to pump sufficient blood). A variety of drugs are available to treat these conditions.

Cholesterol-Lowering Drugs

■ Bile acids are secreted into the small intestine to aid in the digestion of food. They are steroids (Section 15.6) and, like all steroids, are synthesized from cholesterol.

Changes in diet and lifestyle are always the first step to lowering blood cholesterol. The next step is the use of drugs such as *lovastatin* and *cholestyramine.* Lovastatin acts by interfering with cholesterol synthesis in the liver. Cholestyramine acts by binding to bile acids in the intestines and accelerating their excretion. This causes the liver to convert more cholesterol into bile acids, leaving less to enter the circulatory system.

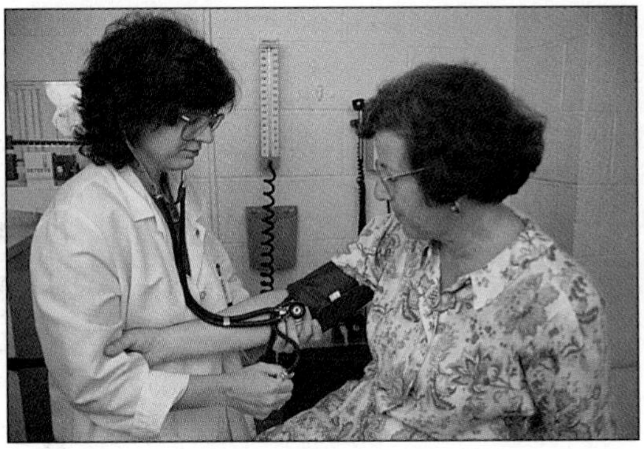

Blood pressure measurement. The pressure on the walls of blood vessels is higher when the heart muscles are pushing blood out (*systolic pressure,* normal values from 115 to 135 in young adults) and lower when the muscles are relaxed (*diastolic pressure,* normal values from 75 to 85 in young adults). It is reported as the two numbers together, for example, 120/80 ("120 over 80"). The cuff is inflated enough to squeeze the artery and prevent blood flow. As the pressure is gradually released, a tapping sound is heard at the systolic pressure and stops when the pressure reaches the diastolic pressure. *(C.D. Winters)*

Diuretics

Treating hypertension early enough can also diminish the risks of heart disease. The first steps are loss of excess weight and restrictions on dietary sodium ions (Na^+) and alcohol. If these measures do not lower blood pressure, drugs are required. The first choice is often a **diuretic,** commonly a *thiazide,* which stimulates the production of urine and excretion of Na^+. With increased urine output, blood volume and, consequently, blood pressure are decreased.

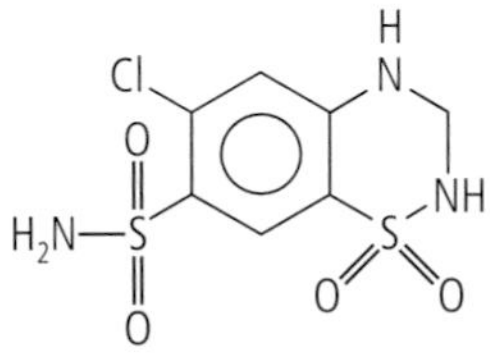

Hydrochlorothiazide, one of the thiazide diuretics

Vasodilators

Angina occurs because of insufficient oxygen delivery to the heart muscle. The attacks are brought on in susceptible individuals by exercise or anxiety, which makes the heart work harder and thus increases its oxygen demand. To treat angina, *vasodilators,* drugs that *dilate* veins (make them open wider), are used. When the veins are dilated, the blood pressure against which the heart must work is reduced. The classic vasodilators are organic nitrates such as nitroglycerin or amyl nitrite. Within tissues they are converted to NO, which acts as described in Chapter 6 (Frontiers in Chemistry: NO News Is Good News).

■ Isn't it interesting that nitroglycerin is also a powerful explosive?

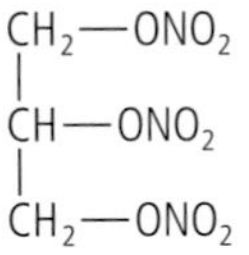

Nitroglycerin

$C_5H_{11}NO_2$

Amyl nitrite

Beta Blockers

Two relatively new classes of drugs, beta blockers and calcium channel blockers, are used to prevent angina and treat other aspects of heart disease. The development of beta blockers illustrates how understanding the biochemistry of disease can lead to the design of drugs.

In the 1960s two types of receptors that are part of the natural regulatory system for heart rate were discovered and named *beta receptors.* The beta-1 receptors are located primarily in the heart—their stimulation speeds up the rate at which the heart beats. The beta-2 receptors are located in the peripheral blood vessels and the bronchial tubes. Stimulation of the beta-2 receptors relaxes muscle fibers, opening up the blood vessels and bronchial tubes so that blood flows more easily and it is easier to breath deeply and quickly.

The beta receptors were found to be activated by epinephrine and norepinephrine. Armed with this information, chemists began to search for

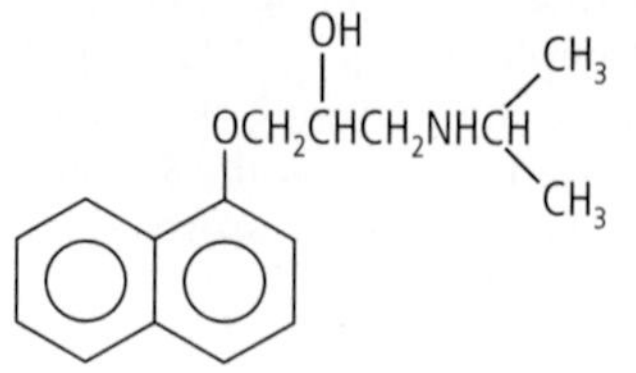

Propranolol (Inderal)

chemicals that would compete with epinephrine and norepinephrine at the beta receptor sites. If these sites could be blocked, stimulation of the heart muscle could be prevented. For a heart already overworked from the buildup of plaque in the arteries, this might produce enough relaxation to avoid an impending heart attack. In addition, these drugs might be able to relieve high blood pressure. The first successful beta blocker was *propranolol (Inderal),* now used to treat cardiac arrhythmias, angina, and hypertension. Look back at the structures in Section 19.5 to see its similarity in structure to the compounds whose action it blocks.

Calcium Channel Blockers

Calcium ions (Ca^{2+}) move into muscle cells through channels that open when the signal calling for muscle contraction arrives. Within the cells, the calcium ions initiate a series of chemical reactions that lead to tightening up the muscle. With the flow of calcium ions blocked, muscles stay in a more relaxed state. When muscles in the walls of coronary arteries relax so that the arteries expand, the supply of blood to the heart increases. In addition, blood pressure is decreased by relaxing muscles in the veins, causing them to dilate. The first calcium channel blockers, verapamil (trade names Isoptin, Calan, Verelan) and nifedipine, were introduced in the United States in 1981. Nifedipine is a powerful dilator of coronary arteries, providing immediate relief from angina, and it has become a widely used drug (see Table 19.1).

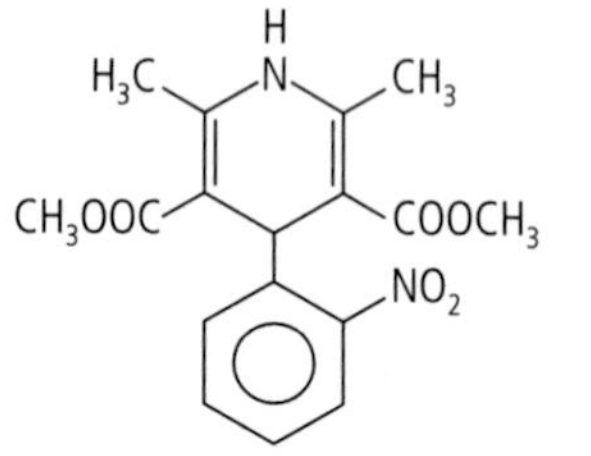

Nifedipine

■ Each of the beta blockers and calcium channel blockers has a specific set of potential side effects that vary for different individuals. Sometimes the drug and dosage must be adjusted several times before the best combination for an individual is found.

Emergency Treatment of Heart Attack Victims

New clot-dissolving drugs given to heart-attack victims in the emergency room (or in ambulances staffed by emergency medical technicians) show promise in reducing the death rate. These drugs are enzymes that act on *plasminogen,* a natural factor in the blood, by converting it to *plasmin.* Once this happens, plasmin proceeds to dissolve blood clots by the body's own natural mechanism. Three enzymes have been developed as drugs that catalyze the plasminogen → plasmin conversion: (1) *urokinase,* a natural enzyme isolated from human urine, (2) *streptokinase,* isolated from a streptococcus bacterium, and (3) *tissue plasminogen activator (TPA),* one of the first drugs produced by recombinant DNA technology to receive government approval.

TPA is made in hamster ovary cells and is identical to the natural human enzyme that activates plasminogen. Intravenous injection of TPA is often enough to halt a heart attack within minutes and to save heart tissue from damage.

Does Aspirin Cut the Risk of a Heart Attack?

Some studies have shown that aspirin can cut the risk of a heart attack even for those who have no overt signs of cardiovascular disease. In a six-year study covering 22,000 male physicians aged 40 to 84 with no history of heart disease, a 325-mg aspirin tablet taken every other day reduced the risk of an initial heart attack by 47%. The aspirin acts by blocking the manufacture of prostaglandins, which are, among their many functions, instrumental in the formation of blood clots. There are potentially serious side effects to routine ingestion of aspirin, however. People susceptible to ulcers or bleeding should not take aspirin frequently. The FDA, which has approved aspirin in low dosage for heart attack prevention, urges that this treatment be undertaken only at the recommendation of a physician who is evaluating all other aspects of a person's health and medications.

■ As of 1994, no comparable large study of the relationship of aspirin and heart attacks in women has yet been completed.

19.10 Cancer and Anticancer Drugs

Cancer is not one but perhaps 100 different diseases caused by a number of factors, including the chemical carcinogens discussed in Section 18.8. The sites of attack for major types of cancer are shown in Figure 19.8. A cancer begins when a cell in the body starts to multiply without restraint and produces descendants that invade other tissues. It seems reasonable, then, that drugs might be able either to stop this undesirable spreading of cancer cells or to prevent cancer from happening at all. A major obstacle to successful drug treatment for cancer is that its biochemistry is not well understood. There is, however, general agreement that cancer is initiated by damage to DNA, which may be done by physical (e.g., ionizing radiation), biological (e.g., viruses), or chemical agents (e.g., compounds in cigarette smoke).

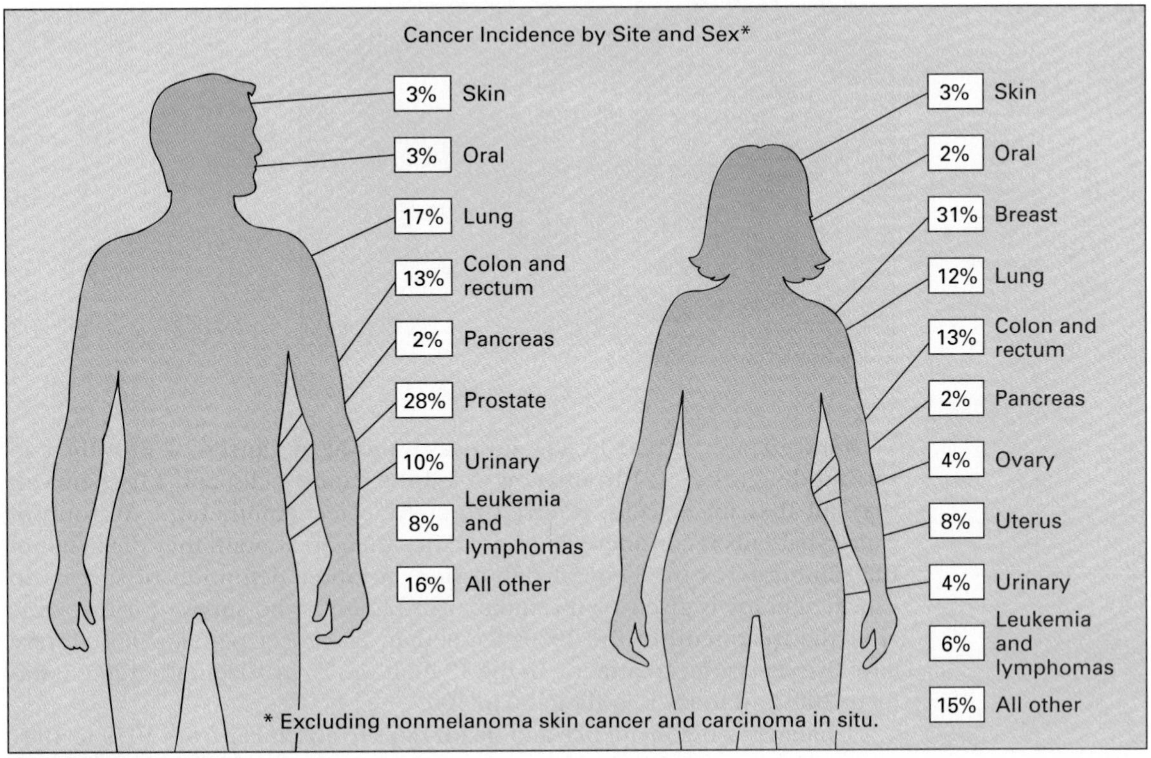

Source: American Cancer Society

Figure 19.8 Estimates of cancer incidence by site and sex for 1993.

Figure 19.9 A status report in the battle against cancer. The bars show percentage changes between 1979 and 1991 in the incidence of and death rate from various cancers.

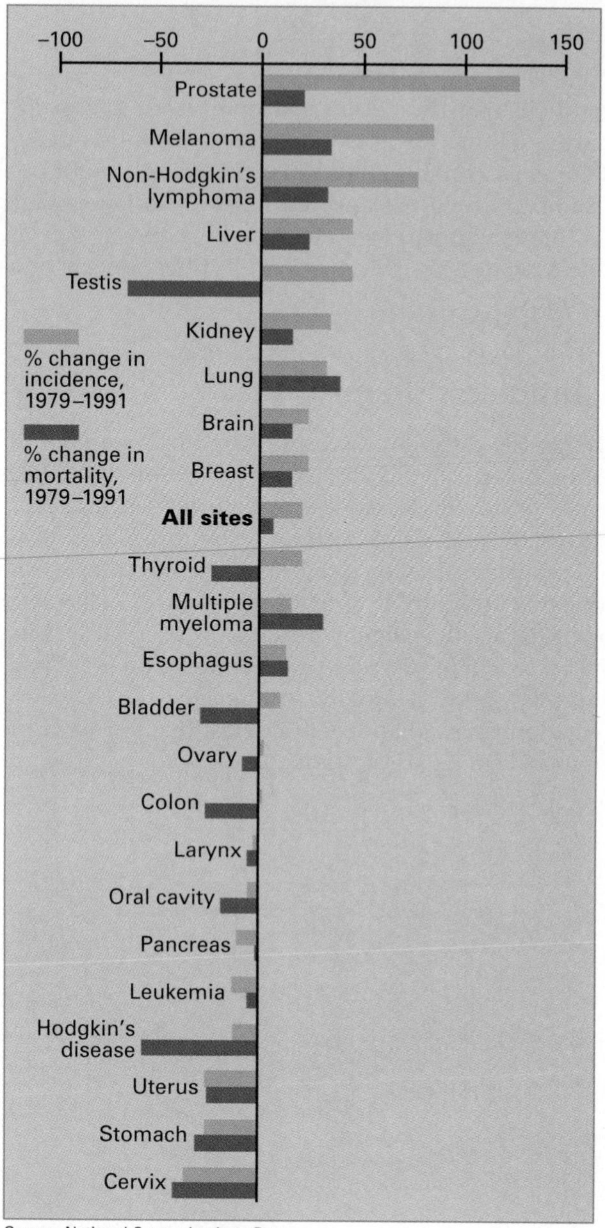

Source: National Cancer Institute Data

Cancers are treated by (1) surgical removal of cancerous growths and surrounding tissue, (2) irradiation to kill the cancer cells, and (3) chemicals that kill the cancer cells, referred to as cancer chemotherapy. A group of cancer patients is considered cured if, after their treatment, they die at about the same rate as the general population. Another definition of success in cancer therapy is given by the number of patients who survive for five years after the treatment. In the 1930s fewer than 20 cancer patients in 100 were alive five years after treatment; in the 1940s it was 25 in 100; in the 1960s it was 33 in 100; and today it is about 50 in 100.

Changes in the incidence and death rates from cancer from 1979 to 1991 (Figure 19.9) vary with the type of cancer, while the *overall* rates continue to increase. Cancers associated with old age (e.g., brain cancer, melanoma) are increasing in incidence, as is lung cancer, which is driven by cigarette smoking. Little change in the death rates from some cancers has been achieved (e.g., breast cancer and pancreatic cancer). Some major decreases in death rates have been achieved, however (e.g., cancer of the testis and cervix, and Hodgkin's disease).

Cancer Chemotherapy

During World War I the toxic effects of the mustard gases were found to include damage to bone marrow and changes in DNA (mutations) that created abnormal offspring. In addition, the so-called *nitrogen mustards* caused cancers in some animals.

■ Chemical warfare is discussed in the *Science and Society* feature in Section 18.5.

■ The name "mustard gas" comes from its mustard-like odor; mustard "gas," however, is not a gas but a high-boiling liquid that was dispersed as a mist of tiny droplets.

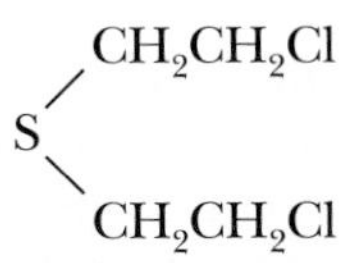

Mustard gas

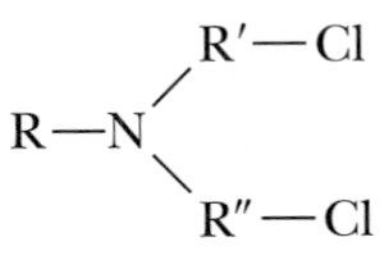

Nitrogen mustard (general formula)

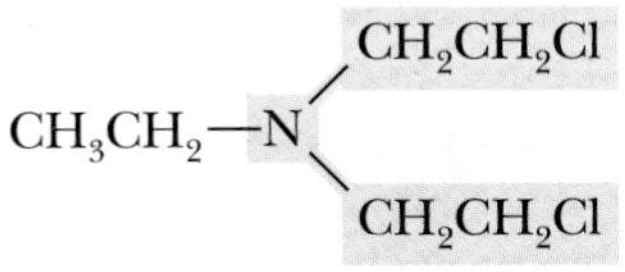

A nitrogen mustard

When the wartime-imposed secrecy surrounding these chemicals ended, it occurred to some researchers that cancers might be treated with similar compounds. The result might be to alter the DNA in cancer cells to the extent that these cells could be destroyed selectively.

One of the most widely used anticancer drugs is now cyclophosphamide, a compound that contains the nitrogen mustard group.

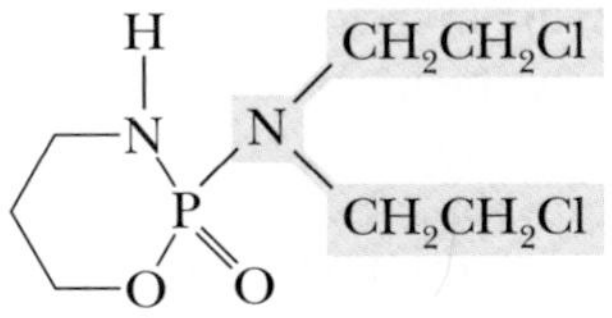

Cyclophosphamide

Cyclophosphamide, and other anticancer drugs that act in the same manner, are **alkylating agents**—reactive organic compounds that transfer alkyl groups in chemical reactions. Their anticancer activity results from the transfer of alkyl groups (e.g., CH_3CH_2—) to the nitrogen bases in DNA, often to guanine. The alkyl group physically gets in the way of base pairing and prevents DNA replication, which stops cell division. Although alkylating agents attack both normal cells and cancer cells, the effect is greater for rapidly dividing cancer cells.

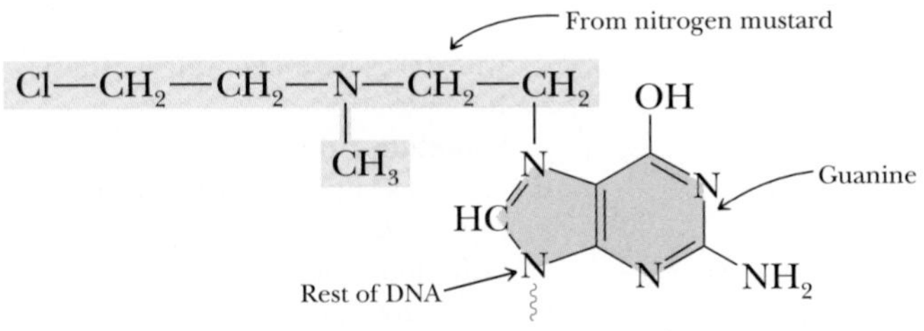

An alkylated guanine in DNA. You can see how the alkyl group alters the size of the guanine residue.

■ Because cells in hair follicles divide rapidly, they are often killed during cancer chemotherapy, resulting in hair loss. (When new cells are generated, hair returns.)

Another class of chemotherapy drug, the **antimetabolites,** interferes with DNA synthesis because they are similar in molecular structure to compounds required for DNA synthesis (metabolites). Methotrexate, for example, is an antimetabolite similar in structure to folic acid (a B vitamin), used in the synthesis of nucleic acids. Methotrexate prevents reduction of folic acid in the first step of nucleic acid synthesis by strongly binding to the enzyme for that reaction. (Methotrexate and folic acid are large molecules. Methotrexate differs from folic acid by the addition of one $-CH_3$ group and replacement of a $C{=}O$ by a $C{-}NH_2$.)

All cancer chemotherapy is tedious and has risks and unpleasant side effects. The problem is that no agent has yet been found that kills only cancer cells, so there is always a balance to be struck between killing healthy cells and killing cancer cells. Because chemotherapy drugs kill actively dividing cells, the goal is to kill many more cancer cells than normal cells (ideally, of course, 100% of the cancer cells). In addition to being highly toxic, most of the cancer chemotherapy drugs are themselves carcinogenic, and very high doses are usually necessary. As a result, single-agent chemotherapy has largely given way to combination chemotherapy because of positive additive, or even synergistic, effects when two or more drugs are used together. Because of their synergistic action, lower doses of each compound can be used when they are given together, and this reduces the harmful side effects. Chemotherapy alone is most successful in treating cancers such as leukemia or lymphoma, in which the cancer cells are widely dispersed in the body.

■ SELF-TEST 19D

1. Heart disease is known medically as ______________.
2. During a heart attack, heart muscle can die from lack of ______________.
3. When beta receptors in the heart are stimulated by epinephrine, the heart beats (a) faster, (b) slower.
4. A drug that blocks calcium ions from flowing into heart muscles has the effect of (exciting/relaxing) the heart muscle.
5. Clot-dissolving drugs used in emergency treatment of heart attacks are classified biochemically as ______________.
6. Which of the following can initiate cancer? (a) Smoking, (b) Viruses, (c) Ionizing radiation.
7. Most anticancer drugs can also cause cancer. (a) True, (b) False.
8. The nitrogen mustards act on cancer cells by blocking ______________ replication.
9. An ideal cancer chemotherapy agent would be able to ______________ cancer cells without doing the same to noncancerous cells.

■ MATCHING SET

___	1. Histamine	a. Interferes with cell wall synthesis
___	2. Penicillin	b. Receptor
___	3. Source of morphine	c. Schedule 1 drug
___	4. Cell surface connector to a messenger	d. Causes symptoms of hay fever
___	5. Heroin	e. Opium poppy
___	6. Sulfa drug	f. Sulfanilamide
		g. Methotrexate

____ 7. Propranolol
____ 8. "Fight or flight" hormone
____ 9. Explosive and heart drug
____ 10. Alkylating drug for cancer
____ 11. Steroid
____ 12. Trade name
____ 13. Generic name
____ 14. Causes heart muscles to contract
____ 15. Antimetabolite drug for cancer

h. Tylenol
i. Testosterone
j. Nitroglycerin
k. Epinephrine
l. Calcium ion
m. Acetaminophen
n. Cyclophosphamide
o. Heart muscle relaxant

QUESTIONS FOR REVIEW AND THOUGHT

1. Give examples of a bacterial disease and a viral disease.
2. Give definitions of the following terms.
   (a) Over-the-counter drug
   (b) Prescription drug
   (c) Generic name for a drug
   (d) Trade name for a drug
3. Name the agency responsible for classifying drugs in the United States.
4. What are antibiotics?
5. Name three major classes of antibiotics.
6. What do the following three acronyms represent?
   (a) AIDS (b) HIV
   (c) AZT
7. What is chemotherapy?
8. Describe how penicillin kills bacteria.
9. What is a retrovirus?
10. For what disease or condition is each of the following classes of drugs used?
    (a) Analgesics (b) Antipyretics
    (c) Antibiotics (d) Antihistamines
11. For what disease or condition is each of the following classes of drugs used?
    (a) Vasodilators (b) Alkylating agents
    (c) Beta blockers (d) Antimetabolites
12. Describe the role of a receptor in biochemistry.
13. What two classes of natural biomolecules require receptors for their action?
14. To what classes of drugs do the following compounds belong?
    (a) Barbiturates (b) Benzodiazepines
    (c) Ethanol (d) Amphetamines
15. To what classes of drugs do the following compounds belong?
    (a) Lovastatins
    (b) Methotrexates
    (c) Chlorphenirimines
    (d) Pseudoephedrines
16. Describe the following and give an example of each.
    (a) Schedule 1 drugs (b) Schedule 2 drugs
    (c) Schedule 3 drugs
17. Which of the following terms apply to codeine?
    (a) Analgesic
    (b) Antibiotic
    (c) Opioid
    (d) A scheduled drug with a potential for abuse
    (e) An antitussive
18. How do antihistamines work in the body? What, if any, side effects do antihistamines have?
19. What symptoms are experienced in
    (a) angina? (b) arrhythmia?
20. Name two drugs that might be used to treat the symptoms described in Question 19.
21. What are prostaglandins and what are some of their functions?
22. What is the physiological effect of excess prostaglandin?
23. What is a barbiturate? What are the physiological effects of barbiturates?
24. What is Reye's syndrome? What common drug is associated with Reye's syndrome?
25. Nitrogen mustards are alkylating agents, drugs that interfere with DNA replication. Explain what this means.
26. What happens when beta receptor sites in heart muscle are stimulated?
27. Name the class of biomolecule that includes dopamine, norepinephrine, and serotonin.
28. The estrogen estradiol has the following structure

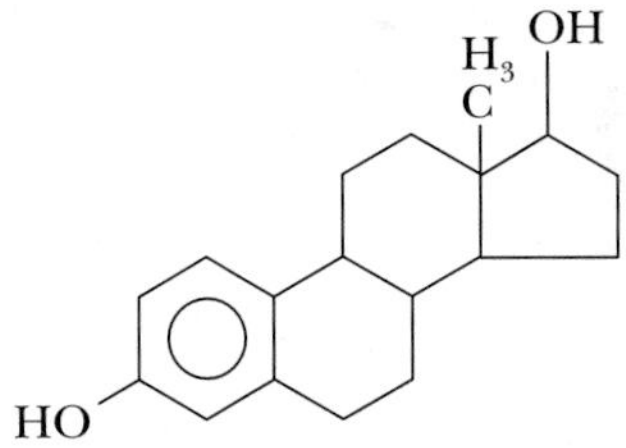

What functional groups are different in ethynyl estradiol?
29. What is the function of premarin? What relationship does it have to osteoporosis?
30. Give the functions for the following.
    (a) Dopamine (b) Epinephrine
31. What are the applications for the following drugs?
    (a) Tylenol (b) Ibuprofen
    (c) Naproxen
32. What are the four classifications of drugs in terms of Drug Enforcement Administration regulations?
33. How are cocaine and crack related?

34. Classify each substance as either an hallucinogen, an antidepressant, or a depressant.
    (a) Mescaline
    (b) Lysergic acid diethylamide, LSD
    (c) Cannabis sativa
    (d) Phencyclidine, PCP
    (e) Barbiturates
    (f) Amphetamines
35. What are the functions of the following over-the-counter drugs?
    (a) Antihistamines (b) Analgesics
    (c) Decongestants (d) Antitussives
    (e) Expectorants
36. What disease is treated with the following drugs? Tell the function of each drug.
    (a) Cholesterol lowering drugs
    (b) Vasodilators
    (c) Diuretics
    (d) Calcium channel blockers
37. What are nitrogen mustards? What was their original purpose? What is their current medical use?
38. What are the three modes of treatment for cancers?
39. Penicillins have the general formula shown below. What is the R group in penicillin G? Why is it necessary to have a number of different penicillins?

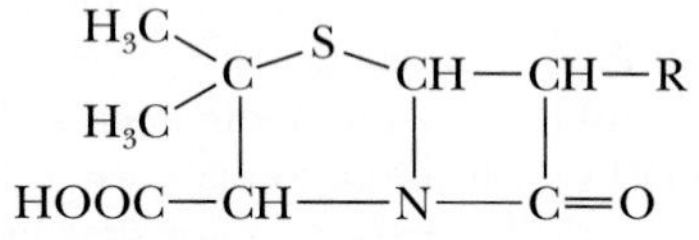

40. How do sulfa drugs work to kill bacteria? Why aren't humans and animals harmed by sulfa drugs?
41. What is the effect of a hallucinogen? Name two examples of hallucinogens.
42. Describe the normal steps in the action of a neurotransmitter at a nerve synapse.
43. Of the three compounds heroin, morphine, and codeine,
    (a) which is the most effective pain killer?
    (b) which is not a natural alkaloid?
    (c) which is so addictive that its sale and use are illegal in the United States?
44. Identify by chemical name the oxygen-containing functional groups in morphine, codeine, and heroin.
45. What is the physiological effect of nitroglycerine? What disease is it used to treat?
46. The FDA requires extensive testing of a prospective drug. The period of development of a new drug can take almost 12 years from initial discovery to final approval. Other countries have much shorter approval processes. Imagine that you have a close relative who is suffering from a serious illness and you have learned that an effective drug for the illness was available only outside the United States. What are the possible courses of action? What would you do?
47. Explain the role of the blood-brain barrier. What does it mean for the design of a drug that must be active in the brain?
48. During the time period represented in Figure 19.7, identify
    (a) the type of cancer whose incidence has decreased the most.
    (b) the type of cancer for which mortality has increased the most.
    (c) the type of cancer for which the incidence has increased the most.
49. Using Figure 19.7, find
    (a) the approximate percentage change in incidence for melanoma, which is associated with exposure to ultraviolet solar radiation.
    (b) the approximate percentage changes for the incidence and mortality of cancer at all sites.
    (c) the approximate percentage changes for the incidence and mortality of lung cancer, which is directly related to cigarette smoking.

The actions of government, industry, and private individuals are all needed if we are to succeed in preventing this kind of sad situation. *(Grant Heilman Photography)*

CHAPTER

20

Water—Plenty of It, But of What Quality?

For a molecule so simple, water, made of two hydrogen atoms covalently bound to an oxygen atom, is extremely important in our world. Our lives and the lives of all the other creatures of the earth depend on the liquid we call water—a collection of these simple H_2O molecules. Too little water and we can die of thirst; too much and we can drown. If an area on the Earth has too little water, it becomes an arid desert. If too much water is present, normal life for land dwellers becomes impossible, and *aquatic* animal life takes over. There is a lot of water on this planet, but much of it is either not in the liquid state—the physical state we find most useful—or contains dissolved substances that make it unfit for most uses. Our largest reserves of water are in the oceans. Most water we come in contact with daily is not *pure;* that is, it is not 100% water molecules and nothing else. Water dissolves all kinds of substances; it is a "universal solvent." To assure a safe water supply, we must limit the dissolved substances in type and quantity to what is safe. In this chapter we shall look at the answers to these questions:

- How can there be a water shortage if water is such an abundant compound?
- How is water used and reused?
- What are the differences between clean water and polluted water?
- How is water pollution measured?
- What are some of the causes of water pollution?
- How is water made pure?
- How can we ensure that our water supplies remain pure?
- Will we be able to use the vast quantities of water found in the oceans?

Assuring the quality of the water supply is a shared responsibility. The federal government enacts laws governing the quality of waste water that can be returned to the environment and also the quality of water that can be

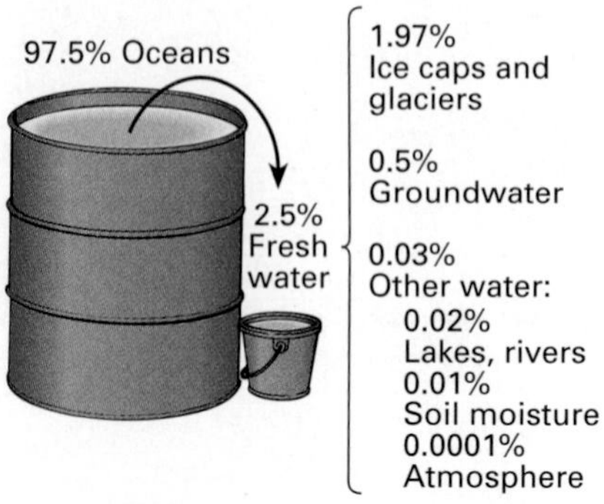

The water supply. Of the 2.5% fresh water, less than 1% is available as groundwater or surface water for human use.

supplied for drinking. The U.S. Environmental Protection Agency (EPA) enforces these laws. State and local governments have their own laws and shoulder the responsibility for enforcing them as well as seeing to it that local industries, sewage treatment plants, and municipal water treatment plants meet federal standards.

The goal of these regulations is to keep water relatively pure so the next user of that water, whether that is a city wanting a water supply or a trout looking for a nice stream to live in, will find it suitable for its intended use.

Individual citizens also have a role to play in keeping the water clean. Voluntary action by an informed public is needed to halt water pollution that originates in our households. Laws alone cannot halt disposal of hazardous waste in the municipal garbage or pouring of potential water pollutants into toilets, sinks, or storm drains.

In this chapter we will look at how water is used and reused, how it becomes polluted, and how it is purified. This chapter will build on what you learned about the properties of water in Section 7.4. As you will see, understanding how water is used, how it becomes polluted, and how it can be purified depends on understanding its properties.

20.1 How Can There Be a Shortage of Something as Abundant as Water?

Water is the most abundant substance on the Earth's surface. Oceans (with an average depth of 2.5 miles) cover about 72% of the Earth. They are the reservoir of 97.5% of the Earth's water. Only 2.5% is fresh water. Water is also the major component of all living things (Table 20.1). For example, the water content of human adults is 70%—the same proportion as for the Earth's surface.

■ The Mississippi River discharges an estimated 400 billion gallons of water daily into the Gulf of Mexico.

An average of 4350 billion gallons of rain and snow fall on the contiguous United States each day (Figure 20.1). Of this amount, 3100 billion gallons return to the atmosphere by evaporation and transpiration. The discharge to the sea and to underground reserves amounts to 800 billion gallons daily, leaving 450 billion gallons of surface water each day for domestic and commercial use. The 48 contiguous states withdrew 40 billion gallons per day from natural sources in 1900, but that rose to over 400 billion gallons by 1993. It is estimated that the demand may be 900 billion gallons per day by the year 2000. You can see from these numbers that the demand for water by our growing population is already as great as the resupply by natural resources. This means that water must be reused on a daily basis. The water molecules you drink from a water fountain may have been in the municipal wastewater of another city just a few days earlier.

■ A shallow well generally goes no deeper than 50 feet. A deep well is generally drilled to between 100 and 200 feet below the surface. The deepest water well in the world is in Montana—it goes 7320 feet into the ground.

The two sources of usable water are **surface water,** such as rivers, lakes, and wetland waters, and **groundwater,** which is beneath the Earth's surface. Figure 20.1 shows our water resources and the flow of groundwater. About 90 billion gallons of the total water withdrawn every day is groundwater drawn from wells drilled into **aquifers,** layers of water-bearing porous rock or sediment held in place by impermeable rock. These kinds of wells, which supply water to many cities in the Great Plains and along the East Coast, are called **artesian wells**.

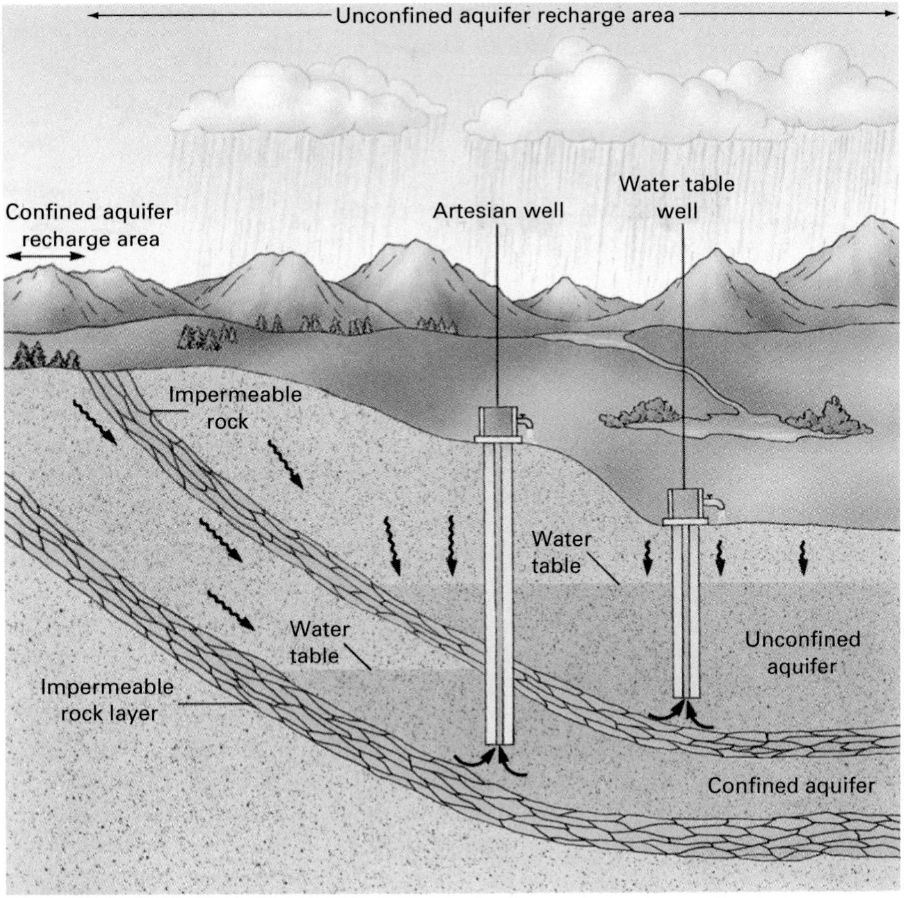

Figure 20.1 Water resources. Surface water includes water that collects in rivers, lakes, and wetlands. Groundwater may be in unconfined aquifers, which are recharged by surface water from directly above, or in confined aquifers, which lie between impermeable rock layers. Wells may tap into either kind of aquifer. Where water in a confined aquifer is under pressure, pumping of the water may not be needed.

TABLE 20.1 ■ Water Content

	Percentage
Marine invertebrates	97%
Human fetus (1 month)	93%
Adult human	70%
body fluids	95%
nerve tissue	84%
muscle	77%
skin	71%
connective tissue	60%
Vegetables	89%
Milk	88%
Fish	82%
Fruit	80%
Lean meat	76%
Potatoes	75%
Cheese	35%

In the arid West, wells used to pump water for irrigation either are going dry or require drilling so deep that irrigation is no longer economically feasible. The huge Ogallala aquifer that stretches through eight states from South Dakota to Texas has 150,000 wells tapping it for irrigation of ten million acres. As a result, the Ogallala aquifer is being drawn down at a rate that has reduced the average thickness of the aquifer from 58 ft in 1930 to 8 ft today.

The depletion of a major aquifer along the Eastern seaboard has caused large sinkholes in Georgia and Florida when the limestone rock strata of the aquifer collapse as the water is withdrawn (Figure 20.2). Many coastal cities are also experiencing problems with **brackish water** that comes from aquifers where the fresh water has been drawn off, causing sea water to flow into the depleted aquifer. Depletion of underground sources has also caused sinkholes in Texas. The entire city of Houston has sunk several feet over the years as the result of extensive use of the underground water sources in that area.

Figure 20.2 A sinkhole in Winter Park, Florida. As the result of aquifer depletion, the top of an underground cavern has collapsed, an event that can happen in the course of just a few minutes. *(M. Timothy O'Keefe/Tom Stack & Associates)*

Steam rising from the cooling towers of a coal-fired power plant. *(Simon Fraser/SPL/Photo Researchers)*

TABLE 20.2 ■ Water Consumption in The United States (Typical) per Day

	Billion Gallons
Households	8.9
Industry	6.7
Steam-electric utilities	4.0
Agriculture	76
	95.6

Source: Statistical Abstracts of the United States, 114th ed., U.S. Department of Commerce, 1994.

20.2 Water Use and Reuse

Who's using the water? Table 20.2 shows the breakdown for water use in the United States. Agriculture is the major user of water, with all other uses being a distant second.

Industrial water usage can be directly related to production of finished products. One gallon of water, for example, is used to produce eight sheets of ordinary typing paper, while 80 gallons of water are needed to produce a gallon of gasoline, and about 25 gallons of water are needed to make a single box of nails. Of course, because none of these products *contain* water, the water used is either recycled into the environment or recycled within the facility. Industrial recycling has not always been the case. For example, the paper industry used more than 400,000 gallons of water to produce a ton of paper in 1900, but modern processes have dropped the water usage to under 40,000 gallons per ton of paper produced.

In many industrial locations the largest single use of water is in plant cooling systems. Water can absorb more heat than any other readily available liquid—it has a heat capacity of 1 cal/g, or 4.18 J/g. Recirculating cooling water is an important means of water conservation. A cooling tower allows the warmed water to lose its heat to the surrounding air and helps to reduce thermal pollution of the river or lake where the used water would be discharged. In addition, the high heat capacity of water enables the industrial user to recycle heat energy captured by the cooling water.

■ Water that is returned to a waterway at a higher temperature than it was withdrawn contributes to thermal pollution of the waterway. Because the solubility of oxygen in water decreases with increased temperature, thermally polluted water cannot support aquatic life as well as cooler water can.

Groundwater and surface water sources each provide half of the more than 44 billion gallons of **potable** (drinkable) water that is used each day in the United States. Table 20.3 gives the average amounts for various personal uses. Only a small fraction of municipal water really needs to be of drinking-water quality. The largest portion of a water supply could be disinfected and made bacteriologically safe, while avoiding the more costly treatment needed to meet drinking-quality standards. Water treated in this way would be suitable for irrigation of parks and golf courses, air conditioning, industrial cooling, and toilet flushing. Consider the data in Table 20.3, which illustrate the inefficiency of residential water systems. We use 33% of residential water for flushing toilets and 25% for bathing. Conventional showers use up to 10 gal/min, which can be reduced by as much as 70% by the installation of inexpensive water-saving shower heads. It is obvious that strong arguments could be made

■ There are newer-design toilets that use about 1.5 gal per flush.

TABLE 20.3 ■ Average Water Usage per Person per Day

Use	Gallons
Flushing toilets	30
Bathing	23
Laundering	11
Drinking and cooking	2
Miscellaneous	10
Dishwashing	6
Other*	86
Total	168†

*Includes car washing, swimming pools, lawn watering, public fountains, losses due to leaks in water mains, etc.

†Based on taking all treated water used per day and dividing by the total population.

Source: American Water Works Association.

Surface level drop in San Joaquin, California. The markers on the utility pole show how the surface level has dropped over the years due to withdrawal of ground water for irrigation. *(Courtesy of U.S. Geological Survey)*

for dual water systems—one system that treats water for drinking and a different system for other uses.

Residential water conservation is a way to cut demand for fresh water supplies. Although Table 20.2 shows that residential use is a small part of the total, home water conservation can be an important step in cutting demand for fresh water supplies in large urban areas.

Almost all water molecules have been recycling since they were formed billions of years ago. Have you ever considered that your next glass of water may include some molecules that Aristotle or Abraham Lincoln drank, or some molecules that flooded into the Titanic when it sank? In spite of these possibilities, the idea of obtaining potable water from wastewater or even sewage, especially wastewater that was *recently* discharged, is psychologically difficult for many people to accept. Nevertheless, the technology has been developed and is currently being used in NASA's space shuttle flights. What this means is if water (pure or not) is available, it should be considered for reuse. In the southwestern United States the rate of depletion of aquifers has led to direct recycling of water from sewage, a process called **groundwater recharge.** For example, in El Paso, Texas, ten million gallons of pure water per day from sewage effluent is pumped into the underground aquifer that is the main source of water for El Paso.

20.3 What Is the Difference Between Clean Water and Polluted Water?

The term **pollution** is used to describe any condition that causes the natural usefulness of air, water, and soil to be diminished. Water that is judged unsuitable for drinking, washing, irrigation, or industrial uses is polluted water. The pollution (Table 20.4) may be heat, radioisotopes, toxic metal ions and anions, organic molecules, acids, alkalis, or organisms that cause disease

TABLE 20.4 ■ U.S. Public Health Service Classes of Pollutants

Pollutant	Example
Oxygen-demanding wastes	Plant and animal material
Infectious agents	Bacteria, viruses
Plant nutrients	Fertilizers such as nitrates and phosphates
Organic chemicals	Solvents, pesticides, detergent molecules
Other minerals and chemicals	Acids from mining operations, inorganic chemicals from metal-working operations
Sediment from land erosion	Clay silt from stream beds
Radioactive substances	Mining wastes, used radioisotopes
Heat from industry	Cooling water from electric generating plants

■ Creeks often have names that indicate a content other than pure water. Here are a few creeks along with the metal compounds they contain. Creeks with these names can be found in many different states. The chemicals are usually the same.

Red Creek	Iron
Sulfur Creek	Sulfur of hydrogen sulfide
Copper Creek	Copper
Red Mountain Creek	Iron
Alum Creek	Aluminum compounds
Bitter Creek	Aluminum
Buttermilk Creek	Colloidal zinc carbonate

(pathogens). Water suitable for some uses might be considered polluted and therefore unsuitable for other uses—you might go swimming in water you would not consider drinking. Water that is unsuitable for use has often been polluted by human activity, but natural processes can also pollute water. For example, water that contacts organic substances such as decaying leaves and animal wastes will pick up numerous organic compounds, many of which impart odors and color to the water and some of which might be pathogenic. Silt, consisting of colloidal-sized particles of dirt and sand, can also pollute water. Table 20.5 lists some of the substances that can be found in "pure" natural water. Looking at this list it is clear that absolutely pure (100%) water is not a common commodity.

As human activities have continued to pollute water, governments have passed laws designed to keep our waters clean. The Clean Water Act of 1977 represented a major change in the thinking of Congress regarding who is

TABLE 20.5 ■ Dissolved Substances Found in "Pure" Water

Name	Formula	Comment
The following come from contact of water with the atmosphere:		
Carbon dioxide	CO_2	Makes water slightly acidic
Dust particles	—	Can be large amounts at times
Nitrogen	N_2	Along with oxygen, causes visible bubbles in hot water
Nitrogen dioxide	NO_2	Formed by lightning
Oxygen	O_2	Supports aquatic life
The following vary considerably, depending on the kinds of rock formations the water has contacted:		
Bicarbonate ions	HCO_3^-	Soils and rocks
Calcium ions	Ca^{2+}	From limestone
Chloride ions	Cl^-	Soils, clays, and rocks
Iron(II) ions	Fe^{2+}	Soils, clays, and rocks
Magnesium ions	Mg^{2+}	Soils, clays, and rocks
Potassium ions	K^+	Soils, clays, and rocks
Sodium ions	Na^+	Soils, clays, and rocks
Sulfate ions	SO_4^{2-}	Soils and rocks

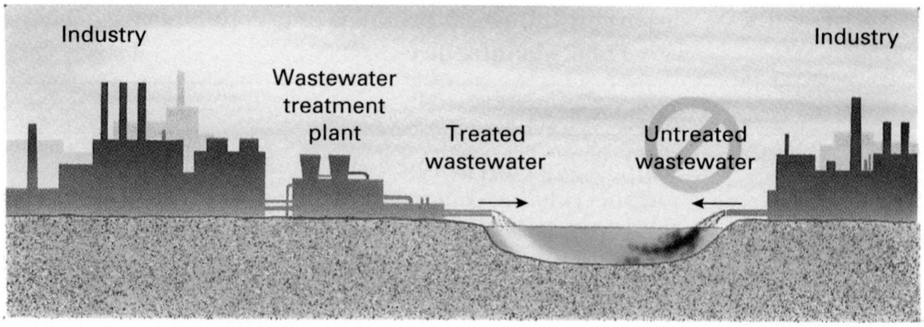

Figure 20.3 The EPA requires that virtually all industrial wastewaters be treated prior to discharge. This water is reasonably pure although it is not as pure as drinking water.

responsible for keeping water clean. This act shifted the burden of producing water suitable for reuse from the user (a municipality, for example) to the wastewater discharger. Because it is easier to clean wastewater prior to dumping than to clean the water after the untreated waste has been discharged (Figure 20.3), the Clean Water Act was a major step in improving the quality of our natural waters. The Act requires the EPA to establish and monitor emission standards—the maximum amounts of water pollutants that can be discharged into natural bodies of water from factories, municipal sewage treatment plants, and other facilities. As it turns out, the wastewater effluent from an industry can now often be clean enough to be used for such purposes as irrigation or industrial cooling.

■ The Clean Water Act is also known as the Federal Water Pollution Control Act. It was revised in 1978, 1980, and 1988.

SELF-TEST 20A

1. Approximately what percentage of the human body is water?
2. What is the major reservoir of water on Earth?
3. Water beneath the Earth's surface is called ____________.
4. Water on the surface of the Earth is called ____________.
5. The average person in the United States uses about ____________ gallons of water per day for all purposes.
6. The actual amount of potable (drinkable) water a person needs is ____________ per day.
7. Three common water pollutants are ____________, ____________, and ____________.
8. A water-bearing stratum of porous rock, sand, or gravel is called a(n) ____________.
9. What happens to most of the water that falls on the United States each day?

20.4 The Impact of Hazardous Industrial Wastes on Water Quality

Industrial processes, whether making paper or automobiles or TV sets, produce waste materials. Table 20.6 (p. 646) lists some of the industrial pollutants that result from the manufacture of products important to us. For many years the disposal of solid wastes in *landfills* was considered good engineering practice. Many of the substances present in those wastes were partially dissolved by rainwater and became part of the groundwater, causing serious pollution of

Corroding waste barrels. *(USDA/Soil Conservation Service)*

TABLE 20.6 ■ Important Industrial Products and Pollutants Associated with Their Manufacture

Products	Pollutants
Plastics	Solvents, organic chlorine compounds
Pesticides	Organic chlorine compounds, organic phosphate compounds
Medicines	Solvents, metals such as mercury and zinc
Paints	Metals, pigments, solvents, organic residues
Petroleum products	Oils, organic solvents, acids, alkalies
Metals	Metals, fluorides, cyanide, acids, oils
Leather	Chromium, zinc
Textiles	Metals, pigments, organic chlorine compounds, solvents

water supplies. Many of these older landfills also allowed surface runoff that contained dissolved substances from the wastes to be carried into natural bodies of water.

It was also common practice in the past to place waste and discarded chemicals into metal drums and bury them directly in the ground. After a few years the drums developed leaks due to corrosion, allowing the drum contents to leak into water that would ultimately become groundwater or surface water. Recognizing this common practice and what it was doing to water quality, the U.S. Congress in 1976 passed the Resource Conservation and Recovery Act (RCRA). In 1980 Congress established the "Superfund," a $1.6 billion program designed to clean up hazardous waste sites that were threatening to contaminate the nation's water supplies. Since 1980, the U.S. EPA has targeted many thousands of waste sites that have the potential of harming our water supplies, and some of these have been cleaned up. In general, the waste site cleanups have seen a lot of tax dollars spent and a lot of litigation involving "Responsible Parties" regarding who will pay for the cleanup. Currently, the EPA has identified more than 1200 sites in the United States where toxic wastes have been stored that should be cleaned up. The cleanup of these remaining waste sites will cost hundreds of billions of dollars.

■ When the Superfund was first established, over 40,000 sites were identified for cleanup.

While the Superfund is dealing with existing hazardous waste sites, the RCRA law and its regulations govern the disposal of newly generated wastes that have the potential to harm the environment. The EPA defines certain industrial wastes as **hazardous wastes** (Table 20.7) and closely regulates how they are generated, stored, transported, and disposed of. The RCRA law was designed to give "cradle-to-grave" (origin to disposal) responsibility to *generators* of hazardous wastes. Before RCRA, an industry could hire almost anyone to haul away its wastes without regard to where it was taken or how it would be disposed of. Today, a generator of a hazardous waste must know who is transporting it, where it is going, and how it will be disposed of. Each shipment of hazardous waste is accompanied by a **manifest,** a document listing the hazardous waste by name, how much is present, and how it will be disposed of. As the load of hazardous waste is transported, temporarily stored, and finally disposed of, parts of the manifests are signed by each responsible party and returned to the generator at each step. The states also receive copies of the

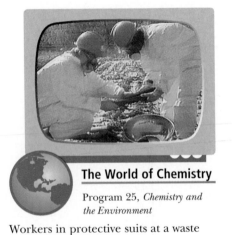

The World of Chemistry

Program 25, *Chemistry and the Environment*

Workers in protective suits at a waste site.

TABLE 20.7 ■ Hazardous Wastes as Defined by EPA (Causing Water Pollution)

Wastes containing the following metals and pesticides:
Arsenic, barium, cadmium, chromium, lead, mercury, selenium, silver, endrin, lindane, methoxychlor, toxaphene, 2,4-D, 2,4,5-TP (Silvex)
Wastes that have the following characteristic properties*:
Ignitible, corrosive, reactive, acutely toxic
Twenty-one wastes from nonspecific sources (such as)
Wastes containing the cyanide ion, distillation residues, used halogenated solvents such as carbon tetrachloride
Eighty-nine wastes from specific sources (such as)
Wastewater sludges from chloride production, wastewater from pesticide manufacture, sludges from the production of petroleum products
A large number of various discarded and off-specification chemicals, many of which are used in the chemical industry to manufacture pharmaceuticals, polymers, paints, dyes, automotive products, cosmetics, and so on.

Note: Shipments of hazardous wastes are carefully monitored by EPA and state governments. In addition, all facilities receiving these wastes must have permits and licenses.

Source: EPA

*Detailed definitions apply to these waste characteristics.

manifests, and annual reports of hazardous waste activities are filed with EPA by everyone who handles hazardous wastes routinely.

Today, industrial hazardous wastes must be placed into *secure* landfills, incinerated, or treated in some way to render them nonhazardous. No hazardous waste is allowed to be disposed of in a way that could pollute the environment. Some hazardous wastes go to secure landfills with plastic linings (Figure 20.4) that prevent their contents from easily reaching surrounding water supplies. These landfills also have carefully spaced monitoring wells so any substances escaping from the landfill's contents may be detected, allowing the

Figure 20.4 A hazardous waste landfill. Underneath, this landfill has several feet of clay covered by three plastic liners. Barrels of hazardous waste are placed above the liners and buried in soil. *(Dennis Barnes)*

FRONTIERS IN THE WORLD OF CHEMISTRY

Arsenic-Eating Microbes

Ordinarily, arsenic is toxic to most organisms (Section 18.4). But a microbe has been discovered that actually thrives on arsenic, using its most highly oxidized form as a source of energy. Dianne Ahmann at the Massachusetts Institute of Technology in Boston and her co-workers isolated a microbe from arsenic-contaminated sediments of the Aberjona watershed in eastern Massachusetts. Until the 1930s arsenic-containing industrial waste had been dumped into the watershed. Normally, when exposed to the environment, arsenic is oxidized to arsenic(V) and, in the form of insoluble arsenate ions, binds to sediment particles by strong electrostatic forces. The arsenic would be expected to remain in the sediment for a long time, all the while having a toxic effect on any organisms that might come in contact with it.

Ahmann and her colleagues found that samples of the Aberjona sediment contained unusually high concentrations of arsenic(III), a less oxidized form of arsenic. After carefully screening samples of the sediment to determine what was producing the arsenic(III), they isolated a microbe, which they named MIT-13. Experiments with the microbe showed that as the concentration of arsenic(V) in its environment decreased, the concentration of arsenic(III) increased, and that at the same time, cell growth increased. The microbe uses arsenic(V) as a source of energy—in other words, as a food! To show just how dependent MIT-13 is on arsenic(V), experiments were done in which everything was kept the same but with no arsenic(V) added. In these studies, minimum cell growth resulted.

What promise does a discovery like this hold for dealing with arsenic-containing wastes? If this microbe can be grown in sufficient quantity, perhaps it could be used routinely to help get rid of persistent arsenic(V) deposits in sediments and perhaps even in soil samples. Of course, there is the arsenic(III) to be dealt with. It is water-soluble, but that property can be used to advantage. If a solution containing arsenic(III) is passed through an organic polymer containing numerous ionically charged sites (an ion-exchange resin), the arsenic(III) can be concentrated and eventually purified and reused. Scientists will continually be on the lookout for microbes like MIT-13 that can thrive on metals and organic compounds that are deadly to most organisms. To turn an old saying around, "one microbe's poison is another microbe's meat."

Dianne Ahmann holding a microbe-filled test tube. (Tom Pantages)

Reference Ahmann, D., et al. "Microbe grows by reducing arsenic," *Nature,* Vol. 371, p. 750, October 27, 1994.

problem to be corrected. Other hazardous wastes may no longer be placed in secure landfills, but must be incinerated or destroyed in some other way. While incineration seems like a logical best choice to dispose of hazardous wastes, current incinerators are operating at near capacity, and it is difficult to get proper permits for new ones due to community opposition. (Incineration of *some* hazardous wastes does produce small quantities of other, even more harmful combustion products.)

States like California that have severe water-shortage problems have taken drastic steps to protect their ground and surface water. In California Proposition 65, the Safe Drinking Water and Toxic Enforcement Act of 1986, lists approximately 200 chemicals or classes of chemicals known to cause cancer or reproductive toxicity and prohibits their discharge into any water that might become a drinking-water supply. Other states have shown interest in creating statutes similar to California's Proposition 65.

■ Proposition 65: "No person in the course of doing business shall knowingly discharge or release a chemical known to cause cancer or reproductive toxicity into water or onto or into land where such chemical passes or probably will pass into any source of drinking water." This proposition resembles the Delaney clause (see Chapter 17, Science and Society) and carries with it similar problems.

TABLE 20.8 ■ Some Common Household Hazardous Wastes and Recommended Disposal

Type of Product	Harmful Ingredients	Disposal*
Bug sprays	Pesticides, organic solvents	Special
Oven cleaner	Caustics	Drain
Bathroom cleaners	Acids or caustics	Drain
Furniture polish	Organic solvents	Special
Aerosol cans (empty)	Solvents, propellants	Trash
Nail-polish remover	Organic solvents	Special
Nail polish	Solvents	Trash
Antifreeze	Organic solvents, metals	Special
Insecticides	Pesticides, solvents	Special
Auto battery	Sulfuric acid, lead	Special
Medicine (expired)	Organic compounds	Drain
Paint (latex)	Organic polymers	Special
Gasoline	Organic solvents	Special
Motor oil	Organic compounds, metals	Special
Drain cleaners	Caustics	Drain
Shoe polish	Waxes, solvents	Trash
Paints (oil-based)	Organic solvents	Special
Mercury batteries	Mercury	Special
Moth balls	Chlorinated organic compound	Special
Batteries	Heavy metals such as Hg	Special

*Special: Professional disposal as a hazardous waste. Drain: disposal down the kitchen or bathroom drain. Trash: Treat as normal trash—no harm to the groundwater. In most households, the items marked special are disposed of as normal trash, which results in groundwater pollution.

Source: "Household Hazardous Waste: What You Should and Shouldn't Do," Water Pollution Control Federation, 1986.

20.5 Household Wastes that Affect Water Quality

Often we do not think about the things we discard in our garbage, but what we throw away and how we do it can affect the quality of natural waters as much as what industry does. Household wastes that are incinerated can contribute to air pollution (see Chapter 21), but because the bulk of our household waste goes to landfills, we too can be responsible for causing pollution of groundwater as well as of rivers, streams, and lakes. Table 20.8 lists some common household products and the kinds of chemicals they contain. Because we are the consumers of industrial products, we can put the very same chemicals into the water as industry can. Although the individual amounts of harmful chemicals used in a household are less than those used in a large industry, the total amounts disposed of daily by all households can be very large, even for a medium-sized city.

■ The EPA estimates that each year 350 million gallons of waste motor oil are poured on the ground or flushed down the drain by individual citizens. That's 35 times more oil than the *Exxon Valdez* spilled in Alaska!

Households have a greater problem disposing of hazardous wastes than industry does. Even where there is an active recycling project for glass, paper, metals, and plastics, there is often no pickup of chemicals that should be separated from the ordinary trash destined for the landfill. If these chemicals are mixed with ordinary garbage, they go to the city landfill or incinerator. If

■ Ordinary garbage costs about \$27/ton for disposal, whereas hazardous waste costs about \$1000/ton for proper disposal.

DISCOVERY EXPERIMENT

Inspection of Your Tap Water

Most of the water you drink probably comes directly from a tap in your home or from a drinking fountain in your school or office. In spite of efforts by municipal water departments, tap water sometimes contains substances that impart color and odor. Here are some simple observations you can make. You will need a clear glass or cup to collect a sample of your tap water. Carefully note the condition of the tap from which you take the water sample. If the tap is on a sink, note the condition of the sink. Now answer these questions:

Does the water sample have an odor? If yes, go to #1.

Does the water have a taste? If yes, go to #2.

Does the water have a slight color? If yes, go to #3.

Has the water stained the tap or the sink, or is there a gray deposit present? If yes, go to #4.

1. *Is the odor like rotten eggs?* If yes, then the water contains hydrogen sulfide (sometimes called "sulfur water." *Is the odor bleach-like?* If yes, then the water contains an excess of chlorine. *Is the odor musty or earthy?* If yes, then the water may contain algae or coliform bacteria that might possibly be a health hazard.

2. *Is the taste metallic?* If yes, then the water may contain iron, zinc, copper, or lead. The pH may also be acidic, meaning the water is dissolving metals from the pipes through which it runs. *Is the taste salty?* If yes, then the water may have a high amount of dissolved solids, including sodium chloride.

3. *Is the color cloudy white?* If yes, then the water is turbid, caused by suspended colloidal particles. *Is the water dark, almost black?* If yes, then there is manganese and possibly iron sulfide present. *Is the water red?* If yes, there is iron present. *Is the water brown or yellow?* If yes, there is possibly iron or tannins from decaying plant materials.

4. *Is the deposit gray to white?* If yes, then the water contains excess Ca^{2+} and Mg^{2+} ions (hardness). *Is there a red or brown stain?* If yes, then the water contains iron. *Is there a black stain?* If yes, then the water contains manganese. *Is there a green or blue stain?* If yes, then the water contains copper.

If your water source is a well and you answered yes to any of the questions above, you might want to call your local health department about getting a water analysis. If you have city or municipal water, contact the water department and request a copy of their Municipal Drinking Water Contaminant Analysis Report.

■ Suspended particles in water are colloidal in size and might include bacteria and viruses, as well as harmless soil particles.

they are poured out in the sink, driveway, or backyard they will eventually reach natural waters.

How can we dispose of hazardous household wastes without danger to the groundwater supply? We can ask our city's municipal waste authorities to provide disposal sites for these wastes or to sponsor periodic household hazardous waste days when these materials can be brought to a central site. In some U.S. cities and some European countries (such as the Netherlands) special trucks

Municipal hazardous waste day. On a designated day, citizens bring their hazardous waste to a site where trained crews sort the materials for disposal. Every community should have such days or have access to a permanent hazardous waste disposal facility. *(Alan Pitcairn /Grant Heilman Photography)*

pick up paints, oil, batteries, and other products for disposal. Another approach is to purchase products with their ingredients in mind. Ordinary alkaline batteries often work just as well as mercury batteries, for example.

20.6 Measuring Water Pollution

Analytical chemistry is that branch of chemistry devoted to the determination of what is present and how much is present in pure chemicals or mixtures. Analytical methods have been developed for every known water pollutant, and these are carefully followed to ensure that our water supplies remain as free from harmful pollutants as possible. To illustrate three of these water analysis methods in a general way, let's first consider how water is analyzed for dissolved organic matter, too much of which can lower the available dissolved oxygen to the point where aquatic life cannot survive. Then we will look at methods that determine the amounts of toxic metals and organic compounds that are commonly found in drinking water.

Biochemical Oxygen Demand

Many organic compounds that find their way into water can easily be oxidized by microorganisms that are also there. This is a natural process that prevents a buildup of organic waste in natural waters. To change this organic material into simple substances (such as CO_2 and H_2O) requires oxygen. The amount of dissolved oxygen required is called the **biochemical oxygen demand (BOD),** and it is a measure of the quantity of dissolved organic matter. The oxygen is necessary so that the bacteria and other microorganisms can metabolize the organic matter that constitutes their food. Ultimately, given near-normal conditions and enough time, the microorganisms will convert huge quantities of organic matter into the following end-products:

$$\text{Organic carbon} \longrightarrow CO_2$$
$$\text{Organic hydrogen} \longrightarrow H_2O$$
$$\text{Organic oxygen} \longrightarrow H_2O$$
$$\text{Organic nitrogen} \longrightarrow NO_3^- \text{ or } N_2$$

The BOD of water leaving sewage treatment plants and industrial wastewater treatment plants is routinely measured to evaluate the effectiveness of the treatment. To do this, the initial oxygen content of a known volume of water is measured. The sample is then mixed with an acid–base buffer, the nutrients needed by the microorganisms (that is, phosphorus, magnesium, calcium), and, if they are not naturally present, a population of bacteria. This mixture is held in the dark in a closed bottle at a constant temperature of 20°C for five days. At the end of this time, the oxygen content of the sample is measured again. The amount of oxygen that has been consumed is taken to be the biochemical oxygen demand.

To illustrate: A stream containing 10 ppm by weight (just 0.001%) of an organic material with the formula $C_6H_{10}O_5$ will contain 0.01 g of this material

■ ppm means parts per million.

■ Characteristic BOD levels (g O_2/L):

Untreated municipal sewage	0.1–0.4
Runoff from barnyards and feed lots	0.1–10
Food-processing wastes	0.1–10

per liter. To oxidize this pollutant to CO_2 and H_2O, the bacteria present will use oxygen according to the equation

$$C_6H_{10}O_5(aq) + 6\ O_2(g) \longrightarrow 6\ CO_2(g) + 5\ H_2O(\ell)$$

From the equation, we can see that for *every mole* of the pollutant that is oxidized, *six moles* of oxygen are required. Since the molar masses of $C_6H_{10}O_5$ and 6 O_2 are 162 g and 192 g, respectively, the grams of oxygen required to oxidize 0.01 g of the pollutant can be calculated.

$$\text{Mass of } O_2 \text{ required} = \frac{0.010\text{ g } C_6H_{10}O_5}{\text{L water}} \times \frac{192\text{ g } O_2}{162\text{ g } C_6H_{10}O_5} = 0.012\text{ g } O_2/\text{L of water}$$

This is the biochemical oxygen demand for this pollutant at this concentration in the water.

At 68°F (20°C) the solubility of O_2 in water under normal pressure of 1 atmosphere is only 0.0092 g O_2/L (Table 20.9). Because the BOD of the waste water (0.012 g/L) is greater than the equilibrium concentration of dissolved oxygen (0.0092 g/L), as the bacteria utilize the dissolved oxygen in a stream or lake with this BOD, the oxygen concentration of the water may soon drop too low to sustain any form of fish life. Whether this happens depends on the opportunities for new oxygen to become dissolved in the water. Life forms can survive in water where the BOD exceeds the dissolved oxygen if the water is flowing vigorously in a shallow stream (this facilitates the absorption of more oxygen from the air via aeration).

BOD values can be greatly reduced by treating industrial wastes and sewage with oxygen and/or ozone. Numerous commercial cleanup operations now being developed and used employ this type of "burning" of the organic wastes. Another benefit of treating waste water with oxygen is that some of the nonbiodegradable material becomes biodegradable as a result of partial oxidation.

TABLE 20.9 ■ Solubility of Oxygen in Water at Various Temperatures

Temperature °C	Solubility of O_2 (g O_2/L H_2O)
0	0.0141
10	0.0109
20	0.0092
25	0.0083
30	0.0077
35	0.0070
40	0.0065

These data are for water in contact with air at 760 mm mercury pressure.

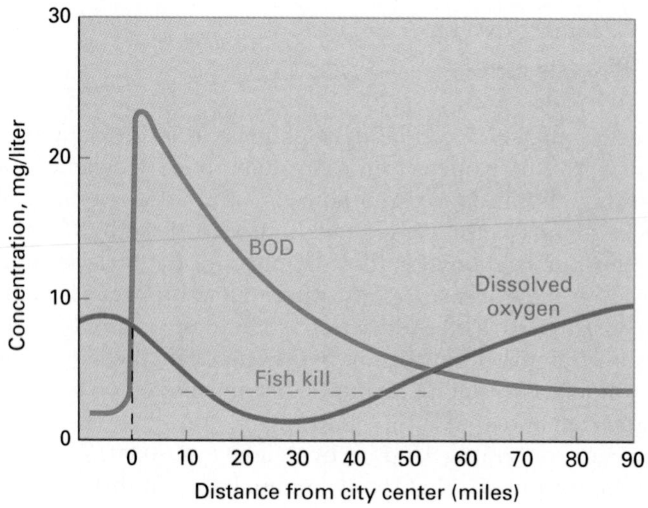

Figure 20.5 Oxygen depletion and rising BOD due to a sewage spill. The plots represent the result of a sewage spill at the center of a city (0 miles) into a river that flows at the rate of 750 gal/s. Near the spill the oxygen content drops and the BOD soars to 23 mg/liter (0.23 g/liter). Note that it takes 70 miles for the stream to recover, and that 15 to 45 miles downstream the oxygen content drops so low that fish will die.

Highly polluted water often has a high concentration of organic material, with resultant large biochemical oxygen demand (Fig. 20.5). In extreme cases, more oxygen is required than is available from the environment. The result is that fish and other aquatic life can no longer survive. The aerobic bacteria (those that require oxygen for the decomposition process) die. As a result, even more lifeless organic matter results, and the BOD soars. Nature, however, has a backup system for such conditions. A whole new set of microorganisms (anaerobic bacteria) takes over; these organisms take oxygen from oxygen-containing compounds to convert organic matter to CO_2 and water. Organic nitrogen is converted to elemental nitrogen by these bacteria. Given enough time, enough oxygen may become available, and aerobic oxidation will then return.

■ Fish cannot live in water that has less than 0.004 g O_2/L (4 ppm).

■ High concentrations of organic pollutants

↓

Low oxygen concentration

↓

Dead organisms

↓

Higher concentrations of organic pollutants

↓

Lower oxygen concentrations

↓

Anaerobic conditions

↓

Oxygen-requiring fish, shellfish, and other aquatic organisms leave or die

Measuring Metal Concentrations in Water

The concentrations of metal ions dissolved in water can be measured in a number of ways, but the most efficient is by the use of **atomic absorption spectrophotometry,** or **AA** for short. As you learned in Section 3.5, electrons in gaseous atoms can be excited to higher energy levels, and when they return to their original energy levels they release energy in the form of light. In an AA instrument the atoms of the metal to be analyzed are excited inside a lamp (Figure 20.6). For example, if a lead analysis is being done, a lamp containing lead atoms is used. The light from the lamp is passed through a flame into which the water sample under analysis is sprayed.

The metal atoms excited in the flame *absorb* the light from the lamp, hence the name atomic absorption. The "spectrophotometry" part of the name means "light measuring." Electrical circuits measure this absorption of light and even average out fluctuations in absorption caused by slight variations in the flame. When the absorption by a sample of *unknown* concentration is compared with that of a sample of *known* concentration (a **standard solu-**

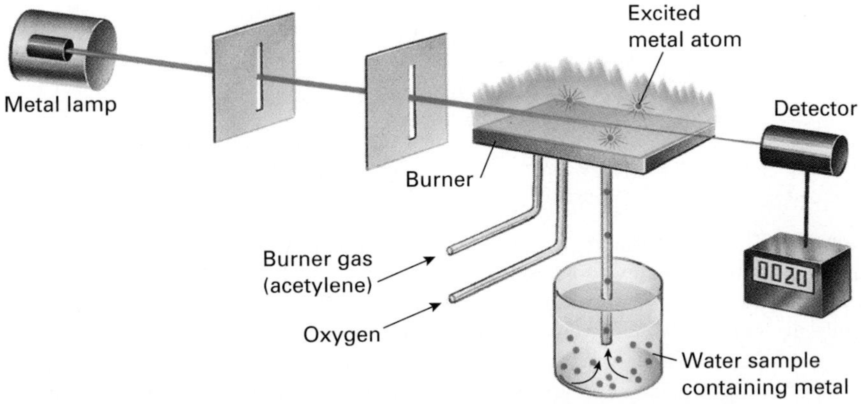

Figure 20.6 An atomic absorption spectrometer. Light from the metal lamp is produced by electrical excitation of the metal under analysis. When this light passes through the metal ions excited in the burner, it is absorbed in proportion to the quantity of the metal that is present.

A water quality laboratory. The large instrument is an atomic absorption spectrometer used to analyze water samples for trace amounts of various metals. *(Courtesy of Metro Water Services, Nashville, Tennessee)*

tion), the concentration of the unknown can be determined. AA analysis is done routinely on water samples in almost all municipalities. Using the proper light sources, it is possible to analyze for any of the metals that commonly occur in both drinking water and waste waters. The EPA has published standard methods for making these analyses of water, and water analysis laboratories must follow these methods.

Detecting and Measuring Organic Compounds in Water

Detecting and measuring trace amounts of organic compounds in water is a difficult task. To do this, the analytical chemist must first separate the organic compounds from the water sample and then separate the organic compounds from one another so they can be identified. One of the best ways to detect and measure organic compounds in water is by using an instrument called a **gas chromatograph–mass spectrometer** or **GC/MS.** In the gas chromatograph part of the instrument, a mixture to be separated is placed inside a metal or glass column containing some chemical that has a slight ability to dissolve the molecules in the mixture. This chemical is called the **stationary** phase because it does not move through the column. As the mixture is heated, its volatile components are vaporized and are swept through the column by a carrier gas such as helium (Figure 20.7). The temperature inside the chromatographic

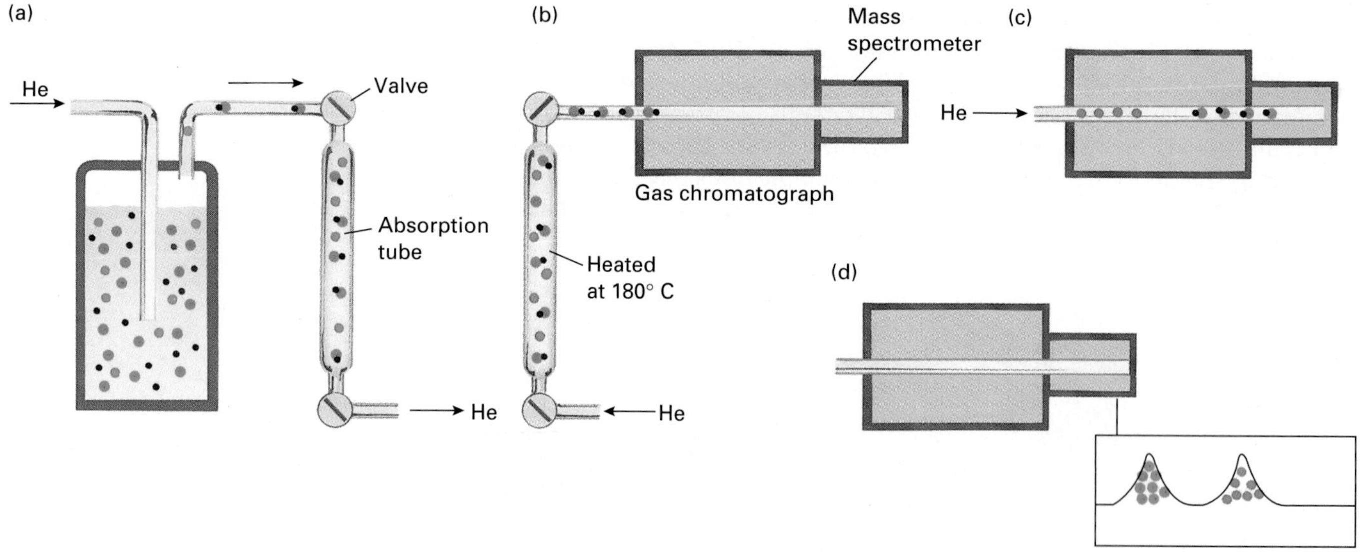

Figure 20.7 A gas chromatograph/mass spectrometer for water analysis. (a) Volatile organic pollutants in a water sample (red and green dots) are first swept out with helium gas into a tube filled with an adsorbent. (b) The tube is then quickly heated while the helium flow is reversed and the valve opens to allow the compounds to be swept into the chromatographic column. (c) As the mixture of volatile compounds passes through the chromatographic column, separation occurs, again illustrated with red and green dots. (d) Finally, the compounds emerge from the chromatographic column and are detected by the mass spectrometer, which makes a positive identification of the compounds. Numerous samples of known composition and known concentration are required to run GC/MS analyses.

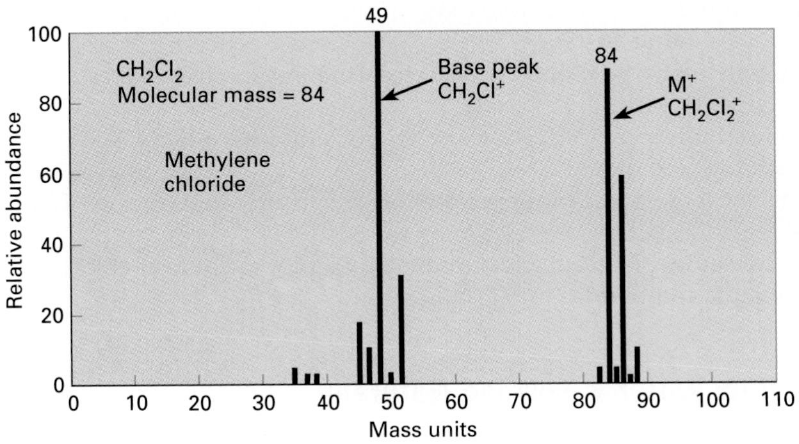

Figure 20.8 The mass spectrum of methylene chloride, a water pollutant and a carcinogen. Inside the mass spectrometer, organic molecules are bombarded with high-energy electrons. The "base peak" is the one most commonly formed when the molecules under analysis are fragmented. A compound can be identified by comparison of its mass spectrum with a library of spectra of known compounds or by careful measurement and analysis of the peaks in the spectrum.

column is kept high enough that the sample molecules would remain in the vapor state and flow through the column along with the carrier gas *if* they had no tendency to dissolve in the stationary phase. By carefully choosing the stationary phase, a mixture of even similar organic molecules can be separated because the stationary phase retains different substances in different degrees as they pass through the column. A detector at the outlet of the chromatographic column produces a **chromatogram** as it responds to the different sample molecules that are separated. In the GC/MS, the detector is a mass spectrometer (see Section 3.4) in which the molecules being analyzed are converted into positive ions and are analyzed according to their mass (Figure 20.8). A computer program then compares the mass spectra obtained against libraries of spectra of known compounds to give positive identification of the compounds. Most large municipalities have one or more GC/MS instruments devoted to the analysis of drinking water and industrial wastewaters.

SELF-TEST 20B

1. The federal law requiring cleanup of hazardous waste sites that can pollute water is called ____________.
2. Hazardous wastes can be indiscriminately placed in landfills. (a) True, (b) False.
3. Name three household wastes that can contaminate groundwater with the same harmful chemicals as industrial wastes. Beside each list the harmful chemical.

Household Waste	**Harmful Chemical**
____________	____________
____________	____________
____________	____________

4. List four household waste types that lend themselves to recycling.
5. The amount of oxygen required to oxidize a given amount of organic material is called the ____________ ____________ ____________, which is abbreviated ____________.

6. Which can hold more dissolved oxygen, (a) 1 L of water at 5°C or (b) 1 L of water at 40°C?
7. A common analytical tool used for measuring metal ions in water is ____________.
8. Name two industrial products whose manufacture introduces heavy metals into groundwater.
9. An analytical tool used for measuring organic compounds dissolved in water is ____________.
10. Name two industrial products whose manufacture introduces chlorinated organic compounds into groundwater.

20.7 How Water Is Purified Naturally

■ Rain water in clean air is very pure, containing only small amounts of dissolved gases such as N_2, O_2, and CO_2.

Water is a natural resource that, within limitations, is continuously renewed. The water cycle offers a number of opportunities for nature to purify its water. The worldwide ***distillation*** process results in rain water containing only traces of nonvolatile impurities, along with gases dissolved from the air. ***Crystallization*** of ice from ocean saltwater results in relatively pure water in the form of icebergs. ***Aeration*** of groundwater as it trickles over rock surfaces, as in a rapidly running brook, allows volatile impurities to be released to the air and allows oxygen to be dissolved. ***Sedimentation*** (or ***settling***) of solid particles occurs in slow-moving streams and lakes. ***Filtration*** of water through sand rids the water of suspended matter such as silt and algae. Of very great importance are the ***oxidation*** processes carried out by bacteria and other microorganisms. Practically all naturally occurring organic materials—plant and animal tissue, as well as their waste materials—can be oxidized in surface waters so long as oxygen is available and their concentration is not too high. Finally, another natural process is ***dilution.*** Most, if not all, pollutants found in nature are rendered harmless if reduced below certain levels of concentration by dilution.

Before the explosion of the human population and the advent of the Industrial Revolution, natural purification processes were quite adequate to provide ample water of very high purity in all but desert regions. Nature's purification processes can be thought of as massive but somewhat delicate.

Today, the activities of humans often push the natural purification processes beyond their limit, and polluted water accumulates. A simple example comes from dragging gravel from stream beds. The excavation leaves large amounts of suspended matter in the water, and for miles downstream, aquatic life is destroyed. Eventually, the solid matter settles, and normal life can be found again in the stream.

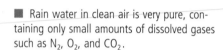

Pumping mud from the bottom to make room for deeper boats in the Mississippi River. When the bottom of a natural body of water is disturbed, the possibility exists of redistributing into the water pollutants that had been trapped in the mud. *(Grant Heilman Photography)*

A more complex example—one that perhaps cannot be solved by relying on natural purification processes—is pollution by organic molecules that cannot be easily oxidized by microorganisms. A **biodegradable** substance is composed of molecules that are broken down to simpler ones by microorganisms. For example, cellulose suspended in water will be converted to carbon dioxide and water. A **nonbiodegradable** substance, on the other hand, cannot be easily converted to simpler molecules by microorganisms. If the conversion process is extremely slow, or if it cannot be done at all by natural microorganisms, nonbiodegradable substances tend to accumulate in the environment.

Some organic compounds, notably some of those produced synthetically, are nonbiodegradable. When these substances are introduced into the environment, they simply stay in the natural waters or are absorbed by life forms and remain intact for a long time. Branched-chain detergent molecules, for example, cannot easily be eaten by microorganisms. The first detergents that contained such molecules accumulated and caused noticeable foaming in rivers and streams.

$$CH_3-\underset{\underset{CH_3}{|}}{CH}-CH_2-\underset{\underset{CH_3}{|}}{CH}-CH_2-\underset{\underset{CH_3}{|}}{CH}-CH_2-\underset{\underset{CH_3}{|}}{CH}-\text{\Large\bigcirc}-SO_3^-\ Na^+$$

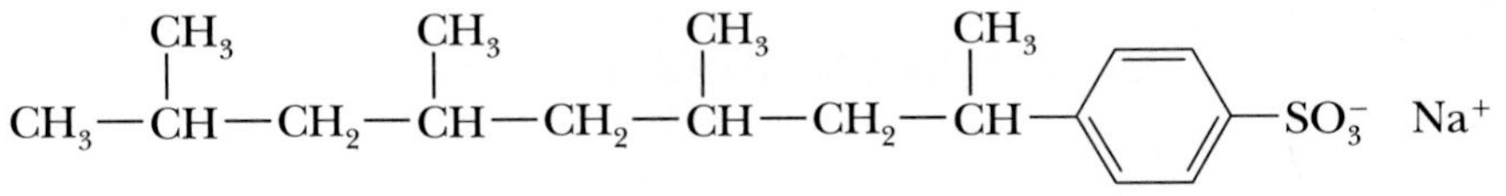

A branched-chain sodium alkylbenzenesulfonate detergent molecule

The branched-chain detergents were soon replaced by linear-chain alkylbenzenesulfonate detergents, which are easily decomposed by microorganisms— they are biodegradable.

$$CH_3CH_2CH_2CH_2CH_2CH_2CH_2CH_2CH_2CH_2CH_2CH_2-\text{\Large\bigcirc}-SO_3^-\ Na^+$$

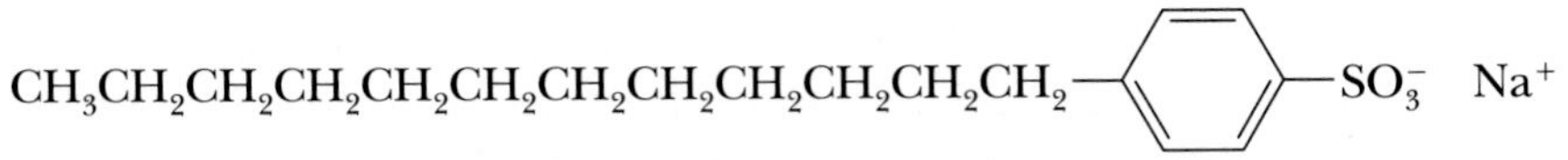

A linear sodium alkylbenzenesulfonate detergent molecule

Other examples of nonbiodegradable organic pollutants are the chlorinated and polychlorinated hydrocarbons. Many of these compounds are used as insecticides. The insect-killing ability of DDT was first recognized in 1939. By the end of World War II its insecticide properties were legendary owing to its ability to kill everything from malaria-causing mosquitoes to lice. This success prompted the introduction of numerous other chlorinated hydrocarbons as insecticides, and by the early 1960s their use was widespread throughout the world. These compounds are broad-spectrum insecticides, killing most insects rather effectively; however, they are also nonbiodegradable, so they tend to accumulate in the environment. This persistence is especially troublesome since their slow biodegradation allows such compounds to accumulate in the food chain. Fish-eating birds, for example, can accumulate large quantities of these insecticides that have accumulated in fish that ate smaller organisms containing these compounds. The populations of falcons, pelicans, bald eagles, ospreys, and other birds have been endangered by persistent pesticides. DDT causes reproductive failure in birds by interfering with the mechanisms that produce strong eggshells. After the use of DDT was banned in the United States in 1972, the numbers of surviving hatchlings increased rather dramatically.

■ DDT is fat soluble.

■ A hypothetical food chain might be

Plant ⟶ insect ⟶ fish ⟶ hawk

20.8 Water Purification Processes: Classical and Modern

The outhouses of some rural dwellers had their counterparts in city cesspools, which were basically holes in the ground into which sewage flowed. In cesspools, organic matter is decomposed by anaerobic bacteria, producing some pretty bad-smelling chemicals such as hydrogen sulfide, which has a character-

■ Cesspools were an early and crude form of the modern activated sludge process.

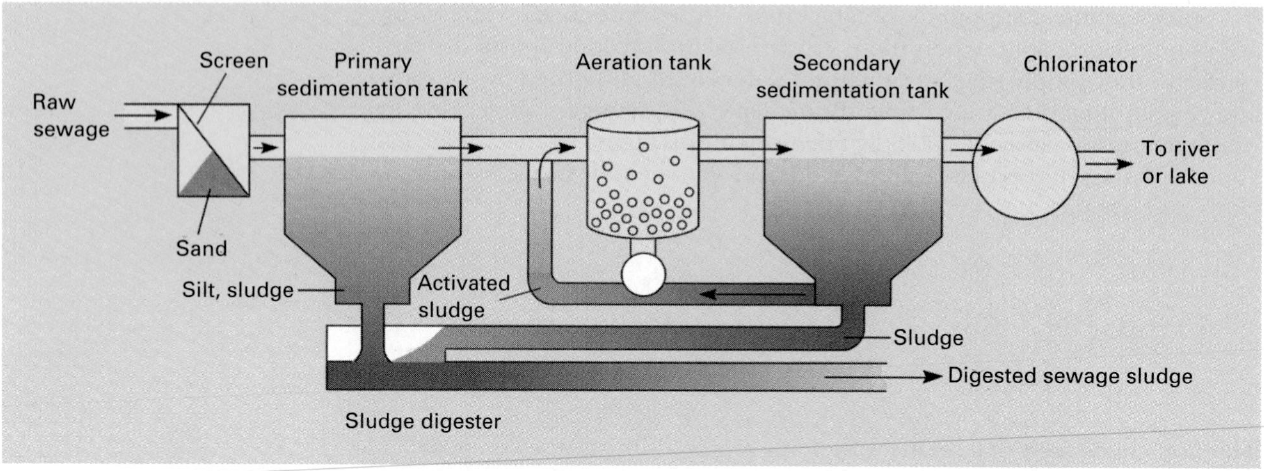

Figure 20.9 The steps in primary and secondary sewage treatment.

■ Sewage includes everything that flows from the sinks, tubs, washing machines, and toilets in our homes, factories, and public buildings. It excludes wastewater treated separately by industrial facilities.

■ Sewage is still 99.9% water!

istic rotten-egg odor. The terrible job of cleaning cesspools inspired the development of cesspools that could be flush-cleaned with water, followed by a connecting series of such pools that could be flushed from time to time. City sewage systems with no holding of the wastes were the next step.

At first, city sewage systems did little but channel sewage water to rivers and streams, where natural purification processes were expected to clean the water for the next users downstream. Today, however, sewage is treated using a combination of methods that can render the treated municipal wastewater almost as clean as the natural waters into which it is being discharged. The simplest treatment method is ***primary wastewater treatment,*** which copies two of nature's purification methods, settling and filtration. In primary treatment, sewage goes from the primary sedimentation tank shown in Figure 20.9 to the chlorinator.

Primary treatment removes 40 to 60% of the solids present in sewage and about 30% of the organic matter present. Calcium hydroxide and aluminum sulfate are added to produce aluminum hydroxide, which is a sticky, gelatinous precipitate that settles out slowly, carrying suspended dirt particles and bacteria with it.

$$3\ Ca(OH)_2(aq) + Al_2(SO_4)_3(aq) \longrightarrow 2\ Al(OH)_3(s) + 3\ CaSO_4(s)$$

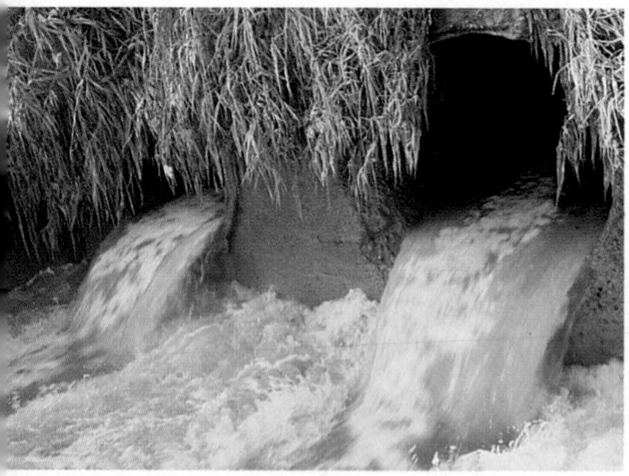

A sewage outfall, where treated water is returned to the environment. This water should be as clean as possible. *(Doug Wechsler)*

For many years municipal sewage-treatment plants had only primary treatment, followed by chlorination of the treated wastewater (see Section 20.10) before it was discharged into a suitable river or stream. Chlorination kills any remaining harmful pathogens. Presumably, natural processes would get rid of the remaining solids and dissolved organic matter.

Realizing that this treatment was not sufficient to protect the public from contaminated water, the writers of the 1972 Clean Water Act required that sewage-treatment plants also provide ***secondary wastewater treatment,*** which revives the old cesspool idea but under more controlled conditions. Modern secondary treatment operates in an oxygen-rich environment (aerobic; Figure 20.10), whereas the cesspool operates in an oxygen-poor environment (anaer-

The World of Chemistry

Program 12, *Water*

Figure 20.10 The aeration tank in a sewage treatment plant.

Chlorine contact tank in a sewage treatment plant. Whatever the level of treatment, the water is chlorinated to kill disease-causing organisms before it is returned to natural waterways. *(Runk/Schoenberger/Grant Heilman Photography)*

obic). The results are the same: The organic molecules that will not settle are consumed by microorganisms and the resulting sludge will settle.

Even a combination of primary and secondary wastewater treatment systems will not remove dissolved inorganic materials such as toxic metal ions, nutrients such as nitrate ions (NO_3^-) or ammonium ions (NH_4^+), or nonbiodegradable organic compounds such as chlorinated hydrocarbons. These materials can be removed by a variety of ***tertiary wastewater treatment*** methods that are selectively introduced where the nature of the wastewater requires them. One obstacle to tertiary treatment is the initial expense of modifying sewage-treatment plants and the ongoing expense of the additional treatment.

Filtration of the water through **carbon black** is a type of tertiary treatment effective for removing soluble organic compounds that are nonbiodegradable and thus remain in the water after secondary treatment. Carbon black consists of finely divided carbon particles with a large surface area on which solutes, including certain potentially toxic substances, can be *adsorbed.*

A different kind of tertiary treatment is needed to remove ammonia or ammonium ion. Because nitrogen is a nutrient for aquatic microorganisms, excessive nitrogen released to natural waters can cause a soaring BOD with the accompanying fish kills and other problems. The water is exposed to **denitrifying bacteria** that convert ammonium ion or ammonia to harmless nitrogen gas.

$$NH_4^+(aq) \text{ or } NH_3(aq) \xrightarrow[\text{bacteria}]{\text{Denitrifying}} N_2(g)$$

■ Definitions of "pure water":
Chemist: "Pure H_2O—no other substance."
Parent: "Nothing harmful to my child."
Game and Fish Commission: "Nothing harmful to animals."
Sunday boater: "Pleasing to the eye and nose, no debris."
Ecologist: "Natural mixture containing necessary nutrients."

■ Carbon black is used in many processes for manufacturing extremely pure organic compounds such as pharmaceuticals or food additives. **Adsorption** is the process by which molecules are attracted and held onto a surface.

20.9 Softening Hard Water

The presence of Ca^{2+}, Mg^{2+}, Fe^{3+}, or Mn^{2+} ions will impart "hardness" to water. Hardness is objectionable because (1) it causes precipitates (scale) to form in boilers and hot-water systems, (2) it causes soaps to form insoluble curds (this reaction does not occur with some synthetic detergents—see Section 16.10), and (3) it can impart a disagreeable taste to the water.

THE WORLD OF CHEMISTRY

Bioremediation of Hazardous Waste Sites

Program 25, *The Environment*

Bioremediation is the microbial detoxification or degradation of wastes. Companies that are designated by the EPA to clean up a hazardous waste site, such as a Superfund site, are faced with high costs to do the cleanup. In the early days of waste-site cleanup contaminated soil was often dug up and transported to a secure landfill. But as transportation costs and landfills began to increase, this method fell out of favor. (When you think about it, it makes more sense to clean up the site and restore it to its former state.) Bioremediation offers the possibility of getting rid of the (at times small amounts of) contaminants or converting them to less toxic substances, and the site is left very close to its original state. It can even possibly be sold to a new owner. (If all the soil is dug up and hauled away, the property might not be as desirable.)

In bioremediation, a cue is taken from the microbes that eat ordinary wastes—if bacteria can be found to eat that, then maybe some exist that can eat other chemicals, including those that are considered hazardous wastes. Bioremediation methods fall into three categories—land treatment, in which the bacteria are spread across the surface of the land; bioreactors, in which soil and liquids are removed from the surface and placed in a reaction vessel for a period of time; and *in situ* treatment, in which bacteria are intimately mixed with the soil or water.

Cleanup from leakage of underground oil storage tanks is one of the most direct applications of bioremediation. The leaked oil, a hydrocarbon, is food for many kinds of bacteria, and finding bacteria that are compatible with the oil at a site is a fairly straightforward task. One of the largest oil bioremediation projects to date has been the cleanup of the March 1989 spill of 11 million gallons of crude oil by the *Exxon Valdez*. Immediately after that spill, it was proposed to use the bacteria already on the beaches there, but one problem arose. There was plenty of food for the bacteria (the oil) but a limited amount of available nutrients such as nitrogen, phosphorus, and trace elements, which they required to thrive. This problem was solved when the EPA allowed the use of an oil-soluble slow-release fertilizer that had been developed a few years earlier by a French company when the tanker *Amoco Cadiz* went aground off the coast of Brittany. The combination of the bacteria and the fertilizer worked, and the shores of Prince William Sound became cleaner much more quickly and probably at much less cost.

According to the EPA, about 135 sites are being considered or planned or are in operation for bioremediation.

The World of Chemistry

Program 12, *Water*

Over time, copper water pipes like this one become coated with deposits of the minerals dissolved in the water.

Hardness due to calcium or magnesium, present as bicarbonates, is produced when water containing carbon dioxide trickles through limestone or dolomite:

$$\underset{\text{Limestone}}{CaCO_3(s)} + CO_2(g) + H_2O(\ell) \longrightarrow Ca^{2+}(aq) + 2\ HCO_3^-(aq)$$

$$\underset{\text{Dolomite}}{CaCO_3 \cdot MgCO_3(s)} + 2\ CO_2(g) + 2\ H_2O(\ell) \longrightarrow$$

$$Ca^{2+}(aq) + Mg^{2+}(aq) + 4\ HCO_3^-(aq)$$

Such "hard water" can be softened by removing these ions. One of the methods for softening water is the lime-soda process. The lime-soda process takes advantage of the facts that calcium carbonate ($CaCO_3$) is much less soluble than calcium bicarbonate ($Ca(HCO_3)_2$) and that magnesium hydroxide is much less soluble than magnesium bicarbonate. The raw materials added to the water in this process are hydrated lime ($Ca(OH)_2$) and soda (Na_2CO_3). In the system, several reactions then take place, which can be summarized as follows:

$$HCO_3^-(aq) + OH^-(aq) \longrightarrow CO_3^{2-}(aq) + H_2O(\ell)$$
$$Ca^{2+}(aq) + CO_3^{2-}(aq) \longrightarrow CaCO_3(s)$$
$$Mg^{2+} + 2\,OH^-(aq) \longrightarrow Mg(OH)_2(s)$$

The overall result of the lime-soda process is to precipitate almost all the calcium and magnesium ions and to leave sodium ions as replacements.

Iron present as Fe^{2+} and manganese present as Mn^{2+} can be removed from water by oxidation with air (aeration) to higher oxidation states. If the pH of the water is 7 or above (either naturally or through the addition of lime), the insoluble compounds $Fe(OH)_3$ and $MnO_2(H_2O)_x$ are produced and precipitate from solution.

■ Hard water contains metal ions that react with soaps and give precipitates.

The desire for and achievement of soft water for domestic use has sparked a rather heated health debate during the past two decades. Soft water is usually acidic and contains Na^+ ions in the place of di- and trivalent metal ions. An increased intake of Na^+ is known to be related to heart disease. The acidic soft water is also more likely to attack metallic pipes, joints, and fixtures, resulting in the dissolution of toxic ions such as Pb^{2+}. One way to avoid sodium ions in drinking water and to use less soap when washing would be to drink only naturally hard water and to do your washing in soft water.

■ Soft water: <65 mg of metal ion/gal
Slightly hard water: 65–228 mg
Moderately hard water: 228–455 mg
Hard water: 455–682 mg
Very hard water: >682 mg

■ For people on low-sodium diets, lime-soda treated water might represent too high a daily dose of sodium.

20.10 Chlorination and Ozone Treatment of Water

With the advent of chlorination of water supplies in the early 1900s, the number of deaths in the United States caused by typhoid and other water-borne diseases dropped from 35 per 100,000 population in 1900 to 3 per 100,000 population in 1930.

Chlorine is introduced into water as the gaseous element (Cl_2), and it acts as a powerful oxidizing agent for the purpose of killing bacteria in water. This process is used in treating both water that will become tap water and wastewater before it is released. Chlorination largely prevents the principal water-borne diseases spread by bacteria, which include cholera, typhoid, paratyphoid, and dysentery.

In spite of chlorination, most city water supplies are not bacteria-free, but only rarely do these surviving bacteria cause disease. Today the most common water-borne bacterial disease is *giardiasis,* a gastrointestinal disorder. Most often this disease comes from surface water, but on occasion it can be traced to city water systems.

Chlorination of industrial wastewater and city water supplies presents a potential threat because of the reaction of chlorine with residual concentrations of organic compounds to produce **disinfection byproducts.**

$$\text{Water containing organic compounds} \xrightarrow{\text{Chlorine}} \text{chlorinated organic disinfection byproducts}$$

These disinfection byproducts, which may be present at levels of a few parts per million or less, include dichloromethane, chloroform, trichloroethylene, and chlorobenzene—all suspected carcinogens. According to the EPA, muta-

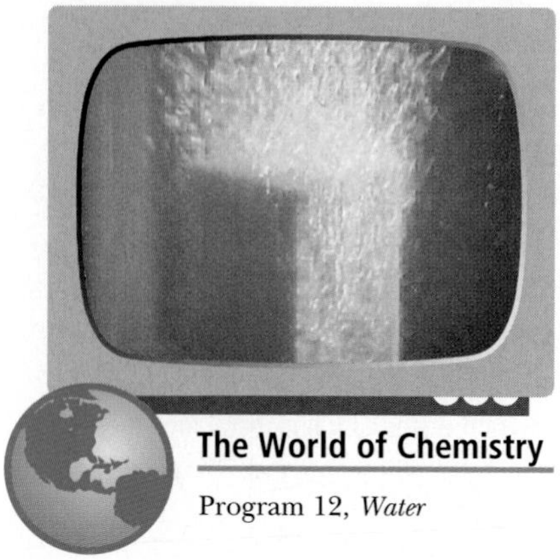

Ozone gas, a disinfectant, bubbling through water.

genic or carcinogenic chemicals have been found in 14 major river basins in the United States. It is estimated that more than 500 water systems in the United States exceed EPA's maximum of 0.1 ppm for chlorinated hydrocarbons. The presence of these chlorinated hydrocarbons can be prevented by more efficient removal of the organic matter that becomes chlorinated, but unfortunately even the best-designed purification systems (including carbon filtration) allow some organic compounds to pass through, only to become chlorinated.

One way to eliminate chlorinated hydrocarbons as disinfection byproducts is to use ozone (O_3) as the disinfectant. Ozone is used in more than 1000 water-treatment plants, mostly in Europe. The ozone is produced on-site by passing oxygen or air through an electric discharge. This process normally gives about a 20% ozone-oxygen mixture that is a very strong oxidizer. Although the use of ozone in the United States has been minimal, 20 of the 25 ozone plants in the United States have been built in the last decade.

Ozonation, like chlorination, is also not without potentially harmful disinfection byproducts. Bromide ion (Br^-), which is found in most natural waters, is oxidized by ozone to bromate ion (BrO_3^-), which is a suspected carcinogen. Generally, this single known harmful disinfection byproduct is considered less of a risk factor than the numerous chlorinated hydrocarbons produced by chlorine disinfection.

20.11 Fresh Water from the Sea

Because sea water covers 72% of the Earth, it is not surprising that it is considered a water source for areas where fresh water supplies aren't sufficient to meet the demand. The oceans contain an average 3.5% (35,000 ppm) dissolved salts by weight, a concentration too high to make ocean water useful for drinking, washing, or agricultural use (Table 20.10). The total of dissolved ions must be reduced to below 500 ppm before the water is suitable for human consumption.

The technology has been developed for the conversion of sea water to fresh water. The extent to which this technology is actually put to use depends on the availability of fresh water and the cost of the energy for the conversion. More than 3000 desalination plants were in operation throughout the world in the early 1990s. Two methods used to purify sea water are reverse osmosis and solar distillation.

TABLE 20.10 ■ **Ions Present in Sea Water at Concentrations Greater Than 0.001 g/kg**

Ion	g/kg Sea Water
Cl^-	19.35
Na^+	10.76
SO_4^{2-}	2.71
Mg^{2+}	1.29
Ca^{2+}	0.41
K^+	0.40
HCO_3^-, CO_3^{2-}	0.106
Br^-	0.067
$H_2BO_3^-$	0.027
Sr^{2+}	0.008
F^-	0.001
Total	35.129

Reverse Osmosis

An extremely thin piece of material such as a sheet of plastic or animal tissue can allow molecules to pass through it. Such a material is called a membrane and is said to be **permeable** to those molecules and ions that can pass through. Permeability is dependent on the presence of tiny passages within the membrane. A membrane permeable to water molecules but not to ions or molecules larger than water molecules is called a **semipermeable** membrane. Many membranes made from synthetic polymers have this characteristic. One such polymer is cellulose acetate. If a semipermeable membrane is placed between sea water and pure water, the pure water will pass through the membrane to

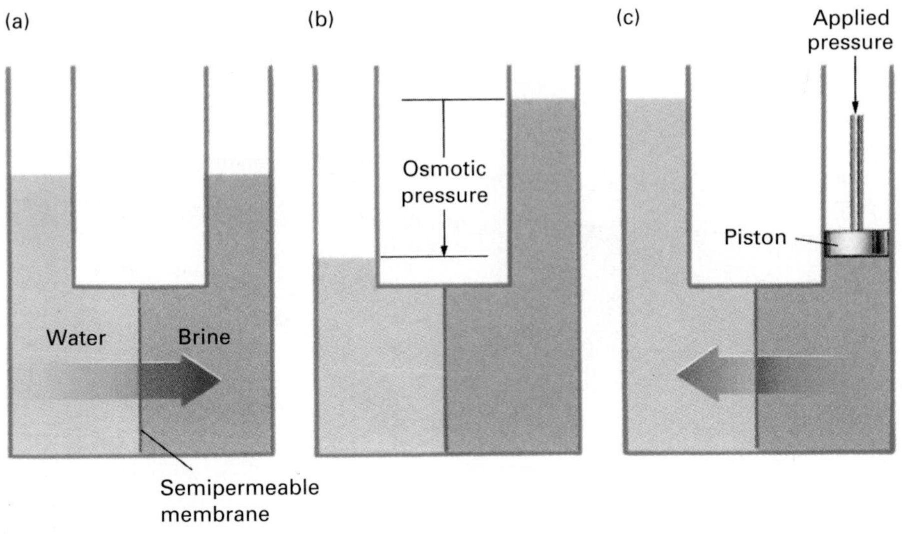

Figure 20.11 Osmosis (a, b) and reverse osmosis (c). (a) In normal osmosis, water molecules pass through the semipermeable membrane from the less concentrated into the more concentrated solution, in this case from pure water into the brine. (b) At a certain height of the brine solution in this apparatus, the pressure of the column of water is equal to the osmotic pressure and the flow stops. (c) In reverse osmosis, the application of external pressure greater than the osmotic pressure forces water molecules to the pure water side.

dilute the sea water. This is a process called **osmosis.** The liquid level on the sea-water side rises as more water molecules enter than leave, and pressure is exerted on the membrane until the rates of diffusion of water molecules in both directions are equal. **Osmotic pressure** is defined as the external pressure required to *prevent* osmosis. Figure 20.11 illustrates the concepts of osmosis and osmotic pressure.

Reverse osmosis is the application of pressure to cause water to pass through the membrane from the aqueous-solution side to the pure-water side (Figure 20.12). The osmotic pressure of normal sea water is 24.8 atm (atmospheres). As a result, pressures greater than 24.8 atm must be applied to cause reverse osmosis. Pressures up to 100 atm are used to provide a reasonable rate of reverse osmosis and to account for the increase in salt concentration that occurs as the process proceeds.

The largest reverse osmosis plant in operation today is the Yuma Desalting Plant in Arizona. This plant, which began operation in the 1980s, can produce 100 million gallons of water per day. The plant was built to reduce the salt concentration of irrigation wastewater in the Colorado River from 3200 ppm to 283 ppm. The project is part of a U.S. commitment to supply Mexico with a sufficient quantity of water suitable for irrigation. The Mediterranean island of

■ Irrigation water of desert fields dissolves about two tons of salt per acre per year. Irrigation wastewater carries the salt back to the Colorado River.

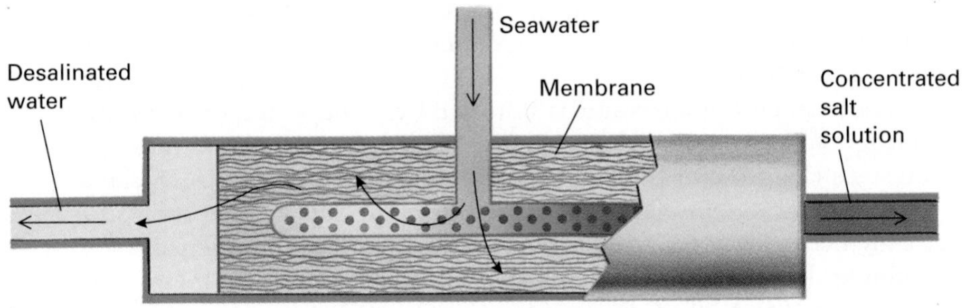

Figure 20.12 An industrial reverse osmosis unit for desalination of sea water.

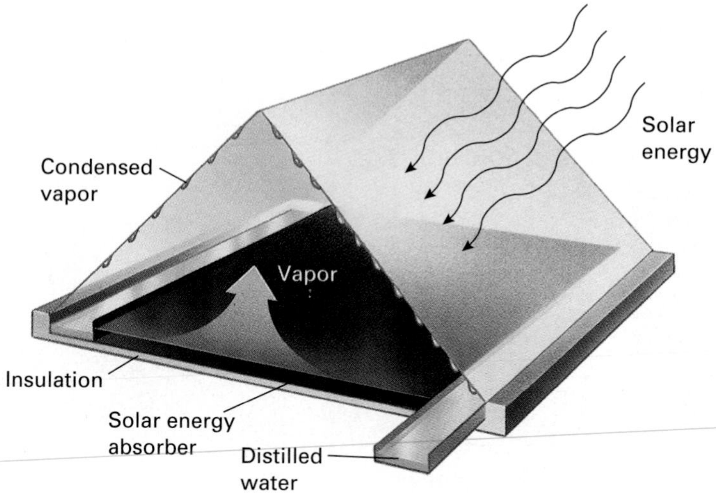

Figure 20.13 A solar still. Solar radiation heats salt water in the black trough. Vapor condenses on the sloping glass surfaces and runs off into the distilled water troughs.

Malta now uses four reverse osmosis plants that produce a total of 12 million gallons of fresh water per day from the sea. On Florida's Sanibel Island, increasing salinity in the well water led to the installation of a reverse osmosis system. This facility has a design capacity of 3.6 million gal/day and has one of the lowest energy-consumption rates per 1000 gal of potable water of any comparably sized system in commercial use.

Solar Distillation

Solar evaporation units can be used to purify sea water in parts of the world that receive a lot of sunlight. The main disadvantage of solar units is the amount of land required to produce appreciable amounts of fresh water. The output of these units is about 3 L/m^2/day or 7000 gal/day for the larger units. Figure 20.13 shows a basic design used for solar stills. Smaller units can provide enough fresh water for homes.

20.12 Pure Drinking Water for the Home

In spite of all the efforts taken to purify public water supplies, many consumers are concerned about the quality of the water that comes out of the taps in their homes, schools, and places of business. Parents of small children are especially worried about chemicals such as lead and carcinogenic organic compounds that are chlorine disinfection byproducts. Many have turned to bottled water or home water-treatment devices that offer some protection from these harmful trace pollutants.

■ Americans spend about $350 million a year on bottled water. Buyer beware: A very wide variety of standards exists for bottled water.

While some bottled water is untreated ground water (sometimes called "spring" water), most bottled water has passed through one or more purification steps (Figure 20.14). The three purification methods (which can also be done at home)—distillation, carbon filtration, and reverse osmosis—have already been discussed as methods used for treating municipal water. The maximum levels of contaminants allowed by the EPA after these treatments are listed in Table 20.11. Each of these methods is expensive and results in a

TABLE 20.11 ■ A Partial List of Maximum Contaminant Levels (MCL) for Drinking Water Allowed by the EPA*

Metals		Volatile Organic Compounds		Herbicides, Pesticides, PCBs		Secondary Contaminants	
Arsenic	0.050	Benzene	0.005	Chlordane	0.002	Iron	0.30
Antimony	0.006	Carbon tetrachloride	0.005	Endrin	0.0002	Manganese	0.05
Barium	2.000	1,4-Dichlorobenzene	0.075	Glyphosate	0.7	Zinc	5.00
Beryllium	0.004	Dichloromethane	0.005	Heptachlor	0.0004	Chloride	250.00
Cadmium	0.005	1,1-Dichloroethylene	0.007	Hexachlorobenzene	0.001	Sulfate	250.00
Chromium	0.100	Vinyl chloride	0.002	Lindane	0.0002	Total dissolved solids	500.00
Copper	1.300	Trichloroethylene	0.005	Methoxychlor	0.040		
Lead	0.015	*o*-Dichlorobenzene	0.600	Toxaphene	0.003		
Nickel	0.1	1,2-Dichloropropane	0.005	PCBs	0.0005		
Selenium	0.050	Hexachlorobenzene	0.001	2,4-D	0.070		
Thallium	0.002	Styrene	0.100	Aldicarb	0.003		
				Atrazine	0.003		
Nonmetals				Dalapon	0.2		
Fluoride	4.00			Ethylene dibromide	0.00005		
Nitrate	10.00			Heptachlor oxide	0.0002		
Nitrite	1.00						

*Values are in milligrams per liter.

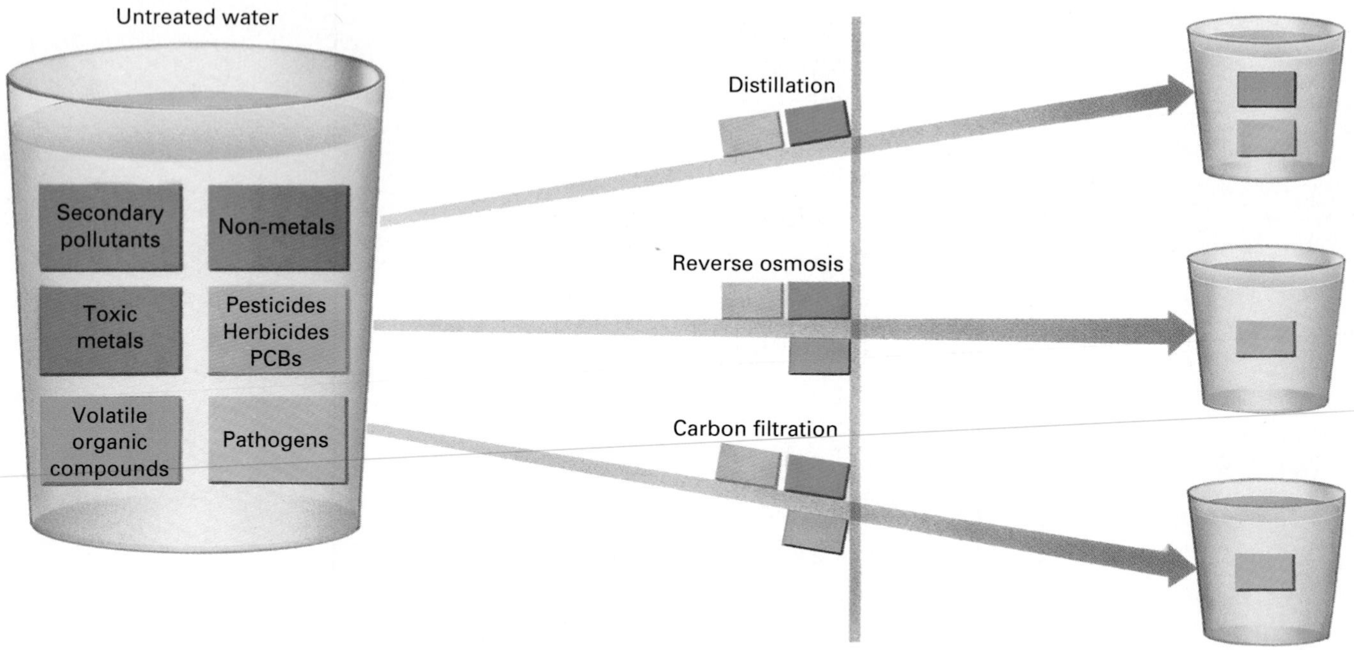

Figure 20.14 Final steps that can be used in water purification. Both bottled water and municipal tap water are purified in these ways. The color code shows which pollutants are removed by each method. Pathogenic bacteria can and do pass through all these methods. This is why municipal tap water must be treated with chlorine or some other disinfectant before release into the system.

high cost per gallon of treated water. In the case of bottled water or home water treatment, the cost of treatment is not the major factor, since only the small amount of water needed for human consumption needs to be specially purified. Figure 20.14 shows which trace pollutants are removed or allowed to remain in the water by each treatment method, color-coded to examples of the types of contaminants in Table 20.11. Thus, the analysis on the label of the bottled water is highly important and should be read with care before purchase.

20.13 What About the Future?

Our water quality in the United States has improved substantially since the passing of the Clean Water Act of 1977 and its subsequent amendments. Municipal water is now monitored more carefully for heavy metals, pesticides, and chlorinated organic molecules than was even thought possible just 30 or 40 years ago. As a result of this continual monitoring, consumers can feel better about the water they drink. Industry is complying more with hazardous waste–reduction and pollution-discharge regulations, and much research is focused on finding ways to totally eliminate releases of harmful chemicals that might find their way into our water supplies.

In spite of this progress, some problems still exist. Household wastes continue to be a major potential source of water pollution. As long as households are not offered inexpensive alternatives to mixing hazardous chemicals such

A paper mill without pollution controls. Prior to the Clean Air Act and the Clean Water Act, paper mills in the United States were among the most notorious of polluters. *(Dan Guravich/Photo Researchers)*

as paints, solvents, and waste oil with ordinary household garbage, groundwater and surface water will remain at risk. **Recycling** of wastes is growing, but the recycling of some items such as automobile batteries, mercury batteries, unused pesticides, solvents, and used lubricating oil is still severely limited by economic factors that overshadow a strong desire by citizens for purer water supplies.

The politics of water protection must improve in the future. Today, many communities are taking rather short-sighted approaches to hazardous-waste disposal. In effect, these communities are saying ''not in my backyard!'' States that allow communities veto power over the location of hazardous-waste sites have been singled out for retaliation by other states with active hazardous-waste disposal sites located within their borders. These states are passing laws effectively banning another state's hazardous wastes if that state doesn't allow hazardous-waste disposal in its own borders or if it allows communities to have veto power over the location of disposal sites within their city or community limits. Political problems like these can only be solved when everyone recognizes the importance of the proper disposal of hazardous industrial and household wastes.

The future will see increased water-conservation measures by everyone who uses water—industry, agriculture, and households. Recently, the city of Boston has encouraged water-efficient fixtures in homes as well as industry water audits and system-wide leak detection. These steps, coupled with public education about how to reduce annual water usage, have resulted in a 16% decrease in water demand. In Texas, farmers drawing water from the Ogallala aquifer (Section 20.1) have begun to use an old-fashioned furrow technique to introduce water more directly to the plant's roots, thus causing less water loss into the ground. We will probably see an expansion of the capacity for desalination of sea water, particularly for industrial and agricultural uses. Most large cities will have to replace leaky water mains and plumbing, which account for up to one third of their water use. This will mean higher water bills in the future.

The discussion in this chapter has focused on water quality in the United States. A combined program of water conservation, protection of water quality, and water recycling will help to alleviate the water crisis in the United States and other industrialized nations. However, contaminated water is still a serious problem for 75% of the world's population. It has been estimated that

The NIMBY response (Not in My Backyard). This protest against a solid waste incinerator in East Liverpool, Ohio, in 1992 is typical of the reaction to plans for incinerators, landfills, and other repositories for waste. Sometimes the protests result from anger at being left out of the decisionmaking process, rather than fact-based conclusions that the site would be harmful to those in the area. *(Visuals Unlimited/Bill Beatty)*

80% of the sickness in the world is caused by contaminated water. For years, many countries and international organizations have provided financial and technical aid to help improve the water quality in developing countries. However, much work remains to be done to reduce sickness caused by contaminated water.

SELF-TEST 20C

1. Which water purification process is not a natural process? (a) Distillation, (b) Aeration, (c) Filtration, (d) Reverse osmosis, (e) Settling.
2. Which word can be used to describe organic pesticides that are not readily biodegradable? (a) Permanent, (b) Persistent, (c) Nonvolatile.
3. Name two water purification methods that are part of primary wastewater treatment.
4. Secondary wastewater treatment operates under (aerobic/anaerobic) conditions.
5. A cesspool operates under (aerobic/anaerobic) conditions.
6. Select the ions that may cause water to be hard: (a) Sodium, (b) Calcium, (c) Magnesium, (d) Potassium.
7. Two methods used to kill harmful microorganisms in water are ____________ and ____________.
8. What are the four metal ions present in sea water at concentrations of 400 ppm or higher?
9. Water flows through a semipermeable membrane from a solution of low concentration to a solution of higher concentration. This process is called ____________.

MATCHING SET

____ 1. Sedimentation
____ 2. Biodegradable
____ 3. Clean Water Act of 1977
____ 4. BOD
____ 5. Proposition 65
____ 6. Ammonium ion
____ 7. Unused paint
____ 8. Reverse osmosis
____ 9. Water hardness
____ 10. Ozone treatment
____ 11. Recycling
____ 12. Superfund
____ 13. Incineration
____ 14. Aeration
____ 15. Carbon adsorption
____ 16. Chlorination
____ 17. Auto battery

a. A measure of dissolved organic material in water
b. Can impart lead to water supplies when improperly disposed of
c. Legislation designed to prevent the discharge of carcinogens and similar chemicals into drinking-water supplies.
d. Caused by metal ions such as Ca^{2+} and Mg^{2+} in solution
e. A nutrient for microorganisms living in water
f. Provides for hazardous-waste site cleanup
g. Alternative to landfills but can contribute to air pollution
h. Disinfection method commonly used for drinking water and wastewater
i. Removes organic compounds from water
j. Relieves dependence on landfills
k. Common household hazardous waste
l. Primary purification process
m. Disinfectant method that can produce bromate ion from bromide ion
n. Secondary purification process
o. Naturally decomposed to simpler compounds
p. Shifted responsibility for water purity to wastewater discharger
q. Uses pressure to purify water

QUESTIONS FOR REVIEW AND THOUGHT

1. Define the following terms.
   (a) Surface water (b) Groundwater
   (c) Aquifer (d) Brackish water
   (e) Pollution
2. Define the following terms.
   (a) ppm (b) Potable
   (c) Dilution (d) Hazardous waste
3. Define the following terms.
   (a) BOD (b) Distillation
   (c) Aeration (d) Sedimentation
4. Define the following terms.
   (a) Biodegradable (b) Nonbiodegradable
   (c) Hard water (d) Disinfection byproducts
   (e) Reverse osmosis (f) Semipermeable
5. Where does most of the water go that falls on the continental United States every day?
6. Explain how rainwater becomes groundwater.
7. Explain how groundwater can become contaminated with pollutants.
8. What is the origin of brackish water?
9. What activity is the largest single user of water?
10. What causes the level of an aquifer to drop? Cite some examples of the effects of aquifers dropping in level.
11. Explain how a "dual" water system would work. What are some advantages and what are some disadvantages of such a system?
12. How much water do you think you use per day? List your uses and include water that might be used "for you," such as in food preparation in a restaurant.
13. What would you expect to find dissolved in "clean" water?
14. What does the term "groundwater recharge"mean? What is the source of water that is used for this purpose?
15. Name five kinds of pollution often found in water. Give a source for each.
16. Prior to the enactment of the Clean Water Act, who was responsible for ensuring that water being used was pure? After passage of the Act, who is now responsible?
17. Both surface water and groundwater (natural waters) often contain dissolved ions. Name three positive and three negative ions that are often found in natural waters.
18. Name two methods of disposal for solid wastes from industry and households. Which one of these has the greater possibility to adversely impact water quality?
19. What is the "Superfund"? Explain how it is used to improve water quality.
20. What is meant by the term "cradle-to-grave" as it applies to hazardous wastes?
21. What is a "manifest" and how does it help insure that hazardous wastes are handled properly?
22. Describe a way by which a landfill can be made more "secure" in terms of water quality protection.
23. What is Proposition 65? How is this proposition similar to the Delaney clause?
24. Name three common household wastes and the kinds of chemicals they contain that might be harmful to water quality.
25. Describe how measuring the biochemical oxygen demand (BOD) of a sample of water indicates something about its purity.
26. How does thermal pollution affect the concentration of dissolved oxygen? How does this relate to aquatic life?
27. Describe how the BOD of wastewater can be lowered.
28. What kind of instrument is commonly used to determine dissolved metals like vanadium (V) and chromium (Cr) in water samples?
29. What kind of instrument is commonly used to detect and measure the amounts of trace organic compounds in water samples?
30. Explain how distillation purifies a sample of water.
31. Explain how aeration purifies a sample of water containing dissolved organic compounds.
32. Explain how settling and filtration purify water samples.
33. What is the difference between "biodegradable" and "nonbiodegradable"? If you had a choice between using a biodegradable and a nonbiodegradable detergent to clean your clothes, which would you choose?
34. Chlorinated hydrocarbons and branched-chain hydrocarbons are nonbiodegradable. What happens to them when they are released into the environment?
35. What is DDT and how does it affect certain species of birds?
36. Name the two methods of primary water purification.
37. Chlorination is used both for the final treatment of sewage and for treating drinking water. What does chlorine do?
38. Why are calcium hydroxide and aluminum sulfate added to sewage during the settling stage of treatment?
39. What is "bioremediation," and how is it similar to secondary water treatment methods?
40. What is "carbon black" and how is it used to treat water?
41. Too-high concentrations of nitrogen compounds like ammonia (NH_3) adversely affect water quality. What tertiary water treatment method gets rid of these compounds?
42. Name the ions commonly present in hard water. What kinds of problems do they cause?
43. What is the soda lime process and how does it affect the concentration of sodium ions in water treated by this process?
44. How is chlorination of water similar to aeration of water? How are these different?
45. What are "disinfection byproducts," and how are these potentially harmful?
46. How is ozonation of water similar to chlorination of water? How are these different?

47. Sea water contains about 35,000 ppm of dissolved solids. What is the dissolved solid present in highest concentration?
48. Explain how reverse osmosis can be used to purify sea water.
49. Explain how solar distillation can be used to purify sea water.
50. Which method of purification of drinking water for the home would most likely get rid of dissolved organic compounds?
51. Explain the meaning of the term "not in my backyard." How does this attitude affect water quality, nationally as well as locally?
52. About how many Superfund sites has the EPA listed for cleanup action? What will these cleanups cost? Read World of Chemistry—Toxic Substances in Perspective in Chapter 18 and tell how you feel about this problem.
53. During 1995, efforts were begun to cancel some requirements of the Clean Water Act. Consult newspapers or magazines in the library to discover the results of these efforts and write a brief summary of your opinions of the outcome.

CHAPTER 21

Air—The Precious Canopy

Planet Earth is enveloped by a few vertical miles of chemicals that compose the gaseous medium in which we exist—the atmosphere. Close to the Earth's surface and near sea level, the atmosphere is mostly nitrogen and life-sustaining oxygen. It is the few little fractions of a percentage point of other chemicals that make a difference in the quality of life in various places on Earth. Urbanization is the main culprit in causing problems in our atmosphere. With its vast number of vehicles and increases in industrialization, urbanization has produced an abnormal increase in some of the naturally occurring "minor" chemicals in the atmosphere—compounds such as nitrogen oxides, sulfur dioxide, carbon monoxide, carbon dioxide, and ozone. Increased amounts of these compounds in the atmosphere can create an unhealthful, unpleasant medium. An atmosphere containing these unwanted and harmful ingredients is called *polluted.* Some questions that will be answered in this chapter are the following:

- What is clean air, and what role does government play in helping to maintain clean air?
- What are the different kinds of air pollution, and what are their sources?
- What kinds of chemical reactions take place in the atmosphere that contribute to air pollution?
- What role does sunlight play in causing air pollution?
- What two kinds of industry-related air pollution have the potential to change our planet?
- What are some of the sources of indoor air pollution?

Air pollution research: measuring the volatile organic compounds given off by trees. *(© Ann States/SABA)*

The atmosphere of the Earth is a fantastically large source of the elements nitrogen (N_2) and oxygen (O_2), with much smaller amounts of certain of the noble gases, including argon (Ar), neon (Ne), and xenon (Xe) (Table 21.1).

TABLE 21.1 ■ Composition of Dry Air at Sea Level

Gas	Percentage by Volume
Nitrogen	78.084
Oxygen	20.948
Argon	0.934
Carbon dioxide	0.033*
Neon	0.00182
Hydrogen	0.0010
Helium	0.00052
Methane	0.0002*
Krypton	0.0001
Carbon monoxide	0.00001*
Xenon	0.000008
Ozone	0.000002*
Ammonia	0.000001
Nitrogen dioxide	0.0000001*
Sulfur dioxide	0.00000002*

*Trace gases of environmental importance discussed in this chapter.

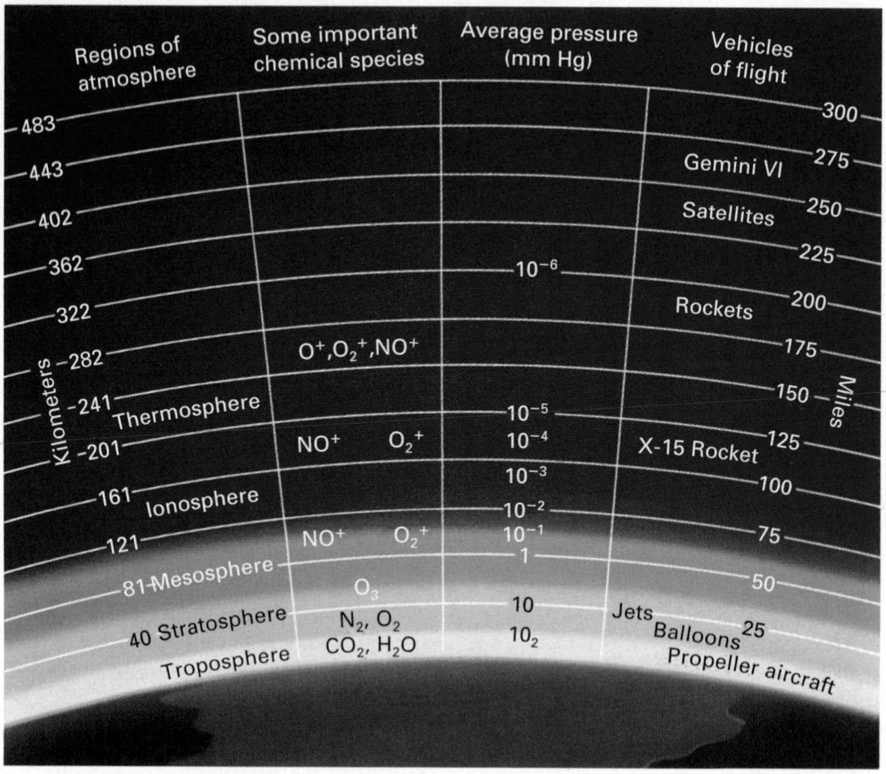

Figure 21.1 Some facts about our Earth's atmosphere.

Air pollutants become mixed with the natural components of the atmosphere. Over time, certain pollutants even react with other components of the atmosphere. Figure 21.1 presents some of the basic facts about our atmosphere, including the naming of the stratified layers that compose the atmosphere and the chemical species present in those layers. Our main concerns are with the layers called the **troposphere** (the air we breathe and where our weather takes place) and the **stratosphere,** where the UV-protective ozone layer is found.

21.1 Doing Something About Polluted Air

Nature pollutes the air on a massive scale with ash, mercury vapor, hydrogen chloride, and hydrogen sulfide from volcanoes, carbon dioxide and chlorinated organic compounds from forest and grassland fires, and reactive organic compounds from coniferous and deciduous plants. Human activity, however, especially in heavily populated (urban) areas, seems to have the most noticeable effects on the quality of the air we breathe. Automobiles, fossil fuel–burning power plants, smelting plants, other metallurgical plants, and petroleum refineries add significant quantities of polluting chemicals to the atmosphere. These atmospheric pollutants, especially in the concentrations found in urban areas, cause people to have burning eyes, coughing, and breathing difficulties. Air pollution is nothing new; Shakespeare wrote about it in the 17th century.

■ *This most excellent canopy, the air, look you, this excellent o'erhanging firmament, this majestical roof fretted with golden fire, why, it appears no other thing to me but a foul and pestilent congregation of vapors.* Hamlet (act II, scene 2)

(a)

(b)

Air pollution by nature and industry. (a) Volcanoes, like the one pictured in Mauna Loa, Hawaii, emit ash and a variety of inorganic chemicals. (b) Industrial smokestacks, like these at a steel plant in Pennsylvania, also emit pollutants. Enforcement of the Clean Air Act has, however, been effective in greatly improving air quality near industrial sites. *(a, Dan McCoy/Rainbow; b, Jack Rosen/Photo Researchers)*

Prior to 1960 there was little concern about air pollution and little effort toward its control in the United States, in spite of some dramatic episodes in which many people suffered as a direct result of polluted air. For example, in October of 1948 the city of Donora, Pennsylvania, was overcome by five days of air pollution that caused almost 6000 residents to become ill and 18 to die. In the past, smoke, carbon monoxide, sulfur dioxide, nitrogen oxides, and organic vapors were emitted into the air from industrial facilities with little apparent thought about their harmful nature as long as they were scattered into the atmosphere and away from human smell and sight.

■ A few decades ago, we operated on the principle that "Dilution is the solution to pollution."

Early in the 1960s, air pollution became generally recognized as a problem in the United States, and this resulted in laws governing emissions of air pollutants by industry. In 1970 the first Clean Air Act was passed. This law helped in controlling air pollution from sources such as industry and automobiles, but it was not very comprehensive. The Clean Air Act was amended in 1977 to add stricter requirements, for example, on emissions from automobiles. In November of 1990 the President signed into law the 1990 Clean Air Act (CAA) amendments, a major overhaul of the earlier Clean Air Act. The 1990 CAA affects almost everything that is manufactured and consumed in this country, all in the name of cleaner, safer air. The substances regulated by the 1990 CAA include those discussed in this chapter—particulates, ozone, carbon monoxide, oxides of nitrogen and sulfur, hydrocarbons, volatile toxic substances, carbon dioxide, and stratospheric ozone-depleting chemicals. Let's begin by looking at the particles that obscure our vision, aggravate respiratory illnesses, and cause regional and global cooling by scattering sunlight.

21.2 Air Pollutants—Particle Size Makes a Difference

One of the most common forms of air pollution occurs as particles. Pollutant particles range in size from fly ash particles, which are big enough to see, down to individual molecules, ions, or atoms. Because of their polar nature, many pollutants are attracted into water droplets and form **aerosols.** Fogs and smoke are common examples of aerosols. Larger solid particles in the atmosphere are called **particulates.** The solids in an aerosol or particulate may be metal oxides, soil particles, sea salt, fly ash from electric generating plants and incinerators, elemental carbon, or even small metal particles. Aerosol particles range upward from a diameter of 1 nm (nanometer) to about 10,000 nm and

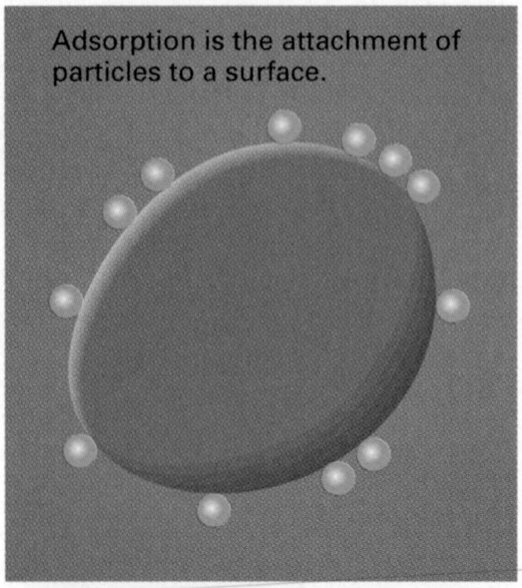

Adsorption is the attachment of particles to a surface.

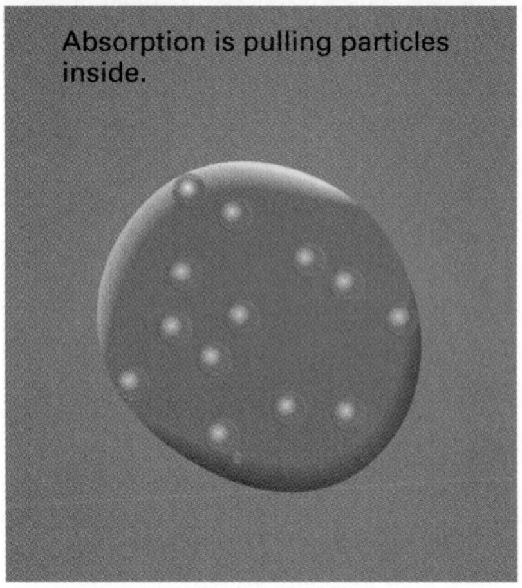

Absorption is pulling particles inside.

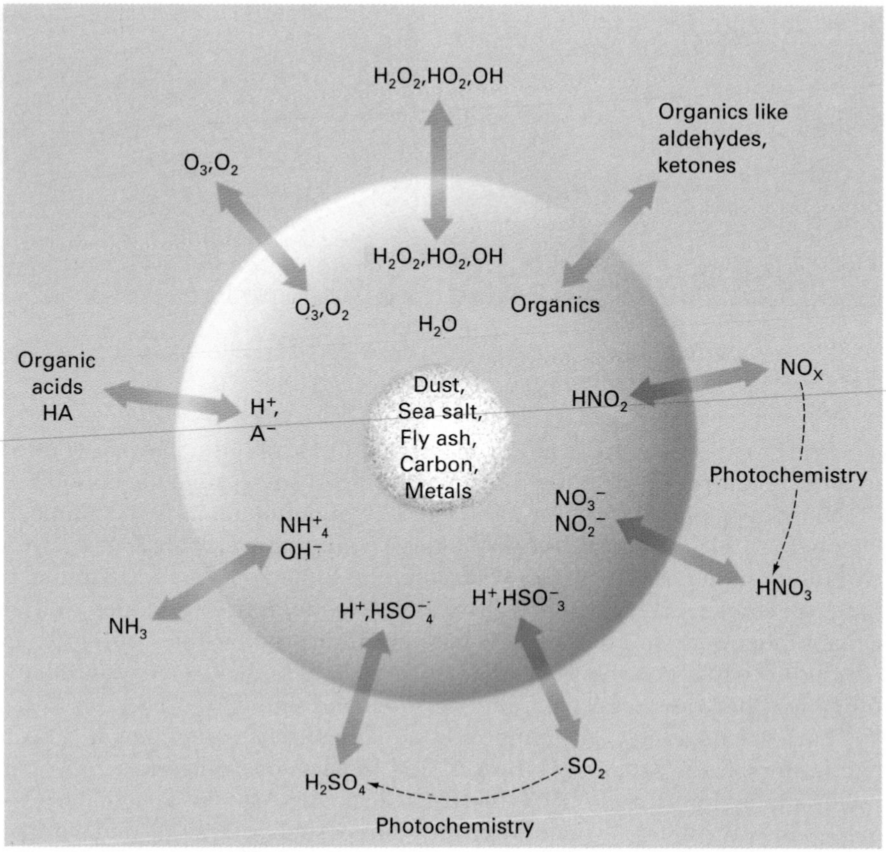

Figure 21.2 A typical urban aerosol particle, showing its composition and some of the chemical reactions of urban air pollutants.

■ 1 μm = 10^{-6} m, or 1000 nm.

■ Major contributors to the amount of atmospheric particulates were volcanic eruptions by Krakatoa, Indonesia, 1883; Mt. Katmai, Alaska, 1912; Hekla, Iceland, 1947; Mt. Spurr, Alaska, 1953; Bezymyannaya, U.S.S.R., 1956; Mt. St. Helens, Washington, 1980; Mt. Pinatubo, 1991.

may contain as many as a trillion (10^{12}) atoms, ions, or small molecules. Particles in the 2000-nm range are largely responsible for the deterioration of visibility.

Aerosol particles are small enough to remain suspended in the atmosphere for long periods. Such particles are easily breathable and can cause lung diseases. They may also contain mutagenic or carcinogenic compounds (see Figure 18.9, Section 18.7). Because of their relatively large surface area, aerosol particles have great capacities to *adsorb* and concentrate chemicals on their surfaces. Liquid aerosols or particles covered with a thin coating of water may *absorb* air pollutants, thereby concentrating them and providing a medium in which reactions may occur. A typical urban aerosol and some of the reactions that can take place there are shown schematically in Figure 21.2.

Millions of tons of soot, dust, and smoke particles are emitted into the atmosphere of the United States each year. The average suspended particulate concentrations in the United States vary from about 0.00001 g/m^3 of air in rural areas to about six times as much in urban locations. In heavily polluted areas, concentrations of particulates may increase to 0.002 g/m^3.

Particulates in the atmosphere can cool the Earth by scattering and partially reflecting light from the Sun. Large volcanic eruptions such as those from Mt. St. Helens in 1980 and Mt. Pinatubo in 1991 had measurable cooling effects on the Earth.

Particulates and aerosols are removed naturally from the atmosphere by gravitational settling and by rain and snow. Industrial emissions of particulates can be prevented by treating the emissions with one or more of a variety of physical methods such as filtration, centrifugal separation, and scrubbing. Another method often used is electrostatic precipitation, which is more than 98% effective in removing aerosols and dust particulates even smaller than 1 μm from exhaust gases. The effects of an efficient electrostatic precipitator can be quite dramatic, as Figure 21.3 shows.

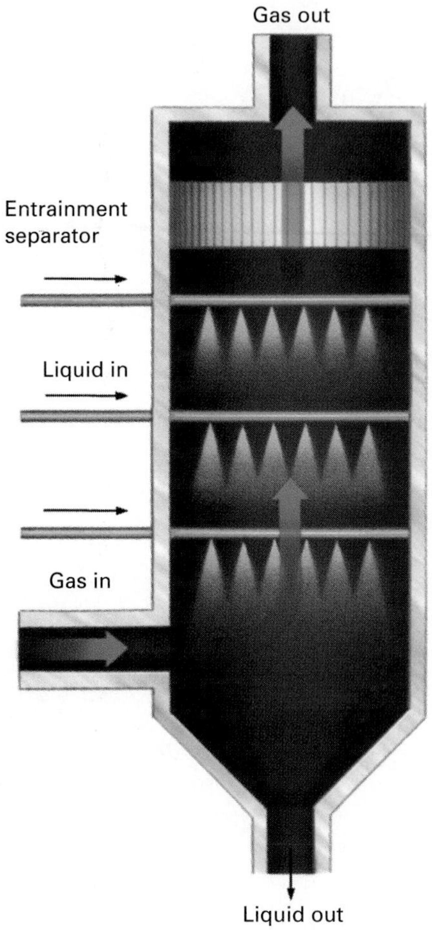

Removing particulate pollutants by scrubbing. The fine mist of water droplets traps particulates entering with the gas stream.

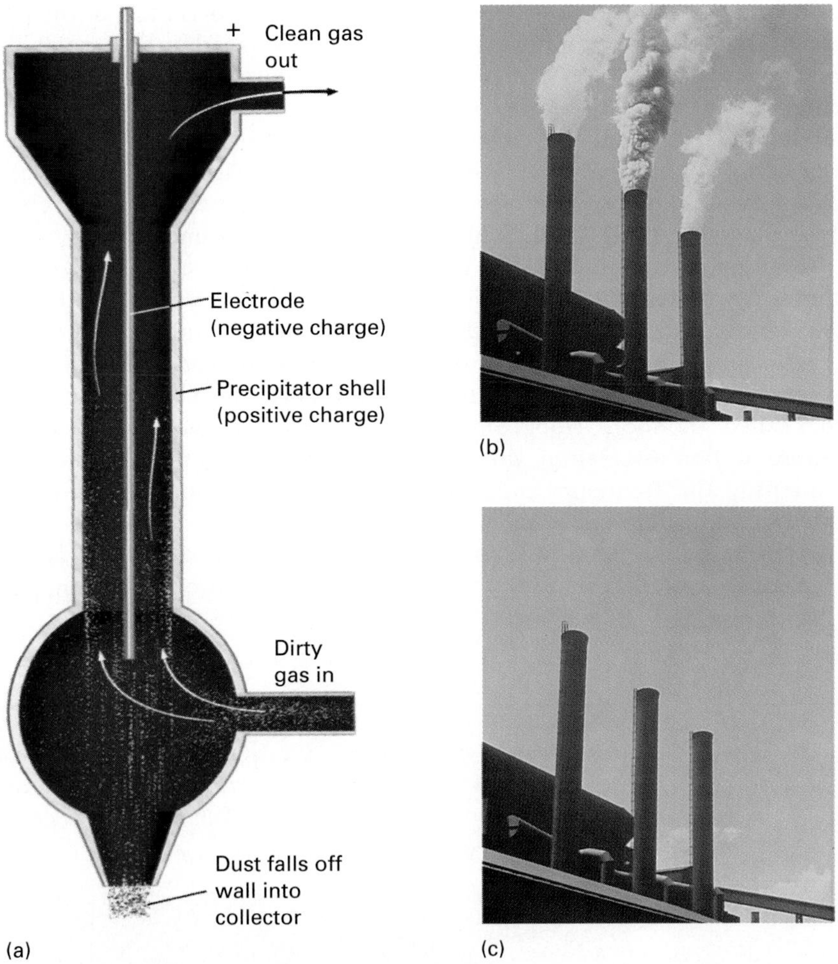

Figure 21.3 Electrostatic precipitation and its effectiveness. (a) An electrostatic precipitator. The central electrode is negatively charged and imparts a negative charge to particles in smoke that pass over it. These charged particles are then attracted to the positively charged walls and fall into the collector. (b) Smokestacks at a steel mill with the electrostatic precipitator turned off. (c) The same smokestacks with the electrostatic precipitator turned on. *(b and c, Visuals Unlimited/John D. Cunningham)*

21.3 Smog

The poisonous mixture of smoke, fog, air, and other chemicals was first called **smog** in 1911 by Dr. Harold de Voeux in his report on a London air-pollution disaster that caused the deaths of 1150 people. Through the years, smog has been a technological plague in many communities and industrial regions.

What general conditions are necessary to produce smog? Although the chemical ingredients of smogs vary depending on the unique sources of the pollutants, certain geographical and meteorological conditions exist in nearly every instance of smog.

■ Thermal inversion: mass of warmer air over a mass of cooler air.

First, there must be a period of windlessness so that pollutants can collect without being dispersed vertically or horizontally. This sets the conditions for a **thermal inversion,** which is an abnormal temperature arrangement for air masses (Figure 21.4). Normally, warmer air is on the bottom, nearer the warm Earth, and this warmer, less dense air rises and transports most of the pollutants to the upper troposphere, where they are dispersed. In a thermal inversion the warmer air is on top, and the cooler, denser air retains its position nearer the Earth. The air becomes stagnated. If the land is bowl-shaped (surrounded by mountains, cliffs, or the like), a stagnant air mass can remain in place for quite some time. When these atmospheric conditions exist, the pollutants supplied by combustion and evaporation in automobiles, electric power plants, and industrial plants accumulate to form smog.

Two general kinds of smog have been identified. One is the *chemically reducing type,* which is derived largely from the combustion of coal and oil, and contains sulfur dioxide mixed with soot, fly ash, smoke, and partially oxidized organic compounds. This type of smog is usually seen near industrial centers. Because it was first characterized in and around the city of London, it is sometimes called "London" smog. Smogs of this kind, also called industrial smogs because of their association with industrial activity, are generally diminishing in intensity and frequency as less coal is burned and more controls are installed on industrial emissions.

■ Industrial smog: fog + SO_2

The main ingredient in industrial smog is sulfur dioxide. Laboratory experiments have shown that sulfur dioxide increases aerosol formation, partic-

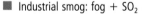

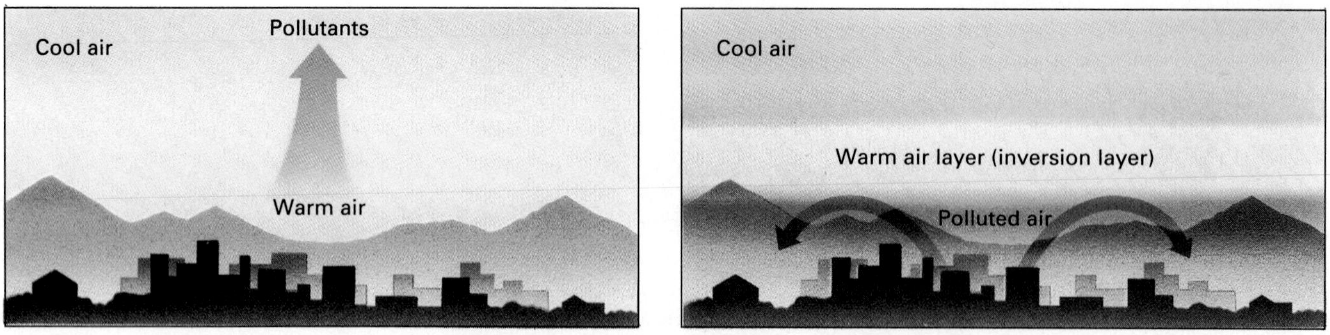

Figure 21.4 A thermal inversion. Normally, air that is warmed near the surface rises, carrying pollutants with it. During a thermal inversion, a blanket of warm air becomes stationary over a layer of cooler, denser air. The result is that pollutants are trapped near the surface.

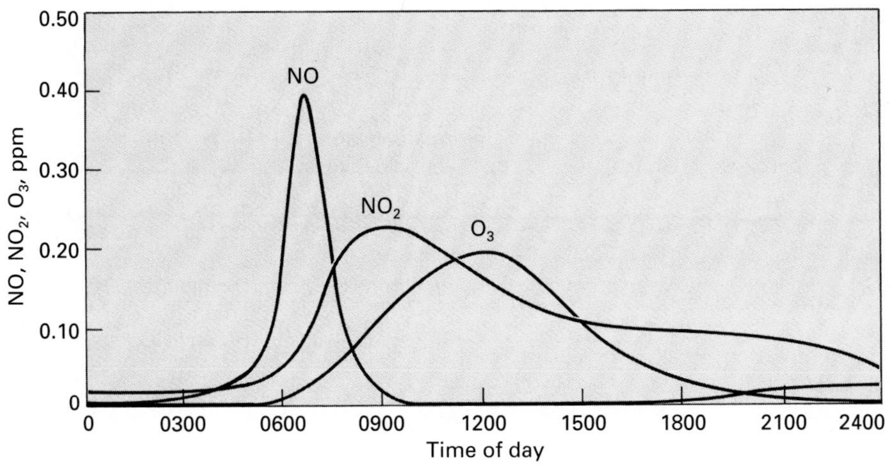

Figure 21.5 The average concentrations of pollutants NO, NO_2, and O_3 on a smoggy day in Los Angeles. The NO concentration builds up during the morning rush hour. Later in the day, the concentrations of NO_2 and O_3 build up.

ularly in the presence of mixtures of hydrocarbons and nitrogen oxides. For example, mixtures of 3 ppm hydrocarbons, 1 ppm NO_2, and 0.5 ppm SO_2 at 50% relative humidity form aerosols that have sulfuric acid as a major product. Breathing a sulfuric acid aerosol is very harmful, especially to people suffering from respiratory diseases such as asthma or emphysema. At a concentration of 5 ppm for one hour, this kind of aerosol can cause constriction of bronchial tubes. A level of 10 ppm for one hour can cause severe breathing distress.

A second type of smog is the *chemically oxidizing type,* typical of Los Angeles and other urban centers where exhaust fumes from internal combustion engines are highly concentrated in the atmosphere. This predominantly urban smog is called **photochemical smog** because light—in this instance sunlight—is important in initiating several chemical reactions that together make the smog harmful. Photochemical smog is practically free of sulfur dioxide but contains substantial amounts of nitrogen oxides, ozone, oxygenated and ozonated hydrocarbons, and organic peroxide compounds, together with unreacted hydrocarbons of varying complexity. The automobile is a direct or indirect source of many of the components of photochemical smog. Consider Figure 21.5. It shows how several components of photochemical smog increase during the rush hour in Los Angeles.

■ Organic peroxides contain the R—O—O—R′ structure and are produced by ozone reacting with organic molecules. Hydrogen peroxide is H—O—O—H.

Many of the chemical reactions that create photochemical smog take place in aerosol particles. These reactions produce **secondary pollutants**—pollutants that are not directly released from some source but are formed by reactions with other components in the air.

■ **Secondary pollutants** are formed in the air by chemical reactions.

The exact reaction scheme by which **primary pollutants** form the secondary pollutants of photochemical smog is still not completely understood (Figure 21.6). One process that is known begins with the absorption of light energy by a molecule of nitrogen dioxide, which causes its breakdown into nitric oxide and atomic oxygen, a highly reactive chemical **free radical.**

■ **Primary pollutants** are emitted directly into the air from a source.

■ A chemical **free radical** is a species with an unpaired valence electron. Both NO and oxygen atoms are free radicals. They are usually very reactive.

$$NO_2(g) \xrightarrow{\text{Light}} NO(g) + O(g)$$

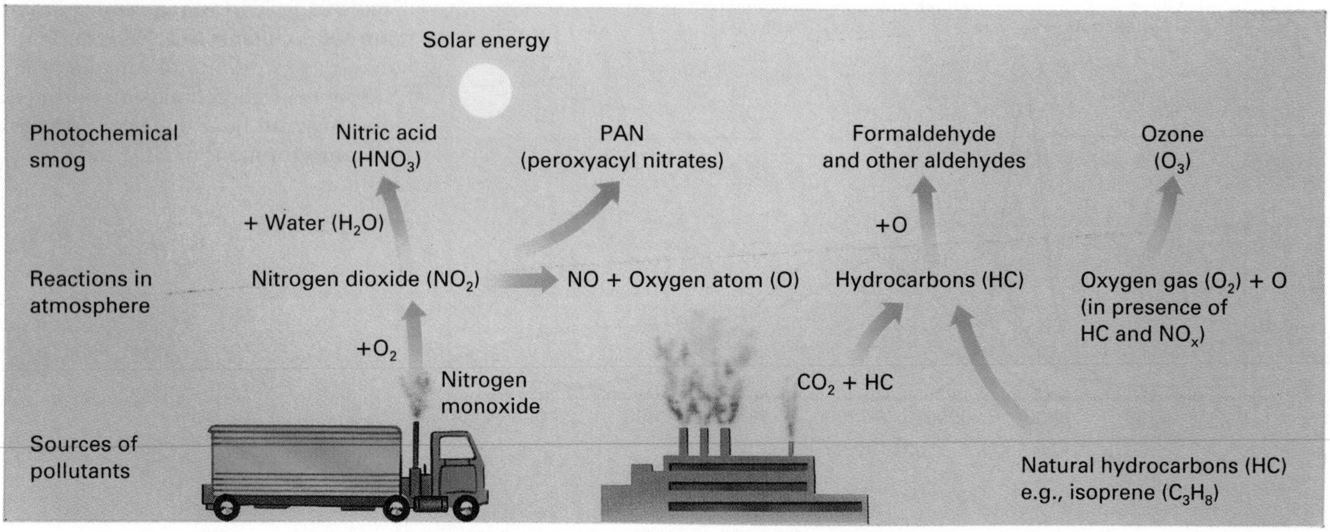

Figure 21.6 The formation of photochemical smog.

The atomic oxygen then reacts with molecular oxygen to form ozone (O_3), which is an important secondary pollutant.

$$O(g) + O_2(g) \longrightarrow \underset{\text{Ozone}}{O_3(g)}$$

Photochemical smog over San Diego. The layer of smog held in place by a thermal inversion layer is clearly visible here. The red-brown color of the smog shows the presence of nitrogen dioxide. *(Alan Pitcairn/Grant Heilman Photography)*

Atomic oxygen that doesn't react with oxygen molecules to form ozone can also react with hydrocarbons, such as olefins (molecules with double bonds) and aromatics, to form other chemicals such as aldehydes and ketones, which are toxic and also impart an odor to the air. On a sunny day only about 0.2 ppm of nitrogen oxides and 1 ppm of reactive hydrocarbons are sufficient to initiate these photochemical smog reactions. The hydrocarbons involved in these reactions come mostly from unburned petroleum products such as gasoline, and the nitrogen oxides come from the exhausts of internal combustion engines.

In the following sections we shall look at the major ingredients of photochemical smog—the primary pollutants, the oxides of nitrogen and hydrocarbons, and the secondary pollutant ozone—to see how they produce urban pollution.

21.4 Nitrogen Oxides

■ Photochemical smog:

Fog + NO_x + hydrocarbons

There are eight known oxides of nitrogen, three of which are recognized as important components of the atmosphere: dinitrogen oxide (N_2O), nitric oxide (NO), and nitrogen dioxide (NO_2). These oxides of nitrogen are collectively known as "NO_x." About 97% of the nitrogen oxides in the atmosphere are naturally produced, and only 3% result from human activity. Certain bacteria can produce N_2O, so this oxide of nitrogen is also commonly found in the atmosphere in trace amounts.

Almost all of the nitrogen chemically bound to oxygen begins as NO (nitric oxide), a colorless reactive gas. Nitrogen, normally a relatively inert gas, readily reacts with oxygen when there is a source of energy to produce a high temperature. The product is NO.

■ Another name for nitric oxide is "nitrogen monoxide."

$$N_2(g) + O_2(g) \xrightarrow{\text{Energy}} \underset{\text{Nitric oxide}}{2\,NO(g)}$$

Nitric oxide is formed in this manner during electrical storms, where lightning can supply the needed energy. Nitric oxide is also formed whenever combustion of fuels in air produces high temperatures. Because the formation of nitric oxide requires heat, it follows that a higher combustion temperature would produce relatively more NO. The combustion of gasoline vapors in an automobile engine is always accompanied by the production of NO because both nitrogen and oxygen are present in the combustion chamber (Figure 21.7).

The NO molecule is short-lived in the atmosphere because it reacts rapidly with atmospheric oxygen to produce NO_2, a brown gas (see Section 3.2).

$$2\,NO(g) + O_2(g) \longrightarrow \underset{\text{Nitrogen dioxide}}{2\,NO_2(g)}$$

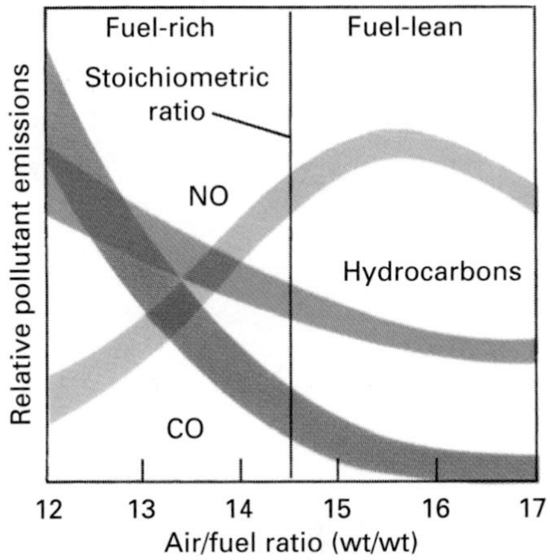

Figure 21.7 The effects of the fuel/air ratio on NO emissions in an automobile internal combustion engine. The stoichiometric ratio is the air/fuel ratio that allows complete combustion of the hydrocarbon fuel to CO_2 and H_2O.

Normally the atmospheric concentration of NO_2 is a few parts per billion (ppb) or less; most of the nitrogen oxides formed during lightning storms are washed out by rain. This is one of the ways nitrogen is made available to plants. Looking at all the sources of oxides of nitrogen (Table 21.2), it is apparent that combustion processes are their primary sources. In the United States, most oxides of nitrogen from sources other than nature are produced from fossil-fuel combustion.

Nitrogen dioxide is a powerful corrosive agent. It will cause severe skin burns in high concentrations. Breathing a concentration of only 3 ppm NO_2 for one hour causes bronchioconstriction in humans, and short exposures at high levels (150–220 ppm) result in corrosive reaction with lung tissue that

TABLE 21.2 ■ Emissions of NO_x

	Emissions (millions of tons)	
Source	*United States*	*Global*
Fossil fuel combustion	66	231
Biomass burning	1.1	132
Lightning	3.3	88
Microbial activity in soil	3.3	88
Input from the stratosphere	0.3	5.5
Total (uncertainty in estimates)	74 (± 1)	544.5 (± 275)

Note: The large uncertainty for global emissions is due to incomplete data for much of the world.
Source: Stanford Research Institute

can be fatal. A seemingly harmless exposure to high concentrations of NO_2 one day can even cause death a few days later.

One of the primary roles of nitrogen dioxide as a pollutant is in the formation of the secondary pollutant ozone. Nitrogen dioxide reacts with light (see Section 3.5), which can be written as $h\nu$, with a wavelength between 280 and 430 nm. This **photodissociation** reaction (*photo,* light; *dissociation,* breaking apart) produces nitric oxide and free oxygen atoms (O, oxygen free radicals) that can react with a molecule of oxygen to produce a molecule of ozone.

$$NO_2(g) + h\nu \longrightarrow NO(g) + O(g)$$
$$O_2(g) + O(g) \longrightarrow O_3(g)$$

The nitric oxide then reacts with oxygen to regenerate NO_2.

Nitrogen dioxide can also react with water to form nitric acid and nitrous acid. This reaction takes place readily in aqueous aerosols, producing acids that help to stabilize the droplet.

$$2\,NO_2(g) + H_2O(\ell) \longrightarrow \underset{\text{Nitric acid}}{HNO_3(aq)} + \underset{\text{Nitrous acid}}{HNO_2(aq)}$$

■ Nitrates are important components of fertilizers.

Of course, breathing air containing these aerosol droplets is harmful because of the corrosive nature of the acids. The acids in turn can react with ammonia or metallic particles in the atmosphere to produce nitrate or nitrite salts. For example,

$$\underset{\text{Ammonia}}{NH_3(g)} + HNO_3(aq) \longrightarrow \underset{\text{Ammonium nitrate (a salt)}}{NH_4NO_3(aq)}$$

Both the acids and the salts stabilize the aerosol particles, which eventually settle from the air or dissolve in larger raindrops. Nitrogen dioxide, besides causing the formation of ozone, is a primary cause of haze in urban or industrial atmospheres because of its participation in the process of aerosol formation.

21.5 Ozone and Its Role in Air Pollution

■ The odor of ozone can be detected by most people at concentrations as low as 0.02 ppm.

Ozone consists of three oxygen atoms bound together in a molecule with the formula O_3. It has a pungent odor that we often smell near sparking electrical appliances or after a thunderstorm when lightning-caused ozone washes out with the rainfall.

As you will see in this chapter, there is "good" and "bad" ozone. The bad ozone is that found in the air we breathe, whereas the good ozone is found in the stratosphere, where it forms a protective blanket, absorbing harmful ultraviolet radiation (see Section 21.10).

Being a secondary pollutant, ozone is one of the most difficult pollutants to control. According to the EPA, in 1990 the upper limit for ozone of 0.12 ppm was exceeded in many of the urban areas of the United States (Table 21.3). These high ozone concentrations were primarily the result of the excess

TABLE 21.3 ■ Urban Areas with the Worst Ozone Air Quality in 1990

Extreme (0.28 ppm O_3)	**Severe (0.18 to 0.19 ppm O_3)**
Los Angeles and south coast air basin, California	Baltimore and the State of Maryland
Very Severe (0.19 to 0.28 ppm O_3)	Philadelphia, Pennsylvania
Chicago and Gary and Lake Counties, Indiana	Wilmington, Delaware
Houston, Galveston, and Brazoria, Texas	Trenton, New Jersey
Milwaukee and Racine, Wisconsin	San Diego, California
New York City and Long Island, New York	Ventura County (between Santa Barbara and Los Angeles), California
Northern New Jersey and Connecticut	
Southeast Desert, California	

nitrogen oxide emissions from automobiles, buses, and trucks. Most major urban areas have vehicle inspection centers for passenger automobiles in an effort to control nitrogen oxide emissions as well as emissions of carbon monoxide and unburned hydrocarbons. In spite of these efforts, large urban centers still have high NO_2 concentrations, which result in too much ozone being formed.

As difficult as it is to attain, the ozone standard that the EPA has set may not be low enough for good health. Exposure to concentrations of ozone at or near 0.12 ppm lowers the volume of air a person breaths out in 1 second (the forced expiratory volume, or FEV_1). Children who were exposed to ozone concentrations close to the EPA standard, but not exceeding it, showed a 16% decrease in the FEV_1. Some scientists have been urging the EPA to lower the standard to 0.08 ppm. If that is done, even some midsized cities would probably fail to meet the EPA standards.

■ Lower FEV_1 accelerates the aging of the lungs.

No matter what the standard becomes, present ozone concentrations in many urban areas represent health hazards to children at play, joggers, others doing outdoor exercise, and older people who may have diminished respiratory capabilities. The only effective way to limit ozone is to limit NO_x emissions. In some areas this means possibly limiting the numbers of automobiles in use on any given day or limiting the number of automobiles with internal combustion engines. This idea has given rise to a state rule in California setting a number of electric-powered automobiles that must be sold (see Chapter 10, Science and Society—Electric Automobiles: Like Them or Not, Here They Come).

Rush hour during a transit strike in New York City. These bicycle commuters are not contributing to air pollution, but they are probably breathing in more of it than is healthy. *(© James H. Karales/Peter Arnold, Inc.)*

21.6 Hydrocarbons and Air Pollution

Hydrocarbons enter the atmosphere from both natural sources and human activities. Certain natural hydrocarbons are produced in large quantities by both coniferous and deciduous trees. Methane gas (CH_4) is produced by such diverse sources as rice-growing, ruminant animals, termites, ants, and decay-causing bacteria acting on dead plants and animals. Human activities such as the use of industrial solvents, petroleum refining and its distribution, and the release of unburned gasoline and diesel fuel components account for a large amount of hydrocarbons in the atmosphere.

■ In 1988 William Chameides, of Georgia Tech in Atlanta, published a report in *Science* magazine, in which he stated that in some cities trees may account for more hydrocarbons in the atmosphere than those produced from human activities. The EPA has since found this to be true. This fact is causing a re-thinking about how to control urban pollution.

FRONTIERS IN THE WORLD OF CHEMISTRY

Smog-Eating Radiators

What could be better? Driving your car *and* destroying more pollutants than you produced! That just might be possible someday if a new catalytic *radiator* being tested by Ford Motor Company proves successful. Radiators found under the hoods of cars allow engine coolant to circulate and be cooled by outside air. Even on hot days, fast-moving air passing through openings in the radiator provide enough cooling to keep a typical engine from overheating. (When the car isn't moving, as in heavy traffic, an electric fan turns on and blows air through the radiator and removes heat from the engine coolant.) The fact that the very air passing through the radiator is also the polluted air we breathe caused some scientists at Engelhard Corporation to see an interesting solution to the air pollution problem. (Engelhard is one of the major world suppliers of platinum-based catalysts for automotive catalytic converters.)

Using their knowledge about how catalysts work, a special platinum-based coating was developed for the automotive radiator. This coating converts ozone, one of the main ingredients of smog, into oxygen. Carbon monoxide is also converted into carbon dioxide as the outside air passes through the radiator. The coating worked so well that about 90% of all the ozone and carbon monoxide passing over it were converted.

Recognizing a potentially good thing, Ford engineers have placed these PremAir catalytic coated radiators on a test fleet of vehicles with the hopes that everyday driving will show them to be net pollution eliminators rather than pollution generators. After determining how well the catalytic radiators hold up under salt, dirt, insects, high altitudes, as well as ice and snow, Ford will determine how much the radiators will cost car buyers. It is expected that the radiators will be "significantly under $1000 per vehicle."

As the new standards of the federal Clean Air Act, requiring significant numbers of non-polluting cars by 1998 go into effect, cars equipped with pollution-reducing radiators might offer buyers a chance to continue to buy gasoline-engine cars instead of models powered by batteries.

Reference Associated Press, June 1995.

■ For every million tons of coal burned, about 750 tons of benzo(α)pyrene can be produced. Coal smoke contains about 300 ppm benzo(α)pyrene.

In addition to simpler hydrocarbons like alkanes, alkenes, and alkynes, a large number of polynuclear aromatic hydrocarbons (PAH) are released into the atmosphere, primarily from motor vehicle exhaust. The greatest danger of these pollutants is their toxic properties. One PAH, benzo(α)pyrene (BAP), is a known carcinogen (see Section 18.8). Concentrations of BAP as high as 60 $\mu g/m^3$ of air have been found in urban air.

Hydrocarbons also contribute to ozone formation. Some of the oxygen atoms formed during the photodecomposition of NO_2 can react with water, forming hydroxyl free radicals (OH).

$$O(g) + H_2O(g) \longrightarrow \underset{\text{Hydroxyl free radical}}{2\,OH(g)}$$

Trees in urban environments may emit as many reactive hydrocarbons as do automobiles.

These hydroxyl radicals in turn react with hydrocarbon molecules, producing a number of compounds including aldehydes and ketones, and NO_2. Of course, NO_2 is easily photodecomposed, producing oxygen atoms, which go on to form ozone.

Although it is practically impossible to control hydrocarbon emissions from living plants and other natural sources, hydrocarbon emissions from automobiles can be controlled. Two means of control are being used at present. First, the spouts and hoses on gasoline pumps have been redesigned to prevent gasoline from entering the air. Second, catalytic converters that reduce emissions of hydrocarbons, CO and NO_x, are now part of every automobile's exhaust system. Careful control of the engine fuel-air ratio is re-

TABLE 21.4 ■ Emission Rates for Hydrocarbons (HC), Carbon Monoxide (CO), and NO_x

1960 (precontrol—no catalytic mufflers installed)		1993 (catalytic mufflers required on all automobiles)		1995 (EPA estimates)	
HC	10.6 g/mile	HC	0.41 g/mile	HC	0.25 g/mile
CO	84.0 g/mile	CO	3.4 g/mile	CO	3.4 g/mile
NO_x	4.1 g/mile	NO_x	1.0 g/mile	NO_x	0.4 g/mile

Note: As of December 31, 1993, there were 189,674,000 motor vehicles registered in the United States. Of these, 75.8% were cars, 23.9% were trucks, and the remainder were buses, motorcycles, and other vehicles.

quired for these catalysts to perform well. This is accomplished by means of an oxygen sensor in the engine. When it operates properly, the fuel-air ratio is correct, but if it malfunctions, some automobiles will not run until the sensor is replaced.

The effectiveness of these catalytic converters, which have been on automobiles sold in the United States since the mid-1970s, can be seen by comparing the emissions of hydrocarbons, CO and NO_x, in grams per vehicle-mile in 1960, before there were controls, to the same values in the 1990s (Table 21.4).

■ Hydrocarbon emissions from vehicles in California in 2003 will have to be much lower than the national standards. By 2003, hydrocarbon emissions for 75% of all vehicles must be no greater than 0.075 g/mile, no greater than 0.04 g/mile for 15% of all vehicles, and 10% of all vehicles must have zero hydrocarbon emissions. That means they will probably be electric vehicles.

■ SELF-TEST 21A

1. What has been the general trend in air pollutants for approximately the past decade? (a) Increase, (b) Decrease.
2. Name a chemical that is considered both an air pollutant and a beneficial chemical.
3. Because of their large surface areas, aerosol particles can (absorb/adsorb) chemicals onto their surfaces.
4. A liquid aerosol particle will probably (adsorb/absorb) a chemical.
5. A thermal inversion occurs when (warm/cool) air is above (warm/cool) air below.
6. Industrial-type smog is often associated with (a) coal burning, (b) sunlight.
7. In all combustion processes in air, some nitrogen ____________ are formed.
8. What are the products of the photodissociation of nitrogen dioxide?
9. What species reacts with molecular oxygen to form ozone?

21.7 Sulfur Dioxide—A Major Primary Pollutant

Sulfur dioxide is produced when sulfur or sulfur-containing compounds are burned in air. While volcanoes put large amounts of SO_2 into the atmosphere annually, human activities probably account for up to 70% of all emissions on a global basis. In the United States, about 21 million tons of sulfur are released annually.

$$S(s) + O_2(g) \longrightarrow SO_2(g)$$

Once formed, SO_2 generally becomes distributed in aerosol droplets, which are numerous enough to contribute to significantly reduced visibility and can affect both global and regional climate by causing the scattering of sunlight that would otherwise warm the Earth. Emissions of SO_2 cause the mean temperature in the United States to be about 1°C cooler than it would be otherwise.

Most of the coal burned in the United States contains sulfur in the form of the mineral pyrite (FeS_2). The weight percent of sulfur in this coal ranges from 1% to 4%. The pyrite is oxidized as the coal is burned, forming SO_2.

$$4\,FeS_2(s) + 11\,O_2(g) \longrightarrow 2\,Fe_2O_3(s) + 8\,SO_2(g)$$

Large amounts of coal are burned in this country to generate electricity. A 1000-megawatt (MW) coal-fired generating plant can burn about 700 tons of coal an hour. If the coal contains 4% sulfur, that equals 56 tons of SO_2 an hour, or 490,560 tons of SO_2 every year. About 800 million tons of coal are burned each year to produce electricity.

Oil-burning electric generating plants can also produce comparable amounts of SO_2 because some fuel oils can contain up to 4% sulfur. The sulfur in the oil is in the form of compounds in which sulfur atoms are bound to carbon and hydrogen atoms.

■ Electric power plants are discussed in Section 13.2.

States that rely mainly on coal for their electricity production and industrial furnaces have the highest SO_2 emissions in the United States. Operators of all coal-fired burners are under EPA orders to eliminate most of the SO_2 before it reaches the stack. The 1990 Clean Air Act requires that by the year 2000, SO_2 emissions from *all* power-generating sources will be no greater than 8.9 million tons per year. That's a 10 million ton per year reduction from 1980 levels.

The removal of sulfur from high-sulfur coal is costly and incomplete. One method is to pulverize the coal to the consistency of talcum powder and remove the pyrite (FeS_2) by magnetic separation. Reducing the sulfur content of fuel oil is also costly. It involves the formation of hydrogen sulfide (H_2S) by bubbling hydrogen through the oil in the presence of metal catalysts.

At present, most sulfur-containing coal is burned without prior treatment, and SO_2 is removed from the exhaust gases. In one method lime reacts with SO_2 to form calcium sulfite, a solid particulate, which can be removed from an exhaust stack by an electrostatic precipitator.

$$\underset{\text{Limestone}}{CaCO_3(s)} \xrightarrow{\text{heat}} \underset{\text{Lime}}{CaO(s)} + CO_2(g)$$

$$CaO(s) + SO_2(g) \longrightarrow \underset{\text{Calcium sulfite}}{CaSO_3(s)}$$

In another method, the exhaust gases containing SO_2 are passed through molten sodium carbonate, and solid sodium sulfite is formed (Figure 21.8).

Notice how both of these methods of SO_2 removal emit additional CO_2, also a pollutant (Section 21.11), into the atmosphere. Newer technology to address this problem is being developed.

A less desirable method of lowering the effects of SO_2 emissions, but one still being used, is sending the smoke up very tall exhaust stacks. Tall stacks emit SO_2 at a high elevation and away from the immediate vicinity, which

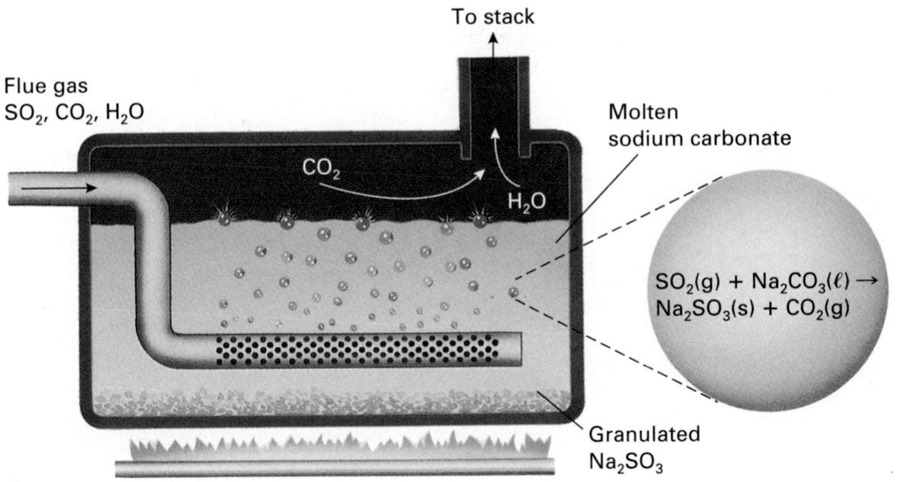

Figure 21.8 Removal of SO_2 from flue gas by reaction with molten sodium carbonate.

allows the SO_2 to be diluted before forming aerosol particles. The fact remains, however, that the longer it stays in the air, the greater chance SO_2 has to become sulfuric acid. A ten-year study in Great Britain showed that although SO_2 emissions from power plants increased by 35%, the construction of tall stacks decreased the ground-level concentrations of SO_2 by as much as 30%. The question is, who got the SO_2? In this case, Britain's solution was others' pollution. In the United States, the EPA may have added to a pollution problem unwittingly with rules in 1970 that caused plants to increase the height of smokestacks and caused pollutants to be carried longer distances by winds. Currently, about 179 stacks in the United States that are 500 ft tall or higher and 20 stacks that are 1000 ft or more tall are in use.

Most of the SO_2 that does get into the atmosphere reacts with oxygen to form sulfur trioxide (SO_3). The SO_3 has a strong affinity for water and dissolves in aqueous aerosol particles, forming sulfuric acid, a strong acid.

$$SO_3(g) + H_2O(\ell) \longrightarrow H_2SO_4(aq)$$

21.8 Acid Rain

The term **acid rain** was first used in 1872 by Robert Angus Smith, an English chemist and climatologist. He used the term to describe the acidic precipitation that fell on Manchester, England, just at the start of the Industrial Revolution. Although neutral water has a pH of 7, rainwater becomes naturally acidified from dissolved carbon dioxide, a normal component of the atmosphere. The carbon dioxide reacts reversibly with water to form a solution of the weak acid carbonic acid.

■ Reversibility and equilibria are discussed in Section 8.4. Weak acids are discussed in Section 9.5. See Figure 9.7 for a review of pH values.

$$2\,H_2O(\ell) + CO_2(g) \rightleftharpoons H_3O^+(aq) + HCO_3^-(aq)$$

At equilibrium, the pH of a solution of CO_2 from the air is 5.6. Any precipitation with a pH below 5.6 is considered to be acid rain.

Figure 21.9 Pollutants can cause acid rain to fall far from where the pollutants are generated.

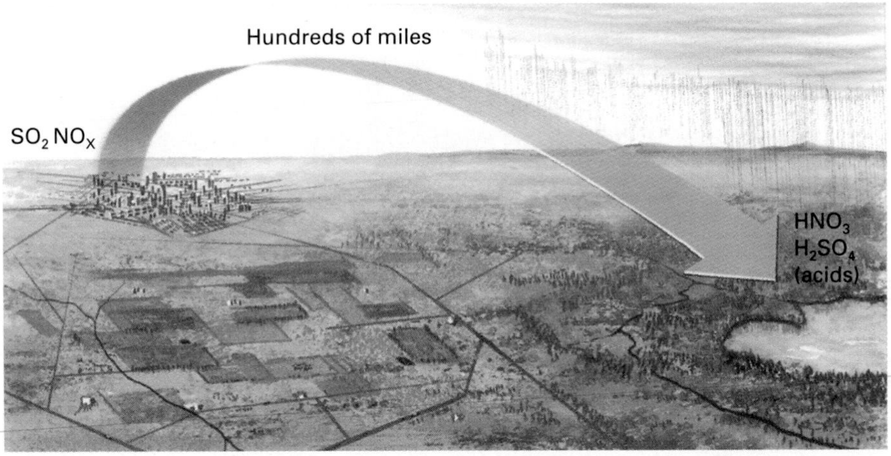

■ A more adequate term for acid rain might be "acid precipitation." Some scientists use "acid deposition."

■ The March 1991 eruption of Mt. Pinatubo in the Philippines injected more than 10^8 kg of SO_2 into the stratosphere. The SO_2 eventually came down as acid rain. During the period from 1991 to 1994 aerosol particles from these eruptions enhanced the beauty of sunrises and sunsets by scattering sunlight more than normal.

■ The leaching of toxic metal ions into groundwater by acid rain may also increase groundwater pollution.

As you have seen, NO_2 and SO_2 can both react with water in the atmosphere to produce acids: NO_2 produces nitric acid (HNO_3) and nitrous acid (HNO_2); SO_2 produces sulfuric acid (H_2SO_4) and sulfurous acid (H_2SO_3). When conditions are favorable, these acidic water droplets precipitate as rain or snow with a low pH. Ice core samples taken in Greenland and dating back to 1900 contain sulfate (SO_4^{2-}) and nitrate (NO_3^-) ions. This indicates that at least from 1900 onward, acid rain has been commonplace.

Acid rain is a problem today owing to the large amounts of these acidic oxides being produced by human activities and put into the atmosphere annually (Figure 21.9). When this precipitation falls on natural areas that cannot easily tolerate such acidity, serious environmental problems occur. The average annual pH of precipitation falling on much of the northeastern United States and northeastern Europe is between 4 and 4.5. Specific rainstorms in some areas where there are numerous sources of SO_2 and NO_x have had pH values as low as 1.5. Further complicating matters is the fact that acid rain is an international problem—rain and snow don't observe borders. Many Canadian residents are offended by the government of the United States because some of the acid rain produced in the United States falls on Canadian cities and forests (Figure 21.10).

The extent of the problems with acid rain can be seen in "dead" (fishless) ponds and lakes, dying or dead forests, and crumbling buildings. Because of wind patterns, Norway and Sweden have received the brunt of western Europe's emission of sulfur oxides and nitrogen oxides as acid rain. As a result, of the 100,000 lakes in Sweden, 4000 have become fishless, and 14,000 other lakes have been acidified to some degree. In the United States, 6% of all ponds and lakes in the Adirondack Mountains of New York are now fishless, and 200 lakes in Michigan are dead. For the most part, these dead lakes are still picturesque, but no fish can live in the acidified water. Lake trout and yellow perch die at pH values below 5.0, and smallmouth bass die at pH values below 6.0. Mussels die when the pH is below 6.5.

Acid rain damages trees in several ways. It disturbs the stomata (openings) in tree leaves and causes increased transpiration and a water deficit in the tree.

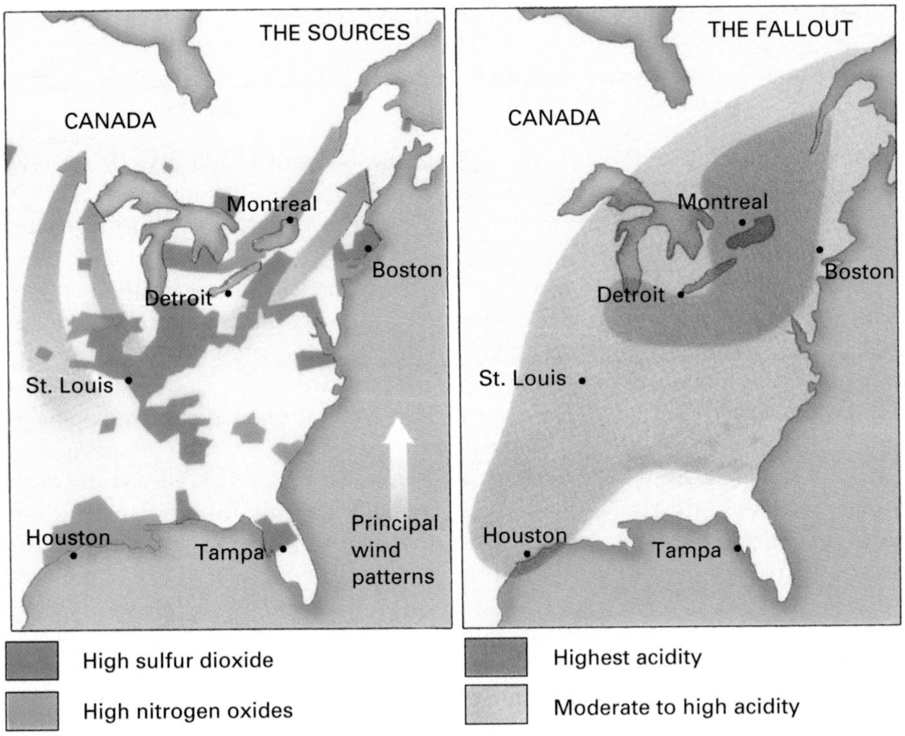

Figure 21.10 Distribution of sulfur dioxide, nitrogen oxides, and acid rain. Prevailing winds carry the acid droplets over the Northeast and into Canada.

The surface structures of the bark and the leaves can also be destroyed by the acid. Acid rainfall can acidify the soil, damaging fine root hairs and thus diminishing nutrient and water uptake. In addition, acid rain dissolves minerals that are insoluble in groundwater and surface waters of normal pH and many of these minerals contain metal ions toxic to plant life. For example, acid rain dissolves aluminum hydroxide in the soil, allowing aluminum ions (Al^{3+}) to be taken up by the roots of plants, where they have toxic effects.

$$Al(OH)_3(s) + 3\,H^+(aq) \longrightarrow Al^{3+}(aq) + 3\,H_2O(\ell)$$

The effects of acid rain and other pollution on stone and metal structures are especially devastating because of their irreversibility. By damaging stone buildings in Europe, acid rain is slowly but surely dissolving the continent's historical heritage. The bas-reliefs on the Cologne (West Germany) cathedral are barely recognizable today. The Tower of London, St. Paul's Cathedral, and the Lincoln Cathedral in London have suffered the same fate. Other beautifully carved statues and bas-reliefs on buildings throughout Europe and the eastern part of the United States and Canada are slowly passing into oblivion by the action of pollutants, in particular, acid rain.

What can be done about acid rain? Obviously, eliminating the emissions of the oxides of nitrogen and sulfur would be the answer. This is not easy, however. Some stopgap measures are being taken, such as spraying hydrated lime,

Effects of acid rain on a tombstone in England that was carved in 1817. *(Bruce F. Molnia/Terraphotographics)*

■ The government of Sweden is spending $40 million a year to neutralize the acid in some of its lakes.

Effects of acid rain on a forest in one of the most polluted parts of Europe. The devastation has been caused by emission of sulfur dioxide and nitrogen oxides from factories in the former East Germany and Czechoslovakia. *(Simon Fraser/SPL/Photo Researchers)*

$Ca(OH)_2$, into acidified lakes to neutralize at least some of the acid and raise the pH toward 7.

$$Ca(OH)_2(s) + 2\,H^+(aq) \longrightarrow Ca^{2+}(aq) + 2\,H_2O(\ell)$$

Some lakes in the problem areas have their own safeguard against acid rain by having limestone-lined bottoms, which supply calcium carbonate ($CaCO_3$) for neutralizing the acid from acid rain (just as an antacid tablet relieves indigestion).

The ultimate answers to acid-rain problems lie with those industries that produce the oxides of sulfur and nitrogen and with the regulatory agencies that govern them. As you read in the earlier section, methods exist for the control of SO_2 emissions, although some of these are costly. In the final analysis, the consumer will bear those costs. The control of oxides of nitrogen is more difficult because there are so many sources of combustion exhaust gases. Catalytic mufflers help control NO_x emissions from automobiles, but most home furnaces, industrial boilers, and even electrical generating plants do not have adequate controls. Fortunately for acid-rain production, NO_2 is so reactive in the troposphere that it is not the major contributor that SO_2 is to acid rain.

21.9 Carbon Monoxide

At least ten times more carbon monoxide enters the atmosphere from natural sources than from all industrial and automotive sources combined. Of the 3.8 billion tons of carbon monoxide emitted every year, about three billion tons are emitted by the oxidation of decaying organic matter in the topsoil. In spite of this fact, carbon monoxide is considered an air pollutant, primarily because so much of it is produced by human activities in the urban environment.

Like ozone, carbon monoxide is one of the most difficult pollutants to control. Cities such as Los Angeles and other highly populated urban centers with their high densities of automobiles tend to be repeatedly cited by the EPA for not attaining the required ambient air quality for carbon monoxide.

Carbon monoxide is always produced when carbon or carbon-containing compounds are oxidized using an insufficient quantity of oxygen.

$$2\,C(s) + O_2(g) \longrightarrow 2\,CO(g)$$

Gasoline engines are notorious sources of CO. This happens because the rapid combustion inside the combustion chamber does not burn all of the carbon to CO_2 before the exhaust gases are swept out. Modern catalytic converters convert much of this carbon monoxide to carbon dioxide, but the amounts that are not converted make being near a heavily traveled street dangerous because of the carbon monoxide concentrations. At peak traffic times, concentrations as high as 50 ppm are common. In the countryside, carbon monoxide levels are closer to the global average of 0.1 ppm. The only effective means of controlling carbon monoxide concentrations in urban air is to control the major emitters—automobiles. Of course, transportation that does not depend on burning hydrocarbon fuels would not emit *any* CO.

A bit of a mystery concerning carbon monoxide is that its global level does not seem to be changing, as is the case with some pollutants. Although polar carbon monoxide molecules dissolve readily in water, they react very slowly with oxygen to form carbon dioxide. The fate of atmospheric carbon monoxide is the subject of ongoing research in atmospheric chemistry.

21.10 Chlorofluorocarbons and the Ozone Layer

Most pollutants are adsorbed onto surfaces, absorbed into water droplets and react, or react in the gas phase with other pollutants in the lower atmosphere (troposphere) and eventually wash out in precipitation. There is one class of industrial pollutants, the halogenated hydrocarbons collectively called **chlorofluorocarbons,** or **CFCs,** that are relatively unreactive and are not easily or quickly eliminated in the troposphere. Being unreactive, CFCs are also virtually nontoxic, so their presence in the troposphere causes none of the problems associated with carbon monoxide or the oxides of sulfur and nitrogen. Instead, as a direct result of their nonreactivity, these compounds have a chance to mix with air in the stratosphere, where they can reside for many years.

■ The low toxicity of CFCs allowed them to be used as propellants for aerosols, dispensing such things as hair sprays, deodorants, and even medicines. Virtually all these uses, except for some medical applications, have been banned.

After the discovery of the refrigerant gas properties of the chlorofluorocarbons (see Section 1.3), their use became widespread in applications such as automotive and home air-conditioning as well as in the manufacture of formed plastics for insulation. During the time CFCs were in common use, little effort was made to prevent them from escaping into the atmosphere. For example, if your automobile air-conditioner needed repair, it was common practice to vent all of the CFC refrigerant gas to the atmosphere before any work was done. In fact, probably most of the CFCs ever manufactured have been released into the atmosphere. The industrialized countries were the first to use these compounds, but by the late 1970s they were also being extensively used in developing countries as well (Figure 21.11). Because of their excellent solvent properties for greases and oils, many of the CFCs also have been used as degreasers during the manufacture of printed circuit boards used in com-

■ Approximately two thirds of U.S. households rely on air-conditioning systems that use CFC refrigerants.

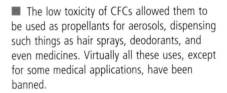

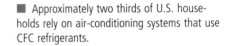

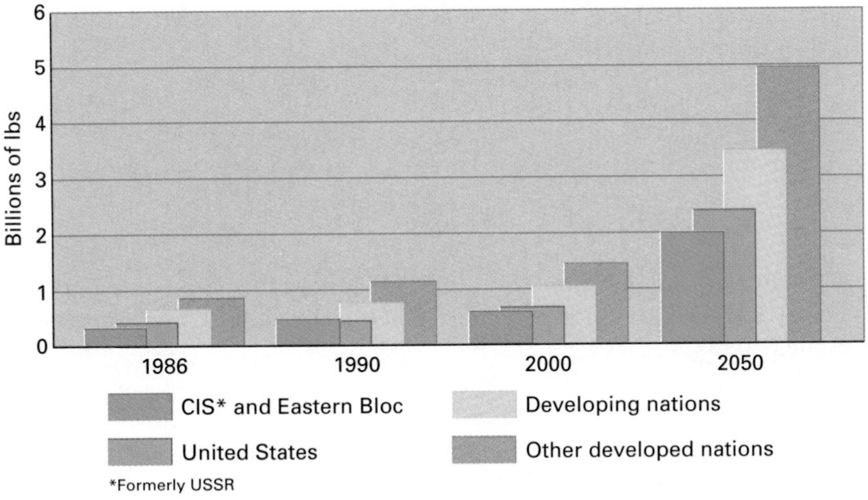

Figure 21.11 Projection of growth in CFC use by geographic region for 1986 to 2050 if no international controls were to be exercised. *(Environmental Protection Agency)*

puters, television sets, and other kinds of appliances. In effect, the increased use of CFCs paralleled the growth of modern, urbanized, industrial society with its indoor climate control, electronics, and electrical appliances.

$$Cl—\overset{\displaystyle Cl}{\underset{\displaystyle Cl}{C}}—F \qquad Cl—\overset{\displaystyle F}{\underset{\displaystyle Cl}{C}}—F$$

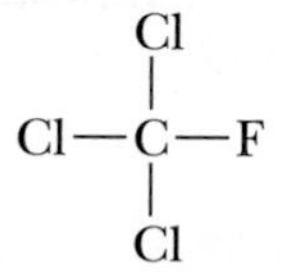

CFC-11
Trichlorofluoromethane

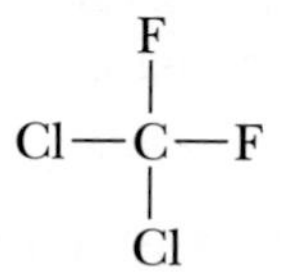

CFC-12
Dichlorodifluoromethane

■ The numbers in the names CFC-11, and CFC-12 are industrial code numbers that identify the compounds without using complicated chemical names.

■ In the United States, California leads all other states in releases of ozone-depleting substances, with more than 10,000,000 pounds released in 1992.

The dangers of CFCs were first announced in 1974, when M. J. Molina and F. S. Rowland of the University of California, Irvine, published a scientific paper in which they predicted that continued use of CFCs would lead to a serious depletion of the Earth's protective stratospheric ozone layer. Depletion of the ozone layer is important because for every 1% decrease in stratospheric ozone, an additional 2% of the sun's most damaging ultraviolet radiation reaches the Earth's surface. The result is increased skin cancer, damage to plants, and possibly other effects we know little about now. Let's examine how these CFCs destroy the ozone layer.

In the stratosphere an abundance of ozone is produced because ultraviolet light with wavelengths below 280 nm readily breaks down oxygen molecules to produce oxygen atoms. These oxygen atoms, in turn, can react with oxygen molecules to produce ozone.

$$O_2 + h\nu \longrightarrow 2\,O$$
$$O + O_2 \longrightarrow O_3$$

Stratospheric ozone formed in this manner is so abundant (about 10 ppm) that it *absorbs between 95 and 99%* of the sunlight in the 200 to 300 nm wavelength range (the ultraviolet range). Light in this wavelength range is especially damaging to living organisms, so the stratospheric ozone layer is highly beneficial.

While the carbon-chlorine bond in a CFC molecule is not easily broken by reactions with acids, bases, or water, it *is* easily broken by ultraviolet light found in the stratosphere. This photodissociation produces a chlorine atom (Cl), for example,

■ Until the 1994 model year, CFC-11 was the refrigerant gas commonly used in automotive air conditioners. It is still used in most older (prior to 1993) automobiles.

$$F—\overset{\displaystyle Cl}{\underset{\displaystyle Cl}{C}}—Cl + h\nu \longrightarrow F—\overset{\displaystyle Cl}{\underset{\displaystyle Cl}{C}} + Cl$$

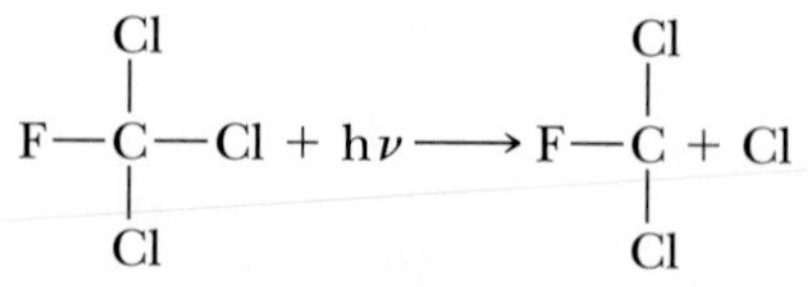

The chlorine atom is quite reactive. If it happens to collide with an ozone molecule, it forms a chlorine oxide (ClO) free radical and an oxygen molecule.

$$Cl + O_3 \longrightarrow ClO + O_2$$

Since there are plenty of ozone molecules present, this reaction is very likely to occur—and each time an ozone molecule is destroyed. If this were the *only* ozone molecule destroyed by the photodissociation of a single CFC, there would be little danger to the ozone layer. However, the ClO free radical can react with oxygen atoms and produce the chlorine atom again, which is ready to react with yet another ozone molecule.

$$ClO + O \longrightarrow O_2 + Cl$$
$$Cl + O_3 \longrightarrow O_2 + ClO$$

These reactions are summarized below, with the overall, or *net* reaction showing the destruction of an ozone molecule for every reaction cycle in which the chlorine atom participates

$$\begin{array}{rl} Cl(g) + O_3(g) & \longrightarrow ClO(g) + O_2(g) \\ ClO(g) + O(g) & \longrightarrow O_2(g) + Cl(g) \\ \hline \text{Net: } O_3(g) + O(g) & \longrightarrow 2\,O_2(g) \end{array}$$

THE PERSONAL SIDE

Susan Soloman (1956–)

Susan Solomon in the Dry Valleys. (Courtesy of S. Solomon)

In 1985 a British team at Halley Bay Station, Antarctica, discovered the existence of a hole in the ozone layer above that continent. This totally unexpected phenomenon needed an explanation, and it was Susan Soloman, a young NOAA (National Oceanic and Atmospheric Administration) scientist, who first proposed a good theory for it. While attending a lecture on polar stratospheric clouds, she realized that ice crystals in the clouds might do more than just scatter light over the Antarctic. Her chemist's intuition told her that the ice crystals could provide a surface on which chemical reactions of CFC compounds could take place.

In 1986 NASA chose Soloman (then 30 years old) to lead a team to Antarctica to sort out the right explanation for the ozone hole. Experiments during that visit to Antarctica showed that her cloud theory was correct, and a second expedition that year added further evidence of its validity. Soloman's team and their experiments led to the first solid proof that there is a connection between CFCs and ozone depletion.

Susan Soloman is one of the youngest members of the National Academy of Sciences. She decided to become a scientist at age 10, having been influenced by watching Jacques Cousteau on TV. At age 16 she won first place in the Chicago Science Fair for a project called "Using Light To Determine Percentage of Oxygen," and went on to place third in the national science fair that year. She said that her winters as a young girl in Chicago prepared her for her visits to Antarctica.

SCIENCE AND SOCIETY

The Montreal Protocol on Substances that Deplete the Ozone Layer

It is rare for many different countries to get together and ban the use of a certain class of chemicals. With the exceptions of bans on chemicals used for warfare and bans on addictive drugs, this kind of activity almost never happens. Probably the reason it doesn't happen often is that most substances are available in such small quantities that the consequences of their use on a global scale would be insignificant. This means that local laws and regulations could control their use. Not so with the CFCs and other chemicals that can deplete the stratospheric ozone layer. These chemicals, as explained in the text, get involved in *cyclic* reactions that actually magnify their effect. That fact alone probably wasn't enough evidence to convince skeptics until observations of a decreasing ozone layer began to become common. By then, 11 years had passed since Roland and Molina first warned of a potential environmental disaster. In 1985 a conference called the Convention for the Protection of the Ozone Layer convened in Vienna. The United States and several Scandinavian countries wanted to freeze the use of ozone-depleters and follow that with a ban on production. Of the attendees, only 27 countries, including the United States, the then Soviet Union, Japan, and the nations of the European community, signed on. Larger developing countries such as India, China, and Brazil were afraid that this action would harm their economic

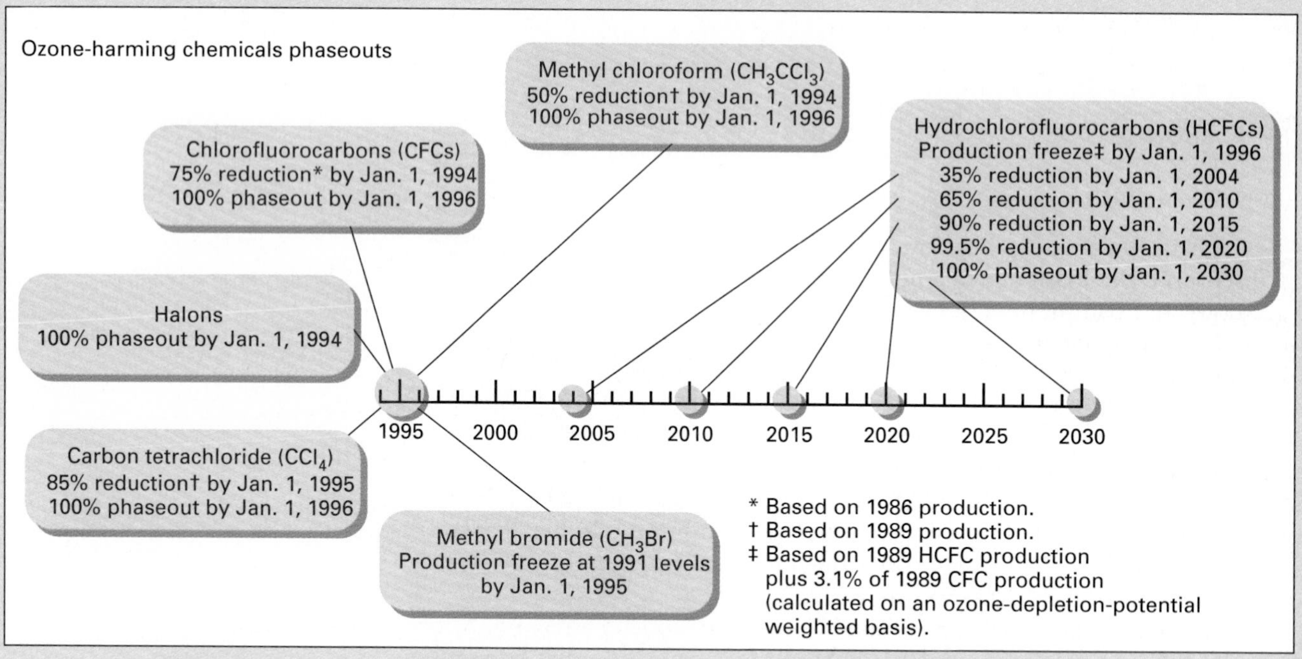

Phaseout schedule for ozone-depleting chemicals according to the Montreal Protocol.

The result of this reaction cycle is that a single chlorine atom may react up to 100,000 times before it eventually reacts with a water molecule to form HCl, which then mixes into the troposphere and washes out in acidic rainfall. This chlorine atom chain is thought to account for about 80% of the known losses of stratospheric ozone.

Molina and Rowland's warning more than 20 years ago regarding CFCs and their potential for depleting the stratospheric ozone layer has proven to

growth. In May of 1985, however, the British Antarctic Survey announced that a 40% loss in ozone was occurring every fall over Halley Bay, Antarctica, and had been since the 1960s! It was shortly after that discovery that the public's attention was drawn to the fact that stratospheric ozone acted as a protective layer, filtering out harmful UV-B radiation. With a growing sense of urgency, nations began signing on to the Vienna Conference pact, and by 1987 another, larger conference was held in Montreal. The agreement, which became effective in January 1989, was called the Montreal Protocol. One of its strongest inducements to sign was trade restrictions that were to be imposed on nations who did not sign the protocol. By 1994, 134 nations had signed on. There have been amendments to the Protocol, but the real enforcement of restrictions on the manufacture, use, and distribution of substances that can deplete the ozone layer is contained in the 1989 agreement.

This agreement on the worldwide control of a class of chemicals that are capable of harming the entire global environment represents an interesting history in the handling of scientific data, its interpretation, drawing conclusions about the implications of those interpretations, and taking drastic steps to correct a problem.

Key Dates in the History of Ozone-Depleting Substances

Dec. 1973:	Rowland and Molina discovery
Oct. 1978:	Use of CFCs in aerosols banned in United States
Oct. 1984:	British team reports 40% loss of ozone over Antarctica during austral (Southern hemisphere) spring.
Sept. 1987:	Montreal Protocol—Representatives from 43 nations agree to CFC reductions of 50% by 2000.
Oct. 1987:	Antarctic expedition verifies huge losses of ozone over Antarctica during austral spring.
Mar. 1988:	United States ratifies Montreal Protocol; large U.S. manufacturers announce they will cease production of CFCs.
Apr. 1988:	Plastic foam manufacturers announce they will stop using CFCs.
Mar. 1989:	Seven hundred representatives from 124 countries attend London conference on saving the ozone layer.
June 1990:	Environment ministers from 93 nations agree to strengthen Montreal Protocol with complete phaseout of CFCs by 2000 (and HCFCs by 2040).
Oct. 1990:	U.S. Congress passes revised Clean Air Act that includes phaseout of CFCs by 2000.
Jan. 1991:	Environment ministers of European Community agree to complete CFC ban by 1997.
Jan. 1992:	Increased concentrations of ozone-depleting chemicals found over populated areas in the Northern Hemisphere.
Feb. 1992:	U.S. president moves target date for phaseout of CFCs from the year 2000 to 1995.
Jan. 1, 1994:	Halon production stops. EPA formally asks some companies to continue production of CFCs through 1995 in order to meet "consumer needs" in automotive air conditioners
Oct. 1994:	Antarctic zone hole appears earlier than normal—covers 23 to 24 million km^2.
Jan. 1995:	NASA satellite data confirm CFC link to ozone hole.

be correct. Satellite and ground-based measurements since 1978 indicate that global concentrations of ozone in the stratosphere have been decreasing. The early stratospheric ozone concentration data showed an average decrease of about 2.5% annually in the decade from 1978 to 1988. Since then, studies over North America in 1993 have shown decreases of 12 to 15% below normal levels. These levels were lower than any measured over the past 35 years. Even larger decreases have been recorded over Europe. In addition, near the North

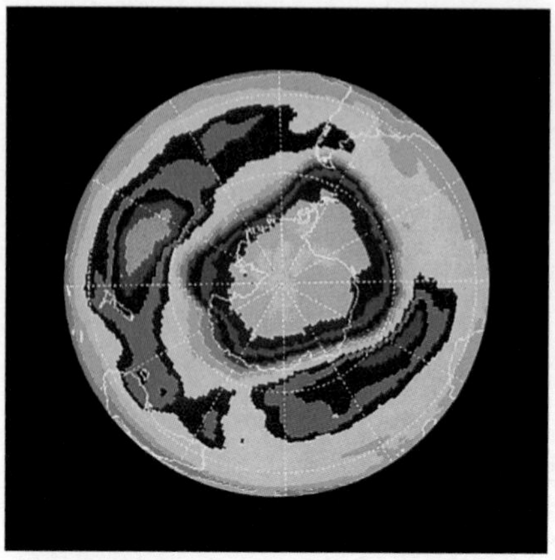

Figure 21.12 Ozone hole over Antarctica in October, 1994. The land is outlined in white, and the "hole" is shown by concentrations of ozone represented by gray for the lowest, followed by pink, and purple. The hole covers 24 million sq km. Higher concentrations of ozone are represented by yellow, green, and brown, in that order. *(NASA/SPL/Photo Researchers)*

and South Poles ozone losses have been so massive that "holes" in the ozone layer have been observed (Figure 21.12). These ozone holes are of special concern because many scientists believe that they may happen at the mid-latitudes in the future.

Alternatives to Ozone-Depleting Chemicals

The alarming drops in ozone concentration along with the regulatory actions taken by countries who signed the Montreal Protocol have led scientists to look for alternatives to compounds that deplete the ozone layer. In the early 1990s, it was believed that finding suitable alternatives for the CFCs most commonly used as refrigerant gases would be almost impossible, yet by 1993 substitutes were found. Today CFC-12, once the most common automotive refrigerant, has been almost totally replaced by HCFC-134a, while CFC-11 has been replaced by HCFC-141b. These compounds, *hydrochlorofluorocarbons* (HCFCs), still have some ozone-depleting capabilities, but they are much more reactive in the lower atmosphere, which lessens their chances of getting into the stratosphere. Under current regulations, the EPA will allow the use of HCFCs as refrigerants until the year 2030. By that time, it is believed that even better alternatives will be found.

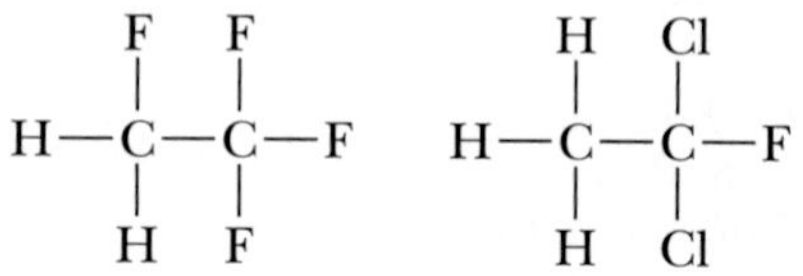

■ Almost all new cars sold in 1994 have air conditioners using HCFC-134a (CH_2FCF_3).

$$
\underset{\text{HCFC-134a}}{H\!-\!\overset{\displaystyle F}{\underset{\displaystyle H}{C}}\!-\!\overset{\displaystyle F}{\underset{\displaystyle F}{C}}\!-\!F}
\qquad
\underset{\text{HCFC-141b}}{H\!-\!\overset{\displaystyle H}{\underset{\displaystyle H}{C}}\!-\!\overset{\displaystyle Cl}{\underset{\displaystyle Cl}{C}}\!-\!F}
$$

When it became apparent that existing automotive air conditioners would have to be "retrofitted" with HCFCs as they were repaired or replaced, fears of high consumer costs sent engineers to the labs searching for solutions. Now, it is known that HCFCs are more compatible with existing air-conditioning systems on older cars and that repairs and retrofits should not cost as much as once feared. Still, retrofitting cars to use HCFC-134a can cost from $200 to $800. (Prices like these will undoubtedly cause some car owners to choose to

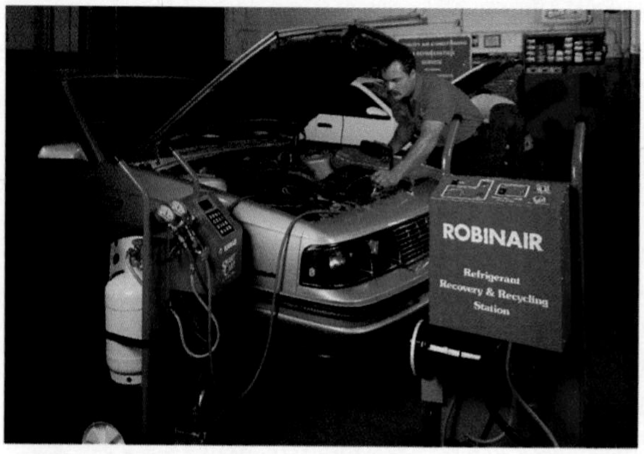

Recycling CFCs during repair of an automotive air conditioner. At many service centers, CFCs have simply been vented into the atmosphere. *(Courtesy of Robinair)*

FRONTIERS IN THE WORLD OF CHEMISTRY

The Role of Iodine in Stratospheric Ozone Depletion

It should not be surprising that iodine, a member of the same chemical family as chlorine and bromine, is also an ozone depleter. Until recently, however, iodine's role in depleting the ozone concentration in the stratosphere was ignored because it was thought that the C—I bond in almost any iodocarbon compound would be broken by photochemical reactions in the lower atmosphere. This would keep the iodine atoms well away from stratospheric ozone.

At altitudes below about 20 km in temperate regions ozone is depleted. Yet chlorine and bromine atoms alone cannot account for all the ozone loss because photodissociation of chlorine and bromine compounds is not great enough at those altitudes. This fact led scientists to suspect iodine, but they were at a loss to explain how the iodine-containing compounds could ever get into the stratosphere. While looking for CFC replacements, researchers studied the compound methyl iodide (CH_3I), and it was shown that the C—I bond was indeed easily broken by sunlight very close to the Earth. This meant that CH_3I would be a good candidate to replace CFCs. In addition, natural production of CH_3I by marine organisms (about 3×10^6 metric tons annually) from iodide ions in sea water meant that much more was being produced naturally than would ever be produced synthetically for commercial use. While studying what happened to all this naturally produced CH_3I, scientists discovered that huge thunderclouds actually sweep pockets of air from the ocean surface into the upper atmosphere in a matter of a few seconds—quick enough to allow the iodine-containing compounds to remain unreacted until they are well aloft. Once the iodine compounds are in the stratosphere they are destroyed by sunlight, and their iodine atoms begin destroying ozone molecules.

The implication of discoveries like this one are uncertain as yet. Since the initial discoveries concerning CFCs and their role in the depletion of stratospheric ozone, it has become clear that many natural sources of halogen atoms play important roles, and in some cases these natural sources of ozone depleters may play a larger role than their synthetic counterparts.

Reference *Chemical and Engineering News,* Nov. 14, 1994, p. 8.

drive with their windows open during the hot months.) In spite of the availability of substitutes, the EPA will allow older cars to be repaired using CFCs. The American Automobile Manufacturers Association has estimated a need for 200,000 metric tons of CFC-12 from 1996 to 2005 to service older vehicles.

The electronics industry has also made significant advances in eliminating CFCs and similar compounds in parts-cleaning operations. One large electronics plant in California had been releasing more than 1.5 million pounds of CFCs annually as part of its normal operations. As a result of the bans on such chemicals, it found a way to get its parts just as clean using soapy water, rinsing, and blow-drying with hot air.

A class of compounds known as **halons** must also be replaced to protect the ozone layer. The halons are structurally similar to CFCs, but contain a carbon-bromine bond. Like the CFCs, the halons photodissociate in the stratosphere, where they produce bromine oxide radicals (BrO) that are destructive to ozone. Finding halon substitutes has presented quite a challenge. The halons are superior fire-fighting agents. They are used on aircraft, for example, and have been credited with saving many lives. Existing halon supplies are being allowed to remain in place, but halon substitutes such as perfluorobutane, bromodifluoromethane, and chlorotetrafluoroethane are being employed where halons were once used and in new systems.

Halons

Name	Formula	Uses
Halon-1211	CF_2BrCl	Portable fire extinguishers
Halon-1301	CF_3Br	Fixed fire extinguishers, aircraft fire extinguishers

The "per-" in perfluorobutane indicates that all of the hydrogens in the butane molecule have been replaced, in this case by fluorine atoms. Another common "per-" compound is perchloroethylene ($CCl_2{=}CCl_2$), also known as "perc," which has been widely used as a dry-cleaning solvent.

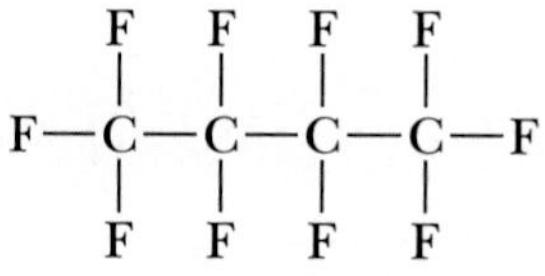

Perfluorobutane

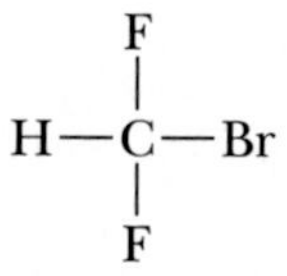

Bromodifluoromethane

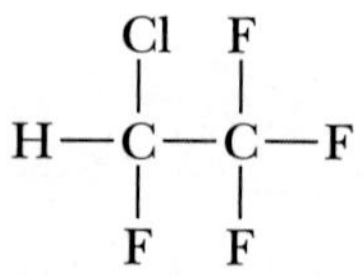

Chlorotetrafluoroethane

These compounds do not have the ozone-depleting potential that the other halons have. Their main disadvantage, and the prime reason they were not used earlier in the place of the other halons, is their higher cost and somewhat lower ability to stop the reactions going on during a fire.

The Future of the Stratospheric Ozone Layer

Unfortunately, a black-market for CFCs has developed with unscrupulous importers smuggling loads of CFCs into the United States and other countries.

It is generally agreed that controls on emissions of CFCs and other compounds that can deplete the stratospheric layer are going to have the desired effect. Indeed, data now indicate that smaller decreases in ozone concentrations are occurring than just a few years ago. Natural processes that produce ozone-depleting compounds will continue, but it appears the stratospheric concentrations of all other ozone depleters will begin to decrease before 1999 and will probably be back to 1978 levels (about 50% over natural levels) by the year 2050. As markets for CFCs disappear, even reserves of those compounds (and hoarded quantities) will become less valuable. As that happens, they will eventually disappear. The present regulatory climate in the United States (Clean Air Act of 1990) as well as in other countries makes it unappealing to illegally manufacture or import CFCs and similar compounds. Industry has certainly indicated that it can "change with the times" and find substitutes even when no substitutes came to mind.

SELF-TEST 21B

1. When coal and fuel oil are burned, what two primary pollutants are formed?
2. When sulfur dioxide reacts with oxygen, what oxide of sulfur is formed? When this oxide reacts with water, what acid is formed?
3. Which chemical would be most likely to react with sulfur dioxide and remove it from combustion gases? (a) Sodium chloride (NaCl), (b) Lime (CaO), (c) Nitric oxide (NO).
4. Name two acids found in acid rain. One must contain sulfur, the other nitrogen.
5. What is the pH of normal rainfall? What dissolved chemical causes this pH to be below pH 7?
6. Approximately when was acid rainfall first observed?
7. Which chemical bond is broken by a photon of light in a typical CFC molecule?
8. Write the reaction producing ozone from oxygen.
9. What is the chemical species containing chlorine that destroys ozone molecules?
10. Over what continent have scientists found an ozone "hole"?
11. If ozone in the stratosphere is destroyed, what form of radiation will pass through to the Earth below?

21.11 Carbon Dioxide and the Greenhouse Effect

How can carbon dioxide (CO_2) be considered a pollutant when it is a natural product of respiration and fossil-fuel burning and is a required reactant for photosynthesis? Actually, CO_2 is not a pollutant in the strictest meaning of the term, but the fact that it is increasing in the Earth's atmosphere is cause for deep concern. Consequently, it is treated as a pollutant. Without human influences, the flow of carbon between the air, plants, animals, and the oceans would be roughly balanced. However, between 1900 and 1970, the global concentration of CO_2 increased from 296 ppm to 318 ppm, an increase of 7.4%. By 1994 the concentration was 360 ppm, and expectations are that the CO_2 concentration will continue to increase (Figure 21.13). For example, since the end of World War II, a world energy growth rate of about 5.3% per year took place until the OPEC oil embargo in the mid-1970s. Rates of energy use have actually decreased since then.

■ OPEC: Oil Producing and Exporting Countries

Population pressures are contributing heavily to increased CO_2 concentrations. In the Amazon region of Brazil, for example, extensive cut-and-burn

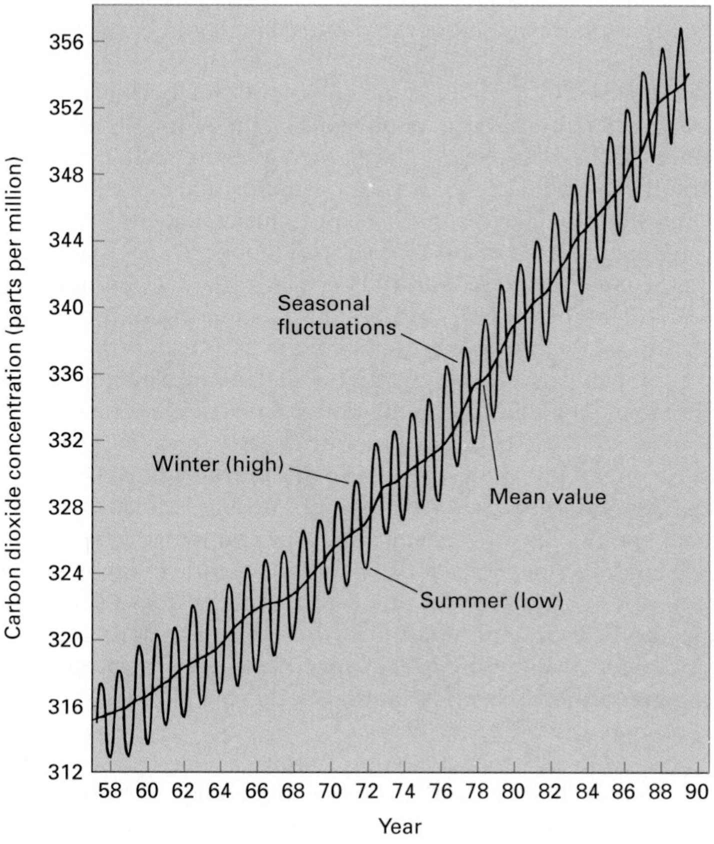

Figure 21.13 Carbon dioxide concentration in the atmosphere, showing the steady increase over the years. The seasonal variation results from high CO_2 in the summer due to greater photosynthesis and low CO_2 in the winter when photosynthesis diminishes. These measurements are taken at the Mauna Loa Observatory in Hawaii, which is located far from urban areas with their industrial CO_2 emissions.

Deforestation in Brazil. This satellite photo taken over the Amazon Basin shows the dramatic extent of the cut-and-burn creation of cropland. The remaining rain forest is dark green and the leveled forest shows in pale greens and browns. The ladder-like pattern at the upper left is a typical result of cut-and-burn in this region. *(Geospace/SPL/Photo Researchers)*

practices are being used to create cropland. This is causing a tremendous burden on the natural CO_2 cycle, because CO_2 is being added to the atmosphere during burning while there are fewer trees present to photosynthesize this additional CO_2 into plant nutrients.

Counting all forms of fossil-fuel combustion worldwide, the amount of CO_2 added to the atmosphere is about 50 billion tons a year. About half of this remains in the atmosphere to increase the global concentration of CO_2. The other half is taken up by plants during photosynthesis and by the oceans, where CO_2 dissolves to form carbonic acid, which then can form bicarbonates and carbonates.

$$CO_2(g) + H_2O(\ell) \rightleftharpoons H^+(aq) + \underset{\text{Bicarbonate ion}}{HCO_3^-(aq)}$$

$$HCO_3^-(aq) \rightleftharpoons H^+(aq) + \underset{\text{Carbonate ion}}{CO_3^{2-}(aq)}$$

To see how easily our everyday activities affect the amount of CO_2 being put into the atmosphere, consider a round-trip flight from New York to Los Angeles. Each passenger pays for about 200 gallons of jet fuel, which weighs 1400 lb. When burned, each pound of jet fuel produces about 3.14 lb of carbon dioxide. So 4400 lb, or 2.2 tons, of carbon dioxide are produced per passenger during that trip. It seems reasonable that if we are rapidly burning fossil fuels that took millions of years to form, we are then going to be adding CO_2 back into the atmosphere at a more rapid pace than it can be used up in natural processes.

What is the problem with increasing atmospheric CO_2? When solar radiation arrives at the Earth's atmosphere, about half of the visible light (400–700 nm) is reflected back into space. (That's a good thing; otherwise, the temperature of the Earth would be far too hot to support life as we know it.) The remainder reaches the Earth's surface and causes warming (Figure 21.14). The warmed surfaces (average temperature about 27°C) then reradiate this energy as heat energy in the infrared portion of the spectrum. Water vapor, CO_2, ozone, and methane (CH_4) readily absorb some of this reradiated energy and in turn *warm* the atmosphere, creating what is called the **greenhouse effect.** A botanical greenhouse works on the same principle. Glass transmits visible light but blocks infrared radiation trying to leave. The effect is a warming of the air inside the greenhouse. In warm weather, the windows of a greenhouse must be opened or the plants inside will overheat and die.

All four of the "greenhouse gases" act as an absorbing blanket that prevents radiation losses and keeps the Earth's atmospheric temperature comfortable (although not in all locations at the same time!). Water vapor in the atmosphere is subject to such vast cycles that human activity doesn't seem to bother it. Because ozone is present in relatively low concentrations, and because methane is produced naturally in vast quantities, our attention is focused on CO_2, the greenhouse gas whose atmospheric concentration is most closely related to human activity.

Recently, Russian scientists took ice core samples dating back 160,000 years. In these ice samples were tiny pockets of air that could be analyzed for CO_2 content. They found a direct correlation between CO_2 and geologic tem-

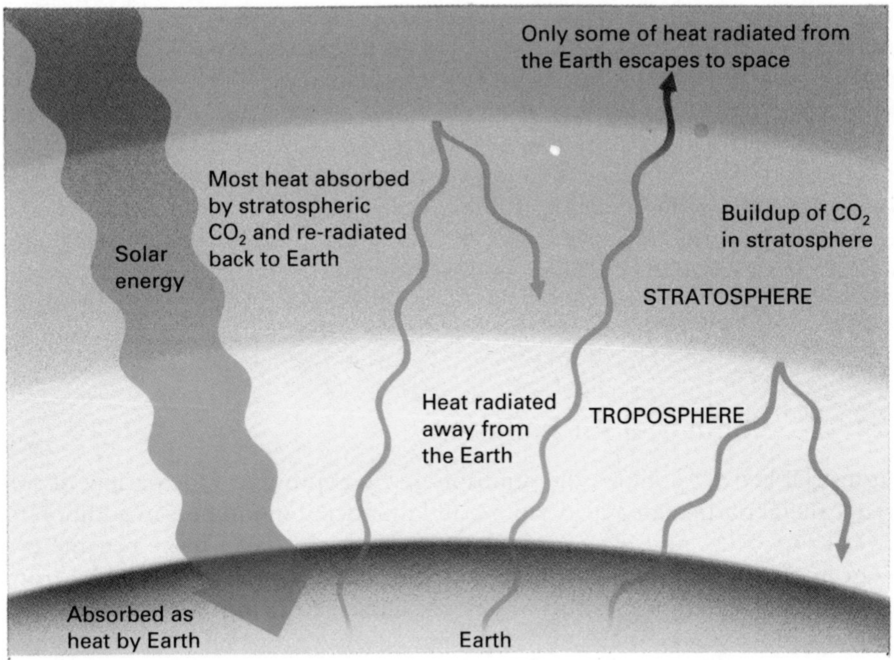

Figure 21.14 The greenhouse effect. Greenhouse gases effectively form a barrier that prevents heat from escaping from the Earth's surface.

peratures known by other means. As the CO_2 increased, the global temperature increased, and as the CO_2 decreased, the global temperature decreased. It is generally agreed that rising CO_2 concentrations will probably lead to increasing global temperatures and corresponding changes in climates. If predictions by the National Academy of Sciences prove correct, when and if the global concentration of CO_2 reaches 600 ppm, the average global temperature will have risen by 1.5° to 4.5°C (2.7° to 8.1°F). Even a 1.5°C warming would produce the warmest climate seen on Earth in the past 6000 years, and a 4.5°C warming would produce world temperatures higher than any since the Mesozoic era—the time of dinosaurs.

Clearly, **global warming** is a major potential problem, and its effects appear to be measurable on a human time scale. It has been calculated that stabilizing the CO_2 concentration in the atmosphere at 360 ppm will require limiting global industrial emissions to something less than 2×10^9 tons CO_2/year, which is well below the current emissions of 6×10^9 tons CO_2/year. Scientists who study the atmosphere fear that at the current rate of emissions, CO_2 concentrations will rise to about 550 ppm before they begin to level off. Controlling CO_2 emissions worldwide will undoubtedly prove to be more difficult than controlling CFCs or the precursors to acid rain. At the 1992 Earth Summit in Rio de Janeiro, Brazil, a number of countries signed a treaty to limit CO_2 emissions by the year 2000 to 1990 levels in their countries. This goal is proving very difficult to achieve. Industrialized countries emit so much CO_2 from so many diverse sources that reducing emissions may be almost impossible unless radical societal changes are made. One of these would be in the ways energy is produced and distributed. Strong arguments can be and are being made for rapid conversion to solar energy as well as to nuclear energy, neither of which contributes significantly to global

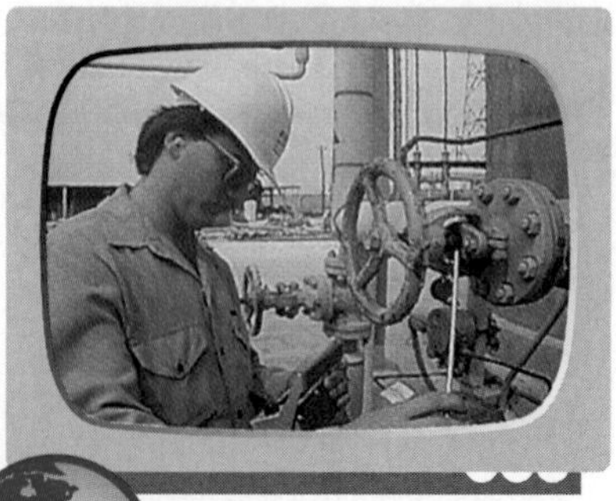

The World of Chemistry

Program 25, *Chemistry and the Environment*

Testing a valve in a chemical plant for leaks.

warming, but which have their own unique negative environmental consequences (see Sections 13.5 and 13.4). Unfortunately, developing countries appear to have no other choice but to continue to emit CO_2 as they try to raise their standards of living by producing food, energy, clothing, and shelter. Most planners agree that global CO_2 emissions will depend on both population growth and world economic growth. Using projections of large population growth (to just over 11.5 billion by the year 2050) and moderate world economic growth (3% per year until 2050), a United Nations panel on climate control has estimated the mean global temperature could increase by almost 3°C.

21.12 Industrial Air Pollution

Industrial activity pollutes the atmosphere by emitting a wide variety of solvents, metal particulates, acid vapors, and unreacted monomers in addition to CO_2, CFCs, NO_x, and SO_2. The extent to which this takes place became evident in 1989 when the first summary of annual releases was published from data received by the EPA. This report was a part of the Superfund Reauthorization and Amendments Act of 1986, regulations that resulted in part from a tragic release of a toxic chemical in Bhopal, India, where more than 4000 people were killed and many thousands more were injured. These release-reporting regulations were placed on manufacturers who use any of a group of about 320 chemicals and classes of compounds representing special health hazards. The reporting was divided into releases to air, water, and land. Recently, the list of reportable chemicals was expanded by the EPA to include a total of 654 compounds in an effort to give citizens an even clearer picture of chemicals that affect their community.

■ Water and land releases directly affect surface and groundwater purity. (See Chapter 20.)

■ These regulations have been called "Community Right-to-Know" regulations because they inform communities about releases of harmful chemicals in their areas (see Chapter 18, Science and Society—Worst Case Scenarios of Toxic Releases).

In a release report, an industrial facility must list all releases of a reportable chemical to the atmosphere, regardless of the type of release. This means that leaky valves and fittings, accidental spills, vapor losses while filling tank trucks and rail tank cars, emissions at stacks, and so forth are all added together on the report. As expected, heavily industrialized states and states with a lot of chemical industry have high releases (Table 21.5), but the amounts of some chemicals released have also been surprisingly large (Table 21.6). These data are now summarized annually and are available by means of publicly accessible computerized databases. Interested persons may call the EPA at 1-800-535-0202 for more information about chemical releases in their community.

TABLE 21.5 ■ Top Five States for Releases of Toxic Substances (1992)

State	Total 10^6 Pounds
Louisiana	464
Texas	419
Tennessee	194
Ohio	143
Indiana	124

Source: U.S. EPA. Toxic Release Reports, 1992.

What do these releases of harmful chemicals into the atmosphere mean? Compared with the vast quantities of matter comprising the atmosphere, industrial releases of chemicals seem low. Even compared with the releases of certain classes of chemicals such as halogenated compounds by marine organisms and methane by ruminant animals, ants, and termites, industrial releases seem small. The problem is that industrial releases are usually concentrated close to home—that is, near population centers where the quality of the atmosphere becomes lowered before the released chemicals ever have a chance to be diluted by all the surrounding atmosphere. (Compare this with the release of some 10^9 tons annually of methyl iodide (CH_3I) by marine organisms over the entire surface of the oceans.) In addition, all of those released compounds represent *financial losses* for the companies involved. If anything, the EPA release rules helped focus industry's attention on the need to reduce emissions.

21.13 Indoor Air Pollution

As if the data about pollutants in the outside air were not enough to concern us, the air inside our homes and workplaces is also contaminated, usually by the same chemicals emitted by industry. Some scientists have concluded that air in our homes may be *more* harmful than the air outdoors, even in heavily industrialized areas. A study by the EPA indicated that indoor pollution levels in rural homes were about the same as in homes in industrialized areas. One cause for this is the emphasis on tighter, more energy-efficient homes, which tend to trap air inside for long periods.

What are the sources of home air pollution (see Figure 21.15)? Tobacco smoke, if present, is an obvious source. Benzene, a known carcinogen, occurs

TABLE 21.6 ■ Top Five Chemicals Released into the Environment—1992

Chemical	Pounds
1. Ammonia	463,000,000
2. Hydrochloric acid	287,000,000
3. Methanol	241,000,000
4. Phosphoric acid	206,000,000
5. Toluene	183,000,000

Source: U.S. EPA Toxic Release Report, 1992

■ We shouldn't be surprised that air in our homes is contaminated by industrial chemicals—after all, we bring industry's products into our homes.

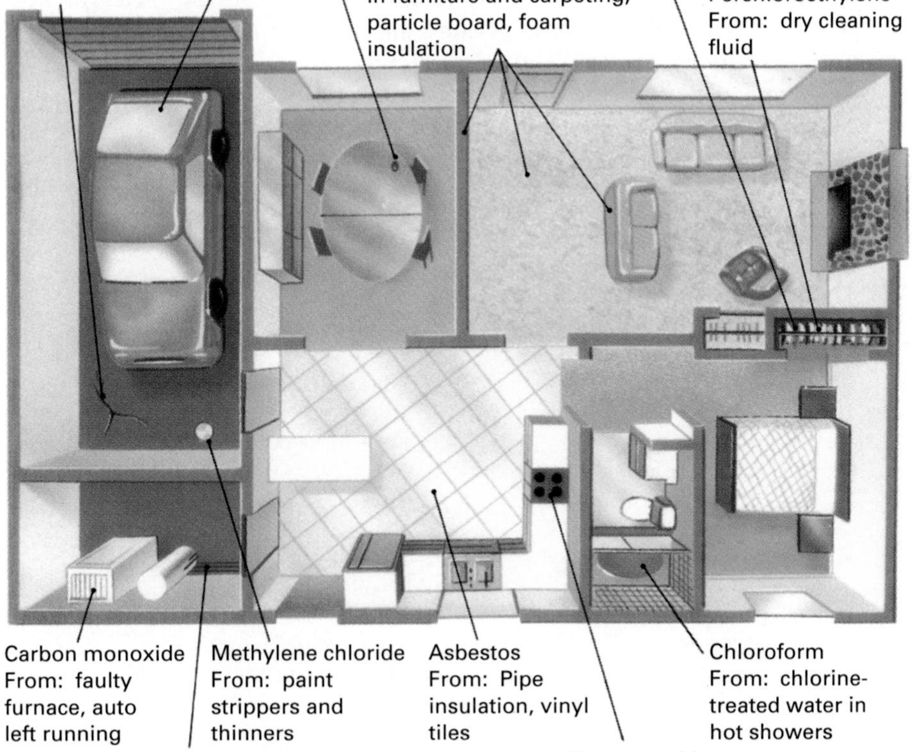

Figure 21.15 Sources of some indoor air pollutants.

at 30% to 50% higher levels in homes of smokers than in homes of nonsmokers. Building materials and other consumer products are also sources of pollutants. Entire buildings can acquire a "sick building syndrome" when a particular chemical or group of chemicals is found in sufficiently high concentration to cause headaches, nausea, stinging eyes, itching nose, or some combination of these symptoms. Usually, the best cure for all forms of indoor air pollution is to limit the introduction of the offending chemicals and to have better exchange with the outside air. Of course, this solution comes at a cost. If you exchange indoor air with outside air when the temperature differences between inside and outside are great, you will be paying a larger heating or cooling bill.

SELF-TEST 21C

1. Name three different human activities that produce large amounts (millions of tons/year) of CO_2.
2. What are two principal processes whereby CO_2 is consumed?
3. Name three major greenhouse gases found in the atmosphere.
4. Which greenhouse gas is most closely associated with human activities?
5. Global temperature seems to follow carbon dioxide concentrations. (a) True, (b) False.
6. Approximately what is the current global CO_2 concentration?
7. What state had the highest air releases of toxic chemicals in 1989, the latest year for which such data are available?
8. What chemical was released in greatest amount nationwide in 1989, according to the EPA release report?
9. What single activity inside the home can account for increased concentrations of benzene, a known human carcinogen?

MATCHING SET

____ 1. Gas with the highest percentage by volume in the atmosphere
____ 2. Consisting mostly of very small water droplets
____ 3. Abnormal temperature arrangement for air masses
____ 4. A mixture of smoke, fog, air, and various other chemicals
____ 5. A pollutant caused by reactions of other chemicals
____ 6. Second most abundant gas in the atmosphere
____ 7. A pollutant that is directly discharged into the atmosphere
____ 8. A chemical species with an unpaired valence electron
____ 9. Main ingredient in industrial smog
____ 10. The product of a reaction between an oxygen atom and an oxygen molecule
____ 11. An oxide of nitrogen produced by certain bacteria
____ 12. The oxide of nitrogen produced by lightning, forest fires, and internal combustion engines
____ 13. "Bad" ozone
____ 14. "Good" ozone
____ 15. A polynuclear aromatic hydrocarbon capable of causing cancer
____ 16. One of the principal producers of SO_2
____ 17. Cause rain to have an abnormally low pH
____ 18. Contribute to the destruction of the ozone layer
____ 19. An alternative to certain ozone-depleting chemicals
____ 20. Ozone-depleting chemicals that contain a carbon-bromine bond
____ 21. The principal greenhouse gas

a. Oxygen
b. NO_2 and SO_2
c. Nitrogen
d. Halons
e. Aerosol
f. Chlorofluorocarbons
g. CO_2
h. Secondary pollutant
i. SO_2
j. HCFC-134a
k. Smog
l. NO
m. Fossil fuel electricity generators
n. Primary pollutant
o. Thermal inversion
p. Ozone in the air we breath
q. N_2O
r. Benzo(α)pyrene
s. Free radical
t. Ozone molecule
u. Stratospheric ozone

QUESTIONS FOR REVIEW AND THOUGHT

1. Define the following terms.
   (a) Polluted air (b) Particulate
   (c) Aerosol (d) Thermal inversion
2. Define the following terms.
   (a) Photochemical smog (b) Reducing type smog
   (c) Secondary pollutant (d) Free radical
   (e) Photodissociation
3. Define the following terms.
   (a) Acid rain (b) CFC
   (c) Ozone hole (d) Greenhouse gas
   (e) Global warming
4. Name three major air pollutants and give a source for each.
5. What federal legislation has the abbreviation CAA? Describe how this legislation has changed over the years.
6. Describe what a thermal inversion is and explain how it can aggravate the problems caused by smog.
7. How is an industrial-type smog different from a photochemical smog?
8. Explain how ozone is formed in the troposphere.
9. Explain how ozone is formed in the stratosphere.
10. What is the difference between "good" ozone and "bad" ozone?
11. What are the main sources of nitrogen oxides in the atmosphere?
12. What is an aerosol and how does it play a role in air pollution?
13. Describe how a volcanic eruption can contribute to global cooling.
14. Name three ingredients necessary for the formation of a chemically oxidizing type smog and explain how they interact.
15. Name two sources of hydrocarbons in the atmosphere. Which one is more readily controlled?
16. Write the equation for the reaction that occurs when lightning causes nitrogen to react with oxygen.
17. Nitrogen dioxide plays a role in the formation of what secondary pollutant?
18. What are the names of the two regions of the atmosphere discussed in this chapter? Which one is closer to the surface of the Earth?
19. What happens when ultraviolet light strikes a molecule of NO_2? Write the reaction.
20. Describe how ozone can be harmful when it is present in the air we breathe.
21. How are the harmful effects of SO_2-containing aerosols, NO_2, and ozone similar?
22. What kinds of processes contribute to the formation of benzo(α)pyrene? How is this substance harmful?
23. How is the control of ozone in the lower atmosphere connected to the control of NO_x emissions?
24. What health risks are posed by breathing air contaminated with particulates?
25. What two air pollutants does an automotive catalytic converter help control?
26. Which state has the most stringent air pollution regulations, often more stringent than those of the federal government? Explain why this is the case.
27. Describe the trends in automobile-related air pollution during the past two decades. Name some factors that have contributed to these trends.
28. What is the product of the combustion of sulfur or sulfur-containing compounds in air? Write the reaction.
29. Pick a source of SO_2 emissions and describe two ways SO_2 emissions from that source can be controlled.
30. Assume you have at your disposal some kind of SO_2 control that produced a mole of CO_2 for every mole of SO_2 removed. Explain how this might have both good and adverse effects on the environment.
31. Explain how rain can have a pH slightly below 7, the pH of pure water. In your explanation be sure to include both natural as well as other sources of compounds that can acidify rain.
32. Briefly describe the adverse effects acid rain produces. How do these effects sometimes become international issues?
33. Write the chemical equation for the reaction between carbon and oxygen when an insufficient amount of oxygen is present. What is the name of the product of this reaction?
34. What are CFCs? Describe some of the uses of this class of compounds.
35. Someone asks you "what's all the fuss about CFCs?" Explain in the simplest terms you can why CFCs are "in the news."
36. Describe briefly the history of the CFC problem and steps that have been taken to address the problem.
37. Write the equation for the chemical reaction that occurs when ultraviolet light strikes an oxygen molecule. Then write the equation for the reaction that produces ozone in the upper atmosphere.
38. Why is a small concentration of ozone in the upper atmosphere important?
39. Name the CFC that was commonly used in air conditioners in new model cars until recently. Why is this CFC no longer used in automotive air conditioners?
40. Explain how a single chlorine atom can cause the destruction of up to 100,000 ozone molecules.

41. Name three ozone-depleting halogenated compounds that are naturally produced by marine organisms.
42. Describe the trend that has been observed in the atmospheric concentration of CO_2. Suggest an explanation for this.
43. Greenhouse gases absorb what kind of radiation? What is the effect on the atmosphere when this radiation is absorbed?
44. Name three greenhouse gases and give a source for each.
45. What two natural processes annually remove vast quantities of CO_2 from the atmosphere?
46. Explain the term "global warming" and explain its significance.
47. Compare the release of industrial chemicals that can pollute the air with the release of chemicals into the atmosphere by natural processes. Explain why the control and lowering of industrial releases is important.
48. Describe some source of indoor air pollution. Why are the chemicals found in indoor air similar to those released into the air by industry?
49. Describe an air pollution problem in your community and discuss how it might be solved.

CHAPTER 22

Feeding the World

When hunting and gathering from nature's bounty were the primary means of obtaining food, only catastrophic events could void the fundamental relationship between effort and a satisfied stomach. Population increases and the concentration of people in cities forced the development of basic agriculture, which has mostly succeeded but sometimes dramatically failed. Now with a world population of 5.7 billion people, scientific and technological advances in agriculture are absolutely necessary. To help grow the enormous amount of food we need, the chemical industry supplies modern scientific agriculture with a large assortment of chemicals—the **agrichemicals,** including fertilizers, medicine for livestock, chemicals to destroy unwanted pests and plant diseases, and food supplements. However, even with the use of new agrichemicals, sound agricultural practices, and biotechnology advances, the world's food supply cannot keep pace with the rate of growth of the world's human population. In order to have sustained food security for the world's population, global cooperation is essential in slowing the rate of population growth and reducing environmental problems that threaten food productivity.

- What are current projections for feeding the growing world population?
- What factors affect the productivity of soil?
- What agricultural practices are essential to sustaining soil productivity?
- How dependent is modern agriculture on insecticides and herbicides?
- What biotechnology advances in agriculture are helping to increase food production?
- What is sustainable agriculture?

Spreading lime. Healthy plants require sunlight, water, and the proper balance of chemicals in the soil. *(Larry Lefever/Grant Heilman Photography)*

Through the 8000 to 10,000 years of recorded human history, food-production techniques have developed enormously. It is generally estimated that 90% of the U.S. population worked to provide food and fiber during most of the 19th century. Although more efficient agricultural methods now allow one U.S. farm worker to feed 100 people, the world's farmers are falling behind in their ability to feed the growing world population. More than one billion people in the world depend upon agricultural lands that are not productive enough to support them adequately. A report released by the United Nations Environment Program in 1992 states that 4.84 billion acres of soil—an area the size of China and India combined—has been degraded to the point where it will be difficult or impossible to reclaim it. The main causes of soil degradation are soil erosion, overgrazing, and deforestation. Global pollution, extinction of wildlife, degradation and loss of natural resources, and depletion of energy reserves in today's world are all related to the increasing world population.

22.1 World Population Growth

How close are we to reaching the Earth's capacity to support the world population? What is the maximum sustainable population level? What factors determine the Earth's capacity? These questions have been studied by organizations such as the Worldwatch Institute and at world conferences such as the United Nations Conference on Environment and Development, commonly known as the Earth Summit, in Rio de Janeiro in 1992 and the United Nations International Conference on Population and Development in Cairo in September 1994. All such studies recognize that control of the rate of growth of the world's population is essential in maintaining adequate food supplies to feed the world. The present world population of 5.7 billion is growing at the rate of about 100 million people per year and is projected to reach 8.9 billion by the year 2030 and level off at 11.5 billion around 2050 (Figure 22.1). Even now, 20% of the world's population (one out of three people in developing countries) are living in extreme poverty, with 700 million of them malnourished.

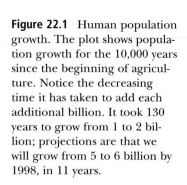

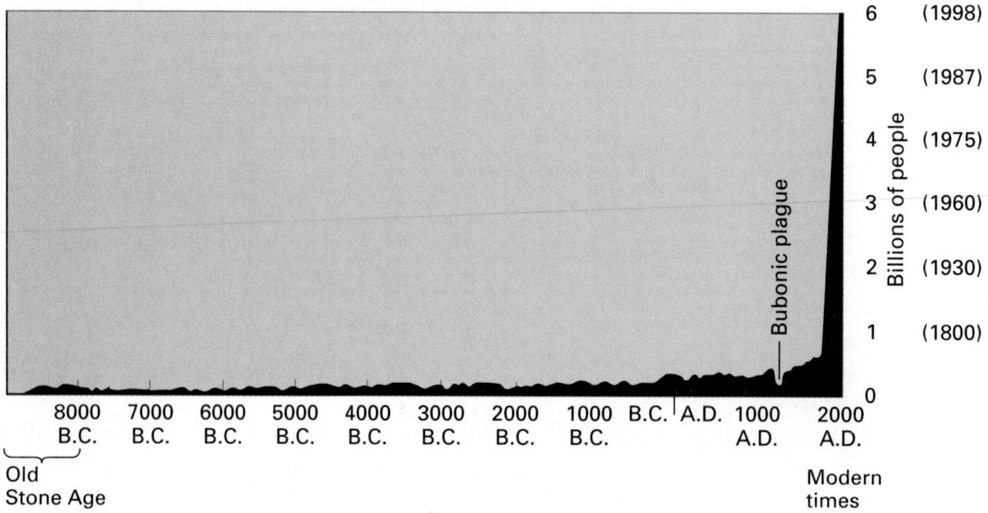

Figure 22.1 Human population growth. The plot shows population growth for the 10,000 years since the beginning of agriculture. Notice the decreasing time it has taken to add each additional billion. It took 130 years to grow from 1 to 2 billion; projections are that we will grow from 5 to 6 billion by 1998, in 11 years.

Our fundamental food is plants. Through photosynthesis powered by energy from the Sun, Earth's green plant population provides the food to sustain life on Earth. We eat either plants or animals that eat plants. In order to grow, plants require the proper temperature, nutrients, air, water, and freedom from disease, weeds, and harmful pests. Agricultural productivity for the past 50 years has relied on agrichemicals to assist nature in giving plants the proper nutrients and freedom from disease and competitive life forms. However, the use of agrichemicals involves risks to the environment and human health; it is important to measure the risks versus benefits in the use of these chemicals.

The World of Chemistry

Program 23, *Proteins: Structure and Function*

Plants and animals, sustained by photosynthesis. Powered by energy from the sun, the Earth's green plant population provides the food to sustain life on earth.

22.2 What is Soil?

Soil is a mixture of four components—mineral particles, organic matter, water, and air (Figure 22.2). Weathering processes in nature over thousands of years break down rock into small mineral particles found in soil. Organic matter in soil is a mixture that includes leaves, twigs, plant and animal parts in various stages of decomposition, and microorganisms. **Humus,** the dark-colored decomposed organic material, is important to a good soil structure. As a source of nutrients for plants, humus is almost like a time-release capsule, slowly releasing its contents.

■ Humus releases its nutrients to plants slowly.

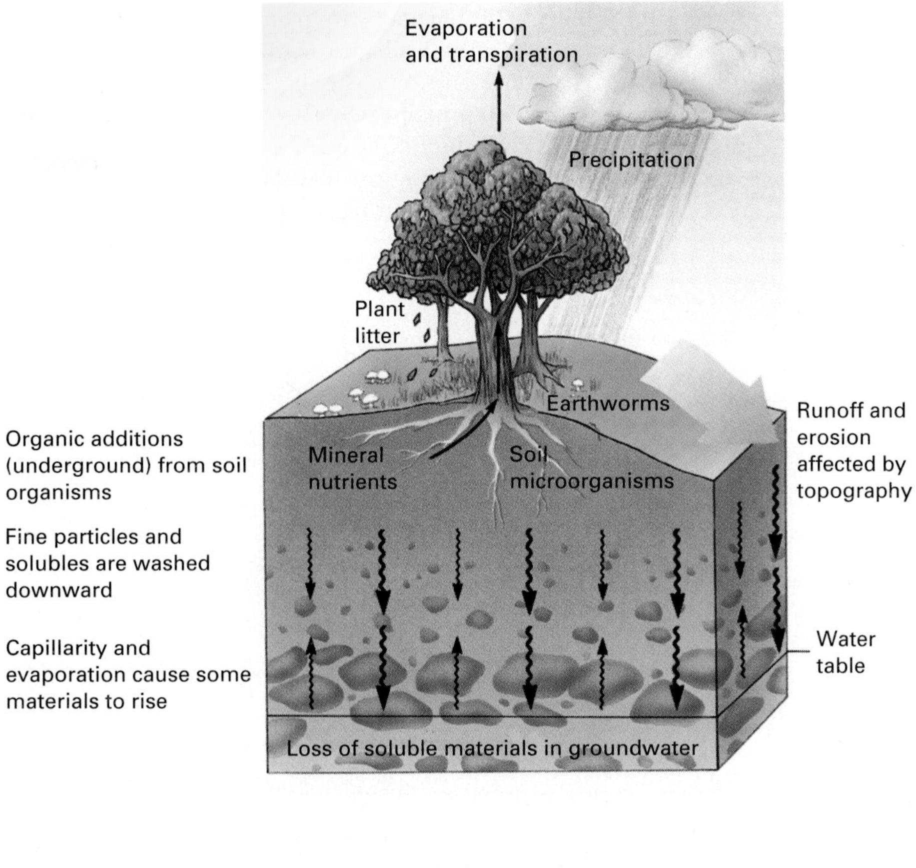

Figure 22.2 Soil formation. Weather, plants and their litter, earthworms and other organisms, and topography interact to produce soil.

Maintaining humus in the soil is of major concern to the agriculturist. Humus such as peat moss or organic fertilizer can be added. However, there is no real substitute for natural plant growth that is returned to the ground for humus formation. Clover is often grown for this purpose and plowed under at the point of its maximum growth. The compost pile of the gardener is another effort to maintain humus for a productive soil.

■ Friable material crumbles easily under slight pressure.

In addition to being a source of plant nutrients, humus is important in maintaining good soil structure, often keeping it *friable.* Soil rich in humus may contain as much as 5% organic matter. Soils in the grasslands of North America have humus to a considerable depth, in contrast to rain-forest regions, where there is only a thin film of humus on the ground surface.

Soil Profile

A handful of humus, which is partially decomposed organic material from plants and animals. Where soil contains ample humus, it is spongy, holds water well, and is a healthy environment for plants and organisms that live in the soil. *(USDA/Soil Conservation Service)*

Layers within the soil are called **horizons** (Figure 22.3). The **topsoil** contains most of the presently living material and humus from dead organisms. Topsoil is usually several inches thick, and in some locations more than 3 ft of topsoil can be found. The **subsoil,** up to several feet in thickness, contains the inorganic materials from the parent rocks as well as organic matter, salts, and clay particles washed out of the topsoil.

Because healthy topsoil has abundant life forms, it must contain an abundant supply of oxygen. Soil that supports vegetative growth and serves as a host for insects, worms, and microbes is typically full of pores; such soil is likely to have as much as 25% of its volume occupied by air. The ability of soil to hold air depends on soil-particle size and how well the particles pack and cling together to form a solid mass. The particle size groups in soils, called **separates,** vary from clays (the finest) through silt and sand to gravel (the coarsest). The particle size of a clay is 0.005 mm or less. The small particles in a clay deposit pack closely together to eliminate essentially all air and thus support little or no life. A typical soil horizon is composed of several separates. A **loam,** for example, is a soil consisting of a friable mixture of varying proportions of clay, sand, and organic matter; a loam has a high air content.

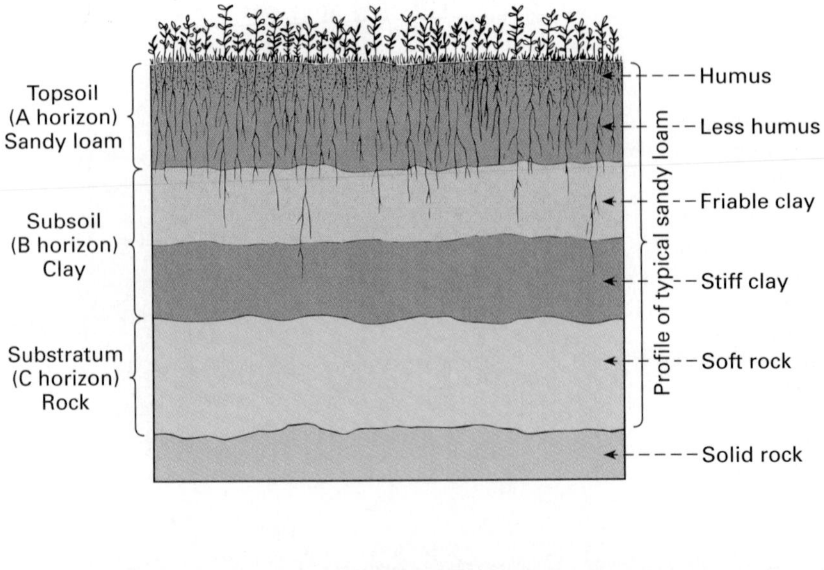

Figure 22.3 Structure of a sandy soil. The layers, known as *horizons,* are built up through weathering of rock and interaction with water, air, plants, and animals as shown in Figure 22.2.

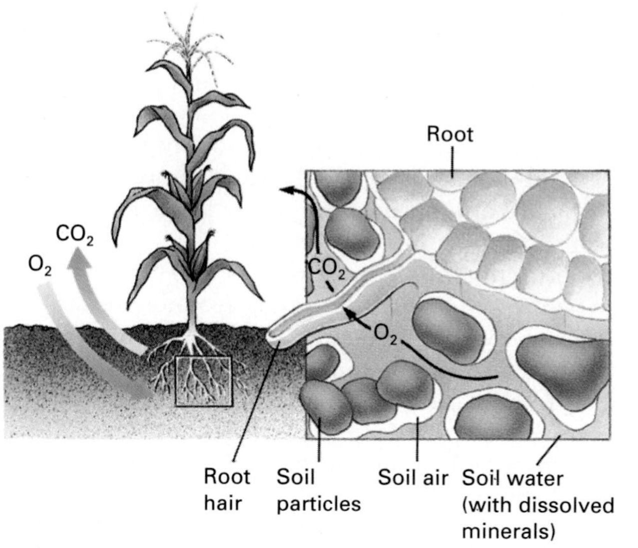

Exchanges between water and air in soil pores. Plant roots absorb oxygen and water, and release carbon dioxide.

Spreading "lime" on the lawn. "Lawn lime" is powdered limestone, which is $CaCO_3$. It increases the pH, that is, makes the soil more alkaline, or "sweeter." Calcium oxide (CaO) and calcium hydroxide ($Ca(OH)_2$), which is made by adding water to CaO, are also both often referred to as "lime." They are too strongly basic for lawns, but can be used in commercial agriculture. *(C.D. Winters)*

Air in soil has a different composition from the air we breathe. Normal dry air at sea level contains about 21% oxygen (O_2) and 0.03% carbon dioxide (CO_2). In soil the percentage of oxygen may drop to as low as 15%, and the percentage of carbon dioxide may rise above 5%. This results from the partial oxidation of organic matter in the closed space. The carbon in the organic material combines with oxygen to form carbon dioxide. This increased concentration of carbon dioxide tends to cause groundwater to become acidic; acidic soils are described as *sour* soils because of the presence of aqueous acids.

$$CO_2(g) + H_2O(\ell) \longrightarrow H^+(aq) + HCO_3^-(aq)$$

Crushed limestone ($CaCO_3$) applied to soil combines with hydrogen ions to form bicarbonate ions, thus raising the pH.

$$H^+(aq) + CO_3^{2-}(aq) \longrightarrow HCO_3^-(aq)$$

If enough limestone is added to neutralize the acid in the soil and leave an excess of limestone, the pH of the soil becomes alkaline (basic).

■ A slightly basic soil is a "sweet" soil.

Water in the Soil—Too Much, Too Little, or Just Right

Water can be held in soil in three ways: it can be *absorbed* into the structure of the particulate material, it can be *adsorbed* onto the surface of the soil particles, and it can occupy the pores ordinarily filled with air.

Water is removed from soil in four ways: plants transpire water while carrying on life processes, soil surfaces evaporate water, water is carried away in plant products, and water moves through the subsoil and rock formations below in a process called **percolation.** Soils with good percolation drain water from all but the small pores in the natural flow of the water.

■ It takes several hundred pounds of water for the typical food crop to make 1 lb of food.

The percolation of a soil depends on the soil-particle size and its chemical composition. Because of their small particle sizes, clays, and to a lesser degree silts, tend to pack together in an impervious mass with little or no percolation. Of course, sand, gravel, and rock pass water readily. Waterlogged soils that do not percolate support few crops because of their lack of air and oxygen. Rice is an important exception. A negative aspect of the massive flow of water through soil is the **leaching effect.** Water, known as the universal solvent because of its ability to dissolve so many different materials, dissolves away, or leaches, many of the chemicals needed to make a soil productive. If the leached material is not replaced, the soil becomes increasingly unproductive.

Soils become acidic, or sour, not only because of the oxidation of organic matter but also because of *selective leaching* by the passing groundwater. Salts of Group IA and IIA metals are more soluble than salts of the Group IIIA and transition metals. For example, a soil containing calcium, magnesium, iron, and aluminum ions is likely to be slightly alkaline, or sweet, before leaching with water. After the selective removal of calcium and magnesium salts, the soil becomes acidic because the iron and aluminum ions each tie up hydroxide ions from water and release hydrogen ions:

$$Fe^{3+} + H_2O \longrightarrow FeOH^{2+} + H^+$$

$$Al^{3+} + H_2O \longrightarrow AlOH^{2+} + H^+$$

22.3 Nutrients

At least 18 known elemental nutrients are required for normal green plant growth (Table 22.1). Three of these, the **nonmineral nutrients**—carbon, hydrogen, and oxygen—are obtained from air and water. The **mineral nutrients** must be absorbed through the plant root system as solutes in water. The 15

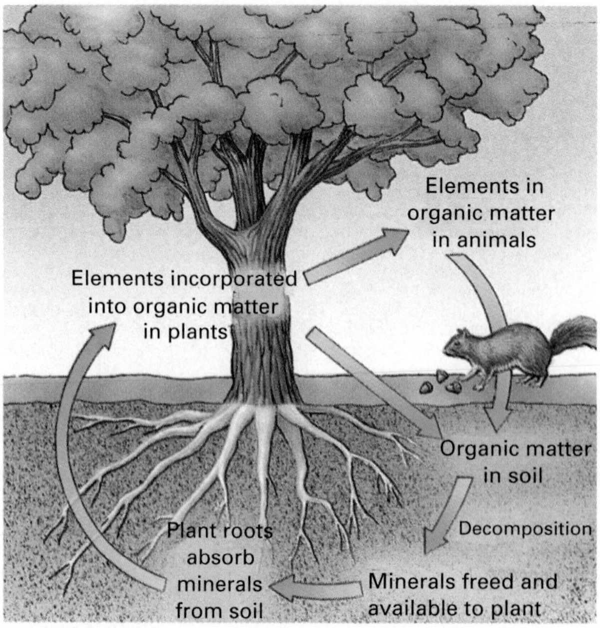

A balanced ecosystem. There is a smooth cycling of nutrients from the soil to plants and animals, and then back to the soil.

TABLE 22.1 ■ Essential Plant Nutrients

Nonmineral	Primary	Secondary	Micronutrients
Carbon	Nitrogen	Calcium	Boron
Hydrogen	Phosphorus	Magnesium	Chlorine
Oxygen	Potassium	Sulfur	Copper
			Iron
			Manganese
			Molybdenum
			Sodium
			Vanadium
			Zinc

A natural method of nitrogen fixation. The energy in a bolt of lightning is sufficient to disrupt the very stable triple bond in a nitrogen molecule (N_2). The result is the reaction with oxygen in the air to produce NO. *(Gordon Garrado/SPL/Photo Researchers)*

known mineral nutrients fall into three groups: **primary nutrients, secondary nutrients,** and **micronutrients,** depending on the amounts necessary for healthy plant growth.

Primary Nutrients

The primary nutrients are nitrogen, phosphorus, and potassium. Although bathed in an atmosphere of nitrogen, most plants are unable to use the air as a supply of this vital element. Nitrogen fixation is the process of changing atmospheric nitrogen into water-soluble compounds that can be absorbed through the plant's roots, and assimilated by the plant (See Section 11.2 and Figure 11.3).

Nature fixes nitrogen in two ways. In the first method, bolts of lightning provide the high energy needed to oxidize nitrogen to nitric oxide (NO). The NO is further oxidized to NO_2 (Section 21.4) which reacts with water to form nitric acid (HNO_3). This method is estimated to provide less than ten percent of the nitrogen fixed by nature. Nitric acid is readily soluble in rain, clouds, or ground moisture and thus increases nitrate concentration in soil.

In the most important method, nitrogen-fixing bacteria that live in the roots of legumes, such as soybeans and alfalfa, use an enzyme, **nitrogenase,** to catalyze a complex series of reactions that convert atmospheric nitrogen into ammonia under normal atmospheric conditions. Legume nitrogen fixation can add more than 100 pounds of nitrogen per acre of soil in one growing season. (See *World of Chemistry* box on nitrogen fixation in Chapter 6).

■ A German, H. Hellriegel, showed in 1886 that leguminous plants such as alfalfa, beans, and soybeans "fix" nitrogen.

■ Another major source of nitrogen replenishment in soil is dead organisms and animal wastes. Even in the absence of legumes, this can be an adequate source of nitrogen.

Like nitrogen, phosphorus must be in a soluble mineral or inorganic form before it can be used by plants. Unlike nitrogen, phosphorus comes totally from the mineral content of the soil. Salts of the dihydrogen phosphate ion ($H_2PO_4^-$) and monohydrogen phosphate ion (HPO_4^{2-}) are the dominant phosphate ions in soils of normal pH (Figure 22.4). Because of the great concentration of electric charge associated with the trivalent phosphate ion (PO_4^{3-}), phosphates are more tightly held to positive ions such as Ca^{2+} and Fe^{3+} and are not as easily leached by groundwater as are nitrate salts, which are generally soluble in water.

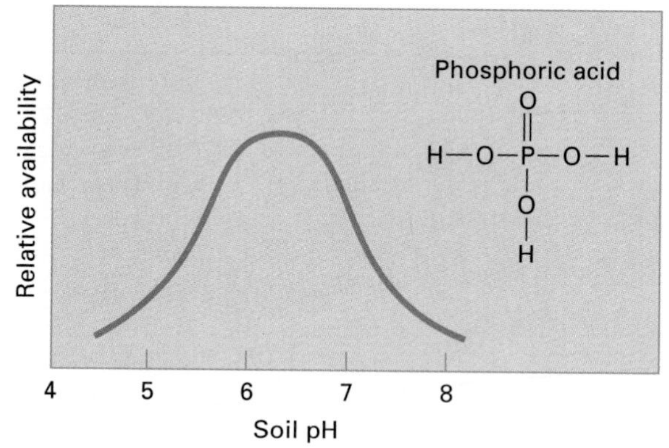

Figure 22.4 Availability of phosphate in the soil as a function of pH. The ions present vary with the pH. At pH 5 to pH 8, $H_2PO_4^-$ and HPO_4^{2-} predominate. At very low pH values, phosphorus is in the form of the nonionized acid H_3PO_4. At a very high pH, all three protons are removed and the phosphorus is in the form of the phosphate ion PO_4^{3-}. Low soil temperatures in temperate regions significantly reduce phosphorus uptake by plants.

■ Potassium is absorbed as the free ion, $K^+(aq)$.

Potassium in the form of K^+ ion is a key element in the enzymatic control of the interchange of sugars, starches, and cellulose. Although potassium is the seventh most abundant element in the Earth's crust, soil used heavily in crop production can be depleted of this important metabolic element, especially if the soil is regularly fertilized with nitrate, with no regard to potassium content.

Secondary Nutrients

Calcium and magnesium are available in small amounts as Ca^{2+} and Mg^{2+} ions as well as in complex ions and crystalline formations. These abundant ele-

FRONTIERS IN THE WORLD OF CHEMISTRY

Trying to Mimic Nature

The process for making ammonia from the reaction of nitrogen and hydrogen at high temperatures and pressures in the presence of a catalyst was invented by Haber in 1913 and is still the major industrial process for the synthesis of ammonia. However, this is far from doing the job of fixing nitrogen as well as nature does. Fixation of nitrogen by bacteria on root nodules of legume crops occurs at atmospheric temperature and pressure. The function of nitrogenase, the biological catalyst for this process, is still not fully understood, but iron and molybdenum are known to be important to the process by which the high-energy triple bond in N_2 is broken during the reduction of N_2 to NH_3 in biological nitrogen fixation. For years, chemists have been studying reactions of transition metal compounds with nitrogen in an attempt to understand the function of metal ions in nitrogenase. Although several compounds were isolated that contained N_2 bonded to the metal center, none involved breaking the triple bond, a necessary step in nitrogen fixation. The recent discovery of a molybdenum compound that breaks the triple bond of N_2 at room temperature and atmospheric pressure to give another molybdenum compound with a single nitrogen atom bonded directly to molybdenum is a major breakthrough in synthetic chemistry. The discovery could lead to both lower energy requirements for industrial nitrogen fixation and a better understanding of the role of molybdenum in the function of nitrogenase in biological nitrogen fixation. H. Hellriegel showed in 1886 that leguminous plants can fix nitrogen, and more than 100 years later, scientists appear to be coming closer to mimicking nature's efficient process for nitrogen fixation.

Reference C.E. Laplaza and C.C. Cummins, "Dinitrogen Cleavage by a Three-Coordinate Molybdenum (III) Complex," *Science,* Vol. 268, pp. 861–863, 12 May, 1995.

ments are bound tightly enough by soil so they are not readily leached yet loosely enough to be available to plants. When in the soil as sulfate (SO_4^{2-}), sulfur is readily available to plants.

Micronutrients

Only very small amounts of micronutrients are required by plants; therefore, unless extensive cropping or other factors deplete the soil of these nutrients, sufficient quantities are usually available.

Iron is also an essential component of the enzyme involved in the formation of chlorophyll. When the soil is iron deficient or when too much lime, $Ca(OH)_2$, is present in the soil, iron availability decreases. Often a gardener or lawn worker will apply phosphate and lime to adjust soil acidity, only to see green plants turn yellow because of **chlorosis.** What happens in such cases is that both phosphate and the hydroxide from the lime tie up the iron and make iron unavailable to the plants:

■ Lime, as CaO or $Ca(OH)_2$, is the number-five commercial chemical in the United States.

$$\underset{}{Fe^{3+}(aq)} + \underset{\text{Phosphate}}{2\,PO_4^{3-}(aq)} \longrightarrow \underset{\text{Tightly bound complex}}{Fe(PO_4)_2^{3-}(aq)}$$

$$Fe^{3+}(aq) + 3\,OH^-(aq) \longrightarrow \underset{\text{Insoluble hydroxide}}{Fe(OH)_3(s)}$$

A leaf on a blackberry plant suffering from chlorosis. Chlorophyll, the green plant pigment, requires nitrogen, magnesium, and iron from the soil. Deficiencies of any of these nutrients cause chlorosis, a condition of low chlorophyll content indicated by yellowing of the leaves. *(C.D. Winters)*

■ SELF-TEST 22A

1. The present world population of ____________ is growing at a rate of ____________ per year.
2. Rank the following types of soils from those with the smallest soil particles to those with the largest: silts, sandy soils, loams, clays
3. Carbon dioxide causes soils to be (a) acidic, (b) basic. Limestone ($CaCO_3$) causes soils to be (a) acidic, (b) basic.
4. The two factors that determine the percolation of a soil are ____________ and ____________.
5. Which is more acidic, a monovalent ion such as Na^+ or a trivalent ion such as Fe^{3+}?
6. A well-decomposed, dark-colored plant residue that is relatively resistant to further decomposition is known as ____________.
7. The primary elemental plant nutrients necessary in the soil for healthy plant growth are ____________, ____________, and ____________.
8. The secondary elemental plant nutrients are ____________, ____________, and ____________.
9. Nitrogen fixation by lightning discharges involves breaking a nitrogen-nitrogen triple bond and combining nitrogen with ____________.

22.4 Fertilizers Supplement Natural Soils

Primitive people raised crops on a cultivated plot until the land lost its fertility; then they moved to a virgin piece of ground where they cut down natural vegetation ("slash") and burned off the stubble to clear the land. In many cases, the slash-burn-cultivate cycle was no more than a year in length, and few found a piece of ground anywhere that could support successful cropping for more than five years without fertilization. Farming villages, developed in ancient times and prevalent throughout the Middle Ages, demanded innovation in fertilization, because they had to use the same land for many years. With the use of legumes in crop rotations, manures, dead fish, or almost any organic matter available, the land was kept in production.

■ Vast amounts of acres have been torched each year in the Amazon basin. Current efforts have curbed this deforestation but have not halted it.

An estimated four billion acres are used worldwide in the cultivation of crops for food, less than 0.8 acre per person. This acreage would probably be sufficient if modern chemical fertilization were employed on all of it. If about \$40 were spent on fertilizer for each cultivated acre, world crop production would increase by 50%, the equivalent of having two billion more acres under cultivation. However, the cost to produce this additional food would approach a prohibitive \$160 billion, and the environmental impact of such a large dispersion of fertilizer chemicals would probably be massive. For example, at the present time, aquifer contamination with nitrates due to corn crop fertilization renders well water from these natural underground basins unfit for drinking in large areas of the U.S. corn belt.

■ Crop yield explosions: (1) U.S. corn—25 bushels per acre in 1800, 110 bushels per acre in the 1980s, 130 bushels per acre in the 1990s; (2) English wheat—less than 10 bushels per acre from A.D. 800 to 1600, more than 75 bushels per acre in the 1980s; (3) rice in Japan, Korea, and Taiwan—fourfold increase in the last 40 years.

Fertilizers that contain only one nutrient are called **straight fertilizers.** Potassium chloride for potassium is an example of a straight fertilizer. Those containing a mixture of the three primary nutrients are called **complete,** or **mixed, fertilizers.** The macronutrients are absorbed by plant roots as simple inorganic ions: nitrogen in the form of nitrates (NO_3^-), phosphorus as phosphates ($H_2PO_4^-$ or HPO_4^{2-}), and potassium as the K^+ ion. Organic fertilizers can supply these ions, but only when used in large quantities over a long time. For example, a manure might be a 0.5–0.24–0.5 fertilizer, in contrast to a typical chemical fertilizer, which might carry the numbers 6–12–6. These numbers indicate the **grade,** or **analysis,** in order, of the percentage of nitrogen as N, phosphorus as P_2O_5, and potassium as K_2O (commonly known as potash) in the fertilizer (Figure 22.5). In addition to containing the desired ions, chemical fertilizers place the ions in the soil in a form that can be absorbed directly by plants. The problem is that these inorganic ions are relatively easily leached from the soil and may pose pollution problems if not contained. The much slower organic fertilizer tends to stay put. **Quick-release fertilizers** are water-soluble, as opposed to **slow-release fertilizers,** which require days or weeks for the material to dissolve completely. Table 22.2 lists the necessary plant nutrients and suitable chemical sources of each.

■ Organic fertilizers contain mostly organic material and very little mineral content. For example, manure as a fertilizer is graded less than 1–2–1.

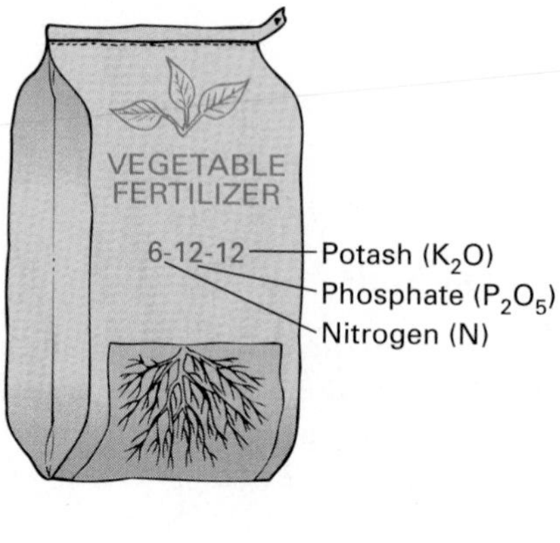

Figure 22.5 How to read a fertilizer label. The three numbers, in order, refer to the percentage by weight of N (nitrogen), P_2O_5 (phosphate), and K_2O (potash). Following the lead of J. von Liebig, his German teacher and the first to suggest adding nutrients to soils, Samuel William Johnson, an American, burned plants and analyzed their ashes. He expressed the nutrient concentrations as the oxides present in the ashes, a practice that has continued to this day.

TABLE 22.2 ■ Some Chemical Sources of Plant Nutrients

Element	Source Compound(s)
	Nonmineral Nutrients
C	CO_2 (carbon dioxide)
H	H_2O (water)
O	H_2O (water)
	Primary Nutrients
N	NH_3 (ammonia), NH_4NO_3 (ammonium nitrate), H_2NCONH_2 (urea)
P	$Ca(H_2PO_4)_2$ (calcium dihydrogen phosphate)
K	KCl (potassium chloride)
	Secondary Nutrients
Ca	$Ca(OH)_2$ (calcium hydroxide, slaked lime), $CaCO_3$ (calcium carbonate, limestone), $CaSO_4$ (calcium sulfate, gypsum)
Mg	$MgCO_3$ (magnesium carbonate), $MgSO_4$ (magnesium sulfate, epsom salts)
S	Elemental sulfur, metallic sulfates
	Micronutrients
B	$Na_2B_4O_7 \cdot 10H_2O$ (borax)
Cl	KCl (potassium chloride)
Cu	$CuSO_4 \cdot 5H_2O$ (copper sulfate pentahydrate)
Fe	$FeSO_4$ (iron(II) sulfate, iron chelates)
Mn	$MnSO_4$ (manganese(II) sulfate, manganese chelates)
Mo	$(NH_4)_2MoO_4$ (ammonium molybdate)
Na	NaCl (sodium chloride)
V	V_2O_5, VO_2 (vanadium oxides)
Zn	$ZnSO_4$ (zinc sulfate, zinc chelates)

Anhydrous ammonia as a fertilizer. At normal temperatures, anhydrous ammonia (ammonia gas containing no water) can be held as a liquid under tank pressure. On release from pressure, the ammonia returns to the gaseous state and is injected into the soil by an "ammonia knife." In slightly acid soil, the ammonia is immediately converted to the water-soluble ammonium ion and enters the natural nitrogen pathways of the soil. *(Grant Heilman/ Grant Heilman Photography)*

Nitrogen Fertilizers

The cheapest source of nitrogen is the air, but it must be combined with relatively expensive hydrogen, obtained from petroleum, to form ammonia in the Haber process (see Section 11.2 and Figure 11.4).

Ammonia can be applied directly to the soil or converted into numerous agrichemicals (Figure 22.6). Two common straight fertilizers for nitrogen are ammonium nitrate (NH_4NO_3) and urea, numbers 14 and 15 in U.S. chemical production in 1994. A slurry of water, urea, and ammonium nitrate is often applied to crops under the name of "liquid nitrogen." Such a solution can contain up to 30% nitrogen and is easy to store and apply.

■ The structural formula of urea is

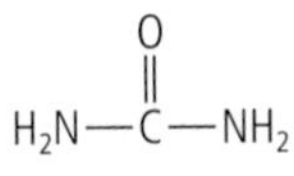

When applied to the surface of the ground around plants, urea is subject to considerable nitrogen loss unless it is washed into the soil by rain or irrigation. When urea decomposes, ammonia is formed, and some ammonia is lost to the air and some is absorbed by moist soil particles. As much as half of the nitrogen applied to the soil can be lost in this way.

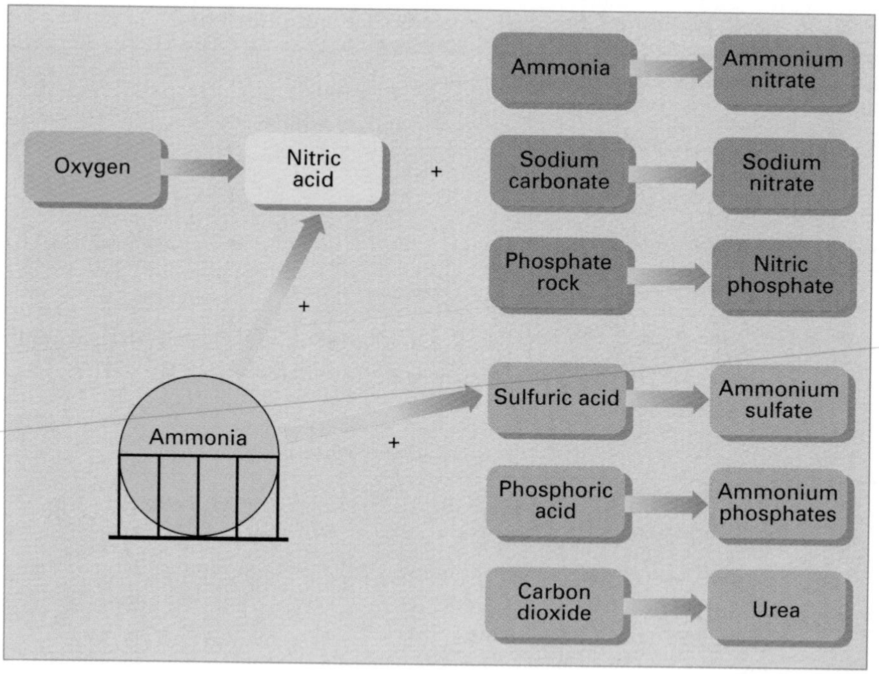

Figure 22.6 Nitrogen fertilizers produced from anhydrous ammonia.

SCIENCE AND SOCIETY

Ammonium Nitrate—A Dilemma

Ammonium nitrate is the number 14 chemical produced in the United States because of its use as both a fertilizer and an explosive. The Oklahoma City bombing on April 19, 1995, that killed 169 people has resulted in a call for regulations to eliminate the ready availability of ammonium nitrate. Regulations on the use of ammonium nitrate as an explosive are feasible, but restrictions on the purchase of ammonium nitrate fertilizer present a major problem. Fifty-pound bags of ammonium nitrate can be purchased at any farm supply store. One approach that is being discussed is to require the addition of a substance that would make fertilizer-grade ammonium nitrate unexplodable.

Samuel J. Porter, a hazardous chemicals consultant, received a patent in 1968 for rendering ammonium nitrate unexplodable by adding 5 to 10% diammonium phosphate to molten ammonium nitrate before the formation of the final product, small spherical pellets. At the time, none of the chemical companies approached by Porter felt that the increased cost of adding a substance to make ammonium nitrate inert was justified. The patent has expired and is now in the public domain.

In May 1995 Representative W. J. Tauzin from Louisiana called on the fertilizer industry to test Porter's process. Gary Myers, president of the Fertilizer Institute, says the fertilizer industry supports studies on how to make fertilizers useless as explosives. The question of whether additives such as diammonium phosphate really work cannot be answered without further testing. Testing is also necessary to determine whether additives could be separated from ammonium nitrate by terrorists. In Northern Ireland and the Republic of Ireland, regulations require the addition of calcium carbonate to ammonium nitrate before it can be sold as fertilizer. The salts in this mixture, referred to as calcium ammonium nitrate, can be separated; however, available information indicates that terrorists do not separate the ammonium nitrate but rather make an explosive mixture by grinding the calcium ammonium nitrate into very fine particles and mixing it with other substances to make a bomb.

The dilemma: How do we retain the availability of ammonium nitrate as an economical fertilizer and eliminate its use by terrorists to make bombs?

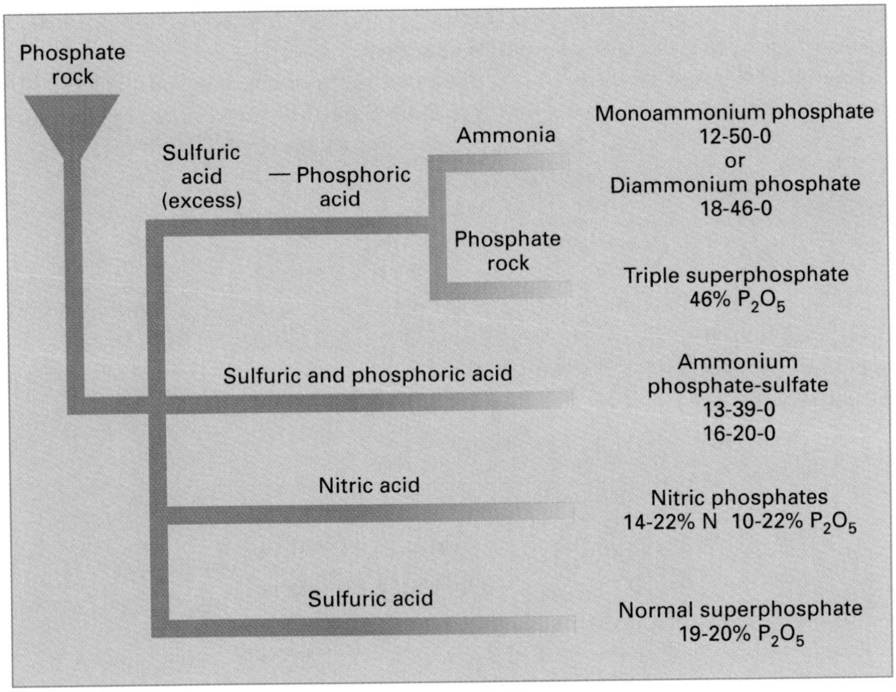

Figure 22.7 Fertilizers produced from phosphate rock.

Phosphate Fertilizers

Phosphorus is readily available in the form of phosphate rock, which can be transformed into the needed fertilizers (Figure 22.7). World deposits of phosphate rock are limited, and costs for supplying phosphorus fertilizers will increase as deposits are depleted. The phosphate rock, $Ca_3(PO_4)_2$, is not very useful because of its very low solubility. When treated with sulfuric acid, however, phosphate rock becomes more soluble, and the resulting mixture of phosphate and sulfate salts of calcium is called "superphosphate."

$$\underset{\text{Phosphate rock}}{Ca_3(PO_4)_2} + 2\ H_2SO_4 \longrightarrow \underset{\text{Superphosphate}}{Ca(H_2PO_4)_2 + 2\ CaSO_4}$$

A phosphate mine in Florida. In the background is a part of this mine that has been restored after mining was completed and is now used as a pasture. *(William Felger/Grant Heilman)*

EXAMPLE 22.1 *Fertilizer Labels*

The label on a fertilizer bag lists the numbers 22-6-8. What do these numbers mean? Calculate the pounds of each nutrient present in a 50.0 pound bag of fertilizer.

SOLUTION

The numbers mean that the fertilizer contains 22.0% nitrogen (N) in the form of some nitrogen-containing compound, 6.00% of a phosphorus-containing compound (calculated as a percentage of P_2O_5), and 8.00% of a potassium-containing compound (calculated as a percentage of K_2O). The pounds of each nutrient present are calculated as follows:

Nitrogen: The pounds of nitrogen present are calculated by multiplying the percentage of nitrogen times 50.0 pounds. Since the unit in this problem is pounds (lb), 22.0% of N can be represented as 22.0 lb N/100 lb fertilizer.

$$\frac{22.0 \text{ lb N}}{100 \text{ lb fertilizer}} \times 50.0 \text{ lb fertilizer} = 11.0 \text{ lb N}$$

Phosphorus: Since the amount of phosphorus present is represented as 6.00% P_2O_5, the actual percentage present as elemental phosphorus is smaller. To calculate the fraction of phosphorus in P_2O_5 requires using the atomic masses of P and O in a stoichiometric factor:

$$\frac{2 \times \text{atomic mass P}}{(2 \times \text{atomic mass P}) + (5 \times \text{atomic mass O})}$$

$$2\text{ P} = 2 \times 31.0 \text{ amu} = 62.0 \text{ amu}$$

$$5\text{ O} = 5 \times 16.0 \text{ amu} = 80.0 \text{ amu}$$

$$\text{Total for } P_2O_5 = 142.0 \text{ amu, so factor is}$$

$$\frac{62.0 \text{ amu P}}{142 \text{ amu } P_2O_5}$$

The relative masses are the same no matter what units are used. Since the unit in this problem is pounds, the stoichiometric factor can be written as

$$\frac{62.0 \text{ lb P}}{142 \text{ lb } P_2O_5}$$

$$\frac{6.00 \text{ lb } P_2O_5}{100 \text{ lb fertilizer}} \times \frac{62.0 \text{ lb P}}{142 \text{ lb } P_2O_5} \times 50.0 \text{ lb fertilizer} = 1.31 \text{ lb P}$$

Potassium: Since the amount of potassium present is represented as 8.00% K_2O, calculating the pounds of potassium present requires using the atomic masses of K and O in a stoichiometric factor:

$$\frac{2 \times \text{atomic mass K}}{(2 \times \text{atomic mass K}) + (\text{atomic mass O})}$$

2 K = 2 × 39.0 amu = 78.0 amu; 1 O = 16.0 amu; total of 2 K and 1 O = 94.0 amu so the stoichiometric factor showing the ratios in pounds is

$$\frac{78.0 \text{ lb K}}{94.0 \text{ lb } K_2O}$$

$$\frac{8.00 \text{ lb } K_2O}{100 \text{ lb fertilizer}} \times \frac{78.0 \text{ lb K}}{94.0 \text{ lb } K_2O} \times 50.00 \text{ lb fertilizer} = 3.32 \text{ lb K}$$

Exercise 22.1 *Fertilizer*

Ammonium nitrate, NH_4NO_3, is widely used as a fertilizer. A bag of ammonium nitrate sold as fertilizer lists the numbers 35-0-0. Use the formula weight for NH_4NO_3 to verify that this is the correct number.

22.5 Protecting Plants in Order to Produce More Food

The natural enemies of plants include more than 80,000 diseases brought on by viruses, bacteria, fungi, algae, and similar organisms; 30,000 species of weeds; 3000 species of nematodes; and about 10,000 species of plant-eating insects. About one third of the food crops in the world are lost to pests each year, with the loss going above 40% in some developing countries. Crop losses to pests amount to $20 billion per year. "Pest" is any organism that in some way reduces crop yields or human health. **Pesticides,** chemicals used to control pests, are classified according to the pests they control: **insecticides** kill insects, **herbicides** kill weeds, **fungicides** kill fungi.

Program 25, *Chemistry and the Environment*

Effect of DDT on egg shells. The massive use of DDT for insect control nearly wiped out several species of birds due to interference in the formation of egg shells.

Pesticides

Pesticides are the chemical answer to pest control. Eighteen common classes of pesticides are fortified with more than 2600 active ingredients to fight the battle with pests. More than five billion pounds of pesticides are produced worldwide each year. In 1993 pesticide sales in the United States totaled $6.8 billion. Although the dollar cost is up and expected to reach $7.5 billion in 1995, the actual poundage use began a decline in 1987. There are three reasons for a slowing in the demand for pesticides in the United States: (1) Cropland planted is less than 330 million acres, down from a high of 383 million acres in 1982. (2) Farming is becoming more cost-effective as farmers learn to use a minimum of pesticide for the desired effect. (3) Farmers are becoming more concerned with environmental and health issues.

Insecticides

Before World War II, the list of insecticides included only a few compounds of arsenic, petroleum oils, nicotine, pyrethrum (obtained from dried chrysanthemum flowers), rotenone (obtained from roots of derris vines), sulfur, hydrogen cyanide gas, and cryolite (the mineral Na_3AlF_6). DDT, the first of the chlorinated organic insecticides, was originally prepared in 1873, but it was not until 1939 that Paul Müller of Geigy Pharmaceutical in Switzerland discovered the effectiveness of DDT as an insecticide.

■ Paul Müller was awarded the Nobel Prize in medicine and physiology in 1948 for his discovery of the insecticidal properties of DDT.

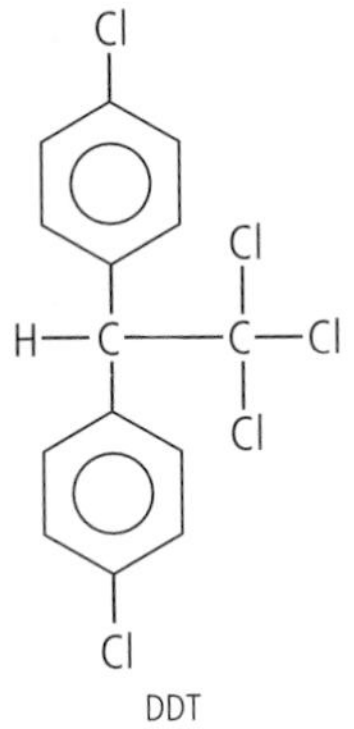

DDT

The use of DDT increased enormously on a worldwide basis after World War II, primarily because of its effectiveness against the mosquito that spreads malaria and lice that carry typhus. The World Health Organization estimates that approximately 25 million lives have been saved through the use of DDT. DDT seemed to be the ideal insecticide—it is cheap and of relatively low toxicity to mammals (oral LD_{50} is 300 to 500 mg/kg). However, problems related to extensive use of DDT began to appear in the late 1940s. Many species of insects developed resistance to DDT, and DDT was also discovered to have a high toxicity toward fish. The chemical stability of DDT and its fat solubility compounded the problem.

DDT is not metabolized very rapidly by animals; instead, it is deposited and stored in the fatty tissues. The biological half-life of DDT is about eight years; that is, it takes about eight years for an animal to metabolize half of the amount it assimilates. If ingestion continues at a steady rate, DDT builds up within the animal over time (see Section 20.7 for a discussion of this problem).

The use of DDT was banned in the United States in 1973, although it is still in use in some other parts of the world. The buildup of DDT in natural waters is a reversible process; the EPA reported a 90% reduction of DDT in Lake Michigan fish by 1978 as a result of the ban on the use of the insecticide in the United States.

The most important insecticides can be grouped in three structure classes—chlorinated hydrocarbons, organophosphorus compounds, and carbamates. Chlorinated hydrocarbons such as DDT are referred to as **persistent pesticides** because they persist in the environment for years after their use.

Malathion, a widely used organophosphorus pesticide that functions as an anticholinesterase poison (see structure on p. 591), is both biodegradable and less toxic than DDT (LD_{50} for female rats is 1000 mg/kg).

The carbamate insecticides are derivatives of carbamic acid:

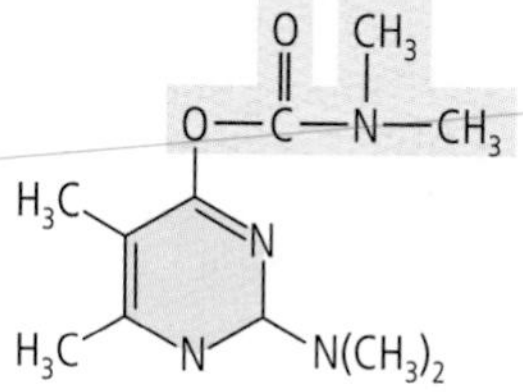

Pirimicarb

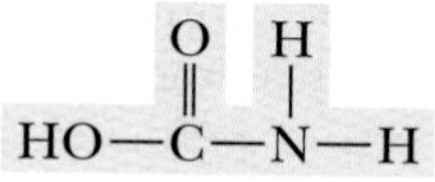

Carbamic acid

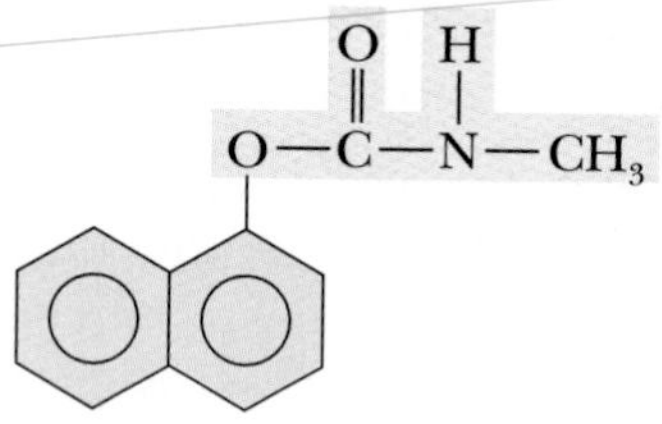

Carbaryl

They are also anticholinesterase poisons and nonpersistent insecticides. The most widely used carbamate is carbaryl, a general-purpose insecticide with a relatively low mammalian toxicity (oral LD_{50} for rats is 250 mg/kg). A serious drawback to the use of carbaryl for spraying crops is its high toxicity toward honeybees. Another carbamate of interest is pirimicarb (oral dose for female rats is 147 mg/kg), which is a selective insecticide for aphids and has the advantage of controling strains that have developed resistance to organic phosphates. Pirimicarb has a low toxicity to predators such as bees. It is rapidly metabolized and leaves no lasting residues in plant materials. Pirimicarb is expensive, however; its preparation requires five or six steps and special handling procedures.

A corn earworm at work. *(Gilbert S. Grant/Photo Researchers)*

The choice of solutions to the problems of insecticide use is not an easy one. Their use introduces trace amounts into our environment and our water supplies. If we fail to use them, we must tolerate malaria, plague, sleeping sickness, and the consumption of a large part of our food supply by insects. Continuing research in the development of more effective and safer insecticides is intense, and new products are introduced each year.

■ The goal of the insecticide quest: a selectively toxic chemical that is quickly biodegradable.

Herbicides

Herbicides kill plants. They may be **selective** and kill only a particular group of plants, such as the broad-leaved plants or the grasses, or they may be **nonselective,** making the ground barren of all plant life.

Selective herbicides act like hormones, very selective biochemicals that control a particular chemical change in a particular type of organism at a particular stage in its development. Most selective herbicides in use today are growth hormones; they cause cells to swell, so that leaves become too thick for chemicals to be transported through them, and roots become too thick to absorb needed water and nutrients. Nonselective herbicides usually interfere

FRONTIERS IN THE WORLD OF CHEMISTRY

Environmentally Friendly Insecticide

Two chemical dyes that are used to color drugs and cosmetics may have an important new use. Field tests have shown that phloxine B and uranine food dyes are effective against Mediterranean fruit flies that currently threaten billions of dollars of crops in California and Mexico. Why are the food dyes safe for use in coloring drugs and cosmetics and yet toxic enough to kill fruit flies? The answer lies in the discovery that the dyes only become toxic after exposure to light. Biochemist James Heitz says "Insects like fruit flies have translucent guts that are easily penetrated by sunlight." As a result, the dyes are activated in the flies' digestive systems to toxic derivatives. This doesn't happen in humans, other mammals, fish, or birds because the dyes pass through the digestive system without being exposed to light.

The hope is that the dyes will eventually replace malathion, the insecticide now being used to spray crops threatened by fruit flies. Before this happens, additional field testing is needed to make sure the decomposition products produced by the light-sensitive dyes do not harm crops.

Reference Robert F. Service, "New Compounds Make Light of Pests," *Science,* Vol. 268, p. 806, 12 May 1995.

with photosynthesis and thereby starve the plant to death. On application, the plant quickly loses its green color, withers, and dies.

The traditional method for the control of weeds in agriculture was **tillage,** plowing the weeds under a layer of soil. Only in the early 1900s was it recognized that some fertilizers were also weed killers. For example, when calcium cyanamide (CaNCN) was used as a source of nitrogen, it was found to retard the growth of weeds. Arsenites, arsenates, sulfates, sulfuric acid, chlorates, and borates have also found use as weed killers. A typical product still in commercial use contains 40% sodium chlorate ($NaClO_3$), 50% sodium metaborate ($NaBO_2$), and 10% inert filler. These herbicides are nonselective and must be used with considerable care to protect desirable plants.

The most widely used herbicide is 2,4-dichlorophenoxyacetic acid (2,4-D). The corresponding trichloro- compound (common name: 2,4,5-T) has also been shown to be highly effective, but it was banned by the EPA because of a number of health problems associated with its use.

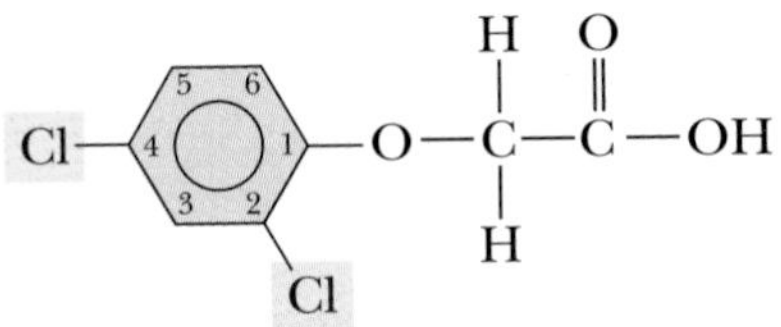

2,4-D
(2,4-dichlorophenoxyacetic acid)

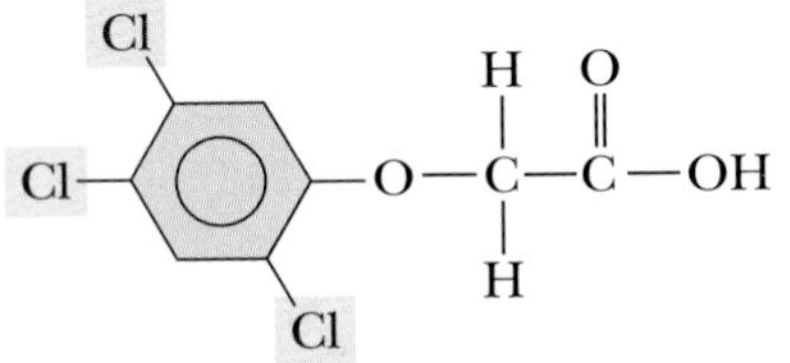

2,4,5-T
(2,4,5-trichlorophenoxyacetic acid)

The only difference between 2,4,5-T and 2,4-D is the additional chlorine atom on the benzene ring in the fifth position. Agent Orange, widely used as a defoliant during the Vietnam War, is a mixture of these two compounds. Many veterans of the Vietnam War have attributed their health problems to exposure to Agent Orange. Dioxin, a highly toxic compound (Section 18.1) produced during the manufacture of 2,4,5-T, was present as an impurity in both

Agent Orange and the 2,4,5-T used for agricultural purposes. Dioxin is regarded as the most likely cause of the health problems associated with the use of Agent Orange and 2,4,5-T. It will be interesting to see whether 2,4,5-T, which is now commercially produced free of dioxin, will be reestablished as a herbicide. Both 2,4-D and 2,4,5-T result in an abnormally high level of RNA (Section 15.7) in the cells of the affected plants, causing the plants to grow themselves to death.

Triazines have been found to be effective as herbicides; the most important one is atrazine.

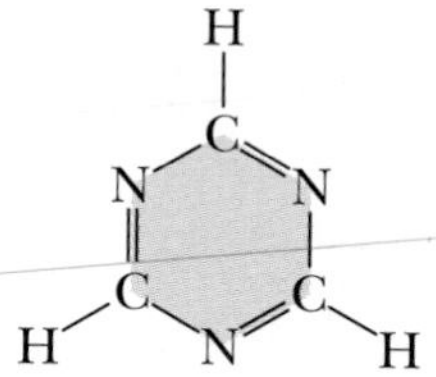

1,3,5-triazine

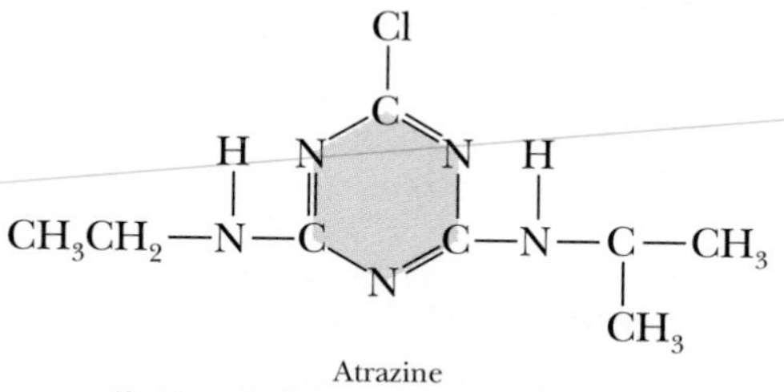

Atrazine
(2-chloro-4-ethylamino-6-isopropyltriazine)

Atrazine is widely used in no-till corn production or for weed control in minimum tillage. Atrazine is a poison to any green plant if it is not quickly metabolized. Corn and certain other crops have the ability to render atrazine harmless, which weeds cannot do. Hence, the weeds die, and the corn shows no ill effect.

Several herbicides work by inhibiting plant enzymes that catalyze the synthesis of amino acids in plants that are essential amino acids in animals (Section 15.2). Unlike animals, plants can synthesize all 20 amino acids. Since animals do not synthesize essential amino acids, developing inhibitors for amino acids that are synthesized by plants but not by animals would produce safer herbicides. Glyphosate and sulfonylureas are herbicides with this mechanism of action. Glyphosate, the active ingredient in Round-up, is a phosphate derivative of glycine that inhibits the synthesis of the essential amino acids tyrosine and phenylalanine. It is used to control perennial grasses.

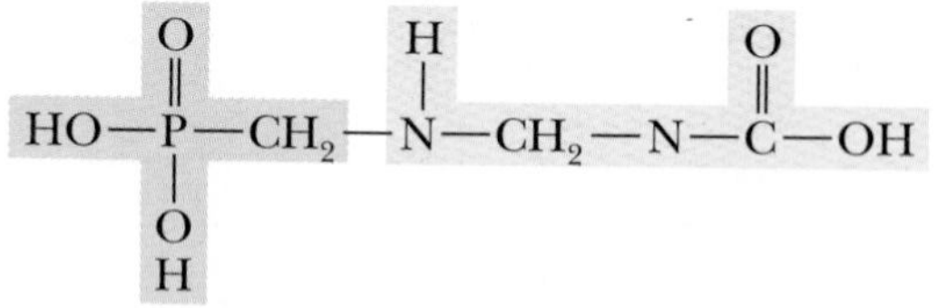

Glyphosate

Sulfmeturon methyl, the active ingredient of Oust, is a sulfonylurea that inhibits the synthesis of the essential amino acids valine, leucine, and isoleucine.

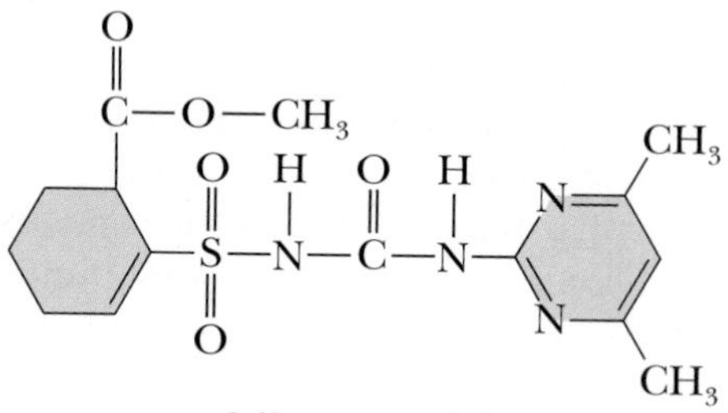

Sulfmeturon methyl

Paraquat is a herbicide and can be used to kill weeds before the crop sprouts. Such herbicides are called "pre-emergent" herbicides. When applied directly to susceptible plants, it quickly causes a frostbitten appearance and death. Like atrazine, paraquat has a nitrogen atom in each aromatic ring of the two-ring system. Paraquat has received considerable attention because it was used to spray illegal poppy and marijuana fields in Mexico and elsewhere, which caused drug users to suffer lung damage.

$$H_3C-{}^+N \bigcirc - \bigcirc N^+ - CH_3$$

Cl⁻ Cl⁻

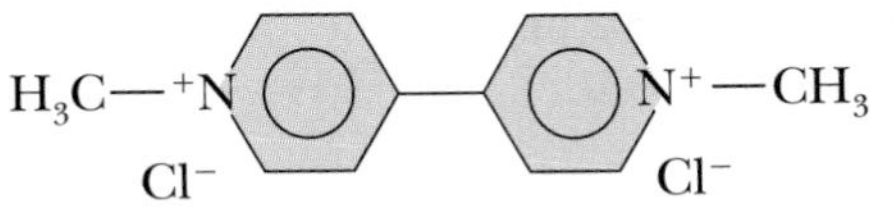

Paraquat
(1,1'-dimethyl-4,4'-bipyridinium dichloride)

The amount of energy saved by herbicides used in no-till farming is enormous. The saving of topsoil is also considerable, because the cover from the previous crop holds the soil against wind and water runoff. However, agriculturists who use herbicides are highly dependent on agricultural research institutions for the selection of herbicides that will do the desired job without harmful side effects. Such selections depend on considerable research, much of which is carried out on a trial-and-error basis on test plots. A procedure that is recommended today may be outdated by the next growing season.

Fungicides

About 200 of the 100,000 classified fungal species are known to cause serious plant disease. Agricultural fungicide application accounts for about 20% of all pesticide use. Most fungicides are applied to the seed or foliage of the growing plant or to harvested produce to prevent storage losses. The earliest fungicides were inorganic substances such as elemental sulfur and compounds of copper and mercury. In the nineteenth century, sulfur was used to control mildew on fruit and grapes.

In recent times, different types of organic compounds have been used. An important class of fungicides is the dithiocarbamates and their derivatives, which are widely used on many crops such as fruits and field vegetables.

$$(CH_3)_2N-\overset{\overset{\displaystyle S}{\|}}{C}-S-S-\overset{\overset{\displaystyle S}{\|}}{C}-N(CH_3)_2$$

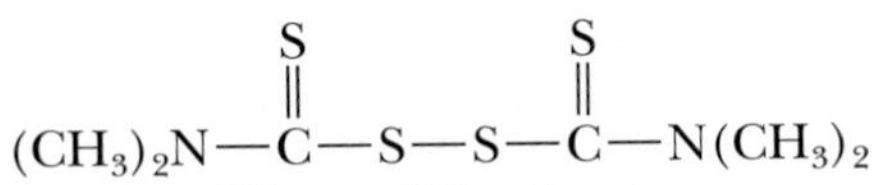

Thiram, a dithiocarbamate

Benomyl (DuPont tradename Benlate) was introduced in 1968, and for 20 years was an effective fungicide for 70 different crops.

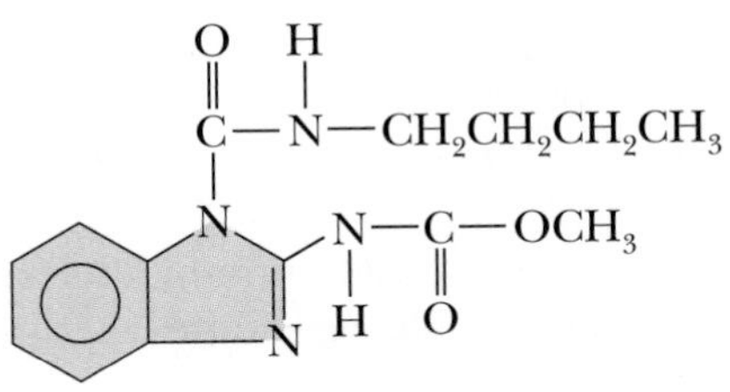

Benomyl

However, in late 1989 DuPont began receiving complaints that plants treated with Benlate had retarded root growth, shriveling leaves, and in some cases died. The problem was most prevalent for ornamental plants raised in hot and humid climates, such as Florida, where high concentrations of benomyl were applied every two months. After several years of extensive testing, over a thousand claims of plant damage, and numerous lawsuits, the cause of the problem is suspected to be a contamination of the fungicide with traces of other herbicides, such as atrazine and sulfonylureas. The symptoms of plant damage agree with this explanation, but the amounts of contaminants were so low that further testing is still being done.

EXAMPLE 22.2 *Pesticide Toxicity*

How much carbaryl could lead to a 50% chance of death in a child that weighs 20 kg if the lethal dose is the same as the oral dose for rats, LD_{50} = 250 mg/kg?

SOLUTION

The lethal dose is given in terms of mg/kg body weight, so

$$20\text{ kg} \times 250\text{ mg/kg} = 5000\text{ mg, or } 5.0\text{ grams could kill the child.}$$

Exercise 22.2

How much 2,4,5-T would it take to kill a farmer who weighs 75 kg if the lethal dose is the same as the oral dose for rats, LD_{50} = 500 mg/kg?

22.6 Sustainable Agriculture

Extensive use of fertilizers and pesticides, and development of higher-yielding varieties of crops have made it possible to increase crop yields dramatically. However, poor farm practices have resulted in soil erosion and loss of soil fertility. Soil erosion is proceeding on the average of 20 times the rate of replenishment. For example, in Iowa, half the topsoil has been lost in the past 150 years. Compaction of soil, which decreases soil fertility, is caused by (1) growing shallow-rooted crops year after year, (2) not incorporating enough organic matter into the soil, (3) practices that alter soil microbial and earthworm populations, and (4) using heavy machinery, particularly on wet soils. Compacted soil, when dry, looks like brick. It restricts root growth and has soil oxygen levels below those necessary for optimum uptake of nutrients.

Soil erosion caused by water runoff, which carries soil, fertilizer, and pesticides with it. *(T. McCabe/Visuals Unlimited)*

Another major problem is the increasing resistance of pests to pesticides. For example, even though there has been a 33-fold increase in the amount of pesticides used since 1945, crop losses from insects, diseases, and weeds have increased from 31% in 1945 to 37% today. About 500 different insect and mite species, 80 fungus species, and 80 weed species are now resistant to commonly used pesticides.

These problems have led to consideration of alternatives such as **organic farming** and **alternative** or **sustainable agriculture** methods.

Organic farming is farming without chemical fertilizers and pesticides. Organic farming uses only about 40% of the energy required for modern farming with synthetic chemicals and produces about 90% of the yield. The costs of energy saved in organic farming are offset by the costs of human labor, which is required by the use of natural fertilizers. Many claims are made that organic farming produces a better product for human consumption. However, there is no real evidence that these claims are generally true. Organic farming does have one clear advantage, however; it is definitely less of a threat to the environment than regular farming if agrichemicals are not very carefully controlled.

"Organic" produce, which is grown without pesticides. Consumer demand for such products is growing. *(Steve Feld)*

Alternative farming, a term that was popularized by a 1989 special report of the National Research Council, is an effort to stake out middle ground between organic farming and the heavy use of agrichemicals. The goals of alternative agriculture are to improve profits, limit the use of agrichemicals, and increase the use of environmentally friendly procedures to fight pests and produce food and fiber. Examples of significant changes in modern farming include (1) expanding crop rotations because the same pests do not afflict every crop, (2) using multiple crops in alternate plantings within a given planting field (Figure 22.8), (3) using as much natural fertilizer as possible before using agrichemicals, (4) increasing the use of biological pest controls, (5) employing renewed efforts at soil and water conservation, and (6) having a diversification in livestock as well as field crops on the same farm.

The main sources of plant nutrients in alternative farming systems are animal and green manures. A green manure crop is a grass or legume that is plowed into the soil or surface-mulched at the end of a growing season to enhance soil productivity.

One broad approach to limiting use of pesticides is **integrated pest management** (IPM), which relies more on disease-resistant crop varieties and

Figure 22.8 Strip farming, in which different crops are grown side by side. Often, some strips contain legumes which produce nitrogen and reduce the need for chemical fertilizers. Soil is preserved by contouring the strips according to the topography of the land. *(USDA)*

Chrysanthemum flowers being harvested in Rwanda, where they are grown as a source of pyrethrum. *(Robert E. Ford/Terraphotographics)*

biological controls such as natural predators or parasites that control pest populations than on agrichemicals. Farmers can select tillage methods, planting times, crop rotations, and plant-residue management practices to optimize the environment for beneficial insects that control pest species. If pesticides are used as a last resort, they are applied when pests are more vulnerable or when any beneficial species and natural predators are least likely to be harmed.

IPM programs have been most effective for cotton, sorghum, peanuts, and fruit orchards, but less effective for corn and soybeans. Some of the biological controls include release of sterilized insect pests, use of insect pheromones to disrupt mating, release of natural predator pests, and use of natural insecticides.

Natural Insecticides

Many plant species contain natural protection against insects. For example, nicotine protects tobacco plants from sucking insects. However, its commercial use is limited by its high mammalian toxicity (oral LD_{50} for rats is 50 mg/kg).

One of the oldest and best-known natural insecticides is pyrethrum, a contact insecticide obtained from dried chrysanthemum flowers by extraction with hydrocarbon solvents. Pyrethrum is very effective in killing flying insects and is relatively nontoxic toward mammals (oral LD_{50} for rats is 129 mg/kg). Pyrethrum aerosol sprays are excellent home insecticides because of their safety for humans and rapid action on pests. However, pyrethrum lacks persistence against agricultural insects because of its instability to air and light, so it must be mixed with small amounts of other insecticides to kill insects that might recover from sublethal doses of pyrethrum.

The insecticidal properties of pyrethrum result from six esters that are collectively called pyrethrins. After the isolation and identification of these compounds, a number of derivatives have been synthesized that are more effective than the natural pyrethrins. For example, dimethrin is effective against mosquito larva and is safe to use (oral LD_{50} for rats is greater than 10,000 mg/kg). (The basic pyrethin structure is shown in black, and the groups that vary in different pyrethins are shown in yellow.)

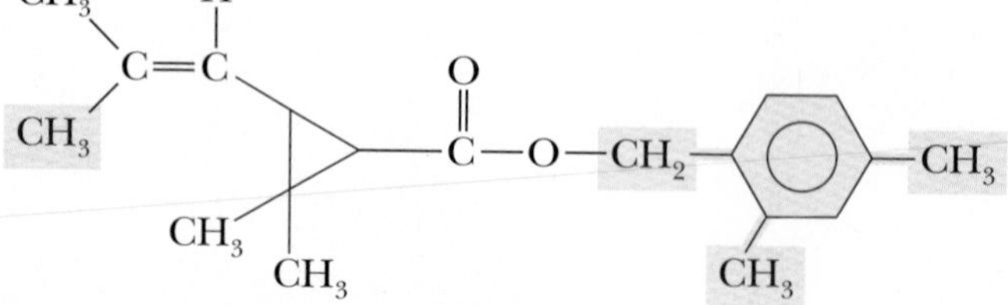

Dimethrin

A class of natural insecticides based on naturally occurring bacterial toxins of *Bacillus thuringiensis (Bt)*, a bacterium found in the soil, have been used widely in the United States in forestry to control gypsy moths and other insects. Since different *Bt* strains are highly selective in killing targeted insects, aerial spraying of *Bt* toxins can be used to control moths or mosquitoes without being hazardous to other beneficial insects or to animals, humans, or plants. The *Bt* toxins also degrade after a few days so they do not contaminate the environment. Two commercial products containing *Bt* are Sharpshooter for weed control in landscape management and recreational turf and DeMoss for controlling moss and algae growth on roofs, buildings, sidewalks, and decks.

However, there is evidence that insects are developing a resistance to *Bt* toxins. Farmers using *Bt* for control of moths in fields of watercress reported a loss of effectiveness. Research has shown that when insects were exposed to high doses that kill 60% to 90% of the insects, their offspring were more resistant to *Bt*. Rotating crops and using *Bt* biopesticides sparingly are recommended ways to slow the development of resistance.

■ Natural insecticides are commonly called *biopesticides.*

Another source of effective natural pesticides is the neem tree that grows widely in Africa and Asia. For centuries, people of India have known of the insect-fighting ability of the neem tree. The oil extracted from neem-tree seeds has been found to be effective against more than 200 species of insects, including locusts, gypsy moths, cockroaches, the California medfly, and aphids. Azadirachtin, an active ingredient of oil from neem-tree seeds, interferes with insect molting, reproduction, and digestion. Tests have shown it is specific to insects without affecting pest predators. For example, use of azadirachtin on an aphid-infested field killed the aphids without harming ladybugs and lacewings, which are aphid predators. Margosan-O is the first commercial neem-based biopesticide, but a large market for neem-based insecticides seems likely in view of their apparently ideal characteristics.

THE WORLD OF CHEMISTRY

Pheromones

Program 10, *Signals from Within*

One important area in insect research is the development of pheromones, the sex attractants used by female insects to entice males. They offer a safe, nontoxic way to lure insects into traps and avoid the use of pesticides. Dr. Meyer Schwartz makes such synthetic insect pheromones in his laboratory.

An important part of Dr. Schwartz's work is confirming that he has made exact copies of the natural pheromones. He uses infrared spectroscopy to verify the structure of the molecules he's tried to duplicate in the lab.

Once an accurate copy is made, it's tried out to see if it works. A miniature wind tunnel lets Agriculture Department scientists see just how tantalizing their creation is. The pheromone is placed at one end of the tunnel, and a love-struck male gypsy moth flies against the wind to get it. It's bad enough that he's going to all this trouble for a synthetic chemical, but the Agriculture Department scientists have another trick up their sleeve. They can control the speed of a striped conveyor belt on the wind tunnel floor. The distracted moth sees the belt moving by and thinks he's flying full tilt toward a female. By carefully adjusting the belt speed, the researchers can stop the moth's forward progress. The speed required to stop the moth gives them a good measure of how strong a sex lure their pheromone is. When he finally is allowed to reach the end, all he finds are synthetic pheromones in a cold steel cage.

A pheromone trap. A male moth is attracted to the trap during pheromone testing. By placing the bait in a wind tunnel with a moving floor, the effort of the moth to reach the "female" can be measured numerically, giving quantitative data that can be used to evaluate the success of the pheromone.

22.7 Agricultural Genetic Engineering

Armed with the ability to insert genes into organisms (see Section 15.9), it follows that we might be able to introduce genes into food plants and animals in order to fight pests, control diseases, and improve the quality of the food being produced. It should also be possible to introduce genes into organisms to cause that organism to produce a particular food or chemical that we desire or to provide a living system in which the effects of a particular gene can be studied. Specific progress has been made in the following areas that relate to agriculture: growing plants that produce their own insecticide, making plants resistant to viral and bacterial infections, matching the chemistry of a plant with a protecting herbicide, and growing animals and plants that produce a better food product.

The number of **transgenic crops** (genetically altered crops) undergoing testing is expanding rapidly. The U.S. Department of Agriculture has authorized tests of more than one thousand genetically engineered fruits, vegetables, and grains, and 486 field tests were in progress in 1994, compared with five in 1987. Figure 22.9 illustrates the series of tests involved. There are concerns that weeds might pick up resistance to herbicides, viruses, and pests from these transgenic crops, but it is unlikely that data based on small test plots will yield any information on hazards associated with transgenic crops. Proponents of transgenic crops feel that safety concerns can be tested in closely monitored large-scale field tests, while opponents want transgenic crops tested for their ability to transfer genes to weeds before large-scale trials are conducted. Data from large-scale field trials already underway in China for transgenic tobacco, tomatoes, and rice may provide answers to the question of transgenic crop safety.

The World of Chemistry

Programs 26, 2, and 25, *Futures, Color, and Genetic Code*

Figure 22.9 The sequence of tests in evaluating a gene-modified plant. (a) After the laboratory work of gene modification, the test plants must be evaluated in (b) greenhouse and (c) field tests to verify that the desired characteristic has been obtained.

Natural breeding methods require up to ten years to produce plants suitable for field testing that show resistance to a particular virus. Genetic engineering requires one year! One of the first examples of a genetically engineered plant was the insertion of DNA segments from the tobacco mosaic virus into the genetic code of tomato plants (see Figure 15.30). The implants caused the tomato plants to be strongly resistant to attack by this virus.

The first genetically engineered food to be sold to consumers is the Flavr Savr tomato developed by Calgene. Ordinary tomatoes are picked green and ripened with ethylene gas in order to avoid excess softening and spoilage before the tomatoes reach the consumer. The Flavr Savr tomato ripens more slowly so it can be picked at a ripened stage, enhancing its flavor and increasing its shelf life. This property was accomplished by inserting a reverse version of a tomato gene for *polygalacturonase* (PG), an enzyme that breaks down cell walls.

The large losses of stored cereal grains, beans, and peas, especially in developing countries where farmers cannot afford to use fungicides, demonstrated a need for genetic engineering of seeds. In 1994, genetic engineering researchers reported the development of the first pest-resistant seeds—a strain of garden pea that resists attack by two weevil species. Techniques learned through this development are expected to pave the way for gene transfer in other legumes.

Experimental tomatoes, genetically engineered for long shelf life. Ordinarily, tomatoes left on the vine until they become this ripe do not last long enough to reach the supermarket. *(© Richard Nowitz/Phototake NYC)*

SELF-TEST 22B

1. Most virgin soils can support crop production for a decade or more before fertilization is needed. (a) True, (b) False.
2. What do the numbers 6-12-8 on a fertilizer mean? The 6 is the percentage of ____________, the 12 is the percentage of ____________, and the 8 is the percentage of ____________ in the fertilizer.
3. Would the nitrogen in ammonia be considered "fixed" nitrogen?
4. Pure ammonia under ordinary conditions is (a) a solid, (b) a liquid, (c) a gas.
5. Approximately what percentage of the food crops of the world is lost to pests each year?
6. The first chlorinated organic insecticide was ____________.
7. Which of the following is a persistent insecticide? (a) DDT, (b) Malathion, (c) Carbaryl, (d) Pirimicarb.
8. ____________ is a natural insecticide extracted from chrysanthemum flowers.
9. Which is more likely to be a hormone, a selective or a nonselective herbicide?
10. The most widely used herbicide today is ____________.
11. Three common classes of pesticides are ____________, ____________, and ____________.

MATCHING SET

____ 1. Legume crops
____ 2. Chlorosis
____ 3. Ammonium nitrate
____ 4. 2,4-D
____ 5. Glyphosate
____ 6. Natural insecticide
____ 7. Pheromones
____ 8. Flavr Savr tomato
____ 9. Fixed nitrogen
____ 10. Carbaryl
____ 11. Selective leaching
____ 12. Limestone
____ 13. Chlorinated hydrocarbon pesticide

a. Most common herbicide
b. *Bacillus thuringiensis*
c. Caused by nutrient deficiency in soil
d. First genetically engineered food on the market
e. Causes basic soil to become acidic
f. Roots contain nitrogen-fixing bacteria
g. Solid added to neutralize acidic soil
h. Nitrogen fertilizer also used as an explosive
i. Anticholinesterase insecticide
j. Herbicide inhibits amino acid synthesis in plants
k. Sex attractants used in integrated pest management
l. DDT
m. Ammonia

QUESTIONS FOR REVIEW AND THOUGHT

1. What factors affect the Earth's carrying capacity?
2. What are the likely consequences if the present rate of human population growth continues?
3. What is an operational definition of "soil"?
4. Give definitions and descriptions for the following.
   (a) Topsoil (b) Subsoil
   (c) Loam (d) Separates
5. What is the structure of a typical soil?
6. What causes soil to be (a) sour? (b) sweet?
7. If crushed limestone is spread on soil, will it raise or lower the pH of the soil? Explain.
8. Give definitions for the following.
   (a) Leaching (b) Selective leaching
9. What are the three nonmineral nutrients obtained from water and air required for normal plant growth?
10. Which is more easily leached from soils, nitrates or phosphates? Why?
11. (a) Which groups of elements are first leached from soils, the Group IA and IIA metals or the Group IIIA and transition metals? (b) What is the effect of this selective leaching on soil pH?
12. (a) What are two important roles of humus in the soil? (b) Do leaves turned into the soil to produce humus raise or lower the soil pH?
13. Give definitions and/or descriptions for the following.
   (a) Nutrients (b) Nonmineral nutrients
   (c) Mineral nutrients
14. What are the three primary mineral plant nutrients that are considered in fertilizer formulations?
15. Which is more likely to be a problem in farming, a soil shortage of N, P, and K, or a shortage of Ca, Mg, and S? Give a reason for your answer.
16. (a) What does the term "nitrogen fixation" mean? (b) Give two natural ways that this process occurs.
17. Write the balanced equation for the Haber process. Why is this process important in the production of fertilizers?
18. What are the three ways water can be held in soil?
19. What are legumes? What is nitrogenase?
20. What are two natural nitrogen-fixation processes? Are either of these controllable? How can this control be exercised?
21. What is chlorosis? What causes it? What are the symptoms of chlorosis?
22. An inorganic fertilizer has a grade of 10-0-0. Is this a complete fertilizer? Explain.
23. Inorganic fertilizers are graded on their content of nitrogen, phosphorus, and potassium. Are any of these actually present as the element? Are any of the reference substances actually in a bag of fertilizer?
24. Phosphate rock is treated with sulfuric acid to make superphosphate. Why isn't the phosphate rock simply used as a phosphorus-containing fertilizer?
25. Urea, NH_2CONH_2, is a common straight fertilizer supplying nitrogen. What happens to urea when it is applied to soil? What nitrogen compound is formed when urea decomposes?
26. What is a straight fertilizer? What is a complete fertilizer?
27. Give definitions or descriptions for the following.

(a) Quick-release fertilizers (b) Slow-release fertilizers

28. Give definitions for the following.
    (a) Pesticide (b) Insecticide
    (c) Herbicide (d) Fungicide
29. Ammonium nitrate, $NH_4NO_3(s)$, is a common straight fertilizer. It decomposes at high temperature to produce N_2, H_2O and O_2. Handling ammonium nitrate has led to accidental explosions. Give one reason for continued use of ammonium nitrate as a fertilizer and one reason why it should be under greater control.
30. What is the reason why the use of DDT is banned in the United States? Why do you suppose that DDT is still used in other parts of the world?
31. Why is DDT fat soluble and not water soluble? What two properties make DDT a problem?
32. Trace the use and fall of the use of DDT in agriculture. Debate whether it has been more good than bad for the human race.
33. What is the approximate percentage of food crops that is lost to pests every year? What is the estimated dollar value of these lost crops?
34. What is the trend in pesticide use? How is the amount of money spent on pesticides in the United States changing? Explain.
35. What are the three structure classes for the most important insecticides?
36. What is a persistent pesticide? Give an example.
37. What is biodegradable pesticide? Give an example.
38. (a) What two herbicides were formulated to produce Agent Orange? (b) Which of these herbicides is currently banned in the United States for agricultural use?
39. (a) Discuss the benefits and possible harms in using pesticides in agriculture. (b) What conclusions can you draw on this controversial issue?
40. In the period after World War II, most farmers fertilized "enough to be sure." Farmers today are likely to have the soil analyzed and have a fertilizer formulated on prescription. (a) What is the cause of this change? (b) How might this change in farming practice affect water pollution?
41. Organic farming saves energy in one area but loses it in another. Explain.
42. Investigate a no-till farming operation. (a) What herbicides were used? (b) How is energy saved and, at the same time, how is additional energy required? (c) What is the effect of no-till farming on the conservation of topsoil?
43. If you grew a garden, would you use chemical fertilizers and pesticides? Why or why not?
44. What is integrated pest management?
45. Two effective natural insecticides are *Bt* toxins and neem-tree seeds. Explain what these are and why they are becoming so popular.
46. (a) What are transgenic crops? (b) Give some examples.
47. What mechanism of action is followed by glyphosate and sulfonylureas? What is the relationship between "essential" amino acids and these herbicides?
48. Why are pheromones interesting compounds for insect control? How are they used?
49. Assume it is possible to use genetic engineering to improve the storage life of a ripe harvested food like oranges. Give two questions you would want answered before you would buy and eat these oranges.
50. What do you think about the connection between population growth and limitations on the amount of cultivable land? Do you see any conflict? Explain.
51. A 1992 United Nations study indicated that 4.84 billion acres of farm land soil were so degraded that it would be impossible to reclaim them. What are the main causes of this soil degradation?
52. What is sustainable agriculture?
53. Transgenic crops are currently being researched and developed. How does this differ from the previous genetic hybrid work on plants and animals? This previous work gave hybrid corn and selectively bred chickens, cows, dogs, horses, etc.

PROBLEMS

1. How many acres of land are estimated to be under cultivation? What is the approximate number of acres cultivated per person? Assume Earth's population is 5.7 billion people.
2. What is the formula for potash? What is the percent composition of potash?
3. An inorganic fertilizer has an analysis of 10-20-20. What is the percent N in this fertilizer? How many pounds of N would be in a 20-pound bag?
4. A fertilizer bag label displays the following grade numbers, 20-10-5. Is this a complete fertilizer? Explain. What is the percent N, the percent P as P_2O_5, and the percent K as K_2O in this fertilizer? How many pounds of N are in a 50-pound bag?
5. Pyrethrum has an LD_{50} of 129 mg/kg for rats. Assuming that this LD_{50} value is valid for humans, what amount of pyrethrum would a 110-pound person need to ingest to create a 50-50 chance of consuming a lethal dose? 1 pound = 0.454 kg.

APPENDIX A

Significant Figures

A temperature of 52.7°C can be read from the thermometer pictured in Figure 2.12 (page 38) of this book by estimating the .7 part of the measurement. This thermometer can only measure a temperature to three **significant figures,** which are the number of digits that can be measured with certainty, plus one more that is estimated. Suppose the temperature of 52.7°C is measured during a chemical experiment and it is decided to try the next experiment at a lower temperature, one third of 52.7°C. Dividing 52.7°C by 3 on a calculator gives an answer of 17.56666667°C. But since 52.7°C has three significant figures, only three digits should be used in the answer to this calculation. The 17.56666667 on the calculator must be rounded off to give a temperature for the new experiment of 17.6°C.

In scientific calculations, it is important to report the results of measurements with only the number of digits that are significant. When these measurements are then used in calculations, the number of digits in the answer must be neither more nor less than are significant. Keeping track of significant figures in calculations requires two skills: (1) counting digits that are significant and (2) applying rules for rounding off the answers to calculations.

Counting Significant Figures

1. **All nonzero digits are significant.**

1 2	1 2 3 4 5
23	15,699
Two significant figures	Five significant figures

2. **Zeroes to the left of the first nonzero digit are not significant**

1 2	1 2
0.23	0.00023
Two significant figures	Two significant figures

3. **Zeroes between nonzero digits are significant**

1 2 3 4	1 2 3
7077	50.2
Four significant figures	Three significant figures

4. Zeroes at the end of a number that includes a decimal point are significant

These zeroes have been added to represent the precision of the measurement.

12 345 50.020	1 23 7.00
Five significant figures	Three significant figures

For a number with zeros at the end and *no* decimal point, it is impossible to know how many of the digits are significant. Does 99,000 have two, three, or four significant figures? To represent the significance of digits in numbers like this, scientists rely on scientific notation, which is discussed in Appendix B.

Counting Significant Figures

Measured Value	Number of Significant Figures
1 3 cm	1
12 34 45.87 mL	4
123 0.223 g	3
1234 4529 m	4
12 0.70 kg	2
12 3 30.6 cm	3
12 34 65.00 m^3	4
1 0.005 mg	1
1 2 3.8×10^3 L	2
1 23 1.05×10^{-10} mm	3

EXAMPLE A.1 *Significant Figures*

How many significant figures are there in each of the following measurements?

(a) 0.007 m (b) 99.00 kg (c) 2.0004°C

SOLUTION

Always start counting with the first nonzero digit. In (a) this is the 7 and there is only one significant figure. In (b), you continue to count past the decimal point—the zeros have been added to show that this measurement has four significant figures. In (c) the zeros in the middle of the number are significant, so this measurement has five significant figures.

(a) 0.007 m (1) (b) 99.00 kg (12 34) (c) 2.0004°C (1 2345)

Exercise A.1

How many significant figures are in each of the following quantities?

(a) 63,332 kg (b) 0.06 mm (c) 407 L (d) 0.0600 sec (e) 0.0101 g

Answers

(a) 5 (b) 1 (c) 3 (d) 3 (e) 3

Calculations with Significant Figures

■ For calculations that have several parts, the best approach is to do the entire calculation first, and then round off the final answer. Rounding off at every separate step can cause errors.

To find the correct number of significant figures for the answer to a calculation requires three steps: (1) Do the arithmetic. (2) Decide how many significant figures the answer can have. (3) Round off the answer.

In **addition and subtraction** an answer should have no more *decimal* places than the number in the calculation with the fewest decimal places. Consider the following addition.

13.672	Three decimal places
4.0	One decimal place
[17.673]	Three decimal places (unrounded answer)

The unrounded answer has two more decimal places than 4.0 and must therefore be rounded off to include only one decimal place as in 4.0. The rules for **rounding off** the answer are simple:

> **If the numbers to be dropped are 5 or greater than 5, then 1 is added to the final digit that is retained.**

> **If the numbers to be dropped are less than five, they are dropped and the final retained digit is unchanged.**

In the addition above, 17.673 must be rounded off to have one decimal place, which means dropping the 73 and increasing the 6 by 1 to give 17.7.

An exception to the need to round off answers occurs with numbers described as *exact.* One source of exact numbers is counting. The number of students in a class or the number of objects in one dozen are exact numbers. Such numbers do not limit the number of digits in the answer to a calculation. Another source of exact numbers is definition. For example, one kilogram is defined as *exactly* equal to 1000 g.

■ Note that in addition and subtraction, rounding off isn't needed if all of the numbers in the calculation are whole numbers.

Rounding-Off Numbers

Original Number*	Number of Significant Figures Desired	Rounded-off Number
1.67	2	1.7
521.1	3	521
4.565	3	4.57
351.53	3	352
1.568	1	2
16.043	3	16.0
8.235×10^3	2	8.2×10^3
7540.916	4	7541

*The first number being dropped is underlined.

EXAMPLE A.2 *Significant Figures in Addition*

What is the total weight of three jars of chemicals weighing 14.57 g, 1835.5 g, and 0.875 g?

SOLUTION

By aligning the numbers on their decimal points for the addition and mentally drawing a vertical line after the number with the smallest number of decimal places, it is easy to see that the answer must be rounded off to one decimal place.

$$\begin{array}{r|l} 14.5 & 7\ \text{g} \\ 1835.5 & \ \ \ \text{g} \\ 0.8 & 75\ \text{g} \\ \hline [1850.9 & 45\ \text{g}] \end{array} \qquad \text{Rounded off answer: 1850.9 g}$$

The total weight of the jars is 1850.9 g.

EXAMPLE A.3 *Significant Figures in Subtraction*

An African Goliath beetle, one of the world's heaviest insects, weighs 99.790 g. What is the difference in weight between this beetle and a hummingbird with a weight of 10.8 g?

SOLUTION

In subtraction as in addition, align the numbers on the decimal point and use a vertical line to find the place to which the answer should be rounded.

$$\begin{array}{r} 99.7|90 \text{ g} \\ -\ 10.6| \ \text{g} \\ \hline [89.1|90 \text{ g}] \end{array} \qquad \text{Rounded off answer: } 89.2 \text{ g}$$

The African Goliath beetle is 89.2 g heavier than a hummingbird.

Exercise A.2

Round off each of the following numbers to three significant figures.

(a) 187.6 (b) 5.281 (c) 1.0334 (d) 0.9265 (e) 30.099

Answers

(a) 188 (b) 5.28 (c) 1.03 (d) 0.927 (e) 30.1

Exercise A.3

Perform the following calculations and give the answers with the proper numbers of significant figures.

(a) $77.2 + 0.531 + 13.27$ (b) $0.815 - 0.2346$
(c) $62 + 49.1$ (d) $1.386 - 1.2$

Answers

(a) 91.0 (b) 0.580 (c) 111 (d) 0.2

Calculations with Significant Figures

Computation	Result
$18.2 + 5.35 + 20.$	44
$15.02 + 0.003 + 700.1$	715.1
$7109.3 - 40.352$	7068.9
$1.521 - 0.81$	0.71
38.2×0.95	36
17.32×1.66	28.8
$182/32.800$	5.55
$0.881/5.2$	0.17
$\dfrac{39.3 + 17.21}{190.}$	0.297
$(58.1 \times 0.82) - 1.19$	46

■ The number of significant figures in the answer is determined by *decimal places* in addition and subtraction, and by total *significant figures* in multiplication and division.

■ To use a percent in a calculation, you must move the decimal point two places to the left:

$$10\% = 0.10 \quad 59.33\% = 0.5933$$
$$230\% = 2.30$$

For **multiplication and division** the answer is limited to the number of digits in the number with the fewest significant figures. When, for example, 75 is multiplied by 0.05 the answer can have only one significant figure (as in 0.05); when 0.084 is divided by 0.00298 the answer can have only two significant figures (as in 0.084).

$$75 \times 0.05 = [3.75] \qquad \text{Rounded off answer: } 4$$

$$\frac{0.084}{0.00298} = [28.18791946] \qquad \text{Rounded off answer: } 28$$

EXAMPLE A.4 ***Significant Figures in Multiplication***

A sample of a metal ore analyzed in the laboratory is found to contain 11.00% of titanium. How much titanium is present in 55 kg of this ore?

SOLUTION

To find the answer requires multiplying 55 kg by 0.1100:

$$0.1100 \times 55 \text{ kg} = [6.05 \text{ kg}] \qquad \text{Rounded off answer: } 6.1 \text{ kg}$$

The answer is limited to two significant figures by 55 kg. The ore contains 6.1 kg of titanium.

EXAMPLE A.5 *Significant Figures in Division*

The Pentagon building in Washington, D.C. has an area of 117,355 square meters, and a football field has an area of 5358 square meters. How many football fields would fit inside the Pentagon?

SOLUTION

To find the answer requires dividing the area of the Pentagon by the area of the football field. On a calculator the answer has 10 digits. It must be rounded off to four digits as in 5358.

$$\frac{117{,}355\ \text{m}^2}{5358\ \text{m}^2} = [21.90276222] \qquad \text{Rounded off answer: } 21.90$$

The Pentagon could hold 21.90, or just about 22 football fields.

Exercise A.5

Perform the following calculations and give the answers with the proper numbers of significant figures.

(a) 68.3×0.92 (b) 3.01×20.225 (c) $1.594/6.23$

(d) $\dfrac{5.0}{25.0}$ (e) $\dfrac{(2.151 - 1.30)}{0.591}$

Answers

(a) 63 (b) 60.9 (c) 0.256 (d) 0.20 (e) 1.40

APPENDIX B

Scientific Notation

Scientific notation, also known as **exponential notation,** is a way of representing large and small numbers as the product of two terms. The first term, the coefficient, is a number between 1 and 10. The second term, the exponential term, is 10 raised to a power—the **exponent.** For example,

Exponent

$$2.5 \times 10^{2}$$

Coefficient

The exponent indicates the number of 10's by which the coefficient is multiplied to give the number represented in scientific notation:

$$2.5 \times 10^{2} = 2.5 \times 10 \times 10 = 2500$$

There are two great advantages to scientific notation. The first, as illustrated in Table B.1, is that the very large and very small numbers often dealt with in the sciences are much less cumbersome in scientific notation. The second is that it removes the ambiguity in the number of significant figures in a number ending with zeroes. We can, for example, write

Two significant figures	2.5×10^{2}
Three significant figures	2.50×10^{2}
Four significant figures	2.500×10^{2}

All of the digits in the coefficient are always considered significant.

For numbers larger than 1, the exponent in scientific notation is a positive whole number, as illustrated above. For numbers less than 1, the exponent is a negative whole number that indicates the number of times the coefficient

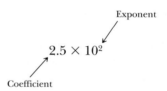

TABLE B.1 ■ Large and Small Numbers in Scientific Notation

Distance from the earth to the sun	150,000,000,000 m	1.5×10^{11} m
Diameter of the Earth	13,000,000 m	1.3×10^{7} m
Height of Mt. Everest	10,000 m	1.0×10^{4} m
Height of Empire State Building	400 m	4.0×10^{2} m
Human height	1.7 m	1.7×10^{0} m
Length of a cockroach	0.040 m	4.0×10^{-2} m
Size of a grain of sand	0.00010 m	1.0×10^{-4} m
Size of a red blood cell	0.0000065 m	6.5×10^{-6} m
Size of a polio virus	0.000000025 m	2.5×10^{-8} m

must be divided by 10 (or multiplied by 0.1) to give the number represented in scientific notation:

$$1.2 \times 10^{-2} = 1.2 \times \frac{1}{10} \times \frac{1}{10} = 0.012$$

■ Reading numbers in scientific notation:

2×10^{4} "Two times ten to the fourth power"

2×10^{-4} "Two times ten to the minus fourth power"

There are two points to remember in converting a number into or out of scientific notation: (1) The value of the exponent is the number of places by which the decimal is shifted. (2) The coefficient should have only one digit before the decimal point. Look through the examples below and count the number of places the decimal points have been moved.

$$10000 = 1 \times 10^{4} \qquad 12345 = 1.2345 \times 10^{4}$$
$$1000 = 1 \times 10^{3} \qquad 1234 = 1.234 \times 10^{3}$$
$$100 = 1 \times 10^{2} \qquad 123 = 1.23 \times 10^{2}$$
$$10 = 1 \times 10^{1} \qquad 12 = 1.2 \times 10^{1}$$
$$1 = 1 \times 10^{0} \qquad \text{(Any number} \times 10^{0} = \text{the number itself.)}$$
$$1/10 = 1 \times 10^{-1} \qquad 0.12 = 1.2 \times 10^{-1}$$
$$1/100 = 1 \times 10^{-2} \qquad 0.012 = 1.2 \times 10^{-2}$$
$$1/1000 = 1 \times 10^{-3} \qquad 0.0012 = 1.2 \times 10^{-3}$$
$$1/10000 = 1 \times 10^{-4} \qquad 0.00012 = 1.2 \times 10^{-4}$$

Some electronic calculators allow you to easily convert numbers to scientific notation. If you have such a calculator, you can change a number shown in the usual form to scientific notation simply by pressing the EE or EXP key and then the "=" key (or in some cases the "=" key and then the exponent key).

EXAMPLE B.1 *Converting Into and Out Of Scientific Notation*

(a) The sun is about 93 million miles from Earth. Express this number in scientific notation. (b) A blue whale has a weight of 1.36×10^{5} kg. Convert this number from scientific notation into a conventional number.

SOLUTION

(a) Writing out the zeros in 93 million gives

93,000,000 miles

To express this distance in scientific notation requires moving the decimal point 7 places to the left. Because the number is larger than 1, the exponent is positive.

$$93{,}000{,}000 \text{ miles} = 9.3 \times 10^7 \text{ miles}$$

7 6 5 4 3 2 1

(b) To go the other way, the positive exponent indicates that the decimal point must be moved to the right, in this case by 5 places.

$$1.36 \times 10^5 \text{ kg} = 136{,}000 \text{ kg}$$

1 2 3 4 5

EXAMPLE B.2 *Converting Into and Out Of Scientific Notation*

(a) Convert the length of a virus from 0.00000002 m to scientific notation. (b) Convert the diameter of a protein molecule from 6.5×10^{-6} m to a decimal number.

SOLUTION

(a) For a number less than 1, the exponent is the negative of the number of places the decimal is moved to the *right*:

$$0.00000002 \text{ m} = 2 \times 10^{-8} \text{ m}$$

1 2 3 4 5 6 7 8

(b) The negative exponent shows that the decimal point must be moved to the left, by 6 places in this case, to convert out of scientific notation.

$$6.5 \times 10^{-6} \text{ m} = 0.0000065 \text{ m}$$

6 5 4 3 2 1

Exercise B.1

Write the numbers below in scientific notation.
(a) 3785 (b) 0.093 (c) 19.2
(d) 552,000 (e) 0.00075

Answers

(a) 3.785×10^3 (b) 9.3×10^{-2} (c) 1.92×10^1
(d) 5.52×10^5 (e) 7.5×10^{-4}

Exercise B.2

Write the numbers below in conventional form.

(a) 4.42×10^2
(b) 8.9×10^{-3}
(c) 3.01×10^{-1}
(d) 6.9×10^4
(e) 7.7×10^0

Answers

(a) 442 (b) 0.0089 (c) 0.301 (d) 69,000 (e) 7.7

Calculations with Numbers in Scientific Notation

For **addition and subtraction** of numbers in scientific notation, the numbers must first be converted to the same powers of 10. The coefficients are then added or subtracted as usual and the exponential term remains the same in the sum or difference.

$$\begin{aligned}(1.234 \times 10^{-3}) + (5.623 \times 10^{-2}) &= (0.1234 \times 10^{-2}) + (5.623 \times 10^{-2}) \\ &= 5.746 \times 10^{-2}\end{aligned}$$

$$\begin{aligned}(6.52 \times 10^{2}) - (1.56 \times 10^{3}) &= (6.52 \times 10^{2}) - (15.6 \times 10^{2}) \\ &= -9.1 \times 10^{2}\end{aligned}$$

For **multiplication** of numbers in scientific notation, the coefficients are multiplied in the usual manner and the exponents are added. If both exponents have the same sign, their values are added and the exponent in the answer has the same sign. If the exponents have different signs, they must be added algebraically, taking into account the + and − signs. The final answer is given with one nonzero digit to the left of the decimal and the correct number of significant figures.

■ To algebraically add a number with a + sign to a number with a − sign, drop the signs, subtract the smaller number from the larger number, and give the answer the sign of the larger number.

$$\begin{aligned}(1.23 \times 10^{3})(7.60 \times 10^{2}) &= (1.23)(7.60) \times 10^{3+2} \\ &= 9.35 \times 10^{5}\end{aligned}$$

$$\begin{aligned}(6.02 \times 10^{23})(2.32 \times 10^{-2}) &= (6.02)(2.32) \times 10^{23-2} \\ &= 13.966 \times 10^{21} \\ &= 1.40 \times 10^{22} \text{ (answer in 3 significant figures)}\end{aligned}$$

$$\begin{aligned}(5.2 \times 10^{-4})(6.1 \times 10^{2}) &= (5.2)(6.1) \times 10^{-4+2} \\ &= 31.72 \times 10^{-2} \\ &= 3.2 \times 10^{-1} \text{ (2 significant figures)}\end{aligned}$$

For **division** of numbers in scientific notation, the coefficients are divided in the usual fashion. The exponent of the divisor is subtracted algebraically from the exponent of the number that is being divided.

■ To algebraically subtract numbers when they have different signs, reverse the sign of the number that is being subtracted, then add the numbers algebraically.

$$\frac{7.60 \times 10^{3}}{1.23 \times 10^{2}} = \frac{7.60}{1.23} \times 10^{3-2} = 6.18 \times 10^{1}$$

$$\frac{6.02 \times 10^{23}}{9.10 \times 10^{-2}} = \frac{6.02}{9.10} \times 10^{(23)-(-2)} = 0.662 \times 10^{25} = 6.62 \times 10^{24}$$

EXAMPLE B.3 *Calculations with Numbers in Scientific Notation*

The distance around the globe from New York to London is 5.6×10^3 km and the distance from London to New Zealand is 1.9×10^4 km. What is the sum of these distances? What is the difference between these distances?

SOLUTION

$$\begin{aligned}(1.9 \times 10^4 \text{ km}) + (5.6 \times 10^3 \text{ km}) &= (1.9 \times 10^4 \text{ km}) + (0.56 \times 10^4)\\ &= 2.5 \times 10^4 \text{ km}\end{aligned}$$

$$\begin{aligned}(1.9 \times 10^4 \text{ km}) - (5.6 \times 10^3 \text{ km}) &= (1.9 \times 10^4 \text{ km}) - (0.56 \times 10^4)\\ &= 1.3 \times 10^4 \text{ km}\end{aligned}$$

It is 2.5×10^4 km from New York to New Zealand via London, and it is 1.3×10^3 km further from London to New Zealand than it is from New York to London.

EXAMPLE B.4 *Calculations with Numbers in Scientific Notation*

To compare their sizes, divide the size of a grain of sand, 1.0×10^{-4} m, by the size of a red blood cell, 6.5×10^{-6}.

SOLUTION

$$\begin{aligned}\frac{1.0 \times 10^{-4} \text{ m}}{6.5 \times 10^{-6} \text{ m}} &= \frac{1.0}{6.5} \times 10^{-4-(-6)}\\ &= 0.15 \times 10^2\\ &= 1.5 \times 10^1 \text{ or } 15\end{aligned}$$

A grain of sand is 15 times larger than a red blood cell.

EXAMPLE B.5 *Calculations with Numbers in Scientific Notation*

In one mole of oxygen gas there are 6.02×10^{23} O_2 molecules. How many O_2 molecules are there in 1.4×10^3 moles of oxygen gas?

$$\begin{aligned}(1.4 \times 10^3 \text{ mol } O_2) \times \left(\frac{6.02 \times 10^{23} \text{ } O_2 \text{ molecules}}{\text{mol } O_2}\right)\\ &= (1.4 \times 6.02) \times 10^{3+23} \text{ molecules}\\ &= 8.4 \times 10^{26} \text{ } O_2 \text{ molecules}\end{aligned}$$

Exercise B.3

Perform the following calculations.

(a) $(8.91 \times 10^{-2}) \times (1.2 \times 10^{5})$

(b) $(1.15 \times 10^{8}) + (6.2 \times 10^{7})$

(c) $\dfrac{5.09 \times 10^{5}}{8.2 \times 10^{-2}}$

(d) $(9.82 \times 10^{4}) - (1.35 \times 10^{3})$

(e) $\dfrac{(2.731 \times 10^{-2}) \times (1.52 \times 10^{9})}{8.3 \times 10^{14}}$

Answers

(a) 1.1×10^{4} (b) 1.77×10^{8} (c) 6.2×10^{6} (d) 9.69×10^{4} (e) 5.0×10^{-8}

APPENDIX

C

Units of Measure, Unit Conversion, and Problem Solving

Units of the SI System

The metric system was begun by the French National Assembly in 1790 and has undergone many modifications. The International System of Units or *Système International* (SI), which represents an extension of the metric system, was adopted by the 11th General Conference of Weights and Measures in 1960. It is constructed from seven base units, each of which represents a particular physical quantity (Table C.1).

The first five units listed in Table C.1 are particularly useful in chemistry. They are defined as follows:

1. The *meter* is the length of the path travelled by light in vacuum during a time interval of 1/299792458 of a second.
2. The *kilogram* represents the mass of a platinum-iridium block kept at the International Bureau of Weights and Measures of Sèvres, France.
3. The *second* was redefined in 1967 as the duration of 9,192,631,770 periods of a certain line in the microwave spectrum of cesium-133.
4. The *kelvin* is 1/273.16 of the temperature interval between absolute zero and the triple point of water (the temperature at which liquid water, ice, and water vapor coexist).
5. The *mole* is the amount of substance that contains as many entities as there are atoms in exactly 0.012 kg of carbon-12 (12 g of ^{12}C atoms).

Decimal fractions and multiples of metric and SI units are designated by using the ***prefixes*** listed in Table C.2. The prefix *kilo*, for example, means a unit is multiplied by 10^3

$$1 \text{ kilogram} = 1 \times 10^3 \text{ grams} = 1000 \text{ grams}$$

and the prefix *centi* means the unit is multiplied by the factor 10^{-2}

$$1 \text{ centigram} = 1 \times 10^{-2} \text{ gram} = 0.01 \text{ gram}$$

TABLE C.1 ■ SI Fundamental Units

Physical Quantity	Name of Unit	Symbol
Length	Meter	m
Mass	Kilogram	kg
Time	Second	s
Temperature	Kelvin	K
Amount of substance	Mole	mol
Electric current	Ampere	A
Luminous intensity	Candela	cd

The prefixes are added to give units of a magnitude appropriate to what is being measured. The distance from New York to London (5.6×10^3 km = 5,600 km) is much easier to comprehend measured in kilometers than in meters (5.6×10^6 m = 5,600,000 m). The following is a list of units for measuring very small and very large distances:

nanometer (nm)	0.000000001 meter
micrometer (μm)	0.000001 meter
millimeter (mm)	0.001 meter
centimeter (cm)	0.01 meter
decimeter (dm)	0.1 meter
meter (m)	1 meter
dekameter (dam)	10 meters
hectometer (hm)	100 meters
kilometer (km)	1,000 meters
megameter (Mm)	1,000,000 meters

In the International System of Units, all physical quantities are represented by appropriate combinations of the base units listed in Table C.1. The result is a derived unit for each kind of measured quantity. The most common derived units are listed in Table C.3. It is easy to see that the derived unit for area is length × length = meter × meter = square meter, m^2, or that the de-

TABLE C.2 ■ Prefixes for Metric and SI Units

Factor	Prefix	Symbol	Factor	Prefix	Symbol
10^{12}	tera	T	10^{-1}	*deci*	d
10^{9}	giga	G	10^{-2}	*centi*	c
10^{6}	mega	M	10^{-3}	*milli*	m
10^{3}	*kilo*	k	10^{-6}	micro	μ
10^{2}	hecto	h	10^{-9}	*nano*	n
10^{1}	deka	da	10^{-12}	*pico*	p
			10^{-15}	femto	f
			10^{-18}	atto	a

*The most common prefixes are shown in italics.

TABLE C.3 ■ Derived SI Units

Physical Quantity	Name of Unit	Symbol	Definition	Symbol
Area	Square meter	m^2	—	
Volume	Cubic meter	m^3	—	
Density	Kilogram per cubic meter	kg/m^3	—	
Force	Newton	N	(kilogram) (meter)/(second)^2	$kg\ m/s^2$
Pressure	Pascal	Pa	Newton/square meter	N/m^2
Energy	Joule	J	(kilogram) (square meter)/(second)^2	$kg\ m^2/s^2$
Electric charge	Coulomb	C	(ampere) (second)	A s
Electric potential difference	Volt	V	joule/(ampere) (second)	J/(A s)

rived unit for volume is length × length × length = meter × meter × meter = cubic meter, m^3. The more complex derivations are arrived at by the same kind of combination of units. Units such as the one for *force* have been given simple names that represent the units by which they are defined.

Conversion of Units for Physical Quantities

The result of a measurement is a **physical quantity,** which consists of a number plus a unit. To convert a physical quantity from one unit of measure to another requires a conversion factor based on equivalences between units of measure like those given in Table C.4. Each equivalence provides two conversion factors that are the inverse of each other. For example, the equivalence between a quart and a liter, 1 quart = 0.9463 liter, gives

$$\frac{1 \text{ quart}}{0.9463 \text{ liter}} \quad \text{There is 1 quart per 0.9463 liter.}$$

$$\frac{0.9463 \text{ liter}}{1 \text{ quart}} \quad \text{There is 0.9463 liter per 1 quart.}$$

The method of cancelling units described in Section 2.8 provides the basis for choosing which conversion factor is needed: It is always the one that allows the unit being converted to be canceled and leaves the new unit uncanceled.

To convert 2 quarts to liters:

$$2 \text{ quarts} \times \frac{0.9463 \text{ liter}}{1 \text{ quart}} = 1.893 \text{ liters}$$

To convert 2 liters to quarts:

$$2 \text{ liters} \times \frac{1 \text{ quart}}{0.9463 \text{ liter}} = 2.113 \text{ quarts}$$

TABLE C.4 ■ Common Units of Measure

Mass and Weight

1 pound = 453.59 grams = 0.45359 kilogram
1 kilogram = 1000 grams = 2.205 pounds
1 gram = 10 decigrams = 100 centigrams = 1000 milligrams
1 gram = 6.022×10^{23} atomic mass units
1 atomic mass unit = 1.6605×10^{-24} gram
1 short ton = 2000 pounds = 907.2 kilograms
1 long ton = 2240 pounds
1 metric tonne = 1000 kilograms = 2205 pounds

Length

1 inch = 2.54 centimeters (exactly)
1 mile = 5280 feet = 1.609 kilometers
1 yard = 36 inches = 0.9144 meter
1 meter = 100 centimeters = 39.37 inches = 3.281 feet = 1.094 yards
1 kilometer = 1000 meters = 1094 yards = 0.6215 mile
1 Ångstrom = 1×10^{-8} centimeter = 0.1 nanometer = 100 picometers
= 1×10^{-10} meter = 3.937×10^{-9} inch

Volume

1 quart = 0.9463 liter
1 liter = 1.0567 quarts
1 liter = 1 cubic decimeter = 1000 cubic centimeters = 0.001 cubic meter
1 milliliter = 1 cubic centimeter = 0.001 liter = 1.056×10^{-3} quart
1 cubic foot = 28.316 liters = 29.924 quarts = 7.481 gallons

Force and Pressure

1 atmosphere = 760 millimeters of mercury (exactly) = 1.013×10^5 pascals
= 14.70 pounds per square inch
1 bar = 10^5 pascals
1 torr = 1 millimeter of mercury
1 pascal = $1\ kg/m\ s^2 = 1\ N/m^2$

Energy

1 joule = 1×10^7 ergs
1 calorie = 4.184 joules (exactly) = 4.184×10^7 ergs
= 4.129×10^{-2} liter-atmospheres
= 2.612×10^{19} electron volts
1 Calorie (food) = 1 kilocalorie
1 erg = 1×10^{-7} joule = 2.3901×10^{-8} calorie
1 electron volt = 1.6022×10^{-19} joule = 1.6022×10^{-12} erg
1 liter-atmosphere = 24.217 calories = 101.32 joules = 1.0132×10^9 ergs
1 British thermal unit = 1055.06 joules = 1.05506×10^{10} ergs = 252.2 calories

EXAMPLE C.1 *Unit Conversion*

Find the weight in (a) kilograms, (b) grams, and (c) decigrams of a person who weighs 115 pounds.

SOLUTION

(a) Table C.4 gives the equivalence between pounds and kilograms in two ways: 1 pound = 0.454 kilogram (rounded off), or 1 kilogram = 2.205 pounds. Either equivalence can be used to convert 115 pounds to kilograms by setting up the problem so that *pounds* cancels out:

$$115\ \cancel{\text{pounds}} \times \frac{0.454\ \text{kilogram}}{1\ \cancel{\text{pound}}} = 52.2\ \text{kg}$$

$$115\ \cancel{\text{pounds}} \times \frac{1\ \text{kilogram}}{2.205\ \cancel{\text{pounds}}} = 52.2\ \text{kg}$$

(b) Using the equivalence 1 kilogram = 1000 grams gives

$$52.2\ \cancel{\text{kilograms}} \times \frac{1000\ \text{grams}}{1\ \cancel{\text{kilogram}}} = 52{,}200\ \text{grams or}\ 5.22 \times 10^4\ \text{grams}$$

(c) Using the equivalence 1 gram = 10 decigrams gives

$$5.22 \times 10^4\ \cancel{\text{grams}} \times \frac{10\ \text{decigrams}}{1\ \cancel{\text{gram}}} = 52.2 \times 10^4\ \text{decigrams or}\ 5.22 \times 10^5\ \text{decigrams}$$

EXAMPLE C.2 *Unit Conversion*

What is the volume of a 5.00×10^2 milliliter flask in (a) liters, (b) quarts, and (c) ounces (1 quart = 32 ounces, exactly)?

(a)

$$5.00 \times 10^2\ \cancel{\text{milliliters}} \times \frac{1\ \text{liter}}{1000\ \cancel{\text{milliliters}}} = 0.500\ \text{liter}$$

(b)

$$0.500\ \cancel{\text{liter}} \times \frac{1.0567\ \text{quart}}{1\ \cancel{\text{liter}}} \times 0.528\ \text{quart}$$

(c)

$$0.528\ \cancel{\text{quart}} \times \frac{32\ \text{ounces}}{1\ \cancel{\text{quart}}} = 16.9\ \text{ounces}$$

Exercise C.1

Use the information in Table C.3 to write the equivalences and two conversion factors for each of the following pairs of units:

(a) grams and milligrams (b) kilometers and miles (c) calories and joules.

Answers

(a) 1 gram = 1000 milligrams

$$\frac{1 \text{ gram}}{1000 \text{ milligrams}} \qquad \frac{1000 \text{ milligrams}}{1 \text{ gram}}$$

(b) 1 kilometer = 0.6215 mile

$$\frac{1 \text{ kilometer}}{0.6215 \text{ mile}} \qquad \frac{0.6215 \text{ mile}}{1 \text{ kilometer}}$$

(c) 1 calorie = 4.184 joules

$$\frac{1 \text{ calorie}}{4.184 \text{ joules}} \qquad \frac{4.184 \text{ joules}}{1 \text{ calorie}}$$

Exercise C.2

Carry out the following unit conversions:

(a) 15 milligrams to grams (b) 453 grams to milligrams (c) 95 kilometers to miles (d) 5 miles to kilometers (e) 325 calories to joules (f) 325 joules to calories.

Answers

(a) 0.015 g (b) 4.53×10^5 milligrams (c) 59 miles (d) 8 kilometers (e) 1360 joules (f) 77.7 calories

Sometimes it is convenient or necessary to use more than one conversion factor to convert a physical quantity into some other unit. Instead of doing one conversion at a time, a string of conversion factors can be put together and unit cancellation checked before the calculation is done. For example, atmospheric pressure is usually given in the weather report in inches of mercury. What is the atmospheric pressure in millimeters of mercury (a customary unit in scientific measurement) on a day when the weatherman reports it to be 29.2 inches of mercury? Since we have Table C.4 handy, it will be most convenient to use conversion factors based on that table. There isn't an equivalence there for millimeters and inches, but the table does give 1 inch = 2.54 centimeters. Since we know from the prefixes that 100 *centi*meters = 1 meter and 1000 *milli*meters = 1 meter, we can apply these equivalences to go from centimeters to millimeters. Combining the factors we know and checking cancellation of units gives

$$29.2 \text{ \sout{inches}} \times \frac{2.54 \text{ \sout{centimeters}}}{1 \text{ \sout{inch}}} \times \frac{1 \text{ \sout{meter}}}{100 \text{ \sout{centimeters}}} \times \frac{1000 \text{ millimeters}}{1 \text{ \sout{meter}}}$$

The cancellations leave only the unit we're looking for—millimeters—so the setup is correct and the answer is

$$29.2 \text{ \sout{inches}} \times \frac{2.54 \text{ \sout{centimeters}}}{1 \text{ \sout{inch}}} \times \frac{1 \text{ \sout{meter}}}{100 \text{ \sout{centimeters}}} \times \frac{1000 \text{ millimeters}}{1 \text{ \sout{meter}}} = 742 \text{ millimeters}$$

Exercise C.3

(a) Is the following setup correct for converting milligrams to pounds?

$$225 \text{ milligrams} \times \frac{1 \text{ gram}}{1000 \text{ milligrams}} \times \frac{1 \text{ pound}}{454 \text{ grams}}$$

(b) Is the following setup correct for converting yards to kilometers?

$$25 \text{ yards} \times \frac{36 \text{ inches}}{1 \text{ yard}} \times \frac{1 \text{ yard}}{3 \text{ feet}} \times \frac{5280 \text{ feet}}{1 \text{ mile}} \times \frac{1 \text{ mile}}{1.609 \text{ kilometers}}$$

Answers

(a) Yes (b) No

Exercise C.4

Carry out the following calculations by using multiple conversion factors. (1 quart = 2 pints; 1 pound = 16 ounces)

(a) What is the volume in milliliters of 1.00 pint of milk?
(b) What is the mass in grams of 15 ounces of cereal?
(c) What is the distance in centimeters of the very small distance of 30.5 picometers measured by using X rays?
(d) What is the volume in cubic meters of a 250-gallon tank?

Answers

(a) 473 milliliters (b) 426 grams (c) 3.05×10^{-9} centimeters
(d) 0.95 cubic meter

Problem Solving by Cancellation of Units

Many kinds of numerical problems in everyday life as well as in science can be solved by extending the use of conversion factors and cancellation of units beyond unit conversion. The use of density expressed as mass per unit volume is a simple example of such an extension. Density provides the *connection* between mass and volume. Given that the density of lead is 11.4 g/cm^3, you can find the mass in grams of a piece of lead of known volume or the volume of a piece of lead of known mass. If the known information is that a piece of lead has a volume of 25.0 cm^3 and the unknown information is its mass, the problem is set up and solved as follows:

$$25.0 \ \cancel{\text{cm}^3} \times \frac{11.4 \text{ g lead}}{1 \ \cancel{\text{cm}^3}} = 285 \text{ g lead}$$

As an example of an everyday problem, consider assigning "units" to the number of people served by a gallon of ice cream. Suppose you know that one gallon of ice cream will serve 16 people—16 servings per gallon. To calculate an unknown (the number of gallons needed to serve 50 people) setting up and solving the problem gives

$$50 \ \cancel{\text{servings}} \times \frac{1 \text{ gallon of ice cream}}{16 \ \cancel{\text{servings}}} = 3.1 \text{ gallons of ice cream}$$

Three gallons should do the job. Applying the method again to calculate the cost, if the ice cream is \$4.35 a gallon,

$$3 \text{ gallons} \times \frac{\$4.35}{1 \text{ gallon}} = \$13.05$$

Notice how the "conversion factors" in the two ice cream calculations are arranged so that the unwanted "units" cancel. That is the secret of what is sometimes called the *dimensional method* of problem solving.

For a more complex example, suppose a manufacturer must figure out how many people are needed to assemble 6000 Snivies each week (6000 Snivies/week). He knows that it takes a person 1 hour to assemble 8 Snivies (8 Snivies/hour) and that each person will put in 30 hours a week on the assembly line (30 hours/person). How many people must be working on the assembly line each week? Using the dimensional method, the "conversion factors" must be arranged so that the "units" cancel to give an answer of "persons per week." In single setup, the calculation can be done as follows:

$$\frac{6000 \text{ Snivies}}{1 \text{ week}} \times \frac{1 \text{ hour}}{8 \text{ Snivies}} \times \frac{1 \text{ person}}{30 \text{ hours}} = 25 \text{ persons/week}$$

The manufacturer will need 25 persons on the assembly line each week.

In Section 8.6, we illustrated dimensional problem solving for calculating the numbers of moles and the masses of reactants and products in chemical reactions. The coefficients in chemical equations and the molar masses of reactants and products provide the factors that must be combined to solve problems of this kind. Consider the equation for the reaction that occurs when copper(II) sulfide is heated in oxygen:

$$Cu_2S(s) + O_2(g) \longrightarrow 2\,Cu(s) + SO_2(g)$$

The molar masses of the reactants and products are

$$159 \text{ g } Cu_2S/1 \text{ mol } Cu_2S \qquad 32 \text{ g } O_2/1 \text{ mol } O_2$$
$$63.5 \text{ g Cu}/1 \text{ mol Cu} \qquad 64 \text{ g } SO_2/1 \text{ mol } SO_2$$

How much copper can be made from 2.50 mol of Cu_2S by this reaction? The connection between Cu_2S and Cu is given by the mole ratio 1 mol Cu_2S/2 mol Cu. This ratio and the molar masses must be combined so that the answer has the units of grams of copper:

$$2.50 \text{ mol } Cu_2S \times \frac{2 \text{ mol Cu}}{1 \text{ mol } Cu_2S} \times \frac{63.5 \text{ g Cu}}{1 \text{ mol Cu}} = 318 \text{ g Cu}$$

The mass of copper that can be produced by roasting 2.50 mol of copper sulfide is 318 g.

EXAMPLE C.4 *Problem Solving*

The Snivie manufacturer has one truck available to deliver his Snivies. The Snivies are packed 50 in a carton and the truck can carry 10 cartons per trip.

How many trips will the truck have to make to deliver the 6000 Snivies produced each week?

SOLUTION

Expressing the known information as "conversion factors" gives 50 Snivies/carton and 10 cartons/trip. The information must be set up to yield the answer in trips per week.

$$\frac{6000 \text{ Snivies}}{1 \text{ week}} \times \frac{1 \text{ carton}}{50 \text{ Snivies}} \times \frac{1 \text{ trip}}{10 \text{ cartons}} = 12 \text{ trips/week}$$

EXAMPLE C.5 *Problem Solving*

Using the information given in the text, find the number of moles of oxygen that will be consumed in the conversion of 683 g of copper sulfide to copper metal.

SOLUTION

The connection here is supplied by the mole ratio, 1 mol O_2/1 mol Cu_2S, and the molar mass of copper sulfide.

$$683 \text{ g } Cu_2S \times \frac{1 \text{ mol } Cu_2S}{159 \text{ g } Cu_2S} \times \frac{1 \text{ mol } O_2}{1 \text{ mol } Cu_2S} = 4.30 \text{ mol } O_2$$

Exercise C.5

(a) If the manufacturer described in the text decides to increase his production to 20,000 Snivies per week, how many persons will he need working on the assembly line each week? (b) If the most trips a truck can make in a week is 20, how many trucks will the manufacturer need to deliver 20,000 Snivies per week?

Answers

(a) 83 persons per week (b) 2 trucks

Exercise C.6

Use the information given in the text about the reaction of copper sulfide with oxygen to answer the following questions:

(a) How many grams of copper can be made from 683 g of copper sulfide?

(b) How many moles of sulfur dioxide will be released during the production of the copper from 683 g of copper sulfide?

Answers

(a) 546 g Cu (b) 4.30 mol SO_2

APPENDIX D

Naming Organic Compounds

Chapters 12 and 14 include both common names and systematic names for organic compounds representing the various classes of hydrocarbons and functional groups. This appendix focuses on the systematic nomenclature for organic compounds, as proposed by the International Union of Pure and Applied Chemistry (IUPAC).

Hydrocarbons

The name of each of the members of the hydrocarbon classes has two parts. The first part, the prefix—*meth-*, *eth-*, *prop-*, *but-*, and so on—reflects the number of carbon atoms. When more than four carbons are present, the Greek or Latin number prefixes are used: *pent-*, *hex-*, *hept-*, *oct-*, *non-*, and *dec-*. The second part of the name, or the suffix, tells the class of hydrocarbon. Alkanes have carbon-carbon single bonds, alkenes have carbon-carbon double bonds, and alkynes have carbon-carbon triple bonds and are indicated by the suffixes *-ane*, *-ene*, and *-yne*, respectively.

Unbranched Alkanes and Alkyl Groups

The names of the first ten unbranched (straight-chain) alkanes are given in Table 12.3. Alkyl groups are named by dropping "-ane" from the parent alkane and adding "-yl." See Table 12.4.

Branched-Chain Alkanes

The rules for naming branched-chain alkanes are as follows:

1. *Find the longest continuous chain of carbon atoms: this chain determines the parent name for the compound.* For example, the following compound has two methyl groups attached to a *heptane* parent.

$$\begin{array}{l}CH_3CH_2CH_2\underset{\displaystyle CH_3}{\underset{|}{C}}HCH_2\underset{\displaystyle CH_3}{\underset{|}{C}}HCH_3\end{array}$$

The longest continuous chain may not be obvious from the way the formula is written, especially for the straight-line format that is commonly used. For example, the longest continuous chain of carbon atoms in the following chain is *eight*, not *four* or *six*.

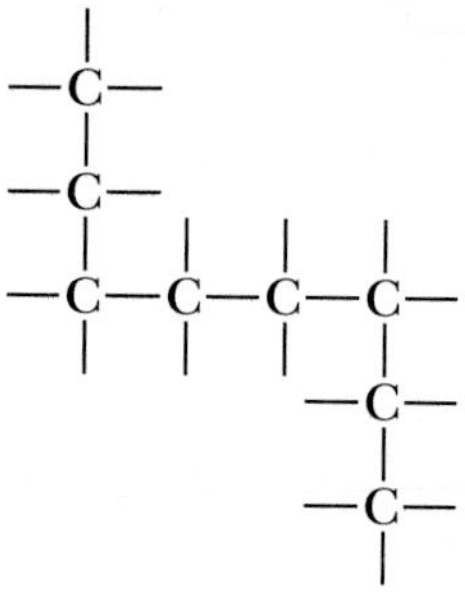

$$
\begin{array}{c}
-\!\!\stackrel{|}{\underset{|}{C}}\!\!- \\[2pt]
-\!\!\stackrel{|}{\underset{|}{C}}\!\!- \\[2pt]
-\!\!\stackrel{|}{\underset{|}{C}}\!\!-\!\!\stackrel{|}{\underset{|}{C}}\!\!-\!\!\stackrel{|}{\underset{|}{C}}\!\!-\!\!\stackrel{|}{\underset{|}{C}}\!\!- \\[2pt]
-\!\!\stackrel{|}{\underset{|}{C}}\!\!- \\[2pt]
-\!\!\stackrel{|}{\underset{|}{C}}\!\!-
\end{array}
\qquad \text{is equivalent to}
$$

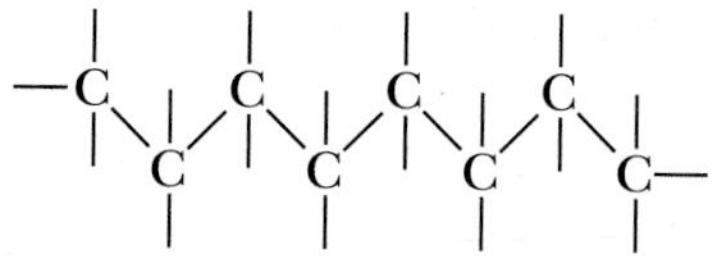

2. *Number the longest chain beginning with the end of the chain nearest the branching. Use these numbers to designate the location of the attached group. When two or more groups are attached to the parent, give each group a number corresponding to its location on the parent chain.* For example, the name of

$$
\overset{7}{C}H_3\overset{6}{C}H_2\overset{5}{C}H_2\overset{4}{C}H\overset{3}{C}H_2\overset{2}{C}H\overset{1}{C}H_3
$$
$$
\qquad\qquad\quad |\qquad\quad |
$$
$$
\qquad\qquad CH_3\quad CH_3
$$

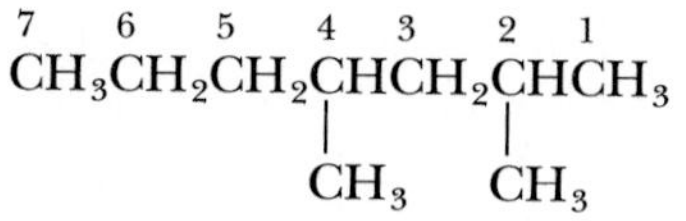

is 2,4-dimethylheptane. The name of the compound below is 3-methylheptane, not 5-methylheptane or 2-ethylhexane.

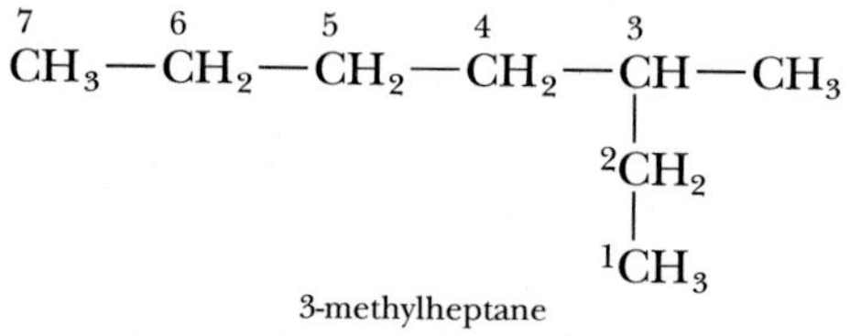

$$
\overset{7}{C}H_3-\overset{6}{C}H_2-\overset{5}{C}H_2-\overset{4}{C}H_2-\overset{3}{C}H-CH_3
$$
$$
\qquad\qquad\qquad\qquad\qquad |
$$
$$
\qquad\qquad\qquad\qquad\quad {}^{2}CH_2
$$
$$
\qquad\qquad\qquad\qquad\qquad |
$$
$$
\qquad\qquad\qquad\qquad\quad {}^{1}CH_3
$$
3-methylheptane

3. *When two or more substituents are identical, indicate this by the use of the prefixes di-, tri-, tetra-, and so on. Positional numbers of the substituents should have the smallest possible sum.*

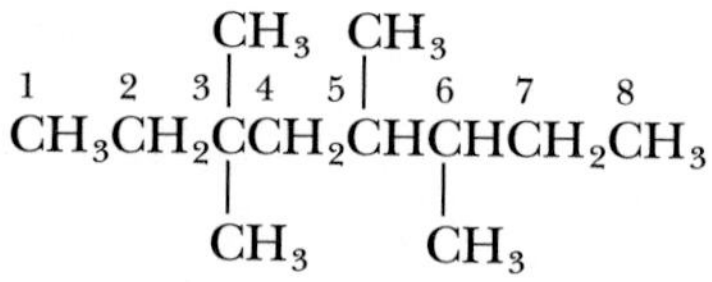

$$
\qquad\qquad\qquad\qquad CH_3\ \ CH_3
$$
$$
\overset{1}{C}H_3\overset{2}{C}H_2\overset{3}{C}\overset{4}{C}H_2\overset{5}{C}H\overset{6}{C}H\overset{7}{C}H_2\overset{8}{C}H_3
$$
$$
\qquad\qquad\quad |\qquad\qquad |
$$
$$
\qquad\qquad CH_3\qquad CH_3
$$

The correct name of the above compound is 3,3,5,6-tetramethyloctane.

4. *If there are two or more different groups, the groups are listed alphabetically.*

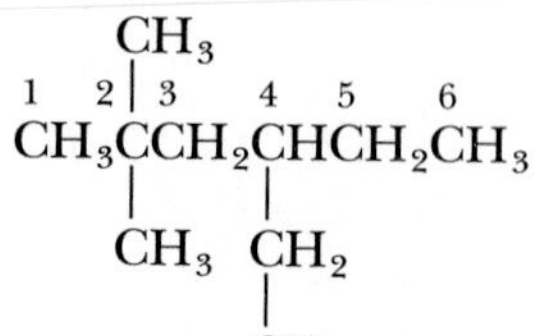

$$
\qquad\qquad\qquad CH_3
$$
$$
\overset{1}{C}H_3\overset{2}{C}\overset{3}{C}H_2\overset{4}{C}H\overset{5}{C}H_2\overset{6}{C}H_3
$$
$$
\qquad\qquad |\qquad\ \ |
$$
$$
\qquad\quad CH_3\ \ CH_2
$$
$$
\qquad\qquad\qquad\quad |
$$
$$
\qquad\qquad\qquad CH_3
$$

The correct name of the above compound is 4-ethyl-2,2-dimethylhexane. Note that the prefix ''di'' is ignored in determining alphabetical order.

Alkenes

Alkenes are named by using the prefix to indicate the number of carbon atoms and the suffix -*ene* to indicate one or more double bonds. The systematic names for the first two members of the alkene series are *ethene* and *propene.*

$$CH_2{=}CH_2 \quad CH_3CH{=}CH_2$$

When groups, such as methyl or ethyl, are attached to carbon atoms in an alkene, the longest hydrocarbon chain is numbered from the end that will give the double bond the lowest number and then numbers are assigned to the attached group. For example, the name of

$$
\begin{array}{c}
CH_3 \\
\overset{5\quad 4|\quad 3\quad\quad 2\quad 1}{CH_3CHCH{=}CHCH_3}
\end{array}
$$

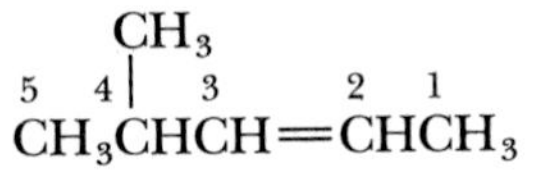

is 4-methyl-2-pentene. See Section 12.3 for a discussion of *cis-trans* isomers of alkenes.

Alkynes

The naming of alkynes is similar to that of alkenes, with the lowest number possible being used to locate the triple bond. For example, the name of

$$
\begin{array}{c}
CH_3 \\
\overset{1\quad 2\quad 3\,4|\quad 5}{CH_3C{\equiv}CCHCH_3}
\end{array}
$$

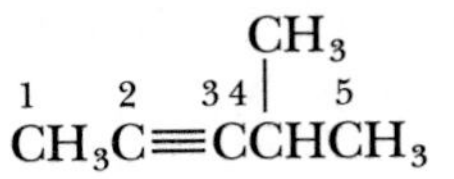

is 4-methyl-2-pentyne.

Benzene Derivatives

Monosubstituted benzene derivatives are named by using a prefix for the substituent. Some examples are

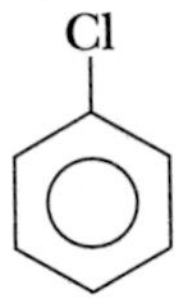

Cl — Chlorobenzene

CH₃ — Methylbenzene (toluene)

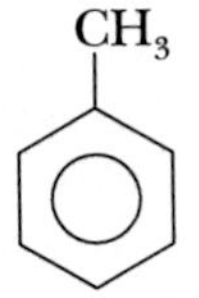

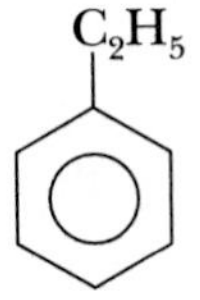

C₂H₅ — Ethylbenzene

Three isomers are possible when two groups are substituted for hydrogen atoms on the benzene ring. The relative positions of the substituents are indicated by either the prefixes *ortho-, meta-, para-* (abbreviated *o-, m-, p-,*), or by numbers. For example,

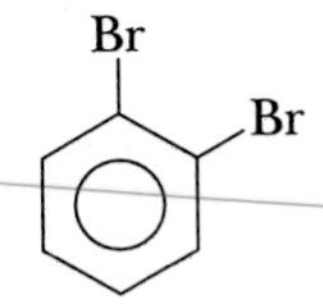

1,2-dibromobenzene (*o*-dibromobenzene) 1,3-dibromobenzene (*m*-dibromobenzene) 1,4-dibromobenzene (*p*-dibromobenzene)

The dimethylbenzenes are called *xylenes.*

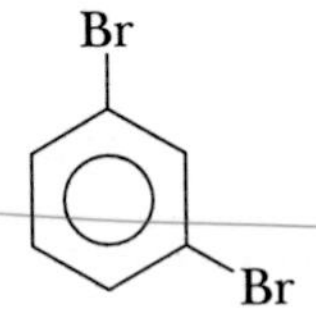

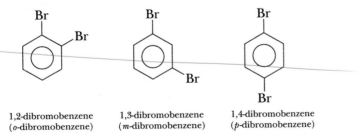

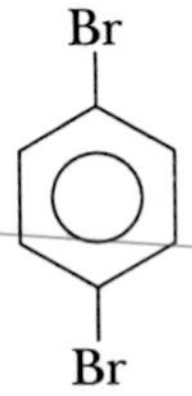

If more than two groups are attached to the benzene ring, numbers must be used to identify the positions. The benzene ring is numbered to give the lowest possible numbers to the substituents.

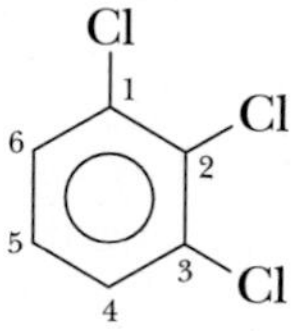

1,2,3-trichlorobenzene 1,2,4-trichlorobenzene 1,3,5-trichlorobenzene

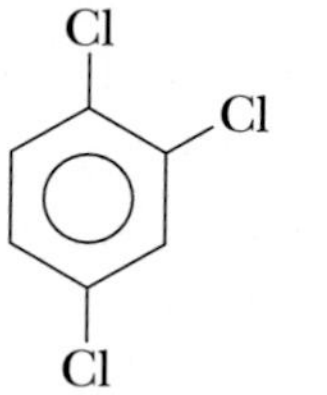

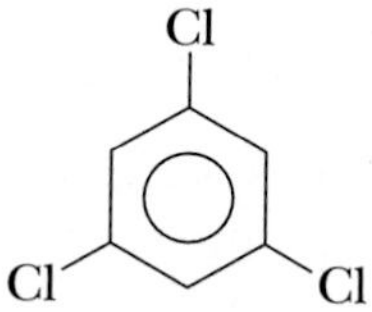

Functional Groups

An atom or group of atoms that defines the structure of a specific class of organic compounds and determines their properties is called a **functional group**. The millions of organic compounds include classes of compounds that are obtained by replacing hydrogen atoms of hydrocarbons with functional groups (Chapter 14). The important functional groups are shown in Table 14.3.

The "R" attached to the functional group represents the nonfunctional, hydrocarbon framework with one hydrogen removed for each functional group added. The IUPAC system provides a systematic method for naming all members of a given class. For example, alcohols end in *-ol* (methan*ol*); aldehydes end in *-al* (methan*al*); carboxylic acids end in *-oic* (ethan*oic* acid); and ketones end in *-one* (propan*one*).

Alcohols

Isomers are also possible for molecules containing functional groups. For example, three different alcohols are obtained when a hydrogen atom in pentane is replaced by —OH, depending on which hydrogen atom is replaced. The rules for naming the "R" or hydrocarbon framework are the same as those for hydrocarbon compounds. (See Section 14.2.)

$$CH_3CH_2CH_2CH_2CH_2OH \qquad CH_3CH_2CH_2\underset{\underset{OH}{|}}{C}HCH_3 \qquad CH_3CH_2\underset{\underset{OH}{|}}{C}HCH_2CH_3$$

1-pentanol 2-pentanol 3-pentanol

Compounds with one or more functional groups and alkyl substituents are named so as to give the functional groups the lowest number. For example, the correct name of

$$\overset{\qquad\qquad CH_3}{\underset{\qquad\quad OH \quad\ CH_3}{\overset{1 \quad 2 \quad 3 \quad 4\,|\,5}{CH_3CHCH_2CCH_3}}}$$

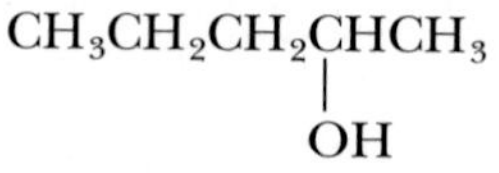

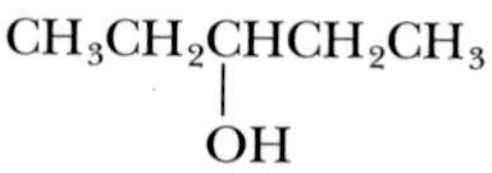

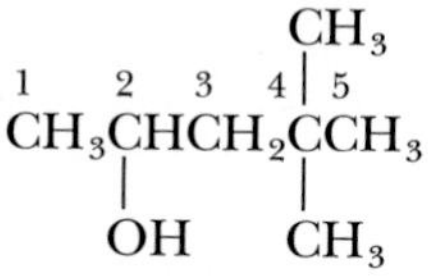

is 4,4-dimethyl-2-pentanol.

Aldehydes and Ketones

The systematic names of the first three aldehydes are methanal, ethanal, and propanal.

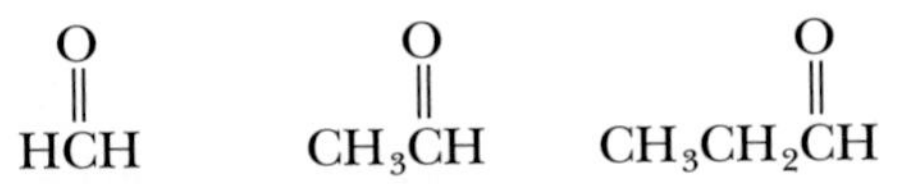

$$\underset{\substack{\text{Methanal}\\\text{(formaldehyde)}}}{\overset{\displaystyle O \atop \|}{HCH}} \qquad \underset{\substack{\text{Ethanal}\\\text{(acetaldehyde)}}}{\overset{\displaystyle O \atop \|}{CH_3CH}} \qquad \underset{\substack{\text{Propanal}\\\text{(propionaldehyde)}}}{\overset{\displaystyle O \atop \|}{CH_3CH_2CH}}$$

For ketones, a number is used to designate the position of the carbonyl group, and the chain is numbered in a way that gives the carbonyl carbon the smallest number.

$$\underset{\substack{\text{2-propanone}\\\text{(acetone)}}}{\overset{\displaystyle O \atop \|}{CH_3CCH_3}} \qquad \underset{\substack{\text{2-butanone}\\\text{(methyl ethyl ketone)}}}{\overset{\displaystyle O \atop \|}{CH_3CH_2CCH_3}} \qquad \underset{\substack{\text{4-penten-2-one}}}{\overset{\displaystyle O \atop \|}{CH_3CCH_2CH{=}CH_2}}$$

Carboxylic Acids

The systematic names of carboxylic acids are obtained by dropping the final *e* of the name of the corresponding alkane and adding *oic acid*. For example, the name of

$$CH_3CH_2CH_2CH_2CH_2COOH$$

is hexanoic acid. The systematic names of the first five carboxylic acids are given in Table 14.7. Other examples are

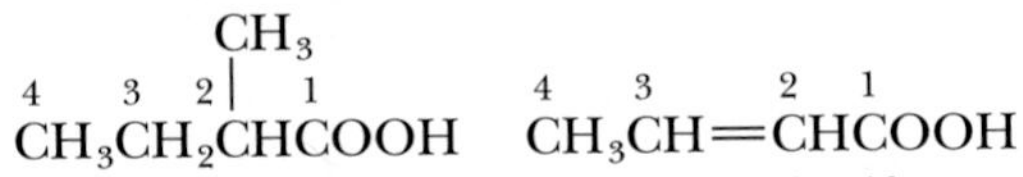

$$\underset{\substack{\text{2-methylbutanoic acid}}}{\overset{\displaystyle CH_3 \atop |}{\underset{4\quad 3\quad 2\quad 1}{CH_3CH_2CHCOOH}}} \qquad \underset{\substack{\text{2-butenoic acid}}}{\underset{4\quad 3\quad 2\quad 1}{CH_3CH{=}CHCOOH}}$$

Esters

The systematic names of esters are derived from the names of the alcohol and the acid used to prepare the ester. The general formula for esters is

$$\overset{\displaystyle O \atop \|}{R-C-OR'}$$

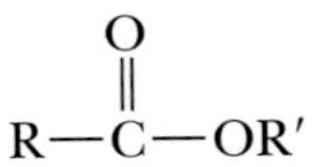

As shown in Section 14.3, the $R-\overset{\displaystyle O \atop \|}{C}$ comes from the acid and the R'O comes from the alcohol. The alcohol part is named first, followed by the name of the acid changed to end in -*ate*. For example,

$$\overset{\displaystyle O \atop \|}{CH_3CH_2C-OCH_3}$$

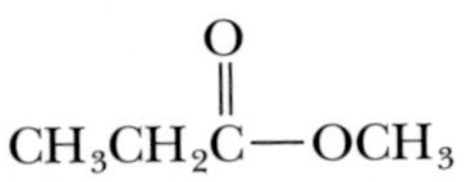

is named methyl propanoate, and

$$\overset{\displaystyle O \atop \|}{CH_3C-OCH{=}CH_2}$$

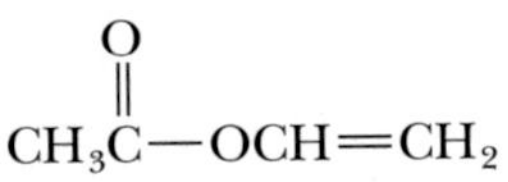

is named ethenyl ethanoate.

APPENDIX E

Answers to Self-Tests and Matching

CHAPTER 2

SELF-TEST 2A

1. (a), (b)
2. an element; cannot be broken down
3. elements and chemical compounds
4. homogeneous
5. (a) True
6. chemical change; changed in identity
7. (a), (b), (c), (d), (e)
8. (a) sugar and water
9. (a) methane and oxygen (b) hydrogen peroxide
10. (a) chemical property (b) physical property (c) chemical property
11. (a) True

SELF-TEST 2B

1. structure of matter
2. (a) copper (b) chlorine
3. solids: carbon, sulfur, phosphorus, iodine, all metals; liquids: mercury and bromine; gases: hydrogen, oxygen, nitrogen, fluorine, chlorine (also neon, argon, krypton, xenon, radon)
4. C, S, P, I, metal symbols in Tables 2.2 and 2.3; Hg, Br; H, O, N, F, Cl
5. Elements: potassium, hydrogen, phosphorus, oxygen; 8 atoms
6. CH_3OH
7. (a) no (b) yes (c) no (d) yes
8. (a) oxygen atom (in a compound)
   (b) two molecules of oxygen
   (c) one molecule of methane (d) yields (e) water
   (f) two molecules of water
9. nonmetal
10. (a) and (d)

SELF-TEST 2C

1. a number, a unit
2. prefixes, multiples of ten
3. cubic centimeter
4. kilo-, milli-
5. mm
6. 2 liters, 67.6 fluid ounces, 2000 milliliters
7. (b)
8. (a) quantitative (b) qualitative

MATCHING SET

1. b	7. i	13. m
2. p (or e)	8. l	14. q
3. d	9. g	15. k
4. a	10. h	16. e (or p)
5. c	11. j	17. o
6. f	12. n	18. r

CHAPTER 3

SELF-TEST 3A

1. (b) by early Greek philosophers
2. (b) philosophy (use of logic)
3. gained, chemical
4. 50%
5. NO and NO_2
6. 5.88%
7. (a) the same (b) atoms
8. (a) Yes. Although the total mass of Ag_2S may change, the percent mass of Ag in Ag_2S remains constant because the elements in a chemical compound are always present in a definite proportion by mass.
9. (a) Yes. Pure methane will have the same composition regardless of source because the percent mass of hydrogen and carbon in CH_4 is constant.

10. NO and NO_2, CO and CO_2, H_2O and H_2O_2, $FeCl_2$ and $FeCl_3$
11. (b) +1
12. (b) electron
13. (a) True
14. (a) nucleus
15. 1800

SELF-TEST 3B

1. 10 protons, 10 electrons, 11 neutrons
2. atomic number, mass number
3. protons, neutrons
4. electrons, protons, neutrons
5. protons and electrons
6. atomic
7. 33 electrons, 33 protons, 42 neutrons
8. 100,000
9. 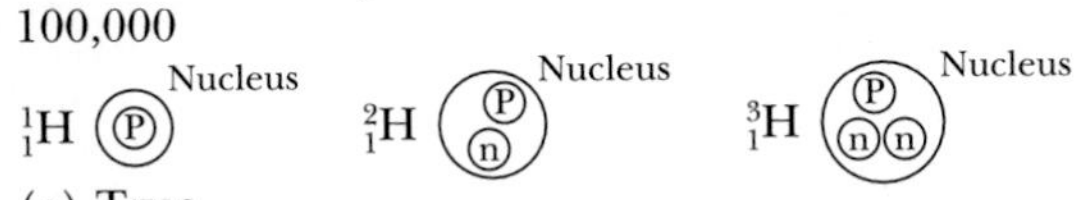

14. (a) True
11. (b) False

SELF-TEST 3C

1. farther from, closer to
2. (d) Infrared light
3. 18
4. 2-8-8-1, 1 valence electron
5. 2-8-7, 7 valence electrons
6. (a) Blue
7. (a) True
8. wave

MATCHING SET

1. d	5. c	9. k
2. a	6. h	10. e
3. j	7. b	11. g
4. m	8. f	

CHAPTER 4

SELF-TEST 4A

1. $\cdot\dot{B}\cdot$
2. 7
3. (a) 1, (b) 2, (c) 3, (d) 3, (e) 8, (f) 7
4. (a) True
5. (b) False
6. (a) True
7. (a) True
8. (b) and (d)
9. $SrCl_2$
10. both have 2 valence electrons
11. Be
12. (a) oxygen (b) calcium (c) fluorine (d) nitrogen (e) sulfur (f) neon
13. (a) True
14. (b) False
15. (a) cesium (b) fluorine (c) calcium (d) oxygen (e) sodium (f) xenon

SELF-TEST 4B

1. HCl (usually hydrogen chemistry is considered separately from other Group IA elements), LiCl, NaCl, KCl, RbCl, CsCl, FrCl
2. CO_2, SiO_2, GeO_2, SnO_2, PbO_2
3. (a) True
4. (a) True
5. Cs
6. (a) True
7. (a) True
8. (a) True
9. Xe
10. (a) True

MATCHING SET

1. d	5. c	9. f
2. e	6. h	10. b
3. a	7. j	
4. g	8. i	

CHAPTER 5

SELF-TEST 5A

1. (a) True
2. (c) Villard
3. alpha decay and positron emission
4. gamma
5. (a) True
6. (b) False
7. $^{212}_{82}Pb$
8. (a) True
9. (a) True
10. 9250 dps (disintegrations per second)

SELF-TEST 5B

1. ^8_4Be
2. $^{239}_{93}Np$
3. (b) False
4. Neptunium
5. (a) True
6. röntgen equivalent man
7. (a) True
8. (a) True
9. Radium
10. (a) True
11. 4 pCi/L of air

SELF-TEST 5C

1. medical diagnosis and treatment, food irradiation, materials testing
2. (a) ^{60}Co
3. Phosphorus-32

4. imaging
5. (c) metastable
6. (a) one eighth of the original dose
7. The one with the half-life of 6 hours. This would expose the patient to less radiation.
8. Increases one's annual dose by about 7 mrem.
9. (a) 180–200 mrem

MATCHING SET

1. b	6. c	11. d
2. i	7. l	12. n
3. a	8. h	13. f
4. k	9. 0	14. g
5. e	10. m	15. j

CHAPTER 6

SELF-TEST 6A

1. ionic
2. (a) losing electrons
3. (a) +1 (b) +3 (c) −2 (d) −1
4. (a) potassium
5. (b) gaining electrons
6. (a) smaller
7. (b) larger
8. (a) Li^+ (b) Ca^{2+} (c) Br (d) Al^{3+}
9. NaCl, sodium chloride
10. Potassium bromide: KBr
    Magnesium oxide: MgO
    Dinitrogen tetroxide: N_2O_4
    Sulfur trioxide: SO_3
    Dichlorine monoxide: Cl_2O
    Sulfur hexafluoride; SF_6

SELF-TEST 6B

1. PCl_5
2. sulfur dioxide
3. SO_3
4. hydrogen bromide
5. Cl_2O
6. SCl_2
7. silicon tetrachloride
8. (a) 4 (b) 6
9. (a) none (b) two (c) none (d) two (e) one
10. (c) C≡C
11. (c) C≡C

SELF-TEST 6C

1. (a) H_2 (b) HCl
2. F
3. (b) decreases down a group
4. 109.5°
5. 120°
6. AX_4
7. (a) H—F (b) C—O (c) H—N
8. (a) H_2O
9. (d) CCl_4
10. (b) nonbonding (c) bonding
11. 2, 2, bent
12. (b) SO_2
13. (d) SO_3 (e) H_2CO

MATCHING SET

1. b	5. j	9. a
2. k	6. c	10. h
3. g	7. d	11. e
4. f	8. i	

CHAPTER 7

SELF-TEST 7A

1. gas, liquid, solid
2. increases
3. liquids and solids
4. gases and liquids
5. gases are compressible, liquids and solids are noncompressible
6. volume
7. decreases
8. (a) True
9. hydrogen bonding
10. oxygen

SELF-TEST 7B

1. (a) True
2. (a) True
3. The 100 g of copper
4. surface tension
5. volatile
6. increase
7. boiling point
8. Miami
9. solute, solvent
10. (b) Corn oil
11. (b) False
12. increases

SELF-TEST 7C

1. (b) False
2. melting; melting point
3. sublimation
4. (a) True
5. (b) False
6. superconductor
7. n-type
8. p-type
9. transistors
10. at or just above the boiling point of nitrogen

SELF-TEST 7D

1. proportional to
2. inversely
3. 250 mL
4. 2.0 L
5. directly
6. (b) False
7. 23 + 273 = 296 kelvins
8. 298.15 − 273.15 = 25°C
9. $V_2 = \frac{(1.75L)(325\ K)}{(1255\ K)} = 0.453\ L$

MATCHING SET

1. p	8. d	15. m
2. k	9. c	16. i
3. q	10. a	17. n
4. l	11. u	18. e
5. r	12. h	19. f
6. b	13. g	20. o
7. t	14. s	21. j

CHAPTER 8

SELF-TEST 8A

Note: When the coefficient is "1" in balanced equations, the "1" is not written. In questions that relate to balancing equations, coefficients of "1" will be placed in parentheses to provide answers for the blanks given in the question.

1. law of conservation of mass
2. mole, dozen
3. (a) (1) Si + 2 Cl_2
   (b) 4 Al + 3 $O_2 \rightarrow$ 2 Al_2O_3
   (c) (1) $(NH_4)_2CO_3$ + (1) $Cu(NO_3)_2 \rightarrow$
   (1) $CuCO_3$ + 2 NH_4NO_3
4. atomic weight
5. (a) 6 CO_2 + 6 $H_2O \rightarrow$ (1) $C_6H_{12}O_6$ + 6 O_2
   (b) 6 molecules (c) 6 mol (d) 180 amu (e) 264 g
6. (c)

SELF-TEST 8B

1. reactant converted to product
2. energy
3. concentration, temperature
4. catalyst
5. activation energy
6. no
7. no
8. yes
9. Le Chatelier's principle

SELF-TEST 8C

1. released
2. absorbed
3. (a) True
4. energy
5. (b)
6. decreases
7. energy, entropy
8. the first law of thermodynamics
9. the second law of thermodynamics

SELF-TEST 8D

1. 2 mol H_2/1 mol O_2, 2 mol H_2O/1 mol O_2
2. 171 g
3. 51.3 g
4. 0.082 mol
5. 125 mol

MATCHING SET

1. h	5. b	9. g
2. d	6. c	10. j
3. a	7. e	11. i
4. k	8. f	12. l

CHAPTER 9

SELF-TEST 9A

1. donor, acceptor
2. base, acid
3. HCO_3^-
4. HPO_4^{2-}
5. bases
6. acids
7. strong
8. weak
9. basic, strong
10. $H^+(aq) + OH^-(aq) \rightarrow H_2O(\ell)$
11. $H^+(aq) + OH^-(aq) \rightarrow H_2O(\ell)$ or
    $H_3O^+(aq) + OH^-(aq) \rightarrow 2\ H_2O(\ell)$

SELF-TEST 9B

1. pH of 2 (d) 10,000
2. (a) high hydronium ion concentration
3. (b) low hydronium ion concentration
4. $H_2PO_4^-$ neutralizes added base, HPO_4^{2-} neutralizes added acid
5. 3
6. 11
7. 1.0×10^{-11} M, 1.0×10^{-3} M
8. 1.0×10^{-4} M
9. 400 g
10. 1
11. 1×10^{-2} M

MATCHING SET

1. m	6. l	11. g
2. k	7. j	12. i
3. b	8. h	13. d
4. a	9. c	14. f
5. n	10. e	

CHAPTER 10

SELF-TEST 10A

1. oxide
2. rust
3. (b) CO_2
4. (a) NO_2
5. H_2O and CO_2
6. (a) True
7. (a) True
8. (b) acetaldehyde
9. (a) Na^+ion
10. (a) True

SELF-TEST 10B

1. (b) reduced
2. (b) reducing agent
3. (b) reduced
4. (b) $CH_3—NH_2$
5. sodium metal
6. magnesium metal
7. (a) Cu
8. oxidizing agent
9. reducing agent

SELF-TEST 10C

1. active
2. (b) False
3. Yes, Ca will reduce Fe^{2+} ions.
4. Yes, Ag^+ will oxidize Ca.
5. No, Ag^+ will not oxidize F^- ion.
6. Rusting requires iron, oxygen, and moisture.
7. galvanizing
8. fluorine
9. lithium

SELF-TEST 10D

1. (a) throw-away battery
2. (b) rechargeable battery
3. (b) zinc
4. (b) decrease
5. (b) reversible
6. (b) Ni-Cad
7. water (H_2O)
8. (b) $Cu^{2+} + 2\,e^- \rightarrow Cu$
9. hydrogen and oxygen

MATCHING SET

1. b	8. m	15. r
2. l	9. g	16. d
3. q	10. h	17. k
4. p	11. i	18. e
5. c	12. s	19. o
6. a	13. n	20. j
7. f	14. t	

CHAPTER 11

SELF-TEST 11A

1. oxygen
2. aluminum
3. (a) i (b) iii (c) ii (d) iii (e) i (f) iii
4. boiling points
5. (b)
6. (a) inert (not reactive) (b) very low boiling points
7. magnesium

SELF-TEST 11B

1. gold, copper
2. minerals
3. ores
4. (b) positive ions
5. oxide
6. limestone
7. slag
8. (i) cast iron: (a), (c), (e) (ii) steel: (b), (d), (f)
9. reduced
10. (a) electrolysis: (i) aluminum
    (b) reaction with carbon: (ii) iron

SELF-TEST 11C

1. silicon and oxygen
2. (i) silica: (a), (b) (ii) silicates: (c) (d) (e)
3. glass is not crystalline (it is amorphous)
4. clay, sand, feldspar
5. temperature, stress
6. cement, sand, aggregate (crushed stone or pebbles)

MATCHING SET

1. i	6. a	11. g
2. j	7. e	12. c
3. n	8. o	13. k
4. h	9. d	14. f
5. b	10. m	15. l

CHAPTER 12

SELF-TEST 12A

1. (c) natural gas
2. (b) hydrogen
3. (a) True
4. oxygen, water
5. tetrahedral
6. ethene or ethylene
7. ethyne or acetylene
8. $—C_2H_5$
9. structural
10. ethylene
11. *cis* and *trans*

SELF-TEST 12B

1. hydrogen
2. 12
3. CO and H_2
4. 3
5. 3
6. (a) ethanol
7. R—O—R′

SELF-TEST 12C

1. fractional distillation
2. methane, CH_4
3. methyl-*tertiary*-butyl ether
4. catalytic re-forming
5. catalytic cracking
6. (c)

MATCHING SET

1. d	6. a
2. f	7. c
3. i	8. b
4. h	9. e
5. j	10. g

CHAPTER 13

SELF-TEST 13A

1. (c) False
2. (b) 80 quads
3. (1) 1000 J
4. (c) 3000 Btu
5. (a) True
6. 36%

SELF-TEST 13B

1. (b) uranium-235
2. $^{236}_{92}U$
3. $^{134}_{52}Te$
4. smaller
5. (a) iron
6. (b) 19%
7. (a) True
8. (b) 24,000 years

SELF-TEST 13C

1. (a) True
2. (a) 50%
3. (c) Wind energy
4. a gardner's greenhouse
5. active
6. (a) True
7. (a) True
8. (a) Solar—aesthetics of having huge solar collectors on the landscape
   (b) Wind—aesthetics of having huge windmills everywhere
   (c) Biomass—dealing with the amounts of waste materials produced
   (d) Geothermal—H_2S produced is toxic and many geothermal streams contain dissolved minerals harmful to aquatic life
9. (b) decay of radioactive elements in the core of the Earth

MATCHING SET

1. g	7. q	13. o
2. f	8. r	14. c
3. i	9. e	15. b
4. j	10. d	16. a
5. n	11. m	17. h
6. p	12. k	18. l

CHAPTER 14

SELF-TEST 14A

1. (a) alcohol (b) carboxylic acid (c) aldehyde (d) ketone
2. aromatic hydrocarbons
3. acetaldehyde
4. rubbing alcohol
5. ethylene glycol
6. 42%
7. denatured
8. 3
9. glycerol
10. CH_3OH

SELF-TEST 14B

1. formaldehyde
2. acetic acid
3. ethanol with acetic acid
4. methyl acetate
5. 3
6. acetone
7. (b) esters

SELF-TEST 14C

1. monomers
2. free radicals, organic peroxides
3. double
4. single
5. (a) True
6. CH_2CHCl
7. *cis*
8. rubbers
9. neoprene
10. (a) low-density polyethylene (b) high-density polyethylene (c) cross-linked polyethylene

SELF-TEST 14D

1. adipic acid, hexamethylenediamine, nylon 66, water
2. condensation

3. water
4. (b) condensation reactions
5. react with moisture
6. (a) True
7. glass fibers
8. polymer

MATCHING SET

1. e	5. l	9. c
2. d	6. h	10. i
3. f	7. j	11. b
4. g	8. k	

CHAPTER 15

SELF-TEST 15A

1. 4
2. glycine
3. (b) False
4. (a) True
5. essential

SELF-TEST 15B

1. 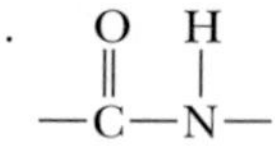
2. (a) 27 (b) 6
3. active site
4. (a) amino acid sequence
 (b) hydrogen bonded structures to form helices or sheets
 (c) how the protein molecule is folded
 (d) shape assumed by all chains in a protein with two or more chains
5. (a) catalyst
6. enzyme
7. (a) True
8. hydrogen

SELF-TEST 15C

1. monosaccharides
2. glucose and fructose
3. α-D-glucose
4. β-D-glucose
5. Hydrogen
6. D-glucose
7. glycerol and fatty acids
8. double bonds. A saturated fat contains only single carbon-carbon bonds while an unsaturated fat contains one or more carbon-carbon double bonds and hence fewer hydrogen atoms than a saturated fat.
9. (a) steroid
10. (a) True
11. long-chain fatty acids and long-chain alcohols

SELF-TEST 15D

1. CO_2 and H_2O, energy
2. ATP
3. ADP, Energy
4. DNA
5. ribose, deoxyribose
6. phosphoric acid group, a sugar (ribose or deoxyribose), a nitrogen heterocyclic base (adenine, guanine, thymine, cytosine, or uracil)
7. double helix
8. hydrogen
9. T, C, A, A
10. three billion

MATCHING SET

1. d	5. g	9. f
2. c	6. i	10. h
3. a	7. j	11. k
4. b	8. e	12. l

CHAPTER 16

SELF-TEST 16A

1. corneal
2. sebum
3. skin and hair
4. ionic, disulfide
5. 10%
6. emulsion
7. (a) True
8. talc
9. (a) True
10. para-aminobenzoic acid (PABA)
11. melanin

SELF-TEST 16B

1. (a) True
2. ionic
3. melanin
4. polymers
5. temporary
6. depilatory
7. 13

SELF-TEST 16C

1. fatty
2. aqueous base
3. more soluble in water
4. trapped air
5. (a) traditional soaps
6. surface active agent
7. they will react with one another
8. visible
9. a surfactant and an abrasive
10. fluorine

MATCHING SET

1. m	7. d	13. n
2. f	8. c	14. j
3. h	9. q	15. b
4. p	10. o	16. e
5. a	11. g	17. k
6. l	12. i	

CHAPTER 17

SELF-TEST 17A

1. simple sugars, amino acids, glycerol + fatty acids, energy
2. basal metabolic rate, weight in pounds
3. 1 kilocalorie
4. carbohydrates, fats, proteins
5. (a) True
6. (b) glucose
7. ATP

SELF-TEST 17B

1. dietary fiber
2. triglycerides
3. high-density lipoproteins (HDLs), transport cholesterol to liver for excretion
4. low-density lipoproteins (LDLs), transport cholesterol to arteries
5. nitrogen
6. fats
7. fruits and vegetables
8. (b)
9. 30%

SELF-TEST 17C

1. (b) False
2. fat-soluble, water-soluble
3. vitamin A
4. antioxidants
5. enzymes
6. enriched
7. ions
8. electrolyte balance, acid-base balance (also fluid balance), nerve
9. anemia, osteoporosis

SELF-TEST 17D

1. drying, salt or sugar
2. Generally Recognized as Safe
3. (b) False
4. antioxidants
5. EDTA
6. (b) False
7. polysaccharides
8. 100 kcal divided by 230 kcal
9. 0.30×2000 kcal

MATCHING SET

A.	1. b	5. d	8. j
	2. c	6. e	9. g
	3. i	7. f	10. a
	4. h		
B.	1. i	5. h	9. d
	2. g	6. b	10. l
	3. c	7. e	11. f
	4. k	8. j	12. a

CHAPTER 18

SELF-TEST 18A

1. dehydrating cellular structures, hydrolysis
2. oxidizing
3. hemoglobin, oxygen
4. (b) False
5. CN^-; cytochrome oxidase, oxygen
6. (a) CO, and (d) CN^-
7. different
8. inhalation, ingestion, skin contact

SELF-TEST 18B

1. mercury
2. arsenic
3. (a) True
4. (a) True
5. 10 μg of Pb/dL of blood
6. arsenic, chelates
7. (a) True
8. (1) drinking water, (2) foods, (3) lead-based paints
9. lead and mercury

SELF-TEST 18C

1. neurotoxins
2. synapses
3. neurotransmitter
4. DNA or chromosomes
5. (b) False
6. chimney sweeping
7. lung cancer and low birth weight for newborns
8. teratogen

MATCHING SET

1. b	5. i	9. c
2. d	6. e	10. j
3. g	7. f	11. a
4. h	8. k	

CHAPTER 19

SELF-TEST 19A

1. generic, trade name
2. pneumonia, influenza, HIV infection

3. pathogens
4. (b) False
5. cross-linking
6. an $—NH_2$ group an —OH group
7. inside, outside
8. retrovirus
9. AIDS, DNA
10. chemotherapy
11. antibiotic-resistant
12. pneumonia and influenza, diabetes mellitus, lung disease

SELF-TEST 19B

1. hormones and neurotransmitters
2. receptors
3. proteins or steroids
4. (a) and (d)
5. dopamine
6. serotonin
7. epinephrine
8. glands
9. very similar

SELF-TEST 19C

1. heroin
2. dose
3. (a) ii (b) i (c) iii
4. acetylsalicylic acid
5. carboxylic acid, anti-inflammatory
6. addictive, an abused drug, medical
7. depressants
8. cocaine, addictive
9. (c) allergic reaction
10. (b) False
11. decongestants, antitussives, expectorants, analgesics, antihistamines
12. the opium poppy
13. an anti-inflammatory

SELF-TEST 19D

1. cardiovascular disease
2. oxygen
3. (a) faster
4. relaxing heart muscle
5. enzymes
6. (a), (b), (c)
7. (a) True
8. DNA
9. kill

MATCHING SET

1. d	6. f	11. i
2. a	7. o	12. h
3. e	8. k	13. m
4. b	9. j	14. l
5. c	10. n	15. g

CHAPTER 20

SELF-TEST 20A

1. 70%
2. oceans
3. groundwater
4. surface water
5. 168 gal
6. suitable for drinking
7. dissolved metals, dissolved organic compounds, microorganisms
8. aquifer
9. runs off into the oceans

SELF-TEST 20B

1. Superfund
2. (b) False
3.

Household Waste	**Harmful Chemical**
automobile batteries	lead and its salts and sulfuric acid
oil-based paint	organic solvents
fluorescent lamps	mercury

4. steel cans, aluminum cans, glass, paper
5. biochemical oxygen demand, BOD
6. (a) a liter of water at 5°C
7. the atomic absorption spectrophotometer (AA)
8. (1) Automotive-battery manufacture introduces lead into groundwater. (2) Consumer-battery manufacture introduces mercury and cadmium into groundwater.
9. gas chromatograph/mass spectrometer
10. (1) chlorinated plastics manufacture, (2) textile manufacture

SELF-TEST 20C

1. (d) Reverse osmosis
2. (b) persistent
3. settling and filtration
4. aerobic
5. anaerobic
6. (b) calcium, (c) magnesium
7. chlorination and ozone treatment
8. Na^+, Mg^{2+}, Ca^{2+}, and K^+
9. osmosis

MATCHING SET

1. l	7. k	13. g
2. o	8. q	14. n
3. p	9. d	15. i
4. a	10. m	16. h
5. c	11. j	17. b
6. e	12. f	

CHAPTER 21

SELF-TEST 21A

1. (b) decrease

2. ozone in the troposphere is harmful, ozone in the stratosphere is beneficial
3. adsorb
4. absorb
5. warm air is above cool air
6. (a) coal burning
7. oxides
8. NO and O
9. oxygen atom

SELF-TEST 21B

1. SO_2 and CO_2
2. SO_3, H_2SO_4
3. (b) Lime
4. sulfuric acid (H_2SO_4) and nitric acid (HNO_3)
5. 5.6 caused by dissolved CO_2
6. 1872
7. the C—Cl bond
8. $O_2(g) + h\nu \rightarrow 2\,O(g)$
   $O(g) + O_2(g) \rightarrow O_3(g)$
9. chlorine atom
10. Antarctica
11. ultraviolet radiation with wavelengths in the 200–300 nm range

SELF-TEST 21C

1. (1) burning coal, oil, and natural gas to produce electricity, (2) burning gasoline to power automobiles, (3) clear-cutting and burning of forests to create cropland
2. (1) photosynthesis, (2) dissolution in the world's oceans
3. (1) carbon dioxide (CO_2), (2) methane (CH_4), and (3) water vapor (H_2O)
4. carbon dioxide
5. (a) True
6. 330 ppm
7. Louisiana
8. ammonia
9. smoking tobacco products

MATCHING SET

1. c	8. s	15. r
2. e	9. i	16. m
3. o	10. t	17. b
4. k	11. q	18. f
5. h	12. l	19. j
6. a	13. p	20. d
7. n	14. u	21. g

CHAPTER 22

SELF-TEST 22A

1. 5.7 billion, 100 million
2. clays, silts, sandy soils, loams
3. (a) acidic, (b) basic
4. particle size and chemical composition
5. a trivalent ion like Fe^{3+} (hydrolyzes more than Na^+)
6. humus
7. nitrogen, phosphorus, potassium
8. calcium, magnesium, sulfur
9. oxygen

SELF-TEST 22B

1. (b) False
2. nitrogen, phosphorus (calculated as P_2O_5), potassium (calculated as K_2O)
3. Yes
4. (c) a gas
5. about 33%
6. DDT
7. (a) DDT
8. Pyrethrum
9. Selective herbicide
10. 2,4-D
11. insecticides, herbicides, fungicides

MATCHING SET

1. f	6. b	11. e
2. c	7. k	12. g
3. h	8. d	13. l
4. a	9. m	
5. j	10. i	

APPENDIX F

Answers to Exercises

CHAPTER 2

2.1. (a) physical change (b) chemical change

2.2. (a) Pb (b) P (c) HCl (d) $AlBr_3$ (e) F_2

2.3. (a) NF_3 (b) 4 atoms in the molecule (c) nitrogen, fluorine

2.4. (a) Hydrogen gas reacts with chlorine gas to form hydrogen chloride gas.
(b) *Reactants*: 2 H atoms in one H_2 molecule, 2 Cl atoms in one Cl_2 molecule; *products*: 2 HCl molecules, which contain 2 H atoms and 2 Cl atoms

2.5. *mega* = 1 million, 20 megabucks = 20 million dollars ($20,000,000)

2.6. to the right. 0.060 g = 60 mg = 60,000 μg; 0.00400 g = 4 mg = 4000 μg

2.7. 225 C (Calories) = 225 kilocalories (kcal)

$$225\text{ kcal} \times \frac{1000\text{ cal}}{1\text{ kcal}} = 225{,}000\text{ kcal}$$

2.8. $$64\text{ fl oz} \times \frac{1\text{ qt}}{32\text{ fl oz}} \times \frac{1\text{ L}}{1.06\text{ qt}} = 1.9\text{ L}$$

2.9. (a) $$16{,}000\text{ g} \times \frac{0.0352\text{ oz}}{1\text{ g}} = 560\text{ oz}$$

(b) $$560\text{ oz} \times \frac{1\text{ lb}}{16\text{ oz}} = 35\text{ lb}$$

CHAPTER 3

3.1. (a) 41.1 g nitrogen and 8.9 g hydrogen are in 50.0 g ammonia
(b) 41.1 tons of nitrogen and 8.9 tons of hydrogen

3.2. $$5.0\text{ cm} \times \frac{1\text{ gold atom}}{2.9 \times 10^{-8}\text{ cm}} = 1.7 \times 10^{8}\text{ atoms}$$

3.3. 28 protons, 28 electrons, and 31 neutrons

3.4. ${}^{107}_{47}Ag$ ${}^{109}_{47}Ag$

3.5. Bohr arrangement for sulfur is 2-8-6. Sulfur has 6 valence electrons.

3.6. Chlorine. 7 valence electrons.

CHAPTER 4

4.1. (a) Group IIA is the only main group that has all metals. Group IA is often regarded as having only metals since hydrogen is a nonmetal. Figure 4.2 shows hydrogen separated from the rest of the members of Group IA for this reason.
(b) Groups VIIA and VIIIA have only nonmetals.
(c) Groups IIIA, IVA, VA, and VIA include metalloids.
The number of valence electrons are: (a) Group IA, 1; Group IIA, 2 (b) Group VIIA, 7; Group VIIIA, 8 (c) Group IIIA, 3; Group IVA, 4; Group VA, 5; Group VIA, 6

4.2. Rb· :B̈r· (with two further dots below Br)

4.3A. (a) Ba (b) Se (c) Si (d) Ga

4.3B. (a) Sr (b) Cl (c) Cs

4.4. Group VA

CHAPTER 5

5.1. ${}^{234}_{91}Pa \rightarrow {}^{0}_{-1}e + {}^{234}_{92}U$

5.2A. ${}^{237}_{93}Np \rightarrow {}^{4}_{2}He + {}^{233}_{91}Pa$

5.2B. ${}^{230}_{90}Th \rightarrow {}^{4}_{2}He + {}^{226}_{88}Ra$
${}^{226}_{88}Ra \rightarrow {}^{4}_{2}He + {}^{222}_{86}Rn$

5.3. First, find the number of half-lives in 244 hr.

$$244\text{ hr} \times \frac{1\text{ half-life}}{61\text{ hr}} = 4.0\text{ half-lives}$$

Four half-lives means the initial quantity of copper-67 will have decreased four times by 1/2, or $(1/2)^4$. Writing this out as a fraction, 21.0 mg × (1/2) × (1/2) × (1/2) × (1/2) = 21.0 mg × $(1/2)^4$ = 21.0 mg × (1/16) = 1.31 mg

CHAPTER 6

6.1. CaF_2

6.2. CoO, Co_2O_3

6.3. (a) rubidium chloride (b) gallium oxide (c) barium bromide (d) iron(II) nitride

6.4. (a) CoS (b) BeF_2 (c) KI

6.5. (a) $MgCO_3$ (b) NaH_2PO_4

6.6. 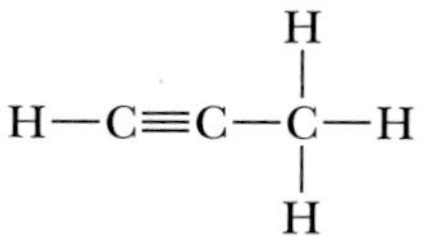

6.7. (a) 24 valence electrons (b) 6 single bonds (c) 1 double bond (d) 8 bonding pairs (e) 4 nonbonding pairs

6.8. Tetraphosphorus trisulfide

6.9. N_2F_4

6.10. 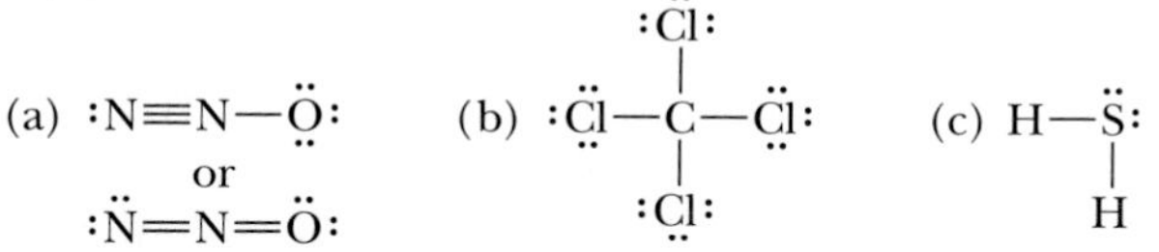

6.11. ClO_3^- has three bonding pairs and one nonbonding pair around the Cl, which gives a tetrahedral electron-pair geometry. The question asks for the predicted molecular geometry, which considers only the shape formed by the atoms in the species, in this case ClO_3^-. The predicted shape of the ion is triangular pyramidal.

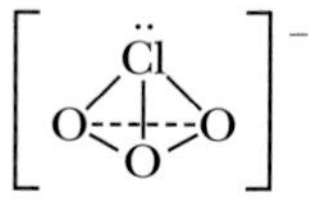

NH_2^- has two bonding pairs and two nonbonding pairs around the N, which gives a tetrahedral electron-pair geometry. The predicted shape of the ion is bent. Refer to the examples in Figure 6.8 for further help on this exercise.

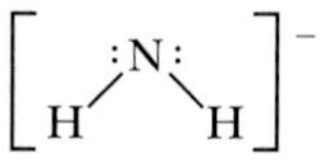

CHAPTER 7

7.1. CH_3CH_2OH will have the higher boiling point because of hydrogen bonding between the molecules using the —OH groups.

7.2. Using $P_1V_1 = P_2V_2$ and substituting,

$$(1\ \text{atm})(400\ \text{L}) = (0.75\ \text{atm})(V_{\text{unknown}})$$

Solving for the unknown volume,

$$V_{\text{unknown}} = \frac{(1\ \text{atm})(400\ \text{L})}{(0.750\ \text{atm})} = 533\ \text{L}$$

7.3. First, convert all temperatures to absolute.

$T_1 = 31 + 273 = 304$ K

$T_2 = 75 + 273 = 348$ K

Using $\frac{T_1}{V_1} = \frac{T_2}{V_2}$ and solving for V_2,

$$V_2 = \frac{V_1T_2}{T_1} = \frac{(765\ \text{mL})(348\ \text{K})}{(304\ \text{K})}$$

$= 875$ mL

7.4. Set up this problem's solution like that in Example 7.4. While doing this, convert all temperatures from Celcius to Kelvins.

$T_1 = 17 + 273 = 290$ K $P_1 = 714$ mm Hg $V_1 = 2125$ mL

$T_2 = ?$ $P_2 = 745$ mm Hg $V_2 = 2333$ mL

Use a combination of Charles' and Boyle's laws.

$$\frac{P_1V_1}{T_1} = \frac{P_2V_2}{T_2}$$

Solve for T_2.

$$T_2 = \frac{P_2V_2T_1}{P_1V_1} = \frac{(745\ \text{mm Hg})(2333\ \text{mL})(290\ \text{K})}{(714\ \text{mm Hg})(2125\ \text{mL})} = 332\ \text{K}$$

or, $T_2 = 332 - 273 = 59$°C

CHAPTER 8

8.1A. $2\ Al(s) + 3\ Cl_2(g) \rightarrow 2\ AlCl_3(s)$

8.1B. (a) yes (b) no

8.2. $Ba(NO_3)_2(aq) + Na_2SO_4(aq) \rightarrow BaSO_4(s) + 2\ NaNO_3(aq)$

8.3.

$CO(g)$ +	$2\ H_2(g) \rightarrow$	$CH_3OH(\ell)$
1 mol	2 mol	1 mol
28 g/mol	2 g/mol	32 g/mol
28 g	2 mol × 2 g/mol = 4 g	32 g

8.4. $$150\ \text{g NaCl} \times \frac{1\ \text{mol NaCl}}{58\ \text{g NaCl}} = 2.6\ \text{mol NaCl}$$

8.5. $$50.\ \text{mol}\ Ba(NO_3)_2 \times \frac{261\ \text{g}\ Ba(NO_3)_2}{1\ \text{mol}\ Ba(NO_3)_2} = 13{,}000\ \text{g}\ Ba(NO_3)_2$$

8.6. $$2\ \text{mol Cr} \times \frac{2\ \text{mol}\ Cr_2O_3}{4\ \text{mol Cr}} = 1\ \text{mol}\ Cr_2O_3$$

8.7. $$550\ \text{g}\ Cr_2O_3 \times \frac{1\ \text{mol}\ Cr_2O_3}{152\ \text{g}\ Cr_2O_3} = 3.6\ \text{mol}\ Cr_2O_3$$

$$3.6\ \text{mol}\ Cr_2O_3 = \frac{4\ \text{mol Cr}}{2\ \text{mol}\ Cr_2O_3} = 7.2\ \text{mol Cr}$$

$$7.2\ \text{mol Cr} \times \frac{52\ \text{g Cr}}{1\ \text{mol Cr}} = 370\ \text{g Cr}$$

■ CHAPTER 9

9.1. $Ca(OH)_2(aq) + 2 HNO_3(aq) \rightarrow$
$$Ca(NO_3)_2(aq) + 2 H_2O(\ell)$$
$H^+(aq) + OH^-(aq) \rightarrow H_2O(\ell)$

9.2. $\underset{\text{Base}}{CO_3^{2-}(aq)} + \underset{\text{Acid}}{H_2O(\ell)} \rightarrow \underset{\text{Conjugate acid}}{HCO_3^-(aq)} + \underset{\text{Conjugate base}}{OH^-(aq)}$

9.3. $\dfrac{0.10 \text{ mol NaOH}}{L} \times \dfrac{40.0 \text{ g NaOH}}{\text{mole}} \times 1.5 \text{ L} = 6.0 \text{ g NaOH}$

9.4. $M_1V_1 = M_2V_2$
$12.1 \text{ M} \times V_1 = 5.0 \text{ L} \times 3.0 \text{ M}$
$V_1 = \dfrac{5.0 \text{ L} \times 3.0 \text{ M}}{12.1 \text{ M}}$ $V_1 = 1.2 \text{ L}$

The solution is prepared by adding 1.2 L of 12.2 M hydrochloric acid to enough water to make 5.0 liters of solution.

9.5. $[H_3O^+] = 0.01$ $[H_3O^+][OH^-] = 1.0 \times 10^{-14}$
$[0.01][OH^-] = 1.0 \times 10^{-14}$
$[OH^-] = 1.0 \times 10^{-12}$

9.6. $[OH^-] = 0.025$ $[H_3O^+][OH^-] = 1.0 \times 10^{-14}$
$[H_3O^+][0.025] = 1.0 \times 10^{-14}$
$[H_3O^+] = \dfrac{1 \times 10^{-14}}{0.025} = 4.0 \times 10^{-13}$

9.7. Small amount of acid added: $CO_3^{2-}(aq) + H_3O^+(aq) \rightarrow HCO_3^-(aq) + H_2O(\ell)$
Small amount of base added: $HCO_3^-(aq) + OH^-(aq) \rightarrow CO_3^{2-}(aq) + H_2O(\ell)$

9.8. Milk of magnesia is more basic than the baking soda solution. Milk of magnesia is 100 times more basic than the baking soda solution.

9.9. (a) 3 (b) 11

9.10. $[H_3O^+] = 1 \times 10^{-4}$

■ CHAPTER 10

10.1A. $2 Ca(s) + O_2(g) \rightarrow 2 CaO(s)$

10.1B. (a) SO_3 (b) Fe_2O_3 (c) PbO_2

10.2. (a) sulfur is oxidized
(c) NO is oxidized

10.3. (a) C_2H_4 is oxidized, hydrogen is lost
(d) $CH_3CH_2OCH_3$ is oxidized, hydrogen is lost

10.4. (a) O_2 is the oxidizing agent
(b) O_2 is the oxidizing agent
(c) F_2 is the oxidizing agent
(d) F_2 is the oxidizing agent
(e) I_2 is the oxidizing agent
The definition of oxidation that says oxidation is the loss of electrons applies to all of the above reactions.

10.5. (a) Cu is not being reduced, it is being oxidized.
(b) CH_3CN is being reduced. Hydrogen could be the reducing agent.
(c) H_2O is being reduced to form H_2. Electricity could do this. (H_2O is also being oxidized to produce O_2.)

(d) SnO is being reduced. Hydrogen could be the reducing agent.

10.6. (a) K is oxidized. H_2O is being reduced.
(b) SnO_2 is being reduced. Carbon is being oxidized.
(c) Cl^- is being oxidized. F_2 is being reduced.

10.7A. *Experiment 1.* Place a piece of iron into a solution of a zinc salt and see what happens. Since zinc is more active than iron, nothing will happen, the zinc atoms have already lost their electrons.
Experiment 2. Place a piece of zinc into a solution of an iron salt like iron(II) nitrate ($Fe(NO_3)_2$). Iron will be observed being formed on the surface of the zinc. The piece of zinc will be observed to dissolve in the iron salt solution.

10.7B. Silver can be seen plating out on the metallic copper. The solution also turns blue as time passes. Silver ions are being reduced: $Ag^+ + e^- \rightarrow Ag$
Copper is being oxidized: $Cu \rightarrow \underset{\text{(Blue)}}{Cu^{2+}} + 2 e^-$

10.8A. Magnesium can reduce Ag^+ to Ag because magnesium is more active than silver. Look at Table 10.6. Magnesium is shown as a stronger reducing agent than Ag. This means that Mg will reduce Ag^+ ions.
$Mg(s) + 2 Ag^+(aq) \rightarrow Mg^{2+}(aq) + 2 Ag(s)$

10.8B. Iron cannot reduce Mg^{2+} ions. Look at Table 10.6. Mg is a stronger reducing agent than iron. This means that Mg^{2+} ions will not gain electrons from iron atoms. The reverse would be true—magnesium atoms would lose electrons to Fe^{2+} ions.

■ CHAPTER 12

12.1. $2 \; H\!-\!\overset{\displaystyle H}{\underset{\displaystyle H}{\overset{|}{\underset{|}{C}}}}\!-\!\overset{\displaystyle H}{\underset{\displaystyle H}{\overset{|}{\underset{|}{C}}}}\!-\!H + 7 \, O\!=\!O \rightarrow$

$4 \, O\!=\!C\!=\!O + 6 \, H\!-\!O\!-\!H$

Breaking bonds: 2 C—C 2×83 kcal per bond = 166 kcal
12 C—H 12×99 kcal per bond = 1188 kcal
7 O=O 7×118 kcal per bond = 826 kcal
Total energy required to break bonds = 2180 kcal

Making Bonds: 8 C=O 8×192 kcal per bond = 1536 kcal
12 O—H 12×111 kcal per bond = 1332 kcal
Total energy released in making bonds = 2868 kcal

Heat of combustion = 2868 kcal − 2180 kcal = 688 kcal for two moles of ethane or *344 kcal per mol ethane*

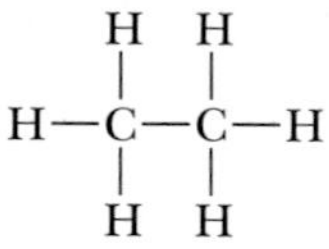

12.2. (a)
$$\underset{\text{2-methyl pentane}}{CH_3\overset{\displaystyle CH_3}{\overset{|}{C}HCH_2CH_2CH_3}}$$
(b)
$$\underset{\text{3-methyl pentane}}{CH_3CH_2\overset{\displaystyle CH_3}{\overset{|}{C}HCH_2CH_3}}$$

(c)
$$\underset{\text{2,2-dimethylbutane}}{CH_3\overset{\displaystyle CH_3}{\underset{\displaystyle CH_3}{\overset{|}{\underset{|}{C}}}CH_2CH_3}}$$
(d)
$$\underset{\text{2,3-dimethylbutane}}{CH_3\overset{\displaystyle CH_3}{\overset{|}{C}H}\overset{\displaystyle CH_3}{\overset{|}{C}HCH_3}}$$

12.3.
$$\underset{\text{3-methyl-1-butene}}{H-\overset{H}{\overset{|}{C}}=\overset{H}{\overset{|}{C}}-\overset{\overset{\displaystyle CH_3}{|}}{\underset{\underset{\displaystyle H}{|}}{C}}-\overset{\overset{\displaystyle H}{|}}{\underset{\underset{\displaystyle H}{|}}{C}}-H}$$

12.4.

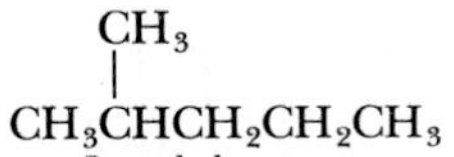

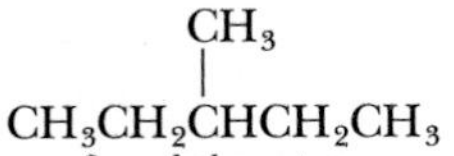

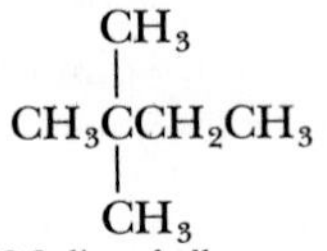

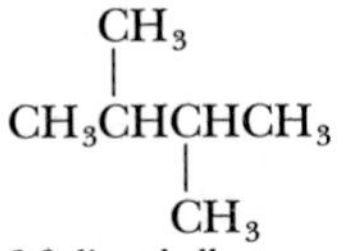

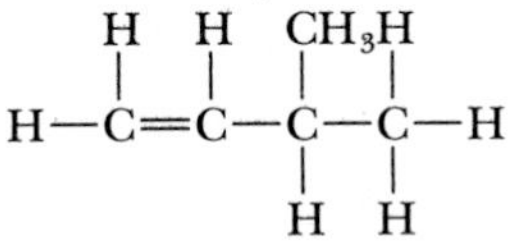

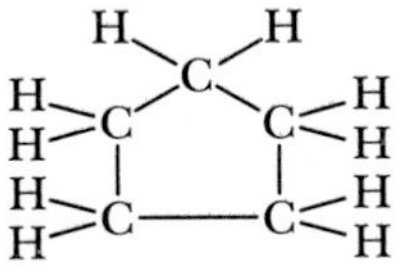

5 C, 10 H

12.5. methyl-*tertiary*-butyl ether, ethanol, benzene, 1-butene, octane

CHAPTER 13

13.1. Find the heating equivalence between these two sources of energy from Table 13.1: 1 million kWh electricity = 614 barrel (bbl) oil. Then use the equivalence as a ratio, as in Example 13.1, allowing the units to cancel.

$$1 \times 10^6 \text{ kWh electricity} \times \frac{614 \text{ bbl oil}}{1 \times 10^6 \text{ kWh electricity}}$$
$$= 614 \text{ bbl oil}$$

13.2. First, convert the food calories to joules.

$$500 \text{ food cal} = \frac{1000 \text{ cal}}{\text{food cal}} = \frac{4.184 \text{ J}}{\text{cal}} = 2.1 \times 10^6 \text{ J}$$

$$\text{Power} = \frac{2.1 \times 10^6 \text{ J}}{10 \text{ min}} \times \frac{1 \text{ min}}{60 \text{ sec}} = 3.5 \times 10^3 \text{ J/sec}$$
$$= 3.5 \times 10^3 \text{ W}$$

13.3. First, convert the electrical energy into a larger value using the 30% efficiency factor.

$$6.5 \times 10^6 \text{ kWh delivered} \times \frac{100 \text{ kWh generated}}{30 \text{ kWh delivered}}$$
$$= 22 \times 10^6 \text{ kWh generated}$$

Next, use Table 13.1 to find the energy equivalents.

$$\text{Tons of coal} = 22 \times 10^6 \text{ kWh} \times \frac{137 \text{ tons coal}}{1 \times 10^6 \text{ kWh}}$$
$$= 3.0 \times 10^3 \text{ tons coal}$$

Cubic feet natural gas

$$= 22 \times 10^6 \text{ kWh} \times \frac{3.41 \times 10^6 \text{ cubic feet}}{1 \times 10^6 \text{ kWh}}$$
$$= 75 \times 10^6 \text{ cubic feet}$$

$$\text{bbl of oil} = 22 \times 10^6 \text{ kWh} \times \frac{614 \text{ bbl oil}}{1 \times 10^6 \text{ kWh}}$$
$$= 14 \times 10^3 \text{ bbl oil}$$

13.4. Energy output $= 100{,}000 \text{ m}^2 \times \dfrac{200 \text{ W}}{\text{m}^2} \times 10 \text{ h}$

$$= 2 \times 10^8 \text{ Wh}$$
$$= 2 \times 10^8 \text{ Wh} \times \frac{1 \text{ kW}}{10^3 \text{ W}} = 2 \times 10^5 \text{ kWh}$$

Using Table 13.1 to find the energy equivalents:

$$\text{Coal} = 2.00 \times 10^5 \text{ kWh} \times \frac{137 \text{ tons coal}}{1 \times 10^6 \text{ kWh}} = 27.4 \text{ tons coal}$$

$$\text{Barrels of oil} = 2.00 \times 10^5 \text{ kWh} \times \frac{614 \text{ bbl oil}}{1 \times 10^6 \text{ kWh}}$$
$$= 123 \text{ bbl oil}$$

kWh electricity $= 2.00 \times 10^5$ kWh (same energy units)

CHAPTER 14

14.1. (a) substitution (b) addition (c) addition

14.2. (a) tertiary (b) secondary (c) primary

14.3. (a)
$$CH_3\overset{\displaystyle O}{\overset{\|}{C}}OCH_3$$

(b)
$$CH_3CH_2\overset{\displaystyle O}{\overset{\|}{C}}OCH_2CH_2O\overset{\displaystyle O}{\overset{\|}{C}}CH_2CH_3$$

14.4. (a)
$$\underset{H}{\overset{H}{C}}=\underset{H}{\overset{H}{C}}$$
(b)
$$\underset{H}{\overset{H}{C}}=\underset{Cl}{\overset{H}{C}}$$

(c)
$$\underset{\text{(benzene ring)}}{\overset{H}{\underset{H}{C}}=\overset{H}{C}}$$

14.5.
$$\left(\overset{\displaystyle O}{\overset{\|}{C}}\text{—}\langle\text{ring}\rangle\text{—}\overset{\displaystyle O}{\overset{\|}{C}}\text{—O—CH}_2\text{—CH}_2\text{—O}\right)_n$$

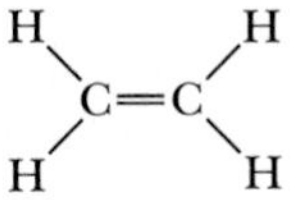

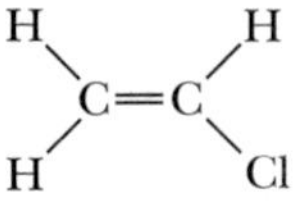

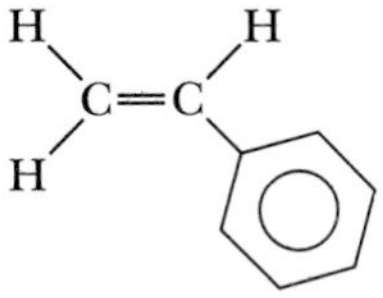

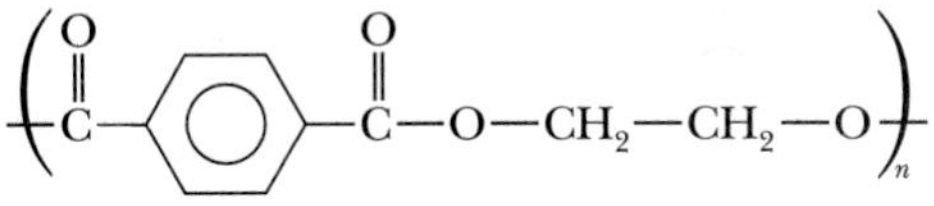

CHAPTER 15

15.1. (b) is chiral.

$$\underset{Cl}{\overset{\displaystyle OH}{\overset{|}{\underset{|}{C}}}}\overset{\quad}{\underset{\quad}{H\cdots C\cdots CH_3}} \qquad \underset{H}{\overset{\displaystyle OH}{\overset{|}{\underset{|}{C}}}}\overset{\quad}{\underset{\quad}{Cl\cdots C\cdots CH_3}}$$

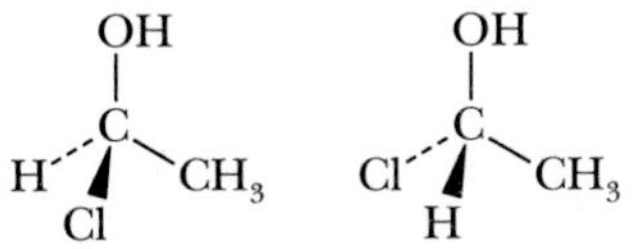

15.2. Amino acids have a zwitterion structure because the basic amino group can accept a proton from the carboxylic acid group.

15.3.
$$CH_3\text{—}\underset{\underset{\displaystyle CH_3}{|}}{CH}\text{—}\underset{\underset{\displaystyle NH_3^{\oplus}}{|}}{CH}\text{—}COO^{\ominus}$$

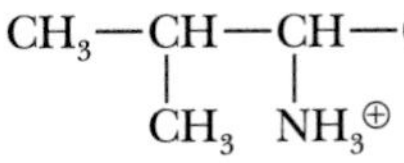

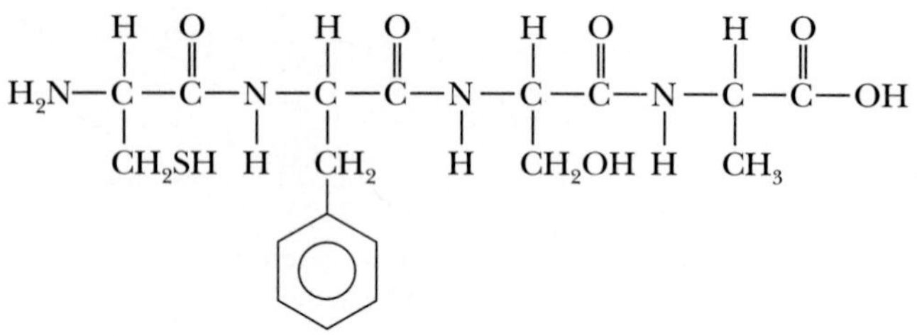

15.4. AGGCTA

CHAPTER 17

17.1. 160. lb × 10 kcal/lb = 1600 kcal
1600 kcal × 1.3 = 2080 kcal

17.2A. Healthy Chicken:

$$\frac{\text{15 Calories from fat}}{\text{90 Calories total}} \times 100 = 17\% \text{ Calories from fat}$$

Cream of Mushroom:

$$\frac{\text{80 Calories from fat}}{\text{140 Calories total}} \times 100 = 57\% \text{ Calories from fat}$$

Cream of Mushroom has more than 30% Calories from fat

17.2B. To decide how many cookies will provide 53 g − 33 g = 20 g of fat requires knowing how many grams of fat per cookie:

$$\frac{\text{45 Calories}}{\text{2 cookies}} \times \frac{\text{1 g fat}}{\text{9 Calories}} = 2.5 \text{ g fat/cookie}$$

$$20 \text{ g fat} \times \frac{\text{1 cookie}}{\text{2.5 g fat}} = 8 \text{ cookies}$$

17.3.

$$3 \text{ g protein} \times \frac{\text{4 kcal}}{\text{g protein}} = 12 \text{ kcal}$$

$$8 \text{ g fat} \times \frac{\text{9 kcal}}{\text{g fat}} = 72 \text{ kcal}$$

$$40 \text{ g carbohydrate} \times \frac{\text{4 kcal}}{\text{g carbohydrate}} = 160 \text{ kcal}$$

$$\text{Total} = 244 \text{ kcal per slice}$$

(b) $$\frac{\text{72 kcal}}{\text{244 kcal}} \times 100 = 30\% \text{ kcal from fat}$$

(c) $$244 \text{ kcal} \times \frac{\text{1 min of lecture}}{\text{1.7 kcal}} = 140 \text{ minutes of lecture}$$

17.4. 1600 kcal × 0.30 = 480 kcal from fat

$$480 \text{ kcal} \times \frac{1 \text{ g}}{9 \text{ kcal}} = 53 \text{ g fat/day}$$

1600 kcal × 0.60 = 960 kcal from carbohydrate

$$960 \text{ kcal} \times \frac{1 \text{ g}}{4 \text{ kcal}} = 240 \text{ g carbohydrate/day}$$

1600 kcal × 0.10 = 160 kcal protein

$$160 \text{ kcal} \times \frac{1 \text{ g}}{4 \text{ kcal}} = 40 \text{ g protein/day}$$

CHAPTER 18

18.1. Dose = (12 mg/kg) × (1.25 kg) = 15 mg

18.2. Use the density of water as 1 g/mL. 1L of water weighs 1000 g.

$$1000 \text{ g water} \times \left(\frac{100 \text{ g Pb}}{10^9 \text{ g water}}\right) = 10^{-4} \text{ g Pb } (100\ \mu\text{g Pb})$$

CHAPTER 22

22.1. Formula weight: 2 N × 14 = 28, 4 H × 1 = 4, 3 O × 16 = 48. Total = 80.

$$\text{Percentage N} = \frac{28}{80} \times 100\% = 35\%$$

There is no K or P in ammonium nitrate, so their values are zero.

22.2. $75 \text{ kg} \times \frac{500 \text{ mg}}{\text{kg}} = 37{,}500$ mg or 37.5 g could kill the farmer.

APPENDIX G

Answers to Selected Questions and Problems

CHAPTER 2

1. Some possibilities: oxygen in the air, water, methane in natural gas, beeswax, wood, etc.
3. No. Two pure substances that have the same set of physical and chemical properties would be the same substance.
5. A spark provides energy to ignite the air and fuel mixture. The oxygen in the air and the hydrocarbon fuel are chemically changed to heat, water vapor, and carbon dioxide (new substances).
7. (a) Density is a physical property because it depends only on the physical properties of mass and volume.
(b) Physical property, because melting temperature is a measure of the amount of energy needed to physically separate the particles in a solid to allow them to move freely in the liquid.
(c) Chemical property, because the substance changes from a compound to two elements, a change in chemical composition.
(d) Physical property. The chemical composition of a piece of copper wire does not change when it conducts electricity.
(e) Chemical property. Anything that refers to the reaction of one substance with another is a chemical property.
(f) Chemical property. It indicates the ease with which the paper reacts with air, resulting in a change in composition.
9. (a) element, contains only Hg atoms
(b) mixture of water, minerals, proteins, fats
(c) compound, contains only one kind of molecule (H_2O)
(d) mixture of cellulose, water. Wood changes weight when dried.
(e) mixture of dye and solvent
(f) mixture of water, caffeine, tea extract
(g) compound, solid pure water containing only one kind of molecule
(h) element, contains only C atoms
(i) element, contains only Sb atoms
11. Properties of iron do not change because all particles in iron are atoms of iron. Steel is a mixture of iron and other atoms. The type of steel depends on what is added to the iron. Properties of steel change with composition.
13. The water can be evaporated or distilled off, trapped, and condensed. The solids will be left behind as a residue.
15. carbon, hydrogen, nitrogen, chlorine
17. One example of each:
(a) BonAmi kitchen and bath cleanser
(b) Coca Cola
(c) Gatorade
(d) Coca Cola
(e) Skippy Peanut Butter
(f) Kraft Grated Parmesan Cheese
(g) Morton's Iodized Salt
(h) Oil of Olay
(i) Mylanta
(j) Kellogg's Frosted Mini-Wheats
19. $N_2 + 3\ H_2 \rightarrow 2\ NH_3$
21. electrical, heat, light, mechanical
23. (a) On the left side of the arrow, "2 Na" means 2 Na atoms; one Cl_2 molecule contains 2 Cl atoms. On the right side, 2 NaCl units contain 2 Na atoms and 2 Cl atoms.
(b) On the left, 1 N_2 molecule contains 2 N atoms and 3 Cl_2 molecules contain 6 Cl atoms. On the right, 2 NCl_3 molecules contain a total of 2 N atoms and 6 Cl atoms.
(c) On the left there are 1 C atom, 2 H atoms, and $2 + 1 = 3$ O atoms. On the right there are 1 C atom, 2 H atoms, and 3 O atoms.
(d) On the left there are 4 H atoms and 4 O atoms in 2 molecules of H_2O_2. On the right there are 4 H atoms in the 2 molecules of water; there are also 2 O atoms in the 2 water

molecules and 2 more O atoms in the O_2 molecule for a total of 4 O atoms.

25. (a) No
    (b) Yes
27. Some examples: baking soda, sugar, flour, water, aluminum foil
29. gram, meter, liter
31. Yes

SELECTED PROBLEMS

1. 96 marbles
2. 50 glasses
3. 200,000 kilobytes, 200,000,000 bytes
4. 17 g
5. 5.5 acres/55 cows
6. 0.200 g, 200,000 μg
7. 1000 cm^3, density
8. 3 g/cm^3, density
9. 3000 kg/m^3
13. (a) 0.04 m
    (b) 43 mg
    (c) 15,500 mm
    (d) 0.328 L
    (e) 980 g
18. 965 g
20. 500 cm^3
22. 0°F

CHAPTER 3

1. Matter is neither created nor destroyed. Examples: flash bulb before and after use, hard boiling an egg, yarn knitted into clothing, melting ice cubes in a glass. There is no change in mass during the chemical changes (first two) or the physical changes.
3. Matter could be divided into smaller and smaller particles with each successive piece duplicating the properties of the previous size particle.
5. (a) Dalton's atoms were indestructible.
   (b) Dalton's model of compounds had specific combinations of atoms.
   (c) Atoms combine in whole-number amounts to form molecules of compound; see answer to question 7.
7. The Law of Multiple Proportions says that in two different compounds consisting of the same two elements, the masses of one element combined with a fixed mass of the second element are related by the ratio of small whole numbers. In SO_2 there are 32 g S with 32 g O, and in SO_3 there are 32 g S with 48 g O; with S fixed at 32 g, the ratio of the O amounts is 48 g/32 g or 3/2 in small whole numbers.
9. Cathode rays are streams of electrons produced in an evacuated tube.
11. Ground state for an atom is the condition in which all electrons are in the lowest possible energy levels. An excited state is a condition in which one or more electrons are in energy levels greater than the lowest ones.
13. (a) the number of protons in the nucleus of an atom
    (b) the total count of protons *plus* neutrons in the nucleus of an atom
    (c) the mass of an average atom of an element compared to an atom of ^{12}C which is assigned a mass of exactly 12 atomic mass units
    (d) atoms with the same number of protons but with different numbers of neutrons
    (e) on Earth, the percentage of all the atoms of the element that are this isotope.
    (f) 1 amu is one-twelfth of the mass of the carbon-12 atom. See also answer to question 13c.
15. (a) No, water evaporating from a puddle merely escapes into the atmosphere to increase water content of the atmosphere.
    (b) No, the apparent decrease in matter occurs because gas products escape into the atmosphere.
17. Atoms exist with 1, 2, . . . , up to 109 protons; each number of protons corresponds to a different element. Most elements exist as isotopes, i.e., atoms with various numbers of neutrons for a given number of protons.
19. (a) ^{50}Ti and ^{50}V are different elements with the same mass number, not isotopes.
    (b) ^{12}C and ^{14}C are isotopes with same proton count (6 each) but with different neutron counts (6 in ^{12}C and 8 in ^{14}C).
    (c) ^{40}Ar and ^{40}Kr are not isotopes; Ar and Kr are different elements.
21. 53 protons and 127 − 53 = 74 neutrons
23. (a) 32, 41, 32, 32, 73
    (b) 14, 14, 14, 14, 28
    (c) 28, 31, 28, 28, 59
    (d) 48, 64, 48, 48, 112
    (e) 77, 115, 77, 77, 192
25. Atomic Number, or number of protons.
27. Bromine-81: 35, 81, 35, 46, 35
    Boron-11: 5, 11, 5, 6, 5
    Chlorine-35: 17, 35, 17, 18, 17
    Chromium-52: 24, 52, 24, 28, 24
    Nickel-60: 28, 60, 28, 32, 28
    Strontium-90: 38, 90, 38, 52, 38
    Lead-206: 82, 206, 82, 124, 82
29. (c) the ratio by weight of the elements in the compound
31. Concepts in science evolve over time and change to account for additional facts.
33. Kr; 36; 83.80 amu; 2-8-18-8
35. (a) 12, 12, 12
    (b) 26, 30, 26
    (c) 49, 66, 49
    (d) 53, 74, 53
    (e) 47, 60, 47
    (f) 86, 136, 86

37. $1s^22s^22p^2$; $1s^22s^22p^6$; $1s^22s^22p^63s^23p^1$; $1s^22s^22p^63s^23p^64s^2$
39. Na, 1; Mg, 2; Al, 3; Si, 4; P, 5; S, 6; Cl, 7; Ar, 8.
41. Be: $1s^22s^2$; B: $1s^22s^22p^1$; C: $1s^22s^22p^2$; N: $1s^22s^22p^3$; O: $1s^22s^22p^4$; F: $1s^22s^22p^5$; Ne: $1s^22s^22p^6$.
43. X-ray, ultraviolet, visible, infrared, microwave, radio wave
45. With the electron starting from n = 1, the smallest jump is to n = 2.
47. Some lines (wavelengths) observed in the solar emission spectrum did not match wavelengths emitted by any of the elements known in 1868. Since each element has its own unique emission spectrum, scientists knew they were looking at emissions from a previously-undiscovered element.

SELECTED PROBLEMS

1. 49.6 g C, 10.4 g H
2. 6.6×10^{-20} C
3. 1000 inches, 25.4 meters
5. Line up more than 1000 marbles to get a length that you can measure.
6. 2.54×10^8 or 254,000,000 taws
8. $3.496 \times 10^8 = 3.50 \times 10^8$ to 3 significant digits
11. (a) 1.2 cm
    (b) 3.9 cm
    The 1.2 cm wave carries more energy.

CHAPTER 4

1. When elements are arranged in order of their atomic numbers, their chemical and physical properties show repeatable trends.
3. Metals are elements that show metallic properties such as the physical properties of malleability, ductility, and good electrical and thermal conductivity; chemically, metals react to lose electrons and form positive ions. Metals are the elements, except for hydrogen, lying to the left and below the diagonal line formed by the metalloids in the periodic table.
5. Metalloids are elements that have properties intermediate between those of metals and nonmetals.
7. elements with atomic numbers 27 and 28, 52 and 53, 90 and 91, 91 and 92, 99 and 100, 101 and 102, 106 and 107
9. Metals: sodium, lead. Nonmetals: hydrogen, chlorine. Metalloids: arsenic, germanium.
11. Metals lose electrons to form positive ions, are malleable and ductile, are good conductors of heat and electricity. Nonmetals gain electrons to form negative ions, are brittle, are insulators not conductors.
13. (a) 7
    (b) 8
    (c) 2, if you ignore H in IA
    (d) Group VIIA and VIIIA
    (e) Yes, period 7.
15. All have the same number of valence electrons and have similar chemical properties.
17. Atomic radius is the distance from the center of the nucleus to the outer surface of the electron cloud around the nucleus. Reactivity is a measure of how readily the substance enters into a chemical reaction.
19. All have one valence electron.
21. (a) 2
    (b) 3
    (c) 5
    (d) 6
    (e) 7
    (f) 1
23. ·Be· :C̤̈l· K· ·Äṡ :K̤̈r:
25. (a) Li
    (b) Cs
    (c) Ba
    (d) Pb
    (e) Al
    (f) Na
27. 8
29. Carbon 4, 2, NM
    Magnesium 2, 3, M
    Chlorine 7, 3, NM
    Rubidium 1, 5, M
    Molybdenum 6, 5, M
    Xenon 8 or 0, 5, NM
31. F, Cl, Br, I, At
33. Beryllium, Be
    Magnesium, Mg
    Calcium, Ca
    Strontium, Sr
    Barium, Ba
    Radium, Ra
35. The smaller the radius of the nonmetal atom, the more reactive it is.
37. Both atomic size and metallic character decrease from left to right across a period; both atomic size and metallic tendency increase from top to bottom of a group.
39. Electronic structure repeats in each level.
41. K (largest), Al, P, S, Cl (smallest)
43. Group IVA; carbon is a nonmetal with nonmetallic properties and lead is a metal with metallic properties.
45. Helium: cooling gas, balloons
    Neon and Argon: lighting, electric discharge tubes
47. Compounds of xenon have now been produced so all group members are not totally unreactive or inert.
49. The number of oxygen atoms per one atom of the other element increases by 1/2 from one group to the next: MgO (1 to 1), Al_2O_3 (1 to 1.5), SiO_2 (1 to 2), etc. We predict a formula of Cl_2O_7.
51. both have 6 outer electrons
53. False
55. 23 amu, 0.7 g/cm³, 122°C, 1044°C
57. First element of period 8, in Group IA.

CHAPTER 5

1. Gamma rays
3. The more hazardous radioisotope is $^{222}_{86}Rn$ with a short half-life of 3 days because more ionizing radiation is emitted over a shorter time period.
5. annihilation, see page 140. $^{0}_{+1}e + ^{0}_{-1}e \rightarrow 2\ ^{0}_{0}\gamma$
7. conservation of matter
9. $^{4}_{2}He$ or $^{1}_{0}n$ projectiles
11. Transuranium elements have atomic numbers greater than 92 and are all artificial.
13. The "actinium series" is a decay series that starts with U-235 and ends with Pb-206.
15. Gamma radiation of food prevents spoilage. Living organisms such as bacteria, molds, and yeasts are killed by the gamma radiation.
17. Cyclotrons accelerate projectile particles to high enough velocities so collisions between projectile and target nuclei result in fusion of the nucleons forming in a new nucleus.
19. The gamma rays can damage DNA sequences necessary for cell reproduction, thereby slowing or stopping cell reproduction. Cancer cells reproduce faster than normal cells and are more sensitive to this.
21. natural radiation in food, water, and air.
23. Gamma irradiation would kill the *E. coli*.
25. This indicated that the radioactive materials gave off ionizing radiation that exposed the film.
27. (a) Rutherford discovered that alpha rays could be stopped by thin layers of paper and that beta rays were more penetrating.
 (b) Irene and Frederic Joliot made the first artificial radioactive isotope, phosphorus-30.
 (c) explosion of the first atomic bomb
 (d) Becquerel discovered that uranium emitted ionizing radiation.
 (e) additional nuclear testing
29. none

SELECTED PROBLEMS

1. (a) $^{64}_{30}Zn$ (b) $^{0}_{-1}e$ (c) $^{0}_{-1}e$
3. (a) $^{218}_{84}Po$ (b) $^{221}_{88}Ra$
4. 18,750
5. 5,000
9. 100.45 mrem, cosmic and ground radiation.
11. alpha particles
12. 3366 disintegrations/minute

CHAPTER 6

1. (a) ion with a positive charge
 (b) ion with a negative charge
 (c) Atoms react to acquire an electron configuration with 8 electrons in the outermost level.
 (d) simplest element ratio for an ionic compound
3. (a) a pair of electrons shared between two atoms
 (b) four electrons (2 pairs) shared between two atoms
 (c) six electrons (3 pairs) shared between two atoms
 (d) a pair of valence electrons on an atom that are not shared with another atom
 (e) two electrons (1 pair) shared between two atoms
 (f) a double or a triple bond
5. (a) bond between two atoms that share electrons evenly or equally
 (b) bond between two atoms where one atom attracts the shared electrons more than does the other atom so electrons are not shared equally
 (c) a measure of an atom's attraction for shared electrons
7. (a) compound consisting only of carbon and hydrogen
 (b) compound consisting only of carbon and hydrogen with only single bonds between carbon atoms
 (c) compound consisting only of carbon and hydrogen with one or more multiple bonds between carbon atoms
 (d) a molecule of carbon with formula C_{60} whose surface has five-membered rings of carbon linked to six-membered rings of carbon (resembles a hollow soccer ball).
 (e) hydrocarbons with one or more carbon-carbon double bonds
9. (a) Br^{1-}
 (b) Al^{3+}
 (c) Na^{1+}
 (d) Ba^{2+}
 (e) Ca^{2+}
 (f) Ga^{3+}
 (g) I^{1-}
 (h) S^{2-}
 (i) 1+ ion
 (j) 1− ion
11. −2, S^{2-}
13. (a) Neutral atom. A cation has more protons than electrons so the electron cloud is pulled in closer to the nucleus.
 (b) Anion. Additional electrons are repelled by other electrons in the electron cloud and there are more electrons than protons so each electron is held more loosely by the nucleus.
15. electrostatic attractions between positive and negative ions in an ionic lattice
17.

Formula	MP	Formula	MP
NaCl	801°C	Mg_3P_2	not listed
Na_2S	1180°C	$AlCl_3$	190°C
Na_3P	decomposes	Al_2S_3	1100°C
$MgCl_2$	708°C	AlP	not listed
MgS	decomposes		

All are solids at room temperature. For a given cation the melting points increase with the charge on the anion.

19. IA, H 1 bond, Na-Fr no covalent bonds
 IIA, no covalent bonds
 IIIA, 3 bonds
 IVA, 4 bonds
 VA, 3 bonds
 VIA, 2 bonds
 VIIA, 1 bond
 VIIIA, 0 bonds
21. $Tb(SO_4)_2$, yes
23. (a) $Al_2(SO_4)_3$
 (b) $Ca(H_2PO_4)_2$
 (c) K_3PO_4
 (d) NH_4NO_3
 (e) Na_2CO_3
 (f) $CaHPO_4$
25. (a) lose 1 electron
 (b) gain 1 electron
 (c) gain 4 electrons
 (d) lose 2 electrons
27. (a) NaH
 (b) MgH_2
 (c) GaH_3
 (d) GeH_4
 (e) AsH_3
 (f) HCl
29. (a) nitrogen monoxide
 (b) sulfur trioxide
 (c) dinitrogen oxide
 (d) nitrogen dioxide
31. (a) 14
 (b) 8
 (c) 8

33. (a) H—N̈—N̈—H (b)
 | |
 H H H
 H—C—Ö—H
 |
 H

 (c) :C̈l:B:C̈l (d) H:P̈:H
 :C̈l: H

 (e) [:Ö:C̈l:Ö:]⁻ (f) [:Ö:S̈:Ö:]²⁻
 :O: :O:

 (g) [H]⁺ (h) :N:::N:
 [H:N̈:H]
 [H]

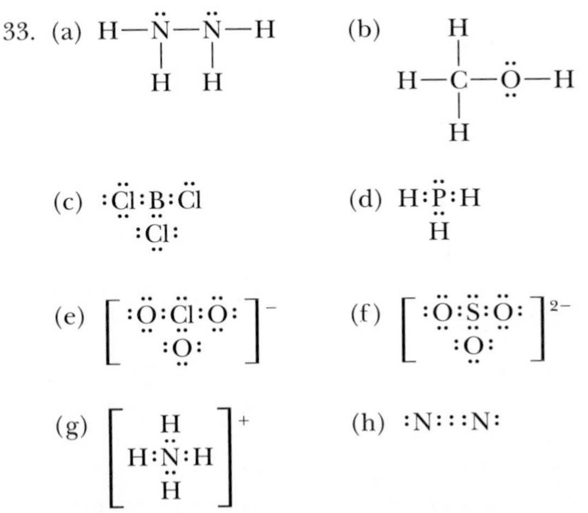

35. The electron-pair geometry depends on the positions of the electron pairs around the center atom in the molecule. The molecular geometry refers to the shape of the molecule; shapes are described in terms of locations of atoms (nuclei) in the molecule. Water: electron-pair geometry is tetrahedral and molecular shape is angular or bent.

37.

Lewis Formula	Electron-Pair Geometry	Molecular Geometry
(a) H:N̈:C̈l: H	tetrahedral	trigonal pyramid
(b) H:Ö:F̈:	tetrahedral	angular
(c) :S::C::S:	linear	linear
(d) [:C̈l:Sn:C̈l:]⁻ :C̈l:	tetrahedral	trigonal pyramid
(e) :B̈r: :B̈r:C̈:B̈r: :B̈r:	tetrahedral	tetrahedral
(f) [:F̈:]⁻ [:F̈:B̈:F̈:] [:F̈:]	tetrahedral	tetrahedral

39. H:C:::N:, linear
41. (a) C—Cl
 (b) C—F
43. $BeCl_2$ is linear so partial charges on the chlorine atoms counteract each other.
45. (a) HF
47. The electron-pair geometry is basically tetrahedral but the lone pair on the N distorts the HNH angle from 109.5° to 106.5°.
49. (a) 24
 (b) 8
 (c) 1
 (d) 10
 (e) 2

 H :O: H
 | ‖ |
 H—C—C—C—H
 | |
 H H

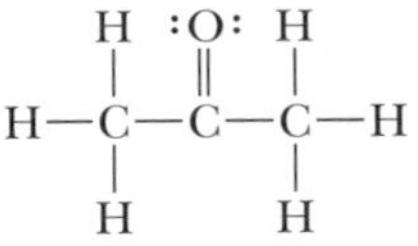

51. (a) (b)

SELECTED PROBLEMS
1. 170 pm, 290 pm
2. The quarter is the anion while the penny is the cation; 1670 pm

■ **CHAPTER 7**

1. (a) The gaseous state has no definite shape or volume; a gas takes the shape and volume of its container.

(b) A liquid has definite volume and assumes the shape of the container.
(c) The solid state has definite shape and definite volume.

3. gas
4. liquid
7. The number of gas molecules in a given volume of gas decreases as you move away from Earth's surface.
9. The pressure exerted by the gas mixture results from the collisions of molecules of both gases acting independently with the container walls.
11. Hydrogen bonds indicated by arrows:

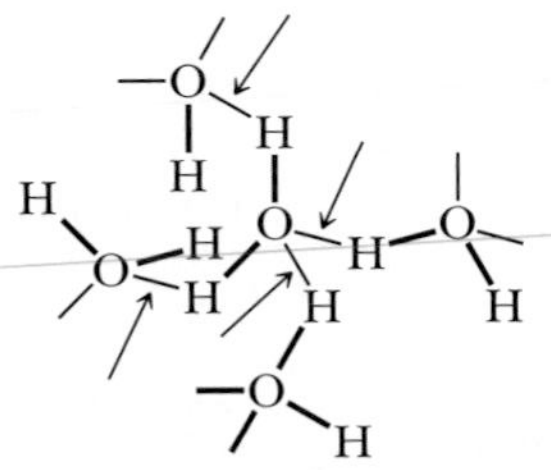

13. (a) Yes. Hydrogen is covalently bonded to oxygen, a small and very electronegative atom.
(b) Yes. Hydrogen is covalently bonded to nitrogen, a small and very electronegative atom.
(f) Yes. Hydrogen is covalently bonded to fluorine, a small and very electronegative atom.
15. Water molecules have strong attractions for one another. The water molecules making up the surface are difficult to separate and the surface seems relatively hard to the "belly flopper."
17. There are great distances between particles in the gas state, while particles in the liquid and solid states are in direct contact.
19. The temperature at which the vapor pressure of the liquid equals 760 mm Hg.
21. Attractive forces between particles in the surface layer are unbalanced because the particles are attracted by the bulk of the liquid below the surface. Water has a high surface tension.
23. about 70°C
25. No, the carbon skeleton is too large.
27. The solubility of oxygen in water decreases when water temperature goes up. Fish may not be able to get enough dissolved oxygen from the warmer water.
29. sublimation
31. water
33. electrical conductivity
35. The n-type have excess electrons while the p-type have a shortage of electrons.
37. Dry air is circulated through the freezer replacing moist air and making any frost sublime to reestablish equilibrium.
39. to find materials that superconduct at relatively high temperatures.

SELECTED PROBLEMS

1. 2200 mL
3. 378 mm Hg
6. 6 L
7. (a) 295 K
(c) 300.3 K
(e) −244°C
(f) 358°C
8. 3.2 atm
9. 497 K = 224°C
13. 300 mL

CHAPTER 8

1. (a) 6 and 6
(b) 7 and 7
(c) 1, 3, 2, 3
(d) The number of atoms of each element is the same on both reactant and product side.
3. (a) $2\ Al + 3\ Cl_2 \rightarrow 2\ AlCl_3$
(b) $3\ Mg + N_2 \rightarrow Mg_3N_2$
(c) $2\ NO + O_2 \rightarrow 2\ NO_2$
(d) $2\ SO_2 + O_2 \rightarrow 2\ SO_3$
(e) $3\ H_2 + N_2 \rightarrow 2\ NH_3$
5. (a) $Ba(s) + 2\ H_2O(\ell) \rightarrow Ba(OH)_2(aq) + H_2(g)$
(b) $3\ Fe(s) + 4\ H_2O(\ell) \rightarrow Fe_3O_4(s) + 4\ H_2(g)$
(c) $2\ Na(s) + 2\ H_2O(\ell) \rightarrow 2\ NaOH(aq) + H_2(g)$
(d) $2\ Li(s) + 2\ H_2O(\ell) \rightarrow 2\ LiOH(aq) + H_2(g)$
7. (a) $Sn(s) + 2\ HBr(aq) \rightarrow SnBr_2(aq) + H_2(g)$
(b) $Mg(s) + 2\ HCl(aq) \rightarrow MgCl_2(aq) + H_2(g)$
(c) $Ca(s) + 2\ H_2O(\ell) \rightarrow Ca(OH)_2(aq) + H_2(g)$
(d) $Zn(s) + 2\ HNO_3(aq) \rightarrow Zn(NO_3)_2(aq) + H_2(g)$
(e) $2\ Cs(s) + 2\ H_2O(\ell) \rightarrow 2\ CsOH(aq) + H_2(g)$
9. the carbon-12 isotope
11. Each is 0.500 mol of that element.
13. 6.02×10^{23}
15. The mole is defined as the number of atoms in exactly 12 g of carbon-12, about 6.02×10^{23} atoms.
17. (a) 6.02×10^{23}
(b) 3.01×10^{23}
(c) 6.02×10^{24}
(d) 1.93×10^{25}
(e) 6.02×10^{20}
19. Some reactions are fast because reactants have weak bonds. These reactions have low activation energies; the reactants have a "low hill to climb" to form products. Other reactions are slow because reactants have strong bonds and a high activation energy; these reactants have a "high energy hill to climb" to form products.
21. Slows reaction rates. Freezing slows reaction rates, slowing spoilage and slowing growth of mold or bacteria.
23. Reactions with high activation energies tend to be slow.

25. The activation energy is high so there is no reaction at room temperature. Once started by the spark (activation energy), the reaction continues with production of energy because the products have lower potential energy than the reactants.
27. Faster because contact with oxygen molecules would be better with the higher concentration of oxygen molecules.
29. Reactions that take place in both forward and reverse directions.
31. (a) shift to form $CaCO_3$
(b) shift to form CaO and CO_2
33. Additional reactants are consumed in the reaction to make more products.
35. Unreacted reactants are essentially zero; all are converted to products.
37. (a) stored energy
(b) a process that gives off heat
(c) a process that consumes heat
(d) a measure of disorder
(e) a process that favors products at the expense of reactants
39. Every spontaneous process occurs with an increase in entropy for the universe. This entropy increase consumes energy that cannot then be used for other purposes. Every spontaneous process has a built-in amount of this "wasted" energy.
41. exothermic
43. No. Photosynthesis requires a continuous input of energy to keep the process going; without this energy the process will stop. Sunlight is the energy source.
45. No, because recycling may organize the recycled materials, but the steps in the process create entropy in other matter.

SELECTED PROBLEMS

1. (a) 18 g
(b) 254 g
(c) 56 g
(d) 17 g
(e) 44 g
(f) 28 g
3. (b) 1 mol Pb
5. (a) 1.0 mol
(c) 10 mol
6. (a) 76.5 g
(b) 0.0782 g
7. 161.8 g HBr
9. $C_6H_{12}O_6(aq) \rightarrow 2\ CH_3CH_2OH + 2\ CO_2$
(a) 12 mol
(b) 10.5 mol
11. $2\ Sn + O_2 \rightarrow 2\ SnO$
(a) 13.5 g
(b) 0.72 g
(c) 6.1 g

CHAPTER 9

1. a substance that donates a proton, H^{1+}, to water molecules
3. H_3O^{1+}
5. (a) Acidic solutions have H_3O^{1+} concentrations greater than the OH^{1-} concentration.
(b) Basic solutions have an OH^{1-} concentration greater than the H_3O^{1+} concentration.
(c) Neutral solutions have pH = 7 and have equal concentrations of hydronium ions, H_3O^{1+}, and hydroxide ions, OH^{1-}.
7. A salt is the substance formed between the anion of an acid and the cation of a base.
9. Spectator ions are essential because they keep electrical "charge balance." They do not participate directly in a reaction or change during the course of a reaction.
11. any molecule or ion that accepts a proton.
13. (a) a molecule or ion that donates an H^{1+} ion (proton)
(b) a molecule or ion that bonds to a proton
(c) a substance that can act as either a proton donor or a proton acceptor
(d) When a base bonds to a proton, i.e., accepts a proton, the particle that results is called the "conjugate acid" of this base.
(e) A conjugate base is what is left of an acid after it gives up its proton.
(f) A conjugate acid-base pair is a set of two particles, one is a proton donor and the other is the particle formed after the proton is lost.
15. aspirin, coffee, vinegar, fruit juice and carbonated beverages
17. (a) acidic
(b) basic
(c) basic
(d) acidic
(e) acidic
19. (a) acidic
(b) acidic
(c) acidic
(d) basic
(e) acidic
(f) basic
(g) basic
(h) basic
(i) acidic
21. (a) acid
(b) base
(c) base
(d) base
(e) acid
23. (a) a hydrogen that can break away easily as H^{1+}, leaving its electron behind
(b) an acid with more than one acidic hydrogen

(c) a concentration measure equal to the number of moles of solute in one liter of solution
(d) measures the relative amount of dissolved solute in a definite amount of solution

25. An acid-base indicator changes color with changes in acidity or pH. Indicator molecules act as both acids and bases. They display one color when protonated in acidic solutions and another color when they lose a proton in basic solutions. Litmus is blue in basic solutions and turns red in acidic solutions. Red cabbage is pink in acid and green in base solutions. Purple grape juice is red in acid and olive green in base.
27. (a) a condition resulting from blood pH levels below 7.35
(b) a condition that occurs when blood pH rises above 7.5
29. $CaO(s) + H_2O(\ell) \rightarrow Ca(OH)_2(aq)$
$Na_2O(s) + H_2O(\ell) \rightarrow 2\ NaOH(aq)$
$MgO(s) + H_2O(\ell) \rightarrow Mg(OH)_2(aq)$
$Al_2O_3(s) + 3\ H_2O(\ell) \rightarrow 2\ Al(OH)_3(aq)$
31. (a) $2\ HBr(aq) + Ca(OH)_2(aq) \rightarrow CaBr_2(aq) + 2\ H_2O(\ell)$
$H^{1+}(aq) + OH^{1-}(aq) \rightarrow H_2O(\ell)$
(b) $3\ HNO_3(aq) + Al(OH)_3(s) \rightarrow 3\ H_2O(\ell) + Al(NO_3)_3(aq)$
$3\ H^{1+}(aq) + Al(OH)_3(s) \rightarrow 3\ H_2O(\ell) + Al^{3+}(aq)$
33. (a) H_3O^{1+}
(b) H_2S
(c) H_2SO_4
(d) H_3PO_4
35. (a) $\underset{\text{Acid}}{HCl(aq)} + \underset{\text{Base}}{H_2O(\ell)} \rightarrow \underset{\text{Conjugate acid}}{H_3O^{1+}(aq)} + \underset{\text{Conjugate base}}{Cl^{1-}(aq)}$
(b) $\underset{\text{Acid}}{H_2SO_4(aq)} + \underset{\text{Base}}{H_2O(\ell)} \rightarrow \underset{\text{Conjugate acid}}{H_3O^{1+}(aq)} + \underset{\text{Conjugate base}}{HSO_4^{1-}(aq)}$
(c) $\underset{\text{Base}}{NH_3(aq)} + \underset{\text{Acid}}{H_2O(\ell)} \rightleftharpoons \underset{\text{Conjugate base}}{OH^{1-}(aq)} + \underset{\text{Conjugate acid}}{NH_4^{1+}(aq)}$
(d) $\underset{\text{Acid}}{CH_3COOH(aq)} + \underset{\text{Base}}{H_2O(\ell)} \rightleftharpoons \underset{\text{Conjugate acid}}{H_3O^{1+}(aq)} + \underset{\text{Conjugate base}}{CH_3COO^{1-}(aq)}$
37. A solution with a pH of 2 has an hydronium ion concentration of $10^{-2} = 0.01$, while in a solution with pH = 10 the $H_3O^{1+}(aq)$ ion concentration is 10^{-10}. The pH 2 solution is much more acidic than the pH 10 solution because it has greater $H_3O^{1+}(aq)$ ion concentration.
39. (a) acidic
(b) basic
(c) acidic
(d) basic
(e) acidic
(f) basic
41. Ammonia solution is 100 times more basic.
43. The acidic HCO_3^{1-} can provide H^{1+} to neutralize added base. The CO_3^{2-} will react with added acid or H^{1+} to produce HCO_3^{1-}. This action will stabilize the H^{1+} concentration and pH when either acid or base is added.
45. The baking soda is a weak base. It will neutralize the acid, but any excess will not produce caustic burns. $NaHCO_3(aq) + HCl(aq) \rightarrow H_2O(\ell) + CO_2(g) + NaCl(aq)$

SELECTED PROBLEMS

1. (a) 0.030 mole HCl
(c) 0.012 mole HNO_3
2. (a) 0.0060 mole HCl
(b) 0.000025 mole KOH
3. (a) 3.2 g NaOH
(b) 31 g NH_3
(c) 2.9 g H_2SO_4
(d) 15 g H_3PO_4
5. (a) 15. g CH_3COOH
(b) 310 g $Al(OH)_3$
7. (a) 0.45 M HBr
(b) 0.18 M H_2SO_4
(c) 2.8 M HNO_3
(d) 0.1 M NaOH
9. (a) 1.88 M NaOH
(b) 0.44 M H_2SO_4
(c) 0.018 M $Ba(OH)_2$
(d) 0.0095 M $Ca(OH)_2$
11. (a) 0.16 M NH_3
(b) 0.30 M HCl
(c) 5.7 M H_2SO_4
(d) 0.031 M KOH
12. (a) pH = 2
(b) pH = 3
(c) pH = 13
(d) pH = 1
13. (d) 7
14. (a) $[H_3O^{1+}] = 1 \times 10^{-1}$ or 0.1
(b) $[H_3O^{1+}] = 1 \times 10^{-0}$ or 1
15. (a) 1×10^{-6}
(b) 1×10^{-4}

CHAPTER 10

1. (a) loss of electrons
(b) gain of electrons
(c) gain of oxygen
(d) gain of hydrogen
(e) substance that accepts electrons, causes oxygen to be gained, or causes hydrogen to be lost
(f) substance that gives up electrons, causes oxygen to be lost, or causes hydrogen to be gained
3. (a) CO_2
(b) NO_2
(c) SO_3
(d) CrO_3
(e) CaO
(f) N_2
5. H_2O
7. Oxidation refers to a reaction in which electrons are lost, hydrogen is lost, or oxygen is gained by another element. Combustion is a reaction in which oxygen combines with another element with heat given off rapidly.

9. (b) CH_3CHO
11. (a) oxidized
13. Oxygen in the atmosphere oxidizes other substances.
15. oxidized
17. oxidized
19. oxidized
21. (a) reducing agent
    (b) reducing agent
    (c) oxidizing agent
    (d) oxidizing agent
    (e) reducing agent
23. CO_2
25. cathode, H_2; anode, O_2.
27. There is less water in the air in Arizona.
29. Oxide coating on aluminum is tightly bonded to the surface and prevents additional oxidation. Oxide coating that makes up the rust on iron is loosely held to the surface and does not protect the underlying iron.
31. Lithium, Li, is the strongest reducing agent. The weakest reducing agent is fluoride ion, F^-.
33. The lead storage battery is used in automobiles; the Ni-Cad battery is used in portable video cameras; and the lithium ion battery is used in cellular phones and laptop computers.
35. anode, salt bridge, cathode
37. Many dry cells use the anode as the reaction container. It disappears as the cell produces electricity and the cell contents can leak out to react with surrounding metals.
39. Ni-Cad batteries contain toxic heavy metals and the batteries have a memory.
41. Both use oxidation-reduction to produce electrical energy. The reactants are added to a fuel cell and products are removed, while a battery is usually self-contained.
43. The reactants and products are still inside the battery. No material was added or removed and no mass was created or lost.
45. Silver in tarnish is present as silver ion while the pan contains aluminum atoms. Silver ion, Ag^+, is a better oxidizing agent and Al is a reducing agent, so silver ions gain electrons from aluminum atoms, forming silver atoms and Al^{3+} ions. The water–baking soda solution allows transfer of electrons.
47. $2\ Al(s) + 3\ Cu(NO_3)_2(aq) \rightarrow 3\ Cu(s) + 2\ Al(NO_3)_3(aq)$

CHAPTER 11

1. (a) layer of gases extending from the Earth's surface out into space
   (b) fresh water and salt water above and below the Earth's surface
   (c) rock formed by solidification of molten rock
   (d) rock formed by deposition of dissolved or suspended substances
3. (a) a substance having a very low boiling point
   (b) use of liquid nitrogen for removal of tissue (or other kinds of surgery)
   (c) process of converting N_2 into nitrogen compounds usable by plants
   (d) condition occurring when nitrogen dissolved in the blood at high pressure escapes at lower pressures to form "bubbles" in the capillaries
   (e) salty water or salt solution
5. (a) controlled cooling of viscous liquid to form a solid such as glass
   (b) describes solids with no regular crystalline order
   (c) materials generally made from clays and then hardened by heat
   (d) substance able to bond mineral fragments into a solid mass
   (e) a hard, noncrystalline substance with random (liquid-like) structure
7. H_2 and He particles have low masses and, therefore, high velocities; their velocities are generally greater than the "escape velocity" needed for them to leave Earth's gravitational field. N_2 and O_2 are much heavier molecules; their velocities are low and they do not escape as easily.
9. (a) liquefaction of air
   (b) liquefaction of air
   (c) separation from natural gas
   (d) liquefaction of air
11. Yes. The atmosphere might be exclusively N_2 and contain no O_2 to breathe.
13. Magnesium. It is used for alloys for auto and aircraft parts, fireworks, flashbulbs.
15. gold, copper, platinum
17. The slag has lower density than the molten iron. The two layers are not soluble in each other so they do not mix.
19. carbon
21. (a) $4\ Al(s) + 3\ O_2(g) \rightarrow 2\ Al_2O_3(s)$
    (b) $3\ Na(s) + AlCl_3(s) \rightarrow Al(s) + 3\ NaCl(s)$
23. bauxite
25. At the time, aluminum was extremely valuable. It has low density so the cap was easily lifted into place. Aluminum reacts with oxygen in the air to form a white oxide coating that would color match the marble color of the rest of the structure and that would protect the rest of the aluminum cap from more corrosion.
27. tetrahedral,

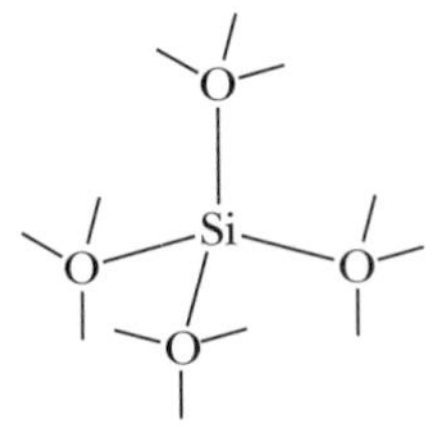

29. Fibers can penetrate lung passages and can cause lung cancer.
31. oxygen and aluminum
33. to produce more uniform bonding forces and minimize strain so glass is stronger
35. Mixture of ceramic materials or ceramic fibers with plastics. This makes a lightweight, heat-tolerant, high-strength material for such things as turbine blades.
37. a substance produced by roasting a powered mix of chalk, sand, clay, iron oxide, and gypsum
39. Glass has the random structure of a liquid; it is an amorphous solid. Sodium chloride is a crystalline solid with an orderly, repetitive structure.
41. window glass
43. Na is oxidized; Al is reduced.
45. Concrete is not uniform and is a mixture of cement, sand, and aggregate. Mortar is uniform and is a mixture of cement, sand, water, and lime. Cement is an ingredient in mortar and concrete.
47. Yes. Separation of useful pure elements would require processing more material and lead to greater environmental disturbance.
49. "Green Design" considers a product's impact on the environment at all stages of its life cycle. Examples: bags of laundry detergent to refill original detergent boxes, concentrated forms of laundry detergent with most inert "fillers" left out, glass bottles returnable for frequent reuse before recycling.

CHAPTER 12

1. fuel formed by decomposition of plant and animal matter
3. compound containing only carbon and hydrogen
5. the energy released when a compound reacts completely with oxygen
7. (a) bond between two atoms sharing either four or six electrons
   (b) bond between two atoms sharing four electrons
   (c) bond between two atoms sharing two electrons
   (d) bond between two atoms sharing six electrons
9. A hydrocarbon fragment formed by removing a hydrogen atom from a hydrocarbon
11.  or  All angles are 109.5°
13. Alkanes have the formula C_nH_{2n+2}, while alkyl groups have the formula C_nH_{2n+1} with one less hydrogen atom.
15. (a) 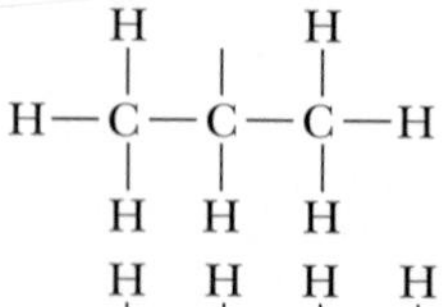 (b) 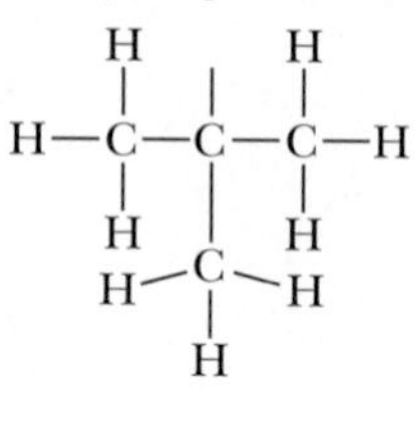
   (c) 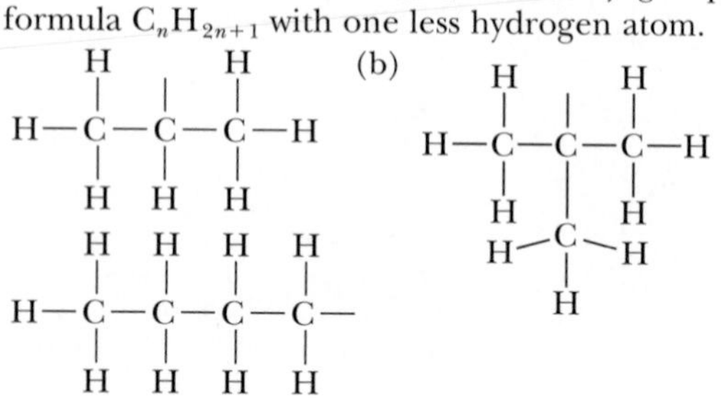 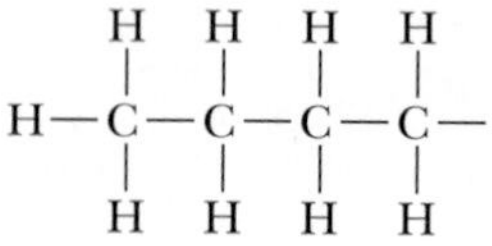
17. (a) 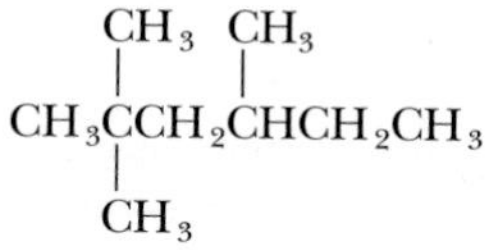 (b) 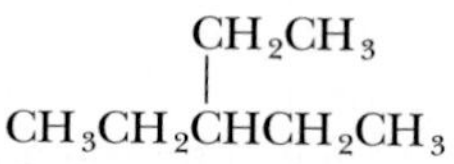
   (c) 
19. 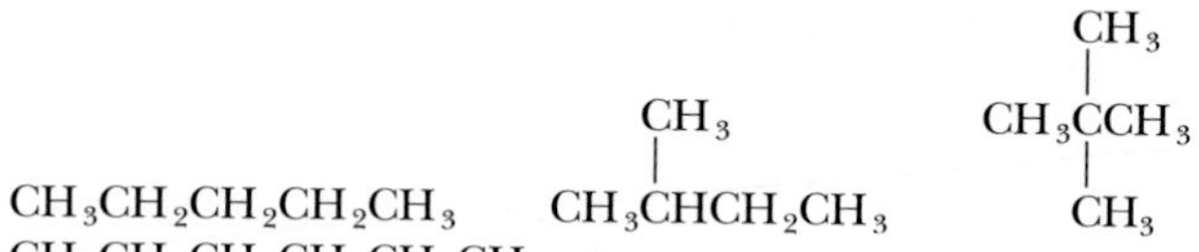
21. $CH_3CH_2CH_2CH_2CH_2CH_3$ hexane
   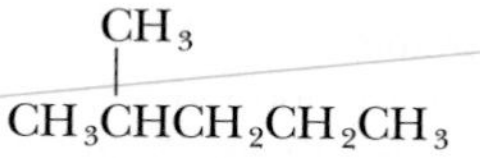 2-methylpentane
   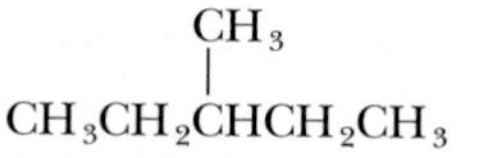 3-methylpentane
   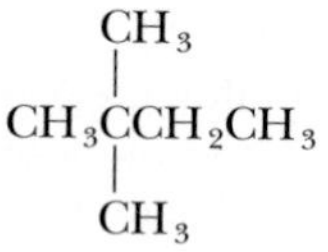 2,2-dimethylbutane
   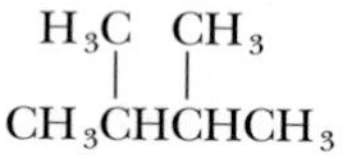 2,3-dimethylbutane
23. 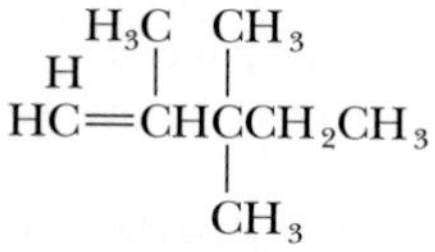
25. 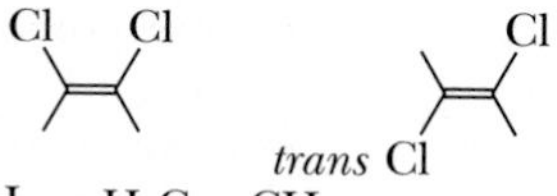
   cis *trans*
27. C_4H_8 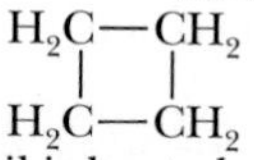
29. Crude oil is heated and fractionally distilled. Desired products are collected based on boiling points.
31. hydrocarbons from fractional distillation that boil in the range 30–200°C
33. gasoline mixture with 90% gasoline and 10% ethanol
35. (a) process used to increase the octane rating of straight-run gasoline
   (b) process used to break large hydrocarbons into shorter ones
   (c) percentage of isooctane in isooctane-heptane mix with the same knocking behavior as the fuel
37. branched chain hydrocarbons and aromatic hydrocarbons
39. Octane ratings are determined by running a standard engine on the fuel and comparing its knocking properties with those of a mix of isooctane and heptane.
41. Oxygenated fuels give more complete combustion.
43. Treating pulverized coal with superheated steam forms either carbon monoxide and hydrogen or, using a catalyst, carbon dioxide and methane.

45. M100 means 100% methanol.
E100 means 100% ethanol.
M85 means 85% methanol, 15% gasoline.
E85 means 85% ethanol, 15% gasoline.
FFV means flexible-fueled vehicle.
47. (a) alkane
(b) alkyne
(c) ether
(d) alcohol
(e) aromatic
49. Lead is a toxic heavy metal. Lead emissions contaminate the atmosphere, soil, and ground water.
51. Aromatics are typically carcinogens. Evaporated fuel and unburned gasoline would contain more carcinogens, causing greater human exposure to these.

SELECTED PROBLEMS

1. (a) $2\ C_4H_{10} + 13\ O_2 \rightarrow 8\ CO_2 + 10\ H_2O$
(b) 640 kcal/mol
3. (a) $2\ C_8H_{18} + 25\ O_2 \rightarrow 16\ CO_2 + 18\ H_2O$
(b) 1331 kcal/mol
(c) 11.7 kcal/gram

CHAPTER 13

1. (a) energy that is replenished by natural processes
(b) rate of energy use or energy per unit time
(c) 10^{15} Btu
(d) a relatively slow neutron with a kinetic energy about the same as that of a gas molecule at room temperature
3. There are energy losses in the mechanical processes used to produce electricity from heat energy. The transmission of electricity incurs additional losses because of the resistance of the power lines and system.
5. (b) fission of uranium-235
7. convenient, clean
9. fission of one mole of uranium-235
11. Uranium is converted to UF_6. The U-235 and U-238 forms of UF_6 have different molar masses; the molecules containing U-235 are lighter and more volatile. This volatility difference is enough to allow separation of the isotopes.
13. binding energy
15. to convert high energy neutrons to thermal neutrons
17. risk of catastrophic accident, plutonium production, disposal of radioactive wastes from plant
19. (a) separation of useful plutonium and fissionable uranium
(b) plutonium may be diverted for weapons
21. Fission and fusion both rely on conversion of mass to energy. Fission requires splitting heavy nuclei into smaller ones, while fusion involves joining small nuclei together to produce a larger nucleus.

Fission

Fuels	*Benefits*	*Problems*	*Current Status*
uranium, plutonium	renewable from breeder reactors	radioactive waste handling	commercially available, proven method
Fusion			
deuterium tritium	unlimited fuel supply	radioactive hardware from power plants	research only, not commercial energy source

23. (a) 24,000 years
(b) It remains radioactive, and therefore hazardous, for a long time.
25. The magnetic fields produced by strong electromagnets are used to contain a plasma by interacting with the magnetic fields generated by the ions in the plasma.
27. warming the atmosphere and land
29. (b) 84 quads
31. Use of parabolic mirror concentrators to produce high temperatures by focusing light on a small area greatly increasing the amount of energy per square inch.
33. molten core of the Earth
35. Special devices must be built to convert them into useful forms.
37. Develop plants that are more efficient biomass producers. These would yield more vegetable oil or other fuel than present crops.
39. Each person may answer differently.

SELECTED PROBLEMS

1. (a) 8×10^{15} Btu
(b) 1×10^{12} Btu
(c) 20.2 kcal
3. 5.2 barrel/cord
6. 2.64×10^{10} Watts
8. (a) 1.3×10^{-2} barrel/day
(b) 4.3×10^{-2} barrel/day (30% efficiency)

CHAPTER 14

1. Carbon can bond to carbon in almost unlimited numbers and in a variety of ways. Introducing other elements allows different molecules due to different atom sequences (functional groups and isomers).

3. (a) aldehyde
   (b) alcohol
   (c) ketone
   (d) carboxylic acid
   (e) ester
   (f) ether
   (g) aldehyde

5. (a)

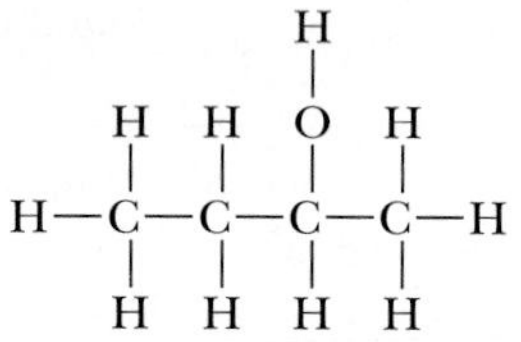

(b)

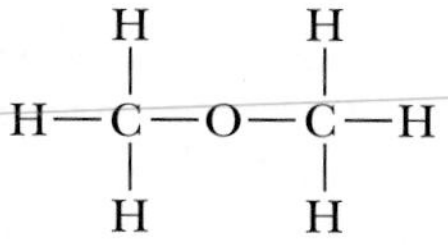

(c)

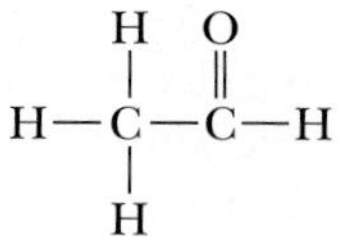

(d)

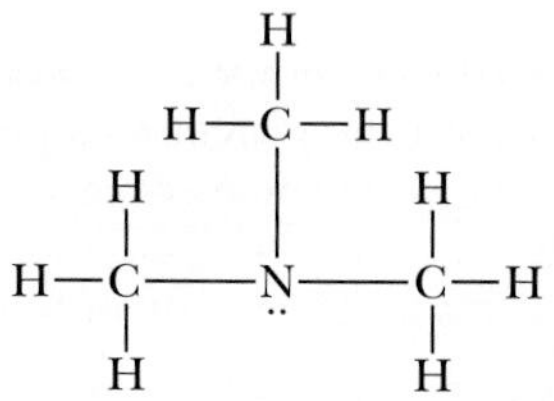

(e)

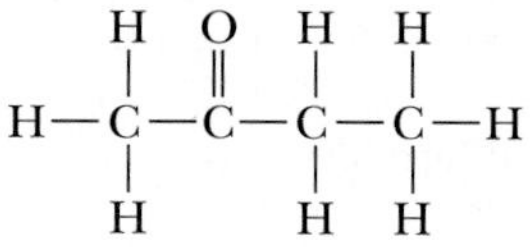

(f)

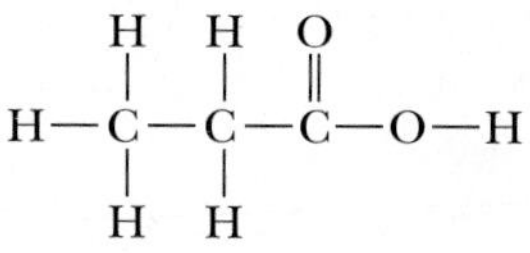

(g)

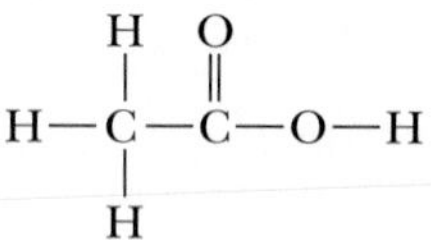

(h)

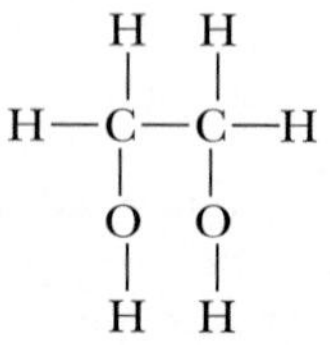

7.

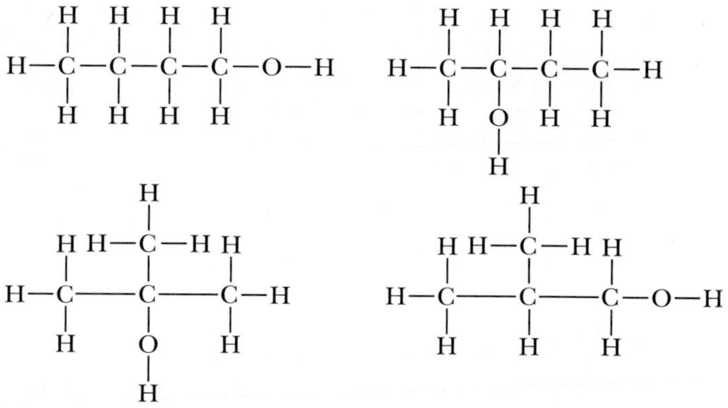

9. (a) pentanoic acid
   (b) 2-pentene
   (c) diethyl ether
   (d) 3-pentanol
   (e) 2-methyl-2-propanol
   (f) acetaldehyde
11. (a) two times the percent alcohol by volume
    (b) ethanol with additives to make it toxic
13. 200
15. an amine
17. (a) isoamyl acetate, bananas; butyl butanoate, pineapple
    (b) formic acid, ants; lactic acid, milk
19. $C + H_2O \rightarrow CO + H_2$
    $CO + 2\,H_2 \rightarrow CH_3OH$
    $CH_3OH + CH_3COOH \rightarrow CH_3COOCH_3$
    $CH_3COOCH_3 + CO \rightarrow (CH_3CO)_2O$
21. many parts
23. (a) small molecule used to make a polymer
    (b) molecule made up of many small molecules (monomers) linked together
25. (a) polymer built from CH_2CH_2
    (b) high-density polyethylene
    (c) low-density polyethylene
    (d) cross-linked polyethylene
27. The difference is in the positions of the CH_2 groups relative to the double bond: *trans* = across, *cis* = same side.

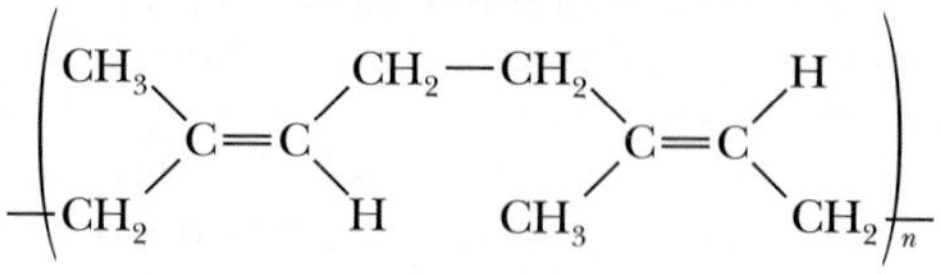

poly-*trans*-isoprene brittle and hard

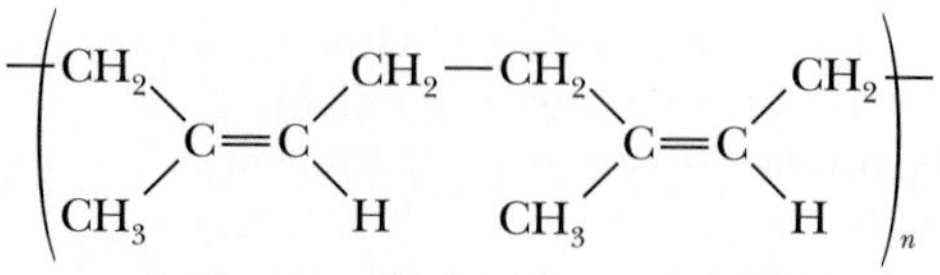

poly-*cis*-isoprene elastic, water-repellent

29. It makes the rubber elastic and water-repellent but not sticky.

31. (a) 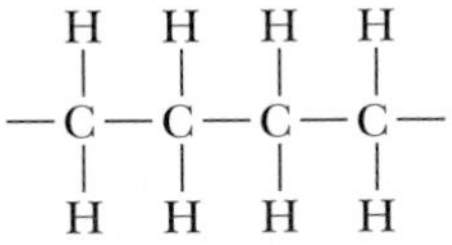

(b) 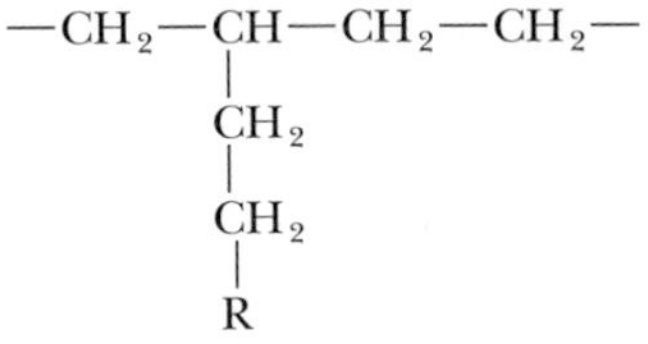

(c) 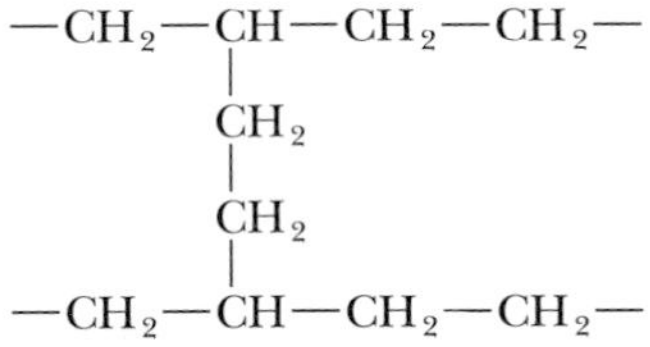

33. (a) 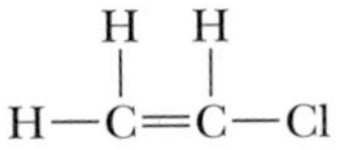

(b) 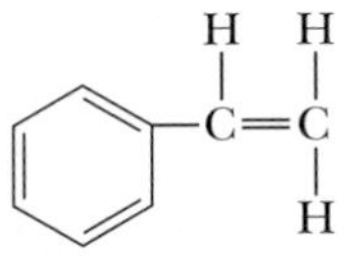

(c) 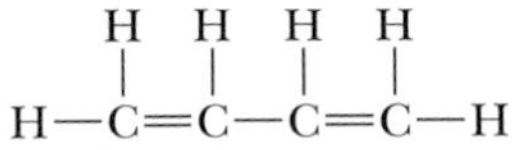

(d) 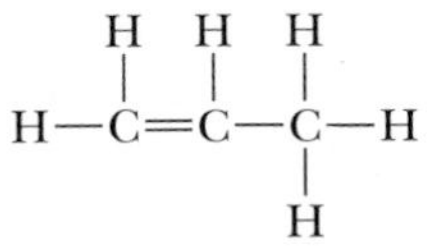

35. (a) Yes, can add to double bond
 (b) Yes, can add to double bond
 (c) No, no multiple bond, cannot add to single bond
37. (a) polymer built from ester monomers
 (b) polymer built from amide monomers
 (c) polymer built from 6-carbon dicarboxylic acid and diamine monomers
 (d) molecule with two carboxylic acid groups
 (e) molecule containing two amine groups
 (f) structural group with a carbonyl adjacent to an amine group
39. compounds with silicon bonded to alkyl groups
41. A small molecule is eliminated.
43. Petroleum. Yes. Plant material sources will, because they are relatively cheaper.
45. collection, sorting, reclamation, end-use. Collected materials will not be sorted or reclaimed.
47. Greater strength. Yes, separation of fibers during reprocessing may be difficult.
49. PET in carpet fibers, HDPE in fencing and drain pipe
51. (a) aldehyde, alcohol
 (b) alcohol, aromatic, amine
 (c) ketone, alkene, alcohol
 (d) aromatic, aldehyde, ether, alcohol
 (e) alcohol, carboxylic acid

SELECTED PROBLEMS

1. 10,000
4. (a) polyester
 (b) polyethylene

CHAPTER 15

1. Greek for "hand." Structures that have nonsuperimposable mirror images.
3. a 50-50 mixture of enantiomers
5. a "double" ion that has a + charge separated from a − charge;

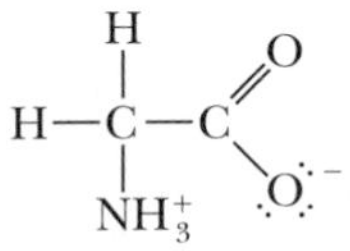

7. an amino acid not produced by the human body
9. (a) two amino acids bonded together
 (b) a protein of 3 or more amino acids
 (c) amino acids linked to form peptide molecule
 (e) very large polypeptide
11. Biomolecules that transmit chemical messages along nerve pathways by connecting with receptors. Examples: methionine-enkephalin, leucine-enkephalin.
13. (a) starch
 (b) cellulose
 (f) proteins
 (g) DNA
 (h) RNA
15. to make it unable to perform its normal function
17. region where substrate interacts with enzyme
19. (a) simple polyhydroxy aldehydes or ketones
 (b) simple sugar
 (c) carbohydrate made from two monosaccharides
 (d) carbohydrate built of many monosaccharides
21. A person with Type I does not produce enough insulin and can be treated by giving insulin injections. A person with Type II produces enough insulin but the insulin receptors on cells do not function properly; this type is treated by altering diet and by oral medication.
23. an organic substance in living systems that is soluble in organic solvents, but is insoluble in water
25 conversion of C=C double bonds to single bonds by adding hydrogen atoms to molecule
27. (a) process in which green plants absorb light to make glucose and oxygen from water and carbon dioxide
 (b) steps in photosynthesis that require light
 (c) photosynthesis steps that occur in the absence of light
 (d) molecule that absorbs light energy for use in photosynthesis

29. DNA and RNA have two structural differences:
    (a) They contain different sugars. DNA contains the sugar α-2-deoxy-D-ribose. RNA contains the sugar α-D-ribose.
    (b) The base uracil occurs only in RNA, whereas the base thymine is found only in DNA. The other bases are found in both DNA and RNA.
31. (a) 24
    (b) Gly-Ala-Ser-Cys
    Gly-Ala-Cys-Ser
    Gly-Ser-Cys-Ala
    Gly-Ser-Ala-Cys
    Gly-Cys-Ser-Ala
    Gly-Cys-Ala-Ser
    Ala-Ser-Cys-Gly
    Ala-Ser-Gly-Cys
    Ala-Cys-Gly-Ser
    Ala-Cys-Ser-Gly
    Ala-Gly-Ser-Cys
    Ala-Gly-Cys-Ser
    Ser-Cys-Gly-Ala
    Ser-Cys-Ala-Gly
    Ser-Gly-Ala-Cys
    Ser-Gly-Cys-Ala
    Ser-Ala-Gly-Cys
    Ser-Ala-Cys-Gly
    Cys-Gly-Ala-Ser
    Cys-Gly-Ser-Ala
    Cys-Ala-Ser-Gly
    Cys-Ala-Gly-Ser
    Cys-Ser-Ala-Gly
    Cys-Ser-Gly-Ala
33. valylcysteyltyrosine
35. Enzymes are stereo specific. The D-glucose fits its active site; the L-glucose will not act as a substrate for the enzyme.
37. nucleotides
39. bases that have sites for matched hydrogen bonding
41. the project intended to completely map the sequence of base pairs in human DNA
43. (a) A C A G U C A C C C G G C G A
    (b) U G U C A G U G G G C C G C U
    (c) Thr-Val-Thr-Arg-Arg
45. (a) $H{-}S{-}CH_2{-}\underset{\underset{\displaystyle NH_2}{|}}{CH}{-}COOH$
    (b) Cysteine
47. hydrogen bonding
49. insulin production by bacteria, TPA production from transgenic animals
51. The amount needed is extremely small and distributing the sweetener uniformly through the product is difficult.
53. (a) Guanine, Adenine, Cytosine
    (b) Uracil
    (c) Thymine
55. There are no wrong answers; each individual will have his or her own opinion.

CHAPTER 16

1. Hydrogen bonding between hydrogens in amide groups in one protein chain and oxygens in amide groups in adjacent strands. The strands are also held together by disulfide bonds, —S—S—, in cystine.
3. The amide group is $-\overset{\overset{\displaystyle O}{\|}}{C}-\overset{\overset{\displaystyle H}{|}}{N}-H$. Amide linkages connect the amino acid monomers in proteins.
5. cost, compatibility with skin, whether oil from animal was obtained without threatening species or harming animal
7. water in oil
9. Tanning results from uv radiation striking the skin and increasing the concentration of melanin pigment. If uv exposure is too great, erythema (reddening of the skin) occurs.
11. PABA absorbs uvB before it can reach skin surface. In contrast, zinc oxide tends to reflect light and not absorb it.
13. The top note is most volatile, middle note is intermediate, and end note is the least volatile.
15. Colognes are diluted perfumes, about one-tenth as strong.
17. Disulfide bonds in the hair are broken by a reducing agent. The hair strand is shaped, and then the disulfide bonds are reformed using an oxidizing agent.
19. to hold hair strand in place
21. any color desired depending on what dye is used
23. to act as an astringent
25. Inhalation can possibly harm nasal passages and lungs.
27. They work because they stabilize small size oil particles so they can remain in suspension in the aqueous layer.
29. Cationic detergents have a positively charged site. Anionic detergents have a negatively charged site. Nonionic detergents are neutral.
31. Nonionic detergents give good foaming characteristics. They stabilize foams and thicken the shampoo to give a smoother-pouring liquid.
33. Abrasives remove unwanted substances from tooth surfaces and detergents assist in suspending unwanted particles in water.
35. Optical brighteners fluoresce. They absorb invisible light of high energy and emit light in the visible range.
37. Toilet bowl cleaners have pH less than 2; drain cleaners have pH greater than 12.
39. Inhalation of caustic oven cleaners can damage nasal passages and lungs; eye exposure can harm the lens of the eye. Skin contact can cause irritation.

CHAPTER 17

1. The process of breaking large molecules in food into substances small enough to be absorbed by the body from the digestive tract.
3. (a) fats, 9 kcal/g (b) proteins, 4 kcal/g
   (c) carbohydrates, 4 kcal/g

5. Basal metabolic rate is the minimum amount of energy required per unit time to keep alive.
7. Triesters of glycerol with three fatty acids.
9. (a) the buildup of fatty deposits on the inner walls of arteries, hardening of the arteries
(b) a steroidal alcohol that contributes to the development of atherosclerosis
(c) the yellowish deposits of cholesterol and lipid-containing materials on artery walls
(d) a complex assemblage of lipids, cholesterol, and proteins
(e) lipoproteins richer in lipid than protein; carry cholesterol to arteries
(f) lipoproteins richer in protein than lipid; carry cholesterol to the liver
11. Empty calories provide energy, while a nutritious calorie provides the same energy and is accompanied by protein, vitamins, or minerals.
13. total calories and calories from fat, the weight and % daily values for total fat, saturated fat, cholesterol, sodium, total carbohydrate, dietary fiber, sugars, protein, vitamins A and C, calcium, iron
15. Nutrients, vitamins, and minerals, needed in small amounts. Iron.
17. Overdoses of vitamins A, D, and B should be avoided. Megadoses can cause varying adverse physiological problems.
19. An electrolyte is a substance that dissolves in water to produce ions. Electrolyte balance is the electrolyte content of the major fluid compartments (extracellular, intracellular, interstitial) of the body during normal health.
21. phosphorus, calcium, magnesium, sodium, potassium, chlorine, and sulfur
23. (a) preservative (b) preservative
(c) antioxidant (d) antioxidant
(e) sequestrant (f) sequestrant
25. (a) flavor enhancer (b) food color
(c) buffer or pH control (d) anticaking agent
(e) thickener (f) antioxidant
27. disaccharide, because it can be broken down into two monosaccharides
29. The volume of these globules would be 9/4 or 2.25 times as bulky.
31. The question each person considers important will differ. Some possibilities: What are the negative effects of the overdose? What percentage of those over 50 will experience negative effects? Are there any benefits to those over 50 in receiving this dosage?
33. The program might benefit individuals who would suffer from osteoporosis. This kind of medication of the entire population might undermine individual responsibility for personal health.
35. oxidation, sequestrants and/or antioxidants
37. There is no single answer for this question.
39. A zero-risk law allows for no risk of harm and does not weigh the risks compared to the benefits.
41. The answer to this question clearly depends on the specific group.

SELECTED PROBLEMS

1. 1760 Calories
3. 20 minutes
7. 15,000 mg
11. 726 Calories
12. 2755 C/day; 92 g fat, 413 g carbohydrates, 69 g protein

CHAPTER 18

1. (a) amount of substance that enters the body of the exposed organism
(b) the dose that is lethal in 50% of a large number of test animals under controlled conditions
(c) substances that are dangerous in small amounts
(d) interaction of a substance with the senses to let person know of exposure
(e) collection of fluid in the lungs
3. (a) a molecule that can bind metal ions to form a water-soluble complex ion
(b) an alloy of mercury and another metal
(c) a craving for non-food items including flaking paint
(d) a poison that can cause illness or death by interfering with metabolic chemical reactions
(e) compound that counteracts the effects of a poison
5. (a) a chemical that can change the hereditary pattern of a cell; one that is capable of altering the structure of DNA
(b) substance that causes cancer
(c) abnormal, out-of-control cell growth
(d) maximum tolerated dose, which is nearly a toxic dose
7. Skin contact, ingestion, inhalation. Inhalation allows contact with the blood stream in the lungs.
9. Hydrochloric acid. Mineral acids dehydrate cells; the water would flow into the lungs from the cells.
11. corrosive
13. Carbon monoxide and cyanide are both metabolic poisons. Carbon monoxide binds to hemoglobin and keeps it from picking up oxygen because CO is more strongly bound than O_2. Cyanide binds to iron in oxidative enzymes like cytochrome oxidase. The use of oxygen for oxidation in cells is disrupted.
15. hemoglobin
17. oxidative enzymes like cytochrome oxidase
19. Carboxyhemoglobin has carbon monoxide bound to the iron in hemoglobin. Normal hemoglobin contains iron with no additional bound species.
21. thiosulfate ($S_2O_3^{2-}$) ion
23. Process in which a molecule uses two or more sites to bond to a metal atom or ion. The chelating agent ties up the metal to keep it from disrupting metabolic processes. Two chelating agents are BAL and EDTA.
25. cyanide
27. (a) immediate, severe response

(b) response or symptoms preceding acute response
(c) prolonged and continuous poisoning

29. British Anti-Lewisite. 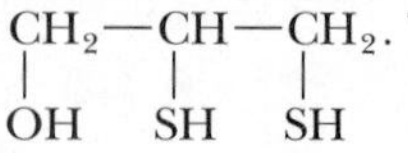 The —SH groups are the active sites.
31. micrograms per deciliter of blood, $\mu g/dL$
33. The solder used in joints in copper pipe contains lead.
35. Acetyl cholinesterase breaks down acetylcholine in a stimulated synapse at a receptor to release choline. This allows the synapse to fire again to send an impulse to the receptor site.
37. parathion, malathion, and carbaryl
39. The compound could interfere with fetal development. Teratogen
41. It provides a quick way to determine whether a compound is a mutagenic hazard. It also is a good indicator of the carcinogenic hazard a substance poses because most mutagens are also carcinogens.
45. Exposure of an organism to the MTD, maximum tolerated dose, for a prolonged period amounts to sustained or chronic wounding. This serves as a promoter of cancer and increases the number of cancers. Many compounds appear carcinogenic because test conditions are wrong.
47. Heavy metals typically are at the bottom of a group or in the transition series.

SELECTED PROBLEMS

1. 3400 mg
3. 300 g, 845 cans

CHAPTER 19

1. bacterial: malaria, tuberculosis; viral: polio, AIDS
3. Food and Drug Administration, FDA
5. penicillins, cephalosporins, tetracyclines
7. Treatment of disease with chemical agents
9. a virus that uses RNA-directed synthesis of DNA instead of DNA-directed synthesis of RNA.
11. (a) asthma
    (b) cancer
    (c) heart disease
    (d) cancer
13. hormones and neurotransmitters
15. (a) cholesterol-lowering drug
    (b) antimetabolite
    (c) antihistamines
    (d) stimulants
17. a, c, d, e
19. (a) chest pain on exertion
    (b) abnormal heart rhythm
21. Lipid derived from 20-carbon carboxylic acid. They control blood clot formation, blood lipid levels, blood vessel contractions, nerve impulses, inflammation response to injury and infection.
23. depressants, inhibit transmission of nerve impulses
25. The mustard alkyl groups attach to the nitrogen bases in DNA, often guanine. This bonding physically blocks base pairing and prevents DNA replication (production of new DNA) and therefore prevents cell division.
27. neurotransmitters
29. Premarin is a female hormone replacement. It slows bone loss that normally occurs with age.
31. (a) analgesic
    (b) analgesic, antiinflammatory
    (c) analgesic, antiinflammatory
33. Crack is a form of cocaine with the usual hydrochloride group removed by heating with sodium bicarbonate.
35. (a) treat allergic reactions
    (b) diminish pain
    (c) shrink nasal passages
    (d) diminish cough
    (e) loosen fluids and allow cough to clear respiratory tract
37. Compounds that have a mustard like odor and a structure like this: 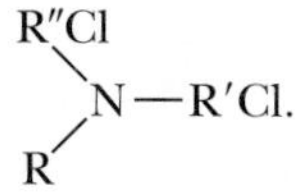
    War gases. Chemotherapy for cancer.
39. $-NH\overset{O}{\overset{\|}{C}}CH_2C_6H_5$. Some are more effective against certain bacteria than others.
41. Cause vivid illusions, fantasies, hallucinations. Mescaline, PCP, LSD.
43. (a) morphine
    (b) heroin
    (c) heroin
45. heart muscle relaxant; heart disease
47. Boundary between blood and brain cells. Drugs must be soluble in lipids (fats or oils) and able to move across the membranes separating brain cells and the circulatory system.
49. (a) +80%
    (b) +25%, +10%
    (c) +35%, +40%

CHAPTER 20

1. (a) rivers, lakes, and streams
   (b) water beneath the earth's surface
   (c) layer of water-bearing porous rock
   (d) fresh water contaminated by sea water
   (e) any condition that causes the natural usefulness of water, air, and soil to be diminished
3. (a) amount of oxygen needed to convert organic material in water to CO_2 and water
   (b) vaporization of water followed by condensation
5. evaporation and transpiration from plants

7. Rainwater passes through materials and dissolves the pollutants, then enters the ground water.
9. Agriculture and industry are nearly equal.
11. Dual water system has a system for potable water and another for water for other uses. Advantages: consumes less potable water, nonpotable water is less expensive; Disadvantages: added cost for second distribution system, dual plumbing needed in homes, potential for cross contamination of the two kinds.
13. CO_2, N_2, NO_2, O_2, HCO_3^-, Ca^{2+}, Mg^{2+}, K^{1+}, Na^+, Cl^-, SO_4^{2-}
15. Pathogens: sewage; Organic chemicals: landfills; Acids: mining; Radioisotopes: mine tailings; Heat: cooling towers for electric generating plants
17. Positive ions: magnesium, calcium, iron(II); Negative ions: chloride, bicarbonate, sulfate
19. Originally a $1.6 billion program for cleaning up hazardous waste sites. Reduces possible contamination of groundwater.
21. A document listing hazardous waste by name, amount present, and procedure for its disposal. Assigns responsibility and specifies disposal method.
23. California law that prohibits businesses from knowingly discharging or releasing 200 different hazardous substances into water that may become part of drinking water. Proposition 65 and the Delaney Clause are both "zero risk" concepts.
25. High BOD indicates high concentration of organic matter in the water.
27. treatment of industrial wastewater and sewage with oxygen or ozone.
29. gas chromatograph–mass spectrometer
31. Dissolved volatile organics can escape into the atmosphere and others are oxidized more readily to smaller, less hazardous substances.
33. Biodegradable substances can be easily broken down into simpler, smaller molecules by microorganisms, while nonbiodegradable substances cannot be easily broken down. Prefer biodegradable.
35. DDT is a persistent nonbiodegradable pesticide. It is implicated in interfering with bird reproduction by causing fragile eggs.
37. kills pathogens
39. The microbial detoxification or degradation of wastes into nonhazardous products. Both are similar in that organic molecules are consumed by microorganisms.
41. use of denitrifying bacteria that convert ammonia or ammonium ion to nitrogen, N_2
43. Soda lime process is used to treat hard water. Sodium ions replace hard water ions. Sodium ion concentrations go up.
45. Chlorination of water containing organic residues can produce chlorinated hydrocarbons that might be hazardous themselves.
47. NaCl
49. Water from salt water is evaporated by solar energy and condensed as pure water. Large surface areas are needed.
51. The philosophy shifts problems from one area to another and avoids the issues of responsibility and ways to prevent or remedy problems.
53. Answer depends on individual interpretation.

CHAPTER 21

1. (a) air that contains unwanted and harmful substances.
(b) solid particles in the air with diameters greater than 10,000 nm
(c) mixtures of water droplets and particulates with diameters in the range 1 nm–10,000 nm
(d) layering of air with cold high-density air near the Earth's surface and warmer low-density air layered above
3. (a) rainwater with a pH lower than 5.6
(b) chlorofluorocarbons
(c) hole in the ozone layer in the stratosphere
(d) gas that absorbs infrared light and radiates it to the atmosphere in the form of heat
(e) worldwide increase in air temperature
5. Clean Air Act in 1970 originally controlled air pollution from cars and industry. Amendments in 1977 imposed stricter auto emission standards. The 1990 CAA extends to manufacturing and commercial activity.
7. Industrial smog is chemically reducing, while photochemical smog is chemically oxidizing.
9. UV light breaks up O_2 to produce oxygen atoms; these react with additional O_2 to form O_3, ozone.
11. electrical storms and combustion in automobile engines
13. Dust particles from the eruption scatters and reflects sunlight into space so the solar energy never reaches the Earth's surface.
15. natural sources and human activities; human activities
17. ozone
19. Nitrogen dioxide dissociates to form an oxygen atom and nitric oxide.

$$NO_2 + h\nu \longrightarrow NO + O$$

21. Ozone, sulfur dioxide, and nitrogen dioxide cause lung damage.
23. Ozone is produced by decomposition of NO_2 so controlling the NO_x emissions will control ozone.
25. Carbon monoxide and NO_x.
27. Increased NO_x emissions occurred with higher engine operating temperatures and more cars on the road. Decreased CO emissions because of catalytic converters; decreased lead emissions through use of unleaded gasolines.
29. Oil-burning electric fuel plants generate SO_2. Using low-sulfur oil or passing plant exhaust gas through molten sodium carbonate to form sodium sulfite.
31. Rain tends to be acidic because it dissolves CO_2 to make H_2CO_3. Human sources of SO_2 and NO_x will yield nitric acid and sulfuric and sulfurous acid when mixed with rain.
33. $2\,C + O_2 \rightarrow 2\,CO$; carbon monoxide

35. CFC's are linked to depletion of ozone concentrations in the stratosphere. These ozone holes could lead to increased ultraviolet light levels at sealevel, resulting in greater rates of skin cancer. Some unforeseen effects on algae, plants, and animals are additional causes for worry.
37. $O_2 + h\nu \rightarrow 2\,O$
$O_2 + O \rightarrow O_3$
39. CFC-12; CFC's reach the stratosphere and break down to produce Cl atoms that deplete ozone concentrations in the stratosphere.
41. methyl iodide, CH_3I
43. infrared radiation; the atmosphere warms up
45. photosynthesis and dissolving in seawater
47. Natural processes release large amounts of pollutants but usually are part of an ongoing natural cycle and equilibrium. Human sources are smaller but, like the proverbial straw that broke the camel's back, can have profound effects.
49. Each community is different so there are many possible right answers.

CHAPTER 22

1. The earth's carrying capacity depends on the amount of productive land, agricultural practices, biotechnology advances, the rate of degradation of farmland, global pollution, and the level of an acceptable quality of life.
3. Soil is a mixture of mineral particles, organic matter, water, and air.
5. Soil has layers of topsoil and subsoil.
7. Raise. Lime is a base.
9. carbon, hydrogen, and oxygen
11. (a) Group IA and Group IIA
(b) The pH drops.
13. (a) 18 elemental substances required for green plant growth
(b) carbon, oxygen, and hydrogen
(c) elements absorbed as solutes by plant roots
15. A shortage of N, P, and K because Ca, Mg, and S are not as easily leached from the soil.
17. $N_2(g) + 3\,H_2(g) \rightleftarrows 2\,NH_3(g)$. This is the synthetic source of ammonia used in fertilizers.
19. Legumes are plants that have nitrogen-fixing bacteria in their roots. Nitrogenase is the enzyme that catalyzes nitrogen fixation in legume root modules.
21. A deficiency of iron, magnesium or nitrogen that leads to low chlorophyll content in plants. Plant leaves are pale yellow instead of green.
23. No. No.
25. Urea decomposes to form ammonia.
27. (a) water soluble, dissolve readily
(b) dissolve very slowly in water
29. Ammonium nitrate is inexpensive and is an excellent source of water-soluble nitrogen compounds. It can be used by terrorists as an explosive as was done in 1995 at the Federal Building in Oklahoma City.
31. DDT is a nonpolar organic compound. It is relatively unreactive and is stored in fat tissue of animals.
33. 33% to 40%
35. chlorinated hydrocarbons, organophosphorus compounds, and carbamates
37. one that is quickly converted to harmless products by microorganisms and natural processes
39. (a) Pesticides can increase crop yields and protect them when stored. Pollution can occur, pesticide residues can contaminate crops. Resistant strains of pests can require higher levels of pesticides.
(b) Pesticides should be used early enough to require smaller amounts and only when absolutely necessary and when alternative methods do not exist.
41. Organic farming uses materials that require only 40% of the energy that is used in farming with synthetic chemical fertilizers. Organic farming uses human labor which has a high energy cost.
43. Answer depends on person's views.
45. Bt toxins are bacterial toxins produced by *Bacillus thuringiensis.* Neem-tree seeds are a source of neem oil which is effective against 200 insect species. Both are specific for pest and do not harm desirable insects.
47. These block synthesis of "essential" (for animals) amino acids in plants by inhibiting enzymes that catalyze their synthesis.
49. No right answer exists. Each person will have his or her own priorities.
51. overgrazing, deforestation, soil erosion
53. Time intervals are tremendously different. Natural breeding methods can take years, while gene splicing can be done within one year. Gene insertion techniques can produce properties that are more specific and do not occur in any existing species.

SELECTED PROBLEMS

2. K_2O; 83% K, 17% O
3. 10% N; 2 pounds

Glossary/Index

Alphabetization is word-for-word. Terms appearing in boldface are followed by a brief definition. Letters i, t, and s following page numbers indicate the entry refers to an illustration, a table, or a structure, respectively.

A

C

E

I

N

O

Q

R

U

V

SOME COMMON CHEMICALS

Common Name or Product Name	Chemical Name[a]	Chemical Formula
Alum	Aluminum potassium sulfate	$KAl(SO_4)_2 \cdot 12\ H_2O$
Antifreeze	Ethylene glycol	$HOCH_2CH_2OH$
Baking soda	Sodium hydrogen carbonate	$NaHCO_3$
Battery acid	Sulfuric acid	H_2SO_4
Borax	Sodium borate	$Na_2B_4O_7$
Calamine lotion	Zinc oxide	ZnO
Caustic potash	Potassium hydroxide	KOH
Chalk	Calcium carbonate	$CaCO_3$
Clorox	Sodium hypochlorite	$NaOCl$
Dry Ice	Carbon dioxide	$CO_2(s)$
Epsom salts	Magnesium sulfate	$MgSO_4$
Glycerine	Glycerol	$HOCH_2CH(OH)CH_2OH$
Grain alcohol	Ethyl alcohol	CH_3CH_2OH
Gypsum	Calcium sulfate	$CaSO_4$
Household ammonia	Ammonia	NH_3
Ice melter	Calcium chloride	$CaCl_2$
Lighter fluid	Butane	$CH_3CH_2CH_2CH_3$
Lime	Calcium oxide	CaO
Limestone	Calcium carbonate	$CaCO_3$
Lye	Sodium hydroxide	$NaOH$
Milk of Magnesia	Magnesium hydroxide	$Mg(OH)_2$
Moth balls	Naphthalene	$C_{10}H_8$
Muriatic acid	Hydrochloric acid	$HCl(aq)$
Nail polish remover	Acetone	CH_3COCH_3
Quicksilver	Mercury	Hg
Rubbing alcohol	Isopropyl alcohol	$CH_3CHOHCH_3$
Sal ammoniac	Ammonium chloride	NH_4Cl
Saltpetre	Potassium nitrate	KNO_3
Lite salt	Potassium chloride	KCl
Slaked lime	Calcium hydroxide	$Ca(OH)_2$
Smelling salts	Ammonium carbonate	$(NH_4)_2CO_3$
Table salt	Sodium chloride	$NaCl$
Table sugar	Sucrose	$C_{12}H_{22}O_{11}$
Vinegar	Acetic acid	CH_3COOH
Washing soda	Sodium carbonate	Na_2CO_3
Wood alcohol	Methyl alcohol	CH_3OH

[a]Chemical name of the active ingredient, which may be present in a mixture.